TO THE STUDENT: Four helpful supplementary study aids are available for this textbook. If these supplements are not stocked in your college bookstore, ask the manager for details on how to order them.

- **Student Guide and Review Manual** by John K. Harris and Dudley W. Curry contains four sections for each chapter in the textbook:

  Main Focus and Objectives - introduces the chapter, provides an overall learning objective, refers to the specific learning objectives in the textbook, and identifies concepts and techniques that deserve special study.

  Review of Key Terms and Concepts - is a comprehensive outline featuring (1) references to specific textbook exhibits and examples and (2) references to the Practice Test Questions and Problems that cover important concepts and techniques.

  Practice Test Questions and Problems - includes an average of 33 objective questions and 4 problems (with check figures). An average of 5 CPA and CMA questions are included per chapter.

  Solutions to Practice Test - provides (1) explanations for all true-false and multiple-choice answers and (2) easy-to-follow solutions to the problems.

  (ISBN: 0-13-184730-9)

- **Student Solutions Manual** by Charles T. Horngren, George Foster, and Srikant M. Datar provides worked-out solutions to all of the even-numbered problems. (ISBN: 0-13-184714-7)

- **Applications in Cost Accounting Using Lotus 1-2-3** by David M. Buehlman and Dennis P. Curtin is a book/disk package that consists of a series of case problems that parallel the concepts in the textbook. The templates on the disk run with Lotus 1-2-3. (ISBN: 1-13-184813-5)

- **Lotus Templates for Selected Problems** by Chitra Rajagopal provides spreadsheets for solving selected problems from the text in a Lotus 1-2-3 format. (ISBN: 1-13-184805-4)

Cost Accounting

PRENTICE HALL SERIES IN ACCOUNTING
Charles T. Horngren, Consulting Editor

AUDITING: AN INTEGRATED APPROACH, 6/E	ARENS/LOEBBECKE
KOHLER'S DICTIONARY FOR ACCOUNTANTS, 6/E	COOPER/IJIRI
FINANCIAL STATEMENT ANALYSIS, 2/E	FOSTER
GOVERNMENTAL AND NONPROFIT ACCOUNTING: THEORY AND PRACTICE, 4/E	FREEMAN/SHOULDERS/LYNN
FINANCIAL ACCOUNTING: PRINCIPLES AND ISSUES, 4/E	GRANOF/BELL
FINANCIAL ACCOUNTING	HARRISON/HORNGREN
COST ACCOUNTING: A MANAGERIAL EMPHASIS, 8/E	HORNGREN/FOSTER/DATAR
ACCOUNTING, 2/E, 1993 ED.	HORNGREN/HARRISON
PRINCIPLES OF FINANCIAL AND MANAGEMENT ACCOUNTING: A SOLE PROPRIETORSHIP APPROACH	HORNGREN/HARRISON/ROBINSON
PRINCIPLES OF FINANCIAL AND MANAGEMENT ACCOUNTING: A CORPORATE APPROACH	HORNGREN/HARRISON/ROBINSON
INTRODUCTION TO FINANCIAL ACCOUNTING, 5/E	HORNGREN/SUNDEM/ELLIOTT
INTRODUCTION TO MANAGEMENT ACCOUNTING, 9/E	HORNGREN/SUNDEM
INTRODUCTORY FINANCIAL ACCOUNTING, 3/E	MUELLER/KELLY
AUDITING PRINCIPLES, 5/E	STETTLER
BUDGETING, 5/E	WELSCH/HILTON/GORDON

Cost Accounting

A Managerial Emphasis

EIGHTH EDITION

Charles T. Horngren
Stanford University

George Foster
Stanford University

Srikant M. Datar
Stanford University

Prentice Hall
Englewood Cliffs, NJ 07632

Library of Congress Cataloging-in-Publication Data

HORNGREN, CHARLES T.,
    Cost accounting : a managerial emphasis / Charles T. Horngren,
George Foster, Srikant Datar . — 8th ed.
        p.   cm.
    Includes bibliographical references and indexes.
    ISBN 0-13-181066-9
    1. Cost accounting.   2. Costs, Industrial.   I. Foster, George.
II. Datar, Srikant. III. Title.
HF5686.C8H59 1994
657. 15'11—dc20
                                          93-31195
                                          CIP

Executive Editor: *Bill Webber*
Editor in Chief: *Joe Heider*
Development Editor: *Stephen Deitmer*
Assistant Editor: *Diane deCastro*
Production Editor: *Carol Burgett*
Managing Editors, Production: *Kris Ann Cappelluti, Joyce Turner*
Copy Editor: *Margo Quinto*
Marketing Manager: *David Gillespie, Judy Perkinson*
Cover Photo: *Walt Urbina/Pittsburgh*
Cover Designer: *Patricia Woszyk*
Interior Designer: *Jayne Conte*
Buyer: *Trudy Pisciotti, Herb Klein*
Editorial Assistant: *John Chillingworth, Renee Pelletier*

 © 1994, 1991, 1987, 1982, 1977, 1967, 1962 by Prentice-Hall, Inc.
A Simon & Schuster Company
Englewood Cliffs, New Jersey 07632

Printed in the United States of America
10 9 8 7 6 5 4 3 2

ISBN   0-13-181066-9

Prentice-Hall International (UK) Limited, *London*
Prentice-Hall of Australia Pty. Limited, *Sydney*
Prentice-Hall Canada Inc., *Toronto*
Prentice-Hall Hispanoamericana, S.A., *Mexico City*
Prentice-Hall of India Private Limited, *New Delhi*
Prentice-Hall of Japan, Inc., *Tokyo*
Simon & Schuster Asia Pte. Ltd., *Singapore*

About the Authors

Charles T. Horngren is the Edmund W. Littlefield Professor of Accounting at Stanford University. A graduate of Marquette University, he received his MBA from Harvard University and his Ph.D. from the University of Chicago. He is also the recipient of honorary doctorates from Marquette University and DePaul University.

A Certified Public Accountant, Horngren served on the Accounting Principles Board for six years, the Financial Accounting Standards Board Advisory Council for five years, and the Council of the American Institute of Certified Public Accountants for three years. For six years, he served as a trustee of the Financial Accounting Foundation, which oversees the Financial Accounting Standards Board and the Government Accounting Standards Board.

In 1990 Horngren was elected to the Accounting Hall of Fame.

A member of the American Accounting Association, Horngren has been its President and its Director of Research. He received the Outstanding Accounting Educator Award in 1973, when the association initiated an annual series of such awards.

The California Certified Public Accountants Foundation gave Horngren its Faculty Excellence Award in 1975 and its Distinguished Professor Award in 1983. He is the first person to have received both awards.

In 1985 the American Institute of Certified Public Accountants presented its first Outstanding Educator Award to Horngren.

In 1993, Horngren was named Accountant of the Year, Education, by the national professional accounting fraternity, Beta Alpha Psi.

Professor Horngren is also a member of the National Association of Accountants, where he was on its research planning committee for three years. He was a member of the Board of Regents, Institute of Management Accounting, which administers the Certified Management Accountant examinations.

Horngren is the coauthor of six other books published by Prentice Hall: *Principles of Financial and Management Accounting: A Sole Proprietorship Approach* and *Principles of Financial and Management Accounting: A Corporate Approach*, 1994 (with Walter T. Harrison, Jr., and Michael A. Robinson), *Introduction to Financial*

Accounting, Fifth Edition, 1993 (with Gary L. Sundem and John Elliott), *Introduction to Management Accounting*, Ninth Edition, 1993 (with Gary L. Sundem), *Financial Accounting*, 1992, and *Accounting*, Second Edition, 1993 (with Walter T. Harrison, Jr.).

Charles T. Horngren is the Consulting Editor for the Prentice Hall Series in Accounting.

George Foster is the Paul L. and Phyllis Wattis Professor of Management at Stanford University. He graduated with a university medal from the University of Sydney and has a Ph.D. from Stanford University. He has been awarded honorary doctorates from the University of Ghent, Belgium, and from the University of Vaasa, Finland. In 1993, he received the Outstanding Educator Award from the American Accounting Association.

Foster has received the Distinguished Teaching Award at Stanford University and the Faculty Excellence Award from the California Society of Certified Public Accountants.

Research awards Foster has received include the Competitive Manuscript Competition Award of the American Accounting Association, the Notable Contribution to Accounting Literature Award of the American Institute of Certified Public Accountants, and the Citation for Meritorious Contribution to Accounting Literature Award of the Australian Society of Accountants.

He is the author of *Financial Statement Analysis*, published by Prentice Hall. He is coauthor of *Security Analyst Multi-Year Earnings Forecasts and The Capital Market* and *Market Microstructure and Capital Market Information Content Research*, both published by the American Accounting Association. Journals publishing his articles include *Abacus, The Accounting Review, Harvard Business Review, Journal of Accounting and Economics, Journal of Accounting Research, Journal of Cost Management*, and *Management Accounting*.

Foster works actively with many companies, including Apple Computer, ARCO, BHP, Digital Equipment Corp., Exxon, Frito-Lay Corp., Hewlett-Packard, McDonald's Corp., Octel Communications, Santa Fe Corp., and Wells Fargo. He also has worked closely with Computer Aided Manufacturing-International (CAM-I) in the development of a framework for modern cost management practices. Foster has presented seminars on new developments in cost accounting in North America, Asia, Australasia, and Europe.

Srikant M. Datar is Professor of Accounting at Stanford University. A graduate with distinction from the University of Bombay, he received gold medals upon graduation from the Indian Institute of Management, Ahmedabad, and the Institute of Cost and Works Accountants of India. A Chartered Accountant, he holds two masters degrees and a Ph.D. from Stanford University.

Cited by his students as a dedicated and innovative teacher, Datar received the George Leland Bach Award for Excellence in the Classroom at Carnegie Mellon University.

Datar has published his research in several journals, including *The Accounting Review, Journal of Accounting and Economics, Journal of Accounting Research, Contemporary Accounting Research*, and *Management Science*. He has served on the editorial board of several journals. He has presented his research to corporate executives and academic audiences in North America, Asia, and Europe.

Datar has worked with many companies, including General Motors, Mellon Bank, Solectron, TRW, and Visa, on field-based projects in management accounting. He is a member of the American Accounting Association and the Institute of Management Accountants.

Brief Contents

Part 1 **COST ACCOUNTING FUNDAMENTALS**

1 The Accountant's Role in the Organization   1
2 An Introduction to Cost Terms and Purposes   25
3 Cost-Volume-Profit Relationships   59
4 Costing Systems in the Service and Merchandising Sectors   97
5 Costing Systems in the Manufacturing Sector   139

Part 2 **BUDGETS AND STANDARDS AS KEYS TO PLANNING AND CONTROL**

6 Master Budget and Responsibility Accounting   181
7 Flexible Budgets, Variances, and Management Control: I   225
8 Flexible Budgets, Variances, and Management Control: II   267
9 Income Effects of Alternative Inventory-Costing Methods   307

Part 3 **COST INFORMATION FOR VARIOUS DECISION AND CONTROL PURPOSES**

10 Determining How Costs Behave   339
11 Relevance, Costs, and the Decision Process   385
12 Pricing Decisions, Product Profitability Decisions, and Cost Management   427
13 Management Control Systems: Choice and Application   465

Part 4 COST ALLOCATION AND MORE ON COSTING SYSTEMS

14 Cost Allocation: I   497
15 Cost Allocation: II   535
16 Cost Allocation: Joint Products and Byproducts   569
17 Process Costing Systems   597
18 Spoilage, Reworked Units, and Scrap   631
19 Operation Costing, Backflush Costing, and Project Control   655

Part 5 DECISION MODELS AND COST INFORMATION

20 Capital Budgeting and Cost Analysis   683
21 Capital Budgeting: A Closer Look   719

Part 6 MORE ON COST ANALYSIS AND COST MANAGEMENT

22 Measuring Mix, Yield, and Productivity   753
23 Cost Management: Quality and Time   793
24 Inventory Management and Just-in-Time   831

Part 7 STRATEGY AND MANAGEMENT CONTROL

25 Systems Choice: Decentralization and Transfer Pricing   859
26 Systems Choice: Performance Measurement and Compensation   889

Appendix A—Surveys of Company Practice   921
Appendix B—Recommended Readings   925
Appendix C—Notes on Compound Interest and Interest Tables   927
Appendix D—Cost Accounting in Professional Examinations   935

Glossary   939
Photo Credits   951
Author Index   953
Company Index   955
Subject Index   957

Contents

Preface   xvi

PART ONE
COST ACCOUNTING FUNDAMENTALS

1
The Accountant's Role
in the Organization   1

Modern Cost Accounting   2
The Major Purposes of Management Accounting
   Systems   4
   Surveys of Company Practice:
     Customer Satisfaction as Priority
     One for CFOs   5
Newly Evolving Management Themes   5
Elements of Management Control   7
   Concepts in Action:
     Customer-Driven Changes in Management
     Accounting Systems   9
Cost-Benefit Approach   10
Organizational Structure and the Management
   Accountant   11
   Surveys of Company Practice:
     Changing Responsibilities of CFOs   14
   Concepts in Action:
     Time as a Key Success Factor for Controllers   15

Professional Ethics   16
Problem for Self-Study   18   *Summary*   19
Terms to Learn   20   *Assignment Material*   20

2
An Introduction to Cost Terms
and Purposes   25

Costs in General   26
Direct Costs and Indirect Costs   27
Cost Drivers and Cost Management   29
Cost Behavior Patterns: Variable Costs
   and Fixed Costs   29
Total Costs and Unit Costs   33
Financial Statements and Cost Information   34
Costs As Assets and Expenses   37
Perpetual and Periodic Inventories   40
Manufacturing Costs   41
   Concepts in Action:
     Harley-Davidson Eliminates
     the Direct Manufacturing Labor
     Cost Category   42
Benefits of Defining Accounting Terms   43
   Surveys of Company Practice:
     Labor Fringe Benefits as a Major Cost Item   44
The Many Meanings of "Product Costs"   45
   Surveys of Company Practice:
     The Relative Sizes of Manufacturing
     Cost Categories   46

Classifications of Costs   46
Problem for Self-Study   47   Summary   49
Terms to Learn   50   Assignment Material   50

3
Cost-Volume-Profit Relationships   59

Cost Drivers and Revenue Drivers:
   The General Case and a Special Case   60
Terminology   61
The Breakeven Point   62
CVP Assumptions   65
Cost Planning and CVP   65
Uncertainty and Sensitivity Analysis   67
The PV Chart   68
Effects of Sales Mix   69
Role of Income Taxes   70
    Concepts in Action:
      How a Jobs-Bank Agreement
      Increased General Motors'
      Breakeven Point   71
Nonprofit Institutions
   and Cost-Volume-Revenue Analysis   72
Contribution Margin and Gross Margin   73
    Surveys of Company Practice:
      Purposes for Companies Distinguishing
      Between Fixed Costs and Variable Costs   73
Problem for Self-Study   75   Summary   76
Appendix: Decision Models and Uncertainty   77
Terms to Learn   82   Assignment Material   82

4
Costing Systems in the Service and Merchandising Sectors   97

Building Block Concepts of Costing Systems   98
Job Costing and Process Costing Systems   99
Job Costing in Service Organizations   101
Tracing Direct Costs to Jobs   104
    Surveys of Company Practice:
      Importance of Labor Compensation
      Costs in Law Firms   104
Allocating Indirect Costs to Jobs   107
Actual, Normal, and Budgeted Costing
   Methods   109
Refinements in Costing Systems   110
Peanut-Butter Costing Approaches   114
Activity-Based Costing   115
Customer Costing   118
    Concepts in Action:
      Customer Profitability Analysis
      at DEC-Europe   123
The Value Chain in Job Costing and Customer
   Costing Systems   124
Problem for Self-Study   124   Summary   126
Terms to Learn   128   Assignment Material   128

5
Costing Systems in the Manufacturing Sector   139

**Part One: Job Costing Systems
   in Manufacturing   140**
General Approach to Job Costing   141
Allocating Indirect Costs to Jobs   144
Actual, Normal, and Budgeted Costing Methods   145
    Surveys of Company Practice:
      Cost Allocation Bases Used
      for Manufacturing Overhead   145
Illustration of a Job Costing System
   in Manufacturing   146
Budgeted Indirect Costs
   and End-of-Period Adjustments   152
**Part Two: Process Costing Systems
   in Manufacturing   155**
Case 1: Process Costing with All Units Fully
   Completed   156
Case 2: Process Costing with Some Units
   Not Fully Completed   156
**Part Three: Activity-Based Costing
   in Manufacturing   159**
Illustration of ABC in Manufacturing   159
    Surveys of Company Practice:
      Growing Interest in Activity-Based
      Costing   161
    Concepts in Action:
      How Clark-Hurth Implemented
      Activity-Based Costing   165
Multipurpose Nature of Costing Systems   166
Problem for Self-Study   167   Summary   169
Terms to Learn   170   Assignment Material   170

PART TWO
BUDGETS AND STANDARDS AS KEYS TO PLANNING AND CONTROL

6
Master Budget and Responsibility Accounting   181

Evolution of Systems   182
Major Features of Budgets   183
Advantages of Budgets   184
Types of Budgets   186
    Surveys of Company Practice: Budget Practices
      Around the Globe   187
Illustration of Master Budget   188
    Concepts in Action: The Paperless Budget at
      Union Pacific Railroad   196
Sales Forecasting—A Difficult Task   198

Activity-Based Budgeting and *Kaizen* Budgeting 199
Computer-Based Financial Planning Models 200
   Concepts in Action: *Kaizen* Budgeting
     at Toyota Motor Company 201
   Concepts in Action:
     The Budgeting Process at Emerson Electric 202
Responsibility Accounting 202
Responsibility and Controllability 206
Human Aspects of Budgeting 208
Problem for Self-Study 208 Summary 208
Appendix: The Cash Budget 209
Terms to Learn 214 Assignment Material 214

7
Flexible Budgets, Variances, and Management Control: I   225

Performance Gaps and Variances 227
Static Budgets and Flexible Budgets 228
Flexible Budgeting without Standard Costs 228
   Surveys of Company Practice:
     The Widespread Use of Standard Costs 232
Flexible Budgeting with Standard Costs 233
Price Variances and Efficiency Variances 238
   Concepts in Action:
     Supplier Performance Evaluation
     at Northrop Aircraft Division 243
Overview of Variance Analysis 245
Performance Measurement Using Standards 245
Impact of Inventories 247
Illustration of General Ledger Entries
   Using Standard Costs 248
Benchmarks and Budgeted Amounts 249
   Concepts in Action:
     An Australian Company Uses
     Benchmarking to Adopt World-Class
     Performance Targets 250
Problem for Self-Study 251 Summary 252
Appendix: Variance Investigation Decisions
   and Uncertainty 253
   Surveys of Company Practice:
     The Decision to Investigate Variances 255
Terms to Learn 256 Assignment Material 256

8
Flexible Budgets, Variances, and Management Control: II   267

Planning of Variable and Fixed Overhead Costs 269
Developing Budgeted Variable Overhead Rates 270
Variable Overhead Cost Variances 271
   Surveys of Company Practice:
     Manufacturing Overhead Allocation Bases
     and Rates in the Electronics Industry 273
Developing Fixed Overhead Rates 276

Fixed Overhead Cost Variances 277
Output Level (Production Volume) Overhead
   Variance 278
Integrated Analysis of Overhead Cost Variances 280
Overhead Cost Variances in Nonmanufacturing
   Settings 281
Different Purposes of Manufacturing
   Overhead Cost Analysis 282
Journal Entries for Overhead Costs
   and Variances 283
Performance Gaps and Variances 285
Actual, Normal, Budgeted,
   and Standard Costing 286
Problem for Self-Study 286 Summary 289
Appendix: Proration of Manufacturing Variances
   Using Standard Costs 290
Terms to Learn 296 Assignment Material 296

9
Income Effects of Alternative Inventory-Costing Methods   307

**Part One: Variable Costing
   and Absorption Costing 308**
Role of Fixed Manufacturing Overhead Costs 308
   Concepts in Action:
     Nothing Beats a Great Deduction:
     The IRS and Packaging Design Costs 313
Comparison of Standard Variable Costing
   and Standard Absorption Costing 314
   Surveys of Company Practice:
     Usage of Variable Costing by Companies 318
Breakeven Points in Variable Costing
   and Absorption Costing 320
Performance Measures and Absorption Costing 321
**Part Two: Role of Various Denominator-Level
   Concepts in Absorption Costing 322**
Alternative Denominator-Level Concepts 323
Effect on Financial Statements 324
Problem for Self-Study 326 Summary 327
Terms to Learn 329 Assignment Material 329

PART THREE
COST INFORMATION FOR VARIOUS DECISION AND CONTROL PURPOSES

10
Determining How Costs Behave   339

General Issues in Estimating Cost Functions 340
Assumptions Underlying Cost Classification 343
Cost Estimation Approaches 344

Surveys of Company Practice:
International Comparison
of Cost Classification by Companies 345
Steps in Estimating a Cost Function 347
Evaluating and Choosing Among Cost Functions 352
Concepts in Action:
Cost Estimation at Neptune Plastics
Manufacturing 353
Data Collection and Adjustment Issues 354
Hierarchy of Costs and Cost Drivers 355
Nonlinearity and Cost Functions 357
Learning Curves and Cost Functions 358
Problem for Self-Study 361 Summary 363
Appendix: Regression Analysis 364
An Example of Choosing Among Cost Functions 369
Terms to Learn 373 Assignment Material 373

11

Relevance, Costs, and the Decision Process 385

Information and the Decision Process 386
The Meaning of Relevance 388
Illustration of Relevance:
Choosing Output Levels 390
Other Illustrations of Relevance 393
Concepts in Action:
Outsourcing Data-Processing Operations
at Eastman Kodak 397
Opportunity Costs, Relevance, and Accounting
Records 397
Customer Profitability and Relevant Costs 400
Concepts in Action:
Opportunity Costs in Energy Transmission 401
Irrelevance of Past Costs 403
How Managers Behave 406
Problem for Self-Study 407 Summary 409
Appendix: Linear Programming 410
Terms to Learn 414 Assignment Material 414

12

Pricing Decisions, Product Profitability Decisions, and Cost Management 427

Major Influences on Pricing 428
Product-Cost Categories 429
Concepts in Action: Competitor Analysis 430
Costing and Pricing for the Short Run 431
Costing and Pricing for the Long Run 434
Alternative Pricing Approaches 437
Target Costing for Target Pricing 438
Concepts in Action:
Target Costing at Mercedes Benz
and Toyota 443

Cost-Plus Pricing 443
Considerations Other Than Costs
in Pricing Decisions 446
Surveys of Company Practice:
Differences in Pricing Practices
and Cost Management Methods
in Various Countries 447
Life-Cycle Product Budgeting and Costing 448
Effects of Antitrust Laws on Pricing 451
Problem for Self-Study 452 Summary 454
Terms to Learn 455 Assignment Material 455

13

Management Control Systems: Choice and Application 465

Management Control Systems 466
Evaluating Management Control Systems 468
An Example of Motivation:
Sales Compensation Plans 470
Concepts in Action:
Management Control Systems Go Green 472
Value-Added/Nonvalue-Added Cost Analysis 473
Engineered, Discretionary,
and Infrastructure Costs 475
Classifying Costs in Business Function Areas
in the Value Chain 476
Budgeting for Discretionary Costs 478
Work Measurement 479
Engineered-Cost versus Discretionary-Cost
Approaches to Budgeting 479
Benchmarking and Cost Management 481
Problem for Self-Study 483 Summary 484
Appendix: Promoting and Monitoring the
Effectiveness or Efficiency of
Discretionary Cost Centers 485
Terms to Learn 487 Assignment Material 487

PART FOUR
COST ALLOCATION AND MORE
ON COSTING SYSTEMS

14

Cost Allocation: I 497

The Terminology of Cost Allocation 498
Purposes of Cost Allocation 499
Criteria to Guide Cost-Allocation Decisions 500
Surveys of Company Practice: Why Allocate
Corporate and Other Support Costs to
Divisions and Departments? 502
Cost Tracing and Cost Allocation 502
Cost Pools and Cost Allocation 504

Allocating Costs from One Department
  to Another   506
    Concepts in Action: How a Quart of Distilled
      Water Cost Cindy Chase $17   506
Allocating Costs of Support Departments   510
Allocation of Common Costs   515
    Surveys of Company Practice: Allocation of
      Support Department Costs   516
Problem for Self-Study   517   Summary   519
Terms to Learn   520   Assignment Material   520

15
Cost Allocation: II   535

Cost Tracing and Cost Allocation   536
Choosing Indirect Cost Pools and Cost Rates   539
Evolving Trends in Cost Assignment   541
    Surveys of Company Practice:
      Is the Product-Costing System Broken?   543
Used and Unused Capacity Distinctions   544
Cost Justification and Reimbursement   546
Cost Assignment and Cost Hierarchies   549
    Concepts in Action:
      Cost Hierarchies Down on the Farm   550
Problem for Self-Study   553   Summary   555
Terms to Learn   556   Assignment Material   556

16
Cost Allocation:
Joint Products and Byproducts   569

Meaning of Terms   570
Why Allocate Joint Costs?   571
Methods of Allocating Joint Costs   572
Irrelevance of Joint Costs for Decision Making   579
    Surveys of Company Practice:
      Joint-Cost Allocation Methods Used
        by U.K. Companies   580
Accounting for Byproducts   582
Problem for Self-Study   584   Summary   586
Terms to Learn   587   Assignment Material   587

17
Process Costing Systems   597

Process Costing
  and Equivalent Unit Computations   598
Process Costing with Equivalent Units:
  Five Key Steps   599
Weighted-Average Method   602
First-In, First-Out Method   606
Comparison of FIFO
  and Weighted-Average Methods   609
Standard Costs and Process Costs   610
    Concepts in Action:
      Process Costing in the Ceramics Industry   611
Transferred-In Costs in Process Costing   614

Additional Aspects of Process Costing   620
Problem for Self-Study   620   Summary   622
Terms to Learn   623   Assignment Material   623

18
Spoilage, Reworked Units, and Scrap   631

Management Effort and Control   632
Terminology   632
Spoilage in General   633
Process Costing and Spoilage   634
Job Costing and Spoilage   640
Reworked Units   641
Accounting for Scrap   642
    Surveys of Company Practice:
      Rejection in the Electronics Industry   642
Comparison of Accounting for Spoilage, Rework,
  and Scrap   644
Problem for Self-Study   645   Summary   645
Appendix: Inspection and Spoilage at Intermediate
  Stages of Completion in Process Costing   646
Terms to Learn   648   Assignment Material   648

19
Operation Costing, Backflush Costing,
and Project Control   655

**Part One: Operation Costing
  and Backflush Costing Systems**   656
Hybrid Costing Systems   656
Operation Costing   657
Just-in-Time (JIT) Production Systems   660
Backflush Costing   661
    Surveys of Company Practice:
      Adopt JIT and Simplify Your Life as a
      Management Accountant   667
Part Two: Control of Projects   668
Project Features   668
Similarity of Control of Jobs and Projects   671
Problem for Self-Study   672   Summary   674
Terms to Learn   675   Assignment Material   675

PART FIVE
DECISION MODELS
AND COST INFORMATION

20
Capital Budgeting and Cost Analysis   683

Contrast in Purposes of Cost Analysis   684
Definition and Stages of Capital Budgeting   685
The Discounted Cash-Flow Methods   687
Sensitivity Analysis   692
Analysis of Selected Items
  Using Discounted Cash Flow   694

The Payback Method 697
Concepts in Action: Environmental Costs
and Capital Budgeting 698
The Accrual Accounting Rate-of-Return Method 701
Surveys of Company Practice:
International Comparison of
Capital Budgeting Methods 703
Complexities in Capital-Budgeting Applications 704
Problem for Self-Study 705 Summary 706
Terms to Learn 708 Assignment Material 708

21
Capital Budgeting: A Closer Look 719

Income Tax Factors 720
Effects of Income Taxes on Cash Flow 724
Capital Budgeting and Inflation 732
Required Rate of Return 736
Surveys of Company Practice:
Risk Adjustment Methods in
Capital Budgeting 737
Applicability to Nonprofit Organizations 738
Implementing the Net Present Value
Decision Rule 738
Implementing the Internal Rate-
of-Return Decision Rule 740
Problem for Self-Study 741 Summary 743
Appendix: Modified Accelerated
Cost Recovery System 744
Terms to Learn 745 Assignment Material 745

PART SIX
MORE ON COST ANALYSIS
AND COST MANAGEMENT

22
Measuring Mix, Yield, and Productivity 753

Part One: Sales Variances 754
Sales-Volume Variances 755
Sales-Quantity and Sales-Mix Variances 758
Market-Size and Market-Share Variances 761
Part Two: Input Variances 764
Direct Materials Yield
and Direct Materials Mix Variances 765
Direct Labor Yield
and Direct Labor Mix Variances 770
Part Three: Productivity Measurement 771
What is Productivity? 773
Partial Productivity Measures 774
Total Factor Productivity 777
Analysis of Annual Cost Changes 780
Service-Sector Productivity 782
Problem for Self-Study 782 Summary 783
Terms to Learn 785 Assignment Material 785

23
Cost Management: Quality and Time 793

Quality as a Competitive Weapon 794
Costs of Quality 795
Methods Used to Identify Quality Problems 798
Relevant Costs and Benefits
of Quality Improvement 800
Concepts in Action: Crysel Wins
Premio Nacional de Calidad—Mexico's
Premier Quality Award 803
Quality and Customer-Satisfaction Measures 804
Quality and Internal Performance Measures 805
Evaluating Quality Performance 806
Time as a Competitive Weapon 807
New Product Development Time 807
Breakeven Time for New Products 808
Operational Measures of Time 811
Costs of Time 814
Theory of Constraints
and Throughput Contribution Analysis 816
Concepts in Action: Throughput Marketing
at Allied-Signal, Skelmersdale, U.K. 818
Problem for Self-Study 820 Summary 820
Terms to Learn 822 Assignment Material 822

24
Inventory Management
and Just-in-Time 831

Managing Goods for Sale in Retail Organizations 832
Difficulties with Accounting Data
for Managing Goods for Sale 838
Just-in-Time Purchasing 840
Managing Inventories in Manufacturing
Organizations 844
Surveys of Company Practice:
JIT Performance Measures
Around the Globe 848
Concepts in Action: Boeing Goes
"Just in Time" Just in Time 849
Problems for Self-Study 849 Summary 851
Terms to Learn 852 Assignment Material 852

PART SEVEN
STRATEGY AND MANAGEMENT CONTROL

25
Systems Choice: Decentralization
and Transfer Pricing 859

Organizational Structure and Decentralization 860
Choices about Responsibility Centers 863
Transfer Pricing 863

Illustration of Transfer Pricing 865
Tax Considerations 867
Multinational Transfer Pricing 868
Market-Based Transfer Prices 869
Cost-Based Transfer Prices 870
  Surveys of Company Practice:
    Domestic and Multinational
    Transfer-Pricing Practices 872
A General Guideline for Transfer-
  Pricing Situations 875
Problem for Self-Study 876   Summary 878
Terms to Learn 879   Assignment Material 879

26
Systems Choice: Performance
Measurement and Compensation 889

Financial and Nonfinancial Performance
  Measures 890
Designing an Accounting-Based Performance
  Measure 891
  Surveys of Company Practice:
    Nonfinancial Measures of Performance 892
Different Performance Measures 892
Alternative Definitions of Investment 897
Measurement Alternatives for Assets 897
Goal Congruence and Performance Measures 900
Distinction between Managers
  and Organization Units 901

Performance Measures at the
  Individual Activity Level 904
Performance Measures at the
  Total Organization Level 905
    Concepts in Action: Sears Auto Repair Shops:
      What Price Quality? 905
Environmental and Ethical Responsibilities 907
Problem for Self-Study 908   Summary 909
Terms to Learn 910   Assignment Material 910

Appendix A—Surveys of Company Practice 921

Appendix B—Recommended Readings 925

Appendix C—Notes on Compound Interest
  and Interest Tables 927

Appendix D—Cost Accounting in Professional
  Examinations 935

Glossary 939

Photo Credits 951

Author Index 953

Company Index 955

Subject Index 957

Preface

Cost accounting provides data for various purposes, including planning, controlling, and the costing of products, services, and customers. We stress our major theme of "different costs for different purposes" throughout this book. The favorable reaction to previous editions is evidence that cost accounting courses can be enriched and relieved of drudgery by broadening the course from coverage of procedures alone to a full-fledged coverage of concepts, analyses, and procedures that pays more than lip service to accounting as a management tool.

STRENGTHS OF THE SEVENTH EDITION RETAINED AND ENHANCED

Reviewers of the seventh edition praised the following features, which have been retained and strengthened in the eighth edition:

◆ Clarity and understandability of the text
◆ Coverage of important topics, including current developments in actual practice
◆ Extensive use of real-world examples
◆ Excellent quantity, quality, and range of assignment material
◆ Helpful Problems for Self-Study for each chapter
◆ Flexible organization through a modular approach

The first sixteen chapters provide the essence of a one-term (quarter or semester) course. There is ample text and assignment material in the book's twenty-six chapters for a two-term course. This book can be used immediately after the student has had an introductory course in financial accounting. Alternatively, this book can build on an introductory course in managerial accounting.

Deciding on the sequence of chapters in a textbook is a challenge. Every instructor has a favorite way of organizing his or her course. Hence, we present a modular, flexible organization that permits a course to be custom-tailored. *Our loosely constrained sequence of chapters facilitates diverse approaches to teaching and learning.*

CHANGES IN CONTENT AND PEDAGOGY
OF THE EIGHTH EDITION

The pace of change in organizations continues to be rapid. This edition has been revised extensively to reflect changes occurring in the role of cost accounting in organizations and in research on cost accounting. Each chapter was scrutinized by knowledgeable critics before a final draft was reached.

1. *Newly evolving management themes.* These themes guided us in choosing the topics to be given increased emphasis:

- ◆ *Focus on customers.* Chapter 1 takes a customer-driven perspective on changes in cost accounting. The customer focus is emphasized in many other chapters. For example, Chapters 4, 11, and 15 include new sections on customer costing and profitability analysis.
- ◆ *Key success factors, such as cost, quality, and time.* Chapters 8 and 12 include expanded discussion on cost planning and cost management. Chapter 23 is a new chapter: Cost Management: Quality and Time.
- ◆ *Total value-chain analysis.* The value chain is now systematically emphasized throughout the book. Chapter 1 introduces the value chain. The text in many areas (such as Chapter 2 on cost drivers, Chapter 6 on budgeting, Chapter 7 on explanations for variances, and Chapter 12 on pricing decisions) has been revised to emphasize the importance of considering all areas of the value chain.
- ◆ *Dual internal/external focus.* Topics such as benchmarking (Chapters 7 and 13) and customer satisfaction (Chapter 23) are now included to reflect the increased attention given to external factors in management control systems. Greater recognition is now given to how cost accounting is expanding its horizons to incorporate environment-related considerations (Chapters 13, 20, and 26).
- ◆ *Continuous improvement.* Topics such as *kaizen* budgeting (Chapter 6) and productivity (Chapter 22) have been added to highlight the heightened emphasis companies now give to continuous improvement.

2. *Increased coverage of the service sector.* Chapter 2 now discusses cost concepts in the service sector as well as in the merchandising and manufacturing sectors. Also, Chapters 4 and 5 have been restructured to illustrate job costing in the service sector first. The service sector is now the single largest sector in the economy, and it is the most straightforward sector to use in teaching job costing because there is no work in process to consider. A new costing method category—budgeted costing—is introduced to describe the system most companies in the service sector use.

3. *Greatly expanded global content of the text.* Our coverage of international business is highly visible in two types of boxed features, new to the eighth edition: Surveys of Company Practice and Concepts in Action. Both types of boxes, described in more detail below, draw on businesses from around the globe. In addition, company examples from many different countries are cited throughout the chapter material.

4. *Professional ethics.* The eighth edition has increased coverage of ethics. The final problem in every chapter of the book now has a component on ethics. This feature gives the instructor the flexibility to reinforce the importance of ethics in as many areas as is deemed appropriate.

5. *Cost management.* The seventh edition expanded its coverage of cost management. The eighth edition continues this expansion. Chapter 3 illustrates how decisions at the planning stage can affect the budgeted breakeven point. Chapter 12 emphasizes the designed-in cost notion as well as offering expanded coverage of target costing and value engineering.

Activity-based management principles are integrated into many chapters of the eighth edition. Chapter 2 discusses cost drivers and value-added costs. Chapters 4 and 5 illustrate how an activity-based costing approach can be used to refine either job costing or process costing systems. Chapters 10 and 15 present and illustrate cost hierarchies. Chapter 12 discusses how companies can reduce costs by re-engineering the design of products or by reducing individual activity costs. Chapter 15 examines capacity costing issues.

6. *Revised placement of material.* Several new appendices contain material that was covered in the later chapters of the seventh edition:

◆ Chapter 3 Appendix, "Decision Models and Uncertainty," incorporates material from Chapter 20 of the seventh edition.
◆ Chapter 7 Appendix, "Variance Investigation Decisions and Uncertainty," incorporates material from Chapter 26 of the seventh edition.
◆ Chapter 10 Appendix, "Regression Analysis," incorporates material from Chapter 25 of the seventh edition.
◆ Chapter 11 Appendix, "Linear Programming," incorporates materials from Chapter 24 of the seventh edition.

The material in Chapter 29 of the seventh edition has been incorporated into sections of Chapters 4 and 13 and into the new Chapter 23, Cost Management: Quality and Time.

ILLUSTRATIONS OF ACTUAL BUSINESSES

Students become more highly motivated to learn cost accounting if they can relate the subject matter to the real world. We have spent considerable time interacting with the management community, investigating new uses of cost accounting data, and gaining insight into how changes in technology are affecting the roles of cost accounting information. Real-world illustrations are found in many parts of the text.

CONCEPTS IN ACTION BOXES.   Found in many chapters, these boxes discuss how cost accounting concepts are applied by individual companies. Examples are drawn from many different countries, including the United States (Harley-Davidson on p. 42, General Motors on p. 71, and Ben & Jerry's on p. 472), Australia (Johnson & Johnson Pacific on p. 9 and Smeltco on p. 250), Canada (Neptune Plastics on p. 353), Japan (Toyota on p. 201), Mexico (Crysel on p. 803), Switzerland (Sandoz on p. 472), and the United Kingdom (Allied-Signal Skelmersdale on p. 818).

SURVEYS OF COMPANY PRACTICE BOXES.   Results from surveys in over 15 countries are cited in the many Surveys of Company Practice boxes found throughout the book. Examples include:

◆ Growing Interest in Activity-Based Costing (p. 161)—cites evidence from the United States, Canada, and the United Kingdom.
◆ Budgeting Practices (p. 187)—cites evidence from the United States, Australia, Netherlands, Japan, and the United Kingdom.
◆ Variable Costing (p. 318)—cites evidence from the United States, Australia, Canada, Japan, Sweden, and the United Kingdom.
◆ Pricing Practices (p. 447)—cites evidence from the United States, Australia, Ireland, Japan, and the United Kingdom.
◆ Purposes of Cost Allocation (p. 502)—cites evidence from the United States, Australia, Canada, and the United Kingdom.
◆ Capital Budgeting Methods (p. 703)—cites evidence from the United States, Australia, Canada, Ireland, Japan, Scotland, South Korea, and the United Kingdom.
◆ Transfer Pricing Methods (p. 872)—cites evidence from the United States, Australia, Canada, India, Japan, and the United Kingdom.

This extensive survey evidence enables students to see that many of the concepts they are learning are widely used across the globe.

PHOTOS FROM ACTUAL COMPANIES.   All chapters open with a photo that illustrates an important concept to be discussed in that chapter. These photos feature companies from many different countries, including the United States (Sun Microsystems on p. 59, Pizza Hut on p. 181, Snapple Beverage on p. 307, Ford Motor Company on p. 427, and Home Depot on p. 465), Australia (Lindsay Brothers on p. 385), Canada (Hudson's Bay Company on p. 889), Japan (Takeda Chemical Industries, Ltd., on p. 139), Netherlands (Rugby Group on p. 631), the United Kingdom (Nestlé Rowntree on p. 225), and Venezuela (Asea Brown Boveri on p. 683).

SUPPLEMENTS TO THE EIGHTH EDITION

A complete package of supplements is available to assist students and instructors in using this book. Supplements available to students are:

◆ *Student Guide and Review Manual* by John K. Harris and Dudley W. Curry. This is a chapter-by-chapter learning aid for students. It reviews key terms and concepts and contains practice test questions and problems, including an average of four CPA/CMA questions per chapter. Solutions and explanations are also included for all questions and problems. We believe this *Student Guide* is a world-class study aid that will be helpful to students at all levels.

◆ *Student Solutions Manual* by Charles T. Horngren, George Foster, and Srikant M. Datar. Designed for student use, this supplement contains solutions for all of the even-numbered questions and problems in the textbook. This may be purchased with your instructor's permission.

◆ *Lotus Templates for Selected Problems* by Chitra Rajagopal. Templates are provided for two problems per chapter, designated by a computer icon in the text.

◆ *Applications in Cost Accounting Using Lotus 1-2-3* by David M. Buehlman and Dennis P. Curtin. Personal computer applications are keyed to selected chapters of the textbook.

Supplements available to instructors are:

◆ *Annotated Instructor's Edition* with annotations by Linda S. Bamber. A great teaching tool, the *Annotated Instructor's Edition* is the regular textbook enhanced through additional comments printed in the margins. Comments are presented in several categories: New in This Edition, Teaching Tips, Examples, Correcting Student Misconceptions, Points to Stress, Curriculum Linkages, and Reinforcing Problems. Check figures appear next to the exercises and problems in the assignment material. (The inside front cover of the *Annotated Instructor's Edition* offers a more detailed look at these features.)

◆ *Prentice Hall Course Manager.* A three-hole punched *Annotated Instructor's Edition*, packaged in a binder, provides instructors with maximum flexibility in organizing course materials. Other instructor supplements are also three-hole punched.

◆ *Instructor's Manual and Media Guide* by William O. Stratton. This supplement offers a chapter overview, chapter outline, additional examples, alternate means of presenting materials, chapter quiz, and suggested readings. The Media Guide section includes a write-up for each video in the *ABC News/PH Video Library for Cost Accounting*, consisting of a brief synopsis of the video and discussion questions that can be used to link material in the video to topics in the text and so stimulate class discussion. A chapter reference, a synopsis, and discussion questions are also provided for each article in the *New York Times Supplement for Management/Cost Accounting*.

◆ *Cases and Extended Problems in Cost Accounting* by Charles T. Horngren, George Foster, and Srikant M. Datar. This supplement contains additional, longer problems for all chapters.

◆ *Test Item File* by James R. Davis. This includes quiz and examination material and is available in both hard copy and diskette form. It offers 20 true/false questions, 50 multiple-choice questions, 10 problems, and 3 essay questions (which require students to use their critical-thinking skills) for each of the text's chapters.

◆ *Solutions Manual* by Charles T. Horngren, George Foster, and Srikant M. Datar. This offers comments on alternative teaching approaches and solutions to all assignment material.

◆ *Transparencies of Solutions*

◆ *Teaching Transparencies.* Full-color transparencies for approximately 50 key exhibits from the text enhance classroom presentation.

Resources from the business world include the following:

- ◆ *ABC News/PH Video Library for Cost Accounting.* Video is the most dynamic of all the supplements you can use to enhance your class. But the quality of the video material and how well it relates to your course can still make all the difference. For these reasons, Prentice Hall and ABC News have decided to work together to bring you the best and most comprehensive video ancillaries available in the college market:

    Through its wide variety of award-winning programs—"Nightline," "Business World," "On Business," "This Week with David Brinkley," "World News Tonight," and "The Health Show"—ABC offers a resource for feature and documentary-style videos related to text concepts and applications. The programs have extremely high production quality, present substantial content, and are hosted by well-versed, well-known anchors. Prentice Hall, its authors, and its editors provide the benefit of having selected videos on topics that will work well with this course and text and give the instructor teaching notes on how to use them in the classroom.

    The ABC News/PH Video Library for Cost Accounting offers video material for almost every chapter in the text. An excellent video guide that is included in the Instructor's Manual carefully and completely integrates the videos into your lecture.

- ◆ *The New York Times* and Prentice Hall are sponsoring "Themes of the Times," a program designed to enhance student access to current information of relevance in the classroom.

    Through this program, the core subject matter provided in the text is supplemented by a collection of time-sensitive articles from one of the world's most distinguished newspapers, *The New York Times*. These articles demonstrate the vital, ongoing connection between what is learned in the classroom and what is happening in the world around us.

    To enjoy the wealth of information of *The New York Times* daily, a reduced subscription rate is available. For information, call toll-free: 1-800-631-1222.

    Prentice Hall and *The New York Times* are proud to co-sponsor "Themes of the Times." We hope it will make the reading of both textbooks and newspapers a more dynamic, involving process.

ACKNOWLEDGMENTS

We are indebted to many for their ideas and assistance. Our primary thanks is to the many academics and practitioners who have advanced our knowledge of cost accounting.

The package of teaching materials we present is the work of many skillful and valued team members. John K. Harris aided us immensely at all stages in the development and production of this book. He critiqued the seventh edition, gave a detailed review of the manuscript of this edition, and assisted in the essential task of proofing. Linda S. Bamber and William O. Stratton gave extensive reviews of the manuscript and suggestions for improvement in addition to working on their supplements. Professor James R. Davis has integrated the new text material into the Test Item File. Computer-related supplements have been skillfully prepared by David M. Buehlmann, Dennis P. Curtin, and Chitra Rajagopal.

Professors and students providing us detailed written reviews of the previous edition or comments on our drafts of this edition include Michael Alles, David Bell, Donald Bedell, Charles Betts, Donald Bostrom, Lisa Brown, Eric Carlsen, Greg Cermignano, Michael Cornick, John Cruickshank, James Emig, Ellen Engel, Zhilong Fang, Sally Foster, Scott Foster, Monojit Ghosal, David Green, Mahendra Gupta, Rachel Hayes, Hal Hoverland, Ze-Kai Hsiau, Zafar Iqbal, Rohit Jain, Diane Janvrin, Sunder Kekre, Steve Kovzan, Alan Larris, Yow-Min Lee, Elizabeth MacLean, Steve Manske, Richard Mayer, Tony Mongkolcheep, V.G. Narayanan, Brian O'Doherty, August Petersen, Chitra Rajagopal, Ratna Sarkar, Michi Sakurai, William Sanders, Michael Schiff, Arnold Schneider, Emanuel Schwarz, Larry Singleton, Leif Sjoblom, Andy Spero, Takao Tanaka, Debbie Then, Howard Toole, Sandra Weber, James Westbrook, Harry Wolk, Tsing Wu, and Ziaofang Zhao. The faculty participating in the many focus groups on the seventh edition provided highly valued feedback. In addition, we have received helpful suggestions from many other users, unfortunately too numerous to be mentioned here. The eighth edition is much improved by the feedback and interest of all these people. We are very appreciative of this support.

Debbie Then managed the collection of photos and the selection of videos. Her enormous energy level and care with detail greatly improved these areas of the eighth edition.

Our association with CAM-I has been a source of much stimulation as well as enjoyment. CAM-I has played a pivotal role in extending the frontiers of knowledge on cost management. We appreciate our extended and continued interaction with Jim Brimson, Callie Berliner, Tom Pryor, Mike Roberts, and Pete Zampino.

We thank the people at Prentice Hall. Carol Burgett (the production editor) and Steve Deitmer (the development editor) went well beyond their call of duty in producing and developing a quality book in a timely fashion. Many others at Prentice Hall gave important assistance, including Linda Albelli, Kris Ann Cappelluti, John Chillingworth, Jayne Conte, Lori Cowen, Terri Daly, Patti Dant, Diane deCastro, Anne DiBisceglie, Annmarie Dunn, Patrice Fraccio, David Gillespie, Daniel Griffin, Joseph Heider, Herb Klein, Trudy Pisciotti, Frances Russello, Judy Perkinson, Lori Sypher, Joyce Turner, Bill Webber, and Pat Wosczyk.

Jiranee Tongudai managed the production aspects of all the manuscript preparation with superb skill and much grace. Her capacity and willingness to handle the many tasks never wavered. We are deeply appreciative of her good spirits, loyalty, and ability to stay calm in the most hectic of times. We also appreciate the help of T. Bush, M. Lonergan, and J. Ochoa.

Appreciation also goes to the American Institute of Certified Public Accountants, the Institute of Management Accountants, the Society of Management Accountants of Canada, the Certified General Accountants' Association of Canada, the Financial Executive Institute of America, and many other publishers and companies for their generous permission to quote from their publications. Problems from the Uniform CPA examinations are designated (CPA); problems from the Certificate in Management Accounting examinations are designated (CMA); problems from the Canadian examinations administered by the Society of Management Accountants are designated (SMA); problems from the Certified General Accountants' Association are designated (CGA). Many of these problems are adapted to highlight particular points.

We are grateful to the professors who contributed assignment material for this edition. Their name are indicated in parentheses at the start of their specific problems.

Comments from users are welcome.

CHARLES T. HORNGREN
GEORGE FOSTER
SRIKANT M. DATAR

INTERNATIONALIZE YOUR EDUCATION: JOIN INTERNATIONAL BUSINESS SEMINARS ON AN OVERSEAS ADVENTURE

EARN COLLEGE CREDIT—GAIN INTERNATIONAL EXPERTISE—INTERACT WITH TOP-LEVEL EXECUTIVES—VISIT THE WORLD'S GREATEST CITIES: MAY 30, 1994–JUNE 23, 1994

Visit organizations such as Procter & Gamble Italia, NATO, The European Parliament, Elektra Breganz, Philip Morris, Allianz Insurance, Deutsche Aerospace, Digital Equipment, Coca-Cola, G.E. International, Ernst & Young, Esso Italiana, Guccio Gucci, Targetti Lighting, University of Innsbruck & British Bankers Association.

Prentice Hall International Business Scholarship 1994

Prentice Hall and International Business Seminars have joined forces to create a scholarship for students to study and travel in Europe in the summer of 1994. We believe that in today's global business environment students should be exposed to as many different cultures as possible. Although many campuses reflect diversity in both their students and faculty, nothing can replace the educational value of learning about a continent, country, or city first hand.

Each professor may sponsor one student to apply for the scholarship.

You can receive more information on the PH Business Scholarship and/or additional travel programs with International Business Seminars by contacting your local Prentice Hall representative or International Business Seminars, P. O. Box 30279, Mesa, Arizona 85275, Telephone: (602) 830-0902; Fax: (602) 924-0527.

1

The Accountant's Role

in the Organization

Modern cost accounting

The major purposes of management accounting systems

Newly evolving management themes

Elements of management control

Cost-benefit approach

Organizational structure and the management accountant

Professional ethics

*A*ccounting provides relevant information to managers in many organizations, including those in the natural-gas marketing group at the Pittsburgh headquarters of Equitable Resources.

Learning Objectives

When you have finished studying this chapter, you should be able to

1. Understand why taking a customer viewpoint assists in the design of an accounting system
2. Describe the set of business functions in a value chain
3. Identify the four broad purposes of an accounting system
4. Describe five evolving management themes that are shaping developments in management accounting systems
5. Understand how accounting can facilitate planning, control, and decision making

6. Describe the difference between line management and staff management
7. Distinguish among the scorekeeping, attention-directing, and problem-solving functions of the controller
8. Understand the importance of professional ethics to management accountants

$\mathbf{A}$s we write this chapter, former accountants are top executives in many large companies, including Nike, PepsiCo, Bell Canada, Cadbury Schweppes, and Nissan Motors. Accounting duties have played a key role in their rise to the management summit. Accounting cuts across all facets of the organization; the management accountant's duties are intertwined with executive planning, control, and decision making.[1]

The study of modern cost accounting yields insight into both the accountant's role and the manager's role in an organization. How are these two roles related? Where may they overlap? How can accounting help managers? This book addresses these questions. In this chapter we look at where the accountant fits in the organization, which gives us a framework for studying the succeeding chapters.

MODERN COST ACCOUNTING

Objective 1

Understand why taking a customer viewpoint assists in the design of an accounting system

Managers as Customers of Accounting

Modern cost accounting is often called *management accounting*. Why? Because cost accountants regard managers within their own organization as the primary users of their accounting information—that is, as their internal customers. Around the globe, managers are becoming increasingly aware of the importance of the quality and timeliness of products and services sold to their external customers. In turn, accountants are becoming increasingly sensitive to the quality and timeliness of accounting information required by managers.

The Value Chain of the Business Functions

Throughout this book we organize our look at organizations by using the value chain of the business functions, which appears as Exhibit 1-1. The **value chain** is the sequence of business functions in which utility (usefulness) is added to the products or services of an organization. These functions are:

---

[1]Robert Half International surveyed 200 executives on the "type of background most suitable to becoming a chief executive officer." Respondents replied sales/marketing 29% and accounting/finance 27%. See "Accountants Make Good CEOs," *Management Accounting* (June 1992), p. 19.

EXHIBIT 1-1
The Value Chain and Management Accounting

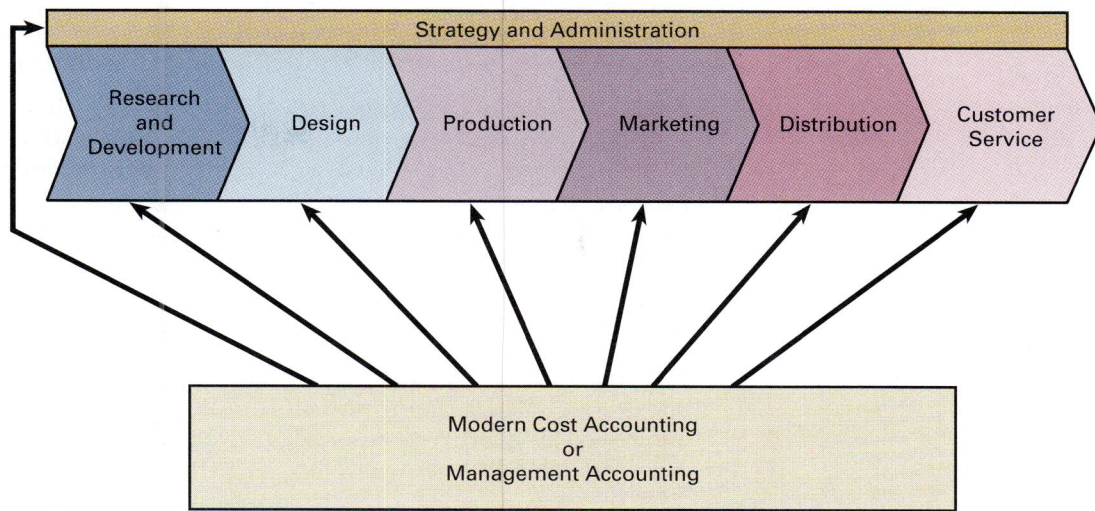

Objective 2
Describe the set of business functions in a value chain

♦ **Research and development**—the generation of, and experimentation with, ideas related to new products, services, or processes.

♦ **Design of products, services, or processes**—the detailed planning and engineering of products, services, or processes.

♦ **Production**—the coordination and assembly of resources to produce a product or deliver a service.

♦ **Marketing**—the process by which individuals or groups (a) learn about and value the attributes of products or services and (b) purchase those products or services.

♦ **Distribution**—the mechanism by which products or services are delivered to the customer.

♦ **Customer service**—the support activities provided to customers.

Exhibit 1-1 also shows a **strategy and administration** function, which spans across all the individual business functions. This category includes senior executives charged with the overall responsibility for the organization. General administrative tasks such as human resource management, legal matters, tax planning, and the like are often included in the strategy and administration function.[2]

Accounting is a major means of helping managers (a) to administer each of the business functions presented in Exhibit 1-1 and (b) to coordinate their activities within the framework of the organization as a whole. This book focuses on how accounting does in fact assist managers in those tasks.

Do not interpret Exhibit 1-1 as implying that managers should proceed sequentially through the value chain. There are important gains to organizations (in terms of, say, cost, quality, and the speed with which new products are developed) from having the various individual parts of the value chain work concurrently as a team.

How long managers enjoy success depends on their pleasing their external customers. Therefore, managers must know their customers' perceptions of company products or services. Similarly, management accountants must understand how managers in their organization—their customers—use accounting information. *The success of management accounting depends on whether managers' decisions are improved by the accounting information provided to them.*

---

[2] For convenience, we will not include the strategy and administration function in the visual presentation of a value chain in subsequent chapters of this book.

THE MAJOR PURPOSES OF MANAGEMENT ACCOUNTING SYSTEMS

Objective 3

Identify the four broad purposes of an accounting system

The accounting system is the principal, and the most credible, quantitative information system in almost every organization. This system should provide information for four broad purposes:

- *Purpose 1:* Internal routine reporting to managers for (a) cost planning and cost control of operations and (b) performance evaluation of people and activities.
- *Purpose 2:* Internal routine reporting to managers on the profitability of products, brand categories, customers, distribution channels, and so on. This information is used in making decisions on resource allocation and in some cases decisions on pricing.
- *Purpose 3:* Internal nonroutine reporting to managers for strategic and tactical decisions on matters such as formulating overall policies and long-range plans, new product development, investing in equipment, and special orders or special situations.
- *Purpose 4:* External reporting through financial statements to investors, government authorities, and other outside parties. To satisfy external purposes, businesses must report income and inventory costs in accordance with the generally accepted accounting principles that guide financial accounting.

Each broad purpose of accounting may require a different way of aggregating or reporting the data. An ideal data base (sometimes called a "data warehouse") will consist of finely granulated bits of information. In turn, accountants combine or adjust ("slice or dice") these data to answer the questions from particular internal or external users. Accountants cannot foresee each and every decision facing these users. Consequently, systems are often designed to fulfill the broadest set of uses that are anticipated among managers.

Management Accounting, Financial Accounting, and Cost Accounting

Management accounting, focusing on internal customers, measures and reports financial and other information that assists managers in fulfilling goals of the organization. It is concerned with the first three purposes of an accounting system listed above. As will be explained in many subsequent chapters, "different costs for different purposes" is a central idea in management accounting. **Financial accounting** focuses on external reporting. It is concerned with the fourth purpose of an accounting system listed above. **Cost accounting** is management accounting plus a part of financial accounting—to the extent that cost accounting provides information that helps the requirements of external reporting. The means by which cost accounting information is reported is called a **cost accounting system** or **costing system.** This relationship can be presented as in the diagram below.

Financial accounting, as mentioned, is constrained by generally accepted accounting principles. These principles restrict the set of revenue and cost measurement rules and the types of items that are classified as assets, liabilities, or owners' equity in balance sheets. In contrast, management accounting is not restricted to those accounting principles acceptable for financial reporting. For example, a consumer

Cost Accounting	
Management Accounting	Financial Accounting
• Purpose 1 • Purpose 2 • Purpose 3	• Purpose 4

products company may present a particular estimated "value" of a brand name (such as the Coca-Cola brand name) in its *internal* financial reports for marketing, although doing so is not in accordance with generally accepted accounting principles.

Financial accounting takes a historical perspective. The reports it generates focus on what has happened in the past. In contrast, management accounting emphasizes the future, providing budgets and other future projections in addition to historical reports.

Cost Management and Accounting Systems

The term *cost management* has become widely used in recent years. Unfortunately, no uniform definition exists. We use **cost management** to describe the actions by managers to satisfy customers while continuously reducing and controlling costs. An important component of cost management is the recognition that prior management decisions often commit the organization to the subsequent incurrence of costs. For example, decisions about plant layout and the extent of physical movement of materials required for production are often made before production begins. Yet it is not until production begins that the actual material-handling costs are incurred. Reductions in material-handling costs are achieved by thoughtful analysis when designing the plant layout as well as by efficient handling of materials on a day-to-day basis while production is occurring.

NEWLY EVOLVING MANAGEMENT THEMES

Objective 4
Describe five evolving management themes that are shaping developments in management accounting systems.

Management accounting exists to help managers make better decisions. Changes in the way managers operate require reevaluating the design and operation of the management accounting systems themselves. Exhibit 1-2 presents key themes in the newly evolving management approach. Subsequent chapters explain how management accounting dovetails with these themes. There are five themes in Exhibit 1-2.

1. Customer satisfaction is priority one. Customers are pivotal to the success of an organization. The number of organizations aiming to be "customer-driven" is large and increasing.

EXHIBIT 1-2
Key Themes in the Newly Evolving Management Approach

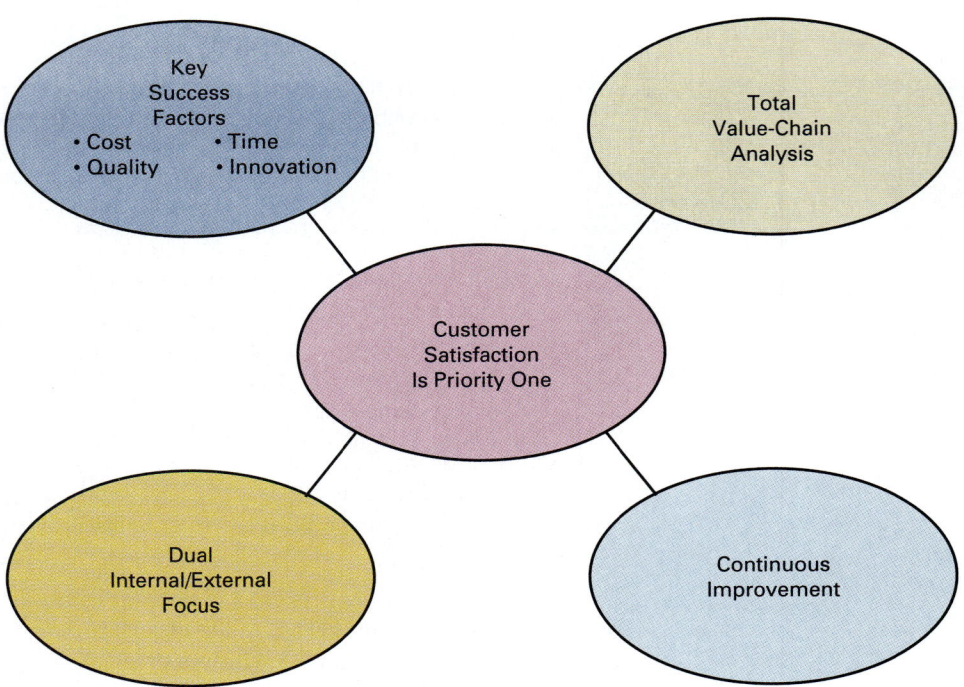

2. Key success factors. Customers are demanding ever-improving levels of performance regarding several (or even all) of the following factors:

◆ *Cost.* Organizations are under continuous pressure to reduce the cost of the products or services they sell to their customers.

◆ *Quality.* The **quality** of a product or service is its conformance with a preannounced or prespecified standard. Customers are expecting higher levels of quality and are less tolerant of low quality than in the past.

◆ *Time.* Time has many components, including the time taken to develop and bring new products to market, the speed at which an organization responds to customer requests, and the reliability with which promised delivery dates are met. Organizations are under pressure to complete activities faster and to meet promised due dates more reliably than in the past in order to increase customer satisfaction.

◆ *Innovation.* There is now heightened recognition that a continuing flow of innovative products or services is a prerequisite for the ongoing success of most organizations.

Factors that directly affect customer satisfaction, such as cost, quality, time, and innovative products and services, are termed **key success factors.**

3. Total value-chain analysis. This theme has two related aspects:

◆ Treating each of the business functions (research and development; design of products, services, or processes; production; marketing; distribution; customer service; and strategy and administration) as an essential and valued contributor and

◆ Integrating and coordinating the efforts of all business functions in addition to developing the capabilities of each individual business function.

The phrase *extended value chain* reinforces the notion that "upstream" parties such as suppliers and "downstream" parties such as customers are essential parts of total value-chain analysis. Indeed, the "customer satisfaction is priority one" theme highlights the importance of taking a broad view of all the parties included in the value chain. See Exhibit 1-3. For example, Pepsi-Cola bottlers work with their materials suppliers to reduce Pepsi's material-handling costs. Similarly, Fujitsu works

EXHIBIT 1-3

An Organization as an Extended Value Chain

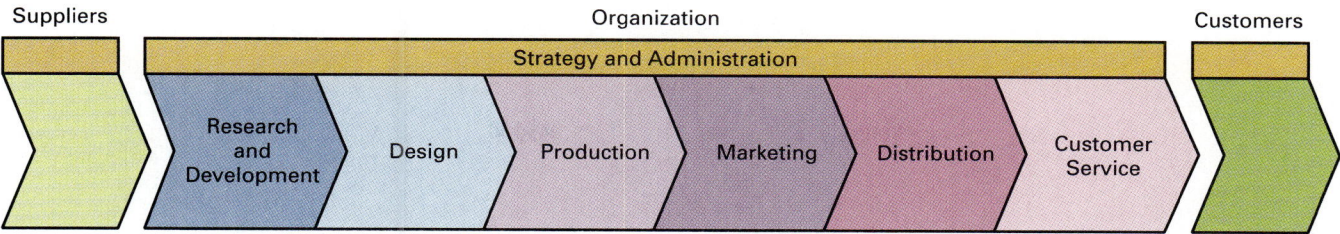

Suppliers	Organization							Customers
	Strategy and Administration							
	Research and Development	Design	Production	Marketing	Distribution	Customer Service		

with the customers of its microchip division to better plan its production scheduling of microchips. An **extended value-chain** analysis focuses on all the business functions related to a product or service from its cradle to grave (womb to tomb), irrespective of whether those functions occur in the same organization or in a set of legally independent organizations.[3]

4. Dual internal/external focus. Managers operate in both an internal and an external environment. The internal environment includes the physical, human, and informational aspects associated with each of the individual business functions and how those functions themselves are coordinated. The external environment includes customers, competitors, suppliers, and government bodies. Successful organizations need to be "nimble" in order to respond to changes in both their internal and external environments.

5. Continuous improvement. Continuous improvement by competitors creates a never-ending search for higher levels of performance within many organizations. Phrases such as the following capture this theme:

- "A journey with no end."
- "We are running harder just to stand still."
- "If you're not going forward, you're going backwards."

The Japanese term for "continuous improvement" is *kaizen*, and Toyota Motor Company uses the phrase *"kaizen* management" to describe its commitment to progress.

The high level of interest managers have in benchmarking also illustrates this theme. **Benchmarking** is the continuous process of measuring products, services, or activities against the best levels of performance that may be found either inside or outside the organization. Chapter 7 discusses benchmarking further.

ELEMENTS OF MANAGEMENT CONTROL

Objective 5
Understand how accounting can facilitate planning, control, and decision making

Is Wal-Mart's management control system better than Sears'? Is Nestlé's better than Cadbury's? This book develops an approach for addressing such questions. This section provides an overview of management control systems.

Planning and Control

There are countless definitions of planning and control. Study the left side of Exhibit 1-4, which uses planning and control at *The Daily Sporting News* (DSN) as an illustration. We define **planning** (the top box) as choosing goals, predicting results under various alternative ways of achieving those goals, and then deciding how to attain the

---

[3] J. Shank and V. Govindarajan, *Strategic Management Accounting* (New York: The Free Press, 1993) provide further discussion of extended value-chain analysis.

EXHIBIT 1-4

How Accounting Facilitates Planning and Control at *The Daily Sporting News*

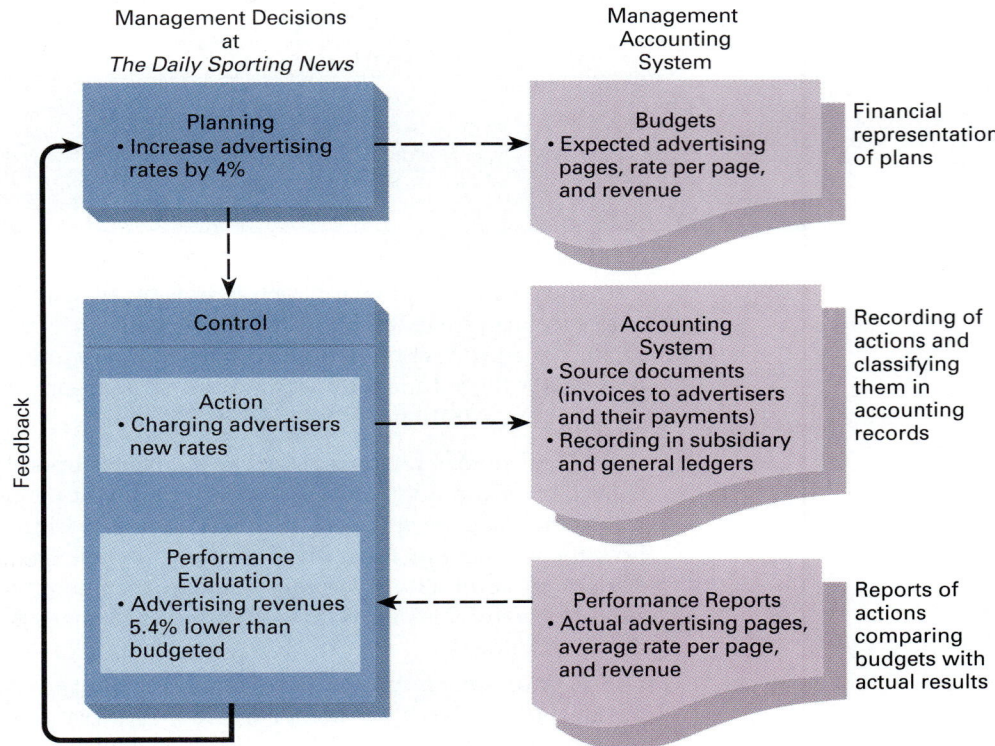

desired goals. For example, one goal of DSN may be to increase operating income. Three main alternatives are considered to achieve this goal:

1. Increase the price per newspaper,
2. Increase the rates per page charged to advertisers, or
3. Reduce labor costs by having fewer workers at DSN's printing facility.

Assume management opts for (2) and increases advertising rates by 4% to $5,200 per page for March 19_5. It budgets advertising revenue to be $4,160,000 ($5,200 × 800 pages predicted to be sold in March 19_5). A **budget** is the quantitative expression of a plan of action and an aid to the coordination and implementation of the plan.

Control (the bottom box in Exhibit 1-4) covers both action that implements the planning decision and performance evaluation of its personnel and its operations. With our DSN example, action would include communicating the new advertising rate schedule to DSN's marketing sales representatives and advertisers. Performance evaluation provides feedback on the actual results.

During March 19_5 DSN sells advertising, sends out invoices, and receives payments. These invoices and receipts are recorded in the accounting system. Exhibit 1-5 shows the March 19_5 advertising revenue performance report for DSN. This report indicates that 760 pages of advertising (40 pages less than the budgeted 800 pages) were sold in March 19_5. The average rate per page was $5,180 compared with the budgeted $5,200 rate, yielding actual advertising revenue in March 19_5 of $3,936,800. The actual advertising revenue in March 19_5 is $223,200 less than the budgeted $4,160,000. Understanding the reasons for any difference between actual results and budgeted amounts is an important part of **management by exception**, which is the practice of concentrating on areas that deserve attention and placing less attention on areas operating as expected. The term *variance* in Exhibit 1-5 refers to the difference between the actual results and the budgeted amount.

Concepts in Action

CUSTOMER-DRIVEN CHANGES IN MANAGEMENT ACCOUNTING SYSTEMS

A growing number of companies are explicitly adopting a customer-driven focus in their management accounting systems. A management accounting group at Johnson & Johnson (the manufacturer of many consumer products, such as Band-Aids) recently developed a vision statement that included the following points among its major goals:[a]

- Delight our customers
- Develop alternative measurement systems
- Keep it simple
- Utilize 20% of time on accounting and 80% on analysis
- Be the best

Examples of changes made at Johnson & Johnson stemming from the statement are:

- Using alternative measurement systems. Johnson & Johnson Pacific has reduced its monthly operations report from in excess of 70 pages to 4 pages. The operations reviews are a series of performance graphs, which are published weekly. A one-page cost-reduction graph, which summarizes the actual cost savings over targets, is published monthly.
- Keeping it simple. We are trying to get all our cost estimates to one-level cost sheets. In the 1993 Original Forecast, we were able to reduce the number of cost reports in one area by 25%.

[a]Adapted from N. Guest, "The Changing Focus of Finance at Johnson & Johnson Pacific," *Johnson & Johnson Financial Management News* (December 1992).

The performance report in Exhibit 1-5 could spur investigation. For example, did other newspapers experience a comparable decline in advertising? Did the marketing department make sufficient efforts to convince advertisers that, even with the new rate of $5,200 per page, advertising in the DSN was a good buy? Why was the actual average rate per page $5,180 instead of the budgeted rate of $5,200? Did some sales representatives offer discounted rates? Answers to these questions could prompt management at DSN to take subsequent actions, including, for example, pushing its marketing people to make renewed efforts to promote advertising to existing and potential advertisers.

A well-conceived plan includes enough flexibility so that managers can seize opportunities unforeseen at the time the plan is drawn up. In no case should control mean

EXHIBIT 1-5

Advertising Revenue Performance Report at *The Daily Sporting News* for March 19_5

	Actual Results	Budgeted Amounts	Variance
Advertising pages sold	760 pages	800 pages	40 pages unfavorable
Average rate per page	$5,180	$5,200	$20 unfavorable
Advertising revenue	$3,936,800	$4,160,000	$223,200 unfavorable

that managers cling to a preexisting plan when unfolding events indicate that actions not encompassed by the original plan would offer the best results to the company.

Planning and control are so strongly intertwined that managers do not spend time drawing artificially rigid distinctions between them. Unless otherwise stated, we use control in its broadest sense to denote the entire management process of both planning and control. For example, instead of referring to a management planning and control system, we will refer to a management control system. Similarly, we will often refer to the control purpose of accounting instead of the awkward planning and control purpose of accounting.

Do not underestimate the role of people in management control systems. Both accountants and managers should always remember that management control systems are not confined exclusively to technical matters such as the type of computer systems used and the frequency with which reports are prepared. Management control is primarily a human activity that should focus on how to help individuals do their jobs better.

Feedback: A Major Key

Exhibit 1-4 (p. 8) shows a feedback loop from control back to planning. Feedback can lead to a variety of responses, including:

Use of Feedback	Example
◆ Changing goals	◆ Based on evaluation of goals, General Electric abandons the home appliance business.
◆ Searching for alternative means of operating	◆ To save costs, the State of California renews drivers' licenses by mail rather than requiring an annual driving test.
◆ Changing methods for making decisions	◆ The University of Illinois bases its maintenance decisions on a comparison of expected repair costs over a five-year period rather than over a one-year period.
◆ Making predictions	◆ Lockheed incorporates average inflation forecasts for wages when predicting future labor costs.
◆ Changing the operating process	◆ Harley-Davidson has materials delivered directly to the assembly floor instead of to a storeroom.
◆ Changing the reward system	◆ Microsoft considers basing its marketing bonuses on profitability of sales rather than on the dollar amount of sales.

Feedback involves managers examining past performance and systematically exploring alternative ways to improve future performance.

COST-BENEFIT APPROACH

Improving Collective Decisions

This book takes a general approach to accounting referred to as the **cost-benefit approach**. That is, the primary criterion for choosing among alternative accounting systems is how well they help achieve organization goals in relation to the costs of those systems.

Accounting systems are economic goods. They cost money, just like bread and milk. The old adage says that if you build a better mousetrap, people will beat a path to your door. But many sellers of mousetraps will testify that when the buyers get to the door they say, "What is the price?" If the price is too high, buyers may be unwilling to pay, and the maker of better mousetraps may face financial ruin.

The costs of buying a new accounting system include user education programs as well as the purchase cost of the new system. The costs of educating users of a new system are frequently substantial, particularly when those users must invest much time in learning it.

As customers, managers buy a more elaborate management accounting system when its perceived expected benefits exceed its perceived expected costs. Although the benefits may take many forms, they can be summarized as the collective set of decisions that will better attain organization goals.

Consider the installation of a company's first budgeting system. Previously the company had probably been using some historical recordkeeping and little formal planning. A major benefit of installing the budgeting system is that it compels managers to plan ahead more formally. They may make a different, more profitable set of decisions than would have been generated by using only a historical system. Thus, the expected benefits exceed the expected costs of the new budgeting system.

Admittedly, the measurement of these costs and benefits is seldom easy. Therefore, you may want to call this approach a conceptual rather than a practical guide. Nevertheless, the cost-benefit approach provides a starting point for analyzing virtually all accounting issues.

The Decision's Dependence on Circumstances

A key question asked in applying the cost-benefit approach is, How much would we be willing to pay for one system versus another? For example, the same concert ticket at a given price may be a "good buy" for one person but a "bad buy" for another person under different circumstances. Similarly, a particular cost accounting system may be a good buy for General Motors but a bad buy for Honda. After all, General Motors and Honda have different plants, processes, and managers.

The choice of an accounting system inherently depends on specific circumstances. Therefore, this book will concentrate on describing alternative systems and on how to go about making the choices among them within the context of those circumstances. It will not present one accounting system as being innately superior to all others. Again, the human element is significant. The same system may work well in one organization but not in another. Why? Because the collective personalities and cultures differ between the two organizations. Systems do not exist in a vacuum. Managers and accountants are integral parts of accounting systems. Costs and benefits cannot be evaluated apart from the managers and accountants who will use the systems.

ORGANIZATIONAL STRUCTURE AND THE MANAGEMENT ACCOUNTANT

Objective 6
Describe the difference between line management and staff management

Exhibit 1-6 shows the most general form of an organizational structure with a corporate headquarters and three divisions. These divisions are most often structured in one of three ways:

1. By business function—for example, a steel company may have a manufacturing division, a marketing division, and a distribution division.
2. By geographic region—for example, an automobile company may have a North American division, an Asia/Pacific division, and a European division.
3. By product groups (or strategic business units)—for example, a media company may have a television division, a radio division, and a magazine division.

These groups are not mutually exclusive. For example, a company with divisions based on product groups may have each product group internally organized along business function lines.

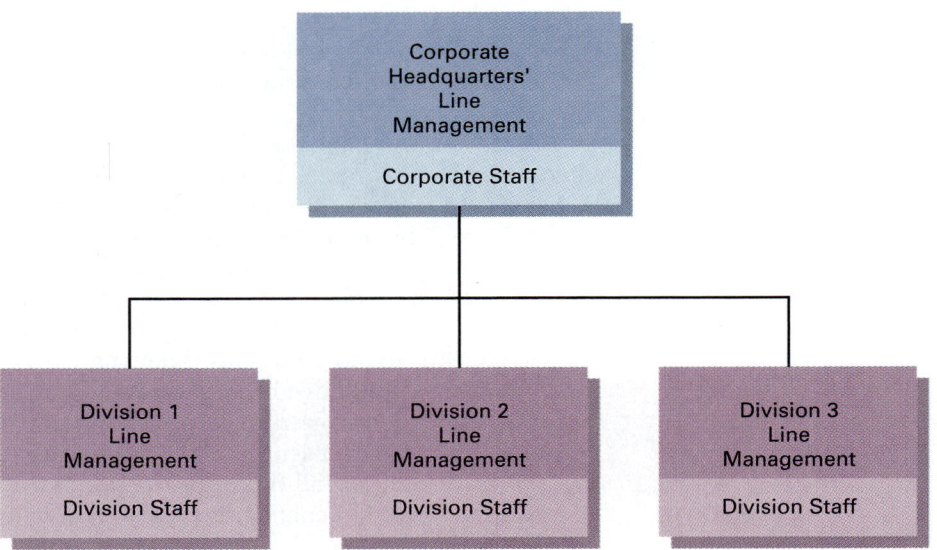

Line and Staff Relationships

Most organizations draw a distinction between line and staff relationships. **Line management** is directly responsible for attaining the objectives of the organization. For example, the manager of a manufacturing division may have as his or her objective an operating income figure as well as targets for product quality, safety, and compliance with environmental laws. **Staff management** exists to provide advice and assistance to line management. For example, a plant manager (a line function) may have responsibility for investing in new plant equipment. A plant financial analyst (a staff function) may prepare detailed operating cost comparisons of alternative pieces of equipment the plant manager is considering purchasing.

Increasingly, organizations are emphasizing the importance of teams in promoting their objectives. These teams can include both line and staff management with the result that the traditional distinctions between line and staff are less clear-cut than they were a decade ago. Line management and staff management designations are best viewed as different ends of a spectrum.

Chief Financial Officer and the Corporate Controller

The **chief financial officer (CFO)**—also called the **finance director**—is the senior officer empowered with oversight of the finance operations of an organization. The responsibilities of the CFO vary among organizations, but they almost always include the following four areas:

- ◆ Controllership/financial control
- ◆ Treasury
- ◆ Taxes
- ◆ Internal audit

In some organizations, the CFO also has responsibility for information systems. In other organizations, an officer of equivalent rank to the CFO—termed chief information officer—has responsibility for information systems.

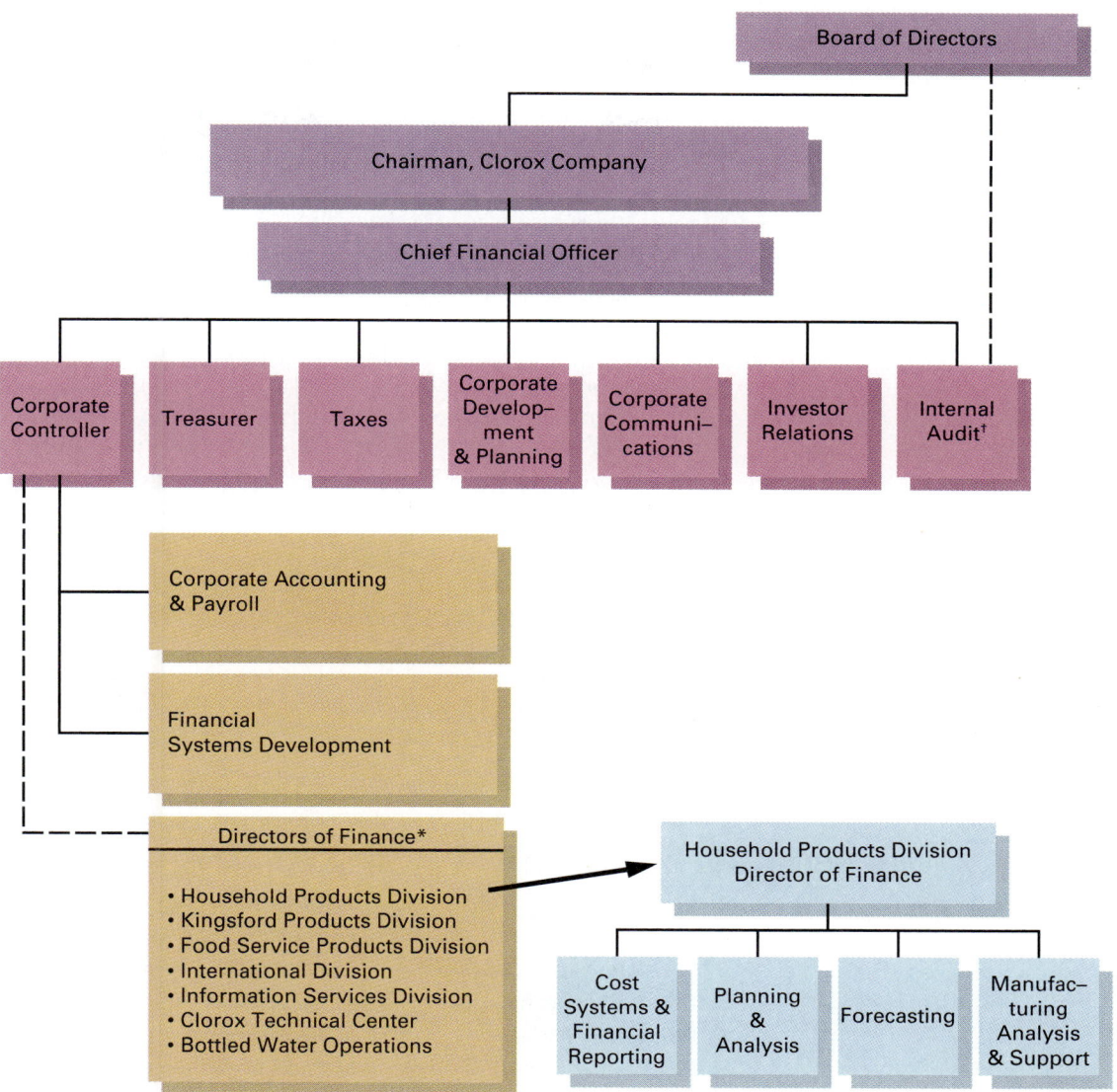

Key: —— = primary (hard line) reporting relationship; — — = secondary (dotted line) reporting relationship.
* The director of finance in each division reports hard line to the division president and dotted line to the corporate controller.
† Internal audit reports hard line to the chief financial officer and dotted line to the board of directors.

Exhibit 1-7 presents the organization charts of the CFO and the Corporate Controller at the Clorox Company. Clorox is a leading bleach-producing company and also has major brands in charcoal and salad dressing. The CFO is a staff management function that reports to the most senior line manager (who in turn reports to the board of directors). As in most organizations, the corporate controller at Clorox reports to the CFO. Organization charts like that in Exhibit 1-7 show formal reporting relationships. In most organizations, informal relationships also exist that are essential to understand when managers attempt to implement their decisions.

Some people confuse the responsibilities of the controller and the treasurer. Their functions include:

Controllership	Treasury
1. Planning and control	1. Provision of capital
2. Internal reporting	2. Short-term financing
3. Evaluation and consulting	3. Banking and custody
4. External reporting	4. Credits and collections
5. Protection of assets	5. Investments
6. Economic appraisal	6. Foreign exchange management

The **controller** is the financial executive primarily responsible for both management accounting and financial accounting. The **treasurer** is the financial executive who is primarily responsible for obtaining investment capital and managing cash.

The Controller: The Chief Management Accountant

In this book, we consider the controller as the chief management accounting executive. The modern controller does not do any controlling in terms of line authority except over his or her own department. Yet the modern concept of controllership maintains that the controller does control in a special sense. That is, by reporting and interpreting relevant data, the controller exerts a force or influence that impels management toward making better informed decisions.

The controller has been compared to a ship's navigator, with the president being the ship's captain. The navigator, with the help of specialized training, assists the captain. Without the navigator, the ship may founder on reefs or miss its destination entirely, but the captain exerts the right to command. The navigator guides and informs the captain as to how well the ship is being steered. This navigator role is especially evident in points 1, 2, and 3 of the controller's functions—planning and control, internal reporting, and evaluation and consulting.

Objective 7

Distinguish among the scorekeeping, attention-directing, and problem-solving functions of the controller

Scorekeeping, Attention-Directing, and Problem-Solving Functions

Management accountants perform three important functions—scorekeeping, attention directing, and problem solving. An example for each function follows.

- ◆ Scorekeeping—recording a sale to a customer
- ◆ Attention directing—computing the difference between an actual result and a budgeted amount for a performance report
- ◆ Problem solving—explaining how to avoid a cost overrun on a government contract

The scorekeeping function involves accumulating data and reporting results to all levels of management. Accountants serving this function are responsible for the integrity (reliability) of the reported information. In this regard, accountants are watchdogs for top management. The scorekeeping function in many organizations requires processing numerous information items (millions of items in some cases). The mechanics of the task should be well understood by those handling it and executed as flawlessly as possible. It is important, however, that the scorekeeping function not become the sole focus of the controller.

Many organizations now have management accountants whose responsibility centers on only the attention-directing or the problem-solving function. The titles of these individuals differ. For example, Clorox has special staff controller positions in its Household Products division for "planning and analysis," "forecasting," and "manufacturing analysis and support." Yoplait Company, the French yogurt company, has staff positions for "operations analysis," "budget analysis and reporting," and "marketing and sales analysis."

Many controllers' departments are actively promoting their attention-directing and problem-solving abilities to their internal customers in the organization. For example, Swissair's corporate controller's group was reorganized in 1988. Each of the 13 staff members in the group was assigned responsibility for assisting an individual Swissair division (such as engineering and maintenance, flight services, and marketing for Europe). Their challenge was to demonstrate to the managers of each division the value of their assistance in areas such as financial analysis, the budgeting process, and cost management.

Alternative Reporting Relationships

Management accountants typically are found in all the areas labeled "staff" in Exhibit 1-6 (p. 12). At the corporate headquarters' staff level, the controller reports directly to the chief financial officer. At the division level, two formal reporting relationships are commonly found:

1. The division controller reports to the division president in a primary ("hard-line") mode and to the corporate headquarters' controller in a secondary ("dotted-line") mode.
2. The division controller reports to the corporate headquarters' controller in a primary ("hard-line") mode and to the division president in a secondary ("dotted-line") mode.

The first reporting relationship emphasizes the helper role of the division controller; the division president will likely view the division controller as part of the division management team. Clorox adopts this first reporting relationship. The second reporting relationship emphasizes the management accountant's role in ensuring the integrity of reported information. A division president will more likely view the division controller as a watchdog for top management in the second reporting relationship.

The reporting relationship of the internal audit function to senior line managers deserves special mention. **Internal auditing** is the function responsible for reviewing and analyzing the financial and other records of an organization to attest to the integrity of its financial statements. In some organizations, internal auditing has a broader responsibility for promoting operating efficiency and effectiveness. At Clorox (see Exhibit 1-7) internal audit reports hard line to the CFO and dotted line to the board of directors. This is the most commonly encountered reporting relationship for internal audit.[4] In contrast, at Hewlett-Packard internal audit reports hard line to the president and chief executive officer and dotted line to the board of directors. This reporting relationship between internal audit and the president gives internal audit the responsibility to report directly to the most senior line manager any failure of Hewlett-Packard personnel to follow audit and other financial responsibility guidelines.

PROFESSIONAL ETHICS

Objective 8
Understand the importance of professional ethics to management accountants

The distinction between management accounting and financial accounting became institutionalized in the United States in 1972 when the **Institute of Management Accountants (IMA)**—formerly the National Association of Accountants, the largest association of internal accountants in the United States—established a program leading to the Certificate in Management Accounting (CMA). **Certified Management Accountant (CMA)** is the professional designation for management accountants and financial executives. It is the internal accountant's counterpart of Certified Public Accountant (CPA). Appendix D, at the end of the text, further discusses professional exams for the CMA and CPA as well as professional associations for management accountants in other countries.

---

[4]One survey reported that the chief internal auditor reports to the CFO for 48% of the respondents, to the chief executive officer 17%, to the controller 15%, the audit committee of the board 10%, and other 10%. J. Schiff, *New Directions in Internal Auditing* (New York: The Conference Board, 1990), p. 13.

EXHIBIT 1-8
Standards of Ethical Conduct for Management Accountants

Management accountants have an obligation to the organizations they serve, their profession, the public, and themselves to maintain the highest standards of ethical conduct. In recognition of this obligation, the Institute of Management Accountants has adopted the following standards of ethical conduct for management accountants. Adherence to these standards is integral to achieving the objectives of management accounting. Management accountants shall not commit acts contrary to these standards nor shall they condone the commission of such acts by others within their organizations.

Competence

Management accountants have a responsibility to:

- Maintain an appropriate level of professional competence by ongoing development of their knowledge and skills.
- Perform their professional duties in accordance with relevant laws, regulations, and technical standards.
- Prepare complete and clear reports and recommendations after appropriate analysis of relevant and reliable information.

Confidentiality

Management accountants have a responsibility to:

- Refrain from disclosing confidential information acquired in the course of their work except when authorized, unless legally obligated to do so.
- Inform subordinates as appropriate regarding the confidentiality of information acquired in the course of their work and monitor their activities to assure the maintenance of that confidentiality.
- Refrain from using or appearing to use confidential information acquired in the course of their work for unethical or illegal advantage either personally or through third parties.

Integrity

Management accountants have a responsibility to:

- Avoid actual or apparent conflicts of interest and advise all appropriate parties of any potential conflict.
- Refrain from engaging in any activity that would prejudice their ability to carry out their duties ethically.
- Refuse any gift, favor, or hospitality that would influence or would appear to influence their actions.
- Refrain from either actively or passively subverting the attainment of the organization's legitimate and ethical objectives.
- Recognize and communicate professional limitations or other constraints that would preclude responsible judgment or successful performance of an activity.
- Communicate unfavorable as well as favorable information and professional judgments or opinions.
- Refrain from engaging in or supporting any activity that would discredit the profession.

Objectivity

Management accountants have a responsibility to:

- Communicate information fairly and objectively.
- Disclose fully all relevant information that could reasonably be expected to influence an intended user's understanding of the reports, comments, and recommendations presented.

Source: Statement on Management Accounting, Standards of Ethical Conduct for Management Accountants (Montvale, NJ: Institute of Management Accountants, 1992—first issued in 1983).

Accountants consistently rank high in public opinion surveys on the ethics exhibited by members of different professions. *CMAs and CPAs are required to adhere to formal codes of ethical conduct. Exhibit 1-8 contains the IMA's "Standards of Ethical Conduct for Management Accountants." Professional accounting organizations have mechanisms for reviewing conduct not consistent with these standards. In addition, many companies and government agencies have their own codes of ethical conduct.* Key employees must sign a statement annually indicating that they have complied with the codes.

Line managers and management accountants share responsibility for both internal and external financial reports. The management accountant must ensure that the underlying accounting systems, procedures, and compilations are reliable and free of manipulation.

PROBLEM FOR SELF-STUDY

(Try to solve the following problem before examining the solution that follows.)

PROBLEM

Campbell Soup Company incurs the following costs:
a. Purchase of tomatoes by canning plant for Campbell's tomato soup products.
b. Materials purchased for redesigning Pepperidge Farm biscuit containers to make biscuits stay fresh longer.
c. Payment to Backer, Spielvogel Bates, the advertising agency for the Healthy Request line of soup products.
d. Salaries of food technologists researching feasibility of a Prego pizza sauce that has zero calories.
e. Cost of legal and investment banking advice for the takeover bid of Arnotts Limited of Australia.
f. Cost of a toll-free telephone line used for customer inquiries about possible product defects in Campbell's Soups.
g. Cost of gloves used by line operators on the Swanson Fiesta breakfast food production line.
h. Cost of hand-held computers used by Pepperidge Farm delivery staff serving major supermarket accounts.

Required
Classify each cost item (a–h) into one of the parts of the value chain shown in Exhibit 1-1 (p. 3).

SOLUTION

a. Production
b. Design of products, services, or processes
c. Marketing
d. Research and development
e. Strategy and administration
f. Customer service
g. Production
h. Distribution

SUMMARY

The following points are linked to the chapter's learning objectives.

1. In its fullest sense, management accounting is well named. It ties management with accounting. Managers are the customers of the management accountant. To maximize their value, accountants must focus on the challenges facing managers as much as on the technical aspects of accounting measurement.

Cost accounting is management accounting plus a part of financial accounting—to the extent that cost accounting satisfies the requirements of external reporting.

2. Managers in all areas of the value chain are customers of accounting information. The business functions in the value chain are research and development; design of products, services, or processes; production; marketing; distribution; customer service; and strategy and administration.

3. Accounting systems provide information for four broad purposes: (a) internal routine reporting for cost planning, cost control, and performance evaluation; (b) internal routine reporting on the profitability of products, brand categories, customers, distribution channels, and the like; (c) internal nonroutine reports for strategic and tactical decisions; and (d) external reporting.

4. Important management themes that are shaping developments in management accounting systems include: (a) the primacy of customer satisfaction, (b) linking planning and control to key success factors, (c) total value-chain analysis, (d) dual internal/external focus, and (e) continuous improvement.

5. Accounting helps facilitate planning, control, and decision making through budgets and other financial benchmarks, its systematic recording of actual results, its analysis of cost behavior, and its role in performance analysis.

6. Management accountants and controllers are staff management in most organizations. Staff managers exist to provide advice and assistance to line managers, who are directly responsible for attaining the objectives of the organization.

7. In most organizations, management accountants perform scorekeeping, attention-directing, and problem-solving functions. The first function emphasizes the importance of the integrity of information, while the other two functions emphasize the helper role of the accountant.

8. Management accountants have important ethical responsibilities that are related to competence, confidentiality, integrity, and objectivity.

TERMS TO LEARN

Each chapter will include this section. Like all technical subjects, accounting contains many terms with precise meanings. Pin down the definitions of new terms when you initially encounter them. The meaning of each of the following terms is explained in this chapter and also in the Glossary at the end of this book.

benchmarking *(p. 7)*   budget *(8)*   Certified Management Accountant (CMA) *(16)*
chief financial officer (CFO) *(12)*   control *(8)*   controller *(14)*   cost accounting *(4)*
cost accounting system *(4)*   cost-benefit approach *(10)*   costing system *(4)*
cost management *(5)*   customer service *(3)*   design of products, services, or processes *(3)*
distribution *(3)*   extended value chain *(7)*   finance director *(12)*
financial accounting *(4)*   Institute of Management Accountants (IMA) *(16)*
internal auditing *(16)*   key success factor *(6)*   line management *(12)*
management accounting *(4)*   management by exception *(8)*   marketing *(3)*
planning *(7)*   production *(3)*   quality *(6)*   research and development *(3)*
staff management *(12)*   strategy and administration *(3)*   treasurer *(14)*
value chain *(2)*

ASSIGNMENT MATERIAL

QUESTIONS

1-1   A leading management observer stated that "the most successful companies are those who have an obsession for their customers." Is this statement pertinent to management accountants? Explain.

1-2   Describe the business functions in the value chain.

1-3   The accounting system should provide information for four broad purposes. Describe them.

1-4   Distinguish between *management accounting* and *financial accounting*.

1-5   "Cost accounting is management accounting plus a part of financial accounting." Explain.

1-6   "Changes in the way managers operate require rethinking the design and operation of management accounting systems." Describe five themes that are affecting both the way managers operate and developments in management accounting.

1-7   Explain the meaning of *cost management*.

1-8   "Our management control system is a technical masterpiece. Our organization would be in much better shape if our line managers would recognize this fact and use the control system more." Comment on this statement made by a controller.

1-9   Feedback may be used for a variety of purposes. Identify at least five uses of feedback.

1-10   Describe three ways the divisions of an organization can be structured.

1-11   Name three areas for which a chief financial officer is frequently responsible.

1-12   As a new controller, reply to this comment by a plant manager: "As I see it, our accountants may be needed to keep records for stockholders and Uncle Sam—but I don't want them sticking their noses in my day-to-day operations. I do the best I know how. No pencil-pushing bean-counter knows enough about my responsibilities to be of any use to me."

1-13   Name three functions performed by the controller.

1-14   As used in accounting, what do IMA and CMA stand for?

1-15   Name the four areas in which standards of ethical conduct exist for management accountants in the United States. What organization sets forth these standards?

EXERCISES AND PROBLEMS

1-16 Financial and management accounting. David Colhane, an able electrical engineer, was informed that he was going to be promoted to assistant plant manager. David was elated but uneasy. In particular, his knowledge of accounting was sparse. He had taken one course in financial accounting but had not been exposed to the management accounting that his superiors found helpful.

Colhane planned to enroll in a management accounting course as soon as possible. Meanwhile, he asked Susan Hansley, an assistant controller, to state two or three of the principal distinctions between financial and management accounting using some concrete examples.

As the assistant controller, prepare a written response to Colhane.

1-17 Value chain and classification of costs, computer company. Apple Computer incurs the following costs:

a. Electricity costs for the plant assembling the Macintosh computer line of products.

b. Transportation costs for shipping Macintosh software to a retail chain.

c. Payment to David Kelley Designs for design of the Powerbook carrying case.

d. Salary of computer scientist working on the next generation of minicomputers.

e. Cost of Apple employee's visit to a major customer to illustrate Apple's ability to interconnect with other computers.

f. Salary of the president of Apple Computer.

g. Payment to television station for running Apple advertisements.

h. Cost of cables purchased from outside supplier to be used with the Macintosh printer.

Required
Classify each of the cost items (a–h) into one of the parts of the value chain shown in Exhibit 1-1 (p. 3)

1-18 Value chain and classification of costs, pharmaceutical company. Merck, a pharmaceutical company, incurs the following costs:

a. Cost of redesigning blister packs to make drug containers more tamper-proof.

b. Cost of videos sent to doctors to promote sales of a new drug.

c. Cost of a toll-free telephone line used for customer inquiries about usage, side effects of drugs, and so on.

d. Equipment purchased by a scientist to conduct experiments on drugs yet to be approved by the government.

e. Cost of fees paid to members of Merck's board of directors.

f. Labor costs of workers in the packaging area of a production facility.

g. Bonus paid to a salesperson for exceeding monthly sales quota.

h. Cost of Federal Express courier service to deliver drugs to hospitals.

Required
Classify each of the cost items (a–h) into one of the parts of the value chain shown in Exhibit 1-1 (p. 3)

1-19 Changes in management and changes in management accounting. A survey on ways organizations are changing their management accounting systems reported the following:

a. Company A now reports for each of the brands it sells a value-chain income statement.

b. Company B now presents in a single report all costs related to achieving high quality levels of its products.

c. Company C now presents in its performance reports estimates of the manufacturing costs of its two most important competitors in addition to its own internal manufacturing costs.

d. Company D reduces by 1% each month the budgeted labor assembly cost of a product when evaluating the performance of a plant manager.

e. Company E now reports profitability and satisfaction measures (as assessed by a third party) on a customer-by-customer basis.

Required
Link each of the above changes to one of the key themes in the newly evolving management approach outlined in Exhibit 1-2 (p. 6).

1-20 Uses of feedback. Six uses of feedback are described in the chapter (p. 10):

1. Changing goals
2. Searching for alternative means of operating
3. Changing methods for making decisions
4. Making predictions
5. Changing the operating process
6. Changing the reward system

Required

Match the appropriate numbers from the above list to each of the following items.

   a. The California State University system subcontracts its gardening operations to a private company instead of hiring its own gardeners.
   b. Sales commissions are to be based on total operating income instead of total revenue.
   c. Ford Motor Company adjusts its elaborate way of forecasting demand for its cars by including the effects of expected changes in the price of crude oil.
   d. The hiring of new sales personnel will include an additional step: an interview and evaluation by the company psychologist.
   e. Quality inspectors at General Motors are now being used in the middle of the production process as well as at the end of the process.
   f. Procter & Gamble enters the telecommunications industry.

1-21 Scorekeeping, attention directing, and problem solving. For each of the activities listed below, identify the major *function* (scorekeeping, attention directing, or problem solving) the accountant is performing.

   1. Preparing a monthly statement of Australian sales for the IBM marketing vice president.
   2. Interpreting differences between actual results and budgeted amounts on a performance report for the customer warranty department of General Electric.
   3. Preparing a schedule of depreciation for forklift trucks in the receiving department of a Hewlett-Packard factory in Scotland.
   4. Analyzing, for a Mitsubishi international manufacturing manager, the desirability of having some auto parts made in Korea.
   5. Interpreting why a Birmingham distribution center did not adhere to its delivery costs budget.
   6. Explaining a Xerox shipping department's performance report.
   7. Preparing, for the manager of production control of a U.S. Steel plant, a cost comparison of two computerized manufacturing control systems.
   8. Preparing a scrap report for the finishing department of a Toyota parts plant.
   9. Preparing the budget for the maintenance department of Mount Sinai Hospital.
   10. Analyzing, for a General Motors product designer, the impact on product costs of some new headlight lamps.

1-22 Scorekeeping, attention directing, and problem solving. For each of the following, identify the major function the accountant is performing—scorekeeping, attention directing, or problem solving.

   1. Interpreting differences between actual results and budgeted amounts on a shipping manager's performance report at a Daewoo distribution center.
   2. Preparing the budget for research and development costs at a DuPont division.
   3. Preparing adjusting journal entries for depreciation on the personnel manager's office equipment at Citibank.
   4. Preparing a customer's monthly statement for a Sears store.
   5. Processing the weekly payroll for the Harvard University maintenance department.
   6. Explaining the product-design manager's performance report at a Chrysler division.
   7. Analyzing the costs of several different ways to blend materials in the foundry of a General Electric plant.
   8. Tallying sales, by branches, for the sales vice president of Unilever.
   9. Analyzing, for the president of Microsoft, the impact of a contemplated new product on net income.
   10. Interpreting why an IBM sales district did not meet its sales quota.

1-23 Responsibility for analysis of performance. Karen Phillipson is the new corporate controller of a multinational company that has just overhauled its organizational structure. The company is now decentralized. Each division is under an operating vice president who, within wide limits, has responsibilities and authority to run the division like a separate company.

Phillipson has a number of bright staff members. One of them, Bob Garrett, is in charge of a newly created performance-analysis staff. Garrett and staff members prepare monthly division performance reports for the company president. These reports are division income statements, showing budgeted performance and actual results, and they are accompanied by detailed written explanations and appraisals of variances. In the past, each of Garrett's staff members was responsible for analyzing one division; each consulted with division line and staff executives and became generally acquainted with the division's operations.

After a few months, Bill Whisler, vice president in charge of Division C, stormed into the controller's office. The gist of his complaint follows:

"Your staff is trying to take over part of my responsibility. They come in, snoop around, ask hundreds of questions, and take up plenty of our time. It's up to me, not you and your detectives, to analyze and explain my division's performance to central headquarters. If you don't stop trying to grab my responsibility, I'll raise the whole issue with the president."

Required
1. What events or relationships may have led to Whisler's outburst?
2. As Phillipson, how would you answer Whisler's contentions?
3. What alternative actions can Phillipson take to improve future relationships?

1-24 Professional ethics and reporting divisional performance. Marcia Miller is division controller and Tom Maloney is division manager of the Ramses Shoe Company. Miller has line responsibility to Maloney, but she also has staff responsibility to the company controller.

Maloney is under severe pressures to achieve budgeted division income for the year. He has asked Miller to book $200,000 of sales on December 31. The customers' orders are firm, but the shoes are still in the production process. They will be shipped on or about January 4. Maloney said to Miller, "The key event is getting the sales order, not shipping of the shoes. You should support me, not obstruct my reaching division goals."

Required
1. Describe Miller's ethical responsibilities.
2. What should Miller do if Maloney gives her a direct order to book the sales?

1-25 Professional ethics, quality control (CMA). Fullrange Inc. produces complex printed circuits for stereo amplifiers. The circuits are sold primarily to major component manufacturers, and any production overruns are sold to small manufacturers. The small manufacturer market segment appears very profitable even though the sales to that segment are made at a substantial discount.

A common product defect that occurs in production is "component failure" caused by precise heat levels not being maintained during the production process. Every printed circuit undergoes testing. Rejects from the testing program can be reworked to acceptable levels if the defect is a component failure. However, in a recent analysis of customer complaints, George Wilson (the cost accountant) and Sarah Young (the quality control engineer) have ascertained that normal rework does not bring the circuits up to specification. Sampling shows that about one-half of the reworked circuits will fail after extended, high-volume amplifier operation. The incidence of failure in the reworked circuits is projected to be about 10% during one to five years' operation.

Unfortunately, there is no way to determine which reworked circuits will fail because testing will not detect this problem. The rework process could be changed to correct the problem, but the cost-benefit analysis for the suggested change in the rework process indicates that it is not feasible. Fullrange's marketing analyst has indicated that this problem will have a significant impact on the company's reputation and customer satisfaction if it is not corrected. Consequently, the board of directors would interpret this problem as having serious negative implications on the company's profitability.

Wilson has included the circuit failure and rework problem in his report that has been prepared for the upcoming quarterly meeting of the board of directors. Due to the potential adverse economic impact, Wilson has followed a long-standing practice of highlighting this information.

After reviewing the reports to be presented, the plant manager and his staff were upset and indicated to the controller that he should control his people better. "We can't upset the board with this kind of material. Tell Wilson to tone that down. Maybe we can get it by this meeting and have some time to work on it. People who buy those cheap systems and play them that loud shouldn't expect them to last forever."

The controller called Wilson into his office and said, "George, you'll have to bury this one. The probable failure of reworks can be referred to briefly in the oral presentation, but it should not be mentioned or highlighted in the advance material mailed to the board."

Wilson feels strongly that the board will be misinformed on a potentially serious loss of income if he follows the controller's orders. Wilson discusses the problem with the quality control engineer, who simply remarks, "That's your problem, George."

Required

1. Discuss the ethical considerations that George Wilson should recognize in deciding how to proceed in this matter.

2. Explain what ethical responsibilities should be accepted in this situation by the (a) controller, (b) quality control engineer, and (c) plant manager and staff.

3. What should George Wilson do in this situation? Explain your answer.

1-26 Professional ethics and "end-of-year games." Janet Taylor is the new division controller of the snack foods division of National Foods. National Foods has reported a minimum 15% growth in annual earnings for each of the past five years. The snack foods division has reported annual earnings growth of over 20% each year in this same period. During the current year, the economy went into a recession. The corporate controller estimates a 10% annual earnings growth rate for National Foods in this year. One month before the December 31, fiscal year end of the current year, Taylor estimates the snack foods division will report an annual earnings growth of only 8%. Warren Ryan, the snack foods division president, is less than happy, but he says with a wry smile, "Let the end-of-year games begin."

Taylor makes some inquiries and is able to compile the following list of end-of-year games that were more-or-less accepted by the prior division controller:

a. Deferring routine monthly maintenance in December on packaging equipment by an independent contractor until January of next year.

b. "Extending" the close of the current fiscal year beyond December 31 so that some sales of next year are included in the current year.

c. Altering dates of shipping documents of next January's sales to record them as sales in December of the current year.

d. Giving salespeople a double bonus to exceed December sales targets.

e. Deferring the current period's advertising by reducing the number of television spots run in December and running more than planned in January of next year.

f. Deferring the current period's reported advertising costs by having National Foods' outside advertising agency delay billing December advertisements until January of next year or having the agency alter invoices to conceal the December date.

g. Persuading carriers to accept merchandise for shipment in December of the current year although they normally would not have done so.

Required

1. Why might the snack foods division president want to "play the end-of-year games" described above?

2. The division controller is deeply troubled and reads the "Standards of Ethical Conduct for Management Accountants" in Exhibit 1-8 (p. 17). Classify each of the end-of-year games (a–g) as (i) acceptable or (ii) unacceptable according to that document.

3. What should Taylor do if Ryan suggests that end-of-year games are played in every division of National Foods and that she would greatly harm the snack foods division if she did not play along and paint the rosiest picture possible of the division's results?

2

An Introduction to

Cost Terms and Purposes

Costs in general

Direct costs and indirect costs

Cost drivers and cost management

Cost behavior patterns: variable costs and fixed costs

Total costs and unit costs

Financial statements and cost information

Costs as assets and expenses

Perpetual and periodic inventories

Manufacturing costs

Benefits of defining accounting terms

The many meanings of "product costs"

Classifications of costs

*T*RW *uses robotics at its Auburn, New York, plant to assemble transmitter units. Companies that make extensive use of robotics often reexamine the way they classify manufacturing labor costs in their costing systems.*

This chapter explains several widely recognized cost concepts and terms. They will help us demonstrate the multiple purposes of cost accounting systems, which we will stress throughout the book.

The cost concepts and terms we discuss are useful in many contexts, including decision making in all areas of the value chain. Illustrative decisions are:

◆ How much should we spend for research and development?

◆ What is the effect of product design changes on manufacturing costs?

◆ Should we replace some production assembly workers with a robotic machine?

◆ Should we spend more of the marketing budget on sales promotion coupons and less on advertising?

◆ Should we distribute from a central warehouse or via regionally dispersed warehouses?

◆ Should we provide a toll-free number for customers to make inquiries about the use of our products?

◆ Should we use our own legal personnel or hire an outside law firm?

The cost concepts and terms we discuss are also useful in making decisions on what products to emphasize, on pricing, in cost management, and in costing inventories for financial reporting purposes.

COSTS IN GENERAL

Cost Objects

Objective 1

Define and illustrate a *cost object*

Accountants usually define **cost** as a resource sacrificed or forgone to achieve a specific objective. For now, consider costs as being measured in the conventional accounting way, as monetary amounts (for example, dollars or pesos) that must be paid to acquire goods and services.

To guide decisions, managers want to know the cost of something. We call this something a **cost object** and define it as anything for which a separate measurement of costs is desired. Examples of cost objects include a product, a service, a project, a customer, a brand category, an activity, a department, and a program. Exhibit 2-1 presents an illustration for each of these cost objects. *Cost objects are chosen not for their own sake but to help decision making.*

EXHIBIT 2-1
Examples of Cost Objects

Cost Object	Illustration
◆ Product	A ten-speed bicycle
◆ Service	Airline flight from Los Angeles to London
◆ Project	Construction of road and rail underwater tunnel from Folkestown, United Kingdom, to Calais, France.
◆ Customer	All products purchased by Safeway (the customer) from General Foods
◆ Brand category	All soft drinks sold by a Pepsi-Cola bottling company with "Pepsi" in their name
◆ Activity	A test to determine the quality level of a television set
◆ Department	A department within a government environmental agency that studies air emissions standards
◆ Program	The athletic program of a university

Cost Accumulation and Cost Assignment

A costing system typically accounts for costs in two broad stages:

1. It *accumulates* costs by some "natural" (often self-descriptive) classification such as materials, labor, fuel, advertising, or shipping, and then
2. It *assigns* these costs to cost objects.

Cost accumulation is the collection of cost data in some organized way through an accounting system. **Cost assignment** is a general term that encompasses both (1) tracing accumulated costs to a cost object and (2) allocating accumulated costs to a cost object. As described below, *direct costs* are those *traced* to a cost object, and *indirect costs* are those *allocated* to a cost object. Nearly all systems accumulate **actual costs**, which are the costs incurred (historical costs), as distinguished from predicted or forecasted costs.

Exhibit 2-2 (p. 28) illustrates the accumulation and assignment of the manufacturing labor costs of assembly-department supervisors at Ford Motor Company. Their costs are initially *accumulated* in an assembly-department supervisory labor cost account. The costs are then *assigned* to one or more cost objects. Exhibit 2-2 shows three such cost objects—a department, a product, and a customer—and a typical decision associated with each cost object. Managers select the appropriate cost object on the basis of the decisions they face.[1]

DIRECT COSTS AND INDIRECT COSTS

Cost Tracing and Cost Allocation

Objective 2
Distinguish between *direct costs* and *indirect costs*

A major question concerning costs is whether they have a direct or an indirect relationship to a particular cost object.

◆ **Direct costs of a cost object:** costs that are related to the cost object and can be traced to it in an economically feasible way

---

[1]Exhibit 2-2 does *not* attempt to present the particular sequence in which assembly-department supervisory labor costs are assigned to cost objects. In some companies, the labor cost may be assigned to the assembly department first, then to products, and then to customers. In other companies, the accounting system may be designed so that a separate allocation is conducted for each cost object.

EXHIBIT 2-2
Decisions as the Guiding Rule for Choosing Cost Objects

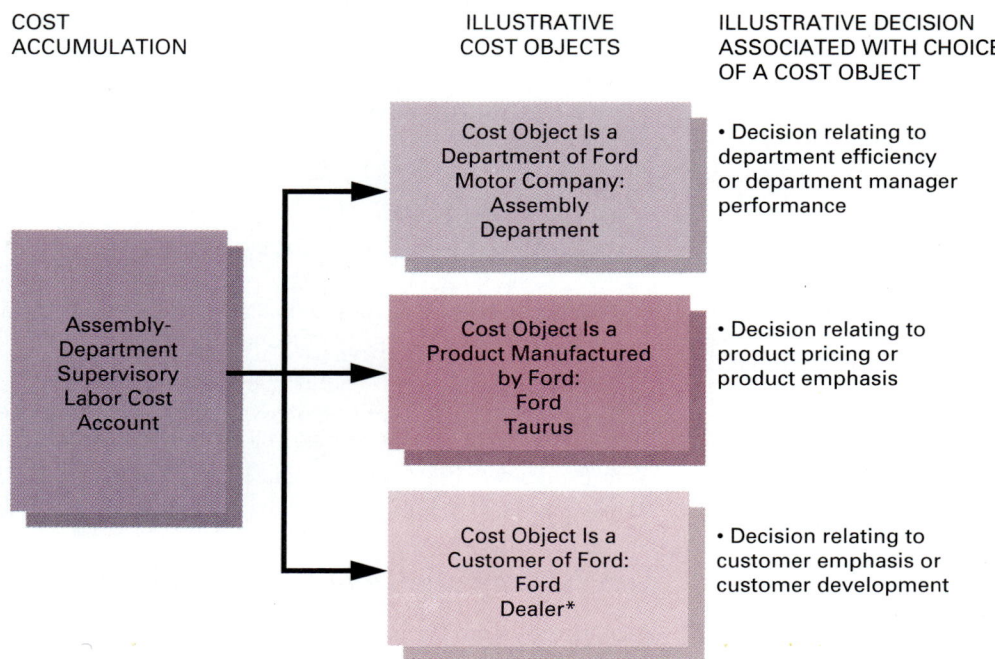

COST ACCUMULATION	ILLUSTRATIVE COST OBJECTS	ILLUSTRATIVE DECISION ASSOCIATED WITH CHOICE OF A COST OBJECT
Assembly-Department Supervisory Labor Cost Account	Cost Object Is a Department of Ford Motor Company: Assembly Department	• Decision relating to department efficiency or department manager performance
	Cost Object Is a Product Manufactured by Ford: Ford Taurus	• Decision relating to product pricing or product emphasis
	Cost Object Is a Customer of Ford: Ford Dealer*	• Decision relating to customer emphasis or customer development

* Ford Motor Company sells automobiles to Ford dealers, who in turn sell to many customers, including individuals, corporations such as Coca–Cola, and rental car companies such as Hertz.

◆ **Indirect costs of a cost object**: costs that are related to the cost object but cannot be traced to it in an economically feasible way. Indirect costs are allocated to the cost object using a cost allocation method.

"Economically feasible" means "cost-effective." The materiality of the cost item affects this cost effectiveness. Consider a mail-order catalog company. It probably would be economically feasible to track the courier charges for delivering a package directly to each customer. In contrast, the cost of the invoice paper included in the package sent to the customer is likely to be classified as an indirect cost because it is not economically feasible to trace the cost of this paper to each customer. Paper towels used in the restroom of a plant are another example of a cost item classified as an indirect cost of a product assembled at the plant.

The available information-gathering technology also affects the economic feasibility of tracing items as direct or indirect costs of cost objects. Consider the recent inclusion of bar codes on materials purchased by many job assembly plants. These bar codes now enable these plants to treat as direct costs of products certain materials previously classified as indirect costs (for example, screws and clips). Managers prefer to make decisions on the basis of direct costs rather than indirect costs. Why? Because they believe direct costs are more accurate than indirect costs.

In summary, the relationship between these terms is :

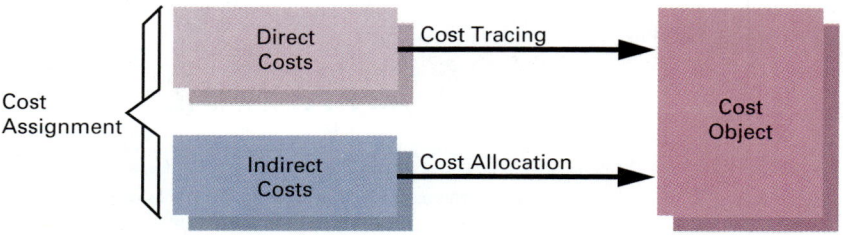

Cost tracing is the assigning of direct costs to the chosen cost object. **Cost allocation** is the assigning of indirect costs to the chosen cost object. *Cost assignment* encompasses both cost tracing and cost allocation.

Classification Depends on Cost Objects

This book examines different ways to assign costs to cost objects. For now, be aware that one particular cost may be both direct and indirect. How? *The direct/indirect classification depends on the choice of the cost object.* For example, the salary of the assembly-department supervisor in Exhibit 2-2 may be a direct cost of the assembly department at Ford but an indirect cost of a product such as the Ford Taurus.

Before proceeding, ponder how direct and indirect costs generally are related to various cost objects. The cost object may be a broadly defined operation (an entire airplane-maintenance facility) or a narrowly defined operation (the painting department within an airplane-maintenance facility). It may be a broadly defined project (designing a new automobile) or a narrowly defined project (designing a door handle for a new automobile). It may be a broadly defined group of products (women's mass-produced suit jackets) or a narrowly defined product (a woman's tailored cashmere suit jacket).

COST DRIVERS AND COST MANAGEMENT

Objective 3
Explain the relationships of *cost drivers*, *variable costs*, and *fixed costs*

Continuous cost reduction by competitors creates a never-ending search by organizations to reduce their own costs. Cost reduction efforts frequently focus on two key areas:

1. Doing only **value-added activities**, that is, those activities that customers perceive as adding utility (usefulness) to the products or services they purchase, and
2. Efficiently managing the use of the cost drivers in those value-added activities.

A **cost driver** is any factor that affects costs. That is, a change in the cost driver will cause a change in the total cost of a related cost object.

Exhibit 2-3 (p. 30) presents examples of cost drivers in each of the business functions of the value chain. Some cost drivers are financial measures found in accounting systems (such as direct manufacturing labor dollars and sales dollars), while others are nonfinancial variables (such as the number of parts per product and the number of service calls). This chapter now discusses the role of cost drivers in describing cost behavior. Chapter 10 further discusses cost drivers and their role in management accounting systems.

Cost management is the set of actions that managers take to satisfy customers while continuously reducing and controlling costs. A caveat on the role of cost drivers in cost management is appropriate. Changes in a particular cost driver do not automatically lead to changes in overall costs. Consider the number of items distributed as a driver of distribution labor costs. Suppose that management reduces the number of items distributed by 25%. This reduction does not automatically translate to a reduction in distribution labor costs. Managers must take steps to reduce distribution labor costs, perhaps by shifting workers out of distribution into other business functions needing additional labor or by laying off some distribution employees.

COST BEHAVIOR PATTERNS: VARIABLE COSTS AND FIXED COSTS

Management accounting systems record the cost of resources acquired and track their subsequent usage. Let us now consider two basic types of cost behavior patterns found in many systems—variable costs and fixed costs. A **variable cost** is a cost that

EXHIBIT 2-3

Examples of Cost Drivers for Business Function Areas in the Value Chain

Business Function	Examples of Cost Drivers
Research and Development	◆ Number of projects ◆ Personnel hours on a project ◆ Technical complexity of projects
Design of Products, Services, and Processes	◆ Number of products ◆ Number of parts per product ◆ Number of engineering hours
Production	◆ Number of units produced ◆ Number of setups ◆ Number of engineering change orders ◆ Direct manufacturing labor costs
Marketing	◆ Number of advertisements run ◆ Number of sales personnel ◆ Sales dollars
Distribution	◆ Number of items distributed ◆ Number of customers ◆ Weight of items distributed
Customer Service	◆ Number of service calls ◆ Number of products serviced ◆ Hours spent servicing products
Strategy and Administration	◆ Number of board of directors members ◆ Number of new government regulations ◆ Hours of legal work subcontracted

changes *in total* in proportion to changes of a cost driver. A **fixed cost** is a cost that does not change *in total* despite changes of a cost driver.

◆ *Variable Costs:* If General Motors buys one type of special clamp at $6 for each of its Saturn cars, then the total cost of clamps should be $6 times the number of cars produced. This is an example of a variable cost, a cost that changes *in total* in proportion to changes in the cost driver but is unchanged per unit of the cost driver. Exhibit 2-4 (Panel A) illustrates this variable cost. A second example of a variable cost is a sales commission plan where a salesperson receives 5% of each sales dollar as a commission. Exhibit 2-4 (Panel B) shows this variable cost example.

◆ *Fixed Costs:* General Motors may incur $20 million in a given year for the property taxes, executive salaries, rent, and insurance of its Saturn plant. These are examples of fixed costs, costs that are unchanged *in total* over a designated range of the cost driver during a given time span, but become progressively smaller on a *per unit* basis as the cost driver increases. For example, if General Motors assembles 10,000 Saturn vehicles at this plant in a year, the fixed cost per vehicle is $2,000 ($20 million ÷ 10,000). In contrast, if 50,000 vehicles are assembled, the fixed cost per vehicle becomes $400.

Major Assumptions

The definitions of variable costs and fixed costs have important underlying assumptions:

1. Costs are defined as variable or fixed with respect to a specific cost object. Products, services, projects, customers, brand categories, activities, departments, and programs are examples of cost objects.

EXHIBIT 2-4

Examples of Variable Costs

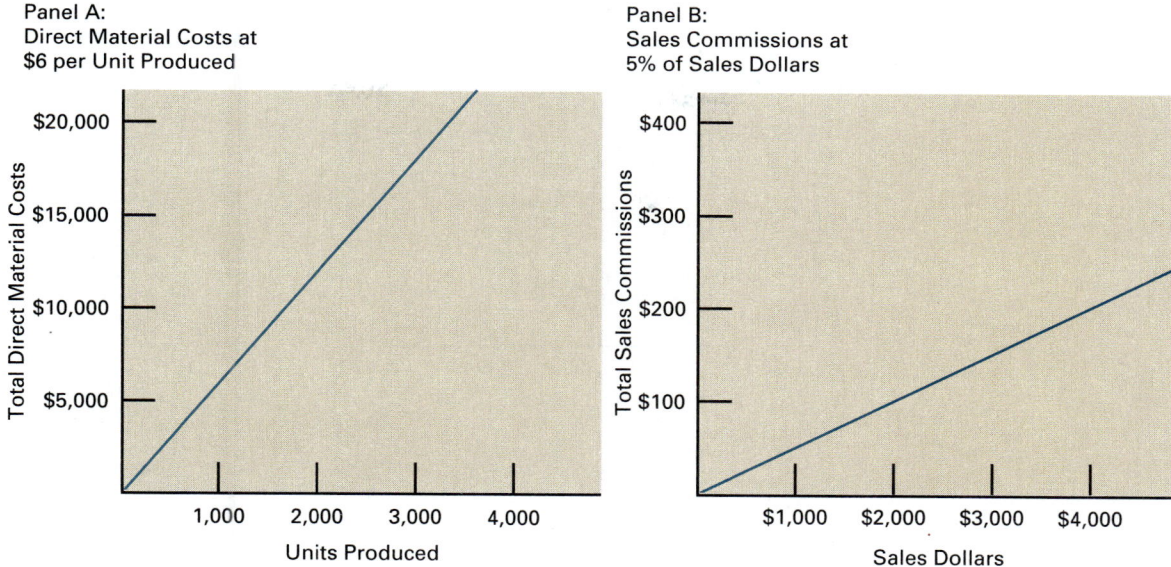

Panel A:
Direct Material Costs at
$6 per Unit Produced

Panel B:
Sales Commissions at
5% of Sales Dollars

2. **The time span must be specified.** Assume that General Electric pays a product-design company $100,000 for the one-year right to use its design of a refrigerator cooling unit in the production of General Electric refrigerators. At the end of the current year the contract will be revised. The $100,000 amount for product design is a fixed cost of the current year. However, if next year the contract is revised so that General Electric pays the product-design company $20 per refrigerator produced, this product design cost would become a variable cost at that time.

3. **Total costs are linear.** That is, when plotted on ordinary graph paper, a total variable cost or a total fixed cost relationship to the cost driver will appear as an unbroken straight line.

4. **There is only one cost driver.** The influences of other possible cost drivers on total costs are held constant or deemed to be insignificant.

5. **Variations in the level of the cost driver are within a relevant range** (which we discuss in the next section).

Variable costs and fixed costs are the two most frequently found cost behavior patterns recognized in existing management accounting systems. Additional cost behavior patterns are discussed in subsequent chapters (see Chapters 10 and 12). These additional patterns recognize contexts where the major assumptions underlying the distinction between variable costs and fixed costs do not adequately hold.

Relevant Range

A **relevant range** is the range of the cost driver in which a specific relationship between cost and the driver is valid. A fixed cost is fixed only in relation to a given relevant range (usually wide) of the cost driver and a given time span (usually a particular budget period). Consider Thomas Transport Company (TTC), which operates two refrigerated trucks that carry agricultural produce to market. Each truck has an annual fixed cost of $40,000 (including an annual insurance cost of $15,000 and an annual registration fee of $8,000) and a variable cost of $2 per mile of hauling. TTC has chosen miles of hauling to be the cost driver. The maximum annual usage of each truck is 120,000 miles. TTC has experienced an annual growth of between 5% and 10% in miles of hauling in each of the last three years. In the current year (19_6), the predicted combined total hauling of the two trucks is 170,000 miles.

Exhibit 2-5
Fixed Cost Behavior at Thomas Transport Company

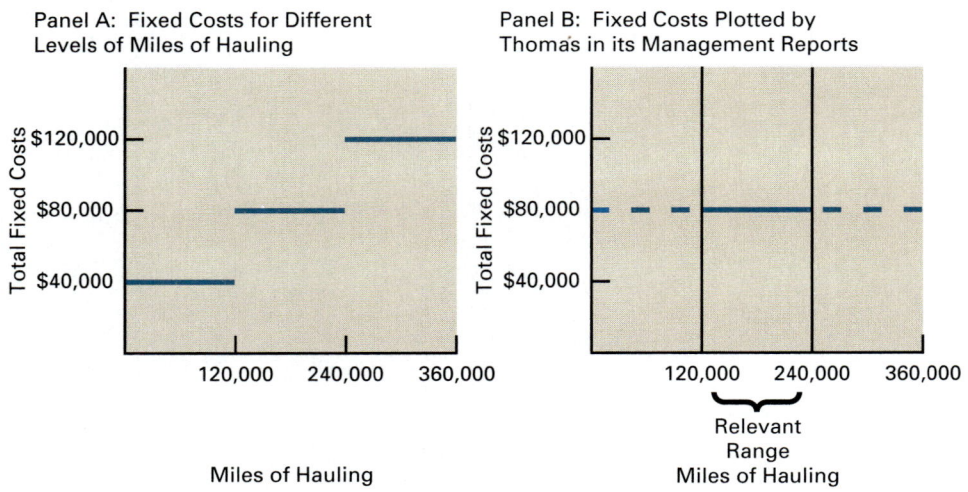

Panel A: Fixed Costs for Different Levels of Miles of Hauling

Panel B: Fixed Costs Plotted by Thomas in its Management Reports

Panel A of Exhibit 2-5 shows how annual fixed costs behave at different levels of miles of hauling. Up to 120,000 miles, TTC can operate with one truck; from 120,001 to 240,000 miles it can operate with two trucks; from 240,001 to 360,000 it can operate with three trucks. This pattern would continue as TTC added trucks to its fleet. Panel B of Exhibit 2-5 shows how TTC graphs fixed costs in its own management documents. The dotted lines signal that the $80,000 fixed-cost level does not apply for miles of hauling less than 120,001 or greater than 240,000. TTC expects to operate in the 120,001 to 240,000 miles range for at least each of the next two years.

Fixed costs may change from one year to the next. For example, if the annual registration fee for refrigerated trucks is increased in 19_7, the total level of fixed costs will increase (unless offset by a reduction in other fixed items).

Exhibit 2-6
Examples of Costs in Combinations of the Direct/Indirect and Variable/Fixed Cost Classifications

Cost Behavior Pattern	Assignment of Costs to Cost Object	
	Direct	**Indirect**
Variable	Cost object: Assembled automobile Example: Tires used in assembly of automobile	Cost object: Assembled automobile Example: Power costs where power usage is metered only to the plant
Fixed	Cost object: Marketing department Example: Annual leasing cost of cars used by sales representatives	Cost object: Marketing department Example: Monthly charge by corporate computer center for marketing's share of corporate computer costs

Relationships of Types of Costs

We have introduced two major classifications of costs: direct/indirect and variable/fixed. Costs may simultaneously be:

- ◆ direct and variable
- ◆ direct and fixed
- ◆ indirect and variable
- ◆ indirect and fixed

Exhibit 2-6 presents examples of costs in each of the four combinations of these two major cost classifications. Be sure to study this important exhibit carefully.

TOTAL COSTS AND UNIT COSTS

Meaning of Unit Costs

Accounting systems typically report both total cost and unit cost numbers. A **unit cost** (also called an **average cost**) is computed by dividing some total cost (the numerator) by some number of units (the denominator). Consider the unit cost of manufacturing a finished good. This unit cost is computed by accumulating manufacturing costs and then dividing the total (assumed to be $980,000) by the number of units manufactured (assumed to be 10,000 units):

$$\frac{\text{Total manufacturing costs}}{\text{Number of units manufactured}} = \frac{\$980,000}{10,000}$$
$$= \$98 \text{ per unit}$$

Suppose 8,000 units are sold and 2,000 units remain in ending inventory. The unit-cost concept helps in the assignment of total costs for financial reporting purposes:

Cost of goods sold, 8,000 units × $98	$784,000
Ending inventory of finished goods, 2,000 units × $98	196,000
Total manufacturing costs of 10,000 units	$980,000

Unit costs are found in all areas of the value chain—for example, unit cost of product design, unit cost of advertisements placed, and unit cost of customer service calls.

Use Unit Costs Cautiously

Objective 4
Understand why unit cost amounts must be interpreted with caution

Unit costs are averages. As we will see, they must be interpreted with caution. For decision making, it is best to think in terms of total costs rather than unit costs. Nevertheless, unit cost numbers are frequently used in many situations. For example, assume the president of a university social club is deciding whether to hire a musical group for an upcoming party. The group charges a fixed fee of $1,000. The president may find it intuitive to compute a unit cost for the group when thinking about an admission price. Given the fixed fee of $1,000, the unit cost is $10 if 100 attend, $2 if 500 attend, and $1 if 1,000 attend. Note, however, that with a fixed fee of $1,000 the *total cost* is unaffected by the attendance level (the size of the denominator), while the *unit cost* is a function of the size of the denominator.

Costs are often neither inherently fixed nor variable. Much depends on the specific context. Consider the $1,000 fixed amount that we assumed was to be paid to the musical group. This is but one way the musical group could be paid. Possible payment schedules that might be considered include:

- Schedule 1: $1,000 fixed fee
- Schedule 2: $1 per person attending + $500 fixed fee
- Schedule 3: $2 per person attending

Under schedules 2 and 3, the dollar amount of the payment to the musical group is not known until after the event.

The effects of these three payment schedules on unit costs and total costs for four different attendance levels are:

Attendance Level	Schedule 1: $1,000 fixed		Schedule 2: $1 per person + $500 fixed		Schedule 3: $2 per person	
	Total Cost	Unit Cost	Total Cost	Unit Cost	Total Cost	Unit Cost
100 people	$1,000	$10	$ 600	$6	$ 200	$2
250 people	1,000	4	750	3	500	2
500 people	1,000	2	1,000	2	1,000	2
1,000 people	1,000	1	1,500	1.50	2,000	2

The unit cost under schedule 1 is computed by dividing the fixed cost of $1,000 by the attendance level. For schedule 2, the unit cost is computed by first determining the total cost for each attendance level and then dividing that amount by that attendance level. Thus, for 250 people, schedule 2 has a total cost of $750 ($500 + 250 x $1), which gives a unit cost of $3 per person. Schedule 3 has a unit cost of $2 per person for any attendance level because the musical group is to be paid $2 per person with no fixed payment.

All three payment schedules would yield the same unit cost of $2 per person only if 500 people attend. The unit cost is not $2 per person under schedule 1 or schedule 2 for any attendance level except 500 people. Thus, it would be incorrect to use the $2 per person amount in schedule 1 or 2 to predict what the total costs would be for 1,000 people. Consider what occurs if 250 people attend and the group is paid a fixed fee of $1,000. The unit cost is then $4 per person. *While unit costs are often useful, they must be interpreted with extreme caution if they include fixed costs per unit.* When estimating total cost, think of variable costs as an amount per unit and fixed costs as a lump sum total amount. The key relationships between total costs and unit costs are summarized in Exhibit 2-7.

EXHIBIT 2-7
Behavior of Total Costs and Unit Costs when Changes Occur in the Level of the Cost Driver

	Total Costs	Unit Costs
Variable Costs	Change	Remain the same
Fixed Costs	Remain the same	Change

FINANCIAL STATEMENTS AND COST INFORMATION

Service-, Merchandising-, and Manufacturing-Sector Companies

Financial statements are an important output of an entity's accounting system. This section presents income statements from three different sectors of the economy—service, merchandising, and manufacturing.

Objective 5

Distinguish among service-sector, merchandising-sector, and manufacturing-sector companies

	Service	**Merchandising**	**Manufacturing**
Examples	Accounting firms Advertising agencies Consulting companies Law firms Television stations	Book stores Department stores Distributors Supermarkets Wholesalers	Automobile makers Computer makers Paper producers Steel mills Food-processing plants
Primary Output as Viewed by Customer	Intangibles (for example, advice and entertainment)	Tangible products in same basic form as purchased from suppliers	Tangible products converted from materials purchased from suppliers
Type of Inventory Held	None	Merchandise	Direct materials Work in process Finished goods

Objective 6

Describe the three categories of inventories commonly found in many manufacturing-sector companies

♦ **Service-sector companies** provide to their customers services or intangible products—for example, legal advice or an audit. These companies do not have any inventory of tangible product at the end of an accounting period.

♦ **Merchandising-sector companies** provide to their customers tangible products they have previously purchased in the same basic form from suppliers. Merchandise purchased from suppliers but not sold at the end of an accounting period is held as inventory. The merchandising sector includes companies engaged in retailing or wholesaling.

♦ **Manufacturing-sector companies** provide to their customers tangible products that have been converted to a different basic form from the materials purchased from suppliers. At the end of an accounting period, inventory of a manufacturer includes direct materials, work in process, and finished goods.

Merchandising companies have a single category for inventory (merchandise inventory). In contrast, manufacturing companies typically adopt a three-part categorization of inventories, each depicting a stage in the conversion of materials and other inputs to the finished product.

1. **Direct materials inventory.** Direct materials in stock and awaiting use in the manufacturing process
2. **Work in process inventory.** Goods partially worked on but not yet fully completed. Also called **work in progress** or **goods in process**
3. **Finished goods inventory.** Goods fully completed but not yet sold.[2]

As noted previously, service companies do not have inventory.

Exhibit 2-8 summarizes some key differences among service-, merchandising-, and manufacturing-sector companies. Several points of caution are appropriate. First, all companies have a service element when interacting with their customers—hence our inclusion of customer service as a function in the value chain is appropriate to any business. Second, many companies have operations in several sectors of the economy. For example, some supermarkets have elements of both retail (such as the resale

---

[2]The cost-benefit criterion introduced in Chapter 1 applies to the choice of the number of inventory categories. Some manufacturing-sector companies find that the three traditional categories of inventories do not pass a cost-benefit test. These companies use only two categories of inventory: direct materials and finished goods. No attempt is made to track work in process under this system, in part because the manufacturing time per unit is very short or the dollar amount of items in work in process is immaterial in amount. Chapter 19 discusses the backflush costing method, which can be used by companies not tracking work in process in their accounting records.

of packaged breakfast cereals) and manufacture (such as in-house bakeries) under the same roof.

Illustrative Income Statements

Companies in the *service sector* provide their customers with an intangible set of products. Consider Elliott and Partners Law Firm, specializing in personal-injury litigation. The customers (clients) of this law firm receive legal advice and representation on their behalf in court and in negotiations. The income statement of Elliott and Partners appears in Panel A of Exhibit 2-9. Labor costs typically are the most significant cost category in a service company. For Elliott and Partners, salaries and wages constitute 67.3% of total operating costs ($970,000 ÷ $1,442,000). The operating-costs line items for service companies will include costs from all areas of the value chain (design of services, marketing, and so on). There is no cost of goods sold line item in the income statement of Elliott and Partners; the business sells only services or intangible products to its customers.

Exhibit 2-9 (Panel B) presents the income statement of Prestige Bathrooms, a merchandiser of bathroom fixtures and furnishings (showers, sinks, hand towels, and the like). The costs of such a retail-sector company fall into one of two categories:

◆ Cost of goods sold
◆ Operating costs

A merchandiser's cost of goods sold consists of the cost of goods purchased for resale adjusted for changes in merchandise inventory:

$$\text{Beginning merchandise inventory} + \text{Purchases} - \text{Ending merchandise inventory} = \text{Cost of goods sold}$$

EXHIBIT 2-9
Illustrative Income Statements of Service-Sector and Merchandising-Sector Companies

PANEL A: Service Sector Elliott and Partners Law Firm Income Statement for the Year Ended December 31, 19_6			PANEL B: Merchandising Sector Prestige Bathrooms Income Statement for the Year Ended December 31, 19_6		
Revenues		$1,600,000	Sales		$1,500,000
Operating costs			Cost of goods sold		
Salaries and wages	$970,000		Beginning merchandise inventory,		
Rent	180,000		January 1, 19_6	$ 95,000	
Depreciation	105,000		Purchases	1,100,000	
Marketing	84,000		Cost of goods available for sale	1,195,000	
Other costs	103,000		Ending merchandise inventory,		
Total operating costs		1,442,000	December 31, 19_6	130,000	
Operating income*		$ 158,000	Cost of goods sold		1,065,000
			Gross margin (or gross profit)		435,000
			Operating costs†		315,000
			Operating income*		$ 120,000

*Note that operating income is determined before deducting income taxes. The relationship between operating income and net income is:

$$\text{Net income} = \text{Operating income} + \text{Nonoperating revenues} - \text{Nonoperating costs} - \text{Income taxes}$$

†Operating costs of a merchandiser will include (1) expensing of noninventoriable capitalized costs and (2) costs from all parts of the value chain, such as design of showroom, marketing, and administration, that are expensed immediately as incurred.

For Prestige Bathrooms in 19_6, the corresponding amounts in Exhibit 2-9 (Panel B) are:

$$\$95{,}000 + \$1{,}100{,}000 - \$130{,}000 = \$1{,}065{,}000$$

Operating costs include costs from all areas of the value chain. For Prestige Bathrooms, these "operating costs" include the costs associated with planning what inventory to carry, the design of the showroom, rent, sales personnel, and advertising.

The income statement of a manufacturer, Cellular Products, is in Exhibit 2-10 (Panel A) (p. 38). This company manufactures telephone systems to be sold to large organizations. Cost of goods sold in a manufacturing company is computed as follows:

$$
\begin{array}{cccc}
\text{Beginning} & \text{Cost of} & \text{Ending} & \text{Cost of} \\
\text{finished goods} + & \text{goods} & - \quad \text{finished goods} = & \text{goods} \\
\text{inventory} & \text{manufactured} & \text{inventory} & \text{sold}
\end{array}
$$

For Cellular Products in 19_6, the corresponding amounts in Exhibit 2-10 (Panel A) are (in thousands):

$$\$22{,}000 + \$104{,}000 - \$18{,}000 = \$108{,}000$$

Cost of goods manufactured refers to the cost of goods brought to completion, whether they were started before or during the current accounting period. In 19_6, these costs amount to $104,000 for Cellular Products (see the Schedule of Cost of Goods Manufactured in Panel B of Exhibit 2-10). A line item in Panel B is "Manufacturing costs incurred during 19_6" of $105,000. This item refers to the "new" direct manufacturing costs and the "new" manufacturing overhead costs that were added to all goods worked on during 19_6, regardless of whether all those goods were fully completed during 19_6.

COSTS AS ASSETS AND EXPENSES

Objective 7
Differentiate among *capitalized costs, inventoriable costs*, and *period costs*

Three cost terms used in many management accounting systems are **capitalized costs, inventoriable costs**, and **period costs**. These terms reflect the influence of financial accounting on management accounting systems.

- **Capitalized costs** are costs that are first recorded as an asset and then subsequently become an expense. Examples are costs incurred to purchase plant, equipment, and computers.

- **Inventoriable costs** are a specific type of capitalized cost. They are costs associated with the purchase of goods for resale (in the case of merchandise inventory) or costs associated with the acquisition and conversion of materials and all other manufacturing inputs into goods for sale (in the case of manufacturing inventories). Under generally accepted accounting principles (GAAP), inventoriable costs are restricted to manufacturing costs for a manufacturing-sector company.

- **Period costs** are costs that are reported as expenses of the period in question. They include costs initially recorded as capitalized costs and costs recorded immediately as expenses as they are incurred.

Exhibit 2-11 (p. 39) illustrates the two classes of capitalized costs and period costs. Service-sector companies hold no inventory and hence have no inventoriable costs. Both merchandising and manufacturing companies, in contrast, have inventoriable costs. Companies in all three sectors in Exhibit 2-11 have both other capitalized costs and period costs.

Exhibit 2-12 (p. 40) shows how inventoriable costs for merchandisers and manufacturers appear in the balance sheet and the income statement. A merchandising-sector company buys goods for resale without changing their basic form. The *only* inventoriable costs are related to the cost of goods purchased for resale. Unsold goods are held as inventory, and their cost is shown as an asset on the balance sheet. As the goods are sold, their costs become expenses in the form of cost of goods sold. A re-

EXHIBIT 2-10
Income Statement and Schedule of Cost of Goods Manufactured of
Manufacturing-Sector Company

PANEL A
Cellular Products
Income Statement for the Year Ended December 31, 19_6
(in thousands)

Sales		$210,000
Cost of goods sold:		
Beginning finished goods, January 1, 19_6	$ 22,000	
Cost of goods manufactured (see Panel B)	104,000	
Cost of goods available for sale	126,000	
Ending finished goods, December 31, 19_6	18,000	
Cost of goods sold		108,000
Gross margin (or gross profit)		102,000
Operating costs*		70,000
Operating income		$ 32,000

PANEL B
Cellular Products
Schedule of Cost of Goods Manufactured†
for the Year Ended December 31, 19_6
(in thousands)

Direct materials costs:		
Beginning inventory, January 1, 19_6	$11,000	
Purchases of direct materials	73,000	
Cost of direct materials available for use	84,000	
Ending inventory, December 31, 19_6	8,000	
Direct materials used		$ 76,000
Direct manufacturing labor costs		17,750
Manufacturing overhead costs		
Indirect manufacturing labor	4,000	
Supplies	1,000	
Heat, light, and power	1,750	
Depreciation—plant building	1,500	
Depreciation—plant equipment	2,500	
Miscellaneous	500	11,250
Manufacturing costs incurred during 19_6		105,000
Add beginning work in process inventory, January 1, 19_6		6,000
Total manufacturing costs to account for		111,000
Deduct ending work in process inventory, December 31, 19_6		7,000
Cost of goods manufactured (to Income Statement)		$104,000

*Operating costs of a manufacturer will include (1) expensing of noninventoriable capitalized costs and (2) costs from all parts of the value chain, such as R&D, product design, marketing, and administration, that are expensed immediately as incurred.

†Note that the term *cost of goods manufactured* refers to the cost of goods brought to completion (finished) during the year, whether they were started before or during the current year. Some of the manufacturing costs incurred are held back as costs of the ending work in process inventory; similarly, the costs of the beginning work in process inventory become part of the cost of goods manufactured for 19_6. Note too that this schedule can become a Schedule of Cost of Goods Manufactured and Sold simply by including the beginning and ending finished goods inventory figures in the supporting schedule rather than directly in the body of the income statement.

EXHIBIT 2-11
Capitalized, Inventoriable, and Period Costs for Financial Reporting under Generally Accepted Accounting Principles

	Service Sector	Merchandising Sector	Manufacturing Sector
1. Capitalized Costs Related to Inventory (Termed Inventoriable Costs)	Services cannot be stored, so there are no inventories and no inventoriable costs	Cost of merchandise purchased for resale (purchase costs, freight-in, sales taxes, and customs duties, for example)	Manufacturing costs associated with converting materials into goods for sale (direct materials, other direct manufacturing costs, and manufacturing overhead costs)
2. Other Capitalized Costs (Termed Noninventoriable Capitalized Costs)	Office equipment, computers, and the like	Fixtures and equipment, computers, and the like	Equipment, computers, and the like *not* used in manufacturing
3. Period Costs	Includes (a) expensing of capitalized costs (such as depreciation on computers) and (b) costs expensed immediately as incurred (design of new services, salaries, wages, marketing costs, and the like)	Includes (a) expensing of noninventoriable capitalized costs (such as depreciation on fixtures) and (b) costs expensed immediately as incurred (salaries, energy-related costs, marketing costs, and so on)	Includes (a) expensing of noninventoriable capitalized costs (such as depreciation on distribution vehicles) and (b) costs expensed immediately as incurred (marketing costs, distribution costs, service warranty costs, and the like)

tailer's other costs include (1) the period expensing of other capitalized costs and (2) the costs of items recorded as an expense immediately as incurred. In the income statement, both (1) and (2) are deducted from sales without ever having been regarded as part of inventoriable cost.

Panel B of Exhibit 2-12 shows how a manufacturing-sector company converts direct materials and other inputs into finished goods. The manufacturing costs of the finished goods include direct materials, other direct manufacturing costs, and manufacturing overhead. All these are inventoriable costs; they are assigned to work in process inventory or finished goods inventory until the goods are sold. Newcomers to cost accounting frequently assume that indirect costs such as rent, telephone, and depreciation are always period costs that are unconnected with inventories. However, if these costs are related to manufacturing per se, they are manufacturing overhead costs and are inventoriable. Other line items in the income statement in Panel B will be (1) the period expensing of capitalized costs not related to manufacturing (such as depreciation on a fleet of delivery vehicles) and (2) the cost of items recorded as an expense immediately as incurred.

EXHIBIT 2-12

Relationship of Inventoriable Costs, Other Capitalized Costs, and Period
Costs for a Merchandising-Sector Company and a Manufacturing-Sector Company

Panel A: Merchandising-Sector Company

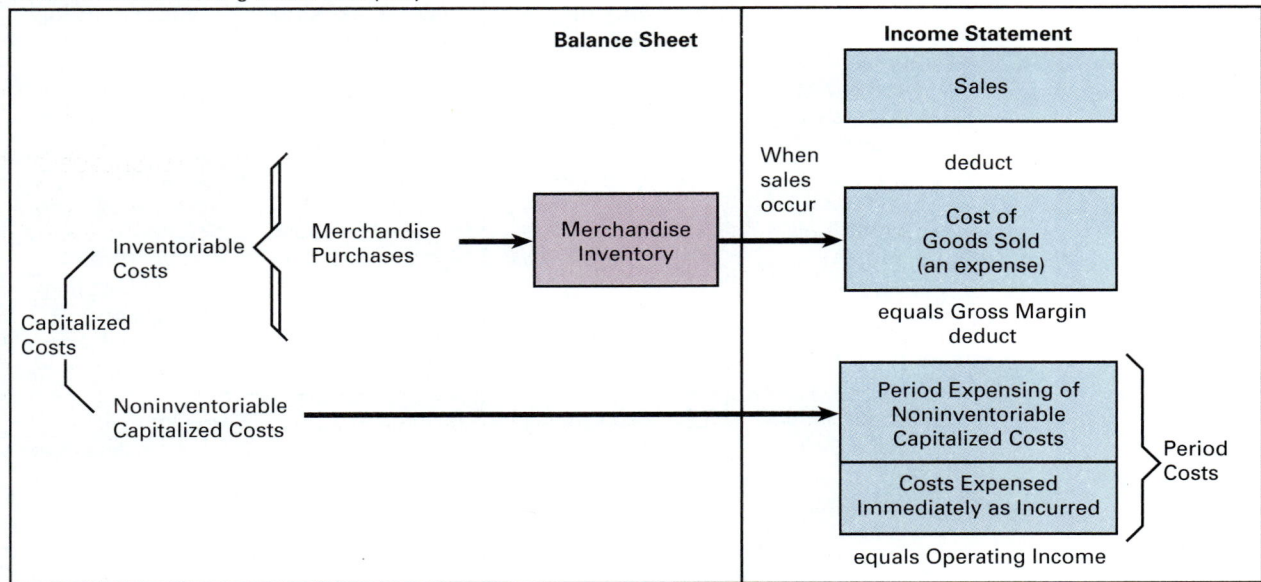

Panel B: Manufacturing-Sector Company

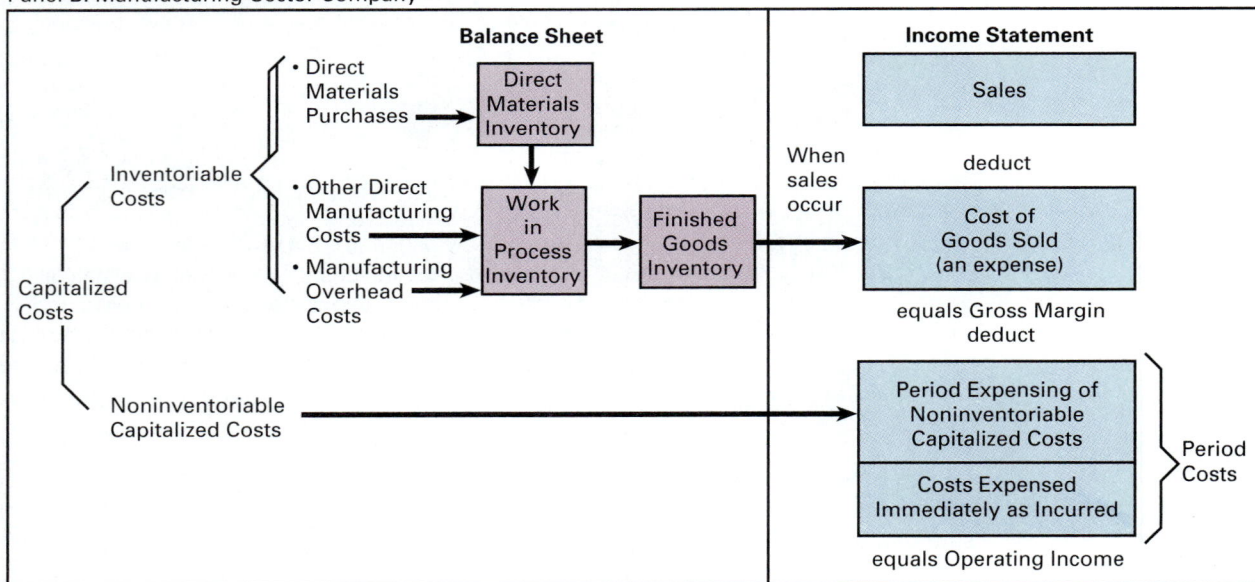

PERPETUAL AND PERIODIC INVENTORIES

There are two fundamental ways of accounting for inventories: perpetual and periodic. The **perpetual inventory method** requires a continuous ("real-time") record of
additions to and reductions from inventory. For a merchandiser, this method entails
a continuous record of additions to and reductions in merchandise for resale. For a
manufacturer, this method entails a continuous record of additions to and reduction
from direct materials, work in process, and finished goods. With this information a
cumulative cost of goods sold can be computed. Such a record helps management

EXHIBIT 2-13

Summary Comparision of Periodic and Perpetual Inventory Methods
for Cellular Products

Periodic Method		Perpetual Method	
Finished goods beginning inventory (by physical count)	$ 22,000	Cost of goods sold (kept on a continuous basis rather than determined periodically)*	$108,000
Add cost of goods manufactured	104,000		
Cost of goods available for sale	126,000		
Deduct finished goods ending inventory (by physical count)	18,000		
Cost of goods sold	$108,000		

*Such a condensed figure does not preclude the presentation of a supplementary schedule showing details of manufacturing costs similar to the one in Exhibit 2-10.

both control inventory and prepare interim financial statements. Physical inventory counts must be taken at least once a year to check on the validity of the clerical records.

Companies using the perpetual inventory method frequently have computer-based information-tracking systems. Consider a manufacturer of television sets. Key components might have a bar code that is machine-read into the computer as the components are used on the assembly line. Through tracking by bar codes, the manufacturer keeps a continuous record of inventory levels for each component. Companies in the merchandising sector (such as supermarkets) also use bar codes to continuously monitor inventory levels. Advances in information-gathering technology are now making it more cost-effective for companies to use perpetual inventory systems.

The **periodic inventory method** does not require a continuous record of inventory changes. Costs of direct materials used or costs of goods sold cannot be computed accurately until ending inventories, determined by physical count, are subtracted from the sum of the beginning inventory, purchases, and other purchasing costs. Costs are recorded by natural classifications, such as direct material purchases, freight-in, and purchase discounts. See Exhibit 2-13 for a comparison of perpetual and periodic inventory methods.

MANUFACTURING COSTS

The language of cost accounting includes specific terms describing manufacturing costs. Three terms with widespread use are direct materials costs, direct manufacturing labor costs, and manufacturing overhead costs.

1. **Direct materials costs.** The acquisition costs of all materials that eventually become part of the cost object (say, units finished or in process) and that can be traced to that cost object in an economically feasible way. Acquisition costs of direct materials include freight-in (inward delivery) charges, sales taxes, and custom duties.

2. **Direct manufacturing labor costs.** The compensation of all manufacturing labor that is considered to be part of the cost object (say, units finished or in process) and that may be traced to the cost object in an economically feasible way.

HARLEY-DAVIDSON ELIMINATES THE DIRECT MANUFACTURING LABOR COST CATEGORY[a]

Harley-Davidson's Motorcycle Division for many years used a three-part cost classification in its manufacturing facilities—direct materials, direct manufacturing labor, and manufacturing overhead. In the mid-1980s a task force of Harley-Davidson managers analyzed how its manufacturing product-cost structure compared with the administrative costs required to "collect, inspect, and report" data in its accounting system:

	Manufacturing Product-Cost Structure	Administrative Cost Effort
Direct materials	54%	25%
Manufacturing overhead	36	13
Direct manufacturing labor	10	62

The administrative costs associated with tracking direct manufacturing labor as a separate cost category included:

◆ operators' time to fill out labor tickets,
◆ supervisors' time to review labor tickets,
◆ timekeepers' time to enter the labor data and review the data output reports for errors, and
◆ cost accountants' time to review the direct labor and variance data.

Harley-Davidson concluded that tracing direct labor to products did not meet a cost-benefit test. Direct labor costs were only 10% of total manufacturing costs but required 62% of the administrative effort used to track all manufacturing costs. The company now includes all manufacturing labor costs as part of manufacturing overhead costs. It uses a two-part classification of direct materials and manufacturing overhead.

———————

[a]Adapted from W. Turk, "Management Accounting Revitalized: The Harley-Davidson Experience," *Journal of Cost Management* (Winter 1990).

Examples include wages and fringe benefits paid to machine operators and assembly-line workers.

3. Manufacturing overhead costs. All manufacturing costs that are considered to be part of the cost object (say, units finished or in process) but that cannot be traced to that cost object in an economically feasible way. Examples of manufacturing overhead include power, supplies, indirect materials, indirect manufacturing labor, plant rent, plant insurance, property taxes on plants, plant depreciation, and the compensation of plant managers. Other terms for this cost category include **indirect manufacturing costs, factory overhead costs,** and **factory burden costs.** We use *manufacturing overhead costs* and i*ndirect manufacturing costs* interchangeably in this book. *Manufacturing overhead costs* are part of *inventoriable costs* and become expenses only when the inventoriable costs are written off as cost of goods sold.

Some accounting systems use the term *manufacturing expenses* interchangeably with *manufacturing overhead costs.* We believe this usage to be confusing. The term *expense* incorrectly implies the item is a cost of the accounting period in which it is incurred.

Three-Part and Two-Part Cost Classifications

Manufacturing cost accounting systems vary among companies. Some use a three-part classification of the above three costs; others use a two-part classification:

Three-Part Classification	Two-Part Classification
◆ Direct materials costs	◆ Direct materials costs
◆ Direct manufacturing labor costs	◆ Manufacturing overhead costs
◆ Manufacturing overhead costs	

Accounting systems of organizations can change over time. For example, a company may change from the three-part classification to the two-part classification if direct labor costs become relatively immaterial in amount because of increased automation. Other alternatives are available. A company may change from the three-part classification to one with two direct cost categories and multiple individual manufacturing overhead cost categories. Chapter 5, which is devoted to a discussion of manufacturing cost accounting systems, provides illustrations of other classification approaches. Managers will choose the classification of costs that best helps them in their planning, control, and decision making.

Prime Costs and Conversion Costs

Two terms found in manufacturing costing systems are *prime costs* and *conversion costs.* **Prime costs** are all direct manufacturing costs. In the three-part classification noted above, prime costs would comprise direct materials costs and direct manufacturing labor costs. In the two-part classification noted above, prime costs would include only direct materials costs. As information-gathering technology improves, companies may add additional direct cost categories. For example, power costs might be metered in various specific areas of a plant that are dedicated totally to the assembly of separate products. In this case, prime costs would include direct materials, direct manufacturing labor, and direct metered power. **Conversion costs** are all manufacturing costs other than direct materials costs. These costs are for converting direct materials into finished goods. In the three-part classification of manufacturing costs, conversion costs would comprise direct manufacturing labor costs and manufacturing overhead costs. In the two-part classification, conversion costs would be only the manufacturing overhead costs.

A summary of the components of prime costs and conversion costs for the three-part and two-part classifications follows:

	Three-Part Classification	Two-Part Classification
Prime costs	Direct materials costs Direct manufacturing labor costs	Direct materials costs
Conversion costs	Direct manufacturing labor costs Manufacturing overhead costs	Manufacturing overhead costs

BENEFITS OF DEFINING ACCOUNTING TERMS

We cannot overemphasize the value of quickly obtaining a thorough understanding of the classifications and cost terms introduced in this chapter and later in this book. Managers, accountants, suppliers, and other people will avoid many misunderstandings if they use the same meaning of technical terms.

Think about the classification of payroll fringe benefit costs of direct manufacturing workers (for example, employer contributions to employees' social security, life insurance, health insurance, and pensions). Some companies classify these costs

Surveys of Company Practice

LABOR FRINGE BENEFITS AS A MAJOR COST ITEM

The labor force in almost all industrialized countries receives sizable fringe benefits—health care, pensions, worker's compensation, and other insurance—as well as wages. The following survey of labor cost per worker in the manufacturing sector of 12 countries illustrates the materiality of the costs of fringe benefits:[a]

	Hourly Wage	Fringe Benefits	Total Hourly Labor Cost	Fringe Benefits as % of Total Hourly Labor Cost
Western Germany	$13.09	$11.30	$24.39	46.3%
Switzerland	15.51	7.88	23.39	33.7
Sweden	12.72	9.58	22.30	43.0
Italy	9.52	9.99	19.51	51.2
Netherlands	10.66	8.69	19.35	44.9
Austria	9.49	9.24	18.73	49.3
Japan	13.60	4.25	17.85	23.8
Canada	12.07	4.35	16.42	26.5
France	8.49	7.61	16.10	47.3
United States	11.18	4.22	15.40	27.4
United Kingdom	9.59	4.12	13.71	30.1
Greece	4.08	2.63	6.71	39.2
Portugal	2.70	2.05	4.75	43.2

These cost figures highlight the significance of fringe benefits and explain why there is much interest now in tracing the costs of fringe benefits as direct costs rather than allocating them as indirect costs.

[a]Adapted from Salowsky, "Labor Costs." Full citation is in Appendix A.

as manufacturing overhead. In many other companies, however, these fringe benefits are charged as an additional direct manufacturing labor cost. Consider a direct laborer, such as a lathe operator or an assembly-line worker, who earns gross wages computed on the basis of a nominal, or stated, wage rate of $20 an hour. This person receives fringe benefits totaling, say, $8 per hour. Some companies classify the $20 as direct manufacturing labor cost and the $8 as manufacturing overhead. Other companies classify the entire $28 as direct manufacturing labor cost. The latter approach is conceptually preferable because these payroll fringe benefit costs are a fundamental part of acquiring manufacturing labor services.[3]

The problem here is to pinpoint what direct labor includes and excludes in a particular situation. Achieving clarity may obviate disputes regarding cost reimbursement contracts, income tax agreements, and labor union matters. For example, some countries offer substantial income tax savings to companies that locate manufacturing plants there. To qualify, the "direct manufacturing labor" costs of these companies in that country must meet a specified minimum percentage of the total manufacturing costs of their products. What incentive does such an income tax agree-

[3] The Institute of Management Accountants (formerly the National Association of Accountants) has issued a series of *Statements on Management Accounting* that discuss objectives, ethics, terminology, and definitions. They are available from the IMA, Montvale, NJ 07645-0433. For example, Statement 4C is "Definition and Measurement of Direct Labor Cost." This statement favors including as many related fringe benefits as feasible as a part of direct labor costs.

ment give managers to classify fringe benefits costs as direct manufacturing labor or manufacturing overhead? Classifying payroll fringe benefit costs as direct manufacturing labor will increase the percentage of direct manufacturing labor costs, thereby making it easier to qualify for the income tax savings. Consider a company with $8 million of payroll fringe benefit costs (figures are assumed, in millions):

Method A			Method B		
Direct materials	$ 40	40%	Direct materials	$ 40	40%
Direct manufacturing labor	20	20	Direct manufacturing labor	28	28
Manufacturing overhead	40	40	Manufacturing overhead	32	32
Total manufacturing costs	$100	100%	Total manufacturing costs	$100	100%

Method A classifies payroll fringe benefit costs as part of manufacturing overhead. In contrast, B classifies payroll fringe benefit costs as part of direct manufacturing labor. If a country set the minimum percentage of direct manufacturing labor costs at 25%, the company would receive a tax break using method B, but not using method A. In addition to fringe benefits, other items subject to different possible classifications include compensation for training time, idle time, vacations, sick leave, and extra compensation for overtime. To prevent disputes, contracts and laws should be as specific as feasible regarding definitions and measurements of accounting terms.

THE MANY MEANINGS OF "PRODUCT COSTS"

Objective 8
Explain how different ways of computing product costs are appropriate for different purposes

An important theme of this book is "different costs for different purposes". This theme can be illustrated with respect to product costing. A **product cost** is the sum of the costs assigned to a product for a specific purpose. Exhibit 2-14 illustrates three different purposes:

1. *Product pricing and product emphasis.* For this purpose, the costs of all those areas of the value chain required to bring a product to a customer should be included.
2. *Contracting with government agencies.* Government agencies frequently provide detailed guidelines on the allowable and nonallowable items in a product-cost amount. For example, some government agencies explicitly exclude "downstream" costs such as marketing costs from reimbursement to contractors and may reimburse only a subset of R&D costs.
3. *External financial reporting.* The focus here is on inventoriable costs for reporting on balance sheets and income statements. Under generally accepted accounting principles, only manufacturing costs are assigned to products for external reporting.

Exhibit 2-14
Different Product Costs for Different Purposes

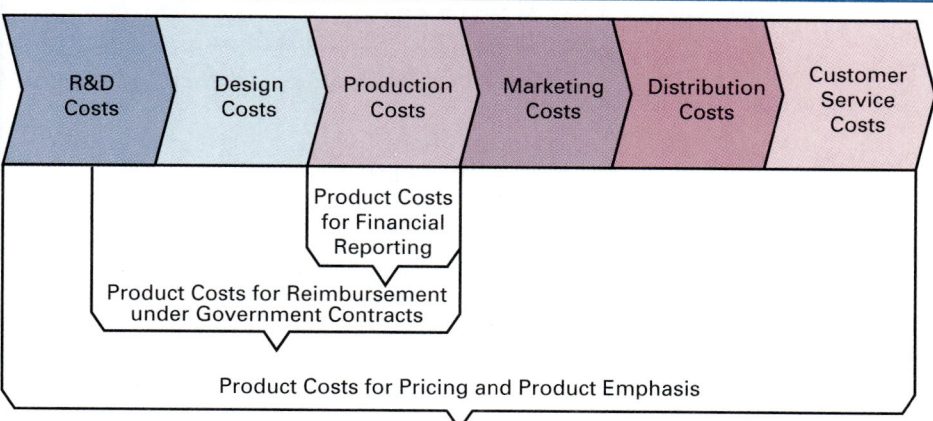

Exhibit 2-14 illustrates how a product cost amount may include inventoriable costs for external financial reporting, a broader set of costs for reimbursement under a contract, and a still broader set of costs for the pricing and product-emphasis purpose.

CLASSIFICATIONS OF COSTS

This chapter has merely hinted at the vast number of classifications of costs that have proved useful for various purposes. Classifications can be made on the basis of:
1. Business function
   a. Research and development
   b. Design of products, services, and processes
   c. Production
   d. Marketing
   e. Distribution
   f. Customer service
   g. Strategy and administration
2. Assignment to a cost object
   a. Direct costs
   b. Indirect costs
3. Behavior pattern in relation to changes of a cost driver
   a. Variable costs
   b. Fixed costs

4. Aggregate or average
    a. Total costs
    b. Unit costs
5. Assets or expenses
    a. Inventoriable costs
    b. Noninventoriable capitalized costs
    c. Period costs

PROBLEM FOR SELF-STUDY

(Try to solve this problem before examining the solution that follows.)

PROBLEM

Consider the following data of the Campbell Company for the year 19_4:

Sandpaper	$ 2,000
Materials-handling costs	70,000
Lubricants and coolants	5,000
Miscellaneous indirect manufacturing labor	40,000
Direct manufacturing labor	300,000
Direct materials, Jan. 1, 19_4	40,000
Finished goods, Jan. 1, 19_4	100,000
Finished goods, Dec. 31, 19_4	150,000
Work in process, Jan. 1, 19_4	10,000
Work in process, Dec. 31, 19_4	14,000
Plant-leasing costs	54,000
Depreciation—plant equipment	36,000
Property taxes on plant equipment	4,000
Fire insurance on plant equipment	3,000
Direct materials purchased	460,000
Direct materials, Dec. 31, 19_4	50,000
Sales	1,260,000
Sales commissions	60,000
Sales salaries	100,000
Shipping costs	70,000
Administration costs	100,000

Required

1. Prepare an income statement with a separate supporting schedule of cost of goods manufactured. For all items except sales, purchases of direct materials, and inventories, indicate by *V* or *F* whether each is basically a variable cost or a fixed cost (where the cost object is a product unit). If in doubt, decide on the basis of whether the total cost will change substantially over a wide range of production output.

2. Suppose that both the direct materials and plant-leasing costs are tied to the production of 900,000 units. What is the unit cost for the direct materials assigned to each unit produced? What is the unit cost of the plant-leasing costs? Assume that the plant-leasing costs are a fixed cost.

3. Repeat the computation in requirement 2 for direct materials and plant-leasing costs, assuming that the costs are being predicted for the manufacturing of 1,000,000 units next year. Assume that the implied cost behavior patterns persist.

4. As a management consultant, explain concisely to the president why the unit costs for direct materials did not change in requirements 2 and 3 but the unit costs for plant-leasing costs did change.

SOLUTION

1.

Campbell Company
Income Statement for the Year Ended December 31, 19_4

Sales		$1,260,000
Cost of goods sold:		
Beginning finished goods, January 1, 19_4	$ 100,000	
Cost of goods manufactured (see schedule below)	960,000	
Cost of goods available for sale	1,060,000	
Ending finished goods, December 31, 19_4	150,000	
Cost of goods sold		910,000
Gross margin (or gross profit)		350,000
Marketing, distribution, customer service,		
and administration costs:		
Sales commissions	60,000 (V)	
Sales salaries	100,000 (F)	
Shipping costs	70,000 (V)	
Administration costs	100,000 (F)*	
Total		330,000
Operating income		$ 20,000

*Probably a mixture of fixed and variable items but most of them would be fixed.

Campbell Company
Schedule of Cost of Goods Manufactured
for the Year Ended December 31, 19_4

Direct materials costs:		
Beginning inventory, January 1, 19_4		$ 40,000
Purchases of direct materials		460,000
Cost of direct materials available for use		500,000
Ending inventory, December 31, 19_4		50,000
Direct materials used		450,000 (V)
Direct manufacturing labor costs:		300,000 (V)
Indirect manufacturing costs:		
Sandpaper	$ 2,000 (V)	
Materials-handling costs	70,000 (V)	
Lubricants and coolants	5,000 (V)	
Miscellaneous indirect manufacturing labor	40,000 (V)	
Plant-leasing costs	54,000 (F)	
Depreciation—plant equipment	36,000 (F)	
Property taxes on plant equipment	4,000 (F)	
Fire insurance on plant equipment	3,000 (F)	214,000
Manufacturing costs incurred during 19_4		964,000
Add beginning work in process inventory,		
January 1, 19_4		10,000
Total manufacturing costs to account for		974,000
Deduct ending work in process inventory,		
December 31, 19_4		14,000
Cost of goods manufactured (to Income Statement)		$960,000

2. Direct materials unit cost = Direct materials costs ÷ Units produced
                     = $450,000 ÷ 900,000 = $0.50
   Plant-leasing unit cost   = Plant-leasing costs ÷ Units produced
                     = $54,000 ÷ 900,000 = $0.06

3. The direct materials costs are variable, so they would increase in total from $450,000 to $500,000 (1,000,000 × $0.50). However, their unit costs would be unaffected: $500,000 ÷ 1,000,000 units = $0.50.

In contrast, the plant-leasing costs of $54,000 are fixed, so they would not increase in total. However, if the plant-leasing costs were assigned to units produced, the unit costs would decline from $0.060 to $0.054: $54,000 ÷ 1,000,000 = $0.054.

4. The explanation would begin with the answer to requirement 3. As a consultant, you should stress that the unitizing (averaging) of costs that have different behavior patterns can be misleading. A common error is to assume that a total unit cost, which is often a sum of variable unit costs and fixed unit costs, is an indicator that *total* costs change in a wholly variable way as production output levels change. The next chapter demonstrates the necessity for distinguishing between cost behavior patterns. You must be wary especially about unit fixed costs. Too often, unit fixed costs are erroneously regarded as being indistinguishable from unit variable costs.

SUMMARY

The following points are linked to the chapter's learning objectives.

1. A cost object is anything for which a separate measurement of costs is desired. Examples include a product, service, project, customer, brand category, activity, department, and program.

2. A direct cost of a cost object is any cost that is related to the cost object and can be traced to that cost object in an economically feasible way. Indirect costs are costs that are related to the cost object but cannot be traced to that cost object in an economically feasible way. A cost may be direct regarding one cost object and indirect regarding other cost objects. This book uses the term *cost tracing* to describe the assignment of direct costs to a cost object and the term *cost allocation* to describe the assignment of indirect costs to a cost object.

3. A cost driver is any factor that affects costs. Examples include the number of setups and direct-labor dollars in manufacturing and the number of sales personnel and sales dollars in marketing. A variable cost is a cost that does change in total in proportion to changes in a cost driver. A fixed cost is a cost that does not change in total despite changes in a cost driver.

4. Unit costs of a cost object should be interpreted with caution when they include a fixed-cost component. When making total cost estimates, think of variable costs as an amount per unit and fixed costs as a total amount.

5. Service-sector companies provide services or intangible products to their customers. In contrast, merchandising- and manufacturing-sector companies provide tangible products to their customers. Merchandising companies do not change the form of the products they acquire and sell. Manufacturing companies convert materials and other inputs into finished goods for sale. These differences are reflected in both the balance sheets and income statements of companies in these sectors.

6. Manufacturing companies typically adopt a three-part classification of inventories that depicts the stage in the conversion process—direct materials, work in process, and finished goods.

7. Capitalized costs are those costs first recorded as an asset and then subsequently becoming an expense. Inventoriable costs are a specific type of capitalized cost. They are costs that are associated with the purchase of goods for resale (in the case of merchandise inventory) or costs associated with the conversion of materials and other inputs into goods for sale (in the case of manufacturing inventories). Period costs are costs that are reported as an expense of the period in question; they include costs initially recorded as capitalized costs and costs recorded immediately as an expense as they are incurred.

8. Managers may assign different costs to the same cost object depending on their purpose. For example, for financial reporting purposes, the (inventoriable) costs of a product include only manufacturing costs. In contrast, costs from all areas of the value chain can be assigned to a product for pricing and product-emphasis decisions.

TERMS TO LEARN

This chapter contains more basic terms than any other in this book. Do not proceed before you check your understanding of the following terms. Both the chapter and the Glossary at the end of the book contain definitions.

actual costs *(p. 27)*   average cost *(33)*   capitalized costs *(37)*   conversion costs *(43)*
cost *(26)*   cost accumulation *(27)*   cost allocation *(29)*   cost assignment *(27)*
cost driver *(29)*   cost object *(26)*   cost tracing *(29)*   direct costs of a cost object *(27)*
direct manufacturing labor costs *(41)*   direct materials costs *(41)*
direct materials inventory *(35)*   factory burden costs *(42)*   factory overhead costs *(42)*
finished goods inventory *(35)*   fixed cost *(30)*   goods in process *(35)*
indirect costs of a cost object *(28)*   indirect manufacturing costs *(42)*
inventoriable costs *(37)*   manufacturing overhead costs *(42)*
manufacturing-sector company *(35)*   merchandising-sector company *(35)*
period costs *(37)*   periodic inventory method *(41)*   perpetual inventory method *(40)*
prime costs *(43)*   product cost *(45)*   relevant range *(31)*
service-sector company *(35)*   unit cost *(33)*   value-added activities *(29)*
variable cost *(29)*   work in process inventory *(35)*   work in progress *(35)*

ASSIGNMENT MATERIAL

QUESTIONS

2-1   Define *cost object* and give three examples.

2-2   Define *cost assignment*, *cost tracing*, and *cost allocation*. How are these terms related?

2-3   Which costs are considered direct? Indirect? Give an example of each.

2-4   Describe how a given cost item can be both a direct cost and an indirect cost.

2-5   What is a *cost driver?* Give one example for each area in the value chain.

2-6   Define *variable cost, fixed cost*, and *relevant range*.

2-7   What are the five assumptions underlying the *variable cost* and *fixed cost* definitions used in this chapter?

2-8   "Total costs on an assembly line can be predicted by multiplying the unit cost per assembly by the units to be assembled." Do you agree? Explain.

2-9   Describe how service-, merchandising-, and manufacturing-sector companies differ from each other.

2-10   What are the three major categories of the inventoriable cost of a manufactured product?

2-11   Distinguish between the *perpetual inventory method* and the *periodic inventory method*.

2-12   Define the following: *direct materials costs, direct manufacturing labor costs, manufacturing overhead costs, prime costs*, and *conversion costs*.

2-13   Give three terms that may be substituted for the term *manufacturing overhead costs*.

2-14   Why is the term *manufacturing expenses* a misnomer for *manufacturing overhead costs*?

2-15   Define *product costs*. Describe three different purposes for computing product costs.

EXERCISES AND PROBLEMS

2-16   Total costs and unit costs.   A student association has hired a musical group for a graduation party. The cost will be a fixed amount of $4,000.

Required

1. Suppose 500 people attend the party. What will be the total cost of the musical group? The unit cost per person?
2. Suppose 2,000 people attend. What will be the total cost of the musical group? The unit cost per person?
3. For prediction of total costs, should the manager of the party use the unit cost in requirement 1? Unit cost in requirement 2? What is the major lesson of this problem?

2-17 Total costs and unit costs. Golden Holidays markets vacation packages to Honolulu from Los Angeles. This package includes a round-trip flight on Global Airways. Golden Holidays pays Global $60,000 for each round-trip flight. The maximum load on a flight is 300 passengers.

Required

1. What is the unit cost to Golden Holidays of each passenger on a Global Airways round-trip flight if there are (a) 200, (b) 250, or (c) 300 passengers?
2. What role can the unit-cost figures per passenger computed in requirement 1 play when Golden Holidays is predicting the total air-flight costs to be paid next month for Global Airways carrying 4,000 passengers on 15 scheduled round-trip flights?

2-18 Variable costs and fixed costs. Consolidated Minerals (CM) owns the rights to extract minerals from beach sands on Fraser Island. CM has costs in three areas:

a. Payment to a mining subcontractor who charges $80 per ton of beach sand mined and returned to the beach (after being processed on the mainland to extract three minerals—ilmenite, rutile, and zircon).
b. Payment of a government mining and environmental tax of $50 per ton of beach sand mined.
c. Payment to a barge operator. This operator charges $150,000 per month to transport each batch of beach sand—up to 100 tons per batch per day—to the mainland and then return to Fraser Island. (That is, 0 to 100 tons per day = $150,000 per month; 101 to 200 tons = $300,000 per month, and so on.) Each barge operates 25 days per month. The $150,000 monthly charge must be paid even if less than 100 tons are transported on any day and even if Consolidated Minerals requires fewer than 25 days of barge transportation in that month.

CM is currently mining 180 tons of beach minerals per day for 25 days per month.

Required

1. What is the variable cost per ton of beach sand mined? What is the fixed cost to CM per month?
2. Make one plot of the variable costs and another plot of the fixed costs of Consolidated Minerals. Your plots should be similar to Exhibits 2-4 and 2-5. Is the concept of relevant range applicable to your plots?
3. What is the unit cost per ton of beach sand mined (a) if 180 tons are mined each day or (b) if 220 tons are mined each day? Explain the difference in the unit cost figures.

2-19 Classification of costs, service sector. Consumer Focus is a marketing research firm that organizes focus groups for consumer product companies. Each focus group has eight individuals who are paid $50 per session to provide comments on new products. These focus groups meet in hotels and are led by a trained independent marketing specialist hired by Consumer Focus. Each specialist is paid a fixed retainer to conduct a minimum number of sessions and a per-session fee of $2,000. A Consumer Focus staff member attends each session to ensure that all the logistical aspects run smoothly.

Required

Classify each of the following cost items as:

a. Direct or indirect (D or I) costs with respect to each individual focus group
b. Variable or fixed (V or F) costs with respect to how the total costs of Consumer Focus change as the number of focus groups conducted changes. (If in doubt, select on the basis of whether the total costs will change substantially if a large number of groups are conducted.)

You will have two answers (D or I; V or F) for each of the following items:

Cost Item	D or I	V or F
A. Payment to individuals in each focus group to provide comments on new products		
B. Annual subscription of Consumer Focus to *Consumer Reports* magazine		
C. Phone calls made by Consumer Focus staff member to confirm individuals will attend a focus group meeting. (Records of individual calls are not kept.)		
D. Retainer paid to focus group leader to conduct up to 20 focus groups per year on new medical products		
E. Hotel meals provided to participants in each focus group		
F. Lease payment by Consumer Focus for corporate office		
G. Cost of tapes used to record comments made by individuals in a focus group session. (These tapes are sent to the company whose products are being tested.)		
H. Gasoline costs of Consumer Focus staff for company-owned vehicles. (Staff members submit monthly bills with no mileage breakdowns.)		

2-20 Classification of costs, merchandising sector. Home Entertainment Center (HEC) operates a large store in San Francisco. The store has both a video section and a musical (compact disks, records, and tapes) section. HEC reports revenues for the video section separately from the musical section.

Required
Classify each of the following cost items as:
a. Direct or indirect (D or I) costs with respect to the video section
b. Variable or fixed (V or F) costs with respect to how the total costs of the video section change as the number of videos sold changes. (If in doubt, select on the basis of whether the total costs will change substantially if a large number of videos are sold.)

You will have two answers (D or I; V or F) for each of the following items:

Cost Item	D or I	V or F
A. Annual retainer paid to a video distributor		
B. Electricity costs of HEC store (single bill covers entire store)		
C. Costs of videos purchased for sale to customers		
D. Subscription to *Video Trends* magazine		
E. Leasing of computer software used for financial budgeting at HEC store		
F. Cost of popcorn provided free to all customers of HEC		
G. Earthquake insurance policy for HEC store		
H. Freight-in costs of videos purchased by HEC for sale		

2-21 Classification of costs, manufacturing sector. The Fremont, California plant of New United Motor Manufacturing Inc. (NUMMI), a joint venture of General Motors and Toyota, assembles two types of cars (Corollas and Geo Prisms). Separate assembly lines are used for each type of car.

Required
Classify each of the following cost items as:
a. Direct or indirect (D or I) costs with respect to the type of car assembled (Corolla or Geo Prism)

b. Variable or fixed (V or F) costs with respect to how the total costs of the plant change as the number of cars assembled changes. (If in doubt, select on the basis of whether the total costs will change substantially if a large number of cars are assembled.)

You will have two answers (D or I; V or F) for each of the following items:

Cost Item	D or I	V or F
A. Cost of tires used on Geo Prisms		
B. Salary of public relations manager for NUMMI plant		
C. Annual awards dinner for Corolla suppliers		
D. Salary of engineer who monitors design changes on Geo Prism		
E. Freight costs of Corolla engines shipped from Toyota City, Japan, to Fremont, California		
F. Electricity costs for NUMMI plant (single bill covers entire plant)		
G. Wages paid to temporary assembly-line workers hired in periods of high production (paid on hourly basis)		
H. Annual fire insurance policy cost for NUMMI plant		

2-22 Income statement, service sector. Creative Advertising (CA) operates a boutique agency specializing in developing television advertisements for consumer products. The following information applies to 19_5.
a. CA's revenues are 15% of the amounts television stations charge companies for running advertisements developed by CA. (That is, if a television station charges a company $10,000 to run an advertisement, CA will be paid $1,500. Thus, the total cost to the company of the advertisement is $11,500.) In 19_5, these companies were charged $10,000,000 for CA-developed advertisements.
b. CA has three partners (paid $150,000 each in salary and benefits in 19_5) and three support staff (paid $50,000 each in salary and benefits in 19_5).
c. Rent costs for CA's office in a prime location were $180,000 in 19_5.
d. Other operating costs of CA in 19_5 were $227,050.

Required
1. Prepare CA's income statement for 19_5.
2. Name one difference between CA's income statement and that of
   a. a merchandising-sector company and
   b. a manufacturing-sector company.

2-23 Income statement, merchandising sector. Morrow Bay Lumber is a retailer of lumber products. It operates a single store that resells lumber products it has bought from over 100 different suppliers. The following account balances pertain to January 1, 19_5, and December 31, 19_5 (in thousands):

	Jan. 1, 19_5	Dec. 31, 19_5
Merchandise inventory	$ 353	$ 303
Purchases of merchandise		2,589
Payroll—nonexecutives		335
Benefits—nonexecutives		133
Executive compensation		70
Depreciation—fixtures and equipment		98
Advertising		94
Rent		70
Repairs and maintenance		79
Taxes: property		51
Office expenses		64
Other expenses		156
Sales		3,916

Required

1. Prepare Morrow Bay's income statement for 19_5.
2. Name one difference between Morrow Bay's income statement and that of:
   a. a service-sector company and
   b. a manufacturing-sector company.

2-24 Cost of goods manufactured. Prepare a schedule of cost of goods manufactured for the Canseco Company from the following account balances (in thousands):

	Beginning of 19_4	End of 19_4
Direct materials inventory	$22,000	$26,000
Work in process inventory	21,000	20,000
Finished goods inventory	18,000	23,000
Purchases of direct materials		75,000
Direct manufacturing labor		25,000
Indirect manufacturing labor		15,000
Plant insurance		9,000
Depreciation—plant building and equipment		11,000
Repairs and maintenance—plant		4,000
Marketing, distribution, and customer-service costs		93,000
General and administrative costs		29,000

2-25 Income statement, manufacturing sector. Prepare an income statement for the company in the preceding problem, assuming sales of $300 million.

2-26 Computing cost of goods manufactured and cost of goods sold. Compute cost of goods manufactured and cost of goods sold from the following account balances relating to 19_5 (in thousands):

Property tax on plant building	$ 3,000
Marketing, distribution, and customer-service costs	37,000
Finished goods inventory, January 1, 19_5	27,000
Plant utilities	17,000
Work in process inventory, December 31, 19_5	26,000
Depreciation of plant building	9,000
General and administrative costs (nonplant)	43,000
Direct materials used	87,000
Finished goods inventory, December 31, 19_5	34,000
Depreciation of plant equipment	11,000
Plant repairs and maintenance	16,000
Work in process inventory, January 1, 19_5	20,000
Direct manufacturing labor	34,000
Indirect manufacturing labor	23,000
Indirect materials used	11,000
Miscellaneous plant overhead	4,000

2-27 Income statement and schedule of cost of goods manufactured. The Howell Corporation has the following account balances (in millions):

For Specific Date		For Year 19_4	
Direct materials, Jan. 1, 19_4	$15	Purchases of direct materials	$325
Work in process, Jan. 1, 19_4	10	Direct manufacturing labor	100
Finished goods, Jan. 1, 19_4	70	Depreciation—plant building and	
Direct materials, Dec. 31, 19_4	20	equipment	80
Work in process, Dec. 31, 19_4	5	Plant supervisory salaries	5
Finished goods, Dec. 31, 19_4	55	Miscellaneous plant overhead	35
		Sales	950
		Marketing, distribution, and	
		customer-service costs	240
		Plant supplies used	10
		Plant utilities	30
		Indirect manufacturing labor	60

Required

Prepare an income statement and a supporting schedule of cost of goods manufactured for the year ended December 31, 19_4. (For additional questions regarding these facts, see the next problem.)

2-28 Interpretation of statements. Refer to the preceding problem.

Required

1. How would the answer to the preceding problem be modified if you were asked for a "schedule of cost of goods manufactured and sold" instead of a "schedule of cost of goods manufactured"? Be specific.

2. Would the sales manager's salary (included in marketing, distribution, and customer–service costs) be accounted for any differently if the Howell Corporation were a merchandising-sector company instead of a manufacturing-sector company? Using the flow of costs outlined in Exhibit 2-12, describe how the wages of an assembler in the plant would be accounted for in this manufacturing company.

3. Plant supervisory salaries are usually regarded as indirect manufacturing costs. When might some of these costs be regarded as direct manufacturing costs? Give an example.

4. Suppose that both the direct materials used and the plant depreciation were related to the manufacture of 1 million units of product. What is the unit cost for the direct materials assigned to those units? What is the unit cost for plant building and equipment depreciation? Assume that yearly plant depreciation is computed on a straight-line basis.

5. Assume that the implied cost behavior patterns in requirement 4 persist. That is, direct materials costs behave as a variable cost and depreciation behaves as a fixed cost. Repeat the computations in requirement 4, assuming that the costs are being predicted for the manufacture of 1.2 million units of product. How would the total costs be affected?

6. As a management accountant, explain concisely to the president why the unit costs differed in requirements 4 and 5.

2-29 Income statement and schedule of cost of goods manufactured. The following items pertain to Chan Corporation (in millions):

For Specific Date		For Year 19_5	
Work in process, Jan. 1, 19_5	$10	Plant utilities	$ 5
Direct materials, Dec. 31, 19_5	5	Indirect manufacturing labor	20
Finished goods, Dec. 31, 19_5	12	Depreciation—plant,	
Accounts payable, Dec. 31, 19_5	20	building, and equipment	9
Accounts receivable,		Sales	350
Jan. 1, 19_5	50	Miscellaneous manufacturing	
Work in process, Dec. 31, 19_5	2	overhead	10
Finished goods, Jan. 1, 19_5	40	Marketing, distribution, and	
Accounts receivable,		customer-service costs	90
Dec. 31, 19_5	30	Direct materials purchased	80
Accounts payable, Jan. 1, 19_5	40	Direct manufacturing labor	40
Direct materials, Jan. 1, 19_5	30	Plant supplies used	6
		Property taxes on plant	1

Required

Prepare an income statement and a supporting schedule of cost of goods manufactured. (For additional questions regarding these facts, see the next problem.)

2-30 Interpretation of statements. Refer to the preceding problem.

Required

1. How would the answer to the preceding problem be modified if you were asked for a "schedule of cost of goods manufactured and sold" instead of a "schedule of cost of goods manufactured"? Be specific.

2. Would the sales manager's salary (included in marketing, distribution and customer-service costs) be accounted for any differently if the Chan Corporation were a merchandising-sector company instead of a manufacturing-sector company? Using the flow of costs out-

lined in Exhibit 2-12, describe how the wages of an assembler in the plant would be accounted for in this manufacturing company.

3. Plant supervisory salaries are usually regarded as indirect manufacturing costs. When might some of these costs be regarded as direct manufacturing costs? Give an example.

4. Suppose that both the direct materials used and the plant depreciation were related to the manufacture of 1 million units of product. What is the unit cost for the direct materials assigned to those units? What is the unit cost for plant building and equipment depreciation? Assume that yearly depreciation is computed on a straight-line basis.

5. Assume that the implied cost behavior patterns in requirement 4 persist. That is, direct materials costs behave as a variable cost and plant depreciation behaves as a fixed cost. Repeat the computations in requirement 4, assuming that the costs are being predicted for the manufacture of 1.5 million units of product. How would the total costs be affected?

6. As a management accountant, explain concisely to the president why the unit costs differed in requirements 4 and 5.

2-31 Finding unknown balances. An auditor for the Internal Revenue Service is trying to reconstruct some partially destroyed records of two taxpayers. For each of the cases in the accompanying list, find the unknowns designated by capital letters.

	Case 1	Case 2
	(in thousands)	
Accounts receivable, 12/31	$ 6,000	$ 2,100
Cost of goods sold	A	20,000
Accounts payable, 1/1	3,000	1,700
Accounts payable, 12/31	1,800	1,500
Finished goods inventory, 12/31	B	5,300
Gross margin	11,300	C
Work in process, 1/1	0	800
Work in process, 12/31	0	3,000
Finished goods inventory, 1/1	4,000	4,000
Direct materials used	8,000	12,000
Direct manufacturing labor	3,000	5,000
Manufacturing overhead	7,000	D
Purchases of direct materials	9,000	7,000
Sales	32,000	31,800
Accounts receivable, 1/1	2,000	1,400

2-32 Fire loss, computing inventory costs. A distraught employee, Fang W. Arson, put a torch to a manufacturing plant on a blustery February 26. The resulting blaze completely destroyed the plant and its contents. Fortunately, certain accounting records were kept in another building. They revealed the following for the period from January 1, 19_5 to February 26, 19_5:

Direct materials purchased	$160,000
Work in process, 1/1/19_5	$34,000
Direct materials, 1/1/19_5	$16,000
Finished goods, 1/1/19_5	$30,000
Manufacturing overhead	40% of conversion costs
Sales	$500,000
Direct manufacturing labor	$180,000
Prime costs	$294,000
Gross margin percentage based on sales	20%
Cost of goods available for sale	$450,000

The loss was fully covered by insurance. The insurance company wants to know the historical cost of the inventories as a basis for negotiating a settlement, although the settlement is actually to be based on replacement cost, not historical cost.

Required
Calculate the cost of

1. Finished goods inventory, 2/26/19_5

2. Work in process inventory, 2/26/19_5

3. Direct materials inventory, 2/26/19_5

2-33 Classification of costs, prime and conversion costs. Tanaka Metal Products reports the following components in its manufacturing costs for April 19_5 (in thousands):

Direct costs:		
Direct materials	¥430	
Direct manufacturing labor salaries	110	
Subcontracting	120	¥660
Manufacturing overhead:		
Fringe benefits on direct manufacturing labor	40	
Production setup	60	
Other manufacturing overhead	240	340
		¥1,000

Subcontracting costs are treated as a direct cost item separate from direct materials.

Required

1. Compute (a) the prime costs and (b) the conversion costs of Tanaka Metal Products using the cost classifications described above.

2. Assume now that Tanaka changes the classification of two items—both benefits for direct manufacturing labor and production setups will now be classified as direct costs. Compute (a) the prime costs and (b) the conversion costs of Tanaka after this change in cost classification, and comment on the results.

3. What information might Tanaka use to change the two items in requirement 2 to be direct cost items rather than manufacturing overhead cost items?

 2-34 Comprehensive problem on unit costs, product costs. Tampa Office Equipment manufactures and sells metal shelving. It began operations on January 1, 19_4. Costs incurred for 19_4 are as follows (V stands for variable; F stands for fixed):

Cost incurred in 19_4:	
Direct materials used	$140,000 V
Direct manufacturing labor costs	30,000 V
Plant energy costs	5,000 V
Indirect manufacturing labor costs	10,000 V
Indirect manufacturing labor costs	16,000 F
Other indirect manufacturing costs	8,000 V
Other indirect manufacturing costs	24,000 F
Marketing, distribution, and customer-service costs	122,850 V
Marketing, distribution, and customer-service costs	40,000 F
Administrative costs	50,000 F

Variable manufacturing costs are variable with respect to units produced. Variable marketing, distribution, and customer-service costs are variable with respect to units sold.
 Inventory data are:

	Beginning, January 1, 19_4	Ending, December 31, 19_4
Direct materials	0 lb	2,000 lb
Work in process	0 units	0 units
Finished goods	0 units	? units

 Production in 19_4 was 100,000 units. Two pounds of direct materials are used to make one unit of finished product.
 Sales in 19_4 were $436,800. The selling price per unit and the purchase price per pound of direct materials were stable throughout the year. The company's ending inventory of finished goods is carried at the average unit manufacturing costs for 19_4. Finished goods inventory at December 31, 19_4, was $20,970.

Required

1. Direct materials inventory, total cost, December 31, 19_4.

2. Finished goods inventory, total units, December 31, 19_4.

3. Selling price per unit, 19_4.

4. Operating income, 19_4. Show computations.

2-35 Budgeted income statement. (Continuation of 2-34) Assume management predicts that the selling price per unit and variable cost per unit each will be the same in 19_5 as in 19_4. Fixed manufacturing costs and marketing, distribution, and customer-service costs in 19_5 are also predicted to be the same as in 19_4. Sales in 19_5 are forecast to be 122,000 units. The desired ending inventory of finished goods, December 31, 19_5, is 12,000 units. Assume zero ending inventories of both direct materials and work in process. The company's ending inventory of finished goods is carried at the average unit manufacturing costs for 19_5. The company uses the first-in, first-out inventory method. Management has asked that you prepare a budgeted income statement for 19_5.

Required

1. Units of finished goods produced in 19_5.

2. Budgeted income statement for 19_5.

2-36 Cost classifications and recording, ethics. Western Outfits manufactures and sells clothing items in many countries. Its headquarters is in Seattle, Washington. Most of the clothing is manufactured in China, Pakistan, Philippines, and Thailand. Five years ago it started manufacturing clothing in Hindstan. This country has a highly skilled labor force. Western Outfits is taxed in Hindstan on income earned in that country. To promote greater employment of Hindstan labor, that country's government allows companies operating there a 150% income tax deduction on labor costs if the total labor costs for Hindstan employed nationals exceeds 30% of total manufacturing costs.

Ray Stevens, a Division Controller of Western Outfits has recently been appointed controller for its Hindstan operations. After two months on the job he returns to Seattle to report to the corporate controller. Stevens notes the following in his report.

a. Western Outfits three years ago set up two separate subsidiaries in Hindstan—Hindstan W.O. Manufacturing and Hindstan W.O. Inc. Only the Hindstan W.O. Manufacturing company qualifies for the 150% tax deduction on labor costs. Western Outfits charges over 90% of its Seattle corporate costs that are related to Hindstan operations to the Hindstan W.O. Inc. subsidiary despite this subsidiary having less than 20% of the total labor force. Several of the charges to Hindstan W.O. Inc. are for patents and design work that are recorded as manufacturing costs in the other non-Hindstan Asian subsidiaries.

b. The accounting manual that guides worldwide division reporting at Western Outfits classifies fringe benefits paid to direct manufacturing labor (for health, insurance, and the like) as a manufacturing overhead cost item. Hindstan W.O. Manufacturing (but not Hindstan W.O. Inc.) departs from this policy and records fringe benefits as part of direct manufacturing labor costs.

c. Hindstan W.O. Manufacturing has many subcontractors. In all its other subsidiaries worldwide, Western Outfits does not classify these subcontractors as direct labor. In Hindstan W.O. Manufacturing, all the subcontractor costs are now classified as direct labor. This situation has created a "nightmare" for accounting, and Stevens knows of at least 10 cases where the subcontract price actually covered materials purchased by the subcontractor as well as labor costs of the subcontractor.

Required

1. Stevens wants to adhere to the "Standards of Ethical Conduct for Management Accountants" (see Exhibit 1-8 on p. 17). What areas in (a), (b), and (c) should cause him concern?

2. What recommendations should Stevens make to the corporate controller of Western Outfits?

3

Cost-Volume-Profit

Relationships

Cost drivers and revenue drivers: The general case and a special case

Terminology

The breakeven point

CVP assumptions

Cost planning and CVP

Uncertainty and sensitivity analysis

The PV chart

Effects of sales mix

Role of income taxes

Nonprofit institutions and cost-volume-revenue analysis

Contribution margin and gross margin

Appendix: Decision models and uncertainty

Trade fairs are an important marketing tool in many industries. Companies such as Sun Microsystems consider cost-volume-profit relationships when deciding at which trade fairs to exhibit their products and the size and location of their booths.

Cost-volume-profit (CVP) analysis naturally appeals to most business students. Why? Because it provides a sweeping financial overview of the planning process. Managers widely use CVP as a tool that helps them answer such questions as, How will costs and revenues be affected if we sell 1,000 more units? If we raise or lower our selling prices? If we increase occupancy levels by 3% in our hotel or hospital? These questions have a common "what-if" theme. CVP is built on simplifying assumptions about cost behavior patterns. This chapter examines CVP analysis and explains how the reasonableness underlying CVP assumptions affects the reliance managers should place on CVP results.

COST DRIVERS AND REVENUE DRIVERS: THE GENERAL CASE AND A SPECIAL CASE

Objective 1

Distinguish between the general case and a special case of CVP

Chapter 2 defined a *cost driver* as any factor that affects costs—that is, a change in the cost driver will cause a change in the total cost of a related cost object. Similarly, a **revenue driver** is any factor (such as units sold) that affects revenues. As Chapter 2 emphasizes, many drivers besides the quantity of units manufactured or sold affect total costs. Exhibit 2-3 (p. 30) lists several such drivers for each business function area. These drivers may simultaneously affect total costs. Similarly, there are many revenue drivers besides the units of output sold. For example, changes in selling price, changes in marketing outlays, and changes in product quality affect total revenues.

The general case for predicting total revenues and total costs would include analyses of how combinations of several revenue drivers and several cost drivers affect total revenues and total costs, respectively. This general case is described in Chapter 10. For now, we focus on a special case where units of output (units manufactured or sold) is assumed to be the only cost and revenue driver.

We focus on the special case of straightforward CVP relationships for two major reasons. First, many companies have found such relationships helpful in decision making. Second, the straightforward relationships provide an excellent base for understanding the more complex relationships that exist in the general case:

General Case	Special Case
Many revenue drivers	Single revenue driver (an output-related driver)
Many cost drivers	Single cost driver (an output-related driver)
Various time spans for decisions (short run, long run, product life cycles)	Short-run decisions (time span, typically less than one year, in which fixed costs do not change within the relevant range)

The term CVP analysis is widely used as representing this special case. **Cost-volume-profit (CVP)** analyzes the behavior of total costs, total revenues, and operating income as changes occur in the output level, selling price, variable costs, or fixed costs; a single revenue driver and a single cost driver are used in this analysis. In CVP the V for "volume" refers to an output-related driver such as units manufactured or units sold. The general case is more accurately labeled as "costs and revenues/costs and revenue drivers/profit analysis." Indeed, some accountants refer to the general case as a major part of strategic profit analysis, which is a tool for strategic planning.

Our restriction that an output-related variable is the sole revenue or cost driver is important to keep in mind. It means that in our CVP model, changes in the level of revenues and costs arise only because the output level changes. This restriction means that we will not consider in this chapter such cost drivers as:

1. The number of product design changes in a manufacturing plant, and
2. The number of complimentary copies of a new magazine distributed in the mail as a marketing promotion.

Changes in the total costs of an organization can arise because of changes in (1) or (2) without any corresponding change in the units manufactured or the units sold. Factors such as (1) and (2) are examples of nonoutput-related cost drivers. This chapter assumes that the only revenue or cost drivers are output-related variables.

Accountants and managers should always assess whether the simplified relationships of the special case generate sufficiently accurate predictions of how total revenues and total costs behave. Otherwise, managers may be misled into making unwise decisions. At the outset, remember that we will be considering simplified versions of the real world. Are these simplifications warranted? The answer depends on the facts in a particular organization. The simpler model is preferable provided that management decisions would not be significantly improved by using a more complicated decision model. More will be said about these simplifications later in the chapter.

TERMINOLOGY

Objective 2

Explain the relationship between operating income and net income

It is important to agree on the meaning of key terms in this chapter. This section defines several terms central to understanding CVP analysis. We use the following synonyms for the key terms:

Key Term	Synonym
Operating revenues	Sales
Operating costs	Operating expenses
Operating income	Operating profit

In this chapter, operating costs are assumed to be made up of only two cost categories—variable operating costs (variable with respect to an output-related cost driver) and fixed operating costs.

$$\text{Operating costs} = \text{Variable operating costs} + \text{Fixed operating costs}$$

Operating income is operating revenues for the accounting period minus all operating costs, including cost of goods sold:

$$\text{Operating income} = \text{Operating revenues} - \text{Operating costs}$$

Net income is operating income plus nonoperating revenues (such as interest revenue) minus nonoperating costs (such as interest cost) minus income taxes. For simplicity, throughout this chapter nonoperating revenues and nonoperating costs are assumed to be zero. Thus, net income will be computed as follows:

$$\text{Net income} = \text{Operating income} - \text{Income taxes}$$

In the examples that follow, the measure of output is the number of units manufactured or units sold. Different industries often use different terminology to describe their measure of output. Examples include:

Industry	Measure of Output
Airlines	Number of passenger miles
Hospitals	Number of patient days
Hotels/motels	Number of rooms occupied
Universities	Number of student credit hours

THE BREAKEVEN POINT

CVP analysis can be used to examine how various alternatives that a decision maker is considering affect operating income. The breakeven point is frequently one point of interest in this analysis. The **breakeven point** is that quantity of output where total revenues and total costs are equal; that is where the operating income is zero.

This section shows three methods for determining the breakeven point: the equation method, the contribution margin method, and the graph method.

> EXAMPLE: Mary Frost plans to sell Do-All, a software package, at a heavily attended two-day computer convention in Chicago. Mary can purchase this software from a computer software wholesaler at $120 per package with the privilege of returning all unsold units. The units (packages) will be sold at $200 each. Frost has already paid $2,000 to Computer Conventions, Inc., for the booth rental for the two-day convention. What quantity of units will she need to sell in order to break even? Assume there are no other costs.

Equation Method

The first approach for computing the breakeven point is the *equation method*. Using the terminology in this chapter, the income statement can be expressed in equation form as follows:

$$\text{Revenues} - \text{Variable costs} - \text{Fixed costs} = \text{Operating income}$$

This equation provides the most general and easy-to-remember approach to any CVP situation. For the example above, let N = number of units sold to break even. Setting operating income equal to zero in the above equation:

$$\$200N - \$120N - \$2,000 = \$0$$
$$\$80N = \$2,000$$
$$N = \$2,000 \div \$80 = 25 \text{ units}$$

If Frost sells fewer than 25 units she will have a loss, if she sells 25 units she will break even, and if she sells more than 25 units she will make a profit. This breakeven point is expressed in units. It can also be expressed in sales dollars: 25 units × $200 selling price = $5,000.

Contribution Margin Method

A second approach is the contribution margin method. **Contribution margin** is equal to revenues minus all costs that vary with respect to an output-related cost driver. This method uses the fact that:

$$\left(\begin{array}{c}\text{Selling} \\ \text{price}\end{array} \times \begin{array}{c}\text{Number} \\ \text{of units}\end{array}\right) - \left(\begin{array}{c}\text{Unit variable} \\ \text{costs}\end{array} \times \begin{array}{c}\text{Number} \\ \text{of units}\end{array}\right) - \begin{array}{c}\text{Fixed} \\ \text{costs}\end{array} = \begin{array}{c}\text{Operating} \\ \text{income}\end{array}$$

$$\left(\begin{array}{c}\text{Selling} \\ \text{price}\end{array} - \begin{array}{c}\text{Unit variable} \\ \text{costs}\end{array}\right) \times \begin{array}{c}\text{Number} \\ \text{of units}\end{array} = \begin{array}{c}\text{Fixed} \\ \text{costs}\end{array} + \begin{array}{c}\text{Operating} \\ \text{income}\end{array}$$

$$\begin{array}{c}\text{Unit contribution} \\ \text{margin}\end{array} \times \begin{array}{c}\text{Number} \\ \text{of units}\end{array} = \begin{array}{c}\text{Fixed} \\ \text{costs}\end{array} + \begin{array}{c}\text{Operating} \\ \text{income}\end{array}$$

Because at the breakeven number of units, operating income is by definition zero, we obtain:

$$\text{Unit contribution margin} \times \text{Breakeven number of units} = \text{Fixed costs}$$

which gives us a general formula for a single product and a single cost driver (based on an output-related factor):

$$\text{Breakeven number of units} = \frac{\text{Fixed costs}}{\text{Unit contribution margin}}$$

In our example, fixed costs are $2,000 and the unit contribution margin is $80:

$$\text{Unit contribution margin} = \text{Selling price} - \text{Unit variable costs}$$
$$= \$200 - \$120 = \$80$$

Therefore,

$$\text{Breakeven number of units} = \$2,000 \div \$80 = 25 \text{ units}$$

If the calculations in the equation method and the contribution margin method appear similar, it is because one is merely a restatement of the other.

The following condensed *contribution income statement* can be used to confirm the breakeven calculations:

	Total
Revenues, $200 × 25 units	$5,000
Variable costs, $120 × 25 units	3,000
Contribution margin, $80 × 25 units	2,000
Fixed costs	2,000
Operating income	$ 0

A **contribution income statement** groups individual line items to highlight the contribution margin, which is the difference between revenues and all costs that vary with respect to an output-related driver.

Graph Method

In the graph method we plot the total costs and total revenues lines to obtain their point of intersection, which is the breakeven point. Exhibit 3-1 illustrates this method for our Do-All example. Given our assumption of linear cost and revenue relationships, we need only two points to plot the total costs and total revenues lines.

1. *Total costs line.* This line is the sum of the fixed costs and the variable costs. Fixed costs are $2,000 at all output levels within the relevant range. To plot fixed costs, measure $2,000 on the vertical axis (point A) and extend a line horizontally. Variable costs are $120 per unit. To plot the total costs line, use as one point the $2,000 fixed costs at 0 output units (point A). Select a second point by choosing any other convenient output level (say, 40 units) and determining its total costs. The total variable costs at this output level are $4,800 (40 × $120). Fixed costs are $2,000 at all output levels within the relevant range. Hence, total costs at 40 units of output are $6,800, which is point B in Exhibit 3-1. The total costs line is the straight line from point A running through point B.

2. *Total revenues line.* One convenient starting point is zero revenue at zero output level, which is point C in Exhibit 3-1. Select a second point by choosing any other convenient output level and determining its total revenues. At 40 units of output, total revenues are $8,000 (40 × $200), which is point D in Exhibit 3-1. The total revenues line is a straight line from point C and passing through point D.

EXHIBIT 3-1
Cost-Volume-Profit Chart

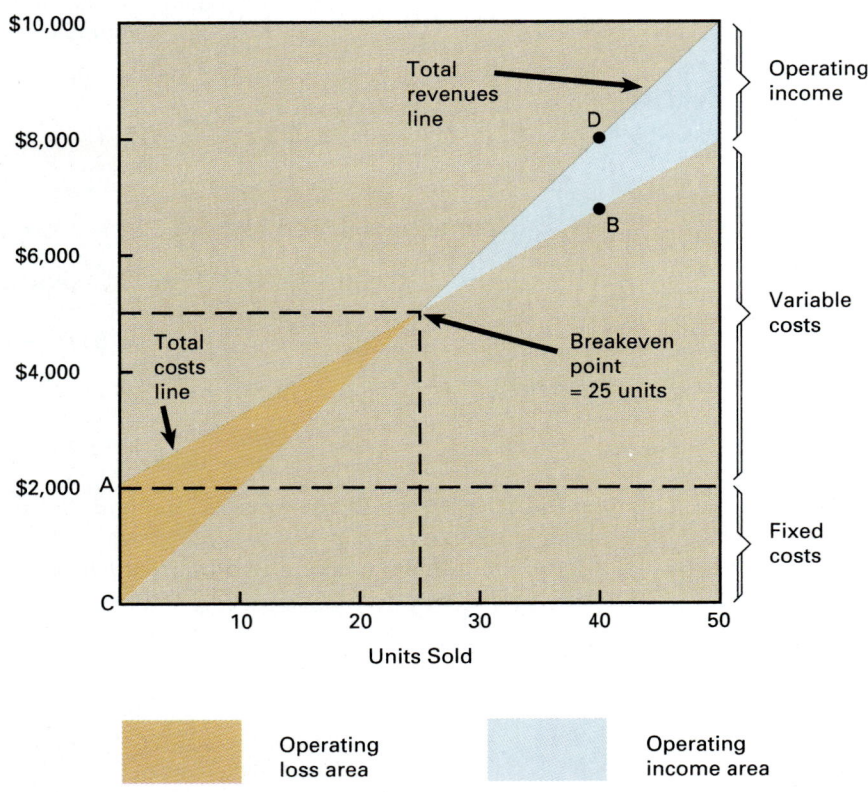

The breakeven point is where the total revenues line and the total costs line intersect. This is where total revenues just equal total costs. But Exhibit 3-1 shows the profit or loss outlook for a wide range of output levels. The confidence we place in any particular CVP chart depends on the relative accuracy of the CVP relationships depicted.

Do not give undue emphasis to the breakeven point. It is but one of many points of interest in CVP analysis. Exhibit 3-1 presents many points (representing operating income at different output levels) that are of interest to managers. Many people describe the topics covered in this chapter as *breakeven analysis*. We prefer to use the phrase *cost-volume-profit analysis* to avoid overemphasizing the single point where total revenues equal total costs.

Target Operating Income

Let us introduce a profit element by asking, How many units must be sold to yield an operating income of $1,200? The equation method provides a straightforward way to answer this question:

Let N = number of units sold to yield target operating income

Revenues – Variable costs – Fixed costs = Target operating income

$$\$200N - \$120N - \$2,000 = \$1,200$$
$$\$80N = \$2,000 + \$1,200$$
$$\$80N = \$3,200$$
$$N = \$3,200 \div \$80 = 40 \text{ units}$$

Proof:	Revenues, $200 × 40	$8,000
	Variable costs, $120 × 40	4,800
	Contribution margin	3,200
	Fixed costs	2,000
	Operating income	$1,200

The graph in Exhibit 3-1 indicates an operating income at the 40-unit volume. The difference between total revenues and total costs at that level of unit sales is the $1,200 operating income.

Alternatively, we could use the contribution margin method. The numerator now consists of fixed costs plus target operating income:

$$N = \frac{\text{Fixed costs + Target operating income}}{\text{Unit contribution margin}}$$

$$N = \frac{\$2,000 + \$1,200}{\$80}$$

$$\$80N = \$3,200$$

$$N = \$3,200 \div \$80 = 40 \text{ units}$$

CVP ASSUMPTIONS

Objective 4

Describe the assumptions underlying CVP

The "special case" CVP analysis that we have been discussing is based on the following assumptions:

1. Total costs can be divided into a fixed component and a component that is variable with respect to an output-related driver (such as units manufactured or units sold).
2. The behavior of total revenues and total costs is linear (straight-line) in relation to output units within the relevant range.[1]
3. There is no uncertainty regarding the cost, revenue, and output quantity used. (This assumption is discussed later in the chapter and in the Appendix.)
4. The analysis either covers a single product or assumes that a given sales mix of products will remain constant as the level of total units sold changes. (This assumption also is discussed later in the chapter.)
5. All revenues and costs can be added and compared without taking into account the time value of money. (Chapters 20 and 21 relax this assumption.)

These CVP assumptions certainly are extreme in the sense that they would rarely match reality. Managers should always question whether a more complicated approach than the special case CVP is warranted. However, managers often find CVP charts a very useful first pass, which helps them understand cost behavior patterns and the interrelationship between revenues and costs over different levels of output.

COST PLANNING AND CVP

Objective 5

Illustrate how CVP can assist cost planning

Alternative Fixed Cost/Variable Cost Structures

CVP is a helpful tool in cost planning. For example, it can highlight the risks that different cost structures pose for a business. Consider again Mary Frost and her booth rental agreement with Computer Conventions, Inc. Our example has Frost paying a

---

[1] For example, one set of conditions in which assumption 2 is descriptive includes: selling prices are constant within the relevant range; productivity is constant; and costs of production inputs are constant. How might nonlinearity arise? On the revenue side, reductions in the selling price may be necessary to spur sales at higher unit levels. On the cost side, variable costs per unit may decline as output increases as employees learn to handle the process more efficiently. The learning curve is discussed in Chapter 10.

EXHIBIT 3-2
Cost-Volume-Profit Charts for Alternative Rental Schedules

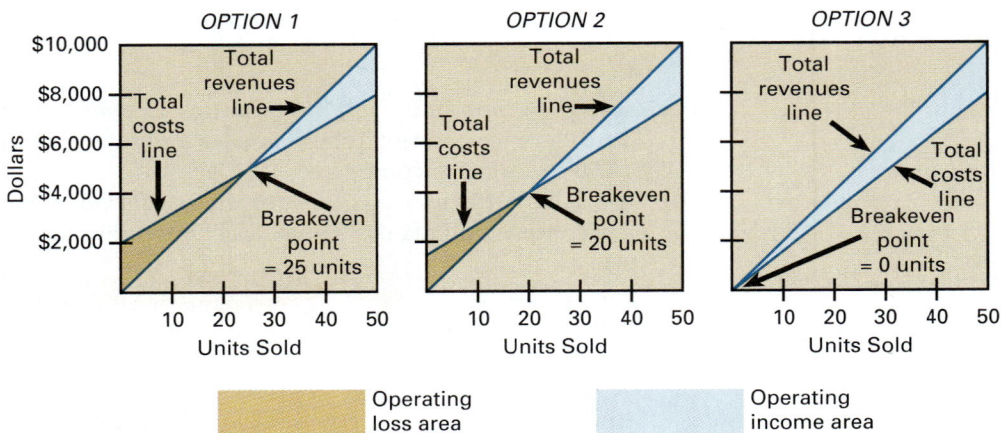

$2,000 booth rental fee. Suppose, however, Computer Conventions offers Frost three rental alternatives:

- ◆ Option 1: $2,000 fixed fee
- ◆ Option 2: $1,400 fixed fee plus 5% of the convention revenues from Do-All sales
- ◆ Option 3: 20% of the convention revenues from Do-All sales with no fixed fee

Frost is interested in how her choice of a rental agreement will affect the risks she faces. Exhibit 3-2 presents these options in the CVP format:

- ◆ *Option 1:* Exposes her to fixed costs of $2,000 and a breakeven point of 25 units. The upside potential is $80 additional operating income for each unit sold above 25 units.
- ◆ *Option 2:* Exposes her to lower fixed costs of $1,400 and a lower breakeven point of 20 units. There is, however, only $70 in additional operating income for each unit sold above 20 units.
- ◆ *Option 3:* Has no fixed costs. Frost makes $40 in additional operating income for each unit sold. This $40 contribution to operating income starts from the first unit sold. This option enables Frost to break even if no units are sold.[2]

CVP highlights the different risks and different returns associated with each option. For example, while option 1 has the most downside risk (a $2,000 fixed up-front payment), it also has the highest contribution margin per unit. The $80 contribution margin per unit translates to high upside potential if Frost is able to generate sales above 25 units. By moving from option 1 to option 2, Frost faces less risk (lowers her fixed costs) if demand is low, but she must accept less upside potential (due to the higher variable costs) if demand is high. The choice among options 1, 2, and 3 will be influenced by her confidence in the level of demand for Do-All software and her willingness to risk money.

Managers increasingly recognize how high levels of fixed costs cause operating income to plummet when large declines in revenues occur. For example, the president of Emery Air Freight (an overnight courier company) commented: "We would prefer to keep out of the airline business and buy space from the existing airlines be-

---

[2] The breakeven point of 25 units for option 1 was computed earlier in this chapter. The breakeven point (N) for option 2 is calculated thusly:

$$\text{Fixed costs} = \$1,400$$
$$\text{Unit variable costs} = \$120 + 0.05\,(\$200) = \$130$$
$$\text{Unit contribution margin} = \$200 - \$130 = \$70 \text{ per unit}$$
$$\$200N - \$130N - \$1,400 = 0$$
$$N = \$1,400 \div \$70 = 20 \text{ units}$$

Option 3 has a breakeven point of zero units because there are no fixed costs. The variable costs per unit are $160 ($120 + 0.20 [$200]). The contribution margin per unit is $40 ($200 − $160).

cause that's a variable cost. If we don't have the need for capacity, we don't buy it, but if we have to own or charter airplanes, the costs become fixed and we're stuck with excess capacity sometimes."

Effect of Time Horizon

Costs are not always neatly classified as fixed or variable, as the examples in this chapter may suggest. In the real world, the shorter the time horizon we consider, the higher the percentage of total costs we may view as fixed. Consider United Airlines. Suppose a United plane will depart from its gate in 30 minutes and there are 20 empty seats. A potential passenger arrives bearing a transferable ticket from a competing airline. What are the variable costs to United of placing one more passenger in an otherwise empty seat? Variable costs (for example, one more meal) would be negligible. Virtually all the costs in that decision situation are fixed. In contrast, suppose United must decide whether to add another flight or to include another city in its routes. Many more costs would be regarded as variable and fewer as fixed in those decisions.

This example underscores the importance of how the decision situation affects the analysis of cost behavior. Of course, total costs are most likely to be affected by long time spans and large changes in output levels. In brief, whether costs are really fixed depends heavily on the relevant range, the length of the time horizon in question, and the specific decision situation.

UNCERTAINTY AND SENSITIVITY ANALYSIS

Objective 6
Explain how sensitivity analysis can help managers cope with uncertainty

Throughout much of this book, we work with single-number "best estimates" in order to emphasize and simplify various important points. For example, our CVP models (this chapter), budget models (Chapter 6), and capital-budgeting models (Chapters 20 and 21) often conveniently assume certainty regarding the levels of variable costs, fixed costs, output, and other factors.

Obviously, our estimates and predictions are subject to varying degrees of **uncertainty**, which is defined here as the possibility that an actual amount will deviate from an expected amount. How do we cope with uncertainty? There are many complex models available that formally analyze expected values in conjunction with probability distributions. But the application of *sensitivity analysis* to an original solution is the most widely used approach. The Appendix to this chapter provides further discussion of uncertainty.

Sensitivity analysis is a "what-if" technique that examines how a result will change if the original predicted data are not achieved or if an underlying assumption changes. In the context of CVP, sensitivity analysis answers such questions as, What will operating income be if the output level decreases by 5% from the original prediction? and What will operating income be if variable costs per unit increase by 10%? The sensitivity to various possible outcomes broadens managers' perspectives as to what might actually occur despite their well-laid plans.

The widespread use of electronic spreadsheets has helped promote the use of CVP in many organizations. Managers can easily conduct CVP-based sensitivity analysis to examine the effect and interaction of changes in selling prices, unit variable costs, fixed costs, and target operating incomes.

Exhibit 3-3 displays a spreadsheet for our Do-All example. Mary Frost can immediately see what she will need to sell to reach particular operating income levels, given alternative levels of fixed costs and variable costs per unit. For example, revenues of $6,000 (30 units at $200 per unit) are required to earn operating income of $1,000 if fixed costs are $2,000 and variable costs per unit are $100. Frost can also use Exhibit 3-3 to assess whether she wants to sell at the Chicago computer convention if, for example, the booth rental is raised to $3,000 or the software supplier raises its price to $140 per unit.

EXHIBIT 3-3

Spreadsheet Analysis of CVP Relationship of Do-All Software

Fixed Costs	Variable Costs Per Unit	Sales Dollars Required at $200 Selling Price to Earn Operating Income of			
		$0	$1,000	$1,500	$2,000
$2,000	$100	$ 4,000	$ 6,000	$ 7,000	$ 8,000
	120	5,000	7,500	8,750	10,000
	140	6,667	10,000	11,667	13,333
$2,500	$100	$ 5,000	$ 7,000	$ 8,000	$ 9,000
	120	6,250	8,750	10,000	11,250
	140	8,333	11,667	13,333	15,000
$3,000	$100	$ 6,000	$ 8,000	$ 9,000	$10,000
	120	7,500	10,000	11,250	12,500
	140	10,000	13,333	15,000	16,667

A tool of sensitivity analysis is the **margin of safety**, which is the excess of budgeted revenues over the breakeven revenues. The margin of safety is the answer to the "what-if" question: If budgeted revenues are above breakeven and drop, how far can they fall below budget before the breakeven point is reached? Assume Mary Frost has fixed costs of $3,000, a selling price of $200, and variable costs per unit of $140. For 75 units sold, the budgeted revenues are $15,000 and the budgeted operating income is $1,500. The breakeven revenue point for this set of assumptions is $10,000 or 50 units. Here the margin of safety is $5,000 ($15,000 – $10,000) or 25 units.

THE PV CHART

We can recast Exhibit 3-1 in the form of a profit-volume (PV) chart. A **PV chart** shows the impact on operating income (profit) of changes in the output level (volume). Exhibit 3-4 (Panel A) presents the PV chart for the original data for our Do-All software example (fixed costs of $2,000, selling price of $200, and variable costs per unit of $120).

Given our linearity assumptions, the PV line can be drawn using two points. One convenient point (X) is the level of fixed costs at zero output—$2,000, which is also the operating loss at this output level. A second convenient point (Y) is the breakeven point—25 units in our example (see p. 64). The PV line is drawn by connecting points X and Y and extending it beyond Y. Each unit sold beyond the breakeven point will add $80 to operating income. At the 35-unit output level, operating income would be $800:

$$(\$200 \times 35) - (\$120 \times 35) - \$2,000 = \$800$$

A comparison of PV charts representing different assumptions can highlight their effect on operating income. Exhibit 3-4 (Panel B) shows the PV chart for our Do-All example assuming fixed costs of $3,300 (compared with $2,000 in Panel A) and variable costs per unit of $90 (compared with $120 in Panel A). The selling price is $200 in both charts. The unit contribution margin in Panel B is $110. The breakeven point in Panel B is 30 units:

EXHIBIT 3-4

The Profit-Volume Chart

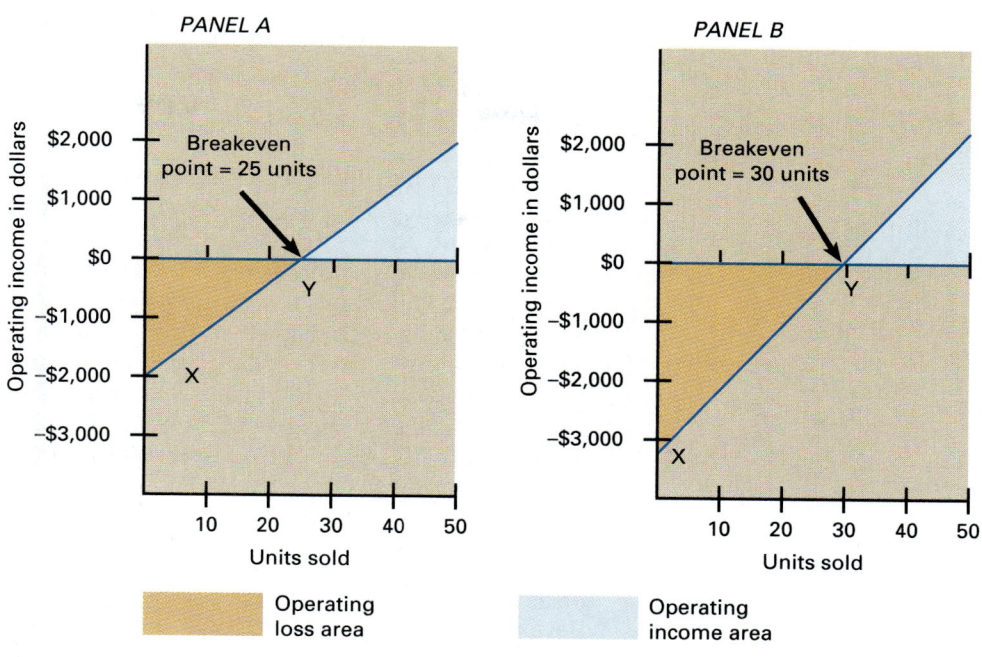

$$\$200N - \$90N - \$3,300 = 0$$
$$N = \$3,300 \div \$110 = 30 \text{ units}$$

Each unit sold beyond the breakeven point will add $110 to operating income. The Panel B PV chart has a steeper slope for its operating income line, which means that the operating income increases at a faster rate as the level of output increases. The Panel B PV chart has higher downside risk and higher upside potential. The Panel A PV chart has lower downside risk and lower upside potential.

EFFECTS OF SALES MIX

Objective 7
Describe the effect of sales mix on operating income

Sales mix is the relative combination of quantities of products or services that constitutes total sales. If the mix changes, overall sales targets may still be achieved. However, the effects on operating income depend on how the original proportions of low-contribution-margin or high-contribution-margin products have shifted.

Suppose Perfect Writer (a two-product pen company) has the following budget:

	Peerless	Fantastic	Total
Sales in units	120,000	40,000	160,000
Revenues, $5 and $10 per unit	$600,000	$400,000	$1,000,000
Variable costs, $4 and $3 per unit	480,000	120,000	600,000
Contribution margin, $1 and $7 per unit	$120,000	$280,000	400,000
Fixed costs			300,000
Operating income			$ 100,000

What is the breakeven point? The usual answer assumes that the budgeted sales mix will not change. That is, three units of Peerless will be sold for each unit of Fantastic:

$$\text{Let } F = \text{Number of units of Fantastic to break even}$$
$$3F = \text{Number of units of Peerless to break even}$$
$$\text{Revenues} - \text{Variable costs} - \text{Fixed costs} = \text{Zero operating income}$$
$$\$5(3F) + \$10(F) - \$4(3F) - \$3(F) - \$300{,}000 = 0$$
$$\$25F - \$15F - \$300{,}000 = 0$$
$$\$10F = \$300{,}000$$
$$F = 30{,}000$$
$$3F = 90{,}000$$

The breakeven point is 120,000 units, consisting of 90,000 Peerless pens and 30,000 Fantastic pens. This is the only breakeven point for a sales mix of three Peerless pens and one Fantastic pen.

But the breakeven point is not a unique number for Perfect Writer. It depends on the sales mix. Obviously, for any given quantity of total unit sales, if the sales mix shifts toward units with higher contribution margins, operating income will be higher. Suppose Perfect Writer had actual sales of 160,000 units—exactly equal to the budgeted total quantity of unit sales—but the sales mix was 100,000 Peerless and 60,000 Fantastic. Operating income would be $220,000, which is a hefty $120,000 higher than the $100,000 budgeted operating income in the original situation, where the sales mix is 3 to 1:

	Peerless	Fantastic	Total
Sales in units	100,000	60,000	160,000
Revenues, $5 and $10 per unit	$500,000	$600,000	$1,100,000
Variable costs, $4 and $3 per unit	400,000	180,000	580,000
Contribution margin, $1 and $7 per unit	$100,000	$420,000	520,000
Fixed costs			300,000
Operating income			$ 220,000

Despite their desire to maximize the sales of all products, managers must frequently cope with limited resources. For instance, additional production capacity may be unavailable. What products should be produced? As Chapter 11 explains in more detail, the best decision is not necessarily to make the product having the highest contribution margin per unit. After all, suppose Perfect Writer can make 1,000 Peerless pens for each hour of capacity instead of 100 Fantastic pens. Then Peerless can generate a contribution margin of $1,000 \times \$1 = \$1,000$ per hour, whereas Fantastic can generate only $100 \times \$7 = \$700$ per hour.

In sum, CVP analysis must be done carefully because one or more initial assumptions may not hold. When the assumed conditions change, the breakeven point and the expected operating incomes at various output levels also change. Of course, the breakeven points are frequently incidental data. Instead, the focus is on the effects on operating income under various production and sales strategies.[3]

ROLE OF INCOME TAXES

Objective 8
Illustrate how CVP can incorporate income taxes

When we introduced a target operating income in our earlier Do-All software example, the following income statement was shown (p. 65):

Revenues, 200×40	$8,000
Variable costs, 120×40	4,800
Contribution margin	3,200
Fixed costs	2,000
Operating income	$1,200

---

[3] Detailed discussion of how CVP analysis can be extended to incorporate uncertainty and multiple products is in M. Schweitzer, E. Trossmann, and G. Lawson, *Break-even Analyses: Basic Model, Variants, Extensions* (Chichester, U.K.: John Wiley, 1992).

Net income of Do-All is operating income minus income taxes. What number of units must Do-All sell to earn net income of $1,200, assuming operating income is taxed at a rate of 40%? The only change in the equation method of CVP analysis is to modify the target operating income to allow for income taxes. Recall our previous equation approach:

$$\text{Revenues} - \text{Variable costs} - \text{Fixed costs} = \text{Operating income}$$

We now introduce income tax effects:

$$\text{Target net income} = (\text{Operating income}) - [(\text{Operating income})(\text{Tax rate})]$$
$$\text{Target net income} = (\text{Operating income})(1 - \text{Tax rate})$$
$$\text{Operating income} = \frac{\text{Target net income}}{1 - \text{Tax rate}}$$

So, taking income taxes into account, the equation method yields:

$$\text{Revenues} - \text{Variable costs} - \text{Fixed costs} = \frac{\text{Target net income}}{1 - \text{Tax rate}}$$

Substituting numbers from our Do-All example, the equation would now be:

$$\$200N - \$120N - \$2,000 = \frac{\text{Target net income}}{1 - \text{Tax rate}}$$
$$\$200N - \$120N - \$2,000 = \frac{\$1,200}{1 - 0.40}$$
$$\$200N - \$120N - \$2,000 = \$2,000$$
$$\$80N = \$4,000$$
$$N = 50 \text{ units}$$

Proof: Revenues, $200 × 50	$10,000
Variable costs, $120 × 50	6,000
Contribution margin	4,000
Fixed costs	2,000
Operating income	2,000
Income taxes, $2,000 × 0.40	800
Net income	$ 1,200

Note again that

$$\text{Operating income} = \frac{\text{Target net income}}{1 - \text{Tax rate}}$$

Suppose the target net income were set at $1,680, instead of $1,200. The required number of unit sales would rise from 50 to 60 units:

$$\$200N - \$120N - \$2,000 = \frac{\$1,680}{1 - 0.40}$$
$$\$80N - \$2,000 = \$2,800$$
$$\$80N = \$4,800$$
$$N = \$4,800 \div \$80 = 60 \text{ units}$$

The presence of income taxes will not change the breakeven point. Why? Because, by definition, operating income at the breakeven point is zero and thus no income taxes will be paid.[4]

NONPROFIT INSTITUTIONS AND COST-VOLUME-REVENUE ANALYSIS

Suppose a social welfare agency has a government budget appropriation (revenue) for 19_4 of $900,000. This nonprofit agency's major purpose is to assist handicapped persons who are seeking employment. On the average, the agency supplements each person's income by $5,000 annually. The agency's fixed costs are $270,000. There are no other costs. The agency manager wants to know how many people could be assisted in 19_4:

Let N = Number of people to be assisted
$$\text{Revenue} - \text{Variable costs} - \text{Fixed costs} = \$0$$
$$\$900,000 - \$5,000N - \$270,000 = \$0$$
$$\$5,000N = \$900,000 - \$270,000$$
$$N = \$630,000 \div \$5,000 = 126 \text{ people}$$

Suppose the manager is concerned that the total budget appropriation for 19_4 will be reduced by 15% to a new amount of $(1 - 0.15) \times \$900,000 = \$765,000$. The manager wants to know how many handicapped persons will be assisted. Assume the same amount of monetary assistance per person:

$$\$765,000 - \$5,000N - \$270,000 = 0$$
$$\$5,000N = \$765,000 - \$270,000$$
$$N = \$495,000 \div \$5,000 = 99 \text{ people}$$

Note the two characteristics of the CVP relationships in this nonprofit situation:

1. The percentage drop in service, $(126 - 99) \div 126$, or 21.4%, is more than the 15% reduction in the budget appropriation. The fixed costs of $270,000 mean that the percentage drop in service exceeds the percentage drop in budget appropriation.
2. If the relationships were graphed, the budgeted appropriation (revenue) amount would be a straight horizontal line of $765,000. The manager could adjust operations in one or more of three major ways: (a) reduce the number of persons assisted; (b) alter the variable costs (the assistance per person); or (c) alter the total fixed costs.

---

[4]Other taxes may affect the breakeven point. For example, an excise tax paid by the seller that is a fixed percentage of sales dollars can be treated as a variable cost and hence will increase the breakeven point.

Two frequently used terms—contribution margin and gross margin—are often confused in management accounting applications in the merchandising and manufacturing sectors:

> **Contribution margin** = Revenues – All costs that vary with respect to an output-related driver
> **Gross margin** = Revenues – Cost of goods sold

The phrase "all costs that vary" refers to variable costs in each of the business function areas of the value chain (research and development; design of products, services, or processes; production; marketing; distribution; and customer service). Cost of goods sold in the merchandising sector is a variable cost made up of goods purchased

Surveys of Company Practice

PURPOSES FOR COMPANIES DISTINGUISHING BETWEEN FIXED COSTS AND VARIABLE COSTS

The distinction between fixed costs and variable costs is fundamental to breakeven analysis. This book illustrates many other contexts where this distinction is also important. One survey of U.S. companies reported the following ranking of purposes for distinguishing between fixed costs and variable costs (1 = most important purpose).[a]

Rank	Purpose	Chapter(s) in This Book Discussing Purpose in Detail
1 (equal)	Pricing decisions	11 and 12
1 (equal)	Budgeting	6
3	Profitability analysis—existing products	11 and 12
4	Profitability analysis—new products	11 and 12
5	Breakeven analysis	3
6	Variance analysis	7, 8, and 22

Surveys of Australian, Japanese, and United Kingdom companies provide additional information on how managers rank the many purposes for distinguishing between fixed costs and variable costs (1 = most important purpose).[b]

Purpose	Ranking By Australian Companies	Ranking By Japanese Companies	Ranking By United Kingdom Companies
Pricing decisions	1	5	1
Budgeting	2	2	3
Making profit plans	3	1	2
Cost reduction	6	3	5 (equal)
Breakeven analysis	4 (equal)	4	4
Cost-benefit analysis	4 (equal)	6	5 (equal)

These two surveys highlight the wide range of decisions for which managers feel an understanding of the relationships among revenues, costs, and operating income is important.

[a]Adapted from Mowen, *Accounting for Costs.*
[b]Blayney and Yokoyama, "Comparative Analysis." Full citations are in Appendix A at the end of the book.

for resale. Cost of goods sold in the manufacturing sector consists of all manufacturing costs (including fixed manufacturing costs).

Service-sector companies can compute a contribution margin figure but not a gross margin figure. Service-sector companies do not have a cost of goods sold line item on their income statement.

Merchandising Sector

The difference between contribution margin and gross margin in the merchandising sector (primarily retailers and wholesalers) affects all variable costs other than cost of goods sold. Contribution margin is computed after all variable costs have been deducted, while gross margin is computed by deducting only cost of goods sold from revenues. Cost of goods sold in the merchandising sector is a variable cost made up of the cost of goods purchased for resale. The following example illustrates this difference (numbers assumed):

Revenues		$200	Revenues		$200
Cost of goods sold	$125		Cost of goods sold		125
Other variable costs	43	168	Gross margin		75
Contribution margin		32	Other costs ($43 + $19)		62
Fixed costs		19	Operating income		$ 13
Operating income		$ 13			

An example of "other variable costs" for a merchandiser is a salesperson's commission that is a percentage of sales dollars.

Manufacturing Sector

The two areas of difference between contribution margin and gross margin for companies in the manufacturing sector are fixed manufacturing costs and variable nonmanufacturing costs. The following example illustrates this difference (numbers assumed):

Revenues		$1,000	Revenues		$1,000
Variable manufacturing costs	$250		Cost of goods sold		
Variable nonmanufacturing			($250 + $160)		410
costs	270	520	Gross margin		590
Contribution margin		480	Nonmanufacturing costs		
Fixed manufacturing costs	160		($270 + $138)		408
Fixed nonmanufacturing costs	138	298	Operating income		$ 182
Operating income		$ 182			

Fixed manufacturing costs are not deducted from sales when computing contribution margin but are deducted when computing gross margin. Cost of goods sold in a manufacturing company includes all manufacturing costs. Variable nonmanufacturing costs are deducted from sales when computing contribution margins but are not deducted when computing gross margins.

Both the *contribution margin* and the *gross margin* can be expressed as totals, as an amount per unit, or as percentages. The **contribution-margin percentage** is the total contribution margin divided by revenues. The **variable-cost percentage** is the total variable costs (with respect to an output-related factor) divided by revenues. The contribution-margin percentage in our manufacturing sector example is 48% ($480 ÷ $1,000), while the variable-cost percentage is 52% ($520 ÷ $1,000). The gross margin percentage is 59% ($590 ÷ $1,000).

PROBLEM FOR SELF-STUDY

PROBLEM

The American-Canadian Chamber of Commerce is planning its July 4 gala ball. There are two possible venues:

a. Toronto Country Golf Club, which has a fixed rental cost of $2,000 plus a charge of $80 per person for its own catering of meals and serving of drinks and hors d'oeuvres.

b. Toronto Town Hall, which has a fixed rental cost of $6,600. The Chamber of Commerce can hire a caterer for meals and waiters and waitresses to serve drinks and hors d'oeuvres at $60 per person.

The Chamber of Commerce budgets $3,500 in costs for administration and marketing. The band will cost a fixed amount of $2,500. Tickets to this prestige event will be $120 per person. All the drinks served and the prizes given away at the ball will be donated.

Required
1. Compute the breakeven point for each venue in terms of tickets sold.
2. Compute the "operating income" on the ball if (a) 150 attend and (b) 300 attend. Comment on the results.
3. At what level of tickets sold will the two venues have the same operating income?

SOLUTION

1. Computation of fixed costs:

	Golf Club	Town Hall
Rental cost of venue	$2,000	$ 6,600
Chamber administration/marketing	3,500	3,500
Band	2,500	2,500
	$8,000	$12,600

Computation of contribution margin per person:

	Golf Club	Town Hall
Selling (ticket) price per person	$120	$120
Catering cost per person	80	60
Contribution margin per person	$ 40	$ 60

$$\text{Breakeven point} = \frac{\text{Fixed costs}}{\text{Unit contribution margin}}$$

$$\text{Breakeven point for Golf Club venue} = \frac{\$8,000}{\$40} = 200 \text{ tickets}$$

$$\text{Breakeven point for Town Hall venue} = \frac{\$12,600}{\$60} = 210 \text{ tickets}$$

2. Operating income = Revenues – Variable costs – Fixed costs

$$\text{Let } X = \text{Number of tickets sold}$$
$$OI = \text{Operating income}$$

Golf Club Venue

$$OI = \$120X - \$80X - \$8,000$$
$$\text{When } X = 150: OI = (\$120 \times 150) - (\$80 \times 150) - \$8,000$$
$$= \$18,000 - \$12,000 - \$8,000$$
$$= -\$2,000$$

$$\text{When } X = 300: OI = (\$120 \times 300) - (\$80 \times 300) - \$8,000$$
$$= \$36,000 - \$24,000 - \$8,000$$
$$= \$4,000$$

Town Hall Venue

$$OI = \$120X - \$60X - \$12,600$$
$$\text{When } X = 150: OI = (\$120 \times 150) - (\$60 \times 150) - \$12,600$$
$$= \$18,000 - \$9,000 - \$12,600$$
$$= -\$3,600$$
$$\text{When } X = 300: OI = (\$120 \times 300) - (\$60 \times 300) - \$12,600$$
$$= \$36,000 - \$18,000 - \$12,600$$
$$= \$5,400$$

The Golf Club venue has higher variable costs per person and lower fixed costs. In contrast, the Town Hall venue has lower variable costs per person and higher fixed costs.

3. Requirement 2 gives the operating income equation for each venue. Setting these two equations equal and solving for X, gives 230 as the level of ticket sales at which the operating incomes for the two venues are equal:

$$\$120X - \$80X - \$8,000 = \$120X - \$60X - \$12,600$$
$$\$40X - \$60X = \$8,000 - \$12,600$$
$$\$20X = \$4,600$$
$$X = 230$$

Above 230, the Town Hall venue will yield higher operating income than the Golf Club venue.

SUMMARY

The following points are linked to the chapter's learning objectives.

1. General profit planning in its full complexity assumes that there are many revenue drivers and many cost drivers. CVP is a special case that, in a restricted number of settings, can assist managers in understanding the behavior of total costs, total revenues, and operating income as changes occur in the output level, selling price, variable costs, or fixed costs.

2. Operating income is computed by subtracting operating costs from operating revenues. Net income is operating income plus nonoperating revenues minus nonoperating costs minus income taxes.

3. The three methods outlined for computing the breakeven point—the equation method, the contribution margin method, and the graph method—are merely restatements of each other. Managers often select the method they find easiest to use in their specific situation.

4. Using CVP requires simplifying assumptions, including that costs are either fixed or variable with respect to an output-related driver (such as units manufactured or units sold) and that total sales and total cost relationships are linear.

5. CVP can highlight to managers the downside risk and upside potential reward of alternatives that differ in their fixed costs and variable costs.

6. Sensitivity analysis, a "what-if" technique, can systematically examine the effect on operating income and net income of different levels of fixed costs, variable costs per unit, selling prices, and output.

7. When CVP is applied to a multiple-product firm, it is assumed that there is a constant sales mix of products as the total quantity of units sold changes.

8. Income taxes can be incorporated into CVP analysis in a straightforward way by adjusting operating income by the income tax rate. The breakeven point is unaffected by the presence of income taxes because no income taxes are paid if there is no operating income.

APPENDIX: DECISION MODELS AND UNCERTAINTY

Managers make predictions and decisions in a world of uncertainty. This Appendix explores the characteristics of uncertainty and describes how managers can cope with it. We also illustrate the additional insights gained when uncertainty is recognized in CVP analysis.

Coping With Uncertainty

ROLE OF A DECISION MODEL. *Uncertainty* is the possibility that an actual amount will deviate from an expected amount. For example, marketing costs might be forecast at $400,000 but actually turn out to be $430,000. A **decision model** helps managers deal with uncertainty; it is a formal method for making a choice that often involves quantitative analysis. It usually includes the following elements:

1. A **choice criterion**, which is an objective that can be quantified. This objective can take many forms. Most often the choice criterion is expressed as a maximization of income or a minimization of cost. The choice criterion, also called an **objective function**, provides a basis for choosing the best alternative action.
2. A set of the alternative actions being considered.
3. A set of all the relevant **events** (also called **states** or **states of nature**) that may occur, where an event is a possible occurrence. This set should be mutually exclusive and collectively exhaustive. Two events are mutually exclusive if they cannot occur at the same time. Events are collectively exhaustive if, taken together, they make up the entire set of possible occurrences (and no other event can occur). Examples are growth or no growth in industry demand and increase or no increase in interest rates. Only one event in a set of mutually exclusive and collectively exhaustive events will actually occur.
4. A set of probabilities, where a **probability** is the likelihood of occurrence of an event.
5. A set of possible **outcomes** (often called **payoffs**) that measures, in terms of the choice criterion, the predicted consequences of the various possible combinations of actions and events.

It is important to distinguish actions from events. *Actions* are choices made by management—for example, the prices it should charge for the firm's products. *Events* are occurrences that management cannot control—for example, a growing or declining economy. The *outcome* is the operating income the company makes, which depends both on the action management selects (pricing strategy) and the event that occurs (how the economy performs). Exhibit 3-5 presents an overview of the link between a decision model, the implementation of the chosen action, its outcome, and subsequent performance evaluation.

EXHIBIT 3-5

Overview of a Decision Model and Its Link to Performance Evaluation

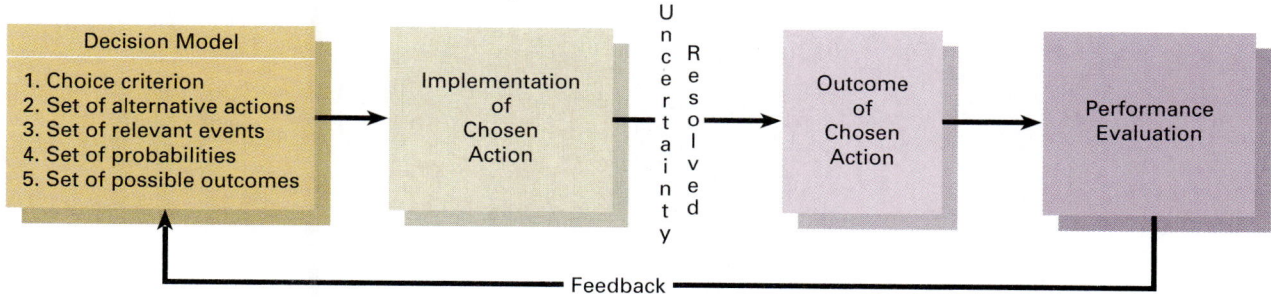

PROBABILITIES. Assigning probabilities is a key aspect of the decision model approach to coping with uncertainty. A **probability distribution** describes the likelihood (or probability) of each of the mutually exclusive and collectively exhaustive sets of events. The probabilities of these events will add to 1.00 because they are collectively exhaustive; that is, one of the events will definitely occur. In some cases, there will be much evidence to guide the assignment of probabilities. For example, the probability of obtaining a head in the toss of a fair coin is 1/2; that of drawing a particular playing card from a standard, well-shuffled deck is 1/52. In business, the probability of having a specified percentage of defective units may be assigned with great confidence, on the basis of production experience with thousands of units. In other cases, there will be little evidence supporting estimated probabilities. For example, how many units of a new pharmaceutical product will be sold next year?

The concept of uncertainty can be illustrated by a decision situation facing a book editor. The editor is deciding between publishing a spy novel and publishing a historical novel. Both book proposals require a $200,000 investment at the beginning of the year. (For simplicity, we ignore the time value of money here. See Chapters 20 and 21 for discussion of that topic.) On the basis of experience, the editor believes that the following *probability distribution* describes the relative likelihood of cash inflows for the next year (assume that the useful sales life of each book is one year):

Proposal A: Spy Novel		Proposal B: Historical Novel	
Probability	Cash Inflows	Probability	Cash Inflows
0.10	$300,000	0.10	$200,000
0.20	350,000	0.25	300,000
0.40	400,000	0.30	400,000
0.20	450,000	0.25	500,000
0.10	500,000	0.10	800,000
1.00		1.00	

Exhibit 3-6 shows a graphical comparison of the probability distributions.

EXPECTED VALUE. An **expected value** is a weighted average of the outcomes with the probability of each outcome serving as the weight. Where the outcomes are measured in monetary terms, *expected value* is often called **expected monetary value**. The expected monetary value of the cash inflows from the spy novel—denoted $E(a_1)$—is $400,000:

EXHIBIT 3-6
Decisions under Uncertainty: Comparison of Probability Distributions

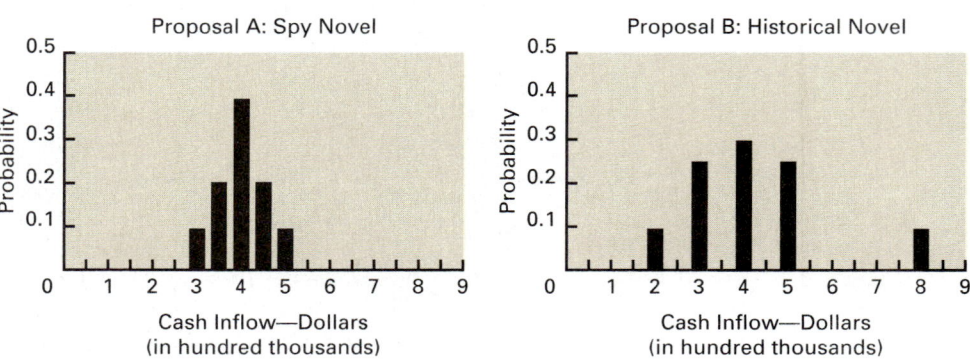

$$E(a_1) = 0.1(\$300{,}000) + 0.2(\$350{,}000) + 0.4(\$400{,}000) + 0.2(\$450{,}000) + 0.1(\$500{,}000) = \$400{,}000$$

The expected monetary value of the cash inflows from the historical novel—denoted $E(a_2)$—is \$420,000:

$$E(a_2) = 0.1(\$200{,}000) + 0.25(\$300{,}000) + 0.3(\$400{,}000) + 0.25(\$500{,}000) + 0.1(\$800{,}000) = \$420{,}000$$

Expected monetary value is widely used as a decision criterion. For a book editor wanting to maximize the expected monetary value, the historical novel is preferable to the spy novel.

To interpret expected value, imagine that the company publishes many historical novels, each with a probability distribution of cash inflows given in proposal B. The expected value of \$420,000 is the average cash inflow per novel that the publisher will receive when averaged across all novels. For a specific novel, the cash inflows will be either \$200,000, \$300,000, \$400,000, \$500,000, or \$800,000. But if the company publishes 100 such novels, it will expect to receive \$42 million in total cash inflows, for an average of \$420,000 per novel.

Many statisticians and accountants favor presenting the entire probability distribution directly to the decision maker. Others present information in three categories: optimistic, most likely, and pessimistic. Both of these presentations remind the user that uncertainty exists in the decision at hand.

Illustrative Problem

Reconsider Mary Frost and the booth rental alternatives offered by Computer Conventions, Inc. to sell Do-All software (p. 66):

- ◆ Option 1: \$2,000 fixed fee
- ◆ Option 2: \$1,400 fixed fee plus 5% of the convention revenues from Do-All sales
- ◆ Option 3: 20% of the convention revenues from Do-All sales (but no fixed fee)

Frost estimates a 0.60 probability that sales will be 40 units and a 0.40 probability that sales will be 70 units. Each Do-All software package will be sold for \$200. Frost will purchase the package from a computer software wholesaler at \$120 per unit with the privilege of returning all unsold units. Which booth rental alternative should Mary Frost choose?

GENERAL APPROACH TO UNCERTAINTY. The construction of a model for decision making consists of five steps that are keyed to the five characteristics described at the beginning of this Appendix.[5]

- ◆ *Step 1. Identify the choice criterion of the decision maker.* Assume that Mary Frost's choice criterion is to maximize expected net cash inflow at the convention.
- ◆ *Step 2. Identify the set of actions under consideration.* The notation for an action is *a*. Mary Frost has three possible actions:

$$a_1 = \text{Pay \$2,000 fixed fee}$$
$$a_2 = \text{Pay \$1,400 fixed fee plus 5\% of convention revenues}$$
$$a_3 = \text{Pay 20\% of convention revenues (but no fixed fee)}$$

- ◆ *Step 3. Identify the set of events that can occur.* The notation for an event is *x*. Mary Frost's only uncertainty is the number of units of Do-All software that she can sell:

$$x_1 = 40 \text{ units}$$
$$x_2 = 70 \text{ units}$$

- ◆ *Step 4. Assign probabilities for the occurrence of each event.* The notation for the probability of an event is $P(x)$. Frost assesses a 60% chance that she will sell 40 units and a 40% chance that she will sell 70 units. Therefore, the probabilities are

---

[5]The presentations here draw (in part) from teaching notes prepared by R. Williamson.

$$P(x_1) = 0.60$$
$$P(x_2) = 0.40$$

◆ *Step 5. Identify the set of possible outcomes that are dependent on specific actions and events.* The outcomes in this example take the form of six possible net cash flows that are displayed in a decision table in Exhibit 3-7. A **decision table** (sometimes called a **payoff table** or **payoff matrix**) is a summary of the contemplated actions, events, outcomes, and probabilities of events.

Mary Frost now can use the information in Exhibit 3-7 to compute the expected net cash inflow of each action as follows:

Pay $2,000 fixed fee: $E(a_1) = 0.60\,(\$1,200) + 0.40\,(\$3,600) = \$2,160$
Pay $1,400 fixed fee
   plus 5% of revenues: $E(a_2) = 0.60\,(\$1,400) + 0.40\,(\$3,500) = \$2,240$
Pay 20% of revenues
   (but no fixed fee): $E(a_3) = 0.60\,(\$1,600) + 0.40\,(\$2,800) = \$2,080$

To maximize expected net cash inflows, Frost should select action a_2—that is, contracting with Computer Conventions to pay $1,400 fixed fee plus 5% of convention revenues.

Exhibit 3-8 illustrates how the five steps can be summarized using a decision tree. The following symbols are widely used in drawing a decision tree:

    ◼ represents a decision node
    ⬤ represents an event node

Attached to each decision node will be all the individual actions being considered by the decision maker. Attached to each event node will be the total set of mutually exclusive and collectively exhaustive events.

Consider the effect of uncertainty on the preferred action choice. If Frost was certain that she would sell only 40 units of Do-All software, she would prefer alternative a_3—pay 20% of revenues and no fixed fee. To follow this reasoning, examine

EXHIBIT 3-7
Decision Table Presentation of Data for Mary Frost

Actions		Probability of Events	
		$X_1 = 40$ units of sales $P(X_1) = 0.60$	$X_2 = 70$ units of sales $P(X_2) = 0.40$
a_1:	Pay $2,000 fixed fee to Computer Conventions, Inc.	$\$1,200^l$	$\$3,600^m$
a_2:	Pay $1,400 fixed fee plus 5% of convention revenues to Computer Conventions, Inc.	$\$1,400^n$	$\$3,500^p$
a_3:	Pay 20% of convention revenues (but no fixed fee) to Computer Conventions, Inc.	$\$1,600^q$	$\$2,800^r$

^{l} Net cash flows = ($200 − $120) (40) − $2,000   = $1,200
^{m} Net cash flows = ($200 − $120) (70) − $2,000   = $3,600
^{n} Net cash flows = ($200 − $120 − $10*) (40) − $1,400 = $1,400
^{p} Net cash flows = ($200 − $120 − $10*) (70) − $1,400 = $3,500
^{q} Net cash flows = ($200 − $120 − $40†) (40)   = $1,600
^{r} Net cash flows = ($200 − $120 − $40†) (70)   = $2,800
*$10 = 5% of selling price of $200
†$40 = 20% of selling price of $200

EXHIBIT 3-8
Decision-Tree Presentation of Data

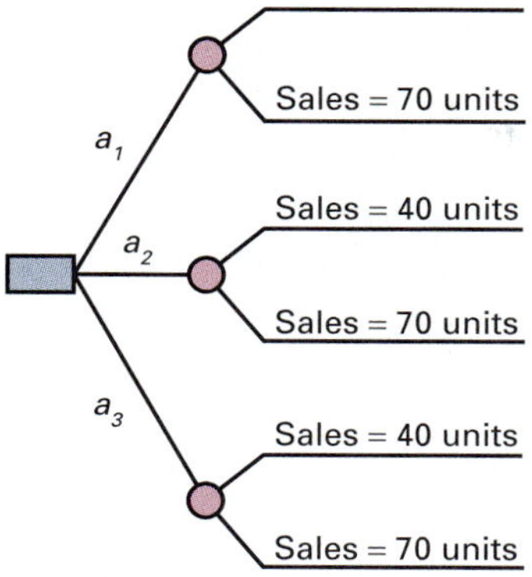

Actions (1)	Events (2)	Outcome (3)	Probability of Events (4)	Expected Cash Flow (5) = (3) × (4)
	Sales = 40 units	$1,200	0.60 =	$ 720
a_1	Sales = 70 units	$3,600	0.40 =	$1,440 $2,160
a_2	Sales = 40 units	$1,400	0.60 =	$ 840
	Sales = 70 units	$3,500	0.40 =	$1,400 $2,240
a_3	Sales = 40 units	$1,600	0.60 =	$ 960
	Sales = 70 units	$2,800	0.40 =	$1,120 $2,080

a_1 = Pay $2,000 fixed fee
a_2 = Pay $1,400 fixed fee plus 5% of revenues
a_3 = Pay 20% of revenues and no fixed fee

Exhibit 3-7. When 40 units are sold, alternative a_3 yields the maximum net cash inflows of $1,600. Because fixed costs are zero, booth rental costs are low when sales are low.

However, if Frost was certain that she would sell 70 units of Do-All software, she would prefer alternative a_1—pay a $2,000 fixed fee. Exhibit 3-7 indicates that when 70 units are sold, alternative a_1 yields the maximum net cash inflows of $3,600. Rental payments under a_2 and a_3 increase with units sold, but are fixed under a_1.

GOOD DECISIONS AND GOOD OUTCOMES. Always distinguish between a good decision and a good outcome. One can exist without the other. By definition, uncertainty rules out guaranteeing, after the fact, that the best outcome will always be obtained. It is possible that bad luck will produce unfavorable consequences even when good decisions have been made.

Suppose you are offered a one-time-only gamble tossing a fair coin. You will win $20 if the event is heads, but you will lose $1 if the event is tails. As a decision maker, you proceed through the logical phases: gathering information, assessing consequences, and making a choice. You accept the bet. Why? Because the expected value is $9.50 [0.5($20) + 0.5(–$1)]. The coin is tossed, and the event is tails. You lose. From your viewpoint, this was a good decision but a bad outcome.

A decision can be made only on the basis of information available at the time of the decision. Hindsight is flawless, but a bad outcome does not necessarily mean that a bad decision was made. Making a good decision is our best protection against a bad outcome.

TERMS TO LEARN

This chapter and the Glossary at the end of the book contain definitions of the following important terms:

breakeven point *(p. 62)*   choice criterion *(77)*   contribution income statement *(63)*
contribution margin *(62)*   contribution-margin percentage *(74)*
cost-volume-profit (CVP) *(61)*   decision model *(77)*   decision table *(80)*
events *(77)*   expected monetary value *(78)*   expected value *(78)*
gross margin *(73)*   margin of safety *(68)*   net income *(61)*
objective function *(77)*   operating income *(61)*   outcomes *(77)*   payoff matrix *(80)*
payoffs *(77)*   payoff table *(80)*   probability *(77)*   probability distribution *(78)*
PV chart *(68)*   revenue driver *(60)*   sales mix *(69)*   sensitivity analysis *(67)*
states *(77)*   states of nature *(77)*   uncertainty *(67)*   variable-cost percentage *(74)*

ASSIGNMENT MATERIAL

QUESTIONS

NOTE: To underscore the basic CVP relationships, the assignment material ignores income taxes unless stated otherwise.

3-1 Describe how the special case labeled CVP is different from the general case for predicting total revenues, total costs, and operating income.

3-2 "There are many cost drivers besides units of output (manufactured or sold)." **Name** two.

3-3 Distinguish between *operating income* and *net income*.

3-4 Identify three methods for determining the breakeven point.

3-5 State a general equation for computing the breakeven point or target operating income in terms of a single product and an output-related cost driver.

3-6 Why is it more accurate to describe the subject matter of this chapter as CVP analysis rather than as breakeven analysis?

3-7 Describe the assumptions underlying CVP analysis.

3-8 "Recognizing income taxes in CVP analysis will increase the breakeven point." Do you agree?

3-9 Define *contribution margin, gross margin, contribution-margin percentage, variable-cost percentage*, and *margin of safety*.

3-10 Describe the approach most widely used by managers to recognize uncertainty in CVP analysis.

3-11 "Gross margin for a manufacturer is contribution margin minus fixed manufacturing overhead costs." Do you agree? Explain.

3-12 "Unlike for merchandising- and manufacturing-sector companies, contribution margin and gross margin are the same for service-sector companies." Do you agree? Explain.

3-13 Give an example of how a manager may decrease variable costs while increasing fixed costs.

3-14 Give an example of how a manager may increase variable costs while decreasing fixed costs.

EXERCISES AND PROBLEMS

3-15 Fill in blanks. In the data following, fill in the information that belongs in the blank spaces for each of the four independent cases.

	Sales	Variable Costs	Fixed Costs	Total Costs	Operating Income	Contribution-Margin%
Case a.	$ —	$500	$ —	$ 800	$1,200	—
Case b.	2,000	—	300	—	200	—
Case c.	1,000	700	—	1,000	—	—
Case d.	1,500	—	300	—	—	0.40

3-16 Fill in blanks. Fill in the blanks for each of the following independent cases.

	Selling Price	Variable Costs per Unit	Units Sold	Contribution Margin	Fixed Costs	Operating Income
Case a.	$30	$20	70,000	$ —	$ —	$15,000
Case b.	25	—	180,000	900,000	800,000	—
Case c.	—	10	150,000	300,000	220,000	—
Case d.	20	14	—	120,000	—	12,000

3-17 CVP, changing cost inputs. Maria Montez is planning to sell a vegetable slicer-dicer for $15 per unit at a county fair. She purchases units from a local distributor for $6 each. She can return any unsold units for full refund. Fixed costs for booth rental, setup, and cleaning are $450.

Required

1. Compute the breakeven point in units sold.
2. Suppose the unit purchase cost is $5 instead of $6, but the selling price is unchanged. Compute the new breakeven point in units sold.

3-18 CVP, margin of safety. Suppose Lattin Corp.'s breakeven point is sales of $1,000,000. Fixed costs are $400,000.

Required

1. Compute the contribution-margin percentage.
2. Compute the selling price if variable costs are $12 per unit.
3. Suppose 80,000 units are sold. Compute the margin of safety.

3-19 CVP, international cost structure differences. Knitwear Inc. is choosing among three countries as the sole site for manufacturing its new sweater—Singapore, Thailand, and the United States. All sweaters are to be sold to retail outlets in the United States at $32 per unit. These retail outlets add their own markup when selling to final customers. The three countries differ in their fixed costs and variable costs per sweater.

	Annual Fixed Costs	Variable Manufacturing Costs per Sweater	Variable Marketing and Distribution Costs per Sweater
Singapore	$ 6.5 million	$ 8.00	$11.00
Thailand	4.5 million	5.50	11.50
United States	12.0 million	13.00	9.00

Required

1. Compute the breakeven point of Knitwear Inc. in both (a) units sold and (b) sales dollars for each of the three countries considered for manufacturing the sweaters.
2. If Knitwear Inc. sells 800,000 sweaters in 19_4, what is the budgeted operating income for each of the three countries considered for manufacturing the sweaters? Comment on the results.

3-20 CVP exercises. The Super Donut owns and operates six donut outlets in and around Kansas City. You are given the following corporate budget data for next year:

Sales	$10,000,000
Fixed costs	1,700,000
Variable costs	8,200,000

Variable costs vary with respect to the number of donuts sold.

Required
Compute budgeted operating income for each of the following deviations from the original budget data. (Consider each case independently.)

1. A 10% increase in contribution margin, holding sales dollars constant
2. A 10% decrease in contribution margin, holding sales dollars constant
3. A 5% increase in fixed costs
4. A 5% decrease in fixed costs
5. An 8% increase in units sold
6. An 8% decrease in units sold
7. A 10% increase in fixed costs and 10% increase in units sold
8. A 5% increase in fixed costs and 5% decrease in variable costs

3-21 CVP exercises. The Doral Company manufactures and sells pens. Present sales output is 5,000,000 units per year at a selling price of $0.50 per unit. Fixed costs are $900,000 per year. Variable costs are $0.30 per unit.

Required
(Consider each case separately.)

1. a. What is the present operating income for a year?
   b. What is the present breakeven point in sales dollars?

Compute the new operating income for each of the following changes:

2. A $0.04 per-unit increase in variable costs
3. A 10% increase in fixed costs and a 10% increase in units sold
4. A 20% decrease in fixed costs, a 20% decrease in selling price, a 10% decrease in variable costs per unit, and a 40% increase in units sold

Compute the new breakeven point in units for each of the following changes:

5. A 10% increase in fixed costs
6. A 10% increase in selling price and a $20,000 increase in fixed costs

3-22 CVP, alternative advertising cost structures. Beverage Products is currently negotiating with Consumer Impact, its prestige advertising agency. Consumer Impact (CI) will design and develop a new set of five advertisements for Diet Super Cola. CI has traditionally been paid a 15% commission on the amount that Beverage Products pays television stations to run its advertisements. In its recent campaign, Beverage Products paid television stations $5,000,000 to run Diet Super Cola advertisements over a three-month period. As a result, CI was paid $750,000 ($5,000,000 × 0.15) for its advertising work. Thus, the total cost of advertising to Beverage Products is $5,750,000.

The new marketing director of Beverage Products argues that the 15% commission is much too high and asks CI for a lower percentage. CI suggests an alternative compensation plan of a $400,000 fixed payment plus a 5% commission on television advertising payments.

Required
1. Graph the advertising agency cost of Beverage Products as a function of television advertising dollars for Diet Super Cola for:
   a. the traditional 15% commission cost structure
   b. the $400,000 plus 5% commission cost structure

Comment on the graphs.

2. What level of television advertising dollars results in Beverage Products paying CI the same amount under the (a) and (b) alternatives in requirement 1?

3. Under the 15% commission plan, what amount does Beverage Products pay CI if television advertising payments for Diet Super Cola are:
   a. $1,000,000
   b. $10,000,000

What rationale could exist for the 15% commission plan if it costs CI no more to develop the advertisements that are repeated many more times in (b) than in (a)?

3-23 CVP, movie production. Royal Rumble Productions has just finished production of *Feature Creatures*, the latest movie directed by Tony Savage and starring Ralph Michaels and Sally Martel. The total production cost to Royal Rumble was $5 million. All the production people and actors on *Feature Creatures* received a fixed salary (included in the $5 million) and will have no "residual" (equity interest) in the revenues or operating income from the movie. Media Productions will handle the marketing of *Feature Creatures*. Media agrees to invest a minimum $3 million of its own money in marketing the movie and will be paid 20% of the revenues Royal Rumble itself receives from the box-office receipts. Royal Rumble receives 62.5% of the total box-office receipts (out of which comes the 20% payment to Media Productions).

Required
1. What is the breakeven point to Royal Rumble for *Feature Creatures* expressed in terms of (a) revenues received by Royal Rumble and (b) total box-office receipts?
2. Assume in its first year of release, *Feature Creatures* has total box-office receipts of $300 million. What is the operating income to Royal Rumble from the movie in its first year?

3-24 CVP, cost structure differences, movie production. (Continuation of 3-23) Royal Rumble is negotiating for *Feature Creatures 2*, a sequel to its mega-blockbuster success *Feature Creatures*. This negotiation is proving more difficult than for the original movie. The budgeted production cost (excluding payments to the director Savage and the stars Michaels and Martel) for *Feature Creatures 2* is $21 million. The agent negotiating for Savage, Michaels, and Martel proposes either of two contracts:

   (A) Fixed salary component of $15 million for Savage, Michaels, and Martel (combined) with no "residual" interest in the revenues from *Feature Creatures 2*, or
   (B) Fixed salary component of $3 million plus a "residual" of 15% of the revenues Royal Rumble receives from *Feature Creatures 2*.

Media Productions will market *Feature Creatures 2*. It agrees to invest a minimum of $10 million of its own money. Because of its major role in the success of *Feature Creatures*, Media Productions will now be paid 25% of the revenues Royal Rumble receives from the total box-office receipts. Royal Rumble receives 62.5% of the total box-office receipts (out of which comes the 25% payment to Media Productions).

Required
1. What is Royal Rumble's breakeven point expressed in terms of:
   a. revenues received by that company
   b. total box-office receipts for *Feature Creatures 2*
   for each of the (A) and (B) contract proposals for Savage, Michaels, and Martel? Explain the difference between the breakeven points for contract proposals (A) and (B).
2. Assume *Feature Creatures 2* achieves the same $300 million in box-office revenues as *Feature Creatures*. What is the operating income to Royal Rumble from *Feature Creatures 2* if it accepts contract proposal (B)? Comment on the difference in operating income between the two films.

3-25 CVP, executive teaching compensation. Brian Smith is an internationally known U.S. professor specializing in consumer marketing. In 19_2 Smith and the United Kingdom Business School (UKBS) agreed to conduct a one-day seminar at UKBS for marketing executives. Each executive would pay £260 to attend the one-day seminar. The dean at UKBS indicates to Smith that the fixed costs for UKBS conducting the seminar would be:

◆ Advertising in magazines	£4,000
◆ Mailing of brochures	£3,000
◆ Administrative labor at UKBS	£2,000
◆ Charge for UKBS lecture auditorium	£1,000

The variable costs to UKBS for each participant attending the seminar would be:

- ◆ Meals and drinks   £25
- ◆ Binders and photocopying   £35

The dean at UKBS initially offered Smith its regular compensation package of (a) business-class airfare and accommodation (£3,000 maximum) and (b) a £2,000 lecture fee. Smith would qualify for the £3,000 maximum allowance. Smith views the £2,000 lecture fee as providing him no upside potential (that is, no sharing in the potential additional operating income that arises if the seminar is highly attended). He suggests he receive 50% of the operating income to UKBS (if positive) from the one-day seminar and no other payments. The dean of UKBS quickly agrees to Smith's proposal after confirming that Smith is willing to pay his own airfare and accommodation and deliver the seminar irrespective of the number of executives signed up to attend.

Required
1. What is UKBS's breakeven point (in number of executives attending) if:
   a. Smith accepts the regular compensation package of £3,000 expenses and a £2,000 lecture fee
   b. Smith receives only 50% of the operating income to UKBS (if positive) from the one-day seminar and no other payments

Comment on the results for (a) and (b).

2. Smith gave the one-day seminar at UKBS in 19_2 (60 attended), 19_3 (90 attended), and 19_4 (180 attended). How much was Smith paid by UKBS for the one-day seminar under the 50% of UKBS's operating income compensation plan in (a) 19_2, (b) 19_3, and (c) 19_4? (Assume that the £260 charge per executive attending and UKBS's fixed and variable costs are the same each year.)

3. After the 19_4 seminar, the dean at UKBS suggested to Smith that the "50%/50% profit-sharing plan was resulting in Smith getting excessive compensation in 19_4 and that a more equitable arrangement to UKBS be used in 19_5." How should Smith respond to this suggestion?

 3-26 CVP, shoe stores. The Walk Rite Shoe Company operates a chain of shoe stores. The stores sell 10 different styles of inexpensive men's shoes with identical unit costs and selling prices. A unit is defined as a pair of shoes. Each store has a store manager who is paid a fixed salary. Individual salespeople receive a fixed salary and a sales commission. Walk Rite is trying to determine the desirability of opening another store, which is expected to have the following revenue and cost relationships:

	Per Pair
Unit variable data:	
Selling price	$ 30.00
Cost of shoes	$ 19.50
Sales commissions	1.50
Total variable costs	$ 21.00
Annual fixed costs:	
Rent	$ 60,000
Salaries	200,000
Advertising	80,000
Other fixed costs	20,000
Total fixed costs	$360,000

Required
(Consider each question independently.)

1. What is the annual breakeven point in (a) units sold and (b) sales dollars?
2. If 35,000 units are sold, what will be the store's operating income (loss)?
3. If sales commissions were discontinued for individual salespeople in favor of an $81,000 increase in fixed salaries, what would be the annual breakeven point in (a) units sold and (b) sales dollars?

4. Refer to the original data. If the store manager were paid $0.30 per unit sold in addition to his current fixed salary, what would be the annual breakeven point in (a) units sold and (b) sales dollars?

5. Refer to the original data. If the store manager were paid $0.30 per unit commission on each unit sold in excess of the breakeven point, what would be the store's operating income if 50,000 units were sold? (This $0.30 is in addition to the commission paid to the sales staff and in addition to the store manager's fixed salary.)

3-27 CVP, shoe stores. (Refer to requirement 3 of 3-26.)

1. Calculate the number of units sold where the operating income under (a) a fixed salary plan and (b) a lower fixed salary and commission plan (for salespeople only) would be equal. Above that number of units sold, one plan would be more profitable than the other; below that number of units sold, the reverse would occur.

2. Compute the operating income or loss under each plan in requirement 1 at sales levels of (a) 50,000 units and (b) 60,000 units.

3. Suppose the target operating income is $168,000. How many units must be sold to reach the target under (a) the fixed salary plan and (b) the lower fixed salary and commission plan?

3-28 Sensitivity and inflation. (Continuation of 3-26)   As president of Walk Rite, you are concerned that inflation may squeeze your profitability. Specifically, you feel committed to the $30 selling price and fear that diluting the quality of the shoes in the face of rising costs would be an unwise marketing move. You expect the cost of shoes to rise by 10% during the coming year. You are tempted to avoid the cost increase by placing a noncancelable order with a large supplier that would provide 50,000 units of the specified quality for each store at $19.50 per unit. (To simplify this analysis, assume that all stores will face identical demands.) These shoes could be acquired and paid for as delivered throughout the year. However, all shoes must be delivered to the stores by the end of the year.

As a shrewd merchandiser, you foresee some risks. If sales were less than 50,000 units, you feel that markdowns of the unsold merchandise would be necessary to sell the goods. You predict that the average selling price of the leftover units would be $18.00. The regular commission of 5% of sales dollars would be paid to salespeople.

Required
1. Suppose that actual sales at $30 for the year is 48,000 units and that you contracted for 50,000 units. What is the operating income for the store?

2. If you had had perfect forecasting ability, you would have contracted for 48,000 units rather than 50,000 units. What would the operating income have been if you had ordered 48,000 units?

3. Given actual sales of 48,000 units, by how much would the average cost per unit have had to rise before you would have been indifferent between having the contract for 50,000 units and not having the contract?

3-29 Target operating incomes and contribution margins.   The Kaplan Company has a maximum capacity of 200,000 units per year. Variable manufacturing costs (with respect to units manufactured) are $12 per unit. Fixed manufacturing overhead is $600,000 per year. Variable marketing, distribution, customer service, and administrative costs (with respect to units sold) are $5 per unit, and fixed marketing, distribution, customer service, and administrative costs are $300,000 per year. The current selling price is $23 per unit. Kaplan has no beginning or ending inventory.

Required
(Consider each situation independently.)

1. What is the breakeven point in (a) units sold and (b) sales dollars?
2. How many units must be sold to earn a target operating income of $240,000 per year?
3. Assume that the company's sales for the year just ended totaled 185,000 units. A strike at a major supplier has caused a materials shortage, so the current year's sales will reach only 160,000 units. Top management is planning to slash fixed costs so that the total for the current year will be $85,000 less than last year. Management is also thinking of increasing the selling price, reducing variable costs, or both, in order to earn a target operating income that will be the same dollar amount as last year's. The company has already sold 30,000 units this

year at a selling price of $23 per unit with variable costs per unit unchanged. What contribution margin per unit is needed on the remaining 130,000 units in order to reach the target operating income?

3-30 Breakeven, CVP. (CMA adapted) Marston Corporation manufactures pharmaceutical products that are sold through a network of independent sales agents located in the United States and Canada. The agents are currently paid an 18% commission on sales. Marston used this percentage in preparing the following budgeted income statement for the fiscal year ending June 30, 19_4 (in thousands):

Sales		$26,000
Cost of goods sold		
Variable	$11,700	
Fixed	2,870	14,570
Gross margin		11,430
Marketing & administrative costs		
Sales commissions	4,680	
Fixed marketing costs	750	
Fixed administrative costs	1,850	7,280
Operating income		4,150
Fixed interest costs		650
Net income before taxes		3,500
Income taxes (40%)		1,400
Net income		$ 2,100

Since preparing this budgeted income statement, Marston has learned that its agents are demanding an increase in the commission rate to 23% for the upcoming year. As a result, Marston's president has decided to investigate the possibility of hiring its own sales staff in place of the network of sales agents and has asked Tom Ross, Marston's controller, to gather information on the costs associated with this alternative.

Ross estimates that Marston will have to hire eight salespeople to cover the current market area. The annual payroll cost of each of these employees will average $80,000, including fringe benefit costs. Travel and entertainment costs are expected to total $600,000 for the year, and the annual cost of a sales manager and sales secretary will be $150,000. In addition to their salary, the eight salespeople will each earn commissions at the rate of 10% on the first $2 million in sales and 15% on all sales over $2 million. Ross expects that all eight salespeople will exceed the $2 million mark and that sales will be at the level originally projected. Ross believes that Marston should also increase its marketing budget by $500,000.

Required
1. Calculate Marston Corporation's breakeven point in sales dollars for the fiscal year ending June 30, 19_4 if the company hires its own sales force and increases its marketing costs.
2. If Marston Corporation continues to sell through its network of sales agents and pays the higher commission rate, determine what the dollar sales for the fiscal year ending June 30, 19_4 would have to be to generate the same net income as projected in the budgeted income statement presented above.

3-31 Service business, cable TV. Cable Vision has been approached by the city of Mirada (population 800,000) to run its cable television operations. Mirada city officials have become tired of reporting losses on the cable television company they have operated for the past five years. Cable Vision currently operates cable television facilities for over 100 other cities or counties.

Cable Vision makes the following assumptions in its planning after negotiations with key parties:

a. A basic set of 10 cable television stations will be offered at a rate of $20 per month per subscriber. These 10 stations include a sports channel, a news channel, and other general-audience channels.

b. The city of Mirada would retain ownership of the physical facilities and would maintain them in working condition. Under a leasing agreement, Cable Vision will pay the city of Mirada $50,000 per month plus 10% of the monthly revenues from the first 10,000 subscribers and 5% of the monthly revenues from additional subscribers.

c. Cable Vision will receive the 10 channels in its basic service from Interlink Cable; Interlink acts as an intermediary between cable television stations and companies such as Cable Vision, which sell to individual subscribers. Interlink charges a monthly fixed fee of $20,000 plus a monthly charge of $8 per subscriber for the first 20,000 subscribers and $6 per additional subscriber.

d. Cable Vision estimates its own operating costs to include both a fixed and a variable component. The fixed component is $60,000 per month. The variable component is $2 per month per subscriber (to cover monthly billing, program news mailings, and so on).

Required

1. How does the contribution margin per subscriber behave over the 0 to 30,000 subscriber range?
2. Calculate the breakeven number of subscribers per month for Cable Vision.
3. What is the operating income per month to Cable Vision with (a) 10,000 subscribers, (b) 20,000 subscribers, and (c) 30,000 subscribers? Comment on the results.
4. You are hired by Cable Vision to give a second opinion on the reliability of the estimate of a breakeven point in requirement 2 above. What seems most certain in Cable Vision's projected costs? What concerns might you have?

3-32 Sales mix, two products. Goldman Company retails two products, a standard and a deluxe version of a luggage carrier. The budgeted income statement follows.

	Standard Carrier	Deluxe Carrier	Total
Sales in units	150,000	50,000	200,000
Sales @ $20 and $30 per unit	$3,000,000	$1,500,000	$4,500,000
Variable costs @ $14 and $18 per unit	2,100,000	900,000	3,000,000
Contribution margins	$ 900,000	$ 600,000	1,500,000
Fixed costs			1,200,000
Operating income			$ 300,000

Required

1. Compute the breakeven point in units, assuming that the planned sales mix is maintained.
2. Compute the breakeven point in units (a) if only standard carriers are sold and (b) if only deluxe carriers are sold.
3. Suppose 200,000 units are sold, but only 20,000 are deluxe. Compute the operating income. Compute the breakeven point if these relationships persist in the next period. Compare your answers with the original plans and the answer in requirement 1. What is the major lesson of this problem?

3-33 Sales mix, three products. The Ronowski Company has three product lines of belts—A, B, and C—having contribution margins of $3, $2, and $1, respectively. The president foresees sales of 200,000 units in the coming period, consisting of 20,000 A, 100,000 B, and 80,000 C. The company's fixed costs for the period are $255,000.

Required

1. What is the company breakeven point in units, assuming that the given sales mix is maintained?
2. If the mix is maintained, what is the total contribution margin when 200,000 units are sold? What is operating income?
3. What would operating income become if 20,000 units of A, 80,000 units of B, and 100,000 units of C were sold? What is the new breakeven point in units if these relationships persist in the next period?

3-34 Sales mix, three products. Mendez Company has three products, tote bags H, J, and K. The president plans to sell 200,000 units during the next period, consisting of 80,000 H, 100,000 J, and 20,000 K. The products have unit contribution margins of $2, $3, and $6, respectively. The company's fixed costs for the period are $406,000.

Required

1. Compute the budgeted operating income. Compute the breakeven point in units, assuming that the given sales mix is maintained.

2. Suppose 80,000 units of H, 80,000 units of J, and 40,000 units of K are sold. Compute the budgeted operating income. Compute the new breakeven point in units if these relationships persist in the next period.

3-35   Nonprofit institution.   The City of Little Rock, Arkansas, makes a $400,000 lump-sum budget appropriation to an agency to conduct a counseling program for drug addicts for a year. All of the appropriation is to be spent. The variable costs for drug prescriptions average $400 per patient per year. Fixed costs are $150,000.

Required
1. Compute the number of patients that could be served in a year.
2. Suppose the total budget for the following year is reduced by 10%. Fixed costs are to remain the same. The same level of service to each patient will be maintained. Compute the number of patients that could be served in a year.
3. As in requirement 2, assume a budget reduction of 10%. Fixed costs are to remain the same. The drug counselor has discretion as to how much in drug prescriptions to give to each patient. She does not want to reduce the number of patients served. On the average, what is the cost of drugs that can be given to each patient? Compute the percentage decline in the annual average cost of drugs per patient.

3-36   CVP, income taxes.   The Bratz Company has fixed costs of $300,000 and a variable-cost percentage of 80%. The company earns net income of $84,000 in 19_4. The income tax rate is 40%.

Required
Compute (1) operating income, (2) contribution margin, (3) total sales, and (4) breakeven point in sales dollars.

3-37   CVP, income taxes.   The Rapid Meal has two restaurants that are open 24 hours per day. Fixed costs for the two restaurants together total $450,000 per year. Service varies from a cup of coffee to full meals. The average sales check for each customer is $8.00. The average cost of food and other variable costs for each customer is $3.20. The income tax rate is 30%. Target net income is $105,000.

Required
1. Compute the total dollar sales needed to obtain the target net income.
2. How many sales checks are needed to earn net income of $105,000? To break even?
3. Compute net income if the number of sales checks is 150,000.

3-38   CVP, income taxes. (CMA)   R. A. Ro and Company, a manufacturer of quality handmade walnut bowls, has experienced a steady growth in sales for the past five years. However, increased competition has led Mr. Ro, the president, to believe that an aggressive marketing campaign will be necessary next year to maintain the company's present growth.

To prepare for next year's marketing campaign, the company's controller has prepared and presented Mr. Ro with the following data for the current year, 19_4:

Variable costs (per bowl):	
Direct manufacturing labor	$ 8.00
Direct materials	3.25
Variable overhead	
(Manufacturing, marketing distribution,	
customer service, & administration)	2.50
Total variable costs	$13.75
Fixed costs:	
Manufacturing	$ 25,000
Marketing, distribution, & customer service	40,000
Administrative	70,000
Total fixed costs	$135,000
Selling price per bowl	$25.00
Expected sales, 19_4 (20,000 units)	$500,000
Income tax rate: 40%	

Required

1. What is the projected net income for 19_4?

2. What is the breakeven point in units for 19_4?

3. Mr. Ro has set the sales target for 19_5 at a level of $550,000 (or 22,000 bowls). He believes an additional marketing cost of $11,250 for advertising in 19_5, with all other costs remaining constant, will be necessary to attain the dollar sales target. What will be the net income for 19_5 if the additional $11,250 is spent and the dollar sales target is met?

4. What will be the breakeven point in sales dollars for 19_5 if the additional $11,250 is spent for advertising?

5. If the additional $11,250 is spent for advertising in 19_5, what is the required 19_5 sales dollars for 19_5's net income to equal 19_4's net income?

6. At a sales level of 22,000 units, what maximum amount can be spent on advertising if a 19_5 net income of $60,000 is desired?

3-39 Review of Chapters 2 and 3. For each of the following independent cases, find the unknowns designated by the capital letters.

	Case 1	Case 2
Direct materials used	$ H	$ 40,000
Direct manufacturing labor costs	30,000	15,000
Variable marketing, distribution, customer service, and administrative costs	K	T
Fixed manufacturing overhead costs	I	20,000
Fixed marketing, distribution, customer service, and administrative costs	J	10,000
Gross margin	25,000	20,000
Finished goods inventory, 1/1	0	5,000
Finished goods inventory, 12/31	0	5,000
Contribution margin	30,000	V
Sales	100,000	100,000
Direct materials inventory, 1/1	12,000	20,000
Direct materials inventory, 12/31	5,000	W
Variable manufacturing overhead costs	5,000	X
Work in process, 1/1	0	9,000
Work in process, 12/31	0	9,000
Purchases of direct materials	15,000	50,000
Breakeven point (in dollars)	66,667	Y
Cost of goods manufactured	G	U
Operating income (loss)	L	(5,000)

3-40 Miscellaneous alternatives, contribution income statement. The income statement of Hall Company follows. Commissions are based on sales dollars. All other variable costs vary in terms of units sold.

The manufacturing plant has a capacity of 150,000 units per year. The results for 19_4 have been disappointing. Top management is sifting through a number of possible ways to make operations profitable in 19_5.

Hall Company
Income Statement
For the Year Ended December 31, 19_4

Sales (90,000 units at $4.00 per unit)			$360,000
Cost of goods sold:			
Direct materials		$90,000	
Direct manufacturing labor		90,000	
Manufacturing overhead:			
Variable	$18,000		
Fixed	80,000	98,000	
Cost of goods sold			278,000
Gross margin			82,000
Operating costs			
Marketing, distribution, & customer service			
Variable:			
Sales commissions*	$18,000		
Shipping	3,600	21,600	
Fixed:			
Advertising, salaries, etc.		40,000	61,600
Administrative costs:			
Variable		4,500	
Fixed		20,400	24,900
Total operating costs			86,500
Operating income (loss)			$ (4,500)

*Based on dollar sales, not units.

Required

Consider requirements 2 through 5 independently.

1. Recast the 19_4 income statement into a contribution format. There will be three major sections: sales, variable costs, and fixed costs. Show costs per unit in an adjacent column. Allow adjacent space for entering your answers to requirement 2.

2. The sales manager is torn between two courses of action:
   a. He has studied the market potential and believes that a 15% slash in selling price would fill the plant to capacity.
   b. He wants to increase selling price by 25%, to increase advertising by $150,000, and to boost sales commissions to 10% of dollar sales. Under these circumstances, he thinks that unit sales will increase by 50%.

Prepare a 19_5 budgeted income statement for each alternative, using a contribution format and two columns, one for per-unit amounts and one for total amounts. Assume that there are no changes in fixed costs other than advertising.

3. The president does not want to change the selling price. How much can advertising be increased to bring production and sales up to 130,000 units and still earn a target operating income of 5% of sales?

4. A mail-order firm is willing to buy 60,000 units of product "if the price is right." Assume that the present market of 90,000 units at a $4 selling price will not be disturbed. Hall Company will not pay any sales commission on these 60,000 units. The mail-order firm will pick up the units at the Hall plant. However, Hall must refund $24,000 of the total dollar sales as a promotional and advertising allowance for the mail-order firm. In addition, special packaging will increase manufacturing costs on these 60,000 units by $0.10 per unit. At what selling price must the mail-order firm business be quoted for Hall to break even on *total* operations in 19_5?

5. The president suspects that an attractive new package will aid consumer awareness and ultimately Hall's sales. Present packaging costs per unit are all variable and consist of $0.05 direct materials and $0.04 direct manufacturing labor; new packaging costs per unit will be $0.30 and $0.13 respectively. Assume no other changes. How many units must be sold to earn a target operating income of $20,000?

3-41 CVP, international expansion alternatives. Dortmund Inc. produces and sells the highly popular Dortmund Lager beer in Europe. In the current year, it will sell 20 million barrels of beer in Europe. For many years it has sold in the North American market through a New York-based importer. The contract with the importer is up for renewal and Dortmund decides to reconsider its North American strategy. After much analysis, it decides that three alternatives warrant further consideration:

- *Alternative A (stay with the importer):* Sell through the current importer who manages all the marketing and distribution of Dortmund Lager in North America. Dortmund receives a net payment of $5 per barrel (net after transportation costs and import duties) sold in the North American market.

- *Alternative B (move to production licensing):* License production of Dortmund Lager to a North American brewer who also will manage its marketing and distribution. This brewer will charge Dortmund a fixed fee of $5 million each year to cover its costs of maintaining the quality of Dortmund products. It will pay Dortmund $10 per barrel of Dortmund Lager it sells in North America.

- *Alternative C (turn to self-production):* Purchase a fully operational brewing plant from a North American brewer with excess capacity. The annual fixed costs of operating the plant are $30 million and the variable costs are $60 per barrel. Dortmund will sell to independent wholesalers in North America at $100 per barrel.

Required
1. Compute the breakeven point for each of the three alternatives that Dortmund Inc. is considering.
2. At which unit sales level (barrels) of Dortmund Lager in North America would Dortmund Inc. report the same operating income under alternatives B and C?
3. What is the operating income to Dortmund Inc. for unit sales of (a) 500,000 barrels and (b) 2,000,000 barrels under each of the three strategy alternatives (A, B, and C) being considered? Comment on the results.

3-42 CVP, service sector, ethics. Home Cleaning Inc. is a franchiser of home cleaning operations. Each franchise operator is given exclusive rights to use the Home Cleaning Inc. name and products in a geographical area. The operator pays Home Cleaning Inc. $2,000 per month plus 20% of its monthly operating income.

Arthur Beetson, the controller of Home Cleaning Inc., has received several letters from former employees of the Tuscany franchise operator. Each letter argued that the Tuscany operator is engaging in questionable practices. Beetson investigates these claims and finds the following:

- **a.** The Tuscany operator employed many illegal immigrants at the rate of $8 per hour and did not pay them any benefits. Approximately 60% of the 2,500 hours billed in October 19_5 was worked by illegal immigrants. The remaining 40% of the 2,500 hours was worked by legal employees who received the standard $12 per hour ($9 per hour salary and $3 per hour fringe benefits). The accounting records accurately recorded the actual labor costs paid to each employee (including employees who were illegal immigrants) in October 19_5.

- **b.** Customers were billed at $20 per hour. The accounting records of the Tuscany operator did not always record all work performed by Tuscany employees. In some cases, customers had requested extra work while the Tuscany employees were on the job. Beetson found several cases where employees agreed to do the extra work for $15 per hour if the payment was made in cash. These cash payments were not recorded as revenues of the Tuscany operator.

- **c.** The fixed costs of the Tuscany operator were the $2,000 monthly fee to Home Cleaning Inc. and $6,000 for office costs. The variable costs were $12 per hour compensation to each employee and $3 per hour for supplies. In addition, Tuscany pays Home Cleaning Inc. 20% of its monthly operating income.

Required
1. Assume that the Tuscany operator paid all employees the standard $12 per hour rate in October 19_5. What is:
   a. the breakeven point for the operator in (i) hours billed to clients and (ii) dollar billing to clients
   b. the operating income to the Tuscany operator for the 2,500 hours billed.

Ignore any "non-reported" cash payments to employees in your answer.

2. Repeat requirement 1 using the actual labor rates and labor mix in October 19_5—that is, 60% of the 2,500 hours at $8 per hour and 40% at $12 per hour. Comment on your results relative to those computed in requirement 1.

3. Repeat requirement 1 taking into account the following change in the way the Tuscany operator pays Home Cleaning Inc. Assume that Tuscany pays $2,000 per month and 5% of monthly revenues. Comment on your results relative to those computed in requirement 1. Why might Home Cleaning Inc. make this change in payments to be made by franchise operators? (Use the $12 rate for all 2,500 hours.)

4. The Tuscany operator was very angry when visited by Beetson. The operator argued that by employing lower-cost employees, his operating income was higher and the 20% operating income franchise fee he paid to Home Cleaning Inc. was also higher. It was a "win-win" situation, he said. The operator also argued that any cash receipts for extra work by Tuscany employees were immaterial and that it was something that everyone in this business had to accept. Customers who were willing to pay for extra work likely were very satisfied, and hence the possibility of this extra work made Home Cleaning Inc. employees work all the harder. How should Beetson respond to these arguments by the Tuscany operator? Refer to the "Standards of Ethical Conduct for Management Accountants" (p. 17) in your answer.

Coverage of the Chapter Appendix

3-43 Purchase order size. Once a day Supervalue, a retailer, stocks fresh-cut carnations. Each flower costs $0.40 and sells for $1.00. Supervalue never reduces the selling price. Unsold carnations at the end of each day are given to a nearby church. Supervalue estimates daily demand characteristics as follows:

Demand	Probability
0	0.05
100	0.20
200	0.40
300	0.25
400	0.10
	1.00

Required
How many carnations should Supervalue purchase each day to maximize expected operating income? Why?

3-44 Purchase of a new lathe, demand uncertainty. (A. Atkinson) The manager of operations at Purcell's Cove Machine Shop is considering leasing equipment for a new flexible manufacturing system (FMS) to replace existing equipment. The FMS lease will increase annual fixed costs by $900,000 per year and will reduce variable costs by $800 per job.

The manager believes that the annual number of jobs processed by the company will be 900, 1,200, or 1,500. The probabilities of these events occurring are:

Annual Number of Jobs	Probability
900	0.25
1,200	0.45
1,500	0.30
	1.00

Required
Should the company lease the FMS? Support your conclusion with appropriate calculations.

3-45 Setting prices and uncertainty. Trent Harris, marketing manager of The Complete Angler, is examining the pricing of its fishing-reel product line. Assume that the unit cost of a fishing reel is known with certainty to be $16. Harris is trying to decide whether to set a selling price of $20 or $22 per unit. The top price has been $20 for the past 30 months. Average monthly sales are forecast as follows:

At a Unit Price of $20		At a Unit Price of $22	
Units	**Probability**	**Units**	**Probability**
1,050,000	0.05	800,000	0.10
1,000,000	0.90	750,000	0.60
950,000	0.05	700,000	0.30
	1.00		1.00

The Complete Angler uses the expected monetary value choice criterion in its decisions.

Required
Determine the optimal selling price of a fishing reel. Show computations.

3-46 CVP under uncertainty. (J. Patell)   In your new position as supervisor of product introduction, you have to decide on a pricing strategy for a talking-doll specialty product with the following cost structure:

Variable costs per unit	$50
Fixed costs	$200,000

The dolls are manufactured upon receipt of orders, so the inventory levels are insignificant. Your market research assistant is very enthusiastic about probability models and has presented the results of his price analysis in the following form:

 a. If you set the selling price at $100 per unit, the probability distribution of total sales is uniform between $300,000 and $600,000. Under this distribution, there is a 0.50 probability of equaling or exceeding sales of $450,000.

 b. If you lower the selling price to $70 per unit, the distribution remains uniform, but it shifts up to the $600,000 to $900,000 range. Under this distribution, there is a 0.50 probability of equaling or exceeding sales of $750,000.

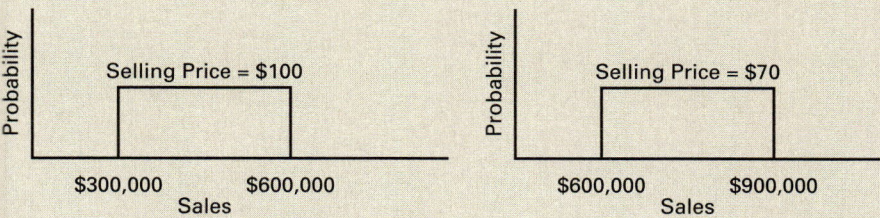

Required
1. This is your first big contract, and, above all, you want to show an operating income. You decide to select the strategy that maximizes the probability of breaking even or earning a positive operating income.
 a. What is the probability of at least breaking even with a selling price of $100 per unit?
 b. What is the probability of at least breaking even with a selling price of $70 per unit?
2. Your assistant suggests that maximum expected operating income might be a better objective to pursue. Which pricing strategy would result in the higher expected operating income? (Use the expected dollar sales under each pricing strategy when making expected operating income computations.)

3-47 CVP under uncertainty. (R. Jaedicke and A. Robichek, adapted)   The Jaro Company is considering two new colors for their umbrella products—emerald green and shocking pink. Either can be produced by using present facilities. Each product requires an increase in annual fixed costs of $400,000. The products have the same selling price ($10) and the same variable costs per unit ($8).
 Management, after studying past experience with similar products, has prepared the following probability distribution:

Event (Units Demanded)	Probability for Emerald Green Umbrella	Probability for Shocking Pink Umbrella
50,000	0.0	0.1
100,000	0.1	0.1
200,000	0.2	0.1
300,000	0.4	0.2
400,000	0.2	0.4
500,000	0.1	0.1
	1.0	1.0

Required
1. What is the breakeven point for each product?
2. Which product should be chosen, assuming the objective is to maximize expected operating income? Why? Show computations.
3. Suppose management was absolutely certain that 300,000 units of shocking pink would be sold, but it still faces the same uncertainty about the demand for emerald green as outlined in the problem. Which product should be chosen? Why? What benefits are available to management from having the complete probability distribution instead of just an expected value?

4

Costing Systems

in the Service and

Merchandising Sectors

Building block concepts of costing systems

Job costing and process costing systems

Job costing in service organizations

Tracing direct costs to jobs

Allocating indirect costs to jobs

Actual, normal, and budgeted costing methods

Refinements in costing systems

Peanut-butter costing approaches

Activity-based costing

Customer costing

The value chain in job costing and customer costing systems

Department stores such as Target carry a broad range of products. These products differ greatly in the resources the company uses in selling them. There is increased interest in cost accounting to measure these differences in resource usage in individual product profitability reports.

How much does it cost Arthur Andersen to audit Federal Express Corporation? How much does it cost Macy's to sell a pair of Levi's blue jeans? How much does it cost Ford Motor Company to manufacture and sell a vehicle to a Ford dealer? This chapter presents a general approach to addressing these questions. Both concepts and techniques are emphasized, with illustrations drawn from the service and merchandising sectors. Chapter 5 extends the discussion with a focus on manufacturing.

Before we explore the details of costing systems, three points are worth noting:

1. The cost-benefit approach is central in designing and choosing management accounting systems. Elaborate systems are expensive in terms of time and money. The costs of educating managers and other personnel must be considered. More sophisticated systems are installed only if managers believe that collective decisions will be sufficiently improved in a cost-benefit sense.

2. Systems should be tailored to the underlying operations and not vice versa. Any significant change in underlying operations is likely to justify a corresponding change in the accompanying accounting systems. The best systems design begins with a careful study of how operations are conducted and a resulting determination of what information to gather and report. The worst systems are those that operating managers perceive as misleading or even useless.

3. Costing systems aim to report cost numbers that reflect the way in which particular cost objects—such as products, services, and customers—use the resources of an organization. Cost information plays a central role in many decisions. For example, which products, services, or customers should the business emphasize? What selling price should be charged? What amount is owed to a contractor who is reimbursed on the basis of incurred costs plus a markup of, say, 15%? Costing systems also include inventoriable cost figures that are reported in the balance sheets of firms in the merchandising and manufacturing sectors.

BUILDING BLOCK CONCEPTS OF COSTING SYSTEMS

Objective 1

Describe the building block concepts of costing systems

Let's review some terms introduced in Chapter 2 that we will use in discussing costing systems:

- ◆ *Cost object:* anything for which a separate measurement of costs is desired
- ◆ *Direct costs of a cost object:* costs that are related to the cost object and can be traced to it in an economically feasible way

♦ *Indirect costs of a cost object:* costs that are related to the cost object but cannot be traced to it in an economically feasible way. Indirect costs are assigned to the cost object using a cost-allocation method.

The relationship among these three terms is:

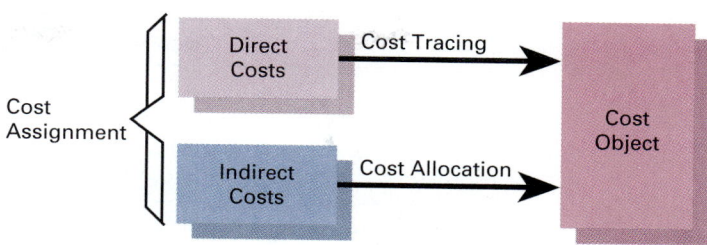

Cost tracing is the assigning of direct costs to the chosen cost object. *Cost allocation* is the assigning of indirect costs to the chosen cost object. *Costs assigned* to a cost object are the direct costs traced plus the indirect costs allocated to the cost object.

Two terms not previously defined are also central to Chapters 4 and 5:

♦ **Cost pool:** a grouping of individual cost items. Cost pools can range from the very broad (such as a companywide total cost pool for telephones and fax machines) to the very narrow (such as the telephone and fax machine cost pool for a sales manager in Iowa City).

♦ **Cost allocation base:** a factor that is the common denominator for systematically linking an indirect cost or group of indirect costs to a cost object. A cost allocation base can be financial (such as sales dollars) or nonfinancial (such as the number of products distributed). Companies often seek to use the cost driver of the indirect costs as the cost allocation base.

These terms are central to understanding how costing systems operate in organizations.

JOB COSTING AND PROCESS COSTING SYSTEMS

Objective 2

Distinguish between job costing and process costing

Companies frequently adopt one of two basic costing systems to assign costs to products or services:

♦ **Job costing system.** In this system, the cost of a product or service is obtained by assigning costs to a distinct, identifiable product or service. A *job* is a task for which resources are expended in bringing a distinct, identifiable product or service to market. Frequently, a job is custom-made for a specific customer.

♦ **Process costing system.** In this system, the cost of a product or service is obtained by assigning costs to masses of similar units and then computing unit costs on an average basis. Frequently, identical items (such as Barbie dolls or nails) are produced for general sale and not for any specific customer.

These two costing systems are best viewed as ends of a spectrum:

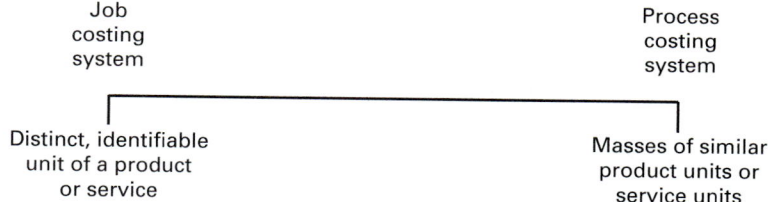

Many companies have costing systems that are neither pure job costing (one end of the spectrum) nor pure process costing (the other end of the spectrum). Rather, they

EXHIBIT 4-1

Examples of Job Costing and Process Costing in the Service, Merchandising, and Manufacturing Sectors

	Service Sector	Merchandising Sector	Manufacturing Sector
Job Costing Used	◆ Accounting firm audits ◆ Advertising campaigns	◆ Mail-order catalog ◆ Bookstore	◆ Aircraft assembly ◆ Manufacture of machining systems
Process Costing Used	◆ Retail banking ◆ Postal delivery (standard items)	◆ Grain dealing ◆ Magazine subscription	◆ Oil refining ◆ Beverage-drink production

have hybrid costing systems that combine elements of both job costing and process costing. For now, we introduce these two systems by focusing on their pure versions. Exhibit 4-1 presents examples of industries in the service, merchandising, and manufacturing sectors that predominantly use either a job costing system or a process costing system.

The particular service or product that companies handle through job costing may differ sizably among customers. For example, an audit is a service that accounting firms provide, and audits can differ markedly in complexity among clients. A mail-order business may define any shipment as a job, and each shipment may be made up of different combinations of any number of items sent to various customers. An aircraft assembly company may have customers who differ in their specifications about electronic equipment, size of restrooms, and so on. In these examples, the service or product is distinct and identifiable. Job costing systems are designed to accommodate the cost accounting for these individual services or products.

Companies that use process costing provide a similar (in many cases identical) product or service to their customers. For example, a retail bank provides the same service to all its customers in processing deposits. A magazine publishing company provides the same product (say, a weekly issue of *Newsweek* or *Time*) to its different customers. An oil-refining company processing crude oil provides the same product to its different customers. Process costing systems average the costs of providing a similar product or service to different customers to obtain a per unit cost.

This chapter discusses job costing in the service and merchandising sectors of the economy. Chapter 5 extends the discussion of job costing to manufacturing. The costing of manufactured products requires additional concepts (associated with work-in-process and finished goods inventories) that are not found in the service and merchandising sectors. Chapters 5, 17, and 18 further discuss process costing.

We cover job costing in the service sector first for two reasons. First, the service sector is now the largest sector in many economies. In the United States, more than 60% of all employment is in the service sector. Second, use of the service sector enables us to present the basic concepts of a costing system before covering the additional complexity added by the existence of inventories.

[THOSE READERS WHO PREFER TO COVER JOB COSTING IN MANUFACTURING FIRST CAN NOW READ PAGES 140–55 WITHOUT LOSS OF CONTINUITY.]

Service sector companies provide their customers with services or intangible products. Within the service sector, jobs often differ considerably as to how long they take, how many resources they consume, and how technically complex they are. Examples include a service call to repair plumbing, an audit engagement, the making of a movie, and a university research project for a government agency.

Job Costing of an Audit Engagement

R. C. Lindsay and Associates is a public accounting firm. Exhibit 4-2 presents the most recent income statement (19_4) and the budgeted income statement for the coming year (19_5). During the year, the costs of each audit job are monitored continuously. There are two main uses of this job cost information:

1. To guide decisions on job pricing and job emphasis (that is, which jobs to actively seek and which to let pass). Lindsay agrees to a fixed fee for each job before doing the work. Actual costs on similar jobs in the recent past are a key input in projecting the costs on future audit jobs.

2. To assist in cost planning and cost management. Information about direct and indirect costs enables Lindsay to manage existing jobs so that work is done efficiently.

EXHIBIT 4-2

Income Statement for Lindsay and Associates: 19_4 Actual Results and 19_5 Budget (in thousands)

	19_4 Actual Results		19_5 Budget	
Operating revenues		$28,020		$29,560
Operating costs				
Professional labor		13,710		14,400*
Other employee costs				
Office staff	$1,840		$2,030	
Information systems	1,530		1,600	
Administrative	500		660	
Other	820	4,690	1,038	5,328
Other (nonlabor) operating costs				
Professional liability				
insurance	1,471		2,160	
Professional				
development	540		880	
Occupancy	1,913		2,000	
Phone/fax/copying	1,330		1,430	
Travel	718		770	
Other	930	6,902	392	7,632
Operating costs		25,302		27,360
Operating income		$ 2,718		$ 2,200

*180 professionals × 1,600 billable hours × $50/hour = $14,400,000

The 19_5 costs of the annual audit of Tracy Transport, a client who is a medium-sized trucking company, are currently being examined. This example will be used to present a general approach to job costing.

General Approach to Job Costing

Objective 3
Outline a five-step approach to job costing

The five steps taken in assigning costs to individual jobs are presented below. These five steps apply equally to job costing in the service, merchandising, or manufacturing sectors.

STEP 1: *IDENTIFY THE JOB THAT IS THE CHOSEN COST OBJECT.* In this example, the job is the annual audit of the financial statements of Tracy Transport.

STEP 2: *IDENTIFY THE DIRECT COST CATEGORIES FOR THE JOB.* Lindsay identifies only one category of direct costs when costing individual audit jobs—professional labor. Each professional keeps a daily time record that is used to trace professional labor-hours to individual audit jobs. Professional labor is charged to individual jobs at $50 per hour. (The calculation of this $50 rate is described later in this chapter.)

STEP 3: *IDENTIFY THE INDIRECT COST POOLS ASSOCIATED WITH THE JOB.* Lindsay groups all its individual indirect costs into a single cost pool called audit support. This cost pool represents all costs in Lindsay's Audit Support Department.

EXHIBIT 4-3
Job Costing Overview at Lindsay and Associates

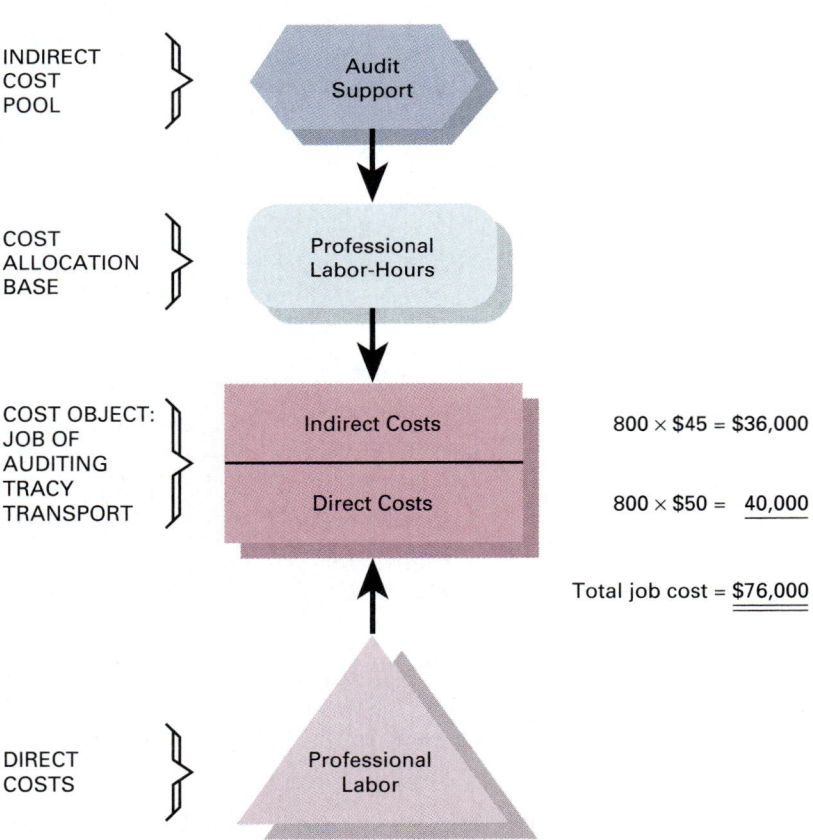

STEP 4: *SELECT THE COST ALLOCATION BASE TO USE IN ASSIGNING EACH INDIRECT COST POOL TO THE JOB.* Lindsay selects the allocation base that has a cause-and-effect relationship between changes in it and changes in the level of indirect costs. The allocation base selected for the audit support indirect cost pool is professional labor-hours. (Chapter 14 further discusses the criteria that guide the choice of a cost allocation base.)

STEP 5: *DEVELOP THE RATE PER UNIT OF THE COST ALLOCATION BASE USED TO ALLOCATE INDIRECT COSTS TO THE JOB.* As explained later in this chapter, Lindsay uses a rate of $45 per professional labor-hour.

Those five steps guide the computation of the job cost of the Tracy Transport audit. Suppose the Tracy Transport audit required 800 professional labor-hours. Its costs would be:

Direct costs	
Professional labor: $50 × 800	$40,000
Indirect costs	
Audit support: $45 × 800	36,000
Total job costs	$76,000

Exhibit 4-3 presents an overview of the job costing system Lindsay uses. This exhibit includes the five building-block concepts of this chapter—cost object, cost pool, direct costs of a cost object, indirect costs of a cost object, and cost allocation base. Costing overview exhibits like Exhibit 4-3 are important learning tools. We urge you to sketch one when you need to understand a costing system. (The symbols in Exhibit 4-3 are used consistently in the costing system overviews presented in this book. For example, a triangle always identifies a direct cost.)

Lindsay uses its job costing system to help manage the costs in its Audit Support Department as well as to determine the costs of individual audit jobs such as Tracy Transport. The Audit Support Department and *each* audit job are important cost objects. The relationship between these two cost objects is as follows:

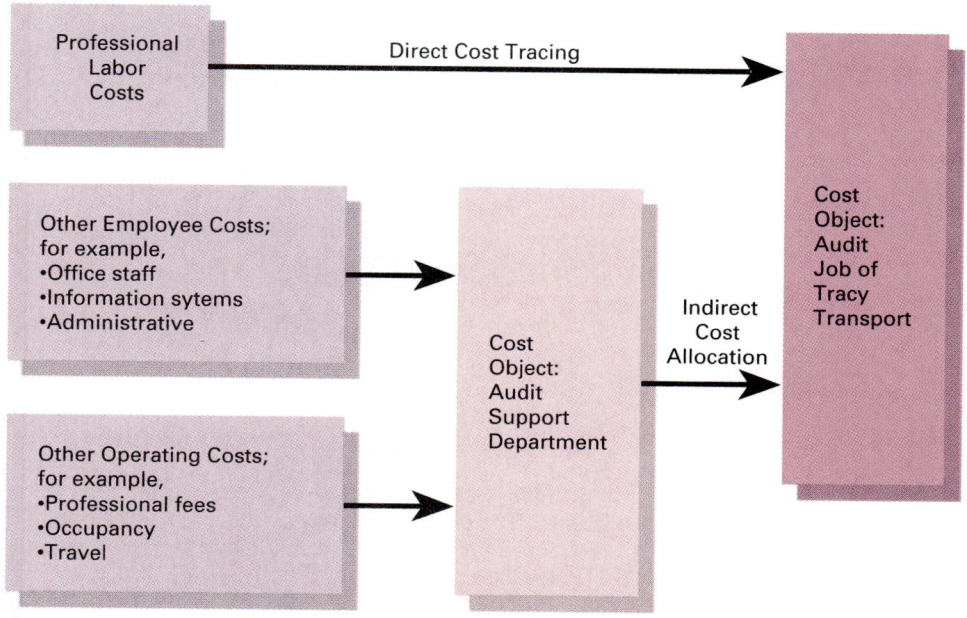

Source Documents

Managers and accountants gather the cost information that goes into their systems through source documents. The source documents Lindsay uses in its job costing system include time records filled out by the professional staff. All professional staff members record (in a computer file) how they spend each half-hour of the day. At the end of each week, the number of total professional labor-hours spent on each job (both for the most recent week and the cumulative total since the start of the job) is tabulated. *The accuracy of information on how employees spend their time is important, especially in service organizations, where labor costs often make up over half of total costs.* Accounting and law firms often impose penalties on personnel who do not submit accurate time records in a timely manner. Computers play an increasingly important role in the recording and preparation of job cost information. *Whatever the source documents that a business uses to collect the information it needs, the object is to trace specific costs to each specific job, to the extent that such tracing is economically feasible.*

TRACING DIRECT COSTS TO JOBS

Two main categories of direct cost items can be distinguished:

1. Resources used exclusively on a specific job. These resources usually can be costed to a job in a straightforward way. For example, if supplies (such as computer disks) are used exclusively on a job, the cost of those materials can be traced to that job. Similarly, if a New York to Toronto round trip is required solely for one job, all the costs of the trip can be traced to that job.

Surveys of Company Practice

IMPORTANCE OF LABOR COMPENSATION COSTS IN LAW FIRMS

How important is it that law firms have complete and timely information from staff members reporting their work billable to clients? To answer this question, consider the following data, drawn up by *The American Lawyer.*[a] These percentages show the ratio of cost item to gross fees for law firms and law partnerships:

Attorney compensation*		52.3%
Other compensation costs		
Secretarial	8.0%	
Legal assistants	2.5	
Word processing	1.2	
Other	12.1	
Other comp. costs		23.8
Other operations costs		
Occupancy	8.8%	
Office operating costs	3.9	
Insurance and taxes	2.4	
Professional activities	1.8	
Other	7.0	
Other oper. costs		23.9
Total costs		100.0%

*Includes salary and profit-sharing component.

Attorney compensation and other compensation costs add up to 76.1% (52.3% + 23.8%) of total gross fees. This high percentage highlights why many law firms give top priority to getting accurate and prompt reports on how their employees spend their working hours.

[a]Adapted from *The American Lawyer* (December 1991).

---

2. Resources used on multiple jobs. Tracing these costs typically requires more work than does tracing the costs in category 1. An example is the compensation of a professional auditor who is hired by an accounting firm—say, on a weekly, monthly, or longer-term basis—to work on the audits of many organizations.

Most direct cost items in service organizations are labor-related and fall into category 2. We now discuss two issues related to category 2 as illustrated by Lindsay's $50 direct cost rate for each hour of professional labor: the use of actual amounts versus budgeted amounts, and the measurement of the capacity of resources that is used on multiple jobs.

Actual Costs versus Budgeted Costs

Objective 4
Compute actual and budgeted direct cost rates

An actual labor cost is the amount that personnel receive in total compensation at the payment date. Actual labor cost rates are often used in costing systems where workers are contracted on an hourly basis or a piece-rate basis. Where workers receive, say, a weekly or monthly salary, the actual labor cost hourly rate is derived by dividing total compensation by actual labor-hours worked.

Many service companies use budgeted direct labor rates in job costing. Lindsay's $50 direct cost rate for professional labor-hours was computed using the following general formula:

$$\frac{\text{Budgeted direct}}{\text{labor cost rate}} = \frac{\text{Budgeted total direct labor compensation}}{\text{Budgeted total direct labor-hours}}$$

$$= \frac{\$14,400,000}{288,000 \text{ hours}} = \$50 \text{ per hour}$$

The $14,400,000 amount is the budgeted professional labor cost in Exhibit 4-2. The 288,000 hours represents 1,600 average annual billable hours per professional times 180 professionals at Lindsay.

Professional service organizations, such as accounting firms and law firms, most often use budgeted costs rather than actual costs for labor in their job costing systems. Why? Three reasons are frequently given:

1. Actual costs will not be available until after a job is completed. However, managers often desire more-timely information. In many organizations, decisions about pricing, bidding, or product emphasis must be made before the job is started. The use of budgeted cost rates enables a costing system to provide managers with information on a continuous basis. Cost reports for individual jobs could not be prepared until the end of the reporting period if *actual* compensation and *actual* hours billed to all clients during that period were used in computing the professional labor cost rate. It is only at the end of the reporting period that the total actual amounts for that period are known.

2. Actual costs may be subject to short-run fluctuations that managers view as "misleading" for individual job costing. For example, actual labor costs can vary if the organization must pay premium rates for overtime work and work done during weekends or on holidays. Consider two identical jobs, each of which takes three days. The first job runs from Monday through Wednesday. The second job, identical to the first, runs from Thursday through Saturday. Premium rates are paid to employees for Saturday work. The first job, handled in regular time, incurs lower labor costs than does the job that stretches into the weekend. In this case, if actual rates are used, the sequence in which jobs are undertaken will affect the actual labor costs assigned to a job. Managers who know that these two jobs are identical will find it misleading that one job has very different cost numbers than the other job. Decisions made on the basis of these numbers alone might not be optimal.

3. When budgeted costs are used, the costs that will be reported on the job cost record are not affected by any change that might occur in the work budgeted for any other job. In contrast, when actual costs are used, the reported job costs of labor can be affected by such a change. Consider Lindsay's 19_5 budgeted direct labor cost rate of $50 per hour. The denominator has 288,000 budgeted hours for 19_5. Assume that half-way through 19_5 Lindsay unexpectedly loses an audit client (with budgeted 19_5 time of 11,000 professional labor-hours) but is unable to reduce its staff. No new clients are engaged for 19_5, so the denominator drops from 288,000 hours to 277,000 hours. The result is this: The actual direct labor cost rate becomes approximately $52 per hour (assuming no change in the numerator: $14,400,000 ÷ 277,000 hours = $51.99). Should the Tracy Transport job have its direct labor cost increased to $52 an hour? The answer is yes if an actual cost rate is used and no if a budgeted cost rate is used. The Lindsay partner in charge of the Tracy Transport audit likely will prefer the budgeted rate. After all, she is not responsible for the loss of the large 11,000-hour audit client.

Measuring Capacity of the Resource

The denominator used to compute the budgeted direct cost rate per professional labor-hour is a measure of labor capacity. We assume that each professional at Lindsay and Associates has 2,000 hours of time available per year, budgeted as follows:

1. Budgeted billable time for clients	1,600 hours
2. Budgeted vacation and sick leave	160 hours
3. Budgeted professional development	240 hours
4. Budgeted unbillable time due to lack of demand	0 hours
	2,000 hours

For simplicity, we assume that there is full demand for the 1,600 hours of billable time of each professional. We defer issues of unused capacity to Chapter 15. Each professional is budgeted to have (on average) a salary of $68,000 per year and fringe benefits (life insurance, medical insurance, retirement plan, and so on) of $12,000. The total compensation is $80,000 per year.

Different denominators for computation of the professional labor rate will affect the resulting rate. Consider including (1) only billable time for clients in the denominator versus (2) billable time, vacation and sick leave, and professional development:

1. *Billable time for clients:* 1,600 hours, which yields a budgeted direct cost rate of $50 per professional labor-hour ($80,000 ÷ 1,600 hours = $50). The numerator of $80,000 includes payments for vacation and sick leave and for professional development time. When the 400 (160 + 240) hours for those items are excluded from the denominator, every $1 of budgeted direct cost will implicitly include an amount to recoup vacation and sick leave and professional development costs, as well an amount to recoup the cost of the hours spent on the job. This approach results in the total compensation amount being billed as a direct cost of jobs worked on in the period (assuming full demand for the available billable time).

2. *Billable time for clients, vacation and sick leave, and professional development:* 2,000 hours, which yields a budgeted direct cost rate of $40 per professional labor-hour ($80,000 ÷ 2,000 hours = $40). This approach results in the vacation/sick leave component and the professional development component each having a rate of $40 per hour, which is not charged for in the professional labor rate. Only $64,000 ($40 × 1,600 hours) of the $80,000 total compensation would be charged as a direct cost of jobs worked on during the period.

Most professional organizations prefer approach 1 over 2 on the grounds that vacation time, sick leave, and professional development time are essential to hiring and retaining the professional staff that provides services to clients. Hence, clients should be charged for that time. Companies that adopt the second approach usually include the vacation and sick leave and professional development components as indirect cost items that are allocated rather than traced to individual jobs.

ALLOCATING INDIRECT COSTS TO JOBS

Objective 5
Compute actual and budgeted indirect cost rates.

Step-by-Step Approach to Computing Budgeted Indirect Cost Allocation Rates

A five-step approach can be used to compute budgeted indirect cost allocation rates such as the $45 rate per professional labor-hour Lindsay uses in its job costing system.

STEP 1: *IDENTIFY THE COSTS TO INCLUDE IN THE INDIRECT COST POOL.* Many organizations have a formal document that describes the chosen set of direct costs and indirect cost pools and how individual items are to be classified. Lindsay groups indirect costs in a single "audit support" cost pool that includes both "other employee costs" (such as the receptionist's wages) and "other (nonlabor) operating costs" (such as the depreciation of a photocopying machine). Lindsay classifies all these indirect costs as "Audit Support Department" costs. Professional labor costs are excluded from this indirect cost pool because they are classified as a direct cost.

STEP 2: *FOR THE BUDGET PERIOD, PREPARE ESTIMATES OF THE COST ITEMS FROM STEP 1.* Lindsay uses an annual budget period. The right-hand column in Exhibit 4-2 presents the budget for 19_5. Budgeted indirect costs are $12,960,000 ($5,328,000 for "other employee costs" and $7,632,000 for "other (nonlabor) operating costs").

STEP 3: *SELECT THE COST ALLOCATION BASE(S).* Professional labor-hours is the allocation base Lindsay uses for its single "audit support" indirect cost pool.

STEP 4: *FOR THE BUDGET PERIOD, ESTIMATE THE QUANTITY OF THE COST ALLOCATION BASE(S) FROM STEP 3.* For 19_5, the budget estimate is 288,000 professional labor-hours to be billed to clients.

STEP 5: *COMPUTE THE BUDGETED INDIRECT COST ALLOCATION RATE USING THE FOLLOWING FORMULA:*

$$\text{Budgeted indirect cost rate} = \frac{\text{Budgeted total costs in indirect cost pool}}{\text{Budgeted total quantity of cost allocation base}}$$

$$= \frac{\$12,960,000}{288,000 \text{ hours}}$$

$$= \$45 \text{ per professional labor-hour}$$

Each hour of professional labor at Lindsay is thus assigned a $50 direct cost rate and a $45 indirect cost rate.

At the end of an accounting period, the actual indirect costs and actual quantity of the cost allocation base likely will differ from their budgeted amounts. Chapter 5 (pp. 152–55) discusses alternative ways of adjusting for differences between the actual and budgeted amounts.

Time Period Used to Set Indirect Cost Rates

Many companies use an annual time period to compute budgeted indirect cost rates. Few organizations use a weekly or monthly time period. Why? There are three main reasons.

1. *The numerator (budgeted indirect costs) reason.* The shorter the time period, the greater the influence of seasonal patterns on the level of costs. For example, if a monthly budget period were used, costs of heating would be charged only to months of winter production. Use of an annual period will incorporate the effect of all four seasons into a single indirect cost rate.

Shorter time periods, such as a month, are also affected by the nonuniform design of the calendar. Some months have 20 workdays, and others have 22 or more. If separate rates are computed each month, jobs undertaken in February, the shortest month, would bear a greater share of indirect costs (such as depreciation and property taxes) than jobs undertaken in March. Many managers believe such results to be "unreasonable." Use of an annual budget period reduces the effect that the number of working days per month has on unit product costs.

2. *The denominator (budgeted quantity of the allocation base) reason.* Some indirect costs (for example, supplies) may be variable with respect to the cost allocation base, whereas other indirect costs are fixed (for example, property taxes and rent). Total variable indirect costs should change in close proportion to variations in output, but total fixed indirect costs will remain unchanged. These relationships mean that total indirect costs based on the monthly quantity of output may differ greatly from month to month solely because of variations in the quantity of output over which fixed indirect costs are spread.

Consider a professional services firm that has a highly seasonal workload. For example, many firms that provide tax advice perform more than 80% of their workload in the four months subsequent to the end of the tax year. Assume the following mix of variable indirect costs (such as phone, fax, and photocopying) and fixed indirect costs (such as insurance and rent):

	Budgeted Indirect Costs			Budgeted Professional Labor-Hours	Allocation Rate per Professional Labor-Hour
	Variable	Fixed	Total		
High-output month	$40,000	$60,000	$100,000	3,200	$31.25
Low-output month	10,000	60,000	70,000	800	87.50

Because of the fixed costs of $60,000, budgeted monthly indirect cost rates can vary sizably—from $31.25 per hour to $87.50 per hour in our example. Few managers believe that identical jobs done in different months should be allocated indirect cost charges per hour that differ so significantly ($87.50 ÷ $31.25 = 280%). Management has committed itself to a specific level of capacity far beyond a mere 30 days. An average, annualized rate based on the relationship of total annual indirect costs to the total annual level of output is more representative of the typical relationship between total indirect costs and total level of output than is a monthly rate.

3. *The cost-benefit reason.* Revising indirect cost rates takes management effort. The shorter the budget period, the more often managers must reconsider cost rates. Most managers believe that a budget period shorter than 6 or 12 months provides few—if any—additional benefits to justify the additional management effort.

Actual Indirect Rates versus Budgeted Indirect Rates

Actual indirect cost rates represent an alternative to the use of budgeted indirect cost rates. Actual rates can be computed only after the end of the accounting period. Consider 19_4 for Lindsay and Associates when actual audit support costs were $11,592,000 ($4,690,000 + $6,902,000) and actual professional hours billed were 276,000 (assumed). The actual indirect cost rate for audit support in 19_4 was $42 per professional labor-hour ($11,592,000 ÷ 276,000 = $42).

Budgeted indirect cost rates are used more frequently than actual indirect cost rates. Why? The reasons given earlier in this chapter for managers preferring budgeted direct cost rates for labor over actual cost rates apply here with equal force.

1. The use of budgeted cost and usage amounts enables an accounting system to provide managers with cost information on a timely basis during the year when decisions about pricing, bidding, or product emphasis must be made.
2. Actual costs may be subject to short-run fluctuations that managers view as "misleading" for individual job costing.
3. Budgeted indirect cost rates are known at the start of a job, whereas actual indirect cost rates are affected by the work done on other jobs in the period (even those jobs started after the job being costed has been finished).

ACTUAL, NORMAL, AND BUDGETED COSTING METHODS

Objective 6

Distinguish among actual costing, normal costing, and budgeted costing methods

The preceding section discussed the computation of actual and budgeted direct cost rates and actual and budgeted indirect cost rates. Three possible combinations of actual and budgeted rates can be used for either a job costing system or a process costing system. The names of costing methods using combinations of these actual and budgeted rates are:

	Actual Costing	Normal Costing	Budgeted Costing
Direct Costs	Actual rate(s)	Actual rate(s)	Budgeted rate(s)
Indirect Costs	Actual rate(s)	Budgeted rate(s)	Budgeted rate(s)

Definitions of these costing methods are:

◆ **Actual costing:** a costing method that traces direct costs to a cost object by using the actual direct cost rate(s) times the actual quantity and allocates indirect costs based on the actual indirect cost rate(s) times the actual quantity.

◆ **Normal costing:** a costing method that traces direct costs to a cost object by using the actual direct cost rate(s) times the actual quantity and allocates indirect costs based on the budgeted indirect cost rate(s) times the actual quantity.

◆ **Budgeted costing:** a costing method that traces direct costs to a cost object by using the budgeted direct cost rate(s) times the actual quantity and allocates indirect costs based on the budgeted indirect cost rate(s) times the actual quantity.

Exhibit 4-4 summarizes the differences in these three methods. Each method in Exhibit 4-4 is represented in its pure version. Individual companies often use hybrid systems. For example, many professional service companies (such as accounting firms and law firms) use budgeted direct cost rates for professional labor (which is often more than 50% of their total costs) but use actual direct cost rates for the costs of travel approved by their clients.

The job costing system for Lindsay and Associates described on pp. 101–103 uses budgeted costing. Both the single direct cost category and the single indirect cost category use budgeted rates when computing the costs of individual jobs such as the Tracy Transport audit.

EXHIBIT 4-4
Actual, Normal, and Budgeted Costing Methods

	Actual Costing	Normal Costing	Budgeted Costing
Direct Costs	Actual rate(s) × actual inputs used	Actual rate(s) × actual inputs used	Budgeted rate(s) × actual inputs used
Indirect Costs	Actual rate(s) × actual inputs used	Budgeted rate(s) × actual inputs used	Budgeted rate(s) × actual inputs used

The costing system overviewed in Exhibit 4-3 reports the cost of the Tracy Transport audit in 19_5 to be $76,000. Suppose Tracy Transport pays Lindsay and Associates $86,000 for the audit. Lindsay earns a reported $10,000 operating income on the job. The ratio of operating income to revenue for this audit is 11.6% ($10,000 ÷ $86,000) compared with the overall 19_5 budgeted rate of 7.4% ($2,200,000 ÷ $29,560,000; see Exhibit 4-2). Near the end of 19_5 the audit firm of Singleton & Partners visits Tracy Transport and notes that it would have bid less than $86,000 for the 19_5 audit. Tracy Transport informs Lindsay that another firm will be bidding for the 19_6 audit and that it expects the other firm to bid less than $86,000.

Lindsay views the Tracy audit as being very important. It was Lindsay's first audit engagement more than 20 years ago, so the historical ties are strong. Moreover, the Tracy job is a very straightforward audit with few difficult areas, and no partner needs to travel out of town. An internal task force at Lindsay concludes that one or more of the following reasons likely explains the willingness of a competitor to underbid the $86,000 figure:

1. The competitor is willing to break even or even lose money on the Tracy Transport audit in the short run to gain a client for the long run.
2. The competitor is more efficient than Lindsay.
3. Lindsay's costing system is overstating the cost of the Tracy Transport audit.

One approach to examining the third factor is to explore ways to refine the existing costing system.

Guidelines for Costing System Refinement

Objective 7
Describe three guidelines for refining a costing system

Costing system refinement means making changes to an existing costing system that result in a better measure of the way that jobs, products, customers, and so on differentially use the resources of the organization. Three guidelines for such refinement are:

◆ *Guideline 1: Direct cost tracing.* Classify as many of the total costs as direct costs as is economically feasible.

◆ *Guideline 2: Indirect cost pools.* Use the notion of a homogeneous cost pool when determining the number of different indirect cost pools. In a *homogeneous cost pool,* all costs have the same or a similar cause-and-effect relationship with the cost allocation base.

◆ *Guideline 3: Cost allocation bases.* Identify the driver of the costs in each indirect cost pool and use it as the cost allocation base. (This guideline assumes that the cause-and-effect criterion guides the choice of a cost allocation base.)

To follow these guidelines, use as many sources of information as are available. Possible sources include (1) interviews with personnel of the organization, (2) personal observation and measurement of activities, (3) analysis of cost and operating records, and (4) feedback about the experiences of other organizations.

Lindsay's costing system, as described in Exhibit 4-3, is very simple. It has a single direct cost category and a single indirect cost pool. We now illustrate how the three guidelines above can be used to refine this costing system. The result will be a costing system that has multiple direct cost categories and multiple indirect cost pools. A word of caution is appropriate. *More complex systems are more costly to implement and operate than are simpler systems. Managers are skeptical about spending time and other resources on highly complex accounting systems. Always keep in mind that the test of a better accounting system is whether it helps managers and others make better decisions. Obtaining "better cost numbers" (however the term is defined) is a means to an end. It should not become an end in itself.*

Refinement via Increased Direct Cost Tracing

Managers prefer costs to be directly traced than to be indirectly allocated, if economically feasible. Why? Because direct tracing increases the accuracy of costs assigned to a cost object.

Suppose Lindsay and Associates examines its activities and collects information about how individual audits vary in their use of resources. Lindsay concludes that individual jobs differ greatly in the way they use the business's resources. Using more direct cost categories would help capture this heterogeneity. An increase in direct cost categories from one to six is possible with a minimal increase in recordkeeping. The six direct cost categories in the refined job costing system are:

1. Professional partner labor: $100 per partner-hour
2. Professional associate labor: $40 per associate-hour
3. Office support labor: $20 per staff-hour
4. Information specialist labor: $35 per specialist-hour
5. Phone/fax/copying: traced on an "as-identified" basis, per monthly billings from third parties or internal cost rates
6. Travel: traced on an "as-identified" basis, per monthly billings from third parties or internal cost rates

The Problem for Self-Study (pp. 124–126) outlines the computation of these 19_5 direct cost rates.

Exhibit 4-5 (Panel B) presents an overview of the direct cost refinement of Lindsay's costing system. The previous system is reproduced in Panel A for comparison. Having professional partner labor and professional associate labor as separate direct-cost pools reflects the different compensation costs of partners and associates; partners average $160,000 total annual compensation, while associates average $64,000 total annual compensation. With professional partners and professional associates as separate direct cost pools, audit jobs that have the same total professional labor time but different mixes of partner and associate time will be costed differently. Direct cost categories 3 to 6 were previously included in the single indirect cost category. The Problem for Self-Study details how these direct cost categories can be traced for a job budgeted at $2,754,000 ($1,872,000 other employee costs and $882,000 other operating costs) that was allocated via the single indirect cost pool in the original system.

Why might the refined costing system in Panel B of Exhibit 4-5 provide more accurate job costs than the one in Panel A? The Panel A costing system assumes much similarity in the way all jobs use the resources of Lindsay. It assumes that:

1. all jobs use the same mix of partner time and associate time, and that
2. all jobs use the items in the single indirect cost pool (including office staff; information systems; phone, fax, and copy machines; and travel) in the same proportion.

On the basis of interviews with its personnel and analysis of its costing records, Lindsay concludes that neither assumption is met. The costing system in Exhibit 4-5, Panel B, relaxes both assumptions. It enables jobs that differ significantly in the way they use the resources of Lindsay to be costed differently.

Refinement via Increased Number of Indirect Cost Pools and Cost Drivers

The job costing systems in both Panels A and B of Exhibit 4-5 have a single indirect cost pool. Guideline 2 for refining a costing system is to use the notion of a homogeneous cost pool to determine how many different indirect cost pools to form. Use of a single indirect cost pool implicitly assumes that total indirect costs are a homoge-

EXHIBIT 4-5

Alternative Job Costing Approaches for Lindsay and Associates

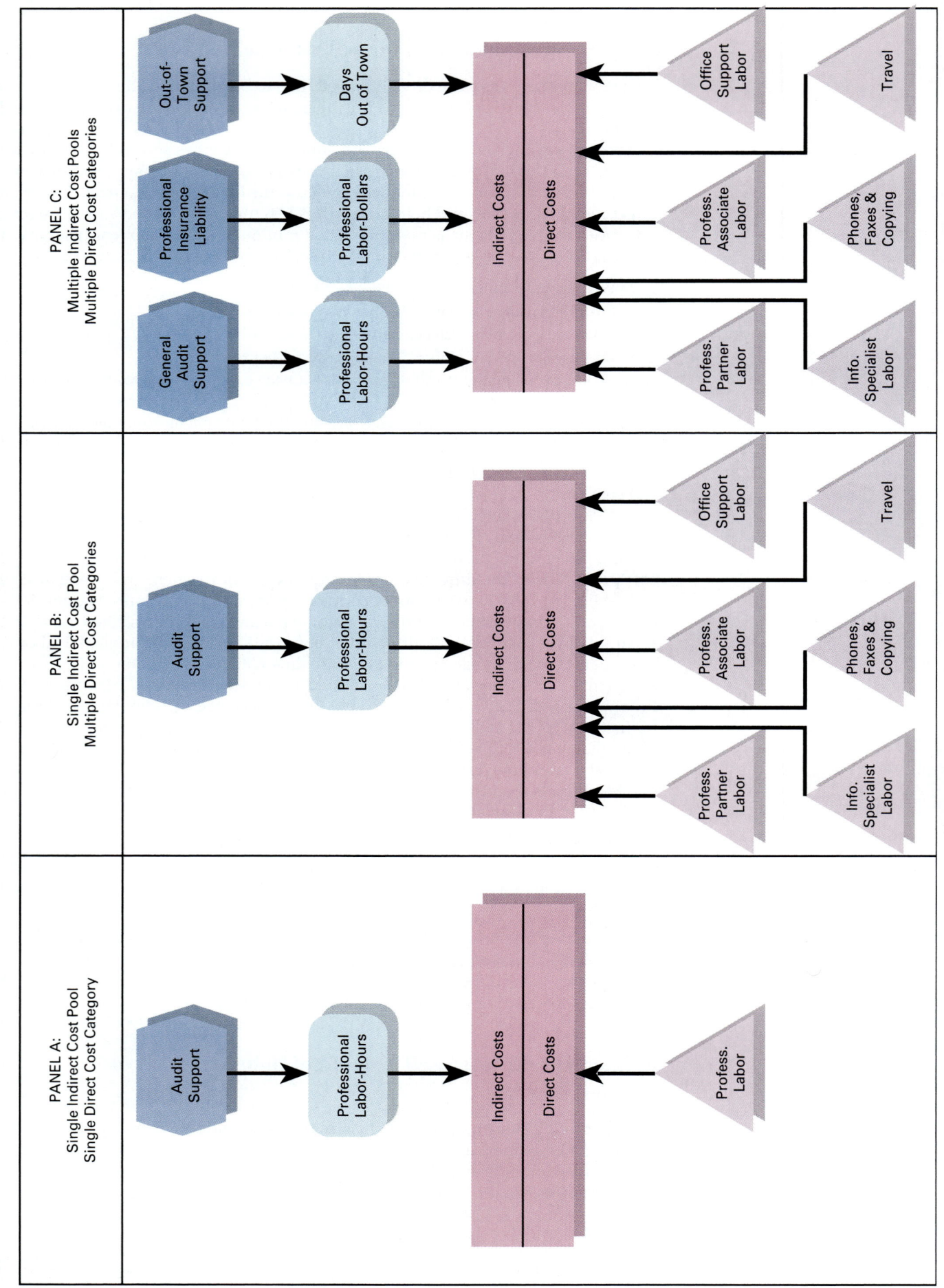

neous cost pool. Suppose Lindsay analyzes the items in the single indirect cost pool. It concludes that the costs in the cost pool are not homogeneous. There is no one cause-and-effect relationship between individual items in this indirect cost pool and the allocation base. Lindsay believes that the following three indirect cost pools provide increased accuracy in job costing:

Indirect Cost Pool	Allocation Base
1. General audit support	$22 per professional labor-hour (same rate for partners and associates)
2. Professional liability insurance	15% per professional labor compensation
3. Out-of-town support	$171 per auditor-day at a nonlocal audit site

The Problem for Self-Study (pp. 124–126) gives details on the computation of these three indirect cost rates.

The use of three indirect cost pools enables Lindsay to cost differently the audit jobs that differ in the way they use the company's support resources. For example, audit jobs with the same number of professional labor-hours but with a different number of audit-days at a nonlocal audit site—that is, with professionals traveling out of town—will now be costed differently in the refined costing system. Panel C of Exhibit 4-5 presents the overview of this further refined job costing system. Panel C illustrates the use of different allocation bases for each of the three indirect cost pools (guideline 3). For example, indirect costs for out-of-town support are now allocated using days out of town, while general support costs are now allocated using professional labor-hours.

The refined job costing system in Panel C of Exhibit 4-5 reports the cost of the Tracy Transport audit job to be $65,020—see Exhibit 4-6. Management at Lindsay believes this $65,020 cost figure is a more accurate estimate of the resources it uses on the audit than the $76,000 estimate from the previous costing system. For example, the Tracy Transport audit job actually uses a ratio of 1 hour of partner time for each 9 hours of associate time. The Panel C system reflects this ratio in its $65,020 job cost estimate. In contrast, the Panel A system assumes that the overall, firmwide ratio of 1 partner hour for every 5 hours of associate time (based on the ratio of 30 partners to 150 associates), applies to the Tracy Transport audit job (and every other audit job). Note also that the Tracy audit requires zero days of out-of-town support (because all

EXHIBIT 4-6

Job Costing of Tracy Transport Audit Using Refined Costing System with Multiple Direct Cost Categories and Multiple Indirect Cost Pools

Direct costs	
Professional partner labor, $100 × 80	$ 8,000
Professional associate labor, $40 × 720	28,800
Office support labor, $20 × 90	1,800
Information specialist labor, $35 × 40	1,400
Phone/fax/copying (as identified)	800
Travel (as identified)	1,100
	41,900
Indirect costs	
General audit support, 800 × $22	17,600
Professional liability insurance, $36,800 × 0.15	5,520
Out-of-town support, $171 × 0	0
	23,120
Total costs	$65,020

its audit sites are near its local office). Hence, it receives no charge for out-of-town support costs in the Panel C system. *The job-costing report in Exhibit 4-6 has been described as "menu-based costing," where the six individual direct cost categories and the three indirect cost pools represent separate items on a restaurant menu. Individual jobs are costed under the Panel C system according to what each job "orders from the menu."*

With its refined costing system in Panel C, Lindsay can reduce its bid for the 19_6 Tracy Transport audit to, say, $69,832 and still make the same ratio of operating income to revenue (7.4%) as the previous accounting system budgeted for 19_5.

PEANUT-BUTTER COSTING APPROACHES

Objective 8
Describe undercosting and overcosting of products

The phrase **peanut-butter costing** describes a costing approach that uniformly assigns ("spreads" or "smoothes out") the cost of resources to cost objects (such as products, services, or customers) when the individual products, services, or customers in fact use those resources in a nonuniform way. The Lindsay costing system in Panel A of Exhibit 4-5 is an example of a peanut-butter costing approach. Why? Because while the individual audit jobs of Lindsay differ greatly in their use of its resources, the Panel A costing system recognizes only one such difference—the use of professional labor-hours. The Panel C system in Exhibit 4-5 was designed after analysis of Lindsay's activities and its audit jobs. It recognizes differences among audit jobs in the areas represented by its six direct cost categories and its three indirect cost pools.

Undercosting and Overcosting

Use of a peanut-butter costing approach can lead to undercosting or overcosting of products (services, customers, and so on):

◆ **Product undercosting:** A product consumes a relatively high level of resources but is reported to have a relatively low cost.

◆ **Product overcosting:** A product consumes a relatively low level of resources but is reported to have a relatively high cost.

Companies that undercost products may actually accept sales that are bringing about losses under the erroneous impression that these sales are profitable. That is, these sales bring in less than they cost in resources. Companies that overcost products run the risk of allowing competitors to enter a market and take market share. Because these products actually cost less than what is reported to management, the company could cut the selling prices to keep competitors out of the market and still make a profit on each sale.

Recall that Lindsay was prompted to refine its job costing system when Tracy Transport informed it that another auditing firm would likely bid less than $86,000 for the 19_6 audit. The refined costing system shows a cost of $65,020. The Tracy Transport audit is an example of an overcosted job; its reported cost under the previous costing system was $76,000.

Product-Cost Cross-Subsidization

Product-cost cross-subsidization means that at least one miscosted product is resulting in the miscosting of other products in the organization. A classic example arises when a cost is uniformly spread ("peanut-buttered") across multiple users without recognition of their different resource demands. Consider the costing of a restaurant bill for four colleagues who meet once a month to discuss business developments. Each diner orders separate entrees, desserts, and drinks. The restaurant bill for the most recent meeting is:

	Entree	Dessert	Drinks	Total
Emma	$11	$ 0	$ 4	$ 15
James	20	8	14	42
Jessica	15	4	8	27
Matthew	14	4	6	24
Total	$60	$16	$32	$108
Average	$15	$ 4	$ 8	$ 27

Suppose the restaurant bill is totaled ($108) and the average cost per dinner is computed ($27). This costing approach treats each diner the same. Emma would probably object to paying $27. Why? Her total bill is only $15. Indeed, she ordered the lowest-cost entree, had no dessert, and had the lowest drink bill. When costs are averaged across all four diners, both Emma and Matthew are overcosted, James is undercosted, and Jessica is accurately costed.

The restaurant example is intuitive, and the amount of cost cross-subsidization can be readily computed given our assumption that all cost items can be traced to an individual diner; that is, all the cost items are direct costs. More complex costing issues arise when items are used simultaneously by two or more individual diners and therefore are indirect costs that require allocation—for example, the cost of a bottle of wine shared by two or more diners. Many subsequent chapters in this book provide discussion of product and customer costing in these more complex settings.

ACTIVITY-BASED COSTING

Pivotal Role of Activities

Many organizations refining their current costing systems are considering **activity-based costing (ABC).** ABC focuses on activities as the fundamental cost objects. It uses the cost of these activities as the basis for assigning costs to other cost objects such as products, services, or customers:

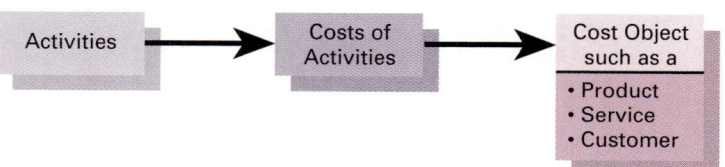

Lindsay and Associates, when refining the job costing system in Panel A of Exhibit 4-5, undertook an analysis of its activities. It interviewed its personnel, observed their audit processes, and analyzed its past costing and operating records. These efforts resulted in a refined system with six direct cost categories and three indirect cost pools. Because of its focus on activities, we would label the refined system in Panel C of Exhibit 4-5 as an ABC job costing system.

ABC is not an alternative costing system to job costing or process costing. Rather, ABC is an approach to developing the cost numbers used in job costing or process costing systems. The distinctive feature of ABC is its focus on activities as the fundamental cost objects. In contrast, more traditional approaches to developing the cost numbers used in job or process costing systems rely on general purpose (generic) accounting systems not tailored to the activities found in individual organizations. The ABC approach is more expensive than traditional approaches. ABC has the potential, however, to provide managers with information they find more useful for costing purposes. Why? Because it provides more accurate costing. For example, Lindsay was able to use it to quote a much lower price to keep the Tracy audit.

Merchandising Sector Application of ABC

We now consider a merchandising-sector application of ABC. A key focus in this sector is the profitability (selling price minus costs assigned to each product) of individual products or product lines. A **product line** is a grouping of similar products. For example, the soft-drink product line at a retail store would include Coca-Cola products, Pepsi products, and many other nonalcoholic beverages.

Family Supermarkets (FS) for many years used a product-profitability system that had a single direct cost category (goods purchased for sale) and a single indirect cost category (marketing, general and administrative: MG&A). MG&A costs were allocated to products at the rate of 30% of the cost of goods sold. Thus, for example, a good sold that cost $2 is allocated a $0.60 MG&A indirect cost charge ($2.00 × 0.30 = $0.60). Panel A of Exhibit 4-7 presents a costing overview of that system, and Panel A of Exhibit 4-8 presents a product-line profitability report from it. FS's cost of goods sold (the goods purchased for resale) makes up 76.92% of total costs ([$200 + $350 + $150] ÷ [$260 + $455 + $195]). This high percentage is typical of many companies in the merchandising sector.

FS managers, at the end of each week, received product-line profitability reports (as in Panel A of Exhibit 4-8). Managers increasingly viewed these reports as overly simplistic. The main criticism was that the system was a classic example of peanut-butter costing. Individual product lines differed greatly in their use of FS's resources, but these differences were not captured in the previous costing system. An analysis

EXHIBIT 4-7
Product-Line Costing Systems at Family Supermarkets

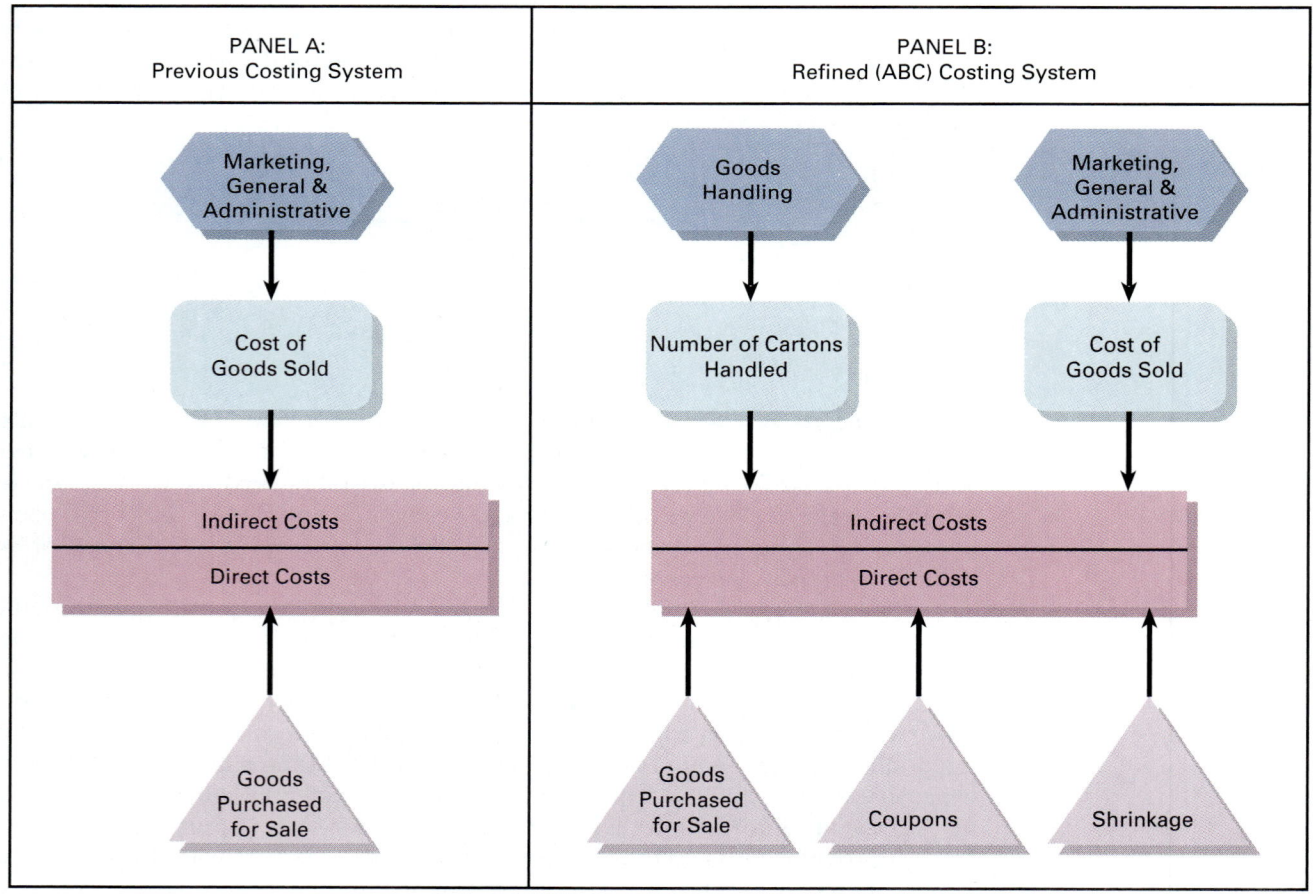

EXHIBIT 4-8

Product-Line Profitability Reports for Family Supermarkets for Week 1, July 19_5 (in thousands)

PANEL A: Previous Costing Systems

	Soft Drinks		Fresh Produce		Bathroom Products	
Sales		$280		$480		$230
Operating costs						
Direct costs						
Cost of goods sold	$200		$350		$150	
Indirect costs						
MG&A (30% of COGS$)	60		105		45	
Operating costs		260		455		195
Operating income		$ 20		$ 25		$ 35
Ratio of operating income to sales		7.14%		5.21%		15.22%

PANEL B: Refined (ABC) Costing System

	Soft Drinks		Fresh Produce		Bathroom Products	
Sales		$280		$480		$230
Operating costs						
Direct costs						
Cost of goods sold	$200		$350		$150	
Coupons	5		0		15	
Shrinkage	1	206	21	371	6	171
Indirect costs						
Goods handling	0		16		6	
MG&A (20% of COGS$)	40	40	70	86	30	36
Operating costs		246		457		207
Operating income		$ 34		$ 23		$ 23
Ratio of operating income to sales		12.14%		4.79%		10.00%

of the activities at FS led managers to make the following changes to the previous costing system:

1. Additions of two direct cost categories:

   ◆ Coupons. Only a subset of the products sold have coupons. FS, on average, covers 25% of the cost of coupons, and its suppliers (such as Procter and Gamble and Lever Brothers) cover the remaining 75% of the cost.

   ◆ **Shrinkage** is the difference between goods purchased for sale and goods actually sold (after making inventory adjustments). This difference can arise from breakages before a sale, theft, perishability, and so on.

   Bar codes are available to trace the cost of coupons to product lines. FS managers also have detailed records documenting shrinkage for specific product lines. Managers use these records to develop specific shrinkage rates for different products. These two additional direct-cost categories had previously been included in MG&A and spread across all products on the basis of a uniform 30% of cost of goods sold.

2. Addition of one indirect cost category. Goods handling (such as incoming handling and stacking of aisles) is a major cost area for supermarkets. The original costing system included all these costs in its single indirect cost pool and spread them across all products on the basis of a uniform 30% of cost of goods sold. FS managers had two problems with that approach:

◆ Some FS suppliers now stack their own products and maintain their shelf appearance. For those products, FS personnel were not involved in goods handling.

◆ The major driver at the materials-handling and shelf-stacking level was believed to be number of cartons handled, not their purchase cost. It took just as much effort to handle low-cost goods as high-cost goods.

In their refined activity-based costing system, FS has a separate goods-handling indirect cost category, with number of cartons handled as its cost allocation base. Goods that are self-stacked by suppliers (such as soft drinks) receive zero indirect cost charge for goods handling.

The refined activity-based costing system reduces the costs in the MG&A indirect cost pool of the original system in two ways. First, some costs are now included in the two new direct cost categories of coupons and shrinkage. Second, another portion of the costs are included in the goods-handling indirect cost pool. As a result, the cost allocation rate for MG&A in the refined system is reduced to 20% of cost of goods sold (from 30% in the original system).

Panel B of Exhibit 4-7 presents a costing overview of the ABC system. Exhibit 4-8 shows an illustrative weekly profitability report under the new system. Managers believe the activity-based system is more credible than the previous system. It better captures the different types of activities at FS. It also better captures how individual products use its resources. Rankings of relative profitability (the percentage of operating income to sales) of the three product lines in Exhibit 4-8 under the original costing system and under the refined costing system follow:

Original Costing System	Refined (Activity-Based) Costing System
1. Bathroom products (15.22%)	1. Soft drinks (12.14%)
2. Soft drinks (7.14%)	2. Bathroom products (10.00%)
3. Fresh produce (5.21%)	3. Fresh produce (4.79%)

Note how the activity-based costing system captures how individual product lines use the resources of FS differently. For example, the increase in the profitability of soft drinks arises primarily because soft drinks have a low shrinkage rate and because suppliers are stacking the shelves themselves.

FS managers believe that the activity-based profitability percentages in the refined costing system provide more reliable guides for deciding how to allocate shelf space and for deciding which products deserve greater emphasis. For example, FS managers might well decide to devote more space to soft drinks and less to bathroom products.

CUSTOMER COSTING

Objective 11
Outline the importance of customer costing

A central theme of the new approach to management is customer satisfaction—see Exhibit 1-2 (p. 6). Accounting systems can help managers focus on customers by reporting revenue and cost information on a customer-by-customer basis. This section discusses general principles of customer costing.

A **customer costing system** reports cost numbers that reflect the way that customers differentially use the resources of a company. We will consider three situations of Rowe & Partners, a law firm with 20 customers (clients).

Situation 1: All customers have only one job (case); all costs are allocated to individual jobs.

Situation 2: Customers can have multiple jobs (cases); all costs are allocated to individual jobs.

Situation 3: Customers can have multiple jobs (cases); not all costs are allocated to individual jobs.

Although customer costing is a relatively new topic in management accounting, it is a vitally important one. Managers need to ensure that customers contributing sizably to the profitability of an organization receive a comparable level of attention from the organization. An accounting system that reports customer profitability helps managers in this task.

Situation 1: All Customers Have Only One Job; All Costs Allocated to Individual Jobs

It is simplest to develop customer-focused accounting information when all customers have only one job, and all costs are allocated to individual jobs (situation 1). Revenue information per customer can be derived from the general ledger. Cost information per customer can be derived from a job costing system of the kind discussed earlier in this chapter. Exhibit 4-9 presents revenue, operating costs, operating income, and the percentage of operating income to revenue for the 20 customers of Rowe & Partners. These customers are ranked on the percentage of operating income to revenue. There are dramatic differences among customers in this percentage. Rowe & Partners might use Exhibit 4-9 to prompt analysis of why these differences exist.

Two alternative presentations of the customer profitability information in Exhibit 4-9 appear in Exhibits 4-10 and 4-11. Exhibit 4-10 shows a ranking of customers on the basis of operating income (column 2). Column 3 reports the cumulative profitability of customers ranked on operating income. Column 4 reports the percentage of cumulative operating income to total operating income of $872. Column 7 reports the percentage of cumulative revenues to total revenues of $10,000. The first four of their 20 customers in Exhibit 4-10 provide 84% of the total operating income, while these same four customers make up only 37% of total revenues. This pattern of a low percentage of customers contributing a high percentage of total operating income is common in many companies. Exhibit 4-11 (p. 122) is a more visual way of highlighting differences among customers in their contribution to total operating income.

Managers find customer profitability analysis useful for several reasons. First, it frequently highlights how vital a small set of customers is to total profitability. Managers need to ensure that the interests of those customers receive high priority. Microsoft uses the phrase "not all revenue dollars are endowed equally in profitability" to stress this key point. Second, when a customer is ranked in the "loss category," managers may focus on ways to make future business with that customer more profitable. For example, an analysis of customer D777 in Exhibit 4-9 revealed that Rowe & Partners quoted a fixed fee for a contract litigation case that took much longer than anticipated. Rowe's normal practice is to bill clients on a "cost basis" using agreed-upon cost rates. The client bears the risk of a case taking longer to settle than anticipated. The customer profitability report in Exhibit 4-9 highlights to Rowe & Partners the dangers of quoting fixed fees in subsequent contract litigation cases.

A caveat on the customer profitability information in Exhibits 4-9, 4-10, and 4-11 is necessary. This information is related to profitability in a single reporting period. A customer may be unprofitable this period, but management may believe that this customer will be profitable in future years. Managers now give high priority to main-

EXHIBIT 4-9

Customer Profitability Analysis with Ranking on Percentage of Operating Income to Revenue for Rowe & Partners (in thousands)

Customer Code	Revenues	Operating Costs	Operating Income	Percentage of Operating Income to Revenues
E091	$ 134	$ 95	$ 39	29%
G001	487	356	131	27
A003	647	492	155	24
C249	805	644	161	20
M764	946	766	180	19
A107	1,333	1,093	240	18
F054	312	259	53	17
B209	381	331	50	13
D333	119	106	13	11
F519	468	445	23	5
C167	365	350	15	4
G641	883	857	26	3
M291	624	618	6	1
N020	129	128	1	1
E288	486	491	−5	−1
N623	611	629	−18	−3
B291	219	226	−7	−3
E764	234	243	−9	−4
L853	126	135	−9	−7
D777	691	864	−173	−25
Total	$10,000	$9,128	$ 872	8.72%*

*$872 ÷ $10,000 = 8.72%

taining long-term ongoing relationships with customers. An extension of the analysis in these exhibits would incorporate a longer time horizon when making profitability computations.

Situation 2: Customers Can Have Multiple Jobs; All Costs Allocated to Individual Jobs

Situation 2 also allocates all costs to individual jobs, but customers can have more than one job. In our example, Rowe & Partners undertakes multiple jobs (such as several litigation cases) for one or more customers. The revenues and costs of these multiple jobs must be added to yield individual customer financial information. For example, assume that the codes in Exhibit 4-9 are job codes not customer codes and that A003 and A107 are jobs for the same customer:

EXHIBIT 4-10
Cumulative Customer Profitability with Individual Customers Ranked on Operating Income for Rowe & Partners (in thousands)

Customer Code (1)	Operating Income (2)	Cumulative Operating Income* (3)	Percentage of Cumulative to Total Operating Income (4) = (3) ÷ $872	Revenues (5)	Cumulative Revenues (6)†	Percentage of Cumulative to Total Revenues (7) = (6) ÷ $10,000
A107	$240	$240	28%	$1,333	$1,333	13%
M764	180	420	48	946	2,279	23
C249	161	581	67	805	3,084	31
A003	155	736	84	647	3,731	37
G001	131	867	99	487	4,218	42
F054	53	920	106	312	4,530	45
B209	50	970	111	381	4,911	49
E091	39	1,009	116	134	5,045	50
G641	26	1,035	119	883	5,928	59
F519	23	1,058	121	468	6,396	64
C167	15	1,073	123	365	6,761	68
D333	13	1,086	125	119	6,880	69
M291	6	1,092	125	624	7,504	75
N020	1	1,093	125	129	7,633	76
E288	-5	1,088	125	486	8,119	81
B291	-7	1,081	124	219	8,338	83
L853	-9	1,072	123	126	8,464	85
E764	-9	1,063	122	234	8,698	87
N623	-18	1,045	120	611	9,309	93
D777	-173	872	100	691	10,000	100

*Column 3 is derived by cumulating the operating income of column 2 of customers ranked on operating income. Thus, row 2 in column 3 has $420, which is the sum of $240 for A107 and $180 for M764 from column 2; row 3 in column 3 has $581, which is the sum of $240 for A107, $180 for M764, and $161 for C249 from column 2.

†Column 6 is derived the same as column 3 using the revenue amounts in column 5.

Customer A revenues		
Job A003	$ 647	
Job A107	1,333	$1,980
Customer A operating costs		
Job A003	492	
Job A107	1,093	1,585
Customer A operating income		$ 395

The percentage of operating income to revenue for customer A is 19.95% ($395 ÷ $1,980). This computation is straightforward at a conceptual level. At a practical level, however, it can be difficult. Large companies have numerous customers, and their

EXHIBIT 4-11

Customer Ranking Based on Operating Income

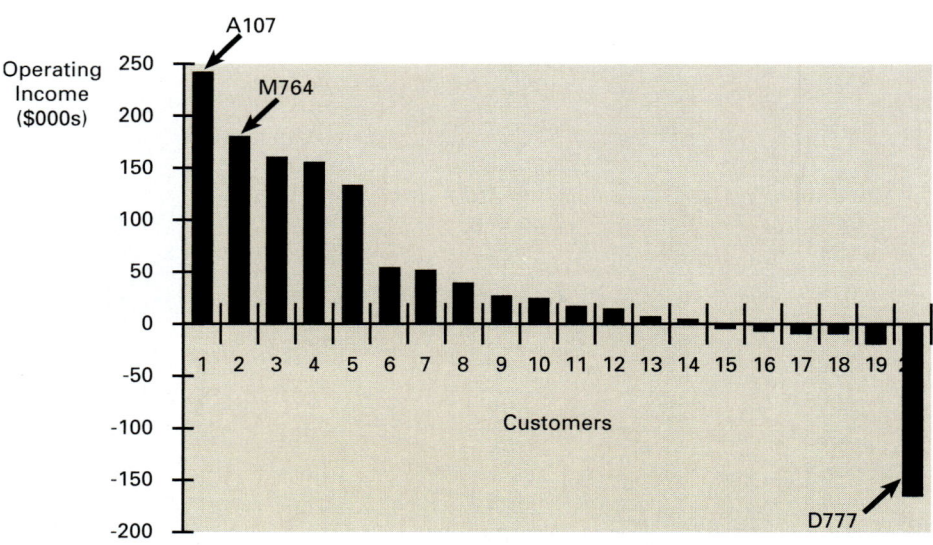

data bases may not consistently code customer names. In addition, the many individual data bases of a company may not be compatible with each other. This situation occurs most often in companies with multiple divisions in different countries. As a result, development of accurate customer-focused data bases (even on the revenue side, let alone the cost side) is a challenging task in many companies.

Situation 3: Customers Can Have Multiple Jobs; Not All Costs Allocated to Individual Jobs

When all costs are allocated to individual jobs, as in situations 1 and 2, customer costing reports can be developed by accumulating cost information collected at the job level. But organizations do not always allocate all costs to individual jobs. This section introduces customer costing in such a context.

For customer-costing purposes, assume that an organization classifies costs into one of two categories:

◆ Job-specific costs (both direct and indirect)
◆ Customer-support costs (both direct and indirect)

Job-specific costs are costs incurred in the process of completing a distinct, identifiable job for a customer. **Customer-support costs** are costs incurred at the individual customer level, regardless of the quantity or mix of individual services or products that the customer buys. Consider Rowe & Partners. Each year the company pays the costs for a representative from each of its five largest clients to attend an American Bar Association seminar on "Making Effective Use of Your Law Firm Partner." Rowe classifies these costs as customer-support costs and does not allocate them to individual jobs of these clients.

Individual customer revenue, cost, and profitability figures can be presented as follows (jobs and amounts assumed):

Customer X revenues		
Job 1 (contract work)	$10,006	
Job 2 (litigation work)	9,487	$19,493
Customer X costs		
Job 1	7,462	
Job 2	5,824	
Customer support	2,812	16,098
Customer X operating income		$ 3,395

The customer-support category is but one type of non-job-specific cost that can be recognized in an accounting system. Chapter 15 provides further examples of other non-job-specific customer costs that are important to managers.

Concepts in Action

CUSTOMER PROFITABILITY ANALYSIS AT DEC-EUROPE[a]

Organizations are increasingly emphasizing the importance of focusing on their customers. Many management accounting systems, however, do not support that focus. Costs are often reported on a departmental basis or a product basis, but not on a customer basis.

So it was for Digital Equipment Company (DEC), which sells computer systems to corporations. The computer system industry is highly competitive; operating margins are low, and new products are continually being introduced. For many years, DEC's management accounting system did not report costs or profitability on an individual-customer basis. In December 1990, the president of DEC-Europe requested that DEC's European operations quickly acquire the capability to report profitability information for each of its 200 largest customers (such as DHL). To increase the credibility of the information, he required that it tie back to the general ledger. The president wanted the infrastructure in place to report customer profitability information within four months and individual customer reports within six months.

After an enormous effort, DEC-Europe developed its customer profitability reports within six months of the president's request. These reports revealed that a small number of DEC-Europe's customers provided the bulk of its operating income, while many customers showed no income or even operating losses. Many other businesses read a similar story in studying their customer profiles.

DEC uses its customer profitability reports in several ways. First, DEC sales and support staff are making extra efforts to ensure that key customers are satisfied with DEC's products and sales support. Second, changes are being made for those customers currently showing operating losses. DEC views dropping those customers as the last resort. Its challenge is to change DEC's internal operations or to work with low-profitability customers to make their accounts more profitable (ideally by increasing revenue, but also by reducing a customer's use of DEC resources).

Many DEC sales representatives were less than comfortable about being informed that customer accounts for which they were responsible were showing operating losses. The reaction of some representatives was to challenge the costs assigned to their customers and to seek to allocate many costs elsewhere within DEC. Other representatives took the approach of rigorously looking at every single cost item in their customer reports to analyze whether the activity that the cost represented added value from the customer's perspective. This second response is leading to major changes in the way DEC interacts with some customers. The result has been an increase in customer satisfaction with DEC-Europe and an increase in its percentage of profitable customers.

[a]Adapted from a presentation by S. Fuller, Manager of Financial Planning and Strategies for DEC-Europe, London, 1992.

THE VALUE CHAIN IN JOB COSTING
AND CUSTOMER COSTING SYSTEMS

A key theme in the new management approach is "total value chain analysis"—see Exhibit 1-2 (p. 6). Chapter 1 presented the following business functions as elements of the value chain:

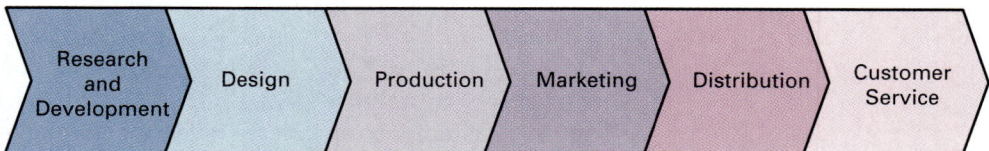

Many organizations do not use these exact labels in their costing systems, nor do they always collect data in terms of separate cost pools for each business function. However, irrespective of the labels used, it is important in both job costing and customer costing to have the capability to combine costs from all areas of the value chain. Consider the pricing strategy of an accounting firm. Accounting firms must invest in developing up-to-date audit technologies on an ongoing basis. This expenditure would be classified in either the research and development or design of products, services, and processes parts of the value chain. Similarly, marketing is now an essential aspect of accounting firms acquiring new clients and retaining existing clients. When developing job-cost estimates for audits, it is essential that the costs of updating audit technologies and the costs of marketing be covered by the total revenues received by the firm. A job costing system or a customer costing system that combines costs from all business function areas of the value chain can greatly assist managers in their decisions on pricing and on which products or customers to emphasize.

PROBLEM FOR SELF-STUDY

PROBLEM

Lindsay and Associates is refining its current costing system, which has a single direct cost pool (professional labor) and a single indirect cost pool (audit support). It conducts extensive interviews with personnel at all levels and examines past audit working papers. It also notes that recent advances in information-gathering technology (such as codes on its phone, fax, and copy machines) enable it to trace directly to jobs some items that currently are classified as indirect costs.

The six direct categories in its refined costing system are:

a. Professional partner labor. Average total annual compensation of $160,000 for each of 30 partners, each of whom has 1,600 hours of budgeted billable time.

b. Professional associate labor. Average total annual compensation of $64,000 for each of 150 associates, each of whom has 1,600 hours of budgeted billable time.

c. Office support labor. Average total annual compensation of $32,000 for each of 27 staff, each of whom has 1,600 hours of budgeted billable time.

d. Information specialist labor. Average total annual compensation of $56,000 for each of 18 specialists, each of whom has 1,600 hours of budgeted billable time.

e. Phone/fax/copying. Costs will be on an "as-identified" basis per monthly bills from third parties or internal cost rates.

f. Travel. Costs will be on an "as-identified" basis per monthly bills from third parties or internal cost rates.

The three indirect cost pools and their chosen allocation bases in the refined costing system are:

g. General audit support (Indirect I). 19_5 budgeted amount is $6,336,000 as shown in column 5 of Exhibit 4-12. The allocation base is budgeted professional labor-hours.

h. Professional liability insurance (Indirect II). 19_5 budgeted amount is $2,160,000 as shown in column 6 of Exhibit 4-12. The allocation base is budgeted professional labor-dollars.

i. Out-of-town support (Indirect III). 19_5 budgeted amount is $1,710,000, as shown in column 7 of Exhibit 4-12. The allocation base is auditor-days at nonlocal audit sites (defined as sites more than 50 miles away from a Lindsay office). In 19_5, there are 10,000 auditor-days budgeted at nonlocal audit sites.

Required

1. Compute the budgeted 19_5 direct cost rate per hour for (i) professional partner labor, (ii) professional associate labor, (iii) office support labor, and (iv) information specialist labor.

2. Compute the budgeted 19_5 indirect cost rate per unit of the allocation base for (i) general audit support, (ii) professional liability insurance, and (iii) out-of-town support.

EXHIBIT 4-12
Budgeted Costing Worksheet for Previous and Refined Costing System of Lindsay and Associates (in thousands)

| Cost Line Items (1) | Previous Costing System | | Refined Costing System | | | |
	Direct (2)	Indirect (3)	Direct (4)	Indirect I (5)	Indirect II (6)	Indirect III (7)
Professional labor						
Partners	$14,400	$ 0	$ 4,800*	$ 0	$ 0	$ 0
Associates		0	9,600†	0	0	0
Total prof. labor	14,400	0	14,400	0	0	0
Other employee costs						
Office staff	0	2,030	864	1,166	0	0
Information specialists	0	1,600	1,008	592	0	0
Administrative	0	660	0	660	0	0
Other	0	1,038	0	1,038	0	0
Total other employee costs	0	5,328	1,872	3,456	0	0
Other (nonlabor) operating costs						
Professional liability insurance	0	2,160	0	0	2,160	0
Professional development	0	880	0	880	0	0
Occupancy	0	2,000	0	2,000	0	0
Phone/fax/copying	0	1,430	600	0	0	830
Travel and allowances	0	770	282	0	0	488
Other	0	392	0	0	0	392
Total other operating costs	0	7,632	882	2,880	2,160	1,710
Total operating costs	$14,400	$12,960	$17,154	$6,336	$2,160	$1,710

*30 partners × 1,600 billable hours × $100   =   48,000 × $100 = $ 4,800,000
†150 associates × 1,600 billable hours × $40 = 240,000 × $ 40 = $ 9,600,000
                                                288,000 $14,400,000

SOLUTION

1. The formula for computing the 19_5 budgeted direct labor cost rates is:

$$\frac{\text{Budgeted direct}}{\text{cost rate}} = \frac{\text{Budgeted total labor compensation}}{\text{Budgeted total labor-hours}}$$

Lindsay uses budgeted billable time for clients as its denominator.

(i)   Professional partner labor: $\frac{\$160,000 \times 30}{1,600 \times 30} = \$100$ per hour

(ii)   Professional associate labor: $\frac{\$64,000 \times 150}{1,600 \times 150} = \$40$ per hour

(iii)   Office support labor: $\frac{\$32,000 \times 27}{1,600 \times 27} = \$20$ per hour

(iv)   Information specialist labor: $\frac{\$56,000 \times 18}{1,600 \times 18} = \$35$ per hour

Column 4 of Exhibit 4-12 shows how $1,872,000 of "other employee costs" and $882,000 of "other (nonlabor) operating costs" are now direct cost items in the refined costing system. In the previous costing system, the $1,872,000 was classified as indirect costs.

2. The formula for computing the 19_5 budgeted indirect cost rates is:

$$\frac{\text{Budgeted indirect}}{\text{cost rate}} = \frac{\text{Budgeted total costs in indirect cost pool}}{\text{Budgeted total quantity of cost allocation base}}$$

(i) General audit support: $\frac{\$6,336,000}{288,000 \text{ hours}} = \$22$ per professional labor-hour

(ii) Professional liability insurance: $\frac{\$2,160,000}{\$14,400,000} = 15\%$ of professional labor compensation

(iii) Out-of-town support: $\frac{\$1,710,000}{10,000 \text{ days}} = \$171$ per day at a nonlocal audit site

SUMMARY

The following points are linked to the chapter's learning objectives.

1. The building blocks of a costing system are *cost object*, *direct costs*, *indirect costs*, *cost pool*, and *cost allocation base*. Costing overview diagrams present these concepts in a systematic way. Costing systems aim to report cost numbers that reflect the way that chosen cost objects (such as products, services, or customers) use the resources of an organization.

2. Job costing systems assign costs to a distinct, identifiable product or service. In contrast, process costing systems assign costs to masses of similar units and compute unit costs on an average basis. These two costing systems are best viewed as opposite ends of a spectrum. The costing systems of many companies combine some elements of both job costing and process costing.

3. A general approach to job costing involves identifying (a) the job, (b) the direct cost categories, (c) the indirect cost categories, (d) the cost allocation base(s), and (e) the cost allocation rate(s).

4. Where labor is compensated on an hourly basis, computing the direct labor rate of a job is straightforward. Where labor is compensated on a longer-term basis, the actual labor cost rate per hour will depend on the actual compensation and the actual hours worked. Computing budgeted direct costs for labor employed on a long-term basis requires estimates of compensation and estimates of how that labor will be used during the period of employment.

5. Actual indirect cost rates cannot be computed until the end of period, when actual indirect costs and actual amounts for the allocation base are known.

Budgeted indirect cost rates, which can be computed at the start of a period, require estimates of total costs in an indirect cost pool and estimates of the total quantity of the cost allocation base for that cost pool. Managers have a strong preference for indirect cost rates based on an annual time period (or a six-month period) rather than on shorter time periods, such as one month or three months.

6. Actual costing, normal costing, and budgeted costing differ in their use of actual or budgeted direct or indirect cost rates:

	Actual Costing	**Normal Costing**	**Budgeted Costing**
Direct Costs	Actual rate(s)	Actual rate(s)	Budgeted rate(s)
Indirect Costs	Actual rate(s)	Budgeted rate(s)	Budgeted rate(s)

All three methods use actual quantities of direct cost inputs and actual quantities of the allocation base(s) for indirect costs.

7. Refining a costing system means making changes that result in cost numbers that better measure the way jobs, products, services, customers, and so on differentially use the resources of the organization. These changes can require additional direct cost tracing, the choice of more indirect cost pools, or the use of different cost allocation bases.

8. Product undercosting (or overcosting) occurs when a product or service consumes a relatively high (low) level of resources, but is reported to have a relatively low (high) cost. Peanut-butter costing, a common cause of under- or overcosting, occurs when an accounting system uniformly assigns ("spreads") the cost of resources to products when the individual products use those resources in a nonuniform way. Product-cost cross-subsidization exists when one or more undercosted (overcosted) products results in one or more other products being overcosted (undercosted).

9. Many organizations seeking to refine their costing systems are giving serious consideration to activity-based costing (ABC). A distinctive feature of ABC is the focus on activities as the fundamental cost objects. The costs of these activities are then assigned to other cost objects such as products, services, or customers.

10. A product line is a grouping of similar products. Product profitability statements report the profitability (revenue minus costs assigned to each product) of individual products or product lines. Cost of goods purchased for resale typically is the largest single cost item for companies in the merchandising sector.

11. Customer costing, a relatively new topic in management accounting, reports cost numbers that reflect the way that customers differentially use the resources of the organization. It helps managers ensure that customers contributing sizably to the profitability of an organization receive a comparable level of attention from the organization.

TERMS TO LEARN

This chapter and the Glossary at the end of the book contain definitions of the following important terms:

activity-based costing (ABC) *(p. 115)*   actual costing *(109)*   budgeted costing *(109)*
cost allocation base *(99)*   cost pool *(99)*   costing system refinement *(110)*
customer costing system *(119)*   customer-support costs *(122)*   job costing system *(99)*
normal costing *(109)*   peanut-butter costing *(114)*   process costing system *(99)*
product-cost cross-subsidization *(114)*   product line *(116)*   product overcosting *(114)*
product undercosting *(114)*   shrinkage *(117)*

ASSIGNMENT MATERIAL

QUESTIONS

4-1 Define *cost pool, cost tracing, cost allocation,* and *cost allocation base.*

4-2 "In the production of services, direct materials are usually the major cost." Is this quotation accurate? Explain.

4-3 How does a *job costing system* differ from a *process costing system?*

4-4 Give an example of a job costing system in the service sector, in the merchandising sector, and in the manufacturing sector.

4-5 What are the five steps in a general approach to job costing?

4-6 Why do most managers use budgeted costs rather than actual costs when computing direct labor rates?

4-7 "Computing the direct labor cost rate in a service organization is far from straightforward." Do you agree? Explain.

4-8 Give two reasons why most organizations use a six-month or annual period rather than a weekly or monthly period to compute budgeted indirect cost rates.

4-9 What is *costing system refinement?* Describe three guidelines for such refinement.

4-10 Define *peanut-butter costing,* and explain how managers may determine if it occurs with their costing system.

4-11 What is an activity-based approach to designing a costing system?

4-12 Why should managers worry about product overcosting or product undercosting?

4-13 What use can managers in a supermarket make of more refined product-line cost information?

4-14 Explain the increased interest by companies in customer costing.

4-15 Comment on the following statement that appeared in a brochure promoting a new costing software package: "The software can be fully operational almost immediately. There are no delays due to tailoring the system to your organization, as it has been found to be highly valuable across a broad range of organizations. Customizing to your organization is not required."

EXERCISES AND PROBLEMS

4-16 Job costing; actual, normal, and budgeted costing. Chirac & Partners is a Quebec-based public accounting partnership specializing in audit services. Its job costing system has a single direct cost category (professional labor) and a single indirect cost pool (audit support, which contains all the costs in the Audit Support Department). Audit support costs are allocated to individual jobs using actual professional labor-hours. Chirac & Partners has 10 professionals who are involved in their auditing services.

Budgeted and actual amounts for 19_5 are:

Budget for 19_5

Professional labor compensation	$960,000
Audit support department costs	$720,000
Professional labor-hours billed to clients	16,000 hours

Actual results for 19_5

Professional labor compensation	$899,000
Audit support department costs	$744,000
Professional labor-hours billed to clients	15,500 hours

Required

1. Compute the direct cost rate per professional labor-hour and the indirect cost rate per professional labor-hour under (a) actual costing, (b) normal costing, and (c) budgeted costing.

2. The 19_5 audit of the Montreal Expos was budgeted to take 110 hours of professional labor time. The actual professional labor time on the audit was 120 hours. Compute the 19_5 job cost using (a) actual costing, (b) normal costing, and (c) budgeted costing. Explain any differences.

3. Chirac, like most professional accounting firms, uses the budgeted costing method in its job costing system. Give two reasons why Chirac may use budgeted costing rather than either actual costing or normal costing.

4-17 Job costing; actual, normal, and budgeted costing. Vista Group provides architectural services for residential and business clients. It has 25 professionals. Its job costing system has a single direct cost category (professional labor) and a single indirect cost pool (client support, which contains all the costs in the Client Support Department). Client support costs are allocated to individual jobs using actual professional labor-hours.

Budgeted and actual amounts for 19_5 are:

Budget for 19_5

Professional labor compensation	$4,000,000
Client support department	$2,600,000
Professional labor-hours billed to clients	40,000 hours

Actual results for 19_5

Professional labor compensation	$4,620,000
Client support department	$2,436,000
Professional labor-hours billed to clients	42,000 hours

Required

1. Compute the direct cost rate per professional labor-hour and the indirect cost rate per professional labor-hour under (a) actual costing, (b) normal costing, and (c) budgeted costing.

2. In 19_5, the Vista Group designed a new retirement village in Tucson, Arizona, for Carefree Years Inc. Vista budgeted to spend 1,500 professional labor-hours on the project. Actual professional labor-hours spent were 1,720. Compute the job cost of the Carefree Years project using (a) actual costing, (b) normal costing, and (c) budgeted costing. Explain any differences.

3. Vista Group, like most architectural firms, uses the budgeted costing method in its job costing system. Give two reasons why Vista may use budgeted costing rather than either actual costing or normal costing.

4-18 Computing direct cost rates, consulting firm. Doherty & Co. is an international consulting firm. Its 19_4 annual budget includes the following for each category of professional labor:

Category	Average Salary	Average Fringe Benefits	Billable Time for Clients	Vacation & Sick Leave	Professional Development	Unbilled Time due to Lack of Demand
Director	$140,000	$60,000	1,600	160	240	0
Partner	105,000	45,000	1,600	160	240	0
Associate	60,000	20,000	1,600	160	240	0
Assistant	38,000	12,000	1,600	160	240	0

Required

1. Compute the budgeted direct cost rate for professional labor (salary and fringe benefits)

per hour for (a) directors, (b) partners, (c) associates, and (d) assistants. Use budgeted bill-able time for clients as the denominator in these computations.

2. Repeat requirement 1 using [budgeted billable time + vacation + sick leave time + professional development time] as the denominator in these calculations.

3. Why are rates different in requirements 1 and 2? How might these differences affect job costing by Doherty & Co.?

4-19 Computing indirect cost rates, job costing.   Mike Rotundo, the president of Tax Assist, is examining alternative ways to compute indirect cost rates. He collects the following information from the budget for 19_5:

◆ Budgeted variable indirect costs: $10 per hour of professional labor time
◆ Budgeted fixed indirect costs: $50,000 per month

The budgeted billable professional labor-hours per quarter are:

January–March	20,000 hours
April–June	10,000 hours
July–September	4,000 hours
October–December	6,000 hours

Rotundo pays all tax professionals employed by Tax Assist on an hourly basis ($30 per hour, including all fringe benefits).

Tax Assist's job costing system has a single direct cost category (professional labor at $30 per hour) and a single indirect cost pool (office support that is allocated using professional labor-hours).

Tax Assist charges clients $65 per professional labor-hour.

Required
1. Compute the budgeted indirect cost rates per professional labor-hour using:
   a. quarterly budgeted billable hours as the denominator
   b. annual budgeted billable hours as the denominator

2. Compute the operating income for the following four customers using:
   a. quarterly-based indirect cost rates
   b. an annual indirect cost rate

   ◆ Stan Hansen: 10 hours work in February
   ◆ Lelani Kai: 6 hours work in March and 4 hours in April
   ◆ Ken Patera: 4 hours in June and 6 hours in August
   ◆ Evelyn Stevens: 5 hours in January, 2 hours in September, and 3 hours in November

3. Comment on the results in requirement 2.

4-20 Job costing, consulting firm.   Taylor & Associates, a consulting firm, has the following condensed budget for 19_5:

Revenues		$20,000,000
Total costs:		
Direct costs		
Professional labor	$ 5,000,000	
Indirect costs		
Client support	13,000,000	18,000,000
Operating income		$ 2,000,000

Taylor has a single direct cost category (professional labor) and a single indirect cost pool (client support). Indirect costs are allocated to jobs on the basis of professional labor costs.

Required
1. Present an overview diagram of the job costing system.
2. Compute the 19_5 budgeted indirect cost rate for Taylor & Associates.
3. The markup rate for pricing jobs is intended to produce a 10% operating income-to-revenue margin. Compute the markup rate as a percentage of professional labor costs.

4. Taylor is bidding on a consulting job for Red Rooster, a fast-food chain specializing in poultry meats. The budgeted breakdown of professional labor on the job is:

Professional Labor Category	Budgeted Rate per Hour	Budgeted Hours
Director	$200	3
Partner	100	16
Associate	50	40
Assistant	30	160

Compute the budgeted cost of the Red Rooster job. How much will Taylor bid for the job if it is to earn its target operating income-to-revenue margin of 10%?

4-21 Job costing, engineering consulting firm. Serra & Co., an engineering consulting firm, specializes in analyzing the structural causes of major building catastrophes. Its job costing system in 19_4 had a single direct cost category (professional labor) and a single indirect cost pool (general support). The allocation base for indirect costs is professional labor-dollars. Actual costs for 19_4 were:

Direct costs	
Professional labor	$10,000,000
Indirect costs	
General support	19,000,000
Total costs	$29,000,000

The following costs were included in the general support indirect cost pool:

Fringe benefits to professional labor	$1,500,000
Secretarial costs	2,700,000
Telephone/fax machine	600,000
Computer time	1,800,000
Photocopying	400,000
	$7,000,000

The firm's data-processing capabilities now make it feasible to trace these costs to individual cases or jobs. The managing partner is considering whether more costs than just professional labor should be traced to each job as a direct cost. In this way, the firm would be better able to justify billings to clients.

In late 19_4, arrangements were made to expand the number of direct cost categories and to trace them to seven client engagements. Two of the case records showed the following:

	Client Case 304	Client Case 308
Professional labor	$20,000	$20,000
Fringe benefits to professional labor	3,000	3,000
Secretarial costs	2,000	6,000
Telephone/fax machine	1,000	2,000
Computer time	2,000	4,000
Photocopying	1,000	2,000
Total direct costs	$29,000	$37,000

Required

1. Present an overview diagram of the 19_4 Serra job costing system. What was the actual indirect cost rate per professional labor-dollar?

2. Assume that the $7 million of costs included in the 19_4 general support indirect cost pool were reclassified as direct costs. The result is a system with six direct cost categories. Compute the revised indirect cost rate as a percentage of:

i. professional labor-dollars (excluding fringe benefits)

ii. total direct costs

3. Compute the total costs of jobs 304 and 308 using:

    a. the 19_4 costing system with a single direct cost category and a single indirect cost pool (professional labor-dollars as the allocation base)

    b. a costing system with six direct cost categories and a single indirect cost pool (professional labor-dollars as the allocation base)

    c. a costing system with six direct cost categories and a single indirect cost pool (total direct costs as the allocation base)

4. Assume that clients are billed at 120% of total job costs (that is, a markup on cost of 20%). Compute the billings in requirement 3 for jobs 304 and 308 for the (a), (b), and (c) costing systems.

5. Which method of job costing in requirement 3 do you favor? Explain.

4-22 Job costing, architectural design firm. Atlas Inc., an architectural design firm, is preparing to bid on two jobs—Brad Armstrong Enterprises and Judy Martin Associates. Both potential clients want a fixed-price quotation. Atlas's policy is to bid 120% of the budgeted total costs for each job. The budgeted direct costs for each job are:

Direct Cost Categories	Brad Armstrong Enterprises	Judy Martin Associates
Professional labor	$ 36,000	$ 25,000
Staff labor	9,000	21,000
Fringe benefits on all direct labor	15,000	14,000
Photocopying	1,000	1,600
Phone/fax machine	400	800
Computer time	3,000	6,400

Budgeted indirect costs of each job are included in a single cost pool (general support) and allocated at the rate of 80% of budgeted total direct costs.

Required

1. Present an overview diagram of the Atlas job costing system.

2. Compute the budgeted total costs of the Armstrong and Martin jobs.

3. What price would Atlas quote on each job?

4. Atlas learns that the Ronald Simmons Group will also bid on the two jobs. Recently, Atlas pirated a Simmons employee, who states that Simmons bases its bids on professional labor costs multiplied by 470%. What price will Simmons bid on each job? (Assume that Simmons has the same cost structure as Atlas.)

5. Compare and comment on the two approaches to bidding on jobs.

4-23 Peanut-butter costing, cross-subsidization. For many years, five former class-mates—Steve Goodyear, Lola Gonzales, Rex King, Elizabeth Poffo, and Gary Young—have had a reunion dinner at the annual meeting of the American Accounting Association. The bill for the most recent dinner at the Seattle Spaceneedle Restaurant was broken down as follows:

Diner	Entree	Dessert	Drinks	Total
Goodyear	$27	$8	$24	$59
Gonzales	24	3	0	27
King	21	6	13	40
Poffo	31	6	12	49
Young	15	4	6	25

For at least the last 10 annual dinners, King put the total restaurant bill on his American Express card. He then mailed to the other four a bill for the same amount (the average cost). They shared the gratuity at the restaurant by paying cash. King continued this practice for the Seattle dinner. However, just before he sent out the bill to the other four diners, Young phoned him to complain. He was livid at Poffo for ordering the steak and lobster entree ("She always does that!") and at Goodyear for having three glasses of imported champagne ("What's wrong with domestic beer?").

Required

1. Why is the average cost approach in the context of the reunion dinner an example of peanut-butter costing?
2. Compute the average cost of the five diners. Who is undercharged and who is overcharged under the average cost approach? Is Young's complaint justified?
3. Give an example of a situation where King would find it more difficult to compute the amount of undercosting or overcosting than in requirement 2.
4. How might the behavior of the diners be affected if each person paid his or her own bill instead of continuing with the average cost approach?

 4-24 Job costing, peanut-butter costing. (Extension of the Problem for Self-Study) Lindsay and Associates (see pp. 124–126) is examining how the refined costing system in Panel C of Exhibit 4-5 will cost three jobs differently from the costing system outlined in Exhibit 4-3. The following information pertains to the 19_5 audit of these three jobs:

	Northern Television	Rooster King	Nambucca Meat Inc.
Professional partner labor	20 hours	24 hours	58 hours
Professional associate labor	140 hours	76 hours	302 hours
Office staff labor	12 hours	18 hours	59 hours
Information specialist labor	8 hours	28 hours	64 hours
Phone/fax/copying	$160	$420	$715
Travel	$ 0	$180	$340
Days out of town	0 days	5 days	8 days

Required

1. Use the job costing system in Exhibit 4-3 to cost the 19_5 audits of Northern Television, Rooster King, and Nambucca Meat Inc. Direct and indirect cost rates are reported on p. 102.
2. Use the job costing system in Panel C of Exhibit 4-5 to cost the 19_5 audits of Northern Television, Rooster King, and Nambucca Meat Inc. Direct and indirect cost rates are reported in the Problem for Self-Study (pp. 124–126).
3. Using the refined costing system from requirement 2 as the benchmark, which 19_5 audit jobs are undercosted or overcosted when the Exhibit 4-3 job costing system is used? Explain reasons for this undercosting or overcosting.
4. How might Lindsay and Associates find the refined costing system in requirement 2 to be more useful than the system in requirement 1 in decisions relating to their audit clients Northern Television, Rooster King, and Nambucca Meat?

4-25 Job costing, law firm. Keating & Associates is a law firm specializing in labor relations and employee-related work. It has 25 professionals who work directly with its clients (5 partners and 20 associates). The average budgeted total compensation per professional for 19_6 is $104,000. Each professional is budgeted to have 1,600 billable hours to clients in 19_6. Keating is a highly respected firm, and all professionals work for clients to their maximum 1,600 billable hours available. All professional labor costs are included in a single direct cost category and are traced to jobs on a per-hour basis.

All costs of Keating & Associates other than professional labor costs are included in a single indirect cost pool (legal support) and are allocated to jobs using professional labor-hours as the allocation base. The budgeted level of indirect costs in 19_6 is $2,200,000.

Required

1. Present an overview diagram of Keating's job costing system.
2. Compute the 19_6 budgeted professional labor-hour direct cost rate.
3. Compute the 19_6 budgeted indirect cost rate per hour of professional labor.
4. Keating & Associates is considering bidding on two jobs:
    a. Litigation work for Richardson Inc. that requires 100 budgeted hours of professional labor

b. Labor contract work for Punch Inc. that requires 150 budgeted hours of professional labor

Prepare a cost estimate for each job.

4-26 Job costing with a refined costing system, law firm. (Continuation of 4-25)   Keating & Associates received feedback from Punch Inc. that its bid for the labor contract work was too high. This feedback prompted Keating to review its work activities and how they are reflected in its job costing system. This review included a detailed analysis of how past jobs used the company's resources and interviews with personnel about what factors drive the level of indirect costs. Management concluded that a system with two direct cost categories (professional partner labor and professional associate labor) and two indirect cost categories (general support and out-of-town support) would yield more accurate job costs. Budgeted information for 19_6 related to the two direct cost categories is:

	Professional Partner Labor	Professional Associate Labor
Number of professionals	5	20
Hours of billable time per professional	1,600 per year	1,600 per year
Total compensation (average per professional)	$200,000	$80,000

Budgeted information for 19_6 relating to the two indirect cost categories is:

	General Support	Out-of-Town Support
Total costs	$1,800,000	$400,000
Cost allocation base	Professional labor-hours	Days out of town

The budgeted total number of days out of town for all professionals in 19_6 is 2,000 days.

Required

1. Present an overview diagram of the refined costing system.
2. Compute the 19_6 budgeted direct cost rates for (a) professional partners and (b) professional associates.
3. Compute the 19_6 budgeted indirect cost rates for (a) general support and (b) out-of-town support.
4. Compute the budgeted job costs for the Richardson Inc. and Punch Inc. jobs, given the following information:

	Richardson Inc.	Punch Inc.
Professional partners	60 hours	30 hours
Professional associates	40 hours	120 hours
Days out of town	6 days	2 days

5. Comment on the results in requirement 4. Why are the job costs different from those computed in requirement 4 of Problem 4-25?

4-27 Job costing with single direct cost category, single indirect cost pool, law firm. Wigan Associates is a recently formed law partnership. Ellery Hanley, the managing partner of Wigan Associates, has just finished a tense phone call with Martin Offiah, president of Widnes Coal. Offiah complained about the price Wigan charged for some conveyancing (drawing up property documents) legal work done for Widnes Coal. He requested a breakdown of the charges. He also indicated to Hanley that a competing law firm, Hull and Kingston, was seeking more business with Widnes Coal and that he was going to ask them to bid for a conveyancing job next month. Offiah ended the phone call by saying that if Wigan bid a price similar to the one charged last month, Wigan would not be hired for next month's job.

Hanley is dismayed by the phone call. He is also puzzled, as he believes that conveyancing is an area where Wigan Associates has much expertise and is highly efficient. The Widnes Coal phone call is the bad news of the week. The good news is that yesterday Hanley received a phone call from its only other client (St. Helens Glass) saying it was very pleased with both the quality of the work (primarily litigation) and the price charged on its most recent case.

Hanley decides to collect data on the Widnes Coal and St. Helens Glass cases. Wigan Associates uses a cost-based approach to pricing (billing) each legal case. Currently it uses a single direct cost category (for professional labor time) and a single indirect cost pool (general support). Indirect costs are allocated to cases on the basis of professional labor-hours per case. The case files show the following:

	Widnes Coal	St. Helens Glass
Professional labor time	104 hours	96 hours

Wigan Associates bills clients for professional labor at $70 an hour. Indirect costs are allocated to cases at $105 an hour. Total indirect costs in the most recent period were $21,000. An operating income component for the partnership is built into the $70 per professional labor-hour billing rate.

Required
1. Why is it important for Wigan Associates to understand the costs associated with individual cases in their pricing decisions?
2. Present an overview diagram of the existing job costing system.
3. Compute the amount Wigan Associates billed Widnes Coal and St. Helens Glass.

4-28 Job costing with multiple direct cost categories, single indirect cost pool, law firm. (Continuation of 4–27) Hanley speaks to the other partners about the pricing of the two cases. Several believe that the relative prices charged seem out of line with their intuition. One partner observes that a useful approach to obtaining more accurate job costs is to increase direct cost tracing.

Hanley asks his assistant to collect details on those costs included in the $21,000 indirect cost pool that can be traced to each individual case. After further analysis, Wigan is able to reclassify $14,000 of the $21,000 as direct costs:

Other Direct Costs Traceable to Cases	Widnes Coal	St. Helens Glass
Support labor	$1,600	$ 3,400
Computer time	500	1,300
Travel	600	4,400
Telephones/fax machine	200	1,000
Photocopying	250	750
Total	$3,150	$10,850

Hanley decides to calculate the price that would have been billed on each case had Wigan used six direct cost pools and a single indirect cost pool. The single indirect cost pool would have $7,000 of costs and would be allocated to each case using the professional labor-hours base.

Required
1. Present an overview diagram of the refined job costing system with its multiple direct cost categories.
2. What is the revised indirect cost allocation rate per professional labor-hour for Wigan Associates when total indirect costs are $7,000?
3. Compute the costs that would have been billed to Widnes and St. Helens if Wigan Associates had used its refined costing system with multiple direct cost categories and one indirect cost pool.
4. Compare the costs billed to Widnes and St. Helens in requirement 3 with those in requirement 3 of Problem 4-27. Comment.

4-29 Job costing with multiple direct cost categories, multiple indirect cost pools, law firm. (Continuation of 4-27 and 4-28) Hanley examines the job costing approaches in Problems 4-27 and 4-28. He questions the use of a single charge-out rate for all professional labor of Wigan Associates. Wigan has two classifications of professional staff—partners and associates.

Hanley asks his assistant to examine the relative use of partners and associates on the recent Widnes Coal and St. Helens cases. The Widnes case used 24 hours of partner time and 80 hours of associate time. The St. Helens case used 56 hours of partner time and 40 hours of associate time.

Hanley decides to examine how the use of separate direct and indirect cost pools for partners and for associates would have affected the amount billed to Widnes and St. Helens. Indirect costs in each cost pool would be allocated on the basis of total hours of that category of professional labor. The rates per category of professional labor are:

Category of Professional Labor	Direct Cost per Hour	Indirect Cost per Hour
Partner	$100.00	$57.50
Associate	50.00	20.00

These indirect cost rates are based on a total indirect cost pool of $7,000; $4,600 of this $7,000 is attributable to the activities of partners, and $2,400 is attributable to the activities of associates. (The indirect cost per hour of $57.50 is calculated by dividing $4,600 by 80 hours of partner time; the indirect cost rate of $20 is calculated by dividing $2,400 by 120 hours of associate time.)

Required

1. Present an overview diagram of the refined job costing system with its multiple direct cost categories and its multiple indirect cost pools.
2. Compute the costs billed to Widnes and St. Helens with Wigan Associates' further refined system, with multiple direct cost categories and multiple indirect cost pools.
3. For what decisions might Wigan Associates find it more useful to use this job costing approach rather than the approach in Problems 4-27 or 4-28?

 4-30 Activity-based costing, merchandising. Figure Four Inc. operates a company that specializes in the distribution of pharmaceutical products. Figure Four buys from pharmaceutical companies and resells to each of three different markets:

A. General supermarket chains
B. Drugstore chains
C. "Ma and Pa" single-store pharmacies

Rick Flair, the new controller of Figure Four, reported the following data for August 19_4:

	General Supermarket Chains	Drugstore Chains	Ma and Pa Single Stores
Average revenue per delivery	$30,900	$10,500	$1,980
Average cost of goods sold per delivery	$30,000	$10,000	$1,800
Number of deliveries	120	300	1,000

For many years, Figure Four has used gross margin percentage ([revenue – cost of goods sold] ÷ revenue) to evaluate the relative profitability of its different distribution outlets.

Flair recently attended a seminar on activity-based costing and decides to consider using it at Figure Four. Flair meets with all the key business area managers and many staff members. People generally agree that there are five key activity areas at Figure Four:

Activity Area	Cost Driver
1. Order processing	Number of orders
2. Line item ordering	Number of line items
3. Store delivery	Number of store deliveries
4. Cartons shipped to stores	Number of cartons shipped to a store per delivery
5. Shelf stacking at customer store	Number of hours of shelf stacking

Each order consists of one or more line items. A line item represents a single product (such as Extra-Strength Tylenol Tablets). Each store delivery entails delivery of 1 or more cartons of products. Each product is delivered in 1 or more separate cartons. Figure Four delivery staff stack cartons directly onto display shelves in a store. Currently there is no charge for this service and not all customers use Figure Four for this activity.

The August 19_4 operating costs (other than cost of goods sold) of Figure Four are $301,080. These operating costs are assigned to the five activity areas. The costs in each area and the number of cost drivers in that area for August 19_4 are:

Activity Area	Total Costs in August 19_4	Total Units of Cost Driver in August 19_4
1. Order processing	$ 80,000	2,000 orders
2. Line item ordering	63,840	21,280 line items
3. Store deliveries	71,000	1,420 store deliveries
4. Carton deliveries	76,000	76,000 cartons
5. Shelf stacking	10,240	640 hours
	$301,080	

Other data for August 19_4 are:

	General Supermarket Chains	Drugstore Chains	Ma and Pa Single Stores
Total number of orders	140	360	1,500
Average number of line items per order	14	12	10
Total number of store deliveries	120	300	1,000
Average number of cartons shipped per store delivery	300	80	16
Average number of hours of shelf stacking per store delivery	3	0.6	0.1

Required

1. Compute the August 19_4 gross-margin percentages for its three distribution markets. What is the operating income of Figure Four?

2. Compute the August 19_4 per unit cost driver rate for each of the five activity areas.

3. Compute the operating income of each distribution market in August 19_4 using the activity-based costing information. Comment on the results. What new insights are available with the activity-based information?

4. Describe four challenging problems Flair would face in assigning the total August 19_4 operating costs of $301,080 to the five activity areas.

4-31 Customer costing, merchandising. (Continuation of 4-30)   Flair decides to further use the activity-based costing information to examine individual customer profitability within each distribution market. He focuses first on the "Ma and Pa Single-Store" distribution market. Two customers are used to highlight the new insights available with the activity-based costing approach. Data pertaining to these two customers in August 19_4 are:

	Charleston Pharmacy	Chapel Hill Pharmacy
Total number of orders	12	10
Average number of line items per order	10	18
Total number of store deliveries	6	10
Average number of cartons shipped per store delivery	24	20
Average number of hours of shelf stacking per store delivery	0	0.5
Average revenue per delivery	$2,400	$1,800
Average cost of goods sold per delivery	$2,100	$1,650

Required

1. Use the activity-based costing information to compute the operating income of each customer in August 19_4. Comment on the results.

2. Flair ranks the individual customers in the "Ma and Pa" single-store distribution market on the basis of operating income. The cumulative operating income of the top 20% of customers is $55,680. Figure Four reports negative operating income of $31,247 for the bottom 40% of its customers. Make four recommendations that you think Figure Four should consider in light of this new customer-profitability information.

4-32 Customer costing, service company. Instant Service (IS) is a repair service company specializing in the rapid repair of photocopying machines. Each of its 10 clients pays a fixed monthly service fee (based on the type of photocopying machines owned by that client and the number of employees at that site). IS keeps records of the time technicians spend at each client as well as the cost of the equipment used to repair each photocopying machine. IS recently decided to compute the profitability of each customer. The following data pertain to May 19_4 (in thousands):

	Customer Revenues	Customer Costs
Avery Group	$260	$182
Duran Systems	180	184
Retail Systems	163	178
Wizard Partners	322	225
Santa Clara College	235	308
Grainger Services	80	74
Software Partners	174	100
Problem Solvers	76	108
Business Systems	137	110
Okie Enterprises	373	231

Required

1. Compute the operating income of each customer. Prepare for Instant Service an exhibit similar to Exhibit 4-10. Comment on the results.

2. What options should Instant Service consider in light of your customer-profitability analysis in requirement 1?

3. What problems might Instant Service encounter in accurately estimating the operating cost of each customer?

4-33 Job costing, accuracy of time records, ethics. Tax Assist provides advice on tax planning and tax return preparation to a broad range of clients. Tax Assist employs its professional staff on an hourly contract basis. The rate is $30 per hour, including all fringe benefits. Most Tax Assist professionals work in their own homes and use the Tax Assist offices only to meet clients. Tax Assist charges clients at $65 per hour of professional labor plus any extra charges for computer time.

Mike Rotundo, the president of Tax Assist, has received complaints from several clients that they are being overcharged. He is dismayed. Tax Assist promotes itself as "Trustworthy Tax Professionals Working For You." Rotundo conducts an investigation. The hours billed to clients by two tax professionals (coded PQR and RST in the investigation) have averaged over 80 hours a week for the last six months.

Required

1. How might Rotundo gain evidence on whether PQR and RST are overstating their hours worked for individual clients?

2. What procedures could Rotundo use to ensure that the time each professional bills to each Tax Assist client is accurate?

3. How should Rotundo respond to a client who argues that the 19_5 hours reported by RST are overstated because his Tax Assist bill for 19_5 is double that of his bill for 19_4?

5

Costing Systems in the Manufacturing Sector

PART ONE: JOB COSTING SYSTEMS IN MANUFACTURING

General approach to job costing

Allocating indirect costs to jobs

Actual, normal, and budgeted costing methods

Illustration of a job costing system in manufacturing

Budgeted indirect costs and end-of-period adjustments

PART TWO: PROCESS COSTING SYSTEMS IN MANUFACTURING

Case 1: Process costing with all units not fully completed

Case 2: Process costing with some units not fully completed

PART THREE: ACTIVITY-BASED COSTING IN MANUFACTURING

Illustration of ABC in manufacturing

Multipurpose nature of costing systems

The manufacture of pharmaceutical products, such as vials of Canferon A for fighting hepatitis, involves the use of high-quality materials, skilled labor, and specialized equipment. Takeda Chemicals has designed its cost accounting system to help improve decisions about cost management and to obtain more accurate product costs.

Learning Objectives

When you have finished studying this chapter, you should be able to

1. Outline a five-step approach to job costing

2. Describe three key source documents used in job costing systems

3. Distinguish between the Work in Process subsidiary ledger and the Work in Process Control account in the general ledger

4. Prepare summary journal entries for typical transactions of a job costing system

5. Describe alternative methods of disposing of period-end underallocated or overallocated indirect costs

6. Identify five key steps in process costing

7. Explain the role of equivalent units in process costing

8. Describe distinctive features of activity-based costing

9. Distinguish between the traditional and the activity-based costing approaches to designing a costing system

Chapter 4 introduced costing systems and distinguished between a job costing system and a process costing system. It then discussed costing systems in the service and merchandising sectors. This chapter focuses on the manufacturing sector. The building block concepts of costing systems (cost object, direct costs, indirect costs, cost pool, and cost allocation base) apply equally to all sectors of the economy. Before reading this chapter, be sure you are comfortable with the contents of pages 98–100.

Costing systems in the manufacturing sector often are more complex than those in the service and merchandising sectors. Manufacturers provide their customers with tangible products that have been converted to a different basic form from the materials and other inputs. Tracking this change in basic form from materials and other inputs through work in process, then to finished goods, and finally to cost of goods sold requires a more complex costing system than is found in most service or merchandising companies.

Part One of this chapter covers job costing in manufacturing. Part Two introduces process costing in manufacturing. Let's review the definitions of a job costing system and a process costing system.

◆ *Job costing system.* The cost of a job or service is obtained by assigning costs to a distinct, identifiable job or service.

◆ *Process costing system.* The cost of a product or service is obtained by assigning costs to masses of similar units and then computing unit costs on an average basis.

Part Three illustrates how an activity-based approach can be incorporated into the design of either a job costing or a process costing system in manufacturing.

◆ PART ONE:
JOB COSTING SYSTEMS IN MANUFACTURING

Jobs undertaken in the manufacturing sector differ considerably in many dimensions—for example, in the time required, the resources expended, and the technical complexity demanded to complete them. The general approach to job costing that follows can be applied to a broad set of different manufacturing jobs. (As illustrated in Chapter 4, this approach also can be used in the service and merchandising sectors.)

Objective 1

Outline a five-step approach to job costing

Step-by-Step Approach

We illustrate a five-step approach to job costing using the Robinson Company, which manufactures specialized machinery for the paper-making industry at its Green Bay, Wisconsin, plant. (See Exhibit 5-1)

STEP 1: *IDENTIFY THE JOB THAT IS THE CHOSEN COST OBJECT.* The job in this case is a pulp machine manufactured for the Western Pulp and Paper Company in 19_4.

STEP 2: *IDENTIFY THE DIRECT COST CATEGORIES FOR THE JOB.* Robinson identifies two direct manufacturing cost categories—direct materials and direct manufacturing labor. Actual cost rates are used for both of these categories. The employees Robinson classifies as direct manufacturing labor are paid on an hourly basis.

STEP 3: *IDENTIFY THE INDIRECT COST POOLS ASSOCIATED WITH THE JOB.* Robinson uses a single indirect manufacturing cost pool termed *manufacturing overhead*. This pool represents the costs of the Green Bay plant's manufacturing department. It includes items such as depreciation on equipment, energy costs, indirect materials, and indirect manufacturing labor and salaries.

STEP 4: *SELECT THE COST ALLOCATION BASE TO USE IN ASSIGNING EACH INDIRECT COST POOL TO THE JOB.* Robinson uses machine-hours as the allocation base for manufacturing overhead.

EXHIBIT 5-1

Job Costing Overview for Manufacturing Costs at Robinson Company

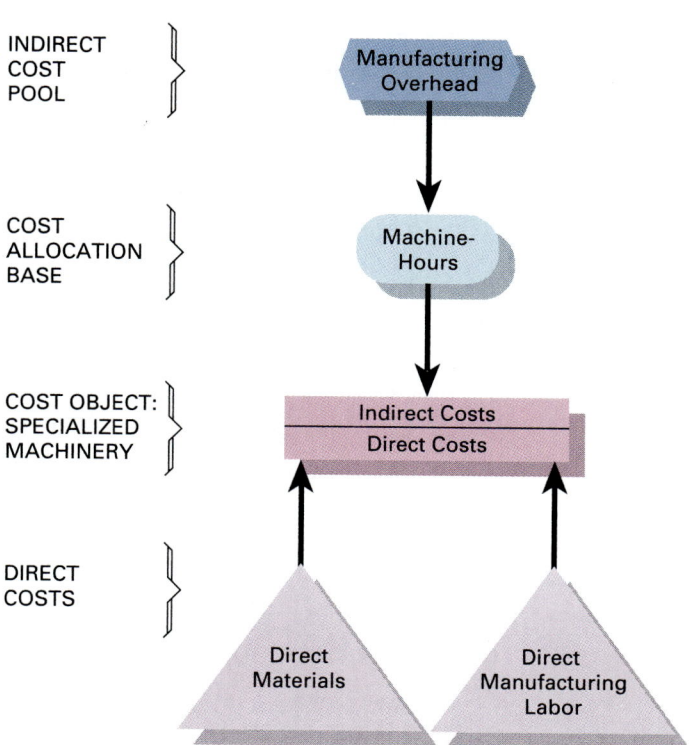

INDIRECT COST POOL — Manufacturing Overhead

COST ALLOCATION BASE — Machine-Hours

COST OBJECT: SPECIALIZED MACHINERY — Indirect Costs / Direct Costs

DIRECT COSTS — Direct Materials / Direct Manufacturing Labor

Robinson uses a budgeted rate of $80 per machine-hour in 19_4. (Details of the computation of this rate are discussed below.)

These five steps guide the computation of the manufacturing job cost of the machine for Western Pulp:

Direct manufacturing costs		
Direct materials	$46,060	
Direct manufacturing labor	13,290	$59,350
Indirect manufacturing costs		
Manufacturing overhead ($80 × 504 machine-hours)		40,320
Total manufacturing job costs		$99,670

Exhibit 5-1 presents an overview of the job costing system for the manufacturing costs of the Robinson Company. Costing overview exhibits are important learning tools. We suggest you sketch one when you need to understand a costing system.

Department Costing and Job Costing

The Green Bay plant uses its job costing system to help manage the costs in its manufacturing department as well as to determine the cost of individual jobs such as the Western Pulp and Paper machine. The manufacturing department is an important cost object, as is each job manufactured. The relationship between these two important cost objects follows:

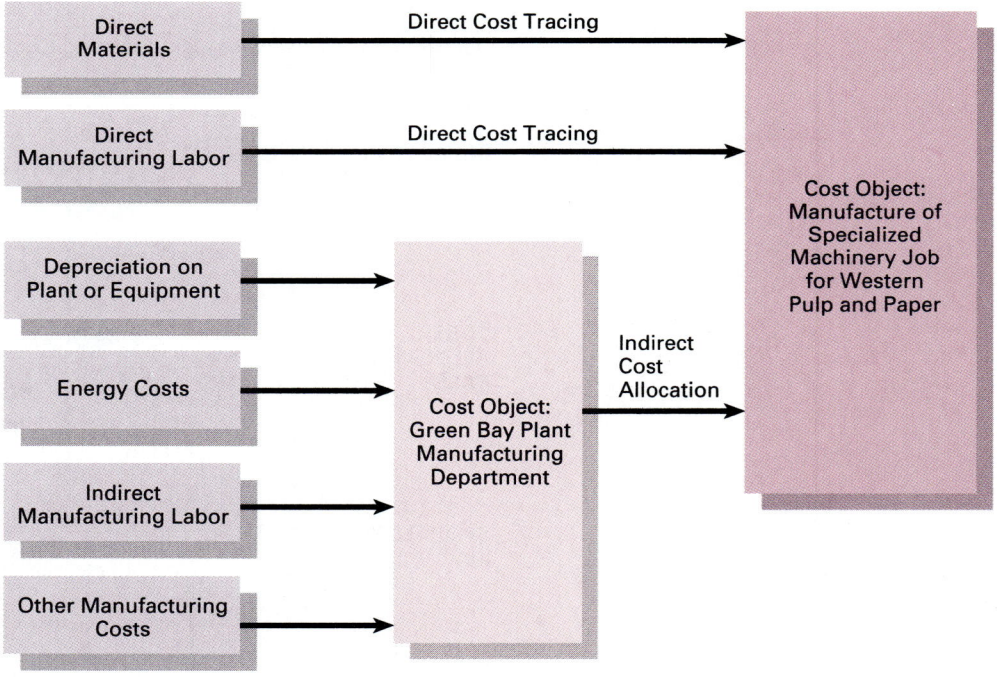

Source Documents

A separate cost record (account) is kept for each job in a job costing system. Exhibit 5-2 presents an illustrative **job cost record** (also called a **job order,** or **job cost, sheet**), which is the key document used to accumulate the costs of a job.

EXHIBIT 5-2

Source Documents at Robinson Company: Job Cost Record*, Materials Requisition Record, and Labor Time Record

JOB COST RECORD

JOB NO:	WPP298		CUSTOMER:	Western Pulp and Paper
Date Started:	Feb.7,19_4		Date Completed:	April 3,19_4

DIRECT MATERIALS

Materials Requisition No.	Part No.	Date Received	Quantity Used	Unit Cost	Billing Amount
MRR: 19_4:198	MB 468-A	Feb. 9, 19_4	8	$14	$112
MRR: 19_4:268	TB 267-F	Feb. 11, 19_4	12	63	756
					•
					•
					•
Total					$46,060

DIRECT MANUFACTURING LABOR

Labor Time Record No.	Employee No.	Period Covered	Hours Used	Hourly Rate	Billing Amount
LTR: 232	551-21-3076	Feb.16-22, 19_4	25	$18	$450
LTR: 247	287-31-4671	Feb.16-22, 19_4	16	18	288
					•
					•
					•
Total					$13,290

MANUFACTURING OVERHEAD

Cost Pool Category	Allocation Base	Allocation Base Units Used	Allocation Base Rate	Billing Amount
Manufacturing	Machine Hours	504	$80	$40,320
				•
				•
				•
Total				$40,320
TOTAL BILLABLE JOB COST				$99,670

MATERIALS REQUISITION RECORD

Materials Requisition Record No:			19_4:198	
Job No: WPP 298	Date:	Feb. 9, 19_4		

Part No.	Part Description	Quantity	Unit Cost	Total Cost
MB468-A	Metal Brackets	8	$14	$112

Issued By: *B. Mackay*   Date: Feb. 9. 19_4
Received By: *J. Hardy*   Date: Feb. 9. 19_4

LABOR TIME RECORD

Labor Time Record No:		LT:232	

Employee Name: G.L. Cook	Employee No: 551-21-3076

Employee Classification Code: Grade 3 Machinist

Week Start:	Feb. 16, 19_4	Week End:	Feb. 22, 19_4

Job. No.	M	T	W	Th	F	S	Su	Total
WPP298	4	8	3	6	4	0	0	25

Supervisor: *S. Gourley*   Date: Feb. 23, 19_4

* Robinson Company uses a single manufacturing overhead cost rate. The use of multiple overhead cost rates would mean multiple entries in the "Manufacturing Overhead" section of its job cost record.

Manufacturers that use job costing usually have several jobs passing through the plant simultaneously. Each job typically requires different kinds of materials and other resources. Thus jobs may have different routings, different operations, and different times required for completion. Standardized source documents provide the information managers need to keep track of transactions and costs that are related to each job on a job cost record. Consider direct materials. The basic source document is a **materials requisition record,** which is a form used to charge departments and job cost records for the cost of the materials used on a specific job. Exhibit 5-2 shows an illustrative materials requisition record for Robinson Company. The basic source document for direct manufacturing labor is a **labor time record,** which is used to charge departments and job cost records for labor time used. Exhibit 5-2 also shows an illustrative labor time record for Robinson Company.

In many costing systems the source documents exist only in the form of computer records. With bar coding and other forms of on-line recording of information, the materials and time used on jobs are recorded routinely without human intervention.

Responsibility and Control

Management should lay out clearly the department responsibility for the use of all resources. Consider direct materials and direct manufacturing labor. Department heads are usually kept informed of direct materials and direct labor performance by hourly, daily, or weekly summaries of requisitions for materials and labor time tickets charged to their departments.

The job cost records also serve a control function. The costs actually assigned to each job are compared with the budgeted costs for that job. Significant deviations between these two amounts can prompt managers to investigate their causes. For example, if the deviation is due to higher actual manufacturing labor time than budgeted, management may conduct a detailed investigation of both its labor budgeting method and its manufacturing labor productivity.

ALLOCATING INDIRECT COSTS TO JOBS

We mentioned that Robinson uses a rate of $80 per machine-hour for indirect costs in 19_4. How does Robinson compute that rate? Like many other companies, Robinson follows a five-step approach to computing indirect cost rates.

STEP 1: *IDENTIFY THE COSTS TO INCLUDE IN THE INDIRECT COST POOL.* Robinson Company has an accounting manual that details the cost items included in its manufacturing overhead cost pool. This pool represents the costs of the manufacturing department at the Green Bay plant.

STEP 2: *FOR THE BUDGET PERIOD, PREPARE ESTIMATES OF THE COST ITEMS FROM STEP 1.* Robinson uses an annual budget period. Budgeted manufacturing overhead costs for 19_4 are $1,280,000.

STEP 3: *SELECT THE COST ALLOCATION BASE(S).* Operating personnel at the Green Bay plant conclude that the number of machine-hours is the major driver of manufacturing overhead. Machine-hours is chosen as the cost allocation base because of this cause-and-effect relationship with manufacturing overhead costs.

STEP 4: *FOR THE BUDGET PERIOD, ESTIMATE THE QUANTITY OF THE COST ALLOCATION BASE(S) FROM STEP 3.* For 19_4, the budgeted quantity of machine-hours is 16,000.

STEP 5: *COMPUTE THE BUDGETED INDIRECT COST ALLOCATION RATE.* Use the following formula:

$$\text{Budgeted indirect cost allocation rate} = \frac{\text{Budgeted total costs in indirect cost pool}}{\text{Budgeted total quantity of cost allocation base}}$$

$$= \frac{\$1,280,000}{16,000 \text{ hours}}$$

$$= \$80 \text{ per machine-hour}$$

Each hour of machine time spent on a job at the Green Bay plant is thus assigned an $80 indirect cost charge.

ACTUAL, NORMAL, AND BUDGETED COSTING METHODS

How does a company use direct rates and indirect rates in computing costs? Three possible combinations of actual and budgeted rates for job costing and process costing systems are used. The following table shows how the *actual, normal,* and *budgeted costing* methods use actual and budgeted rates.

	Actual Costing	Normal Costing	Budgeted Costing
Direct Costs	Actual rate(s) or price(s) x Actual input(s) used	Actual rate(s) or price(s) x Actual input(s) used	Budgeted rate(s) or price(s) x Actual input(s) used
Indirect Costs	Actual rate(s) x Actual input(s) used	Budgeted rate(s) x Actual input(s) used	Budgeted rate(s) x Actual input(s) used

The phrase "actual direct cost rate" is most often used to describe direct labor costs. The phrase "actual direct materials prices" is most often used to describe direct materials costs.

Chapter 2 discussed the actual costing method. The Robinson Company's example in this chapter illustrates normal costing. Robinson uses actual cost rates times actual quantity when computing both the direct materials costs and the direct manufacturing labor costs of each job. It uses the budgeted indirect cost rate times the actual number of machine-hours when computing the indirect costs of each job. Chapter 4 (pp. 101–109) illustrates use of the budgeted costing method by a public accounting firm.

Surveys of Company Practice

COST ALLOCATION BASES USED FOR MANUFACTURING OVERHEAD

How do companies around the world allocate manufacturing overhead costs to products? The percentages in the table below indicate how frequently particular cost allocation bases are used in management accounting systems in five countries. The reported percentages exceed 100% because many companies surveyed use more than one cost allocation base.

	United States[a]	Australia[b]	Ireland[c]	Japan[b]	United Kingdom[b]
Direct labor-hours	31%	36%	30%	50%	31%
Direct labor-dollars	31	21	22	7	29
Machine-hours	12	19	19	12	27
Direct materials dollars	4	12	10	11	17
Units of production	5	20	28	16	22
Prime cost (%)	—	1	—	21	10
Other	17	—	9	—	—

[a]Adapted from Cohen and Paquette, "Management Accounting." [b]Adapted from Blayney and Yokoyama, "A Comparative Analysis." [c]Adapted from Clarke, "Survey." Full citations are in Appendix A.

Why do companies use budgeted direct or indirect cost rates when actual cost rates are more accurate? A detailed discussion of this question appears in Chapter 4 (pp. 105–106). The three main reasons Robinson uses budgeted indirect cost rates match the points made in that discussion.

1. An accounting system using budgeted rates can provide managers with cost information on a timely basis during the year, when decisions about pricing, bidding, or product emphasis must be made. The budgeted indirect rate is known at the start of 19_4. In contrast, the actual indirect cost rate is not known until the end of 19_4, when both the actual total indirect costs and the actual total machine hours are known. (This delay in obtaining actual rates does not cause a problem with Robinson's two direct cost categories. For example, its direct manufacturing labor rates are known at the start of a job because direct labor is paid a precontracted hourly rate. The actual direct materials used on a job are also known when the job is completed.)

2. Actual costs may be subject to short-run fluctuations that managers view as misleading for individual job costs. For example, short-run variations in energy cost rates may occur at Robinson's Green Bay plant because of the pricing policy of the utility company that provides the electricity and gas. Robinson believes that decisions about pricing, bidding, or product emphasis are best made if these short-run variations in energy cost rates are averaged in a single budgeted energy cost rate.

3. Actual costs are affected by changes from the budgeted work done on other jobs in the period. In contrast, once the budgeted rate has been set, it will not be affected by any change or work done on other jobs.

At the end of each year, Robinson's actual total indirect costs and actual total machine-hours likely will differ from their budgeted amounts. A subsequent section of this chapter discusses alternative ways of adjusting for differences between the actual and budgeted amounts.

ILLUSTRATION OF A JOB COSTING SYSTEM IN MANUFACTURING

The Robinson Company illustrates how a job costing system operates in manufacturing. Recall that its job costing system has two direct cost categories (direct materials and direct manufacturing labor) and one indirect cost pool (manufacturing overhead). See Exhibit 5-1. The following example looks at events that took place in September 19_4.

General Ledger and Subsidiary Ledgers

Objective 3

Distinguish between the Work in Process subsidiary ledger and the Work in Process Control account in the general ledger

As we have noted, a job costing system has a separate job cost record for each job. This record typically is found in a subsidiary ledger. The general ledger combines these separate job cost records in the Work in Process Control account, which pertains to all jobs within the organization.

Exhibit 5-3 shows T-account relationships for the Robinson Company's general ledger and illustrative records in the subsidiary ledgers. The first part (Panel A) of Exhibit 5-3 has the general ledger section that gives a bird's-eye view of the costing system. The amounts shown are based on the illustration that follows. The second part (Panel B) of Exhibit 5-3 has the subsidiary ledgers and the basic source documents that contain the underlying details—the worm's-eye view. General ledger accounts that have the word "Control" in their titles (such as "Materials Control" and "Accounts Payable Control") are supported by underlying subsidiary ledgers.

Software programs guide the processing of transactions in most accounting systems. Some programs make general-ledger entries simultaneously with entries in the subsidiary-ledger accounts. Other software programs make general-ledger entries at, say, weekly or monthly intervals, with entries in the subsidiary-ledger accounts on a more frequent basis. Robinson Company makes entries in its subsidiary ledger when transactions occur and then makes entries in its general ledger on a monthly basis.

A general ledger should be viewed as one of many tools that assist management in planning and control. To control operations, managers frequently use the source documents in the subsidiary ledgers and study nonfinancial variables, such as the percentage of jobs requiring any rework.

Explanations of Transactions

Objective 4

Prepare summary journal entries for typical transactions of a job costing system

The following transaction-by-transaction summary analysis explains how a job costing system serves the twin goals of (1) departmental responsibility and control and (2) product costing. These transactions track (a) the purchases of materials and other inputs, (b) their conversion into work in process, (c) their conversion into finished goods, and (d) the sale of finished goods.

1. *Transaction:* Purchases of materials (direct and indirect), $89,000 on account.

   Analysis: The asset Materials Control is increased. The liability Accounts Payable Control is increased. Both accounts have the word *control* in their title in the general ledger because they are supported by records in the subsidiary ledger. The subsidiary records for materials at the Robinson Company form a perpetual inventory record called *Materials Records.* At a minimum, these records would contain columns for quantity received, issuance to jobs, and balance. There is a separate subsidiary materials record for each type of material in the subsidiary ledger. The following journal entry summarizes all the September 19_4 entries in the materials subsidiary ledgers.

   Journal Entry: Materials Control $89,000
                            Accounts Payable Control $89,000

   Post to the General Ledger:

Materials Control		**Accounts Payable Control**	
① 89,000			① 89,000

2. *Transaction:* Materials sent to manufacturing plant floor: direct materials, $81,000, and indirect materials, $4,000.

   Analysis: The accounts Work in Process Control and Manufacturing Overhead Control are increased. The account Materials Control is decreased. Responsibility is fixed by using *materials requisitions records* as a basis for charging departments for the materials issued to them. Requisitions are accumulated and posted monthly to the general ledger at the Robinson Company. As direct materials are used, they are charged to individual job records, which are the subsidiary ledger accounts to the Work in Process Control general ledger account. Indirect materials are charged to individual manufacturing departments' overhead cost records, which are the subsidiary ledger for Manufacturing Overhead Control at the Robinson Company. Department managers are responsible for monitoring costs, item by item.

   Each indirect cost pool in a job costing system will have its own account in the general ledger. Robinson has only one indirect cost pool—manufacturing overhead.

   Journal Entry: Work in Process Control $81,000
                         Manufacturing Dept. Overhead Control   4,000
                               Materials Control $85,000

   Post to the General Ledger:

Materials Control		**Work in Process Control**	
① 89,000	② 85,000	② 81,000	

Manufacturing Department Overhead Control	
② 4,000	

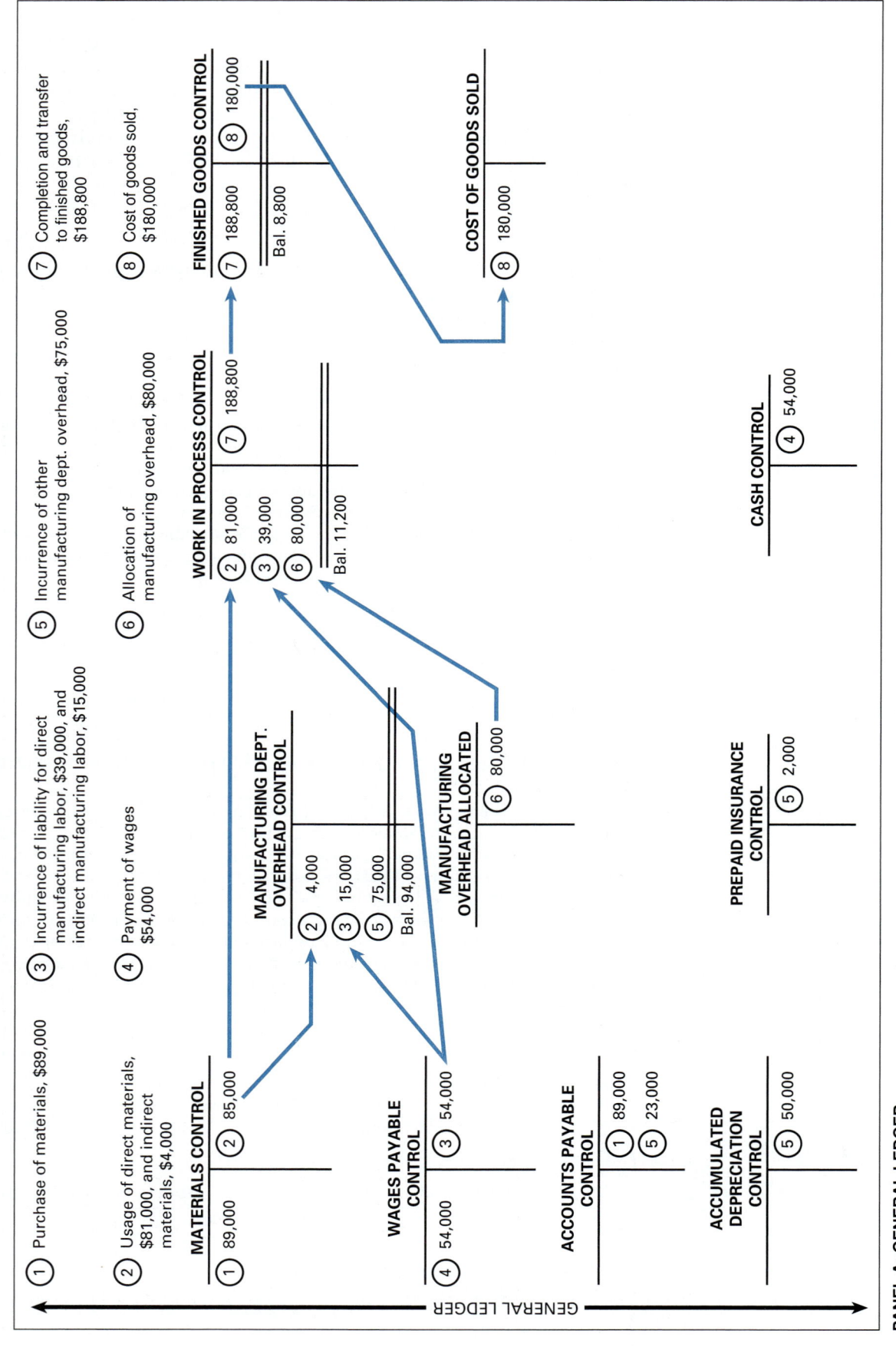

① Purchase of materials, $89,000

② Usage of direct materials, $81,000, and indirect materials, $4,000

③ Incurrence of liability for direct manufacturing labor, $39,000, and indirect manufacturing labor, $15,000

④ Payment of wages $54,000

⑤ Incurrence of other manufacturing dept. overhead, $75,000

⑥ Allocation of manufacturing overhead, $80,000

⑦ Completion and transfer to finished goods, $188,800

⑧ Cost of goods sold, $180,000

PANEL A: GENERAL LEDGER

ILLUSTRATIVE SUBSIDIARY LEDGERS

MATERIALS RECORDS

Received	Issued	Balance
①	②	

Copy of invoice or receiving reports → ①

Copy of materials requisition records → ②

WORK IN PROCESS: JOB RECORD No. XYZ

Direct Materials	Direct Manuf. Labor	Allocated Manuf. Overhead
②	③	⑥

Copy of materials requisition records → ②

Manuf. labor time records → ③

Budgeted rate based on machine-hours → ⑥

FINISHED GOODS RECORDS

Received	Issued	Balance
⑦	⑧	

Completed job cost records → ⑦

Costed sales invoices → ⑧

MANUFACTURING DEPT. OVERHEAD RECORDS

Indirect Materials	Indirect Manuf. Labor	Utilities	Depreciation	Insurance
②	③	⑤	⑤	⑤

Copy of materials requisition → ②

Manuf. labor time record or payroll analysis → ③

Invoices, special authorizations → ⑤

NOTE: Type of source document used is designated at the tail of the arrow. These entries would be shown in much more detail in the subsidiary ledger than in the general ledger.

3. *Transaction:* Manufacturing labor wages liability incurred, direct ($39,000) and indirect ($15,000).

Analysis: The accounts Work in Process Control and Manufacturing Department Overhead Control are increased. Wages Payable Control is also increased. Labor time records are used to trace direct manufacturing labor to Work in Process Control and to accumulate the indirect manufacturing labor in Manufacturing Department Overhead Control. The indirect manufacturing labor is, by definition, difficult to trace to the individual job. Department managers have responsibility for making efficient use of available labor. The percentage of labor time available that is assigned (either directly via tracing or indirectly via allocation) to individual jobs is often used as a measure of labor efficiency. The higher this percentage, the more efficiently labor is being managed at the department level.

Journal Entry: Work in Process Control $39,000
      Manufacturing Dept. Overhead Control   15,000
          Wages Payable Control $54,000

Post to the General Ledger:

Wages Payable Control		**Work in Process Control**		
	③ 54,000	②	81,000	
		③	39,000	

Manufacturing Department Overhead Control	
② 4,000	
③ 15,000	

4. *Transaction:* Payment of total manufacturing payroll for the month, $54,000. (Payroll withholdings from employees are ignored in this example.)

Analysis: The liability Wages Payable Control is decreased. The asset Cash Control is decreased.

Journal Entry: Wages Payable Control $54,000
      Cash Control $54,000

Post to the General Ledger:

Wages Payable Control			**Cash Control**	
④ 54,000	③	54,000	④	54,000

For convenience here, wages payable for the month is assumed to be completely paid at month-end.

5. *Transaction:* Additional manufacturing department overhead costs incurred during the month, $75,000. These costs consist of utilities and repairs, $23,000; insurance expired, $2,000; and depreciation on equipment, $50,000.

Analysis: The Manufacturing Department Overhead Control account is increased. The liability Accounts Payable Control is increased, the asset Prepaid Insurance Control is decreased, and the asset Equipment is decreased by means of a related contra asset account Accumulated Depreciation Control. The detail of these costs is entered in the appropriate columns of the individual manufacturing overhead cost records that make up the subsidiary ledger for Manufacturing Department Overhead Control. The source documents for these distributions include invoices (for example, a utility bill) and special schedules (for example, a depreciation schedule) from the responsible accounting officer.

Journal Entry: Manufacturing Dept. Overhead Control $75,000
      Accounts Payable Control $23,000
      Accumulated Depreciation Control   50,000
      Prepaid Insurance Control   2,000

Post to the General Ledger:

Accounts Payable Control			Manufacturing Department Overhead Control		
	①	89,000	②	4,000	
	⑤	23,000	③	15,000	
			⑤	75,000	

Accumulated Depreciation Control			Prepaid Insurance Control		
	⑤	50,000		⑤	2,000

6. *Transaction:* Allocation of manufacturing overhead to products, $80,000.

   Analysis: The asset Work in Process Control is increased. The Manufacturing Department Overhead Control account is, in effect, decreased by means of its contra account, called Manufacturing Overhead Allocated. **Manufacturing overhead allocated** is made up of all manufacturing costs that are assigned to a product (or service) using a cost allocation base because they cannot be traced to a product (or service) in an economically feasible way. The 19_4 budgeted overhead rate used by Robinson is $80 per machine-hour. The overhead cost allocated to each job depends on the machine-hours used on that job. The job record for each individual job in the subsidiary ledger will include a debit item for manufacturing overhead allocated. It is assumed that 1,000 machine-hours were used for all jobs, resulting in a total manufacturing overhead allocation of 1,000 × $80 = $80,000.

   Note that the time when a subsidiary entry is made for Manufacturing Overhead Allocated is when machine-hours are used on a job. In contrast, the time when subsidiary entries are made for Manufacturing Overhead Control is when actual transactions occur during the period.

   Journal Entry: Work in Process Control $80,000
                    Manufacturing Overhead Allocated $80,000

   Post to the General Ledger:

Manufacturing Overhead Allocated			Work in Process Control		
	⑥	80,000	②	81,000	
			③	39,000	
			⑥	80,000	

7. *Transaction:* Completion and transfer to finished goods of eight individual jobs, $188,800.

   Analysis: The asset Finished Goods Control is increased. The asset Work in Process Control is decreased. The total costs of each job are computed in the subsidiary ledger as each job is completed. Given Robinson's use of normal costing, this total will consist of *actual* direct materials, *actual* direct manufacturing labor, and *budgeted* manufacturing overhead that is allocated to each job.

   Journal Entry: Finished Goods Control $188,800
                   Work in Process Control $188,800

   Post to the General Ledger:

Work in Process Control				Finished Goods Control		
②	81,000	⑦	188,800	⑦	188,800	
③	39,000					
⑥	80,000					

8. *Transaction:* Cost of Goods Sold, $180,000.

   Analysis: The expense Cost of Goods Sold is increased. The asset Finished Goods Control is decreased.

   Journal Entry: Cost of Goods Sold $180,000
                   Finished Goods Control $180,000

Post to the General Ledger:

Finished Goods Control		Cost of Goods Sold	
⑦ 188,800	⑧ 180,000	⑧ 180,000	

At this point, please pause and review all eight entries in the illustration. Be sure to trace each journal entry, step by step, to the general-ledger accounts in the general-ledger section in Panel A of Exhibit 5-3.

BUDGETED INDIRECT COSTS
AND END-OF-PERIOD ADJUSTMENTS

The use of budgeted indirect (overhead) cost rates enables job costs to be computed on an ongoing basis during the accounting period. At the end of the accounting period, the actual costs in the indirect cost pool and/or the actual quantity of the cost allocation base will almost always differ from their respective budgeted amounts. This difference, of course, is to be expected. The result is that indirect costs will be underallocated or overallocated to individual jobs during the period.

Underallocated indirect (overhead) costs occur when the allocated amount of indirect costs is less than the actual (incurred) amount. **Overallocated indirect (overhead) costs** occur when the allocated amount of indirect costs exceeds the actual (incurred) amount.

$$\begin{matrix} \text{Underallocated or} \\ \text{overallocated} \\ \text{manufacturing overhead} \end{matrix} = \begin{matrix} \text{Manufacturing} \\ \text{overhead} \\ \text{incurred} \end{matrix} - \begin{matrix} \text{Manufacturing} \\ \text{overhead} \\ \text{allocated} \end{matrix}$$

Equivalent terms are **underapplied** (or **overapplied**) **indirect (overhead) costs** and **underabsorbed** (or **overabsorbed**) **indirect (overhead) costs.**

Robinson Company has a single indirect cost pool (Manufacturing Department Overhead) in its job costing system. There are two manufacturing overhead accounts in its general ledger:

◆ Manufacturing Department Overhead Control, which is the record of the *actual* costs in all the individual overhead categories (such as indirect materials, indirect manufacturing labor, power, rent, and so on).

◆ Manufacturing Overhead Allocated, which is the record of the *budgeted* manufacturing overhead allocated to individual jobs on the basis of actual machine-hours.

Assume the following annual data for Robinson Company:

Manufacturing Department Overhead Control		Manufacturing Overhead Allocated	
Bal. Dec. 31, 19_4 1,200,000			Bal. Dec. 31, 19_4 1,000,000

The $1,200,000 debit balance in Manufacturing Department Overhead Control is the sum of all the actual costs incurred for manufacturing overhead in 19_4. The $1,000,000 credit balance in Manufacturing Department Overhead Allocated results from 12,500 actual machine-hours worked on all the jobs in 19_4 times the budgeted rate of $80 per hour. Note the $200,000 difference (a net debit) between the two amounts. It is an *underallocated* amount because actual overhead costs exceed the allocated amount. This $200,000 difference in 19_4 arises from two reasons:

1. *Numerator reason.* Actual manufacturing overhead cost of $1,200,000 is less than the budgeted $1,280,000.

2. *Denominator reason.* Actual machine-hours of 12,500 are less than the budgeted 16,000 hours.

We now discuss the two main approaches to disposing of this $200,000 underallocation of manufacturing overhead in Robinson's costing system: (1) the restated allocation rate approach and (2) the end-of-period account(s) approach.

Restated Allocation Rate Approach

The restated allocation rate approach, in effect, restates all entries in the general ledger by using actual cost rates rather than budgeted cost rates. First, the actual indirect (overhead) cost rate is computed at the end of each period. Then, every cost object to which indirect costs were allocated during the period has its account recomputed using the actual indirect cost rate (rather than the budgeted indirect cost rate). Finally, end-of-period closing entries are made. The result is that every single job cost record—as well as the ending work in process, finished goods, and cost of goods sold amounts—accurately represents actual manufacturing overhead cost incurrence. Although this method is the most accurate approach to disposing of variances, it is also the most costly. Very few firms believe that the restated allocation rate approach passes the cost-benefit test.

Still, the widespread adoption of computerized accounting systems has greatly reduced the cost of using the restated allocated rate approach. Consider the Robinson Company example. The actual manufacturing overhead ($1,200,000) exceeds the manufacturing overhead allocated ($1,000,000) by 20%. At year-end Robinson could increase the 19_4 manufacturing overhead allocated to each job in that year by 20% using a single software directive. The directive would apply to the subsidiary ledger as well as to the general ledger. This approach increases the accuracy of each individual product cost and the accuracy of the end-of-year account balance for work in process, finished goods, and cost of goods sold. This increase in accuracy is an important benefit. After-the-fact analysis of individual product profitability can provide managers with useful insights for future decisions about product pricing and about which products to emphasize. It is important to have accurate product-profitability numbers underlying such decisions.

End-of-Period Account(s) Approach

In the end-of-period account(s) approach, underallocated or overallocated overhead is written off entirely to COGS or is prorated among ending work in process, finished goods, and cost of goods sold. **Proration** is the spreading of underallocated or overallocated overhead among ending work in process, finished goods, and cost of goods sold. Assume the following actual results for Robinson Company in 19_4:

	End-of-Year Balances (before Proration)	Manufacturing Overhead Allocated Component of End-of-Year Balances (before Proration)
Work in process	$ 50,000	$ 13,000
Finished goods	75,000	25,000
Cost of goods sold	2,375,000	962,000
	$2,500,000	$1,000,000

The three main alternatives for disposing of the underallocated $200,000 manufacturing overhead into the costing system at the end of 19_4 follow.

Alternative 1. Immediate write-off to Cost of Goods Sold (COGS). Here the total under- or overallocated overhead is included in this year's cost of goods sold. In our Robinson Company example, the journal entry would be:

Cost of Goods Sold	$ 200,000	
Manufacturing Overhead Allocated	1,000,000	
Manufacturing Dept. Overhead Control		$1,200,000

Robinson's two manufacturing overhead accounts are closed out with all the difference between them now included in cost of goods sold. The closing cost of goods sold (after proration) is: $2,375,000 before proration + $200,000 under-allocated overhead amount = $2,575,000.

Alternative 2. Proration based on total ending balances (before proration) in work in process, finished goods, and cost of goods sold. In our Robinson Company example, the $200,000 underallocated overhead is prorated over the three pertinent accounts in proportion to their total ending balances (before proration) in column 2, resulting in the ending balances (after proration) in column 4:

(1)	Balance (before Proration) (2)	Proration of $200,000 Manufacturing Overhead Underallocated (3)	Balance (after Proration) (4) = (2) + (3)
Work in process	$ 50,000 (2%)	0.02 × $200,000 = $ 4,000	$ 54,000
Finished goods	75,000 (3%)	0.03 × 200,000 = 6,000	81,000
Cost of goods sold	2,375,000 (95%)	0.95 × 200,000 = 190,000	2,565,000
	$2,500,000 100%	1.00 $200,000	$2,700,000

For example, work in process is 2% of the $2,500,000 total, so we assign 2% of the underallocated amount (0.02 × $200,000 = $4,000) to the work in process account. The journal entry for this proration would be:

Work in Process Control	$ 4,000	
Finished Goods Control	6,000	
Cost of Goods Sold	190,000	
Manufacturing Overhead Allocated	1,000,000	
Manufacturing Overhead Control		$1,200,000

This alternative (as well as the next) prorates the underallocated overhead to those accounts that receive allocations of manufacturing overhead. There is no proration to materials inventory. Why? Because Robinson uses machine-hours to allocate manufacturing overhead to products, and it is only when direct materials become work in process that machining is started.

Alternative 3. Proration based on the total amount of allocated overhead (before proration) in the ending balances of work in process, finished goods, and cost of goods sold.

(1)	Balance (before Proration) (2)	Overhead Allocated Component in the Balance in Column (2) (3)	Proration of $200,000 Manufacturing Overhead Underallocated (4)	Balance (after Proration) (5) = (2) + (4)
Work in process	$ 50,000	$ 13,000 (1.3%)	0.013 × $200,000 = $ 2,600	$ 52,600
Finished goods	75,000	25,000 (2.5%)	0.025 × 200,000 = 5,000	80,000
Cost of goods sold	2,375,000	962,000 (96.2%)	0.962 × 200,000 = 192,400	2,567,400
	$2,500,000	$1,000,000 100.0%	$200,000	$2,700,000

The journal entry for this proration would be:

Work in Process Control	$ 2,600	
Finished Goods Control	5,000	
Cost of Goods Sold	192,400	
Manufacturing Overhead Allocated	1,000,000	
Manufacturing Overhead Control		$1,200,000

This journal entry results in the 19_4 ending balances for work in process, finished goods, and cost of goods sold being restated to what they would have been had actual cost rates rather than budgeted cost rates been used.

Choice Among Methods

The reported account balances under each of the three alternatives are:

	Alternative 1	Alternative 2	Alternative 3
			Proration Based on Manufacturing
	Write-off to Cost of Goods Sold	**Proration Based on Total Ending Balances**	**Overhead Allocated Component of Ending Balances**
Work in process	$ 50,000	$ 54,000	$ 52,600
Finished goods	75,000	81,000	80,000
Cost of goods sold	2,575,000	2,565,000	2,567,400
	$2,700,000	$2,700,000	$2,700,000

Alternative 3 is the theoretically preferred one. It yields the same ending balances of work in process, finished goods, and cost of goods sold that would have been reported had actual indirect cost rates been used. (The restated allocated rate approach results in the same ending balances as does alternative 3.) Alternative 2 is frequently justified as being a lower-cost way of approximating the results from alternative 3. The implicit assumption in alternative 2 is that the ratio of manufacturing overhead costs allocated to total manufacturing costs is similar for the jobs in work in process, finished goods, and cost of goods sold. Where this assumption is not appropriate, alternative 2 can yield numbers quite different from those alternative 3 gives.

Many companies use alternative 1 for several reasons. First, it is the simplest. Second, the three alternatives often yield similiar amounts for ending work in process, finished goods, and cost of goods sold.

This section has examined end-of-period adjustments for underallocated or overallocated indirect manufacturing costs. The same issues also arise when budgeted direct manufacturing cost rates are used and end-of-period adjustments must be made. End-of-period adjustment issues also occur when budgeted cost rates are used for business function areas other than manufacturing.

◆ PART TWO:
PROCESS COSTING SYSTEMS IN MANUFACTURING

The principal difference between process costing and job costing is the extent of averaging used to compute unit costs of products or services. The cost object in a job costing system is a job that constitutes a distinctly identifiable product or service. Costs are assigned to each cost object (job) with minimal averaging. In contrast, in a process costing system, the cost object is a process that produces a mass of similar units of a product or service. Individual unit costs are computed by averaging total costs of the process over the total number of similar units.

In this section we provide an introduction to process costing. Two cases are illustrated:

◆ *Case 1*. Process costing with all units fully completed at the end of the reporting period.
◆ *Case 2*. Process costing with some units incomplete at the end of the reporting period.

Both cases assume no beginning inventory of work in process. Chapters 17 and 18 provide more detailed discussion of process costing systems. Industries using process costing in their manufacturing area include chemical processing, oil refining, beverages, and breakfast cereals. Each of these industries computes individual unit costs by averaging total costs of the process over the total number of similar units manufactured.

CASE 1: PROCESS COSTING WITH ALL UNITS FULLY COMPLETED

Suppose Advanced Electronics manufactures special microchips for high-speed computers. During April, 2,500 identical microchips were placed into production by the Fremont plant. There was no beginning inventory. The plant's manufacturing costs for April were:

Direct materials	$ 825,000
Conversion costs	425,000
Costs to account for	$1,250,000

Advanced Electronics' process costing system has one direct cost category (direct materials) and one indirect cost pool (conversion costs). *Conversion costs* are all manufacturing costs other than direct materials costs. At Advanced Electronics, conversion costs include manufacturing labor, indirect materials, energy, depreciation, plant leasing costs, and so on. All microchips placed in production in April were fully completed at the end of the month. The unit cost of goods completed would simply be $1,250,000 ÷ 2,500 = $500. An itemization would show:

Direct materials ($825,000 ÷ 2,500)	$ 330
Conversion costs ($425,000 ÷ 2,500)	170
Manufacturing unit cost of a completed unit	$ 500

This process costing system does not keep a job cost record for each individual microchip.

Case 1 applies to all types of organizations that use process costing and have no incomplete units when each reporting period ends. Service-sector organizations that use a process costing system would adopt this approach to compute the unit cost of masses of similar services. For example, case 1 would apply to a bank computing the unit cost of 100,000 customer deposits made in a month.

CASE 2: PROCESS COSTING WITH SOME UNITS NOT FULLY COMPLETED

Objective 6
Identify five key steps in process costing

Suppose the same facts as in case 1 with the exception that at the end of April not all 2,500 microchips were fully completed at Advanced Electronics' Fremont plant. Assume that 500 units were still in process at the end of April; only 2,000 were started and fully completed in April. All direct materials had been added to each microchip

EXHIBIT 5-4

Microchip Output in Equivalent Units for the Month Ended April 30, 19_4 for Advanced Electronics; Fremont Plant

	(Step 1)	(Step 2) Equivalent Units	
Flow of Production	Physical Units	Direct Materials	Conversion Costs
Started and completed	2,000	2,000	2,000
Work in process, ending*	500	500	125
Accounted for	2,500		
Work done to date		2,500	2,125

* Degree of completion: direct materials, 100%; conversion costs, 25%.

still in process, but on average only 25% of the conversion costs had been allocated to the 500 units in ending inventory. Advanced Electronics has an extensive testing procedure for each microchip, and 500 units were not yet fully tested. How should the Fremont plant calculate (a) the cost of completed units in April and (b) the cost of work in process not yet completed at the end of April?

A process costing system yields answers to (a) and (b) using the following five key steps:

- ◆ *Step 1:* Summarize the flow of physical units of a product or output.
- ◆ *Step 2:* Compute output in terms of equivalent units.
- ◆ *Step 3:* Summarize the total costs to account for, which are the total of the costs charged (debited) to Work in Process.
- ◆ *Step 4:* Compute equivalent unit costs.
- ◆ *Step 5:* Assign costs to units completed and to units in ending Work in Process.

Physical Units and Equivalent Units (Steps 1 and 2)

Step 1, as the physical units column in Exhibit 5-4 shows, tracks the physical units of product or output. In step 2, how should the output for April be measured? The output was 2,000 *fully* completed units and 500 *partially* completed units. A partially completed unit is certainly not the same as a fully completed unit. Accordingly, output in step 2 is stated in *equivalent units,* not physical units.

Equivalent units measure the output in terms of the quantities of each of the factors of production that have been consumed by the units. An equivalent unit is the collection of inputs necessary to produce one fully complete unit of product or output. In our example, as step 2 in Exhibit 5-4 shows, the output would be measured as 2,500 equivalent units of direct materials costs and 2,125 equivalent units of conversion costs. There are 2,500 equivalent units of direct materials costs because all 2,500 units are fully complete with respect to materials. There are 2,125 equivalent units of conversion costs because 2,000 units are fully completed, while the 500 units in ending work in process inventory are only 25% complete as to conversion costs— 2,000 + (500 × 0.25) = 2,125.

Calculation of Product Costs (Steps 3, 4, and 5)

Exhibit 5-5 is a production cost worksheet. It shows steps 3, 4, and 5. Step 3 summarizes the total costs to account for (that is, the total costs charged or debited to Work in Process). Step 4 obtains equivalent unit costs by dividing each category of total costs by the related measure of equivalent units. The unit cost of a completed unit is $330 + $200 = $530. Why is the unit cost $530 instead of the $500 calculated earlier in case 1? Because the $425,000 conversion costs are spread over only 2,125 equivalent units instead of 2,500 equivalent units. Step 5 then uses these unit costs to assign costs to products.

In Exhibit 5-5 notice how the costs are assigned to obtain an ending Work in Process of $190,000. The 500 physical units are fully completed regarding direct materials. Therefore, direct materials costs are 500 equivalent units times $330, which equals $165,000. In contrast, the 500 physical units are 25% completed regarding conversion costs. Therefore, the conversion costs assigned are 125 equivalent units (25% of 500 physical units) times $200, which equals $25,000.

Exhibit 5-6 presents an overview of the process costing system at Advanced Electronics. Conversion costs are allocated to masses of similar units using equivalent units as the allocation base. We have assumed that the company uses the actual costing method—that is, it uses actual cost rates for both direct materials and conversion costs. If Advanced Electronics had used either the normal costing (combination of ac-

EXHIBIT 5-5

Production Cost Worksheet for the Month Ended April 30, 19_4 for Advanced Electronics, Fremont Plant

	Cost	Totals	Direct Materials	Conversion Costs
			Details	
(Step 3)	Total costs to account for	$ 1,250,000	$825,000	$425,000
	Divide by equivalent units		÷ 2,500	÷ 2,125
(Step 4)	Equivalent unit costs		$ 330	$ 200
(Step 5)	Assignment of costs:			
	Completed and transferred out (2,000 units)	$1,060,000	(2,000* × $ 530†)	
	Work in process, ending (500 units)			
	Direct materials	165,000	(500* × $330)	
	Conversion costs	25,000		(125* × $200)
	Total work in process	190,000		
	Total costs accounted for	$1,250,000		

*From Exhibit 5-4

†Cost per equivalent unit = $330 + $200 = $530.

EXHIBIT 5-6

Process Costing Overview at Advanced Electronics

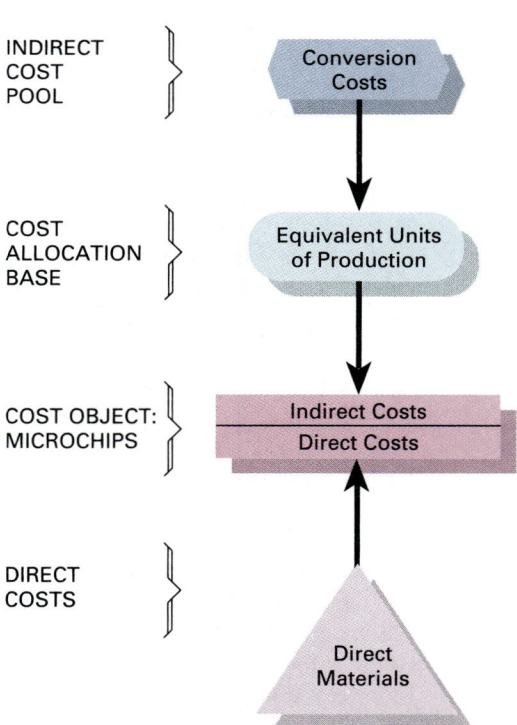

tual and budgeted costs) or the budgeted costing (budgeted costs only) method, end-of-period adjustments for underallocated or overallocated conversion costs would have been required.

Journal Entries

The summary journal entries for the data in our Advanced Electronics illustration would be:

1. Work in Process $ 825,000
          Materials Control $ 825,000
       Direct materials used in production in April.

2. Work in Process $ 425,000
          Various accounts $ 425,000
       To record conversion costs; examples include energy, all manufacturing labor, and pertinent depreciation.

3. Finished Goods $1,060,000
          Work in Process $1,060,000
       Cost of goods completed and transferred
          from work in process to finished goods in April.

The key T-account for Work in Process would show:

Work in Process			
1. Direct materials	825,000	3. Transferred out to	
2. Conversion costs	425,000	finished goods	1,060,000
Costs to account for	1,250,000		
Bal.: April 30	190,000		

[CHAPTERS 17 AND 18 DESCRIBE PROCESS COSTING IN MORE DETAIL. THEY CAN BE STUDIED NOW, IF DESIRED, WITHOUT LOSS OF CONTINUITY.]

◆ PART THREE:
ACTIVITY-BASED COSTING IN MANUFACTURING

Objective 8
Describe distinctive features of activity-based costing

Chapter 4 introduced and illustrated the use of an *activity-based costing (ABC)* approach to designing cost systems in the service and merchandising sectors. We now illustrate the use of ABC in designing costing systems in the manufacturing sector. Manufacturing is where most ABC implementation has occurred. The distinctive feature of ABC is its focus on activities as the fundamental cost objects. The costs of these activities are built up to compute the costs of products, services, and customers. An ABC approach can be used in designing either a job costing system or a process costing system.

ILLUSTRATION OF ABC IN MANUFACTURING

Our illustration of ABC summarizes the experience of an actual manufacturing facility, here called Instruments Inc., that assembles and tests more than 800 electronic instrument products, including printed-circuit boards. Every printed-circuit (PC) board has various parts (diodes, capacitors, and integrated circuits) inserted on it. The prior job costing system was typical of many systems worldwide. Job costing was based on a system with two direct cost categories and two indirect manufacturing cost pools:

- ◆ Direct manufacturing costs
    - Direct materials
    - Direct manufacturing labor
- ◆ Indirect manufacturing costs
    - Procurement (purchasing) department—allocated to products on the basis of their direct materials costs
    - Production department—allocated to products on the basis of their direct labor costs

As business became increasingly competitive, managers in product design, manufacturing, and marketing became more skeptical about the accuracy of Instruments Inc.'s cost accounting system. A common complaint was that the costing system did not produce numbers that reflected the way various products differed in use of Instruments Inc.'s resources. For example, one product design manager commented:

> Why is it when I use a $0.20 capacitor part, the procurement overhead charge is $0.02, but when I use a $100 co-processor part, the procurement overhead charge is $10? Procuring and handling a co-processor does not consume 500 times ($0.02 × 500 = $10.00) the resources used to procure and handle a capacitor. This overhead costing approach is from looney-toon land.

Managers in manufacturing believed that different factors were causing or driving costs in individual activity areas, but the costing system did not reflect information about those differences. These managers found the numbers in the existing costing system to be of limited use or even a detriment in their decisions. Managers in marketing perceived that the costing system tended to "overcost" the intensely competitive high-volume products. How? By loading too much of the indirect manufacturing costs on high-volume products and too little on low-volume products.

Refinement of a Job Costing System Using ABC

Representatives from product design, manufacturing, and accounting worked as a cross-functional team to refine the job costing system using an activity-based focus. Our description of Instruments Inc.'s activity-based job costing system will follow the five-step general approach to job costing.

STEP 1: *IDENTIFY THE JOB THAT IS THE CHOSEN COST OBJECT.* A job at Instruments Inc. is an order of any size for one of its over 800 electronic instrument products.

STEP 2: *IDENTIFY THE DIRECT COST CATEGORIES FOR THE JOB.* Instruments Inc. decided to retain its existing two direct cost categories in its refined job costing system—direct materials and direct manufacturing labor.

STEP 3: *IDENTIFY THE INDIRECT COST POOLS ASSOCIATED WITH THE JOB.* The refined system has six indirect cost pools. These indirect cost pools represent individual activity areas at Instrument Inc.'s manufacturing facility.

1. *Materials handling.* All the parts necessary for manufacturing the PC board are combined into a kit.
2. *Start station.* Instructions for manufacturing the PC board are entered into a computer. The software program tells the automated equipment which parts to insert where.
3. *Machine insertion of parts.* Automated and semiautomated equipment insert components on the board.
4. *Manual insertion of parts.* Skilled workers insert those components that are not machine-inserted (because of their shape, weight, location on the board, and so on).
5. *Wave soldering.* All parts inserted on the board are simultaneously soldered to ensure that they remain attached.

6. *Quality testing*. Tests are made to check that all components are inserted and in the right place and that the final product performs to specification.

STEP 4: *SELECT THE COST ALLOCATION BASE TO USE IN ASSIGNING EACH INDIRECT COST POOL TO THE JOB.* The cause-and-effect criterion guided Instruments Inc. to choose cost allocation bases that are cost drivers. The team refining the costing system interviewed operating personnel, observed operations at the plant, and analyzed operating data at each activity area. The chosen allocation bases are presented together with their rates in step 5.

STEP 5: *DEVELOP THE RATE PER UNIT OF EACH COST ALLOCATION BASE USED TO ALLOCATE INDIRECT COSTS TO THE JOB.* Use the general approach to computing indirect cost allocation rates outlined in an earlier part of this chapter (pp. 144–146) and in Chapter 4 (pp. 107–109).

$$\text{Budgeted indirect cost allocation rate} = \frac{\text{Budgeted total costs in indirect cost pool}}{\text{Budgeted total quantity of cost allocation base}}$$

Consider the indirect cost area for the machine insertion of parts. For 19_5, the budgeted total costs at this activity area are $2,000,000. The budgeted number of machine-inserted parts on PC boards in 19_5 is 4,000,000. Thus, the 19_5 budgeted indirect cost allocation rate for the machine insertion of parts activity area is:

$$\frac{\$2,000,000}{4,000,000 \text{ insertions}} = \$0.50 \text{ per machine insertion}$$

A similar procedure was used to compute each of the following 19_5 budgeted indirect cost allocation rates in each activity area.

Activity Area	Cost Driver Used as Cost Allocation Base	Indirect Cost Allocation Rate
1. Materials handling	Number of parts	$2 per part
2. Start station	Number of PC boards	$20 per board
3. Machine insertion of parts	Number of machine-inserted parts	$0.50 per insertion
4. Manual insertion of parts	Number of manually inserted parts	$4 per insertion
5. Wave soldering	Number of PC boards	$30 per board
6. Quality testing	Hours of test time	$50 per test hour

Exhibit 5-7 presents an overview of the activity-based job costing system of Instruments Inc.

EXHIBIT 5-7

Overview of Activity-Based Job Costing System of Instruments Inc.

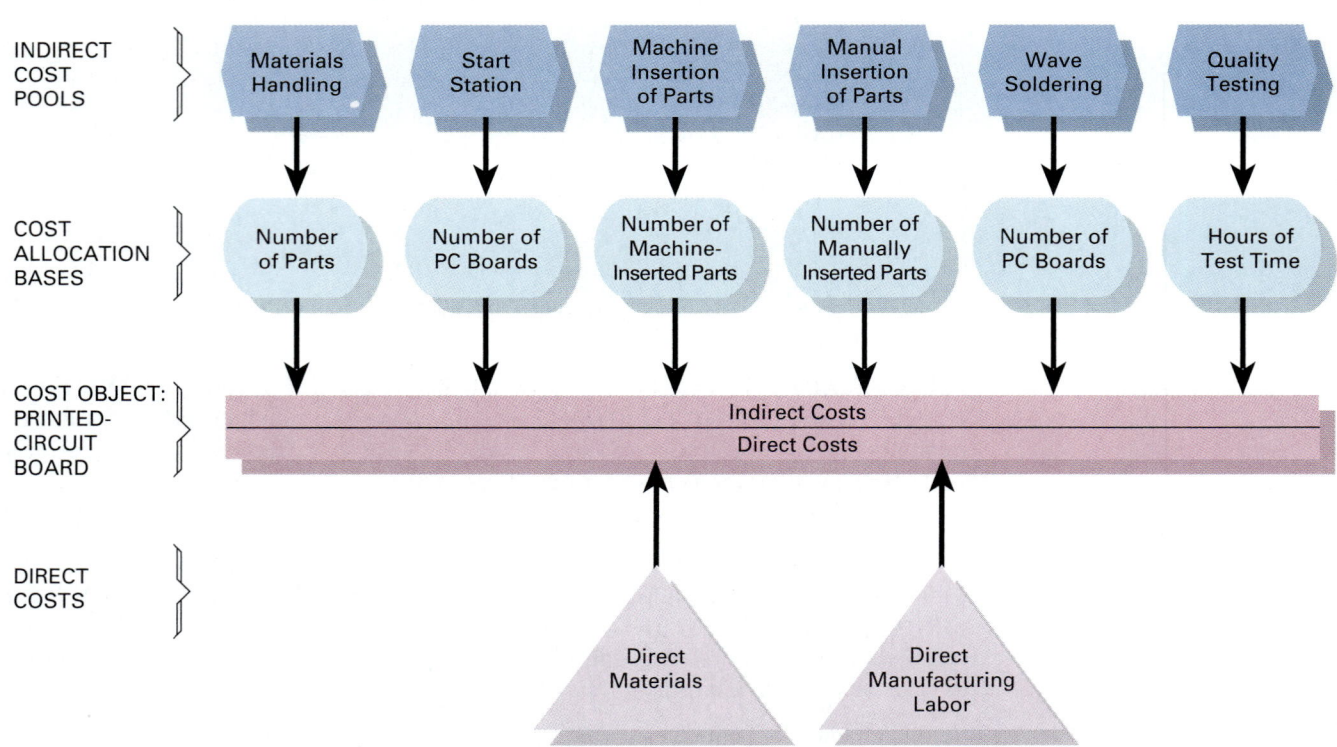

Numerical Example

Exhibit 5-8 presents the product costing of two PC boards using Instruments Inc.'s activity-based job costing system. Note the following points.

1. The activity-based job costing system pinpoints opportunities for cost reductions. Managers might ask, why does board A cost less? ABC costing reveals three reasons:

   a. Board A has fewer parts (81 for A versus 121 for B)
   b. On board A, a higher percentage of total insertions are made by machine, which are cheaper than insertions made by hand. (Board A has 70 out of its 80 component parts—87.5%—inserted by machine, while board B has 90 out of its 120 component parts—75%—inserted by machine.)
   c. Board A requires less test time (1.5 hours for board A versus 6.5 hours for board B).

The activity-based-job costing system explicitly signals that (a), (b), and (c) each lead to a lower cost for assembling a PC board. In fact, board A is a standard, no-frills board that Instruments Inc. produces in large quantities.

2. Manufacturing personnel at Instruments Inc. are using ABC information in their cost-reduction efforts. Each activity area at Instruments Inc. has its own supervisor. The measures used to evaluate the supervisor's performance now include the achievement of cost-reduction targets. For example, the supervisor of the activity area machine insertion of parts receives a bonus if the actual cost rate per machine-

EXHIBIT 5-8

Product Costing of PC Boards A and B Using Instruments Inc.'s
Activity-Based Job Costing System

	PC Board A	PC Board B
Direct manufacturing costs		
Direct materials	$600	$ 280
Direct manufacturing labor	32	96
	632	376
Indirect manufacturing costs		
Materials handling*		
(A, 81 parts; B, 121 parts) × $2	162	242
Start station		
(A, 1 board; B, 1 board) × $20	20	20
Machine insertion of parts		
(A, 70 insertions; B, 90 insertions) × $0.50	35	45
Manual insertion of parts		
(A, 10 insertions; B, 30 insertions) × $4	40	120
Wave soldering		
(A, 1 board; B, 1 board) × $30	30	30
Quality testing		
(A, 1.5 hours; B, 6.5 hours) × $50	75	325
Total	362	782
Total manufacturing costs	$994	$1,158

*The number of parts includes the raw printed-circuit board (counted as one part) plus the number of component parts to be inserted on the board.

inserted part is reduced by 10% or more. (Recall that manufacturing personnel found the numbers from the previous costing system of limited use in their decisions.)

3. The reported costs of the standard (no-frills) PC boards, such as board A, are lower with the activity-based job costing system than with the previous system. In contrast, PC boards that require the manual insertion of a large number of parts or extended test time (as does PC board B) have higher reported costs under the activity-based job costing system.

4. Each of the indirect cost allocation bases in Exhibit 5-8 is a nonfinancial variable (number of parts, hours of test time, and so on). One manufacturing manager used the phrase "Let's get physical" to introduce the cost allocation bases for the activity-based job costing system. Controlling physical items such as hours, parts, and defects is often the most fundamental way that operating personnel manage costs. For example, the product design manager who criticized the previous costing system believes that number of parts rather than direct materials dollars is the main driver of procurement cost at Instruments Inc.

5. The differences in the product costs of PC boards A and B reflect how these products use different amounts of the resources of each activity area. Consider differences in the relative use of the following four activity areas:

	PC Board A	PC Board B
Materials handling	81 parts	121 parts
Machine insertion of parts	70 insertions	90 insertions
Manual insertion of parts	10 insertions	30 insertions
Quality testing	1.5 hours	6.5 hours

The activity-based job costing system, a more refined costing system than the prior system, reports cost numbers that better measure the way jobs, products, customers, and so on differently use the resources of the organization.

Traditional versus ABC Approach to Designing a Costing System

The Instruments Inc. example above, the Lindsay and Associates example in Panel C of Exhibit 4-5 (p. 112), and the Family Supermarkets example in Panel B of Exhibit 4-7 (p. 116) each illustrate how an ABC approach can be used to refine a costing system. The major differences between the traditional approach and an ABC approach follow:

Objective 9

Distinguish between the traditional and the activity-based costing approaches to designing a costing system

Traditional Approach

◆ One or a few indirect cost pools for each department or entire plant, usually with little homogeneity (that is, there is lack of a cause-and-effect relationship between the cost allocation bases and the indirect cost pools).

◆ Indirect cost allocation bases may or may not be cost drivers.

◆ Indirect cost allocation bases are often financial variables, such as direct labor costs or direct materials costs.

ABC Approach

◆ Many homogeneous indirect cost pools because of many activity areas. Operating personnel play a key role in designating which activity areas need to be considered.

◆ Indirect cost allocation bases are highly likely to be cost drivers.

◆ Indirect cost allocation bases are often nonfinancial variables, such as number of parts in a product or hours of test time.

The traditional approach often uses too few pools of indirect costs, so cost allocations have overly broad averages. The resulting costs may lead managers to make erroneous decisions about activities, products, or customers. For example, a product that is overcosted by the traditional approach may be priced too high, resulting in loss

Concepts in Action

HOW CLARK-HURTH IMPLEMENTED ACTIVITY-BASED COSTING[a]

Clark-Hurth (C-H), a division of Clark Equipment Company, manufactures a broad range of axle and transmission products. As a supplier to many off-highway and equipment manufacturers, C-H has been put under great pressure in recent years to provide ever higher quality products at ever lower costs. To obtain more accurate product cost information, C-H managers believed that they had to abandon their costing system and move to ABC. The company had used direct materials and direct manufacturing labor as direct cost categories and manufacturing overhead (allocated using direct manufacturing labor-hours) as the indirect cost pool. But this system, managers felt, provided few insights into how different products were differently using C-H resources. ABC promised to improve C-H's accounting system.

Stage One in implementing ABC at C-H meant surveying every salaried and indirect worker at the company to get information on the activities they performed. Over 170 activities were listed. C-H then ranked these activities in order of frequency. Stage Two in the ABC implementation was determining those activities that a C-H customer would view as valuable. This effort led C-H to discontinue several activities. At Stage Three, managers selected cost drivers for the 40 most frequent activities. The cost drivers chosen included both traditional measures (such as direct manufacturing labor-hours and machine-hours) and nontraditional measures (such as the number of parts in a product).

At Stage Four, managers estimated the costs per driver unit at each of the 40 activities. ABC-based product costs were then developed at Stage Five.

The revised product costs showed several patterns. One pattern that emerged was that many products with low sales volume were being undercosted and that C-H was actually losing money on them. Several high-sales-volume products that required no "bells and whistles" on the production line were being significantly overcosted. C-H has used this revised activity-based product cost information in bidding for work from off-highway companies.

C-H now is also using ABC for cost management. For example, in Stage One C-H found that the large number of different parts it purchased was consuming much of its time and other resources. C-H has undertaken a "parts de-proliferation" program. One step is to have a standard part used on many products, which reduces the number of different parts to be purchased. A second step is to require procurement personnel to justify placing an order for a new part when a part on hand at C-H may be adequate.

The president of C-H notes that the company has enjoyed three major benefits from its switch to ABC:

1. A better understanding of what C-H people are doing,
2. A better understanding of "real costs," and
3. A better understanding of the opportunities available for reducing costs.

[a]Adapted from a presentation by Clark-Hurth at Computer Aided Manufacturing-International (CAM-I); December 1992 meeting, Clearwater, Florida.

of market share. Similarly, managers may be setting selling prices for some products that are below the cost of the resources used to produce them. The dangers are especially pronounced when hundreds of diverse products are manufactured in various annual output levels ranging from a few units of, say, one or two kinds of motors or computers to thousands of units of other kinds. Case studies have shown that the broad averages in the traditional approach can load indirect manufacturing costs too heavily on high-output-level (volume) products and too lightly on low-output-level (volume) products.

Most fundamentally, managers manage costs by overseeing activities rather than products. The pooling of costs by activities or activity areas provides information that may help managers to better plan and control costs throughout the value chain, from research and development to customer service.

MULTIPURPOSE NATURE OF COSTING SYSTEMS

Costing systems serve multiple purposes. Chapters 4 and 5 have emphasized the costing of products, services, and customers. Consider now (1) the planning and control purpose and (2) inventory costing for the financial reporting purpose.

Managers plan and control business functions through personal observation and through systems that contain actual costs, budgets, and variances. These business functions are often divided into departments and then into activity areas. Exhibit 5-9 shows how cost objects become ever more finely granulated, from a particular business function to departments to activities. These progressively more granulated cost objects can be specific responsibility centers in a control system.

Consider now the inventory costing for financial reporting. Exhibit 5-9 (Panel A) highlights that only manufacturing costs are inventoriable. Panels B and C show that activity areas in a costing system become more finely granulated. Traditional costing systems typically stay at the Panel B level when developing information for the product cost or inventory cost for financial reporting purposes. Activity-based costing systems typically move to the more granular Panel C level as is illustrated by Instruments Inc.'s ABC job costing system in Exhibits 5-7 and 5-8.

EXHIBIT 5-9
Overview of Business Functions, Departments, and Activities

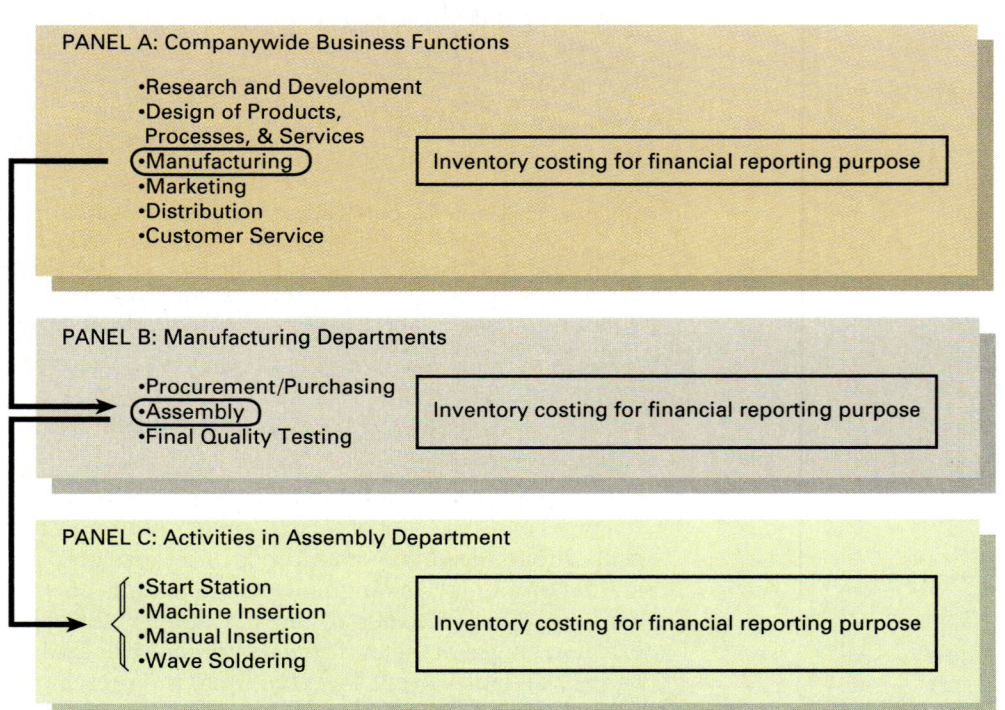

PROBLEM FOR SELF-STUDY

Restudy the Exhibit 5-3 illustration of a job costing system. Then try to solve the following problem, which requires consideration of many of this chapter's important points.

PROBLEM

You are asked to bring the following incomplete accounts of Endeavor Printing Inc. up to date through January 31, 19_5. Consider the data that appear in the T-accounts as well as information in items (a) through (i) below.

Endeavour's job costing system has two direct cost categories (direct materials and direct manufacturing labor) and one indirect cost pool (manufacturing overhead, which is allocated using direct manufacturing labor costs).

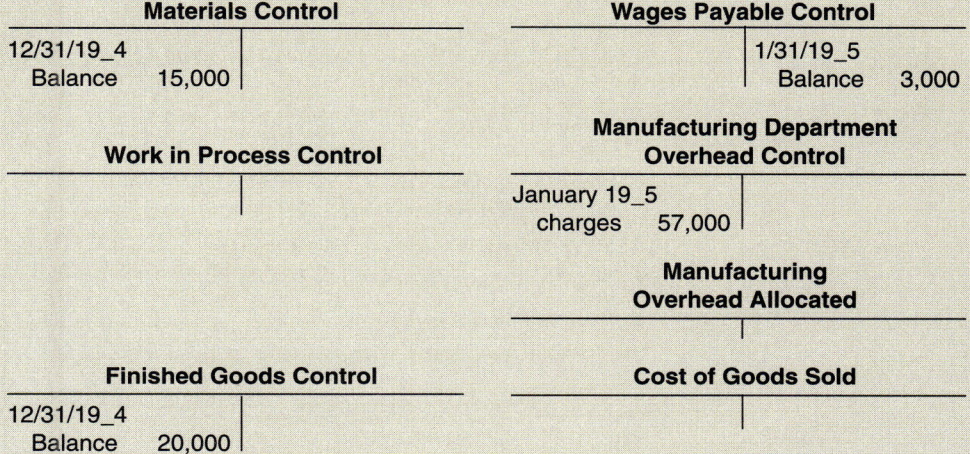

Materials Control	
12/31/19_4	
Balance 15,000	

Wages Payable Control	
	1/31/19_5
	Balance 3,000

Work in Process Control	

Manufacturing Department Overhead Control	
January 19_5	
charges 57,000	

Manufacturing Overhead Allocated	

Finished Goods Control	
12/31/19_4	
Balance 20,000	

Cost of Goods Sold	

Additional Information:

a. Manufacturing department overhead is allocated using a budgeted rate set every December. Management forecasts next year's overhead and next year's direct manufacturing labor costs. The budget for 19_5 is $400,000 of direct manufacturing labor and $600,000 of manufacturing overhead.

b. The only job unfinished on January 31, 19_5, is No. 419, on which direct manufacturing labor costs were $2,000 (125 direct manufacturing labor-hours) and direct materials costs are $8,000.

c. Total materials placed into production during January are $90,000.

d. Cost of goods completed during January is $180,000.

e. Materials inventory as of January 31, 19_5 is $20,000.

f. Finished goods inventory as of January 31, 19_5 is $15,000.

g. All plant workers earn the same wage rate. Direct manufacturing labor-hours for January totals 2,500. Other labor and supervision totals $10,000.

h. The gross plant payroll on January paydays totals $52,000. Ignore withholdings. All personnel are paid on a weekly basis.

i. All "actual" manufacturing department overhead incurred during January has already been posted.

Required
1. Materials purchased during January
2. Cost of Goods Sold during January
3. Direct Manufacturing Labor Costs incurred during January
4. Manufacturing Overhead Allocated during January
5. Balance, Wages Payable Control, December 31, 19_4

6. Balance, Work in Process Control, January 31, 19_5
7. Balance, Work in Process Control, December 31, 19_4
8. Balance, Finished Goods Control, January 31, 19_5
9. Manufacturing Overhead Underallocated or Overallocated for January

SOLUTION

Amounts from the T-accounts are labeled (T).

1. Materials purchased: $90,000 + $20,000 − $15,000 (T) = $95,000
2. Cost of Goods Sold: $20,000 (T) + $180,000 − $15,000 = $185,000
3. Direct manufacturing wage rate: $2,000 ÷ 125 hours = $16 per hour
   Direct manufacturing labor costs: 2,500 hours × $16 = $40,000 (see item g)
4. Manufacturing overhead rate: $600,000 ÷ $400,000 = 150%
   Manufacturing overhead allocated: 150% × $40,000 (see 3) = $60,000
5. Wages Payable Control, December 31, 19_4: $52,000 + $3,000 (T) − $40,000 (see 3) − $10,000
   = $5,000
6. Work in Process Control, January 31, 19_5: $8,000 + $2,000 + 150% of $2,000 = $13,000.
   (This answer is used in 7 below.)
7. Work in Process Control, December 31, 19_4: $180,000 + $13,000 (see 6) − $90,000 − $40,000
   (see 3) − $60,000 (see 4) = $3,000
8. Finished Goods Control, January 31, 19_5: $20,000 (T) + $180,000 − $185,000 = $15,000
9. Manufacturing overhead overallocated: $60,000 (see 4) − $57,000 (T) = $3,000.

Entries in T-accounts are lettered in accordance with the Additional Information in
the problem and are numbered in accordance with the requirements.

Materials Control

12/31/19_4 Bal. (given)		15,000		(c)	90,000
	(1)	95,000*			
1/31/19_5 Bal.	(e)	20,000			

Work in Process Control

12/31/19_4 Bal.	(7)	3,000		(d)	180,000
Direct materials	(c)	90,000			
Direct manuf. labor	(b) (g)	(3) 40,000			
Manuf. overhead alloc.	(g) (a)	(4) 60,000			
1/31/19_5 Bal.	(b) (6)	13,000			

Finished Goods Control

12/31/19_4 Bal. (given)		20,000		(f) (2)	185,000
	(d)	180,000			
1/31/19_5 Bal.	(9)	15,000			

Wages Payable Control

	(h)	52,000	12/31/19_4	(5)	5,000
				(g)	{ 40,000
					{ 10,000
1/31/19_5 Bal. (given) 3,000					

Manufacturing Department Overhead Control

Total January charges (given)	57,000	

Manufacturing Overhead Allocated

(g)	(a)	(4)	60,000

Cost of Goods Sold

(f)	(2)	185,000

*Can be computed only after all other postings in the account have been found.

SUMMARY

The following points are linked to the chapter's learning objectives.

1. Costing systems aim to report cost numbers that reflect the way chosen cost objects (such as products, services, or customers) use the resources of an organization. Designing a job costing system involves identifying (a) the job, (b) the direct cost categories, (c) the indirect cost categories, (d) the cost allocation base(s), and (e) the cost allocation rate(s).

2. Three key source documents in a job costing system are a job cost record, a materials requisition record, and a labor time record.

3. Subsidiary ledgers contain the underlying details of a costing system. The general ledger contains the summary information. Subsidiary ledgers provide a worm's-eye view, while the general ledger provides a bird's-eye view of the costing system.

4. The transactions in a job costing system in manufacturing track: (a) the acquisition of materials and other inputs, (b) their conversion into work in process, (c) their eventual conversion into finished goods, and (d) the sale of finished goods.

5. The theoretically correct alternative to disposing of underallocated or overallocated indirect costs is to prorate that amount on the basis of the already total amount of the allocated indirect cost in the ending balances of work in process, finished goods, and cost of goods sold. Many organizations simply write off any such amount to cost of goods sold.

6. Process costing systems assign costs to processes that produce masses of similar units and then compute unit costs on an average basis.

7. Equivalent unit calculations are required in process costing systems where all units are not fully completed at the beginning or end of the accounting period. An equivalent unit is the collection of inputs necessary to produce one fully complete unit of a product or output.

8. Activity-based costing (ABC) focuses on activities as the fundamental cost objects. The costs of these objects are then assigned to other cost objects such as products, services, or customers.

9. An ABC approach can be used in the design of a job costing system or a process costing system. It differs from the traditional approach by its fundamental focus on activities. An ABC approach typically results in (a) more indirect cost pools than the traditional approach, (b) more cost drivers used as cost allocation bases, and (c) more frequent use of nonfinancial variables as cost allocation bases.

TERMS TO LEARN

This chapter and the Glossary at the end of the book contain definitions of the following important terms:

equivalent units *(p. 157)*   job cost record *(142)*   job cost sheet *(142)*
job order sheet *(142)*   labor time record *(143)*
manufacturing overhead allocated *(151)*   materials requisition record *(143)*
overabsorbed indirect (overhead) costs *(152)*
overallocated indirect (overhead) costs *(152)*
overapplied indirect (overhead) costs *(152)*   proration *(153)*
underabsorbed indirect (overhead) costs *(152)*
underallocated indirect (overhead) costs *(152)*
underapplied indirect (overhead) costs *(152)*

ASSIGNMENT MATERIAL

QUESTIONS

5-1 Why are costing systems in the manufacturing sector often more complex than those in the service and merchandising sectors?

5-2 Give two ways that jobs undertaken in the manufacturing sector can differ.

5-3 Give three examples of specific manufacturing overhead cost items allocated to products.

5-4 Describe two source documents used in assigning manufacturing costs to products in a job costing system.

5-5 Distinguish among *actual costing, normal costing,* and *budgeted costing.*

5-6 What are two major goals of a job costing system?

5-7 Describe the role of a manufacturing overhead allocation base in job costing.

5-8 What are the limitations of the general ledger as a management tool?

5-9 Give two reasons for underallocation or overallocation of indirect (overhead) costs at the end of an accounting period.

5-10 Describe three alternative ways to make end-of-period adjustments for underallocated or overallocated indirect costs.

5-11 Give three examples of industries that often use process costing systems.

5-12 "There are five key steps in process costing when all units are not fully completed." What are they?

5-13 Why might a company have different numbers of equivalent units for direct materials and conversion costs?

5-14 How is an activity-based approach different from a traditional approach to designing a job costing system?

5-15 "Activity-based costing is the wave of the present and the future. All companies should adopt it." Do you agree? Explain.

EXERCISES AND PROBLEMS

5-16 Job costing, accounting for manufacturing overhead, budgeted rates.   The Lynn Company uses a job costing system at its Minneapolis plant. The plant has a machining department and an assembly department. Its job costing system has two direct cost categories (direct materials and direct manufacturing labor) and two manufacturing overhead cost pools (the machining department, allocated using actual machine-hours, and the assembly department, allocated using actual direct manufacturing labor cost). The 19_4 budget for the plant is:

	Machining Department	Assembly Department
Manufacturing overhead	$1,800,000	$3,600,000
Direct manufacturing labor costs	$1,400,000	$2,000,000
Direct manufacturing labor-hours	100,000	200,000
Machine-hours	50,000	200,000

The company uses a budgeted overhead rate for allocating overhead to production orders on a machine-hour basis in Machining and on a direct manufacturing labor-cost basis in Assembly.

Required
1. Present an overview diagram of Lynn's job costing system.
2. Compute the budgeted manufacturing overhead rate for each department.
3. During February the cost record for Job 494 contained the following:

	Machining Department	Assembly Department
Direct materials used	$45,000	$70,000
Direct manufacturing labor cost	$14,000	$15,000
Direct manufacturing labor-hours	1,000	1,500
Machine-hours	2,000	1,000

Compute the total manufacturing overhead costs of Job 494.
4. At the end of 19_4, the actual manufacturing overhead costs were $2,100,000 in Machining and $3,700,000 in Assembly. Assume that 55,000 actual machine-hours were used in Machining and that actual direct manufacturing labor costs in Assembly was $2,200,000. Compute the overallocated or underallocated manufacturing overhead for each department.

5-17 Job costing, accounting for manufacturing overhead, budgeted rates. The Solomon Company uses a job costing system at its Dover, Delaware, plant. The plant has a machining department and a finishing department. Its job costing system has two direct cost categories (direct materials and direct manufacturing labor) and two manufacturing overhead cost pools (the machining department, allocated using actual machine-hours and the finishing department, allocated using actual manufacturing labor costs). The 19_4 budget for the plant is:

	Machining Department	Finishing Department
Manufacturing overhead	$10,000,000	$8,000,000
Direct manufacturing labor costs	$ 900,000	$4,000,000
Direct manufacturing labor-hours	30,000	160,000
Machine-hours	200,000	33,000

Required
1. Present an overview diagram of Solomon's job costing system.
2. What is the budgeted overhead rate that should be used in the machining department? In the finishing department?
3. During the month of January, the cost record for Job 431 shows the following:

	Machining Department	Finishing Department
Direct materials used	$14,000	$3,000
Direct manufacturing labor costs	$ 600	$1,250
Direct manufacturing labor-hours	30	50
Machine-hours	130	10

What is the total manufacturing overhead allocated to Job 431?

4. Assuming that Job 431 consisted of 200 units of product, what is the unit product cost of Job 431?

5. Balances at the end of 19_4 are as follows:

	Machining Department	Finishing Department
Manufacturing overhead incurred	$11,200,000	$7,900,000
Direct manufacturing labor costs	$ 950,000	$4,100,000
Machine-hours	220,000	32,000

Compute the underallocated or overallocated manufacturing overhead for each department and for the Dover plant as a whole.

6. Why might Solomon use two different manufacturing overhead cost pools in its job costing system?

5-18 Job costing, journal entries. The University of Chicago Press is wholly owned by the university. It performs the bulk of its work for other university departments, which pay as though the Press were an outside business enterprise. The Press also publishes and maintains a stock of books for general sale. A job costing system is used to cost each job. There are two direct cost categories (direct materials and direct manufacturing labor) and one indirect cost pool (manufacturing overhead, allocated on the basis of direct labor costs).

The following data pertain to 19_5 (in thousands):

Direct materials and supplies purchased on account	$ 800
Direct materials used	710
Indirect materials issued to various production departments	100
Direct manufacturing labor	1,300
Indirect manufacturing labor incurred by various departments	900
Depreciation on building and manufacturing equipment	400
Miscellaneous manufacturing overhead* incurred by various departments (ordinarily would be detailed as repairs, photocopying, utilities, etc.)	550
Manufacturing overhead allocated at 160% of direct manufacturing labor costs	?
Cost of goods manufactured	4,120
Sales	8,000
Costs of goods sold	4,020
Inventories, December 31, 19_4 (not 19_5):	
Materials control	100
Work in process control	60
Finished goods control	500

*The term *manufacturing overhead* is not used uniformly. Other terms that are often encountered in printing companies include *job overhead* and *shop overhead*.

Required

1. Present an overview diagram of the University of Chicago Press's job costing system.

2. Prepare general journal entries to summarize 19_5 transactions. As your final entry, dispose of the year-end overallocated or underallocated manufacturing overhead as a direct write-off to Cost of Goods Sold. Number your entries. Explanations for each entry may be omitted.

3. Show posted T-accounts for all inventories, Cost of Goods Sold, Manufacturing Overhead Control, and Manufacturing Overhead Allocated.

4. Sketch how the subsidiary ledger would appear for Manufacturing Overhead Control. Assume that there are three departments: art, photo, and printing. You need not show any numbers.

For more details concerning these data, see Problem 5-19.

5-19 Job costing, journal entries and source documents. (Continuation of 5-18) For each journal entry in your answer to Problem 5-18, (a) indicate the source document that would most likely authorize the entry and (b) give a description of the entry into the subsidiary ledgers, if any entry needs to be made there.

5-20   Job costing, journal entries.   Donnell Transport assembles prestige mobile homes. Its job costing system has two direct cost categories (direct materials and direct manufacturing labor) and one indirect cost pool (manufacturing overhead allocated at a budgeted $30 per machine-hour in 19_5). The following data pertain to operations for the year 19_5 (in millions):

Materials Control, December 31, 19_4	$ 12
Work in Process Control, December 31, 19_4	2
Finished Goods Control, December 31, 19_4	6
Materials and supplies purchased on account	150
Direct materials used	145
Indirect materials (supplies) issued to various production departments	10
Direct manufacturing labor	90
Indirect manufacturing labor incurred by various departments	30
Depreciation on plant and manufacturing equipment	19
Miscellaneous manufacturing overhead incurred (credit Various Liabilities, ordinarily would be detailed as repairs, utilities, etc.)	9
Manufacturing overhead allocated, 2,100,000 actual machine-hours	?
Cost of goods manufactured	294
Sales	400
Cost of goods sold	292

Required
1. Present an overview diagram of Donnell Transport's job costing system.
2. Prepare general journal entries. Number your entries.
3. Post to T-accounts. What is the ending balance of Work in Process Control?
4. Sketch how the subsidiary ledger would appear for Manufacturing Overhead Control, assuming that there are four departments. You need not show any numbers.
5. Show the journal entry for disposing of overallocated or underallocated manufacturing overhead directly as a year-end write-off to Cost of Goods Sold. Post the entry to T-accounts.

For more details concerning these data, see Problem 5-21.

5-21   Journal entries and source documents.   (Continuation of 5-20) For each journal entry in your answer to Problem 5-20, (a) indicate the source documents that would most likely authorize the entry and (b) give a description of the entry into the subsidiary ledgers, if any.

5-22   Accounting for manufacturing overhead.   Consider the following selected cost data for the Pittsburgh Forging Company for 19_2.

Budgeted manufacturing overhead	$7,000,000
Budgeted machine-hours	200,000
Actual manufacturing overhead	$6,800,000
Actual machine-hours	195,000

Pittsburgh's job costing system has a single manufacturing overhead cost pool (allocated using a budgeted rate based on actual machine-hours). Any amount of underallocation or overallocation is immediately written off to cost of goods sold.

Required
1. Compute the budgeted manufacturing overhead rate.
2. Journalize the allocation of manufacturing overhead.
3. Compute the amount of underallocation or overallocation of manufacturing overhead. Is the amount significant? Journalize the disposition of this amount on the basis of the ending balances in the relevant accounts.

5-23   Proration of overhead. (Z. Iqbal, adapted)   The Zaf Radiator Company uses a single manufacturing overhead cost pool in its job costing system. The following data are for 19_5:

Budgeted manufacturing overhead		$4,800,000
Overhead allocation base		Machine-hours
Budgeted machine-hours		80,000
Manufacturing overhead incurred		$4,900,000
Actual machine-hours		75,000

Machine-hours data and the ending balances (before proration of underallocated or overallocated overhead) follow:

	Actual Machine-Hours	19_5 End-of-Year Balance
Cost of goods sold	60,000	$8,000,000
Finished goods	11,000	$1,250,000
Work in process	4,000	$ 750,000

Required
1. Compute the budgeted manufacturing overhead rate for 19_5.
2. Compute the underallocated or overallocated manufacturing overhead of Zaf Radiator in 19_5.
3. Prorate the underallocated or overallocated amount from requirement 2 using:
   a. Immediate write-off to cost of goods sold
   b. Proration based on ending balances (before proration) in work in process, finished goods, and cost of goods sold
   c. Proration based on the allocated overhead amount (before proration) in the ending balances of work in process, finished goods, and cost of goods sold
4. Which proration method do you prefer in requirement 3? Explain.

5-24 Meaning of overallocated overhead. The Umberto Company had budgeted the following performance for 19_4 at its engineering products plant:

Machine-hours	30,000
Beginning inventories	None
Sales	$4,000,000
Total variable costs	3,000,000
Total fixed costs	800,000
Operating income	200,000
Manufacturing overhead	
Variable	300,000
Fixed	600,000

It is now the end of 19_4. A manufacturing overhead rate of $30 per machine-hour was used in 19_4 for allocating budgeted costs in the single manufacturing overhead cost pool to products. Total manufacturing overhead incurred in 19_4 was $900,000. Overallocated manufacturing overhead at the end of 19_4 was $54,000. There was no ending work in process.

Required
1. Explain why Umberto might prefer to use a budgeted rather than an actual manufacturing overhead rate for job costing.
2. Why might manufacturing overhead be underallocated or overallocated at the end of 19_4?
3. How many machine-hours were used by Umberto in 19_4?

5-25 Allocation and proration of manufacturing overhead. (SMA, heavily adapted) Nicole Limited is a company that produces machinery to customer order. Its job costing system has two direct cost categories (direct materials and direct manufacturing labor) and one indirect cost pool (manufacturing overhead, allocated using a budgeted rate based on direct manufacturing labor costs). The budget for 19_5 was:

Direct manufacturing labor	$420,000
Manufacturing overhead	$252,000

At the end of 19_5, two jobs were incomplete: No. 1768B (total direct manufacturing labor costs were $11,000) and No. 1819C (total direct manufacturing labor costs were $39,000). Machine time totaled 287 hours for No. 1768B and 647 hours for No. 1819C. Direct materials issued to No. 1768B amounted to $22,000. Direct material for No. 1819C came to $42,000.

Total charges to the Manufacturing Overhead Control account for the year were $186,840. Direct manufacturing labor charges made to all jobs were $400,000, representing 20,000 direct manufacturing labor-hours.

There were no beginning inventories. In addition to the ending work in process described above, the ending finished goods showed a balance of $156,000 (including a direct manufacturing labor cost component of $40,000). Sales for 19_5 totaled $2,700,680, cost of goods sold were $1,600,000, and marketing costs were $857,870.

Required

1. Prepare a detailed schedule showing the ending balances in the inventories and cost of goods sold (before considering any underallocated or overallocated manufacturing overhead). Show also the manufacturing overhead allocated to these ending balances.
2. Compute the underallocated or overallocated manufacturing overhead for 19_5.
3. Prorate the amount computed in requirement 2 on the basis of:
   a. the ending balances (before proration) of work in process, finished goods, and cost of goods sold
   b. the allocated overhead amount (before proration) in the ending balances of work in process, finished goods, and cost of goods sold
4. Assume that Nicole decides to immediately write off to cost of goods sold any underallocated or overallocated manufacturing overhead. Will operating income be higher or lower than the operating income that would have resulted from the proration in requirements 3(a) and 3(b) above?

5-26 Overview of general-ledger relationships. The Blakely Company is a small machine shop that uses highly skilled labor and a job costing system. The total debits and credits in certain accounts *just before* year-end are:

	December 30, 19_6	
	Total Debits	**Total Credits**
Direct materials control	$100,000	$ 70,000
Work in process control	320,000	305,000
Manufacturing department overhead control	85,000	—
Finished goods control	325,000	300,000
Cost of goods sold	300,000	—
Manufacturing overhead allocated	—	90,000

Note that "total debits" in the inventory accounts would include beginning inventory balances, if any.

The above accounts *do not* include the following:

a. The manufacturing labor costs recapitulation for the December 31 working day: direct manufacturing labor, $5,000, and indirect manufacturing labor, $1,000
b. Miscellaneous manufacturing overhead incurred on December 30 and December 31: $1,000

Additional Information

◆ Manufacturing overhead has been allocated as a percentage of direct manufacturing labor costs through December 30.
◆ Direct materials purchased during 19_6 were $85,000.
◆ There were no returns to suppliers.
◆ Direct manufacturing labor costs during 19_6 totaled $150,000, not including the December 31 working day described above.

Required

1. Compute the inventories (December 31, 19_5) of direct materials control, work in process control, and finished goods control. Show T-accounts.

2. Prepare all adjusting and closing journal entries for the above accounts. Assume that all underallocated or overallocated manufacturing overhead is closed directly to Cost of Goods Sold.

3. Compute the ending inventories (December 31, 19_6), after adjustments and closing, of direct materials control, work in process control, and finished goods control.

5-27 Process costing.   International Electronics manufactures microchips in large quantities. Each microchip undergoes assembly and testing. The assembly costs during January 19_5 were:

Direct materials used	$ 720,000
Conversion costs	760,000
Total manufacturing costs	$1,480,000

There was no beginning inventory on January 1, 19_5. During January, 10,000 microchips were placed into production.

Required

1. Assume that all 10,000 microchips are fully completed at the end of January. What is the unit cost of a completed microchip in January 19_5?

2. Assume now that only 9,000 microchips were fully completed at the end of January. All direct materials had been added to the remaining 1,000 microchips. However, on average, these remaining 1,000 microchips were only 50% complete as to conversion costs. (a) What are the equivalent units for direct materials and conversion costs and their respective equivalent unit costs for January? (b) What is the cost of a completed unit in January 19_5?

3. Explain the difference in your answers to requirements 1 and 2.

5-28 Journal entries.   Refer to the International Electronics data in requirement 2 of Problem 5-27. Prepare summary journal entries for the use of direct materials and conversion costs. Also prepare a journal entry to transfer out the cost of goods completed. Show the postings to the Work in Process—Assembly account.

5-29 Process costing.   Vought Company produces a standard missile for the Department of Defense at its Rhode Island plant. The manufacturing costs of the plant for March 19_4 were:

Direct materials used	$11,875,000
Conversion costs	9,000,000
Total manufacturing costs	$20,875,000

There is no beginning inventory on March 1, 19_4. Suppose 100 missiles were put into production in March.

Required

1. Assume that all 100 missiles are fully completed in March 19_4. What is the unit cost of a completed missile in March 19_4?

2. Assume now that only 80 missiles are fully completed by March 31, 19_4. The remaining 20 missiles were 75% complete as to direct materials and 50% complete as to conversion costs. (a) What are the equivalent units for direct materials and conversion costs and their respective equivalent unit costs for March 19_4? (b) What is the cost of a completed missile in March 19_4?

3. Explain the difference in your answers to requirements 1 and 2.

5-30 Journal entries.   Refer to the Vought plant data in requirement 2 of Problem 5-29. Prepare summary journal entries for the use of direct materials and conversion costs. Also prepare a journal entry to transfer out the cost of goods completed. Show the postings to the Work in Process account.

 5-31 Process costing, single department.   Vaasa Chemicals has a mixing department and a refining department. Its process costing system in the mixing department has two direct materials cost categories (Chemical P and Chemical Q) and one conversion costs pool. The following data pertain to the mixing department for July 19_5:

Units	
Work in process, July 1	0
Units started	50,000
Completed and transferred to refining department	35,000
Costs	
Chemical P	$250,000
Chemical Q	70,000
Conversion costs	135,000

Chemical P is introduced at the start of operations in the mixing department, and Chemical Q is added when the product reaches the three-fourths stage of completion in the mixing department. Conversion costs are incurred uniformly throughout the process. The ending work in process in the mixing department is two-thirds completed.

Required
1. Compute the equivalent units in the mixing department for July 19_5.
2. Compute (a) the cost of goods completed and transferred to the refining department during July and (b) the cost of work in process as of July 31, 19_5.

5-32   Journal entries.   Refer to Problem 5-31. Prepare journal entries. Assume that the completed goods are transferred to the refining department.

5-33   Activity-based job costing system.   The Denver Company manufactures and sells packaging machines. It recently used an activity-based approach to refine the job costing system at its Denver plant. The resulting job costing system has one direct cost category (direct materials) and four indirect manufacturing cost pools. These four indirect cost pools and their allocation bases were chosen by a team of product designers, manufacturing personnel, and marketing personnel:

Indirect Manufacturing Cost Pool	Cost Allocation Base	Budgeted Cost Allocation Rate
1. Materials handling	Number of component parts	$ 8 per part
2. Machining	Machine-hours	$ 68 per hour
3. Assembly	Assembly line hours	$ 75 per hour
4. Inspection	Inspection hours	$104 per hour

Cola Supreme recently purchased 50 can-packaging machines from the Denver Company. Each machine has direct materials costs of $3,000, requires 50 component parts, 12 machine-hours, 15 assembly line hours, and 4 inspection hours.

Denver's prior costing system had one direct cost category (direct materials) and one indirect cost category (manufacturing overhead, allocated using assembly line hours).

Required
1. Present overview diagrams of the prior job costing system and the refined activity-based job costing system.
2. Compute the unit manufacturing costs of each machine and the total manufacturing cost of the Cola Supreme job.
3. The activity-based job costing system of Denver has only one manufacturing direct cost category—direct materials. A competitor of Denver Company has two direct cost categories at its manufacturing plant—direct materials and direct manufacturing labor. Why might Denver not have a direct manufacturing labor costs category in its job costing system? Where are the manufacturing labor costs included in the Denver costing system?
4. What information might members of the team that refined the prior costing system find useful in the activity-based job costing system?

5-34   Activity-based costing, job costing system.   The Hewlett-Packard (HP) plant in Roseville, California, assembles and tests printed-circuit (PC) boards. The job costing system at this plant has two direct cost categories (direct materials and direct manufacturing labor) and eight indirect cost pools. These eight indirect cost pools represent the eight activity areas that operating personnel at the plant determined were sufficiently different (in terms of cost behavior patterns or in terms of individual products being assembled) to warrant separate cost pools. The cost allocation base chosen for each activity area is the cost driver at that activity area.

Debbie Berlant, a newly appointed marketing manager at HP, attends a training session that describes how an activity-based costing approach has been used to design the Roseville plant's job costing system. Berlant is provided with the following incomplete information for a specific job (an order for a single PC board, No. A82):

Direct materials	$75.00	
Direct manufacturing labor	15.00	$90.00
Manufacturing overhead (see below)		?
Total manufacturing costs		$?

Manufacturing Overhead Cost Pool	Cost Allocation Base	Cost Allocation Rate	Units of Base Used on Job No. A82	Manufacturing Overhead Allocated to Job
1. Start station	No. of raw PC boards	$ 1.10	1	$ 1.10
2. Axial insertion	No. of axial insertions	0.08	45	?
3. Dip insertion	No. of dip insertions	0.25	?	6.00
4. Manual insertion	No. of manual insertions	?	11	5.50
5. Wave solder	No. of boards soldered	3.50	?	3.50
6. Backload	No. of backload insertions	?	6	4.20
7. Test	Budgeted time board is in test activity	90.00	0.25	?
8. Defect analysis	Budgeted time for defect analysis and repair	?	0.10	8.00

Required
1. Present an overview exhibit of the activity-based job costing system at the Roseville plant.
2. Fill in the blanks (signaled by a question mark) in the cost information provided to Berlant for job No. A82.
3. Why might manufacturing managers and marketing managers favor this ABC job costing system over the prior costing system, which had the same two direct cost categories but only a single indirect cost pool (manufacturing overhead allocated using direct labor cost)?

5-35 Activity-based job costing. The Schramka Company manufactures a variety of prestige boardroom chairs. Its job costing system was designed using an activity-based approach. There are two direct cost categories (direct materials and direct manufacturing labor) and three indirect cost pools. These three cost pools represent three activity areas at the plant.

Manufacturing Activity Area	Budgeted Costs for 19_5	Cost Driver Used as Allocation Base	Cost Allocation Rate
Materials handling	$ 200,000	Number of parts	$ 0.25
Cutting	2,160,000	Number of parts	2.50
Assembly	2,000,000	Direct manufacturing labor-hours	25.00

Two styles of chairs were produced in March, the executive chair and the chairman chair. Their quantities, direct material costs, and other data for March 19_5 follow:

	Units Produced	Direct Material Costs	Number of Parts	Direct Manufacturing Labor-Hours
Executive chair	5,000	$600,000	100,000	7,500
Chairman chair	100	25,000	3,500	500

The direct manufacturing labor rate is $20 per hour. Assume no beginning or ending inventory.

Required

1. Compute the March 19_5 total manufacturing costs and unit costs of the executive chair and the chairman chair.

2. Suppose that the upstream activities to manufacturing (R&D and design) and the downstream activities (marketing, distribution, and customer service) were analyzed. The unit costs in 19_5 were budgeted to be:

	Upstream Activities	Downstream Activities
Executive chair	$ 60	$110
Chairman chair	146	236

Compute the full product costs per unit of each line of chairs. (Full product costs are the sum of the costs in all business function areas.)

5-36 Activity-based job costing, unit cost comparisons. Tracy Corporation has a machining facility specializing in jobs for the aircraft components market. The prior job costing system had two direct cost categories (direct materials and direct manufacturing labor) and a single indirect cost pool (manufacturing overhead, allocated using direct labor-hours). The indirect cost allocation rate of the prior system for 19_5 would have been $115 per direct labor-hour.

Recently a team with members from product design, manufacturing, and accounting used an activity-based approach to refine its job costing system. The two direct cost categories were retained. The team decided to substitute the single indirect cost pool with five indirect cost pools. These five cost pools represent five activity areas at the facility, each with its own supervisor and budget responsibility. Pertinent data follow:

Activity Area	Cost Driver Used as Allocation Base	Cost Allocation Rate
Materials handling	Number of parts	$ 0.40
Lathe work	Number of turns	0.20
Milling	Number of machine-hours	20.00
Grinding	Number of parts	0.80
Testing	Number of units tested	15.00

Information-gathering technology has advanced to the point where all the data necessary for budgeting in these five activity areas are automatically collected.

The two jobs processed under the new system at the facility in the most recent period had the following characteristics:

	Job 410	Job 411
Direct materials cost per job	$ 9,700	$59,900
Direct manufacturing labor cost per job	$ 750	$11,250
Number of direct manufacturing labor-hours per job	25	375
Number of parts per job	500	2,000
Number of turns per job	20,000	60,000
Number of machine-hours per job	150	1,050
Number of units per job (all tested)	10	200

Required

1. Compute the per-unit manufacturing costs of each job under the prior job costing system.

2. Compute the per-unit manufacturing costs of each job under the activity-based job costing system.

3. Compare the per-unit cost figures for Jobs 410 and 411 computed in requirements 1 and 2. Why do the prior and the activity-based costing systems differ in their job cost estimates for each job? Why might these differences be important to Tracy Corporation?

5-37   Job costing, contracting, ethics.   Jack Halpern is the owner and CEO of Aerospace Comfort, a firm specializing in the manufacture of seats for air transport. He has just received a copy of a letter written to the General Audit Section of the U.S. Navy. He believes it is from an ex-employee of Aerospace.

Dear Sir,

Aerospace Comfort in 19_5 manufactured 100 X7 seats for the Navy. You may be interested to know the following:

1. Direct materials costs billed for the 100 X7 seats was $25,000.
2. Direct manufacturing labor costs billed for 100 X7 seats was $6,000. This cost includes 16 hours of set up labor at $25 per hour, an amount included in the manufacturing overhead cost pool as well. The $6,000 also includes 12 hours of design time at $50 an hour. Design time was explicitly identified as a cost the Navy was not to reimburse.
3. Manufacturing overhead costs billed for 100 X7 seats was $9,000 (150% of direct manufacturing labor costs). This amount includes the 16 hours of setup labor at $25 per hour that is incorrectly included as part of direct manufacturing labor costs.

You may also want to know that over 40% of the direct materials is purchased from Frontier Technology, a company that is 51% owned by Jack Halpern's brother.

For obvious reasons, this letter will not be signed.

c.c.: *The Wall Street Journal*
Jack Halpern, CEO of Aerospace Comfort

Aerospace Comfort's contract states that the Navy reimburses Aerospace at 130% of manufacturing costs.

Required
Assume that the facts in the letter are correct in answering the following questions.
1. What is the cost amount per X7 seat that Aerospace Comfort billed the Navy? Assume that the actual direct materials costs are $25,000.
2. What is the amount per X7 seat that Aerospace Comfort should have billed the Navy? Assume that the actual direct materials costs are $25,000.
3. What should the Navy do to tighten its procurement procedures to reduce the likelihood of such situations recurring?

6

Master Budget and

Responsibility Accounting

Evolution of systems

Major features of budgets

Advantages of budgets

Types of budgets

Illustration of master budget

Sales forecasting—a difficult task

Activity-based budgeting and *kaizen* budgeting

Computer-based financial planning models

Responsibility accounting

Responsibility and controllability

Human aspects of budgeting

Appendix: The cash budget

*I*n Pizza Hut's budgeting system, unit managers budget and are responsible for sales and for operating costs within the general guidelines of Pizza Hut pricing policy. Favorable performance is often a combination of effective promotion of new products and efficient management of costs.

Learning Objectives

When you have finished studying this chapter, you should be able to

1. Define *master budget* and describe its major benefits to an organization
2. Distinguish various components of the master budget
3. Construct the budgeted income statement and its supporting schedules
4. Define *activity-based budgeting* and *kaizen budgeting*
5. Describe the uses of computer-based financial planning models
6. Describe responsibility centers and responsibility accounting
7. Explain how controllability is related to responsibility accounting

Budgets quantify future plans of action. A budgeting system builds on historical (actual) performance and expands to include consideration of future (expected) performance. Budgeting systems turn managers' perspectives forward. Managers prepare both financial and nonfinancial budgets. Financial budgets detail the expected revenue and cost impacts that the organization's plans will have. Nonfinancial, or physical, budgets focus on machinery, space, equipment, workers, and the like. This chapter examines the master, or comprehensive, budget as a planning and coordinating tool. The succeeding chapters, especially Chapters 7 and 8, examine various aspects of budgeting decisions and their implementation.

Chapter 1 described some newly evolving themes in management accounting. Budgets give financial expression to many of these themes. For example, budgets quantify the planned financial effects of activities aimed at continuous improvement and cost reduction. The budget also expresses management's spending plans for various value-chain functions (research and development, design production, marketing, distribution, and customer service).

EVOLUTION OF SYSTEMS

Consider the evolution of control systems. In a small, new organization, personal observation is usually the dominant means of control. A manager sees, touches, and hears the relationship between inputs and outputs; he or she oversees the work and actions of various personnel.

Over time, managers add historical records to their personal observations. Historical records allow managers to compare current performance with past performance. How did a department perform in 19_6 compared with 19_5? Analyses of past performance may help improve future performance. Managers must deal with a series of periods, not just one.

As the organization matures, budgeting becomes an important step in the growth and improvement of the accounting system. A manager would find it helpful to compare *actual performance* in 19_6 with the *plans* that had been prepared for 19_6. Budgeting systems help promote this future perspective.

Do budgeting systems meet the cost-benefit test? Evidently they do. Typically, organizations purchase budgeting systems voluntarily rather than as a result of outside forces (such as regulatory or government mandates). Why? Because budgeting systems change human behavior—and decisions—in ways sought by top management. For example, budgeting systems may prompt managers to use longer planning horizons. As a result, many potential difficulties are foreseen and avoided.

MAJOR FEATURES OF BUDGETS

Definition and Role of Budgets

A **budget** is a quantitative expression of a plan of action and an aid to the coordination and implementation of this plan. Managers formulate budgets for the organization as a whole or for any subunit. The **master budget** summarizes the financial projections of all the organization's budgets and plans. It describes the financial plans for all value-chain functions. The budget quantifies management's expectations regarding future income, cash flows, and financial position. These expectations arise from a careful look at the organization's future. Where does management want the company to be a year from now? Five years from now?

Consider the diagram in Exhibit 6-1. It shows how budgets and performance reports help managers. Note the key role that budgets play throughout the process. Our focus in this chapter will be on the planning of operations. Budgets, however, serve a variety of additional functions: coordinating activities, implementing plans, communicating, authorizing actions, motivating, controlling, and evaluating performance. The authorization function is especially important in government budgeting and nonprofit budgeting, where budget appropriations serve as approvals and ceilings for management expenditures.

Well-managed organizations usually have the following budgeting cycle:

1. Planning the performance of the organization as a whole as well as its subunits. The entire management team agrees as to what is expected.
2. Providing a frame of reference, a set of specific expectations against which actual results can be compared.
3. Investigating variations from plans. If necessary, corrective action follows investigation.
4. Planning again, considering feedback and changed conditions.

The master budget embraces the impact of both *operating* decisions and *financing* decisions. Operating decisions center on the acquisition and use of scarce resources. Financing decisions center on how to get the funds to acquire resources. This book concentrates on how accounting helps managers make operating decisions; this chapter emphasizes operating budgets. Financing decisions and the role of cash budgets are covered in many finance texts. The chapter Appendix explains cash budgets.

Are Budgets Widely Used?

Surveys of companies in many countries indicate that over 90% of responding firms use budgets (see Surveys of Company Practice on p. 187). Surveys of Japanese companies indicate that budgetary planning and control is the most important cost man-

EXHIBIT 6-1
Budgets and Performance Reports

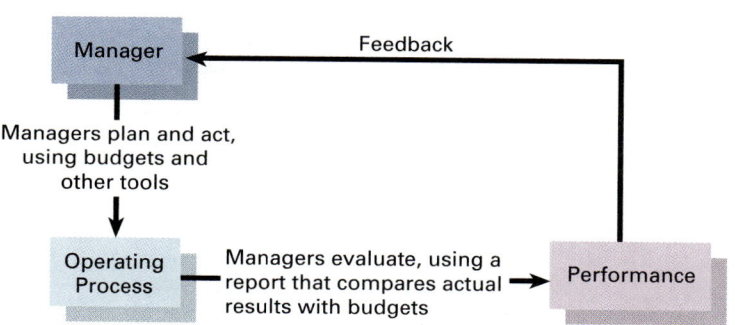

agement technique for "costing down" (continuously reducing product costs.)[1] Budgeting systems are more common in larger companies, and their systems are more formalized. Still, small businesses also use budgets but in a less structured way. Budgets force entrepreneurs to quantify their dreams and directly face the uncertainties of their ventures. As one observer said: "Few businesses plan to fail, but many of those that flop failed to plan."[2]

ADVANTAGES OF BUDGETS

Budgets are a major feature of most management control systems. When administered intelligently, budgets (1) compel planning, (2) provide performance criteria, and (3) promote communication and coordination within the organization.

Strategy and Plans

Too often, executives practice "management by crisis." Organizations or individuals are caught in undesirable situations that should have been anticipated and avoided. Forced planning for changing conditions is by far the greatest contribution of budgeting to management.

Budgeting is an integral part of strategy and tactics. **Strategy** is a broad term that usually means selection of overall objectives. **Tactics** are the general means for attaining strategic goals. Strategy analysis includes consideration of such questions as:

1. What are the overall goals or objectives of our organization?
2. Are the markets for our products local, regional, national, or global? What trends will affect our markets? How are we affected by the economy, the industry, and our competitors?
3. What forms of organization and financial structures serve us best?
4. What are the risks of alternative strategies, and what are our contingency plans if our preferred plan fails?

Consider the diagram in Exhibit 6-2. Strategy analysis underlies both long-run and short-run planning.[3] In turn, these plans lead to the formulation of budgets.[4] The arrowheads in the diagram are pointing in two directions. Why? Because strategy, plans, and budgets are interrelated and affect one another. Budgets provide feedback to managers about the likely effects of their strategic plans. Managers use this feedback to revise their strategic plans. Apple Computer's strategic decision to reduce the prices of its personal computers provides an example of the interrelation between strategy analysis and budgets. By reducing its prices, Apple expected to increase the demand for its computers. The budget, however, indicated that even at the predicted higher sales quantities Apple would be unable to meet its financial targets. For the strategy to succeed, Apple would need to reduce operating costs by streamlining operations and moving facilities to lower-cost areas.

---

[1]See S. Inoue, "JIT Production and Its Influence on Cost Management in Japan," *Annals* (Kagawa University) 1990; and M. Gietzmann and S. Inoue, "The Adaptation of Management Accounting Systems to Changing Market Conditions," *British Journal of Management*, (vol. 2, 1991).

[2]Dun & Bradstreet, *Business Failure Record* (1992), reports that the major causes of business failure include industry weakness, insufficient capital, and heavy operating costs.

[3]Not all organizations undertake strategy analysis before or during the development of a budget. For example, some budgets in the nonprofit sector are based merely on last year's expenditures and then are adjusted for an inflation factor.

[4]See J. Shank and V. Govindarajan, *Strategic Cost Management: Three Key Themes for Managing Costs Effectively* (New York: Free Press, 1993).

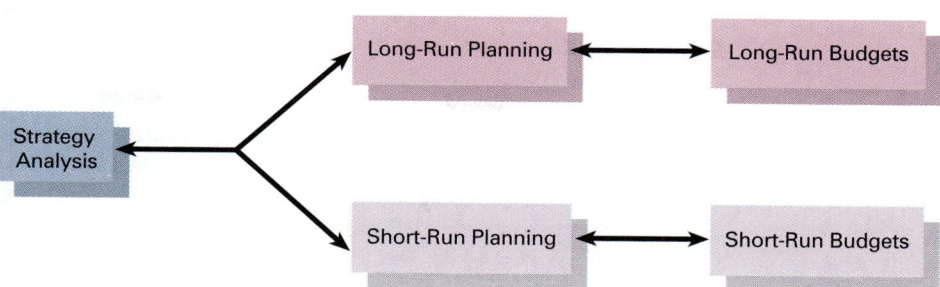

Framework for Judging Performance

As a basis for judging actual results, budgeted performance is generally a better criterion than past performance. For example, sales this year may be higher or direct materials costs may be lower, which may be encouraging—but by no means conclusive as a measure of success. Assume that sales rose from 90,000 units to 100,000 units. But suppose that the entire market has grown. Anticipating this growth, the company had budgeted sales of 112,000 units this year, on the basis of its current market share. Actual sales of 100,000 units, then, is a disappointing performance. Management would likely give the personnel who are responsible for the lackluster sales a negative evaluation. The key point: Using a budget, management evaluates employees on targeted amounts, not on whim or past performance. Employees know what their performance is expected to be.

One major weakness of using historical data for judging performance is that inefficiencies may be buried in past actions. A good budgeting system forces managers to examine the business as they plan, so managers may detect inefficiencies that would otherwise go unnoticed. Management would alert individual employees to these inefficiencies and monitor performance for improvement. Also, the usefulness of comparisons with the past may be hampered by intervening changes in technology, personnel, products, competition, general economic conditions, and other factors.

A second weakness of using historical data is that opportunities in the future, which did not exist in the past, may be ignored. Budgets alert managers to possible revenues, costs, and profits arising from new opportunities. Budgets, based on these new opportunities, are also useful for evaluating management performance. Consider, for example, programming innovations in the cable television industry, such as pay-per-view special events for live concerts and championship boxing. In the year when these innovations are introduced, a manager's performance will likely exceed the prior year's performance. But performance relative to the prior year is not a good basis for evaluating the manager. A better measure is how well the manager performed relative to the budget for the year.

Coordination and Communication

Coordination is the meshing and balancing of all factors of production or service and of all the departments and functions so that the company can meet organization objectives. *Communication* is getting the company's plans across to all personnel in departments and functions.

Coordination forces executives to think of relationships among individual operations, departments, and the company as a whole. Coordination implies, for example, that purchasing officers make material purchase plans based on production requirements. Also, production managers plan personnel and machinery needs to produce the number of products necessary to meet sales forecasts. How does a budget lead to coordination? Consider the following example. Production managers, evaluated on the basis of output produced per machine (machine efficiency), would like to increase efficiency by producing more output. But if the output produced cannot be sold, costly inventory will build up. The budget achieves coordination by constraining production managers to produce only what is expected to sell. Top managers design budgeting and performance evaluation systems so that the self-interests of managers do not conflict with the interests of the organization.

For coordination to succeed, communication is essential. The production manager must know the sales plan. The purchasing manager must know the production plan, and so on. The budget is an effective way to communicate a consistent set of plans to the organization as a whole.

Management Support and Administration

Budgets help managers, but budgets need help. *That is, top management must understand and enthusiastically support the budget and all aspects of the management control system.* Top management support is critical for obtaining active line participation in the formulation of budgets and for successful administration of the budget. If line managers feel that top management does not "believe" in the budget, line managers are unlikely to be active participants in the budget process. To increase participation in and acceptance of budgets, companies tend to begin building the budget at relatively low organization levels. This participation also means that line managers' expertise and knowledge are reflected in the budget.

Budgets should not be administered rigidly. Changing conditions call for changes in plans. A manager may commit to the budget, but matters might develop so that some special repairs or a special advertising outlay would best serve the interests of the organization. The manager should not defer the repairs or the advertising in order to meet the budget, if such actions will hurt the organization in the long run. Instead, the manager should feel comfortable enough in his or her position to request permission for such outlays, or the budget itself should provide enough flexibility to permit reasonable discretion in deciding how best to get the job done. The budget is a means to achieve strategic goals, not an end in itself.

TYPES OF BUDGETS

Time Coverage

Budgets often span a period of one year or less, but in cases of plant, equipment, and product changes, the budget may span five or more years. The usual planning-and-control budget period is one year. The annual budget is often broken down by months for the first quarter and by quarters for the remainder of the year. The budgeted data for a year are frequently revised as the year unfolds. For example, at the end of the first quarter, the budget for the next three quarters is changed in light of new information.

Businesses are increasingly using *rolling budgets*. A **rolling budget** is a budget or plan that is always available for a specified future period by adding a month, quarter, or year in the future as the month, quarter, or year just ended is dropped. Rolling budgets constantly force management to think concretely about a forthcoming span of time. Arizona Public Service Company has a budget that looks ahead two years and is updated every month. NEC Corporation of Japan has a one-year operating budget that is updated each month. Companies also frequently use rolling budgets when developing five-year budgets for long-range planning. For example, NEC Corporation also has a five-year budget that is updated each year.

Surveys of Company Practice

BUDGET PRACTICES AROUND THE GLOBE[a]

Surveys of financial officers of the largest industrial companies in the United States, Japan, Australia, the United Kingdom, and Holland indicate that the master budget is widespread, but that differences arise with respect to other dimensions of budgeting. U.S. controllers and managers prefer more participation and regard return on investment as the most important budget goal. In comparison, Japanese controllers and managers, for example, prefer less participation and regard sales revenues as the most important budget goal.

	United States	Japan	Australia	United Kingdom	Holland
1. Percentage of firms that prepare complete master budget	91%	93%	95%	100%	100%

Survey information for points 2 and 3 is available for the United States and Japan only.

	United States	Japan
2. Percentage of firms that report division manager's participation in budget committee discussions	78%	67%

	United States	Japan
3. Ranking of the most important budget goals for division managers (1 = most important)		
Return on investment	1	4
Operating income	2	2
Sales revenues	3	1
Production costs	4	3

[a]Adapted from Asada, Bailes, and Amano, "An Empirical Study," Blayney and Yokoyama, "Comparative Analysis," and de With and Ijskes "Current Budgeting." Full citations are in Appendix A.

The choice of budget period largely flows from the objectives, uses, and dependability of the budget data. For example, British Petroleum, when budgeting for exploration and development costs for its oil and gas properties, uses budgets that span several years. By contrast, manufacturers of children's toys, such as Toys "Я" Us, use relatively short budget periods. Why? Because the company faces constantly changing consumer fads.

Classification of Budgets

Terminology used to describe budgets varies among organizations. For example, budgeted financial statements are sometimes called **pro forma statements**. The budgeted financial statements of many companies include the budgeted income statement, the budgeted balance sheet, and the budgeted statement of cash flows. Some organizations, such as Hewlett-Packard, refer to budgeting as *targeting*. Indeed, to give a more positive thrust to budgeting, many organizations—for example, Nissan Motor Company and Owens-Corning—describe the *budget* as a *profit plan*.

There are countless forms of budgets. Many special budgets and related reports are prepared, including:

◆ Comparisons of budgets with actual results (performance reports)
◆ Flexible budgets (see Chapter 7)
◆ Product life cycle budgets (see Chapter 12)
◆ Long-term budgets, often called "capital," "capital expenditures," "facilities," or "project" budgets (see Chapters 20 and 21)
◆ Reports for specific management needs—for example, cost-volume-profit projections

ILLUSTRATION OF MASTER BUDGET

Objective 2
Distinguish various components of the master budget

We use Halifax Engineering, a manufacturer of aircraft replacement parts, to illustrate a master budget. Its job costing system for manufacturing costs has two direct cost categories (direct materials and direct manufacturing labor) and one indirect cost pool (manufacturing overhead). Manufacturing overhead (both variable and fixed) is allocated to products using direct manufacturing labor hours as the allocation base.

Exhibit 6-3 shows a simplified diagram of the various parts of the *master budget* for Halifax Engineering. The master budget summarizes the financial projections of all the organization's budgets and plans. The master budget results in a set of detailed financial statements for short periods, usually a year. The bulk of the diagram presents various elements that together are often called the **operating budget,** which is the budgeted income statement and its supporting schedules. The supporting schedules cut across different categories of the value chain, from research and development to customer service. The **financial budget** is that part of the master budget that comprises the capital budget, cash budget, budgeted balance sheet, and budgeted statement of cash flows. It focuses on the impact that cash has on operations and on other factors, such as planned capital outlays for equipment.

Once the sales budget is completed, then research and development, design, production, marketing, distribution, customer-service, and administration functions can often be working on their budgets *simultaneously*. Similarly, the various ingredients of the financial budget are often prepared simultaneously. In addition to dollars, managers budget physical resource needs such as personnel, machines, and space. The master budget, however, describes the financial impact of all the organization's other budgets and plans.

The following discussion focuses on the master budget itself, but note that key top-management decisions regarding pricing, product lines, production scheduling, capital expenditures, research and development, management assignments, and so on affect the master budget. Developing a budget is an iterative process; each draft of the budget almost always leads to consideration of choices and decisions that prompt further revisions before a final budget is chosen.

Basic Data and Requirements

Halifax Engineering is a machine shop that uses skilled labor and metal alloys to manufacture two types of aircraft replacement parts—regular (R) and heavy-duty (HD). Its managers are ready to prepare a master budget for the year 19_5. To keep our illustration manageable for clarifying basic relationships, we make the following assumptions:

1. Work in process inventories are negligible and are ignored.
2. Direct materials inventory and finished goods inventory are costed using the first-in-first-out (FIFO) method.
3. Unit costs of direct materials purchased and finished goods sold remain unchanged throughout the budget year (19_5).

Schedule 3A : Direct Materials Usage Budget in Kilograms and Dollars
For the Year Ended December 31, 19_5

	Material 111 Alloy	Material 112 Alloy	Total
Kilograms of direct materials to be used for production of regular (6,000 units × 12 and 6 kilograms)	72,000	36,000	
Kilograms of direct materials to be used for production of heavy-duty (1,000 units × 12 and 8 kilograms)	12,000	8,000	
Total direct materials to be used (in kilograms)	84,000	44,000	
Kilograms of direct materials to be used from beginning inventory (assuming a FIFO cost flow assumption)	7,000	6,000	
Multiply by cost per kilogram of beginning inventory	$ 7	$ 10	
Cost of direct materials to be used from beginning inventory (a)	$ 49,000	$ 60,000	$ 109,000
Kilograms of direct materials to be used from purchases (84,000 − 7,000; 44,000 − 6,000)	77,000	38,000	
Multiply by cost per kilogram of purchased materials	$ 7	$ 10	
Cost of direct materials to be used from purchases (b)	$539,000	$380,000	$ 919,000
Total costs of direct materials to be used (a + b)	$588,000	$440,000	$1,028,000

Schedule 3B computes the budgeted direct materials purchases. The budget for direct materials purchases depends on the budgeted direct materials to be used, the beginning inventory of direct materials, and the target ending inventory of direct materials:

Purchases of direct materials	=	Usage of direct materials	+	Target ending inventory of direct materials	−	Beginning inventory of direct materials

Schedule 3B : Direct Materials Purchases Budget
For the Year Ended December 31, 19_5

	Material 111 Alloy	Material 112 Alloy	Total
Direct materials to be used in production (in kilograms) from schedule 3A	84,000	44,000	
Add target ending direct materials inventory in kilograms	8,000	2,000	
Total requirements in kilograms	92,000	46,000	
Deduct beginning direct materials inventory in kilograms	7,000	6,000	
Kilograms of direct materials to be purchased	85,000	40,000	
Multiply by cost/kg of purchased materials	$ 7	$ 10	
Direct materials purchases costs	$595,000	$400,000	$995,000

STEP 4: *DIRECT MANUFACTURING LABOR BUDGET.* These costs depend on the types of products produced, labor rates, production methods, and hiring plans. The computations of budgeted direct manufacturing labor costs appear in schedule 4.

Schedule 4 : Direct Manufacturing Labor Budget
For the Year Ended December 31, 19_5

	Output Units Produced (Schedule 2)	Direct Manufacturing Labor Hours per Unit	Total Hours	Hourly Rate	Total
Regular	6,000	4	24,000	$20	$480,000
Heavy-duty	1,000	6	6,000	$20	120,000
Total			30,000		$600,000

STEP 5: *MANUFACTURING OVERHEAD COSTS BUDGET.* The total of these costs depends on how individual overhead costs vary with the cost driver, direct manufacturing labor hours. The calculations of budgeted manufacturing overhead costs appear in schedule 5.

Schedule 5 : Manufacturing Overhead Costs Budget*
For the Year Ended December 31, 19_5

	At Budgeted Level of 30,000 Direct Manufacturing Labor Hours
Variable manufacturing overhead costs:	
Supplies	$ 90,000
Indirect manufacturing labor	200,000
Direct and indirect manufacturing labor	
fringe costs	320,000
Power	90,000
Maintenance	70,000
Variable manufacturing overhead costs	$ 770,000
Fixed manufacturing overhead costs:	
Depreciation	230,000
Property taxes	50,000
Property insurance	10,000
Supervision	100,000
Power	20,000
Maintenance	20,000
Fixed manufacturing overhead costs	430,000
Total manufacturing overhead costs	$1,200,000

*Data are from page 190.

   Halifax treats both variable and fixed manufacturing overhead as inventoriable costs. (This inventory costing method is termed *absorption costing;* see Chapter 9 for further discussion.) The manufacturing overhead will be inventoried at the rate of $40 per direct manufacturing labor hour (total manufacturing overhead, $1,200,000 ÷ 30,000 budgeted direct manufacturing labor hours).

STEP 6: *ENDING INVENTORY BUDGET.* Schedule 6A shows the calculations of the dollar amounts of target ending inventories of direct materials and finished goods. This information is required not only for the production budget and the direct materials purchases budget but also for details on the budgeted income statement and the budgeted balance sheet.

Schedule 6A : Ending Inventory Budget
December 31, 19_5

	Kilograms	Cost per Kilogram	Total	
Direct materials:				
111 Alloy	8,000*	$ 7	$56,000	
112 Alloy	2,000*	10	20,000	$ 76,000

	Units	Cost Per Unit		
Finished goods:				
Regular	1,100†	$384‡	422,400	
Heavy-duty	50†	524‡	26,200	448,600
Total ending inventory				$524,600

*Data are from page 190.

†Data are from page 190.

‡From schedule 6B below based on 19_5 costs of manufacturing finished goods because under the FIFO costing method, the units in finished goods ending inventory consist of units that are produced during 19_5.

Schedule 6B: Computation of Unit Costs of Manufacturing Finished Goods in 19_5

	Cost Per Kilogram or Hour of Input	Regular		Heavy-Duty	
		Inputs in Kilograms or Hours	Amount	Inputs in Kilograms or Hours	Amount
Direct materials 111 Alloy	$ 7	12	$ 84	12	$ 84
Direct materials 112 Alloy	10	6	60	8	80
Direct manufacturing labor	20*	4	80	6	120
Manufacturing overhead	40†	4	160	6	240
Total			$384		$524

*Data are from page 190.

†Direct manufacturing labor hours is the sole allocation base for manufacturing overhead (both variable and fixed). The manufacturing overhead rate per direct manufacturing labor hour of $40 was calculated in step 5.

STEP 7: *COST OF GOODS SOLD BUDGET.* The information gathered in schedules 3 through 6 leads to schedule 7:

Schedule 7 : Cost of Goods Sold Budget
For the Year Ended December 31, 19_5

	From Schedule		Total
Beginning finished goods inventory, January 1, 19_5	Given*		$ 64,600
Direct materials used	3A	$1,028,000	
Direct manufacturing labor	4	600,000	
Manufacturing overhead	5	1,200,000	
Cost of goods manufactured			2,828,000
Cost of goods available for sale			2,892,600
Deduct ending finished goods inventory, December 31, 19_5	6A		448,600
Cost of goods sold			$2,444,000

*Given in description of basic data and requirements (regular $38,400; heavy-duty $26,200).

Note that,

$$\begin{array}{c} \text{Cost of goods} \\ \text{sold} \end{array} = \begin{array}{c} \text{Beginning} \\ \text{finished goods} \\ \text{inventory} \end{array} + \begin{array}{c} \text{Cost of goods} \\ \text{manufactured} \end{array} - \begin{array}{c} \text{Ending} \\ \text{finished goods} \\ \text{inventory} \end{array}$$

STEP 8: *RESEARCH AND DEVELOPMENT/DESIGN COSTS BUDGET.* These costs are budgeted at lump-sum amounts judged by top management to be adequate for the research and development/design plans of Halifax Engineering.

Schedule 8: Research and Development/Design Costs Budget*
For the Year Ended December 31, 19_5

		Total
Fixed costs:		
Research and development/design salaries	$ 105,000	
Prototype and materials costs	31,000	$136,000

*Data are from page 191.

STEP 9: *MARKETING COSTS BUDGET.* Marketing costs are budgeted expenditures determined by top management as the "correct" amounts to spend on marketing activities.

Schedule 9 : Marketing Costs Budget*
For the Year Ended December 31, 19_5

		Total
Fixed costs:		
Advertising and promotion	$ 30,000	
Marketing salaries	130,000	
Travel and miscellaneous	40,000	$200,000

* Data are from page 191.

Concepts in Action

THE PAPERLESS BUDGET AT UNION PACIFIC RAILROAD[a]

How does Union Pacific Railroad, a rail transportation company with 5,000 cost centers, 1,500 cost codes, and operations extending over 22,000 far-flung route miles do its budgeting? The answer: without any paper. Instead, Union Pacific gets all managers who incur costs or control costs to prepare their budgets using an on-line, computerized budgeting system

The budgeting system is "turned-on" for three weeks for annual and quarterly budget preparation. During this period, responsibility-center managers input revenue and cost budgets into the system. The system is then turned off (managers can no longer input data), and all the individual budgets are combined. Analyses, adjustments, and negotiations follow. The system is then turned back on for a week to finalize the budget.

[a]Draws on Union-Pacific, Harvard Business School case 9-186-176, and discussions with Union Pacific Railroad management.

STEP 10: *DISTRIBUTION COSTS BUDGET.* The distribution budget is deter-mined by management on the basis of product distribution decisions and warehous-ing needs.

Schedule 10 : Distribution Costs Budget*
For the Year Ended December 31, 19_5

		Total
Fixed costs:		
Distribution salaries and wages	$ 60,000	
Rent, power, maintenance, and taxes	40,000	$100,000

*Data are from page 191.

STEP 11: *CUSTOMER-SERVICE COSTS BUDGET.* These costs reflect top man-agement's lump-sum budget allocations for customer-service activities.

Schedule 11 : Customer-Service Costs Budget*
For the Year Ended December 31, 19_5

		Total
Fixed costs:		
Customer-service salaries	$ 40,000	
Travel and miscellaneous	20,000	$60,000

*Data are from page 191.

STEP 12: *ADMINISTRATION COSTS BUDGET.* These costs are amounts identi-fied by top management to support administration activities during 19_5.

Schedule 12 : Administration Costs Budget*
For the Year Ended December 31, 19_5

		Total
Fixed costs:		
Officers' salaries	$ 165,000	
Clerical wages	80,000	
Office costs and supplies	75,000	
Miscellaneous	54,000	$374,000

*Data are from page 191.

STEP 13: *BUDGETED INCOME STATEMENT.* Schedules 1, 7, 8, 9, 10, 11, and 12 provide the necessary information to complete the budgeted income statement, shown in Exhibit 6-4. Of course, more details could be included in the income state-ment, and then fewer supporting schedules would be needed.

Top management's strategies for achieving sales and operating income goals in-fluence the costs planned for the different elements of the value chain. As strategies change, the budget allocations for different areas of the value chain will also change. For example, a shift in strategy toward emphasizing product development and cus-tomer service will result in increased resources being allocated to those areas in the master budget. The actual data resulting from this strategy will be compared with budgeted results. Management can then evaluate whether the focus on product de-velopment and customer service has been successful. This feedback forms the basis for subsequent plans.

EXHIBIT 6-4

Budgeted Income Statement for Halifax Engineering for the Year Ended December 31, 19_5

Sales	Schedule 1		$3,800,000
Cost of goods sold	Schedule 7		2,444,000
Gross margin			1,356,000
Operating costs			
Research and development/Design costs	Schedule 8	$136,000	
Marketing costs	Schedule 9	200,000	
Distribution costs	Schedule 10	100,000	
Customer-service costs	Schedule 11	60,000	
Administration costs	Schedule 12	374,000	
Operating costs			870,000
Operating income			$ 486,000

SALES FORECASTING—A DIFFICULT TASK

Factors in Sales Forecasting

The term *sales forecast* is distinguished from *sales budget* as follows: The forecast is the estimate—the prediction—that may or may not become the sales budget. The forecast often leads to adjustments of management plans, so the final sales budget may differ from the original sales forecast.

Many factors influence the sales forecast, including the following:

◆ Past sales volume
◆ General economic and industry conditions
◆ Relationship of sales to economic indicators such as gross domestic product, personal income, employment, prices, and industrial production
◆ Relative product profitability
◆ Market research studies
◆ Pricing policies

◆ Advertising and other promotion
◆ Quality of sales force
◆ Competition
◆ Seasonal variations
◆ Production capacity
◆ Long-run sales trends for various products
◆ Regulatory policies and restrictions

Forecasting Procedures

Budgets based on ill-conceived sales projections have little value for planning, coordination, and performance evaluation. Because budgets can be effective only if the information that underlies them is accurate, the best available information should be used to shape the sales forecast. An effective aid to accurate forecasting is to approach the task by several different methods. Each forecast acts as a check on the others. The three methods described below are usually combined in some fashion that is suitable for a specific company.

INFORMATION FROM SALES STAFF.   As is the case for all budgets, those responsible for carrying out the budget should have an active role in budget formulation. Sales personnel work closely with customers and have a more detailed understanding of market potential, customer needs, and competitors' products than top management has. The sales staff's superior expertise and information can yield more accurate and realistic sales forecasts. Better sales forecasting leads to more reliable production plans and better resource-acquisition decisions. The result is improved planning and coordination.

One problem in obtaining accurate sales forecasts occurs when the firm uses differences between actual sales and budgeted amounts to evaluate salespersons' performance. Salespersons may then respond by underreporting achievable sales. **Padding** the budget or introducing **budgetary slack** refers to the practice of underestimating budgeted revenues (or overestimating budgeted costs) in order to make budgeted targets more easily achievable. With budgetary slack it is more likely that actual sales will exceed budgeted amounts and make salespersons' performance look good. From the sales staff's standpoint, budgetary slack hedges against unexpected adverse circumstances.

The key point is that the manner in which a firm evaluates the performance of its sales staff will affect the sales staff's reporting incentives. Some companies, such as IBM and Kodak, have designed innovative performance evaluation methods that create incentives for sales staff to report what they know while maintaining incentives for the sales staff to meet and exceed budgeted targets.[6]

STATISTICAL APPROACHES.   Regression analysis and trend analysis are useful techniques for sales forecasting. Correlations between sales and economic indicators help make sales forecasts more reliable, especially if fluctuations in certain economic indicators precede fluctuations in company sales. However, no organization should depend entirely on this approach. The relationship between sales and specific economic indicators may change over time. Trend analysis examines past sales data to forecast future sales. Seasonal and month-to-month patterns in past sales are used to predict future sales. Too much reliance on statistical analysis is dangerous, however, because chance variations in statistical data may completely distort a prediction. Statistical analysis can provide help but not flawless answers.

GROUP EXECUTIVE JUDGMENT.   All top officers, including those from product development, purchasing, manufacturing, marketing, distribution, finance, and administration, may use their collective experience and knowledge to project sales on the basis of group opinion. Often, the group will obtain and analyze market research data before forecasting sales.

Surveys of firms in many countries show consistency among methods used for sales forecasting. Judgment of marketing staff and estimates based on last year's sales are the most popular methods. Statistical techniques are least used.[7]

ACTIVITY-BASED BUDGETING AND *KAIZEN* BUDGETING

Activity-Based Budgets

Objective 4
Define *activity-based budgeting* and *kaizen budgeting*

In Chapters 4 and 5 we looked at how activity-based costing can lead to improved decision making because it provides more detailed information that more accurately describes operations than information used in traditional costing systems. The fundamental advantage of activity-based costing is that a business can identify multiple cost drivers. Activity-based costing principles extend to budgeting. **Activity-based budgeting** is an approach to budgeting that focuses on the costs of activities necessary to produce and sell products and services. Activity-based budgeting is especially valuable in the case of indirect costs. *Activity-based budgeting* partitions indirect costs into separate homogeneous activity cost pools. Management uses the cause-and-effect criterion to identify the cost drivers for the separate cost pools.

---

[6]See also S. Reichelstein, "Constructing Incentive Schemes for Government Contracts: An Application of Agency Theory," *The Accounting Review* (October, 1992).

[7]See, for example, E. de With and E. Ijskes, "Current Budgeting Practices in Dutch Companies," Working Paper, University of Amsterdam, 1992; and P. Clarke, "Management Accounting Practices and Techniques in Irish Manufacturing Firms," Working Paper, University College, Dublin, 1992.

Four key steps in activity-based budgeting are:

1. To determine the budgeted cost of performing each unit of activity at each activity area.
2. To determine the demand for each individual activity on the basis of sales and production targets.
3. To compute the budgeted cost of performing each activity.
4. To describe the budget as costs of performing various activities (rather than budgeted costs of functional or spending categories).

Kaizen Budgeting

Kaizen is the Japanese term for continuous improvement. Customer satisfaction demands continuous improvements in internal work processes. ***Kaizen* budgeting** is a budgeting approach that projects costs on the basis of future improvements rather than current practices and methods. Under *kaizen* budgeting, existing practices are analyzed to identify potential product and process improvements. These improvements can come, for example, from changes in operating procedures, improved work processes, and faster setup times. The budget staff estimates the financial impact of the changes and also determines the cost of implementing them. The costs included in the master budget and activity-based budget are then based on process improvements that are *yet to be implemented*. The key point is that the budget cannot be achieved unless improvements are made.

Suggestions for improvements can come from employees in all parts of the organization. Successful organizations create environments that encourage all employees to make suggestions. Employees of Toyota Corporation, for example, offered 2 million suggestions in 1991 (35 per employee), and 97% were adopted.[8]

COMPUTER-BASED FINANCIAL PLANNING MODELS

Objective 5
Describe the uses of computer-based financial planning models

A master budget can be a comprehensive planning model for the organization. As the budget is formulated, it is frequently altered as executives exchange views on various aspects of expected activities and ask "what-if" questions. A "what-if" (sensitivity) analysis examines the impact on the master budget if conditions differ from the master budget assumptions. Departures from budgeted assumptions can occur, for example, because of changes in the competitive environment, general economic conditions, and availability of direct materials. Sensitivity analysis enables managers to explore the effects of uncertainty on their businesses and to plan for them.

Computer-based **financial planning models** often use the master budget as their structural base. Financial planning models are mathematical statements of the relationships among all operating activities, financial activities, and the other major internal and external factors that may affect decisions. The models are used for budgeting, for revising budgets, for conducting "what-if" analysis, and for comparing a variety of decision alternatives.

Consider Halifax Engineering. Their financial planning model assumes:

◆ Direct materials and direct manufacturing labor costs vary proportionately with the quantities of heavy-duty and regular parts produced.
◆ Variable manufacturing overhead varies with direct manufacturing labor hours.
◆ All other costs are fixed.
◆ Target ending inventories remain unchanged.

---

[8]See T. Tanaka, "*Kaizen* Budgeting: Toyota's Cost Control System under Total Quality Control," Working Paper, Tokyo Keizai University, April, 1992.

EXHIBIT 6-5

Effect of Changes in Budget Assumptions on Budgeted Income for Halifax Engineering

What-If Scenario	Units Sold		Selling Price		Direct Materials Costs per Kilogram		Budgeted Operating Income	
	Regular	Heavy-Duty	Regular	Heavy-Duty	111 Alloy	112 Alloy	Amount	Percent Change from Master Budget
Master budget	5,000	1,000	$600	$800	$7.00	$10.00	$486,000	—
Scenario 1	5,000	1,000	582	776	7.00	10.00	372,000	23% decrease
Scenario 2	4,800	960	600	800	7.00	10.00	419,273	14% decrease
Scenario 3	5,000	1,000	600	800	7.35	10.50	448,380	9% decrease

Exhibit 6-5 presents the budgeted operating income (without supporting details) for three "what-if" scenarios for Halifax Engineering:

◆ Scenario 1: 3% decrease in the selling price of the regular part and 3% decrease in the selling price of the heavy-duty part.

◆ Scenario 2: 4% decrease in units sold of the regular part and 4% decrease in units sold of the heavy-duty part.

◆ Scenario 3: 5% increase in the price per kilogram of 111 alloy and 5% increase in the price per kilogram of 112 alloy.

Exhibit 6-5 indicates that, relative to the master budget, budgeted operating income decreases by 23% under scenario 1, by 14% under scenario 2, and by 9% under scenario 3. Managers can use this information to plan actions that they may need to take to improve the situation if confronted with one of these scenarios.

Concepts in Action

KAIZEN BUDGETING AT TOYOTA MOTOR COMPANY[a]

The *kaizen* budgeting process at Toyota starts when top management determines a *target* operating income for the coming year and calculates an *estimated* operating income based on the company's existing revenue and cost structures. The target operating income is generally more than the estimated operating income. Toyota refers to the difference between the two as "kaizen value." Under the Toyota system, roughly half the *kaizen* value must come from higher sales and the other half from reductions in costs. Toyota applies *kaizen* budgeting only to variable costs.

Toyota management determines the *kaizen* value that each division (for example, transmission, engine, and chassis) must obtain. Division managers and the plant managers who report to them then submit detailed *kaizen* activity plans for attaining these goals. Toyota regards *kaizen* budgeting as central to successful "costing down," which is reducing the expected cost of making and selling cars.

[a]Draws on T. Tanaka "*Kaizen* Budgeting: Toyota's Cost-Control System under Total Quality Control," Working Paper, Tokyo Keizai University, April, 1992, and discussions with M. Sakurai and T. Tanaka.

RESPONSIBILITY ACCOUNTING

Objective 6
Describe responsibility centers and responsibility accounting

Organizational Structure

To attain the goals described in the master budget, an organization must coordinate the efforts of all its employees—from the top executive through all levels of management to every supervised worker. Coordinating the organization's efforts means assigning responsibility to managers who are accountable for their actions in planning and controlling human and physical resources. Management is essentially a human activity. Budgets exist not for their own sake, but to help managers.

Organizational structure is an arrangement of lines of responsibility within the entity. A company such as Shell Canada, a petroleum company, may be organized primarily by business function: exploration, refining, and marketing. Another company, such as Procter & Gamble, a household products giant, may be organized by product line. If so, the managers of the individual divisions (toothpaste, soap, and so on) would each have decision-making authority concerning all the functions (manufacturing, marketing, and so on) within that division.

Definition of Responsibility Accounting

Each manager, regardless of level, is in charge of a *responsibility center*. A **responsibility center** is a part, segment, or subunit of an organization, whose manager is accountable for a specified set of activities. The higher the manager's level, the broader the responsibility center he or she manages, and generally, the larger the number of subordinates who report to him or her. **Responsibility accounting** is a system that measures the plans (by budgets) and actions (by actual results) of each responsibility center. Four major types of responsibility centers are:

1. **Cost center**—manager accountable for costs only
2. **Revenue center**—manager accountable for revenues only
3. **Profit center**—manager accountable for revenues and costs
4. **Investment center**—manager accountable for investments, revenues, and costs

The maintenance department of a Marriott hotel would be a cost center because the maintenance manager is responsible only for costs. The sales department of the hotel would be a revenue center because the sales manager is responsible only for revenues. The hotel manager would be in charge of a profit center because the hotel manager is accountable for both revenues and costs. The regional manager responsible for investments in new hotel projects and for revenues and costs would be in charge of an investment center.

Responsibility accounting affects behavior. For example, consider the following incident:

> The sales department requests a rush production run. The plant scheduler argues that it will disrupt his production and cost a substantial though not clearly determined amount of money. The answer coming from sales is: "Do you want to take the responsibility of losing the X Company as a customer?" Of course the production scheduler does not want to take such a responsibility, and he gives up, but not before a heavy exchange of arguments and the accumulation of a substantial backlog of ill feeling. The controller proposes an innovative solution. He analyzes the payroll in the assembly department to determine the costs involved in getting out rush orders. This information eliminates the cause for argument. Henceforth, any rush order would be accepted by the production scheduler, "no questions asked." The extra cost would be duly recorded and charged to the sales department.
>
> As a result, the tension created by rush orders disappeared completely, and, somehow, the number of rush orders requested by the sales department was progressively reduced to an insignificant level.[9]

The responsibility accounting approach traces costs to either (1) the individual who has the best knowledge about why the costs arose or (2) the activity that caused the costs. In this incident, the cause was the sales activity, and the resulting costs were charged to the sales department. If rush orders occur regularly, the sales department might have a budget for such costs, and the department's actual performance would then be compared against the budget.

Illustration of Responsibility Accounting

The simplified organization chart of Pizza Hut in Exhibit 6-6 illustrates how companies may use responsibility accounting in a service industry, the fast-food industry. At the top level, a regional manager oversees the area managers, who supervise the managers of the individual units. Unit managers have limited freedom to make operating decisions. They may decide on how to handle local advertising, the number of employees and their schedules, and the hours that the unit is open. Area managers oversee several units, evaluate unit managers' performance, and set unit managers' compensation levels. In turn, regional managers oversee several areas, evaluate area managers' performance and compensation, and decide on regional prices and sales promotions. Regional managers are accountable to state managers, who answer to home-office vice-presidents.

Exhibit 6-7 provides a more detailed view of how responsibility accounting is used to evaluate profit centers at Pizza Hut. Examine the lowest level first (at the bottom of the exhibit) and then move upward. Follow how the reports are related through the three levels of responsibility.

[9]R. Villers, "Control and Freedom in a Decentralized Company," *Harvard Business Review*, 32, no. 2, p. 95.

EXHIBIT 6-6

Simplified Partial Organizational Chart for Pizza Hut

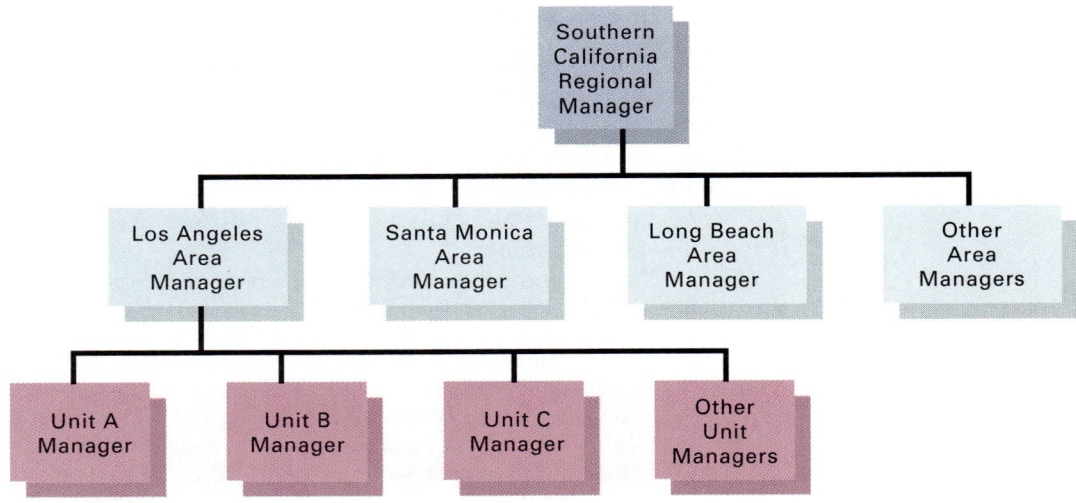

Trace the $3,000 operating income from the Unit B manager's report to the Los Angeles area manager's report. The area manager's report summarizes the actual results of the stores under his supervision. In addition, charges incurred by the area manager's office are included in this report.

Trace the $58,000 total from the Los Angeles area manager's report to the southern California regional manager's report. The report of the regional manager includes data for her own regional office plus a summary of the entire region's operating income performance. All variances, whatever the level, may be subdivided for further analysis, either in these reports or in supporting schedules. (A variance is the difference between an actual result and a budgeted amount, when that budgeted amount is found in an accounting system. Chapters 7, 8, and 22 present further discussion of variances.)

Performance Report Format

Exhibit 6-7 stresses variances from the budget. This focus is a highlight of *management by exception,* the practice of concentrating on areas that deserve attention and placing less attention on areas that are operating as expected. For example, the Los Angeles area's operating income lagged behind budgeted operating income, while the operating incomes of other areas exceeded their budgets during the current month and for the year to date. The regional manager would concentrate her efforts on improving the Los Angeles area's operation. Managers spend less time investigating smoothly running operations.

A complete performance report would likely include line-by-line presentations of other strategic data, some of which are nonfinancial. For example, a report for a restaurant will show the number of customers served and the average sales per customer. In the hotel industry, performance measures include the percentage of rooms occupied and the average daily rate per room. The performance measures used are key success factors in their respective industries. Informative performance reports, in many cases, include accounting information plus key strategic factors.

Southern California Regional Manager — Monthly Responsibility Report

Operating income of areas and regional manager's office costs:	Budget — This Month	Budget — Year to Date	Variance from Budget Favorable (Unfavorable) — This Month	Variance from Budget Favorable (Unfavorable) — Year to Date
Regional manager's office costs	$ (35)	$ (150)	$ (4)	$ (12)
Los Angeles area	58	224	(5)	(4)
Santa Monica area	38	158	9	15
Long Beach area	36	152	3	13
Other areas	97	377	7	30
Operating income	$194	$ 761	$10	$ 42

Los Angeles Area Manager — Monthly Responsibility Report

Operating income of stores and area manager's office costs:	Budget — This Month	Budget — Year to Date	Variance from Budget Favorable (Unfavorable) — This Month	Variance from Budget Favorable (Unfavorable) — Year to Date
Area manager's office costs	$ (6)	$ (26)	$(1)	$(2)
Unit A	5	21	(1)	(1)
Unit B	3	14	1	3
Unit C	4	17	—	—
Other units	52	198	(4)	(4)
Operating income	$58	$ 224	$(5)	$(4)

Unit B Manager — Monthly Responsibility Report

Sales and costs:	Budget — This Month	Budget — Year to Date	Variance from Budget Favorable (Unfavorable) — This Month	Variance from Budget Favorable (Unfavorable) — Year to Date
Sales	$26	$110	$3	$6
Food costs	7	32	(1)	(2)
Supplies	1	4	—	—
Wages	8	33	—	(2)
Utilities, depreciation, and maintenance	4	17	(1)	(1)
Advertising	2	6	—	2
Delivery	1	4	—	—
Total costs	23	96	(2)	(3)
Operating income	$ 3	$ 14	$1	$3

Feedback and Fixing Blame

Budgets coupled with responsibility accounting provide systematic help for managers, particularly if managers interpret the feedback carefully. Managers, accountants, and students of cost accounting repeatedly tend to "play the blame game"—using variances appearing in the responsibility accounting system to pinpoint fault for operating problems. In looking at revenues, costs or variances, managers should determine whom they should talk to in that specific situation to obtain information and seek solutions—not whom they should blame. Variances only suggest questions or direct attention to persons who should have the relevant information. Nevertheless, variances, properly used, can be helpful in evaluating managers' performance.

RESPONSIBILITY AND CONTROLLABILITY

Objective 7
Explain how controllability is related to responsibility accounting

Definition of Controllability

Controllability is the degree of influence that a specific manager has over costs, revenues, or other items in question. A **controllable cost** is any cost that is primarily subject to the influence of a given *manager* of a given *responsibility center* for a given *time span*. Ideally, responsibility accounting systems either exclude all uncontrollable costs from a manager's performance report or else segregate such costs from the controllable costs. For example, a machining supervisor's performance report might be confined to quantities (not prices) of direct materials, direct manufacturing labor, power, and supplies.

In practice, controllability is difficult to pinpoint:

1. Few costs are clearly under the sole influence of one manager. For example, prices of direct materials may be influenced by a purchasing manager, but *prices* also depend on market conditions beyond the manager's control. *Quantities* used may be influenced by a production manager, but quantities used also depend on the quality of materials purchased. Moreover, managers often work in groups or teams. How can individual responsibility be evaluated in a group decision?

2. With a long enough time span, all costs will come under somebody's control. However, most performance reports focus on periods of a year or less. A current manager may have inherited problems and inefficiencies from his or her predecessor. For example, the present manager may have to work under undesirable contracts with suppliers or labor unions that were negotiated before he or she became manager. How can we separate what the current manager actually controls from the results of decisions made by others? Exactly what is the current manager accountable for? In practice, answers to such questions may not be clear-cut.

Research on controllability has led to the following observation:

> When profit center managers are held accountable for individual items which they do not fully control, the managers are taking on risk; their rewards depend on factors they cannot fully influence. The firm must compensate managers for taking this risk. If the firm fails to compensate managers for the additional risk, the firm will bear the costs of frustration, lower motivation, and possibly management turnover. [10]

Emphasis on Information and Behavior

Avoid overemphasizing controllability. Responsibility accounting is more far-reaching. It focuses on *information and knowledge,* not on control. The key question is, Who is the best informed? Put another way, Who is the person who can tell us the most

---

[10]K. Merchant, *Rewarding Results* (Boston: Harvard Business School Press, 1989), p. 5.

about the specific item in question, regardless of that person's ability to exert personal control? For instance, purchasing managers may be held accountable for total purchase costs, not because of their ability to affect market prices, but because of their ability to predict uncontrollable prices and explain uncontrollable price changes. Similarly, unit managers at Pizza Hut may be held responsible for operating income of the unit, even though they do not fully control sales and costs. Why? Because unit managers are in the best position to explain differences between their actual operating income and their budgeted operating income.

Performance reports for responsibility centers also may include uncontrollable items because such inclusion could change behavior in the directions top management desires. For example, some companies have changed the accountability of a cost center to a profit center. Why? Because the manager will probably behave differently. A cost-center manager may emphasize production efficiency and deemphasize the pleas of sales personnel for faster service and rush orders. In a profit center, the manager is responsible for both costs and revenues. Thus, even though the manager still has no control over sales personnel, the manager will now more likely weigh the impact of his or her decisions on costs and revenues, rather than solely on costs.

Importance of Budget and Management by Objectives

Many organizations have successfully used some form of **management by objectives (MBO).** Under this procedure, a subordinate and his or her manager jointly formulate the subordinate's set of objectives and the plans for attaining those objectives for a subsequent period. Subordinates are not asked to maximize a single measure such as operating income. Instead, they are asked to attain a *budgeted* operating income and simultaneously achieve additional objectives, such as a targeted share of a market. Focusing on a single measure such as operating income, sales, costs of materials, or some other number tends to lead people to neglect other important aspects of their jobs.

The plans are frequently crystallized in responsibility accounting and a budget. In addition, various subgoals may be measured, such as levels of product innovation, quality, market share, and management training. The subordinate's performance is then evaluated in comparison to these agreed-upon budgeted goals.

MBO is also often used by nonprofit organizations. For example, in the late 1980s the Episcopal Diocese of Newark, New Jersey, started paying ministers according to their performance. Under the merit-pay plan, ministers can qualify for salary increases based on such goals as parish growth, education, and quality of sermons.

Responsibility and Uncertainty

Managers live in an uncertain world. Actions and events outside and inside the organization affect the measures of their performance. In evaluation of managers, distinctions between controllable and uncontrollable items should be made whenever feasible. However, financial responsibility may be assigned to managers even though controllability may be minimal. The final test is what information and what behavior will be generated by the measurement system. Put another way:

1. What costs do you, as a top manager, want the responsibility-center managers to control? Assign those costs to the manager. Why? Because only then will the responsibility-center manager pay attention to those costs.
2. How much attention do you want responsibility-center managers to pay to the assigned costs? The more you want your managers to focus on these costs, the greater the control managers should be able to exercise over them.

HUMAN ASPECTS OF BUDGETING

Why did we cover the two major topics, master budgets and responsibility accounting, in a single chapter? Primarily to emphasize that human factors are crucial parts of budgeting. Too often, students study budgeting as though it were a mechanical tool.

The budgeting techniques themselves are free of emotion; however, their administration requires education, persuasion, and intelligent interpretation. Many managers regard budgets negatively. To them, the word *budget* is about as popular as, say, *strike* or *layoff*. Top managers must convince their subordinates that the budget is a positive tool designed to help them choose and reach goals. But budgets are not cure-alls. They are not remedies for weak management talent, faulty organization, or a poor accounting system.

PROBLEM FOR SELF-STUDY

Before trying to solve the homework problems, review the illustration of the master budget, p. 188.

PROBLEM

Prepare a budgeted income statement, including all necessary detailed supporting schedules. Use the data given in the illustration of the master budget to prepare your own budget schedules. (See the "Basic Data and Requirements" section on pp. 188, 190, 191.)

SUMMARY

The following points are linked to the chapter's learning objectives.

1. The master budget summarizes the financial projections of all the oganization's budgets and plans. It expresses management's comprehensive operating and financial plan—the formalized outline of the organization's financial objectives and their means of attainment. When administered wisely, budgets compel management planning, provide definite expectations that are an appropriate framework for judging subsequent performance, and promote communication and coordination among the various subunits of the organization.

2. Two major parts of the master budget are the operating budget, which is the budgeted income statement and its supporting budgets, and the financial budget, which comprises the capital budget, cash budget, budgeted balance sheet, and budgeted statement of cash flows.

3. The foundation for the operating budget is generally the sales budget. The following supporting schedules are geared to the sales budget: production budget in units, direct materials usage budget, direct materials purchases budget, direct manufacturing labor budget, manufacturing overhead costs budget, ending inventory budget, cost of goods sold budget, research and development/design costs budget, marketing costs budget, distribution costs budget, customer-service costs budget, and administration costs budget. The operating budget ends with the budgeted income statement.

4. Activity-based budgeting is an approach to budgeting that focuses on the costs of activities necessary to produce and sell products and services. In *kaizen* budgeting, costs are based on future improvements that are yet to be implemented rather than on current practices or methods.

5. Computer-based financial planning models are mathematical statements of the relationships among operating activities, financial activities, and other factors that affect the budget. These models allow management to conduct "what-if" (sensitivity) analyses of the effects on the master budget of changes in the original predicted data or changes in budget assumptions.

6. A responsibility center is a part, segment, or subunit of an organization, whose manager is accountable for a specified set of activities. Four major types of responsibility centers are cost centers, revenue centers, profit centers, and investment centers. Responsibility accounting systems measure the plans (by budgets) and actions (by actual results) of each responsibility center.

7. Controllable costs are costs that are primarily subject to the influence of a given manager of a given responsibility center for a given time span. Performance reports of responsibility-center managers, however, often include costs, revenues, and investments that the managers do not control. Responsibility accounting associates financial items with managers on the basis of which manager has the most knowledge and information about the specific items, regardless of the manager's ability to exercise full control. The important question is who should be asked, *not* who should be blamed.

APPENDIX: THE CASH BUDGET

The major illustration in the chapter features the operating budget. The other major part of the master budget is the financial budget, which includes the capital budget, cash budget, budgeted balance sheet, and budgeted statement of cash flows. This Appendix focuses on the cash budget and the budgeted balance sheet. Capital budgeting is covered in Chapters 20 and 21; coverage of the budgeted statement of cash flows is beyond the scope of this book.

Suppose Halifax Engineering in our chapter illustration had the following balance sheet for the year ended December 31, 19_4:

HALIFAX ENGINEERING
Balance Sheet,
December 31, 19_4

Assets

Current assets:		
Cash	$ 30,000	
Accounts receivable	400,000	
Direct materials	109,000	
Finished goods	64,600	$ 603,600
Property, plant, and equipment:		
Land	200,000	
Building and equipment	2,200,000	
Accumulated depreciation	(690,000)	1,710,000
Total		$2,313,600

Liabilities and Stockholders' Equity

Current liabilities:		
Accounts payable	$ 150,000	
Income taxes payable	50,000	$ 200,000
Stockholders' equity:		
Common stock, no-par,		
25,000 shares outstanding	350,000	
Retained earnings	1,763,600	2,113,600
Total		$2,313,600

Budgeted cash flows for 19_5 are:

| | Quarter | | | |
	1	2	3	4
Collections from customers	$913,700	$984,600	$976,500	$918,400
Disbursements:				
For direct materials	314,360	283,700	227,880	213,800
For payroll	557,520	432,080	409,680	400,720
For income taxes	50,000	46,986	46,986	46,986
For other costs	184,000	156,000	151,000	149,000
For machinery purchase	—	—	—	35,080

The quarterly data are based on the cash effects of the operations formulated in schedules 1 through 12 in the chapter, but the details of that formulation are not shown here in order to keep the illustration relatively brief and focused.

The company wants to maintain a $35,000 minimum cash balance at the end of each quarter. The company can borrow or repay money in multiples of $1,000 at an interest rate of 12% per year. Management does not want to borrow any more cash than is necessary and wants to repay as promptly as possible. By special arrangement, interest is computed and paid when the principal is repaid. Assume that borrowings take place at the beginning and repayments occur at the end of the quarters in question. Interest is computed to the nearest dollar.

An accountant at Halifax Engineering was given the above data and the other data contained in the budgets in the chapter (pp. 191–98). He was instructed as follows:

1. Prepare a cash budget. That is, prepare a statement of cash receipts and disbursements by quarters, including details of borrowing, repayment, and interest expense.
2. Prepare a budgeted balance sheet.
3. Prepare a budgeted income statement, including the effects of interest expense and income taxes. Assume that income taxes for 19_5 (at a tax rate of 40%) are $187,944.

Preparation of Budgets

1. The **cash budget** (Exhibit 6-8) is a schedule of expected cash receipts and disbursements. It predicts the effects on the cash position at the given level of operations. Exhibit 6-8 presents the cash budget by quarters to show the impact of cash-flow timing on bank loans and their repayment. In practice, monthly—and sometimes weekly—cash budgets are very helpful for cash planning and control. Cash budgets help avoid unnecessary idle cash and unexpected cash deficiencies. Cash balances are kept in line with needs. Ordinarily, the cash budget has the following main sections:

a. The beginning cash balance plus cash receipts equals the total cash available before financing. Cash receipts depend on collections of accounts receivable, cash sales, and miscellaneous recurring sources such as rental or royalty receipts. Information on the prospective collectibility of accounts receivable is needed for accurate predictions. Key factors include bad-debt (uncollectible account) experience and average time lag between sales and collections.

b. Cash disbursements:
   (i) Direct material purchases—depends on credit terms extended by suppliers and bill-paying patterns of the buyer.
   (ii) Direct labor and other wage and salary outlays—depends on payroll dates.
   (iii) Other costs—depends on timing and credit terms. *Note that depreciation does not require a cash outlay.*

(iv) Other disbursements—outlays for property, plant, and equipment and long-term investments.

c. Financing requirements depend on how the total cash available for needs, keyed as (a) in Exhibit 6-8, compares with the total cash needed, keyed as (c). Total cash needed includes total disbursements, keyed as (b), plus the minimum ending cash balance desired. The financing plans will depend on the relationship between total cash available for needs and total cash needed. If there is excess cash, loans may be repaid or temporary investments made. The outlays for interest expense are usually shown in this section of the cash budget.

d. The ending cash balance. The total effect of the financing decisions on the cash budget, keyed as (d) in Exhibit 6-8, may be positive (borrowing) or negative (repayment), and the ending cash balance is (a) − (b) + (d).

The cash budget in Exhibit 6-8 shows the pattern of short-term "self-liquidating cash loans." Seasonal peaks of production or sales often result in heavy cash disbursements for purchases, payroll, and other operating outlays as the products are produced and sold. Cash receipts from customers typically lag behind sales. The loan is *self-liquidating* in the sense that the borrowed money is used to acquire resources that are combined for sale, and the proceeds from sales are used to repay the loan. This **self-liquidating cycle**—sometimes called the **working-capital, cash, or operating cycle**—is the movement from cash to inventories to receivables and back to cash.

EXHIBIT 6-8

Cash Budget for Halifax Engineering
for the Year Ended December 31, 19_5

| | Quarter | | | | Year as a Whole |
	1	2	3	4	
Cash balance, beginning	$ 30,000	$ 35,820	$ 35,934	$ 35,188	$ 30,000
Add receipts:					
Collections from customers	913,700	984,600	976,500	918,400	3,793,200
(a) Total cash available for needs	943,700	1,020,420	1,012,434	953,588	3,823,200
Deduct disbursements:					
For direct materials	314,360	283,700	227,880	213,800	1,039,740
For payroll	557,520	432,080	409,680	400,720	1,800,000
For income taxes	50,000	46,986	46,986	46,986	190,958
For other costs	184,000	156,000	151,000	149,000	640,000
For machinery purchase	0	0	0	35,080	35,080
(b) Total disbursements	1,105,880	918,766	835,546	845,586	3,705,778
Minimum cash balance desired	35,000	35,000	35,000	35,000	35,000
(c) Total cash needed	1,140,880	953,766	870,546	880,586	3,740,778
Cash excess (deficiency) (a − c)*	$ (197,180)	$ 66,654	$ 141,888	$ 73,002	$ 82,422
Financing:					
Borrowing (at beginning)	$ 198,000	$ 0	$ 0	$ 0	$ 198,000
Repayment (at end)	—	(62,000)	(130,000)	(6,000)	(198,000)
Interest (at 12% per year)†	—	(3,720)	(11,700)	(720)	(16,140)
(d) Total effects of financing	$ 198,000	$ (65,720)	$ (141,700)	$ (6,720)	$ (16,140)
Cash balance, ending (a − b + d)	$ 35,820	$ 35,934	$ 35,188	$101,282	$ 101,282

*Excess of total cash available over total cash needed before current financing.
†The interest payments pertain only to the amount of principal being repaid at the end of a given quarter. Also note that *depreciation does not require a cash outlay.* The specific computations regarding interest are: $62,000 × 0.12 × 2/4 = $3,720; $130,000 × 0.12 × 3/4 = $11,700; and $6,000 × 0.12 × 4/4 = $720.

2. The budgeted balance sheet is in Exhibit 6-9. Each item is projected in light of the details of the business plan as expressed in the previous schedules. For example, the ending balance of Accounts Receivable of $406,800 is computed by adding the budgeted sales of $3,800,000 (from schedule 1) to the beginning balance of $400,000 (given) and subtracting cash receipts of $3,793,200 (given and in Exhibit 6-8).

3. The budgeted income statement is in Exhibit 6-10. It is merely the budgeted operating income statement in Exhibit 6-4, page 198, expanded to include interest expense and income taxes.

For simplicity, the cash receipts and disbursements were given explicitly in this illustration. Frequently, there are lags between the items reported on the accrual basis in an income statement and their related cash receipts and disbursements. In the Halifax Engineering example, collections from customers are derived under the

EXHIBIT 6-9
Halifax Engineering: Budgeted Balance Sheet
December 31, 19_5

Assets			
Current assets:			
Cash (from Exhibit 6-8)		$ 101,282	
Accounts receivable (1)		406,800	
Direct materials (2)		76,000	
Finished goods (2)		448,600	$1,032,682
Property, plant, and equipment:			
Land (3)		200,000	
Building and equipment (4)	$2,235,080		
Accumulated depreciation (5)	(920,000)	1,315,080	1,515,080
Total			$2,547,762
Liabilities and Stockholders' Equity			
Current liabilities:			
Accounts payable (6)		$ 105,260	
Income taxes payable (7)		46,986	$ 152,246
Stockholders' equity:			
Common stock, no-par, 25,000 shares			
outstanding (8)		350,000	
Retained earnings (9)		2,045,516	2,395,516
Total			$2,547,762

Notes:
Beginning balances are used as the starting point for most of the following computations:
(1) $400,000 + $3,800,000 sales – $3,793,200 receipts = $406,800.
(2) From schedule 6A, page 195.
(3) From beginning balance sheet, page 211.
(4) $2,200,000 + $35,080 purchases = $2,235,080.
(5) $690,000 + $230,000 depreciation from schedule 5, page 194, = $920,000.
(6) $150,000 + $995,000 (schedule 3B) – $1,039,740 (Exhibit 6-8) = $105,260.
    There are no wages payable. The detailed payroll consists of $600,000 direct manufacturing labor (schedule 4) + $620,000 manufacturing overhead salaries ($200,000 indirect manufacturing labor + $320,000 direct and indirect manufacturing labor fringe cost + $100,000 supervision from schedule 5) + research and development/design salaries $105,000 (schedule 8) + marketing salaries $130,000 (schedule 9) + distribution salaries and wages $60,000 (schedule 10) + customer-service salaries $40,000 (schedule 11) + administration salaries and clerical wages $245,000 (schedule 12) = $1,800,000, all of which was disbursed per Exhibit 6-8.
(7) $50,000 + $187,944 current year – $190,958 payment = $46,986.
(8) From beginning balance sheet.
(9) $1,763,600 + $281,916 net income per Exhibit 6-10 = $2,045,516.

EXHIBIT 6-10

Budgeted Income Statement for Halifax Engineering for the Year Ended December 31, 19_5

Sales	Schedule 1		$3,800,000
Cost of goods sold	Schedule 7		2,444,000
Gross margin			1,356,000
Operating costs			
Research and development/Design costs	Schedule 8	$136,000	
Marketing costs	Schedule 9	200,000	
Distribution costs	Schedule 10	100,000	
Customer-service costs	Schedule 11	60,000	
Administration costs	Schedule 12	374,000	
Total operating costs			870,000
Operating income			486,000
Interest expense	Exhibit 6-8		16,140
Income before income taxes			469,860
Income taxes	Given		187,944
Net income			$ 281,916

following assumptions: (1) In any month, 10% of sales are cash sales and 90% of sales are on credit and (2) half the total credit sales are collected in each of the two months subsequent to the sale, as the following table shows:

	May	June	July	August	September	Cash Collections in Third Quarter as a Whole
Monthly Sales Budget for Halifax (Given)						
Credit sales, 90%	$307,800	$307,800	$280,800	$280,800	$ 280,800	
Cash sales, 10%	34,200	34,200	31,200	31,200	31,200	
Total sales, 100%	$342,000	$342,000	$312,000	$312,000	$ 312,000	
Cash Collections from						
Cash sales this month			$ 31,200	$ 31,200	$ 31,200	
Credit sales last month			153,900*	140,400‡	140,400§	
Credit sales two months ago			153,900†	153,900*	140,400‡	
Total collections			$339,000	$325,500	$ 312,000	$976,500

* 0.50 × $307,800 (June sales) = $153,900   ‡ 0.50 × $280,800 (July sales) = $140,400
† 0.50 × $307,800 (May sales) = $153,900   § 0.50 × $280,800 (August sales) = $140,400

Of course, such schedules of cash collections depend on credit terms, collection histories, and expected bad debts. Similar schedules can be prepared for operating costs and their related cash disbursements.

Computer-based financial planning models help to examine the impact of changes in master-budget assumptions on budgeted cash flows. The computer model for Halifax Engineering was introduced on page 201. Under scenario 1, when selling prices of regular and heavy-duty aircraft parts each decreased by 3%, the computer model indicated that Halifax would need to borrow $213,000 instead of $198,000 at the beginning of the first quarter (computation not shown)

TERMS TO LEARN

The chapter and Glossary contain definitions of the following important terms:

activity-based budgeting *(p. 199)*   budget *(183)*   budgetary slack *(199)*
cash budget *(210)*   cash cycle *(211)*   controllability *(206)*   controllable cost *(206)*
cost center *(203)*   financial budget *(188)*   financial planning models *(200)*
investment center *(203)*   *kaizen* budgeting *(200)*   master budget *(183)*
management by objectives (MBO) *(207)*   operating budget *(188)*
operating cycle *(211)*   organizational structure *(202)*   padding *(199)*
pro forma statements *(187)*   profit center *(203)*   responsibility accounting *(202)*
responsibility center *(202)*   revenue center *(203)*   rolling budget *(186)*
self-liquidating cycle *(211)*   strategy *(184)*   tactics *(184)*
working-capital cycle *(211)*

ASSIGNMENT MATERIAL

QUESTIONS

6-1 Identify three major steps in the evolution of management control systems.

6-2 Define *master budget*.

6-3 What are the elements of the budgeting cycle?

6-4 "Strategy, plans, and budgets are unrelated to one another." Do you agree? Explain.

6-5 "Budgets meet the cost-benefit test. They force managers to act differently." Do you agree? Explain.

6-6 "Budgeted performance is a better criterion than past performance for judging managers." Do you agree? Explain.

6-7 "Budgets are wonderful vehicles for communication." Comment.

6-8 "Operating plans are contracts, and I want them met. If your revenue is below expectations, you should cut your costs accordingly." Do you agree? Explain.

6-9 Define *rolling budget* and *pro forma statements*.

6-10 Define *operating budget* and *financial budget*.

6-11 Distinguish between *sales forecast* and *sales budget*.

6-12 "The sales forecast is the cornerstone for budgeting." Why?

6-13 Define *activity-based budgeting*. Describe the four key steps in activity-based budgeting.

6-14 Define *kaizen budgeting*.

6-15 Define *responsibility accounting*.

EXERCISES AND PROBLEMS

6-16 Sales and production budget. The Mendez Company expects 19_5 sales of 100,000 units of serving trays. Mendez's beginning inventory for 19_5 is 7,000 trays; target ending inventory, 11,000 trays. Compute the number of trays budgeted for production in 19_5.

6-17 Sales and production budget. The Gallo Company had a target ending inventory of 70,000 four-liter bottles of burgundy wine. Gallo's beginning inventory was 60,000 bottles, and its budgeted production was 900,000 bottles. Compute the budgeted sales in number of bottles.

6-18 Direct materials budget. Inglenook Co. produces wine. The company expects to produce 1,500,000 two-liter bottles of chablis in 19_6. Inglenook purchases empty glass bottles from an outside vendor. Its target ending inventory of such bottles is 50,000; its beginning inventory is 20,000. For simplicity, ignore breakage. Compute the number of bottles to be purchased in 19_6.

6-19 Budgeted manufacturing costs. The Weber Company has budgeted sales of 100,000 units of its product for 19_4. Expected unit costs, based on past experience, should be $60 for direct materials, $40 for direct manufacturing labor, and $30 for manufacturing overhead. Assume no beginning or ending inventory in process. Weber begins the year with 30,000 finished units on hand but budgets the ending finished goods inventory at only 10,000 units. Compute the budgeted costs of production for 19_4.

6-20 Budgeting material purchases. The Mahoney Company has prepared a sales budget of 42,000 finished units for a three-month period. The company has an inventory of 22,000 units of finished goods on hand at December 31 and has a target finished goods inventory of 24,000 units at the end of the succeeding quarter.

It takes three gallons of direct materials to make one unit of finished product. The company has an inventory of 90,000 gallons of direct materials at December 31 and has a target ending inventory of 110,000 gallons. How many gallons of direct materials should be purchased during the three months ending March 31?

6-21 Budgeting sales, cost of goods sold, and gross margin. Janet Grossman operates Centrum Gift Shop. She expects cash sales of $10,000 for October, $11,000 for November, and $16,000 for December. Grossman expects credit card sales of $7,000 during October and $8,000 and $12,000 during November and December, respectively. Sales returns and allowances can be ignored. Credit card companies like VISA and MasterCard charge 4% on credit card sales, so Centrum's net sales will be 96%. Cost of goods sold averages 40% of net sales.

Grossman asks you to prepare a schedule of budgeted sales, cost of goods sold, and gross margin for each month of the last quarter. Also show totals for the quarter.

6-22 Sales, production, and purchases budget. The Suzuki Co. in Japan has a division that manufactures two-wheel motorcycles. Its budgeted sales for Model G in 19_6 are 800,000 units. Suzuki's target ending inventory is 100,000 units, and its beginning inventory is 120,000 units. The company's budgeted selling price to its distributors and dealers is 400,000 yen (¥) per motorcycle.

Suzuki buys all its wheels from an outside supplier. No defective wheels are accepted. (Suzuki's needs for extra wheels for replacement parts are ordered by a separate division of the company.) The company's target ending inventory is 30,000 wheels, and its beginning inventory is 20,000 wheels. The budgeted purchase price is 16,000¥ per wheel.

Required
1. Compute the budgeted sales in yen.
2. Compute the number of motorcycles to be produced.
3. Compute the budgeted purchases of wheels in units and in yen.

6-23 Budget for production and direct manufacturing labor. (CMA, adapted)   Roletter Company makes and sells artistic frames for pictures of weddings, graduations, and other special events. Bob Anderson, controller, is responsible for preparing Roletter's master budget and has accumulated the information below for 19_5.

	January	February	March	April	May
			19_5		
Estimated sales in units	10,000	12,000	8,000	9,000	9,000
Selling price	$54.00	$51.50	$51.50	$51.50	$51.50
Direct manufacturing labor hours per unit	2.0	2.0	1.5	1.5	1.5
Wages per direct manufacturing labor hour	$10.00	$10.00	$10.00	$11.00	$11.00

Besides wages, direct manufacturing labor-related costs include pension contributions of $0.50 per hour, worker's compensation insurance of $0.15 per hour, employee medical insurance of $0.40 per hour, and social security taxes. Assume that as of January 1, 19_5, the social security tax rates are 7.5% for employers and 7.5% for employees. The cost of employee benefits paid by Roletter on its employees is treated as a direct manufacturing labor cost.

Roletter has a labor contract that calls for a wage increase to $11.00 per hour on April 1, 19_5. New labor-saving machinery has been installed and will be fully operational by March 1, 19_5.

Roletter expects to have 16,000 frames on hand at December 31, 19_4, and has a policy of carrying an end-of-month inventory of 100% of the following month's sales plus 50% of the second following month's sales.

Required

Prepare a production budget and a direct manufacturing labor budget for Roletter Company by month and for the first quarter of 19_5. Both budgets may be combined in one schedule. The direct manufacturing labor budget should include labor hours and show the detail for each direct manufacturing labor cost category.

6-24 Sales and production budgets. (CPA, adapted) The Scarborough Corporation in Australia manufactures and sells two products, Thingone and Thingtwo. In July 19_7, Scarborough's budget department gathered the following data in order to prepare budgets for 19_8:

19_8 Projected Sales

Product	Units	Price
Thingone	60,000	$165
Thingtwo	40,000	$250

19_8 Inventories, in Units

Product	Expected January 1, 19_8	Target December 31, 19_8
Thingone	20,000	25,000
Thingtwo	8,000	9,000

To produce one unit of Thingone and Thingtwo, the following direct materials are used:

		Amount Used per Unit	
Direct Material	Unit	Thingone	Thingtwo
A	lb	4	5
B	lb	2	3
C	each	0	1

Projected data for 19_8 with respect to direct materials are as follows:

Direct Material	Anticipated Purchase Price	Expected Inventories, January 1, 19_8	Target Inventories, December 31, 19_8
A	$12	32,000 lb	36,000 lb
B	$ 5	29,000 lb	32,000 lb
C	$ 3	6,000 units	7,000 units

Projected direct manufacturing labor requirements and rates for 19_8 are as follows:

Product	Hours per Unit	Rate per Hour
Thingone	2	$12
Thingtwo	3	$16

Manufacturing overhead is allocated at the rate of $20 per direct manufacturing labor hour.

Required

On the basis of the above projections and budget requirements for Thingone and Thingtwo, prepare the following budgets for 19_8:

1. Sales budget (in dollars)
2. Production budget (in units)
3. Direct materials purchases budget (in quantities)
4. Direct materials purchases budget (in dollars)
5. Direct manufacturing labor budget (in dollars)
6. Budgeted finished goods inventory at December 31, 19_8 (in dollars)

6-25 Budgeted income statement. (CMA, adapted) Easecom Company is a manufacturer of video-conferencing products. Regular units are manufactured to meet marketing projections, and specialized units are made after an order is received. Maintaining the video-conferencing equipment is an important area of customer satisfaction. With the recent downturn in the computer industry, the video-conferencing equipment segment has suffered, leading to a decline in Easecom's financial performance. The following income statement shows results for the year 19_4.

Income Statement for Easecom Company
for the Year Ended December 31, 19_4
(in thousands)

Net sales		
Equipment	$ 6,000	
Maintenance contracts	1,800	
Net sales		$7,800
Cost of goods sold		4,600
Gross margin		3,200
Operating costs		
Marketing	600	
Distribution	150	
Customer maintenance	1,000	
Administration	900	
Operating costs		2,650
Operating income		$ 550

Easecom's management team is in the process of preparing the 19_5 budget and is studying the following information.

a. Selling prices of equipment are expected to increase by 10% as the economic recovery begins. The selling price of each maintenance contract is unchanged from 19_4.

b. Equipment sales in units are expected to increase by 6%, with a corresponding 6% growth in units of maintenance contracts.

c. Cost of each unit sold is expected to increase by 3% to pay for the necessary technology and quality improvements.

d. Marketing costs are expected to increase by $250,000, but administration costs are expected to be held at 19_4 levels.

e. Distribution costs vary in proportion to the number of units of equipment sold.

f. Two maintenance technicians are to be added at a total cost of $130,000, which covers wages and travel costs. The objective is to improve customer service and shorten response time.

g. There is no beginning or ending inventory of equipment.

Required

Prepare a budgeted income statement for 19_5.

6-26 *Kaizen*-budgeted income statement and supporting schedules. (Refer to the Halifax Engineering example in the chapter.) Assume the following expected sales quantities, selling prices, and target ending inventories for regular and heavy-duty aircraft parts.

	Sales Quantity	Selling Price	Target Ending Inventory
Regular	4,500 units	$585 per unit	800 units
Heavy-duty	900 units	$780 per unit	50 units

Expected direct materials prices are:

111 alloy:	$6.50 per kilogram
112 alloy:	$9.50 per kilogram

According to its current practices, Halifax needs 12 kilograms of 111 alloy to produce one unit of regular aircraft part and 12 kilograms of 111 alloy to produce one unit of heavy-duty aircraft part. Management estimates that through *kaizen* activities, the amount of 111 alloy required to produce one regular unit can be reduced from 12 kilograms to 11 kilograms. Similarly, the amount of 111 alloy required to produce one heavy-duty unit can be reduced from 12 kilograms to 11 kilograms. Retain all other assumptions from the Halifax Engineering chapter illustration.

Required

1. Prepare a *kaizen*-budgeted income statement and supporting schedules for Halifax Engineering for the year 19_5.

2. What is the dollar effect of the *kaizen* activities on the 19_5 budgeted income statement?

6-27 Responsibility of purchasing agent. (Adapted from a description by R. Villers) Mark Richards is the purchasing agent for the Hart Manufacturing Company. Kent Sampson is head of the production planning and control department. Every six months, Sampson gives Richards a general purchasing program. Richards gets specifications from the engineering department. He then selects suppliers and negotiates prices. When he took this job, Richards was informed very clearly that he bore responsibility for meeting the general purchasing program once he accepted it from Sampson.

During week 24, Richards was advised that Part No. 1234—a critical part—would be needed for assembly on Tuesday morning of week 32. He found that the regular supplier could not deliver. He called everywhere and finally found a supplier in the Midwest and accepted the commitment.

He followed up by mail. Yes, the supplier assured him, the part would be ready. The matter was so important that on Thursday of week 31, Richards checked by phone. Yes, the shipment had left in time. Richards was reassured and did not check further. But on Tuesday of week 32, the part had not arrived. Inquiry revealed that the shipment had been misdirected by the railroad and was still in Chicago.

Required

What department should bear the costs of time lost in the plant? Why? As purchasing agent, do you think it fair that such costs be charged to your department?

6-28 A study in responsibility accounting. The David Machine Tool Company is in the doldrums. Production output has fallen to a 10-year low. The company has a nucleus of skilled tool-and-die workers who could find employment elsewhere if they were laid off. Three of these workers have been transferred temporarily to the building and grounds department, where they have been doing menial tasks such as washing walls and sweeping floors for the past month. They have earned their regular rate of $13 per hour. Their wages have been charged to the building and grounds department. The supervisor of building and grounds has just confronted the controller as follows: "Look at the cockeyed performance report you pencil pushers have given me." The helpers' line reads:

	Actual Results	Budget	Variance
Wages of helpers	$6,552	$4,032	$2,520 Unfavorable

"This is just another example of how unrealistic you bookkeepers are! Those tool-and-die people are loafing on the job because they know we won't lay them off. The regular hourly rate for my three helpers is $8 each. Now that my regular helpers are laid off, my work is piling up, so that when they return they'll either have to put in overtime or I'll have to get part-time help to catch up with things. Instead of charging me at $13 per hour, you should charge about $6—that's all those tool-and-die slobs are worth at their best."

Required

As the controller, what would you do *now*? Would you handle the accounting for these wages any differently?

6-29 Fixing responsibility. (Adapted from a description by Harold Bierman, Jr.) The city of Mountainvale hired its first city manager four years ago. She favored a "management by objectives" philosophy and accordingly set up many profit responsibility centers, including a sanitation department, a city utility, and a repair shop.

For many months, the sanitation manager had been complaining to the utility manager about wires being too low at one point in the road. There was barely clearance for large sanitation trucks. The sanitation manager asked the repair shop to make changes in the clearance. The repair shop manager asked, "Should I charge the sanitation or the utility department for the $20,000 cost of making the adjustment?" Both departments refused to accept the charge, so the repair department refused to do the work.

Late one day the top of a sanitation truck caught the wires and ripped them down. The repair department made an emergency repair at a cost of $26,000. Moreover, the city lost $10,000 of utility income because of the disruption of service.

Investigation disclosed that the sanitation truck had failed to clamp down its top properly. The extra two inches of height caused the wire to be caught.

Both the sanitation and utility managers argued strenuously about who should bear the $26,000 cost. Moreover, the utility manager demanded reimbursement from the sanitation department of the $10,000 of lost utility income.

Required

As the city controller in charge of the responsibility accounting system, how would you favor accounting for these costs? Specifically, what would you do next? What is the proper role of responsibility accounting in fixing the blame for this situation?

6-30 Comprehensive review of budgeting. British Beverages bottles two soft drinks under license to Cadbury Schweppes at its Sydney plant. Bottling at this plant is a highly repetitive, automated process. Empty bottles are removed from their carton, placed on a conveyor, and cleaned, rinsed, dried, filled, capped, and heated (to reduce condensation). All inventory is in direct materials and finished goods at the end of each working day. There is no work in process inventory.

The two soft drinks bottled by British Beverages are lemonade and diet lemonade. The syrup for both soft drinks is purchased from Cadbury Schweppes. Syrup for the regular lemonade contains a higher sugar content than the syrup for the diet lemonade.

British Beverages uses a lot size of 1,000 cases as the unit of analysis in its budgeting. (Each case contains 24 bottles.) Direct materials are expressed in terms of lots, where one lot of direct materials is the input necessary to yield one lot (1,000 cases) of beverage. In 19_6, the following purchase prices are forecast for direct materials:

	Lemonade	Diet Lemonade
Syrup	$1,200 per lot	$1,100 per lot
Containers (bottles, caps, etc.)	$1,000 per lot	$1,000 per lot
Packaging	$ 800 per lot	$ 800 per lot

The two soft drinks are bottled using the same equipment. The equipment is sanitized daily, but it is rinsed only when a switch is made during the day between diet lemonade and lemonade. Diet lemonade is always bottled first each day to reduce the risk of sugar contamination. The only difference in the bottling process for the two soft drinks is the syrup.

Summary data used in developing budgets for 19_6 are:

1. Sales
   ◆ Lemonade, 1,080 lots at $9,000 selling price per lot
   ◆ Diet lemonade, 540 lots at $8,500 selling price per lot

2. Beginning (January 1, 19_6) inventory of direct materials
   ◆ Syrup for lemonade: 80 lots at $1,100 purchase price per lot
   ◆ Syrup for diet lemonade: 70 lots at $1,000 purchase price per lot
   ◆ Containers: 200 lots at $950 purchase price per lot
   ◆ Packaging: 400 lots at $900 purchase price per lot

3. Beginning (January 1, 19_6) inventory of finished goods
   ◆ Lemonade: 100 lots at $5,300 per lot
   ◆ Diet lemonade: 50 lots at $5,200 per lot

4. Target ending (December 31, 19_6) inventory of direct materials
   ◆ Syrup for lemonade: 30 lots
   ◆ Syrup for diet lemonade: 20 lots
   ◆ Containers: 100 lots
   ◆ Packaging: 200 lots

5. Target ending (December 31, 19_6) inventory of finished goods
   ◆ Lemonade: 20 lots
   ◆ Diet lemonade: 10 lots

6. Each lot requires 20 direct manufacturing labor hours at the 19_6 budgeted rate of $25 per hour. Indirect manufacturing labor costs are included in the manufacturing overhead forecast.

7. Variable manufacturing overhead is forecast to be $600 per hour of bottling time; bottling time is the time the filling equipment is in operation. It takes two hours to bottle one lot of lemonade and two hours to bottle one lot of diet lemonade.

   Fixed manufacturing overhead is forecast to be $1,200,000 for 19_6.

8. Hours of budgeted bottling time is the sole allocation base for all fixed manufacturing overhead.

9. Administration costs are forecast to be 10% of the cost of goods manufactured for 19_6. Marketing costs are forecast to be 12% of dollar sales for 19_6. Distribution costs are forecast to be 8% of dollar sales for 19_6.

Required
Assume that British Beverages uses the first-in, first-out method for costing all inventories. On the basis of the above data, prepare the following budgets for 19_6:
   a. Sales budget (in dollars)
   b. Production budget (in units)
   c. Direct materials usage budget (in units and dollars)
   d. Direct materials purchases budget (in units and dollars)
   e. Direct manufacturing labor budget
   f. Manufacturing overhead costs budget
   g. Ending inventory budget
   h. Cost of goods sold budget
   i. Marketing costs budget
   j. Distribution costs budget
   k. Administration costs budget
   l. Budgeted income statement

6-31 Budgetary slack and ethics. (CMA, adapted)   Marge Atkins, the budget manager at Norton Company, a manufacturer of infant furniture and carriages, was working on the 19_5 annual budget. In discussions with Scott Ford, the sales manager, Atkins discovered that Ford's sales projections were lower than what Ford actually believed are achievable. When Atkins asked Ford about this, Ford said: "Well, we don't want to fall short of the sales projections, so we generally give ourselves a little breathing room by lowering the sales projections anywhere from 5% to 10%." Atkins also found that Pete Granger, the production manager, makes similar adjustments. He pads budgeted costs, adding 10% to estimated costs.

Required
As a management accountant, should Marge Atkins take the position that Scott Ford's and Pete Granger's behavior is unethical? Refer to the "Standards of Ethical Conduct for Management Accountants" described in Chapter 1 (p. 17).

Coverage of the Chapter Appendix

6-32 Collections and disbursements. (CPA) The following accrual accounting information was available from Montero Corporation's books:

19_5	Purchases (before discounts)	Sales
January	$42,000	$72,000
February	48,000	66,000
March	36,000	60,000
April	54,000	78,000

Collections from customers are normally 70% in the month of sale, 20% in the month following the sale, and 9% in the second month following the sale. The balance is expected to be uncollectible. Montero takes full advantage of the 2% discount allowed on purchases paid for by the 10th of the following month. Purchases for May are budgeted at $60,000 (before discounts), while sales for May are forecasted at $66,000. Cash disbursements for costs are expected to be $14,400 for the month of May. Montero's cash balance at May 1 was $20,000.

Required
Prepare the following schedules:

1. Expected cash collections during May
2. Expected cash disbursements during May
3. Expected cash balance at May 31

6-33 Preparing a cash budget. Dulcet Tones, a family-owned stereo store, began October with $10,000 cash. Management forecasts that collections from credit customers will be $90,000 in October and $122,000 in November. The store is scheduled to receive $40,000 cash on a note receivable in November. Projected cash disbursements include inventory purchases ($102,000 in October and $121,000 in November) and operating costs ($30,000 each month).

The store's bank requires a $7,500 minimum balance in the store's checking account. At the end of any month when the account balance goes below $7,500, the bank automatically extends credit to the store in multiples of $1,000 so that the balance in the checking account is at least $7,500. Dulcet Tones borrows as little as possible and repays these loans as rapidly as possible in multiples of $1,000 plus 1.5% monthly interest on the entire unpaid principal. The first payment occurs at the end of the month following the loan.

Required
Prepare the store's cash budget for October and November. Compute the amount owed to the bank, if any, on November 30.

6-34 Cash budgeting. On December 1, 19_5, the Itami Wholesale Co. is attempting to project cash receipts and disbursements through January 31, 19_6. On this latter date, a note will be payable in the amount of $100,000. This amount was borrowed in September to carry the company through the seasonal peak in November and December.

The trial balance on December 1 shows in part:

Cash	$ 10,000	
Accounts receivable	280,000	
Allowance for bad debts		$15,800
Inventory	87,500	
Accounts payable		92,000

Sales terms call for a 2% discount if payment is made within the first 10 days of the month after purchase, with the balance due by the end of the month after purchase. Experience has shown that 70% of the billings will be collected within the discount period, 20% by the end of the month after purchase, 8% in the following month, and that 2% will be uncollectible. There are no cash sales.

The average selling price of the company's products is $100 per unit. Actual and projected sales are:

October actual	$ 180,000
November actual	250,000
December estimated	300,000
January estimated	150,000
February estimated	120,000
Total estimated for year ended June 30, 19_6	1,500,000

All purchases are payable within 15 days. Thus approximately 50% of the purchases in a month are due and payable in the next month. The average unit purchase cost is $70. Target ending inventories are 500 units plus 25% of the next month's unit sales.

Total budgeted marketing, distribution, and customer-service costs for the year are $400,000. Of this amount, $150,000 is considered fixed (and includes depreciation of $30,000). The remainder varies with sales. Both fixed and variable marketing, distribution, and customer-service costs are paid as incurred.

Required
Prepare a cash budget for December and January. Supply supporting schedules for collections of receivables, payments for merchandise, and marketing, distribution, and customer-service costs.

6-35 Comprehensive budget; fill in schedules. The following information pertains to Newport Stationery Store.

Balance sheet information as of Sept. 30:
Current assets	
Cash	$ 12,000
Accounts receivable	10,000
Inventory	63,600
Equipment, net	100,000
Liabilities	None

Recent and anticipated sales:
September	$ 40,000
October	48,000
November	60,000
December	80,000
January	36,000

Credit sales: Sales are 75% for cash and 25% on credit. Assume that credit accounts are all collected within 30 days from sale. The accounts receivable on September 30 are the result of the credit sales for September (25% of $40,000).

Gross margin averages 30% of sales. Newport treats cash discounts on purchases in the income statement as "other income."

Operating costs: Salaries and wages average 15% of monthly sales; rent, 5%; other operating costs, excluding depreciation, 4%. Assume that these costs are disbursed each month. Depreciation is $1,000 per month.

Purchases: Newport keeps a minimum inventory of $30,000. The policy is to purchase each month additional inventory in the amount necessary to provide for the following month's sales. Terms on purchases are 2/10, n/30. (Payments on purchases are to be made in 30 days; a 2% discount is available if the payment is made within 10 days after purchase.) Assume that payments are made in the month of purchase and that all discounts are taken.

Light fixtures: In October, $600 is spent for light fixtures, and in November, $400 is to be expended for this purpose. These amounts are to be capitalized.

Assume that a minimum cash balance of $8,000 must be maintained. Assume also that all borrowing is effective at the beginning of the month and all repayments are made at the end of the month of repayment. Loans are repaid when sufficient cash is available. Interest is paid only at the time of repaying principal. The interest rate is 18% per year. Management does not want to borrow any more cash than is necessary and wants to repay as soon as cash is available.

Required

On the basis of the facts as given above:

1. Complete schedule A.

Schedule A
Budgeted Monthly Cash Receipts

Item	September	October	November	December
Total sales	$40,000	$48,000	$60,000	$80,000
Credit sales	10,000	12,000		
Cash sales				
Receipts:				
Cash sales		$36,000		
Collections on accounts receivable		10,000		
Total		$46,000		

2. Complete schedule B.

Schedule B
Budgeted Monthly Cash Disbursements for Purchases

Item	October	November	December	4th Quarter
Purchases	$42,000			
Deduct 2% cash discount	840			
Disbursements	$41,160			

3. Complete schedule C.

Schedule C
Budgeted Monthly Cash Disbursements for Operating Costs

Item	October	November	December	4th Quarter
Salaries and wages	$7,200			
Rent	2,400			
Other cash operating costs	1,920			
Total	$11,520			

4. Complete schedule D.

Schedule D
Budgeted Total Monthly Cash Disbursements

Item	October	November	December	4th Quarter
Purchases	$41,160			
Cash operating costs	11,520			
Light fixtures	600			
Total	$53,280			

5. Complete schedule E.

Schedule E
Budgeted Cash Receipts and Disbursements

Item	October	November	December	4th Quarter
Receipts	$46,000			
Disbursements	53,280	_____	_____	_____
Net cash increase				
Net cash decrease	$ 7,280	======	======	======

6. Complete schedule F (assume that borrowings must be made in multiples of $1,000).

Schedule F
Financing Required

Item	October	November	December	4th Quarter
Beginning cash balance	$12,000			
Net cash increase				
Net cash decrease	7,280	_____	_____	_____
Cash position before borrowing	4,720			
Minimum cash balance required	8,000	_____	_____	_____
Excess (deficiency)	(3,280)			
Borrowing required	4,000			
Interest payments				
Borrowing repaid		_____	_____	_____
Ending cash balance	$ 8,720	======	======	======

7. What do you think is the most logical type of loan needed by Newport? Explain your reasoning.

8. Prepare a budgeted income statement for the fourth quarter and a budgeted balance sheet as of December 31. Ignore income taxes.

9. Some simplifications have been introduced in this problem. What complicating factors would be met in a typical business situation?

7

Flexible Budgets, Variances, and Management Control: I

Performance gaps and variances

Static budgets and flexible budgets

Flexible budgeting without standard costs

Flexible budgeting with standard costs

Price variances and efficiency variances

Overview of variance analysis

Performance measurement using standards

Impact of inventories

Illustration of general ledger entries using standard costs

Benchmarks and budgeted amounts

Appendix: Variance investigation decisions and uncertainty

*M*aterials costs for items such as cocoa, sugar, and packaging are sizable for confectionary companies. Nestlé-Rowntree uses comparisons of actual costs with budgeted costs to help manage costs of the KitKat production line at its York, United Kingdom, plant.

Learning Objectives

When you have finished studying this chapter, you should be able to

1. Distinguish between a performance gap and a variance
2. Describe the difference between a static budget and a flexible budget
3. Illustrate how a flexible budget can be developed
4. Use the flexible-budget approach to compute flexible-budget variances and sales-volume variances
5. Describe two different definitions of standards used in computing a standard cost
6. Compute the price and efficiency variances for direct cost categories
7. Explain why purchasing performance measures should focus on more factors than just price variances
8. Describe how the continuous improvement theme can be integrated into a standard costing system
9. Describe a cost benchmark used for cost management

We have learned that managers quantify their plans in the form of budgets. This chapter focuses on how flexible budgets and variances can play a key role in management planning and control. Recall from Chapter 1 that feedback enables managers to compare the actual results with the planned performance. Flexible budgets and variances help managers gain insights into why the actual results differ from the planned performance. It is this insight into "why" that makes the topics covered in this chapter and the next important ones to master.

Our discussion of flexible budgets and variances in this chapter uses two examples. Goodride Company manufactures tires, and Webb company manufactures jackets. Two terms used frequently in our discussion of these companies are *budgeted* and *standard*. All amounts in a budget are budgeted amounts, but not all budgeted amounts are standard amounts. A standard amount is a specific type of budgeted amount:

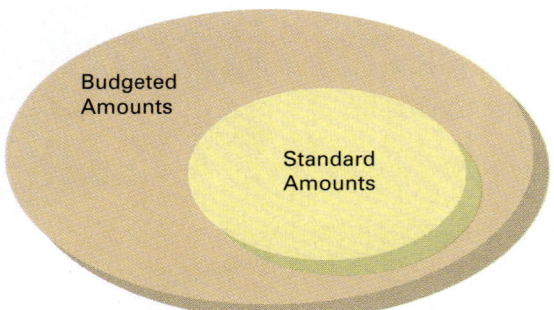

A **standard** represents a good level of performance or the best level of performance. A standard usually is developed from a careful study of the specific operations and is expressed on a per-unit basis. For example, a labor time standard for a delivery operator (such as a delivery driver for Federal Express) could be based on a detailed time-and-motion study of how long it takes to perform every step involved in picking up and delivering a package. As a second example, materials standards in manufacturing are frequently based on detailed engineering studies at the specific plant of interest. A **standard input** is the allowed quantity of inputs (such as hours of labor time or pounds of materials) for one unit of output, given a good level or the best level of performance. A **standard cost** is the per-unit cost of output for a good or the best level of performance. Standard costs are most frequently found in the manufacturing and distribution parts of the value chain.

Why are budgeted costs not necessarily the same as standard costs? Because not all budgeted costs are based on a careful study of the specific operations of an organization. Rather, many budgeted costs are based on past cost relationships, without a detailed analysis of whether those past costs represent good or best levels of performance. As illustrated in this chapter, flexible budgets can use any budgeted amounts for individual cost items, irrespective of whether or not those amounts are based on standard costs.

PERFORMANCE GAPS AND VARIANCES

Objective 1

Distinguish between a performance gap and a variance

Performance gaps and variances are two key terms used in this book. A **performance gap** is the difference between an actual result and a benchmark amount. A *benchmark* amount is a best level of performance that can be found inside or outside the organization. A **variance** is the difference between an actual result and a budgeted amount when that budgeted amount is a financial variable reported by the accounting system. It may or may not be a benchmark amount or a standard amount.

This book uses the term *performance gap* as the general case and *variance* as a special case. That is, a variance is only one type of performance gap. The relationship between a variance and a performance gap can be illustrated by distinguishing among three types of benchmarks:

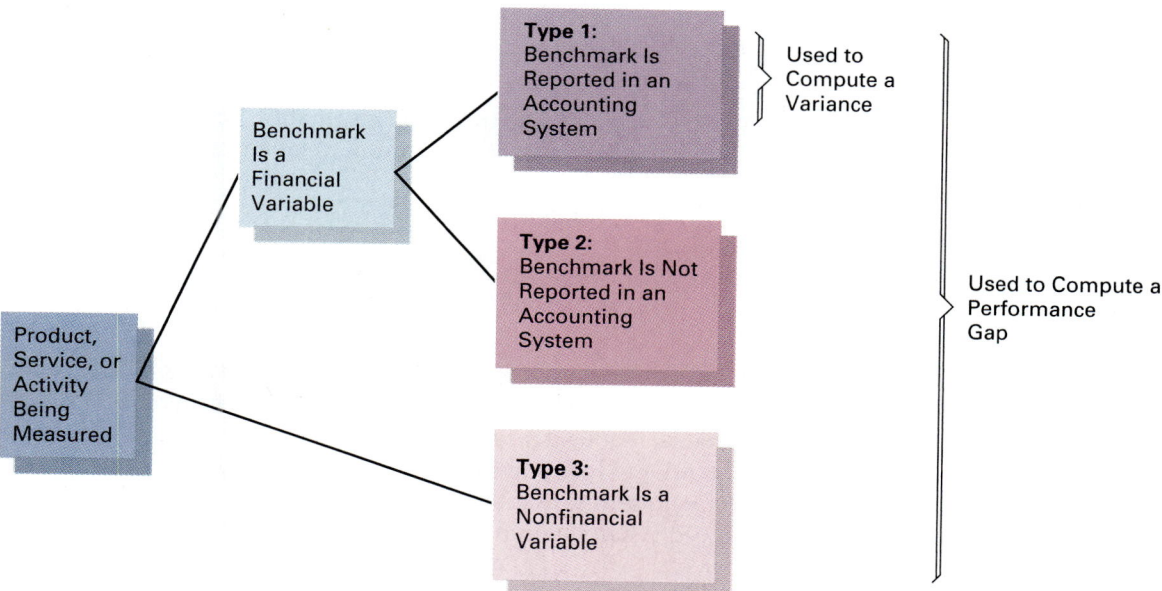

Managers are interested in all three types of benchmarks. Consider American Airlines. An example of a type 1 benchmark is the compensation it pays its own pilots. Comparison of its actual compensation with its budgeted compensation provides useful insights to American when budgeting future costs and when evaluating the performance of its managers. This book uses the term *variance analysis* to describe such comparisons.

Performance gap comparisons involving benchmarks 2 and 3 are also important to American Airlines. An example of a type 2 benchmark (a financial variable not reported in an accounting system) is the compensation other airlines (such as Delta Airlines and United Airlines) pay their own pilots and managers. An example of a type 3 benchmark is the percentage of American Airlines' flights that arrive on time. While this chapter focuses on benchmarks found in accounting systems, keep in mind that the resultant variance analysis is only one of several types of information managers use in their planning and control decisions.

STATIC BUDGETS AND FLEXIBLE BUDGETS

Objective 2
Describe the difference between a static budget and a flexible budget

This chapter illustrates both static budgets and flexible budgets. A **static budget** is a budget that is based on one level of output and is not adjusted or altered after it is finalized. A **flexible budget** is a budget that is developed using budgeted revenue or cost amounts; it is adjusted (flexed) to the actual level of output achieved or expected to be achieved during the budget period. As we will see, a flexible budget enables managers to compute a richer set of variances than does a static budget.

Budgets, both static and flexible, can differ in the level of detail they report. Increasingly, organizations are developing approaches to budgeting that report summary figures with the capability to display on a computer screen more detailed breakdowns of these figures. Level 0 reports the least detail, Level 1 offers more information, and so on. Different terms are used to describe the variances that occur at these various levels. We will look at three budget levels and their associated variances using data from Goodride Company, a tire manufacturer:

Variances Illustrated

◆ Level 0: based on a *static budget*
◆ Level 1: based on a *static budget* with more detail than in Level 0

} Static-budget variance

◆ Level 2: based on a *flexible budget*

} Flexible-budget variance and sales-volume variance

The Webb Company data includes information about input prices and input quantities that enable it to undertake a more detailed Level 3 analysis yielding two additional variances:

Variances Illustrated

◆ Level 3: based on a flexible budget with more detail than in Level 2

} Price variance and Efficiency variance

A **favorable variance**—denoted F in the exhibits—is a variance that increases operating income relative to the budgeted amount. An **unfavorable variance**—denoted U—is a variance that decreases operating income.

In order to focus on the key concepts, both the Goodride and Webb examples in this chapter assume that there is no beginning or ending inventory. The complexities introduced when beginning or ending inventory exists are covered in Part One of Chapter 9.

FLEXIBLE BUDGETING WITHOUT STANDARD COSTS

The Goodride Company manufactures tires for sale to automobile companies. The 19_4 static budget operating income, based on a sales (and production) budget of 800,000 units, is $3.200 million. In 19_4 actual operating income was $4.734 million.

Static Budget: Level 0 and Level 1 Analysis

Exhibit 7-1 presents the Level 0 and Level 1 variance analyses. Level 0 is the most general comparison of the actual and budgeted operating income. The variance of $1.534 million is simply the result of subtracting the budgeted operating income of $3.200 million from the actual operating income of $4.734 million:

EXHIBIT 7-1

Static Budget Based Variance Analysis
for Goodride Company for 19_4*
($ in millions)

Level 0 Analysis

Actual operating income	$4.734
Budgeted operating income	3.200
Static budget variance of operating income	$1.534 F

	Actual Results (1)	Static-Budget Variances (2) = (1) -- (3)	Static Budget (3)
Level 1 Analysis			
Units sold	840,000	40,000	800,000
Revenues (sales)	$42.840	2.840 F	$40.000
Variable costs	24.024	0.824 U	23.200
Contribution margin	18.816	2.016 F	16.800
Fixed costs	14.082	0.482 U	13.600
Operating income	$ 4.734	$1.534 F	$ 3.200
		$1.534 F	

Total static-budget variance

*F = favorable effect on operating income; U = unfavorable effect on operating income.

$$\text{Static budget variance of operating income} = \left(\begin{array}{c}\text{Actual} \\ \text{results}\end{array}\right) - \left(\begin{array}{c}\text{Static budget} \\ \text{amount}\end{array}\right)$$
$$= \$4.734 - \$3.200$$
$$= \$1.534 \text{ million F}$$

This $1.534 million F amount is often called a static-budget variance because the number used for the budgeted amount is taken from a static budget.

Level 1 analysis in Exhibit 7-1 provides managers with more detailed information on the operating income variance of $1.534 million F. The additional information added in Level 1 pertains to units sold, revenues, variable costs, and fixed costs. Note that 840,000 units were sold instead of the budgeted 800,000 units. The contribution margin of 42% ($16.800 ÷ $40.000) in the static budget increases to 43.3% ($18.564 ÷ $42.840) for the actual results.

While Level 1 contains more information than Level 0, more insights can be gained by incorporating a flexible budget into the computation of variances.

Flexible Budget: Level 2 Analysis

Goodride Company uses a flexible-budget approach based on output units. The flexible budget is the budget that would have been formulated if Goodride had perfect foresight about the actual level of output units. The steps in this approach to developing a flexible budget follow:

Objective 3
Illustrate how a flexible budget can be developed

STEP 1: *DETERMINE THE BUDGETED SELLING PRICE, THE BUDGETED VARIABLE COST PER OUTPUT UNIT, AND THE BUDGETED FIXED COSTS.* These budgeted amounts could be developed in several ways: for example, through (a) analyzing past costs and then making an inflation adjustment for the budget period or (b) analyzing past costs and interviewing operating personnel to incorporate any expected improvement for the budgeted period. Column 2 of Exhibit 7-2 shows the 19_4 budgeted selling price ($50) and the budgeted variable costs per output unit ($29) for Goodride Company. Budgeted fixed costs for 19_4 are $13.600 million. Budgeted sales output is 800,000 units.

STEP 2: *DETERMINE THE ACTUAL QUANTITY OF OUTPUT UNITS.* In 19_4, Goodride actually sold 840,000 units.

STEP 3: *DETERMINE THE FLEXIBLE BUDGET BASED ON THE ACTUAL QUANTITY OF OUTPUT UNITS.* Columns 3 to 5 of Exhibit 7-2 show the flexible budget for illustrative output levels of 760,000, 800,000, and 840,000 units respectively. That is, these budgeted amounts are flexed to illustrate what the company *would expect* if actual output level is 760,000, 800,000, or 840,000 units. For the actual sales output of 840,000 units, the flexible budget is:

$$\text{Flexible-budget revenues} = \$50 \times 840{,}000 = \$42.000 \text{ million}$$
$$\text{Flexible-budget variable costs} = \$29 \times 840{,}000 = \$24.360 \text{ million}$$
$$\text{Flexible-budget fixed costs} = \$13.600 \text{ million}$$

Column 6 of Exhibit 7-2 shows the actual results for Goodride Company for 19_4.

Exhibit 7-3 presents the Level 2 flexible budget based variance analysis for Goodride Company. Note that the $1.534 million favorable static-budget variance of operating income is now split into two categories—a flexible-budget variance and a sales-volume variance.

Objective 4

Use the flexible-budget approach to compute flexible-budget variances and sales-volume variances

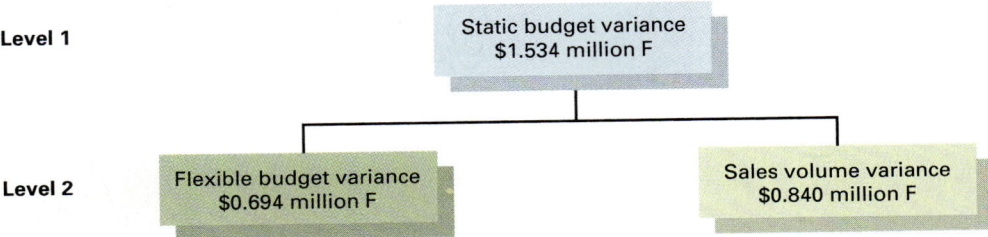

Level 1

Static budget variance
$1.534 million F

Level 2

Flexible budget variance
$0.694 million F

Sales volume variance
$0.840 million F

The **flexible-budget variance** is the difference between the actual result and the flexible budget amount for the actual output achieved. The **sales-volume variance** is the difference between the flexible-budget amount and the static-budget amount; unit selling prices, unit variable costs, and fixed costs are held constant. Information on these two variances is unattainable except through a Level 2 analysis.

Flexible-Budget Variances

The first three columns of Exhibit 7-3 compare the actual results with the flexible budget amounts. Flexible-budget variances are reported in column 2 for each line item in the summary income statement:

$$\frac{\text{Flexible budget}}{\text{variance}} = \binom{\text{Actual}}{\text{results}} - \binom{\text{Flexible budget}}{\text{amount}}$$

Thus, for example, the flexible-budget variance for operating income is $0.694 million F ($4.734 – $4.040). Most managers want a line-by-line listing of revenues and various costs and their flexible-budget variances. Consider the revenue line item in Exhibit 7-3, where the flexible-budget variance is $0.840 million F. This variance came about because management raised the actual selling price by an average $1 per unit (from the budgeted $50 to $51; that is, $42.840 million ÷ 840,000 units = $51) after the static budget was prepared.

EXHIBIT 7-2

**Flexible Budget Data
for Goodride Company for 19_4**

Line Items (1)	Budgeted Amount per Unit (2)	Flexible Budget Amounts for Alternative Levels of Output Units Sold			Actual Results for 840,000 Units (6) (in millions)
		760,000 (3) (in millions)	800,000 (4) (in millions)	840,000 (5) (in millions)	
Revenue (sales)	$50	$38.000	$40.000	$42.000	$42.840
Variable costs					
Direct materials	15	11.400	12.000	12.600	12.684
Direct manufacturing labor	7	5.320	5.600	5.880	5.460
Variable manufacturing overhead	4	3.040	3.200	3.360	3.276
Variable marketing overhead	3	2.280	2.400	2.520	2.604
Total variable costs	29	22.040	23.200	24.360	24.024
Contribution margin	$21	15.960	16.800	17.640	18.816
Fixed costs					
Manufacturing overhead		6.400	6.400	6.400	6.720
Marketing overhead		7.200	7.200	7.200	7.362
Total fixed costs		13.600	13.600	13.600	14.082
Operating income		$ 2.360	$ 3.200	$ 4.040	$ 4.734

EXHIBIT 7-3

**Flexible Budget Based Variance Analysis
for Goodride Company for 19_4*
(in millions)**

Level 2 Analysis	Actual Results (1)	Flexible-Budget Variances (2) = (1) – (3)	Flexible Budget† (3)	Sales-Volume Variances (4) = (3) – (5)	Static Budget‡ (5)
Units sold	840,000	0	840,000	40,000 F	800,000
Revenues (sales)	$42.840	$0.840 F	$42.000	$2.000 F	$40.000
Variable costs	24.024	0.336 F	24.360	1.160 U	23.200
Contribution margin	18.816	1.176 F	17.640	0.840 F	16.800
Fixed costs	14.082	0.482 U	13.600	0.000	13.600
Operating income	$ 4.734	$0.694 F	$ 4.040	$0.840 F	$ 3.200
	↑	$0.694 F	↑	$0.840 F	↑

Total flexible-budget variance   Total sales-volume variance

$1.534 F

Total static-budget variance

*F = favorable effect on operating income; U = unfavorable effect on operating income.

†Flexible budget, based on actual units sold and budgeted selling price, variable costs per output unit, and fixed costs.

‡Static budget, based on budgeted units sold and budgeted selling price, variable costs per output unit, and fixed costs.

Sales-Volume Variances

The flexible budget amount in column 3 of Exhibit 7-3 and the static-budget amount are also computed using the budgeted selling prices and budgeted costs. The sales-volume variance arises solely because the actual number of output units sold differs from the budgeted number of output units sold. The number of units sold is frequently called sales volume—hence the label "sales-volume variance."

$$\text{Sales volume variance} = \left(\begin{array}{c}\text{Flexible-budget}\\\text{amount}\end{array}\right) - \left(\begin{array}{c}\text{Static-budget}\\\text{amount}\end{array}\right)$$

Thus, for example, the sales-volume variance for operating income is $0.840 million F ($4.040 − $3.200).

Who has responsibility for the sales volume variance? Fluctuations in unit sales are attributable to many factors, but the senior manager of marketing is usually in the best position to explain why the actual level of units sold differs from the budgeted level in the static budget. Chapter 22 explores sales-volume variances in more detail and discusses sales mix, market share, and market size variances.

Surveys of Company Practice

THE WIDESPREAD USE OF STANDARD COSTS

Surveys of company practice across the globe report widespread use of standard costs by manufacturers. The following data are representative of surveys conducted in five countries:

	Percentage of Respondents Using Standard Costs in Their Accounting System
United States[a]	86%
Ireland[b]	85%
United Kingdom[c]	76%
Sweden[d]	73%
Japan[e]	65%

What explains the popularity of standard costs? Companies based in four countries report the following reasons, ranked 1 for most important down to 4 for least important[f]:

	United States	Canada	Japan	United Kingdom
Cost management	1	1	1	2
Price making and price policy	2	3	2	1
Budgetary planning and control	3	2	3	3
Financial statement preparation	4	4	4	4

The materials price and efficiency variances discussed in this chapter illustrate the use of standard costs in promoting cost management.

---

[a]Cornick, Cooper, and Wilson, "How Do Companies." [b]Clarke, "Management Accounting." [c]Drury, Braund, Osborne, and Tayles, "A Survey." [d]Ask and Ax, "Trends." [e]Scarbrough, Nanni, and Sakurai, "Japanese Management." [f]Inoue, "A Comparative Study." Full citations are in Appendix A.

The Level 2 analysis provides insight into the $1.534 million F total static-budget variance that is not available from the Level 0 or Level 1 analyses. Goodride management performed well in 19_4 on two dimensions. First, it sold 40,000 more tires than budgeted; holding everything else constant, this 40,000 increase in unit sales would have increased operating income by $0.840 million in 19_4. Second, it obtained a higher selling price than budgeted ($51 versus $50 budgeted) and a lower variable cost per output unit than budgeted ($28.60 versus $29 budgeted). In contrast, the actual fixed costs of $14.082 million exceeded the budgeted amount of $13.600 million. The net effect of these differences in selling prices, variable costs per output unit, and fixed costs is an increase in operating income of $0.694 million above that budgeted for an output level of 840,000 units.

Information Required for Level 2 Analysis

The Level 2 analysis in Exhibit 7-3 requires information about budgeted and actual output units, selling price (per output unit), variable cost per output unit, and fixed costs. This information is available in many organizations for many parts of their value chain. Thus, Level 2 analysis can be undertaken in a broad range of areas.

The Goodride Company example used *output* amounts and assumed no information about budgeted or actual *input* prices or *input* quantities. In contrast, the Webb Company example in the next section uses input price and input quantity information in its flexible budgeting process. This input price and quantity information enables Webb to compute additional variances not discussed in our Goodride example.

FLEXIBLE BUDGETING WITH STANDARD COSTS

We now consider the Webb Company, which manufactures and sells a single product, a distinctive jacket that requires many materials, tailoring, and hand operations. Sales are made to independent clothing stores and retail chains. Webb sets budgeted revenues (budgeted selling price × budgeted units sold) after discussion with its marketing personnel and analysis of general and industry economic conditions.

The costing system at Webb includes both manufacturing costs and marketing costs. There are direct and indirect costs in each category:

	Direct Costs	**Indirect Costs**
Manufacturing	1. Direct materials 2. Direct manufacturing labor	4. Variable manufacturing overhead 5. Fixed manufacturing overhead
Marketing	3. Direct marketing (distribution) labor	6. Variable marketing overhead 7. Fixed marketing overhead

Webb's manufacturing costs include direct materials, direct manufacturing labor, and manufacturing overhead (both variable and fixed). Its marketing (distribution and customer-service as well as advertising) costs are made up of direct marketing labor (primarily distribution personnel) and marketing overhead (both variable and fixed). The cost driver for direct materials, direct manufacturing labor, and variable manufacturing overhead is the number of units manufactured. The cost driver for direct marketing labor and variable marketing overhead is the number of units sold. The relevant range for the $180 selling price per jacket and for the cost drivers in both manufacturing and marketing is from 8,000 to 16,000 units.

Webb's budgeted unit costs are taken from its standard costing system. This use of standard costs enables Webb to compute the price and efficiency variances discussed in the next section, with good measures of performance as the benchmark.

Alternative Definitions of Standards

As noted in the introduction to this chapter, *standards* represent the results arising from a good or the best level of performance. They usually are developed from a careful study of the operations and are expressed on a per-unit basis. For example, standard labor hours of input per unit of output could be based on detailed time-and-motion studies at a plant. Organizations vary in what they mean by "good or best level of performance." Examples include:

- ◆ **Perfection standards:** The best level of performance under the best conceivable conditions. No provision is made for spoilage, waste, and the like.
- ◆ **Currently attainable standards:** A good level of performance taking into account normal spoilage, waste, and the like.

Most organizations use the "currently attainable" meaning of standards in their accounting systems. However, a growing number are now using perfection standards because they harmonize with the recent management trend toward continuous improvement of quality and operating efficiency.

Webb Company uses currently attainable standards when computing its standard costs.

Static Budget: Level 0 and Level 1 Analysis

Exhibit 7-4 presents the Level 0 and Level 1 analyses for Webb Company. The Level 0 analysis shows a static-budget variance of operating income of $237,000 U.

$$\begin{aligned}
\text{Static-budget variance} \atop \text{of operating income} &= \left(\text{Actual} \atop \text{results}\right) - \left(\text{Static budgeted} \atop \text{amount}\right) \\
&= \$25,000 - \$262,000 \\
&= \$237,000 \text{ U}
\end{aligned}$$

EXHIBIT 7-4

Static Budget Based Variance Analysis for Webb Company for April 19_4*

Level 0 Analysis

Actual operating income	$ 25,000
Budgeted operating income	262,000
Static-budget variance of operating income	$237,000 U

Level 1 Analysis	Actual Results (1)	Static-Budget Variances (2) = (1) -- (3)	Static Budget (3)
Units sold	10,000	2,000 U	12,000
Revenues (sales)	$1,850,000	$310,000 U	$2,160,000
Variable costs	1,120,000	68,000 F	1,188,000
Contribution margin	730,000	242,000 U	972,000
Fixed costs	705,000	5,000 F	710,000
Operating income	$ 25,000	$237,000 U	$ 262,000
		$237,000 U	

Total static-budget variance

*F = favorable effect on operating income; U = unfavorable effect on operating income.

Managers at Webb quickly proceed to Level 1 analysis to understand better why the unfavorable static-budget variance arose. As noted in the Goodride example, the information added in Level 1 pertains to units sold, revenues, variable costs, and fixed costs. Note that only 10,000 units were sold instead of the target of 12,000. This shortfall, however, was not accompanied by a proportional decline in variable costs. Variable costs decreased by only $68,000 and amounted to 60.5% of actual revenue ($1,120,000 ÷ 1,850,000) instead of the 55.0% budgeted amount ($1,188,000 ÷ $2,160,000). Thus, the contribution margin was only 39.5% of sales instead of the 45.0% budgeted.

Flexible Budget: Level 2 Analysis

Webb Company uses a standard costing system to develop its flexible budget. The variable costs per output unit for each of its five variable-cost items are computed using the following formula:

$$\left(\begin{array}{c}\text{Standard inputs allowed} \\ \text{for one output unit}\end{array}\right) \times \left(\begin{array}{c}\text{Standard cost} \\ \text{per input unit}\end{array}\right)$$

These standards are based on engineering time-and-motion studies in Webb's manufacturing and marketing areas. The marketing, distribution, and customer-service business functions are included in the marketing cost items. The direct marketing labor category is for distribution activities.

The standard costs for each of the five variable-cost items are:

◆ *Direct materials:* 2.00 square yards of cloth input allowed per output unit (jacket) manufactured, at $30 standard cost per square yard

Standard cost = 2.00 × $30 = $60.00 per output unit manufactured

◆ *Direct manufacturing labor:* 0.80 manufacturing labor-hours of input allowed per output unit manufactured, at $20 standard cost per hour

Standard cost = 0.80 × $20 = $16.00 per output unit manufactured

◆ *Direct marketing labor:* 0.25 marketing labor-hours of input allowed per output unit sold, at $24 standard cost per hour

Standard cost = 0.25 × $24 = $6.00 per output unit sold

◆ *Variable manufacturing overhead:* assigned on the basis of 1.20 machine-hours per output unit manufactured, at $10 standard cost per machine-hour

Standard cost = 1.20 × $10 = $12.00 per output unit manufactured

◆ *Variable marketing overhead:* assigned on the basis of 0.125 direct marketing labor-hours per output unit sold, at $40 standard cost per hour

Standard cost = 0.125 × $40 = $5.00 per output unit sold

The total standard cost per output unit is $99 ($60 + $16 + $6 + $12 + $5). This $99 standard cost is the budgeted amount used in Webb's flexible budget.

The steps in Webb's approach to developing a flexible budget follow:

STEP 1: DETERMINE THE BUDGETED SELLING PRICE, THE BUDGETED VARIABLE COSTS PER OUTPUT UNIT, AND THE BUDGETED FIXED COSTS. Exhibit 7-5 (Column 2) shows the budgeted selling price of $180 per jacket and the budgeted variable costs of $99 per jacket. Budgeted fixed costs are $710,000. Budgeted output is 12,000 jackets.

STEP 2: DETERMINE THE ACTUAL QUANTITY OF OUTPUT UNITS. In April 19_4, Webb sold 10,000 jackets.

EXHIBIT 7-5

Flexible Budget Data for Webb Company
for the Month April 19_4

Line Items (1)	Standard Cost Amount per Unit (2)	Flexible-Budget Amounts for Alternative Levels of Output Units Sold			Actual Results for 10,000 Units (6)
		10,000 (3)	12,000 (4)	15,000 (5)	
Revenue (sales)	$180	$1,800,000	$2,160,000	$2,700,000	$1,850,000
Variable costs:					
Direct materials	60	600,000	720,000	900,000	688,200
Direct manufacturing labor	16	160,000	192,000	240,000	198,000
Direct marketing labor	6	60,000	72,000	90,000	57,600
Variable manufacuturing overhead	12	120,000	144,000	180,000	130,500
Variable marketing overhead	5	50,000	60,000	75,000	45,700
Total variable costs	99	990,000	1,188,000	1,485,000	1,120,000
Contribution margin	$ 81	810,000	972,000	1,215,000	730,000
Fixed costs:					
Manufacturing overhead		276,000	276,000	276,000	285,000
Marketing overhead		434,000	434,000	434,000	420,000
Total fixed costs		710,000	710,000	710,000	705,000
Total costs		1,700,000	1,898,000	2,195,000	1,825,000
Operating income		$ 100,000	$ 262,000	$ 505,000	$ 25,000

STEP 3: *DETERMINE THE FLEXIBLE BUDGET BASED ON THE ACTUAL QUAN-TITY OF OUTPUT UNITS.* Columns 3 to 5 of Exhibit 7-5 show the flexible budget for illustrative output levels of 10,000, 12,000, and 15,000 units, respectively. For the actual sales output of 10,000 units, the flexible budget is:

Flexible-budget revenues $= \$180 \times 10,000 = \$1,800,000$
Flexible-budget variable costs $= \$99 \times 10,000 = \$ \ \ 990,000$
Flexible-budget fixed costs $= \$710,000$

The flexible budget is the budget that would have been formulated if Webb Company had perfect foresight that the actual number of jackets sold would be 10,000 jackets. Column 6 of Exhibit 7-5 reports the actual results for the 10,000 units sold in April 19_4.

Exhibit 7-6 presents the Level 2 flexible budget based variance analysis for Webb. The $237,000 unfavorable static-budget variance is now split into the flexible-budget variance and the sales-volume variance:

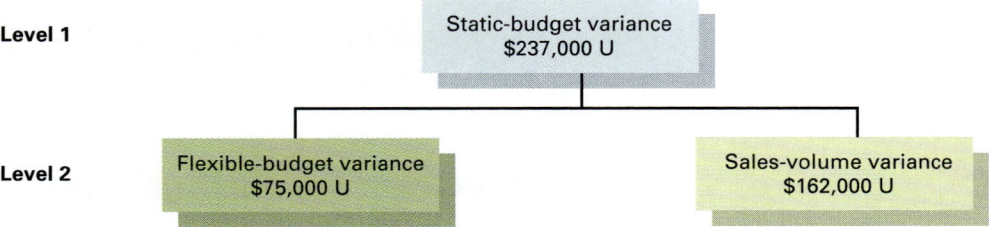

EXHIBIT 7-6

Flexible Budget Based Variance Analysis
for Webb Company for the Month April 19_4*

Level 2 Analysis	Actual Results (1)	Flexible Budget Variances (2) = (1) – (3)	Flexible Budget (3)	Sales-Volume Variances (4) = (3) – (5)	Static Budget (5)
Units sold	10,000	0	10,000	2,000 U	12,000
Revenues (sales)	$1,850,000	$ 50,000 F	$1,800,000	$ 360,000 U	$2,160,000
Variable costs	1,120,000	130,000 U	990,000	198,000 F	1,188,000
Contribution margin	730,000	80,000 U	810,000	162,000 U	972,000
Fixed costs	705,000	5,000 F	710,000	0	710,000
Operating income	$ 25,000	$ 75,000 U	$ 100,000	$ 162,000 U	$ 262,000
		$75,000 U		$162,000 U	

Total flexible-budget variance   Total sales-volume variance

$237,000 U

Total static-budget variance

*F = favorable effect on operating income; U = unfavorable effect on operating income.

The sales-volume variance of $162,000 U arises solely because Webb sold 2,000 fewer jackets in April 19_4 than were budgeted:

Sales-volume variance = (Flexible-budget amount) – (Static-budget amount)
= $100,000 – $262,000
= $162,000 U

The flexible-budget variance of $75,000 U arises because the actual selling price, unit variable costs, and fixed costs differ from the budgeted amounts. Consider the revenue line item in Exhibit 7-6, where the flexible-budget variance is $50,000 F. This variance comes about because management raised the actual selling price by an average $5 per unit (from the budgeted $180 to $185; that is, $1,850,000 ÷ 10,000 units = $185) after the static budget was prepared. Consider also the fixed costs line item. The $5,000 F variance comes about because the actual fixed costs of $705,000 are $5,000 lower than the budgeted $710,000.

Effectiveness and Efficiency

Managers may use variance analysis to evaluate performance. Two attributes of performance are commonly measured:

- **Effectiveness:** the degree to which a predetermined objective or target is met
- **Efficiency:** the amount of inputs used to achieve a given level of output

For example, Webb Company set a static-budget target of manufacturing and selling 12,000 units. Only 10,000 units were manufactured and sold. Sales performance would be judged as not fully effective. Whether sales performance was efficient is a separate question. In Webb's case, performance was also inefficient, in sales and especially in the manufacturing area. Webb required a higher quantity of manufacturing inputs than was budgeted to yield the actual 10,000 output units. (Further analysis is presented later in this chapter.)

How are these ideas of effectiveness and efficiency related to variances? The sales-volume variance of operating income is a measure of effectiveness. The flexible-budget variance of operating income is often a measure of efficiency, but it is also affected by changes in selling prices and unit costs. Of course, all variances and hence inferences about effectiveness or efficiency are affected by the care used in formulating credible budgeted amounts.

PRICE VARIANCES AND EFFICIENCY VARIANCES

Objective 6

Compute the price and efficiency variances for direct cost categories

The Level 2 flexible-budget amounts for Webb's variable-costs line items in Exhibit 7-5 are computed by multiplying the budgeted variable costs per output unit of each line item times the actual quantity of output units. The Level 3 variance analysis we now discuss requires information on budgeted input prices and input quantities and actual input prices and input quantities. Both price variances and efficiency variances are computed in Level 3 analysis. These two variances are subdivisions of the Level 2 flexible-budget variance (which is itself a subdivision of the Level 1 static-budget variance).

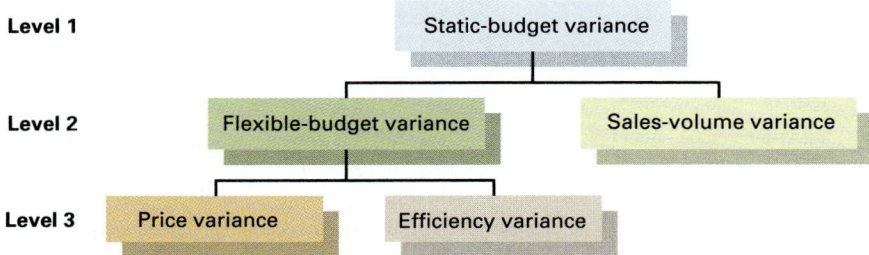

The definitions are:

◆ **Price variance:** the difference between actual price and budgeted price multiplied by the actual quantity of input in question (for example, direct materials purchased, or used). *Price variances* are sometimes called **rate variances,** especially when those variances are for direct labor categories.

◆ **Efficiency variance:** the difference between the actual quantity of input used (such as yards of materials) and the budgeted quantity of input that should have been used, multiplied by the budgeted price. *Efficiency variances* are sometimes called **quantity variances** or **usage variances.**

Definitions of these variances can differ. Always check their precise meaning when you encounter these terms.

The breakdown of the flexible-budget variance into its price- and efficiency-variance components is important information for responsible managers. One manager (say, a purchasing manager) may be responsible for price variances. Another manager (say, a production manager) may be responsible for efficiency variances. The separate computation of price variances enables the efficiency variances to be computed using budgeted input prices. Thus, judgment about efficiency (the quantity of inputs used to produce a given level of output) is not affected by whether actual input prices differ from their budgeted input prices. A word of caution, however, is appropriate. As will be discussed below, the causes of price variances and efficiency variances can be interrelated. For this reason, do not interpret these variances in isolation of each other.

Illustration of Price and Efficiency Variances

Consider Webb's three direct cost categories. The standard cost for each of these three categories is:

Direct Cost Category	(Standard inputs allowed for one output unit)	×	(Standard cost per input unit)	=	Standard cost per output unit
Direct materials	2.00	×	$30	=	$60
Direct manufacturing labor	0.80	×	20	=	16
Direct marketing labor	0.25	×	24	=	6

The following data were compiled regarding *actual* performance at Webb in April 19_4:

Actual units of output	10,000
Direct materials used (and purchased)	
Direct materials costs	$688,200
Square yards of input purchased and used	22,200
Actual price per yard	$31
Direct manufacturing labor	
Direct manufacturing labor costs	$198,000
Manufacturing labor-hours of input	9,000
Actual price per hour	$22
Direct marketing labor	
Direct marketing labor costs	$ 57,600
Marketing labor-hours of input	2,304
Actual price per hour	$25

The actual results and flexible-budget amounts for each category of direct costs for 10,000 actual output units are reported respectively in columns 6 and 3 of Exhibit 7-5 (p. 236).

	Actual Results	Flexible-Budget Amount	Flexible-Budget Variances
Direct materials	$688,200	$600,000	$88,200 U
Direct manufacturing labor	198,000	160,000	38,000 U
Direct marketing labor	57,600	60,000	2,400 F
Total	$943,800	$820,000	$123,800 U

We now use this Webb Company data to illustrate the price and efficiency variances computed in Level 3. Consider first the price variance.

Price Variances

The formula for computing a price variance is:

$$\text{Price variance} = \left(\begin{array}{c}\text{Actual price} \\ \text{of input}\end{array} - \begin{array}{c}\text{Budgeted price} \\ \text{of input}\end{array}\right) \times \left(\begin{array}{c}\text{Actual quantity} \\ \text{of input}\end{array}\right)$$

Price variances for each of Webb's three direct cost categories are:

Direct Cost Category	(Actual price of input − Budgeted price of input)	×	(Actual quantity of input)	=	Price Variance
Direct materials	($31 − $30)	×	22,200	=	$22,200 U
Direct manufacturing labor	($22 − $20)	×	9,000	=	18,000 U
Direct marketing labor	($25 − $24)	×	2,304	=	2,304 U

All three price variances are unfavorable (they reduce operating income) because the actual price of each direct cost input exceeds the budgeted price; that is, Webb paid more per input unit than was budgeted.

Always consider a broad range of possible reasons for price variances arising. For example, Webb's direct materials unfavorable price variance could be due to one or more of the following reasons.

- ◆ Webb's purchasing officer negotiating less skillfully than was assumed in the budget
- ◆ Webb purchasing in smaller lot sizes than budgeted when quantity discounts are available for the larger lot sizes
- ◆ unexpected increase in materials prices due to, say, weather conditions or strikes
- ◆ budgeted purchase prices being set without careful analysis of the market for Webb's materials

Changes in Webb's relationships with its suppliers should be considered when interpreting price variances. For example, assume that Webb moves to a long-term relationship with a single supplier of a material. Webb and the supplier agree to a single purchase price per unit for all purchases in the next six months. It is likely that price variances will be minimal for this material because all purchases will be made from this supplier.

Efficiency Variances

Consider now the efficiency variance. Measurement of efficiency variances requires measurement of *inputs* for a given level of output. For any actual level of output, the efficiency variance is the difference between the input that was actually used and the input that should have been used to achieve that actual output, holding input price constant at the budgeted level:

$$\text{Efficiency variance} = \left(\begin{array}{c}\text{Actual quantity} \\ \text{of input used}\end{array} - \begin{array}{c}\text{Budgeted quantity} \\ \text{of input allowed} \\ \text{for actual output units}\end{array}\right) \times \left(\begin{array}{c}\text{Budgeted price} \\ \text{of input}\end{array}\right)$$

The intuition here is that an organization is inefficient (or efficient) if it uses more (less) inputs than budgeted for the actual output units achieved.

The efficiency variances for each of Webb's direct cost categories are:

Direct Cost Category	(Actual quantity of input used − Budgeted quantity of input allowed for actual output units)	×	(Budgeted price of input) =	Efficiency variance
Direct materials	[22,200 yards − (10,000 units × 2.00 yards)] (22,200 yards − 20,000 yards)	× ×	$30 $30	= $66,000 U
Direct manufacturing labor	[9,000 hours − (10,000 units × 0.80 hours)] (9,000 hours − 8,000 hours)	× ×	$20 $20	= $20,000 U
Direct marketing labor	[2,304 hours − (10,000 units × 0.25 hours)] (2,304 hours − 2,500 hours)	× ×	$24 $24	= $ 4,704 F

The two manufacturing efficiency variances (direct materials and direct manufacturing labor) are each unfavorable because more input was used than was budgeted, resulting in a decrease in operating income. The marketing efficiency variance is favorable because this amount increases operating income.

As with price variances, always consider a broad range of possible reasons for efficiency variances arising. For example, Webb's unfavorable direct manufacturing labor variance could be due to one or more of the following reasons.

- ◆ Webb's personnel officer hired underskilled workers.
- ◆ Webb's production scheduler inefficiently scheduled work, resulting in longer labor time per jacket.
- ◆ Webb's maintenance department did not maintain machines in good condition, resulting in longer labor time per jacket.
- ◆ Budgeted time standards were set without careful analysis of the operating conditions and the employees' skills.

Presentation of Price and Efficiency Variances

Note how the sum of the price variance and the efficiency variance equals the flexible-budget variance:

	Price Variance	Efficiency Variance	Flexible-Budget Variance
Direct materials	$22,200 U	$66,000 U	$88,200 U
Direct manufacturing labor	18,000 U	20,000 U	38,000 U
Direct marketing labor	2,304 U	4,704 F	2,400 F

A columnar presentation of variance analysis of the direct materials cost appears in Exhibit 7-7. This columnar presentation will also be used throughout this chapter, Chapter 8, and Chapter 22.

The price variance as computed in this chapter uses the *actual quantity of inputs* rather than the *budgeted quantity of inputs* allowed for *actual output units*. The result is that inefficiencies in usage affect the price-variance computation. Some companies further refine the price variance into a pure price variance component and a joint price-efficiency variance component. We illustrate this approach for direct materials:

$$\text{Pure price variance} = \left(\begin{array}{c}\text{Actual} \\ \text{price}\end{array} - \begin{array}{c}\text{Budgeted} \\ \text{price}\end{array}\right) \times \left(\begin{array}{c}\text{Budgeted quantity of input} \\ \text{allowed for actual output units}\end{array}\right)$$

$$= (\$31 - \$30) \times (10,000 \text{ units} \times 2 \text{ yards})$$
$$= \$20,000 \text{ U}$$

$$\text{Joint price efficiency variance} = \left(\begin{array}{c}\text{Actual} \\ \text{price}\end{array} - \begin{array}{c}\text{Budgeted} \\ \text{price}\end{array}\right) \times \left(\begin{array}{c}\text{Actual quantity of} \\ \text{input used}\end{array} - \begin{array}{c}\text{Budgeted quantity of} \\ \text{input allowed for} \\ \text{actual output units}\end{array}\right)$$

$$= (\$31 - \$30) \times [22,200 - (10,000 \text{ units} \times 2 \text{ yards})]$$
$$= \$1 \times 2,200 = \$2,200 \text{ U}$$

The pure price variance enables managers to analyze the effect of price changes without considering efficiency issues.

EXHIBIT 7-7

Columnar Presentation of Variance Analysis:
Direct Materials Cost for Webb Company*

Actual Costs Incurred (Actual Input × Actual Price) (1)	Actual Input × Budgeted Prices (2)	Flexible Budget (Budgeted Input Allowed for Actual Output Achieved × Budgeted Price) (3)
(22,200 × $31)	(22,200 × $30)	(20,000 × $30)
$688,200	$666,000	$600,000

    ↑ $22,200 U ↑ $66,000 U ↑
      Price variance   Efficiency variance

    ↑ $88,200 U   ↑
         Flexible-budget variance

*F = favorable effect on operating income; U = unfavorable effect on operating income.

EXHIBIT 7-8
Variance Analysis of Direct Materials

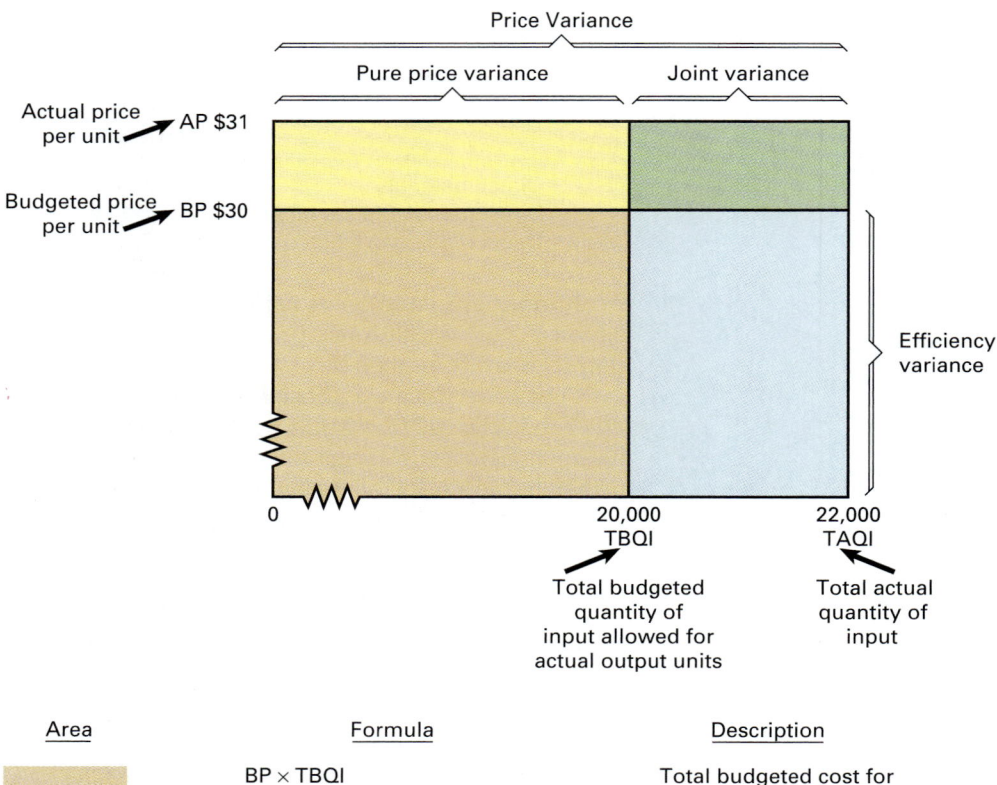

Area	Formula	Description	
	BP × TBQI $30 × 20,000	Total budgeted cost for actual output units	
	(TAQI – TBQI) × BP (22,000 – 20,000) × $30	Efficiency variance	
	(AP – BP) × TBQI ($31 – $30) × 20,000	Pure price variance	Price variance
	(AP – BP) × (TAQI – TBQI) ($31 – $30) × (22,000 – 20,000)	Joint price- efficiency variance	

Exhibit 7-8 is a graphical representation of variances for direct materials. This graphical representation is easiest to interpret when:

a. the actual price (AP) exceeds the budgeted price (BP), and

b. the total actual quantity of input (TAQI) exceeds the total budgeted quantity of input (TBQI).

As Exhibit 7-8 shows for direct materials, four nonoverlapping areas can be distinguished when both (a) and (b) hold. When either (a) or (b) does not hold, the graphical representation is more difficult to understand because the four areas can overlap.

Information Required for Level 3 Analysis

Webb's Level 3 analysis of price and efficiency variances in Exhibit 7-7 requires information about budgeted and actual output units, input prices and input quantities for variable cost line items, and fixed costs. This information is most commonly found in the production and distribution areas of organizations. Hence, analysis of price and efficiency variances is most frequently found in these two areas of the value chain.

Concepts in Action

SUPPLIER PERFORMANCE EVALUATION AT NORTHROP AIRCRAFT DIVISION

Northrop, like many other companies in the aerospace industry, is under great pressure to reduce its costs. Materials costs are a major part of the cost buildup of an aircraft. The Northrop Aircraft Division (NAD) is now evaluating its materials suppliers using a supplier performance index (SPI) to rate individual suppliers:

$$SPI = \frac{\text{Nonconformance costs} + \text{Purchase costs}}{\text{Purchase costs}}$$

Purchase costs are the "total invoiced costs of goods purchased from that supplier during the rating period." The nonconformance costs are based on the number of nonconformance occurrences for each of seven categories times the "standard cost" of resolving the problem (based on an industrial engineering study). These seven categories (with assumed standard hour and cost numbers) are:

Nonconformance Event	Standard Hours to Correct	Standard Costs (hours × $50)
Documentation problem	3	$150.00
Materials review board time required	12	600.00
Return of materials to supplier	6	300.00
Rework required	15	750.00
Undershipment of materials	7	350.00
Overshipment of materials	2	100.00
Late delivery of materials	10	500.00

The following illustrates the computation of an SPI measure for a supplier:

Purchase costs for month		$250,000
Nonconformance costs		
Return of materials to supplier		
(2 units returned at $300 each)	$600	
Undershipment of materials		
(5 shipments undershipped at $350 each)	1,750	
Late delivery of materials		
(3 shipments late at $500 each)	1,500	
Total nonconformance costs		$3,850

The $250,000 and $3,850 amounts are then combined in the SPI measure:

$$SPI = \frac{\$3,850 + \$250,000}{\$250,000} = 1.015$$

NAD uses the SPI rating to evaluate its materials suppliers. The lower the SPI, the more valued the supplier. NAD also used the SPI rating to signal to some suppliers the need to improve their performance.

Source: L. Carr and C. Ittner, "Measuring the Cost of Ownership," *Journal of Cost Management* (Fall 1992).

Although Webb Company uses a standard costing system, this is not a requirement for firms to be able to compute price and efficiency variances. The key requirement is that budgeted and actual input prices and input quantities for individual cost categories be available. While standard costs and standard quantities are one source of these budgeted amounts, other sources exist. For example, the budgeted amounts for April 19_4 could have been based on the actual input prices and actual input quantities in March 19_4.

EXHIBIT 7-9

Overview of Variance Analysis
for Webb Company for April 19_4*

Level 0 Analysis

Actual operating income	$ 25,000
Budgeted operating income	262,000
Static-budget variance of operating income	$237,000 U

Level 1 Analysis

	Actual Results (1)	Static-Budget Variances (2) = (1) − (3)	Static Budget (3)
Units sold	10,000	2,000 U	12,000
Revenues (sales)	$1,850,000	$310,000 U	$2,160,000
Variable costs	1,120,000	68,000 F	1,188,000
Contribution margin	730,000	242,000 U	972,000
Fixed costs	705,000	5,000 F	710,000
Operating income	$ 25,000	$237,000 U	$ 262,000
		$237,000 U	

Total static-budget variance

Level 2 Analysis

	Actual Results (1)	Flexible-Budget Variances (2) = (1) − (3)	Flexible Budget (3)	Sales-Volume Variances (4) = (3) − (5)	Static Budget (5)
Units sold	10,000	0	10,000	2,000 U	12,000
Revenues (sales)	$1,850,000	$ 50,000 F	$1,800,000	$360,000 U	$2,160,000
Variable costs	1,120,000	130,000 U	990,000	198,000 F	1,188,000
Contribution margin	730,000	80,000 U	810,000	162,000 U	972,000
Fixed costs	705,000	5,000 F	710,000	0	710,000
Operating income	$ 25,000	$ 75,000 U	$ 100,000	$162,000 U	$ 262,000
		$ 75,000 U		$162,000 U	

Total flexible-budget variance   Total sales-volume variance

$237,000 U

Total static-budget variance

Level 3 Analysis

Detailed Sales-Price Variances by Customers		Detailed Cost Variances		Detailed Sales-Volume Variances

Detailed Sales-Price Variances by Customers				Detailed Cost Variances		Detailed Sales-Volume Variances
Retail chains	$ 0			Price Variance	Efficiency Variance	Sales-quantity variances
Independent stores	50,000 F					Sales-mix variances
Sales price variances	$50,000 F					(These aspects are
		Direct materials		$22,000 U	$66,000 U	covered in Chapter 22.)
		Direct manufacturing				
		labor		18,000 U	20,000 U	
		Direct marketing				
		labor		2,304 U	4,704 F	
		(Chapter 8 contains further discussion				
		of Level 3 cost-variance analysis.)				

*F = favorable effect on operating income; U = unfavorable effect on operating income.

OVERVIEW OF VARIANCE ANALYSIS

Exhibit 7-9 presents a comprehensive road map of where we have been. Level 0 and Level 1 are reproductions of Exhibit 7-4. Level 2 is a reproduction of Exhibit 7-6. We have just discussed price and efficiency variances, which are Level 3. Two other examples of Level 3 analysis are also presented in Exhibit 7-9—detailed sales-price variances by customers, and detailed sales-volume variances. Consider a detailed sales-price variance analysis by customers. In our Webb Company illustration, suppose half the unit sales were to retail chain customers and half were to independent stores. The average $5 increase in selling price over the budgeted $180 amount may have been due to a zero increase to retail chains and a $10 increase to independent stores:

Level 3 Analysis of Customer Categories	Detailed Sales-Price Variances
Retail chains	$ 0
Independent stores	50,000 F
Total sales-price variance	$50,000 F

Exhibit 7-9 also refers to a Level 3 analysis of detailed sales-volume variances. Webb could, for example, analyze differences between actual unit sales and budgeted unit sales by geographic area, by season of the year, by customer, and so on. Chapter 22 presents detailed analysis of sales-quantity variances and sales-mix variances, which provide important information for managers.

Do not attach undue importance to the specific Level 0, Level 1, Level 2, and so on. By themselves, these levels have no universal meaning in the field of management accounting. The importance of the labels is that they represent increasingly detailed analyses that can assist managers in their planning and control decisions.

PERFORMANCE MEASUREMENT USING STANDARDS

Objective 7

Explain why purchasing performance measures should focus on more factors than just price variances

Price variances and efficiency variances are important performance measures for many manufacturers. For example, direct materials costs constitute over 50% of total costs for some computer companies. Materials price variances for such companies provide information on a key aspect of management performance. Be careful, however, to understand the cause(s) of a variance when using it as a performance measure.

Assume that a purchasing manager has just negotiated a deal that results in a favorable price variance. The deal could have succeeded for any or all of three reasons:

1. The purchasing manager bargained effectively with suppliers.
2. The purchasing manager accepted lower-quality materials.
3. The purchasing manager secured a discount for buying in bulk. However, he bought higher quantities than necessary for the short run, which resulted in excessive inventories.

If the purchasing manager's performance is evaluated solely on price variances, then only reason 1 is considered, and the evaluation will be positive. However, reasons 2 and 3 will likely cause the company to incur additional costs, such as higher material scrap costs and higher storage costs, respectively.

Performance measures increasingly focus on reducing the total costs of the company as a whole. Such a focus is central to the total value-chain-analysis theme in the new management approach. In the purchasing manager example, the company may ultimately lose more because of reasons 2 and 3 than it gains from reason 1. Conversely, manufacturing costs may be deliberately increased (for instance, because higher costs are paid for better materials or more manufacturing labor time) in order to obtain better product quality. In turn, the costs of the better product quality may be more than offset by reductions in the costs of customer service.

If any single performance measure (for example, a labor efficiency cost variance or a consumer rating report) receives excessive emphasis, managers tend to make decisions that maximize their own reported performance in terms of that single performance measure. In turn, managers' actions may conflict with the organization achieving its overall goals. This faulty perspective on performance arises because top management has designed a performance measurement and reward system that does not adequately emphasize total organization objectives.

Continuous Improvement Management Theme

Objective 8
Describe how the continuous improvement theme can be integrated into a standard costing system

Continuous improvement is one of the newly evolving management themes highlighted in this book. See, for example, Exhibit 1-2 (p. 6) and the discussion of *kaizen* in Chapter 6 (pp. 199–200). This theme can be incorporated into the topics covered in this chapter by the use of a **continuous improvement standard cost.** This is a standard cost that is successively reduced over succeeding time periods. A synonym is a **moving cost reduction standard cost.** The standard direct materials cost for each jacket that Webb Company manufactured in April 19_4 is $60 per unit. The standard cost used in variance analysis for subsequent periods could be based on a targeted 1% reduction each period:

Month	Prior Month's Standard Cost	Reduction in Standard Cost	Revised Standard Cost
April 19_4	—	—	$60.00
May 19_4	$60.00	$0.600 (0.01 × $60.00)	59.40
June 19_4	59.40	0.594 (0.01 × $59.40)	58.81
July 19_4	58.81	0.588 (0.01 × $58.81)	58.22

The source of this reduction in standard cost most likely would be efficiency improvements rather than input price reductions. By using continuous improvement standard costs, an organization signals the importance of constantly seeking ways to reduce total costs. For example, one way managers could avoid unfavorable materials efficiency variances with these continuously revised standards is to continually reduce materials waste.

Responsibility and Variances

Organizations adopting a total value chain approach to analyzing variances are recognizing the diversity of the possible sources of variances. Consider an unfavorable materials efficiency variance in the manufacturing area of an organization. The possible sources of this variance include:

◆ poor design of products or processes,
◆ problems with the quality or availability of materials from vendors,
◆ problems arising from poor work in the manufacturing area,
◆ problems arising from inadequate training of the labor force,
◆ problems arising from inappropriate assignment of labor or machines to specific jobs, and
◆ scheduling congestion due to a large number of rush orders required by marketing.

This list is far from exhaustive. However, it does suggest that areas outside manufacturing can cause unfavorable manufacturing efficiency variances.

The most important task in variance analysis is to understand why variances arise and then to use that knowledge to promote learning and continuous improvement. For instance, in our list of examples above, we may seek improvements in product design, in the timeliness of vendor deliveries, in the commitment of the manufacturing labor force to do the job right the first time, and so on. *Variance analysis should not be a tool to "play the blame game." Rather, it should be an essential ingredient that helps promote learning in the organization.*

When to Investigate Variances

When should variances be investigated? Frequently, managers base their answer on subjective judgments, or rules of thumb. The most troublesome aspect of feedback is deciding when a variance (either favorable or unfavorable) is significant enough to warrant management's attention. For critical items, a small deviation may prompt follow-up. For other items, a minimum dollar amount or a certain percentage of deviation from budget may prompt investigations. Of course, a 4% variance in a $1 million materials cost may deserve more attention than a 20% variance in a $10,000 repair cost. Therefore, rules such as "Investigate all variances exceeding $5,000 or 25% of budgeted cost, whichever is lower" are common.

Variance analysis is subject to the same cost-benefit test as other phases of a management control system. The trouble with using rules of thumb is that they are too frequently based on subjective assessments, guesses, or hunches. The field of statistics offers tools that can help managers make decisions regarding variance analysis. These tools help answer the cost-benefit question, and they help separate variances caused by random events from variances that are controllable.

Accounting systems have traditionally implied that a standard is a single acceptable measure. Practically, accountants (and everybody else) realize that the standard is a *band* or *range* of possible acceptable outcomes. Consequently, accountants expect variances to fluctuate randomly within some normal limits.

By definition, a random variance per se is within this band or range. It calls for no corrective action to an existing process. Random variances are attributable to chance rather than to management's implementation decisions.

Further discussion of the decision to investigate variances is in the Appendix to this chapter.

IMPACT OF INVENTORIES

Both our Goodride Company and Webb Company illustrations assumed the following:

1. All units are manufactured and sold in the same accounting period. There are no work-in-process or finished-goods inventories at either the beginning or the end of the period.
2. All direct materials are purchased and used in the same accounting period. There is no direct materials inventory at either the beginning or the end of the period.

Both assumptions can be relaxed without changing the key concepts introduced in this chapter. However, changes in the computation or interpretation of variances may be required when beginning or ending inventories exist.

Suppose direct materials are purchased sometime prior to their use and that direct materials inventories can exist at the beginning or the end of the accounting period. Managers typically want to pinpoint variances closest to when decisions can be informed by those variances. For direct materials price variances, the purchase date almost always will be closer to this time than the usage date. As a result, many organizations compute direct materials price variances using quantities purchased in an accounting period. The Problem for Self-Study at the end of the chapter illustrates this approach, including discussion of how the columnar presentation in Exhibit 7-7 can be adapted to the use of two different times to pinpoint direct materials variances. Chapter 9 provides more extensive discussion of the effect of inventories on a standard costing system.

ILLUSTRATION OF GENERAL LEDGER ENTRIES USING STANDARD COSTS

Control Feature of Standard Costs

Chapter 5 illustrated general-ledger accounting when normal costs are used. We now illustrate general ledger entries when standard costs are used. For illustrative purposes, we focus on direct materials and direct manufacturing labor.

We will continue with the data in the Webb Company illustration with one exception. Assume that during April 19_4 Webb purchases 25,000 square yards of materials. Recall that the actual quantity used is 22,200 yards and that the standard quantity allowed for the actual output achieved is 20,000 yards. The actual purchase price was $31 per square yard, while the budgeted price was $30.

Note that in each of the following entries unfavorable variances are always debits and favorable variances are always credits.

Entry 1a: Isolate the direct materials price variance at the time of purchase by debiting Materials Control at standard prices. This is the earliest date possible to isolate this variance.

1a.	Materials Control		
	(25,000 yards × $30)	$750,000	
	Direct Materials Price Variance		
	(25,000 yards × $1)	25,000	
	Accounts Payable Control		
	(25,000 yards × $31)		$775,000
	To record direct materials purchased.		

Entry 1b: Isolate the direct materials efficiency variance at the time of usage by debiting Work in Process Control at standard quantities allowed for actual output units achieved at standard prices.

1b.	Work in Process Control		
	(20,000 yards × $30)	$ 600,000	
	Direct Materials Efficiency Variance		
	(2,200 yards × $30)	66,000	
	Materials Control		
	(22,200 yards × $30)		$666,000
	To record direct materials used.		

Entry 2: Isolate the direct manufacturing labor price and efficiency variances at the time of usage by debiting Work in Process Control at standard quantities allowed for actual output units achieved at standard rates. Note that Wages Payable Control is always at *actual* wage rates because this is the payroll liability.

2.	Work in Process Control		
	(8,000 hours × $20)	$160,000	
	Direct Manufacturing Labor Price Variance		
	(9,000 hours × $2)	18,000	
	Direct Manufacturing Labor Efficiency Variance		
	(1,000 hours × $20)	20,000	
	Wages Payable Control		
	(9,000 hours × $22)		$198,000
	To record liability for direct manufacturing labor costs.		

The major advantage of this standard costing system is its emphasis on the control feature of standard costs. All variances are isolated at the earliest possible time, when managers can make informed decisions centered on those variances.

End-of-Period Adjustments

Chapter 5 discussed two main approaches to recognizing the underallocated or over-allocated manufacturing overhead at the end of a period:

◆ The restated allocation rate approach, which restates every job cost record using the actual manufacturing overhead rate

◆ The end-of-period account(s) approach, which makes adjustments to one or more of the following end-of-period account balances: work in process, finished goods, and cost of goods sold

Price and efficiency variances can also be disposed of using these same two approaches. The Appendix to Chapter 8 (pp. 290–95) contains further discussion of this topic.

BENCHMARKS AND BUDGETED AMOUNTS

Objective 9
Describe a cost benchmark used for cost management

Another theme in the newly evolving approach to management is a dual internal/external focus (see Exhibit 1-2 on p. 6). Both this theme and the continuous improvement theme have implications for the topics covered in this chapter.

The continuous improvement theme stresses the never-ending search for higher levels of performance. When the benchmark used to evaluate performance is continuously increased, managers can use variances to assess whether continuous improvement is, in fact, occurring in an organization. Earlier in this chapter we illustrated this approach with the concept of a continuous improvement standard cost.

The dual internal/external focus theme stresses the importance of managers' keeping abreast of both changes within their own organization and changes in the external environment with their customers, competitors, suppliers, and so on. Different benchmarks can highlight either an internal or an external focus for managers. Consider the following benchmarks that companies can use to compute performance measures or variances:

◆ Internally generated actual costs from the most recent period.
◆ Internally generated actual costs from the most recent period adjusted for expected improvement.
◆ Internally generated standard costs based on perfection standards.
◆ Internally generated standard costs based on currently attainable standards.
◆ Internally generated costs from other plants or divisions of a company that are designated as "most efficient." For example, the company Yamazaki Mazak compares actual costs at its U.S. and U.K. plants against the actual costs of its Japanese plant that produces similar products.
◆ Externally generated *target cost* numbers based on an analysis of the cost structure of a leading competitor in an industry. (See Chapter 12 for a discussion of target cost.)

Not all of these benchmarks would be found in most accounting systems.

Organizations often use different benchmarks depending on the purpose at hand. For example, one food-processing company uses as its benchmark for each plant a measure of the best level of performance derived from analysis of all its plants. Every six months it ranks its 20 worldwide plants on actual costs per ton of a specific type of food processed. The average actual costs of the two most efficient plants becomes the benchmark in cost management decisions for each of the 20 plants. This is an example of a process now widely called benchmarking. **Benchmarking** is the continuous process of measuring products, services, and activities against best levels of performance, which may be found either inside or outside the organization.

Financial and Nonfinancial Benchmarks

Almost all organizations use a combination of financial and nonfinancial benchmarks rather than relying exclusively on either type. Consider our Webb Company illustration. In its cutting room, fabric is laid out and cut into pieces, which are then matched together and assembled. Control is often exercised at the floor level by focusing on nonfinancial benchmarks such as the number of square yards of cloth used to produce 1,000 jackets or the percentage of jackets started and completed without requiring any rework. Production managers at Webb also likely will use financial benchmarks to evaluate the overall cost efficiency with which operations are being run and to help guide decisions about, say, changing the mix of inputs used in manufacturing jackets. Financial benchmarks are often critical in an organization because they summarize the economic impact of diverse physical activities in a way managers readily understand. Moreover, these managers are often evaluated on results measured against financial benchmarks.

Concepts in Action

AN AUSTRALIAN COMPANY USES BENCHMARKING TO ADOPT WORLD-CLASS PERFORMANCE TARGETS[a]

Smeltco[b], an Australian-headquartered natural resources company, was proud of the major decreases in cost and increases in productivity it had achieved in the 1980s relative to past cost and productivity measures at its smelting plants. However, management believed there was still much room for continued improvement. Smeltco teamed up with the consulting firm of McKinsey & Company to undertake a benchmarking study. The plan was to identify key economic variables at each of its plants and then to determine the gap between each site's performance and the world's best practice in that area. This benchmarking study had six steps:

Step 1	Step 2	Step 3	Step 4	Step 5	Step 6
Develop economic model for each smelter	Identify key variables & activities at smelter	Select benchmark sites around the globe	Define benchmark questions	Gather information about benchmark sites	Determine and interpret performance gap

Feedback

The key activities identified in step 2 included materials purchasing, capacity utilization, labor productivity, and overhead cost control. The benchmark sites visited in step 5 included smelting and other plants in the United States, Japan, Belgium, France, and Germany. The focus was on specific activities (such as materials purchasing) for which world-class standards could be observed. Not all companies were willing to provide Smeltco with the information it sought or even to allow Smeltco to visit their plants. But enough businesses opened up their doors for Smeltco to gather the information it needed to set benchmarks.

The benchmarking study took three months. Smeltco learned it had performance gaps in key areas, such as labor use, energy use, and overhead costs. The company is now using these performance gaps to help focus all members of the organization on the need to improve to achieve world-class operations.

[a]Adapted from A. Walleck, D. O'Halloran, and C. Leader, "'A Race with No Finish': Benchmarking World-Class Performance" (Sydney, Australia: McKinsey & Company, 1991).
[b]Smeltco is a disguised name for the actual company.

PROBLEM FOR SELF-STUDY

PROBLEM

O'Shea Company manufactures ceramic vases. It uses its standard costing system when developing its flexible-budget amounts. In April 19_5, 2,000 finished units were produced. The following information is related to its two direct manufacturing cost categories of direct materials and direct manufacturing labor.

Direct materials used were 4,400 pounds. The standard direct materials input allowed for one output unit is two pounds at $15 per pound. Six thousand pounds were purchased at $16.50 per pound, a total of $99,000.

Actual direct manufacturing labor-hours were 3,250 at a total cost of $40,300. Standard manufacturing labor time allowed is 1.5 hours per output unit, and the standard direct manufacturing labor cost is $12 per hour.

Required

1. Calculate the direct materials price and efficiency variances and the direct manufacturing labor price and efficiency variances. The direct materials price variance will be based on a flexible budget for actual quantities purchased, but the efficiency variance will be based on a flexible budget for actual quantities used.

2. Journal entries for a standard-cost system that isolates variances as early as feasible.

SOLUTION

1. Exhibit 7-10 shows how the columnar presentation of variances introduced in Exhibit 7-7 can be adjusted for the difference in timing between the purchase and use of materials. In particular, note the two sets of computations in column 2 for direct materials. The $90,000 pertains to the direct materials *purchased*; the $66,000 pertains to the direct materials *used*.

EXHIBIT 7-10
Columnar Presentation of Variance Analysis:
Direct Materials and Direct Manufacturing Labor*

	Actual Costs Incurred (Actual Input × Actual price) (1)		Actual Input × Budgeted Price (2)	Flexible Budget (Budgeted Input Allowed for Actual Output Achieved × Budgeted Price) (3)
Direct Materials	(6,000 × $16.50) $99,000	(6,000 × $15.00) $90,000	(4,400 × $15.00) $66,000	(4,000 × $15.00) $60,000
	⤒ $9,000 U ⤒		⤒ $6,000 U ⤒	
	Price variance		Efficiency variance	
Direct Manufacturing Labor	(3,250 × $12.40) $40,300		(3,250 × $12.00) $39,000	(3,000 × $12.00) $36,000
	⤒ $1,300 U ⤒		⤒ $3,000 U ⤒	
	Price variance		Efficiency variance	

*F = favorable effect on operating income; U = unfavorable effect on operating income.

2. Materials Control
   (6,000 pounds × $15) — $90,000
   Direct Materials Price Variance
   (6,000 pounds × $1.50) — 9,000
   Accounts Payable Control
   (6,000 pounds × $16.50) — $99,000

Work in Process Control
   (4,000 pounds × $15) — $60,000
Direct Materials Efficiency Variance
   (400 pounds × $15) — 6,000
Materials Control
   (4,400 pounds × $15) — $66,000

Work in Process Control
   (3,000 hours × $12) — $36,000
Direct Manufacturing Labor Efficiency Variance
   (250 hours × $12) — 3,000
Direct Manufacturing Labor Price Variance
   (3,250 hours × $0.40) — 1,300
Wages Payable Control
   (3,250 hours × 12.40) — $40,300

SUMMARY

The following points are linked to the chapter's learning objectives.

1. Managers find that understanding the causes of *performance gaps* (actual results minus benchmark amounts) enables them to plan better and to manage costs better. A *variance* is a special case of a performance gap where the benchmark used is a financial amount reported in an accounting system.

2. A *static budget* is a budget that is based on one level of output and is not adjusted or altered after it has been finalized. A *flexible budget* is a budget that is developed using budgeted revenue or cost amounts; it is adjusted (flexed) to the actual level of output achieved or expected to be achieved during the budget period. Flexible budgets help managers gain more insight into the causes of variances than do static budgets.

3. A flexible budget can be developed by expressing revenues and variable costs on a budgeted per-output-unit basis and then multiplying these amounts by the appropriate output level. Alternatively, the flexible budget can be developed, first by computing standard costs for its inputs, then summing those amounts to obtain the standard cost per output unit, and finally by multiplying these standard costs per output unit by the appropriate output level.

4. The static-budget variance can be broken into a flexible-budget variance (the difference between the actual result and the flexible-budget amount) and a sales-volume variance. The sales-volume variance arises solely because the actual output units differ from the budgeted output units.

5. Standards represent good or best levels of performance; they usually are developed from a careful study of the operations. Alternative definitions of standards used in computing a standard cost are perfection standards and currently attainable standards.

6. The computation of price variances and efficiency variances helps managers gain insight into two different (but not independent) aspects of performance. Price variances focus on the difference between actual and budgeted input prices. Efficiency variances focus on the difference between actual inputs used and the budgeted inputs allowed for the actual output.

7. Price variances capture only one aspect of a manager's performance. Other aspects include the quality of the inputs the manager purchases and his or her ability to get suppliers to deliver on time.

8. Managers can use continuous improvement standard costs in their accounting system to highlight to all employees the importance of continuously seeking ways to reduce total costs.

9. Increased competition is causing managers to consider external benchmarks as well as internal benchmarks. The framework presented in this chapter can be adapted to incorporate external benchmarks. For example, a cost benchmark could be based on the cost structure of a leading competitor or the cost structure of the most efficient organization for a specific business function area.

APPENDIX: VARIANCE INVESTIGATION DECISIONS AND UNCERTAINTY

The approach to decision making under uncertainty explained in the Appendix to Chapter 3 can be used in variance investigation decisions. Exhibit 7-11 presents an example involving the glass-bottle filling line of a soft-drink bottling plant. For simplicity, we assume two actions and two events:

Actions: a_1 = investigate the process
a_2 = do not investigate the process

Events: x_1 = in-control process (that is, the bottle-filling process is functioning properly)
x_2 = out-of-control process (that is, the bottle-filling process is not functioning properly; for example, it may not be filling bottles to their required level)

Several categories of costs are important in the decision analysis approach summarized in Exhibit 7-11:

C = cost of investigating = $3,000
D = cost per period of the process being out of control = $5,000
M = cost of correcting, if an out-of-control process is discovered = $2,000
L = cost of a process being out of control this period and not corrected this period. This cost includes an estimate of the number of future periods the process wll be out of control without a correction and the cost of correcting it when a decision is finally made to correct it = $37,000.

EXHIBIT 7-11
Decision Table Presentation of Data
for Variance Investigation Decision

Actions	Probability of Event*	
	x_1 = In-Control Process $P(x_1)$ = 0.80	x_2 = Out-of-Control Process $P(x_2)$ = 0.20
a_1 = Investigate the process	C = $3,000	$C + D + M$ = $10,000
a_1 = Do not investigate the process	$ 0	L = $37,000

*C = Cost of investigating = $3,000.
D = Cost per period of being out of control = $5,000.
M = Cost of correcting, if out-of-control process discovered = $2,000.
L = Cost of a process being out of control this period and not corrected this period = $37,000.

We assume that it takes one period to determine whether a process is out of control and then to correct it. Our simple example also assumes (1) that a process that goes out of control stays out of control for the planning period and (2) that when an out-of-control process is corrected, it stays in control until the end of the planning period. The length of each time period in Exhibit 7-11 is one day. The decision analysis framework we outline here, however, is a general one that can be applied to much longer time periods.

The probability of the process being in control is 0.80, and the probability of its being out of control is 0.20:

$$P(x_1) = 0.80 \quad P(x_2) = 0.20$$

The expected costs from the investigate (a_1) and do not investigate (a_2) actions are:

Investigate:

$$E(a_1) = [P(x_1) \times C] + [P(x_2) \times (C + D + M)]$$
$$= (0.80 \times \$3,000) + (0.20 \times \$10,000) = \$4,400$$

Do not investigate:

$$E(a_2) = [P(x_1) \times \$0] + [P(x_2) \times L]$$
$$= (0.80 \times \$0) + (0.20 \times \$37,000) = \$7,400$$

The optimal action is investigate the process, because the expected cost is $3,000 less than the do not investigate alternative ($4,400 – $7,400).

Role of Probabilities

The preceding example illustrates the critical role of probabilities in a manager's deciding whether to investigate. A lower probability—say, 0.05—that the process is out of control will reduce the desirability of conducting an investigation:

Investigate:

$$E(a_1) = (0.95 \times \$3,000) + (0.05 \times \$10,000) = \$3,350$$

Do not investigate:

$$E(a_2) = (0.95 \times \$0) + (0.05 \times \$37,000) = \$1,850$$

When $P(x_2) = 0.05$, the optimal action is do not investigate.

The probability level at which a manager would be indifferent between investigating and not investigating whether a process is out of control is[1]:

$$P(x_2) = \frac{C}{L - (D + M)}$$

Given $C = \$3,000$, $D = \$5,000$, $M = \$2,000$, and $L = \$37,000$, then

$$P(x_2) = \frac{\$3,000}{\$37,000 - (\$5,000 + \$2,000)} = 0.10$$

The expected cost when $P(x_2) = 0.10$ for both a_1 and a_2 is the same, $3,700:

Investigate:

$$E(a_1) = (0.90 \times \$3,000) + (0.10 \times \$10,000) = \$3,700$$

---

[1]This formula is derived as follows:

$$\text{Given} \quad P(x_1) + P(x_2) = 1$$
$$\text{then} \quad P(x_1) = 1 - P(x_2)$$

Equating the expected costs of a_1 and a_2 and substituting $1 - P(x_2)$ for $P(x_1)$ yields:

$$\{[1 - P(x_2)] \times C\} + [P(x_2) \times (C + D + M)] = P(x_2) \times L$$
$$C - [C \times P(x_2)] + [C \times P(x_2)] + [D \times P(x_2)] + [M \times P(x_2)] = L \times P(x_2)$$
$$C = P(x_2)[L - (D + M)]$$
$$P(x_2) = \frac{C}{L - (D + M)}$$

Do not investigate:

$$E(a_2) = (0.90 \times \$0) + (0.10 \times \$37,000) = \$3,700$$

Investigation is desirable in the Exhibit 7-11 example only if the probability of the process being out of control exceeds 0.10.

Estimation of Costs and Benefits

The usefulness of our cost-benefit calculation critically depends on having accurate estimates of $P(x_1)$ and $P(x_2)$ and of the C, D, M, and L parameters. Much care should be taken in estimating them.

Consider the estimation of C, the cost of investigation. Signals that a process is potentially out of control can often occur well before the cost-variance report is prepared. For example, line operators on the soft-drink bottling production line may complain about the quality of the glass bottles if they encounter more breakages than expected. The line manager may immediately direct that a different batch of bottles be used on the production line. The investigation of the first batch begins. By the time the line manager receives a routine materials yield variance report, the preliminary investigation may already be completed. In this case, estimation of C for an investigation prompted by a materials yield variance report should include *only* the incremental investigation costs beyond those already incurred at the time the report is received. This example illustrates how the timelines of cost accounting data can affect the values of parameters used in cost-variance investigation models.

Surveys of Company Practice

THE DECISION TO INVESTIGATE VARIANCES

A survey of U.S. managers reported the following approaches to investigating direct materials and direct labor variances[a]:

	Direct Materials (%)	Direct Labor (%)
1. All variances investigated	6.9	5.3
2. Variances over prescribed dollar limits investigated	34.8	31.0
3. Variances over prescribed percentage limits investigated	12.2	14.1
4. Statistical procedures used to select cases for investigation	0.9	0.9
5. Variances never investigated	0.0	0.9
6. Judgment used to decide if investigation is needed	45.2	47.8
	100.0%	100.0%

Investigating all variances may be justified if the cost of the process being out of control is extremely high. An example is the manufacture of a door lock for a space shuttle.

[a]Adapted from Gaumnitz and Kollaritsch, "Manufacturing Variances." Full citation is in Appendix A.

TERMS TO LEARN

This chapter and the Glossary at the end of the book contain definitions of the following important terms:

benchmarking *(p. 249)*   continuous improvement standard cost *(246)*
currently attainable standard *(234)*   effectiveness *(237)*   efficiency *(237)*
efficiency variance *(238)*   favorable variance *(228)*   flexible budget *(228)*
flexible-budget variance *(230)*   moving cost reduction standard cost *(246)*
perfection standard *(234)*   performance gap *(227)*   price variance *(238)*
quantity variance *(238)*   rate variance *(238)*   sales-volume variance *(230)*
standard *(226)*   standard cost *(227)*   standard input *(226)*   static budget *(228)*
unfavorable variance *(228)*   usage variance *(238)*   variance *(227)*

ASSIGNMENT MATERIAL

QUESTIONS

7-1   Describe the relationship between a performance gap and a variance.

7-2   Give an example of the three types of benchmarks of interest to managers.

7-3   What is the key question in deciding which variances should be computed and analyzed?

7-4   Why might managers find a Level 2 flexible-budget analysis more informative than a Level 1 static-budget analysis?

7-5   Comment on the following statement made by a manufacturing plant controller: "I would prefer to use a flexible budget rather than a static budget, but I cannot, as we don't have a standard costing system."

7-6   "Performance may be both effective and efficient, but either condition can occur without the other." Give an example of effectiveness. Give an example of efficiency.

7-7   What is the relationship between a budgeted amount and a standard amount?

7-8   List four purposes of standard costs.

7-9   How might a manager gain insight into the causes of a flexible-budget variance for direct materials?

7-10   List four sources of an unfavorable materials efficiency variance.

7-11   Describe why direct materials price and direct materials efficiency variances may be computed with reference to different points in time.

7-12   List and briefly describe two different types of standard costs.

7-13   How might the continuous improvement theme be incorporated into the process of setting standard costs?

7-14   Give two examples of performance gaps based on benchmarks not found in accounting systems.

7-15   Comment on the following statement made by a plant supervisor: "Meetings with my plant accountant are frustrating. All he wants to do is pin the blame for the many variances he reports."

EXERCISES AND PROBLEMS

7-16   **Flexible budget.**   The budgeted prices for direct materials, direct manufacturing labor, and direct marketing (distribution) labor per attaché case are $40, $8, and $12, respectively. The president is pleased with the following performance report:

	Actual Results	Static Budget	Variance
Direct materials	$364,000	$400,000	$36,000 F
Direct manufacturing labor	78,000	80,000	2,000 F
Direct marketing (distribution) labor	110,000	120,000	10,000 F

Required

Actual output was 8,800 attaché cases. Is the president's pleasure justified? Prepare a revised performance report that uses a flexible budget and a static budget. Assume all three direct costs items above are variable costs.

7-17 Materials and manufacturing labor variances. Consider the following data collected for Great Homes, Inc.:

	Direct Materials	Direct Manufacturing Labor
Costs incurred (actual input × actual price)	$200,000	$90,000
Actual input × budgeted price	214,000	86,000
Budgeted input allowed for actual output × budgeted price	225,000	80,000

Required

Compute the price, efficiency, and flexible-budget variances for direct materials and direct manufacturing labor.

7-18 Flexible-budget preparation. The managing partner of Roan Music Box Fabricators has become aware of the disadvantages of static budgets. She asks you to prepare a flexible budget for October 19_4 for the main style of music box. The following partial data are available for the actual operations in August 19_4 (a recent typical month):

Boxes produced and sold	4,500
Direct materials costs	$90,000
Direct manufacturing labor costs	$67,500
Depreciation and other fixed manufacturing costs	$50,700
Average selling price per box	$ 70
Fixed marketing costs	$81,350

Assume no beginning or ending inventory of music boxes.

A 10% increase in the selling price is expected in October. The only variable marketing cost is a commission of $5.50 per unit paid to the manufacturer's representatives, who bear all their own costs of traveling, entertaining customers, and so on. A patent royalty of $2 per box manufactured is paid to an independent design firm. Salary increases that will become effective in October are $12,000 per year for the production superintendent and $15,000 per year for the sales manager. A 10% increase in direct materials prices is expected to become effective in October. No changes are expected in direct manufacturing labor wage rates or in the productivity of the direct manufacturing labor personnel. Roan uses a normal costing system and does not have standard costs for any of its inputs.

Required

1. Prepare a flexible budget for October 19_4, showing budgeted amounts at each of three output levels of music boxes: 4,000 units, 5,000 units, and 6,000 units. (Use the flexible-budget approach of developing budgeted revenue and variable costs on a budgeted per-output-unit basis.)
2. Why might Roan Music Box Fabricators find a flexible budget more useful than a static budget? Explain.

7-19 Flexible budgets, variance analysis. You have been hired as a consultant by Mary Flanagan, the president of a small manufacturing company that makes automobile parts. Flanagan is an excellent engineer, but she has been frustrated by working with inadequate cost data.

You helped install flexible budgeting and standard costs. Flanagan has asked you to consider the following May data and recommend how variances might be computed and presented in performance reports:

Static budget in output units	20,000
Actual output units produced and sold	23,000
Budgeted selling price per output unit	$ 40
Budgeted variable costs per output unit	$ 25
Budgeted total fixed costs per month	$200,000
Actual revenues	$874,000
Actual variable costs	$630,000
Favorable variance in fixed costs	$ 5,000

Flanagan was disappointed. Although output units sold exceeded expectations, operating income did not. Assume that there was no beginning or ending inventory.

Required
1. You decide to present Flanagan with alternative ways to analyze variances so that she can decide what level of detail she prefers. The reporting system can then be designed accordingly. Prepare an analysis similar to Levels 0, 1, and 2 in Exhibit 7-9, page 244.
2. What are some likely causes for the variances you report in requirement 1?

7-20 Flexible budget preparation and analysis. Bank Management Printers, Inc. produces luxury checkbooks with three checks and stubs per page. Each checkbook is designed for an individual customer and is ordered through the customer's bank. The company's operating budget for September 19_4 included these data:

Number of checkbooks	15,000
Selling price per book	$ 20
Variable costs per book	$ 8
Total fixed costs for the month	$145,000

The actual results for September 19_4 were:

Number of checkbooks produced and sold	12,000
Average selling price per book	$ 21
Average variable costs per book	$ 7
Total fixed costs for the month	$150,000

The executive vice-president of the company observed that the operating income for September was much less than anticipated, despite a higher-than-budgeted selling price and a lower-than-budgeted variable cost per unit. You have been asked to provide explanations for the disappointing September results.

Bank Management develops its flexible budget on the basis of budgeted per-output-unit revenue and per-output variable costs without detailed analysis of budgeted inputs.

Required
1. Prepare a Level 1 analysis of the September performance.
2. Prepare a Level 2 analysis of the September performance.
3. Why might Bank Management find the Level 2 analysis more informative than the Level 1 analysis? Explain your answer.

7-21 Flexible budget preparation, service sector. Meridian Finance helps prospective homeowners of substantial means to find low-cost financing and assists existing homeowners in refinancing their current loans at lower interest rates. Meridian works only for customers with excellent borrowing capacity. Hence, Meridian is able to obtain a loan for every customer with whom it decides to work.

Meridian charges clients 1/2% of the loan amount it arranges. In 19_3, the average loan amount per customer was $199,000. In 19_4, the average loan amount was $200,210. In its 19_5 flexible budgeting system, Meridian assumes the average loan amount will be $200,000. Budgeted cost data per loan application for 19_5 are:

- Professional labor: 6 budgeted hours at a budgeted rate of $40 per hour
- Fees for filing: budgeted at $100
- Checks on credit worthiness: budgeted at $120
- Courier mailings: budgeted at $50

Office support (the costs of leases, secretarial workers, and others) is budgeted to be $31,000 per month. Meridian Finance views this amount as a fixed cost.

Required

1. Prepare a static budget for November 19_5 assuming 90 loan applications.
2. Actual loan applications in November 19_5 were 120. Other actual data for November 19_5 were:

- Professional labor: 7.2 hours per loan application at $42 per hour
- Loan filing fees: $100 per loan application
- Credit-worthiness checks: $125 per loan application
- Courier mailings: $54 per loan application

Office support costs for November 19_5 were $33,500. The average loan amount for November 19_5 was $224,000. Meridian received its 1/2% fee on all loans. Prepare a Level 2 variance analysis of Meridian Finance for November 19_5.

7-22 Professional labor efficiency and effectiveness. (Continuation of 7-21) Meridian Finance is analyzing the efficiency and effectiveness of its professional labor staff.

Required

1. Compute professional labor price and efficiency variances for November 19_5. (Compute labor price on a per-hour basis.)
2. What factors would you consider in evaluating the effectiveness of professional labor in November 19_5?

7-23 Flexible and static budgets, service company. Avanti Transportation Company executives have had trouble interpreting operating performance for a number of years. The company has used a budget based on detailed expectations for the forthcoming quarter. For example, the condensed performance report for a midwestern branch for the most recent quarter was:

	Actual Results	Budget	Variance*
Revenue	$9,500,000	$10,000,000	$500,000 U
Variable costs			
Fuel	986,000	1,000,000	14,000 F
Repairs and maintenance	98,000	100,000	2,000 F
Supplies and miscellaneous	196,000	200,000	4,000 F
Variable labor payroll	5,500,000	5,700,000	200,000 F
Total variable costs†	6,780,000	7,000,000	220,000 F
Fixed costs			
Supervision	200,000	200,000	0
Rent	200,000	200,000	0
Depreciation	1,600,000	1,600,000	0
Other fixed costs	200,000	200,000	0
Total fixed costs	2,200,000	2,200,000	0
Total costs	8,980,000	9,200,000	220,000 F
Operating income	$ 520,000	$ 800,000	$280,000 U

*U = unfavorable; F = favorable.

†For purposes of this analysis, assume that all these costs are purely variable (in relation to revenue dollars). In practice, many have to be subdivided into variable and fixed components before a meaningful analysis can be made. Also assume that the prices and mix of services sold remain unchanged.

Although the branch manager was upset about the unfavorable revenue variance, he was happy that his cost performance was favorable; otherwise his operating income would have been even lower.

His immediate superior, the vice-president for operations, was totally unhappy and remarked: "I can see some merit in comparing actual performance with budgeted performance, because we can see whether actual revenue coincided with our best guess for budget purposes. But I can't see how this performance report helps us evaluate the cost control performance of the branch manager."

Required

1. Prepare a columnar flexible budget for Avanti at revenue levels of $9,000,000, $10,000,000, and $11,000,000. Use the format of Exhibit 7-2, page 231. Assume that the prices and mix of products sold are equal to the budgeted prices and mix.

2. Express the flexible budget for costs in formula form.

3. Prepare a condensed contribution-format income statement showing the static-budget variance, the sales-volume variance, and the flexible-budget variance. Use the format of Exhibit 7-3, page 231.

7-24 Direct materials variances, long-term agreement with supplier. Yamazaki Mazak manufactures large-scale machining systems that are sold to other industrial companies. Each machining system has a sizable direct materials cost, consisting primarily of the purchase price for a metal compound. For its Lexington, Kentucky, manufacturing facility, Mazak has a long-term contract with Fuji Metals. Fuji will supply to Mazak up to 2,400 pounds of metal per month at a fixed purchase price of $120 per pound for each month in 19_5. For purchases above 2,400 pounds in any month, Mazak renegotiates the price for the additional amount with Fuji Metals (or another supplier). The standard price per pound is $120 for each month in the January to December 19_5 period.

Production data, direct materials actual usage in dollars, and direct materials actual price per pound for the January to May 19_5 period are:

	Number of Machining Systems Produced	Total Actual Direct Materials Usage	Average Actual Direct Materials Purchase Price per Pound of Metal
January	10	$242,400	$120
February	12	286,560	120
March	18	442,260	126
April	16	395,264	128
May	11	253,440	120

The average actual direct materials purchase price is for all units purchased in that month. Assume that (a) the direct materials purchased in each month are all used in that month and (b) each machining system is started and completed in the same month.

The Lexington facility is one of three plants that Mazak operates to manufacture large-scale machining systems. The other plants are in Worcester, the United Kingdom, and Tokyo, Japan.

Required

1. Assume that Mazak's standard materials input per machining system is 198 pounds of metal. Compute the direct materials price variance and direct materials efficiency variance for each month of the January to May 19_5 period.

2. How does the signing of a long-term agreement with a supplier—an agreement that includes a fixed-purchase-price clause—affect the interpretation of a materials price variance?

7-25 Continuous improvement standards. Assume in Problem 7-24 that Mazak uses the following continuous improvement standards for the direct materials input per machining system:

January	200 pounds of metal
February	198 pounds of metal
March	196 pounds of metal
April	194 pounds of metal
May	192 pounds of metal

Required

1. Using these standards, compute the direct materials efficiency variance for each month of the January to May 19_5 period.
2. Outline two basic ways that Mazak might develop continuous improvement standards for the direct materials input per machining system.

7-26 Direct manufacturing labor variances. (CMA, adapted) Day-Mold was founded several years ago by two designers who had developed several popular lines of living room, dining room, and bedroom furniture for other companies. The designers believed that their design for dinette sets could be standardized and would sell well. They formed their own company and soon had all the orders they could complete in their small plant in Dayton, Ohio.

From this strong beginning, the firm continued to succeed. The owners bought a microcomputer and software that produced financial statements, which an employee prepared. The owners thought that the information they needed was contained in these statements.

Recently, however, the employees have been requesting raises. The owners wonder how to evaluate the employees' requests. The owners have hired a consultant who is a CMA to implement a standard costing system. The consultant believes that the calculation of variances will aid management in setting responsibility for manufacturing labor's performance.

The floor supervisors believe that under normal conditions the dinette set can be assembled with five hours of direct manufacturing labor at a cost of $20 per hour. The consultant has assembled manufacturing labor cost information for the most recent month and would like your advice in calculating variances.

During the month, the actual direct manufacturing labor wages were $127,600 for 5,760 hours worked. The plant produced 1,200 dinette sets during the month.

Required

1. Compute price and efficiency variances for direct manufacturing labor.
2. Comment on the variances in requirement 1.

7-27 Professional labor variances, efficiency comparisons.   Norma Tuck is manager of The Tax Experts, a firm that provides assistance in the preparation of individual tax returns. Because of the highly seasonal nature of her business, Tuck hires staff on a monthly basis from two accounting placement firms—Professional Assist (PA) and Office Support (OS). In February 19_5, The Tax Experts hired 12 staff members from PA and 10 from OS. PA is the prestige firm in its area. OS is a recently formed firm.

Tuck budgets the following for February 19_5:

	PA Staff	OS Staff
Budgeted hourly rate	$45	$40
Budgeted hours per tax return	0.40	0.50

Actual results for February 19_5 were:

	PA Staff	OS Staff
Actual hourly rate	$48	$42
Actual hours per tax return	0.42	0.46
Number of tax returns completed	4,608	3,600

Required

1. Compute professional labor price and efficiency variances for (a) the 12 PA staff and (b) the 10 OS staff hired in February 19_5.

2. Comment on the efficiency of the PA and OS staff The Tax Experts hired.

3. What factors other than efficiency might Tuck consider in deciding whether to hire staff from PA or OS?

7-28 Direct labor variances, solving for unknowns. The city of Chicago has a maintenance shop where truck repairs are performed. Through the years, various labor standards have been developed to judge performance. However, during a March strike, some labor records vanished. The actual hours of input were 1,000. The direct labor flexible-budget variance was $1,700, favorable. The standard labor price was $14 per hour; however, a recent labor shortage had necessitated using higher-paid workers for some jobs and had produced a labor price variance for March of $400, unfavorable.

Required

1. Compute the actual direct labor price per hour.

2. What were the standard hours allowed for actual output achieved?

7-29 Materials and manufacturing labor variances, nonfinancial measures. The production report of Illawarra Office Equipment for April 19_5 included the following information pertaining to the manufacture of a line of tables:

	Direct Materials	Direct Manufacturing Labor
Actual price per unit of input (board-feet and hours)	$14	$18
Standard price per unit of input	$12	$20
Standard inputs allowed per unit of output	5	2
Actual units of input	48,000	22,000
Actual units of output (product)	10,000	10,000

Required

1. Compute the price, efficiency, and flexible-budget variances for direct materials and direct manufacturing labor.

2. Give a plausible explanation for the performance.

3. What type of data would the production manager probably watch most closely on a day-to-day basis?

7-30 Price and efficiency variances, journal entries. Chemical, Inc., has set up the following standards per finished unit for direct materials and direct manufacturing labor:

Direct materials: 10 lb at $3.00 per lb	$30.00
Direct manufacturing labor: 0.5 hours at $20.00 per hour	10.00

The number of finished units budgeted for February 19_5 was 10,000; 9,810 units were actually produced.
    Actual results in February 19_5 were:

Direct materials: 98,073 lb used	
Direct manufacturing labor: 4,900 hours	$102,900

Assume that there was no beginning inventory of either direct materials or finished units.
    During the month, materials purchases amounted to 100,000 lb, at a total cost of $310,000. Price variances are isolated upon purchase. Efficiency variances are isolated at the time of usage.

Required

1. Compute the February 19_5 price and efficiency variances of direct materials and direct manufacturing labor.

2. Prepare journal entries to record the variances in requirement 1.

3. Comment on the February 19_5 price and efficiency variances of Chemical, Inc.

4. Why might Chemical, Inc., calculate materials price variances and materials efficiency variances with reference to different points in time?

7-31 Materials and manufacturing labor variances, nonfinancial measures. (SMA, adapted) The Carberg Co. manufactures and sells a single product for $90. The company uses a standard-costing system, isolating all variances as soon as possible. The standards for one finished output unit include:

Direct materials (1.0 kg at $20 per kg)	$20
Direct manufacturing labor (0.6 hour at $20 per hour)	12

Actual results for November were:

Output units produced	5,100
Direct materials purchased at $21.00 per kg	5,200 kg
Direct materials used in production	5,300 kg
Direct manufacturing labor at $20.40 per hour	3,200 hr

Required

1. Compute the price and efficiency variances for direct materials and direct manufacturing labor.

2. What type of data would the production manager probably watch most closely on a day-to-day basis?

7-32 Materials and manufacturing labor variances, standard costs. Consider the following selected data regarding the manufacture of a line of upholstered chairs:

	Standards per Chair
Direct materials	Two square yards of input at $10 per square yard
Direct manufacturing labor	One-half hour of input at $20 per hour

The following data were compiled regarding actual performance: actual output units (chairs) produced, 20,000; square yards of input purchased and used, 37,000; price per square yard, $10.20; direct manufacturing labor costs, $176,400; actual hours of input, 9,000; labor price per hour, $19.60.

Required

1. Show computations of price and efficiency variances for direct materials and for direct manufacturing labor. Give a plausible explanation of why the variances occurred.

2. Suppose 60,000 square yards of materials were purchased (at $10.20 per square yard) even though only 37,000 square yards were used. Suppose further that variances are identified with their most likely control point; accordingly, direct materials price variances are isolated and traced to the purchasing department rather than to the production department. Compute the price and efficiency variances under this approach.

7-33 Journal entries and T-accounts. Prepare journal entries and post them to T-accounts for all transactions in Problem 7-33, including requirement 2. Summarize in three sentences how these journal entries differ from the normal costing entries described in Chapter 5, page 145.

7-34 Flexible budget. Refer to Problem 7-32. Suppose the static budget was for 24,000 units of output. The general manager is thrilled about the following report:

	Actual Results	Static Budget	Variance
Direct materials	$377,400	$480,000	$102,600 F
Direct manufacturing labor	$176,400	$240,000	$ 63,600 F

Required
Is the manager's glee warranted? Prepare a report that provides a more detailed explanation of why the static budget was not achieved. Actual output was 20,000 units.

7-35 Direct materials and manufacturing labor variances, nonfinancial measures.
Consider some data for the Cobb Company for April 19_5, a manufacturer of specialty rainjackets:

Output units produced, 10,700
Standard input amounts per output unit
    Direct materials (4 square yards at $5.00 per square yard) $20
    Direct manufacturing labor (2 hours at $16.00 per hour)   32

Actual data:

Direct materials costs	$270,000	Direct manufacturing labor costs	$343,200
Square yards of input purchased and used	50,000	Direct manufacturing labor hours of input	22,000
Price per square yard	$ 5.40	Direct manufacturing labor price per hour	$ 15.60

Required
1. Compute (a) price and efficiency variances for direct materials and (b) price and efficiency variances for direct manufacturing labor. Present your answer in a format similar to the analysis that appears in Exhibit 7-7, page 241.
2. Suppose that the Cobb Company control system was designed to isolate materials price variances at the time of purchase rather than when usage of the materials occurs. Suppose further that 60,000 square yards of materials were purchased during April and that only 50,000 square yards were issued to production. Compute the price variance that would be reported by the control system.
3. What type of data would the production manager probably watch most closely on a day-to-day basis?

7-36 Direct materials and manufacturing labor variances, solving unknowns. (CPA adapted) On May 1, 19_5, Bovar Company began the manufacture of a new paging machine known as "Dandy." The company installed a standard costing system to account for manufacturing costs. The standard costs for a unit of Dandy follow:

Direct materials (3 lb at $5 per lb)	$15.00
Direct manufacturing labor (1/2 hour at $20 per hour)	10.00
Manufacturing overhead (75% of direct manufacturing labor costs)	7.50
	$32.50

The following data were obtained from Bovar's records for the month of May:

	Debit	Credit
Sales		$125,000
Accounts payable control (for May's purchases of direct materials)	$68,250	
Direct materials price variance	3,250	
Direct materials efficiency variance	2,500	
Direct manufacturing labor price variance	1,900	
Direct manufacturing labor efficiency variance		2,000

Actual production in May was 4,000 units of Dandy, and actual sales in May was 2,500 units.

The amount shown above for direct materials price variance applies to materials purchased during May. There was no beginning inventory of materials on May 1, 19_5.

Required

1. Compute each of the following items for Bovar for the month of May. Show computations.
   a. Standard direct manufacturing labor hours allowed for actual output achieved.
   b. Actual direct manufacturing labor hours worked.
   c. Actual direct manufacturing labor wage rate.
   d. Standard quantity of direct materials allowed (in pounds).
   e. Actual quantity of direct materials used (in pounds).
   f. Actual quantity of direct materials purchased (in pounds).
   g. Actual direct materials price per pound.

7-37 Direct materials and direct manufacturing labor variances, responsibility reports.
The Chester Company uses standard costs for metalworking operations. The purchasing manager, Amy Strotz, is responsible for materials price variances, and the production manager, Juan Morales, is responsible for materials efficiency variances and direct manufacturing labor price and efficiency variances.

The standard price for metal used as a principal material was $4.00 per pound. The standard allowance was six pounds per finished unit of product.

The standard price for direct manufacturing labor was $14 per hour. The standard allowance was one-half hour per finished unit of product.

During the past week, 10,000 output units were produced. However, labor trouble caused the production manager to use temporary workers. Actual direct manufacturing labor costs were $78,000 for 6,500 actual hours. Also, 80,000 pounds of metal were purchased for $3.60 per pound; and 71,000 pounds of metal were consumed during production.

Required

1. Compute the materials price variance, materials efficiency variance, direct manufacturing labor price variance, and direct manufacturing labor efficiency variance.
2. As a supervisor of both the purchasing manager and the production manager, how would you interpret the feedback provided by the variances computed in requirement 1?
3. What are the flexible-budget allowances for the production manager for direct materials and direct manufacturing labor? What would they be if production were 7,000 output units?
4. Prepare a short performance report for the production manager if 10,000 output units were produced. Show three columns: charges to department, flexible budget, and variance.
5. Describe how the managers probably control their day-to-day operations.

7-38 Price and efficiency variances, problems in standard setting, ethics. Savannah Fashions manufactures shirts for retail chains. Jorge Andersen, the controller, is becoming increasingly disenchanted with Savannah's six-month-old standard costing system. The budgeted amounts for both its direct materials and direct manufacturing labor are drawn from its standard costing system. The budgeted and actual amounts for July 19_4 were:

	Budgeted	Actual
Shirts manufactured	4,000	4,488
Direct materials cost	$20,000	$20,196
Direct materials units (rolls of cloth)	400	408
Direct manufacturing labor costs	$18,000	$18,462
Direct manufacturing labor-hours	1,000	1,020

There was no beginning or ending inventory of materials.

Andersen observes that in the last six months he has rarely seen an unfavorable variance, of any magnitude. The standard costing system is based on a study of the operations conducted by an independent consultant. Andersen decides to play "detective" and makes some unobtrusive observations of the workforce at the plant. He notes that even at their current output levels, the workers seemed to have much time for discussion about baseball, sitcoms, and the local hot fishing spots.

At a recent industry conference, Andersen had a discussion with Mary Blanchard, the controller of Winston Fabrics. Blanchard told him that Winston had employed the same independent consultant to design a standard costing system. However, the company dismissed him after two weeks. The Winston employees quickly became aware of the consultant observing their work. They immediately took actions to increase the time taken for making each shirt and to reduce the number of shirts made from each roll of cloth. Blanchard said she had decided to use the last month's actual input prices and input quantities as benchmarks when making inferences about efficiency in the current month.

Required

1. Compute the price and efficiency variances of Savannah Fashions for direct materials and direct manufacturing labor in July 19_4.
2. Describe the types of actions the employees at Winston Fabrics may have taken to reduce the accuracy of the standards set by the independent consultant. Why would employees take those actions? Is this behavior ethical?
3. How might Andersen determine if the standards currently being used at Savannah Fashions are demanding enough?
4. Why do so many firms use standard costs if there is the possibility that employees will attempt to undermine their reliability?

Coverage of the Chapter Appendix

7-39 Cost-benefit analysis of variance investigation decision, timing of variance reports. Southern Beverages bottles mineral waters. Nancy Daus, the newly appointed production manager, has just received the weekly direct materials efficiency variance report. Daus's predecessor used the following estimates in his variance investigation decisions:

$$C = \text{cost of investigating} = \$10,000$$
$$D = \text{cost per period of the process being out of control} = \$12,000$$
$$M = \text{cost of correcting, if an out-of-control process is discovered} = \$3,000$$

The file containing this information also included a report on "L, the cost of a process being out of control this period and not being corrected this period." No estimate on L, however, was included in the file. Daus remembered that her predecessor had been indifferent about conducting an investigation when there was a probability of 0.60 that the process was in control.

Required

1. What is the estimate of L that Daus's predecessor used in deciding whether to investigate a variance?
2. Daus decides to have a daily direct materials efficiency report. Would you recommend she use the same estimates of $C, D, M,$ and L as her predecessor in deciding whether to investigate a variance? Explain.

7-40 Cost-benefit analysis of variance investigation decision. You are the manager of a manufacturing process. A materials efficiency variance of $10,000 has been reported for the past week's operations. You are trying to decide whether to investigate this variance. You feel that if you do not investigate and the process is out of control, the cost savings (L) over the planning horizon is $3,800. The cost to investigate is $500. The cost per period of being out of control (D) is $700. If an out-of-control process is discovered, the cost of correcting it (M) is $300. You assess the probability that the process is out of control at 0.30.

Required

1. Should the process be investigated? What are the expected costs of investigating and of not investigating?
2. At what level of probability that the process is out of control would the expected costs of each action be the same?
3. If the cost variance is $10,000, why is L only $3,800?

8

Flexible Budgets, Variances, and Management Control: II

- Planning of variable and fixed overhead costs
- Developing budgeted variable overhead rates
- Variable overhead cost variances
- Developing fixed overhead rates
- Fixed overhead cost variances
- Output level (production volume) overhead variance
- Integrated analysis of overhead cost variances
- Overhead cost variances in nonmanufacturing settings
- Different purposes of manufacturing overhead cost analysis
- Journal entries for overhead costs and variances
- Performance gaps and variances
- Actual, normal, budgeted, and standard costing
- Appendix: Proration of manufacturing variances using standard costs

*M*anufacturing overhead items such as energy, depreciation, and maintenance are large cost items for thread manufacturers. American & Efird uses variance analysis to plan and control variable overhead and fixed overhead costs at its Mt. Holly, North Carolina, plant.

Overhead costs are a major component of the total costs in most organizations. This chapter covers some popular methods of planning and controlling overhead costs, allocating these indirect costs to products, and analyzing overhead variances. We continue the Chapter 7 analysis of the Webb Company in several ways. First, we continue the Level 1, 2, and 3 approach to variance analysis. Second, we continue to stress that variances are only part of the information managers use to plan and control overhead costs.

Exhibit 8-1 presents an overview of Webb's costing system. Chapter 7 discussed flexible budgets and variances for Webb's three direct cost categories. This chapter discusses flexible budgets and variances for Webb's four indirect cost pools. The text analyzes the variable and fixed manufacturing overhead cost pools. The Problem for Self-Study illustrates how the same analysis applies to the variable and fixed marketing overhead cost pools.

Chapter 7 illustrated how a static-budget variance can be divided into a flexible-budget variance and a sales-volume variance. This chapter focuses on understanding flexible-budget variances for overhead costs and their causes. Chapter 22 presents further discussion of sales-volume variances.

Please proceed slowly as you study this chapter. Trace the data to the analysis in a systematic way. In particular, note how fixed manufacturing overhead is accounted for in one way for the planning and control purpose and in a different way for the inventory costing purpose. Accounting for fixed manufacturing overhead is usually the most puzzling aspect of the study of flexible budgets and variances.

The following data from Exhibit 7-5 (p. 236) for the actual results (column 6), the flexible budget for 10,000 units (column 3), and the flexible budget for 12,000 units (column 4) are central to this chapter. (The flexible budget for 12,000 units is Webb's static budget for April 19_4.)

EXHIBIT 8-1
Costing Overview for Webb Company

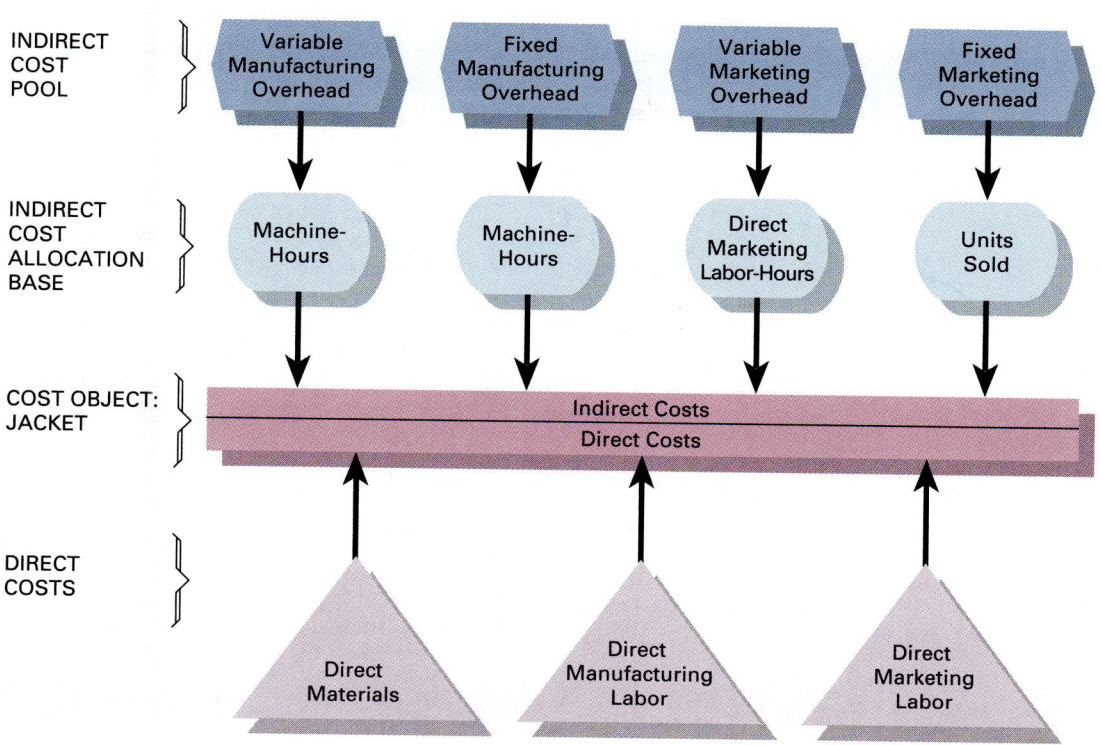

Overhead Category	Actual Results	Flexible Budget Amount for 10,000 Output Units	Static Budget (Based on 12,000 Output Units)
Variable manufacturing overhead	$130,500	$120,000	$144,000
Fixed manufacturing overhead	285,000	276,000	276,000
Variable marketing overhead	45,700	50,000	60,000
Fixed marketing overhead	420,000	434,000	434,000

PLANNING OF VARIABLE AND FIXED OVERHEAD COSTS

Objective 1
Explain differences in the planning and control of variable overhead costs and fixed overhead costs

Webb's cost structure illustrates why the planning and control of overhead costs are important. The following percentages of total budgeted costs are based on Webb's budget for 12,000 output units for April 19_4 (see Exhibit 7-5, p. 236):

	Variable Costs	Fixed Costs	Total Costs
Manufacturing overhead	7.59%	14.54%	22.13%
Marketing overhead	3.16	22.87	26.03

Budgeted manufacturing overhead costs are 22.13% of budgeted total costs, while budgeted marketing overhead costs are 26.03% of that amount. Clearly, Webb can greatly improve its profitability by effective planning and control of its overhead costs, both variable and fixed.

Planning Variable Overhead Costs

Among Webb's variable manufacturing overhead costs are energy, engineering support, indirect materials, and indirect manufacturing labor. Effective planning of variable overhead costs involves planning to undertake only those variable overhead activities that add value and then planning to use the drivers of costs in those activities in the most efficient way.

A **value-added cost** is one that customers perceive as adding value to a product or service. A **non-value-added cost** is one that customers perceive as not adding value to a product or service. Consider the cost of electricity used in the sewing of jackets manufactured by Webb. Customers view sewing as an essential element of manufacturing a jacket. Hence, costs associated with sewing (such as electricity) would be classified as "adding value." In contrast, consider the cost of a warehouse used to store rolls of cloth to be used in an emergency (if, say, a supplier fails to deliver). Customers typically do not view a jacket sewn from cloth stored in a warehouse as being different from a jacket sewn from cloth delivered by a supplier directly to the production floor. Hence, costs associated with warehousing are likely viewed as "not adding value." As discussed further in Chapter 13, there is a continuum between value-added costs and non-value-added costs. Many overhead cost items are in a gray, uncertain area between adding value and not adding value.

The analysis and understanding of the causes of variances computed in this chapter assist managers both in focusing on only value-added variable overhead costs and on using the drivers of these costs in the most efficient way.

Planning Fixed Overhead Costs

Effective planning of fixed overhead costs involves planning to undertake only those fixed overhead activities that add value and then planning the appropriate level for those activities. Examples in Webb's manufacturing area include depreciation or leasing costs on plant and equipment, some administrative costs (such as the plant manager's salary), and property taxes.

The decision of what capacity level to provide for many items that are fixed costs is one of the most challenging facing managers. Consider Webb's leasing of loom machines. These machines offer different capacity levels and are leased on an annual basis. Failure to lease sufficient machine capacity will result in an inability to meet demand and thus in lost sales of jackets. In contrast, if Webb greatly overestimates demand, it will incur additional leasing costs on machines that are not being fully utilized during the year.

At the start of an accounting period, management likely will have made most of the key decisions that determine the level of actual fixed overhead costs to be incurred. In contrast, day-to-day, ongoing management decisions play a larger role in determining the level of actual variable overhead costs incurred in that period. We turn now to a discussion of variable overhead rates.

DEVELOPING BUDGETED VARIABLE OVERHEAD RATES

Webb uses a three-step approach when developing its variable overhead rate(s):

STEP 1: *IDENTIFY THE COSTS TO INCLUDE IN THE VARIABLE OVERHEAD COST POOL(S).* Webb groups all its variable manufacturing overhead costs in a single cost pool. Costs in this pool include energy, engineering support, indirect materials, and indirect manufacturing labor.

STEP 2: *SELECT THE COST ALLOCATION BASE(S).* Webb's operating managers believe that machine-hours is the cost driver of variable manufacturing overhead and decide to use this measure as the cost allocation base.

Several approaches can be used in this step. One approach is to make an adjustment to the past *actual* variable manufacturing overhead cost rate per unit of the allocation base—for example, an adjustment to take account of expected inflation in items in the cost pools, or to take account of expected improvements in the efficiency with which particular items are used. A second approach is to develop the budgeted variable overhead cost rate on the basis of standards set for the individual line items in the overhead cost pool per unit of the allocation base.

Webb uses the standard costing approach to develop its April 19_4 budgeted variable overhead cost rate of $30 per machine-hour. Its budgeted machine-hour rate per output unit of 0.40 hours also is developed in Webb's standard-setting process:

◆ Budgeted variable manufacturing overhead per machine-hour $30
◆ Budgeted machine-hours allowed per actual output unit 0.40

These input amounts combine to yield the budgeted variable manufacturing overhead rate per output unit:

$$\begin{pmatrix} \text{Budgeted inputs allowed} \\ \text{for one output unit} \end{pmatrix} \times \begin{pmatrix} \text{Budgeted cost} \\ \text{per input unit} \end{pmatrix} = 0.40 \times \$30$$
$$= \$12 \text{ per output unit}$$

VARIABLE OVERHEAD COST VARIANCES

We now illustrate how a variable manufacturing overhead rate is used in computing Webb's variable manufacturing overhead cost variances. The following data are for April 19_4:

Budgeted output units (static budget)	12,000 jackets
Budgeted machine-hours (static budget)	4,800
Budgeted variable manufacturing overhead costs (static budget)	$144,000
Budgeted variable manufacturing overhead costs per machine-hour	$30
Budgeted variable manufacturing overhead costs per output unit ($30.00 per machine hour × 0.4 machine-hours)	$12
Actual output units produced	10,000 jackets
Actual machine-hours used	4,500
Actual variable manufacturing overhead costs	$130,500
Actual variable manufacturing overhead costs per machine-hour ($130,500 ÷ 4,500)	$29
Actual variable manufacturing overhead costs per output unit ($130,500 ÷ 10,000)	$13.05

The relevant range for these budgeted cost relationships (both variable and fixed) is from 8,000 to 16,000 output units.

Static-Budget and Flexible-Budget Analysis

The Level 1 and Level 2 analyses discussed in Chapter 7 are presented in Exhibit 8-2 for variable manufacturing overhead costs. Level 1 static-budget variance is $13,500 F:

$$\begin{matrix} \text{Variable manufacturing overhead} \\ \text{static budget variance} \end{matrix} = \begin{pmatrix} \text{Actual} \\ \text{results} \end{pmatrix} - \begin{pmatrix} \text{Static budgeted} \\ \text{amount} \end{pmatrix}$$
$$= \$130,500 - \$144,000$$
$$= \$13,500 \text{ F}$$

EXHIBIT 8-2

Static-Budget and Flexible-Budget Analysis
for Variable Manufacturing Overhead Costs for
Webb Company for April 19_4*

Level 1 Analysis

	Actual Results (1)	Static-Budget Variance (2) = (1) − (3)	Static Budget (3)
Units produced (sold)	10,000	2,000 U	12,000
Variable manufacturing overhead	$130,500		$144,000

$13,500 F
Static-budget variance

Level 2 Analysis

	Actual Results (1)	Flexible-Budget Variance (2) = (1) − (3)	Flexible Budget (3)	Sales-Volume Variance (4) = (3) − (5)	Static Budget (5)
Units produced (sold)	10,000	0	10,000	2,000 U	12,000
Variable manufacturing overhead	$130,500		$120,000		$144,000

$10,500 U $24,000 F
Flexible-budget variance Sales-volume variance

$13,500 F
Static-budget variance

*F = favorable effect on operating income; U = unfavorable effect on operating income.

Additional insight into the ability of Webb's managers to control variable manufacturing overhead can be gained by moving to a Level 2 analysis. Level 2 analysis uses a flexible budget to restate ("flex") Webb's variable manufacturing overhead to include the fact that 10,000 output units were sold instead of the budgeted 12,000 output units. The flexible budget for variable manufacturing overhead in April 19_4 is $120,000 (10,000 × $12).

The variable manufacturing overhead sales-volume variance arises solely because the actual number of output units sold by Webb differs from the budgeted number of output units sold:

$$\text{Variable manufacturing overhead sales-volume variance} = \left(\begin{array}{c}\text{Flexible-budget}\\ \text{amount}\end{array}\right) - \left(\begin{array}{c}\text{Static-budget}\\ \text{amount}\end{array}\right)$$
$$= \$120,000 - \$144,000$$
$$= \$24,000 \text{ F}$$

The variable manufacturing overhead flexible budget variance arises because Webb's actual variable manufacturing overhead differs from that budgeted for the actual output units sold:

$$\text{Variable manufacturing overhead flexible-budget variance} = \left(\begin{array}{c}\text{Actual}\\ \text{results}\end{array}\right) - \left(\begin{array}{c}\text{Flexible-budget}\\ \text{amount}\end{array}\right)$$
$$= \$130,500 - \$120,000$$
$$= \$10,500 \text{ U}$$

MANUFACTURING OVERHEAD ALLOCATION BASES AND RATES IN THE ELECTRONICS INDUSTRY

Manufacturing overhead costs are the second most important category of manufacturing costs for electronics companies such as Apple Computer, Hewlett-Packard, Hitachi, Philips, Siemens, and Toshiba. Two frequently used allocation bases for manufacturing overhead costs in this industry are direct manufacturing labor dollars (or hours) and direct materials dollars. Many individual companies use both allocation bases (and, in addition, other bases such as hours of equipment testing). Individual segments of this industry differ in their overhead rates, in part, because of differences in their cost structures. The following industry data for four segments of the electronics industry are drawn from companies that are members of the American Electronics Association.[a] The four segments are:

◆ components: includes capacitors, amplifiers, oscillators, and wire and cable
◆ computers: includes mainframes, minicomputers, and microcomputers
◆ peripherals: includes disc drives, printers, and keyboards
◆ instruments: includes test, analytical, and scientific equipment and medical instruments

	Components	Computers	Peripherals	Instruments
COST STRUCTURE				
Revenues	100.0%	100.0%	100.0%	100.0%
Research and development costs	6.3%	11.4%	8.4%	9.6%
Manufacturing costs				
◆ Materials and subcontracts	28.3%	36.5%	40.2%	28.6%
◆ Direct manuf. labor	10.6	2.8	3.2	4.3
◆ Manufacturing overhead	22.6	10.3	12.2	17.0
Total manufacturing costs	61.5	49.6	55.6	49.9
Marketing costs	12.4	17.9	15.9	20.5
General & administrative costs	11.9	7.8	9.6	11.4
Other costs, taxes, and profits	7.9	13.3	10.5	8.6
	100.0%	100.0%	100.0%	100.0%
AVERAGE MANUFACTURING OVERHEAD RATES				
◆ Allocation base is direct manufacturing labor dollars	214.5%	440.0%	277.0%	262.0%
◆ Allocation base is materials dollars	17.0%	15.5%	12.5%	15.0%

The ratio of manufacturing overhead costs to direct manufacturing labor costs ranges from 3.95 for companies manufacturing instruments (17.0% ÷ 4.3%) to 2.13 for companies manufacturing components (22.6% ÷ 10.6%). Clearly, planning and controlling manufacturing overhead costs is a high priority for managers in the electronics industry.

[a]Adapted from *Operating Ratios Survey*, American Electronics Association. Full citation is in Appendix A.

This $10,500 unfavorable flexible-budget variance shows that Webb's actual variable manufacturing overhead exceeded the budgeted amount by $10,500 for the 10,000 jackets actually produced in April 19_4.

We now discuss how managers can gain additional insight by dividing the Level 2 variable manufacturing overhead flexible-budget variance into its Level 3 spending and price variances.

Variable Overhead Spending Variance

The **variable overhead spending variance** is the difference between the actual amount of variable overhead incurred and the budgeted amount allowed for actual output achieved. Focusing on units of the cost allocation base (machine-hours), the formula for the variable overhead (VOH) spending variance is:

$$\begin{pmatrix} \text{Variable overhead-} \\ \text{spending variance} \end{pmatrix} = \begin{pmatrix} \text{Actual VOH costs} \\ \text{per unit of} \\ \text{cost allocation base} \end{pmatrix} - \begin{pmatrix} \text{Budgeted VOH costs} \\ \text{per unit of} \\ \text{cost allocation base} \end{pmatrix} \times \begin{pmatrix} \text{Actual quantity of} \\ \text{VOH cost allocation} \\ \text{base for actual} \\ \text{output units achieved} \end{pmatrix}$$

$$= (\$29 - \$30) \times 4{,}500$$
$$= -\$1 \times 4{,}500 = \$4{,}500 \text{ F}$$

Webb operated in April 19_4 with a lower than budgeted variable overhead cost per machine hour. Hence, there is a favorable variable overhead spending variance.

The variable overhead spending variance is computed similarly to the price variance described in Chapter 7 (p. 240) for direct cost items. Do not assume, however, that the causes of these two variances are the same. Two main reasons explain a variable overhead spending variance:

1. The actual prices of individual items included in variable overhead differ from the budgeted prices, and
2. The actual usage of individual items included in variable overhead differs from the budgeted usage.

The $4,500 favorable variable overhead spending variance for Webb could have arisen because, say, the April 19_4 purchase price of energy, indirect materials, or indirect manufacturing labor was less than budgeted prices. This variance could also have arisen because the actual usage of these three items was less than the budgeted usage assumed in setting the $30 budgeted variable manufacturing overhead rate per machine-hour.

Variable Overhead Efficiency Variance

The formula for the **variable overhead efficiency variance** is:

$$\begin{pmatrix} \text{Variable overhead} \\ \text{efficiency variance} \end{pmatrix} = \begin{pmatrix} \text{Actual quantity of} \\ \text{VOH cost allocation} \\ \text{base for actual} \\ \text{output achieved} \end{pmatrix} - \begin{pmatrix} \text{Budgeted quantity of} \\ \text{VOH cost allocation} \\ \text{base allowed for actual} \\ \text{output achieved} \end{pmatrix} \times \begin{pmatrix} \text{Budgeted} \\ \text{VOH cost} \\ \text{allocation rate} \end{pmatrix}$$

$$= [4{,}500 - (10{,}000 \times 0.40)] \times \$30$$
$$= (4{,}500 - 4{,}000) \times \$30 = 500 \times \$30$$
$$= \$15{,}000 \text{ U}$$

The variable overhead efficiency variance is computed similarly to the efficiency variance described in Chapter 7 (p. 240) for direct cost items. But the interpretations of the Chapter 7 and Chapter 8 efficiency variances differ. In Chapter 7, efficiency variances for direct cost items reflect differences between actual inputs used and the budgeted inputs allowed for actual outputs achieved. In Chapter 8, efficiency variances for variable overhead costs reflect the efficiency with which the cost allocation base is used. Webb's unfavorable variable overhead efficiency variance of $15,000 means that actual machine-hours (the cost allocation base) was higher than the budgeted machine-hours allowed to manufacture 10,000 jackets.

Use of cotton thread for sewing jackets serves as another example illustrating the difference between the efficiency variance discussed in Chapter 7 and the efficiency variance under investigation in Chapter 8. If Webb classifies cotton thread as a direct cost item, the direct materials efficiency variance will reflect whether more or less cotton thread per jacket is used than was budgeted for the actual output achieved. In contrast, if Webb classifies cotton thread as an indirect cost item, the variable manufacturing overhead efficiency variance will reflect whether Webb used more or less machine-hours (the cost allocation base for variable manufacturing overhead) than were budgeted for the actual output achieved. Any variation in cotton thread usage that differs from that budgeted to vary with respect to machine-hours will be reflected in the variable manufacturing overhead spending variance.

A summary of the variable manufacturing overhead variances computed in this section follows:

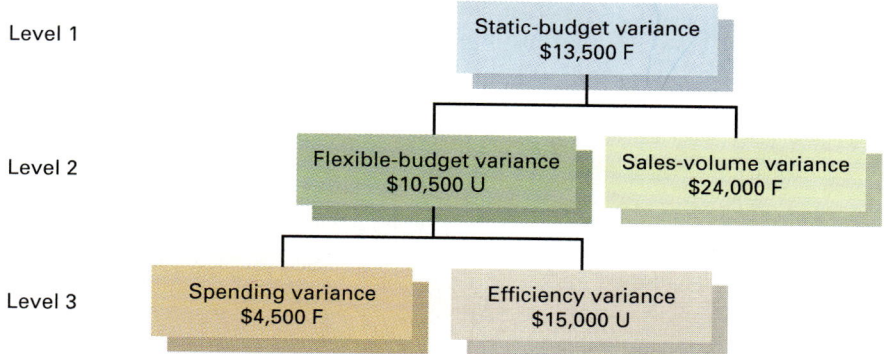

Exhibit 8-3 shows the columnar presentation of the Level 3 spending and efficiency variances. The key cause of Webb's unfavorable flexible-budget variance is that the actual use of machine-hours is higher than budgeted.

EXHIBIT 8-3

Columnar Presentation of Variance Analysis:
Variable Manufacturing Overhead for Webb Company*

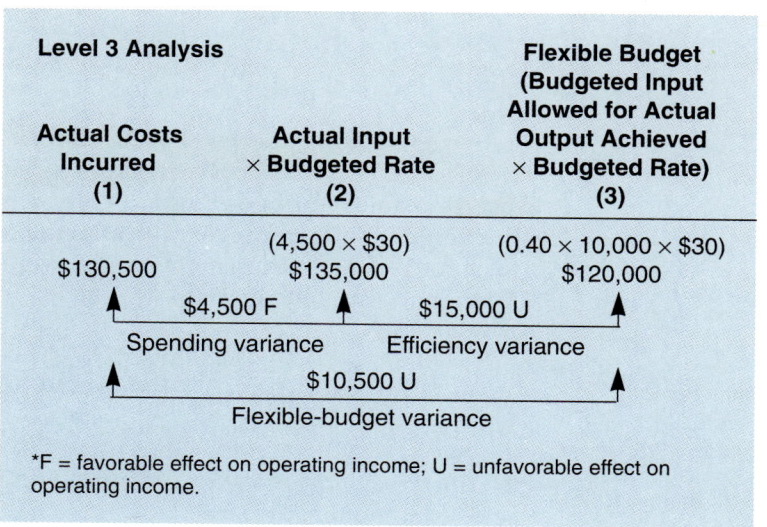

DEVELOPING FIXED OVERHEAD RATES

Objective 3

Illustrate how to compute the budgeted fixed manufacturing overhead rate

Fixed overhead costs are, by definition, a lump sum that does not change in total despite changes in a cost driver. While total fixed costs are frequently included in flexible budgets, they remain the same total amount regardless of the output level chosen to "flex" the variable costs and revenues. Webb Company has budgeted fixed manufacturing overhead of $276,000 in April 19_4. The steps in developing its budgeted fixed overhead rate are:

STEP 1: *IDENTIFY THE COSTS IN THE FIXED OVERHEAD COST POOL.*   For Webb, fixed manufacturing overhead costs include depreciation, plant-leasing costs, property taxes, salaries, and some administrative costs, all of which are included in a single cost pool. Webb's budget is $276,000 for April 19_4.

STEP 2: *CHOOSE THE DENOMINATOR LEVEL OF THE BUDGETED FIXED OVERHEAD RATE.*   This denominator level may be expressed in input units (such as machine-hours) or output units (jackets manufactured). Webb budgets to manufacture 12,000 jackets in April 19_4. The budgeted machine-hours to manufacture 12,000 jackets is 4,800 (12,000 × 0.40 budgeted machine-hours per output unit).

STEP 3: *COMPUTE THE BUDGETED FIXED OVERHEAD RATE.*   This step uses information from steps 1 and 2:

$$\begin{array}{l} \text{Budgeted fixed} \\ \text{overhead rate} \\ \text{per output unit} \end{array} = \frac{\text{Budgeted fixed overhead costs}}{\text{Denominator level in output units}}$$

$$= \frac{\$276,000}{12,000 \text{ output units}}$$

$$= \$23 \text{ per output unit (jacket)}$$

Alternatively, the fixed overhead rate can be expressed in terms of input units (machine-hours for Webb):

$$\begin{array}{l} \text{Budgeted fixed} \\ \text{overhead rate} \\ \text{per input unit} \end{array} = \frac{\text{Budgeted fixed overhead costs}}{\text{Denominator level in input units}}$$

$$= \frac{\$276,000}{4,800 \text{ machine-hours}}$$

$$= \$57.50 \text{ per machine-hour}$$

The relationship between the budgeted fixed overhead rates is the ratio of machine-hour input to each output unit:

$$\left(\begin{array}{c} \$57.50 \\ \text{per machine-hour} \end{array} \right) \times \left(\begin{array}{c} 0.40 \text{ machine-hours} \\ \text{per jacket} \end{array} \right) = \begin{array}{c} \$23.00 \\ \text{per jacket} \end{array}$$

The **denominator level** is the preselected level of the cost allocation base used to set a budgeted fixed overhead rate for allocating fixed overhead costs to a cost object. In manufacturing settings, the *denominator level* is commonly termed the **production denominator level** or **production denominator volume.** The denominator level for Webb's fixed manufacturing overhead costs is 12,000 output units or 4,800 input units (machine-hours).

Objective 4

Describe how per-unit inventory costs are affected by the denominator level chosen for fixed manufacturing overhead

Unit Fixed Manufacturing Overhead Costs with Different Denominator Levels

The total fixed manufacturing overhead for planning and control is unaffected by the selected denominator level. However, the fixed manufacturing overhead rate (and thus the inventoriable cost per unit) will be affected by the choice of the denominator level. Consider how the budgeted fixed overhead rate changes with different denominator levels in the following table.

Total Budgeted Monthly Fixed Manufacturing Overhead (1)	Total Monthly Denominator Level in Output Units (2)	Budgeted Fixed Manufacturing Overhead Rate per Output Unit (3) = (1) ÷ (2)	Total Monthly Denominator Level in Input Units (Machine-Hours) (4)	Budgeted Fixed Manufacturing Overhead Rate per Input Unit (Machine-Hours) (5) = (1) ÷ (4)
$276,000	8,000	$34.50	3,200	$86.25
$276,000	10,000	$27.60	4,000	$69.00
$276,000	12,000	$23.00	4,800	$57.50
$276,000	14,000	$19.71	5,600	$49.29
$276,000	16,000	$17.25	6,400	$43.13

Unitizing fixed manufacturing costs seemingly transforms a fixed cost into a variable cost. However, by definition fixed costs do not change as output levels change. Chapters 9 and 15 further discuss the choice of the denominator level when computing budgeted overhead rates.

FIXED OVERHEAD COST VARIANCES

Level 1 static-budget variance for Webb's fixed manufacturing overhead is $9,000 U:

$$\text{Fixed overhead static-budget variance} = \begin{pmatrix}\text{Actual}\\\text{results}\end{pmatrix} - \begin{pmatrix}\text{Budgeted amount}\\\text{for fixed overhead}\end{pmatrix}$$
$$= \$285{,}000 - \$276{,}000$$
$$= \$9{,}000 \text{ U}$$

Budgeted fixed overhead costs remain the same across different output levels. Hence, there is no "flexing" of fixed costs.

A summary of the Level 1, Level 2, and Level 3 variance analysis for Webb's fixed manufacturing overhead in April 19_4 is as follows:

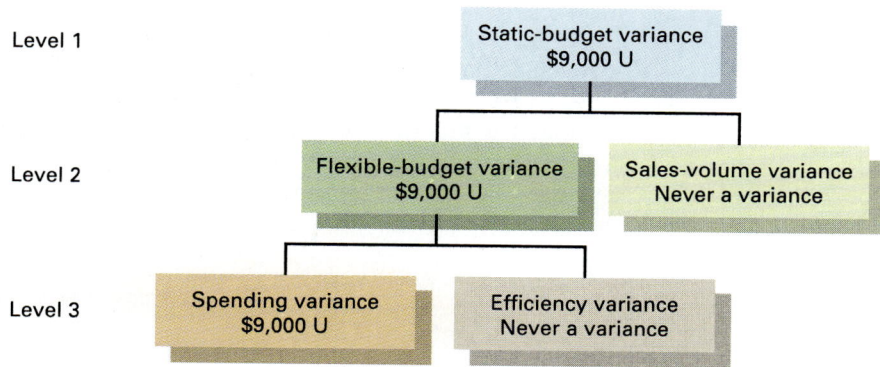

There is never a sales-volume variance in Level 2 for fixed overhead costs. Budgeted fixed costs are, by definition, unaffected by sales-volume changes. Similarly, there is never an efficiency variance in Level 3 for fixed overhead costs. After all, a manager cannot be more or less efficient in dealing with a fixed amount.

The formulas for the fixed manufacturing overhead flexible budget variance and the spending variance are the same as those we presented in Chapter 7:

$$\text{Fixed manufacturing overhead flexible-budget variance} = \left(\begin{array}{c}\text{Actual} \\ \text{results}\end{array}\right) - \left(\begin{array}{c}\text{Budgeted} \\ \text{amount}\end{array}\right)$$
$$= \$285{,}000 - \$276{,}000$$
$$= \$9{,}000 \text{ U}$$

$$\text{Fixed manufacturing overhead spending variance} = \left(\begin{array}{c}\text{Actual} \\ \text{results}\end{array}\right) - \left(\begin{array}{c}\text{Budgeted} \\ \text{amount}\end{array}\right)$$
$$= \$285{,}000 - \$276{,}000$$
$$= \$9{,}000 \text{ U}$$

The $9,000 unfavorable variance simply means that Webb spent more on fixed manufacturing overhead in April 19_4 than it budgeted.

OUTPUT LEVEL (PRODUCTION VOLUME) OVERHEAD VARIANCE

Objective 5

Explain why the output level (production volume) overhead variance applies to fixed overhead but not to variable overhead

The analysis of fixed overhead generates a new variance not previously introduced in Chapter 7 or in prior sections of this chapter. In its most general setting, this variance is termed an **output level overhead variance.** It is the difference between budgeted fixed overhead and the fixed overhead allocated to actual output units achieved. In a manufacturing setting, the output level variance is commonly termed a **production volume variance** or a **production level variance.** It is the difference between budgeted fixed manufacturing overhead and the fixed manufacturing overhead allocated to actual output units achieved.

Computing an Output Level Variance

The formula for an output level (production volume) overhead variance expressed in terms of a rate per output unit is:

$$\text{Output level overhead variance} = \left(\begin{array}{c}\text{Denominator} \\ \text{level in} \\ \text{output units}\end{array} - \begin{array}{c}\text{Actual} \\ \text{output units} \\ \text{achieved}\end{array}\right) \times \left(\begin{array}{c}\text{Budgeted fixed} \\ \text{overhead rate} \\ \text{per output unit}\end{array}\right)$$
$$= (12{,}000 - 10{,}000) \times \$23$$
$$= \$46{,}000 \text{ U}$$

When actual output units are less than the denominator level, the output level overhead variance is unfavorable. Why? This variance indicates that the facilities were not utilized as much as the denominator level had specified.

The formula for an output level (production volume) overhead variance expressed in terms of input units (machine-hours for Webb) is:

$$\text{Output level overhead variance} = \left(\begin{array}{c}\text{Budgeted} \\ \text{fixed} \\ \text{overhead}\end{array} - \begin{array}{c}\text{Fixed overhead allocated using} \\ \text{budgeted input allowed for} \\ \text{actual output units achieved}\end{array}\right)$$
$$= \$276{,}000 - (0.40 \times 10{,}000 \times \$57.50)$$
$$= \$276{,}000 - (4{,}000 \times \$57.50)$$
$$= \$276{,}000 - \$230{,}000$$
$$= \$46{,}000 \text{ U}$$

The amount used for budgeted fixed overhead will be the same amount shown in the static budget and also in any flexible budget in the relevant range.

Exhibit 8-4 (Panel B) is a columnar presentation for Webb's analysis of its fixed manufacturing variances. Panel A of Exhibit 8-4 is exactly like Exhibit 8-3 except that column 4 has been added to parallel Panel B.

EXHIBIT 8-4

EXHIBIT 8-4

Columnar Presentation of Variance Analysis:
Variable and Fixed Manufacturing Overhead for Webb Company*

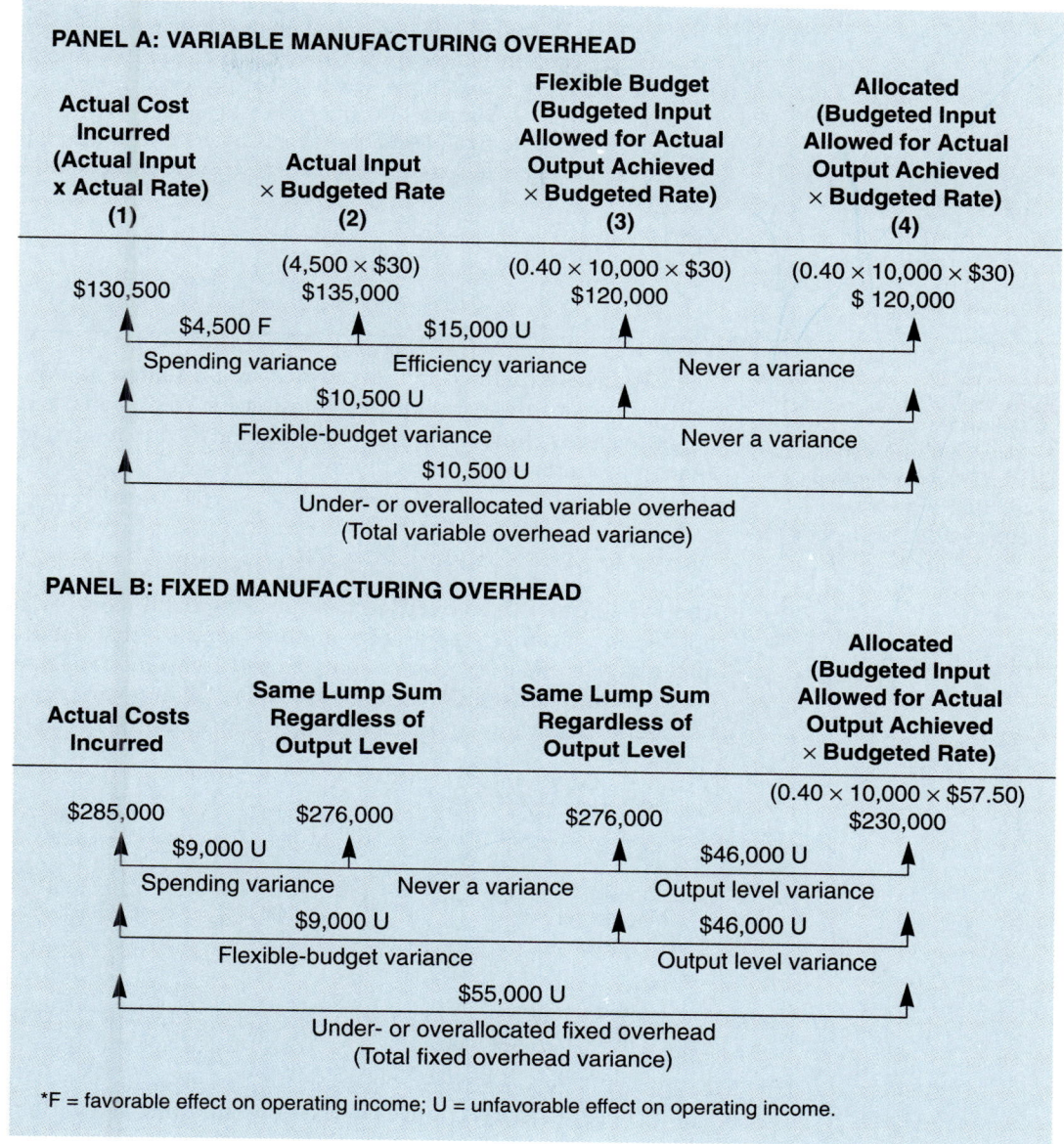

PANEL A: VARIABLE MANUFACTURING OVERHEAD

Actual Cost Incurred (Actual Input x Actual Rate) (1)	Actual Input × Budgeted Rate (2)	Flexible Budget (Budgeted Input Allowed for Actual Output Achieved × Budgeted Rate) (3)	Allocated (Budgeted Input Allowed for Actual Output Achieved × Budgeted Rate) (4)
$130,500	(4,500 × $30) $135,000	(0.40 × 10,000 × $30) $120,000	(0.40 × 10,000 × $30) $ 120,000

$4,500 F ← Spending variance
$15,000 U ← Efficiency variance
Never a variance

$10,500 U ← Flexible-budget variance
Never a variance

$10,500 U
Under- or overallocated variable overhead
(Total variable overhead variance)

PANEL B: FIXED MANUFACTURING OVERHEAD

Actual Costs Incurred	Same Lump Sum Regardless of Output Level	Same Lump Sum Regardless of Output Level	Allocated (Budgeted Input Allowed for Actual Output Achieved × Budgeted Rate)
$285,000	$276,000	$276,000	(0.40 × 10,000 × $57.50) $230,000

$9,000 U ← Spending variance
Never a variance
$46,000 U ← Output level variance

$9,000 U ← Flexible-budget variance
$46,000 U ← Output level variance

$55,000 U
Under- or overallocated fixed overhead
(Total fixed overhead variance)

*F = favorable effect on operating income; U = unfavorable effect on operating income.

Interpreting Output Level Overhead Variances

The output level variance arises because the actual output level differs from the output level used as the denominator to calculate the budgeted fixed overhead rate. We compute this rate because inventory costing and some contracts require fixed overhead costs to be expressed on a unit-of-output basis. There are two reasons why the output level variance is not a good measure of the opportunity cost (that is, the income forgone) of unused capacity.

1. The denominator output level may be less than the plant capacity output level. Consider our Webb Company example and assume that 18,000 units of production output is the capacity level. Webb uses 12,000 output units as its denominator level. Hence, the output

level variance calculated above does not take into account budgeted unused capacity. See Chapters 9 (p. 322) and 15 (p. 544) for further discussion.

2. The economic costs of not fully using capacity should reflect both revenue and cost factors. The output level overhead variance includes only cost factors. Incorporating revenue factors requires analysis of whether the selling price received on the actual units sold could also have been received on the units that could have been produced with the unused capacity but were not. Often, the selling price would not stay the same across additional units. For example, Webb sold 10,000 units in April 19_4 instead of the budgeted 12,000 units. A competitor could have taken market share from the Webb Company by price cutting to two of Webb's key customers. Webb may have decided not to quote a lower price to retain those customers and as a result reduced actual production by 2,000 units.

INTEGRATED ANALYSIS OF OVERHEAD COST VARIANCES

Objective 7
Explain how a 4-variance analysis can provide an integrated overview of overhead cost variances

The variances explained in this chapter are presented in Exhibit 8-4. The Problem for Self-Study requires a similar presentation for Webb's marketing overhead variances. We now discuss how the variances in Exhibit 8-4 are related to the level of detail managers may find useful.

4-, 3-, 2-, 1-Variance Analysis

The variable and fixed overhead variances in Exhibit 8-4 can be presented in most detail as four variances (4-variance analysis) and in least detail as one variance (1-variance analysis). The 3-variance analysis and 2-variance analysis offer intermediate levels of detail. The "4" in 4-variance analysis ("3" in 3-variance, and so on) indicates the number of separate variances managers can examine in that format.

4-Variance Analysis

	Spending Variance	Efficiency Variance	Output Level Variance
Variable Manufacturing Overhead	$4,500 F	$15,000 U	Never a variance
Fixed Manufacturing Overhead	$9,000 U	Never a variance	$46,000 U

There are four variances in this presentation (two variable manufacturing overhead variances and two fixed manufacturing overhead variances). There is never an output level overhead variance for variable overhead. Why? Because the concept of output level overhead variance arises from our attempt to treat a fixed cost (for inventory costing and, in some cases, for contract reimbursement) as if it were a variable cost, but in fact it is not. Also, there is never an efficiency variance for fixed overhead. There can be no efficiency variance for fixed manufacturing overhead because this amount is a lump sum regardless of the output level.

3-Variance Analysis

	Spending Variance	Efficiency Variance	Output Level Variance
Total Manufacturing Overhead	$4,500 U	$15,000 U	$46,000 U

The two spending variances from the 4-variance analysis have been combined in the 3-variance analysis. Given this change, note that it is not necessary to separate variable and fixed manufacturing overhead incurred.

2-Variance Analysis

	Flexible-Budget Variance	Output Level Variance
Total Manufacturing Overhead	$19,500 U	$46,000 U

The spending and efficiency variances from the 3-variance analysis have been combined under 2-variance analysis.

1-Variance Analysis

	Total-Overhead Variance
Total Manufacturing Overhead	$65,500 U

The single variance of $65,500 U in 1-variance analysis is simply the difference between the total actual manufacturing overhead incurred ($130,500 + $285,000 = $415,500) and the manufacturing overhead allocated ($120,000 + $230,000 = $350,000) to the actual output units produced. Clearly, the large drop in output level, from 12,000 budgeted units to 10,000 actual units, has resulted in a large amount of fixed manufacturing overhead costs not being allocated to actual units produced.

Interdependencies among Variances

The $65,500 unfavorable total manufacturing overhead variance for Webb Company in April 19_4 is largely the result of the $46,000 unfavorable output level variance, which reflects the 2,000 unit shortfall in production output from the denominator level. Using the 4-variance analysis presentation, the next largest amount (after the $46,000) is the $15,000 unfavorable variable overhead efficiency variance. This amount arises from the additional 500 machine-hours used in April 19_4 above the 4,000 machine-hours allowed to manufacture the 10,000 jackets. The two spending variances ($4,500 F and $9,000 U) partially offset each other.

The variances in Webb's 4-variance analysis are not always independent of each other. For example, Webb may be *spending* less than the fixed budgeted amount on equipment maintenance, with the result that individual sewing machines are operating less *efficiently* than assumed in the budget.

OVERHEAD COST VARIANCES IN NONMANUFACTURING SETTINGS

Our Webb Company example examines variable and fixed manufacturing overhead costs. Under generally accepted accounting principles, both variable and fixed manufacturing overhead costs are inventoriable costs for financial reporting purposes. In contrast the overhead costs of nonmanufacturing areas of the value chain (such as research and development and marketing) are not inventoriable costs under generally accepted accounting principles. Should the overhead costs of nonmanufacturing areas be examined using the variance analysis framework discussed in this chapter?

Many decisions that are related to product pricing or product emphasis use variable-cost information. For these decisions, *all* variable costs (nonmanufacturing as well as manufacturing) should be included in computing product profitability. Variance analysis of variable overhead costs in all areas of the value chain provides useful information to managers making decisions.

Few companies conduct detailed variance analysis of fixed nonmanufacturing overhead costs. Many managers believe little information is gained by computing flexible-budget variances or spending variances for those costs. One exception is where a company is performing a contract on a "full-cost-plus" basis—that is, it is reimbursed for its "full costs" plus an additional percentage of those full costs. Here, full costs typically will include an allocated amount of fixed overhead costs (nonmanufacturing and manufacturing) as well as direct costs and allocated variable overhead costs.

The Problem for Self-Study at the end of this chapter illustrates how the framework used in Exhibit 8-4 for Webb's manufacturing overhead costs can also be used to examine Webb's marketing overhead costs.

DIFFERENT PURPOSES OF MANUFACTURING OVERHEAD COST ANALYSIS

Different types of cost analysis may be appropriate for different purposes. Consider planning and control and inventory costing for financial reporting. Panel A of Exhibit 8-5 depicts variable manufacturing overhead for each purpose; Panel B depicts fixed manufacturing overhead for each purpose.

Variable Manufacturing Overhead Costs

Webb's variable manufacturing overhead is shown in Panel A of Exhibit 8-5 as being variable with respect to output units (jackets) for both the planning and control purpose and the inventory costing purpose. The greater the number of output units manufactured, the higher the budgeted total variable manufacturing overhead costs and the higher the total variable manufacturing overhead costs allocated to output units.

Panel A of Exhibit 8-5 presents an overall picture of how total variable overhead might behave. Of course, variable overhead consists of many items, including energy costs, repairs, indirect labor, and so on. Managers help control variable overhead costs by budgeting each line item and then investigating possible causes for any significant variances.

Some variable manufacturing overhead costs are greatly affected by decisions made by other business function areas. For example, equipment repair costs may be higher or lower because of a combination of decisions in the product design department and the equipment purchasing department. The interpretation of variable manufacturing overhead variances requires considerable knowledge of the diverse factors affecting the line items included in those variable manufacturing overhead cost pools.

Fixed Manufacturing Overhead Costs

Panel B of Exhibit 8-5 (the left-hand graph) shows that for the planning and control purpose, fixed overhead costs do not change in the 8,000 to 16,000 unit output range. Consider a monthly leasing cost of $20,000 for a building under a three-year leasing agreement. Managers control this fixed leasing cost at the time the lease is signed. During any month in the leasing period, there is little management can do to change this $20,000 lump-sum payment. Contrast this description of fixed overhead with how these costs are depicted for the inventory costing purpose, the right-hand graph of Panel B. Under generally accepted accounting principles, fixed manufacturing costs are capitalized as part of inventory on a unit-of-output basis. Every output unit Webb manufactures will increase the fixed overhead allocated to products by $23. Managers should not use this unitization of fixed manufacturing overhead costs for their planning and control.

EXHIBIT 8-5

EXHIBIT 8-5

Behavior of Variable and Fixed Manufacturing Overhead Costs for Planning and Control and for Inventory Costing

PANEL A: Variable Manufacturing Overhead Costs

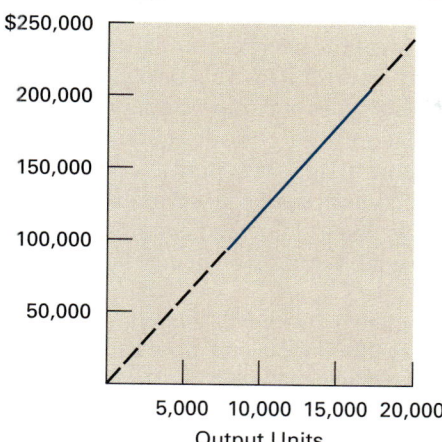

PLANNING AND CONTROL PURPOSE

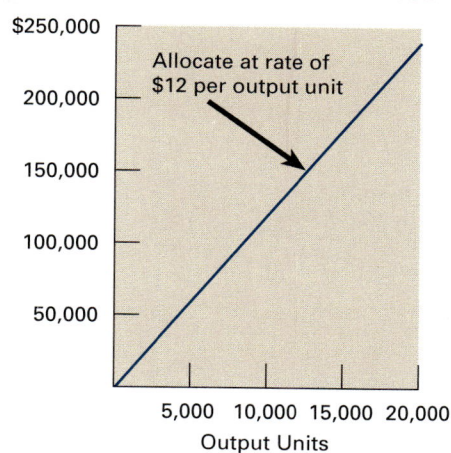

INVENTORY COSTING PURPOSE

Allocate at rate of $12 per output unit

PANEL B: Fixed Manufacturing Overhead Costs

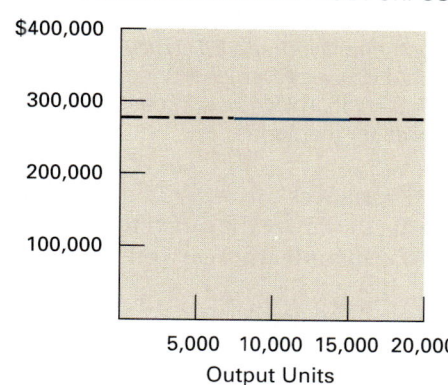

PLANNING AND CONTROL PURPOSE

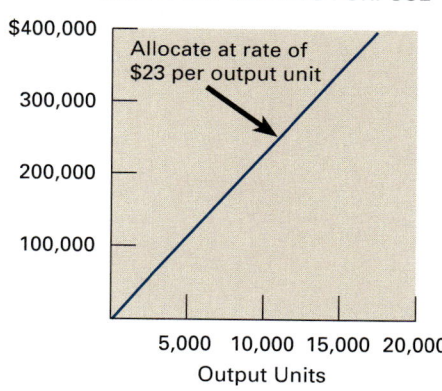

INVENTORY COSTING PURPOSE

Allocate at rate of $23 per output unit

The denominator level in each graph in Exhibit 8-5 is expressed in output units. Alternatively, we could also have expressed this denominator in terms of input units. For Webb, machine-hours would be the chosen denominator, as this is the allocation base for both variable and fixed manufacturing overhead costs.

As noted above, some contracts provide for companies to be reimbursed on a full-cost basis for individual products or services. The computation of product costs per unit under these contracts will require fixed overhead to be unitized. Again, this unitization should not be interpreted as meaning that the level of fixed costs increases as more units are produced.

JOURNAL ENTRIES FOR OVERHEAD COSTS AND VARIANCES

Objective 8

Prepare journal entries for variable and fixed overhead variances

Recording Overhead Costs

The Chapter 5 job costing example (Robinson Company; pp. 141–55) used a single manufacturing overhead control account. This chapter illustrates separate variable overhead and fixed overhead control accounts. Each overhead control account requires its own overhead allocated account.

Consider the journal entries for the Webb Company. Recall that for April 19_4:

	Actual	Budgeted
Variable manufacturing overhead	$130,500	$120,000
Fixed manufacturing overhead	$285,000	$276,000

The budgeted variable overhead rate is $12 per output unit ($30 per machine-hour). The denominator level for fixed manufacturing overhead is 12,000 units of output (or 4,800 machine-hours of input) with a budgeted rate of $23 per output unit ($57.50 per machine-hour). Webb uses 4-variance analysis.

During the accounting period, actual variable overhead and actual fixed overhead costs are accumulated in separate control accounts. As each unit is manufactured, the budgeted variable and fixed overhead rates are used to record the amounts in the respective overhead allocated accounts.

Entries for variable manufacturing overhead for April 19_4 are:

1. Variable Manufacturing Overhead Control $130,500
    Accounts Payable Control and other accounts $130,500
   To record actual variable manufacturing overhead
   costs incurred.

2. Work in Process Control $120,000
    Variable Manufacturing Overhead Allocated $120,000
   To record allocation of variable manufacturing overhead
   costs. (10,000 × $12) or (0.40 × 10,000 × $30).

3. Variable Manufacturing Overhead Allocated $120,000
   Variable Manufacturing Overhead Efficiency Variance 15,000
       Variable Manufacturing Overhead Control $130,500
       Variable Manufacturing Overhead Spending Variance 4,500
   To isolate variances for the period.

As individual variances can differ in their causes, isolating each one helps managers see those areas that differ the most from their budgeted amounts.
Entries for fixed manufacturing overhead are:

1. Fixed Manufacturing Overhead Control $285,000
    Wages Payable Control, Accumulated Depreciation
    Control, etc. $285,000
   To record actual fixed overhead costs incurred.

2. Work in Process Control $230,000
    Fixed Manufacturing Overhead Allocated $230,000
   To record allocation of fixed manufacturing overhead costs:
   (10,000 × $23) or (0.40 × 10,000 × $57.50).

3. Fixed Manufacturing Overhead Allocated $230,000
   Fixed Manufacturing Overhead Spending Variance 9,000
   Fixed Manufacturing Output Level Overhead Variance 46,000
       Fixed Manufacturing Overhead Control $285,000
   To isolate variances for the period.

The disposition of these overhead variances is discussed next.

Disposition of Overhead Variances

Chapter 5 outlined the restated allocation rate and the end-of-period account approaches to handling the end-of-period difference between manufacturing overhead incurred and manufacturing overhead allocated. Consider Webb's variable manufacturing overhead:

	Budgeted	Actual
Variable manufacturing overhead costs	$144,000	$130,500
Machine-hours	4,800	4,500
Variable manufacturing overhead rate per machine-hour	$30	$29

Under the restated allocation rate approach, Webb would adjust every job record worked on during the year. This adjustment would entail using the actual rate per machine-hour of $29 instead of the budgeted rate of $30. Then, Webb would accordingly recompute the work in process, finished goods, and cost of goods sold amounts for the period. This approach has several benefits. Individual job records are restated accurately. Also, ending work in process, finished goods, and cost of goods sold amounts accurately reflect actual variable overhead incurrence. A similar approach could be used to restate the fixed manufacturing overhead.

If managers view the restated allocation rate approach as not being cost-effective, their firms can use the end-of-period account approach. The three main variants of this approach are:

◆ immediate writeoff to Cost of Goods Sold

◆ proration based on total ending balances (before proration) in Work in Process, Finished Goods, and Cost of Goods Sold

◆ proration based on the allocated overhead amount (before proration) in the ending balances of Work in Process, Finished Goods, and Cost of Goods Sold

Webb can use any one of these variants when disposing of the overhead variances. Further discussion of these variants appears in the Appendix to this chapter.

PERFORMANCE GAPS AND VARIANCES

Objective 9

Explain why managers frequently use both financial and nonfinancial variables to plan and control overhead costs

Chapter 7 noted that variances are one type of performance gap. A *performance gap* is the difference between actual results and a benchmark amount. A *variance* is the difference between an actual result and a budgeted amount where that budgeted amount is a financial variable reported by the accounting system. Managers find that many nonfinancial benchmarks provide useful information in their planning and control decisions. Examples of performance gaps based on nonfinancial benchmarks that Webb likely would find useful in planning and controlling its manufacturing overhead costs are:

1. actual indirect materials usage in yards per machine-hour, relative to budgeted indirect materials usage in yards per machine-hour,

2. actual energy usage per machine-hour, relative to budgeted energy usage per machine-hour, and

3. actual machining time per job, relative to budgeted machining time per job.

These performance gaps, like the variances discussed in this chapter, are best viewed as attention directors, not problem solvers. These three performance gaps probably would be reported on the manufacturing floor on a daily, or even hourly, basis. The manufacturing overhead variances we discuss in this chapter capture the financial effects of items such as 1, 2, and 3, which in many cases first appear as nonfinancial performance gaps.

Many companies use both financial measures and nonfinancial measures when evaluating the performance of their managers. In Chapter 7 (pp. 245–47), we noted that it is inappropriate to overemphasize individual price or efficiency variances for direct cost categories such as direct materials or direct manufacturing labor. The same caution applies to overemphasizing the manufacturing overhead variances presented in this chapter.

Many subsequent chapters discuss performance analysis. For example, Chapters 9, 13, and 26 note potential conflicts between financial and nonfinancial performance measures. There are no easy solutions to designing an appropriate management control system. Focusing only on nonfinancial measures or focusing only on financial measures is nearly always too simplistic.

ACTUAL, NORMAL, BUDGETED, AND STANDARD COSTING

Chapter 4 presented three possible combinations of actual and budgeted direct cost rates and actual and budgeted indirect cost rates. Exhibit 8-6 presents these three costing systems along with a fourth system—standard costing—discussed in Chapters 7 and 8. **Standard costing** is a costing method that traces direct costs to a cost object by using the standard price(s) or rate(s) times the standard inputs allowed for actual outputs achieved and allocates indirect costs on the basis of the budgeted indirect rate(s) times the standard inputs allowed for the actual outputs achieved. (Recall from Chapter 7 that a standard amount, which represents a good level of performance or the best level of performance, is a special case of a budgeted amount. Standard costing is a special case of budgeted costing.)

As with the actual, normal, and budgeted costing methods, the standard costing method is presented in Exhibit 8-6 in its pure version. Individual companies may use some combination of these four methods when recording costs in their different business function areas.

EXHIBIT 8-6
Actual, Normal, Budgeted, and Standard Costing Methods

	Actual Costing	Normal Costing	Budgeted Costing	Standard Costing
Direct costs	Actual direct prices or rates × Actual quantity	Actual direct prices or rates × Actual quantity	Budgeted direct prices or rates × Actual quantity	Standard prices or rates × Standard inputs allowed for actual outputs achieved
Overhead (indirect) costs	Actual indirect rate(s) × Actual quantity	Budgeted indirect rate(s) × Actual quantity	Budgeted indirect rate(s) × Actual quantity	Budgeted indirect rate(s) × Standard inputs allowed for actual outputs achieved

PROBLEM FOR SELF-STUDY

PROBLEM

Webb Company is analyzing its marketing overhead costs. It uses a 4-variance analysis of its marketing overhead costs. The following information is collected for April 19_4.

a. Variable marketing overhead is allocated to products using budgeted direct marketing labor-hours per jacket. Fixed marketing overhead is allocated to products on a per-jacket basis.

b. Budgeted amounts for April 19_4 are:
  (i) Direct marketing labor-hours: 0.25 hours per jacket
  (ii) Variable marketing overhead rate is $20 per direct marketing labor-hour
  (iii) Fixed marketing overhead is budgeted to be $434,000
  (iv) Budgeted output is 12,000 jackets, which is used as the denominator level of output

c. Actual amounts for April 19_4 are:
  (i) Variable marketing overhead: $45,700
  (ii) Fixed marketing overhead: $420,000
  (iii) Direct marketing labor-hours: 2,304
  (iv) Actual output: 10,000 jackets

Required

1. Describe how Webb Company might plan and control (a) its variable marketing overhead costs and (b) its fixed marketing overhead costs.

2. Present an analysis of the April 19_4 marketing overhead costs using the format shown in Panels A and B of Exhibit 8-4.

3. Provide at least one explanation for each of the four variances in Webb's 4-variance analysis of its marketing overhead costs.

SOLUTION

1. a. The main approaches to planning and controlling variable marketing overhead costs are:

  (i) Working creatively at the design stage to avoid non-value-added activities (such as double-checking marketing mailings).
  (ii) Working on reducing the rate per cost driver or the number of cost driver units per output. For example, marketing managers could exert tighter control over the price of department purchases.
  (iii) Monitoring variances on an ongoing basis.
  (iv) Assigning responsibilities for the marketing variance to managers who will promote the productivity of the marketing staff.

  b. The main approaches to planning and controlling fixed marketing overhead costs are:

  (i) Planning capacity needs in detail, including providing incentives for managers to estimate their budgeted usage in an unbiased way.
  (ii) Having marketing managers make careful, cost-conscious planning decisions on individual line items.

Day-to-day monitoring of variances is not likely to play an important role in the control of fixed marketing overhead costs.

2. Exhibit 8-7 offers the columnar presentation of variances. The budgeted fixed marketing overhead rate is $36.1667 (rounded) per jacket. A summary of these variances follows:

	Spending Variance	Efficiency Variance	Output Level Variance
Variable Marketing Overhead	$380 F	$3,920 F	Never a variance
Fixed Marketing Overhead	$14,000 F	Never a variance	$72,333 U

EXHIBIT 8-7
Columnar Presentation of Variance Analysis:
Variable and Fixed Marketing Overhead for Webb Company

PANEL A: VARIABLE MARKETING OVERHEAD

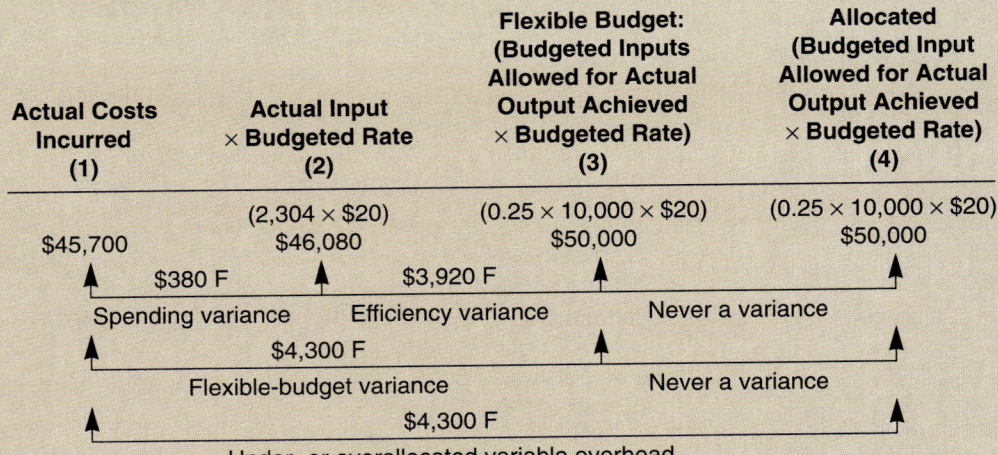

PANEL B: FIXED MARKETING OVERHEAD

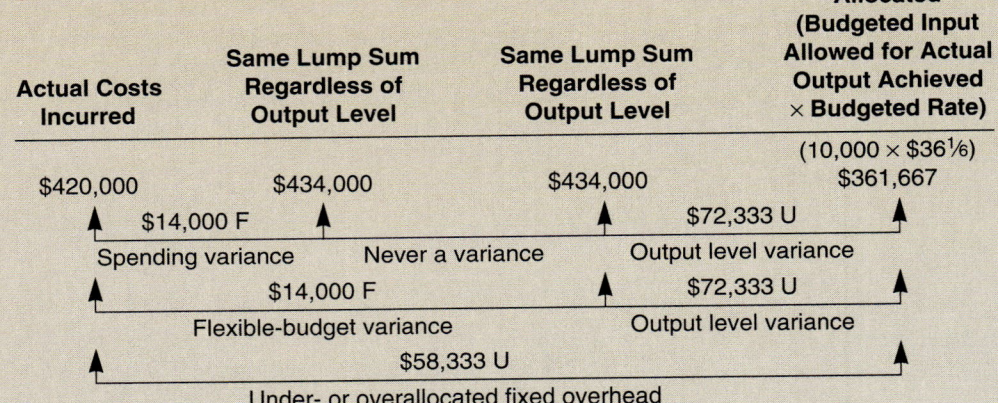

*F = favorable effect on operating income; U = unfavorable effect on operating income.

3. *Variable marketing overhead spending variance* ($380 F). The reasons for this variance include:

   a. lower than expected prices for line items in the variable marketing overhead budget, such as lower wage rates for marketing support staff, lower prices for long-distance phone calls, and lower prices for gasoline used by salespeople

   b. lower than expected usage of line items in the variable marketing overhead budget, such as fewer marketing support staff, fewer long-distance phone calls, and fewer gallons of gasoline per direct marketing labor-hour

Variable marketing overhead efficiency variance ($3,920 F). The reason for this variance is more productive use of the cost allocation base (direct marketing labor-hours); 2,304 direct marketing labor-hours were used, compared with a budgeted 2,500 hours. Perhaps marketing personnel classified as direct labor were more efficient, maybe because of a new incentive plan, a better training program, or greater than expected continuous improvement.

Fixed marketing overhead spending variance ($14,000 F). The reasons for this variance include:

a. lower than expected prices for line items in the fixed marketing overhead budget. Perhaps a marketing department supervisor resigned and was replaced by a lower-paid supervisor, or maybe a building lease was renegotiated at a lower than budgeted amount.

b. lower than expected usage of line items in the fixed marketing overhead budget, such as the marketing sales force leasing five rather than six cars and thus reducing the fixed monthly auto-leasing payment.

Output level overhead variance ($72,333 U). This variance arises because the output level (sales) was 10,000 jackets rather than 12,000 jackets. One explanation is that marketing sales personnel were much less effective than budgeted. Other explanations include a downturn in the economy, poor quality work at Webb's manufacturing plant, and a reduction in tariffs, resulting in the import of lower-cost jackets.

SUMMARY

The following points are linked to the chapter's learning objectives.

1. Planning and control of variable overhead costs emphasize undertaking only value-added activities and then efficiently managing the use of the cost drivers of those activities. Planning and control of fixed overhead costs emphasize undertaking only value-added activities and then choosing the appropriate capacity level, given the expected demand and the level of uncertainty pertaining to that demand.

2. When the flexible budget for variable overhead is developed, a spending overhead variance and an efficiency variance can be computed. The variable overhead spending variance is the difference between the actual amount of variable overhead incurred and the budgeted amount that is allowed for the actual output level achieved. The variable overhead efficiency variance measures the efficiency with which the cost allocation base is used; this is a different type of efficiency variance than that calculated in Chapter 7 for direct cost items, such as direct materials.

3. The budgeted fixed manufacturing overhead rate is calculated by dividing the budgeted fixed overhead costs by the chosen denominator level (expressed in either output units or input units). This rate is calculated for inventory costing, and in some cases, for contract reimbursement.

4. The output level overhead variance arises because we treat a fixed cost as if it were variable when computing inventoriable costs. Variable overhead costs are, by definition, not fixed, and hence there is no output level variance for them.

5. Fixed manufacturing overhead is a lump sum that is unaffected by the selected denominator level. In contrast, the unitized fixed manufacturing overhead rate will be affected by the chosen denominator level.

6. Output level overhead variances are rarely a good measure of the opportunity cost of unused capacity. For example, the plant capacity level may exceed the chosen denominator level; hence, some unused capacity may not be included in the denominator. Moreover, the output level variance focuses only on costs. It does not take into account any price changes necessary to spur extra demand that would in turn make use of any idle capacity.

7. A 4-variance analysis presents spending and efficiency variances for variable overhead costs and spending and output level variances for fixed overhead costs. By analyzing these four variances together, managers can consider possible interrelationships among them. These four variances collectively measure differences between actual and budgeted amounts for output level, selling prices, variable costs, and fixed costs.

8. The separate analysis of variable and fixed overhead costs requires the use of separate variable and fixed overhead control accounts and separate variable and fixed overhead allocated accounts. At the end of each accounting period, any under- or overallocated amount for variable or fixed overhead costs can be disposed as illustrated in Chapter 5.

9. Managers use both financial and nonfinancial measures to plan and control overhead costs. In many cases overhead variances initially appear as nonfinancial measures. Expressing these measures in financial terms can highlight to managers the relative importance of different types of nonfinancial performance gaps.

APPENDIX: PRORATION OF MANUFACTURING VARIANCES USING STANDARD COSTS

This appendix explores the proration of variances in a standard-costing system. We also illustrate how the end-of-period account approach discussed in Chapter 5 can be refined by a more detailed analysis of direct materials price variances.

Morales Company makes one uniform product, a specialty plastic container. The company uses a standard costing system. Both variable and fixed costs are inventoried. Morales distributes monthly income statements to its internal managers. The following results pertain to October 19_4. (To keep the calculations manageable, the following numbers are deliberately small.)

Direct materials purchased (charged to Materials Control at standard prices):	
200,000 lb × $0.50	$100,000
Direct materials price variance: 200,000 lb × $0.05	10,000 U
Direct materials, assigned at standard prices: 160,000 lb × $0.50	80,000
Direct materials efficiency variance: 8,000 lb × $0.50	4,000 U
Direct manufacturing labor costs incurred: 2,200 hours × $20.4545	45,000
Direct manufacturing labor costs, assigned at standard rate:	
2,000 hours × $20.00	40,000
Direct manufacturing labor price variance: 2,200 hours × $0.4545	1,000 U
Direct manufacturing labor efficiency variance: 200 hours × $20.00	4,000 U
Manufacturing overhead allocated: at budgeted rate per machine-hour	70,000
Manufacturing overhead incurred	75,000
Underallocated manufacturing overhead	5,000
Sales	273,000
Marketing, distribution, and customer-service costs	130,000

Materials price variances are measured when the material is purchased rather than when it is used.

Assume that 40% of the October 19_4 production is in the ending inventory of finished goods and that 60% of the October production has been sold in that month. There is no ending work in process on October 31, 19_4. All manufacturing variances are unfavorable. There were no beginning inventories on October 1, 19_4. There were no variances for marketing, distribution, and customer-service costs. The balances (before proration) at the end of the month are based on the data given above:

	Dollars	Percentage
Work in process	$ 0	0%
Finished goods, 40% of the total standard manufacturing costs assigned for materials, manufacturing labor, and manufacturing overhead ($80,000 + $40,000 + $70,000):		
40% of $190,000	76,000	40
Cost of goods sold: 60% of $190,000	114,000	60
Total	$190,000	100%

Morales purchases 200,000 pounds of direct materials, of which 160,000 pounds is the standard amount used for October's production. The standard amounts of direct materials in ending work in process, ending finished goods, and cost of goods sold in October are 0 pounds (0% × 160,000), 64,000 pounds (40% × 160,000), and 96,000 pounds (60% × 160,000), respectively. Given purchases of 200,000 pounds, standard usage in production of 160,000 pounds, and an unfavorable direct materials efficiency quantity of 8,000 pounds, there are 32,000 pounds in ending direct materials inventories:

	Pounds	Total Costs at $0.50 Standard Cost per Pound	Percentage
To account for	200,000	$100,000	100%
Now present in:			
Direct materials efficiency variance	8,000	$ 4,000	4%
Work in process	0	0	0
Finished goods	64,000	32,000	32
Cost of goods sold	96,000	48,000	48
Remainder, in direct materials inventory	32,000	16,000	16
Accounted for	200,000	$100,000	100%

Managers are considering two different approaches to handling end-of-period accounting variances:

1. Morales prorates direct manufacturing labor and manufacturing overhead variances based on the total end-of-period account balances of Work in Process, Finished Goods, and Cost of Goods Sold. It prorates direct materials variances based on the total end-of-period account balances of Direct Materials, Work in Process, Finished Goods, and Cost of Goods Sold.

2. Morales immediately writes off any end-of-period account balances to Cost of Goods Sold.

What effect does each approach have on operating income for October 19_4?

Proration of Direct Manufacturing Labor and Manufacturing Overhead Variances

We will present the analysis using T-accounts. The following entries are numbered in accordance with the logical flow of the amounts through the accounts for direct manufacturing labor and manufacturing overhead.

Work in Process

① Direct manufacturing labor 40,000	③ Transferred* 110,000
② Manufacturing overhead allocated 70,000	

Finished Goods

③ 110,000	④ Sold (60% of 110,000*) 66,000

Cost of Goods Sold

④ 66,000	

Manufacturing Overhead Control

⑤ Incurred 75,000	

Manufacturing Overhead Allocated

	② 70,000

Direct Manufacturing Labor Price Variance

① 1,000	

	Cash, Current Liabilities, Etc.	Direct Manufacturing Labor Efficiency Variance

① Direct manufacturing labor 45,000	① 4,000
⑤ Manufacturing overhead 75,000	

* Transferred as a part of the total standard costs transferred; direct manufacturing labor of $40,000 + manufacturing overhead allocated of $70,000 = $110,000.

In our example, there is a zero ending Work in Process balance. Finished Goods represents 40% of total manufacturing costs, Cost of Goods Sold, 60%. All direct manufacturing labor and manufacturing overhead variances may be prorated accordingly. See the final three prorations in Exhibit 8-8.

	Total Variance	Work in Process 0%	Finished Goods 40%	Cost of Goods Sold 60%
Direct manufacturing labor price variance	$ 1,000 U	$ 0	$ 400	$ 600
Direct manufacturing labor efficiency variance	4,000 U	0	1,600	2,400
Manufacturing overhead variance	5,000 U	0	2,000	3,000
Totals	$10,000 U	$ 0	$4,000	$6,000

For simplicity, the manufacturing overhead variance of $5,000 has not been subdivided into spending, efficiency, or output-level (production-volume) variances.

Based on the proration of these variances, the following journal entry would be made:

EXHIBIT 8-8

Comprehensive Schedule of Proration of Variances:
Morales Company for October 19_4 (All Manufacturing Variances Are Unfavorable)

Type of Variance	Total Variance (1)	To Direct Materials Inventory (2)	To Direct Materials Efficiency Variance (3)	To Work in Process (4)	To Finished Goods (5)	To Cost of Goods Sold (6)
Direct materials price	$10,000*	$1,600	$ 400	$0	$3,200	$ 4,800
Direct materials efficiency						
Balance before proration	4,000		4,000			
Balance after proration			$4,400†	0	1,760	2,640
Direct manufacturing labor price	1,000†			0	400	600
Direct manufacturing labor efficiency	4,000†			0	1,600	2,400
Manufacturing overhead	5,000†			0	2,000	3,000
Total variances prorated	$24,000	$1,600		$0	$8,960	$13,440
*Percentage used for proration	100%	16%	4%	0%	32%	48%
†Percentage used for proration	100%			0%	40%	60%

Finished Goods Control	$ 4,000	
Cost of Goods Sold	6,000	
Manufacturing Overhead Allocated	70,000	
Direct Manufacturing Labor Price Variance		$ 1,000
Direct Manufacturing Labor Efficiency Variance		4,000
Manufacturing Overhead Control		75,000

To prorate variances and to close manufacturing overhead accounts.

Proration of Direct Materials Variances

Direct materials is inventoried when purchased. In contrast, other manufacturing costs are not inventoried until used. This difference requires an extra step when prorating direct materials variances (assuming that materials price variances are measured when the material is purchased rather than when it is used). Some key T-accounts follow regarding the flow of the direct materials cost (only) through the accounts:

Direct Materials Inventory

① Purchased		② Issued	84,000
100,000			

Work in Process Control†

②	80,000	③ Transferred*	
			80,000

Finished Goods Control†

③	80,000	④ Sold*	48,000
Bal.	32,000		

Cost of Goods Sold†

④*	48,000	

Accounts Payable Control

	①	110,000

Direct Materials Price Variance

①	10,000	

Direct Materials Efficiency Variance

②	4,000	

*Transferred as a part of the total standard costs transferred.

†Direct materials cost component only; standard cost of direct materials in cost of goods sold is $48,000 (96,000 pounds times the $0.50 standard cost per pound).

The most complex proration is the direct materials price variance. To be most accurate, its proration in our illustration should be traced at $10,000 ÷ 200,000 = $0.05 per pound to wherever the 200,000 pounds have been charged at standard cost. The pounds are not only in Work in Process, Finished Goods, and Cost of Goods Sold. They are also in Direct Materials Inventory and in the Direct Materials Efficiency Variance account. Hence, we begin with a proration of the materials price variance to five accounts, using the percentages shown for direct materials in the Morales data on page 290.

Direct materials price variance	$10,000
Allocated to:	
Direct materials efficiency variance, 4%	$ 400
Work in process inventory, 0%	0
Finished goods inventory, 32%	3,200
Cost of goods sold, 48%	4,800
Direct materials inventories, 16%	1,600
Total allocated	$10,000

The following compound journal entry prorates the direct materials price variance:

Direct Materials Efficiency Variance	$ 400	
Finished Goods Control	3,200	
Cost of Goods Sold	4,800	
Direct Materials Inventory	1,600	
Direct Materials Price Variance		$10,000

After the proration of the direct materials price variance is posted, the Direct Materials Efficiency Variance account will be:

Direct Materials Efficiency Variance

Balance before proration	4,000	
Proration of unfavorable direct materials price variance	400	
Balance after proration	4,400	

In turn, the direct materials efficiency variance after proration is allocated to:

Work in process inventory, 0%	$ 0
Finished goods inventory, 40%	1,760
Cost of goods sold, 60%	2,640
Total allocated	$4,400

The following journal entry prorates the direct materials efficiency variance:

Finished Goods	$1,760	
Cost of Goods Sold	2,640	
Direct Materials Efficiency Variance		$4,400
To prorate the direct materials efficiency variance.		

Exhibit 8-8 is a comprehensive schedule of all the variance prorations explained here. The T-accounts for inventories and cost of goods sold after proration are:

Direct Materials Inventory

Purchased	100,000	Issued	84,000
Proration of direct materials price variance	1,600		
Balance	17,600		

Work in Process

Direct materials	80,000	Transferred	190,000
Direct manufacturing labor	40,000		
Manufacturing overhead allocated	70,000		

Finished Goods Control

Transferred	190,000	Sold	114,000
Proration of direct manufacturing labor and manufacturing overhead variances	4,000		
Proration of direct materials price variance	3,200		
Proration of direct materials efficiency variance	1,760		
Balance	84,960		

Cost of Goods Sold

Sold	114,000	
Proration of direct manufacturing labor and manufacturing overhead variances	6,000	
Proration of direct materials price variance	4,800	
Proration of direct materials efficiency variance	2,640	
Balance	127,440	

Exhibit 8-9 compares how immediate write-off and proration affect operating income. All the variances are unfavorable, so writing off all the variances to cost of goods sold reduces operating income by $24,000. When prorated, however, the variances reduce income by only $13,440. The difference between these two approaches—$10,560—shows up in the operating income amounts. Immediately writing off all the

EXHIBIT 8-9

Effects of Disposal of Manufacturing Variances on Operating Income:
Morales Company for October 19_4

	Standard Absorption Costing	
	All Variances Cost of Goods Sold	Variance Proration
Sales	$273,000	$273,000
Cost of goods sold (at standard costs)	114,000	114,000
Gross margin (at standard costs)	159,000	159,000
Total variances (from column 1, Exhibit 8-8)	24,000	0
Prorated variances (from column 6, Exhibit 8-8)	0	13,440
Gross margin	135,000	145,560
Marketing, distribution, and customer-service costs	130,000	130,000
Operating income	$ 5,000	$ 15,560

variance to cost of goods sold will result in operating income of $5,000; prorating the variances will result in operating income of $15,560. Managers at Morales likely are very interested in these differences, especially if operating income is used as one of their performance measures.

Alternative Views on Proration

Accountants who support proration of variances among end-of-period account balances frequently view the method as a means of approximating the actual costs, which are assumed to be the most accurate costs to report. An alternative viewpoint is that actual costs are not the appropriate costs to report in a balance sheet for inventories. For example, with proration, an unfavorable materials efficiency variance is capitalized in inventory even though that variance may be due to poor work or inadequate maintenance of plant and equipment. Advocates of writing off all variances to cost of goods sold typically argue that standard costs best represent the appropriate cost for an inventory asset item in a balance sheet.

Generally accepted accounting principles and income tax laws typically require that financial statements show actual costs, not standard costs, of inventories and cost of goods sold. Consequently, proration of manufacturing variances is required if it results in a material change in inventories or operating income.

TERMS TO LEARN

This chapter and the Glossary at the end of the book contain definitions of the following important terms:

denominator level *(p. 276)* non-value-added cost *(270)*
output level overhead variance *(278)* production denominator level *(276)*
production denominator volume *(276)* production level variance *(278)*
production volume variance *(278)* standard costing *(286)* value-added cost *(270)*
variable overhead efficiency variance *(274)* variable overhead spending variance *(274)*

ASSIGNMENT MATERIAL

QUESTIONS

8-1 What are the steps in planning variable overhead costs?

8-2 How does the planning of fixed overhead costs differ from the planning of variable overhead costs?

8-3 "Budgeting for variable manufacturing overhead requires knowledge of cost drivers." Name three possible cost drivers.

8-4 Both financial and nonfinancial measures are used to control variable manufacturing overhead. Give two examples of each type of measure.

8-5 "The spending variance for variable manufacturing overhead is affected by several factors." Explain.

8-6 Describe the difference between a direct materials efficiency variance and a variable manufacturing overhead efficiency variance.

8-7 What are the steps in developing a budgeted fixed overhead rate?

8-8 Why is the flexible-budget variance the same amount as the spending variance for fixed manufacturing overhead?

8-9 Give two terms commonly used to describe an output level overhead variance in a manufacturing context.

8-10 Give two reasons why an output level overhead variance may not be a good measure of the cost of unused capacity.

8-11 Explain how 4-variance analysis differs from 1-variance, 2-variance, and 3-variance analyses.

8-12 Give an example of how the variances in a 4-variance analysis might be interdependent.

8-13 Describe a situation in which fixed marketing overhead costs might be unitized and allocated to products.

8-14 Explain how the analysis of fixed overhead costs differs for (a) planning and control on the one hand and (b) inventory costing for financial reporting on the other hand.

8-15 Give two examples of nonfinancial measures used to control manufacturing overhead.

EXERCISES AND PROBLEMS

8-16 Manufacturing overhead, under- or overallocated amounts. (Z. Iqbal) Peach Company estimated variable manufacturing overhead of $800,000 and fixed manufacturing overhead of $500,000 for the year. The company uses a standard costing system. It allocates manufacturing overhead on the basis of machine-hours. The estimates were based on the budgeted usage of 100,000 machine-hours. During the period, Peach actually used 110,000 machine-hours.

According to the standard costing system, 95,000 machine-hours should have been used for actual output achieved. Actual costs incurred for variable and fixed manufacturing overhead were $815,000 and $470,000, respectively.

Required
1. Compute the under- or overallocated variable manufacturing overhead amount.
2. Compute the under- or overallocated fixed manufacturing overhead amount.
3. Compute the under- or overallocated total manufacturing overhead amount.

8-17 4-variance analysis, fill in the blanks. Use the given manufacturing overhead data to fill in the blanks.

	Variable	Fixed
Actual costs incurred	$11,900	$6,000
Allocated to products	9,000	4,500
Flexible budget (Budgeted input allowed for actual output achieved × Budgeted rate)	9,000	5,000
Actual input × Budgeted rate	10,000	5,000

Use *F* for favorable and *U* for unfavorable:

	Variable	Fixed
1. Spending variance	$_____	$_____
2. Efficiency variance	_____	_____
3. Output level (production volume) variance	_____	_____
4. Flexible-budget variance	_____	_____
5. Underallocated (overallocated) manufacturing overhead	_____	_____

8-18 Spending and efficiency overhead variances, service sector. Meals on Wheels (MOW) operates a home meal delivery service. It has agreements with 20 restaurants to pick up and deliver meals to customers who phone or fax orders to MOW. MOW is currently examining its overhead costs for May 19_5.

Variable overhead costs for May 19_5 were budgeted at $2 per hour of home delivery time. Fixed overhead costs were budgeted at $24,000. The budgeted number of home deliveries in May 19_5 was 8,000. Delivery time, the allocation base for variable and fixed overhead costs, is budgeted to be 0.80 hours per delivery.

Actual results for May 19_5 were:

Variable overhead	$14,174
Fixed overhead	$27,600
Number of home deliveries	7,460
Hours of delivery time	5,595

Customers are charged $12 per delivery. The delivery driver is paid $7 per delivery.

MOW receives a 10% commission on the meal costs that the restaurants charge the customers who use MOW.

Required
1. Compute spending and efficiency variances for MOW's variable overhead and fixed overhead in May 19_5. Comment on the results.
2. How might MOW manage its variable overhead costs differently from how it manages its fixed overhead costs?

8-19 Spending and efficiency overhead variances, distribution. Package Postal Service (PPS) operates a parcel delivery service. PPS's costing system has one direct cost category (delivery driver payments) and two overhead categories—variable delivery overhead and fixed delivery overhead. In 19_5 it charged retail companies and mail-order catalog companies $15 per delivery. Delivery drivers in 19_5 were contracted at $5 per delivery. Variable delivery overhead for September 19_5 was budgeted at $2 per hour of delivery time. Budgeted fixed delivery overhead in September 19_5 was $120,000. PPS budgeted 100,000 deliveries for September 19_5. Delivery time, the allocation base for variable and fixed overhead costs, is budgeted to be 0.25 hours per delivery.

Actual results for September 19_5 were:

Variable delivery overhead	$ 60,000
Fixed delivery overhead	$128,400
Number of deliveries	96,000
Hours of delivery time	28,800

Required
1. Compute spending and efficiency variances for PPS's two categories of delivery overhead costs in September 19_5. Comment on the results.
2. What problems might PPS face in managing (a) its direct costs, (b) its variable delivery overhead costs, and (c) its fixed delivery overhead costs?

8-20 Graphs and overhead variances. The Carvelli Company is a manufacturer of housewares. In its job costing system, manufacturing overhead (both variable and fixed) is allocated to products on the basis of budgeted machine-hours. The budgeted amounts are taken from Carvelli's standard costing system. The budget for 19_4 included:

Variable manufacturing overhead	$9 per machine-hour
Fixed manufacturing overhead	$72,000,000
Denominator level	4,000,000 machine-hours

Required
1. Prepare four graphs, two for variable manufacturing overhead and two for fixed manufacturing overhead. Each pair of graphs should display how total manufacturing overhead costs of Carvelli will be depicted for the purpose of (a) planning and control and (b) inventory costing.
2. Suppose that 3,500,000 machine-hours were allowed for actual output achieved in 19_4, but 3,800,000 actual machine-hours were used. Actual manufacturing overhead was: variable, $36,100,000; fixed, $72,200,000. Compute (a) variable manufacturing overhead spending and efficiency variances and (b) the fixed manufacturing overhead spending and output level (production volume) variances. Use the columnar presentation illustrated in Exhibit 8-4 (p. 279).
3. What is the amount of the underallocated or overallocated variable manufacturing overhead? Of the underallocated or overallocated fixed manufacturing overhead? Why are the flexible-budget variance and the underallocated or overallocated overhead amount always the same for variable manufacturing overhead but rarely the same for fixed manufacturing overhead?
4. Suppose the denominator level was 3,000,000 rather than 4,000,000 machine-hours. What variances in requirement 2 would be affected? Recompute them.

8-21 Journal entries. Refer to the preceding problem, requirement 2. Consider variable manufacturing overhead and then fixed manufacturing overhead. Prepare the journal entries for (a) the incurrence of overhead, (b) the allocation of overhead, and (c) the isolation and closing of overhead variances to Cost of Goods Sold for the year.

8-22 Straightforward 4-variance overhead analysis. The Lopez Company uses a standard-cost system in its manufacturing plant for auto parts. Its standard cost of an auto part, based on a denominator level of 4,000 output units per year, included 6 machine-hours of variable manufacturing overhead at $8 per hour and 6 machine-hours of fixed manufacturing overhead at $15 per hour. Actual output achieved was 4,400 units. Variable manufacturing overhead incurred was $245,000. Fixed manufacturing overhead incurred was $373,000. Actual incurred machine-hours were 28,400.

Required

1. Prepare an analysis of all variable manufacturing overhead and fixed manufacturing overhead variances, using the 4-variance analysis on p. 279.
2. Prepare journal entries using the 4-variance analysis.
3. Describe how individual variable manufacturing overhead items are controlled from day to day. Also, describe how individual fixed manufacturing overhead items are controlled.

8-23 Total manufacturing-overhead variances, 2-variance analysis. (SMA, adapted) The Weser Company uses a budgeted total manufacturing overhead rate of $40 per output unit based on a denominator level of 60,000 units a year, or 5,000 units a month.

During October, the company produced 5,200 units and used 2-variance analysis to compute the following manufacturing overhead variances:

Total overhead flexible budget variance	$2,500 U
Output level (production volume) variance	$5,000 F

During November, production was 4,900 units, and the total manufacturing overhead cost incurred was $2,000 less than October's amount.

Required

Compute the manufacturing overhead flexible-budget variance and the output level (production volume) variance for the month of November.

8-24 Straightforward coverage of manufacturing overhead, standard-cost system. The Singapore division of a Canadian telecommunications company uses a standard-cost system for its machine-paced production of telephone equipment. Data regarding production during June follow:

Variable manufacturing overhead costs incurred: $155,100
Variable manufacturing overhead costs allocated at $12 per
  standard machine-hour allowed for actual output achieved
Fixed manufacturing overhead costs incurred: $401,000
Fixed manufacturing overhead budgeted: $390,000
Denominator level in machine-hours: 13,000
Standard machine-hours allowed per unit of output: 0.30
Units of output: 41,000
Actual machine-hours used: 13,300
Ending work in process inventory: 0

Required

1. Prepare an analysis of all manufacturing overhead variances. Use the 4-variance analysis framework illustrated in Exhibit 8-4, (p. 279).
2. Prepare journal entries for manufacturing overhead without explanations.
3. Describe how individual variable manufacturing overhead items are controlled from day to day. Also, describe how individual fixed manufacturing overhead items are controlled.

8-25 Total overhead, 3-variance analysis. The Wright-Patterson Air Force Base has an extensive repair facility for jet engines. It developed standard costing and flexible budgets to account for this activity. Budgeted variable overhead at a level of 8,000 standard monthly direct labor-hours was $64,000; budgeted total overhead at 10,000 standard direct labor-hours was $197,600. The standard cost allocated to repair output included a total overhead rate of 120% of standard direct labor cost.

Total overhead incurred for October was $249,000. Direct labor costs incurred were $202,440. The direct labor price variance was $9,640, unfavorable. The direct labor flexible-budget variance was $14,440, unfavorable. The standard labor price was $16 per hour. The output level variance was $14,000, favorable.

Required

1. Compute the direct labor efficiency variance and the spending, efficiency, and output level variances for overhead. Also, compute the denominator level.
2. Describe how individual variable manufacturing overhead items are controlled from day to day. Also, describe how individual fixed manufacturing overhead items are controlled.

8-26 Price (spending) and efficiency variances. The Favaro Company used a flexible budget and standard costs for controlling its customer services. In March, the company serviced 7,000 product units. Assume that 15,000 actual hours were used at an actual rate of $16.40 per hour. Two direct labor-hours is the standard allowance for servicing one product unit. The standard labor rate is $16 per hour. The flexible budget for miscellaneous supplies (classified as part of manufacturing overhead) is based on a formula of $2.40 per product unit serviced, which can also be expressed as $1.20 per standard direct labor-hour. The actual cost of supplies was $18,800.

Required
1. Compute the price and efficiency variances for direct labor.
2. Compute the spending and efficiency variances for supplies.
3. The plant manager has made the following comments: "I have been troubled by the waste of supplies for months. I know that the prices are exactly equal to the budgeted prices for every supply item, because a two-year stock of supplies was bought a year ago in anticipation of prolonged inflation. Consequently, we used known prices for preparing our budgets. Given these facts, please explain how a spending variance can arise for miscellaneous supplies. Why doesn't all waste appear as an efficiency variance?" Respond to the manager's comments. Be clear and specific; the manager is impatient with muddled explanations, and your next pay raise will be affected by her appraisal of your explanation.
4. If prices rise, the two-year stock of supplies will result in savings of overall purchase costs. Give two arguments against acquiring such a large stock of supplies.

8-27 4-variance analysis, find the unknowns. Consider each of the following situations—cases A, B, and C—independently. Data refer to operations of April 19_5. For each situation, assume a standard-cost system. Also assume the use of a flexible budget for control of variable and fixed manufacturing overhead based on machine-hours.

	Case A	Case B	Case C
(1) Fixed manufacturing overhead incurred	$10,600	—	$12,000
(2) Variable manufacturing overhead incurred	7,000	—	—
(3) Denominator level in machine-hours	500	—	1,100
(4) Standard machine-hours allowed for actual output achieved	—	650	—
Flexible-budget data:			
(5) Fixed manufacturing overhead	—	—	—
(6) Variable manufacturing overhead (per standard machine-hour)	—	8.50	5.00
(7) Budgeted fixed manufacturing overhead	10,000	—	11,000
(8) Budgeted variable manufacturing overhead*	—	—	—
(9) Total budgeted manufacturing overhead*	—	12,525	—
Additional data:			
(10) Standard variable manufacturing overhead allocated	7,500	—	—
(11) Standard fixed manufacturing overhead allocated	10,000	—	—
(12) Output level (production volume) variance	—	500 U	500 F
(13) Variable manufacturing overhead spending variance	950 F	0	350 U
(14) Variable manufacturing overhead efficiency variance	—	0	100 U
(15) Fixed manufacturing overhead spending variance	—	300 F	—
(16) Actual machine-hours used	—	—	—

*For standard machine-hours allowed for actual output achieved.

Required
Fill in the blanks under each case. (*Hint:* Prepare a worksheet similar to that in Exhibit 8-4, p. 279. Fill in the knowns and then solve for the unknowns.)

8-28 Hospital overhead variances, 4-variance analysis. The Sharon Hospital, a large metropolitan health-care complex, has had much trouble in controlling its accounts receivable. Bills for patients, for various government agencies, and for private insurance companies have frequently been inaccurate and late. This situation has led to intolerable levels of bad debts and investments in receivables.

You were employed by the hospital as a consultant on this matter. After conducting a careful study of the billing operation, you developed some currently attainable standards that were implemented in conjunction with a flexible budget. You had divided costs into fixed and variable categories. You regarded the bill as the product, the unit of output.

You have reasonable confidence that the underlying source documents for compiling the results have been accurately tallied. However, the accountant has had some trouble summarizing the data and provided the following:

Variable overhead costs, allowance per standard hour	$ 10
Fixed overhead flexible-budget variance, favorable	200
Total budgeted overhead costs for the bills prepared	22,500
Output level (production volume) variance, favorable	900
Variable cost spending variance, unfavorable	2,000
Variable cost efficiency variance, favorable	2,000
Standard hours allowed for the bills prepared, 1,800	

Required
1. Actual hours of input used.
2. Fixed overhead budget.
3. Fixed overhead allocated.
4. Budgeted fixed overhead rate per hour.
5. Denominator level in hours.

8-29 Working backward from given variances. The Mancuso Company uses a flexible budget and standard costs to aid planning and control of its manufacturing operations. At a 60,000 direct manufacturing labor-hour level, budgeted variable overhead is $120,000 and budgeted direct labor is $480,000.

The following actual results are for August:

Variable manufacturing overhead flexible budget variance	$ 10,500 U
Variable manufacturing overhead efficiency variance	20,000 U
Direct manufacturing labor costs incurred	574,000
Materials price variance (based on purchases)	16,000 F
Materials efficiency variance	9,000 U
Fixed manufacturing overhead incurred	50,000
Fixed manufacturing overhead spending variance	2,000 U

The standard cost per pound of direct materials is $1.50. The standard allowance is one pound of direct materials for each unit of product. Ninety thousand units of product were made during August. There was no beginning or ending work in process. In July, the materials efficiency variance was $1,000, favorable, and the price variance was $0.20 per pound, unfavorable. In August, the price variance was $0.10 per pound.

In July, labor troubles caused a major slowdown in the pace of production, resulting in an unfavorable direct manufacturing labor efficiency variance of $60,000. There was no manufacturing labor price variance. These troubles persisted into August. Some workers quit. Their replacements had to be hired at higher rates, which had to be extended to all workers. The actual average wage rate in August exceeded the standard average wage rate by $0.20 per hour.

Required
1. Total pounds of direct materials purchased during August.
2. Total number of pounds of excess materials used.
3. Variable manufacturing overhead spending variance.
4. Total number of actual hours of direct manufacturing labor-hours used.

5. Total number of standard direct manufacturing labor-hours allowed for the units produced.
6. Describe how individual variable manufacturing overhead items are controlled from day to day. Also, describe how individual fixed manufacturing overhead items are controlled.

8-30 Flexible budgets, 4-variance analysis. (CMA, adapted)   Nolton Products uses a standard costing system. It allocates manufacturing overhead (both variable and fixed) to products on the basis of standard direct manufacturing labor-hours (DLH). Nolton develops its manufacturing overhead rate from the current annual budget. The manufacturing overhead budget for 19_4 is based on budgeted output of 720,000 units requiring 3,600,000 direct manufacturing labor-hours. The company is able to schedule production uniformly throughout the year.

A total of 66,000 output units requiring 315,000 direct labor-hours was produced during May 19_4. Manufacturing overhead (MOH) costs incurred for May amounted to $375,000. The actual costs as compared with the annual budget and 1/12 of the annual budget are shown below.

Annual Manufacturing Overhead Budget 19_4

	Total Amount	Per Output Unit	Per DLH Input Unit	Monthly MOH Budget May 19_4	Actual MOH Costs for May 19_4
VARIABLE MOH					
Indirect manufacturing labor	$ 900,000	$1.25	$ 0.25	$ 75,000	$ 75,000
Supplies	1,224,000	1.70	0.34	102,000	111,000
FIXED MOH					
Supervision	648,000	0.90	0.18	54,000	51,000
Utilities	540,000	0.75	0.15	45,000	54,000
Depreciation	1,008,000	1.40	0.28	84,000	84,000
Total	$4,320,000	$6.00	$ 1.20	$360,000	$375,000

Required
Calculate the following amounts for Nolton Products for May 19_4:

1. Total manufacturing overhead costs allocated.
2. Variable manufacturing overhead spending variance.
3. Fixed manufacturing overhead spending variance.
4. Variable manufacturing overhead efficiency variance.
5. Output level (production volume) variance.

Be sure to identify each variance as favorable (F) or unfavorable (U).

8-31 Variance analysis from fragmentary evidence, 4-variance analysis.   Being a bright young person, you have just landed a wonderful job as assistant controller of Gyp-Clip, a new and promising Singapore division of Croding Metals Corporation. The Gyp-Clip Division has been formed to produce a single product, a new-model paper clip. Croding Laboratories has developed an extremely springy and lightweight new alloy, Clypton, which is expected to revolutionize the paper-clip industry.

It is your first day on the job. Gyp-Clip has been in business one month. The controller takes you on a tour of the plant and explains the operation in detail: Clypton wire is received on two-mile spools from the Croding mill at a fixed price of $40 a spool, which is not subject to change. Clips are bent, cut, and shipped in bulk to the Croding packaging plant. Plant rent, depreciation, and all other items of fixed manufacturing overhead are handled by the home office at a set amount of $100,000 per month. Ten thousand tons of paper clips have been produced, which is only 75% of the budgeted denominator level.

The controller has just computed the month's variances. She is looking for a method of presenting them in clear, logical form to top management at the home office. You say that you know of just the method and promise to have the analysis ready the next morning.

Filled with zeal and enthusiasm, feeling that your future as a rising star in this growing company is secure, you decide to take your spouse out to dinner to celebrate the trust and confidence that your superior has placed in you.

Upon returning home, you are horrified to discover your dog happily devouring the controller's figure sheet. You manage to salvage only the accompanying fragments.

Miles used.. 5,300
Variable manufacturing overhead incurred....$50,000

Manufacturing overhead efficiency variance....$8,500 U
variance...$7,000 U

dard Costs per ton:

bor hours............................. 0.6
age rate.................................... $20
al overhead............................. $12.60
Clypton wire........................... 1/2 mile

You remember that the $7,000 unfavorable variance did not represent the grand total of all variances. You also recall that the company allocates manufacturing overhead on the basis of direct manufacturing labor-hours.

Required
Don't let the controller down. Go ahead and reconstruct your analysis of all variances.

8-32 Review of Chapters 7 and 8, 3-variance analysis. (CPA, adapted)   The Beal Manufacturing Company's job costing system has two direct cost categories—direct materials and direct manufacturing labor. Manufacturing overhead (both variable and fixed) is allocated to products on the basis of standard direct manufacturing labor-hours (DLH). At the beginning of 19_5, Beal adopted the following standards for its manufacturing costs:

	Input	Cost per Output Unit
Direct materials	3 lb at $5.00 per lb	$ 15.00
Direct manufacturing labor	5 hr at $15.00 per hr	75.00
Manufacturing overhead:		
Variable	$6.00 per DLH	30.00
Fixed	$8.00 per DLH	40.00
Standard manufacturing cost per output unit		$160.00

The denominator level for total manufacturing overhead per month in 19_5 is 40,000 direct manufacturing labor-hours. Beal's flexible budget for January 19_5 was based on this denominator level. The records for January indicated the following:

Direct materials purchased	25,000 lb at $5.20 per lb
Direct materials used	23,100 lb
Direct manufacturing labor	40,100 hr, at $14.60 per hr
Total actual manufacturing overhead (variable and fixed)	$600,000
Actual production	7,800 output units

Required
1. Prepare a schedule of total standard manufacturing costs for the 7,800 output units in January, 19_5.
2. For the month of January 19_5, compute the following variances, indicating whether each is favorable (F) or unfavorable (U):
   a. Direct materials price variance, based on purchases
   b. Direct materials efficiency variance
   c. Direct manufacturing labor price variance
   d. Direct manufacturing labor efficiency variance

e. Total manufacturing overhead spending variance
f. Variable manufacturing overhead efficiency variance
g. Output level (production volume) variance

8-33 Working backward, comprehensive variance analysis. (H. Hoverland)   Durrell Company manufactures a single product name, "Preston." It uses a standard costing system for both its two direct costs (direct materials and direct manufacturing labor) and its variable manufacturing overhead costs. The standard variable cost for one unit of Preston in 19_4 is:

Direct materials	
(10 lb at $0.50 per lb)	$ 5
Direct manufacturing labor	
(2 hr × $10 per hr)	20
Variable manufacturing overhead	
(3 machine-hours at $5 per hour)	15
Total variable cost per unit	$40

Fixed manufacturing costs for 19_4 are budgeted to be $400,000.
Selected actual results and variances for 19_4 are:

Production output	100,000 units
Direct materials inventory, 1/1/19_4	0
Direct materials inventory, 12/31/19_4	100,000 lb
Materials price variance (MPV)	$ 24,000 U
Materials efficiency variance (MEV)	$ 50,000 U
Manufacturing labor price variance	$ 50,000 F
Manufacturing labor efficiency variance	$100,000 U
Variable manufacturing overhead spending variance (VMOSV)	$ 50,000 F
Variable manufacturing overhead efficiency variance (VMOEV)	$150,000 U
Fixed manufacturing overhead spending variance (FMOSV)	$ 30,000 U
Output level (production volume) variance	$ 0

Required
Compute the following items:

1. Direct materials purchased ($).
2. Direct materials purchased (pounds).
3. Average price paid per pound of direct material.
4. Direct materials used ($).
5. Direct materials used (pounds).
6. Direct manufacturing labor used ($).
7. Direct manufacturing labor used (DLH).
8. Average rate paid per DLH.
9. Actual variable overhead ($).
10. Machine-hours used.
11. Average variable manufacturing overhead rate per machine-hour.
12. Actual fixed manufacturing overhead ($).
13. Planned production level (units).
14. Planned production level (machine-hours).

8-34 Joint venture, 4-variance analysis of marketing costs, ethics.   London Life Insurance sells life insurance policies to citizens of the United Kingdom (U.K.). London Life has a joint-venture agreement with Georgia Life Agents (GLA). GLA markets London Life policies to U.K. citizens living in the United States, and over 5% of London Life policyholders currently live in the United States. London Life pays GLA for the marketing cost of each new life insurance policy signed by a GLA sales representative. The joint-venture agreement specifies that at the start of each year GLA determine its budgeted marketing cost per new policy sold. This budgeted amount consists of:

- ◆ Direct marketing labor costs (costs of sales representatives)
- ◆ Variable marketing overhead costs, expressed on a sales-call basis, which is the cost driver
- ◆ Fixed marketing overhead costs, expressed on the basis of the number of new policies budgeted to be sold

For simplicity, assume a sales representative makes only one sales call to each new potential customer.

At the end of each month, London Life pays GLA the budgeted marketing cost per new London Life policy sold times the number of new policies GLA has sold. James Barnes, the former finance director of London Life, stressed that the use of a budgeted rate successfully protected London Life from any inefficiencies in GLA's marketing.

Jeanette Smith, the new finance director, is more skeptical of the joint-venture agreement than her predecessor was. The agreement provides that, in the event of a dispute, an independent arbitrator can investigate and make a ruling binding on both parties. Smith requests that an independent arbitrator examine whether the amount London Life is paying for GLA's marketing of its policies in the United States is excessive. The arbitrator collects the following information pertaining to GLA for 19_5:

	Budgeted Amount	Actual Results
Direct marketing labor costs per sales call	$50	$45
Variable marketing overhead costs per sales call	$20	$16
Fixed marketing overhead costs	$1,800,000	$1,600,000
Number of sales calls	60,000	48,000
Number of new policies sold	10,000	12,000

Required
1. What were GLA's budgeted marketing costs for each new London Life insurance policy sold in 19_5? What was the actual marketing cost?
2. Use the 4-variance analysis framework to analyze GLA's variable and fixed marketing costs in 19_5. Comment on the results. Assume a unit of output is a new policy sold.
3. The arbitrator suspects that GLA may be systematically misrepresenting key numbers in its budget. How would you gain evidence on whether systematic misrepresentation may have occurred?

8-35 Joint venture, incentives for cost misrepresentation, ethics. (Continuation of 8-34) The arbitrator reports that the 19_5 budgeted rate GLA charged was indeed the budgeted rate GLA computed for 19_5, but that budgeted rate differed greatly from the actual rate for GLA in 19_5. Jeanette Smith, the finance director of London Life, seeks your advice on redesigning the contract. She is concerned about reducing the incentives for GLA to misrepresent the costs of its marketing of London Life policies in the United States.

Required
Recommend an alternative payment plan for the joint venture between London Life and GLA. Comment on its strengths and weaknesses relative to the existing plan described in Problem 8-34.

Coverage of the Chapter Appendix

8-36 Straightforward proration of manufacturing variances. Consider the following balances of standard costs before proration at the end of the year: work in process, $180,000; finished goods, $720,000; cost of goods sold, $900,000. The output-level (production-volume) variance was $50,000, favorable. All other manufacturing variances were $330,000, unfavorable. Management has decided to prorate all variances in proportion to the ending balances in work in process, finished goods, and cost of goods sold before proration.

Required
1. Prepare a schedule that prorates the manufacturing variances.
2. Prepare a journal entry that closes all variance accounts.

3. Assume that the major justification for proration is the attempt to approximate the "actual" cost of the units produced. Name two likely sources of inaccuracy in the proration here. Explain.

8-37 Proration of manufacturing variances and income effects of standard costs. The Stefano Company uses a standard costing system, which shows the following account balances (before proration of any variances) at December 31, 19_4:

Direct materials, ending inventory	$175,000
Work in process, ending inventory	100,000
Finished goods, ending inventory	300,000
Cost of goods sold	600,000
Direct materials price variance	64,000
Direct materials efficiency variance	25,000
Direct manufacturing labor price variance	5,000
Direct manufacturing labor efficiency variance	25,000
Manufacturing overhead incurred	210,000
Manufacturing overhead allocated, at standard rate	170,000
Sales	900,000
Marketing and administrative costs	180,000

Materials price variances are measured when the material is purchased rather than when it is used. Assume that Work in Process, Finished Goods, and Cost of Goods Sold contain standard costs in uniform proportions of direct materials, direct manufacturing labor, and manufacturing overhead. The direct materials component represented 60% of the ending balance in Work in Process, Finished Goods, and Cost of Goods Sold. All manufacturing variances are unfavorable. There are no beginning inventories. There are no variances for marketing and administrative costs. Both variable and fixed manufacturing costs are inventoried.

Required
1. Prepare a comprehensive schedule showing the proration of all manufacturing variances.
2. Prepare a journal entry for the proration.
3. Prepare comparative summary income statements based on a standard costing system:
   a. Without proration of any manufacturing variances (all variances written off to Cost of Goods Sold)
   b. With proration of all manufacturing variances based on total ending balances
4. Compute the amount of direct manufacturing labor cost included in the balance of finished goods ending inventory (before proration).

9

Income Effects of Alternative Inventory-Costing Methods

PART ONE: VARIABLE COSTING
    AND ABSORPTION COSTING

Role of fixed manufacturing
    overhead costs

Comparison of standard variable
    costing and standard absorption
    costing

Breakeven points in variable costing
    and absorption costing

Performance measures and
    absorption costing

PART TWO: ROLE OF VARIOUS
    DENOMINATOR-LEVEL
    CONCEPTS IN ABSORPTION
    COSTING

Alternative denominator-level
    concepts

Effect on financial statements

*T*he bottling lines at Snapple Beverage Corp. can operate
at varying degrees of speed. Management considers
these different operating speeds, as well as demand factors, when determining the denominator level to use in setting fixed overhead cost rates.

Learning Objectives

When you have finished studying this chapter, you should be able to

1. Identify the fundamental feature that distinguishes variable costing from absorption costing
2. Construct income statements using absorption costing and variable costing
3. Compare and contrast eight alternative inventory-costing systems
4. Explain differences in operating income under absorption costing and variable costing
5. Describe the effects of absorption costing on computing the breakeven point
6. Understand how absorption costing influences performance-evaluation decisions
7. Describe the various denominator-level concepts that can be used in absorption costing
8. Explain how the choice of the denominator level affects reported operating income and inventory amounts

Managers and accountants face the decision of which inventory costing method to use. This decision will affect reported income, which is a principal measure of managers' performance as viewed from both inside and outside the organization. In this chapter we examine two topics related to inventory costing of manufacturing companies:

1. variable costing and absorption costing, and
2. choice of the denominator-level concept in absorption costing.

These two topics are related closely enough to warrant inclusion in a single chapter. *However, they may be studied independently.*

We concentrate on how different approaches to measuring inventories and income affect reported balance sheet and income statement numbers. These approaches, however, have implications for many management decisions, including choices of product mix, pricing, and making or buying components. For instance, the computation of the full product cost for guiding pricing not only includes manufacturing costs (commonly regarded as inventoriable costs) but also includes the costs of the other business function areas of the value chain, such as product design, distribution, and customer service.

◆ PART ONE: VARIABLE COSTING AND ABSORPTION COSTING

ROLE OF FIXED MANUFACTURING OVERHEAD COSTS

Impact of Fixed Manufacturing Overhead

Objective 1
Identify the fundamental feature that distinguishes variable costing from absorption costing

Two major methods of costing the inventories of manufacturing companies are variable costing and absorption costing. These methods differ in only one conceptual respect: whether fixed manufacturing overhead (indirect manufacturing costs) is an inventoriable cost. Recall from Chapter 2 that *inventoriable costs* for a manufacturing company are costs associated with the acquisition and conversion of materials and all other manufacturing inputs into goods for sale; these costs are first recorded as an asset and then subsequently become an expense when the goods are sold. Definitions for the two inventory costing methods are:

- ◆ **Variable costing** is a method of inventory costing in which all direct manufacturing costs and variable manufacturing overhead costs are included as inventoriable costs; fixed manufacturing overhead costs are excluded from the inventoriable costs and are costs of the period in which they are incurred.
- ◆ **Absorption costing** is a method of inventory costing in which all direct manufacturing costs and all manufacturing overhead costs—both variable and fixed—are considered as inventoriable costs. That is, inventory "absorbs" all these costs.

Throughout this chapter, to emphasize underlying concepts, we assume that the chosen denominator level for computing the variable and fixed manufacturing overhead allocation rates is a production, output-related variable. Examples include direct labor-hours, direct machine-hours, and units of production output.

Radius Company, which manufactures specialty belts, illustrates the difference between variable costing and absorption costing. Radius uses a normal costing system. That is, its direct costs are traced to products using the actual prices times the actual inputs used, and indirect (overhead) costs are allocated using the budgeted indirect cost rate times the actual inputs used. Company management has chosen budgeted production output units as the denominator for allocating fixed manufacturing overhead to products. To keep the example simple, we assume that:

1. the 19_4 actual fixed manufacturing overhead of $2,200,000 is the same as the budgeted amount, and
2. the 19_4 actual production level of 1,100,000 units is the same as the budgeted denominator level.

The fixed manufacturing overhead rate is $2 per unit ($2,200,000 ÷ 1,100,000 units). Although 1,100,000 units were produced in 19_4, only 1,000,000 units were sold that year, leaving an ending inventory of 100,000 units. Radius had no beginning inventory.

Working with these data for the variable-costing method and for the absorption-costing method results in the following chart:

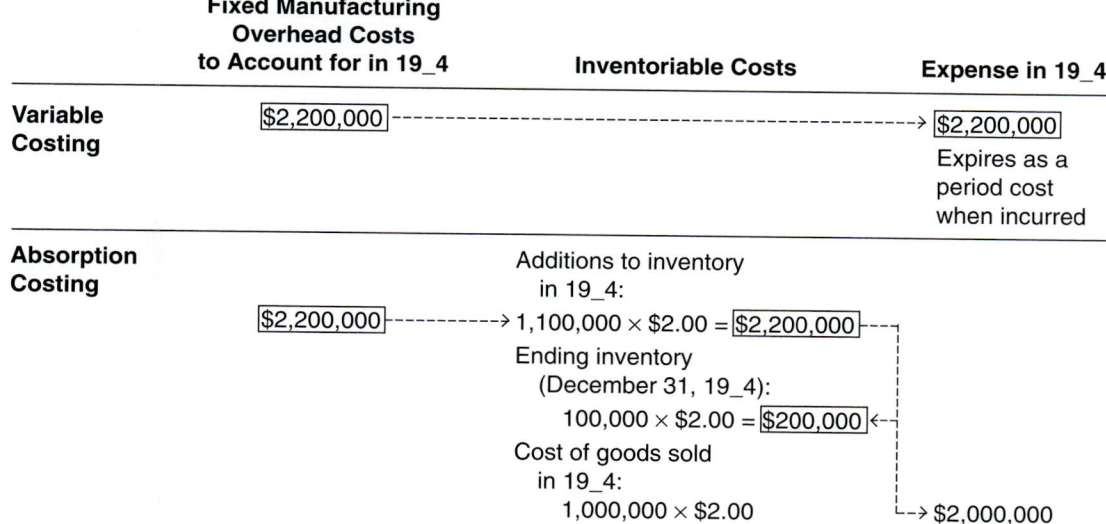

This chart contains only one cost item—fixed manufacturing overhead. All other manufacturing costs—direct manufacturing costs and variable manufacturing overhead costs—are accounted for exactly alike under both variable costing and absorption costing. Moreover, all nonmanufacturing costs are also accounted for exactly alike under both variable costing and absorption costing. Thus, fixed manufacturing overhead is the only difference between variable costing and absorption costing.

EXHIBIT 9-1

Comparison of Variable Costing and Absorption Costing
Income Statements for the Year Ended December 31, 19_4
for Radius Company (in thousands)

PANEL A: VARIABLE COSTING

Sales: $17.00 × 1,000,000 units		$17,000
Variable costs		
Beginning inventory	$ 0	
Variable cost of goods manufactured:		
$6.00 × 1,100,000	6,600	
Cost of goods available for sale	6,600	
Ending inventory: $6.00 × 100,000	600	
Variable manufacturing cost of goods sold	6,000	
Variable marketing and administrative costs	2,400	
Adjustment for variances	0	
Total variable costs		8,400
Contribution margin		8,600
Fixed costs		
Fixed manufacturing overhead costs	2,200	
Fixed marketing and administrative costs	5,500	
Total fixed costs		7,700
Operating income		$ 900

PANEL B: ABSORPTION COSTING

Sales: $17.00 × 1,000,000 units		$17,000
Cost of goods sold		
Beginning inventory	$ 0	
Variable manufacturing costs:		
$6.00 × 1,100,000	6,600	
Fixed manufacturing costs allocated:		
$2.00 x 1,100,000	2,200	
Cost of goods available for sale	8,800	
Ending inventory: $8.00 × 100,000	800	
Adjustment for variances	0	
Total cost of goods sold		8,000
Gross margin		9,000
Marketing and administrative		
Variable marketing and administrative costs	2,400	
Fixed marketing and administrative costs	5,500	
Adjustment for variances	0	
Total marketing and administrative costs		7,900
Operating income		$ 1,100

Objective 2

Construct income statements using absorption costing and variable costing

Exhibit 9-1 presents the variable-costing and absorption-costing income statements for Radius Company in 19_4. The variable-costing income statement uses the contribution-margin format introduced in Chapter 2. The absorption-costing income statement uses the gross-margin format, also introduced in Chapter 2. Why these differences in format? The distinction between variable costs and fixed costs is central to variable costing; the contribution-margin format highlights this distinction. The distinction between manufacturing and nonmanufacturing costs is central to absorption costing; the gross margin format highlights this distinction. Companies using absorption costing sometimes do not make any distinction between variable costs and fixed costs in their accounting system.

We assume the following information for Radius Company for 19_4 (see Exhibit 9-1):

Selling price	$17.00
Variable costs per unit	
Direct materials	3.50
Direct manufacturing labor	1.60
Variable manufacturing overhead costs	0.90
Total variable manufacturing costs	6.00
Variable marketing and administrative costs	2.40
Total variable costs per unit	$ 8.40
Fixed costs	
Fixed manufacturing overhead costs	$2,200,000
Fixed marketing and administrative costs	$5,500,000
Total fixed costs	$7,700,000
No beginning inventory of finished goods	
No variances	

Trace the $2,200,000 fixed manufacturing overhead amount in Exhibit 9-1. The income statement under variable costing deducts the $2,200,000 lump sum as a period cost in 19_4. In contrast, the income statement under absorption costing regards each finished unit as bearing $2 of fixed manufacturing overhead. Under absorption costing the $2,200,000 is initially capitalized as an inventory cost in 19_4. Subsequently, $2,000,000 becomes an expense in 19_4, and $200,000 remains an asset—part of ending inventory, 100,000 units × $2—at December 31, 19_4. The variable manufacturing costs are accounted for in the same way in both income statements in Exhibit 9-1.

Never overlook the heart of the matter. The differences between variable costing and absorption costing center on how to account for fixed manufacturing overhead. If inventory levels change, income will differ because of the different accounting for fixed manufacturing overhead. Compare sales of 900,000, 1,000,000, and 1,100,000 units by Radius Company in 19_4. Fixed manufacturing overhead would be included in the 19_4 expense as follows:

	Fixed Manufacturing Overhead Costs Treated as an Expense in 19_4
Variable costing, whether	
sales are 900,000, 1,000,000, or 1,100,000 units	$2,200,000
Absorption costing, where:	
◆ sales are 900,000 units, $400,000 (200,000 × $2)	
held back in inventory	$1,800,000
◆ sales are 1,000,000 units, $200,000 (100,000 × $2)	
held back in inventory	$2,000,000
◆ sales are 1,100,000 units, $0 held back in inventory	$2,200,000

Some companies use the term **direct costing** to describe the inventory costing method we call *variable costing*. Direct costing is an unfortunate choice of terms for two reasons: (1) Variable costing in Exhibit 9-1 does not include all direct costs as inventoriable costs. Only direct manufacturing costs are included, while any direct marketing and any direct administrative costs are excluded from inventoriable costs. (2) Variable costing in Exhibit 9-1 includes as inventoriable costs not only direct manufacturing costs but also some indirect costs (variable manufacturing costs).

Indeed, the term *variable costing* could be improved if it were called *variable manufacturing costing*. Why? Because the distinction between the variable cost and absorption costing approaches centers on how to account for *manufacturing* costs. Variable costs of other business function areas in the value chain—such as marketing, distribution, and customer service—are written off as period costs under both variable costing and absorption costing.

Capsule Comparisons of Inventory-Costing Methods

Exhibit 9-2 presents a capsule comparison of the Work in Process account under eight alternative inventory-costing systems:

Variable Costing	Absorption Costing
1. Actual costing	5. Actual costing
2. Normal costing	6. Normal costing
3. Budgeted costing	7. Budgeted costing
4. Standard costing	8. Standard costing

The boxes in Exhibit 9-2 represent the debits to Work in Process (that is, the amounts assigned to product) under alternative inventory-costing systems. The major systems in use for manufacturing costs are standard costing systems, which were introduced in Chapters 7 and 8, and normal costing systems, which were introduced in Chapters 4 and 5. As the boxes indicate, variable costing or absorption costing may be combined with actual, normal, budgeted, or standard costing.

Variable costing has been a controversial subject among accountants—not so much because there is disagreement about the need for delineating between variable costs and fixed costs for management planning and control, but because there is a question about using variable costing for *external* reporting. Those favoring variable costing for external reporting maintain that the fixed portion of manufacturing overhead is more closely related to the capacity to produce than to the production of specific units. Those opposing variable costing maintain that inventories should carry a fixed manufacturing overhead cost component. Why? Because both variable and fixed manufacturing costs are necessary to produce goods; both types of costs should be inventoriable, regardless of their differences in behavior patterns.

Variety of Practices Concerning "Variable" Costs

Surveys indicate that many companies use a variable-costing format in reporting to top management (see Surveys of Company Practice on p. 318). Various interpretations of variable costing will be found for internal purposes because management can

EXHIBIT 9-2

Capsule Comparison of Alternative Inventory-Costing Systems

		Work in Process Inventory			
		Actual Costing	**Normal Costing**	**Budgeted Costing**	**Standard Costing**
ABSORPTION COSTING — VARIABLE COSTING	**Direct Manufacturing Costs**	Actual prices × Actual inputs used	Actual prices × Actual inputs used	Budgeted prices × Actual inputs used	Standard prices × Standard inputs allowed for actual output achieved
	Variable Manufacturing Overhead Costs	Actual variable overhead rates × Actual inputs used	Budgeted variable overhead rates × Actual inputs used	Budgeted variable overhead rates × Actual inputs used	Budgeted variable overhead rates × Standard inputs allowed for actual output achieved
	Fixed Manufacturing Overhead Costs	Actual fixed overhead rates × Actual inputs used	Budgeted fixed overhead rates × Actual inputs used	Budgeted fixed overhead rates × Actual inputs used	Budgeted fixed overhead rates × Standard inputs allowed for actual output achieved

decide how to apply the basic concepts without worrying about the inventoriable costs to report to external parties.

Regarding internal reporting, some managers advocate "super-variable costing." This interpretation assumes that in the short run, only direct materials are variable costs; all other costs are regarded as fixed and not identifiable with products. The "throughput contribution" measure discussed in Chapter 23 is an example of this approach. Conversely, some opponents of variable costing favor "super-absorption costing." They take a long-run view and maintain that almost all costs (including research, design, marketing, administrative, and customer-service costs) are variable and must be identified with the products. Consider Radius Company. For the budgeted 19_4 production output of 1,100,000 units, the budgeted per-unit costs are:

Direct materials	$3.50
Direct manufacturing labor	1.60
Variable manufacturing overhead costs	0.90
Variable marketing and administrative costs	2.40
Fixed manufacturing overhead costs	2.00
Fixed marketing and administrative costs	5.00
	$15.40

Under super-variable costing, Radius's per-unit cost would be $3.50 per unit. Under super-absorption costs, the per-unit cost would be $15.40.

For reporting to the U.S. Internal Revenue Service, all manufacturing costs plus some design and administrative costs must be included as inventoriable costs. For example, legal department costs must be allocated between those related to manufacturing activities (inventoriable costs) and those not related to manufacturing activities (period costs).

For external reporting to shareholders, companies around the globe tend to follow the generally accepted accounting principle that all manufacturing overhead is inventoriable. In practice, however, some companies do not inventory all manufacturing overhead. Instead, they charge some manufacturing overhead, such as depreciation on plant and equipment, to period expense immediately, as it is incurred. These depreciation methods are unlikely to materially affect reported operating income if:

1. The consistent application of such depreciation is coupled with minimal changes in beginning and ending inventory levels.
2. Depreciation is not a significant part of the manufacturing costs.

In this book, the term *inventoriable cost* for a manufacturing company refers to the costs associated with the acquisition and conversion of materials and all other manufacturing inputs into goods for sale.

COMPARISON OF STANDARD VARIABLE COSTING AND STANDARD ABSORPTION COSTING

Our next example explores the implications of accounting for fixed manufacturing overhead in more detail. Stassen Company manufactures and markets telescopes. It uses a standard costing system for both its manufacturing and marketing/administrative costs.[1] It began business on January 1, 19_4. It is now March 19_4. The president asks you to prepare comparative income statements for January 19_4 and February 19_4. The following simplified unit data are available:

	January 19_4	February 19_4
Beginning inventory	0	200
Production	600	650
Sales	400	750
Ending inventory	200	100

Other data:

Selling price	$99
Standard variable manufacturing costs per unit	$20
Standard variable marketing and administrative costs per unit	$19
Standard fixed manufacturing overhead costs per month	$12,800
Standard fixed marketing and administrative costs per month	$10,400
Budgeted denominator level per month	800 output units

There were no beginning or ending inventories of materials or work in process. There were no price, efficiency, or spending variances for any costs in either January or February of 19_4. The standard fixed manufacturing overhead cost per unit is $16 ($12,800 ÷ 800). Thus the key unit standard cost data are:

---

[1]For ease of exposition, we assume that Stassen Company uses a standard costing system for all its operating costs—that is, it uses standards for both variable and fixed costs in both its manufacturing and marketing/administrative cost areas. Companies frequently use hybrids of the four costing systems in Exhibit 9-2. For example, many companies use a standard costing system for their variable manufacturing costs and a budgeted costing system for their fixed manufacturing and variable and fixed marketing/administrative costs.

Unit variable costs	
Standard variable manufacturing costs	$20
Standard variable marketing/administrative costs	19
Total variable costs	$39
Unit manufacturing costs	
Standard variable manufacturing costs	$20
Standard fixed manufacturing costs	16
Total manufacturing costs	$36

Stassen immediately expenses all variances to cost of goods sold.

Let's turn now to Stassen's comparative income statements under variable costing and absorption costing.

Comparative Income Statements

Exhibit 9-3 contains the comparative income statements under variable costing (Panel A) and absorption costing (Panel B) for Stassen Company in January 19_4 and February 19_4.

In Panel A, Variable Costing, all variable-cost line items are at standard cost except the last line item, "Adjustment for variances." It includes all price, spending, and efficiency variances related to variable-cost items.

In Panel B, Absorption Costing, all cost of goods sold line items are at standard cost except the last line item, "Adjustment for variances." It includes all manufacturing cost variances—price, spending, efficiency, and output-level (production-volume) variances.[2]

Keep the following points in mind about absorption costing as you study Panel B of Exhibit 9-3:

a. The inventoriable cost is $36 per unit, not $20, because fixed manufacturing overhead ($16) as well as variable manufacturing costs ($20), is allocated to each unit of product.

b. The $16 fixed manufacturing overhead rate was based on a denominator level of 800 units per month ($12,800 ÷ 800 = $16). Whenever *production* (not sales) deviates from the denominator level, an output-level (production-volume) variance arises. The measure of the variance is $16 multiplied by the difference between the actual level of production and the denominator level.

c. The output-level variance exists only under absorption costing, not under variable costing. All other variances exist under both absorption costing and variable costing.

d. The absorption-costing income statement classifies costs primarily by *business function*, such as manufacturing and marketing. In contrast, the variable-costing income statement features *cost behavior* (variable or fixed) as the basis of classification. Absorption costing income statements need not differentiate between the variable and fixed costs. (Exhibit 9-3 does make this differentiation for Stassen Company in order to highlight how individual line items are classified differently in our example in the variable-costing and the absorption-costing formats.)

---

[2]Companies differ in how they present the "adjustment for variances" information in their income statements. For example, two alternative approaches to that shown in Exhibit 9-3 are:

♦ Adjusting each single line item in the income statement (for example, direct materials) for the variances applicable to that line item

♦ Computing the operating income amount using standard costs and then making a single one-line "adjustment for variances"

Managers who examine summary percentages, such as the contribution-margin percentage and the gross-margin percentage, should attempt to ensure that differences in reporting format are minimal when they make comparisons among companies using these percentages. Changes in reporting format can lead to changes in reported percentages for either the contribution-margin percentage or the gross-margin percentage.

EXHIBIT 9-3
Comparison of Variable Costing and Absorption Costing
Income Statements for January 19_4 and February 19_4 for Stassen Company

PANEL A: VARIABLE COSTING

	Jan. 19_4	Feb. 19_4
Sales*	$39,600	$74,250
Variable costs		
Beginning inventory	0	4,000
Variable cost of goods manufactured[†]	12,000	13,000
Cost of goods available for sale	12,000	17,000
Ending inventory[‡]	4,000	2,000
Variable manufacturing cost of goods sold	8,000	15,000
Variable marketing and administrative costs[§]	7,600	14,250
Total variable costs (at standard costs)	15,600	29,250
Contribution margin (at standard costs)	24,000	45,000
Adjustment for variances	0	0
Contribution margin	24,000	45,000
Fixed costs		
Fixed manufacturing overhead	12,800	12,800
Fixed marketing and administrative	10,400	10,400
Total fixed costs (at standard costs)	23,200	23,200
Adjustment for variances	0	0
Total fixed costs	23,200	23,200
Operating income	$ 800	$21,800

*400 × $99 = $39,600; 750 × $99 = $74,250.
[†]600 × $20 = $12,000; 650 × $20 = $13,000.
[‡]200 × $20 = $4,000; 100 × $20 = $2,000.
[§]400 × $19 = $7,600; 750 × $19 = $14,250.

PANEL B: ABSORPTION COSTING

	Jan. 19_4	Feb. 19_4
Sales*	$39,600	$74,250
Cost of goods sold		
Beginning inventory	0	7,200
Variable manufacturing costs[†]	12,000	13,000
Fixed manufacturing costs[‡]	9,600	10,400
Cost of goods available for sale	21,600	30,600
Ending inventory[§]	7,200	3,600
Total cost of goods sold (at standard costs)	14,400	27,000
Gross margin (at standard costs)	25,200	47,250
Adjustment for variances[ǁ]	3,200 U	2,400 U
Gross margin	22,000	44,850
Marketing and administrative costs		
Variable marketing and administrative costs[#]	7,600	14,250
Fixed marketing/administrative costs	10,400	10,400
Total marketing/administrative (at standard costs)	18,000	24,650
Adjustment for variances	0	0
Total marketing/administrative costs	18,000	24,650
Operating income	$ 4,000	$20,200

*400 × $99 = $39,600; 750 × $99 = $74,250.
[†]600 × $20 = $12,000; 650 × $20 = $13,000.
[‡]600 × $16 = $9,600; 650 × $16 = $10,400.
[§]200 × ($20 + $16) = $7,200; 100 × ($20 + $16) = $3,600.
[ǁ]Jan. 19_4 has $3,200 unfavorable output-level (production-volume) variance
    (600 − 800) × $16 = $3,200 unfavorable.
 Feb. 19_4 has $2,400 unfavorable output-level (production-volume) variance
    (650 − 800) × $16.00 = $2,400 unfavorable.
[#]400 × $19 = $7,600; 750 × $19 = $14,250.

Explaining Differences in Operating Income

In addition to points (a) through (d) in the preceding section, let's consider another important point:

e. If the inventory level increases during a period, variable costing will generally report less operating income than will absorption costing; when the inventory level decreases, variable costing will generally report more operating income than will absorption costing. These differences in operating income are due *solely* to moving fixed manufacturing overhead into inventories as they increase and out of inventories as they decrease.

The difference between the operating incomes under absorption and variable costing can be computed by formula 1:

FORMULA 1

$$\begin{pmatrix} \text{Absorption} \\ \text{costing} \\ \text{operating} \\ \text{income} \end{pmatrix} - \begin{pmatrix} \text{Variable} \\ \text{costing} \\ \text{operating} \\ \text{income} \end{pmatrix} = \begin{pmatrix} \text{Fixed manufacturing} \\ \text{overhead in} \\ \text{ending inventory} \end{pmatrix} - \begin{pmatrix} \text{Fixed manufacturing} \\ \text{overhead in} \\ \text{beginning inventory} \end{pmatrix}$$

Fixed manufacturing overhead in ending inventory (FMOH:EI) is the current-period expense under variable costing that absorption costing defers to the future period. Fixed manufacturing overhead in beginning inventory (FMOH:BI) is a prior-period expense under variable costing that becomes a current-period expense under absorption costing. Thus, if FMOH:EI > FMOH:BI—that is, if the amount deferred to the future period exceeds the amount that is placed in the current period—the absorption-costing income will exceed the variable-costing income.[3]

Two alternative formulas can be used if we assume that all manufacturing variances are written off as period costs, that no change occurs in work in process inventory, and that no change occurs in the budgeted fixed manufacturing overhead rate between periods:

FORMULA 2

$$\begin{pmatrix} \text{Absorption} \\ \text{costing} \\ \text{operating} \\ \text{income} \end{pmatrix} - \begin{pmatrix} \text{Variable} \\ \text{costing} \\ \text{operating} \\ \text{income} \end{pmatrix} = \begin{pmatrix} \text{Units} \\ \text{produced} \\ \text{minus} \\ \text{units sold} \end{pmatrix} \times \begin{pmatrix} \text{Budgeted fixed} \\ \text{manufacturing} \\ \text{overhead rate} \end{pmatrix}$$

or

FORMULA 3

$$\begin{pmatrix} \text{Absorption} \\ \text{costing} \\ \text{operating} \\ \text{income} \end{pmatrix} - \begin{pmatrix} \text{Variable} \\ \text{costing} \\ \text{operating} \\ \text{income} \end{pmatrix} = \begin{pmatrix} \text{Increase or} \\ \text{decrease in} \\ \text{inventory units} \end{pmatrix} \times \begin{pmatrix} \text{Budgeted fixed} \\ \text{manufacturing} \\ \text{overhead rate} \end{pmatrix}$$

Let's use the data from Exhibit 9-3 to illustrate these formulas:

USING FORMULA 1
Jan. 19_4 $4,000 – $800 = (200 × $16) – (0 × $16)
= $3,200
Feb. 19_4 $20,200 – $21,800 = (100 × $16) – (200 × $16)
= –$1,600

USING FORMULA 2
Jan. 19_4 $4,000 – $800 = (600 – 400) × $16
= $3,200
Feb. 19_4 $20,200 – $21,800 = (650 – 750) × $16
= –$1,600

---

[3]This formula assumes that the amounts used for ending and beginning inventory are after proration of manufacturing overhead variances.

USING FORMULA 3

Jan. 19_4 $\$4,000 - \$800$ $= (200 - 0) \times \$16$
$= \$3,200$

Feb. 19_4 $\$20,200 - \$21,800 = (100 - 200) \times \$16$
$= -\$1,600$

Effects of Sales and Production on Operating Income

A final important point, in addition to the previous points (a) through (e), deals with cost-volume-profit relationships:

f. The period-to-period change in operating income under variable costing is driven by changes in the unit sales level given a constant contribution margin per unit. Consider the February 19_4 variable-costing operating income of Stassen versus the January 19_4 operating income:

Surveys of Company Practice

USAGE OF VARIABLE COSTING BY COMPANIES

Surveys of company practice in many countries report that approximately 30% to 50% of companies use variable costing in their internal accounting system.

	United States[a]	Canada[a]	Australia[b]	Japan[b]	Sweden[c]	United Kingdom[b]
Variable costing used	31%	48%	33%	31%	42%	52%
Absorption costing used	65	52	} 67	} 69	} 58	} 48
Other	4	0				

Many companies that use variable costing for internal reporting also use absorption costing for external reporting or tax reporting. How do companies that use variable costing treat fixed manufacturing overhead (MOH) in their internal reporting system?

	Australia[b]	Japan[b]	United Kingdom[b]
Prorate fixed MOH to inventory/ cost of goods sold at period end	41%	39%	25%
Use variable costing for monthly costing, and adjust to absorption costing once a year	11	8	4
Use both variable costing and absorption costing as dual systems	23	33	31
Treat fixed MOH as a period cost	25	3	35
Other	0	17	5

The most common problem reported by companies that use variable costing was the difficulty of classifying costs into fixed or variable categories.

---

[a]Adapted from Inoue, "A Comparative Study." [b]Adapted from Blayney and Yokoyama, "A Comparative Analysis." [c]Adapted from Ask and Ax, "Trends." Full citations are in Appendix A.

$$\begin{bmatrix} \text{Change in} \\ \text{operating income} \end{bmatrix} = \begin{pmatrix} \text{Contribution} \\ \text{margin} \end{pmatrix} \times \begin{pmatrix} \text{Change in unit} \\ \text{sales level} \end{pmatrix}$$

$$\$21,800 - \$800 = (\$99 - \$39) \times (750 - 400)$$
$$= \$60 \times 350$$
$$= \$21,000$$

Under absorption costing, however, period-to-period change in operating income is driven by variations in both the unit sales level and the unit production level. Exhibit 9-4 illustrates this point. The exhibit shows how absorption-costing operating income for February 19_4 changes as the production level in February 19_4 changes. Beginning inventory in February 19_4 of 200 units and February sales of 750 units are unchanged. Exhibit 9-4 shows that production of only 550 units meets February 19_4 sales of 750. Operating income at this production level is $18,600. By producing more than 550 units in February 19_4, Stassen increases absorption-costing operating income. Each unit in February 19_4 ending inventory will increase February operating income by $16. For example, if 800 units are produced, ending inventory will be 250 units and operating income $22,600. This amount is $4,000 more than what operating income is with zero ending inventory (250 units × $16 = $4,000) on February 28, 19_4.

Exhibit 9-5 is a concise comparison of the preceding six points regarding variable costing and absorption costing.

EXHIBIT 9-4

Effect on Absorption-Costing Operating Income of Different Production Levels Holding the Unit Sales Level Constant: Stassen Company Data for February 19_4 with Sales of 750 Units

	February 19_4 Production Level					
	550	600	650	700	750	800
UNIT DATA						
Beginning inventory	200	200	200	200	200	200
Production	550	600	650	700	750	800
Goods available for sale	750	800	850	900	950	1,000
Sales	750	750	750	750	750	750
Ending inventory	0	50	100	150	200	250
INCOME STATEMENT						
Sales	$74,250	$74,250	$74,250	$74,250	$74,250	$74,250
Beginning inventory	7,200	7,200	7,200	7,200	7,200	7,200
Variable manufacturing costs*	11,000	12,000	13,000	14,000	15,000	16,000
Fixed manufacturing costs†	8,800	9,600	10,400	11,200	12,000	12,800
Cost of goods available for sale	27,000	28,800	30,600	32,400	34,200	36,000
Ending inventory‡	0	1,800	3,600	5,400	7,200	9,000
Cost of goods sold (at standard cost)	27,000	27,000	27,000	27,000	27,000	27,000
Adjustment for variances§	4,000 U	3,200 U	2,400 U	1,600 U	800 U	0
Total cost of goods sold	31,000	30,200	29,400	28,600	27,800	27,000
Gross margin	43,250	44,050	44,850	45,650	46,450	47,250
Total marketing and administrative costs	24,650	24,650	24,650	24,650	24,650	24,650
Operating income	$18,600	$19,400	$20,200	$21,000	$21,800	$22,600

*$20 per unit.
†Allocated at $16 per unit.
‡$36 per unit.
§(Production in units − 800) × $16.

EXHIBIT 9-5

Comparative Income Effects of Variable Costing and Absorption Costing

Question	Variable Costing	Absorption Costing	Comment
a. Is fixed manufacturing overhead inventoried?	No	Yes	Basic theoretical question of timing—that is, when this cost should become an expense.
b. Is there an output-level (production-volume) variance?	No	Yes	Choice of denominator level affects measurement of operating income under absorption costing only.
c. How are the other variances treated?	Same	Same	Highlights that the basic difference is the accounting for fixed manufacturing overhead, not the accounting for any variable manufacturing costs.
d. Are classifications between variable and fixed costs routinely made?	Yes	Not always	Absorption costing can be easily modified to obtain subclassifications for variable and fixed costs, if desired.
e. How do changes in inventory levels affect operating income?			Differences are attributable to timing of the transformation for fixed manufacturing overhead into expense.
Production = sales	Equal	Equal	
Production > sales	Lower*	Higher†	
Production < sales	Higher	Lower	
f. What are the effects on cost-volume-profit relationships?	Driven by sales level	Driven by sales level *and* production level	Management control benefit: Effects of changes in production level on operating income are easier to understand under variable costing.

*That is, lower operating income than under absorption costing.
†That is, higher operating income than under variable costing.

BREAKEVEN POINTS IN VARIABLE COSTING AND ABSORPTION COSTING

Objective 5

Describe the effects of absorption costing on computing the breakeven point

Chapter 3 introduced cost-volume-profit analysis. If variable costing is used, the breakeven point as reported in an income statement is computed in the usual manner. It is unique; there is only one breakeven point. Moreover, operating income is a function of the unit sales level. As the level of unit sales rises, operating income rises, and vice versa. In our Stassen illustration for February 19_4:

$$\text{Breakeven point in units} = \frac{\text{Total fixed costs}}{\text{Unit contribution margin}}$$

$$\text{Let } N = \text{Number of units sold to break even}$$

$$N = \frac{\$12,800 + \$10,400}{\$99 - (\$20 + \$19)} = \frac{\$23,200}{\$60}$$

$$N = 387 \text{ units (rounded)}[4]$$

[4] Proof:

Sales: $99 × 387		$38,313
Variable costs: $39 × 387		15,093
Contribution margin		23,220
Fixed costs		23,200
Operating income		$ 20 (rounding error)

If absorption costing is used, the breakeven point as reported in an income statement is not unique. The following formula, which can be used to compute the breakeven point under absorption costing, highlights the multiple factors that will affect the breakeven point:

$$\text{Breakeven sales in units} = \frac{\text{Total fixed costs} + \left[\text{Fixed manufacturing overhead rate}\left(\text{Breakeven sales in units} - \text{Units produced}\right)\right]}{\text{Unit contribution margin}}$$

Let N = the number of units sold to break even. Consider Stassen Company in February 19_4:

$$N = \frac{\$12{,}800 + \$10{,}400 + [\$16\,(N - 650)]}{\$99 - (\$20 + \$19)}$$

$$N = \frac{\$23{,}200 + \$16N - \$10{,}400}{\$60}$$

$$\$60N = \$12{,}800 + \$16N$$

$$\$44N = \$12{,}800$$

$$N = 291 \text{ units (rounded)}[5]$$

The breakeven point under absorption costing depends on three factors: (1) the units sold, (2) the units produced, and (3) the denominator level chosen to set the fixed manufacturing overhead rate. For Stassen in February 19_4, a combination of 291 units sold (rounded), 650 units produced, and an 800-unit denominator level yields an operating income of zero. Note, however, there are many combinations of these three factors that yield an operating income of zero.

We see that variable costing dovetails with cost-volume-profit analysis. The user of variable costing can easily compute the breakeven point or any effects that changes in the unit sales level may have on operating income. In contrast, the user of absorption costing must consider both the unit sales level and the unit production level before making such computations.

Suppose in our illustration that actual production in February 19_4 were equal to the denominator level, 800 units. Also suppose that there were no sales and no fixed marketing/administrative costs. All the production would be placed in inventory, and so all the fixed manufacturing overhead would be included in inventory. There would be no output-level variance. Thus, the company could break even with no sales whatsoever! In contrast, under variable costing the operating loss would be equal to the fixed costs of $12,800.

PERFORMANCE MEASURES AND ABSORPTION COSTING

Objective 6

Understand how absorption costing influences performance-evaluation decisions

Undesirable Buildups of Inventories

Absorption costing enables managers to increase operating income in the short run by increasing the production schedule independent of customer demands for products. Exhibit 9-4 shows how a manager could increase February 19_4 operating income from $18,600 to $22,600 by producing for inventory an additional 250 units. Such an increase in the production schedule can increase the costs of doing business without any attendant increase in sales. For example, a manager who is evaluated on

---

[5] Proof:

Sales: $99 × 291		$28,809
Cost of goods sold		
Unit cost of goods sold: $36 × 291	$10,476	
Output level variance: (800 – 650) × $16	2,400	
Total cost of goods sold		12,876
Gross margin		15,933
Marketing/administrative costs		
Variable marketing/administrative: $19 × 291	5,529	
Fixed marketing/administrative	10,400	
Total marketing/administrative		15,929
Operating income		$ 4 (rounding error)

the basis of absorption-costing income may increase production at the end of a performance review period solely to increase reported income. Each additional unit produced absorbs fixed manufacturing overhead that would otherwise have been an expense of the period.

The undesirable effects of such an increase in production may be sizable, and they can arise in several ways. For example:

1. A plant manager may switch production to those orders that absorb the highest amount of manufacturing overhead, irrespective of their demand by customers (called "cherry picking" the production line). Some difficult-to-manufacture items may be delayed, resulting in missed promises on customer delivery dates.

2. A plant manager may accept a particular order to increase his or her production even though another plant in the same company is better suited to handle that order.

3. To meet increased production, a manager may defer maintenance beyond the current period. Although operating income may increase now, future income will probably decrease because of increased repairs and less-efficient equipment.

Early criticisms of absorption costing concentrated on whether fixed manufacturing overhead qualified as an asset under generally accepted accounting principles. Criticisms of absorption costing have increasingly emphasized its potential undesirable incentives for managers. Indeed, one critic labels absorption costing as "one of the black holes of cost accounting," in part because it may induce managers to make decisions "against the long-run interests" of the company.[6]

Proposals for Revising Performance Evaluation

Critics of absorption costing are making proposals for revising how the performance of managers is evaluated: (a) reduce the types of costs that are inventoried and (b) emphasize nonfinancial performance measures.

By accounting for fixed manufacturing overhead as period costs, variable costing reduces the incentives for managers to build up inventory levels. Some authors argue that even more costs should be expensed. For example, consider "super-variable costing" in which all costs other than direct materials are expensed to the period in which they are incurred. By expensing direct manufacturing labor and variable manufacturing overhead in the current period, managers will have less incentive to build inventory levels beyond those required to meet customer demand. (See Chapter 23 for an example of this approach.)

Nonfinancial performance measures can be used to monitor key variables such as attaining but not exceeding inventory levels, meeting promised customer delivery dates, and abiding by plant maintenance schedules. By including such nonfinancial performance measures in performance reviews, managers are given explicit signals about the factors top management views as important.

Designing an effective performance measurement and reward system is a challenging task. Chapters 13 and 26 provide further discussion of the potential limitations of using single-performance benchmarks, whether financial or nonfinancial in nature.

◆ PART TWO:
ROLE OF VARIOUS DENOMINATOR-LEVEL CONCEPTS
IN ABSORPTION COSTING

The effect that the choice of a denominator-level concept has on budgeted overhead or indirect cost rates is discussed in various parts of this book—for example, Chapters 4 (pp. 107–9), 5 (pp. 144–45), 8 (pp. 270–78), and 15 (pp. 544–46). This part of

---

[6]R. Schmenner, "Escaping the Black Holes of Cost Accounting," *Business Horizons* (January-February 1988), p. 66.

Chapter 9 examines how alternative denominator-level concepts affect fixed manufacturing overhead rates and reported operating income under absorption costing.

ALTERNATIVE DENOMINATOR-LEVEL CONCEPTS

We use an iced-tea bottling plant to illustrate alternative concepts. Bushells Company produces 12-ounce bottles of iced tea. The variable manufacturing costs of each bottle are $0.35. The fixed monthly manufacturing costs of the bottling plant are $50,000. Bushells uses absorption costing for its monthly internal reporting system and for the financial reports it sends to its shareholders. Bushells could use any one of at least four different denominator-level concepts for computing the fixed manufacturing overhead rate—theoretical capacity, practical capacity, normal utilization, and master-budget utilization. Whatever the denominator-level concept, Bushells defines its denominator in output units (12-ounce bottles of iced tea).

Theoretical Capacity and Practical Capacity

The term *capacity* means "constraint," an "upper limit." **Theoretical capacity** (also called maximum capacity) is the denominator-level concept that is based on the production of output at maximum efficiency for 100% of the time. Bushells can produce 2,400 bottles an hour when the bottling lines are operating at maximum speed. There is a maximum of two eight-hour shifts a day due to labor union agreements. Thus, the theoretical monthly capacity would be:

$$2,400 \text{ per hour} \times 16 \text{ hours} \times 30 \text{ days} = 1,152,000 \text{ bottles}$$

Theoretical capacity is theoretical in the sense that it does not allow for any plant maintenance, any interruptions because of bottle breakages on the filling lines, or a host of other factors.

Practical capacity is the denominator-level concept that reduces theoretical capacity for unavoidable operating interruptions such as scheduled maintenance time, shutdowns for holidays and other days, and so on. Assume that the practical hourly production rate is 2,000 bottles an hour and that the plant can operate 25 days a month. The practical monthly capacity is thus:

$$2,000 \text{ per hour} \times 16 \text{ hours} \times 25 \text{ days} = 800,000 \text{ bottles}$$

To estimate theoretical or practical capacity, a manager must consider both engineering and economic factors. Engineers at the Bushells plant understand well the technical capabilities of machines for filling bottles, the machines that cap bottles, and the machines that glue labels on bottles. In some cases, an increase in capacity may be technically possible but not economically sound. For example, the labor union may actually permit a third shift per day but only at burdensome penalty wage rates that clearly do not make financial sense in the iced-tea market.

Normal Utilization and Master-Budget Utilization

Both theoretical capacity and practical capacity measure the denominator level in terms of what a plant can supply. In contrast, normal utilization and master-budget utilization measure the denominator level in terms of the demand for the output of the plant. In many cases, the budgeted demand is well below the supply available (productive capacity).

Normal utilization is the denominator-level concept based on the level of capacity utilization that satisfies average customer demand over a period (say, two to three years) that includes seasonal, cyclical, or other trend factors. **Master-budget**

utilization is the denominator-level concept based on the anticipated level of capacity utilization for the coming budget period. These two denominator levels can differ—for example, when an industry has cyclical periods of high and low demand or when management believes that the budgeted production for the coming period is unrepresentative of "long-term" demand.

Consider our Bushells example of iced-tea production. The master budget of Bushells for 19_5 is based on production of 400,000 bottles per month. Hence the master-budget denominator level is 400,000 bottles. However, Bushells' senior management believes that over the next one to three years the normal monthly production level will be 500,000 bottles. These people view the 19_5 budgeted production level of 400,000 bottles to be too low. Why? A major competitor has been heavily discounting its iced-tea selling prices and also spending enormous amounts on advertising. Bushells expects that this price discounting and advertising blitz will be a short-run phenomenon and that in 19_6 and 19_7 the market share Bushells has lost to this competitor will be regained.

A major reason for choosing master-budget utilization over normal utilization is the difficulty of forecasting normal utilization in many industries with long-run cyclical patterns. For example, many U.S. steel companies in the 1980s believed that they were in the downturn of a demand cycle that would have an upturn in a two- to three-year period. Unfortunately, the cycle did not turn up for many years, and plants closed. A similar problem occurs when estimating "normal" demand. Some marketing managers are prone to overestimate their ability to regain lost market share. Their estimate of "normal" demand for their product may be based on an overly optimistic outlook (anticipating roses when all that exists are thorns).

EFFECT ON FINANCIAL STATEMENTS

Objective 8
Explain how the choice of the denominator level affects reported operating income and inventory amounts

Bushells has budgeted fixed manufacturing overhead costs of $50,000 per month. Assume that actual costs are also $50,000. The budgeted manufacturing overhead rates in May 19_5 for each of the alternative denominator-level concepts discussed are:

Denominator-Level Concept (1)	Budgeted Fixed Manufacturing Overhead per Month (2)	Budgeted Denominator Level (3)	Budgeted Fixed Manufacturing Overhead Cost Rate (4) = (2) ÷ (3)
Theoretical capacity	$50,000	1,152,000	$0.0434
Practical capacity	50,000	800,000	0.0625
Normal utilization	50,000	500,000	0.1000
Master-budget utilization	50,000	400,000	0.1250

The budgeted fixed manufacturing overhead rate based on master-budget utilization ($0.1250) is more than 180% above the rate based on theoretical capacity ($0.0434).

Assume now that Bushells' actual production in May 19_5 is 460,000 bottles of iced tea and that actual sales are 420,000 bottles. Also assume no beginning inventory on May 1, 19_5 and no price, spending, or efficiency variances in manufacturing for May 19_5. Bushells' manufacturing plant sells bottles of iced tea to another division for $0.50 per bottle. Its only costs are variable manufacturing costs of $0.35 per bottle and $50,000 per month for fixed manufacturing overhead. Bushells writes off to cost of goods sold all variances on a monthly basis.

The budgeted manufacturing cost per bottle of iced tea for each denominator-level concept is the sum of $0.35 in variable manufacturing costs and the budgeted fixed manufacturing costs (shown from the preceding table).

Denominator-Level Concept (1)	Variable Manufacturing Costs (2)	Budgeted Fixed Manufacturing Overhead Cost Rate (3)	Total Manufacturing Costs (4) = (2) ÷ (3)
Theoretical capacity	$0.3500	$0.0434	$0.3934
Practical capacity	0.3500	0.0625	0.4125
Normal utilization	0.3500	0.1000	0.4500
Master-budget utilization	0.3500	0.1250	0.4750

Each denominator-level concept will result in a different output-level (production-volume) variance.

$$\text{Output-level variance} = \begin{pmatrix} \text{Denominator} \\ \text{level in} \\ \text{output units} \end{pmatrix} - \begin{pmatrix} \text{Actual} \\ \text{output} \\ \text{units} \end{pmatrix} \times \begin{pmatrix} \text{Budgeted fixed} \\ \text{manufacturing} \\ \text{overhead rate} \\ \text{per output unit} \end{pmatrix}$$

Theoretical capacity = (1,152,000 − 460,000) × $0.0434
= $30,033 U (rounded up)

Practical capacity = (800,000 − 460,000) × $0.0625
= $21,250 U

Normal utilization = (500,000 − 460,000) × $0.1000
= $4,000 U

Master-budget utilization = (400,000 − 460,000) × $0.1250
= $7,500 F

Exhibit 9-6 shows how the choice of a denominator affects the May 19_5 operating income at the Bushells plant. Using the master-budget denominator results in assigning the highest amount of fixed manufacturing overhead cost per bottle to the 40,000 bottles in ending inventory. As a result, operating income is highest using the master-budget denominator. Recall that Bushells had no beginning inventory on May 1, 19_5, production in May of 460,000 bottles, and sales in May of 420,000 bottles. Hence, the ending inventory on May 31 is 40,000 bottles. The differences among the operating income for the four denominator-level concepts in Exhibit 9-6 are due to different amounts of fixed manufacturing overhead being inventoried:

	Fixed Manufacturing Overhead in May 31, 19_5 Inventory
Theoretical capacity	$1,736(40,000 × $0.0434)
Practical capacity	2,500(40,000 × 0.0625)
Normal utilization	4,000(40,000 × 0.1000)
Master-budget utilization	5,000(40,000 × 0.1250)

Thus, the difference in operating income between the master-budget utilization concept and the normal utilization concept of $1,000 ($18,000 − $17,000) is due to the $1,000 difference in fixed manufacturing overhead inventoried ($5,000 − $4,000).

There is no requirement that U.S. companies use the same denominator-level concept for management purposes and for income tax purposes. Nevertheless, the costs of recordkeeping and the desire for simplicity often lead companies to choose the same denominator for internal reporting and tax purposes.

Income taxation rulings by the U.S. Internal Revenue Service effectively prohibit the use of the theoretical capacity or practical capacity denominator-level concepts. Both these concepts would typically result in companies taking write-offs of fixed manufacturing overhead as tax deductions more quickly than the IRS wants. These write-offs are taken more slowly under the normal utilization or master-budget utilization concept. The IRS requires companies to use the master-budget denominator level measure (along with full proration of variances between inventories and cost of goods sold) for income tax reporting. It requires use of master-budget level rather than

EXHIBIT 9-6

Income Statement Effects of Using Alternative Denominator-Level Concepts:
Bushells Company for May 19_5

	Theoretical Capacity	Practical Capacity	Normal Utilization	Master-Budget Utilization
Sales: $0.50 × 420,000	$210,000	$210,000	$210,000	$210,000
Cost of goods sold				
Beginning inventory	0	0	0	0
Variable manufacturing costs	161,000	161,000	161,000	161,000
Fixed manufacturing overhead costs*	19,964	28,750	46,000	57,500
Cost of goods available for sale	180,964	189,750	207,000	218,500
Ending inventory†	15,736	16,500	18,000	19,000
Total cost of goods sold (at standard costs)	165,228	173,250	189,000	199,500
Adjustment for variances‡	30,033 U	21,250 U	4,000 U	7,500 F
Total cost of goods sold	195,261	194,500	193,000	192,000
Gross margin	14,739	15,500	17,000	18,000
Other costs	0	0	0	0
Operating income	$ 14,739	$ 15,500	$ 17,000	$ 18,000

*Fixed manufacturing overhead costs:
$0.0434 × 460,000 = $19,964
$0.0625 × 460,000 = $28,750
$0.1000 × 460,000 = $46,000
$0.1250 × 460,000 = $57,500

†Ending inventory costs:
($0.3500 + $0.0434) × (460,000 − 420,000) = $15,736
($0.3500 + $0.0625) × (460,000 − 420,000) = $16,500
($0.3500 + $0.1000) × (460,000 − 420,000) = $18,000
($0.3500 + $0.1250) × (460,000 − 420,000) = $19,000

‡The only variance for Bushells in May 19_5 is the output-level (production-volume) variance.
See text (p. 325) for computation of amounts.

normal budget level, in part because the concept of a normal utilization level is not as well defined (or auditable) as the master-budget denominator-level concept.

Many of the issues discussed in this section arise because managers typically prefer the use of budgeted rather than actual overhead rates. Chapter 15 (pp. 544–46) discusses additional issues relating to the choice of the denominator level that arise when either budgeted or actual overhead rates are used.

PROBLEM FOR SELF-STUDY

PROBLEM

Suppose that Bushells Company in our example is computing the operating income for May 19_6. This month is identical to May 19_5, the results of which are in Exhibit 9-6, except that master-budget utilization for 19_6 is 600,000 bottles per month instead of 400,000 bottles. There was no beginning inventory on May 1, 19_6, and no variances other than the output-level variance. Bushells writes off this variance to cost of goods sold on a monthly basis.

Required
How would the results in Exhibit 9-6 for Bushells Company be different if the month is May 19_6 rather than May 19_5? Show computations.

SOLUTION

The only change in the Exhibit 9-6 results will be for the master-budget utilization level. The budgeted fixed manufacturing overhead cost rate in May 19_6 is:

$$\frac{\$50,000}{600,000 \text{ bottles}} = \$0.0833 \text{ per bottle}$$

The manufacturing cost per bottle becomes $0.4333 ($0.3500 + $0.0833). In turn, the output-level (production-volume) variance for May 19_6 becomes:

$$(600,000 - 460,000) \times (\$0.0833) = \$11,662 \text{ U}$$

The income statement for May 19_6 is now:

Sales	$210,000
Cost of goods sold	
Beginning inventory	0
Variable manufacturing costs:	
$0.35 × 460,000	161,000
Fixed manufacturing costs:	
$0.0833 × 460,000	38,318
Cost of goods available for sale	199,318
Ending inventory:	
$0.4333 × (460,000 − 420,000)	17,332
Total cost of goods sold (at standard costs)	181,986
Adjustment for variances	11,662 U
Total cost of goods sold	193,648
Gross margin	16,352
Other costs	0
Operating income	$ 16,352

The higher denominator level in the 19_6 master budget means that less fixed manufacturing overhead costs are inventoried in May 19_6 than in May 19_5 given identical sales and production levels.

SUMMARY

The following points are linked to the chapter's learning objectives.

1. Variable costing and absorption costing differ in only one respect—how to account for fixed manufacturing overhead costs. Under variable costing, fixed manufacturing overhead costs are excluded from inventoriable costs and are a cost of the period in which they are incurred. Under absorption costs, these costs are inventoriable and become expenses only when a sale occurs.

2. The variable-costing income statement is based on the contribution-margin format. The absorption-costing income statement is based on the gross-margin format.

3. Earlier chapters of this book described four alternative costing systems—actual costing, normal costing, budgeted costing, and standard costing. Each of these four systems can use variable costing or absorption costing. Hence, eight alternative costing systems can be used to cost inventory and to compute operating income. Exhibit 9-2 (p. 312) presents an overview of how these eight systems differ when costing work in process inventory.

4. Under variable costing, reported operating income is driven by variations in unit sales levels. Under absorption costing, reported operating income is driven by variations in unit production levels as well as by variations in unit sales levels.

5. There is only one breakeven point with a variable costing income statement. In contrast, there can be multiple breakeven points with an absorption costing income statement; there are multiple combinations of unit sales, unit production, and denominator level that yield an operating income of zero.

6. Managers can increase operating income when absorption costing is used by producing more inventory. Critics of absorption costing label this as the major negative consequence of treating fixed manufacturing overhead as an inventoriable cost.

7. The denominator level chosen for fixed manufacturing overhead can greatly affect reported inventory and operating income amounts. Denominator levels focusing on the capacity of a plant to supply product are theoretical capacity and practical capacity. Denominator levels focusing on the demand for the products a plant can manufacture are normal utilization and master-budget utilization.

8. The smaller the denominator level chosen, the higher the fixed manufacturing overhead cost per output unit that is inventoriable. The IRS's requirement that the master-budget utilization concept be used typically results in higher operating income amounts being reported compared with the operating income reported using the practical capacity or theoretical capacity denominator-level concepts.

TERMS TO LEARN

This chapter and the Glossary at the end of the book contain definitions of the following important terms:

absorption costing (p. 309)   direct costing (311)   master-budget utilization (323)
normal utilization (323)   practical capacity (323)   theoretical capacity (323)
variable costing (309)

ASSIGNMENT MATERIAL

QUESTIONS

9-1 "Differences in operating income between variable costing and absorption costing are due solely to accounting for fixed costs." Do you agree? Explain.

9-2 Why is the term *direct costing* a misnomer?

9-3 "The term *variable costing* could be improved if it were called *variable manufacturing costing*." Do you agree? Why?

9-4 Explain the main conceptual issue under variable costing and absorption costing regarding the proper timing for the release of fixed manufacturing overhead as expense.

9-5 What is the key difference between a variable-costing income statement and an absorption-costing income statement?

9-6 Describe how eight alternative inventory costing systems differ from each other.

9-7 "The main trouble with variable costing is that it ignores the increasing importance of fixed costs in modern manufacturing." Do you agree? Why?

9-8 Give an example of how, under absorption costing, operating income could fall even though the unit sales level rises.

9-9 Why might the IRS require package design costs spent before production commences to be inventoriable?

9-10 List the three factors that affect the breakeven point under absorption costing.

9-11 "Critics of absorption costing have increasingly emphasized its potential for leading to undesirable incentives for managers." Give an example.

9-12 What are two ways of reducing the negative aspects associated with using absorption costing to evaluate the performance of a plant manager?

9-13 List four different concepts of denominator level.

9-14 Name one reason why many companies use the master-budget utilization-level concept rather than the normal utilization-level concept.

9-15 "The IRS should not specify what denominator-level concept a company selects." Why might an IRS officer disagree with this statement by the corporate controller of a steel company?

EXERCISES AND PROBLEMS

Coverage of Part One of the Chapter

9-16 **Straightforward variable-costing income statement.** Prepare a variable-costing income statement (through operating income) for O'Mara Company. Use the following data: No beginning inventories of work in process or finished goods; no ending inventories of work in process. Production was 500,000 units, of which 400,000 were sold for $50 each. Direct material cost was $6 per unit; direct manufacturing labor cost was $8 per unit; variable manufacturing overhead costs were $1 per unit; fixed manufacturing overhead costs were $2,000,000; variable marketing and administrative costs were $5 per unit sold; and fixed marketing and administrative costs were $7,500,000.

9-17 Absorption and variable costing. (CMA)   Osawa Inc. planned and actually manufactured 200,000 units of its single product in 19_5, its first year of operation. Variable manufacturing costs were $20 per unit produced. Variable marketing and administrative costs were $10 per unit sold. Planned and actual fixed manufacturing costs were $600,000. Planned and actual marketing and administrative costs totaled $400,000 in 19_5. Osawa sold 120,000 units of product in 19_5 at a selling price of $40 per unit.

Required
1. Osawa's 19_5 operating income using absorption costing is
   (a) $440,000, (b) $200,000, (c) $600,000, (d) $840,000, (e) none of these.
2. Osawa's 19_5 operating income using variable costing is
   (a) $800,000, (b) $440,000, (c) $200,000, (d) $600,000, (e) none of these.

9-18 Comparison of actual costing methods.   The Rehe Company sells its razors at $3 per unit. The company uses a first-in, first-out actual-costing system. A new fixed manufacturing overhead allocation rate is computed each year by dividing the actual fixed manufacturing overhead cost by the actual production units. The following simplified data are related to its first two years of operation:

	Year 1	Year 2
Sales	1,000 units	1,200 units
Production	1,400 units	1,000 units
Costs		
Variable manufacturing	$ 700	$ 500
Fixed manufacturing	700	700
Variable marketing and administrative	1,000	1,200
Fixed marketing and administrative	400	400

Required
1. Prepare income statements based on variable costing for each of the years.
2. Prepare income statements based on absorption costing for each of the years.
3. Prepare a reconciliation and explanation of the difference in the operating income for each year resulting from the use of absorption costing and variable costing.
4. Critics have claimed that a widely used accounting system had led to undesirable buildups of inventory levels. (a) Is variable costing or absorption costing more likely to lead to such buildups? Why? (b) What can be done to counteract undesirable inventory buildups?

9-19 Income statements. (SMA)   The Mass Company manufactures and sells a single product. The following data cover the two latest years of operations:

		19_3	19_4
Selling price		$ 40	$ 40
Sales in units		25,000	25,000
Beginning inventory in units		1,000	1,000
Ending inventory in units		1,000	5,000
Fixed manufacturing overhead costs		$120,000	$120,000
Fixed marketing and administrative costs		$190,000	$190,000
Standard variable costs per unit			
Direct materials	$10.50		
Direct manufacturing labor	9.50		
Variable manufacturing overhead	4.00		
Variable marketing and administrative	1.20		

The denominator level is 30,000 output units per year. Mass Company's accounting records produce variable-costing information, and year-end adjustments are made to produce external reports showing absorption-costing information. Any variances are charged to cost of goods sold.

Required
1. Prepare two income statements for 19_4, one under variable costing and one under absorption costing.

2. Explain briefly why the operating income figures computed in requirement 1 agree or do not agree.

3. Give two advantages and two disadvantages of using variable costing for internal reporting.

 9-20 Variable costing versus absorption costing. The Mavis Company uses an absorption-costing system based on standard costs. Total variable manufacturing costs, including direct materials costs, were $3 per unit; the standard production rate was 10 units per machine-hour. Total budgeted and actual fixed manufacturing overhead costs were $420,000. Fixed manufacturing overhead was allocated at $7 per machine-hour ($420,000 ÷ 60,000 machine-hours of denominator level). Selling price is $5 per unit. Variable marketing and administrative costs, which are driven by units sold, were $1 per unit. Fixed marketing and administrative costs were $120,000. Beginning inventory in 19_5 was 30,000 units; ending inventory was 40,000 units. Sales in 19_5 were 540,000 units. The same standard unit costs persisted throughout 19_4 and 19_5. For simplicity, assume that there were no price, spending, or efficiency variances.

Required

1. Prepare an income statement for 19_5 assuming that all underallocated or overallocated overhead is written off directly at year-end as an adjustment to Cost of Goods Sold.

2. The president has heard about variable costing. She asks you to recast the 19_5 statement as it would appear under variable costing.

3. Explain the difference in operating income as calculated in requirements 1 and 2.

4. Graph how fixed manufacturing overhead is accounted for under absorption costing. That is, there will be two lines—one for the budgeted fixed overhead (which is equal to the actual fixed manufacturing overhead in this case) and one for the fixed overhead allocated. Show how the overallocated or underallocated manufacturing overhead might be indicated on the graph.

5. Critics have claimed that a widely used accounting system has led to undesirable buildups of inventory levels. (a) Is variable costing or absorption costing more likely to lead to such buildups? Why? (b) What can be done to counteract undesirable inventory buildups?

9-21 Breakeven under absorption costing. Refer to Problem 9-20.

Required

1. Compute the breakeven point (in units) under variable costing.

2. Compute the breakeven point (in units) under absorption costing.

3. Suppose that production was exactly equal to the denominator level, but no units were sold. Fixed manufacturing costs are unaffected. Assume, however, that *all* marketing and administrative costs were avoided. Compute operating income under (a) variable costing and (b) absorption costing. Explain the difference between your answers.

9-22 The All-Fixed Company in 19_6. (R. Marple, adapted) It is the end of 19_6. The All-Fixed Company began operations in January 19_5. The company is so named because it has no variable costs. All its costs are fixed; they do not vary with output.

The All-Fixed Company is located on the bank of a river and has its own hydroelectric plant to supply power, light, and heat. The company manufactures a synthetic fertilizer from air and river water and sells its product at a price that is not expected to change. It has a small staff of employees, all hired on a fixed annual salary. The output of the plant can be increased or decreased by adjusting a few dials on a control panel.

The following are data regarding the operations of the All-Fixed Company:

	19_5	19_6*
Sales	10,000 tons	10,000 tons
Production	20,000 tons	—
Selling price	$30 per ton	$30 per ton
Costs (all fixed)		
Manufacturing	$280,000	$280,000
Marketing and administrative	40,000	40,000

*Management adopted the policy, effective January 1, 19_6, of producing only as much product as was needed to fill sales orders. During 19_6, sales were the same as for 19_5 and were filled entirely from inventory at the start of 19_6.

Required

1. Prepare income statements with one column for 19_5, one column for 19_6, and one column for the two years together, using
   a. Variable costing
   b. Absorption costing
2. What is the breakeven point under (a) variable costing and (b) absorption costing?
3. What inventory costs would be carried on the balance sheets at December 31, 19_5 and 19_6, under each method?
4. Assume that the performance of the top manager of the company is evaluated and rewarded largely on the basis of reported operating income. Which costing method would the manager prefer? Why?

9-23 The Semi-Fixed Company in 19_6. The Semi-Fixed Company began operations in 19_5 and differs from the All-Fixed Company (described in Problem 9-22) in only one respect: It has both variable and fixed manufacturing costs. Its variable manufacturing costs are $7 per ton and its fixed manufacturing costs $140,000 per year. Denominator level is 20,000 tons per year.

Required

1. Using the same data as in Problem 9-22 except for the change in manufacturing cost behavior, prepare income statements with adjacent columns for 19_5, 19_6, and the two years together, under
   a. Variable costing
   b. Absorption costing
2. Why did the Semi-Fixed Company have operating income for the two-year period when the All-Fixed Company in Problem 9-22 suffered an operating loss?
3. What inventory costs would be carried on the balance sheets at December 31, 19_5 and 19_6, under each method?
4. Assume that the performance of the top manager of the company is evaluated and rewarded largely on the basis of reported operating income. Which costing method would the manager prefer? Why?

9-24 Comparison of variable costing and absorption costing. Consider the following data:

Hinkle Company
Income Statements for the Year Ended December 31, 19_5

	Variable Costing	Absorption Costing
Sales	$7,000,000	$7,000,000
Cost of goods sold (at standard)	3,660,000	4,575,000
Fixed manufacturing overhead	1,000,000	—
Manufacturing variances (all unfavorable)		
Direct materials price and efficiency	50,000	50,000
Direct manufacturing labor price		
and efficiency	60,000	60,000
Variable manufacturing overhead spending		
and efficiency	30,000	30,000
Fixed manufacturing overhead		
Spending	100,000	100,000
Output level (production volume)	—	400,000
Total marketing costs (all fixed)	1,000,000	1,000,000
Total administrative costs (all fixed)	500,000	500,000
Total costs	6,400,000	6,715,000
Operating income	$ 600,000	$ 285,000

The inventories, carried at standard costs, were:

	Variable Costing	Absorption Costing
December 31, 19_4	$1,320,000	$1,650,000
December 31, 19_5	60,000	75,000

Required

1. Tim Hinkle, president of the Hinkle Company, has asked you to explain why the operating income for 19_5 is less than for 19_4, even though sales have increased 40% over last year. What will you tell him?

2. At what percentage of denominator level was the plant operating during 19_5?

3. Prepare a numerical reconciliation and explanation of the difference between the operating incomes under absorption costing and variable costing.

4. Critics have claimed that a widely used accounting system has led to undesirable buildups of inventory levels. (a) Is variable costing or absorption costing more likely to lead to such buildups? Why? (b) What can be done to counteract undesirable inventory buildups?

9-25 Inventory costing and management planning. It is November 30, 19_4. Consider the income statement for Industrial Products Inc.'s operations for January through November, 19_4.

Industrial Products Inc.
Income Statement for Eleven Months Ended November 30, 19_4

	Units	Dollars	
Sales @ $1,000	1,000		1,000,000
Cost of goods sold			
Beginning inventory, December 31, 19_3,			
@ $800	50	40,000	
Manufacturing costs @ $800, including			
$600 per unit for fixed manufacturing overhead	1,100	880,000	
Total standard cost of goods available			
for sale	1,150	920,000	
Ending inventory, November 30, 19_4,			
@ $800	150	120,000	
Standard cost of goods sold*	1,000		800,000
Gross margin			200,000
Marketing, distribution and customer-service costs			
Variable, 1,000 units @ $50		50,000	
Fixed, @ $10,000 monthly		110,000	160,000
Operating income			40,000

*There are no variances for the 11-month period considered as a whole.

Production in the past three months has been 100 units monthly. Practical capacity is 125 units monthly. To retain a stable nucleus of key employees, management never schedules monthly production at less than 40 units.

Maximum available storage space for inventory is regarded as 200 units. The sales outlook for the next four months is 70 units monthly. Inventory is never to be less than 50 units.

The company uses a standard absorption-costing system. Denominator production level is 1,200 units annually. All variances are disposed of at year-end as an adjustment to Cost of Goods Sold—at standard.

Required

1. The division manager is given an annual bonus that is geared to operating income. Assume that the manager wants to maximize the company's operating income for 19_4. How many units should the manager schedule for production in December? Note carefully that you do not have to (nor should you) compute the operating income for 19_4 in this or in subsequent parts of this problem.

2. Assume that standard variable costing is in use rather than standard absorption costing. Would variable-costing operating income for 19_4 be higher, lower, or the same as standard absorption-costing income, assuming that production for December is 80 units and sales are 70 units? Why?

3. If standard variable costing were used, what production schedule should the division manager set? Why?

4. Assume that the manager is interested in maximizing his performance over the long run and that performance is being judged on the basis of net income. Assume that the company's income tax rate will be cut in half in 19_5 and that the year-end write-offs of variances are acceptable for income tax purposes. Assume that standard absorption costing is used. How many units should be scheduled for production in December? Why?

5. Assume that the total production and total sales for 19_4 and 19_5, taken together, will be unchanged by the specific decision in requirement 4. Assume also that the standards will be unchanged in 19_5. Suppose the decision in requirement 4 is to schedule 50 units instead of an originally scheduled 120 units. By how much will operating income in 19_5 be affected by the decision to schedule 50 units in December 19_4? (That is, how much operating income is shifted from 19_4 to 19_5?)

9-26 Some additional requirements to Problem 9-25; absorption costing and output-level variances. Refer to Problem 9-25.

Required

1. What operating income will be reported for 19_4 as a whole, assuming that the implied cost-behavior patterns will continue in December as they did in January through November and assuming—without regard to your answer to requirement 1 in Problem 9-25—that production for December is 80 units and sales are 70 units?

2. Assume the same conditions as in requirement 1 except that a monthly denominator level of 125 units (practical capacity) was used in setting fixed manufacturing overhead rates for inventory costing throughout 19_4. What output-level (production-volume) variance would be reported for 19_4?

9-27 Variable costing versus absorption costing. (CMA, adapted) Portland Optics Inc. specializes in manufacturing lenses for large telescopes and cameras used in space exploration. Customers determine the specifications for their lenses, so the lenses vary considerably. Portland Optics uses a job costing system. Manufacturing overhead is allocated to jobs on the basis of direct manufacturing labor-hours, utilizing the absorption-costing method. Portland's budgeted overhead rates for 19_4 and 19_5 were based on the following estimates.

	19_4	19_5
Direct manufacturing labor-hours	32,500	44,000
Direct manufacturing labor costs	$325,000	$462,000
Fixed manufacturing overhead costs	$130,000	$176,000
Variable manufacturing overhead costs	$162,500	$198,000

Jim Bradford, Portland's controller, would like to use variable costing for internal reporting purposes. He believes statements prepared using variable costing are more appropriate for making product decisions. To illustrate the benefits of variable costing to the other members of Portland's management team, Bradford plans to convert the company's income statement from the absorption-costing method to the variable-costing method. He has gathered the following information for this purpose, along with a copy of Portland's 19_4/19_5 comparative income statement (following).

Portland's actual manufacturing data for the two years are:

	19_4	19_5
Direct manufacturing labor-hours	30,000	42,000
Direct manufacturing labor costs	$300,000	$435,000
Direct materials used	$140,000	$210,000
Fixed manufacturing overhead costs	$132,000	$175,000

The company's actual inventory balances were:

	12/31/19_3	12/31/19_4	12/31/19_5
Direct materials	$32,000	$36,000	$18,000
Work in process			
Costs	$44,000	$34,000	$60,000
Direct manufacturing			
labor-hours	1,800	1,400	2,500
Finished goods			
Costs	$16,000	$25,000	$14,000
Direct manufacturing			
labor-hours	700	1,080	550

For 19_4 and 19_5, all administrative costs were fixed. A portion of the marketing costs resulting from an 8% commission on net sales was variable. Portland reports any over- or underallocated manufacturing overhead as an adjustment to the cost of goods sold.

Portland Optics Inc.
Comparative Income Statement
for the Years 19_4/19_5

	19_4	19_5
Sales	$1,140,000	$1,520,000
Cost of goods sold		
Beginning finished goods	16,000	25,000
Cost of goods manufactured	720,000	976,000
Cost of goods available for sale	736,000	1,001,000
Ending finished goods	25,000	14,000
Cost of goods sold before		
overhead adjustment	711,000	987,000
Overhead variance adjustment	12,000 U	7,000 U
Cost of goods sold	723,000	994,000
Gross margin	417,000	526,000
Other operating costs		
Marketing costs	150,000	190,000
Administrative costs	160,000	187,000
Total other operating costs	310,000	377,000
Operating income	$ 107,000	$ 149,000

Required

1. For the year ended December 31, 19_5, prepare the revised income statement for Portland Optics Inc. under variable costing.

2. Describe two advantages of using variable costing over absorption costing that Bradford might cite.

Coverage of Part Two of the Chapter

9-28 Overhead rates in cyclical businesses. It is a time of severe business recession throughout the capital-goods industries. A manager of a truck-assembly division of a large corporation in a heavy-machinery industry is confused and unhappy. The manager is distressed with the controller, whose cost accounting department keeps reporting costs to the manager that are of little comfort because they are higher than ever before. At the same time, the manager has to quote lower prices than before in order to get any business.

Required

1. What measure of denominator-level concept is probably being used for allocation of manufacturing overhead?

2. How might the manufacturing overhead be allocated in order to make the cost data more useful in making price quotations?

3. Would the product costs being furnished by the cost accounting department be satisfactory for the costing of the annual inventory?

9-29   Alternative denominator-level concepts.   Lucky Larger recently purchased a brewing plant from a bankrupt company. The brewery is in Austin, Texas. It was constructed only two years ago. The plant has budgeted fixed manufacturing overhead of $42 million ($3.5 million each month) in 19_5. Paul Vautin, the controller of the brewery, must decide on the denominator-level concept to use in its absorption costing system for 19_5. The options available to him are:

a. Theoretical capacity: 600 barrels an hour for 24 hours a day × 365 days = 5,256,000 barrels

b. Practical capacity: 500 barrels an hour for 20 hours a day × 350 days = 3,500,000 barrels

c. Normal utilization for 19_5: 400 barrels an hour for 20 hours a day × 350 days = 2,800,000 barrels

d. Master-budget utilization for 19_5 (separate rates computed for each half-year)

   ◆ January to June 19_5 budget: 320 barrels an hour for 20 hours a day × 175 days = 1,120,000 barrels

   ◆ July to December 19_5 budget: 480 barrels an hour for 20 hours a day × 175 days = 1,680,000 barrels

Variable standard manufacturing costs per barrel are $45 (variable direct materials, $32; variable manufacturing labor, $6; and variable manufacturing overhead, $7). The Austin brewery "sells" its output to the sales division of Lucky Larger at a budgeted price of $68 per barrel.

Required

1. Compute the budgeted fixed manufacturing overhead rate using each of the four denominator-level concepts for (a) beer produced in March 19_5 and (b) beer produced in September 19_5. Explain why any differences arise.
2. Explain why the theoretical capacity and practical capacity concepts are different.
3. Which denominator-level concept would the plant manager of the Austin brewery prefer when senior management of Lucky Larger is judging plant manager performance during 19_5? Explain.

9-30   Operating income effects of alternative denominator-level concepts. (Continuation of 9-29)   In 19_5 the Austin brewery of Lucky Larger showed these results:

Beginning inventory, January 1, 19_5	0 barrels
Production	2,600,000 barrels
Ending inventory, December 31, 19_5	200,000 barrels

The Austin brewery had actual costs of:

Variable manufacturing costs	$120,380,000
Fixed manufacturing overhead costs	$ 40,632,000

The sales division of Lucky Larger purchased 2,400,000 barrels in 19_5 at the $68 per barrel rate.
All manufacturing variances are written off to cost of goods sold in the period in which they are incurred.

Required

1. Compute the operating income of the Austin brewery using (a) theoretical capacity, (b) practical capacity, and (c) normal utilization denominator-level concepts. Explain any differences among (a), (b), and (c).
2. What denominator-level concept would Lucky Larger prefer for income tax reporting? Explain.
3. Explain the ways in which the Internal Revenue Service might restrict the flexibility of a company like Lucky Larger, which uses absorption costing, to reduce its reported taxable income.

9-31 Role of denominator-level concepts. The Moroso Company incurs fixed manufacturing overhead of $2,700,000 annually. Master-budget utilization level is 37,500 machine-hours; normal utilization level, 45,000 hours; practical capacity, 50,000 hours allowed; and theoretical capacity, 54,000 hours. In 19_4, 37,500 units were produced and 30,000 units were sold. One standard machine-hour is allowed for each unit produced. There was no beginning inventory and no fixed manufacturing overhead spending variance.

Required
1. Prepare a four-column comparison (similar to Exhibit 9-6) using the four denominator-level concepts for allocating fixed manufacturing overhead to product. Designate which denominator-level concept would result in the highest and which in the lowest operating income. For each method, show the amounts that would be charged to

   ◆ Cost of goods sold
   ◆ Output-level (production-volume) variance
   ◆ Ending inventory

2. What are the advantages of using the master-budget utilization as the denominator level rather than normal utilization, practical capacity, or theoretical capacity for judging current operating performance?
3. If your bonus as a manager for the current period were affected by the selection of the denominator-level concept, which concept would you prefer? Why?

9-32 Effects of denominator-level concept choice. The Wong Company installed standard costs and a flexible budget on January 1, 19_4. The president had been pondering how fixed manufacturing overhead should be allocated to products. Machine-hours had been chosen as the allocation base. Her remaining uncertainty was the denominator-level concept for machine-hours. She decided to wait for the first month's results before making a final choice of what denominator-level concept should be used from that day forward.

In January 19_4, the actual units of output had a standard of 70,000 machine-hours allowed. If the company used practical capacity as the denominator-level concept, the fixed manufacturing overhead spending variance would be $10,000, unfavorable, and the output-level (production-volume) variance would be $36,000, unfavorable. If the company used normal utilization as the denominator-level concept, the output-level variance would be $20,000, favorable. Budgeted fixed manufacturing overhead was $120,000 for the month.

Required
1. Compute the denominator level, assuming that the normal utilization concept is chosen.
2. Compute the denominator level, assuming that the practical capacity concept is chosen.
3. Suppose you are the executive vice-president. You want to maximize your 19_4 bonus, which depends on 19_4 operating income. Assume that the output-level variance is charged or credited to income at year-end. Which denominator-level concept would you favor? Why?

9-33 Absorption costing, standard costs, management ethics. Industrial Equity Company (IEC) is a multinational business selling metal products used in the assembly of many cars, trucks, and planes. IEC has over 50 manufacturing divisions worldwide and is listed on the New York Stock Exchange. IEC has consistently reported annual earnings growth rates of 15% or more for each of the last 10 years.

Division managers at IEC receive an annual bonus of 30% of their annual salary if the plant operating income increases 15% or more over the previous year's operating income. Division managers who increase operating income more than 10% but less than 15% receive a bonus of 5% of their annual salary. Division managers who do not achieve a 10% increase in operating income receive no bonus. Instead, they receive a visit from the "IEC corporate consulting team."

Bob Wood is manager of the Flint, Michigan, division, which manufactures crankshafts for sale to automobile manufacturers. Wood has just received a 30% bonus for 19_4. Mary Easson, head of the IEC corporate consulting team, is less than impressed by Wood's performance. She suspects him of "producing for inventory" and collects the following information on the Flint division for 19_4:

Beginning inventory	0 crankshafts
Production	480,000 crankshafts
Ending inventory	30,000 crankshafts
Sales	450,000 crankshafts
Selling price	$66 per unit
Standard variable costs per crankshaft	
Direct materials	$20
Direct manufacturing labor	5
Manufacturing overhead	12
Variable marketing	4
Standard fixed costs	
Manufacturing overhead	$9,000,000
Marketing	1,000,000

Manufacturing overhead is allocated to each crankshaft on the basis of standard machine-hours. Each crankshaft has a standard machining time of 30 minutes. The denominator level in 19_4 was the master-budget utilization for the Flint plant, 500,000 crankshafts. A standard absorption-costing system is used for each IEC plant. All variances are recorded as a cost of the period in which they are incurred.

All auto companies require suppliers to deliver on a just-in-time basis (that is, just before the crankshafts are required for assembly). The last four months of 19_4 saw a reduction in the orders auto companies placed for crankshafts.

The price, spending, and efficiency manufacturing variances for 19_4 were $300,000 unfavorable. The total marketing variances were $156,000 favorable (variable $130,000 favorable and fixed $26,000 favorable).

Operating income for the Flint division in 19_3 was $1,427,010.

Required

1. Compute the absorption-costing operating income for the Flint division in 19_4.
2. Why might Easson believe that in 19_4 Wood engaged in behavior not in the best interests of IEC? How might Wood respond to any charges Easson might make about "producing for inventory"?
3. Is the problem Easson raised likely to be eliminated by her talking to Wood about management ethics? Explain.

9-34 Absorption costing, management ethics. (Continuation of 9-33) Mary Easson decides to undertake a systematic investigation of how the combination of the existing division manager bonus plan and absorption costing may be causing division managers to make decisions not in the best interests of Industrial Equity Company (IEC). She will first visit the Morristown division of IEC, which manufactures more than 100 different metal products.

Required

1. Name three types of behavior that Easson should look for that would suggest problems for IEC with the existing bonus plan and accounting system.
2. What possible changes might Easson consider if her investigation produces widespread evidence of systematic poor decision making by division managers at IEC?

10

Determining

How Costs Behave

General issues in estimating cost functions

Assumptions underlying cost classification

Cost estimation approaches

Steps in estimating a cost function

Evaluating and choosing among cost functions

Data collection and adjustment issues

Hierarchy of costs and cost drivers

Nonlinearity and cost functions

Learning curves and cost functions

Appendix: Regression analysis

*A*ircraft assembly companies report that learning-curve effects cause unit variable costs to decrease as the number of planes assembled increases. Raytheon has observed such learning-curve effects when assembling Beechjets at its Wichita, Kansas, plant.

Knowing how costs behave and distinguishing fixed from variable costs are frequently the keys to making good decisions. Cost behavior is a central idea in budgeting and in making other planning decisions such as, What price should we charge? What bid should we make on that contract? Do we make the item or buy it? What effect will a 20% increase in sales have on operating income? Understanding cost behavior is important in predicting the change in the level of costs as the quantity of different products produced changes. Decisions in the control area, such as the interpretation of variances, similarly rely heavily on knowledge of cost behavior. This chapter focuses on how to determine cost behavior.

GENERAL ISSUES IN ESTIMATING COST FUNCTIONS

Basic Terms and Assumptions

Objective 1
State the two assumptions frequently used in cost-behavior estimation

Cost estimation is the attempt to measure *past* cost relationships. Managers are interested in estimating past cost behavior primarily because these estimates may help them make more accurate **cost predictions,** or forecasts, about *future* costs. Better cost predictions help managers make more informed planning decisions.

Two assumptions are frequently used in the estimation of cost functions:

◆ *Assumption 1:* Variations in a *single* cost driver explain variations in total costs. A *cost driver* is any factor that affects costs—that is, a change in the cost driver will cause a change in the total cost of a related cost object. A *cost object* is anything for which a separate measurement of costs is desired. Examples of cost drivers when the cost object is a product include machine-hours in manufacturing and the weight of items to be shipped in distribution.

◆ *Assumption 2:* A linear function adequately approximates cost behavior within the relevant range of the cost driver. The *relevant range*, described in Chapter 2, is the range of the cost driver in which a specific relationship between total costs and the driver is valid.

These two assumptions are used throughout much of this chapter. Later sections give examples of nonlinear cost behavior. The last section in the Appendix describes how variations in two or more cost drivers can explain variations in the level of total costs.

Given the assumptions of linearity and a single cost driver, each cost has some underlying cost-behavior pattern, more technically described as a *cost function*. The expected value of cost, $E(C)$ has the form

$$E(C) = G + WD$$

where

- C is total costs
- G is an underlying (but unknown) parameter that represents that component of total costs that, within the relevant range, does not vary with changes in the level of the cost driver (D)
- D is the quantity of the cost driver
- W is another underlying (but unknown) parameter that indicates how much total costs (C) change for each unit change in the cost driver (D) within the relevant range

A **parameter** is a constant, such as G, or a coefficient, such as W, in the model equation.

$E(\cdot)$ is called the *expectations operator*. The expectation symbol E indicates the average or expected amount of cost for each level of the cost driver. The expected cost is used because random factors may cause the *actual* observed cost to differ from what would be predicted on the basis of the relationship between cost and the quantity of the cost driver. Suppose, for example, that producing one unit of the cost driver requires one pound of materials costing $10 per pound. Then, on average, producing 100 units of the product (the amount of the cost driver) requires $1,000 of materials costs (the expected cost). This average amount does not mean, however, that exactly $1,000 of materials costs will be incurred every time 100 units of the product are produced. Actual materials costs for a specific batch of 100 units of the product may be more or less than expected materials costs of $1,000 because of random variations in the quality of the materials or performance of the equipment.

Working with historical data, consisting of a set of observed values for C and D, and recognizing that the observed values of C measure the relationship between C and D with error, the cost analyst can develop a formula to *estimate* total costs as:

$$c = g + wD$$

where terms in lower case letters represent estimates:

- ◆ c is the estimated value (as distinguished from the observed value, C)
- ◆ g (termed the **constant**, or **intercept**) and w (termed the **slope coefficient**) are estimates of the underlying but unknown G and W parameters.

Note that the estimation equation computes estimated total costs for actual values of the cost driver D.

Examples of Cost Functions

Objective 2
Describe three types of linear cost functions

This section illustrates three types of linear cost functions. A **linear cost function** is a function in which a single constant (g) and a single slope coefficient (w) describe the behavior of total costs for all changes in the level of the cost driver within the relevant range.

Assume Cannon Services is negotiating with World Wide Communications (WWC) for exclusive use of a telephone line between New York and Paris. WWC offers Cannon Services three alternative cost structures:

- ◆ *Alternative 1:* $5 per minute of phone use. This is a *strictly variable cost* for Cannon Services, with the number of phone-minutes as the cost driver, D. In terms of our equation,

$$c = \$5D$$
$$\text{with } g = 0 \text{ and } w = \$5 \text{ per minute}$$

Graph 1 in Exhibit 10-1 presents the *strictly variable* or *proportionately variable cost*. Total costs change in direct proportion to the number of phone-minutes used within the relevant range. There are no fixed costs, and so the intercept is zero. Every additional minute used adds $5 to total costs.

EXHIBIT 10-1
Examples of Linear Cost Functions

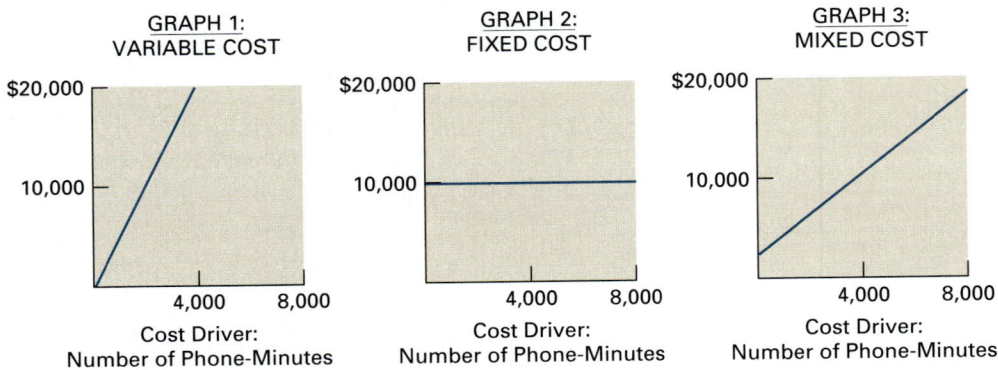

GRAPH 1:
VARIABLE COST
Cost Driver:
Number of Phone-Minutes

GRAPH 2:
FIXED COST
Cost Driver:
Number of Phone-Minutes

GRAPH 3:
MIXED COST
Cost Driver:
Number of Phone-Minutes

◆ *Alternative 2:* $10,000 per month. Under this alternative, Cannon Services has a *fixed cost* of $10,000. The total costs will be $10,000 per month regardless of the number of phone-minutes used. In terms of our equation,

$$c = \$10,000$$
with $g = \$10,000$ and $w = 0$

Graph 2 in Exhibit 10-1 presents the *fixed cost*.

◆ *Alternative 3:* $3,000 per month plus $2 per minute of phone use. This is an example of a *mixed cost*. A **mixed cost** (or **semivariable cost**) is a cost that has both fixed and variable elements. Alternative 3 has one component that is fixed with respect to the number of phone-minutes ($3,000 per month) and another component that is variable with respect to the number of phone-minutes ($2 per minute of phone use). In our equation

$$c = \$3,000 + \$2D$$
with $g = \$3,000$ and $w = \$2$ per minute

Graph 3 in Exhibit 10-1 presents the *mixed cost*. The total cost in the relevant range increases as the number of phone-minutes used increases. But the average cost per phone-minute differs at different levels of use. For example, when 4,000 phone-minutes are used, the average cost is [$3,000 + ($2 × 4,000)] ÷ 4,000 = $2.75 per phone-minute, but when 8,000 phone-minutes are used, the average cost is [$3,000 + ($2 × 8,000)] ÷ 8,000 = $2.375 per phone-minute.

Cause-and-Effect Criterion in Choosing Cost Drivers

The most important issue in estimating a cost function is to determine whether a cause-and-effect relationship exists between the cost driver and the resulting costs. The cause-and-effect relationship might arise in several ways.

1. It may be due to a physical relationship between costs and the cost driver. An example of a physical relationship is when units of production is used as the cost driver of materials costs. To produce more units requires more materials, which results in higher materials costs.
2. Cause and effect can arise from a contractual arrangement, as in the Cannon Services example described above.
3. Cause and effect can be implicitly established by logic and knowledge of operations. For example, it seems intuitively clear that a complex product design with many component parts that must fit together precisely will require higher design costs than a simple product with few parts.

Be careful not to interpret a high correlation, or connection, between two variables to mean that either variable causes the other. A high correlation between two variables,

x and y, indicates merely that the two variables move together. It is possible that x may cause y, y may cause x, x and y may interact, both may be affected by a third variable z, or the correlation may be due to chance. *No conclusions about cause and effect are warranted by high correlations.* For example, higher production, generally, results in higher materials costs and higher labor costs. Materials costs and labor costs are highly correlated, but neither causes the other.

Consider the high correlation between the original league of the winning team in the Super Bowl and security returns on the New York Stock Exchange (NYSE). Over the past 25 years, when an original National Football League team (such as the San Francisco 49ers) has won the Super Bowl, the NYSE stock index has almost always increased during that year. Conversely, in a year when an original American Football League team (such as the Miami Dolphins) has won, the NYSE index has almost always decreased.[1] It is difficult to think of any cause-and-effect reason that makes this high correlation plausible. Even though a high correlation has existed in the past between the original league of the Super Bowl winner and stock prices, it is doubtful that this correlation will continue to hold in the future.

Before estimating any cost functions, the cost analyst must first be satisfied that the relationship between the cost driver and costs is economically plausible. Without any plausibile economic rationale for a relationship, it is unlikely that a high level of correlation observed between the cost driver and costs in one set of data will be found in other similar sets of data. Economic plausibility is established if a cause-and-effect relationship exists between the cost driver and costs.

ASSUMPTIONS UNDERLYING COST CLASSIFICATION

When classifying costs into variable or fixed costs, we need to specify

- the cost object,
- the time span, and
- the relevant range.

We discuss each in turn.

Choice of Cost Object

Whether a cost is fixed or variable depends on the cost object. A particular cost item could be fixed with respect to one cost object and variable with respect to another. Consider a company that owns and operates a chain of restaurants. If the cost object is companywide equipment, depreciation on ovens would be a variable cost with respect to the number of restaurants in the chain. If the cost object is the meal served in a particular restaurant, depreciation on that restaurant oven would be a fixed cost with respect to the number of meals served.

Time Span

Whether a cost is variable or fixed with respect to a particular driver is affected by the time span considered in the decision situation. The longer the time span, other things being equal, the more variable the cost. For example, inspection salaries and costs are typically fixed in the short run. That is, total inspection costs will not change even if the amount of inspection actually required changes. But in the long run, total inspection costs will vary with the inspection time required. If more inspection is needed, inspection costs will increase (more inspectors will be hired); if less inspection is needed, inspection costs will decrease (some inspectors will be reassigned to other tasks).

---

[1] J. Granelli and T. Petruno, "You Can Take Heart From The January Gain, or You Can Punt," *Los Angeles Times* (February 1, 1993).

Relevant Range

Costs can be described as fixed or variable only within a relevant range. For example, direct manufacturing labor cost is proportionately variable with respect to output quantity produced only as long as no overtime premium rates (such as wages at one and a half times the normal hourly rate) need to be paid. If 100,000 units of output can be produced without requiring overtime work, then the relevant range over which direct manufacturing labor is a strictly variable cost is 0 to 100,000.

Likewise, depreciation cost of plant and equipment is a fixed cost only if the output quantity produced is within the capacity level of the plant. If plant capacity is 120,000 units and output required to be produced is 150,000 units, additional investment in plant and equipment will be necessary. This investment will increase depreciation cost. Depreciation cost is fixed only for output levels of fewer than 120,000 units.

COST ESTIMATION APPROACHES

Objective 3
Describe four approaches to cost estimation

There are four approaches to cost estimation:

1. Industrial-engineering method
2. Conference method
3. Account analysis method
4. Quantitative analysis of current or past cost relationships

These approaches differ in the costs of conducting the analysis, the assumptions they make, and the evidence they yield about the accuracy of the estimated cost function. They are not mutually exclusive. Many organizations use a combination of these approaches.

Industrial-Engineering Method

The **industrial-engineering method**, also called the **work-measurement method**, first analyzes the relationship between inputs and outputs in physical terms. For example, inputs for a carpet manufacturer include cotton, wool, dyes, direct labor, machine time, and power. Output is square yards of carpet. Time and motion studies analyze the time and materials required to perform the various operations to produce the carpet. For example, the time and motion study may conclude that to produce 20 square yards of carpet requires 2 bales of cotton and 3 gallons of dye. Standards and budgets transform these physical input and output measures into costs.

The industrial-engineering method is very time-consuming. Some government contracts mandate its use. Many firms, however, find it too costly for analyzing their entire cost structure. More frequently, firms use this approach for direct cost categories such as materials and labor but not for indirect cost categories such as manufacturing overhead. Physical relationships between inputs and outputs may be difficult to specify for individual overhead cost items.

Conference Method

The **conference method** develops cost estimates on the basis of analysis and opinions gathered from various departments of an organization (purchasing, process engineering, manufacturing, employee relations, and so on). One company has a cost-estimating department that develops product costs on the basis of a consensus of the relevant departments. In another company, representatives of all value-chain areas (R&D, design, production, marketing, distribution, and customer service) provide individual cost estimates that are then combined into a product cost estimate.

The conference method allows cost estimates to be developed quickly. The pooling of expert knowledge from each value-chain area gives the conference method credibility. The accuracy of the cost estimates largely depends on the care and detail taken by the people providing the inputs.[2]

Account Analysis Method

The **account analysis method** classifies cost accounts in the ledger as variable, fixed, or mixed with respect to the cost driver. Typically, managers use qualitative rather than quantitative analysis when making these cost-classification decisions. The account analysis approach is widely used in practice.[3]

Organizations differ with respect to the care taken in implementing account analysis. In some organizations, individuals thoroughly knowledgeable about the operations make the cost-classification decisions. For example, manufacturing personnel may classify costs such as machine lubricants and materials-handling labor, while marketing personnel may classify costs such as advertising brochures and sales salaries. In other organizations, only cursory analysis is conducted, sometimes by individuals with limited knowledge of operations, before cost-classification decisions are made. Clearly, the former approach would provide more reliable cost classifica-

---

[2]The conference method is further described in W. Winchell, *Realistic Cost Estimating for Manufacturing*, 2nd ed. (Dearborn, MI: Society for Manufacturing Engineers, 1991).

[3]Survey evidence on the widespread use of the account analysis approach is in M. M. Mowen, *Accounting for Costs as Fixed and Variable* (Montvale, NJ: National Association of Accountants, 1986).

tions than the latter. The account analysis approach may be helpful as a first step in cost classification and estimation. Supplementing this analysis by the conference method improves its credibility.

The account analysis method is often adequate for simple cost structures. Exhibit 10-2 shows the budgeted costs for Colorado Skis, a small retail ski store, for the year 19_5. Colorado Skis budgets sales of 2,000 skis (that is, 2,000 pairs of skis).

The management of Colorado Skis uses the account analysis method to identify fixed and variable costs with respect to the number of skis sold. Management treats costs of skis sold as a variable cost, and the store manager's salary, cashier's salary, and store operating costs as fixed costs. Management considers salespersons' salaries and commissions and also advertising and promotion costs as mixed costs. These costs have a fixed and a variable component that management estimates on the basis of experience. Exhibit 10-2 separates total budgeted costs of $376,000 into budgeted fixed costs of $118,000 and budgeted variable costs of $258,000. Budgeted variable costs vary with the number of skis sold, and column 5 calculates the budgeted variable cost per ski sold. The total budgeted variable cost per ski sold is $258,000 ÷ 2,000 = $129. The general cost equation is expressed as $c = g + wD$. Total budgeted costs = $118,000 + ($129 × number of skis sold). The total budgeted cost per ski sold for 19_5 is $376,000 ÷ 2,000 = $188.

What if the management of Colorado Skis wants to estimate the cost of selling 2,200 skis? The account analysis method provides useful information:

$$\text{Estimated total costs} = \$118,000 + (\$129 \times 2,200) = \$401,800$$

The estimated total cost per ski sold decreases to $401,800 ÷ 2,200 = $182.64. Unit costs decrease because fixed costs are spread over a greater number of units sold.

Quantitative Analysis of Cost Relationships

Cost functions are often estimated using data on past cost relationships. These data may be time-series data or cross-sectional data. *Time-series data* pertain to the same entity (firm, plant, activity area, and so on) over a sequence of past time periods. For example, monthly observations of manufacturing overhead and machine-hours for a particular plant for the most recent year would yield time-series data. *Cross-sectional data* pertain to different entities for the same time period. For example, studies of per-

EXHIBIT 10-2
Budgeted Costs for Colorado Skis for 19_5

Account (1)	Total Costs (2) = (3) + (4)	Fixed Costs (3)	Variable Costs (4)	Variable Costs per Ski Sold (5) = (4) ÷ 2,000
Cost of skis sold	$240,000		$240,000	$120
Store manager's salary	30,000	$ 30,000		
Cashier's salary	10,000	10,000		
Store operating costs: Rent, taxes, insurance, depreciation, utilities	62,000	62,000		
Salespersons' salaries and commissions	24,000	14,000	10,000	5
Advertising and promotion	10,000	2,000	8,000	4
Total costs	$376,000	$118,000	$258,000	$129

EXHIBIT 10-3

Weekly Indirect Manufacturing Labor Costs
and Machine-Hours for Elegant Rugs

Week (1)	Indirect Manufacturing Labor Costs (2)	Cost Driver: Machine-Hours (3)
1	$1,190	68
2	1,211	88
3	1,004	62
4	917	72
5	770	60
6	1,456	96
7	1,180	78
8	710	46
9	1,316	82
10	1,032	94
11	752	68
12	963	48

sonnel costs and loans processed at 50 individual branches of a bank during March would produce cross-sectional data.

Let us examine indirect manufacturing labor costs consisting of machine maintenance costs and setup labor costs at Elegant Rugs, which weaves carpets for homes and offices. Machine maintenance costs are the costs of adjusting and repairing machines during production. Machine maintenance costs also include the costs of servicing and cleaning machines after each production run. Setup labor costs are the costs of preparing equipment to produce a particular pattern of carpets. The manufacturing operation is highly automated with state-of-the-art weaving machines.

Exhibit 10-3 shows weekly (time-series) data for the most recent 12-week period. Note that the data are paired. Each week, managers report data on the indirect manufacturing labor costs and the quantity of cost driver used (machine-hours). For example, week 12 shows indirect manufacturing labor costs of $963 and 48 machine-hours. The next section uses the data in Exhibit 10-3 to illustrate two different quantitative ways to estimate a cost function: the high-low method and regression analysis.

STEPS IN ESTIMATING A COST FUNCTION

Objective 4
Outline six steps in estimating a cost function on the basis of current or past cost relationships

There are six steps in estimating a cost function on the basis of an analysis of current or past cost relationships: (1) choose the dependent variable (the variable to be predicted, which is some type of cost), (2) identify the cost driver(s), (3) collect data on the dependent variable and the cost driver(s), (4) plot the data, (5) estimate the cost function, and (6) evaluate the estimated cost function. Frequently the cost analyst will cycle through these six steps several times before identifying an acceptable cost function.

STEP 1: *CHOOSE THE DEPENDENT VARIABLE.* Choice of the **dependent variable** (the cost variable to be predicted), usually called C, will depend on the purpose for estimating a cost function. For example, if the purpose is to determine indirect manufacturing costs for a production line, then C should incorporate all costs that are classified as indirect with respect to the production line. Ideally, all the individual items included in the dependent variable, C, will have a similar relationship with the cost driver(s), chosen in Step 2. Where a single relationship does not exist, the cost analyst should investigate the possibility of estimating more than one cost function.

Consider several types of fringe benefits paid to employees and their cost drivers:

Fringe Benefit	Cost Driver
1. Health benefits	Number of employees
2. Meals in cafeteria	Number of employees
3. Pension benefits	Salaries of employees
4. Life insurance	Salaries of employees

The costs of fringe benefits 1 and 2 can be grouped or pooled into one dependent variable because they both have the same cost driver. The costs of fringe benefits 3 and 4 should not be included in a dependent variable containing 1 and 2 because 3 and 4 are not driven by the number of employees. Instead, the costs of 3 and 4 should be pooled into a separate dependent variable and estimated using salaries of employees as the cost driver.

STEP 2: *IDENTIFY THE COST DRIVER(S).* Examples of cost drivers when the cost object is a product include the number of parts in a product and the hours of test time for a product. Examples of cost drivers when the cost object is an activity include power and labor consumed in the maintenance activity. Ideally, the chosen cost driver should be economically plausible and accurately measurable.

STEP 3: *COLLECT DATA ON THE DEPENDENT VARIABLE AND THE COST DRIVER(S).* This step is usually the most difficult one in cost analysis. Cost analysts obtain data from company documents, interviews with managers, and through special studies. The ideal database would contain numerous observations for a firm whose operations have not been affected by economic or technological change. Stable technology ensures that data collected in the estimation period represent the same underlying relationship between the dependent variable and the cost driver(s). Moreover, the time periods (for example, daily, weekly, or monthly) used to measure the dependent variable and the cost driver(s) should be identical. A later section of this chapter gives examples of how data frequently do not fit this "ideal" description.

STEP 4: *PLOT THE DATA.* This step is important. The expression "a picture is worth a thousand words" conveys the benefits of plotting the data. The general relation between the dependent variable and the cost driver (often called **correlation**) can readily be observed in a plot of the data. Moreover, the plot highlights extreme observations that analysts should check. Was there an error in recording the data or an unusual event, such as a labor strike, that makes these observations unrepresentative of the normal relationship between the dependent variable and the cost driver? Plotting the data can also provide insight into whether the relation is approximately linear and what the relevant range of the cost function is.

Exhibit 10-4 plots the weekly data from Exhibit 10-3. There is strong visual evidence of a positive relation between indirect manufacturing labor costs and machine-hours (that is, when machine-hours go up, so do costs). There do not appear to be any extreme observations in Exhibit 10-4. The relevant range is from 46 to 96 machine-hours per week.

STEP 5: *ESTIMATE THE COST FUNCTION.* The next section of this chapter illustrates the use of both the high-low method and regression analysis to estimate the indirect manufacturing labor cost function at Elegant Rugs.

STEP 6: *EVALUATE THE ESTIMATED COST FUNCTION.* We describe criteria for evaluating a cost function after illustrating the high-low method and regression analysis.

High-Low Method

Managers occasionally use very simplified ways to estimate cost functions. An example is the **high-low method**, which entails using only the highest and lowest observed values of the *cost driver* within the relevant range. The line connecting these two points becomes the estimated cost function.

EXHIBIT 10-4

Plot of Weekly Indirect Manufacturing Labor Costs
and Machine-Hours for Elegant Rugs

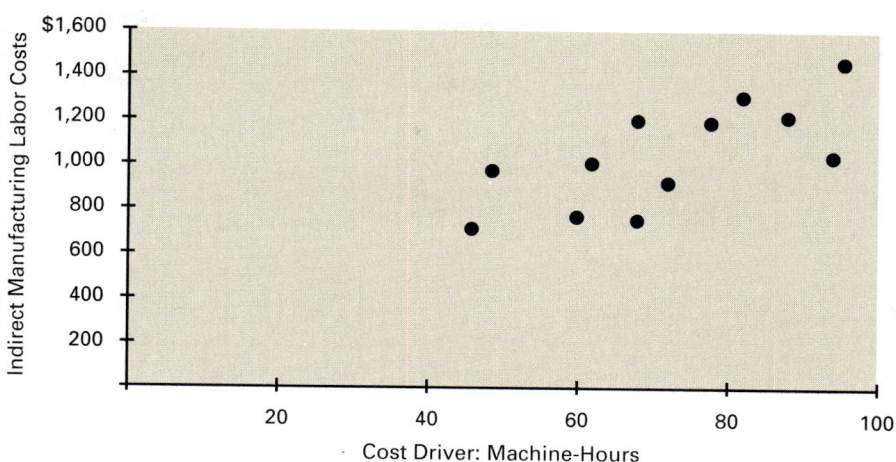

We illustrate the high-low method using data from Exhibit 10-3.

	Cost Driver: Machine-Hours	Indirect Manufacturing Labor Costs
Highest observation of cost driver	96	$1,456
Lowest observation of cost driver	46	710
Difference	50	$ 746

$$\text{Slope coefficient, } w = \frac{\text{Difference between costs associated with highest and lowest observations of the cost driver}}{\text{Difference between highest and lowest observations of the cost driver}}$$

$$\text{Slope coefficient, } w = \frac{\$746}{50} = \$14.92 \text{ per machine-hour}$$

$$\text{Total indirect manufacturing labor costs} = \text{Constant} + (\text{Slope coefficient} \times \text{Quantity of cost driver})$$

That is,

$$\text{Constant} = \text{Total costs} - (\text{Slope coefficient} \times \text{Quantity of cost driver})$$

The constant, or intercept, term does not serve as an estimate of the fixed cost of Elegant Rugs if no machines were run. Why? Because running no machines and shutting down the plant is outside the relevant range. The intercept term is the constant component of the equation that provides the best (linear) approximation of how a cost behaves within the relevant range, based on the high-low method.

To compute the constant, we can use either the highest or the lowest observation of the cost driver. Both calculations yield the same answer (because the solution technique solves two linear equations with two unknown parameters, the slope coefficient and the constant).

At the highest observation of the cost driver:

$$\text{Constant, } g = \$1,456 - (\$14.92 \times 96) = \$23.68$$

At the lowest observation of the cost driver:

$$\text{Constant, } g = \$710 - (\$14.92 \times 46) = \$23.68$$

Therefore, the high-low estimate of the cost function is

$$c = g + wD$$
$$= \$23.68 + \$14.92 \text{ (machine-hours)}$$

EXHIBIT 10-5
Danger of High-Low Method Using Nonrepresentative
Observations for Gemco Distributors

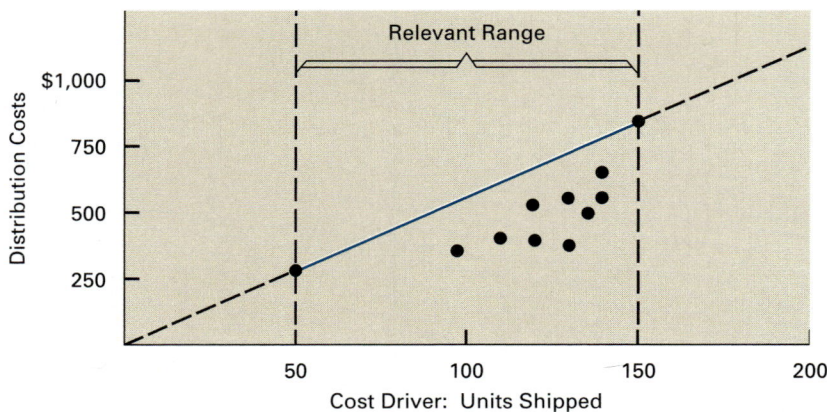

In some cases, the highest (lowest) observation of the cost driver will not coincide with highest (lowest) observation of the dependent variable (cost). Given that causality runs *from* the cost driver to the dependent variable in a cost function, choosing the highest observation and the lowest observation of the cost driver is appropriate.

There is an obvious danger of relying on only two observations. They may not be representative of all the observations. Always plot all the data. Consider a second example of Gemco Distributors. The graph in Exhibit 10-5 plots data for units shipped and distribution costs for Gemco Distributors. The exhibit illustrates the danger of mechanically applying the high-low method. It shows how picking the highest and lowest observations for the units-shipped variable can result in an estimated cost function that poorly describes the underlying (linear) cost relationship between distribution costs and units shipped.

Sometimes the high-low method is modified so that the two observations chosen are a "representative high" and a "representative low." The reason is that management wants to avoid having extreme observations, which arise from abnormal events, affect the cost function. Even with such a modification, this method ignores information on all but two observations when estimating the cost function.

Regression Analysis Method

Regression analysis is a statistical model that measures the average amount of change in the dependent variable that is associated with a unit change in one or more independent variables. Unlike the high-low method, regression analysis uses *all* available data to estimate the cost function. The *dependent variable*, denoted C, is the variable estimated by the regression model. In the Elegant Rugs example, C is total indirect manufacturing labor costs. The *independent variable*, denoted D, is the variable used to estimate the dependent variable in the regression model. In the Elegant Rugs example, D is machine-hours. **Simple regression** analysis uses only one independent variable to estimate the dependent variable; **multiple regression** analysis uses more than one independent variable.

We emphasize the interpretation and use of output from computer software programs for regression analysis. Commonly available programs (such as SPSS, SAS, Lotus, and Excel) on mainframes and personal computers calculate almost all the statistics referred to in this chapter.

Exhibit 10-6 shows the line developed using regression analysis that best fits the data in Exhibit 10-3. The estimated cost function is

$$c = \$300.98 + \$10.31D$$

EXHIBIT 10-6

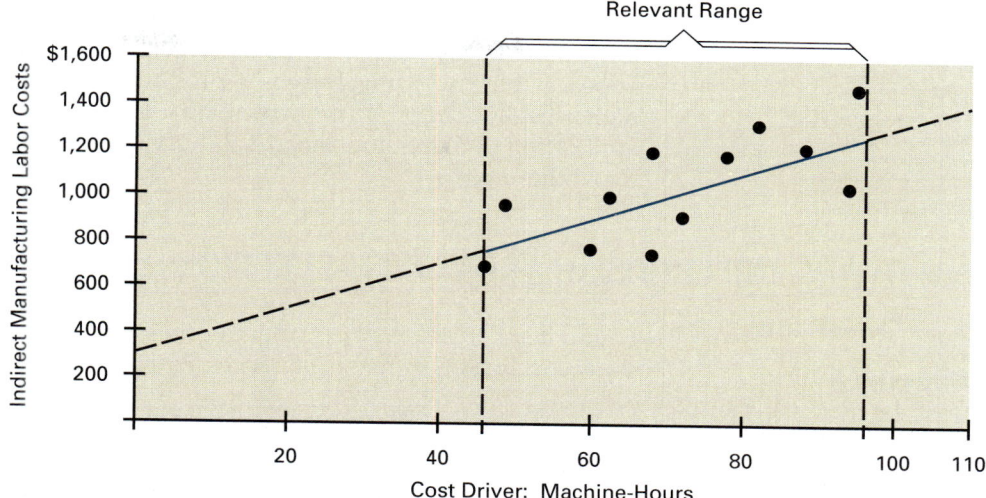

where c is the predicted indirect manufacturing labor costs for any level of machine-hours (D). The constant, or intercept, term of the regression (g) is $300.98, and the slope coefficient (w) is $10.31 per machine-hour.

How do we derive the regression equation and regression line in Exhibit 10-6? We use the least-squares technique. We draw the regression line to minimize the sum of the squared vertical distances from the data points to the regression line. Vertical differences measure distance between actual cost (C) and the predicted cost (c) for each observation. The difference between actual and predicted cost is called the **residual term.** The chapter Appendix gives detailed computations for deriving the regression line.

The vertical dashed lines in Exhibit 10-6 indicate the relevant range. The manager's concern is typically with cost levels *within the relevant range*, not with cost levels outside the relevant range. *The constant component of the equation $300.98, is not* an estimate of the fixed cost that Elegant Rugs incurs if it shuts down all machines. Shutting down the plant is outside the relevant range. The estimated cost function is the combination of g and w that provides the best available estimate of how a cost behaves in the relevant range.

The estimate of the slope coefficient w indicates that the average indirect manufacturing labor costs vary at the rate of $10.31 for every machine-hour. Management can use this equation when budgeting for future indirect manufacturing labor costs. For instance, if management budgeted 90 machine-hours for the upcoming week, the predicted indirect manufacturing labor costs would be

$$c = \$300.98 + \$10.31(90) = \$1,228.88$$

Compare the regression equation with the high-low equation in the preceding section, which was $23.68 + $14.92 per machine-hour. For 90 machine-hours, the predicted cost based on the high-low equation is $23.68 + $14.92(90) = $1,366.48. The difference of $137.60 between the two predictions is 11% of the regression prediction. This difference may be significant to a particular decision.

Intelligent application of regression analysis requires knowledge of both operations and cost accounting. Consider repair costs. Scheduling repairs when production is at a low level has the advantage of taking machines out of service when they are needed least. In this case, regression analysis of production output and repair costs would indicate that the higher the level of production, the lower the repair costs, and vice versa. The engineering link between repairs and production, however, is usually clear-cut. Over time there is a cause-and-effect relation; the *higher* the level of production, the *higher* the repair costs. To estimate the relation better, the analyst must recognize that repair costs may tend to lag behind periods of high production.

Similarly, pooling repair costs with other overhead costs understates the estimated variability of overhead costs with changes in production. At high production levels, low repair costs offset (other) high overhead costs; at low production levels, low overhead costs balance high repair costs. The net effect is that total overhead costs appear not to vary with changes in production levels. In fact, the relation is clear: the higher the level of production, the higher the total overhead costs. An indiscriminate application of regression analysis, however, might mask the true extent of variation in overhead costs.

EVALUATING AND CHOOSING AMONG COST FUNCTIONS

Objective 5

Describe four criteria for evaluating and choosing among cost functions

Four criteria are relevant in evaluating an estimated cost function before managers can draw conclusions or make inferences about the impact of cost driver(s) on total costs.

1. *Economic plausibility.* As we pointed out earlier, the basic relationship between the dependent variable and the independent variable(s) should make economic sense and be intuitive to both the operating manager and the accountant. For example, managers at General Motors' lamp division believe that complex lamps that perform multiple functions demand greater testing and inspection costs than simple single-function lamps.

2. *Goodness of fit.* Goodness of fit measures how well the predicted values, c, based on the cost driver, D, match actual cost observations, C. The regression analysis method computes a formal measure of goodness of fit, called the coefficient of determination. The **coefficient of determination,** r^2, measures the percentage of variation in C explained by D (the independent variable). When the predicted values exactly equal the actual cost values, the independent variable, D, has perfectly explained variations in actual costs, C, and $r^2 = 1$. The range of r^2 is from 0 (implying no explanatory ability) to 1 (implying perfect explanatory ability).[4] Generally, an r^2 of 0.30 or higher passes the goodness-of-fit test. The Appendix describes details of the computation of r^2. Relying exclusively on the goodness-of-fit criterion can be dangerous. It can lead to the indiscriminate inclusion of independent variables that increase r^2 but have no economic plausibility as cost driver(s).

Economic plausibility and goodness of fit serve as checks on one another. For example, data may show that a clerical overhead cost is more highly related to changes in the cost of electricity than to changes in the number of documents processed. But there may be no logical cause-and-effect relationship that supports such goodness of fit. In this situation, a manager should be reluctant to use electricity costs as the cost driver to predict how clerical overhead costs will behave. Managers have greater confidence that an observed statistical relationship will continue in the future if that relationship is economically plausible.

3. *Significance of independent variable(s).* A key question that managers ask is, Do changes in the independent variable(s) significantly affect total costs? Regression analysis provides a formal way to test for significance. The key statistic is the t-value of the slope coefficient(s), w, which measures whether changes in the independent variable result in significant changes in the dependent variable. If the t-value is low (less than 2.00), the manager can conclude that the independent variable is not a driver of cost. The manager may

---

[4] Computer programs frequently report an r^2 and an adjusted r^2. The adjusted r^2 is calculated as follows:

$$\text{Adjusted } r^2 = r^2 - \frac{(k-1)}{(n-k)} (1-r^2)$$

where k is the number of independent variables and n is the number of observations. The adjusted r^2 reduces the unadjusted r^2 as extra independent variables are added to the regression. The adjusted r^2 measure can in some cases be negative (when there is a low r^2 and many independent variables in the regression). Only r^2 values are reported in this chapter. For further details on adjusted r^2 see J. Johnston, Econometric Methods (New York: McGraw-Hill, 1984), pp. 177–178. Many econometric textbooks discuss, in much detail, issues that arise in regression analysis. See, for example, E. Berndt, The Practice of Econometrics (Reading, MA: Addison-Wesley, 1991).

Concepts in Action

COST ESTIMATION AT NEPTUNE PLASTICS MANUFACTURING[a]

Neptune Plastics Manufacturing Company of Canada makes plastic products using the plastic-injection molding process. Neptune was considering a bid to supply milk crates to a large dairy. In coming to a decision, Neptune wanted to know the variable costs of manufacturing and distributing the crates.

Neptune uses the account analysis method to identify some of the variable costs—direct materials, direct manufacturing labor, and shipping. But using account analysis to estimate variable overhead costs is much harder. Instead, Neptune applies regression analysis. In this case, it used hours of machine molding time as the independent variable and total overhead costs as the dependent variable. The results indicated no apparent relationship between the level of activity (molding-machine-hours) and overhead costs. Neptune concluded that overhead costs are entirely fixed within the relevant range with respect to short-term changes in the activity levels at the plant.

[a] Adapted from A. Atkinson, *Cost Estimation in Management Accounting-Six Case Studies* (Hamilton, Ontario: Society of Management Accountants of Canada, 1987).

then examine other economically plausible independent variables to see if they satisfy the t-test. The chapter Appendix explains the calculation of the t-value.

4. *Specification analysis of estimation assumptions.* **Specification analysis** refers to testing the assumptions underlying regression analysis. The chapter Appendix discusses these assumptions: linearity within the relevant range, constant variance of residuals, independence of residuals, and normality of residuals. If the statistical model satisfies these assumptions, users can be more confident in the estimates of cost behavior from the simplest regression procedures. Consider that a violation of these assumptions might result in mistakenly labeling an independent variable nonsignificant when, in fact, it is significant. Many software regression packages include systematic tests of whether these and other assumptions hold.

Cost drivers are those independent variables in a regression model that satisfy these four criteria. Regression analysis offers a structured approach, based on past data relationships, for identifying these cost drivers.

Managers and accountants must often choose among various cost functions that use different combinations of independent variables. For example, one cost function may use machine-hours as the independent variable, a second may use direct manufacturing labor-hours, and a third may use both as independent variables. How do managers select among these alternative cost functions? They apply the four criteria we have described. Economic plausibility and goodness of fit are especially important criteria in guiding managers' choice of cost functions.

Exhibit 10-7 presents a convenient format for summarizing the regression results. The Appendix to this chapter elaborates on the statistics presented in Exhibit 10-7.

EXHIBIT 10-7

Regression Results for Simple Regression with Indirect Manufacturing Labor Costs as Dependent Variable and Machine-Hours as Independent Variable for Elegant Rugs

Variable (1)	Coefficient (2)	Standard Error (3)	t-Value (4) = (2) ÷ (3)
Constant	$300.98	$170.54	1.76
Independent variable 1: machine-hours	10.31	3.12	3.30
$r^2 = 0.52$; Durbin-Watson statistic = 2.05			

DATA COLLECTION AND ADJUSTMENT ISSUES

The ideal data base to be used in estimating cost functions quantitatively has two characteristics:

1. It contains numerous reliably measured observations of the cost driver(s) and the dependent variable. Errors in measuring the cost driver(s) are particularly serious. They result in inaccurate estimates of the effect of the cost driver(s) on costs.

2. It includes considerable variation in the values of the cost driver(s). Variations in the values of the cost driver give greater confidence in the estimates obtained using regression techniques.

Cost analysts typically do not have the advantage of working with a data base having both characteristics. This section outlines some frequently encountered data problems.

Frequently Encountered Data Problems

Objective 6
Understand data problems encountered in estimating cost functions

In most cases, a cost analyst will come across one or more of the following seven problems.

1. The time period for measuring the dependent variable (for example, indirect manufacturing labor costs) does not properly match the period for measuring the cost driver(s). This problem often arises when accounting records are not kept on an accrual basis. Consider a cost function with machine-lubricant costs as the dependent variable and machine-hours as the cost driver. Assume that the lubricant is purchased sporadically and stored for later use. Records maintained on a cash basis will indicate no lubricant consumption in many months and sizable lubricant consumption in other months. This is an obviously inaccurate picture of what is actually taking place. Accrual accounting would better match costs with the cost driver in this example.

2. Fixed costs are allocated as if they are variable. For example, costs such as depreciation, insurance, or rent may be allocated on a per-unit-of-output basis. *The danger is to regard these costs as variable rather than as fixed. They seem to be variable because of the allocation methods used.* The analyst needs to distinguish carefully between fixed and variable costs.

3. Time periods differ for items included in the dependent variable and the cost driver(s). For example, labor costs could be accumulated on a monthly basis, whereas the quantity of output could be accumulated on a weekly basis.

4. Data are either not available for all observations or not uniformly reliable. Missing cost observations often arise from a failure to record a cost or from classifying a cost incorrectly. Data on cost drivers often originate outside the internal accounting system. For example, the accounting department may get data on testing times for medical instruments from the company's manufacturing department and data on the number of items shipped to customers from the distribution department. The reliability of such data varies greatly among organizations. In some systems, data are still recorded manually rather than entered electronically. Manually recorded data typically have a higher percentage of missing observations and erroneously entered observations than electronically entered data.

5. Extreme values of observations occur from errors in recording costs (for instance, a misplaced decimal point), from nonrepresentative time periods (for instance, from a period in which a major machine breakdown occurred or from a period in which delay in delivery of materials from an international supplier curtailed production), or from observations made outside the relevant range. Analysts should adjust or eliminate unusual observations before estimating a cost relationship; otherwise, an incorrect estimate may result.

6. There is no homogeneous relationship between the individual cost items in the dependent variable pool and the cost driver(s). A homogeneous relationship exists when each activity whose costs are included in the dependent variable has the same cause-and-effect relationship with the cost driver. Consider materials procurement overhead cost. This overhead cost account can include a diverse set of activities (for example, new ven-

dor negotiations, materials ordering, incoming inspection, and materials handling). A cost analyst has to decide whether to estimate a separate cost function for each activity or to combine two or more of the cost pools associated with these activities before estimating a cost function for materials procurement overhead.

7. Inflation has occurred in the dependent variable, a cost driver, or both. For example, inflation may cause costs to change even when there is no change in the cost driver. To study the underlying cause-and-effect relationship between the cost driver and costs, the analyst removes purely inflationary price effects from the data.

In many cases, a cost analyst must expend much effort to reduce the effect of these seven problems before estimating a cost function on the basis of past data.

HIERARCHY OF COSTS AND COST DRIVERS[5]

Objective 7
Describe the hierarchy of costs and cost drivers

Recall that indirect manufacturing labor costs at Elegant Rugs consist of machine maintenance costs and setup labor costs. Setup activity is performed to prepare and reconfigure the automatic weaving machines to produce a carpet pattern different from the previous carpet pattern. Once the machine is set up, Elegant Rugs chooses the length of the run—say, 1,000 square yards. Setup activity does not depend on either how much carpet is subsequently produced or how long the weaving machines are run after they are set up. Setup activity occurs every time a new batch of carpets is produced regardless of the size of each batch. If the management of Elegant Rugs is interested in understanding the behavior of setup costs, the appropriate cost driver to consider is the number of batches of carpets produced (or the setup time taken), not the quantity of carpet produced.

Contrast the analysis of setup costs with our previous description and analysis of machine maintenance costs at Elegant Rugs. Machine maintenance activity is influenced by the number of hours the machines are run and the quantity of carpet produced rather than the number of batches in which the carpet is produced. The notion of a hierarchy of cost drivers comes from the observation that different levels of cost drivers (unit versus batch) influence different costs. In the Elegant Rugs example, machine maintenance costs are affected by the individual quantity of carpets produced, while setup costs are related to the number of batches in which the carpets are produced.

We identify four hierarchical levels of costs:

1. output unit-level costs,
2. batch-level costs,
3. product-sustaining costs, and
4. facility-sustaining costs.

Output unit-level costs are resources sacrificed on activities performed on each unit of product or service. Direct materials costs of producing carpets are examples of *output unit-level costs*. If direct materials costs of producing 1 square yard of carpet is $8 and Elegant Rugs produces 10,000 square yards of carpet, then direct materials costs are expected to equal $80,000 ($8 × 10,000), whether the carpets are made in batch sizes of 1,000 square yards or 5,000 square yards.

Machine maintenance costs at Elegant Rugs is another example of unit output-level costs. Why? Because machine-hours and output quantities influence maintenance costs. Recognize one key point, however: Machine maintenance costs, unlike our example of direct materials costs, may not change proportionately with changes in machine-hours in the short-run. That is, machine maintenance cost may be a mixed cost with both fixed and variable components.

[5]Further discussion of cost hierarchies is in R. Cooper and R. S. Kaplan, *The Design of Cost Management Systems* (Englewood Cliffs, N J: Prentice-Hall, 1991).

Batch-level costs are resources sacrificed on activities that are related to a group of units of product(s) or service(s) rather than to each individual unit of product or service. Setup costs are an example of batch-level costs.

To see the distinction between batch-level costs and output unit-level costs, consider the following situation. Elegant Rugs produces 200,000 square yards of carpet in both January and February 19_5. Can we conclude, on the basis of this information, that setup costs are the same in each of the two months? The answer is no. Why? Because setup costs are batch-level costs that depend on the number of batches produced and the setup time taken (batch-level cost drivers). Suppose further that batch sizes in February were half the batch sizes in January and that setup costs varied proportionately with the number of batches. Even though the output quantities produced in the two months were the same, the setup costs in February would be double the setup costs in January, because there were twice as many setups in February.

Other examples of batch-level costs are purchase-order costs and materials-handling costs. These costs are affected by changes in batch-level cost drivers such as number of orders placed and number of batches handled rather than the total unit quantities ordered or handled.

Product-sustaining (or **service-sustaining**) **costs** are resources sacrificed on activities undertaken to support specific products or services. Product-sustaining costs cannot be traced to individual batches or units of products. For example, the costs incurred to design a new carpet are product-sustaining costs. These costs are not related to the number of carpets that are subsequently made or to the number of batches in which the carpets are made. Another example of a product-sustaining cost is the cost of preparing process routings. Process routings describe the steps in making a carpet—the manner in which the operations should be performed and the sequence in which the operations should be done.

What are the drivers of product-sustaining costs? Why might one product incur more product-sustaining costs than another? The most common reason is complexity—complexity of the product and complexity of the manufacturing process. The quantity of output produced and the number of batches started have no impact on product-sustaining costs.

Facility-sustaining costs are resources sacrificed on activities that cannot be traced to specific products or services but support the organization as a whole. Examples of facility-sustaining costs are general administration and corporate-image advertising.

Facility-sustaining costs bear no relation to output units, number of batches, or complexity of individual products. Identifying cost drivers for facility-sustaining costs is extremely difficult.

The hierarchy of costs and cost drivers is helpful to managers in understanding cost behavior. Consider, for example, a retail store manager interested in studying the behavior of purchase-order costs. These costs arise from the activity of processing individual purchase orders rather than the number of units of products entered on each purchase order. An economically plausible relation exists between purchase-order costs and the number of purchase orders placed. It is less plausible that purchase-order costs are related to the number of units of products purchased, an output unit-level cost driver.

We emphasize that purchase-order costs may not be strictly variable with respect to the number of purchase-orders placed. These costs have fixed *and* variable components. Various cost estimation methods described in this chapter can be used to determine the relationship between purchase-order costs and the number of purchase orders placed.

Exhibit 2-3 (p. 30) in Chapter 2 presented examples of variables that managers find to be important cost drivers in different business functions of the value chain. Some cost drivers are output unit-level cost drivers (number of units produced or number of total items distributed); others are batch-level drivers (number of machine setups or number of service calls); and still others are product-sustaining drivers (number of engineering change orders). Some costs may have both output unit-level and batch-level cost drivers. For example, the Appendix to this chapter illustrates the use of regression analysis to test whether both machine-hours and production batches are cost drivers of indirect manufacturing labor costs at Elegant Rugs.

NONLINEARITY AND COST FUNCTIONS

Objective 8
Explain and give examples of nonlinear cost functions

In practice, cost functions are not always linear. A **nonlinear cost function** is a cost function in which a single constant (g) and a single slope coefficient (w) do not adequately describe the behavior of the costs for all changes in the level of the cost driver. For example, economies of scale in advertising may enable an agency to double the number of advertisements for less than double the costs. Even direct materials costs are not always linear variable costs. Consider quantity discounts on direct materials purchases. As shown in Exhibit 10-8, the total cost rises, but it rises more slowly as the cost driver increases because of the quantity discounts. The cost function in Exhibit 10-8 has $w = \$25$ for 1 to 1,000 units purchased, $w = \$15$ for 1,001 to 2,000 units purchased, and $w = \$10$ for 2,000 or more units purchased ($g = \$0$ for all ranges of the units purchased). The cost per unit falls at each price break—that is, the cost per unit decreases with larger orders.

Step functions are also examples of nonlinear cost functions. The graph in Exhibit 10-9 shows a step-variable cost function. The cost of the input is constant over various small ranges of the cost driver, but the cost increases by discrete amounts (that is, in steps) as the cost driver moves from one relevant range to the next. This step-pattern behavior occurs when inputs such as setup labor, production scheduling, product design labor, process engineering labor, and machines and equipment are acquired in discrete quantities but used in fractional quantities.

The graph in Exhibit 10-10 shows a step-fixed cost function. The main difference relative to Exhibit 10-9 is that the cost of the input is constant over large ranges of the cost driver in each relevant range. Consider, for example a firm that operates two large heat-treatment furnaces to harden steel parts. The steps represent costs of operating each furnace ($300,000 per furnace). The relevant range indicates that the firm expects to operate with two furnaces at a cost of $600,000. Management considers the cost of operating furnaces as a fixed cost within the relevant range of operation.

EXHIBIT 10-8
Effects of Quantity Discounts on Slope of Total Cost Function

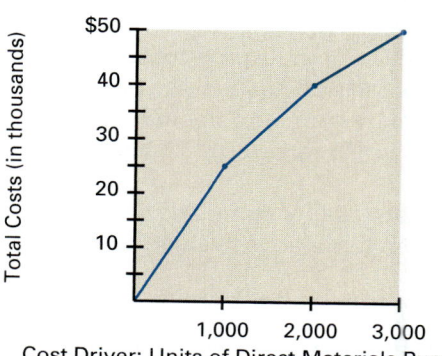

EXHIBIT 10-9
Step-Variable Cost Function

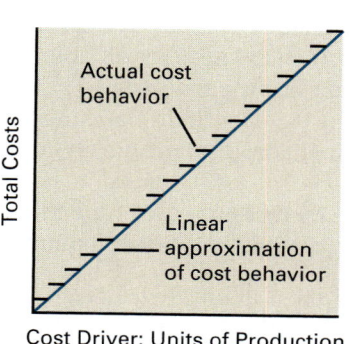

EXHIBIT 10-10
Step-Fixed Cost Function

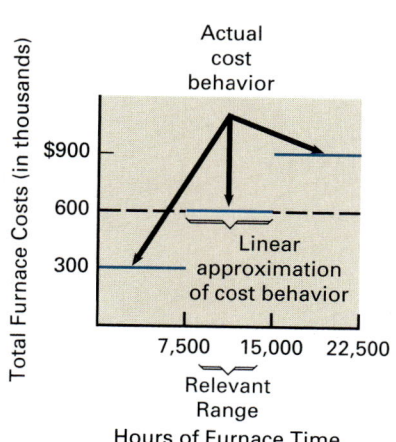

The aircraft-assembly industry first documented the effect that learning has on efficiency. As workers become more familiar with their tasks their efficiency improves. Managers learn how to improve the scheduling of work shifts. Plant operators learn how best to use the operating facility. Unit costs decrease as productivity increases, which means that the unit cost function behaves nonlinearly.

A **learning curve** is a function that shows how labor-hours per unit decline as units of output increase. Managers use learning curves to predict how labor-hours (or labor costs) will change as more units are produced.

Managers are now extending the learning-curve notion to include other cost areas in the value chain, such as marketing, distribution, and customer service. The term *experience curve* describes this broader application of the learning curve. An **experience curve** is a function that shows how full product costs per unit (including manufacturing, distribution, marketing, and so on) decline as units of output increase.

We now describe two learning-curve models: the cumulative average-time learning model and the incremental unit-time learning model.[6]

Cumulative Average-Time Learning Model

Objective 9
Distinguish between cumulative average-time learning model and incremental unit-time learning model

In the **cumulative average-time learning model**, the cumulative average time per unit declines by a constant percentage each time the cumulative quantity of units produced is doubled. Exhibit 10-11 illustrates the cumulative average-time learning model with an 80% learning curve. The 80% means that when the quantity of units produced is doubled from X to $2X$, the cumulative average time *per unit* for the $2X$ units is 80% of the cumulative average time *per unit* for the X units. The left graph in Exhibit 10-11 shows the average time *per unit* as a function of units produced. The right graph in Exhibit 10-11 shows the *total* number of labor-hours as a function of units produced. The observations underlying Exhibit 10-11, and the details of their calculation, are presented in Exhibit 10-12. To obtain the cumulative total time, multiply the cumulative average time per unit by the cumulative number of units produced. For example, to produce 4 cumulative units would require 256.00 hours (4×64).

Incremental Unit-Time Learning Model

In the **incremental unit-time learning model**, the incremental unit time (the time needed to produce the last unit) declines by a constant percentage each time the cumulative quantity of units produced is doubled. Exhibit 10-13 illustrates the incremental unit-time learning model with an 80% learning curve. The 80% here means that when the quantity of units produced is doubled from X to $2X$, the time needed to produce the *last unit* at the $2X$ production level is 80% of the time needed to produce the *last unit* at the X production level. The left graph in Exhibit 10-13 shows the average time *per unit* as a function of units produced. The right graph in Exhibit 10-13 shows the *total* number of labor-hours as a function of units produced. The observations underlying Exhibit 10-13, and the details of their calculation, are presented in Exhibit 10-14. We obtain the cumulative total time by summing the individual unit times. For example, to produce 4 cumulative units would require 314.21 hours ($100.00 + 80.00 + 70.21 + 64.00$).

The incremental unit-time model predicts that a higher cumulative total time is required to produce two or more units than does the cumulative average-time model,

---

[6]For further discussion, see J. Chen and R. Manes, "Distinguishing the Two Forms of the Constant Percentage Learning Curve Model," *Contemporary Accounting Research* (Spring 1985), pp. 242–252. See also the Northern Aerospace Manufacturing case study in A. A. Atkinson, *Cost Estimation in Management Accounting—Six Case Studies* (Hamilton, Ontario: *Society of Management Accountants of Canada*, 1987).

EXHIBIT 10-11

Plots for Cumulative Average-Time Learning Model

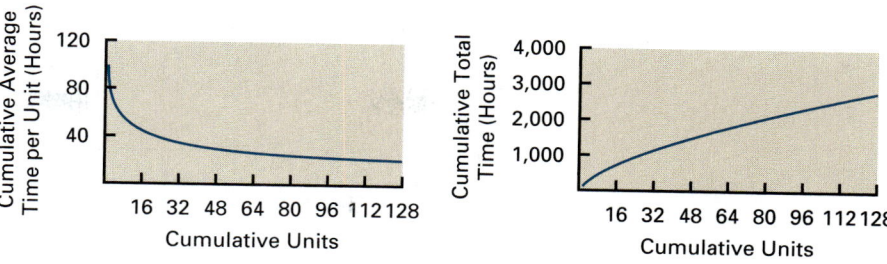

EXHIBIT 10-12

Cumulative Average-Time Learning Model

Cumulative Number of Units (1)	Cumulative Average Time per Unit (y): Hours (2)	Cumulative Total Time: Hours (3) = (1) × (2)	Individual Unit Time for Xth Unit: Hours (4)
1	100.00	100.00	100.00
2	80.00 (100 × 0.8)	160.00	60.00
3	70.21	210.63	50.63
4	64.00 (80 × 0.8)	256.00	45.37
5	59.57	297.85	41.85
6	56.17	337.02	39.17
7	53.45	374.15	37.13
8	51.20 (64 × 0.8)	409.60	35.45
⋮	⋮	⋮	⋮
16	40.96 (51.2 × 0.8)	655.36	28.06

Note: The mathematical relationship underlying the cumulative average-time learning model is

$$y = pX^q$$

where y = cumulative average time (hours) per unit
  X = cumulative number of units produced
  p = time (hours) required to produce the first unit
  q = the index of learning

The value of q is calculated as

$$q = \frac{\ln (\% \text{ learning})}{\ln 2}$$

For an 80% learning index

$$q = \frac{-0.2231}{0.6931} = -0.3219$$

As an illustration, when $X = 3$, $p = 100$, and $q = -0.3219$

$$y = 100 \times 3^{-0.3219} = 70.21 \text{ hours}$$

The cumulative total time when $X = 3$ is 70.21 × 3 = 210.63 hours.
    The individual unit times in column 4 are calculated using the data in column 3. For example, the individual unit time of 50.63 hours for the third unit is calculated as 210.63 minus 160.00.

assuming the same learning rate for the two models (compare results in Exhibit 10-12 with results in Exhibit 10-14). For example, to produce 4 cumulative units, the 80% incremental unit-time learning model predicts 314.21 hours versus 256.00 hours predicted by the 80% cumulative average-time learning model.

EXHIBIT 10-13

Plots for Incremental Unit-Time Learning Model

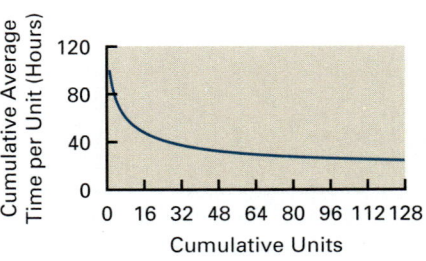 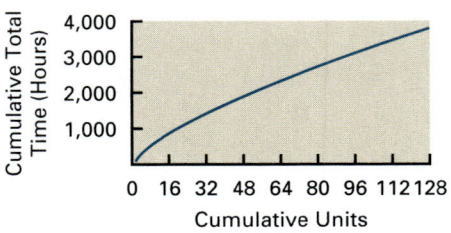

EXHIBIT 10-14

Incremental Unit-Time Learning Model

Cumulative Number of Units (1)	Individual Unit Time for xth Unit (m): Hours (2)	Cumulative Total Time: Hours (3)	Cumulative Average time per Unit: Hours (4) = (3) ÷ (1)
1	100.00	100.00	100.00
2	80.00 (100 × 0.8)	180.00	90.00
3	70.21	250.21	83.40
4	64.00 (80 × 0.8)	314.21	78.55
5	59.57	373.78	74.76
6	56.17	429.95	71.66
7	53.45	483.40	69.06
8	51.20 (64 × 0.8)	534.60	66.82
•	•	•	•
•	•	•	•
•	•	•	•
16	40.96 (51.2 × 0.8)	892.00	55.75

Note: The mathematical relationship underlying the incremental unit-time learning model is

$$m = pX^q$$

where m = time (hours) taken to produce the last single unit
X = cumulative number of units produced
p = time (hours) required to produce the first unit
q = the index of learning

The value of q is calculated as

$$q = \frac{\ln (\% \text{ learning})}{\ln 2}$$

For an 80% learning curve,

$$q = -0.3219$$

As an illustration, when $X = 3$, $p = 100$, and $q = -0.3219$

$$m = 100 \times 3^{-0.3219} = 70.21 \text{ hours}$$

The cumulative total time when $X = 3$ is $100 + 80 + 70.21 = 250.21$ hours.

Which of these two models is preferable? The one that more accurately approximates the behavior of labor-hour usage as output levels increase. The choice can be decided only on a case-by-case basis. Engineers, plant managers, and workers are good sources of information on the amount of learning actually occurring as output increases. Plotting this information is helpful in selecting the appropriate model.

EXHIBIT 10-15
Predicting Costs Using Learning Curves

Cumulative Number of Units	Cumulative Total Labor-Hours*	Cumulative Costs	Additions to Cumulative Costs
1	100.00	$ 5,000 (100.00 × $50)	$5,000
2	160.00	8,000 (160.00 × $50)	3,000
4	256.00	12,800 (256.00 × $50)	4,800
8	409.60	20,480 (409.60 × $50)	7,680
16	655.36	32,768 (655.36 × $50)	12,288

*Based on cumulative average-time learning model. See Exhibit 10-12 for computation of these amounts.

The Problem for Self-Study that follows this section illustrates the cumulative average-time learning model and the incremental unit-time learning model in a job-costing situation.

Setting Prices, Budgets, and Standards

Predictions of costs should allow for learning. Consider the data in Exhibit 10-12 for the cumulative average-time learning model. Suppose the variable costs subject to learning effects consist of direct labor ($20 per hour) and related overhead ($30 per hour). Management could predict the costs shown in Exhibit 10-15.

These data on the effects of the learning curve could have a major influence on decisions. For example, a company might set an extremely low selling price on its product in order to generate high demand. As the company's output increases to meet this growing demand, the cost per unit drops. The company "rides the product down the learning curve" as it establishes a higher market share. Although the company may not have earned much on its first sale—it may actually have lost money—the company gains more profit per unit as output increases.

Alternatively, subject to legal and other considerations, the company might set a low price on just the final eight units. After all, the labor and related overhead cost per unit is predicted to be only $12,288 for these final eight units ($32,768 – $20,480). The per-unit cost of $1,536 on these final eight units ($12,288 ÷ 8) is much lower than the $5,000 cost per unit of the first unit produced.

The learning-curve models examined in Exhibits 10-11 to 10-14 assume that learning is driven by a single variable (production output) and is product-related. Other models of learning focus on how quality (rather than labor-hours) will change over time (rather than as more units are produced). Some recent studies suggest that factors other than production output—such as job rotation and organizing workers into teams—contribute to learning that improves quality.

PROBLEM FOR SELF-STUDY

PROBLEM

The Helicopter Division of Aerospatiale is examining helicopter assembly costs at its plant in Marseilles, France. It has received an initial order for eight of its new land-surveying helicopters. Aerospatiale can adopt one of two methods of assembling the helicopters:

Labor-Intensive Assembly Method

1. Direct materials costs of $40,000 per helicopter.
2. Direct assembly labor time for the first helicopter will be 2,000 hours. Assembly labor time per helicopter follows an 85% cumulative average-time learning curve model.

   (An 85% learning curve is expressed mathematically as $q = -0.2345$.) Direct assembly labor costs $30 per hour.
3. Indirect manufacturing costs are predicted using two separate cost functions, each with its own cost driver.
   a. Equipment-related indirect costs at the rate of $12 per direct assembly labor-hour.
   b. Materials-handling related indirect costs at the rate of 50% of direct materials costs.

Machine-Intensive Assembly Method

1. Direct materials costs of $36,000 per helicopter.
2. Direct assembly labor time for the first helicopter will be 800 hours. Assembly labor time per helicopter follows a 90% incremental unit-time learning curve model. (A 90% learning curve is expressed mathematically as $q = -0.1520$.) Direct assembly labor costs $30 per hour.
3. Indirect manufacturing costs are predicted using two separate cost pools, each with its own cost driver:
   a. Equipment-related indirect costs at the rate of $45 per direct assembly labor-hour.
   b. Materials-handling related indirect costs at the rate of 50% of direct materials costs.

The capital-intensive equipment used in this alternative is operated by workers who control the speed of production and hence the utilization of the equipment.

Required

1. What is the number of direct assembly labor-hours required to assemble the first eight helicopters under
   a. the labor-intensive method?
   b. the machine-intensive method?
2. What is the cost of assembling the first eight helicopters under
   a. the labor-intensive method?
   b. the machine-intensive method?

SOLUTION

1. a. Cumulative average-time learning model (85% learning)

Cumulative Number of Units (1)	Cumulative Average Time per Unit (y): Hours (2)	Cumulative Total Time: Hours (3) = (1) × (2)	Individual Unit Time for Xth Unit: Hours (4)
1	2,000	2,000	2,000
2	1,700 (2,000 × 0.85)	3,400	1,400
3	1,546	4,638	1,238
4	1,445 (1,700 × 0.85)	5,780	1,142
5	1,371	6,855	1,075
6	1,314	7,884	1,029
7	1,267	8,869	985
8	1,228.25 (1,445 × 0.85)	9,826	957

The cumulative average time per unit for the Xth unit in column 2 is calculated as $y = pX^q$; see Exhibit 10-12, p. 359. For example, when $X = 3$, $y = 2,000 \times 3^{-0.2345} = 1,546$ hours.

   b. Incremental unit-time learning model (90% learning)

Cumulative Number of Units (1)	Individual Unit Time for Xth Unit (m): Hours (2)	Cumulative Total Time: Hours (3)	Cumulative Average Time per Unit: Hours (4) = (3) ÷ (1)
1	800	800	800
2	720 (800 × 0.9)	1,520	760
3	677	2,197	732
4	648 (720 × 0.9)	2,845	711
5	626	3,471	694
6	609	4,080	680
7	595	4,675	668
8	583 (648 × 0.9)	5,258	657

The individual unit time for the Xth unit in column 2 is calculated as $m = pX^q$; see Exhibit 10-14, p. 360. For example, when $X = 3$, $m = 800 \times 3^{-0.1520} = 677$ hours.

2. Costs of assembling the first eight helicopters are:
   a. With cumulative average-time learning model (85% learning):

Direct materials, 8 × $40,000	$320,000
Direct assembly labor, 9,826 × $30	294,780
Indirect manufacturing costs:	
Equipment-related, 9,826 × $12	117,912
Materials-related, 0.50 × $320,000	160,000
	$892,692

   b. With incremental unit-time learning model (90% learning):

Direct materials, 8 × $36,000	$288,000
Direct assembly labor, 5,258 × $30	157,740
Indirect manufacturing costs:	
Equipment-related, 5,258 × $45	236,610
Materials-related, 0.50 × $288,000	144,000
	$826,350

The machine-intensive method has an assembly cost $66,342 lower than the labor-intensive method ($892,692 − $826,350).

SUMMARY

The following points are linked to the chapter's learning objectives.

1. Two assumptions frequently made in cost-behavior estimation are (a) a single cost driver can explain variations in total costs and (b) a linear function adequately approximates cost behavior within the relevant range.

2. A linear cost function is a function in which a single constant (g) and a single slope coefficient (w) describe the behavior of total costs for all changes in the level of total the cost driver within the relevant range. The constant (g) represents the estimate of the total cost component that does not vary with changes in the level of the cost driver. The slope coefficient (w) represents the estimate of the amount by which total costs do change for each unit change in the level of the cost driver. Three types of linear cost functions are variable, fixed, and mixed or semivariable cost functions.

3. Many decisions by managers use predictions of future costs. A better understanding of how past costs behave helps managers make more accurate predictions of future costs. Cost estimation attempts to measure past cost relationships. Four broad approaches to estimating cost functions are the industrial-engineering method, the conference method, the account analysis method, and quantitative analysis of cost relationships (the high-low method and regression analysis). Regression analysis is a systematic approach to estimating a cost function on the basis of identified cost drivers. Regression analysis can incorporate single and multiple cost driver variables. Ideally, the cost analyst applies more than one approach; each approach serves as a check on the others.

4. The six steps in estimating a cost function on the basis of an analysis of current or past cost relationships are: (a) choose the dependent variable, (b) identify the cost driver(s), (c) collect data on the dependent variable and the cost driver(s), (d) plot the data, (e) estimate the cost function, and (f) evaluate the estimated cost function. In most applications, the cost analyst will cycle through these steps several times before identifying an acceptable cost function.

5. Four criteria for evaluating and choosing among cost functions are: (a) economic plausibility, (b) goodness of fit, (c) significance of independent variable(s), and (d) specification analysis of estimation assumptions.

6. The most difficult task in cost estimation is collecting high-quality, reliably measured data on the dependent variable and the cost driver(s). Common problems include missing data, extreme values of observations, and distortions resulting from inflation.

7. Four hierarchical levels of costs are (a) output unit-level costs, (b) batch-level costs, (c) product-sustaining costs, and (d) facility-sustaining costs. Output unit-level costs are resources sacrificed on activities performed on each unit of product or service. Batch-level costs are resources sacrificed on activities that relate to a group of units of product(s) or service(s) rather than to each individual unit of product or service. Product-sustaining (or service-sustaining) costs are resources sacrificed on activities undertaken to support specific products or services. Facility-sustaining costs are resources sacrificed on activities that cannot be traced to specific products or services but support the business as a whole.

8. A nonlinear cost function is a cost function in which a single constant, (g), and a single slope coefficient, (w), do not adequately describe the behavior of total costs for all changes in the level of the cost drivers. Nonlinear costs can arise due to economies of scale, quantity discounts, step-cost functions, and learning-curve effects.

9. The learning curve is an example of a nonlinear cost function. Labor-hours per unit decline as units of output increase. In the cumulative average-time learning model, the cumulative average-time per unit declines by a constant percentage each time the cumulative quantity of units produced doubles. In the incremental unit-time learning model, the incremental unit time (the time needed to produce the last unit) declines by a constant percentage each time the cumulative quantity of units produced doubles.

APPENDIX: REGRESSION ANALYSIS

This Appendix describes formulas for estimating the regression equation and several commonly used statistics. We use the data for Elegant Rugs presented in Exhibit 10-3. The Appendix also discusses goodness of fit, significance of independent variables, and specification analysis when using regression analysis.

Estimating the Regression Line

The least-squares technique for estimating the regression line minimizes the sum of the squares of the vertical deviations (distances) from the data points to the estimated regression line.

The object is to find the values of g and w in the predicting equation $c = g + wD$, where c is the predicted value as distinguished from the observed value, C. We wish to find the numerical values of g and w that minimize $\Sigma(C - c)^2$. This calculation is accomplished by using two equations, usually called the normal equations:

$$\Sigma C = ng + w(\Sigma D)$$
$$\Sigma DC = g(\Sigma D) + w(\Sigma D^2)$$

where n is the number of data points; ΣD and ΣC are, respectively, the sums of the given D values and C values; ΣD^2 is the sum of squares of the D values; and ΣDC is the sum of the amounts obtained by multiplying each of the given D values by the associated observed C value.

Exhibit 10-16 shows the calculations required for obtaining the line that best fits the data for Elegant Rugs. Substituting into the two linear equations simultaneously we obtain

$$12{,}501 = 12g + 862w$$
$$928{,}716 = 862\,g + 64{,}900w$$

The solution is $g = \$300.98$ and $w = \$10.31$.

Placing the amounts for g and w in the equation of the least-squares line, we have

$$c = \$300.98 + \$10.31D$$

where c is the predicted indirect manufacturing labor costs for any specified number of machine-hours within the relevant range. Generally, these computations are done using computer software packages.

Goodness of Fit

The coefficient of determination (r^2) indicates the proportion of the variance of C, $\Sigma(C - \bar{C})^2 \div n$, that is explained by the independent variable D (where $\bar{C} = (\Sigma C) \div n$, and n = the number of data points). It is more convenient to express the coefficient of de-

EXHIBIT 10-16

Computation for Least-Squares Regression between Indirect Manufacturing Labor Costs and Machine-Hours for Elegant Rugs

Week (1)	Machine-Hours* D (2)	Indirect Manufacturing Labor Costs* C (3)	D^2 (4)	DC (5)	c (6)	Total Variance of C $(C - \bar{C})^2$ (7)	Unexplained Variance $(C - c)^2$ (8)	Total Variance of D $(D - \bar{D})^2$ (9)
1	68	1,190	4,624	80,920	1,002.06	21,978	35,321	15
2	88	1,211	7,744	106,568	1,208.26	28,646	8	261
3	62	1,004	3,844	62,248	940.20	1,425	4,070	97
4	72	917	5,184	66,024	1,043.30	15,563	15,952	0
5	60	770	3,600	46,200	919.58	73,848	22,374	140
6	96	1,456	9,216	139,776	1,290.74	171,603	27,311	584
7	78	1,180	6,084	92,040	1,105.16	19,113	5,601	38
8	46	710	2,116	32,660	775.24	110,058	4,256	667
9	82	1,316	6,724	107,912	1,146.40	75,213	28,764	103
10	94	1,032	8,836	97,008	1,270.12	95	56,701	491
11	68	752	4,624	51,136	1,002.06	83,955	62,530	15
12	48	963	2,304	46,224	795.86	6,202	27,936	568
Total	862	12,501	64,900	928,716	≈12,501	607,699	290,824	2,980

*Same data as in Exhibit 10-3.

termination as 1 minus the proportion of total variance that is *not* explained. The unexplained variance arises because of differences between the actual values C and the predicted value, c.

$$r^2 = 1 - \frac{\text{Unexplained variation}}{\text{Total variation}} = 1 - \frac{\Sigma(C - c)^2}{\Sigma(C - \bar{C})^2}$$

From Exhibit 10-16, $\Sigma C = 12,501$ and $\bar{C} = 12,501 \div 12 = 1,041.75$
Therefore,

$$\Sigma(C - \bar{C})^2 = (1,190 - 1,041.75)^2 + (1,211 - 1,041.75)^2 + \ldots (963 - 1,041.75)^2 = 607,699$$

Each value of D generates a prediction, c. For example in week 1, $c = \$300.98 + \$10.31(68) = \$1002.06$.

$$\Sigma(C - c)^2 = (1,190 - 1,002.06)^2 + (1,211 - 1,208.26)^2 \ldots (963 - 795.86)^2$$
$$= 290,824$$
$$r^2 = 1 - \frac{290,824}{607,699} = 0.52$$

The calculations indicate that r^2 increases as the predicted values, c, more closely approximate the actual observations, C.

Significance of Independent Variables

The estimates, g and w, of the true underlying population parameters, G and W, are sample-specific. They are subject to random factors, as are all sample statistics. The **standard error of the estimated coefficient** indicates how much the estimated value, w, is likely to be affected by random factors. The t-value of the w coefficient measures how large the value of the estimated coefficient is relative to its standard error. A t-value with an absolute value greater than 2.00 suggests that the w coefficient is significantly different from zero.[7] In other words, a relationship exists between the independent variable and the dependent variable that cannot be attributed to chance alone.

For the regression reported for Elegant Rugs in Exhibit 10-7, the t-value for the slope coefficient (w) is $\$10.31 \div \$3.12 = 3.30$, which exceeds the benchmark of 2.00. The coefficient of the machine-hours variable is significantly different from zero. The probability is low that random factors could have caused the coefficient (w) to be positive. Therefore, we can conclude that changes in machine-hours affect indirect manufacturing labor costs. Similarly, the t-value for the constant term (g) is $\$300.98 \div \$170.54 = 1.76$, which is less than 2.00. This value indicates that, within the relevant range, the constant term is not significantly different from zero.

Specification Analysis of Estimation Assumptions

Specification analysis is the testing of the assumptions of regression analysis. If the assumptions of (1) linearity within the relevant range, (2) constant variance of residu-

---

[7]The benchmark for inferring that a w coefficient is significantly different from zero is a function of the degrees of freedom in a regression. The benchmark of 2.00 assumes a sample size of 60 observations. The number of degrees of freedom is calculated as the sample size minus the number of g and w parameters estimated in the regression. For a simple regression, the benchmark values for the t-values are:

Sample Size	Benchmark*
10	$\lvert t \rvert > 2.31$
15	$\lvert t \rvert > 2.16$
20	$\lvert t \rvert > 2.10$
30	$\lvert t \rvert > 2.05$
60	$\lvert t \rvert > 2.00$

*$\lvert t \rvert$ denotes the absolute value of the t-statistic.

als, (3) independence of residuals, and (4) normality of residuals hold, the simplest regression procedures yield reliable estimates of unknown coefficient values. This section provides a brief overview of specification analysis. When these assumptions are not satisfied, more complex regression procedures are necessary to obtain the best estimates. For details of these procedures, see any standard econometric text.

1. LINEARITY WITHIN THE RELEVANT RANGE. A common assumption is that a linear relationship exists between the dependent variable C and the independent variable D within the relevant range. If a linear model is used to estimate a fundamentally nonlinear relationship, however, the coefficient estimates obtained will be inaccurate (often referred to as biased).

Where there is one independent variable, the easiest way to check for linearity is by studying the data on a scatter diagram, a step that often is unwisely skipped. Exhibit 10-6 (p. 351) presents a scatter diagram for the indirect manufacturing labor costs and machine-hours variables of Elegant Rugs in Exhibit 10-3. Linearity appears to be a reasonable assumption for these data.

The learning-curve models discussed in the chapter (pp. 358–61) are examples of nonlinear cost functions; costs increase when the level of production increases, but by lesser amounts than would occur with a linear cost function. In this case, the analyst should estimate a nonlinear cost function that explicitly incorporates learning effects.

2. CONSTANT VARIANCE OF RESIDUALS. The vertical deviation of the *observed* value, C, from the regression line estimate, c, is called the *residual term*, **disturbance term**, or **error term**, $u = C - c$. The assumption of constant variance implies that the residual terms are unaffected by the level of the independent variable. The assumption also implies that there is a uniform scatter, or dispersion, of the data points about the regression line. The scatter diagram is the easiest way to check for *constant variance*. This assumption holds for the graph in Panel A of Exhibit 10-17 but not for the graph in Panel B. Constant variance is also known as homoscedasticity. Violation of this assumption is called heteroscedasticity.

Violation of the assumption of constant variance does not affect the accuracy of the regression estimates, g and w. It does, however, reduce the reliability of the estimates of the standard errors, and thus affects the precision with which inferences can be drawn.

EXHIBIT 10-17
Constant Variance of Residuals Assumption

PANEL A: Example of Constant Variance (Uniform Scatter of Data Points around Regression Line)

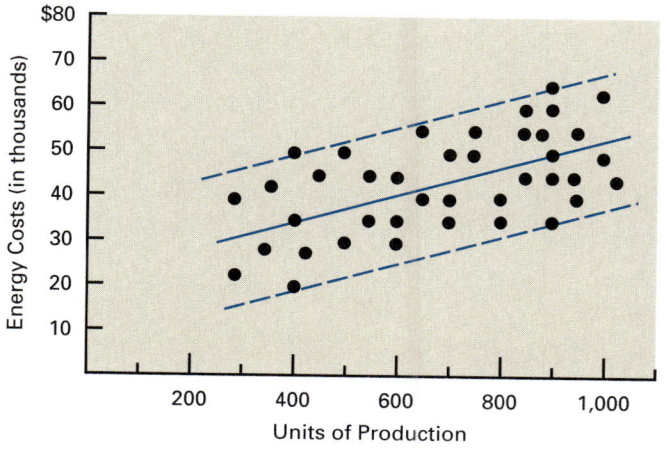

PANEL B: Example of Nonconstant Variance (Higher Outputs Have Larger Residual)

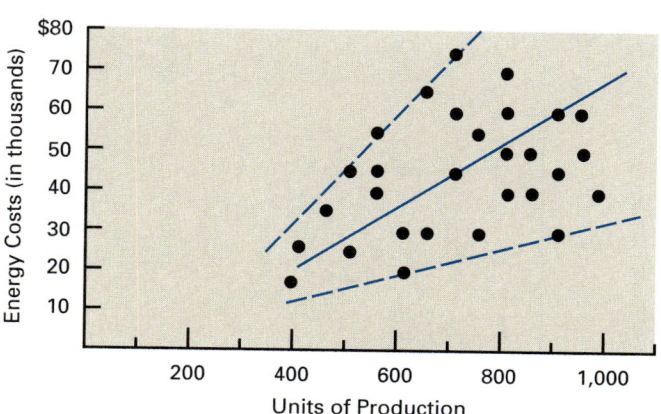

3. INDEPENDENCE OF RESIDUALS. The assumption of the independence of residuals is that the residual term for any one observation is not related to the residual term for any other observation. The problem of **serial correlation** in the residuals (also called **autocorrelation**) arises when the residuals are not independent. Serial correlation means that there is a systematic pattern in the sequence of residuals such that the residual in period t conveys information about the residuals in periods $t + 1, t + 2$, and so on. The scatter diagram helps in identifying autocorrelation. Autocorrelation does not exist in the graph in Panel A of Exhibit 10-18 but exists in the graph in Panel B. Observe the systematic pattern of the residuals in the Panel B graph—negative residuals for low quantities of direct materials costs and positive residuals as direct materials costs increase over time. No such systematic pattern prevails for the Panel A graph.

Like nonconstant variance in the residuals, serial correlation does not affect the accuracy of the regression estimates g and w. But it affects the standard errors of the coefficients, which in turn affect the precision with which inferences about the population parameters can be drawn from the regression estimates.

The Durbin-Watson statistic is one measure of serial correlation in the estimated residuals. For samples of 10 to 20 observations, a Durbin-Watson statistic in the 1.30 to 2.70 range suggests that the residuals are independent. The Durbin-Watson statistic for the regression results of Elegant Rugs in Exhibit 10-7 is 2.05. Therefore, an assumption of independence in the estimated residuals seems reasonable for this regression model.

4. NORMALITY OF RESIDUALS. The normality of residuals assumption means that the residuals, u, are distributed normally around the regression line. This assumption is necessary for making inferences about c, g, and w.

A major limitation of regression analysis is the assumption that the relationship in the equation will continue in the future—that is, that there is an ongoing, stable relationship between the dependent variable and the independent variable(s). For example, for many years manufacturing operations at Elegant Rugs were heavily labor-intensive with minimal use of automated machinery. In the last 2 years Elegant Rugs has invested in highly automated weaving machines. The cost function estimated with data from the labor-intensive manufacturing period would be different from that reported in Exhibit 10-7, which is based on data from the automated-machinery period.

EXHIBIT 10-18
Independence of Residuals

PANEL A: Example of Independence of Residuals (No Pattern in Residuals)

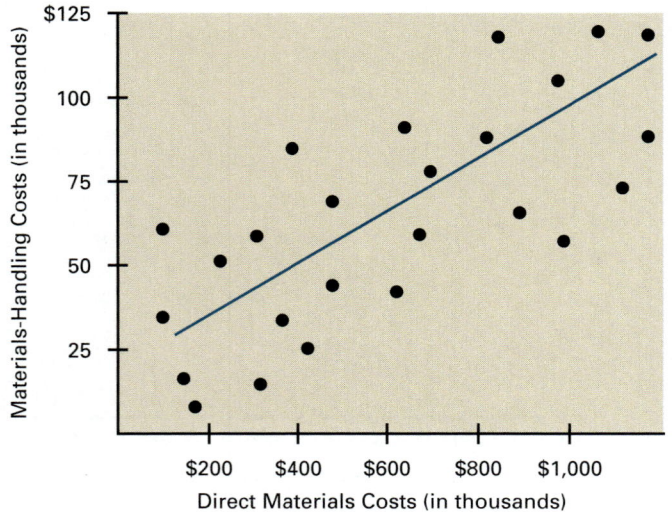

PANEL B: Example of Serial Correlation in Residuals (Negative Residuals for Low DirectMaterials Costs; Positive Residuals for High Direct Materials Costs)

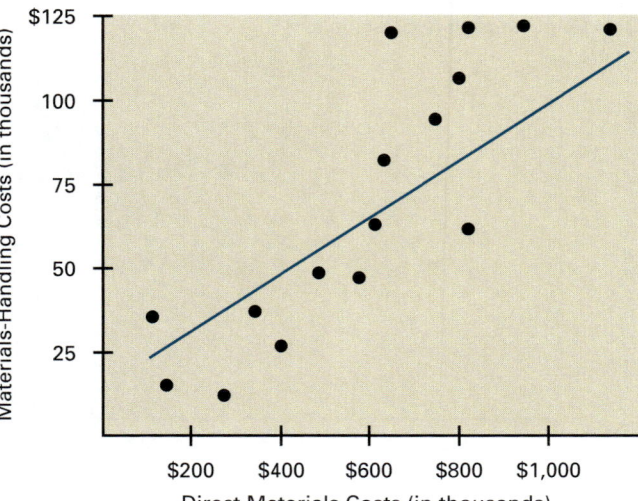

AN EXAMPLE OF CHOOSING AMONG COST FUNCTIONS

Exhibit 10-19 presents additional data for Elegant Rugs. We use Exhibit 10-19 to illustrate the previously outlined criteria for choosing among alternative cost functions. Consider two cost functions:

$$c = g + w \text{ (machine-hours)}$$
$$c = g + w \text{ (direct manufacturing labor-hours)}$$

Exhibits 10-6 and 10-7 present a plot of the data and regression results, respectively, for the cost function using machine-hours as the independent variable. Exhibits 10-20 and 10-21 present comparable information for the cost function using direct manufacturing labor-hours as the independent variable.

EXHIBIT 10-19

Weekly Indirect Manufacturing Labor Costs, Machine-Hours, Direct Manufacturing Labor-Hours, and Number of Production Batches for Elegant Rugs

Week (1)	Indirect Manufacturing Labor Costs (2)	Machine-Hours (3)	Direct Manufacturing Labor-Hours (4)	Number of Production Batches (5)
1	1,190	68	30	12
2	1,211	88	35	15
3	1,004	62	36	13
4	917	72	20	11
5	770	60	47	10
6	1,456	96	45	17
7	1,180	78	44	17
8	710	46	38	7
9	1,316	82	70	14
10	1,032	94	30	12
11	752	68	29	7
12	963	48	38	14

EXHIBIT 10-20

Plot of Weekly Indirect Manufacturing Labor Costs and Direct Manufacturing Labor-Hours for Elegant Rugs

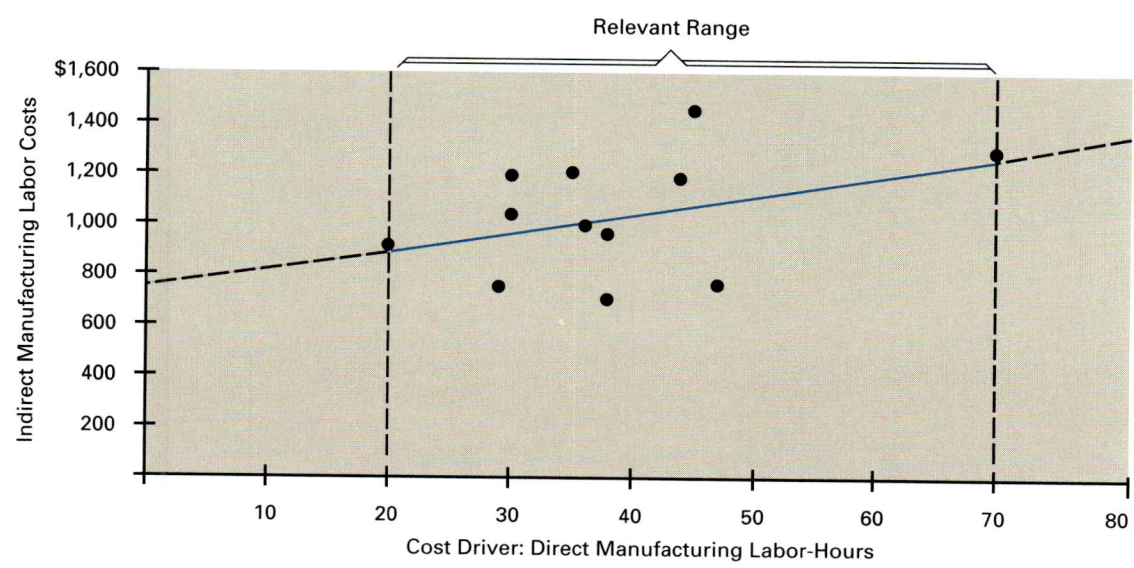

EXHIBIT 10-21

Regression Results for Simple Regression with Direct Manufacturing Labor-Hours
as Independent Variable for Elegant Rugs

Variable (1)	Coefficient (2)	Standard Error (3)	t-Value (4) = (2) ÷ (3)
Constant	$744.67	$224.61	3.32
Independent variable: Direct manufacturing labor-hours	7.72	5.40	1.43

$r^2 = 0.17$; Durbin-Watson statistic = 2.26

On the basis of the material in this Appendix, which regression is better? Exhibit 10-22 compares these two cost functions. For several criteria, the cost function based on machine-hours is preferable to the cost function based on direct manufacturing labor-hours. The economic plausibility criterion is especially important. Machine repair, maintenance, and service costs are important components of indirect manufacturing labor costs. These costs are incurred to support the smooth functioning of the automatic weaving machines. Operating personnel at Elegant Rugs would probably have greater confidence in machine-hours than in direct manufacturing labor-hours as the driver of these indirect manufacturing labor costs. The information in Exhibit

EXHIBIT 10-22

Comparison of Alternative Cost Functions for Indirect Manufacturing Labor Costs
Estimated with Simple Regression for Elegant Rugs

Criterion	Cost Function 1: Machine-Hours as Independent Variable	Cost Function 2: Direct Manufacturing Labor-Hours as Independent Variable
1. Economic plausibility	Positive relationship between indirect manufacturing labor costs (technical support labor) and machine-hours is economically plausible in a highly automated plant.	Positive relationship between indirect manufacturing labor costs and direct manufacturing labor-hours is economically plausible, but less so than machine-hours on a week-to-week basis.
2. Goodness of fit	$r^2 = 0.52$ Excellent goodness of fit.	$r^2 = 0.17$ Poor goodness of fit.
3. Significance of independent variables	t-value on machine-hours of 3.30 is significant.	t-value on direct manufacturing labor-hours of 1.43 is not significant.
4. Specification analysis	Plot of the data indicates that assumptions of linearity, constant variance, and independence of residuals hold, but inferences drawn from only 12 observations are not reliable; Durbin-Watson statistic = 2.05.	Plot of the data indicates that assumptions of linearity, constant variance, and independence of residuals hold, but inferences drawn from only 12 observations are not reliable; Durbin-Watson statistic = 2.26.

10-22 identifies machine-hours as a significant cost driver of monthly indirect manufacturing labor costs.

Do not always assume that any one cost function will perfectly satisfy all the criteria in Exhibit 10-22. A cost analyst must often make a choice among "imperfect" cost functions, in the sense that the data of any particular cost function will not perfectly meet one or more of the assumptions underlying regression analysis.

Multiple Regression

In some cases, satisfactory predictions of a cost may be based on only one independent variable, such as machine-hours. In many cases, however, basing the prediction on more than one independent variable improves accuracy. The most widely used equations to express relationships between a dependent variable and two or more independent variables are linear in the form

$$C = g + w_1 D_1 + w_2 D_2 + \cdots + u$$

where

$$
\begin{aligned}
C &= \text{the cost variable to be predicted} \\
D_1, D_2 &= \text{the independent variables on which the prediction is to be based} \\
g, w_1, w_2 &= \text{the estimated coefficients of the regression model} \\
u &= \text{the residual term that includes the net effect of other factors not} \\
&\quad \text{in the model and measurement errors in the dependent and} \\
&\quad \text{independent variables}
\end{aligned}
$$

EXAMPLE: Consider again the Elegant Rugs data in Exhibit 10-19 (p. 369), which include weekly observations on three potential independent variables in a cost function—machine-hours, direct manufacturing labor-hours, and number of production batches. The production-batch variable measures the number of separate carpet jobs worked on during the week; the same line of carpet worked on multiple times (say, on different days) is counted as multiple batches. Operating personnel at Elegant Rugs report sizable changeover costs when production on one carpet batch is stopped and production on another batch is started. For example, materials on the weaving looms must be changed. Indirect manufacturing labor costs include the cost of setup and changeover labor. Management believes there is a positive relationship between the indirect manufacturing labor costs and the number of production batches.

Exhibit 10-23 presents results for the following multiple regression model, using data in columns 2, 3, and 5 of Exhibit 10-19:

$$c = \$95.13 + \$4.99D_1 + \$47.38D_2$$

where D_1 is the number of machine-hours and D_2 is the number of production batches. It is economically plausible that both machine-hours and production batches would help explain variations in indirect manufacturing labor costs at Elegant Rugs. The r^2 of 0.52 for

EXHIBIT 10-23

Regression Results for Multiple Regression with Two Independent Variables (Machine-Hours and Production Batches) for Elegant Rugs

Variable (1)	Coefficient (2)	Standard Error (3)	t-Value (4) = (2) ÷ (3)
Constant	$95.13	$107.42	0.89
Independent variable 1: machine-hours	4.99	2.37	2.11
Independent variable 2: production batches	47.38	11.77	4.03

$r^2 = 0.83$; Durbin-Watson statistic = 2.15

the simple regression using machine-hours (Exhibit 10-7) increases to 0.83 with the multiple regression in Exhibit 10-23. The t-values suggest that the independent variable coefficients of both machine-hours and production batches are significantly different from zero ($t = 2.10$ for the coefficient on machine-hours, and $t = 4.03$ for the coefficient on production batches). The multiple regression model in Exhibit 10-23 satisfies both economic and statistical criteria, and it explains much greater variation in indirect manufacturing labor costs than does the simple regression model using machine-hours as the independent variable. The information in Exhibit 10-23 indicates that both machine-hours and production batches are important cost drivers of monthly indirect manufacturing labor costs at Elegant Rugs.

In Exhibit 10-23, the slope coefficients—$4.99 for machine-hours and $47.38 for production batches—measure the change in indirect manufacturing labor costs associated with a unit change in an independent variable (assuming that the other independent variable is held constant). For example, indirect manufacturing labor costs increase by $47.38 when one more production batch is added, assuming that the number of machine-hours is held constant.

An alternative approach would create two separate cost pools—one for costs tied to machine-hours and another for costs tied to production batches. Elegant Rugs would then estimate the relationship between the cost driver and overhead costs separately for each cost pool. The difficult task, under that approach, would be dividing overhead costs into the two cost pools.

Multicollinearity

A major concern that arises with multiple regression is multicollinearity. **Multicollinearity** exists when two or more independent variables are highly correlated with each other. Generally, users of regression analysis believe that a coefficient of correlation—that is, r—greater than 0.70 indicates multicollinearity. Multicollinearity increases the standard errors of the coefficients of the individual variables. The result is that there is greater uncertainty about the underlying value of the coefficients of the individual independent variables.

The coefficients of correlation between the potential independent variables for Elegant Rugs in Exhibit 10-19 are:

Pairwise Combinations	Coefficient of Correlation
Machine-hours and direct manufacturing labor-hours	0.11
Machine-hours and production batches	0.56
Direct manufacturing labor-hours and production batches	0.38

These results suggest that multiple regressions using any pair of the independent variables in Exhibit 10-23 are not likely to encounter multicollinearity problems.

TERMS TO LEARN

This chapter and the Glossary at the end of the book contain definitions of the following important terms:

account analysis method *(p. 345)*   autocorrelation *(368)*   batch-level costs *(356)*
coefficient of determination (r^2) *(352)*   conference method *(344)*   constant *(341)*
correlation *(348)*   cost estimation *(340)*   cost predictions *(340)*
cumulative average-time learning model *(358)*   dependent variable *(347)*
disturbance term *(367)*   error term *(367)*   experience curve *(358)*
facility-sustaining costs *(356)*   high-low method *(348)*
incremental unit-time learning model *(358)*   industrial-engineering method *(344)*
intercept *(341)*   learning curve *(358)*   linear cost function *(341)*
mixed cost *(342)*   multicollinearity *(372)*   multiple regression *(350)*
nonlinear cost function *(357)*   output unit-level costs *(355)*   parameter *(341)*
product-sustaining costs *(356)*   regression analysis *(350)*   residual term *(351)*
semivariable cost *(342)*   serial correlation *(368)*   service-sustaining costs *(356)*
simple regression *(350)*   slope coefficient *(341)*   specification analysis *(353)*
standard error of the estimated coefficient *(366)*   work-measurement method *(344)*

ASSIGNMENT MATERIAL

QUESTIONS

10-1   What two assumptions are frequently made when estimating a cost function?

10-2   What is a cost driver? Give two examples of cost drivers.

10-3   What is the difference between a *linear* and a *nonlinear* cost function? Give an example of each type of cost function.

10-4   "High correlation between two variables means that one is the cause and the other is the effect." Do you agree? Explain.

10-5   Name the four approaches to estimating a cost function.

10-6   Describe the conference-method approach to estimating a cost function. What are two advantages of this method?

10-7   List the six steps in estimating a cost function on the basis of an analysis of current or past cost relationships. Which step is typically the most difficult for a cost analyst?

10-8   When using the high-low method, should you base the high and low observations on the dependent variable or on the cost driver?

10-9   Describe four criteria for evaluating and choosing among cost functions.

10-10   Discuss four frequently encountered problems when collecting cost data on variables included in a cost function.

10-11   Describe four hierarchical levels of costs.

10-12   Define *learning curve*. Outline two models that can be used when incorporating learning into the estimation of cost functions.

10-13   What are the four key assumptions examined in specification analysis in the case of simple regression?

10-14   "All the independent variables in a cost function estimated with regression analysis are cost drivers." Do you agree? Explain.

10-15   "Multicollinearity exists when the dependent variable and the independent variable are highly correlated." Do you agree? Explain.

EXERCISES AND PROBLEMS

10-16 Estimating a cost function. The controller of the Ijiri Co. wants you to estimate a cost function from the following two observations in a general-ledger account called Maintenance:

Monthly Machine-Hours	Monthly Maintenance Costs Incurred
4,000	$3,000
7,000	3,900

Required
1. Estimate the cost function for maintenance.
2. Can the constant in the cost function be used as an estimate of fixed maintenance cost per month? Explain.

10-17 Identifying variable, fixed, and mixed cost functions. Pacific Corp. operates car rental agencies at over 20 airports. Customers can choose from one of three contracts for car rentals of one day or less:

◆ Contract 1: $50 for the day
◆ Contract 2: $30 for the day plus $0.20 per mile traveled
◆ Contract 3: $1.00 per mile traveled

Required
1. Present separate plots of each of the three contracts, with costs on the vertical axis and miles traveled on the horizontal axis.
2. Describe each contract as a linear cost function of the form $c = g + wD$.
3. Describe each contract as a variable, fixed, or mixed cost function.

10-18 Various cost-behavior patterns. (CPA, adapted) Select the graph below that matches the numbered manufacturing-cost data. Indicate by letter which of the graphs best fits each of the situations or items described.

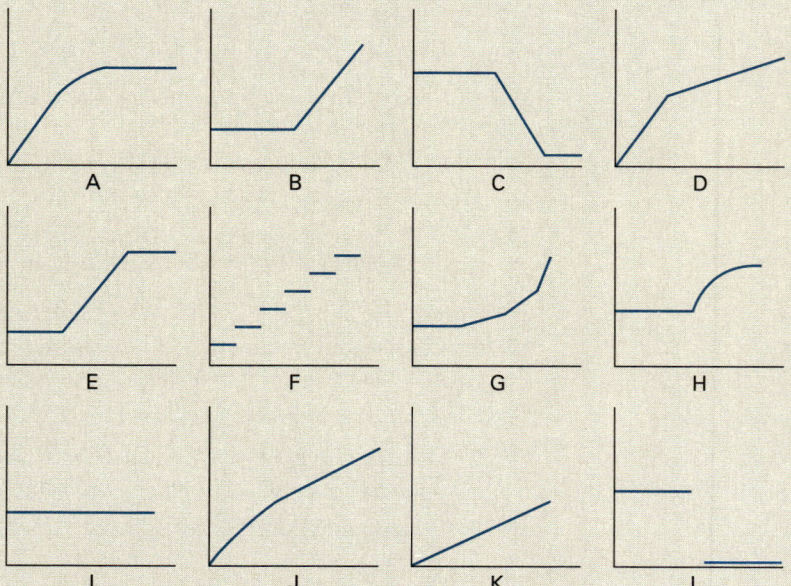

The vertical axes of the graphs represent *total* dollars of cost, and the horizontal axes represent production output during a calendar year. In each case, the zero point of dollars and production is at the intersection of the two axes. The graphs may be used more than once.

1. Annual depreciation of equipment, where the amount of depreciation charged is computed by the machine-hours method.

2. Electricity bill—a flat fixed charge, plus a variable cost after a certain number of kilowatt-hours are used, where the quantity of kilowatt-hours used varies proportionately with quantity of production output.

3. City water bill, which is computed as follows:

First 1,000,000 gallons or less	$1,000 flat fee
Next 10,000 gallons	0.003 per gallon used
Next 10,000 gallons	0.006 per gallon used
Next 10,000 gallons	0.009 per gallon used
and so on	and so on

The gallons of water used vary proportionately with the quantity of production output.

4. Cost of lubricant for machines, where cost per unit decreases with each pound of lubricant used (for example, if 1 pound is used, the cost is $10; if 2 pounds are used, the cost is $19.98; if 3 pounds are used, the cost is $29.94) with a minimum cost per pound of $9.20.

5. Annual depreciation of equipment, where the amount is computed by the straight-line method. When the depreciation rate was established, it was anticipated that the obsolescence factor would be greater than the wear-and-tear factor.

6. Rent on a manufacturing plant donated by the city, where the agreement calls for a fixed-fee payment unless 200,000 labor-hours are worked, in which case no rent need be paid.

7. Salaries of repair personnel, where one person is needed for every 1,000 machine-hours or less (that is, 0 to 1,000 hours requires one person, 1,001 to 2,000 hours requires two people, and so forth).

8. Cost of direct materials used (assume no quantity discounts).

9. Rent on a manufacturing plant donated by the county, where the agreement calls for rent of $100,000 reduced by $1 for each direct manufacturing labor-hour worked in excess of 200,000 hours, but minimum rental payment of $20,000 must be paid.

10-19 Matching graphs with descriptions of cost behavior. (D. Green)   Given below are a number of charts, each indicating some relationship between cost and a cost driver. No attempt has been made to draw these charts to any particular scale; the absolute numbers on each axis may be closely or widely spaced.

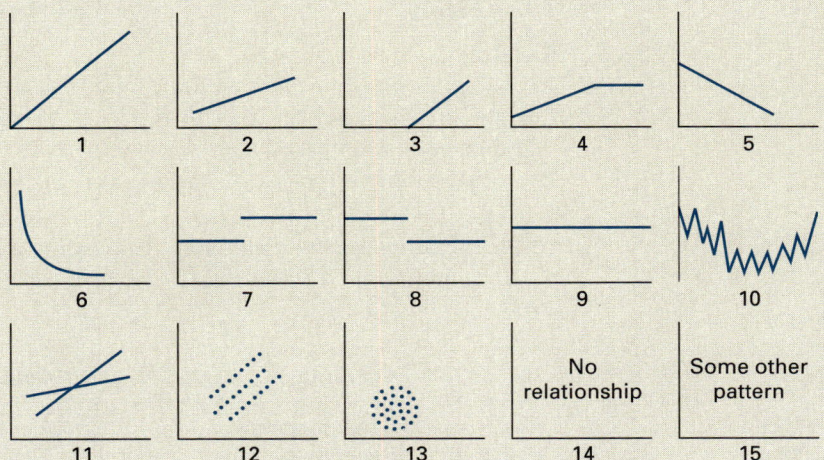

Indicate by number which one of the charts best fits each of the situations or items described. Each situation or item is independent of all the others; all factors not stated are assumed to be irrelevant. Some charts will be used more than once; some may not apply to any of the situations. Note that category 14 "No relationship," is not the same as 15, "Some other pattern."

I. If the horizontal axis represents the production output over the year and the vertical axis represents *total cost* or *revenue*, indicate the one best pattern or relationship for:
    a. Direct materials costs.
    b. Supervisors' salaries.
    c. A breakeven chart.
    d. Mixed costs—for example, fixed electrical power *demand* charge plus variable usage rate.
    e. Depreciation of plant, computed on a straight-line basis.
    f. Data supporting the use of a variable cost rate, such as manufacturing labor cost of $14 per unit produced.
    g. Incentive bonus plan that pays managers $0.10 for every unit produced above some level of production.
    h. Interest charges on money borrowed at a fixed rate of interest to finance the acquisition of a plant, before any payments on principal.

II. If the horizontal axis represents the production output over the year and the vertical axis represents the *cost per unit*, indicate the one best pattern or relationship for:
    i. Average total unit cost, assuming total cost consists of a fixed and a variable cost.
    j. Variable costs *per unit* of production output.

III. If the horizontal axis represents a *time series of weeks* during a year and the vertical axis represents *total cost per week*, match the following with the relationship in the charts:
    k. Direct manufacturing labor cost under stable production output.
    l. Declining production output over the year.
    m. Weekly underallocated or overallocated manufacturing overhead, when output varies widely and the cost rate is assumed to be correct.

10-20   Account analysis method.   Lorenzo operates a brushless car wash. Incoming cars are put on an automatic, continuously moving conveyor belt. Cars are washed as the conveyor belt carries the car from the start station to the finish station. After the car moves off the conveyor belt, the car is dried manually. Workers then clean and vacuum the inside of the car. Workers are managed by a single supervisor. Lorenzo serviced 80,000 cars in 19_5. Lorenzo reports the following costs for 19_5.

Account Description	Costs
Car wash labor	$240,000
Soap, cloth, and supplies	32,000
Water	28,000
Power to move conveyor belt	72,000
Depreciation	64,000
Supervision	30,000
Cashier	16,000

Required
1. Classify each account as variable or fixed with respect to cars washed. Explain.
2. Lorenzo expects to wash 90,000 cars in 19_6. Use the cost classification you developed in requirement 1 to estimate Lorenzo's total costs in 19_6.
3. Calculate the average cost of washing a car in 19_5 and 19_6. (Use the expected 90,000 car-wash level for 19_6.)

10-21   Account analysis method.   Gower Inc., a manufacturer of plastic products, reports the following manufacturing costs and account analysis classification for the year ended December 31, 19_5.

Account	Classification	Amount
Direct materials	All variable	$300,000
Direct manufacturing labor	All variable	225,000
Power	All variable	37,500
Supervision labor	20% variable	56,250
Materials-handling labor	50% variable	60,000
Maintenance labor	40% variable	75,000
Depreciation	0% variable	95,000
Rent, property taxes, and administration	0% variable	100,000

Gower Inc. produced 75,000 units of product in 19_5. Gower's management is estimating costs for 19_6 on the basis of 19_5 numbers. The following additional information is available for 19_6.

a. Direct material prices in 19_6 are expected to increase by 5% compared with 19_5.
b. Under the terms of the labor contract, direct manufacturing labor wage rates are expected to increase by 10% in 19_6 compared with 19_5.
c. Power rates and wage rates for supervision, materials handling, and maintenance are not expected to change from 19_5 to 19_6.
d. Depreciation costs are expected to increase by 5%, and rent, property taxes, and administration costs are expected to increase by 7%.
e. Gower Inc. expects to manufacture and sell 80,000 units in 19_6.

Required
1. Prepare a schedule of variable, fixed, and total manufacturing costs for each account category in 19_6. Calculate total manufacturing costs for 19_6.
2. Calculate Gower's unit total manufacturing costs in 19_5 and 19_6.
3. How can you get better estimates of fixed and variable costs? Why would these better estimates be useful to Gower?

10-22   Estimating a cost function, high-low method.   Laurie Daley is examining customer-service costs at the Southern Region of Capitol Products. Capitol Products has over 200 separate electrical products that are sold with a six-month guarantee of full repair or replacement with a new product. When a product is returned by a customer, a service report is made. This service report includes details of the problem and the time and cost of resolving the problem.
Weekly data for the most recent 10-week period are:

Week	Customer-Service Department Costs	Number of Service Reports
1	$13,845	201
2	20,624	276
3	12,941	122
4	18,452	386
5	14,843	274
6	21,890	436
7	16,831	321
8	21,429	328
9	18,267	243
10	16,832	161

Required
1. Plot the relationship between customer-service costs and number of service reports. Is the relationship economically plausible?
2. Use the high-low method to compute the cost function, relating customer-service costs to the number of service reports.
3. What variables, in addition to number of service reports, might be cost drivers of monthly customer-service costs of Capitol Products?

10-23   Linear cost approximation.   Terry Lawler, managing director of the Memphis Consulting Group, is examining how overhead costs behave with variations in monthly professional labor-hours billed to clients. Assume the following historical data:

Total Overhead Costs	Professional Labor-Hours Billed to Clients
$340,000	3,000
400,000	4,000
435,000	5,000
477,000	6,000
529,000	7,000
587,000	8,000

Required

1. Compute the linear cost function, relating total overhead cost to professional labor-hours, using the representative observations of 4,000 hours and 7,000 hours. Plot the linear cost function.

2. What would be the predicted total overhead costs for (a) 5,000 hours and (b) 8,000 hours using the cost function estimated in requirement 1? Plot the predicted costs and actual costs for 5,000 and 8,000 hours.

3. Lawler had a chance to accept a special job that would have boosted professional labor-hours from 4,000 to 5,000 hours. Suppose Lawler, guided by the linear cost function, rejected this job because it would have brought a total increase in contribution margin of $38,000, before deducting the predicted increase in total overhead cost, $43,000. What is the total contribution margin actually forgone?

4. Does the constant component of the cost function represent the fixed overhead costs of the Memphis Consulting Group? Why?

10-24 Cost hierarchies. Graham Electronics of England makes a radio-cassette player, model CE100, with 80 components. The chief engineer at Graham has proposed a new design that would decrease the number of components to 50. This design would reduce materials purchase costs and the time necessary for testing. Graham produces 7,000 players each month. In order to estimate the costs of the new design, Graham needs to understand how the following five categories of costs will behave. A brief description for each cost follows.

a. Repairs and maintenance costs. These costs appear to be most closely related to the amount of time that the machines are operating.

b. Ordering costs. Graham places two purchase orders each month with each component supplier.

c. Testing costs. Graham tests each radio-cassette player individually.

d. Engineering costs. These costs are incurred to handle different problems that develop from time to time while the cassette players are being manufactured. The demand for engineering services seems to be most closely related to the number of components in the cassette player.

e. Costs of reworking defective output. On average, Graham has had to rework 12 players out of every 100 manufactured.

Graham's management has identified the following potential cost drivers for these costs:

1. Dollar value of purchases
2. Units of output produced
3. Manufacturing labor-hours
4. Machine-hours
5. Number of purchase orders
6. Number of components in the player
7. Number of setups
8. Number of units reworked

Required

1. Classify each of the costs as output unit-level, batch-level, product-sustaining, or facility-sustaining. Explain your answer.

2. Identify the cost driver for each of the costs described.

3. What is the advantage of classifying costs in a hierarchy in this case? That is, how does this classification help Graham's management in estimating the cost of the new design?

10-25 Cost hierarchies. Boris Printing Works runs a small printing press in Germany. Budgeted costs in deutsche marks (DM) for October 19_5 for selected cost categories follow:

Paper and supplies	DM 8,000
Printing labor	4,000
Setup labor	6,000
Rent	3,500
Press manager's salary	3,000

The following additional information is available.

a. Boris prints 400,000 pages each month.

b. Paper and supplies costs and printing labor costs vary with the number of pages printed.

c. Printing machines need to be configured and set up for running each new customer order. Indirect setup labor costs vary with the number of setups. Boris anticipates receiving 400 new orders in October 19_5.

d. Rent costs and the press manager's salary are fixed costs of operating the printing press. These costs are independent of the number of pages printed, the number of customer orders, or any other activity measure in the press.

Required

1. Classify each of the costs as output unit-level, batch-level, product-sustaining, or facility-sustaining costs. Explain your answer.

2. Identify the cost driver for each of the costs described. (Recall that facility-sustaining costs do not have cost drivers.)

3. Boris anticipates printing 500,000 pages in November 19_5 but expects that this higher output will come from only 300 customer orders. Use the budgeted cost structures from October 19_5 to estimate Boris's total costs for November 19_5.

10-26 Cost estimation, cumulative average-time learning curve. The Nautilus Company, which is under contract to the U.S. Navy, assembles troop deployment boats. As part of its research program, it completes the assembly of the first of a new model (PT109) of deployment boats. The Navy is impressed with the PT109. It requests that Nautilus submit a proposal on the cost of producing another seven PT109s.

The accounting department at Nautilus reports the following cost information for the first PT109 assembled by Nautilus:

Direct materials	$100,000
Direct manufacturing labor (10,000 hours @ $30)	300,000
Tooling cost*	50,000
Variable manufacturing overhead†	200,000
Other manufacturing overhead‡	75,000
	$725,000

*Tooling can be reused at no extra cost, since all of its cost has been assigned to the first deployment boat.

†Variable overhead incurred is directly affected by direct manufacturing labor-hours; a rate of $20 per hour is used for purposes of bidding on contracts.

‡Other overhead is allocated at a flat rate of 25% of direct manufacturing labor costs for purposes of bidding on contracts.

Nautilus uses an 85% cumulative average-time learning curve as a basis for forecasting direct manufacturing labor-hours on its assembling operations. (An 85% learning curve implies $q = -0.2345$.)

Required

1. Prepare a prediction of the total costs for producing the seven PT109s for the Navy. (Nautilus will keep the first deployment boat assembled, costed at $725,000, as a demonstration model for other potential customers.)

2. What is the difference between (a) the predicted total costs for producing the seven PT109s in requirement 1 and (b) the predicted total costs for producing the seven PT109s assuming there is no learning curve for direct manufacturing labor—that is, for (b) assume a linear function for direct labor-hours and units produced.

10-27 Cost estimation, incremental unit-time learning curve. Assume the same information for the Nautilus Company as that in requirement 1 of Problem 10-26 with one exception. This exception is that Nautilus uses an 85% incremental unit-time learning curve as a basis for forecasting direct manufacturing labor-hours on its assembling operations. (An 85% learning curve implies $q = -0.2345$.)

Required
1. Prepare a prediction of the total expected costs for producing the seven PT109s for the Navy.
2. If you solved requirement 1 of Problem 10-26, compare your cost prediction there with the one you made here. Why are the predictions different?

10-28   Cost estimation, cumulative average-time learning model. (CMA, adapted)   Cooper Corporation has received a contract to supply 240 units of new telecommunication equipment. The direct materials costs are $60,000 per unit. The average direct manufacturing labor costs for each unit (in the first lot of 30 units) was estimated to be $40,000. Direct manufacturing labor on a per-lot basis is subject to a 90% cumulative average-time learning model. (A 90% learning curve implies $q = -0.1520$.) Variable manufacturing overhead was estimated to be 60% of direct manufacturing labor cost. Cooper's price includes a markup of 25% on total variable manufacturing costs.

The cumulative average-time learning model has proven to be accurate over the first two lots, totaling 60 units. Maximum efficiency is expected to be achieved with the production of 240 units.

Required
1. Determine Cooper Corporation's cumulative average unit cost of manufacturing labor for production of the 240 units contracted for.
2. Determine the total variable manufacturing costs for producing the 240 units of the new telecommunication equipment.
3. Assume Cooper Corporation is asked to produce additional telecommunication equipment beyond the 240 units currently under contract. Calculate the unit price Cooper should bid, employing the same markup that was used in the original bid.

10-29   Cost estimation, incremental unit-time learning curve.   Assume the same information for Cooper Corporation as that in requirement 1 of Problem 10-28 with one exception. This exception is that Cooper uses a 90% incremental unit-time learning curve as a basis for forecasting direct manufacturing labor-hours. (A 90% learning curve implies $q = -0.1520$)

Required
Determine the total variable manufacturing costs for producing the 240 units of the new telecommunication equipment. If you solved requirement 2 of Problem 10-28, compare your cost prediction there with the one you make here. Why are the predictions different?

10-30   Data collection issues, use of high-low method.   Robin Green, financial analyst at Central Railroad, is examining the behavior of monthly transportation costs for budgeting purposes. Transportation costs at Central Railroad are the sum of two types of costs: (a) operating costs (labor, fuel, and so on), and (b) maintenance costs (overhaul of engines and track, and so on).

Green collects monthly data on (a), (b), and track miles hauled. Track miles hauled are the miles clocked by the engine that pulls the rail cars. Monthly observations for the most recent year are:

Month (1)	Operating Costs (2)	Maintenance Costs (3)	Total Transportation Costs (4) = (2) + (3)	Track Miles Hauled (5)
January	$471	$437	$ 908	3,420
February	504	388	892	5,310
March	609	343	952	5,410
April	690	347	1,037	8,440
May	742	294	1,036	9,320
June	774	211	985	8,910
July	784	176	960	8,870
August	986	210	1,196	10,980
September	895	282	1,177	4,980
October	651	394	1,045	5,220
November	481	381	862	4,480
December	386	514	900	2,980

Central Railroad earns its greatest revenues carrying agricultural commodities such as wheat and barley.

Required

1. Present plots of the monthly data underlying each of the following cost functions:
   a. Operating costs = $g + w$ (track miles hauled)
   b. Maintenance costs = $g + w$ (track miles hauled)
   c. Total transportation costs = $g + w$ (track miles hauled)
   Comment on the patterns in the three plots.

2. Compute estimates of the three cost functions in requirement 1 using the high-low method. Comment on the estimated cost functions.

3. Green anticipates 6,000 track miles hauled each month next year. What total transportation costs should Green budget for next year?

4. Outline three limitations of the high-low method for estimating a cost function.

10-31 Data analysis and ethics. Comdex Electronics makes video cassette recorders (VCRs). Jack Gibbs, the manager of the department that makes the head mechanism for the VCR, is keen on introducing robots into the department. Gibbs believes that investment in robots is necessary to improve the VCR quality. To obtain funding, Gibbs knows that he will need to justify the investment in terms of cost savings. Most of these cost savings will have to come from savings in labor costs. Gibbs uses historical data to estimate the average labor costs for making heads, and his analysis shows that savings in labor costs would be large enough to justify the investment in robots. Gibbs asks Joan Hansen, the management accountant, to review his calculations before he submits the robot proposal to senior management.

Joan finds two problems with Gibbs's analysis. First, Gibbs had used a very long time period for estimating direct manufacturing labor costs. The most recent direct manufacturing labor costs per unit are much lower than the direct manufacturing labor costs in earlier periods because of learning curve effects. By considering a long time period, Gibbs computes higher savings in direct manufacturing labor costs because average labor costs over the longer period are higher than current labor costs. Second, Gibbs's analysis included savings in indirect manufacturing labor costs that are unlikely to occur.

Hansen knew that Gibbs would be unhappy with these findings. She also felt that the robot investment was good for the company. She tried to redo the analysis in a way that might show larger cost savings, even though she knew that the assumptions she was using were not appropriate. Nothing she tried could change the conclusion that the cost savings are not large enough to justify the investment in robots. Gibbs is upset when he sees Hansen's report. He tells Hansen, "Try something else. I am sure you can come up with a set of assumptions under which this investment can be justified. You and I both know this is a good investment for the company to make. Quality is essential if we are to compete."

Required

1. Referring to the Standards of Ethical Conduct for Management Accountants described in Chapter 1 (p. 17), explain whether Joan Hansen's initial attempts to redo the data analysis to justify the robot investment were ethical.

2. Identify the steps that Joan Hansen should follow in attempting to resolve this situation.

Coverage of the Chapter Appendix

10-32 Maintenance department costs, high-low and regression approaches. (H. Nurnberg, adapted) Bristol Engineering wishes to set flexible budgets for each of its operating departments. A separate maintenance department performs all routine and major repair work on the corporation's equipment and facilities. It has been determined that maintenance cost is primarily a function of machine-hours worked in the various production departments. The maintenance costs incurred and the actual machine-hours worked during the first 4 months of 19_5 are as follows:

	Machine-Hours in Production Departments (D)	Maintenance Department Costs (C)
January	800	$350
February	1,200	350
March	400	150
April	1,600	550

Required

1. Plot the relationship between monthly maintenance department costs and machine-hours in production departments at Bristol Engineering.

2. Compute the constant (g) and slope coefficient (w) of the following cost function using (a) the high-low approach and (b) the regression approach:

$$c = g + wD$$

3. Compute the coefficient of determination, r^2, of the cost function in requirement 2 when the regression approach is used.

10-33 Data collection issues, regression analysis. (Continuation of 10-30) Robin Green, financial analyst at Central Railroad, is again examining the behavior of monthly transportation costs for budgeting purposes. Monthly observations for the most recent year are shown in Problem 10-30. Green examines the results of the following three regressions:

Regression 1:
Operating costs = $g + w$ (track miles hauled)

Variable	Coefficient	Standard Error	t-value
Constant	$309.19	$96.05	3.22
Independent var.: Track miles hauled	0.054	0.014	3.96
$r^2 = 0.61$; Durbin-Watson statistic = 1.20			

Regression 2:
Maintenance costs = $g + w$ (Track miles hauled)

Variable	Coefficient	Standard Error	t-value
Constant	$531.55	$46.95	11.32
Independent var.: Track miles hauled	−0.031	0.007	−4.66
$r^2 = 0.68$; Durbin-Watson statistic = 2.02			

Regression 3:
Total transportation costs = $g + w$ (Track miles hauled)

Variable	Coefficient	Standard Error	t-value
Constant	$840.73	$80.25	10.48
Independent var.: Track miles hauled	0.023	0.011	2.02
$r^2 = 0.29$; Durbin-Watson statistic = 1.28			

Required

1. Evaluate the three regressions using the economic plausibility, goodness of fit, significance of independent variables, and specification analysis criteria.

2. Green expects 6,000 track miles hauled each month next year. Given your results in requirement 1, what total transportation costs should Green budget for next year?

3. Name three variables, other than track miles hauled, that could be important cost drivers of railroad operating costs.

4. Describe an alternative data base Green might want to use to examine the cost drivers of railroad maintenance costs.

10-34 Purchasing department cost drivers, simple regression analysis. Fashion Flair operates a chain of 10 retail department stores. Each department store makes its own purchasing decisions. Barry Lee, assistant to the president of Fashion Flair, is interested in better understanding the drivers of purchasing department costs at each store. For many years, Fashion Flair has allocated purchasing department costs to products on the basis of the dollar value of merchandise purchased. An item costing $100 is allocated 10 times as much overhead costs associated with the purchasing department as an item costing $10 is allocated.

Lee recently attended a seminar titled "Cost Drivers in the Retail Industry." In a presentation at the seminar, Couture Fabrics, a leading competitor that has 30 retail outlets, reported the number of purchase orders and the number of suppliers to be the two most important cost drivers of purchasing department costs. The dollar value of merchandise purchased on each purchase order was reported to be not a significant cost driver by Couture Fabrics. Lee interviewed several members of the purchasing department at the Fashion Flair store in Miami. These people told Lee that they believed the Couture Fabrics conclusions also applied to their purchasing department.

Lee collects the following data for the most recent year for the 10 retail department stores of Fashion Flair:

Department Store	Purchasing Department Costs (PDC)	Dollar Value of Merchandise Purchased (MPS)	Number of Purchase Orders (No. of POs)	Number of Suppliers (No. of Ss)
Baltimore	$1,523,000	$ 68,315,000	4,357	132
Chicago	1,100,000	33,456,000	2,550	222
Los Angeles	547,000	121,160,000	1,433	11
Miami	2,049,000	119,566,000	5,944	190
New York	1,056,000	33,505,000	2,793	23
Phoenix	529,000	29,854,000	1,327	33
Seattle	1,538,000	102,875,000	7,586	104
St. Louis	1,754,000	38,674,000	3,617	119
Toronto	1,612,000	139,312,000	1,707	208
Vancouver	1,257,000	130,944,000	4,731	201

Lee decides to use simple regression analysis to examine whether one or more of three variables (the last three columns in the table) are cost drivers of purchasing department costs. Summary results for these regressions follow:

Regression 1:
PDC = $g + w$ (MP$)

Variable	Coefficient	Standard Error	t-value
Constant	$1,039,061	$343,439	3.03
Independent variable 1: MP$	0.0031	0.0037	0.85
$r^2 = 0.08$; Durbin-Watson statistic = 2.41			

Regression 2:
PDC = $g + w$ (No. of POs)

Variable	Coefficient	Standard Error	t-value
Constant	$730,716	$265,419	2.75
Independent variable 1: No. of POs	156.97	64.69	2.43
$r^2 = 0.42$; Durbin-Watson statistic = 1.98			

Regression 3:
PDC = $g + w$ (No. of Ss)

Variable	Coefficient	Standard Error	t-value
Constant	$814,862	$247,821	3.29
Independent variable 1: No. of Ss	3,875	1,697	2.28
$r^2 = 0.39$; Durbin-Watson statistic = 1.97			

Required

1. Compare and evaluate the three simple regression models estimated by Lee. Graph each one. Also, use the format employed in Exhibit 10-22 (p. 370) to evaluate the information.

2. Do the regression results support the Couture Fabrics presentation about purchasing department cost drivers?

3. How might Lee gain additional evidence on drivers of purchasing department costs at each store of Fashion Flair?

10-35 Purchasing department cost drivers, multiple regression analysis. (Continuation of 10-34) Barry Lee decides that the simple regression analysis reported in Problem 10-34 could be extended to a multiple regression analysis. He finds the following results for several multiple regressions:

Regression 4:
PDC = $g + w_1$(No. of POs) + w_2(No. of Ss)

Variable	Coefficient	Standard Error	*t*-value
Constant	$485,384	$257,477	1.89
Independent variable 1: No. of POs	123.22	57.69	2.14
Independent variable 2: No. of Ss	2,952	1,476	2.00

$r^2 = 0.63$; Durbin-Watson statistic = 1.90

Regression 5:
PDC = $g + w_1$ (No. of POs) + w_2(No. of Ss) + w_3(MP$)

Variable	Coefficient	Standard Error	*t*-value
Constant	$494,684	$310,205	1.59
Independent variable 1: No. of POs	124.05	63.49	1.95
Independent variable 2: No. of Ss	2,984	1,622	1.80
Independent variable 3: MP$	−0.0002	.0030	−0.07

$r^2 = 0.63$; Durbin-Watson statistic = 1.90

The coefficient of correlation (*r*) between pairwise combinations of the variables is:

	PDC	MP$	No. of POs
MP$	0.29		
No. of POs	0.65	0.27	
No. of Ss	0.63	0.34	0.29

Required
1. Evaluate regression 4 using the economic plausibility, goodness of fit, significance of independent variables, and specification analysis criteria. Compare regression 4 with Regressions 2 and 3 in Problem 10-34. Which model would you recommend that Lee use? Why?
2. Compare regression 5 with regression 4. Which model would you recommend that Lee use? Why?
3. Lee estimates the following data for the Baltimore store for next year: dollar value of merchandise purchased (MP$), $75,000,000; number of purchase orders (No. of POs), 3,900; number of suppliers (No. of Ss), 110. How much should Lee budget for purchasing department costs for the Baltimore store for next year?
4. What difficulties may arise in multiple regressions that do not arise in simple regressions? Is there evidence of such difficulties in either of the multiple regressions presented in this problem?
5. Give two examples of decisions where the regression results reported here (and in Problem 10-34) could be informative.

11

Relevance, Costs, and the Decision Process

Information and the decision process

The meaning of relevance

Illustration of relevance: choosing output levels

Other illustrations of relevance

Opportunity costs, relevance, and accounting records

Customer profitability and relevant costs

Irrelevance of past costs

How managers behave

Appendix: Linear programming

*L*indsay Brothers faces many special-order decisions in its East Coast Australian trucking operations. An analysis of relevant revenues and relevant costs assists Lindsay Brothers in making informed special-order decisions.

Cost data play an important role in many decisions made by managers. This chapter illustrates how a solid understanding of cost behavior can help managers make decisions. We focus on decisions such as accepting or rejecting a one-time-only special order, insourcing or outsourcing products or services, and replacing or keeping equipment. Subsequent chapters will discuss pricing decisions (Chapter 12) and capital budgeting decisions (Chapters 20 and 21).

The accountant serves as a technical expert to line personnel responsible for making decisions. Managers should be supplied with relevant data for guiding their decisions. The abilities to distinguish relevant from irrelevant items and to analyze cost-behavior patterns (as explained in Chapter 10) together form the basis for making many decisions. The various decisions presented in this chapter are all based on an analysis of relevant costs and relevant revenues.

INFORMATION AND THE DECISION PROCESS

The manager has a method, often called a decision model, for deciding among courses of action. A **decision model** is a formal method for making a choice, frequently involving quantitative and qualitative analyses. For now, let us focus on accounting information as an input into decision models.

Predictions and Models

Consider a decision Home Appliances faces: Should it rearrange a manufacturing assembly line to reduce manufacturing labor costs? Assume that the only alternatives are "do not rearrange" and "rearrange." The rearrangement will eliminate manual handling of materials. The current manufacturing line uses 20 workers—15 workers operate machines, and 5 workers handle materials. Each worker puts in 2,000 hours annually. The rearrangement is predicted to cost $90,000. The predicted output of 25,000 units for the next year will be unaffected by the decision. Also unaffected by the decision are the predicted selling price per unit of $250, direct materials costs per unit of $50, other manufacturing overhead of $750,000, and marketing costs of $2,000,000. The cost driver is units of production.

Study Exhibit 11-1. It outlines a five-step sequence that highlights the role of accounting information in predicting the manufacturing labor-cost savings. The histor-

Objective 1
Describe the five-step sequence in a decision process

ical manufacturing labor cost of $14 per hour is the starting point for predicting total manufacturing labor costs under both alternatives. Manufacturing labor costs are expected to increase to $16 per hour following a recently negotiated increase in employee benefits. Predicted manufacturing labor costs under the "do not rearrange" alternative are 20 workers × 2,000 hours × $16 per hour = $640,000.

Predicted manufacturing labor costs under the "rearrange" alternative are 15 workers × 2,000 hours × $16 per hour = $480,000. Rearranging the manufacturing assembly line eliminates the materials-handling activities. The predicted savings come from eliminating materials-handling labor costs (5 workers × 2,000 hours × $16 per hour = $160,000).

Models and Feedback

Assume that management chooses the "rearrange" alternative. This decision is implemented, and the subsequent evaluation of actual performance provides feedback. In turn, the feedback might affect future predictions, the prediction method itself, the decision model, or the implementation.

In our illustration, the actual results of the plant rearrangement may show that the new manufacturing labor costs are $550,000 (due to, say, lower-than-expected manufacturing labor productivity) rather than the predicted $480,000. This feedback may lead to better implementation (step 4 in Exhibit 11-1)—through, for example, a change in supervisory behavior, employee training, or personnel—so that the $480,000 target is achieved in subsequent periods. However, the feedback may convince the decision maker that the prediction method, rather than the implementation,

EXHIBIT 11-1
Accounting Information and the Decision Process

FIVE-STEP SEQUENCE   AN ILLUSTRATION

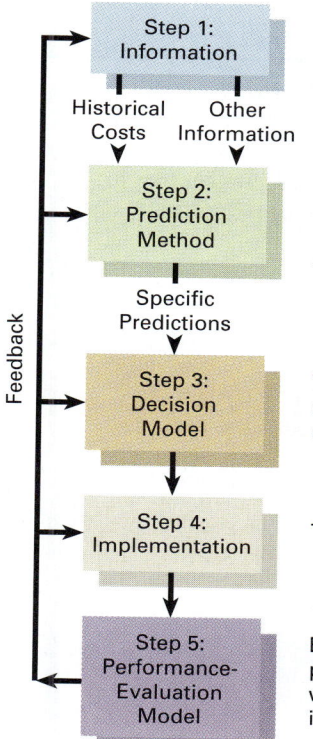

Step 1: Information — Historical labor costs were $14 per hour. A recently negotiated increase in employee benefits of $2 per hour will apply in the future. The rearrangement of the production line is expected to reduce materials-handling costs.

Historical Costs   Other Information

Step 2: Prediction Method — Use the information from step 1 together with an assessment of probability as a basis for predicting the future labor costs of $640,000 and $480,000, respectively, for the "do not rearrange" and "rearrange" alternatives. The rearrangement is predicted to cost $90,000.

Specific Predictions

Step 3: Decision Model — The predicted benefits and costs from step 2 are compared and are related to the size of the required investment together with other considerations (such as likely effects on employee morale).

Step 4: Implementation — The manager implements the decision reached in step 3.

Step 5: Performance-Evaluation Model — Evaluation of performance of the implemented decision in step 4 provides the feedback as the five-step sequence is repeated in whole or in part. This historical information can help managers in making subsequent predictions.

Feedback

was faulty. Perhaps the prediction method for similar decisions in the future should be modified to allow for worker training or learning time.

To highlight and simplify various points throughout this chapter, we assume that dollar amounts of future revenues and future costs occur with certainty. In practice, the forecasting of these figures is generally the most difficult aspect of the decision process. The Appendix to Chapter 3 discusses uncertainty and ways to incorporate it into decision making.

THE MEANING OF RELEVANCE

Historical Data and Predictions

Objective 2
Outline the meaning of relevant cost, and describe its two key aspects

Exhibit 11-1 shows that *every decision deals with the future*—whether it be 20 seconds ahead (the decision to adjust a dial) or 20 years ahead (the decision to plant and harvest pine trees). A decision (step 3) always involves a prediction (step 2). Therefore, the function of decision making is to select courses of action for the future. *Nothing can be done to alter the past.*

Relevant costs are those *expected future costs* that differ among alternative courses of action. Note the two key aspects to this definition:

1. expected future costs
2. that differ among alternative courses of action.

Relevant revenues are those expected future revenues that differ among alternative courses of action.

In Exhibit 11-2, the $640,000 and $480,000 manufacturing labor costs are relevant costs—they are expected future costs that differ between the two alternatives. The past manufacturing labor rate of $14 per hour and total past manufacturing labor costs of $560,000 ($14 per hour × 2,000 hours × 20 workers) are not relevant, al-

EXHIBIT 11-2

Determining Relevant Revenues and Relevant Costs for Home Appliances

	All Data		Relevant Data	
	Alternative 1: Do Not Rearrange	**Alternative 2: Rearrange**	**Alternative 1: Do Not Rearrange**	**Alternative 2: Rearrange**
Revenues*	$6,250,000	$6,250,000	$ —	$ —
Costs				
Direct materials†	1,250,000	1,250,000	—	—
Manufacturing labor	640,000‡	480,000§	640,000‡	480,000§
Manufacturing overhead	750,000	750,000	—	—
Marketing	2,000,000	2,000,000	—	—
Rearrangement costs	—	90,000	—	90,000
Total costs	$4,640,000	$4,570,000	$640,000	$570,000
		$70,000 difference		$70,000 difference

*25,000 × $250.
†25,000 × $50.
‡$16 × 2,000 × 20.
§$16 × 2,000 × 15.

though they may play a role in preparing the $640,000 and $480,000 labor cost predictions. *Historical costs in themselves are irrelevant to a decision, although they may be a useful basis for predicting future costs.*

The quantitative data underlying the choice between the "do not rearrange" and the "rearrange" alternatives are presented in Exhibit 11-2. The first two columns present all data. The last two columns present only expected future costs or revenues that differ between the two alternatives. The revenues, direct materials, manufacturing overhead, and marketing items can be ignored. Why? Because although they are expected future costs, they do not differ between the alternatives. The data in Exhibit 11-2 indicate that rearranging the production line will decrease next year's total predicted costs by $70,000. Note that we reach the same conclusion whether we use all data or include only the relevant data in the analysis. By confining the analysis to only the relevant data, we can focus on the items that really make the difference.

The difference in total cost between two alternatives is an **incremental cost.** The incremental cost between alternative 2 and alternative 1 in Exhibit 11-2 is $70,000. Synonyms for incremental cost are **differential cost** and **net relevant cost.**

The definition of relevance is the major conceptual lesson in this chapter. The remainder of this chapter will show how to apply the concept of relevance to some commonly encountered decisions.

Time Value of Money and Income Taxes

Exhibit 11-2 does not take into account the time value of money or income taxes. We defer discussion of those factors to Chapters 20 and 21. When interest costs associated with the time value of money are recognized, expected future costs and future revenues with the same magnitude but different timing become relevant. Similarly, when income tax issues are introduced, expected future pretax costs and future pretax revenues with the same magnitude may affect income tax costs differently; hence, income tax costs would become relevant.

Qualitative Factors Can Be Relevant

Objective 3
Distinguish between quantitative factors and qualitative factors in decisions

We divide the consequences of alternatives into two broad categories: *quantitative and qualitative.* **Quantitative factors** are outcomes that are measured in numerical terms. Some quantitative factors are financial—that is, they can be easily expressed in financial terms. Examples include the costs of direct materials, direct manufacturing labor, and marketing. Other quantitative factors are nonfinancial—that is, they can be measured numerically, but they are difficult to express in financial terms. Reduction in product-development time for a manufacturing company and the percentage of on-time flight arrivals for an airline company are examples of quantitative, nonfinancial factors. **Qualitative factors** are outcomes that cannot be measured in numerical terms. Employees' morale is an example.

Cost analysis generally emphasizes quantitative factors that can be expressed in financial terms. But just because qualitative factors and nonfinancial quantitative factors cannot be easily measured in financial terms does not make them unimportant. Managers recognize and respond to this problem in two ways.

One approach managers use is to evaluate nonfinancial and qualitative factors separately after first analyzing financial outcomes. In fact, managers must at times give more weight to a qualitative factor than to a quantitative factor. For example, Home Appliances may use a single supplier for a variety of important subassemblies. To preserve good business relationships—a qualitative factor—Home Appliances might decide to continue to buy all subassemblies from that supplier even though it could manufacture one of those subassemblies itself at a cost—a quantitative factor—

lower than it pays to the supplier. The difficulty with this approach is determining how, precisely, to trade off nonfinancial and financial considerations.

A second approach, and one that is increasingly gaining favor among managers, is to attempt to financially quantify factors traditionally considered as nonfinancial or qualitative. One example of this approach is Hewlett-Packard's attempts to measure the financial benefits from faster product-development time (see Chapter 23). The difficulty here is obtaining an accurate financial measurement.

ILLUSTRATION OF RELEVANCE: CHOOSING OUTPUT LEVELS

Managers often make decisions that affect output levels. Many such decisions are essentially short-run in nature, but they have long-run consequences that should never be overlooked. Managers are rewarded for making decisions that maximize the chosen objective of the organization (typically operating income in our illustrations). Cost analysis is essential in determining how operating income will change with changes in output levels. Intelligent analysis of costs often depends on explicit distinctions among cost-behavior patterns—for example, whether costs are fixed or variable with respect to output produced (the cost driver).

One-Time-Only Special Order

Management sometimes faces the decision of accepting or rejecting one-time-only special orders when there is idle production capacity and where the order has no long-run implications. In this situation, costs can be classified as either variable with respect to a single driver (units of output) or fixed. The following example illustrates how the contribution-margin approach can provide key information for decisions about the choice of output level. The example also indicates how reliance on fully allocated unit-cost numbers can mislead managers about the effect that increasing output has on operating income.

> EXAMPLE: Fancy Fabrics manufactures quality bath towels at its highly automated Charlotte plant. The plant has a production capacity of 48,000 towels each month. Current monthly production is 30,000 towels. Retail department stores account for all existing sales. Expected results for the coming month (August) are in Exhibit 11-3. (Note that these amounts are predictions.) The manufacturing costs per unit of $12 comprise direct materials $6 (all variable), direct manufacturing labor $2 ($0.50 of which is variable), and manufacturing overhead $4 ($1 of which is variable). The marketing costs per unit are $7 ($5 of which is variable). Fancy Fabrics has no research and development costs or product-design costs. Marketing costs include distribution costs and customer-service costs.
>
> A four-star hotel chain offers to buy 5,000 towels per month at $11 a towel for each of the next three months. No subsequent sales to this customer are anticipated. No marketing costs will be necessary for the 5,000-unit one-time-only special order. The acceptance of this special order is not expected to affect the selling price or the quantity of regular sales. Should Fancy Fabrics accept the hotel chain's offer?

Exhibit 11-3 presents data in an absorption-costing format, when fixed costs are included in unit product costs (see Chapter 9). The unit manufacturing cost is $12 ($7.50 of which is variable), which is above the $11 price offered by the hotel chain. Using the $12 full-absorption unit cost as a guide in deciding, a manager might reject the offer.

Exhibit 11-4 presents data in a contribution income statement format. The relevant costs are the expected future costs that differ between the alternatives—the variable manufacturing costs of $7.50 per unit (direct materials $6 + direct manufacturing labor $0.50 + variable manufacturing overhead $1). The fixed manufacturing costs and all marketing costs (including variable marketing costs) are irrelevant in this case; they will not change in total whether or not the special order is accepted. Accepting the order will not increase fixed costs—for example, costs of setup and pro-

EXHIBIT 11-3

Budgeted Income Statement for August,
Absorption-Costing Format for Fancy Fabrics

	Total	Per Unit
Sales—30,000 towels × $20	$600,000	$20
Cost of goods sold	360,000	12
Gross margin (gross profit)	240,000	8
Marketing costs	210,000	7
Operating income	$ 30,000	$ 1

EXHIBIT 11-4

Comparative Income Statements for August,
Contribution Income Statement Format for Fancy Fabrics

	Without One-Time-Only Special Order, 30,000 Units		With One-Time-Only Special Order, 35,000 Units	Difference
	Per unit*	Total	Total	
Sales	$20.00	$600,000	$655,000	$55,000†
Variable costs				
Manufacturing	7.50	225,000	262,500	37,500‡
Marketing	5.00	150,000	150,000	— §
Total variable costs	12.50	375,000	412,500	37,500
Contribution margin	7.50	225,000	242,500	17,500
Fixed costs				
Manufacturing	4.50	135,000	135,000	— §
Marketing	2.00	60,000	60,000	— §
Total fixed costs	6.50	195,000	195,000	—
Operating income	$ 1.00	$ 30,000	$ 47,500	$17,500

*Based on analysis of costs per unit shown below.
†5,000 × $11.00.
‡5,000 × $7.50.
§No variable marketing costs will be necessary for the 5,000-unit one-time-only special order. Fixed
manufacturing costs and fixed marketing costs are also unaffected by the special order.

Computation of Cost per Unit for Fancy Fabrics

	Variable Costs	Fixed Costs	Business Function Costs*
Direct materials	$ 6.00	$ —	
Direct manufacturing labor	0.50	1.50	
Manufacturing overhead	1.00	3.00	
Manufacturing cost	7.50	4.50	$12.00
Marketing cost	5.00	2.00	7.00
Full product cost	$12.50	$6.50	$19.00

*Business function costs are the sum of all the costs (variable costs and fixed costs) in a particular
business function.

duction scheduling—because there is surplus capacity of those resources. Therefore, the only relevant items here are sales revenues and variable manufacturing costs. Given the $11 relevant revenue per unit (the special-order price) and the $7.50 relevant cost per unit, Fancy Fabrics gains an additional $17,500 ([$11.00 – $7.50] × 5,000]) in operating income per month by accepting the special order. This example illustrates how comparisons based on either total amounts or relevant amounts (Exhibit 11-4) avoid the misleading implication of the absorption unit-cost data (Exhibit 11-3).

The additional costs of $7.50 that Fancy Fabrics will incur if it accepts the special order for 5,000 towels are sometimes called out-of-pocket costs or outlay costs. **Out-of-pocket,** or **outlay, costs** are current or near-future cash disbursements made to meet costs incurred because of a specific decision. Fancy Fabrics could avoid these costs if it did not accept the special order. Fixed manufacturing costs (including depreciation on the machinery and equipment used in the job) of $4.50 per unit are not out-of-pocket costs; those costs will occur whether or not the special order is accepted.

The analysis in Exhibit 11-4 assumes that the 5,000-towel special order will use otherwise idle capacity "for each of the next three months." Given this assumption, a focus on short-run revenues and short-run costs is appropriate. Chapter 12 explores additional considerations that arise when special-order decisions have long-run revenue or long-run cost implications.

Pitfalls in Relevant-Cost Analysis

Objective 4
Identify two common pitfalls in relevant-cost analysis

One pitfall in relevant-cost analysis is to assume that all variable costs are relevant. In the Fancy Fabrics example, the marketing costs are variable but not relevant. Why? Because for the special-order decision, Fancy Fabrics incurs no extra marketing costs.

A second pitfall is to assume that all fixed costs are irrelevant. Consider fixed manufacturing costs. In our example, we assume that the extra 5,000-towel production per month does not affect fixed manufacturing costs. (In terms of the Chapter 3 terminology, we assume that the relevant range is at least from 30,000 to 35,000 towels per month.) In some cases, however, the extra 5,000 towels might increase fixed manufacturing costs. Assume that Fancy Fabrics would have to run three shifts of 16,000 towels per shift to achieve full capacity of 48,000 towels per month. Increasing the monthly production from 30,000 to 35,000 would require a partial third shift because two shifts alone could produce only 32,000 towels. This extra shift would probably increase fixed manufacturing costs, meaning fixed manufacturing costs are relevant for this decision.

The best way to avoid these two pitfalls is to focus first and foremost on the relevance concept. Always require each item included in the analysis *both* (1) to be an expected future revenue or cost and (2) to differ among the alternatives.

How Unit Costs Can Mislead

Unit-cost data can mislead decision makers in two major ways:

1. When irrelevant costs are included. Consider the $4.50 per unit allocation of fixed direct manufacturing labor and manufacturing overhead costs in the one-time-only special-order decision for Fancy Fabrics. This $4.50 per unit cost is irrelevant given the assumptions of our example.

2. When unit costs not computed at the same output level are compared. Generally, use total costs rather than unit costs. Then, if desired, the total costs can be unitized. Machinery sales personnel, for example, may brag about the low unit costs of using their new machines. Sometimes they neglect to say that the unit costs are based on outputs far in excess of their prospective customer's current and anticipated production levels. Consider, for example, a new machine that costs $100,000 and is capable of producing 100,000 units over its useful life. The salesperson may represent the machine-related costs of pro-

Objective 5
Indicate two ways in which per-unit cost data can mislead decision makers

ducing each unit to be $1. This amount is incorrect if the firm anticipates a total demand of say, only 50,000 units over the useful life of the machine (unit cost is $100,000 ÷ 50,000 = $2). Unitizing fixed costs over different production levels can be particularly misleading.

Confusing Terminology

Many different terms are used to describe the costs of specific products and services. Exhibit 11-5 presents several different unit-cost numbers using the data from column 1 of Exhibit 11-4. **Business function costs** are the sum of all the costs (variable costs and fixed costs) in a particular business function. Manufacturing costs are $12 per unit, and marketing costs are $7 per unit. For inventory costing purposes, *absorption cost* is often used as a synonym for manufacturing cost.

Full product costs refer to the sum of all the costs in all the business functions (research and development, design, production marketing, distribution, and customer service). Full product costs in Exhibit 11-5 are $19 per unit.

Managers use terms such as *business function cost* and *full product cost* differently. In a given situation, be sure to understand their exact meaning.

OTHER ILLUSTRATIONS OF RELEVANCE

Product Mix Decisions

When a multiproduct plant operates at full capacity, managers must often make decisions regarding which products to emphasize. These decisions frequently have a short-run focus. For example, some food-processing companies continually change their product mix in response to short-run fluctuations in materials costs or the selling price of their finished goods. We assume that the only costs that change with

EXHIBIT 11-5

Variety of Cost Terms for Fancy Fabrics*

		Unit Costs		
Variable manufacturing costs	$ 7.50	$ 7.50		$ 7.50
Variable marketing costs	5.00			5.00
Variable product costs	$12.50			
Fixed manufacturing costs		4.50	$4.50	4.50
Manufacturing (absorption) costs		$12.00		
Fixed marketing costs			2.00	2.00
Fixed product costs			$6.50	
Full product costs				$19.00

*In this example, note that marketing costs include distribution costs and customer-service costs.

Unit-cost data from Exhibit 11-4

	Variable Costs	Fixed Costs	Business Function Costs*
Manufacturing costs	$ 7.50	$4.50	$12.00
Marketing costs	5.00	2.00	7.00
Full product costs	$12.50	$6.50	$ 19.00

short-run changes in product mix are costs that are variable with respect to the number of units produced (and sold).

Analysis of individual product contribution margins provides insight into the product mix that maximizes operating income. Consider Power Engines, a company that manufactures engines for a broad range of commercial and consumer products. At its Lexington, Kentucky, plant, it assembles two engine models—a snowmobile engine and an outboard boat engine. Information on these products follows:

	Snowmobile Engine	Boat Engine
Selling price	$800	$950
Unit variable costs	600	700
Unit contribution margin	$200	$250
Contribution-margin percentage	25%	26.3%

At first glance, boat engines appear more profitable than snowmobile engines. The product to be emphasized, however, is not necessarily the product with the higher individual unit contribution margin or contribution-margin percentage. Rather, managers should aim for the highest contribution margin per unit of the constraining factor—that is, the scarce, limiting, or critical factor. The constraining factor restricts or limits the production or sale of a given product. (See also Chapter 23, the theory of constraints.)

Assume that only 600 hours of machine capacity are available for assembling engines. Additional capacity cannot be obtained in the short run. Power Engines can sell as many engines as it produces. The constraining factor, then, is machine-hours. If it takes 2 machine-hours to produce one snowmobile engine and 5 machine-hours to produce one boat engine, choosing to emphasize snowmobile engines is the correct decision. Producing snowmobile engines contributes more margin per machine-hour, which is the constraining factor in this example.

	Snowmobile Engine	Boat Engine
Contribution margin per engine	$ 200	$ 250
Machine-hours required to produce one engine	2	5
Contribution margin per machine-hour ($200 ÷ 2; $250 ÷ 5)	$ 100	$ 50
Total contribution margin for 600 machine-hours ($100 × 600; $50 × 600)	$60,000	$30,000

The constraining factor in this example is machine-hours. Other constraints in manufacturing settings can be the availability of direct materials, components, or skilled labor, as well as financial and sales considerations. In a retail department store, the constraining factor may be cubic feet of display space. The greatest possible contribution margin per unit of the constraining factor yields the maximum operating income.

As you can imagine, in many cases a manufacturer or retailer must meet the challenge of trying to maximize total operating income for a variety of products, each with more than one constraining factor. The problem of formulating the most profitable production schedules and the most profitable product mix is essentially that of maximizing the total contribution margin in the face of many constraints. Optimization techniques, such as the linear-programming technique discussed in the Appendix to this chapter, help solve these complicated problems.

Outsourcing and Idle Facilities

Outsourcing is the process of purchasing goods and services from outside vendors rather than producing the same goods or providing the same services within the firm. Producing the same goods or providing the same services within the firm is called **in-**

sourcing. A manufacturer may stop making component parts and assemblies within its plant in favor of purchasing the parts and assemblies from outside vendors. A company may decide to hire an outside firm to process its payroll rather than handle payroll internally. Another organization may choose to retain the services of an outside legal firm rather than keep its own in-house staff of lawyers.

Decisions about whether a manufacturer will produce its own parts or purchase them from outside suppliers are also called make-or-buy decisions. Sometimes qualitative factors dictate management's response to the make-or-buy decision. For example, the production of parts may require special know-how, unusually skilled labor, or scarce material that the company needing the part lacks. If so, the manufacturer must buy the parts. Sometimes, a company may prefer to produce parts and assemblies to retain greater control of product design and process technology. Surveys of company practice indicate that quality considerations and dependability of supplies rank with cost as the most important factors in the make-or-buy decision.[1]

In the El Cerrito Company example described below, assume that financial factors predominate in the make-or-buy decision. The question we address is, What financial factors are relevant to the make-or-buy decision?

El Cerrito Company reports the following costs for making part no. 300:

	Total Costs of Producing 10,000 Units	Costs per Unit
Direct materials	$ 80,000	$ 8
Direct manufacturing labor	10,000	1
Variable manufacturing overhead costs for power and utilities	40,000	4
Fixed manufacturing overhead costs of purchasing, receiving, and setups	20,000	2
Fixed manufacturing overhead costs of depreciation, plant insurance, and administration	30,000	3
Total costs	$180,000	$18

Another manufacturer offers to sell El Cerrito Company the same part for $16 per unit. Should El Cerrito Company make or buy the part?

The unit cost of $18 seemingly indicates that the company should buy. A make-or-buy decision, however, is rarely obvious. A key question for management to ask is, What is the difference in relevant costs between the alternatives? Consider the $30,000 in depreciation, plant insurance, and plant administration costs. These costs represent fixed manufacturing overhead that will not vary regardless of the decision made. However, the $20,000 in setup, receiving, and purchasing costs that are fixed over short-run fluctuations in output become variable in the long run. The decision to discontinue part no. 300 is a long-run decision. The $20,000 of fixed manufacturing costs will be saved if the parts are bought instead of made. We see that $30,000 (which amounts to $3 per unit for 10,000 units of production) is therefore irrelevant, but the $20,000 ($2 per unit) that may be avoided in the future is relevant.

For the moment, suppose the capacity now used to make the parts will become idle if the parts are purchased. Exhibit 11-6 presents the relevant-cost computations. Making the part is the preferred alternative. El Cerrito saves $1 per part by making the part rather than buying it from the external manufacturer.

More generally, the choice in our example is not fundamentally whether to make or buy; it is how best to use available facilities. If the component part is bought from the external manufacturer, the released facilities can potentially be used for other, more profitable purposes. The figures in Exhibit 11-6 are valid only if the released facilities remain idle.

---

[1]P. J. Clarke, "Management Accounting Practices and Techniques in Irish Manufacturing Firms," Working Paper, University College, Dublin, 1992.

EXHIBIT 11-6

Relevant Items for Make-or-Buy Decision for Part No. 300
at El Cerrito Company

	Total Costs		Per-Unit Costs	
Relevant Item	**Make**	**Buy**	**Make**	**Buy**
Outside purchase of parts		$160,000		$16
Direct materials	$ 80,000		$ 8	
Direct manufacturing labor	10,000		1	
Variable manufacturing overhead	40,000		4	
Fixed manufacturing overhead that can be avoided by not making	20,000		2	
Total relevant costs	$150,000	$160,000	$15	$16
Difference (in favor of making)	$10,000		$1	

Recognition of how the use of otherwise idle resources can increase profitability occurred at a Chinese machine-repairing plant of Beijing Engineering, where the decision was whether to drop or keep a product. The *China Daily* noted that workers were "busy producing electric plaster-spraying machines" even though the unit cost exceeded the selling price. According to the prevailing method of calculating its cost, each sprayer costs 1,230 yuan to make. However, each sprayer sells for only 985 yuan, leading to a loss of 245 yuan per sprayer. Still, to meet market demand the plant continues to produce sprayers. Workers and machines would otherwise be idle, and the plant would still have to pay 759 yuan even if no sprayer were made. The production of sprayers, even at a loss, then, actually helps cut the company's deficit.

Relevant-Cost Analysis and Contribution-Margin Analysis

Several examples in this chapter illustrate how contribution-margin analysis can assist managers in predicting relevant costs used in decision making. However, contribution-margin analysis is not always synonymous with relevant-cost analysis. In the El Cerrito case, for example, $20,000 of fixed overhead costs of purchasing, receiving, and setup were relevant for the make-or-buy decision. In the Fancy Fabrics example, earlier in this chapter, variable marketing costs of $5 per unit were irrelevant to the decision of whether or not to accept the one-time-only special order of 5,000 towels per month. Relevant-cost analysis focuses on a specific decision with a given set of alternatives. This focus is not necessary when conducting contribution-margin analysis. Many companies use a contribution-margin format in their internal reporting system.

When using contribution-margin data in relevant-cost analysis, examine whether the assumptions underlying the fixed-cost and variable-cost functions are appropriate for the decision at hand. Chapter 3 noted several key assumptions of contribution-margin analysis:

◆ a stipulated (typically short) time span,
◆ a specific relevant range of the cost driver, and
◆ one cost driver is sufficient to explain variations in the variable-cost function.

For decisions such as accepting or rejecting a one-time-only special order or making a short-run change in the product mix, data from a routine contribution income statement may be a helpful starting point. In other cases, decisions will affect a relatively long time span and will involve multiple cost drivers for each business function cost.

Concepts in Action

OUTSOURCING DATA-PROCESSING OPERATIONS AT EASTMAN KODAK[a]

Kodak Corporation recently outsourced its data-processing operations to a data-processing facility that IBM built and runs for Kodak. Some experts estimate the savings to Kodak from outsourcing at several hundred million dollars over the next decade.

How does Kodak save costs by buying this outside service rather than handling data processing internally? Kodak's managers argue that using a firm specializing in data-processing is more efficient and economical than using in-house personnel. Data-processing firms have access to better technology, enjoy economies of scale, and use innovative software practices (such as structured design and analysis techniques, reusable code, and object-oriented programming). Outsourcing has enabled Kodak to keep up with dramatic technological change.

Also, specialized firms can more easily overcome the enormous difficulty of recruiting software professionals. Analysts estimate the shortage of information-technology and business-systems professionals to be between two and five million globally. Overall, Kodak's managers believe that the greater effectiveness of the external data-processing operations has resulted in better quality and faster response time to data-processing questions.

Are there disadvantages of outsourcing data-processing? Kodak's managers suggest that some degree of customization is probably lost, but the gains apparently are worth this price.

[a]Adapted from J. Teresko, "Make or Buy? Now It's a Data-Processing Question Too," *Industry Week* (July 16, 1990) and discussions with Kodak managers.

Activity-based costing is helpful in making such decisions because this approach is based on multiple activities and multiple cost drivers that typically consider longer time horizons.

The El Cerrito example illustrates the application of activity-based costing analysis to the decision to make or buy part no. 300. The cost of part no. 300 is based on the costs of activities undertaken to produce that part. Part no. 300 requires purchasing, receiving, and setup activities costing $20,000. Discontinuing the part reduces the demand for those activities. The relevant costs of not making part no. 300 are the savings in costs from performing fewer activities. The full absorption cost of part no. 300 also includes machine-related and plant-management activities costing $30,000. Depreciation costs and the cost of plant-management activities are unaffected by the decision to stop manufacturing part no. 300. The costs of those activities are irrelevant to the make-or-buy decision.

OPPORTUNITY COSTS, RELEVANCE, AND ACCOUNTING RECORDS

Objective 7
Describe the opportunity cost concept; explain why it is used in decision making

Ideally, a decision maker should be able to make an exhaustive list of alternatives and then compute the expected results under each, giving full consideration to interdependent effects. Practically, the decision maker sifts among the alternatives, quickly discards many as being obviously unattractive, probably overlooks some attractive possibilities, and concentrates on a limited number. The idea of an opportunity cost arises because some alternatives are not selected.

Consider again the El Cerrito Company example from the previous section. The example assumed that capacity currently used to make part no. 300 became idle if the parts were purchased. Suppose instead that El Cerrito has alternative uses for the extra capacity. The best available alternative is for El Cerrito to use the capacity to produce 5,000 units (each year) of its component part A603 to be sold to Terrence Corporation. John Marquez, the accountant at El Cerrito, estimates the following future revenues and future costs if A603 is manufactured and sold.

Expected incremental future revenues		$80,000
Expected incremental future costs:		
Direct materials	$30,000	
Direct manufacturing labor	5,000	
Variable overhead (power, utilities)	20,000	
Total expected incremental future costs		55,000
Excess of expected incremental future revenues		
over expected incremental future costs		$25,000

The three alternatives available to the management of El Cerrito are:

1. make part no. 300
2. buy part no. 300 and do not make A603 for Terrence
3. buy part no. 300 and use surplus capacity to make and sell A603 to Terrence

Exhibit 11-7, Panel A, summarizes the "total-alternatives" approach—the total expected future costs and expected future revenues for *all* alternatives. Buying part no. 300 and using the surplus capacity to make A603 and sell it to Terrence is the preferred alternative. Buying part no. 300 from outside vendors costs more than making

EXHIBIT 11-7

Total-Alternatives Approach and Opportunity-Costs
Approach to Make-or-Buy Decisions for El Cerrito

PANEL A: TOTAL-ALTERNATIVES APPROACH TO MAKE-OR-BUY DECISIONS

	Choices for El Cerrito		
	Make Part No. 300	**Buy Part No. 300 Do Not Make A603**	**Buy Part No. 300 Make A603**
Total relevant costs of making/ buying part no. 300 (from Exhibit 11-6)	$150,000	$160,000	$160,000
Excess of future revenues over future costs for A603	0	0	(25,000)
Total incremental costs	$150,000	$160,000	$135,000

PANEL B: OPPORTUNITY-COSTS APPROACH TO MAKE-OR-BUY DECISIONS

	Choices for El Cerrito	
	Make Part No. 300	**Buy Part No. 300**
Total relevant costs of making/ buying part no. 300 (from Exhibit 11-6)	$150,000	$160,000
Opportunity cost: profit contribution forgone because capacity cannot be used to make A603, the best alternative	25,000	0
Total incremental costs	$175,000	$160,000

part no. 300 in-house ($160,000 to buy versus $150,000 to make). But the capacity freed up by buying part no. 300 from outside suppliers enables El Cerrito to gain $25,000 in operating income (expected incremental future revenues of $80,000 minus expected incremental future costs of $55,000) by making A603 and selling it to Terrence. The total incremental cost of buying part no. 300 (and making and selling A603) is $160,000 – $25,000 = $135,000.

In many decision-making situations, managers are confronted with too many alternatives to analyze thoroughly. Because managers exclude some alternatives from explicit consideration, the concept of opportunity cost comes about. **Opportunity cost** is the contribution to income that is forgone (rejected) by not using a limited resource in its best alternative use.

Exhibit 11-7, Panel B, displays the opportunity-costs approach for analyzing the alternatives faced by El Cerrito. This format does not explicitly consider making and selling A603 in the analysis. The problem confronting management is whether to make or buy part no. 300, and these are the only alternatives considered. The key point, however, is that if management chooses the make alternative, El Cerrito will forgo the operating income it could have made by using its limited capacity to make and sell A603 (the best alternative use of the capacity). This lost operating income of $25,000 is the opportunity cost of using El Cerrito's capacity to make part no. 300. From Exhibit 11-7, Panel B, the cost (including the opportunity cost of $25,000) to make part no. 300 is $175,000. The cost to buy part no. 300 is $160,000. There is zero opportunity cost under the buy alternative because the capacity is still available to make and sell A603, and therefore no operating income is forgone. Exhibit 11-7, Panel B, leads management to the same conclusion as Exhibit 11-7, Panel A does—buying part no. 300 is the preferred alternative by an amount of $15,000.

Opportunity costs are seldom incorporated into formal financial accounting reports. Such costs represent contributions forgone if the limited resources are not used in the best alternative way. Therefore, opportunity costs do not entail cash receipts or disbursements. Accountants usually confine their systematic recording of costs to the outlay or out-of-pocket costs—costs that currently or in the near future require cash disbursements. Accountants confine their historical recordkeeping to alternatives selected rather than those rejected, primarily because of the infeasibility of accumulating appropriate data on what might have been. A second reason is that opportunity costs are forever changing as circumstances change.

Exhibit 11-7, Panel B, demonstrates the importance of recognizing opportunity costs in relevant-cost analysis. On the basis of outlay costs alone, it is less costly for El Cerrito to make part no. 300. Recognizing the opportunity cost of $25,000 when El Cerrito makes part no. 300 leads to a different conclusion. It is preferable to buy part no. 300.

Our analysis emphasizes purely quantitative considerations. Final conclusions, however, will often consider qualitative factors as well. For example, before deciding to buy part no. 300 from outside vendors, El Cerrito management will consider such qualitative factors as the vendor's reputation for quality and the vendor's dependability for on-time delivery.

Suppose El Cerrito has sufficient excess capacity to make A603 (and indeed any other part) even if it makes part no. 300. Under these assumptions, the opportunity cost of making part no. 300 is zero. Why? Because El Cerrito gives up nothing even if it chooses to manufacture part no. 300. It follows from Exhibit 11-7, Panel B (substituting opportunity costs equal to zero), that under these conditions El Cerrito will prefer to make part no. 300.

Carrying Costs of Inventory

The notion of opportunity cost can also be illustrated with a direct-materials purchase-order decision:

Estimated direct-materials requirements for the year	120,000 pounds
Cost per pound for purchase orders below 120,000 pounds	$ 10.00
Cost per pound for purchase orders equal to or greater than 120,000 pounds; $10 minus 2% discount	$ 9.80
Two alternatives under consideration:	
A: Buy 120,000 pounds at start of year	
B: Buy 10,000 pounds per month	
Average investment in inventory:	
A: (120,000 pounds × $9.80) ÷ 2*	$588,000
B: (10,000 pounds × $10) ÷ 2*	$ 50,000

*The example assumes that the direct materials purchased will be used up uniformly at the rate of 10,000 pounds per month. If direct materials are purchased at the start of the year (month), the average investment in inventory during the year is the cost of the inventory at the beginning of the year (month) plus the cost of inventory at the end of the year (month) divided by 2.

The savings in outlay costs under alternative A (purchasing 120,000 pounds at the beginning of the year) compared with alternative B (purchasing 10,000 pounds at the beginning of each month) are $24,000 ([$10.00 − $9.80] per pound × 120,000 pounds). But alternative A requires a substantially higher average investment in inventory than alternative B: $588,000 − $50,000 = $538,000.

The difference of $538,000 could be invested in risk-free government bonds and return, say, 0.06 × $538,000 = $32,280 for the year. This $32,280 is an opportunity forgone when alternative A is chosen; it is the opportunity cost of the 120,000-pound purchase order. The $32,280 would not be recorded in the accounting system because it is an opportunity cost rather than an outlay cost. In fact, the opportunity cost of $32,280 exceeds the benefit from the outlay cost savings of $24,000 under alternative A. Therefore, purchasing 10,000 pounds per month is the preferred alternative.

Imputed Costs

Imputed costs are those costs recognized in particular situations that are not regularly recognized by accrual accounting procedures. Imputed costs attempt to dovetail accounting with economic reality. Suppose a firm lends its supplier $100,000 for one year at an interest rate of 2% when loans of comparable risk in the marketplace carry a 10% interest rate. The firm may extend the 2% loan in consideration for receiving the supplier's products under a purchase contract at prices lower than the prevailing market price. A very refined accounting of the transaction requires recognizing that the interest of $8,000 ($100,000 × [10% − 2%]) forgone over the course of the year is really an additional cost to the buyer for the products purchased during the contract term. The interest forgone is *imputed* as a cost of the products purchased.[2]

CUSTOMER PROFITABILITY AND RELEVANT COSTS

This section illustrates relevant revenue and relevant-cost analysis when different cost drivers are identified for different activities, as in activity-based costing. The cost object in our example is customers. The analysis focuses on customer profitability at Allied West, the West Coast sales office of Allied Furniture, a wholesaler of specialized furniture.

Allied West supplies furniture to three local retailers, Vogel, Brenner, and Wisk. Exhibit 11-8 presents Allied West's revenues and costs by customer. The information in Exhibit 11-8 is representative of Allied West's financial relationship with these customers over the last few years. Additional information on Allied West's costs follows:

---

[2]The technical treatment is to create an initial loan discount amount and then amortize the discount as a part of the cost of products purchased.

1. Materials-handling labor costs vary with the number of units of furniture shipped to customers.

2. Different areas of the warehouse stock furniture for different customers. Materials-handling equipment in an area and depreciation costs on the equipment are identified with individual customer accounts. Any equipment not used remains idle. The equipment has a zero disposal price.

3. Allied West allocates rent to each customer account on the basis of the area of the warehouse space occupied by the products to be shipped to that customer.

4. Marketing costs vary with the number of sales visits made to customers.

5. Purchase-order costs vary with the number of purchase orders placed; delivery-processing costs vary with the number of shipments.

6. Allied West allocates fixed general administration costs to customers on the basis of dollar sales made to each customer.

The Allied West example illustrates cost hierarchies, described in Chapter 10 (pp. 355–57), in a nonmanufacturing, customer-focused setting. If you have *not* studied cost hierarchies in Chapter 10, you can skip the next paragraph without disrupting the flow of the subsequent analysis.

We now describe examples of output unit-level costs, batch-level costs, customer-sustaining (as distinguished from product-sustaining) costs, and facility-sustaining costs in Allied West's cost structure. Materials-handling labor costs are output unit-level costs because these costs vary with individual units of furniture shipped. Delivery-processing costs are batch-level costs because they vary with the number of shipments (the number of batches of furniture shipped) rather than with each individual piece of furniture shipped. The costs of materials-handling equipment are customer-sustaining costs because these costs are related to specific customers and not to

Concepts in Action

OPPORTUNITY COSTS IN ENERGY TRANSMISSION[a]

Electric utility companies own and operate transmission lines that carry power to their customers. The Federal Energy Regulatory Commission (FERC) regulates electric utilities in the United States. In approving the merger between Northeast Utilities and the Public Service Company of New Hampshire, FERC restricted the amount of Northeast's existing transmission capacity that the utility could reserve for its own use to the capacity needed to provide reliable service to its own customers. The excess transmission capacity was to be made available to third parties (other utilities and users), a setup called open-access, which cost Northeast Utilities the opportunity to use the surplus transmission capacity for its own uses—for example, to supply extra power to its customers at cheaper rates. The question at issue: What cost will third parties pay to use the transmission capacity?

In its original ruling, FERC required third-party users to pay only for the actual out-of-pocket costs that Northeast Utilities incurs for the transmissions. Under this ruling, third-party users captured the benefit of the excess transmission capacity, which was built for and paid by Northeast Utilities customers. FERC quickly realized, however that transmission capacity is limited, so the cost of third-party transmission is not limited to the out-of-pocket cost of transmission. Northeast Utilities also incurs an opportunity cost because it is not able to use the capacity for the benefit of its customers. That is, the cost of the transmission capacity third parties take up includes the "lost profits" the utility could have earned by using the transmission capacity for its own use.

The question is open. FERC must ensure that utilities do not take advantage of their monopoly over transmission capacity, but how much of the burden of transmission costs for third parties should Northeast Utilities bear?

[a]Adapted from C. M. Studness, "The FERC Edges toward Opportunity-Cost Pricing for Transmission," *Public Utilities Fortnightly* (March 15, 1992).

EXHIBIT 11-8

Customer Profitability Analysis at Allied West
(in thousands)

	Vogel	Brenner	Wisk	Total
Sales	$500	$300	$400	$1,200
Operating costs				
Cost of goods sold (variable)	370	220	330	920
Materials-handling labor	42	20	30	92
Materials-handling equipment (depreciation)	10	6	8	24
Rent	14	8	14	36
Marketing support	11	9	10	30
Purchase orders and delivery processing	13	7	12	32
General administration	20	12	16	48
Total operating costs	480	282	420	1,182
Operating income	$ 20	$ 18	$ (20)	$ 18

the individual units or batches of furniture shipped. General administration costs are facility-sustaining costs because these costs cannot be identified with specific units of furniture, batches, or customers. These costs support the business as a whole. Identifying the level at which a cost varies is helpful in relevant-cost analysis. For example, when computing the savings in delivery-processing costs from dropping a customer account (say, Wisk), the cost analyst needs to determine the number of shipments that Allied West dispatched to Wisk (a batch-level activity) and not the number of individual pieces of furniture shipped to Wisk.

Exhibit 11-8 indicates a loss of $20,000 on sales to Wisk. Allied West's manager believes this loss occurred because Wisk places many low-volume orders with Allied. Should Allied West discontinue the Wisk account?

The key question is, What are the relevant costs and relevant revenues? The following information is available.

1. Dropping the Wisk account will save cost of goods sold, materials-handling labor, marketing support, purchase-order, and delivery-processing costs incurred on the Wisk account.

2. Dropping the Wisk account will mean that the warehouse space currently occupied by Wisk products and the materials-handling equipment used to move Wisk products will become idle.

3. Dropping the Wisk account will have no effect on fixed general administration costs.

Exhibit 11-9 presents the relevant-cost computations. Allied West's operating income will be $18,000 higher if it keeps the Wisk account. The last column in Exhibit 11-9 shows that the cost savings from dropping the Wisk account, $382,000, is not enough to offset the loss of $400,000 in revenue. The key reason is that depreciation, rent, and general administration costs will not decrease if the Wisk account is dropped.

The conclusion would be quite different if, after dropping the Wisk account, Allied could lease the extra warehouse space for $20,000 per year. The $20,000 that Allied would receive is the opportunity cost of continuing to use the warehouse to service Wisk. Allied would gain $2,000 ($20,000 from lease revenue minus $18,000 loss from dropping the Wisk account) by dropping the Wisk account and leasing out the warehouse. Before coming to a final decision, however, Allied must consider the possibility of changing Wisk into a profitable customer and also consider qualitative factors, such as the effect the decision might have on Allied's reputation for developing stable, long-run business relationships.

EXHIBIT 11-9

Relevant-Cost Analysis for Dropping Wisk Account
(in thousands)

	Amount of Total Revenues and Total Costs		Difference: (Loss in Revenue) and Savings in Costs from Dropping Wisk Account
	Keep Wisk Account	Drop Wisk Account	
Sales	$1,200	$ 800	$(400)
Operating costs			
Cost of goods sold (variable)	920	590	330
Materials-handling labor	92	62	30
Materials-handling equipment (depreciation)	24	24	0
Rent	36	36	0
Marketing support	30	20	10
Purchase orders and delivery processing	32	20	12
General administration	48	48	0
Total operating costs	1,182	800	382
Operating income (Loss)	$ 18	$ 0	$ (18)

IRRELEVANCE OF PAST COSTS

Objective 8

Explain why the book value of equipment is irrelevant in equipment-replacement decisions

As defined earlier, a relevant cost is (1) an expected future cost that (2) will differ among alternatives. The illustrations in this chapter have shown that expected future costs that do not differ among alternatives are irrelevant. Now we return to the idea that all past costs are irrelevant.

Consider an example of equipment replacement. The irrelevant cost illustrated here is the **book value** (original cost minus accumulated depreciation) of the existing equipment. Assume that the Toledo Company is considering replacing a metal-cutting machine for aircraft parts with a more technically advanced model. The new machine has an automatic quality-testing capability and is more efficient than the old machine. The new machine, however, has a shorter life. Toledo Company uses the straight-line depreciation method. Revenue from aircraft parts ($1.1 million per year) will be unaffected by the replacement decision. Summary data on the existing machine and the replacement machine follow:

	Existing Machine	Replacement Machine
Original cost	$1,000,000	$600,000
Useful life in years	5	2
Current age in years	3	0
Useful life remaining in years	2	2
Accumulated depreciation	$ 600,000	$ 0
Book value	$ 400,000	Not acquired yet
Current disposal price (in cash now)	$ 40,000	Not acquired yet
Terminal disposal price (in cash 2 years from now)	$ 0	$ 0
Annual cash operating costs (maintenance, energy, repairs, coolants, and so on)	$ 800,000	$460,000

To focus on the main concept of relevance, we ignore the time value of money in this illustration.

EXHIBIT 11-10

Cost Comparison—Replacement of Machinery, Including Relevant and Irrelevant Items
(in thousands)

	Two Years Together		
	Keep	Replace	Difference
Sales	$2,200	$2,200	$—
Costs			
Cash operating costs	1,600	920	680
Old machine book value:			
Periodic write-off as depreciation, or	400	—	
Lump-sum write-off	—	400*	—
Current disposal price of old machine	—	(40)*	40
New machine, written off periodically as depreciation	—	600	(600)
Total costs	2,000	1,880	120
Operating income	$ 200	$ 320	$120

*In a formal income statement, these two items would be combined as "loss on disposal of machine" of $360,000.

Exhibit 11-10 presents a cost comparison of the two machines. Some managers would not replace the old machine because it would entail recognizing a $360,000 "loss on disposal"($400,000 book value minus $40,000 current disposal price); retention would allow spreading the $400,000 book value over the next two years in the form of "depreciation expense" (a term more appealing than "loss on disposal").

We can apply our definition of relevance to four commonly encountered items in equipment-replacement decisions, such as the one facing Toledo Company:

1. *Book value of old machine.* Irrelevant, because it is a past (historical) cost. All past costs are down the drain. Nothing can change what has already been spent or what has already happened.
2. *Current disposal price of old machine.* Relevant, because it is an expected future cash inflow that differs between alternatives.
3. *Gain or loss on disposal.* This is the algebraic difference between items 1 and 2. It is a meaningless combination blurring the distinction between the irrelevant book value and the relevant disposal price. Each item should be considered separately.
4. *Cost of new machine.* Relevant, because it is an expected future cash outflow that will differ between alternatives.

Exhibit 11-10 should clarify these four assertions. The difference column in Exhibit 11-10 shows that the book value of the old machine is not an element of difference between alternatives and could be completely ignored for decision-making purposes. *No matter what the timing of the charge against revenue, the amount charged is still $400,000 regardless of the alternative chosen.* Note that the advantage of replacing is $120,000 for the two years together.

In either event, the undepreciated cost will be written off with the same ultimate effect on operating income. The $400,000 enters into the income statement either as a $400,000 offset against the $40,000 proceeds to obtain the $360,000 loss on disposal in the current year or as $200,000 depreciation in each of the next two years. But how it appears is irrelevant to the replacement decision. In contrast, the $600,000 cost of the new machine is relevant because it can be avoided by deciding not to replace.

Exhibit 11-11 concentrates on relevant items only. Note that the same answer (the $120,000 net difference) will be produced even though the book value is completely omitted from the calculations. The only relevant items are the cash operating costs, the disposal price of the old machine, and the cost of the new machine (represented as depreciation in Exhibit 11-11).

EXHIBIT 11-11

Cost Comparison—Replacement of Machinery, Relevant Items Only
(in thousands)

		Two Years Together	
	Keep	Replace	Difference
Cash operating costs	$1,600	$ 920	$680
Current disposal price of old machine	—	(40)	40
New machine, written off periodically as depreciation	—	600	(600)
Total relevant costs	$1,600	$1,480	$120

Decision makers can vary in their preference between the formats in Exhibits 11-10 and 11-11. Some prefer the format used in Exhibit 11-10 because it illustrates why some items are irrelevant to the decision. Other managers prefer the format used in Exhibit 11-11 because it is concise.

Past costs that are unavoidable because they cannot be changed no matter what action is taken are sometimes described as **sunk costs**. In our example, old equipment has a book value of $400,000 and current disposal price of $40,000. What are the sunk costs in this case? The entire $400,000 is sunk because it represents an outlay made in the past that cannot be changed. Because all past costs are irrelevant, we do not emphasize the distinction between past costs and sunk costs.

Consider the Allied West example again (p. 400). *Suppose Allied West has only Vogel and Brenner as its customers.* Allied is evaluating the profitability of adding Wisk as a customer. Allied is already paying rent of $36,000 for the warehouse and is incurring general administration costs of $48,000. Suppose Allied predicts other revenues and costs of doing business with Wisk as described under the Wisk column of Exhibit 11-8. Should Allied West add Wisk as a customer? Exhibit 11-12 presents a relevant revenue and relevant cost comparison. Relevant revenues exceed relevant costs by $10,000. If the only available alternatives are to maintain the status quo or to add the

EXHIBIT 11-12

Relevant-Cost Analysis for Adding Wisk Account
(in thousands)

	Amount of Total Revenues and Total Costs		
	Do Not Add Wisk Account	Add Wisk Account	Difference
Sales	$800	$1,200	$400
Operating costs			
Cost of goods sold (variable)	590	920	330
Materials-handling labor	62	92	30
Materials-handling equipment (depreciation)	16	24	8
Rent	36	36	0
Marketing support	20	30	10
Purchase orders and delivery processing	20	32	12
General administration	48	48	0
Total operating costs	792	1,182	390
Operating income	$ 8	$ 18	$ 10

Wisk account (that is, there are no other opportunities), adding Wisk as a customer is the preferred alternative. The key point to note is that the cost of acquiring new equipment to support the Wisk order (represented by depreciation of $8,000 in Exhibit 11-12) is included as a relevant cost. Why? Because this cost can be avoided if Allied decides not to do business with Wisk. Note the critical distinction here. Depreciation cost is irrelevant in deciding whether to drop Wisk as a customer (because it is a past cost), but it is relevant in deciding whether to add Wisk as a new customer (because it is a future cost of acquiring the necessary equipment that arises as a result of the decision to accept Wisk as a customer).

HOW MANAGERS BEHAVE

Objective 9
Explain how conflicts can arise between the decision model used by a manager and the performance model used to evaluate the manager

Consider our equipment-replacement example (p. 403) in light of the five-step sequence in Exhibit 11-1 (p. 387).

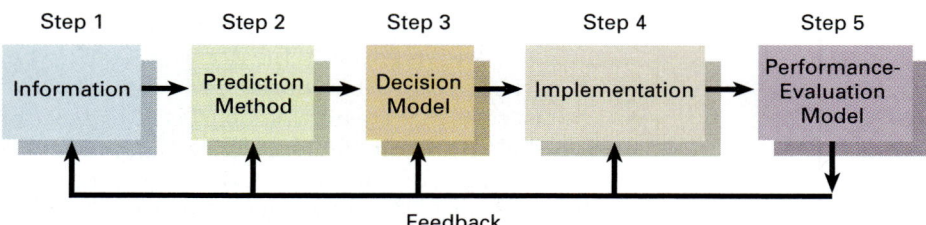

Step 1	Step 2	Step 3	Step 4	Step 5
Information	Prediction Method	Decision Model	Implementation	Performance-Evaluation Model

Feedback

Impact of Reported Loss

If the decision model (step 3) demands choosing the alternative that will minimize total costs over the life span of the equipment, then the analysis in Exhibits 11-10 and 11-11 dictates replacing rather than keeping. In the real world, however, would the manager replace? The answer depends on the manager's perceptions of whether the decision model is consistent with the performance-evaluation model (step 5). The performance-evaluation model describes the basis on which the manager's performance is judged.

Managers tend to favor the alternative that makes their performance look best, so they focus on the measures used in the performance-evaluation model. If the performance-evaluation model conflicts with the decision model, the performance-evaluation model often prevails in influencing a manager's behavior. For example, the decision model in Exhibit 11-10, based on a relevant-cost analysis over the life of the two machines, favors replacing the machine. But if the subordinate manager's promotion or bonus hinges on the first year's operating income performance under accrual accounting, the subordinate's temptation *not* to replace will be overwhelming. Why? Because the accrual accounting model for measuring performance will show a first-year operating loss if the old machine is replaced:

	First Year			
	Keep		**Replace**	
Revenues		$1,100		$1,100
Costs				
Operating costs	$800		$460	
Depreciation	200		300	
Loss on disposal	—		360	
Total costs		1,000		1,120
Operating income (loss)		$ 100		$ (20)

On the basis of these data, the subordinate's performance will look much better in the first year by keeping the old machine rather than by replacing it and showing a loss. Even if top management's goals are long-run (and consistent with the decision model), the subordinate manager's concern is more likely to be short-run if the manager's evaluation is based on short-run measures.

Synchronizing the Models

Resolving the conflict between the decision model and the performance-evaluation model is frequently a baffling problem in practice. In theory, resolving the difficulty seems obvious—merely design consistent models. Consider our replacement example. Year-by-year effects on operating income of replacement can be budgeted over the planning horizon of two years. The first year would be expected to be poor, the next year much better. If the subordinate manager knows that top management will evaluate his or her performance over the longer run, the subordinate is more likely to make decisions that have a long-run orientation.

The practical difficulty is that accounting systems rarely track each decision separately. Performance evaluation focuses on responsibility centers for a specific time period, not on projects or individual items of equipment for their entire useful lives. Therefore, the impacts of many different decisions are combined in a single performance report. Top management, through the reporting system, is rarely aware of particular desirable alternatives *not* chosen by subordinates. Chapter 26 further discusses problems of synchronizing decision models and performance-evaluation models.

PROBLEM FOR SELF-STUDY

PROBLEM

Wally Lewis is manager of the engineering development division of Goldcoast Products, Inc. This division provides contract research for the operating divisions of Goldcoast Products. Lewis has just received a proposal signed by all 10 of his engineers to replace the existing mainframe computing system with 10 workstations. Workstations are minicomputers with extensive memory capacity and work capabilities. Lewis is not enthusiastic about the proposal. The mainframe was purchased only two years ago for $300,000 and has a remaining useful life of three years.

The workstations will cost $13,500 each and have a useful life of three years. Straight-line depreciation is used for all computer equipment at Goldcoast Products. Given the pace of technology, Lewis believes both the mainframe and the workstations will have zero disposal prices in three years' time. Annual cash operating costs for the mainframe are $40,000. Annual cash operating costs for the 10 workstations are $10,000 (10 × $1,000). The current disposal price of the mainframe is $95,000.

Annual revenues of the engineering development division of $1,000,000 and non-computer-related costs of $880,000 are both expected to be unaffected by the computer-equipment replacement decision.

Lewis's annual bonus includes a component based on division operating income. He is keen to maintain his track record of the last three years of increasing division operating income each year. He has a promotion possibility next year that would make him a group vice-president of Goldcoast Products.

Required
1. Summarize the financial data for the two alternatives: (a) keep the mainframe and (b) replace the mainframe with workstations.
2. Tabulate a comparison that includes both relevant and irrelevant items. Combine the three years together in a format similar to Exhibit 11-10.
3. Tabulate a comparison of all relevant items for the next three years together.
4. Why might Lewis be reluctant to purchase the 10 workstations?

SOLUTION

1.

	Keep Mainframe	Proposed Workstations
Original cost	$ 300,000	$ 135,000
Useful life in years	5	3
Current age in years	2	0
Remaining useful life in years	3	3
Accumulated depreciation	$ 120,000	Not acquired yet
Current book value	$ 180,000	Not acquired yet
Current disposal price (in cash now)	$ 95,000	Not acquired yet
Terminal disposal price (in cash 3 years from now)	$ 0	$ 0
Annual computer-related cash operating costs	$ 40,000	$ 10,000
Annual revenues	$1,000,000	$ 1,000,000
Annual non-computer-related operating costs	$ 880,000	$ 880,000

This table includes two types of irrelevant items: (a) past costs (for example, the original cost of the mainframe) and (b) future revenues or costs that are expected to be the same whether the mainframe is retained or replaced with workstations (for example, the annual revenues).

2.

	Three Years Together		
Relevant and Irrelevant Items	Mainframe	Workstations	Difference
Revenues	$3,000,000	$3,000,000	—
Costs			
Non-computer-related operating costs	2,640,000	2,640,000	—
Computer-related cash operating costs	120,000	30,000	$ 90,000
Mainframe book value:			
Periodic write-off as depreciation, or	180,000	—	
Lump-sum write-off	—	180,000	—
Current disposal price of old machine	—	(95,000)	95,000
New workstations, written off periodically as depreciation	—	135,000	(135,000)
Total costs	2,940,000	2,890,000	50,000
Operating Income	$ 60,000	$ 110,000	$ 50,000

This table includes two types of irrelevant items: (a) past costs (for example, the original cost of the mainframe), and (b) future revenues or costs that are expected to be the same whether the mainframe is retained or replaced with workstations (for example, the annual revenues).

3.

	Three Years Together		
Relevant Items	Mainframe	Workstations	Difference
Computer-related cash operating costs	$120,000	$ 30,000	$ 90,000
Current disposal price of mainframe	—	(95,000)	95,000
Depreciation—workstations	—	135,000	(135,000)
Total relevant costs	$120,000	$ 70,000	$ 50,000

The cost comparison in requirement 3 highlights the three relevant cost items for Goldcoast Products, Inc. The basic issue is whether an investment of $40,000 now (the $135,000 cost of the workstations minus the $95,000 current disposal price of the mainframe) to gain a reduction in annual computer-related operating costs of $30,000 for the next three years is justified.

4. If Goldcoast Products replaces the mainframe with the workstations, the operating income of Lewis's division will include an $85,000 operating loss (book value of $180,000 minus $95,000 current disposal price) on the disposal of the mainframe. The operating loss for the current year would be $20,000:

Revenues		$1,000,000
Costs		
Non-computer-related operating costs	$880,000	
Computer-related cash operating costs	10,000	
Depreciation—workstations	45,000	
Loss on disposal of mainframe	85,000	
Total costs		1,020,000
Operating income (loss)		$ (20,000)

Lewis would probably react negatively to the expected operating loss of $20,000 because it would eliminate the component of his bonus based on operating income. He might also perceive that the $20,000 operating loss would reduce his chances of a promotion to a group vice-president.

SUMMARY

The following points are linked to the chapter's learning objectives.

1. The five steps in a decision process are (a) obtaining information (b) making predictions (c) building decision models (d) implementing decisions, and (e) evaluating performance.

2. Accountants can help managers make better decisions by distinguishing relevant from irrelevant revenues and costs. To be relevant to a particular decision, a revenue or cost must meet two criteria: (a) It must be an expected future revenue or cost, and (b) it must differ among alternative courses of action.

3. The consequences of alternative actions can be quantitative and qualitative. Quantitative factors are outcomes that are measured in numerical terms. Some quantitative factors can be easily expressed in financial terms. Qualitative factors, such as employee morale, cannot be measured in numerical terms. Due weight must be given to both types of factors in making decisions.

4. There are two common pitfalls in relevant cost analysis. To avoid them (a) do not assume all variable costs are relevant, and (b) do not assume all fixed costs are irrelevant.

5. Unit-cost data can mislead decision makers in two major ways: (a) when costs that are irrelevant to a particular decision are included in unit costs, and (b) when unit costs that are computed at different output levels are used to choose among alternatives. Unitized fixed costs are often erroneously interpreted as if they behave like unit variable costs. Generally, use total costs rather than unit costs in relevant-cost analysis.

6. In choosing which of multiple products to produce when capacity is constrained, managers should emphasize the product that yields the highest contribution margin per unit of the constraining, or limiting, factor.

7. Opportunity cost is the maximum available contribution to income that is forgone (rejected) by not using a limited resource in its best alternative use. The idea of an opportunity cost arises when there are multiple uses for resources and some alternatives are not selected. It is included in decision making because it represents the best alternative way in which a firm may have used its resources had it not made the decision it did.

8. Past revenues and costs, though irrelevant, are useful because they provide evidence that often helps sharpen predictions of future relevant revenues and costs. But expected future revenues and costs are the only revenue and cost ingredients in any decision model. Book value of existing equipment in equipment-replacement decisions represents past (historical) cost and is therefore irrelevant. Only expected future revenues and costs are relevant to any decision about the future.

9. Top management faces a persistent challenge: making sure that the performance-evaluation model is consistent with the decision model. A common inconsistency is to tell subordinate managers to take a multiple-year view in their decision models but to judge their performance only on the basis of the current year's (operating income) results.

APPENDIX: LINEAR PROGRAMMING

Linear programming (LP) is a solution technique used to maximize total contribution margin (the objective function), given multiple constraints. LP models typically assume that all costs can be classified as either variable or fixed with respect to a single driver (units of output). LP models also require certain other linear assumptions to hold. When these assumptions fail, consider other decision models.[3]

Consider the Power Engines example described in the chapter (p. 394). Suppose that both the snowmobile engine and the boat engine must be tested on a very expensive machine before the engines are shipped to customers. The testing machine resources are limited. Production data follow:

Department	Available Capacity in Hours	Use of Capacity in Hours per Unit of Product		Daily Maximum Production in Units	
		Snowmobile Engine	Boat Engine	Snowmobile Engine	Boat Engine
Department 1: Assembly	600 machine-hours	2.0	5.0	300	120
Department 2: Testing	120 testing hours	1.0	0.5	120	240

Suppose a department works exclusively on a single product (engine type). The table indicates that the assembly department can assemble a maximum of 300 snowmobile engines (600 machine-hours ÷ 2.0 machine-hours per unit = 300 units) or 120 boat engines (600 machine-hours ÷ 5.0 machine-hours per unit = 120 units). Similarly, the testing department can test 120 snowmobile engines (120 testing hours ÷ 1.0 testing-hour per unit = 120 units) or 240 boat engines (120 testing hours ÷ 0.5 testing hours per unit = 240 units).

Exhibit 11-13 summarizes these and other relevant data. Note that snowmobile engines have a contribution margin of $200 per engine and that boat engines have a contribution margin of $250 per engine. Material shortages for boat engines will limit production to 110 boat engines per day. How many engines of each type should be produced daily to maximize operating income?

Steps in Solving an LP Problem

We use the data in Exhibit 11-13 to illustrate the three steps in solving an LP problem. Throughout this discussion, S equals the number of units of snowmobiles produced and B equals the number of units of boat engines produced.

STEP 1: *DETERMINE THE OBJECTIVE.* The **objective function** of a linear program expresses the objective or goal to be maximized (for example, operating income) or minimized (for example, operating costs). In our example, the objective is to find the combination of products that maximizes total contribution margin in the short run.

---

[3]Other decision models are described in G. Eppen, F. Gould, and C. Schmidt, *Quantitative Concepts for Management* (Englewood Cliffs, NJ: Prentice Hall, 1991); and S. Nahmias, *Production and Operations Analysis* (Homewood, IL: Irwin, 1993).

EXHIBIT 11-13

Power Engines: Operating Data

| | Capacities (per Day) in Product Units | | | | |
Product	Department 1: Assembly	Department 2: Testing	Selling Price per Unit	Variable Costs per Unit	Contribution Margin per Unit
If only snowmobile engines produced	300	120	$800	$600	$200
If only boat engines produced	120	240	$950	$700	$250

Fixed costs remain the same regardless of the product mix chosen and are, therefore, irrelevant. The linear function expressing the objective is:

$$\text{Total contribution margin (TCM)} = \$200S + \$250B$$

STEP 2: *DETERMINE THE BASIC RELATIONSHIPS.* These relationships include constraints expressed in linear functions. A **constraint** is a mathematical inequality or equality that must be satisfied by the variables in a mathematical model. The following inequalities depict the relationships in our example:

Department 1 constraint (assembly)	$2S + 5B \leq 600$
Department 2 constraint (testing)	$1S + 0.5B \leq 120$
Material shortage constraint for product B	$B \leq 110$
Negative production is impossible	$S \geq 0$ and $B \geq 0$

The coefficients of the constraints are often called technical coefficients. For example, in the assembly department the technical coefficient for snowmobile engines is 2 machine-hours, and for boat engines is 5 machine-hours.

The three solid lines on the graph in Exhibit 11-14 show the existing constraints for departments 1 (assembly) and 2 (testing) and the material shortage constraint.[4] The feasible alternatives are those combinations of quantities of snowmobile engines and boat engines that satisfy all the constraining factors. The "area of feasible solutions" in Exhibit 11-14 shows the boundaries of those product combinations that are feasible, or technically possible.

STEP 3: *COMPUTE THE OPTIMAL SOLUTION.* In step 2, we concentrate on physical relationships. We now return to the economic relationships expressed as the objective function in step 1. We present the trial-and-error solution approach and the graphic solution approach. These approaches are easy to use in our example because there are only two variables in the objective function and a small number of constraints. An understanding of these two approaches provides insight into LP modeling. In most real-world LP applications, managers use computer software packages to calculate the optimal solution.

TRIAL-AND-ERROR SOLUTION APPROACH. The optimal solution can be found by trial and error, by working with coordinates of the corners of the area of feasible solutions. The approach is simple.

First, select any set of corner points and compute the total contribution margin (TCM). Five corner points appear in Exhibit 11-14. It is helpful to use simultaneous

---

[4]As an example of how the lines are plotted in Exhibit 11-14, use equal signs instead of inequality signs and assume for department 1 (assembly) that $B = 0$; then $S = 300$ (600 machine-hours ÷ 2 machine-hours per snowmobile engine). Assume that $S = 0$; then $B = 120$ (600 machine-hours ÷ 5 machine-hours per boat engine). Connect those two points with a straight line.

EXHIBIT 11-14
Power Engines: Linear Programming—Graphic Solution

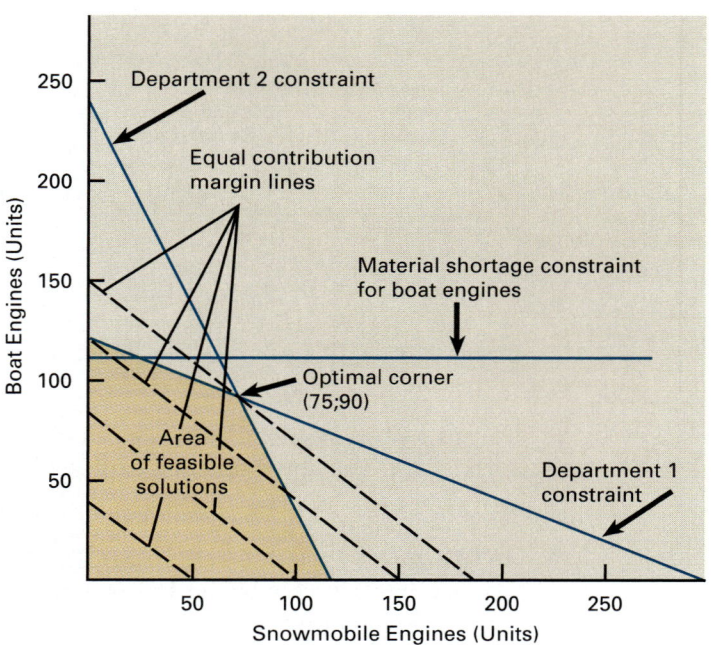

equations to obtain the graph coordinates. To illustrate, the point $(S = 75; B = 90)$ can be derived by solving the two pertinent constraint inequalities as simultaneous equations:

(1) $\qquad 2S + 5B = 600$

(2) $\qquad 1S + 0.5B = 120$

Multiply (2) by 2.0:

(3) $\qquad 2S + 1B = 240$

Subtract (3) from (1):

$$4B = 360$$
$$B = 360 \div 4 = 90$$

Substitute for B in (2):

$$1S + 0.5 (90) = 120$$
$$S = 120 - 45 = 75$$

Given $S = 75$ and $B = 90$:

$$TCM = \$200(75) + \$250(90) = \$37,500$$

Second, move from corner point to corner point, computing the total contribution margin at each corner point. The total contribution margin, at each corner point follows:

Trial	Corner Point (S;B)	Snowmobile Engines	Boat Engines	Total Contribution Margin		
1	(0;0)	0	0	$200(0)	+ $250(0)	= $ 0
2	(0;110)	0	110	$200(0)	+ $250(110)	= 27,500
3	(25;110)	25	110	$200(25)	+ $250(110)	= 32,500
4	(75;90)	75	90	$200(75)	+ $250(90)	= 37,500*
5	(120;0)	120	0	$200(120)	+ $250(0)	= 24,000

*Indicates optimal solution.

The optimal product mix is the mix that yields the highest total contribution—75 snowmobile engines and 90 boat engines.

GRAPHIC SOLUTION APPROACH. Consider all possible combinations that will produce an equal total contribution margin of, say, $10,000. That is:

$$\$200S + \$250B = \$10,000$$

This set of $10,000 contribution margins is a straight dashed line through $(S = 50; B = 0)$ and $(S = 0; B = 40)$. Other equal total contribution margins can be represented by lines parallel to this one. In Exhibit 11-14, we show four dashed lines. The equal total contribution margins increase as the lines get farther from the origin because lines drawn farther from the origin represent more sales of both snowmobile engines and boat engines.

The optimal line is the one farthest from the origin but still at a point in the area of feasible solutions. This line represents the highest contribution margin. The optimal solution is the point at the corner $(S = 75; B = 90)$. This solution will become apparent if you put a ruler on the graph and move it outward from the origin and parallel with the $10,000 line. The idea is to move the ruler as far away from the origin as possible (that is, to increase the total contribution margin) without leaving the area of feasible solutions. In general, the optimal solution in a maximization problem lies at the corner where the dashed line intersects an extreme point of the area of feasible solutions. Moving the ruler out any farther puts it outside the feasible region.

The slope of the objective function (the dashed line representing the equal total contribution margin) can be computed from the equation

$$TCM = \$200S + \$250B$$

To find the slope (the rate of change of B for one additional unit of S), divide by the coefficient of B and write the equation for B:

$$\frac{TCM}{\$250} = \frac{\$200}{\$250}S + B$$

$$B = \frac{TCM}{\$250} - \frac{\$200}{\$250}S$$

$$B = \frac{TCM}{\$250} - \frac{4}{5}S$$

The slope of the objective function is a negative $200 ÷ $250, or − 4 ÷ 5. The graphic solution approach provides insight into the computation of the optimal solution. Its use, however, is restricted to LP problems with two products in the objective function (so that the solution can be represented on a two-dimensional graph).[5]

[5]Although the trial-and-error and graphic approaches can be useful for two or possibly three variables, they are impractical when many variables exist. Standard computer software packages rely on the simplex method. The *simplex method* is an iterative step-by-step procedure for determining the optimal solution to an LP problem. It starts with a specific feasible solution and then tests it by substitution to see if the result can be improved. These substitutions continue until no further improvement is possible, and thus the optimal solution is obtained.

TERMS TO LEARN

This chapter and the Glossary at the end of the book contain definitions of the following important terms:

book value *(p. 403)*   business function costs *(393)*   constraint *(411)*
decision model *(386)*   differential cost *(389)*   full product cost *(393)*
imputed costs *(400)*   incremental cost *(389)*   insourcing *(394)*
net relevant cost *(389)*   objective function *(410)*   opportunity cost *(399)*
outlay costs *(392)*   out-of-pocket costs *(392)*   outsourcing *(394)*
qualitative factors *(389)*   quantitative factors *(389)*   relevant costs *(388)*
relevant revenues *(388)*   sunk costs *(405)*

ASSIGNMENT MATERIAL

QUESTIONS

11-1 Define *decision model.*

11-2 Outline the five-step sequence in a decision process.

11-3 Define *relevant cost.* Why are historical costs irrelevant?

11-4 "All future costs are relevant." Do you agree? Why?

11-5 Distinguish between *quantitative* and *qualitative* factors in decision making.

11-6 "Variable costs are always relevant, and fixed costs are always irrelevant." Do you agree? Why?

11-7 Name two ways unit-cost data can mislead a decision maker.

11-8 "Management should always maximize sales of the product with the highest contribution margin per unit." Do you agree? Why?

11-9 "A component part should be purchased whenever the purchase price is less than the total unit cost to manufacture the component part." Do you agree? Why?

11-10 Define *opportunity cost.*

11-11 Suppose you are a senior manager with many years of experience. A new employee makes the following comment during a heated disagreement over the purchase of a new machine: "No amount of rhetoric is going to change the fact that all your experience and knowledge is about the past and all your decisions are about the future." How would you respond?

11-12 "Managers will always choose the alternative that maximizes operating income or minimizes costs in the decision model." Do you agree? Why?

11-13 Describe the three steps in solving a linear programming problem.

11-14 State the key assumption made when using the linear programming model.

11-15 How might the optimal solution of a linear programming problem be determined?

EXERCISES AND PROBLEMS

11-16 Disposal of assets.

1. A company has an inventory of 1,000 assorted missile parts for a line of missiles that has been discontinued. The inventory cost is $80,000. The parts can be either (a) remachined at total additional costs of $30,000 and then sold for $35,000 or (b) scrapped for $2,000. Which action should be taken?

2. A truck, costing $100,000 and uninsured, is wrecked the first day in use. It can be either (a) disposed of for $10,000 cash and replaced with a similar truck costing $102,000 or (b) rebuilt for $85,000 and be brand-new as far as operating characteristics and looks are concerned. What should be done?

11-17   The careening personal computer. (W. A. Paton)   An employee in the accounting department of a certain business was moving a personal computer from one room to another. As he came alongside an open stairway, he carelessly slipped and let the computer get away from him. It went careening down the stairs with a great racket and wound up at the bottom, completely wrecked. Hearing the crash, the office manager came rushing out and turned rather pale when he saw what had happened. "Someone tell me quickly," the manager yelled, "if that is one of our fully depreciated items." A check of the accounting records showed that the smashed computer was, indeed, one of those items that had been written off. "Thank God!" said the manager.

Required
Explain and comment on the point of this anecdote.

11-18   Inventory decision, opportunity cost.   A manufacturer of lawn mowers predicts that 240,000 spark plugs will have to be purchased during the next year. The manufacturer estimates that 20,000 spark plugs will be required each month. A supplier quotes a price of $8 per spark plug. The supplier also offers a special discount option: If all 240,000 spark plugs are purchased at the start of the year, a discount of 5% off the $8 price will be given. The manufacturer can invest its cash at 8% per year.

Required
1. What is the opportunity cost of interest forgone from purchasing all 240,000 units at the start of the year instead of in 12 monthly purchases of 20,000 units per order?
2. Would this opportunity cost ordinarily be recorded in the accounting system? Why?

11-19   Multiple choice. (CPA)   Choose the best answer:

1. Woody Company, which manufactures slippers, has enough idle capacity available to accept a one-time-only special order of 20,000 pairs of slippers at $6 a pair. The normal selling price is $10 a pair. Variable manufacturing costs are $4.50 a pair, and fixed manufacturing costs are $1.50 a pair. Woody will not incur any marketing costs as a result of the special order. What would the effect on operating income be if the special order could be accepted without affecting normal sales? (a) $0, (b) $30,000 increase, (c) $90,000 increase, (d) $120,000 increase.

2. The Reno Company manufactures part no. 498 for use in its production line. The manufacturing costs per unit for 20,000 units of part no. 498 are as follows:

Direct materials	$ 6
Direct manufacturing labor	30
Variable overhead	12
Fixed overhead allocated	16
	$64

The Tray Company has offered to sell 20,000 units of part no. 498 to Reno for $60 per unit. Reno will make the decision to buy the part from Tray if there is an overall savings of $25,000 for Reno. If Reno accepts Tray's offer, $9 per unit of the fixed overhead allocated would be totally eliminated. Furthermore, Reno has determined that the released facilities could be used to save relevant costs in the manufacture of part no. 575. For Reno to have an overall savings of $25,000, the amount of relevant costs that would have to be saved by using the released facilities in the manufacture of part no. 575 would have to be (a) $80,000, (b) $85,000, (c) $125,000, (d) $140,000.

11-20   Relevant costs, contribution margin, product emphasis.   The Beach Comber is a take-out food store at a popular beach resort. Susan Sexton, owner of the Beach Comber, is deciding how much refrigerator space to devote to four different drinks. Pertinent data on these four drinks follow:

	Cola	Lemonade	Punch	Natural Orange Juice
Selling price per case	$18.00	$19.20	$26.40	$38.40
Variable costs per case	$13.50	$15.20	$20.10	$30.20
Cases sold per foot of shelf space per day	25	24	4	5

Sexton has a maximum front shelf space of 12 feet to devote to the four drinks. She wants a minimum of 1 foot and a maximum of 6 feet of front shelf for each drink.

Required
1. What is the contribution margin per case of each type of drink?
2. A co-worker of Sexton's recommends that she maximize the shelf space devoted to those drinks with the highest contribution margin per case. Evaluate this recommendation.
3. What shelf-space allocation for the four drinks would you recommend for the Beach Comber?

11-21 Selection of most profitable product. Body-Builders Inc. produces two basic types of weight-lifting equipment, Model 9 and Model 14. Pertinent data follow:

	Per Unit	
	Model 9	**Model 14**
Selling price	$100.00	$70.00
Costs		
Direct materials	28.00	13.00
Direct manufacturing labor	15.00	25.00
Variable manufacturing overhead*	25.00	12.50
Fixed manufacturing overhead*	10.00	5.00
Marketing costs (all variable)	14.00	10.00
Total costs	92.00	65.50
Operating income	$ 8.00	$ 4.50

*Allocated on the basis of machine-hours.

The weight-lifting craze is such that enough of either Model 9 or Model 14 can be sold to keep the plant operating at full capacity. Both products are processed through the same production departments.

Required
Which product should be produced? If both should be produced, indicate the proportions of each. Briefly explain your answer.

11-22 Relevance of equipment costs. The Auto Wash Company has just today paid for and installed a special machine for polishing cars at one of its several outlets. It is the first day of the company's fiscal year. The machine cost $20,000. Its annual operating costs total $15,000, exclusive of depreciation. The machine will have a four-year useful life and a zero terminal disposal price.

After the machine has been used one day, a machine salesperson offers a different machine that promises to do the same job at a yearly operating cost of $9,000, exclusive of depreciation. The new machine will cost $24,000 cash, installed. The "old" machine is unique and can be sold outright for only $10,000, minus $2,000 removal cost. The new machine, like the old one, will have a four-year useful life and zero terminal disposal price.

Sales, all in cash, will be $150,000 annually, and other cash costs will be $110,000 annually, regardless of this decision.

For simplicity, ignore income taxes, interest, and present value considerations.

Required
1. Prepare a statement of cash receipts and disbursements for each of the four years under both alternatives. What is the cumulative difference in cash flow for the four years taken together?
2. Prepare income statements for each of the four years under both alternatives. Assume straight-line depreciation. What is the cumulative difference in operating income for the four years taken together?
3. What are the irrelevant items in your presentations in requirements 1 and 2? Why are they irrelevant?
4. Suppose the cost of the "old" machine was $1 million rather than $20,000. Nevertheless, the old machine can be sold outright for only $10,000, minus $2,000 removal cost. Would the net differences in requirements 1 and 2 change? Explain.

5. "To avoid a loss, we should keep the old machine." What is the role of book value in decisions about replacement of machines?

11-23 Opportunity cost. (H. Schaefer)   Wolverine Corp. is working at full production capacity producing 10,000 units of a unique product, Rosebo. Manufacturing costs per unit for Rosebo follow:

Direct materials	$ 2
Direct manufacturing labor	3
Manufacturing overhead	5
	$10

The unit manufacturing overhead cost is based on a variable cost per unit of $2 and fixed costs of $30,000 (at full capacity of 10,000 units). The nonmanufacturing costs, all variable, are $4 per unit, and the selling price is $20 per unit.

A customer, the Miami Co., has asked Wolverine to produce 2,000 units of a modification of Rosebo to be called Orangebo. Orangebo would require the same manufacturing processes as Rosebo, and the Miami Co. has offered to share equally the nonmanufacturing costs with Wolverine. Orangebo will sell at $15 per unit.

Required

1. What is the opportunity cost to Wolverine of producing the 2,000 units of Orangebo? (Assume that no overtime is worked.)
2. The Buckeye Corp. has offered to produce 2,000 units of Rosebo for Wolverine so that Wolverine may accept the Orangebo offer. Buckeye would charge Wolverine $14 per unit for the Rosebo. Should Wolverine accept the Buckeye offer? (Support with specific analysis.)
3. Suppose Wolverine had been working at less than full capacity, producing 8,000 units of Rosebo at the time the Orangebo offer was made. What is the *minimum* price Wolverine should accept for Orangebo under these conditions? (Ignore the $15 price above.)

11-24 Contribution approach, relevant costs.   Air Frisco owns a single jet aircraft and operates between San Francisco and the Fiji Islands. Flights leave San Francisco on Mondays and Thursdays and depart from Fiji on Wednesdays and Saturdays. Air Frisco cannot offer any more flights between San Francisco and Fiji. Only tourist-class seats are available on its planes. An analyst has collected the following information.

Seating capacity per plane	360
Average passengers per flight	200
Flights per week	4
Flights per year	208
Average one-way fare	$500
Variable fuel costs	$14,000 per flight
Food and beverage service costs (no charge to passenger)	$20 per passenger
Commission to travel agents paid by Air Frisco on each ticket booked on Air Frisco. Assume that all of Air Frisco's tickets are booked by travel agents.	8% of fare
Fixed annual lease costs allocated to each flight	$53,000 per flight
Fixed ground services (maintenance, check in, baggage handling) cost allocated to each flight	$7,000 per flight
Fixed salaries of flight crew allocated to each flight	$4,000 per flight

For simplicity, assume that fuel costs are unaffected by the actual number of passengers on a flight.

Required

1. What is the operating income that Air Frisco makes on each one-way flight between San Francisco and Fiji?
2. The market research department of Air Frisco indicates that lowering the average one-way fare to $480 will increase the average number of passengers per flight to 212. Should Air Frisco lower its fare?

3. Travel International, a tour operator, approaches Air Frisco to charter (rent out) its jet aircraft twice each month, first to take Travel International's tourists from San Francisco to Fiji and then to bring the tourists back from Fiji to San Francisco. If Air Frisco accepts Travel International's offer, Air Frisco will be able to offer only 184 (208 − 24) of its own flights each year. The terms of the charter: (a) For each one-way flight Travel International will pay Air Frisco $75,000 to charter the plane and to use its flight crew and ground service staff; (b) Travel International will pay for fuel costs; (c) Travel International will pay for all food costs. On purely financial considerations, should Air Frisco accept Travel International's offer? What other considerations should Air Frisco consider in deciding whether or not to charter its plane to Travel International?

11-25 Make or buy, unknown level of volume. (A. Atkinson)   Oxford Engineering manufactures small engines. The engines are sold to manufacturers who install them in products such as lawn mowers. The company currently manufactures all the parts used in these engines but is considering a proposal from an external supplier who wishes to supply the starter assembly used in these engines.

The starter assembly is currently manufactured in Division 3 of Oxford Engineering. The costs relating to Division 3 for the last 12 months were as follows:

Direct materials	$200,000
Direct manufacturing labor	150,000
Manufacturing overhead	400,000
Total	$750,000

Over the last year, Division 3 manufactured 150,000 starter assemblies; the average cost for the starter assembly is computed as $5 ($750,000 ÷ 150,000).

Further analysis of manufacturing overhead revealed the following information. Of the total manufacturing overhead reported, only 25% is considered variable. Of the fixed portion, $150,000 is an allocation of general overhead that would remain unchanged for the company as a whole if production of the starter assembly is discontinued. A further $100,000 of the fixed overhead is avoidable if self-manufacture of the starter assembly is discontinued. The balance of the current fixed overhead, $50,000, is the division manager's salary. If self-manufacture of the starter assembly is discontinued, the manager of Division 3 will be transferred to Division 2 at the same salary. This move will allow the company to save the $40,000 salary that would otherwise be paid to attract an outsider to this position.

Required

1. Tidnish Electronics, a reliable supplier, has offered to supply starter assembly units at $4 per unit. Since this price is less than the current average cost of $5 per unit, the vice-president of manufacturing is eager to accept this offer. Should the outside offer be accepted? (*Hint:* Production output in the coming year may be different from production output in the last year.)

2. How, if at all, would your response to requirement 1 change if the company could use the vacated plant space for storage and, in so doing, avoid $50,000 of outside storage charges currently incurred? Why is this information relevant or irrelevant?

11-26 Service company, renting capacity, subcontracting, opportunity costs. Roadway Lines is a transport company that moves goods between New York (NY) and Los Angeles (LA). It owns only one truck. Roadway schedules a trip from NY to LA on the 1st of each month and the return trip from LA to NY on the 15th of each month. Roadway reports the following cost information:

a. Variable costs for a one-way trip

(1)	Fuel costs	$ 500
(2)	Driver and helper wages	2,000
(3)	Repairs and maintenance	300
	Total variable costs	$2,800

b. Allocated fixed costs for a one-way trip

(1)	General manager's salary	$1,000
(2)	Depreciation on truck	500
	Total allocated fixed costs	$1,500

c. The cost driver is a single one-way trip.

Roadway is currently accepting bookings for its NY–LA trip on October 1, 19_6. It already has a confirmed booking for October 15, 19_6, for the LA–NY trip. On September 18, 19_6, Roadway receives a request from Brooklyn Enterprises to transport a full truckload of goods on the October 1, 19_6 trip, at a price of $4,500. Brooklyn needs to know immediately if Roadway will accept. The following additional facts are available:

a. If Roadway accepts the offer and decides to use its own truck to transport the goods, it cannot accept any more business for its NY–LA trip on October 1, 19_6.

b. Roadway can rent a truck and driver from Easy Hauling for $3,500. Under this alternative, Roadway will also incur fuel costs of $500.

c. Roadway can subcontract the Brooklyn order to Countrywide Trucking. Under the subcontract terms, Countrywide will transport the goods for a price of $4,200.

d. On the basis of past business, Roadway is certain that it will receive an order at a price of $5,000 from Manhattan Corporation to transport goods from New York to Los Angeles on October 1, 19_6. If Roadway accepts Manhattan's order, Roadway will have to transport Manhattan's goods in its own truck. Manhattan requires the truck to be configured in a special way, and only Roadway's truck meets these requirements.

Required
1. What alternatives are available to Roadway for transporting Brooklyn Enterprises' order?
2. What are the future revenues and future outlay costs under each alternative?
3. Should Roadway Lines accept Brooklyn Enterprises' order? If Roadway accepts Brooklyn's order, how should Roadway transport Brooklyn's goods from NY to LA?

11-27 Which bases to close, relevant-cost analysis, opportunity costs. The U.S. Defense Department has the difficult decision of deciding which military bases to close down. Military and political factors obviously matter, but cost savings are also an important factor. Consider two naval bases located on the West Coast—one in Alameda, northern California, and the other in Everett, Washington. The Navy has decided that it needs permanently only one of those two bases, so one must be shut down. The decision regarding which base to shut down will be made on cost considerations alone. The following information is available.

a. The Alameda base was built at a cost of $10 million. The operating costs of the base are $400 million per year. The base is built on land owned by the federal government, so the Defense Department pays nothing for the use of the property. If the base is closed, the land will be sold to developers for $500 million.

b. The Everett base was built by the U.S. Government at a cost of $15 million on land leased by the Defense Department. The federal government can choose to lease the land for the next 50 years at an annual lease payment of $3 million per year. The land and buildings will immediately revert back to the owner if the base is closed before the term of the lease expires. The operating costs of the base, excluding lease payments, are $300 million.

c. If the Alameda base is closed down, the Defense Department will have to transfer some personnel to the Everett facility. As a result, the yearly operating costs at Everett will increase by $100 million per year. If the Everett facility is closed down, no extra costs will be incurred to operate the Alameda facility.

Required
The California delegation in Congress argues that it is cheaper to close down the Everett base for two reasons: (a) It would save $100 million per year in additional costs required to operate the Everett base, and (b) it would save $3 million dollars per year in lease payments over the next 50 years. (Recall that the Alameda base requires no cash payments for use of the land because the land is owned by the Defense Department.) Do you agree with the California delegation's arguments and conclusions? In your answer, identify and explain all costs that you consider relevant and all costs that you consider irrelevant for the base-closing decision.

11-28 Adding and dropping business segments. (Continuation of the Allied West example described in the chapter)   Refer to the information in Exhibit 11-8 (p. 402) and the information on Allied West's costs on p. 401. Further assume these facts.

a. General administration costs of Allied West of $48,000 include depreciation of $10,000 on furniture and office equipment that has zero terminal disposal price.

b. Allied Furniture has decided to allocate corporate head-office costs to all its warehouse locations, including $25,000 to Allied West. Corporate head-office costs consist of top management salaries and corporate-image advertising costs. Allied West shows a loss of $7,000 after corporate allocations—$18,000 operating income (Exhibit 11-8) minus $25,000 corporate costs. Because Allied West's operating income does not cover corporate costs, Allied Furniture is contemplating closing Allied West.

Required
Show the relevant revenues and relevant costs that help answer the following questions.

1. Should Allied West be shut down?
2. Suppose Allied Furniture had the opportunity to open a new office whose revenues and costs were identical to Allied West. Should Allied Furniture open such an office?

11-29 Make versus buy, activity-based costing, opportunity costs. (N. Melumad and S. Reichelstein, adapted) Ace Bicycle Company produces bicycles. This year's expected production is 10,000 units. Currently, Ace also makes the chains for its bicycles. Ace's accountant reports the following costs for making the 10,000 bicycle chains:

	Per-Unit Costs	Costs for 10,000 Units
Direct materials	$ 4.00	$ 40,000
Direct manufacturing labor	2.00	20,000
Variable manufacturing overhead (power and utilities)	1.50	15,000
Inspection, setup, materials handling		2,000
Machine lease		3,000
Allocated fixed plant administration, taxes, and insurance		30,000
Total costs		$110,000

Ace has received an offer from an outside vendor to supply any number of chains Ace requires at $8.20 per chain. The following additional information is available:

a. Inspection, setup, and materials-handling costs vary with the number of batches in which the chains are produced. Ace produces chains in batch sizes of 1,000 units. Ace estimates that it will produce the 10,000 units in 10 batches.

b. The costs for the machine lease are the payments Ace makes for renting the equipment used in making the chains. If Ace buys all its chains from the outside vendor, it does not need this machine.

Required
1. Assume that if Ace purchases the chains from the outside supplier, the facility where the chains are currently made will remain idle. Should Ace accept the outside supplier's offer at the anticipated production (and sales) volume of 10,000 units?
2. For this question, assume that if the chains are purchased outside, the facilities where the chains are currently made will be used to upgrade the bicycles by adding mud flaps and reflector bars. As a consequence, the selling price on bicycles will be raised by $20. The variable per-unit cost of the upgrade would be $18, and additional tooling costs of $16,000 would be incurred. Should Ace make or buy the chains, assuming that 10,000 units are produced (and sold)?
3. The sales manager at Ace is concerned that the estimate of 10,000 units may be high and believes that only 6,200 units will be sold. Production is cut back, which opens up more work space. Now there is room to add the mud flaps and reflectors whether Ace goes outside for the chains or makes them in-house. At this lower output, Ace will produce the chains in 8 batches of 775 units each. Should Ace purchase the chains from the outside vendor?

11-30 Considering three alternatives. (CMA) Auer Company had received an order for a piece of special machinery from Jay Company. Just as Auer Company completed the machine, Jay Company declared bankruptcy, defaulted on the order, and forfeited the 10% deposit paid on the selling price of $72,500.

Auer's manufacturing manager identified the costs already incurred in the production of the special machinery for Jay as follows:

Direct materials used		$16,600
Direct manufacturing labor		21,400
Overhead allocated:		
Manufacturing		
Variable	$10,700	
Fixed	5,350	16,050
Fixed marketing and administrative		5,405
Total costs		$59,455

Another company, Kaytell Corp., would be interested in buying the special machinery if it is reworked to Kaytell's specifications. Auer offered to sell the reworked special machinery to Kaytell as a special order for a net price of $68,400. Kaytell has agreed to pay the net price when it takes delivery in two months. The additional traceable costs to rework the machinery to the specifications of Kaytell follow:

Direct materials	$ 6,200
Direct manufacturing labor	4,200
	$10,400

A second alternative available to Auer is to convert the special machinery to the standard model. The standard model lists for $62,500. The additional traceable costs to convert the special machinery to the standard model are:

Direct materials	$ 2,850
Direct manufacturing labor	3,300
	$ 6,150

A third alternative for the Auer Company is to sell, as a special order, the machine as is (that is, without modification) for a net price of $52,000. However, the potential buyer of the unmodified machine does not want it for 60 days. The buyer offers a $7,000 down payment with final payment upon delivery.

The following additional information is available regarding Auer's operations:

◆ Sales commission rate on sales of standard models is 2%, and the sales commission rate on special orders is 3%. All sales commissions are calculated on net selling price (that is, list price minus cash discount, if any).

◆ Normal credit terms for sales of standard models are 2/10, n/30 (2/10 means a discount of 2% is given if payment is made within 10 days; n/30 means full amount is due within 30 days). Customers take the discounts except in rare instances. Credit terms for special orders are negotiated with the customer.

◆ The allocation rates for manufacturing overhead and the fixed marketing and administrative costs are as follows:

Manufacturing:	
Variable	50% of direct manufacturing labor cost
Fixed	25% of direct manufacturing labor cost
Marketing and administrative:	
Fixed	10% of the total of direct manufacturing materials, direct manufacturing labor, and manufacturing overhead costs

◆ Normal time required for rework is one month.

◆ A surcharge of 5% of the selling price is placed on all customer requests for minor modifications of standard models.

◆ Auer normally sells a sufficient number of standard models for the company to operate at a volume in excess of the breakeven point.

Auer does not consider the time value of money in analyses of special orders and projects whenever the time period is less than one year, because the effect is not significant.

Required

1. Determine the dollar contribution that each of the three alternatives will add to the Auer Company's operating income.
2. If Kaytell makes Auer a counteroffer, what is the lowest price Auer should accept for the reworked machinery from Kaytell? Explain your answer.
3. Discuss the influence that fixed manufacturing overhead costs should have on the selling prices Auer quotes for special orders when (a) the firm is operating at or below the breakeven point; (b) the firm's special orders constitute efficient utilization of unused capacity above the breakeven point.

11-31 Multiple choice; comprehensive problem on relevant costs. The following are the Class Company's *unit* costs of manufacturing and marketing a high-style pen at a level of 20,000 units per month:

Manufacturing costs:	
Direct materials	$1.00
Direct manufacturing labor	1.20
Variable manufacturing indirect costs	0.80
Fixed manufacturing indirect costs	0.50
Marketing costs:	
Variable	1.50
Fixed	0.90

The following situations refer only to the data given above; there is *no connection* between the situations. Unless stated otherwise, assume a regular selling price of $6 per unit.

Required

Choose the best answer to each of the seven questions. Support each answer with summarized computations.

1. In an inventory of 10,000 items presented on the balance sheet, the unit cost used is (a) $3.00, (b) $3.50, (c) $5.00, (d) $2.20, (e) $5.90.
2. The pen is usually produced and sold at the rate of 240,000 units per year (an average of 20,000 per month). The selling price is $6 per unit, which yields total annual sales of $1,440,000. Total costs are $1,416,000, and operating income is $24,000, or $0.10 per unit. Market research estimates that unit sales could be increased by 10% if prices were cut to $5.80. Assuming the implied cost-behavior patterns to be correct, this action, if taken, would (a) decrease operating income by a net of $7,200; (b) decrease operating income by $0.20 per unit ($48,000) but increase operating income by 10% of sales ($144,000) for a net increase of $96,000; (c) decrease unit fixed costs by 10%, or $0.14, per unit, and thus decrease operating income by $0.06 ($0.20 − $0.14) per unit; (d) increase unit sales to 264,000 units, which at the $5.80 price would give total sales of $1,531,200; costs of $5.90 per unit for 264,000 units would be $1,557,600, and a loss of $26,400 would result; (e) none of these.
3. A cost contract with the government (for 5,000 units) calls for the reimbursement of all costs of manufacturing plus a fixed fee of $1,000. No variable marketing costs are incurred on the government contract. You are required to compare the following two alternatives:

	Alternative A	Alternative B
Sales each month to regular customers	15,000 units	15,000 units
Sales each month to government	0 units	5,000 units

Operating income under alternative B is greater than operating income under alternative A by (a) $1,000, (b) $2,500, (c) $3,500, (d) $300, (e) none of these.

4. Assume the same data with respect to the government contract as in requirement 3 above except that the two alternatives to be compared are:

	Alternative A	Alternative B
Sales each month to regular customers	20,000 units	15,000 units
Sales each month to government	0 units	5,000 units

Operating income under alternative B is greater (less) than operating income under alternative A by (a) $4,000 less, (b) $3,000 greater, (c) $6,500 less, (d) $500 greater, (e) none of these.

5. The company wants to enter a foreign market in which price competition is keen. The company seeks a one-time-only special order for 10,000 units on a minimum-unit-price basis. It expects that shipping costs for this order will amount to only $0.75 per unit, but the fixed costs of obtaining the contract will be $4,000. The company incurs no variable marketing costs other than shipping costs. Domestic business will be unaffected. The selling price to break even is (a) $3.50, (b) $4.15, (c) $4.25, (d) $3.00, (e) $5.00.

6. The company has an inventory of 1,000 units of pens that must be sold immediately at reduced prices. Otherwise the inventory will be worthless. The unit cost that is relevant for establishing the minimum selling price would be (a) $4.50, (b) $4.00, (c) $3.00, (d) $5.90, (e) $1.50.

7. A proposal is received from an outside supplier who will make and ship these pens directly to the Class Company's customers as sales orders are forwarded from Class's sales staff. Class's fixed marketing costs will be unaffected, but its variable marketing costs will be slashed by 20%. Class's plant will be idle, but its fixed manufacturing overhead will continue at 50% of present levels. How much per unit would the company be able to pay the supplier without decreasing operating income? (a) $4.75, (b) $3.95, (c) $2.95, (d) $5.35, (e) none of these.

11-32 Make or buy. (Continuation of 11-31)   Assume that, as in requirement 7 of Problem 11-31, a proposal is received from an outside supplier who will make and ship pens directly to Class Company's customers as sales orders are forwarded from Class's sales staff. If the supplier's offer is accepted, the present plant facilities will be used to make a new pen whose unit costs will be:

Variable manufacturing costs	$5.00
Fixed manufacturing costs	1.00
Variable marketing costs	2.00
Fixed marketing costs	0.50

Total fixed manufacturing overhead will be unchanged from the original level given at the beginning of Problem 11-31. Fixed marketing costs for the new pens are over and above the fixed marketing costs incurred for marketing the old pens at the beginning of Problem 11-31. The new product will sell for $9. The minimum desired operating income on the two products taken together is $50,000 per year.

Required
What is the maximum purchase cost per unit that the Class Company should be willing to pay for subcontracting the production of the old pens?

11-33 Relevant costs, opportunity costs.   Larry Miller, the general manager of Basil Software, scheduled a meeting on June 2, 19_5 with Sally Shields, sales manager, Andy Ashby, accountant, and Ellen Eisner, software operations manager, to discuss the development and release of Basil Software's new version of its spreadsheet package, Easyspread 2.0. Other software firms are working on similar software, but no one has as yet developed a spreadsheet package with the advanced features of Easyspread 2.0. However, it is only a question of time before other software firms have a package that matches Easyspread 2.0. Sally Shields, the sales manager, could hardly control her enthusiasm for the new product: "This product is exactly what the market has been waiting for. We should not delay, by even a single day, the introduction of this product. Let's make July 1, 19_5, the sales release date."

Ellen Eisner: "I don't disagree with Sally's assessment of the market potential for this product, but I have a problem. The threatened strike by our printers caused us to purchase large quantities of User Manuals for Easyspread 1.0. We don't like to store the manuals separately, so we also got extra diskettes duplicated. The manuals and diskettes were then packaged and shrink-wrapped in our colorful boxes. We are currently holding 60,000 completed packages, which equals the expected sales for July, August, and September 19_5 of Easyspread 1.0. I think we should make October 1, 19_5, the expected release date of Easyspread 2.0. This date would enable us to sell all of our inventory of Easyspread 1.0."

Larry Miller: "Sally, do you see any problem with Ellen's suggestion? The inventory of Easyspread 1.0 seems rather large for us to ignore. If we introduced Easyspread 2.0 on July 1, what can we do with the inventory of Easyspread 1.0 that we currently hold?"

Sally Shields: "We currently sell Easyspread 1.0 to our wholesalers and distributors for $150 each. The additional optimization features in Easyspread 2.0 means that we should be able to sell Easyspread 2.0 to our distributors for about $200. We should not ignore the higher profit margins from Easyspread 2.0. It is true, though, that each time we sell one unit of Easyspread 2.0, we forgo the sale of one unit of Easyspread 1.0. Since the expected demand for Easyspread 2.0 is at least as large as the demand for Easyspread 1.0, we may have to throw away the existing inventory of Easyspread 1.0 once we introduce Easyspread 2.0."

Larry Miller: "Andy, you've heard what Sally and Ellen have to say. I would like you to do a detailed analysis of the alternatives, and let me know within a week what you come up with. We need to make a decision on this one way or another and we need to do so soon."

When Andy returned to his office he pulled out the cost records he had developed for Easyspread 1.0 and Easyspread 2.0. The unit costs for the two products are summarized below.

	Easyspread 1.0	Easyspread 2.0
Manuals, diskettes	$ 20	$ 25
Development costs	75	105
Marketing and administration costs	25	30
Total unit costs	$120	$160

The following additional facts are available:

a. Basil contracts with outside vendors to print manuals and duplicate diskettes.

b. Development costs are allocated on the basis of the total costs of developing the software and the anticipated unit sales over the life of the software.

c. Marketing and administration costs are fixed costs in 19_5, incurred to support all activities of Basil Software. Marketing and administration costs are allocated to products on the basis of the budgeted revenues from each of the products.

Required
1. On the basis of financial considerations only, is Basil Software better off introducing Easyspread 2.0 immediately instead of waiting? Explain your conclusion, clearly identifying relevant and irrelevant costs.
2. What other factors might affect Miller's decision?

 11-34 Choice among alternative product designs, demand uncertainty, expected monetary value. (CMA, adapted)   This problem relies on the discussion of uncertainty in Chapter 3, Appendix. Steven Company has been producing component parts and assemblies for use in the manufacture of microcomputers. The company plans to introduce a magnetic-tape-cartridge-backup unit for IBM-compatible microcomputers in the near future.

Steven's Research and Development (R&D) and Market Research departments have been working on this project for an extended period. The development costs of the two departments incurred to date amount to $1,500,000. R&D created several alternative designs for the backup units. Three of the designs were approved for development into prototypes, and from these only one will be manufactured and sold. Market Research has determined that the appropriate selling price would be $540 per unit, regardless of the model selected.

There is uncertainty about demand. Three alternative levels of demand are possible—light, moderate, and heavy. Steven can meet all demand levels because its production facility is currently operating below full capacity.

Level of Demand	Unit Sales	Probability of Event
Light	20,000	0.25
Moderate	80,000	0.60
Heavy	120,000	0.15
		1.00

Variable manufacturing overhead is allocated to Steven's products using a plantwide rate of 250% of direct manufacturing labor costs. Steven's engineering and accounting staffs have worked together to develop manufacturing cost estimates for each of the three model designs:

	Model A	Model B	Model C
Unit Variable Costs			
Direct materials	$150	$100	$114
Direct manufacturing labor	40	50	48
Manufacturing overhead	100	125	120
Marketing	140	140	140
Total unit variable costs	$430	$415	$422
Other Costs			
Fixed manufacturing overhead	$1,000,000	$1,400,000	$1,300,000
Fixed marketing costs	2,000,000	3,100,000	2,800,000
Development costs (already incurred)	1,500,000	1,500,000	1,500,000

Steven has decided to use an expected monetary value choice criterion in its analysis of which of the three prototypes it will manufacture and sell.

Required

1. Compute the unit contribution margins of Models A, B, and C.
2. Which prototype should Steven Company manufacture and sell?

11-35 Ethics and relevant costs. Pastel Company must reach a make-or-buy decision with respect to a high-volume, easily made part, RG1. Sarah Gray, the cost analyst, is responsible for preparing the formal analysis. The buy alternative appears to have a lower cost. Jim Berry, the controller, is very concerned that purchasing RG1 from an outside supplier will mean that some of his close friends who work on the RG1 line will be laid off. He asks Sarah to review all her assumptions and calculations, with the comment, "The yield assumptions you made are very low. I think this plant can achieve much better quality than we have in the past. Better quality will reduce our material and conversion costs. That will make our costs quite competitive with the outside purchase price."

Sarah rechecks her calculations. She believes it is highly unlikely that the plant can achieve the quality levels it would take for the make alternative to be superior to the buy alternative. However, she knows that Jim will be extremely upset if she does not change her conclusion to favor the make alternative.

Required

Refer to the "Standards of Ethical Conduct for Management Accountants" in Chapter 1 (p. 17). Evaluate whether Jim Berry's suggestion to Sarah to use unrealistic estimates is unethical. Will it be unethical for Sarah to change her analysis to support the make alternative? What steps should Sarah take to resolve this situation?

Coverage of Chapter Appendix

11-36 Optimal production plan, computer manufacturer. Information Technology, Inc., assembles and sells two products: printers and desktop computers. Customers can purchase either (a) a computer or (b) a computer plus a printer. The printers are *not* sold without the computer. The result is that the quantity of printers sold is equal to or less than the quantity of desktop computers sold. The contribution margins are $200 per printer and $100 per computer.

Each printer requires 6 hours assembly time on production line 1 and 10 hours assembly time on production line 2. Each computer requires 4 hours assembly time on production line 1 only. (Many of the components of each computer are preassembled by external vendors.) Production line 1 has 24 hours of available time per day. Production line 2 has 20 hours of available time per day.

Let X represent units of printers and Y represent units of desktop computers.

Required

1. Express the relationships in an LP format.
2. Which combination of printers and computers will maximize the operating income of Information Technology? Use both the trial-and-error and the graphic approaches.

11-37 Minimum cost combination, fertilizer mix. The local agricultural center has advised Sam Bowers to spread at least 4,800 pounds of a special nitrogen fertilizer ingredient and at least 5,000 pounds of a special phosphate fertilizer ingredient in order to increase his crop yield. Neither ingredient is available in pure form.

A dealer has offered 100-pound bags of VIM at $10 each. Each 100-pound VIM bag contains the equivalent of 20 pounds of nitrogen and 80 pounds of phosphate. VOOM is also available in 100-pound bags, at $30 each. Each 100-pound VOOM bag contains the equivalent of 75 pounds of nitrogen and 25 pounds of phosphate.

Let X represent bags of VIM and Y represent bags of VOOM.

Required

How many bags of VIM and VOOM should Bowers buy in order to obtain the required fertilizer at minimum cost? Solve graphically.

11-38 Optimal sales mix for a retailer. Always Open, Inc. operates a chain of food stores open 24 hours a day. Each store has a standard 40,000 square feet of floor space available for merchandise. Merchandise is grouped in two categories: grocery products and dairy products. Always Open requires each store to devote a minimum of 10,000 square feet to grocery products and a minimum of 8,000 square feet to dairy products. Within these restrictions, each store manager can choose the mix of products to carry.

The manager of the Winnipeg store estimates the following weekly contribution margins per square foot: grocery products, $10; dairy products, 3.

Required

1. Formulate the decision facing the store manager as an LP model. Use G to represent square feet of floor space for grocery products and D to represent square feet of floor space for dairy products.
2. Why might Always Open set minimum bounds on the floor space devoted to each line of products?
3. Compute the optimal mix of grocery products and dairy products to carry at the Winnipeg store. In the graphic solution approach, use the horizontal axis for grocery products and the vertical axis for dairy products. Use both the graphic solution approach and the trial-and-error solution approach.

11-39 Optimal production plan, fixed and variable overhead costs. (A. Atkinson, adapted) Coldbrook Manufacturing produces two special concentrates that are derived from processing apples. These two products, Delicious (D) and Tasty (T), are used for flavoring in the food-manufacturing industry. Because of the high quality of these products, Coldbrook Manufacturing can sell all the output it can produce at the existing prices.

Characteristics of the two products follow:

	Product D	Product T
Selling price	$20	$35
Direct materials used	$12.40	$17
Direct manufacturing labor-hours used	0.2	0.5
Machine-hours used	1	2

The company allocates manufacturing overhead to products at the rate of $18 per direct manufacturing labor hour. The fixed component of the overhead cost per direct manufacturing labor-hour was estimated by dividing the budgeted fixed manufacturing overhead of $450,000 by the total direct manufacturing labor-hours (45,000) available for production during the year.

Direct manufacturing labor is paid $10 per hour. Coldbrook has 200,000 machine-hours available for production during the year. Fixed marketing and administrative costs are expected to be $300,000 during the year.

Required

1. Determine the optimal production mix for the year.
2. The company controller decides that a cost function with two cost drivers will yield better estimates of variable overhead costs. The new cost function used to estimate variable overhead (V.OVH) is:

$$\text{V.OVH per unit} = \$1.00 \text{ (direct labor-hours)} + \$2.50 \text{ (machine-hours)}$$

How will this approach to variable overhead cost estimation affect the optimal production mix determined in requirement 1?

12

Pricing Decisions, Product Profitability Decisions, and Cost Management

- Major influences on pricing
- Product-cost categories
- Costing and pricing for the short run
- Costing and pricing for the long run
- Alternative pricing approaches
- Target costing for target pricing
- Cost-plus pricing
- Considerations other than costs in pricing decisions
- Life-cycle product budgeting and costing
- Effects of antitrust laws on pricing

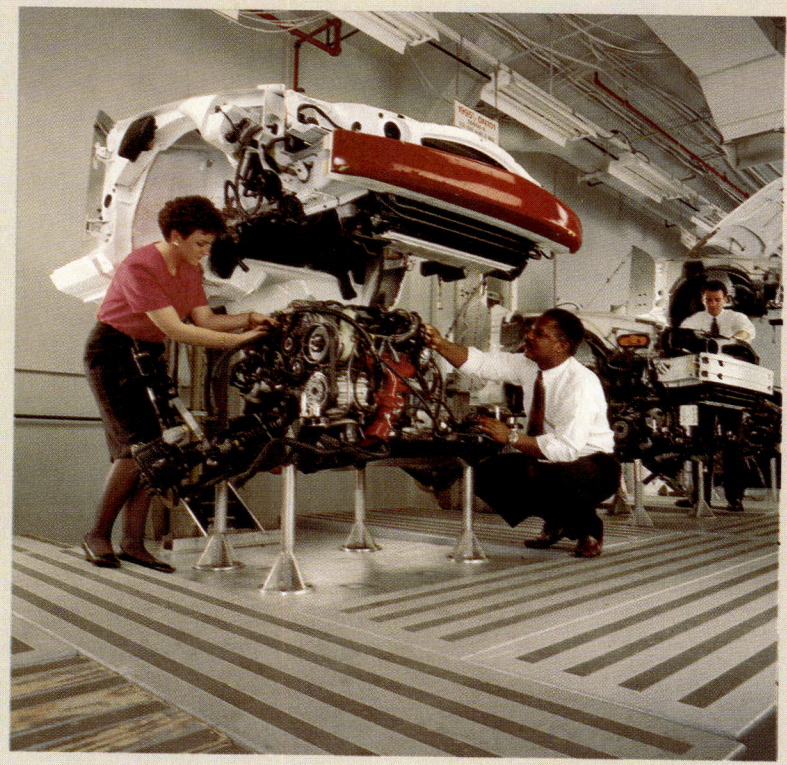

Cost management in the automotive industry requires emphasis on where costs are locked in as well as where costs are incurred. Ford Motor Company uses multi-function teams when designing new models that simultaneously satisfy customers on quality and are competitive on cost.

Learning Objectives

When you have finished studying this chapter, you should be able to

1. Discuss the three major influences on pricing decisions
2. Describe six important categories of costs relevant in pricing decisions
3. Distinguish between pricing decisions for the short run and for the long run
4. Describe the target-costing approach to pricing
5. Distinguish between cost incurrence and locked-in costs
6. Describe the cost-plus approach to pricing
7. Describe two pricing practices in which noncost reasons are important when setting price
8. Explain how life-cycle product budgeting and costing assist in pricing decisions
9. Explain the effects of antitrust laws on pricing

Managers frequently face decisions on the pricing and relative profitability of their products. This chapter illustrates the important role that cost data can play in making pricing decisions.[1] We emphasize how an understanding of cost-behavior patterns and cost drivers can lead to better decisions. This chapter applies the relevant-revenue and relevant-cost framework outlined in Chapter 11.

Pricing decisions differ greatly in both their time horizons and their contexts. Consequently, no single way of computing a product-cost figure is universally relevant for all pricing decisions. This chapter illustrates how managers may find different measures of product cost helpful in making various pricing decisions.

MAJOR INFLUENCES ON PRICING

Objective 1

Discuss the three major influences on pricing decisions

There are three major influences on pricing decisions: customers, competitors, and costs.

CUSTOMERS. Managers must always examine pricing problems through the eyes of their customers. A price increase may cause a customer to reject the company's product and choose one from a competitor. Alternatively, a price increase may cause a customer to choose a substitute product that fits desired specifications in a more cost-effective way. For example, increases in the price of glass drove customers for containers to substitute aluminum cans for bottles.

COMPETITORS. Competitors' reactions influence pricing decisions. At one extreme, a rival's intense drive may force a business to lower its prices to be competitive. At the other extreme, a business without a rival in a given situation can set higher prices. A business with knowledge of its rivals' technology, plant capacity, and operating policies is able to estimate its rivals' costs, which is valuable information in setting competitive prices.

Competition spans international borders. For example, when firms have overcapacity in their domestic markets, they often take an aggressive pricing policy in their export markets. Increasingly, managers are taking a global viewpoint and consider both domestic and international rivals in making pricing decisions. The concept of *target costs*, described later in this chapter, plays a key role in competitor analysis.

---

[1]For brevity, the term *pricing decision* is used in this chapter to encompass decisions about the relative profitability of products.

COSTS. The study of cost-behavior patterns yields insight into the income that results from different combinations of price and output quantities sold for a particular product. A product consistently priced below its costs can drain sizable amounts of resources from an organization.

Surveys of how executives make pricing decisions reveal that companies weigh customers, competitors, and costs differently. Firms selling commodity-type products in highly competitive markets must accept the price determined by market forces. For example, the gold market has many competitors, each offering the identical product at the same price. The market sets the price, but cost data can help managers in the gold market to decide, say, on the output level that best meets a company's particular objective.

In less competitive markets, managers have some discretion in setting prices. The pricing decision depends on how much customers value the product, the pricing strategies of competitors, and the costs of the product. Economic models of pricing describe the price of a product or service as the outcome of the interaction between *demand* for the product or service and *supply* of the product or service. Customers influence prices through their effect on demand. Costs and competitors influence prices because they affect supply.

Chapter 1 described customer satisfaction, continuous improvement, and the dual internal/external focus as important, newly evolving themes in management accounting. Pricing is an area where many of these themes explicitly come together. For example, charging lower prices for quality products is important for customer satisfaction, an external focus. But when prices are lower, costs must be reduced as well. Continuous improvement, an internal focus, is the key to "costing down."

PRODUCT-COST CATEGORIES

Objective 2

Describe six important categories of costs relevant in pricing decisions

Exhibit 12-1 presents six basic categories of business function costs that can be included in the cost buildup of a product or service. Managers in the manufacturing sector frequently group costs in these business functions as follows.

Premanufacturing (upstream) costs are all costs incurred in the value-chain business functions of research and development and the design of products or processes. **Postmanufacturing (downstream) costs** are all costs incurred in the value-chain business functions of marketing, distribution, and customer service.

The Importance of Relevance in Pricing Decisions

What costs are relevant to pricing decisions? As always, the key question is, What difference does it make? Expressed another way, How will future *total* costs be affected by a specific pricing decision? As we explained in Chapters 4 and 5, the full cost of a product or service includes the direct and indirect costs in all categories of the value chain. A *direct product cost* is a cost that can be traced to the product in an economi-

EXHIBIT 12-1
Value Chain Cost Buildup for a Manufacturing Company

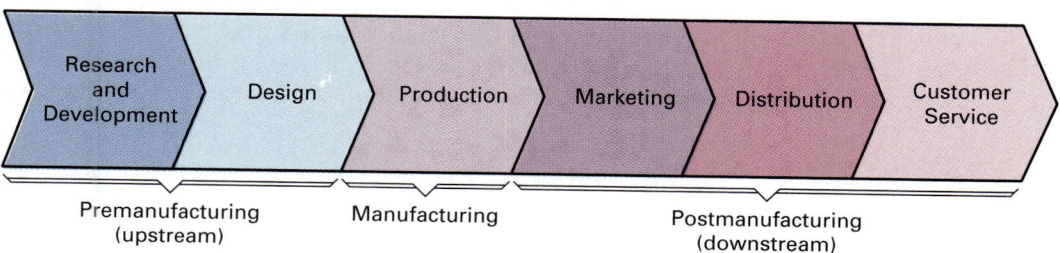

Research and Development | Design | Production | Marketing | Distribution | Customer Service

Premanufacturing (upstream)   Manufacturing   Postmanufacturing (downstream)

cally feasible way. An *indirect product cost* is assigned to products using a cost allocation base.

The relevant cost buildup for a pricing decision may include all six categories in Exhibit 12-1 or only some of them. Consider a firm that manufactures office furniture and that receives a special order for office desks. Marketing costs would be irrelevant in determining the product cost buildup. Why? Because the firm would incur no marketing costs for the order.

The Importance of the Time Horizon

Objective 3
Distinguish between pricing decisions for the short run and for the long run

In computations of the relevant costs in a pricing decision, the time horizon of the decision is critical. The two ends of the time-horizon spectrum are:

Short-Run	Long-Run
Pricing Decisions	Pricing Decisions

Short-run decisions include (1) pricing for a one-time-only special order with no long-term implications and (2) adjusting product mix and volume in a competitive market. The time horizon used to compute those costs that differ among the alternatives for short-run decisions is typically six months or less but sometimes as long as a year. Long-run decisions include pricing a main product in a major market where price setting has considerable leeway. A time horizon of longer than a year is often used when computing relevant costs for these long-run decisions.

This chapter presents examples of both short-run and long-run pricing decisions. Short-run pricing decisions and long-run pricing decisions are best viewed as ends of the spectrum. Most pricing decisions will not fit neatly at either end of this spectrum.

One-Time-Only Special Order

This section illustrates pricing decisions for which the time horizon for computing relevant costs is relatively short. Consider a one-time-only special order from a customer to supply products for the next four months. Acceptance or rejection of the order will not affect the revenues (output quantity sold or the price per unit) from existing sales outlets. The customer is unlikely to place any more sales orders in the future.

EXAMPLE: National Brewery Group (NBG) operates a brewery with a monthly capacity of 1 million barrels of a nonalcoholic beer product (Champion) that has gained significant market share in recent years. Current production and sales are 600,000 barrels per month. The selling price is $90 per barrel. Research and development and design of product and process costs at NBG are negligible. Customer-service costs are also small and are included with marketing costs. The variable cost per barrel and the fixed cost per barrel (based on a production quantity of 600,000 barrels per month) follow:

	Variable Cost per Barrel	Fixed Cost per Barrel	Variable and Fixed Cost per Barrel
Manufacturing costs			
Direct materials (barley, hops, etc.) costs	$ 7	$ 0	$ 7
Packaging (bottles, cans, etc.) costs	18	0	18
Direct manufacturing labor costs	4	0	4
Manufacturing overhead costs	6	13	19
Marketing costs	5	16	21
Distribution costs	9	8	17
Full product costs	$49	$37	$86

Variable manufacturing overhead of $6 per barrel is the cost of power and utilities. The details for total fixed manufacturing overhead costs and their per-barrel unitized costs (based on a production quantity of 600,000 barrels per month) follow:

	Total Fixed Manufacturing Overhead	Per-Barrel Fixed Manufacturing Overhead
Depreciation and production support costs	$3,000,000	$ 5
Materials procurement costs	600,000	1
Salaries paid for process changeover	1,800,000	3
Product and process engineering costs	2,400,000	4
Total fixed manufacturing overhead costs	$7,800,000	$13

Variable marketing costs ($5 per barrel) are direct product costs consisting of salespersons' commissions. Fixed marketing costs consist of advertising costs and salaries of marketing staff. Variable distribution costs (distribution labor and fuel costs for equipment) and fixed distribution costs (warehouse rent and equipment costs) are indirect product costs.

Canadian Brewery (CB) is constructing a new brewery in Toronto. Brewery operations will not begin for four months. Management, however, wants to start marketing immediately. It decides to buy from another brewery 250,000 barrels of nonalcoholic beer for each of the next four months to sell in Canada. CB has asked NBG and two other brewing companies to bid on this special order. From a production-cost viewpoint, the beer to be brewed for CB is identical to that currently brewed by NBG.

If NBG brews the extra 250,000 barrels, an additional $300,000 in manufacturing costs (materials-procurement costs of $100,000 and process-changeover costs of $200,000) will be required each month. This additional $300,000 is not driven by the quantity of barrels in the special order; it is an additional monthly setup cost of the special order. No additional costs will be required for research and development, design, marketing, distribution, or customer service. The 250,000 barrels will be marketed by CB in Canada, where NBG does not sell its Champion brand or any other nonalcoholic beers. CB will assume all costs associated with marketing, distribution, and customer service.

NBG's indirect manufacturing costs include variable manufacturing overhead and fixed manufacturing overhead. These costs are allocated on a per-barrel basis. Total fixed manufacturing overhead ($7,800,000 per month) would be unaffected by the additional manufacturing of 250,000 barrels per month.

A vice-president of CB notifies each potential bidder that a bid above $45 per barrel will probably be "noncompetitive." NBG knows that one of its competitors, with a highly efficient plant, has sizable idle capacity and will definitely bid for the contract. What price should NBG bid for the 250,000 barrel contract?

The relevant costs for the price-bidding decision would be computed by systematically analyzing the direct and indirect costs of each category of the value chain. In this case, only manufacturing costs are relevant. Note that all marketing costs, whether direct or indirect, will be unaffected if the special order is accepted, so they are irrelevant.

Exhibit 12-2 presents an analysis of the relevant costs. They include all manufacturing costs that will change in total if the special order is obtained: all direct and indirect variable manufacturing costs plus procurement and changeover costs related to the special order. Fixed manufacturing overhead costs are irrelevant. Why? Because these costs will not change if the special order is accepted.

Exhibit 12-2 shows the total relevant costs of $9,050,000 per month (or $36.20 per barrel) for the 250,000-barrel special order. Any bid above $9,050,000 per month will increase the profitability of NBG. For example, a successful bid of $40 per barrel, well under CB's ceiling of $45 per barrel, will add $950,000 to NBG's monthly operating income: $250,000 \times (\$40 - \$36.20) = \$950,000$.

Cost data, though key information in NBG's decision on the price to bid, are not the only inputs. NBG must consider business rivals and their likely bids. Bidding by competitors would probably cause NBG to offer a lower price than it otherwise would.

The relevant costs computed in Exhibit 12-2 are developed specifically for price bidding on the one-time-only special order of 250,000 barrels. Exhibit 12-3 presents the June 19_5 monthly income statement of NBG under an absorption-costing format

EXHIBIT 12-2
Relevant Costs for NBG: The 250,000-Barrel
One-Time-Only Special Order

Direct costs:		
Direct materials (250,000 × $7)	$1,750,000	
Packaging (250,000 × $18)	4,500,000	
Direct manufacturing labor (250,000 × $4)	1,000,000	
Materials procurement	100,000	
Process changeover salaries	200,000	
Total direct costs		$7,550,000
Indirect costs:		
Variable manufacturing overhead		
(250,000 × $6)	1,500,000	
Total indirect costs		1,500,000
Total relevant costs		$9,050,000
Per-barrel relevant costs: $9,050,000 ÷ 250,000 = $36.20		

EXHIBIT 12-3

June 19_5 Income Statement for NBG
(in thousands)

PANEL A: ABSORPTION-COSTING FORMAT*

		Total	Per Unit
Revenues		$54,000	$90
Cost of goods sold:			
Direct materials		$ 4,200	$ 7
Packaging		10,800	18
Direct manufacturing labor		2,400	4
Manufacturing overhead:			
Power and utilities	$3,600		
Depreciation and production support	3,000		
Material procurement	600		
Process-changeover salaries	1,800		
Product and process engineering	2,400		
Total manufacturing overhead		11,400	19
Total cost of goods sold		28,800	48
Gross margin		25,200	42
Marketing costs		12,600	21
Distribution costs		10,200	17
Operating income		$ 2,400	$ 4

PANEL B: VARIABLE-COSTING (CONTRIBUTION INCOME STATEMENT) FORMAT*

		Total	Per Unit
Revenues		$54,000	$90
Variable costs:			
Manufacturing			
Direct materials	$ 4,200		$ 7
Packaging	10,800		18
Direct manufacturing labor	2,400		4
Power and utilities	3,600		6
Total variable manufacturing costs		21,000	35
Marketing		3,000	5
Distribution		5,400	9
Total variable costs		29,400	49
Contribution margin		24,600	41
Fixed costs			
Manufacturing:			
Depreciation and production support	3,000		5
Material procurement	600		1
Process-changeover salaries	1,800		3
Product and process engineering	2,400		4
Total fixed manufacturing costs		7,800	13
Marketing		9,600	16
Distribution		4,800	8
Total fixed costs		22,200	37
Operating income		$ 2,400	$ 4

*Production and sales in June = 600,000 barrels.

(Panel A) and a variable-costing (contribution) format (Panel B) at an output level of 600,000 barrels. The absorption-costing income statement reports the total manufacturing cost to be $48 per barrel. This income statement erroneously implies that a bid of $45 per barrel to Canadian Breweries will result in NBG sustaining a $3 per-barrel loss on the contract. Why erroneous? Because the absorption-costing format incorporates $13 per barrel of irrelevant costs with regard to the special order—the $13 fixed manufacturing cost per barrel that will not be incurred on the 250,000-barrel special order. To determine relevant costs for the one-time-only special order using the absorption-costing format, the analyst must examine how *total* costs in each category will change if the special order is accepted.

The variable-costing (contribution) income statement in Exhibit 12-3 (Panel B) highlights the variable manufacturing costs and fixed costs for various functions and activities. Relevant costs for analyzing the one-time-only special order are the variable manufacturing costs and the additional $300,000 costs for materials-procurement and process-changeover salaries.

Both the absorption-costing format and the variable-costing format in Exhibit 12-3 do not explicitly indicate the relevant $300,000 additional monthly manufacturing costs for materials procurement and manufacturing process changeover. But by listing the costs of individual activities, both the absorption-costing format and the variable-costing format can highlight potential activities and costs that might be affected by accepting CB's special offer.

Setup Costs and Short-Run Pricing Decisions

The relevant manufacturing costs in the special-order decision facing National Brewery Group include both the manufacturing costs that vary with each unit of output produced and the additional monthly setup cost of $300,000. The buildup of relevant costs for some product-mix decisions may include setup costs incurred on a daily basis or even for a single-shift time period. For example, a mining company may incur setup costs when it processes several different types of minerals rather than one type of mineral in a given shift. Processing 100 tons of lead and then 180 tons of zinc rather than 300 tons of lead alone during a shift may require that some pieces of equipment be cleaned or repositioned. The setup costs of cleaning or repositioning equipment are relevant short-run costs in deciding on the relative profitability of alternative combinations of minerals to process.

The nature of the decisions that managers must make should help determine what costs are collected in an internal reporting system on an ongoing basis. For example, when setup costs are important for regularly recurring decisions, separately tracking those costs in the internal reporting system is likely to be cost-effective.

COSTING AND PRICING FOR THE LONG RUN

Many pricing decisions are made for the long run. Buyers, whether an individual buying a box of Wheaties or a big-time construction outfit buying a fleet of tractors, prefer stable, steady prices over an extended time horizon. A stable price has several advantages. It reduces the need for buyers to continually monitor prices offered by different suppliers. Greater price stability helps companies plan better and builds long-run buyer-seller relationships.

Calculating Product Costs

When competitive forces set the price for a product, knowledge of long-run product costs can help guide decisions about entering or remaining in the market for that product. When managers have some control over the price charged for a product, long-run product costs can act as a base for setting that price. Determining accurate product-cost information is essential to a manager making a pricing decision.

Environmental costs are becoming increasingly important when computing product costs. The enactment of strict environmental laws (for example, the Clean Air Act and the Superfund Amendment and Reauthorization Act) has introduced tougher environmental standards and increased the penalties and fines for polluting the air and contaminating subsurface soil and ground water. To avoid these liabilities, companies must incur costs to prevent and reduce pollution. The computer manufacturers Compaq and Apple, for example, have recently introduced costly recycling programs to ensure that nickel-cadmium batteries (used to run portable computers) are disposed of in an environmentally safe way.

Consider Astel Computer. Astel manufactures two brands of personal computers (PCs)—Deskpoint and Provalue. Deskpoint is Astel's top-of-the-line product, a 486 DX chip-based PC sold through computer dealers to large organizations and government accounts. Our analysis focuses on Provalue, a less powerful 386 DX chip-based machine sold through catalogs and mass merchandisers to individual consumers and small organizations.

The manufacturing costs of Provalue are calculated using the activity-based costing (ABC) approach described in Chapters 4 and 5. Astel has three direct manufacturing cost categories (direct materials, direct manufacturing labor, and direct machining costs) and three indirect manufacturing cost pools in its accounting system. Astel treats machining costs as a direct cost of Provalue because Provalue is manufactured on machines that are used exclusively for its manufacture. The table below summarizes the activity cost pools, the cost driver for each activity, and the cost per unit of cost driver used by Astel to allocate manufacturing overhead costs to products.

Manufacturing Activity	Description of Activity	Cost Driver	Cost per Unit of Cost Driver
1. Ordering and receiving	Placing of purchase orders for components and receiving and paying for ordered components	Number of orders	$80 per order
2. Testing and inspection	Testing components and final product	Testing-hours	$2 per testing-hour
3. Rework	Correcting and fixing errors and defects	Units reworked	$100 per unit reworked

Astel is using a long-run time horizon to price Deskpoint and Provalue. Over this horizon, Astel's management views direct materials costs and direct manufacturing labor costs as variable with respect to the units of Provalue produced. Direct machining costs do not vary over this time horizon; they are fixed long-run costs. Manufacturing overhead costs vary with respect to their chosen cost drivers. For example, ordering and receiving costs vary with the number of orders. Staff members responsible for placing orders can be reassigned or laid off in the long run if fewer orders need to be placed.

Note that the cost drivers are chosen in an attempt to measure the cause-and-effect relationship between an activity's cost driver and costs in the activity's cost pool. Consider the ordering and receiving activity. Ordering and receiving is a batch-level cost; that is, purchase orders are placed to buy a group of components, not separately for each individual Provalue produced. Over a long-run time horizon, the costs of this activity will vary with the number of orders placed and not with the individual units of Provalue produced. For example, will ordering and receiving costs increase if the number of Provalues produced increases? Not if Astel places the same number of orders and increases the quantity of components in each order.

How does Astel calculate the costs of manufacturing Provalue? It uses the following information, which indicates the resources used to manufacture Provalue in 19_4:

1. Direct materials costs per unit of Provalue are $460.
2. Direct manufacturing labor costs per unit of Provalue are $60.
3. Direct fixed costs of machines used exclusively for the manufacture of Provalue are $12,000,000.
4. Number of orders placed to purchase components required for the manufacture of Provalue is 22,500.
5. Number of testing-hours used for Provalue is 4,500,000.
6. Number of units of Provalue reworked during the year is 12,000.

The detailed calculations underlying each of these numbers are shown in Exhibit 12-4. This exhibit indicates that the total costs of manufacturing Provalue are $102 million and the unit manufacturing costs of Provalue are $680. Manufacturing, however, is just one function in the value chain. For setting long-run prices and for managing costs, Astel must determine the full product costs of Provalue. *Full product costs* refer to the sum of all costs of all business functions (research and development, design, production, marketing, distribution, and customer service).

For brevity, we do not present any analyses or calculations for the other value-chain functions. Astel chooses cost drivers and cost pools in each value-chain function to measure the cause-and-effect relationship between activities and costs within

EXHIBIT 12-4
Manufacturing Costs of Provalue in 19_4 Based on Activity Analysis of Manufacturing Costs

	Total Manufacturing Costs for 150,000 Units (1)	Unit Manufacturing Costs (2) = (1) ÷ 150,000
Direct materials costs (150,000 × $460)	$ 69,000,000	$460
Direct manufacturing labor costs (150,000 × $60)	9,000,000	60
Direct machining costs (fixed costs of $12,000,000)	12,000,000	80
Manufacturing overhead costs:		
Ordering and receiving costs (22,500* × $80)	1,800,000	12
Testing and inspection costs (4,500,000† × $2)	9,000,000	60
Rework costs (12,000‡ × $100)	1,200,000	8
Total manufacturing overhead costs	12,000,000	80
Total manufacturing costs	$102,000,000	$680

*Provalue has 450 components, and 50 orders are placed for each component for a total of 22,500 (450 × 50) orders.

†Each Provalue unit is tested and inspected for 30 hours; total time spent testing Provalue is 4,500,000 (30 × 150,000) hours.

‡8% of Provalue units manufactured require rework; total units of Provalue reworked are 12,000 (8% × 150,000).

EXHIBIT 12-5

Product Profitability of Provalue in 19_4
Based on Value-Chain Activity Analysis

	Total Revenues and Costs for 150,000 Units (1)	Unit Revenue and Unit Costs (2) = (1) ÷ 150,000
Revenue	$150,000,000	$1,000
Cost of goods sold (first four items from Exhibit 12-4)		
Direct materials costs	69,000,000	460
Direct manufacturing labor costs	9,000,000	60
Direct machining costs	12,000,000	80
Manufacturing overhead costs	12,000,000	80
Total cost of goods sold	102,000,000	680
Gross margin	48,000,000	320
Research and development costs	5,400,000	36
Design of products and processes costs	6,000,000	40
Marketing costs	15,000,000	100
Distribution costs	3,600,000	24
Customer-service costs	3,000,000	20
Operating income	$ 15,000,000	$ 100

each activity's cost pool. Costs are allocated to Provalue on the basis of the quantity of cost driver units that Provalue requires. For example, the cost driver for design costs is the complexity of the PC and the number of different features and functions available on it (not the number of units of Provalue produced). Provalue, a simple machine, requires few design resources and is allocated very little in design costs. If you are familiar with the hierarchy of costs presented in Chapter 10 (p. 355), you will recognize design costs as *product-sustaining costs*. Exhibit 12-5 summarizes the product operating income statement for Provalue for the year 19_4 based on an activity analysis of costs in all value-chain functions (supporting calculations for nonmanufacturing value-chain functions are not given). Astel earned $15 million from Provalue, or $100 per unit sold; there is no beginning or ending inventory.

ALTERNATIVE PRICING APPROACHES

The starting point for pricing decisions can be

1. market-based or
2. cost-based (also called cost-plus).

The market-based approach starts by asking, "Given what our customers want and how our competitors will react to what we do, what price should we charge?" The cost-based approach starts by asking, "What does it cost us to make this product, and so what price should we charge?"

In very competitive markets, such as the oil-and-gas and the mining industries, the market-based approach is logical. The items produced or mined by any firm are very similar to those produced by other firms, and the firm has no influence over the price to charge. In other industries, such as appliances and automobiles, a company will have some discretion over prices and products. The company chooses product features and price on the basis of anticipated customer and competitor reactions. A final decision on the product and the price is made after combining these external influences on pricing with the costs to produce and sell the product.

Under the cost-plus approach, price is first computed on the basis of the costs to produce and sell a product. Typically, a markup, representing a reasonable return, is added to cost. Often, the price is then modified on the basis of anticipated customer reaction to alternative price levels and the prices charged by competitors for similar products.

TARGET COSTING FOR TARGET PRICING

Objective 4
Describe the target-costing approach to pricing

An important form of market-based price is the *target price*. A **target price** is the estimated price for a product (or service) that potential customers will be willing to pay. This estimate is based on an understanding of customers' perceived value for a product and the responses of competitors. The target price leads to a *target cost*. A **target cost** is the estimated long-run cost of a product (or service) that when sold enables the company to achieve the targeted income. Target cost is derived by subtracting the target profit margin from the target price. Target cost is often lower than the existing full product cost of making and selling the product. To achieve target cost and the desired profit margin, then, the organization must improve its products and processes. Target costing is widely used among industries in different countries. Mercedes, Toyota, Nissan, and Daihatsu in the automobile industry, Matsushita, Panasonic, and Sharp in the electronics industry, and Apple, Compaq, and Toshiba in the personal computer industry all use target pricing and target costing.

Implementing Target Pricing and Target Costing

Developing target prices and target costs requires the following four steps:

- ◆ *Step 1:* Develop a product that satisfies the needs of potential customers.
- ◆ *Step 2:* Choose a *target price* based on customers' perceived value for the product and the prices competitors charge.
- ◆ *Step 3:* Derive a *target cost* by subtracting the desired profit margin from target price.
- ◆ *Step 4:* Perform *value engineering* to achieve target costs. **Value engineering** is a systematic evaluation of all aspects of research and development, design of products and processes, production, marketing, distribution, and customer service, with the objective of reducing costs while satisfying customer needs. Value engineering can result in improvements in product designs, changes in materials specifications, or modifications in process methods.

We illustrate the four steps for target pricing and target costing using the Astel Computers example.

STEP 1: *PRODUCT PLANNING FOR PROVALUE.* Astel is in the process of planning product changes and design modifications for Provalue, its low-end personal computer (PC), which is sold to individuals and small organizations. Astel is very concerned about severe price competition from several competitors.

STEP 2: *TARGET PRICE OF PROVALUE.* Astel expects its competitors to lower the prices of PCs that compete against Provalue by 15%. Astel's management believes that it must respond aggressively by reducing Provalue's price by 20%, from $1,000 per unit to $800 per unit. At this lower price, Astel's marketing manager forecasts an increase in annual sales from 150,000 to 200,000 units.

STEP 3: *TARGET COST OF PROVALUE.* Astel's management wants a 10% target operating income on sales revenues.

Total target sales revenues = $800 × 200,000 units = $160 million.
Total target operating income = 10% × $160 million = $16 million.
Target operating income per unit = $16 million ÷ 2,000,000 units = $80 per unit.
Target cost per unit = Target price – Target operating income per unit = $800 – $80 = $720.
Total current operating costs of Provalue = $135 million (from Exhibit 12-5)
Current operating costs per unit of Provalue = $135 million ÷ 150,000 units = $900 per unit.

The target cost is substantially lower than Provalue's existing product costs. The goal is to find ways to reduce the cost per unit of Provalue by $180, from $900 to $720. The challenge in step 4 is to achieve target cost through value engineering.

STEP 4: ***VALUE ENGINEERING FOR PROVALUE.*** An important element of Astel's value engineering is determining the kind of low-end PC that will meet the needs of potential customers. For example, the current design of Provalue accommodates various upgrades that can make the PC run faster and handle calculations more quickly. It comes with special audio features. An essential first step in the value-engineering process is evaluating whether potential customers will be willing to pay the price for those features. Customer feedback indicates that customers do not value Provalue's extra features. They want Astel to redesign Provalue into a no-frills PC and sell it at a much lower price. Value engineering at Astel then proceeds with teams consisting of marketing managers, product designers, manufacturing engineers, and production supervisors making suggestions for design improvements and process modifications. Cost accountants estimate the savings in costs that would result from the proposed changes.

Two key cost-related concepts important in value engineering are *cost incurrence* and *locked-in costs*. Successful value engineering requires drawing a careful distinction between when costs are incurred and when costs are locked in.

Cost Incurrence and Locked-in Costs

Objective 5
Distinguish between cost incurrence and locked-in costs

Cost incurrence occurs when resources are actually sacrificed. Cost systems emphasize cost incurrence. They recognize and record costs only when costs are incurred. Astel's cost system, for example, recognizes the direct materials costs of Provalue when Provalue is assembled and sold. But Provalue's direct materials costs are determined much earlier. When? At the time the design of Provalue is finalized. Direct materials costs of Provalue are *locked in* (or *designed in*) at the product design stage. **Locked-in costs (designed-in costs)** are those costs that have not yet been incurred but that will be incurred in the future on the basis of decisions that have already been made.

Why is it important to distinguish between when costs are locked in and when costs are incurred? Because it is difficult to alter or reduce costs that have already been locked in. For example, Astel's ability to reduce the direct materials costs of Provalue are considerably limited once the design of Provalue is finalized.

Examples of how Astel's design decisions affect costs follow:

1. Design decisions influence direct materials costs through the choices of printed circuit boards and add-on features used in Provalue. Better designs also help control defects, rework, and scrap generated during manufacturing and reduce product failures at customer sites.
2. Designing Provalue so that it is easy to manufacture and easy to assemble decreases direct manufacturing labor costs. For example, designing Provalue so that various parts snap-fit together (rather than have various parts soldered together) saves manufacturing labor time.
3. Designing Provalue with fewer components reduces ordering and materials handling costs.
4. Simplifying the Provalue design decreases the time required for testing and inspection.
5. Design decisions also affect customer service. A good design can significantly reduce the need for repairs and the time it takes to service and repair Provalue at customer sites.

Exhibit 12-6 graphically illustrates how the locked-in cost curve and the cost incurrence curve might appear in the case of Provalue. (The numbers underlying the graph are assumed.) The bottom curve plots the cumulative costs per unit incurred in different business functions. The top curve plots the cumulative costs locked in. The point of the graph is to emphasize the wide divergence between when costs are locked in and when those costs are incurred. In our example, once the product and processes are designed, more than 86% ($780, say, ÷ $900) of the unit costs of Provalue are locked in when only about 8% ($72, say, ÷ $900) of the unit costs are actually incurred.

What are the lessons from Exhibit 12-6? First, most costs are locked in well before costs are actually incurred. Second, once design is set, costs are difficult to influence. Third, the key to cost reduction is often to understand when and how costs get locked in; when and how costs are incurred is less important.

We caution that it is not always the case that costs are locked in early in the design stage. In process industries, such as chemicals, oil, steel, and paper, many costs are locked in and incurred at about the same time. When costs are not locked in early, cost-reduction activities can be successful right up to the time that costs are incurred. In these industries the key to lowering costs is improved efficiency and productivity rather than better design.

Achieving the Target Cost for Provalue

Astel's value-engineering team focused its cost-reduction efforts on analyzing the Provalue design. Several points quickly became apparent. Customers rarely used Provalue's capabilities. Provalue was overengineered and overpriced. To achieve target cost, the value-engineering team looked to design and build a high-quality, highly reliable machine with fewer features.

Provalue was discontinued. In its place, Astel introduced Provalue II. Provalue II has 425 components compared with 450 components in Provalue. How can

EXHIBIT 12-6

Pattern of Cost Incurrence and Locked-in Costs for Provalue

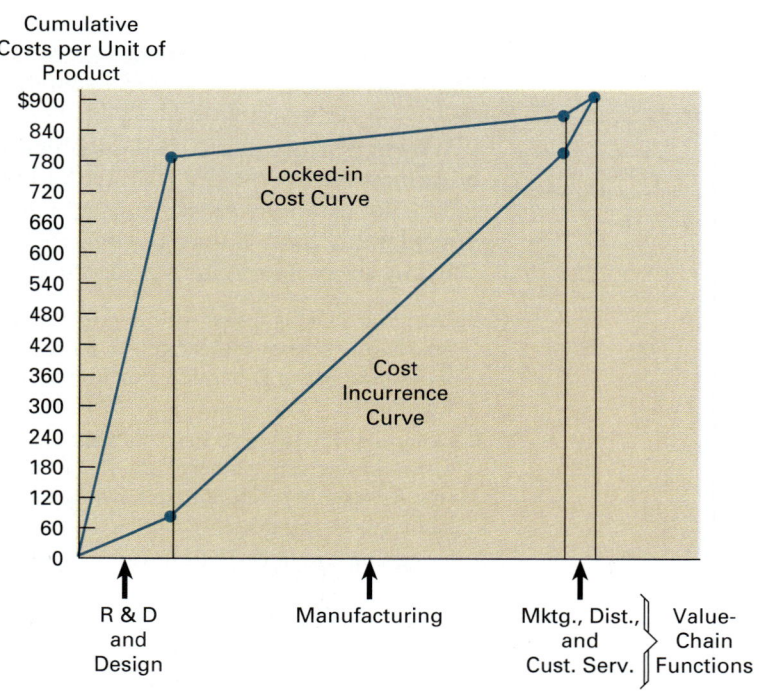

Provalue II be sold at a much lower price? To understand how, look at the following tables, which compare the direct costs incurred and the manufacturing overhead cost drivers used by Provalue and Provalue II. In place of the 150,000 Provalue units manufactured and sold, Astel expects to make and sell 200,000 Provalue II units.

DIRECT COSTS

Cost Category	Costs per Unit		Explanation of Costs for Provalue II
	Provalue	Provalue II	
Direct materials	$460	$385	The Provalue II design uses a simplified main printed-circuit board, fewer components, and no audio features.
Direct manufacturing labor	$ 60	$ 50	The Provalue II requires less assembly time.
Direct machining costs	$ 80	$ 60	Machining costs are fixed at $12 million. Astel can use the capacity to produce 200,000 units of Provalue II. The new design enables Astel to manufacture each unit of Provalue II in less time than a unit of Provalue. Direct machining costs per unit of Provalue II are $60 ($12,000,000 ÷ 200,000).

MANUFACTURING OVERHEAD COSTS

Cost Driver	Quantity of Cost Driver		Explanation for Quantity of Cost Driver Used by Provalue II
	Provalue (from Exhibit 12-4)	Provalue II	
1. Number of orders	22,500	21,250	Astel will place 50 orders for each of the 425 components in Provalue II. Total orders for Provalue II equal 21,250 (425 × 50).
2. Testing-hours	4,500,000	3,000,000	Provalue II is easier to test and requires 15 hours of testing per unit. Total number of expected testing-hours are 3,000,000 (15 × 200,000).
3. Units reworked	12,000	13,000	Provalue II has a lower rework rate of 6.5% because it is easier to manufacture and hence has fewer defects. Total units expected to be reworked are 13,000 (6.5% × 200,000).

Exhibit 12-7 presents the target manufacturing costs of Provalue II (assuming no change in the cost per unit of the cost drivers). To help comparison, Exhibit 12-7 also reproduces the unit manufacturing costs of Provalue from Exhibit 12-4. Exhibit 12-7 shows that the new design is expected to reduce the per-unit manufacturing cost by $140 to $540 from $680. A similar analysis (not presented) estimates the expected effect of the new design on costs in other value-chain functions. Exhibit 12-8 shows that full product costs per unit equal $720—the target cost for the Provalue II.

EXHIBIT 12-7
Target Manufacturing Costs of Provalue II

	PROVALUE II		PROVALUE
	Total Manufacturing Costs for 200,000 Units (1)	Estimated Unit Manufacturing Costs (2) = (1) ÷ 200,000	Unit Manufacturing Costs (Exhibit 12-4, Column 2) (3)
Direct materials costs (200,000 × $385)	$ 77,000,000	$385.00	$460
Direct manufacturing labor costs (200,000 × $50)	10,000,000	50.00	60
Direct machining costs	12,000,000	60.00	80
Total direct manufacturing costs	99,000,000	495.00	600
Manufacturing overhead costs:			
Ordering and receiving costs (21,250 orders × $80)	1,700,000	8.50	12
Testing and inspection costs (3,000,000 × $2)	6,000,000	30.00	60
Rework costs (13,000 × $100)	1,300,000	6.50	8
Total manufacturing overhead costs	9,000,000	45.00	80
Total manufacturing costs	$108,000,000	$540.00	$680

EXHIBIT 12-8
Target Product Profitability of Provalue II

	Total Revenue and Costs for 200,000 Units (1)	Unit Revenue and Unit Costs (2) = (1) ÷ 200,000
Revenue	$160,000,000	$800
Cost of goods sold (from Exhibit 12-7)		
Direct materials costs	77,000,000	385
Direct manufacturing labor costs	10,000,000	50
Direct machining costs	12,000,000	60
Manufacturing overhead costs	9,000,000	45
Total cost of goods sold	108,000,000	540
Operating costs		
Research and development costs	4,000,000	20
Design of products and processes costs	6,000,000	30
Marketing costs	18,000,000	90
Distribution costs	5,000,000	25
Customer-service costs	3,000,000	15
Total Operating costs	36,000,000	180
Full product costs	144,000,000	720
Operating income	$ 16,000,000	$ 80

Concepts in Action

TARGET COSTING AT MERCEDES BENZ AND TOYOTA[a]

Within three weeks of taking over as the chief of Mercedes Benz, Germany's largest industrial company, Helmut Werner made a dramatic announcement. Mercedes Benz would do away with its engineering-dominated cost-plus approach to pricing cars. In the future, the price of Mercedes cars would be set by a market-driven target price. The first test case is the 1994 190E, the Baby Benz. The car is loaded with $5,000 more in features than the prior year's model had, but the 1994 Baby Benz is priced below the 1993 car's $25,000 sticker. To achieve the desired operating margins, Werner's target is to cut costs by 30%. "No one in the world is prepared to pay for German complacency on the cost front," he warns.

Like Mercedes, Toyota Motor Company also chooses target prices for its cars. Toyota believes that a substantial fraction of its costs are locked in at the design stage. It views costs as inflexible once the first prototype is built, so all value engineering is done at the design stage. For example, Toyota's design engineers are trained to recognize the influence that design choices have on materials consumption, yield, and machining methods. Specially trained cost estimators then estimate the manufacturing costs of alternative designs. Toyota often simultaneously designs a family of cars to take advantage of common parts and processes. Cars in the Celica line, consisting of the Celica itself, the Corona Exsiv, and the Carina ED, have much in common in the engine and chassis but are differentiated along external styling dimensions.

[a] Adapted from J. Templeman, "Mercedes is Downsizing—and That Includes the Sticker," *Business Week* (February 8, 1993); and T. Tanaka, "Target Costing at Toyota," *Journal of Cost Management* (Spring 1993), pp. 4-11.

COST-PLUS PRICING

Objective 6
Describe the cost-plus approach to pricing

Astel uses an external market-based approach in its long-run pricing decisions. An alternative approach is to determine a cost-based price. Managers can turn to numerous pricing formulas based on cost. The general formula for setting a price is adding a markup to the cost:

Cost base	$ X
Markup component	Y
Prospective selling price	$X + Y

Consider a cost-based pricing formula that Astel could use for Provalue II. Assume that Astel's engineers have redesigned Provalue into Provalue II as described earlier and that Astel uses full product costs plus a 12% markup on costs in developing the prospective selling price.

Cost base (full product costs, from Exhibit 12-8)	$720.00
Markup component (12% × $720)	86.40
Prospective selling price	$806.40

How is the markup percentage of 12% determined? One approach is to choose a markup to earn a *target return on investment*. The **target return on investment** is the target operating income that an organization expects to earn divided by invested capital.

Suppose that Astel's capital investment to make and sell Provalue II is $96 million. Astel's (pretax) target return on investment for capital invested in its business is 18%. The desired operating income can then be calculated as follows:

Invested capital	$96.00 million
Target return on investment	18%
Total target operating income (18% × $96)	$17.28 million
Target operating income per unit of Provalue II ($17.28 million ÷ 200,000 units)	$86.40

The target return on investment calculation indicates that Astel would like to earn a target operating income of $86.40 on each unit of Provalue II. What markup does this return amount to? Markups are often expressed as a percentage of costs. Since the full product cost is $720, the markup is equal to 12% ($86.40 ÷ $720). Do not confuse the 18% target return on investment with the 12% markup percentage. The 18% target return on investment expresses Astel's expected operating income as a percentage of *investment*. The 12% markup expresses operating income as a percentage of *costs*. Working through the target return on investment, we determine the markup percentage.

Astel could use alternative cost bases and markup percentages to set the price. To use these alternatives, Astel needs to separate the costs per unit for each business function into their variable and fixed components. Exhibit 12-9 presents this information (without providing details of the calculations). Now, let's consider the options available to Astel. The following table illustrates some alternative cost bases and markup percentages.

Prospective Cost Base Provalue II (1)	Unit Cost for Provalue II (2)	Markup Percentage (3)	Markup Component for Provalue II (4) = (2) × (3)	Prospective Selling Price for Provalue II (5) = (2) + (4)
Variable manufacturing costs	$480.00	70%	$336.00	$816.00
Variable product costs	544.00	45	244.80	788.80
Manufacturing function costs	540.00	55	297.00	837.00
Full product costs	720.00	12	86.40	806.40

To illustrate the markup calculations, we have assumed (but not derived) the markup percentages in the table. In practice, firms choose cost bases and markup percentages on the basis of their experiences in pricing products to earn a desired rate of return on investment. The different cost bases and markup percentages that we use in the table give prospective selling prices that are relatively close to one another.

The markup percentages in the table, however, vary a great deal, from a low of 12% on full product costs to a high of 70% on variable manufacturing costs. Why? The markup based on variable manufacturing costs takes into account the need to cover

EXHIBIT 12-9

Estimated Cost Structure for Provalue II

Business Function	Variable Costs per Unit	Fixed Costs per Unit*	Business Function Cost per Unit
Research and development	$ 8.00	$ 12.00	$ 20.00
Design of product/process	10.00	20.00	30.00
Production	480.00	60.00	540.00
Marketing	25.00	65.00	90.00
Distribution	15.00	10.00	25.00
Customer service	6.00	9.00	15.00
Product costs	$544.00	$176.00	$720.00
	↑	↑	↑
	Variable product costs	Fixed product costs	Full product costs

*Based on budgeted annual production quantity of 200,000 units.

fixed manufacturing costs and other business function costs such as research and development, marketing, and distribution. These costs are already included in full product costs, so the markup percentage can be much lower.

Surveys indicate that most managers use full product costs (see Surveys of Company Practice on p. 447)—that is, they include both unit fixed costs and unit variable costs in the cost base in making their pricing decisions. The advantages cited for including unit fixed costs for pricing decisions include the following.

1. *Fixed-cost recovery.* In the long run, fixed costs must be recovered to stay in business. That is, in the long run, fixed costs are relevant costs. Why? Because if a company charges prices that cannot generate sufficient revenues to *meet all of its costs in the long run,* then the best alternative for the company is to shut down. Relative to the shutting-down alternative, *all costs*—whether fixed or variable—are relevant. Some managers believe that fixed-cost recovery is best achieved by having every product priced above its full cost (in the case of Provalue II, $720 at an annual production quantity of 200,000 units).

Some managers are concerned that if prospective prices are based on variable costs, there will be a temptation to engage in excessive long-run price cutting. Why? Because managers pricing on the basis of variable costs are often unaware of full product costs. Fixed costs are often considered irrelevant. Short-run price cutting is viewed as acceptable as long as it yields a positive contribution margin. But short-run price cutting often becomes prolonged as competitors react to the lower prices. Provalue II's variable product costs are $544 per unit. At a markup percentage of 45%, Provalue's selling price, $788.80, covers the $720 full product costs. If Astel sells 200,000 units of Provalue II, but sets long-run prices with a markup percentage of, say, 25% of variable costs, the resulting $680 ($544 + [25% × $544]) prospective price leads to long-run revenues being lower than long-run costs.

2. *Price stability.* Managers believe that full-cost-formula pricing promotes price stability, because it limits the ability of managers to cut prices. Managers prefer price stability because it facilitates planning.

3. *Simplicity.* A full-cost formula for pricing does not require a detailed analysis of cost-behavior patterns to determine fixed and variable costs for each product. Calculating variable costs for each product is expensive and prone to errors. For these reasons, many managers believe that full-cost-formula pricing meets the cost-benefit test.

4. *Price justification.* From a legal standpoint, full-cost-formula pricing reduces the likelihood that other parties could prove that the prices are predatory or anti-competitive. This topic is discussed more fully in "Effects of Antitrust Laws on Pricing" on p. 451.

Including unit fixed costs for pricing is not without its problems. Allocating common and unavoidable fixed costs to products can be somewhat arbitrary. Calculating fixed costs per unit requires an estimate of expected future sales quantities. If actual sales fall short of this estimate, the actual full product costs can exceed price.

Note that the selling prices we have computed are *prospective* prices. *Other factors beyond the cost base and a markup component enter into managers' pricing decisions.* These inputs include expected customer reaction to alternative price levels and the prices of similar competing products. For example, the greater the demand for the firm's product, the higher the markup a firm can charge. When demand is low or when competitors price aggressively, a firm may be forced to reduce its markup. The target-costing approach for pricing described earlier in this chapter more explicitly considers customer reactions and competitor responses in making pricing decisions.

Refined cost driver and cost information plays an important role in both target costing and cost-plus pricing. Identification of cost drivers is critical as managers do value engineering to "cost down" their products. *Cost down* refers to reducing the cost of a product while still satisfying customer expectations. If a firm uses cost-plus pricing, refined cost data reduces the possibility of miscosting and hence mispricing of products and services. Both underpricing and overpricing of products and services hurt the firm. Products that are priced too low earn less income than they should or even no income (if actual costs are higher than price). Products that are priced too high suffer decreases in sales quantities and revenues and invite more competition.

CONSIDERATIONS OTHER THAN COSTS IN PRICING DECISIONS

Objective 7
Describe two pricing practices in which noncost reasons are important when setting price

Consider the prices that publishers charge for hardcover and softcover versions of the same book. The price difference is sometimes in excess of $20. Can this price difference be explained by the difference in the cost of producing these books? No, it costs roughly a dollar or two more to produce a hardcover book than it does to produce a softcover one. How can we explain this difference in price? We must recognize the potential for price discrimination.

Price discrimination is the practice of charging some customers a higher price than is charged to other customers. Why and how does price discrimination work? We continue with our book example. The demand for books comes from two main sources, (1) libraries (and other institutions) and (2) individuals. Libraries value hardcover books very highly. Hardcover books are more durable than softcover books and hence are less likely to tear and fall apart from repeated use. Libraries also need to ensure that they have a wide range of books on hand, to satisfy the reading and research needs of their customers. The twofold responsibility to have books on hand and to have books that last a long time makes libraries relatively insensitive to the prices of hardcover books. This insensitivity of demand to price changes is called demand inelasticity. Publishers can charge libraries higher prices for hardcover books because higher prices have little effect on demand and earn higher operating income.

Individuals, too, value hardcover books, but only slightly more than they value softcover books. Therefore, generally, individuals are less willing to pay a higher price for the hardcover book when a cheaper softcover version is available. The demand among individuals for books is much more sensitive to price than the demand among institutions. In selling to individuals, it is profitable for the publisher to keep the prices of softcover books low to stimulate the demand. The publisher is discriminating in its pricing among different market segments to take advantage of the different sensitivities to prices exhibited by different segments of the market. Pricing in these cases is largely divorced from the cost of producing the books.

Managers often make pricing decisions after taking capacity constraints into account. **Peak-load pricing** is the practice of charging a higher price for the same product or service when demand approaches physical capacity limits. That is, the prices charged during busy periods (when loads on the system are high) are greater than the prices charged when slack or excess capacity is available. Peak-load pricing can be found in the telephone, telecommunications, electric utility, and airline industries. The following table illustrates the prices per minute charged by MCI Communications in August 1993 for long-distance telephone calls at different times during the week.

	San Francisco to Las Vegas	San Francisco to New York
Peak period (8 A.M. to 5 P.M. Monday through Friday)	$0.22	$0.23
Evenings (5 P.M. to 11 P.M.) Sunday through Friday	$0.13	$0.14
All other times	$0.12	$0.12

DIFFERENCES IN PRICING PRACTICES AND COST MANAGEMENT METHODS IN VARIOUS COUNTRIES

Surveys[a] of financial officers of the largest industrial companies in several countries indicate similarities and differences in pricing practices across the globe. The use of cost-based pricing appears to be more prevalent in the United States than in Ireland, Japan, and the United Kingdom.

Some Japanese survey data indicate that market-based target pricing practices vary considerably among industries. For example, while a majority of Japanese companies in assembly-type operations (for example, electronics and automobiles) use target costing for pricing, it is far less prevalent in Japanese process-type industries (for example, chemicals, oil, and steel). Japanese companies use value engineering more frequently and involve designers more often when estimating costs. When costs are used for pricing decisions, the pattern is consistent—overwhelmingly, companies across the globe use full costs rather than variable costs.

Ranking of factors that are primarily used to price products (1 is most important)

	United States	Japan	Ireland	United Kingdom
Market based	2	1	1	1
Cost based	1	2	2	2

	Australia	Japan	United Kingdom
Percentage of companies that use value engineering or analysis for cost reduction	24	58	29
Percentage of companies in which designers are involved in estimating costs	25	46	32

Ranking of cost methods used in pricing decisions (1 is most important)

	United States	United Kingdom	Ireland
Full product cost based	1	1	1
Variable cost based	2	2	2

[a] Adapted from Management Accounting Research Group, "Investigation"; Blayney and Yokoyama, "Comparative Analysis"; *Grant Thornton Survey*; Cornick, Cooper, and Wilson, "How Do Companies"; Mills and Sweeting, *Pricing Decisions*; and Drury, Braund, Osborne, and Tayles, "A Survey." Full citations are in Appendix A.

The table indicates that although telephone prices increase with distance, they vary even more with the time of the day or week when the call is placed.

What explains the differences in prices? Are these differences based on differences in costs? We offer two separate, but related, explanations. One explanation is that when capacity is limited, as during weekdays, the telephone company raises prices to levels that the market will bear.

A second explanation is that these prices are a form of price discrimination. By their very nature, business calls have to be made during business hours. As a result, the demand for telephone services by businesses is relatively insensitive to prices. Charging higher prices during the day is profitable because it has little effect on demand. In contrast, personal calls by individuals are more price-sensitive. Lower prices stimulate demand and increase the telephone company's operating income.

Whatever the explanation, the pricing decision is not driven by cost considerations. The actual outlay costs incurred by the phone company to service the calls are roughly the same every day and during the day, evening, or night. But management prices the telephone calls as high as it can, given capacity constraints and the sensitivity to price that different callers (business versus individual) have.

LIFE-CYCLE PRODUCT BUDGETING AND COSTING

Objective 8
Explain how life-cycle product budgeting and costing assist in pricing decisions

The **product life cycle** spans the time from initial research and development to the time at which support to customers is withdrawn. For motor vehicles, this time span may range from five to ten years. For some pharmaceutical products, the time span may be three to five years. For fashion clothing products, the time span may be less than one year.

Using **life-cycle budgeting**, managers estimate the revenues and costs attributable to each product from its initial research and development to its final customer servicing and support in the marketplace. **Life-cycle costing** tracks and accumulates the actual costs attributable to each product from start to finish. The terms "cradle-to-grave costing" and "womb-to-tomb costing" convey the sense of capturing fully all costs associated with the product.

Life-Cycle Budgeting and Pricing Decisions

Life-cycle budgeted costs can provide important information for pricing decisions. For some products, the development period is relatively long, and many costs are incurred prior to manufacturing. Consider Insight, Inc., a computer software company developing a new accounting package, "General Ledger." Assume the following budgeted amounts for the "General Ledger" software package:

Years 1 and 2	
Research and development costs	$240,000
Design costs	$160,000

Years 3 through 6		
	One-Time Setup Costs	Costs per Package Unit
Manufacturing costs	$100,000	$40
Marketing costs	70,000	24
Distribution costs	50,000	16
Customer service costs	80,000	30

A product life-cycle budget highlights to managers the importance of setting prices and budgeting revenues to cover costs in *all* the value-chain categories (Exhibit 12-1, p. 429) rather than just those costs in some of the categories (such as production). To be profitable, Insight must generate revenue to cover costs in all six categories. The life-cycle budget also indicates the costs to be incurred over the life of the product.

Exhibit 12-10 presents the life-cycle budget for the "General Ledger" software package of Insight, Inc. Three combinations of the selling price per package and predicted demand are shown. The high premanufacturing and postmanufacturing costs at Insight are readily apparent in Exhibit 12-10. For example, premanufacturing costs

EXHIBIT 12-10

Budgeted Life-Cycle Revenues and Costs for "General Ledger" Software Package of Insight, Inc.*

	Alternative Selling Price/ Sales-Volume Combinations		
	1	2	3
Selling price per package	$400	$480	$600
Sales quantity in units	5,000	4,000	2,500
Life-cycle revenues:			
$400 × 5,000; $480 × 4,000; $600 × 2,500	$2,000,000	$1,920,000	$1,500,000
Life-cycle costs:			
Research and development costs	240,000	240,000	240,000
Design of product/process costs	160,000	160,000	160,000
Manufacturing costs:			
$100,000 + ($40 × 5,000); $100,000 + ($40 × 4,000); $100,000 + ($40 × 2,500)	300,000	260,000	200,000
Marketing costs:			
$70,000 + ($24 × 5,000); $70,000 + ($24 × 4,000); $70,000 + ($24 × 2,500)	190,000	166,000	130,000
Distribution costs:			
$50,000 + ($16 × 5,000); $50,000 + ($16 × 4,000); $50,000 + ($16 × 2,500)	130,000	114,000	90,000
Customer-service costs:			
$80,000 + ($30 × 5,000); $80,000 + ($30 × 4,000); $80,000 + ($30 × 2,500)	230,000	200,000	155,000
Total life-cycle costs	1,250,000	1,140,000	975,000
Life-cycle operating income	$ 750,000	$ 780,000	$ 525,000

*This exhibit does not take into consideration the time value of money when summing life-cycle revenue or life-cycle costs. Chapters 20 and 21 outline how this important factor can be incorporated into such summations.

(research and development and product design) constitute over 30% of total costs for each of the three combinations of selling price and predicted sales quantity. Insight should put a premium on having as accurate a set of revenue and cost predictions for the "General Ledger" package as possible, given the high percentage of total life-cycle costs paid out before any manufacturing begins and before any revenue is received.

Exhibit 12-10 presents the summary life-cycle revenues and life-cycle costs for the "General Ledger" product, which has a six-year time horizon from "cradle to grave." Exhibit 12-10 does not take into account the time value of money when computing life-cycle revenues or costs. Chapters 20 and 21 outline how the time value of money can affect these amounts.

The Insight illustration presents three pricing alternatives for the "General Ledger" software package independent of other packages that Insight sells. In fact, strategic considerations may sometimes affect the prices that firms charge for individual products. For example, a firm that is successfully selling quality products at high prices may not want to sell a somewhat lower quality product at a lower price even if profitable opportunities for a lower-priced product exist. Why? Because doing so may negatively affect the sales of the high-quality products.[2]

[2]J. Shank and V. Govindarajan, *Strategic Cost Management: Three Key Themes for Managing Costs Effectively* (New York: The Free Press, 1993).

Developing Life-Cycle Reports

Most accounting systems emphasize reporting on a calendar basis—monthly, quarterly, and annually. In contrast, product-life-cycle reporting does not have this calendar-based focus. Consider four Insight products:

	Year 1	Year 2	Year 3	Year 4	Year 5
Accounting Package (General Ledger)					
Law Package					
Payroll Package					
Engineering Package					

Each product spans more than one calendar year.

Developing life-cycle reports for each product requires tracking costs and revenues on a product-by-product basis over several calendar periods. The numbers in these life-cycle reports may differ from those in traditional calendar-based accounting reports. For example, based on external reporting requirements, research and development costs are expensed to the period in which they are incurred. The R&D expenses in each period are the total of R&D costs for all products. *Individual product-by-product identification of R&D costs typically is not shown in calendar reports.* In contrast, the R&D costs included in a product-life-cycle cost report are often incurred in different calendar years. When R&D costs are tracked over the entire life cycle, the total magnitude of these costs for each individual product can be computed and analyzed.

A product-life-cycle reporting format offers at least three important benefits:

1. The full set of revenues and costs associated with each product becomes visible. Manufacturing costs are highly visible in most accounting systems. However, the costs associated with upstream areas (for example, research and development) and downstream areas (for example, customer service) are frequently less visible at a product-by-product level in organizations.

2. Differences among products in the percentage of total costs incurred at early stages in the life cycle are highlighted. The higher this percentage, the more important it is for managers to develop, as early as possible, accurate predictions of the revenues for that product.

3. Interrelationships among cost categories are highlighted. For example, several companies that sizably cut back their R&D and product-design cost categories experienced major increases in customer-service-related costs in subsequent years. Those costs arose because products failed to meet promised quality-performance levels. A life-cycle revenue and cost report prevents such causally related changes among cost categories from being hidden ("buried") as they are in the calendar income statements.

Life-cycle costs further reinforce the importance of locked-in costs, target costing, and value engineering in pricing and cost management. For products with long life cycles, a very small fraction of the total life-cycle costs are actually incurred at the point in time when costs are locked in. *But locked-in costs will determine how costs will be incurred over several years.* Automobile companies combine target costing with life-cycle budgeting. For example, Mercedes, Nissan, and Toyota determine target prices and target costs for their car models on the basis of estimated costs and revenues over a multiyear horizon.

A different notion of life-cycle costs—customer life-cycle costs—has recently been proposed. **Customer life-cycle costs** focus on the total costs to a customer of acquiring and using a product or service until it is replaced. Customer life-cycle costs for a car, for example, include the cost of the car itself plus the costs of operating and maintaining the car minus the disposal price of the car. Customer life-cycle costs can be an important consideration in the pricing decision. For example, a company can sell a car that has low operation and maintenance costs and retains a significant portion of its value for a higher price than it can charge for a similar car with higher maintenance costs and a lower resale value.

EFFECTS OF ANTITRUST LAWS ON PRICING

Objective 9
Explain the effects of antitrust laws on pricing

To comply with U.S. antitrust laws, such as the Sherman Act, the Clayton Act, the Federal Trade Commission Act, and the Robinson-Patman Act, pricing must not be predatory.[3] A business engages in **predatory pricing** when it deliberately prices below its costs in an effort to drive out competitors and restrict supply and then raises prices rather than enlarge demand or meet competition.[4]

Recent court decisions have been influenced by a classic article by Areeda and Turner,[5] which argues that:

◆ A price at or above reasonably anticipated short-run marginal and average variable costs should be deemed nonpredatory.

◆ Unless at or above average cost, a price below reasonably anticipated (1) short-run marginal costs or (2) average variable costs should be deemed predatory.[6]

The case of *Adjustor's Replace-a-Car* v. *Agency Rent-a-Car* illustrates how the courts are willing to use variable-cost information in decisions concerning predatory pricing.[7] Agency Rent-a-Car (the defendant) used selective price cuts to enter the Austin and San Antonio, Texas, car-rental market; cars were rented to customers for extended time periods. Adjustor's (the plaintiff) claimed that it was forced to depart from these markets because Agency had engaged in predatory pricing. A circuit court judge reaffirmed a lower-court decision that Agency "did not predatorily price its service by underselling a competing service in light of the facts that its charges were above average variable cost and there were no significant entry barriers to the market." Evidence presented by Adjustor's included income statements of Agency showing that it had operated its outlets in Austin and San Antonio at a "net operating loss." These statements included an allocated portion of the overhead cost of Agency's headquarters.

The circuit court judge ruled that it was sufficient (in regard to cost justification) for Agency "to demonstrate that the price it charged for a rental car never dropped below its average variable cost." The judge noted:

Agency's expert testified that Agency's average variable costs in San Antonio and Austin during the relevant time periods fluctuated between approximately $3.65 and approximately

---

[3]Discussion of the Sherman Act and the Clayton Act is in A. Barkman and J. Jolley, "Cost Defenses for Antitrust Cases," *Management Accounting* 67, no. 10, pp. 37–40.

[4]See W. Viscusi, J. Vernon, and J. Harrington, *Economics of Regulation and Antitrust* (Lexington: D. C. Heath, 1992), p. 213; and D. Greer, *Industrial Organization and Public Policy* (New York: Macmillan, 1984), pp. 316–17.

[5]P. Areeda and D. Turner, "Predatory Pricing and Related Practices under Section 2 of the Sherman Act," *Harvard Law Review* 88 (1975), pp. 697–733. See also F. Scherer, "Predatory Pricing and the Sherman Act: A Comment," *Harvard Law Review* 89 (1976), pp. 869–903. For an overview of case law, see W. Viscusi, J. Vernon, J. Harrington, *Economics of Regulation and Antitrust* (Lexington: D. C. Heath, 1992). See also the "Legal Developments" section of the *Journal of Marketing* for summaries of court cases.

[6]Areeda and Turner, "Predatory Pricing," p. 733.

[7]Adjustor's Replace-a-Car, Inc. v. Agency Rent-a-Car, 735 2nd 884 (1984).

$5.00 (a day). Thus, Agency's price was always at least 40% greater than its average variable cost. The expert also testified that Agency's average variable costs in Austin were $5.23 when Agency went to a $9.00 price; the price was thus 72% above average variable cost. This testimony cut to the quick of plaintiff's predatory pricing claim.

The circuit court judge rejected Adjustor's claim that "a net loss from operations" on an income statement including an allocation of Agency's headquarters' overhead was "effectively an admission of predatory pricing."

The variable-cost guidelines proposed by Areeda and Turner, although having the support of several court decisions, are not explicitly incorporated into the statute law or supported by the U.S. Supreme Court. Caution should always be used when generalizing from individual legal cases. Managers and accountants who are concerned with their conformance to antitrust laws would be prudent to have a system that incorporates the following procedures:

1. Collect data in such a manner as to permit relatively easy compilation of variable costs.
2. Review all proposed prices below variable costs in advance, with a presumption of claims of predatory intent.
3. Keep as detailed a set of records as feasible, not only of manufacturing costs but also of (a) upstream costs such as research and development and design of products and processes and (b) downstream costs such as marketing, distribution, and customer service.

A company that follows these guidelines will be prepared to respond effectively to inquiries by regulatory agencies.

Closely related to predatory pricing is dumping. Under United States laws **dumping** occurs when a foreign company sells a product in the U.S. at a price below the market value in the country of its creation, and this action materially injures or threatens to materially injure an industry in the U.S. If dumping is proven, an antidumping duty can be imposed under U.S. tariff laws equal to the amount by which the foreign market value exceeds the U.S. price. Cases related to dumping have occurred in the steel, semiconductor, and sweater industries. For example, in 1990, the Commerce Department ruled that manufacturers in the Far East were dumping sweaters on the U.S. market and that importers should pay an antidumping tariff of 36.9% on imported cost.[8]

Collusive pricing occurs when companies in an industry conspire in their pricing and output decisions to achieve a price above the competitive price. Collusive pricing violates the antitrust laws of the United States because it restrains trade. In 1990, for example, the Justice Department charged that the use of a common computer reservation system enabled airlines to collude on maintaining noncompetitive prices. The airlines involved—American, Continental, Delta, Midway, Northwest, PanAm, TWA, United, and U.S. Air—have agreed to reimburse customers under the terms of the settlement.

---

[8]"Ruling Could Stretch Some Sweater Prices," *San Jose Mercury News*, April 24, 1990.

PROBLEM FOR SELF-STUDY

PROBLEM

Continuing Education Programs (CEP) markets teaching packages for companies to use in their own internal training programs. It currently has teaching packages in over 20 different business areas (accounting, personnel management, and so on). The existing internal accounting system at CEP emphasizes overall revenues and costs, year by year. There is no systematized reporting of an individual teaching package's profitability over its life cycle.

Jamie West, the recently appointed editor, is evaluating a proposed teaching package on *Successful Commercial Lending* by Terry Funk. Funk was the author of a prior CEP teaching package, *Successful Consumer Lending*. Funk is requesting a development grant of $350,000 for *Successful Commercial Lending*. West is reluctant to recommend this grant. The highest amount West previously gave was the $250,000 grant to Funk for his prior teaching package. An analysis of the annual financial statements of CEP provides few insights to West in her evaluation of Funk's request. Columns 2 to 7 of Exhibit 12-11 present year-by-year revenues and costs associated with the *Successful Consumer Lending* training package over its six-year life cycle.

Required

1. Did CEP earn operating income on the *Successful Consumer Lending* package? (Do not adjust the numbers in your answer for the time value of money.)

2. Why might Jamie West find a product-life-cycle report informative in decisions relating to Funk's proposed *Successful Commercial Lending* teaching package?

SOLUTION

1. Column 8 of Exhibit 12-11 summarizes the life-cycle revenues and life-cycle costs of the *Successful Consumer Lending* training package. Total life-cycle operating income was $1,092,000. (No adjustment is made for the time value of money.)

2. West makes many decisions on individual training packages. The product life cycle of each package extends over several calendar years. Reports that include costs and revenues for the entire life cycle highlight the overall profitability of an individual training package. Such reports also highlight the magnitude and timing of development costs on each training package. Among the costs for the *Successful Consumer Lending* package, development costs were relatively high, but these were more than offset by relatively high revenues in years 3 through 6. Funk could well argue that "you have to spend money to make money" in the training package development business.

    A year-by-year reporting format captures only part of the costs and revenues associated with each package. Moreover, year-by-year income statements typically report total costs and revenues for CEP rather than product-by-product costs and revenues for that year.

    From Jamie West's perspective, the issue is not a year-by-year set of reports versus a product-by-product life-cycle set of reports. Rather, the issue is whether, given the existing year-by-year reporting system, her collective decisions will be improved by reporting budgets and actual results on a product-by-product life-cycle basis.

EXHIBIT 12-11

Life-Cycle Revenues and Costs of *Successful Consumer Lending* Training Package (in thousands)

Revenues/Costs (1)	Year 1 (2)	Year 2 (3)	Year 3 (4)	Year 4 (5)	Year 5 (6)	Year 6 (7)	Total* (8)
Life-cycle revenues	$ 0	$ 0	$300	$1,030	$820	$370	$2,520
Life-cycle costs:							
Development payments to author	100	150	0	0	0	0	250
Development costs of CEP	2	98	40	0	0	0	140
Production costs of CEP	0	0	75	96	135	80	386
Marketing costs of CEP	0	15	120	100	35	28	298
Distribution and customer-service costs of CEP	0	5	40	30	15	12	102
Royalty payments to author	0	0	30	103	82	37	252
Total life-cycle costs	$102	$268	$305	$329	$267	$157	1,428
Life-cycle operating income							$1,092

*The time value of money is not taken into account when computing life-cycle revenues or life-cycle costs in this exhibit. Chapters 20 and 21 outline how this important factor can be incorporated into such calculations.

SUMMARY

The following points are linked to the chapter's learning objectives.

1. Three major influences in pricing decisions are customers, competitors, and costs.

2. Six important business function cost categories to be considered in pricing decisions are research and development, design, production, marketing, distribution, and customer service.

3. Short-run pricing decisions focus on a period of a year or less and have no long-run implications. Long-run pricing decisions focus on a main product in a major market with a time horizon of one year or longer. The time horizon appropriate to a decision on pricing dictates which costs are relevant.

4. One approach to pricing is to use a target price. Target price is the estimated price that potential customers are willing to pay for a product (or service). Management bases this estimate on an understanding of customers' perceived value for a product and the responses of competitors. A desired profit margin is subtracted from target price to determine target cost. The target cost is the estimated long-run cost of a product (or service) that when sold enables the firm to achieve the targeted income. The challenge for the organization is to make the cost improvements necessary through value engineering methods to achieve target cost. Value engineering is a systematic evaluation of all aspects of the production, marketing, distribution, and customer-service processes, with the objective of reducing costs while satisfying customer needs.

5. Cost incurrence arises when resources are actually sacrificed. Locked-in costs refer to costs that have not yet been incurred but which, based on decisions that have already been made, will be incurred in the future.

6. The cost-plus approach to pricing sets prices by adding a markup to a cost base. Many different costs (such as full product costs or manufacturing costs) can serve as the cost base in applying the cost-plus formula.

7. Price discrimination is the practice of charging some customers a higher price than is charged to other customers. Peak-load pricing is the practice of charging a higher price for the same product or service when demand approaches physical capacity limits. Under price discrimination and peak-load pricing, prices differ among market segments even though the outlay costs of providing the product or service are approximately the same.

8. Life-cycle budgeting and life-cycle costing estimate, track, and accumulate the costs (and revenues) attributable to each product from its initial research and development to its final customer servicing and support in the marketplace. Life-cycle costing offers three important benefits: (a) the full set of costs associated with each product become visible; (b) differences among products in the percentage of total costs incurred at early stages in the life cycle are highlighted; and (c) interrelationships among cost categories are also highlighted.

9. To comply with antitrust laws, a company must not engage in predatory pricing, dumping, or collusive pricing, which lessens competition or puts another firm at a competitive disadvantage.

TERMS TO LEARN

This chapter and the Glossary at the end of the book contain definitions of the following important terms:

collusive pricing *(p. 452)*   cost incurrence *(439)*   customer life-cycle costs *(451)*
designed-in costs *(439)*   downstream costs *(429)*   dumping *(452)*
life-cycle budgeting *(448)*   life-cycle costing *(448)*   locked-in costs *(439)*
peak-load pricing *(446)*   postmanufacturing costs *(429)*   predatory pricing *(451)*
premanufacturing costs *(429)*   price discrimination *(446)*   product life cycle *(448)*
target cost *(438)*   target price *(438)*   target return on investment *(443)*
upstream costs *(429)*   value engineering *(438)*

ASSIGNMENT MATERIAL

QUESTIONS

12-1   What are the three major influences on pricing decisions?

12-2   Name the six business function cost categories that can be included in the cost buildup of a product or service.

12-3   The relevant costs for pricing decisions are full product costs. Comment.

12-4   Give two examples of pricing decisions with a short-run focus.

12-5   What is undercosting or overcosting of products? How does undercosting or overcosting affect the pricing of a product?

12-6   Describe two alternative approaches to pricing.

12-7   What is a target cost?

12-8   What are two ways a company can manufacture products that are competitively priced?

12-9   "It is not important for a firm to distinguish between cost incurrence and locked-in costs." Do you agree? Explain.

12-10   What is cost-plus pricing?

12-11   Give two examples where the difference in the costs of two products or services is much smaller than the difference in their prices.

12-12   What is the product life cycle?

12-13   What are three benefits of using a product-life-cycle reporting format?

12-14   Define predatory pricing, dumping, and collusive pricing.

EXERCISES AND PROBLEMS

12-15   Peak-load pricing of transportation services.   Examples of round-trip ticket prices on the London Underground follow:

	Piccadilly to Wembley Park	Heathrow to Trafalgar Square	Edgware to Wimbledon
Peak hours (for travel starting between 5:30 A.M. and 9:30 A.M. Monday through Friday)	£4.20	£5.60	£5.20
Off-peak hours	£3.10	£3.50	£3.50

Required

1. Are there differences in outlay costs for London Underground to carry passengers in peak hours versus off-peak hours?

2. Why do you think London Underground charges different rates for peak-hour and off-peak-hour travel?

12-16 Peak-load pricing of telephone services. Examples of prices charged per minute by AT&T for long-distance telephone calls within the United States at different times of the day and week follow:

	Washington, D.C. to Philadelphia	Washington, D.C. to St. Louis	Washington, D.C. to Los Angeles
Peak period (8 A.M. to 5 P.M., Monday through Friday)	$0.21	$0.22	$0.24
Evenings (5 P.M. to 11 P.M. Monday through Friday)	$0.13	$0.13	$0.14
Nights and weekends	$0.11	$0.11	$0.12

Required

1. Are there differences in outlay costs for AT&T to connect telephone calls during peak hours compared to telephone calls made at other times of the day?

2. Why do you think AT&T charges different prices per minute for telephone calls made during peak hours compared to telephone calls made at other times of the day?

12-17 Pricing of hotel rooms on weekends. Paul Diamond is the owner of the Galaxy chain of four-star prestige hotels. These hotels are in Chicago, London, Los Angeles, Montreal, New York, Seattle, San Francisco, and Tokyo. Diamond is currently struggling to set weekend rates for the San Francisco hotel (the San Francisco Galaxy). From Sunday through Thursday, the Galaxy has an average occupancy rate of 90%. On Friday and Saturday nights, however, average occupancy declines to less than 30%. Galaxy's major customers are business travelers who stay mainly Sunday through Thursday.

The current room rate at the Galaxy is $150 a night for single occupancy and $180 a night for double occupancy. These rates apply seven nights a week. For many years, Diamond has resisted having rates for Friday and Saturday nights that are different from rates for the remainder of the week. Diamond has long believed that price reductions convey a "nonprestige" impression to his guests. The San Francisco Galaxy values highly its reputation for treating its guests as "royalty."

Most room costs at the Galaxy are fixed on a short-stay (per-night) basis. Diamond estimates the variable costs of servicing each room to be $20 a night per single occupancy and $22 a night per double occupancy.

Many prestige hotels in San Francisco offer special weekend rate reductions (Friday and/or Saturday) of up to 50% of their Sunday-through-Thursday rates. These weekend rates also include additional items such as a breakfast for two, a bottle of champagne, and discounted theater tickets.

Required

1. Would you recommend that Diamond reduce room rates at the San Francisco Galaxy on Friday and Saturday nights? What factors should be considered in his decision?

2. In six months' time the Super Bowl is to be held in San Francisco. Diamond observes that several four-star prestige hotels have already advertised a Friday-through-Sunday rate for Super Bowl weekend of $300 a night. Should Diamond charge extra for the Super Bowl weekend? Explain.

12-18 Apparent price anomalies. Postal prices for within-country mail in New Zealand vary with the weight of a letter (or parcel). New Zealand Post (NZP) charges NZ$0.45 (NZ$ is New Zealand dollars) for delivering a letter of "standard" weight. Postal prices are not related to the distance from sender to receiver.

Independent of its pricing policy, the costs NZP incurs for delivering a letter are affected by both distance and weight. For example, it costs NZP very little to sort and deliver intracity mail (mail picked up and delivered within one city). Intercity mail (mail picked up in one city and delivered to another city) is more expensive to handle. Even more costly is delivering or picking up mail from the rural areas of New Zealand.

Private businesses have begun entering the mail-delivery business for intracity mail. For example, private mail carriers are approaching electric utilities and local telephone companies with offers to deliver their monthly bills to customers, all of whom are located within the city, at rates that are 30% lower than the regular mail prices. Private mail carriers can afford to charge a lower price because of the low cost of delivering intracity mail. NZP's pricing structure, conversely, is to charge a constant price that represents the average cost of delivering mail all over New Zealand. The price is not related to the cost of delivery. If NZP loses the low-cost intracity mail to private carriers, it will be left with only the more costly intercity and rural mail. This loss would in turn result in still higher postal prices in New Zealand.

Required

1. Why do you believe NZP, which incurs very different costs for handling intracity mail compared with intercity and rural mail, charges the same price for all letter deliveries?
2. How should NZP react to the threat to its business in the cities? Should NZP change its practice of charging the same price for a letter independent of the cost of handling it?

12-19   Relevant-cost approach to pricing decisions, special order.   The following financial data apply to the videotape production plant of the Dill Company:

October 19_4	Budgeted Manufacturing Costs per Video Tape
Direct materials	$1.50
Direct manufacturing labor	0.80
Variable manufacturing overhead	0.70
Fixed manufacturing overhead	1.00
Total manufacturing costs	$4.00

Variable manufacturing overhead varies with respect to units produced. Fixed manufacturing overhead of $1 per tape is based on budgeted fixed manufacturing overhead of $150,000 per month and budgeted production of 150,000 tapes per month. Dill Company sells each tape for $5.
  Marketing costs have two components:

◆ Variable marketing costs (sales commissions) of 5% of dollar sales
◆ Fixed monthly costs of $65,000

  During October 19_4, Lyn Randell, a Dill Company salesperson, asked the president for permission to sell 1,000 tapes at $3.80 per tape to a customer not in its normal marketing channels. The president refused this special order on the grounds that the selling price was below the total budgeted manufacturing cost.

Required

1. What would have been the effect on monthly operating income of accepting the special order?
2. Comment on the president's "below manufacturing costs" reasoning for rejecting the special order.
3. What factors would you recommend that the president consider when deciding whether to accept or reject the special order?

12-20   Relevant-cost approach to short-run pricing decisions.   The San Carlos Company is an electronics business with eight product lines. Income data for one of the products (XT-107) for the month just ended (June 19_5) follow:

Sales: 200,000 units at average price of $100		$20,000,000
Variable costs:		
Direct materials at $35 per unit	$7,000,000	
Direct manufacturing labor at $10 per unit	2,000,000	
Variable manufacturing overhead at $5 per unit	1,000,000	
Sales commissions at 15% of sales	3,000,000	
Other variable costs at $5 per unit	1,000,000	
Total variable costs		14,000,000
Contribution margin		6,000,000
Fixed costs		5,000,000
Operating income		$ 1,000,000

  Abrams Inc., an instruments company, has a problem with its preferred supplier of XT-107 component products. This supplier has had a three-week strike by its employees and will not be able to supply Abrams 3,000 units next month. Abrams approaches the sales representative, Sarah Holtz, of the San Carlos Company about providing 3,000 units of XT-107 at a price of $80 per unit. Holtz informs the XT-107 product manager, Jim McMahon, that she would accept a flat commission of $6,000 rather than the usual 15% if this special order were accepted. San Carlos has the capacity to produce 300,000 units of XT-107 each month, but demand has not exceeded 200,000 in any month in the last year.

Required
1. If the 3,000-unit order from Abrams Inc. is accepted, what will be the effect on monthly operating income? (Assume the same cost structure as occurred in June 19_5.)
2. McMahon ponders whether to accept the 3,000-unit special order. He is afraid of the precedent that might be set by cutting the price. He said, "The price is below our full cost of $95 per unit. I think we should quote a full price, or Abrams will expect favored treatment again and again if we continue to do business with them." Do you agree with McMahon? Explain.

12-21 Relevant-cost approach to pricing decisions: contribution versus absorption-costing income statement. Stardom Inc. cans peaches for sale to food distributors. All costs are classified as either manufacturing or marketing. Stardom prepares monthly budgets. The March 19_6 budgeted absorption-costing income statement follows:

Revenues (1,000 crates at $100 a crate)	$100,000	100%
Cost of goods sold	60,000	60
Gross margin	40,000	40
Marketing costs	30,000	30
Operating income	$ 10,000	10%
Normal markup percentage:		
$40,000 ÷ $60,000 = 66.7% of absorption cost		

Monthly costs are classified as fixed or variable (with respect to the cans produced for manufacturing costs and with respect to the cans sold for marketing costs):

	Fixed	Variable
Manufacturing	$20,000	$40,000
Marketing	16,000	14,000

Stardom has the capacity to can 1,500 crates per month. The relevant range in which monthly fixed manufacturing costs will be "fixed" is from 500 crates to 1,500 crates per month.

Required
1. Recast the income statement in a variable-costing (contribution) format. Indicate the normal markup percentage based on total variable costs.
2. Assume that a new customer approaches Stardom to buy 200 crates at $55 per crate. The customer does not require additional marketing effort. Additional manufacturing costs of $2,000 (for special packaging) will be required. Stardom believes that this is a one-time-only special order, because the customer is discontinuing business in six weeks' time. Stardom is reluctant to accept this 200-crate special order because the $55 per-crate price is below the $60 per-crate absorption cost. Do you agree with this reasoning? Explain.
3. Assume that the new customer decides to remain in business. How would this longevity affect your willingness to accept the $55 per-crate offer? Explain.

12-22 Target prices, target costs, value engineering, incurrence, and locked-in cost. Cutler Electronics makes a radio-cassette player, CE100, which has 80 components. Cutler sells 7,000 units each month for $70 each. The costs of manufacturing CE100 are $45 per unit, or $315,000 per month. Monthly manufacturing costs incurred follow:

Direct materials costs	$182,000
Direct manufacturing labor costs	28,000
Machining costs (fixed)	31,500
Testing costs	35,000
Rework costs	14,000
Ordering costs	3,360
Engineering costs (fixed)	21,140
Total manufacturing costs	$315,000

Cutler's management identifies the activity cost pools, the cost drivers for each activity, and the cost per unit of cost driver for each overhead cost pool as follows:

Manufacturing Activity	Description of Activity	Cost Driver	Cost per Unit of Cost Driver
1. Machining costs	Machining of components.	Fixed costs	No cost driver
2. Testing costs	Testing components and final product. Each unit of CE100 is tested individually.	Testing-hours	$2 per testing-hour
3. Rework costs	Correcting and fixing errors and defects.	Units of CE100 reworked	$20 per unit
4. Ordering costs	Ordering of components.	Number of orders	$21 per order
5. Engineering costs	Designing and managing of products and processes.	Fixed costs	No cost driver

Over a long-run time horizon, Cutler's management views direct materials costs and indirect manufacturing labor costs as variable with respect to the units of CE100 manufactured. Each of the overhead costs described in the above table varies, as described, with the chosen cost drivers.

The following additional information describes the existing design.

a. Testing and inspection time per unit is 2.5 hours.

b. 10% of the CE100s manufactured are reworked.

c. Cutler places two orders with each component supplier each month. Each component is supplied by a different supplier. It takes 1 hour to place an order.

To respond to competitive pressures, Cutler must reduce its price to $62 per unit and reduce its costs by $8 per unit. No additional sales are anticipated at this lower price. However, Cutler stands to lose significant sales if it does not cut its price. Manufacturing has been asked to reduce its costs by $6 per unit. Improvements in manufacturing efficiency are expected to yield net savings of $1.50 per radio-cassette player but that is not enough. The chief engineer has proposed a new modular design that reduces the number of components to 50 and also simplifies testing. The newly designed radio-cassette player, called "New CE100" will replace CE100. The expected effects of the new design follow:

d. Direct materials costs for the New CE100 are expected to be lower by $2.20 per unit.

e. Direct manufacturing labor costs for the New CE100 are expected to be lower by $0.50 per unit.

f. Machining time required to manufacture the New CE100 is expected to be 20% less. It currently takes 1 hour to manufacture one unit of CE100.

g. Time required for testing the New CE100 is expected to be lower by 20%.

h. Rework is expected to decline to 4% of New CE100s manufactured.

Assume that the cost per unit of the cost driver for CE100 continues to apply to New CE100.

Required

1. Calculate Cutler's cost to manufacture one unit of New CE100.

2. Will the new design achieve the per unit cost reduction targets that have been set for manufacturing costs?

3. The problem describes two strategies to reduce costs: (a) improving manufacturing efficiency and (b) modifying the design. Which strategy has a bigger impact on costs? Why? Explain briefly.

 **12-23 Product costs, activity-based costing systems.** Executive Power (EP) manufactures and sells computers and computer peripherals to several nationwide retail chains. John Farnham is the manager of the printer division. Its two largest-selling printers are P-41 and P-63.

The manufacturing cost of each printer is calculated using EP's activity-based costing system. EP has one direct-manufacturing cost category (direct materials) and the following five indirect-manufacturing cost pools:

Indirect Manufacturing Cost Pool	Allocation Base	Allocation Rate
1. Materials handling	Number of parts	$1.20 per part
2. Assembly management	Hours of assembly time	$40 per hour of assembly time
3. Machine insertion of parts	Number of machine-inserted parts	$0.70 per machine-inserted part
4. Manual insertion of parts	Number of manually inserted parts	$2.10 per manually inserted part
5. Quality testing	Hours of quality testing time	$25 per testing-hour

Product characteristics of P41 and P63 follow:

	P41	P63
Direct materials costs	$407.50	$292.10
Number of parts	85	46
Hours of assembly time	3.2	1.9
Number of machine-inserted parts	49	31
Number of manually inserted parts	36	15
Hours of quality testing time	1.4	1.1

Required

What is the manufacturing cost of P-41? Of P-63?

12-24 Target cost, activity-based costing systems. (Continuation of 12-23)   Assume all the information in Problem 12-23. Farnham has just received some bad news. A foreign competitor has introduced products very similar to P-41 and P-63. Given their announced selling prices, Farnham estimates the P-41 clone to have a manufacturing cost of approximately $680 and the P-63 clone to have a manufacturing cost of approximately $390. He calls a meeting of product designers and manufacturing personnel at the printer division. They all agree to have the $680 and $390 figures become target costs for redesigned versions of EP's P-41 and P-63, respectively. Product designers examine alternative ways of designing printers with comparable performance but lower cost. They come up with the following revised designs for P-41 and P-63 (termed P-41 REV and P-63 REV, respectively):

	P41-REV	P63-REV
Direct materials cost	$381.20	$263.10
Number of parts	71	39
Hours of assembly time	2.1	1.6
Number of machine-inserted parts	59	29
Number of manually inserted parts	12	10
Hours of quality testing time	1.2	0.9

Required

1. What is a target cost?
2. Using the activity-based costing system outlined in Problem 12-23, compute the manufacturing costs of P-41 REV and P-63 REV. How do they compare with the $680 and $390 target costs?
3. Explain the differences between P-41 and P-41 REV and between P-63 and P-63 REV.
4. Assume now that John Farnham has achieved major cost reductions in one of the activity areas. As a consequence, the allocation rate in the assembly-management activity area will be reduced from $40 to $28 per assembly-hour. How will this activity-area cost reduction affect the manufacturing costs of P-41 REV and P-63 REV? Comment on the results.

12-25 Cost-plus pricing. (CMA, adapted)   The Sommers Company, located in southern Wisconsin, manufactures a variety of industrial valves and pipe fittings that are sold to customers in nearby states. Currently, the company is operating at about 70% capacity and is earning a satisfactory return on investment.

Sommers management has been approached by Glascow Industries Ltd. of Scotland with an offer to buy 120,000 units of a pressure valve. Glascow Industries manufactures a valve that is almost identical to Sommers' pressure valve; however, a fire in Glascow Industries' valve plant has temporarily shut down its manufacturing operations. Glascow needs the 120,000 valves over the next four months to meet commitments to its regular customers. The company is prepared to pay $19 each for the valves. Glascow will arrange to pick up the valves from the Sommers plant.

Sommers' manufacturing costs, based on currently attainable standards, for the pressure valve are:

Direct materials costs	$ 5.00
Direct manufacturing labor costs	6.00
Manufacturing overhead costs	9.00
Total manufacturing costs	$20.00

Manufacturing overhead is allocated to production at the rate of $18 per standard direct labor-hour. This overhead rate is made up of the following components.

Variable manufacturing overhead	$ 6.00
Fixed manufacturing overhead	12.00
Budgeted manufacturing overhead rate	$18.00

Additional costs incurred in connection with sales of the pressure valve include sales commissions of 5% and freight costs of $1 per unit. However, the company does not pay sales commissions on special orders that come directly to management.

In determining selling prices, Sommers adds a 40% markup to manufacturing costs. This approach provides a $28 suggested selling price for the pressure valve. The marketing department, however, has set the current selling price at $27 in order to maintain market share.

Production management believes that it can handle the Glascow Industries order without disrupting its scheduled production. The order would, however, require additional fixed manufacturing overhead of $12,000 per month in increased supervision and clerical costs.

If management accepts the order, 30,000 pressure valves will be manufactured and picked up by Glascow Industries each month for the next four months. Shipments will be made weekly.

Required:

1. Determine how many additional direct manufacturing labor-hours would be required each month to fill the Glascow Industries order.

2. Prepare an incremental analysis showing the impact on revenues and costs of accepting the Glascow Industries order.

3. Calculate the minimum unit price that Sommers management could accept for the Glascow Industries order without changing operating income.

4. Identify the factors, other than price, that the Sommers Company should consider before accepting the Glascow Industries order.

12-26 Cost-plus pricing. (CMA, adapted)   Marcus Fibers Inc. specializes in manufacturing synthetic fibers, which the company uses in many products, such as blankets, coats, and uniforms for police and firefighters. Marcus has been in business since 1975 and has been profitable each year since 1988. The company uses a standard-costing system and allocates overhead costs on the basis of direct manufacturing labor-hours.

Marcus has recently received a request to bid on the manufacture of 800,000 blankets scheduled for delivery to several military bases. The bid must be stated at full cost per unit plus a return on full cost of no more than 9% after income taxes. Full cost has been defined as including all variable costs of manufacturing the product, a reasonable amount of fixed manufacturing overhead, and a reasonable amount of incremental administrative costs associated with the manufacture and sale of the product. The contractor has indicated that bids in excess of $25 per blanket are not likely to be considered.

In order to prepare the bid for the 800,000 blankets, Andrea Lightner, cost accountant, has gathered information about the costs associated with the production of the blankets.

Direct materials costs	$1.50 per pound of fibers
Direct manufacturing labor costs	$7.00 per hour
Direct machine costs*	$10 per blanket
Variable manufacturing overhead costs	$3.00 per direct manufacturing labor-hour
Fixed manufacturing overhead costs	$8.00 per direct manufacturing labor-hour
Incremental administrative costs	$2,500 per 1,000 blankets
Special fee†	$0.50 per blanket
Materials usage	6 pounds per blanket
Production rate	4 blankets per direct manufacturing labor-hour
Income tax rate	40%

*Direct machine costs consist of items such as special lubricants, replacement of needles used in stitching, and maintenance costs. These costs are not included in the budgeted overhead rates.

† Marcus recently developed a new blanket fiber at a cost of $750,000. In an effort to recover this cost, Marcus has instituted a policy of adding a $0.50 fee to the cost of each blanket that uses the new fiber. To date, the company has recovered $125,000. Lightner knows that this fee does not fit within the definition of full cost because it is not a cost of manufacturing the product.

Required

1. Calculate the minimum price per blanket that Marcus Fibers Inc. could bid without changing the company's net income.

2. Using the full cost criterion and the maximum allowable return specified, calculate Marcus Fibers Inc.'s bid price per blanket.

3. Without considering your answer to requirement 2, assume that the price per blanket that Marcus Fibers Inc. calculated using the cost-plus criterion is greater than the maximum bid of $25 per blanket allowed. Discuss the factors that Marcus Fibers Inc. should consider before deciding whether or not to submit a bid at the maximum acceptable price of $25 per blanket.

 12-27 Cost-plus target return on investment pricing. John Beck is the managing partner of a partnership that has just finished building a 60-room motel. Beck estimates the following operating costs for next year:

Variable operating costs	$3 per room
Fixed costs	
Salaries and wages	$175,000
Maintenance of building and pool	37,000
Other operating and administration costs	140,000
Total fixed costs	$352,000

Beck anticipates that he will rent 16,000 rooms next year. All rooms are similar and will rent for the same price. The capital invested in the motel is $960,000. The partnership's target return on investment is 25%. Beck expects demand for rooms to be about uniform throughout the year. He plans to price the rooms at cost plus a markup to earn the target return on investment.

Required

1. What price should Beck charge for a room? What is the markup over the full cost of a room?

2. Beck's market research indicates that if the price of a room determined in requirement 1 was reduced by 10%, the expected number of rooms Beck could rent would increase by 10%. Should Beck make the 10% cut?

12-28 Airline pricing, price discrimination. Air Americo is about to introduce a daily round-trip flight from New York (NY) to Los Angeles (LA). Air Americo offers only one class of seats—Comfort Class, which allows more leg room for passengers—on all its flights. No other airline offers this kind of seat. Air Americo is in the process of determining how it should price its round-trip tickets. The following information is available:

Seating capacity per plane	360
Maximum demand for seats on any flight	300
Food and beverage service costs for a round-trip (no charge to passenger)	$40 per passenger

Commission to travel agents paid by Air Americo on each ticket booked on Air Americo. (Assume all of Air Americo's tickets are booked by travel agents.)		8% of fare
Fuel costs for a round-trip flight		$24,000
Fixed annual lease costs allocated to a round-trip flight		$100,000
Fixed ground services (maintenance, check-in, baggage-handling) costs allocated to a round-trip flight		$10,000
Fixed flight crew salaries allocated to a round-trip flight		$8,000

For simplicity, assume that fuel costs are not affected by the actual number of passengers on a flight.

The market research group at Air Americo segments the market into two types of travelers—business and pleasure—and provides the following information on the effect of two different prices on the estimated number of seats sold:

	Price Charged	Number of Seats Expected to Be Sold
Business travelers	$500	200
	$2,000	190
Pleasure travelers	$500	100
	$2,000	20

Assume that these two prices are the only two choices available to Air America. The market research team offers one additional piece of information. Pleasure travelers start their travel in one week, stay over at least one weekend at the place they are visiting, and return in some following week. Business travelers usually start and complete their travel within the week. They do not stay away from their homes over weekends.

Required

1. If you could charge different prices to business travelers and pleasure travelers, would you? Show all computations.
2. Explain the key factor (or factors) that drives your answer in requirement 1.
3. How might Air America implement price discrimination? That is, what scheme could the airline devise so that business travelers pay the price the airline would like business travelers to pay, and pleasure travelers pay the price the airline would like pleasure travelers to pay?

12-29 Life-cycle product costing, product emphasis. Decision Support Systems (DSS) is examining the profitability and pricing policies of its software division. The DSS software division develops software packages for engineers. DSS collects data on three of its more recent packages:

◆ EE-46: package for electrical engineers
◆ ME-83: package for mechanical engineers
◆ IE-17: package for industrial engineers

Summary details on each package over their two-year "cradle-to-grave" product lives follow:

		Number of Units Sold	
Package	Selling Price	Year 1	Year 2
EE-46	$250	2,000	8,000
ME-83	300	2,000	3,000
IE-17	200	5,000	3,000

Assume that no inventory remains on hand at the end of year 2.

DSS is deciding which product lines in its software division to emphasize. In the past two years, the profitability of this division has been mediocre. DSS is particularly concerned with the increase in research and development costs in several of its divisions. An analyst at the software division pointed out that for one of its most recent packages (IE-17), major efforts had been made to cut back research and development costs.

Last week Nancy Sullivan, the software division manager, attended a seminar on product-life-cycle management. The topic of life-cycle reporting was discussed. Sullivan decides to use this approach in her own division. She collects the following life-cycle revenue and cost information for the EE-46, ME-83, and IE-17 packages (in thousands).

	EE-46		ME-83		IE-17	
	Year 1	Year 2	Year 1	Year 2	Year 1	Year 2
Revenues	$500	$2,000	$600	$900	$1,000	$600
Costs						
Research and development	700	0	450	0	240	0
Design	185	15	110	10	80	16
Manufacturing	75	225	105	105	143	65
Marketing	140	360	120	150	240	208
Distribution	15	60	24	36	60	36
Customer service	50	325	45	105	220	388

Required

1. How does a product-life-cycle income statement differ from a calendar-based income statement? What are the benefits of using a product-life-cycle reporting format?

2. Present a product-life-cycle income statement for each software package. Which package is the most profitable, and which is the least profitable?

3. How do the three software packages differ in their cost structure (the percentage of total costs in each cost category)?

12-30 Ethics and pricing. Baker Inc. manufactures ball bearings. Baker is preparing to submit a bid for a new ball-bearings order. Greg Lazarus, controller of the Bearings Division of Baker Inc., has asked John Decker, the cost analyst, to prepare the bid. Baker determines price on the basis of full product costs plus a markup of 10%. Lazarus tells Decker that he is keen on winning the bid and that the price he calculates should be competitive.

Decker completes his calculations and shows them to Lazarus. Lazarus reviews the numbers and says, "As usual your costs are way too high. You have allocated a lot of overhead costs to this job. You know our fixed overhead is not going to change if we win this order and manufacture the bearings. Ever since we installed this new activity-based costing system, we never seem to be able to come up with reasonable product and job costs. Rework your numbers. The costs have got to be lower."

On returning to his office, Decker rechecks his numbers. He knows that Lazarus wants this order because the additional revenue from the order would lead to a big bonus for Lazarus and the senior division managers. Decker wonders if he can adjust the costs downward, perhaps by excluding some fixed costs. He knows that if he does not come up with a lower bid, Lazarus will be very upset.

Required

Refer to the Standards of Ethical Conduct for Management Accountants described in Chapter 1 (p. 17). Evaluate whether Lazarus's suggestion to Decker to use lower cost numbers is unethical. Will it be unethical for Decker to change his analysis so that a lower price can be bid? What steps should Decker take to resolve this situation?

13

Management Control Systems:

Choice and Application

Management control systems

Evaluating management control systems

An example of motivation: sales compensation plans

Value-added/nonvalue-added cost analysis

Engineered, discretionary, and infrastructure costs

Classifying costs in business function areas in the value chain

Budgeting for discretionary costs

Work measurement

Engineered-cost versus discretionary-cost approaches to budgeting

Benchmarking and cost management

Appendix: Promoting and monitoring the effectiveness or efficiency of discretionary-cost centers

*M*anagers in organizations such as Home Depot have many balls to juggle. The management control system at Home Depot helps managers monitor financial aspects, such as operating income and sales dollars per transaction, and nonfinancial aspects, such as customer satisfaction, safety, and the environment.

Managers and accountants do not choose a particular management control system on the basis of its technical aspects alone. It is also essential to consider the behavior of people who will use that system. This chapter presents an overview of management control systems. We emphasize behavioral issues in this overview. Because of that emphasis the material in this chapter is "softer" (there is less number crunching) than in most other chapters. Often there will not be a pat answer or, in some cases, even a systematic method of studying the issue. Nevertheless, knowing how to identify the central issue(s) when choosing systems is a skill in itself. This chapter provides a method for identifying central issues and weighing how alternative management control systems may help solve associated problems.

MANAGEMENT CONTROL SYSTEMS

Objective 1
Describe a management control system

A **management control system** is a means of gathering information to aid and coordinate the process of making planning and control decisions throughout the organization. The system improves the collective decisions within an organization.

Exhibit 13-1 shows four types of information a management control system should include:

- ◆ Financial/internal
- ◆ Financial/external
- ◆ Nonfinancial/internal
- ◆ Nonfinancial/external

The examples included in Exhibit 13-1 pertain to Home Depot, Inc., a home improvement retailer.[1]

Companies differ on who has the responsibility for collecting, reporting and ensuring the reliability of information in each of the four areas in Exhibit 13-1. The controller/senior management accountant is the executive responsible for most financial/internal information. The chief financial officer is responsible for monitoring

---

[1]Additional examples of information in the four categories in Exhibit 13-1 are in S. Hronec, *Vital Signs* (New York: American Management Association, 1993).

Financial/Internal	Financial/External
◆ Sales dollars per transaction ◆ Sales dollars per employee ◆ Operating income ◆ Total assets	◆ Stock price of Home Depot on the New York Stock Exchange ◆ Stock price to earnings per share ratio of Home Depot
Nonfinancial/Internal	Nonfinancial/External
◆ Number of sales transactions per employee ◆ Number of employee accidents ◆ Shrinkage on shelves due to breakages, theft, or other factors ◆ Customer time in check-out lines	◆ Customer satisfaction measures from follow-up phone interviews ◆ Third-party ratings on environmental responsibility (such as disposal of waste)

financial/external information. The responsibility for nonfinancial information is often spread among executives in the business function areas of the value chain (such as manufacturing, distribution, and customer service).

Many Balls to Juggle

Objective 2

Identify different areas where information may be gathered in a management control system

Financial/internal information has long been recognized as pivotal to an effective management control system. There is now increased recognition of the importance of the other areas in Exhibit 13-1. The following extracts from a recent Home Depot annual report highlight how areas such as customer satisfaction, employee teamwork, the environment, and ethics are important concerns for all its managers and employees:

◆ "Home Depot is a leading innovator in retailing by combining the economies of scale of warehouse-format stores with a high level of customer service."

◆ "We make sure our employees know what is expected of them and what they should expect—even demand— from their managers. We value free expression, individuality, and self-reliance within the context of teamwork, experimentation, and calculated risk taking."

◆ "We have become recognized as a leader in environmental performance in the retail industry."

◆ "We are committed to dealing only with ethical trading partners."

Today's managers have many balls to juggle. Companies are now being evaluated in many different areas, including profitability, customer satisfaction, labor relations and safety, environmental aspects, and ethical behavior. Moreover, greater computer power has made it cost-effective for more and more companies to expand the documentation that they process and thereby expand the role of the manager. Because of this increased diversity of management responsibility, a well-functioning management control system assumes even greater importance to the company's success today than it did yesterday.

Interrelated Purpose of Many Cost Outlays

Chapter 12 (p. 439) noted how effective cost management requires a focus on the R&D and design stages of the value chain as well as on the production and subsequent stages. This point is especially true in the environmental and safety area. For example, consider ICI, the international chemical company headquartered in London. ICI conducts a "Hazard Study for Safety, Health and Environment Protection" at each of the six stages in its new project program (exploratory concept, process definition, design, construction, commissioning, and full operation). Note that with this integrated approach to project planning, it is difficult to uniquely identify cost outlays as related to any single purpose (whether that purpose be, say, productivity, quality, safety, or the environment).

EVALUATING MANAGEMENT CONTROL SYSTEMS

What is a useful starting point for evaluating a management control system? Obtain a specification of top management's goals for the organization as a whole. Examples of goals include:

◆ Maximizing five-year reported net income
◆ Boosting the net-income-to-shareholders'-equity ratio into the top 20% of all publicly traded companies
◆ Maximizing stock return to shareholders
◆ Attaining the largest market share in the industry
◆ Reaching the highest consumer-satisfaction rating in the industry

An observer may disagree with the goals specified by top management, but a control system should be judged in light of how well it helps achieve a given set of goals. For example, the main goal might be maximization of short-run net income. Although many people may regard this goal as unworthy, if top management chooses it, the system should be appraised in relation to it.

Cost, Congruence, Effort

The primary criterion for comparing management control system A against management control system B is how each system promotes the attainment of top management's goals in a cost-effective manner. Control systems are commodities. They benefit an organization by helping collective decision making. They cost money. The system with the largest favorable difference between benefits and costs should be chosen. Determining the benefits and costs of individual systems can be an imposing task. Two secondary criteria useful in making the cost-benefit criterion more specific are *goal congruence* and *effort*.

Goal congruence exists when individuals and groups work toward the organization goals that top management desires. Goal congruence occurs when managers, working in their own perceived best interests, make decisions that further the overall goals of top management.

Effort is defined here as exertion toward a goal. The concept of effort is not confined to its common meaning of physical exertion, such as a worker producing at a faster rate; it embraces all conscientious actions (physical and mental) that accompany the behavior of individuals. This effort can be guided toward the attainment of tangible goals such as cash, cars, and clothing or intangible items such as enjoyment, self-esteem, and power. In the context of management control systems, we desire effort that is exertion toward attaining top management's goals.

Motivation

Considered together, goal congruence and effort are really subparts of *motivation*.

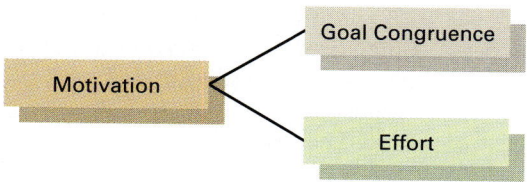

Motivation is the desire to attain a selected goal (the goal-congruence aspect) combined with the resulting drive or pursuit toward that goal (the effort aspect).

Goal congruence and effort are often reinforcing aspects of motivation, but each can exist without the other. A manager can share an organizational goal of reaching a target sales level or a product quality level. But the manager may pursue such goals unenthusiastically. Similarly, a manager may avidly pursue a goal that is incongruent with organizational goals. For example, a manager may increase advertising in an effort to gain sales when the organizational goal is to reduce advertising and instead build up the sales force.

The primary criterion for evaluating a system is how it promotes the attainment of top management's goals in a cost-effective manner. To apply that criterion, the motivational effects of system A and of system B deserve center attention. Why? Because the measurement of benefits and costs is based on how each system under consideration motivates individuals.

Formal and Informal Control Systems

The formal control system of an organization includes those explicit rules, procedures, performance-evaluation measures, and incentive plans that guide the behavior of its managers and employees. The management accounting system is one of several information systems that collectively constitute a formal control system. Other information systems pertain to employee relations, product quality, and compliance with environmental regulations.

The informal part of the management control system includes such aspects as shared values, loyalties, and mutual commitments among members of the organization and the unwritten norms about acceptable behavior for promotion. Companies have developed slogans, readily understandable to individuals in diverse parts of the organization, to reinforce those values and loyalties: "At Ford, Quality is Job 1," and "At Home Depot, low prices are just the beginning," for example. The idea is to get people throughout the organization to feel that they all belong to the same team and to pull together for the company's goals.

This book emphasizes one part of the formal control system—the cost and management accounting system—in much detail. However, this accounting system is only one of many factors that affect the value of the overall formal control system. Other factors include the stage of development of the organization, the business strategies, and the management style of the key senior executives.

Systems Goals and Risk Sharing

What is the role of risk in the design of a management control system? The potential of bearing risk changes behavior. Consider Daley and Langé, an accounting firm with offices in many different European cities. Assume that the executive committee in Brussels has decided that the firm's best long-run interest calls for aggressively seeking European banking clients. The company learns that a British bank seeks competitive bids for its audit. The London managing partner of Daley and Langé may agree with the accounting firm's goal but be unwilling to make a competitive bid for the audit. The internal cost to the London office of thoroughly preparing a bid is £50,000.

The partner views the likelihood of winning the audit engagement to be no more than 25%, even with a billing rate set 10% below normal. How can the executive committee encourage the London office to take the 25% chance of winning the new audit client?

The executive committee in Brussels can take several steps to reduce the risks facing the London office. For example:

1. It could fully reimburse the London office for the £50,000 cost of preparing the bid, regardless of whether it is awarded the audit engagement.
2. It could award the London office its normal billing rate through a subsidy, even though the actual billing rate to the banking client is below normal.

The effect of items 1 and 2 is to shift a major portion of the risk of bidding on the audit engagement from the London office to the accounting firm as a whole. Ideally, the control system should encourage the local partner to accept or avoid the risks that top management desires.

General Guideline

A favorite pastime of accountants, managers, and professors regarding management control systems is "fault finding" and "system bashing." But the heart of judging a system is not in assessing its degree of imperfection. The central question should be, How can the existing system, with all its imperfections, be improved? A system exists to help managers in their collective decision making in organizations. Top managers should predict how system A and system B would affect the collective actions of managers and compare the predicted results. In making such predictions, managers must consider the likely motivational effects (goal congruence and management effort) of each system.

AN EXAMPLE OF MOTIVATION: SALES COMPENSATION PLANS

Compensation plans for sales personnel illustrate the importance of motivation, goal congruence, and effort. Organizations that ignore these factors can find themselves losing super performers and retaining poor performers.

Pay-for-Performance Plans

Consider Horderns, a major department-store chain with more than 50 individual stores. Each store is located in a large shopping mall. The longstanding policy at Horderns has been to pay sales personnel a fixed salary based on number of years with the company: $1,250 per month for sales personnel in their first year, increasing $125 per year to a maximum of $2,500 per month in the eleventh year. Top management at Horderns emphasizes increasing the sales of each store, but sales growth in recent years has been flat. In fact, the business has begun losing market share to Grace Brothers.

Six months ago, 20 of the top-performing salespeople at the Horderns store in Seattle resigned and joined Grace Brothers. Grace Brothers pays all personnel a base salary of $800 per month and a sales commission of 4% of net sales made by each employee. Horderns decided to examine how the sales plan of Grace Brothers would have compensated two super-performing ex-employees—Stephen Kearny and Nina Landis—who left to join Grace Brothers. The company also wanted to know how the sales plan would compensate the average salesperson (total net sales divided by the total number of salespeople).

Panel A of Exhibit 13-2 presents net sales figures for the last three months for Kearney, Landis, and the average Horderns salesperson at Seattle. There are dramatic

EXHIBIT 13-2

Net Sales by and Compensation Paid to Sales Staff at the Horderns Seattle Store

	April	May	June
PANEL A: NET SALES PER MONTH AT HORDERNS			
Stephen Kearney	$49,300	$56,700	$61,200
Nina Landis	54,200	67,000	69,000
Average Seattle salesperson	22,600	23,200	21,200
PANEL B: COMPENSATION			
Current plan at Horderns:			
Stephen Kearney	$1,500	$1,500	$1,500
Nina Landis	1,250	1,250	1,250
Average salesperson	2,100	2,095	2,090
Assuming Grace Brothers plan applied at Horderns:			
Stephen Kearney*	$2,772	$3,068	$3,248
Nina Landis†	2,968	3,480	3,560
Average salesperson‡	1,704	1,728	1,648

	April	May	June
*Stephen Kearney			
Salary base	$ 800	$ 800	$ 800
Commission (4%)	1,972	2,268	2,448
	$2,772	$3,068	$3,248
†Nina Landis			
Salary base	$ 800	$ 800	$ 800
Commission (4%)	2,168	2,680	2,760
	$2,968	$3,480	$3,560
‡Average salesperson			
Salary base	$ 800	$ 800	$ 800
Commission (4%)	904	928	848
	$1,704	$1,728	$1,648

differences in the sales made by individuals at Horderns' Seattle store. Such differences are common in sales organizations.

Panel B of Exhibit 13-2 shows compensation levels under the current Horderns plan and under the Grace Brothers plan. The current plan at Horderns rewards seniority. It does not highlight the importance of making sales. In contrast, the Grace Brothers plan explicitly rewards successful sales efforts. Consider Nina Landis, whose June sales were more than three times as high as those of the average Horderns salesperson. Under the Grace Brothers plan, Landis would have been paid $3,560 ($800 + 4% of $69,000). Under the existing plan at Horderns, she received $1,250.

There is no upper limit to the sales commission at Grace Brothers. Super performers are continually motivated to seek additional sales. It is not surprising that both Stephen Kearney and Nina Landis find the Grace Brothers plan more attractive. At Grace Brothers, goal congruence and effort reinforce each other to promote a high level of motivation toward top management's goal of sales growth.

Caveats about Incentive Plans

It is critical to ensure correspondence between an organization's goals and how its incentive plan is designed. Grace Brothers' goal is to expand sales aggressively. Hence, a compensation plan related to sales (or net sales) is appropriate. If the goal shifts to promoting the sale of high-margin products, Grace Brothers should consider including costs as well as sales informaton in its compensation plan.

Concepts in Action

MANAGEMENT CONTROL SYSTEMS GO GREEN

The "green movement" is having a marked impact on the information collected in management control systems. Government environmental and safety regulations now require detailed records on ozone-destroying chlorofluorocarbons, hazardous waste disposal, worker fatalities, and the like. Government regulations, however, set only minimum requirements. Some companies now systematically collect financial and nonfinancial data pertaining to safety and the environment (S&E) beyond what legislation demands. For example, Sandoz, the Swiss-based pharmaceutical company, requires each of its facilities worldwide (more than 350) to report the following 10 data items on a regular basis:[a]

1. Total S&E investments
2. Total S&E costs
3. Lost time accident rate
4. Lost workday rate
5. Total energy consumption
6. Total water consumption
7. Total liquid and solid waste
8. Total S&E personnel
9. Total production
10. Total personnel

By monitoring trends in these variables and differences in them among facilities, Sandoz can highlight areas where individual facilities should reexamine the adequacy of their existing S&E policies.

Ben & Jerry's, the ice cream manufacturing company, is expanding the environmental measures reported in its control system. These measures track financial as well as nonfinancial variables. A company official noted:[b]

We are trying to do detailed cost accounting, which includes the cost of managing our waste stream. In the Waterbury plant, we have a special unit to treat the waste stream. How much is it costing in terms of money, energy, chemical, and man-hours to run that? Then there are disposal costs, to separate water from "float" and send that remainder to a landfill or compost, or to make animal feed.

Many companies do not now have a cost accounting system that tracks environment-related costs. Other companies are only beginning to develop such a system. Heightened concerns over environmental issues—and over the costs of complying with environmental regulations—have added one more ball that the designer of a management control system must juggle.

[a]Adapted from C. Fitzgerald, "Selecting Measures for Corporate Environmental Quality: Examples from TQEM Companies," *Total Quality Environmental Management* (Summer 1992).
[b]Adapted from Cutler Information Corp., *Newsletter on Business and Environment* (January 1993).

Chapter 26 presents further discussion of incentive plans. Be cautious when designing these plans. Sears Corporation experienced much negative publicity when some employees in its automotive repair shops recommended excessive amounts of repair work for its customers. Short-run bonuses for repair employees increased, but in the end the media learned what these Sears employees were doing and ran many unfavorable stories. It is no surprise that the level of customer dissatisfaction grew and business fell.

VALUE-ADDED/NONVALUE-ADDED COST ANALYSIS

Objective 4

Explain how information on value-added/nonvalue-added costs is useful to managers

The concept of *value-added activity* is a key one in cost management. Value-added activities are those that customers perceive as increasing the utility (usefulness) of the products or services they purchase. Some companies are adopting this concept to help identify which activities to keep and which to eliminate.

General Electric Example

Consider General Electric (GE), which uses the value-added activity concept for cost management at a plant that assembles medical equipment. GE implemented this approach in three steps:[2]

STEP 1: *IDENTIFY THE ATTRIBUTES OF PRODUCTS THAT CUSTOMERS PERCEIVE TO BE IMPORTANT.* For the GE plant, these attributes included quality, dependability, and price.

STEP 2: *IDENTIFY THOSE ACTIVITIES THAT CAUSE WORK IN THE OPERATING PROCESS (SUCH AS THE PRODUCTION LINE), AND ASSESS WHETHER EACH ACTIVITY ADDS VALUE.* GE identified 30 activities. Operating personnel then classified each activity as value-added, nonvalue-added (waste), or in the gray area in between. Examples follow:

Example	Category
Assembly time	Value-added
Rework time	Nonvalue-added
Engineering change orders[3]	Gray area

The work force as a whole spent 35% of its time on value-added activities, 51% on nonvalue-added activities, and 14% on activities in the gray area.

STEP 3: *IDENTIFY AND ELIMINATE THE CAUSES OF NONVALUE-ADDED ACTIVITIES.* The first column in Exhibit 13-3 presents the five highest (in labor-hours) nonvalue-added activities identified by General Electric; the second column presents variables that operating personnel believed were the cost drivers of those activities. Management then devoted considerable effort to reducing the nonvalue-added cost drivers. For example, changes in plant layout reduced materials movements. Similarly, changes in the design of products reduced the number of different parts in a product. The results of these and similar efforts over the next 12 months included:

♦ a 21% reduction in payroll cost per unit,
♦ a 50% reduction in work-in-process inventory, and
♦ a 50% reduction in defects detected in final testing.

The value-added/nonvalue-added classifications of individual activities depend on context. That is, activities viewed as value-added at a medical instrument facility may be viewed as nonvalue-added at a home-appliance assembly plant. Also, as managers gain experience, they may reclassify an activity. Adopting a value-

---

[2]This discussion draws on a visit to the General Electric plant and on H. T. Johnson, "Performance Measurement for Competitive Excellence," *Measures for Manufacturing Excellence*, ed. R. S. Kaplan (Boston, MA: Harvard Business School Press, 1990).

[3]An engineering change order (ECO) is a design change in a product or a process made after production has begun. In some cases, ECOs are value-added (such as a product redesign based on customer feedback to make a product more marketable to a larger market). In other cases, ECOs are nonvalue-added (such as a redesign arising because product designers and manufacturing personnel did not interact with each other during the preproduction stage of a product).

EXHIBIT 13-3

Nonvalue-Added Activities and Their Cost Drivers
at a General Electric Medical Equipment Assembly Plant

Nonvalue-Added Activity	Examples of Cost Drivers
1. Accumulated materials (materials and work-in-process inventories on shop floor)	Stock location (layout) Stocking procedures Assembly sequencing Number of separate part numbers in a product
2. Special delivery (expediting) of materials	Stock balance errors Ordering errors Variation in material yields Schedule changes
3. Movement of materials	Plant layout Handling equipment
4. Rework	Assembly errors Handling damages Supplier quality
5. Testing/verifying	Quality problems Design of products

EXHIBIT 13-4

Value-Added/Nonvalue-Added Reporting Format:
Job Cost Record for Trim-Axle Assembly

Activity Number	Primary Activity Description	Value-Added Cost	Nonvalue-Added (NVA) Cost	Activity Cost of Product
1	Process sales order	$ 164	$ 53	$ 217
2	Procure direct materials	451	162	613
3	Nontraceable, office	588	228	816
4	Move product	62	37	99
5	Ship product	47	21	68
6	Repair tooling, unplanned	33	46	79
7	Design/make tool, planned	23	0	23
8	Process engineering change orders	94	67	161
9	Investigate supplier	0	47	47
10	Inspect supplier for certification	10	0	10
11	Build pallets for WIP inventory	0	69	69
	Total per-unit activity costs	$1,472	$730	$2,202

Source: Adapted from Integrated Cost Management Systems software (Arlington, TX; April 1993).

added/nonvalued-added classification of costs in an internal accounting system probably requires updating the system frequently.

Value-Added Cost Reporting System

Exhibit 13-4 shows a costing record for the trim-axle assembly activity of an automotive component manufacturer. Using an approach similar to steps 1 to 3 for General Electric, we learn that only $1,472 (66.85%) of this cost is classified as value-added. Given the high amount of nonvalue-added costs in this example (33.15%), the benefits from eliminating the causes of those costs is considerable.

ENGINEERED, DISCRETIONARY, AND INFRASTRUCTURE COSTS

Objective 5
Describe how engineered costs, discretionary costs, and infrastructure costs differ

Costs can differ (1) as to the time between acquisition of resources and when the resources are budgeted to be fully used and (2) as to the major accounting technique used to control them. The three main categories of costs are engineered, discretionary, and infrastructure:

Type of Cost	Time between Acquisition of Resource and Its Budgeted Complete Use	Major Accounting Techniques for Control
1. Engineered	Short	Flexible budgets and standards
2. Discretionary	Short to longer	Negotiated static budgets
3. Infrastructure	Longest	Capital-expenditure budgets

Differences among these three cost categories imply that a different mix of approaches to promoting effectiveness and efficiency is appropriate in each category.

Engineered Costs

Engineered costs are costs that result specifically from a clear cause-and-effect relationship between inputs and outputs. This relationship is usually personally observable. Examples of inputs include direct materials costs, direct labor costs, and energy costs. Examples of outputs include cars, computers, and telephones. Inputs of the engineered-cost kind are often fully used soon after being acquired. For example, direct materials now are frequently delivered directly to the assembly floor and then assembled into finished goods within hours or days of delivery. Many engineered costs are locked in at the design area of the value chain (see Chapter 12, p. 439).

The major accounting techniques for controlling engineered costs are flexible budgets and standards. Chapters 7 and 8 provide a detailed discussion of those techniques. The feedback time is short, and nonfinancial measures are often used as the foundation of the control systems. For example, direct materials waste may be monitored as each unit is produced.

Discretionary Costs

Discretionary costs have two important features: (1) They arise from periodic (usually yearly) decisions regarding the maximum outlay to be incurred, and (2) they are not tied to a clear cause-and-effect relationship between inputs and outputs. Examples include advertising, public relations, executive training, teaching, research, health care, and management consulting services. Subsequent sections of this chapter discuss techniques used to control discretionary costs.

The most noteworthy aspect of discretionary costs is that managers are seldom confident that the "correct" amounts are being spent. The founder of Lever Brothers, an international consumer-products company, once noted, "Half the money I spend on advertising is wasted; the trouble is, I don't know which half."

Infrastructure Costs

Infrastructure costs are costs that arise from having property, plant, equipment, and a functioning organization; little can be done in the short run to change infrastructure costs. Capital-expenditure budgeting is the principal accounting technique for controlling these costs. The planning horizon is long, as is the feedback time. Infrastructure costs include depreciation, long-run lease rental, and the acquisition of long-run technical capabilities. There is often a high level of uncertainty about the benefits or cash inflows associated with long-run capital-expenditure decisions.

Chapters 20 and 21 outline the formal decision models (such as discounted cash flow) that managers use in making choices about infrastructure costs. These costs are relatively difficult to influence by short-run actions. Careful long-range planning, rather than day-to-day monitoring, is the key to managing infrastructure costs. Once a building has been erected or a telecommunications satellite launched into space, little can be done in day-to-day operations to affect the *total level* of infrastructure costs. From a control standpoint, the objective is usually to increase current utilization of facilities because this result will ordinarily increase operating income.

CLASSIFYING COSTS IN BUSINESS FUNCTION AREAS IN THE VALUE CHAIN

The following business function cost categories in the value chain have been discussed in earlier chapters:

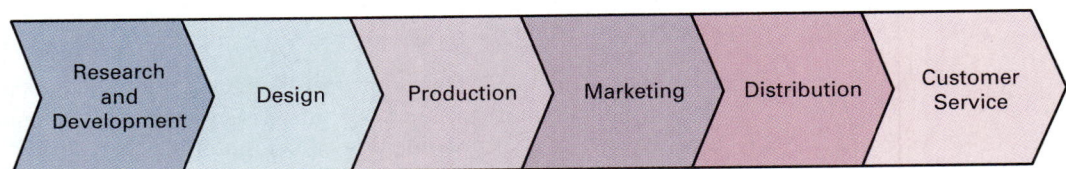

Engineered costs are most frequently found in production and, to a lesser extent, in distribution. Discretionary costs are typically found in research and development, design, marketing, and customer service.[4] Infrastructure costs, in the form of property, plant, and equipment, occur in each of the functional-cost categories.

Exhibit 13-5 compares engineered costs with discretionary costs. The exhibit focuses on four areas: inputs, process, outputs, and the level of uncertainty. Further discussion of infrastructure costs is deferred to Chapters 20 and 21.

Inputs

As Exhibit 13-5 indicates, the primary inputs that affect engineered costs are both physical and human resources. Examples are direct materials and direct labor in any manufacturing facility that assembles machines. In contrast, the primary inputs that affect discretionary costs are human resources. Examples are salaries paid to research engineers, product designers, and sales representatives.

---

[4]In a survey of marketing managers in the food industry, 76.4% reported using static (fixed) budgets while 40% reported using flexible budgets. J. Ratnatunga, "Management Accounting and Marketing in Practice," in J. Ratnatunga, J. Miller, N. Mudalige, and A. Sohal, *Issues in Strategic Management Accounting* (Sydney, Australia: Harcourt Brace Jovanovich, 1993).

EXHIBIT 13-5

Comparison of Engineered and Discretionary Costs*

	Engineered Costs	Discretionary Costs
1. Primary inputs	Physical and human resources (for example, direct materials)	Human resources (for example, time of a research scientist)
2. Process	a. Detailed and physically observable b. Repetitive	a. Black box (that is knowledge of process is sketchy or unavailable) b. Often nonrepetitive or nonroutine
3. Primary outputs	a. Products or quantifiable services b. Value easy to determine c. Quality easy to ascertain	a. Information b. Value difficult to determine c. Quality difficult to ascertain
4. Level of uncertainty	Moderate or small (for example, manufacturing setting)	Great (for example, advertising)

*This exhibit is a modification of one suggested by H. Itami.

Process

Exhibit 13-5 shows that processes that are detailed and repetitive are prime candidates for using engineered-cost techniques. An example is the number of deliveries made by drivers from a distribution center. In contrast, discretionary costs are associated with processes that are sometimes labeled as *black boxes.* That is, the processes are not easily understood. Knowledge of the precise process may be sketchy, unmeasurable, or unavailable. Examples are research and development and some aspects of product design.

Outputs

The contrast between engineered costs and discretionary costs is also evident when outputs are considered. In particular, Exhibit 13-5 points out that the outputs linked with many discretionary-cost centers are types of information. Examples are written or oral reports from the research, legal, or personnel department. The value and quality of such outputs are more difficult to measure than are the value and quality of outputs generated by engineered costs.

Level of Uncertainty

The level of uncertainty also helps to determine whether a cost may be engineered or discretionary. *Uncertainty* is defined here as the possibility that an actual amount will deviate from an expected amount. The higher the level of uncertainty about the relationship between inputs and outputs, the more likely a cost will be classified as a discretionary cost rather than as an engineered cost.

The marketing costs of footwear companies such as Nike and Reebok are typically classified as discretionary costs. There is a high level of uncertainty of the resulting sales when one of these companies signs a multimillion-dollar contract with a sports star or recording star. In contrast, materials costs of footwear companies are classified as engineered costs. There is a low level of uncertainty regarding the direct materials costs for a particular type of sports shoe.

BUDGETING FOR DISCRETIONARY COSTS

Effectiveness and Efficiency

Budgets are an important part of a management control system. They promote effectiveness and efficiency in discretionary-cost centers. Recall that Chapter 7 introduced the following definitions:

- *Effectiveness*—the degree to which a predetermined objective or target is met.
- *Efficiency*—the amount of inputs used to achieve a given level of output. The fewer the inputs used to obtain a given output, the greater the efficiency.

One manager expressed the difference as follows: "Effectiveness is doing the right thing, while efficiency is doing things right." We apply these two terms in the discussion that follows.

The most common accounting technique for controlling discretionary costs is a **negotiated static budget.** In such a budget, a fixed amount of costs is established through negotiations before the start of the budget period. Recall from Chapter 7 that a static budget is not adjusted after it is finalized, regardless of changes in the level of output.

Classifying Negotiated Static Budgets

Objective 6
Explain how ordinary incremental budgets, priority incremental budgets, and zero-base budgets differ

Negotiated static budgets can be classified as ordinary incremental, priority incremental, and zero-base. Each will be discussed in turn.

Ordinary incremental budgets consider the previous period's budget and the actual results. A comparison of these figures may lead to a change to a more realistic budgeted amount for the coming period. Also, expectations for the new period may lead to changes from the previous period's budgeted amounts. For instance, a budget for a research department might be increased because of salary raises, the addition of personnel, or the introduction of a new project.

Priority incremental budgets are similar to ordinary incremental budgets. However, priority incremental budgets also include a description of what incremental activities or changes would occur (1) if the budget were increased by, say, 10% and (2) if the budget were decreased by the same percentage. For example, a university sports department may decide that if its budget is cut by 10%, it will drop scholarships for sports that do not attract large numbers of spectators. This plan establishes priorities and forces the head of the sports department to think more carefully and concretely about operations. In a way, priority budgeting is a simple and economical compromise between ordinary incremental budgeting and zero-base budgeting.

Zero-base budgeting (ZBB) is budgeting from the ground up, as though the budget were being prepared for the first time. Every proposed expenditure comes under review. ZBB gets at bedrock questions by requiring managers to take the following major steps and to document them when possible:

1. Determine objectives, operations, and costs of all activities under the manager's jurisdiction.
2. Explore alternative means of conducting each activity.

3. Evaluate alternative budget amounts for various levels of effort for each activity.
4. Establish measures of workload and performance.
5. Rank all activities in the order of their importance to the organization.

ZBB requires much more extensive work than other forms of budgeting. Because of the paperwork and copying required, some people refer to it as "Xerox-base budgeting." However, it does force managers to better justify their outlays on a more regular and a more systematic basis.

Very few firms adopt ZBB on an annual basis for all departments. Most firms adopt ordinary incremental or priority incremental budgets on an annual basis. To the extent that zero-base budgeting is undertaken, it is typically on a less regular basis and for only a subset of responsibility centers at any one time.

WORK MEASUREMENT

Objective 7

Define work measurement; describe its use in an engineered-cost approach to controlling costs

Work measurement is the careful analysis of a task, its size, the method used in its performance, and the efficiency with which it is performed. The objective of work measurement is to determine the workload in an operation and the number of workers needed to perform that work efficiently. Work measurement is one source of the standards used in cost-management approaches that rely heavily on engineered-cost relationships.

Work-measurement techniques include:[5]

- *Micromotion study.* Film or videotape records of what a job entails and how long it takes. This technique is used most frequently in studying high-volume settings, such as a workstation on an assembly line.
- *Work sampling.* A large number of random observations are made of the job. These observations are used to determine the number and type of steps for the job in its "normal" operating mode.

A **control-factor unit** is the measure of workload used in work measurement. Examples of control-factor units include the following:

Activity	Control-Factor Unit
◆ Computer software writing	◆ Lines of computer code written
◆ Materials handling	◆ Number of incoming materials items handled
◆ Delivery	◆ Number of individual delivery stops
◆ Customer service	◆ Number of customer complaints processed
◆ Accounts receivable (payable)	◆ Number of accounts receivable (payable) posted per hour

Dayton-Hudson, a U.S. retailer, is an enthusiastic advocate of work-measurement techniques. This company measures cartons handled, invoices processed, and store items ticketed on a per-hour basis. Productivity programs in its administrative and distribution areas have been aided by detailed work measures.

ENGINEERED-COST VERSUS DISCRETIONARY-COST APPROACHES TO BUDGETING

Budgeting for engineered costs using work measurement treats the costs of individual control-factor units as variable. In contrast, budgeting for discretionary costs often treats those costs as fixed. We now compare and contrast these two budgeting approaches.

---

[5]For extensive discussion, see R. Failing, J. Janzen, and L. Blevins, *Improving Productivity Through Work Measurement: A Cooperative Approach* (New York: AICPA, 1988).

Assume that Family Farm employs five people (called customer representatives) to process orders for its mail-order catalog business. Each employee earns $1,800 per month and should, if operating efficiently, process 1,000 orders per month.

Family Farm has two categories of decisions to make in planning and control of personnel costs at its customer-order processing department:

1. *Personnel planning.* How many customer-order personnel are required? Should the business hire and lay off personnel as the quantity of work fluctuates?
2. *Control.* How does the business effectively and efficiently control personnel resources on a day-to-day basis?

In June, 4,700 individual orders were processed. The variances shown using the engineered-cost approach and the discretionary-cost approach are tabulated in the next section.

Engineered-Cost Approach

The engineered-cost approach to control of the customer-order payroll costs bases the budget formula on the unit costs of an individual order processed: $1,800 ÷ 1,000 orders, or $1.80 per order. Therefore, the flexible-budget allowance for customer-order personnel would be $8,460 (4,700 × $1.80). Assuming that the five people worked throughout the month at $1,800 per month (giving actual costs of $9,000), the following performance report would be prepared:

Actual Costs	Flexible Budget: Total Budgeted Costs Allowed for Actual Output Achieved	Flexible-Budget Variance
$9,000	$8,460	$540 U

The graphic representation of this example appears in Exhibit 13-6. For simplicity, the flexible-budget variance is not divided into price and efficiency variances here.

The work-measurement approach to day-to-day control assumes a variable-cost budget and the complete divisibility of the workload into small units. Note that the *budget line* on the engineered-cost graph in Exhibit 13-6 is purely variable, even though the costs are actually incurred in steps.

The unfavorable variance of $540 alerts management to the possibilities of over-staffing or inefficiency; the step of the cost-behavior pattern from $7,200 to $9,000 on the graph in Exhibit 13-6 was only partially utilized. The workload capability was 5,000 customer orders, in contrast to the 4,700 actually processed.

Discretionary-Cost Approach

The budget line in the discretionary-cost approach in Exhibit 13-6 is flat from 4,000 to 5,000. Looking at the right side of Exhibit 13-6 provides no insight into how to control costs between 4,000 and 5,000 customer orders processed. The primary means of control with a discretionary-cost approach is personal observation. That is, the department manager uses his or her experience to judge the size of the work force needed to carry out the department's functions. The manager is reluctant to overhire because of the corresponding reluctance to discharge or lay off people when volume slackens. Consequently, occasional peak loads are often met by hiring temporary workers or by having the regular employees work overtime.

Some managers regard a cost as a discretionary cost for *cash-planning purposes* in the preparation of the master budget but may use the engineered-cost approach for *control purposes* in preparing flexible budgets for performance evaluation. These two views may be reconciled within the overall system. In our example, a master budget for Family Farm could conceivably include the following items:

	Engineered-Cost Approach	Discretionary-Cost Approach
Actual costs incurred	$9,000	$9,000
Budget allowance	8,460*	9,000
Variance	540 U	0

*Rate = $1,800 ÷ 1,000 orders = $1.80 per order.
Total = 4,700 orders × $1.80 per order = $8,460.

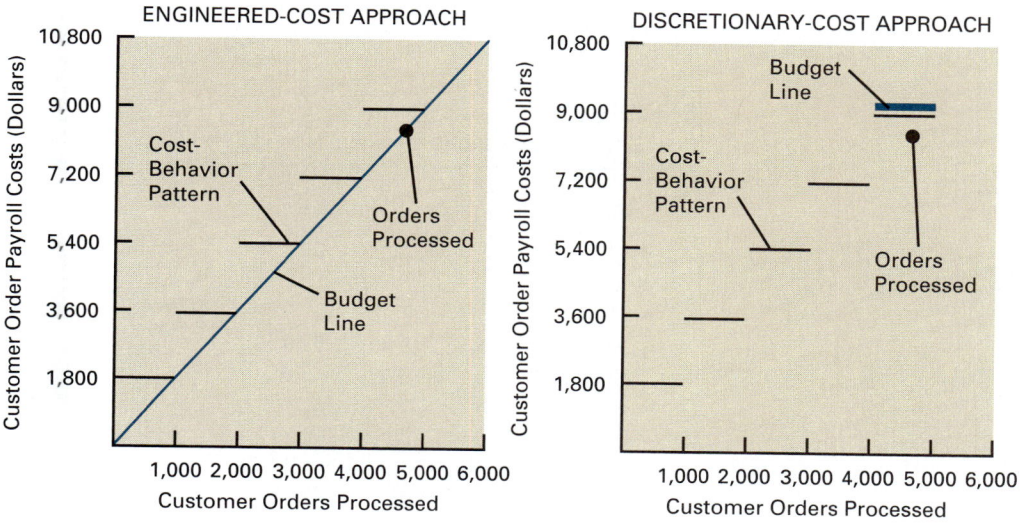

Customer-order processing personnel costs:

Flexible-budget allowance for control	$8,460
Expected unfavorable flexible-budget variance (due to conscious overstaffing)	540
Total budget allowance for cash planning	$9,000

A discretionary-cost approach stresses planning and de-emphasizes daily control through a formal work-measurement system. The company relies more on hiring capable people and less on monitoring their everyday performance.

Whether the engineered-cost approach is "better" than the discretionary-cost approach must be decided on a case-by-case basis, using a cost-benefit approach that focuses on how much operating decisions will be improved. Of course, the consultants who want to install work-measurement systems will typically try to demonstrate that their more costly formal systems will generate net benefits in the form of cost savings and better customer service.

BENCHMARKING AND COST MANAGEMENT

Objective 8

Outline how benchmarking can highlight areas where potential improvements in cost efficiency may exist in an organization

A benchmark study can provide important information on the comparative cost efficiency of an organization. Benchmarking is the continuous process of measuring products, services, and activities against the best levels of performance. Benchmarks may be found inside or outside the organization. We now illustrate benchmarking for hospital cost management.

Market Insights (MI), based in San Francisco, California, analyzes cost information that hospitals submit to various U.S. regulatory bodies. MI develops benchmark reports that show how the cost level at one hospital compares with that at numerous other U.S. hospitals. Reports can be prepared at the total hospital level (such as cost per patient-day or at a specific diagnostic-group level (such as cardiology, orthopedics, and gynecology).

Exhibit 13-7 illustrates an MI report for a client hospital. Panel A shows that the client hospital's cost per case is 10% above the average for comparable hospitals. Panel B shows an extract of an MI report at the diagnostic-group level. For example, the client hospital has a cost per stroke patient of $33,700 compared with a market average among all hospitals of $31,300; the twenty-fifth percentile is $21,900, and the average for the best quartile (0 to 25th percentile) is $20,500.[6] The cost level at this client

---

[6] The cost amounts refer to the insurance premium per month (× 1,000) that an insuree would have to pay to the client hospital for the hospital to break even.

EXHIBIT 13-7

Cost Benchmark Reports for Client Hospital by Market Insights

PANEL A: Total Hospital Cost Comparisons

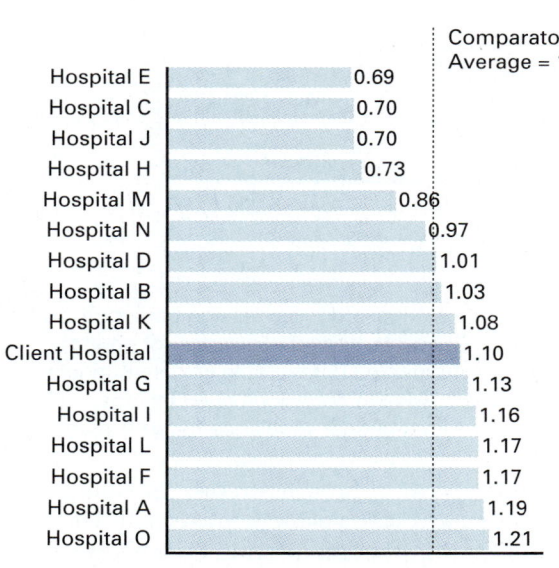

Cost per case index

HOW TO READ THIS CHART:

This chart shows your hospital's relative case cost performance versus comparator hospitals. Your hospital is benchmarked with the average case costs of your peers at the DRG level. The findings from this analysis can help assess the need for making cost reduction an institutional priority.

EXAMPLE:

On an overall cost per case basis, after adjusting for case mix, the client hospital is 10% higher cost than the comparator group average. There are nine hospitals that have lower overall case costs than the client hospital.

PANEL B: Diagnostic–Group Cost Comparisons

Diagnostic Group	Client Hospital	Market Average	25th Percentile	Average of Lowest–Cost Quartile (0–25th)
Stroke	$33,700	$31,300	$21,900	$20,500
Respiratory disorders	66,800	53,700	44,400	38,400
Simple pneumonia	37,100	29,500	23,300	22,000
Heart catherization	24,800	21,200	20,100	17,100

Source: Market Insights (San Francisco, CA).

hospital is well above the lowest quartile diagnostic-group level for patients inflicted with a stroke.

Cost benchmark reports are an attention-directing mechanism. An individual hospital administrator may well be able to justify an above-average cost level by documenting above-average quality levels or revenue levels. However, in many cases, hospitals with above-average costs have no documentable superiority in service quality levels or success in operations.

Exhibit 13-7 highlights how hospitals can differ sizably on costs. An administrator of a hospital with above-average costs potentially has much to learn from administrators at hospitals with below-average costs. Be cautious, however, in using benchmark reports such as Exhibit 13-7. The reliability of individual hospital cost data used in benchmark reports is highly variable. Many hospitals have not invested heavily in refining their cost accounting system. In addition, cost figures for individual diagnostic groups require numerous cost allocations, which vary quite a bit in reliability.

Cost reports like Exhibit 13-7 provide an external benchmark that forces the administrator to ask *why* cost levels differ among hospitals and *how* best practices can be transferred from the more-efficient to the less-efficient hospitals. These reports are especially beneficial in the medical sector, where there is not always a well-defined relationship between inputs and outputs.

Evaluating the overall performance of a hospital or hospital personnel requires analysis of other factors in addition to costs. These include the perceived quality of service to patients, the success rate of operations (how many patients with strokes survive?), and the morale of the doctors, nurses, and other staff. In many cases, however, cost factors have been given too little weighting in the past, in part because of the lack of reliable information on cost relationships in this sector of the economy.

PROBLEM FOR SELF-STUDY

PROBLEM

The Bombay Co. has many small customers. Work measurement of billing activities has shown that each billing clerk can process 4,000 customer accounts per month. The company employs 15 billing clerks at a monthly salary of $1,600. The predicted number of customers for the next month is 56,300; the company had 59,900 in the current month.

Required
1. Assume that management has decided to continue to employ the 15 clerks despite the expected drop in billings. Show two approaches—(a) the engineered-cost approach and (b) the discretionary-cost approach—to budgeting billing-clerk labor. Show how the performance report for the next month would appear under each approach.
2. Some managers favor using tight budgets (based on the engineered-cost approach) as motivating devices for controlling operations. In these cases, managers really expect an unfavorable variance and must allow, in cash planning, for such a variance so that adequate cash will be available as needed. Explain how the flexible-budget variance under the engineered-cost approach could be used in cash planning.
3. Assume that the workers are reasonably efficient. (a) Interpret the flexible-budget variances under the engineered-cost approach and under the discretionary-cost approach. (b) What should management do to exert better control over the cost of billing-clerk labor?

SOLUTION

1. a. Engineered-cost approach:

$$\frac{\text{Budgeted}}{\text{unit rate}} = \frac{\$1,600}{4,000}$$

$$= \$0.40 \text{ per customer account processed}$$

	Actual Costs	Flexible Budget: Total Budgeted Costs Allowed for Actual Output Achieved	Flexible-Budget Variance
Billing-clerk labor	(15 × $1,600) $24,000	(56,300 × $0.40) $22,520	$1,480 U

b. Discretionary-cost approach:

	Actual Costs	Flexible Budget: Total Budgeted Costs Allowed for Actual Output Achieved	Flexible-Budget Variance
Billing-clerk labor	(15 × $1,600) $24,000	$24,000	$0

2. From 1(a), the flexible-budget variance is $1,480 unfavorable. The master budget must provide for monthly labor costs of $24,000. Therefore, if the engineered-cost approach were being used for control, the master budget might specify:

Billing-clerk labor costs	
Flexible-budget allowance	$22,520
Expected unfavorable flexible-budget variance	1,480
Total budget allowance for cash planning	$24,000

3. Management decisions and policies are often crucial in classifying a cost as fixed or variable. If management refuses, as in this case, to control costs rigidly in accordance with short-run fluctuations in customer-account levels, these costs are discretionary. The unfavorable flexible-budget variance of $1,480 represents the cost that management is willing to incur in order to maintain a stable work force geared to "normal needs."

Management should be given an approximation of the extra cost of maintaining a stable workforce. There is no single "right way" to keep managers informed on such matters. Two approaches were demonstrated in the previous parts of this problem. The important point is that clerical workloads and capability must be measured before effective control can be exerted. Such measures may be formal or informal. The latter is often achieved through a supervisor's regular observation of how efficiently work is being performed.

SUMMARY

The following points are linked to the chapter's learning objectives.

1. A management control system is a means of gathering information to aid and coordinate the process of making planning and control decisions throughout the organization.

2. Information included in a control system can be (a) financial or nonfinancial and (b) internal or external. A well-functioning control system will include information from all four combinations of (a) and (b): financial/internal, financial/external, nonfinancial/internal, and nonfinancial/external.

3. The primary criterion for judging a control system is how well it promotes the attainment of top management's goals in a cost-effective manner. Two secondary criteria useful in making this criterion more specific are goal congruence and effort.

4. Value-added activities are those that customers perceive as increasing the utility (usefulness) of the products or services they purchase. Managers can often achieve substantial reductions in costs by explicitly examining whether each activity currently undertaken is adding value and then taking steps to eliminate the costs associated with activities that do not add value.

5. Cost items can be classified into one of three general categories—engineered, discretionary, or infrastructure. These cost categories differ as to the time horizon over which the resources acquired are used and as to the major accounting techniques for their control.

6. The most frequently used budgeting approaches for discretionary costs are ordinary incremental budgets and priority incremental budgets. Zero-base budgeting (budgeting from the ground up) is used less frequently.

7. The budgeting of engineered costs is often based on work measurement, which is the careful analysis of a task, its size, the method used in its performance, and the efficiency with which it is performed.

8. Benchmarking is the continuous process of measuring products, services, and activities against best levels of performance. A comparison of costs per process (such as processing an order) or per output (such as per product sold) among many organizations can highlight which organizations are most cost-efficient and which are least cost-efficient. An analysis of the activities that the most efficient organizations undertake can help pinpoint ways in which the least efficient organizations can improve their cost performance.

APPENDIX: PROMOTING AND MONITORING THE EFFECTIVENESS OR EFFICIENCY OF DISCRETIONARY-COST CENTERS

Exhibit 13-8 summarizes nine approaches used in various combinations to promote the effectiveness and efficiency of cost management in discretionary-cost centers. The text of this chapter discussed the first two approaches in Exhibit 13-8—the use of budgets and the monitoring of financial and nonfinancial performance measures.

Subsequent chapters cover topics related to several other approaches given in Exhibit 13-8. Chapter 25 discusses organizational design and the competitive forces of the market. Chapter 26 covers rewards for performance. Several of the approaches in Exhibit 13-8—for example, leadership and organization culture—typically use minimal accounting information, but this in no way diminishes their importance. Indeed, leadership is likely to be the single most important approach of the nine listed in Exhibit 13-8. However, it is dangerous to rely on leadership alone. What if a leader resigns to join a competitor? What if the leader is transferred within the company? What if the leader suffers health problems and cannot continue at the job?

EXHIBIT 13-8
Approaches to Promoting and Monitoring the Effectiveness or Efficiency of Discretionary-Cost Centers

Type of Approach	Example
1. Detailed analysis when preparing financial budgets	Government departments using zero-base budgeting
2. Monitoring financial and nonfinancial measures of effectiveness or efficiency on an ongoing and systematic basis	Accounts receivable department benchmarking their performance against "best-in-class" worldwide for management of the accounts receivable function
3. Monitoring customer opinion and satisfaction indicators on an ongoing and systematic basis	Wall Street brokerage firm using an independent magazine survey of how investment managers rate the advice of security analysts
4. Organization design	Keeping corporate (headquarters) functions to a minimum by, for example, allowing each division to have its own computer service center that reports to the president of the division rather than to a vice-president of computer services at headquarters
5. Exposing discretionary-cost centers to competitive forces of the marketplace	Prices charged to divisions for company-run executive-development programs being limited by the prices charged by external programs
6. Rewards for performance	Running "fraction of the action" programs (for example, R&D staff having a percentage of the operating income associated with the commercial development of their R&D activity)
7. Leadership	Head of a project team in an advertising agency being a charismatic and creative individual who promotes a high team spirit and a willingness in team members to work long and productive hours
8. Administrative approval mechanisms	Detailed approval procedures checking the hiring of new personnel, upgrading of classifications, working of overtime, and so on
9. Promotion of organization culture	High-technology firms promoting shared norms of strong loyalty, teamwork, and a commitment to technological leadership

TERMS TO LEARN

This chapter and the Glossary contain definitions of the following important terms:

control-factor unit *(p. 479)*   discretionary costs *(475)*   effort *(468)*
engineered costs *(475)*   goal congruence *(468)*   infrastructure costs *(476)*
management control system *(466)*   motivation *(469)*   negotiated static budget *(478)*
ordinary incremental budgets *(478)*   priority incremental budgets *(478)*
work measurement *(479)*   zero-base budgeting (ZBB) *(478)*

ASSIGNMENT MATERIAL

QUESTIONS

13-1   What is a *management control system*?

13-2   What are four types of information that can be included in a management control system? Give an example for each type.

13-3   What is the primary criterion for comparing management control system A against management control system B?

13-4   What is the relationship among motivation, goal congruence, and effort?

13-5   "The potential of bearing risk affects behavior." Do you agree? If yes, give an example.

13-6   "Advances in information technology permit the design of sales compensation plans previously not possible." Give an example.

13-7   Give two examples each of a *value-added cost* and a *nonvalue-added cost*.

13-8   Give two examples each of an *engineered cost*, a *discretionary cost*, and an *infrastructure cost*.

13-9   In which business function areas of the value chain are (a) engineered costs most often found and (b) discretionary costs most often found?

13-10   What are the most common outputs linked with discretionary costs?

13-11   Distinguish among an *ordinary incremental budget*, a *priority incremental budget*, and a *zero-base budget*.

13-12   What is *work measurement*? Describe two work-measurement techniques.

13-13   Give an example of how a management control system can motivate a manager to behave against the best interests of the company as a whole.

13-14   How might benchmark cost reports assist a hospital administrator in promoting cost efficiency?

13-15   "I'm majoring in accounting. This study of human behavior is fruitless. You've got to be born with a flair for getting along with others. You can't learn it!" Do you agree? Why?

EXERCISES AND PROBLEMS

13-16 Goals of public accounting firms. All personnel, including partners, of public accounting firms must usually turn in biweekly time reports, showing how many hours were devoted to their various duties. These firms have traditionally looked unfavorably on idle or unassigned staff time. They have looked favorably on heavy percentages of chargeable (billable) time because this maximizes revenue.

Required
What effect is such a policy likely to have on the behavior of the firm's personnel? Can you relate this practice to the problem of goal congruence that was discussed in the chapter? How?

13-17 Management control systems, changing scope. The following information is collected on two large retailing companies.

		HandyMart			HomeMart	
Variable	19_2	19_3	19_4	19_2	19_3	19_4
1. Square footage of retail sales space at year end (millions)	13.3	16.4	20.8	24.6	22.2	19.7
2. Number of customer transactions (millions)	112.4	146.2	188.5	201.2	189.1	161.6
3. Net earnings to sales (%)	4.3	4.8	5.1	1.3	0.2	1.1
4. Net earnings (millions)	$163.4	$249.1	$362.8	$247.6	$210.7	$212.1
5. Stock price to earnings per share ratio	34.1	37.2	39.6	10.6	11.9	9.8
6. Greenspotlight* score on waste disposal (1 = lowest, 10 = excellent)	7	6	5	8	7	8

*Greenspotlight is an independent environmental group that monitors how companies comply with various environmental and safety standards.

Required

1. Classify each of the six variables into the four categories of information in a management control system in Exhibit 13-1.

2. Who has primary responsibility for collecting, reporting, and ensuring the reliability of each of the above six variables?

3. Why are there growing demands for management control systems to include many items that are not of a financial/internal kind?

4. Compare and contrast the performance of HandyMart and HomeMart over the 19_2 to 19_4 period using these six variables.

13-18   Budgets and motivation.   You are working as a supervisor in a distribution department that has substantial amounts of personnel and equipment costs. You are paid a "base" salary that is actually low for this type of work. The firm has a very liberal bonus plan, which pays you another $1,000 each month that you "make the budget" plus 2% of the amount you are under budget.

Data on your past experiences follow:

Month	Jan.	Feb.	Mar.	Apr.	May	June
Actual result	$41,000	$39,500	$37,000	$37,000	$36,500	$36,000
Budget	40,000	40,000	39,000	36,000	36,000	36,250
Variance	1,000 U	500 F	2,000 F	1,000 U	500 U	250 F

Required

1. What would you do if you were starting the job all over again from January and had the above information?

2. What would you recommend, if anything, be done to the bonus plan if you receive a promotion and are now required to supervise the bonus plan in this department?

13-19   Budgets and motivation.   Christine Everette operates a chain of health centers. One of the largest is in the heart of London. Everette uses the services of three managers, Dave, Nick, and Sarah. Each is in charge of a group of exercise and apparatus rooms similar in all regards. When she hired them, she offered three methods of weekly payment:

◆ *Method X.* A base rate of £6 per hour and 30% of all reductions in costs below a "norm" of £600 per week

◆ *Method Y.* A flat wage of £7 per hour

◆ *Method Z.* No base rate, but a bonus of £300 for meeting the "norm," plus 10% of all reductions in costs below the norm

The managers chose their method of compensation at the time they started employment. Assume a 40-hour week. The record for the past six weeks for the three areas follows. All data are in pounds.

			Week			
	1	2	3	4	5	6
Dave:						
Utilities	250	250	250	250	250	250
Supplies	180	20	20	260	100	20
Repairs and misc.	305	250	220	265	260	200
Total	735	520	490	775	610	470
Nick:						
Utilities	250	250	250	250	250	250
Supplies	100	100	100	100	100	100
Repairs and misc.	250	200	265	220	260	305
Total	600	550	615	570	610	655
Sarah:						
Utilities	250	250	250	250	250	250
Supplies	100	100	100	100	100	100
Repairs and misc.	250	250	250	250	250	250
Total	600	600	600	600	600	600

Required

Which payment method most benefits Dave, Nick, and Sarah? Assume that each manager picked a different method. Briefly explain your choices.

13-20 Incentive systems, motivation. Consider the following excerpt from the company employee newsletter of Superior Security Guards. It announces the creation of a Perpetual Trophy, which will be awarded to the branch with the "most outstanding performance" during each fiscal year.

The trophy will be several feet high and will travel from branch to branch each year. In addition, a small replica of the trophy will be awarded to the manager of the winning branch.

The trophy will be awarded to the branch with the best combination of percentage increases in sales and operating income. The actual calculation will be to add the sales percentage increase over the prior year and the operating income percentage increase over the prior year and divide the result by two, which gives equal weight to both sales and operating income percentage growth. Only branches achieving a minimum 15% sales increase will be eligible.

Superior Security Guards has branches in four cities. Details on sales and operating income of the four branches for the most recent four years are (in millions):

	19_1	19_2	19_3	19_4
Sales				
Iowa City branch	$62.4	$75.6	$ 94.3	$110.4
Kansas City branch	23.7	25.9	28.9	32.8
New Orleans branch	60.5	85.1	122.7	153.8
St. Louis branch	80.7	86.2	99.8	111.3
Operating Income				
Iowa City branch	$ 9.9	$10.5	$ 15.1	$ 18.7
Kansas City branch	5.7	5.4	2.6	7.2
New Orleans branch	6.2	8.3	15.9	21.5
St. Louis branch	14.5	14.6	20.9	21.1

Required

1. Which of the four branches won the trophy for 19_4?

2. Comment on the way Superior Security Guards measures "most outstanding performance."

 13-21 Incentive systems for used-car sales personnel. Ted Dexter is owner of Regal Motors, a used-car sales company. Regal Motors has sales lots in five different cities. For many years, Dexter has paid his salespeople a monthly salary plus a fixed dollar bonus for each car sold. In March 19_5, the monthly salary was $1,600 and the fixed dollar bonus was $25 per car sold.

In April 19_5, Dexter received information that Fred Truman, one of his best salespersons, was about to resign and join Cowdrey Motors. Cowdrey Motors is owned in part by

Michael Cowdrey, a recently retired racing driver superstar. Dexter called Truman to discuss his pending resignation. Truman told him the decision was a "tough personal one" but felt Cowdrey offered him a better deal. Salespeople at Cowdrey Motors receive a base monthly salary of $1,200 plus a bonus of 6% of the gross margin on each car sold. *Gross margin* is defined as the selling price minus the purchase cost and all other directly traceable costs incurred before that car is put on the lot for sale.

Dexter collected the following information on Truman, two other salespersons, and the average for all 48 salespeople of Regal Motors for March 19_5:

	Fred Truman	Louise Hutton	Brian Statham	Average for 48 Salespeople
Cars sold	37	16	46	21
Average selling price	$9,000	$10,000	$8,000	$8,400
Average purchase cost and other traceable costs	$7,900	$9,200	$7,800	$7,950

Required

1. Compute the March 19_5 salary and bonus payments Regal Motors paid to Truman, Hutton, Statham, and the average salesperson.

2. Compute the March 19_5 salary and bonus payments that Regal Motors would have paid if it had used Cowdrey's payment structure (a base monthly salary of $1,200 plus a bonus of 6% of the gross margin on each car sold).

3. Compare and contrast the Regal Motors and the Cowdrey Motors compensation plans for salespeople.

4. What other factors might be important for Ted Dexter to consider in rewarding the salespeople at Regal Motors?

13-22 Risk sharing, motivation, new clients for advertising agency. The Media Impact Group (MIG) is the second largest advertising agency in Canada, with offices in Vancouver, Calgary, Edmonton, Regina, Winnipeg, Toronto, Montreal, Quebec City, and Halifax. The head office is in Toronto. The partners of the firm have agreed to place high priority on developing a larger client base in the lodging industry.

Quality Inns is a major motel chain with over 100 motels across Canada. Its headquarters is in Vancouver. The managing director has called for bids for its advertising account. Quality Inns' current advertising has been described in advertising circles as "uninspiring, nonmemorable, and without focus."

Trent Natham, the partner in charge of MIG's Vancouver office, is unenthusiastic about devoting large amounts of the resources of the Vancouver office to bidding on the Quality Inns account. He views the minimum cost of preparing a credible bid to be $100,000. Moreover, Quality Inns has informed all potential bidders that it will not consider any bid with an average billing rate higher than $80 an hour. The billing rate of the Vancouver office of MIG currently averages $110 per hour. Moreover, Natham estimates that the Vancouver office would receive only 20% of the revenues of the Quality Inns account. The Toronto and Montreal offices would jointly receive over 50% of the revenues of the Quality Inns account. Competition for the account will be intense, with at least six other advertising agencies preparing bids.

The head office views each MIG office as a profit center. Over 50% of Natham's annual bonus is based on profitability for that year at the Vancouver office. The current year has been a highly successful one for the Vancouver office. Many partners have worked 60 hours each week for months at a time.

Required

1. Why is Natham unenthusiastic about making a serious bid for the Quality Inns advertising account?

2. How might Natham be encouraged by the head office in Toronto to make a serious bid for the Quality Inns advertising account?

13-23 Management control systems and the environment. ICI, a multinational chemical company with its head office in London, publishes a section in its *Annual Report* on "Safety, Health and the Environment." It now also publishes a separate *Environmental Report* that is sent to its shareholders.

Examples of information in this *Environmental Report* include:

1. Details of environment-related expenditures. For instance, at ICI's Huddersfield plant, a new boiler plant was constructed using "novel technology for capturing sulfur dioxide. The result has been a halving of sulphur dioxide emissions in the flue gases."

ICI stated that all major new projects they undertake now are assessed for their impact on the environment at all stages in their development. Environmental expenditures, total revenues, and net earnings of ICI were reported to be:

	19_2	19_3	19_4
Environmental expenditures (£ billions)			
Capital equipment related	0.081	0.132	0.164
Operating related	0.194	0.187	0.197
Revenues (£ billions)	12.906	12.488	12.061
Net earnings (£ billions)	0.919	0.789	0.565

2. Number of fines and prosecutions for noncompliance with environmental laws and regulations:

19_2	19_3	19_4
36	26	21

3. Total waste emissions to land, air, and water (in millions of tons):

	19_2	19_3	19_4
Nonhazardous	5.334	5.205	4.817
Hazardous	0.678	0.475	0.350
	6.012	5.680	5.167

4. Number of reportable accidents per 100,000 working hours:

19_2	19_3	19_4
0.28	0.23	0.18

Required

1. Why might ICI send an *Environmental Report* to its shareholders in addition to sending its *Annual Report*?

2. One commentator argued that ICI's *Environmental Report* should not have been sent to shareholders with its *Annual Report*. His argument was: "The financial information in the *Annual Report* is objective and audited by KPMG Peat Marwick. Information about safety, health, and the environment is subjective and nonaudited. I object to this pandering to the greenies. We should not waste ICI money responding to every social pressure group." How would you respond to this commentator?

3. Should the data in items 2 (fines), 3 (emissions), and 4 (accidents) be included in a management control system, or should a management control system focus only on financial/internal information? Explain.

4. What problems might arise in ICI determining what amount of its expenditures are related to "safety, health, or the environment."

5. Comment on trends in ICI's data reported in this question. Does an increase in environment-related expenditures mean an improvement in environment performance?

13-24 Value-added versus nonvalue-added cost classifications. Olivia Johns is manager of the Home Appliance plant of Newton Products. Johns decides to experiment with the value-added/nonvalue-added classification of costs in the accounting records. She selects the clothes dryer-assembly product line for her test. She asks your advice on classifying items of labor time (and cost) in the plant:

a. Moving component parts from warehouse to assembly line
b. Assembling the tumbler unit
c. Expediting materials to the door-assembly area because of stock-balance error
d. Assembling the dial-presentation component
e. Inserting the owner's manual and instruction guide in the dryer package
f. Reworking faulty latches on doors
g. Testing the operating capabilities of the assembled unit
h. Packaging the clothes dryer in a breakage-resistant box

Required

1. What is the distinction between a value-added cost and a nonvalue-added cost?
2. Classify each of the eight items (a–h) as (i) a value-added cost, (ii) in the gray area, or (iii) a nonvalue-added cost.
3. How can Johns use your classifications in requirement 2 in making decisions at the plant?
4. Johns attends a conference where a well-known cost-accounting writer expresses a great deal of cynicism about the value-added/nonvalue-added distinction. He calls it a "fad with a shelf life shorter than freshly cut roses." Johns has the chance to meet the cost-accounting writer. What question should she pose to him about her proposed experiment with the value-added/nonvalue-added cost distinction?

13-25 Work measurement, cost control in a warehouse. The manager of a warehouse for a mail-order firm is concerned with the control of her fixed costs. She has recently applied work-measurement techniques and an engineered-cost approach to the staff of order clerks and is wondering if a similar technique could be applied to the workers who collect merchandise in the warehouse and bring it to the area where orders are assembled for shipment.

The warehouse assistant contends that this action should not be taken, because the present work force of 20 persons should be viewed as a fixed cost necessary to handle the usual volume of orders with a minimum of delay. These employees work a 40-hour week at $15 per hour.

Preliminary studies show that it takes an average of 12 minutes for a worker to locate an article and move it to the order-assembly area and that the average order is for two different articles. At present 1,800 orders are processed per week.

Required
1. For the present volume of orders, prepare the weekly performance report using (a) a discretionary-cost approach and (b) an engineered-cost approach for the weekly performance report.
2. Repeat requirement 1 for volume levels of 1,600 and 1,400 orders per week.
3. What other factors should be considered along with the cost variances found in requirements 1 and 2 in order to make a decision on the size of the work force?

13-26 Work measurement, cost control in customer service. Venture Vision is a major cable television operating company. Three months ago (March 19_4) it acquired Cable Galore, the largest cable television operator in California. Cable Galore had 120,000 subscribers, many of whom were not happy with their service. Local newspapers had run several articles in which subscribers voiced their complaints about service under the Cable Galore ownership. One article was titled "Cable Galore Becomes Cable Screw-Up." One customer reported "reading a novel" while waiting for his telephone call to come to the "top of the list."

Venture Vision decided to work fast and effectively on the customer complaint problem. It immediately hired more customer-service representatives. It also sent troubleshooters to eliminate the causes of the customer complaints and to give personnel additional training.

Venture Vision hired six full-time customer-service representatives, paying each of them a salary of $2,000 a month to handle customer complaints. The two customer-service representatives employed by Cable Galore resigned in March 19_4. The budget for the next three months (April to June) assumed that each representative would handle 5 calls (complaints) an hour, work 8 hours a day (8 A.M. to 5 P.M. with 1 hour of breaks) for 20 working days (Monday through Friday) per month. The actual number of calls in the April to June 19_4 period were:

Month	Number of Calls
April	4,900
May	4,200
June	3,300

Required

1. Show the performance report each month (April, May, and June) under (a) the engineered-cost approach and (b) the discretionary-cost approach. Comment on the different inferences that might be drawn from (a) and (b).

2. What changes should Venture Vision consider in the staffing of its customer-service area?

13-27 Discretionary-cost center, research and development function of a computer software company Susan Teece, president of Software Advance Inc., has a set of problems many of her competitors would love to have. Teece has a large cash fund of $50 million that could be allocated to research and development. Her twin problems are (a) how much of the $50 million to spend on R&D and (b) how to ensure that the amount budgeted for R&D is used effectively and efficiently.

The sales, operating income, and R&D expenditures of Software Advance over the last four years (19_1 to 19_4) are (in millions):

	19_1	19_2	19_3	19_4
Sales	$ 2	$12	$47	$86
R&D	0.5	2	8	15
Operating income	(1)	3	11	16

Software Advance's success is based almost exclusively on Insight 1-2-3, a spreadsheet program for personal computers. The Insight program was developed by Jeff Moore, a major shareholder in Software Advance and head of its R&D department. Over 30% of the R&D budget in 19_3 and 19_4 was paid to independent contractors who operate out of their own homes or private offices. Despite $23 million being spent on R&D in 19_3 and 19_4, Software Advance has not been successful in developing a second generation of products to continue its recent high growth rate of sales.

The 19_3 and 19_4 R&D budgets were based on the 19_2 percentage of R&D to sales, the year in which the Insight 1-2-3 program was developed. Teece used this fixed percentage-of-sales approach because it seemed a convenient way to reach a decision quickly. Moreover, the percentage approach allowed Teece to focus on her many other pressing marketing problems in 19_3 and 19_4.

The electronics industry trade association reports the following 19_4 summary data for 33 software-development companies, classified by firm size:

	Firm Size Category by Sales ($ millions)			
	1–5	5–10	10–20	20–100
Number of firms	18	5	5	5
R&D as a percentage of sales	25.9%	18.4%	15.0%	7.4%

Required

1. Why might the R&D department of Software Advance be viewed as a discretionary-cost center?

2. What limitations arise from use of the fixed percentage-of-sales approach to decide the size of the annual R&D budget?

3. How might the industry trade association data be useful in the R&D decisions facing Teece?

4. How might Teece and Moore help ensure that the R&D budget of Software Advance is used effectively and efficiently?

13-28 Discretionary cost center, promoting effectiveness and efficiency in an engineering department. The Sharp Company develops, manufactures, and markets several product lines of low-cost consumer goods. Top management of the company is attempting to evaluate the present method and a new method of charging the different production departments for the services they receive from one of the engineering departments, which is called Manufacturing Engineering Services (MES).

The function of MES, which consists of about 30 engineers and 10 drafters, is to reduce the costs of producing the different products of the company by improving machine and manufacturing process design while maintaining the required level of quality. The MES manager reports to the engineering supervisor, who reports to the vice-president, manufacturing. The MES manager, George Amershi, may increase or decrease the number of engineers under him. He is evaluated on the basis of several variables, one of which is the difference between the budgeted costs and the actual costs of the department. These costs consist of actual salaries, a share of corporate overhead, the cost of office supplies his department uses, and a cost of capital charge. An individual engineer is evaluated on the basis of the annual savings he or she brings about in relation to annual salary. The salary range of an engineer is defined by personnel classification. There are four classifications, and promotion from one classification to another depends on the approval of a panel that includes both production and engineering personnel.

Production department managers report to a production supervisor, who reports to the vice-president, manufacturing. The production department for each product line is treated as a profit center, and engineering services are provided at a cost, according to the following plan. When a production department manager and an engineer agree on a possible project to improve production efficiency, they sign a contract that specifies the scope of the project, the estimated savings to be realized, the probability of success, and the number of engineering hours of each personnel classification required. The charge to the particular production department is the number of hours required multiplied by the "classification rate" for each personnel classification. This rate depends on the average salary for the classification involved and a share of the engineering department's other costs. An engineer is expected to spend at least 85% of his or her time on specific, contracted projects; the remainder may be used for preliminary investigations of potential cost-saving projects or self-improvement study. A recent survey showed that production managers have a high degree of confidence in the MES engineers.

A new plan has been proposed to top management. No charge will be made to production departments for engineering services. In all other respects, the new system will be identical to the present one. Production managers will continue to request engineering services, as under the present plan. Proponents of the new plan believe that production managers will take greater advantage of existing engineering talent. Regardless of how engineering services are accounted for, the company is committed to the idea of production departments being profit centers.

Required
Evaluate the strong and weak points of the present and proposed plans. Will the company tend to hire the optimal quantity of engineering talent? Will this engineering talent be used as effectively as possible?

13-29 Benchmarking, hospital cost comparisons. Julie Smith is the newly appointed President of National University. National University Hospital (NUH) is a major problem for her. It is running large deficits. Sam Horn, the chairman of the hospital, tells Smith that he and his staff have cut costs to the bare bone. Any further cost cutting, he argues, would "destroy the culture" of the hospital. He also argues that the use of detailed cost studies is totally inappropriate for a medical institution because of (a) the inability to have well-defined relationships between inputs and outputs and (b) the problem of even defining what is a "good output" for a hospital. He notes that he is "fed up with people equating continuous improvement at NUH with continued cost reduction. This is only a cost accountant's view of the world. The point is to help doctors save lives and to help people recover their health."

Smith hears about a new benchmark cost analysis service offered by Market Insights. She asks Horn to hire Market Insights to provide a benchmark cost report that pertains to NUH. Horn is not enthusiastic about doing so, but he complies with her request. The report includes the following:

1. Aggregate Hospital Cost Comparison (Average = 1.00)

Hospital E	0.69
Hospital C	0.70
Hospital J	0.70
•	
•	
•	
Hospital A	1.19
National University Hospital	1.20
Hospital O	1.21

2. Diagnostic Group Cost Comparison

Diagnostic Group	National University Hospital	Market Average	25th Percentile	Average of Best Quartile (0–25th)
Angina, chest pain	$23,000	$20,500	$17,300	$15,300
Asthma, bronchitis	15,400	13,100	10,400	9,000
Skin disorders, cellulitis	9,600	9,200	6,500	5,800
Renal failure and dialysis	7,600	5,500	4,200	3,600
Diabetes	6,700	5,100	3,700	3,100
Gastroenteritis	12,000	18,500	16,000	12,800

Required

1. Do you agree with Horn that the use of detailed cost studies at NUH is totally inappropriate? Explain your answer and comment on Horn's reasoning.
2. What inferences do you draw from the MI benchmark cost report on NUH?
3. What use might Smith make of the MI benchmark cost report?
4. What criticisms might you anticipate Horn would make of the MI benchmark cost report?
5. What factors other than cost might Smith consider in evaluating Horn's performance and that of NUH?

13-30 Incentives, compensation plans, ethics. Lisa Vachon is the founder and, up to two months ago, the sole owner of Sports Gear Inc. Vachon's compensation includes a base salary ($50,000) and an annual bonus of 4% of Sports Gear's operating income. Sports Gear has grown rapidly. Sales have grown to over $100 million. Vachon, a former Olympic gold medal athlete, is now a national celebrity a second time around for her business success.

Ralph Smith is the controller of Sports Gear. He has been with the company since its outset. Vachon has always paid her managers very well. She sets aside an annual bonus pool equal to 5% of operating income, which is shared among the top 20 managers (including Smith). Since its inception, Sports Gear has used accounting principles in its executive bonus computation that differ from those used in its income tax reporting:

◆ Goods available for shipment to customers but not yet shipped are recorded as sales (for tax, these goods are recorded as inventory)

◆ Advertising costs, which total over 15% of operating costs, are capitalized and expensed over a four-year period (for tax, these costs are expensed in the period in which they are incurred).

◆ The cost of bonuses paid to managers is excluded from operating income computations (for tax, bonuses are expensed in the period in which they are accrued).

Smith makes the bonus "operating income" calculations at the end of each year by adjusting information in the management accounting system. Smith's share of the 5% bonus pool for senior managers has been 4% for each of the last four years. The operating income numbers Smith used in computing Sports Gear bonuses in the 19_1 to 19_4 period are (in millions):

	19_1	19_2	19_3	19_4
Sales	$30.1	$46.7	$91.3	$140.6
Operating income	2.4	4.1	6.7	11.5

In February 19_5, L'Express—a Paris-based company—acquired a 49% interest in Sports Gear by paying Vachon $150 million. In March 19_5, she distributed $20 million of this amount among her top 20 managers. Smith received $500,000. Vachon had long promised her top managers that she would share any such amount, but the exact formula had never been put in writing.

It is now May 19_5. In one week's time, a group from L'Express' senior management team is due to visit Sports Gear. Vachon is reluctant to let L'Express have full access to the way it computes executive bonuses. Vachon had expected L'Express to ask many questions about how operating income was computed during their January 19_5 negotiations, but the French firm did not. Vachon implied during a January 14, 19_5, lunch that she and Smith had with the president and marketing manager of L'Express that the computations were based on the management accounting system. However, she gave no further details, and L'Express asked no further questions. After lunch Smith told Vachon that fuller disclosure should have been made to L'Express. She responded that she never made any inaccurate statements. Smith wanted Vachon to fax L'Express full details of the bonus plan computations. She became very angry and accused him of trying to derail the L'Express investment in their company. Vachon argued that she was in charge of the negotiations and that if Smith was unhappy, he should resign. He had never seen Vachon so angry and decided not to push her further on this issue. Smith decided to stay with Sports Gear and now is very unsure how to handle the forthcoming visit from L'Express. Vachon is adamant that he should not volunteer any information about how he computes operating income for bonus calculations.

Required

1. What are the pros and cons to Sports Gear that Vachon receives a low base salary and an annual bonus of 4% of operating income?

2. Compute the bonuses paid to Vachon and Smith each year in the 19_1 to 19_4 period.

3. Explain the likely effect that the three noted accounting differences (between the operating income computations and the taxable income computations) had in the computation of the amount used in bonus calculations. Why might Sports Gear use different computations for bonus determinations than for income tax reporting?

4. What should Smith have done after the January 14, 19_5 lunch? Should he have resigned?

5. What advice would you now give Smith about how to approach his forthcoming meetings with the L'Express senior management team? Refer to the "Standards of Ethical Conduct for Management Accountants" (p. 17) in your answer.

14

Cost Allocation: I

- The terminology of cost allocation
- Purposes of cost allocation
- Criteria to guide cost-allocation decisions
- Cost tracing and cost allocation
- Cost pools and cost allocation
- Allocating costs from one department to another
- Allocating costs of support departments
- Allocation of common costs

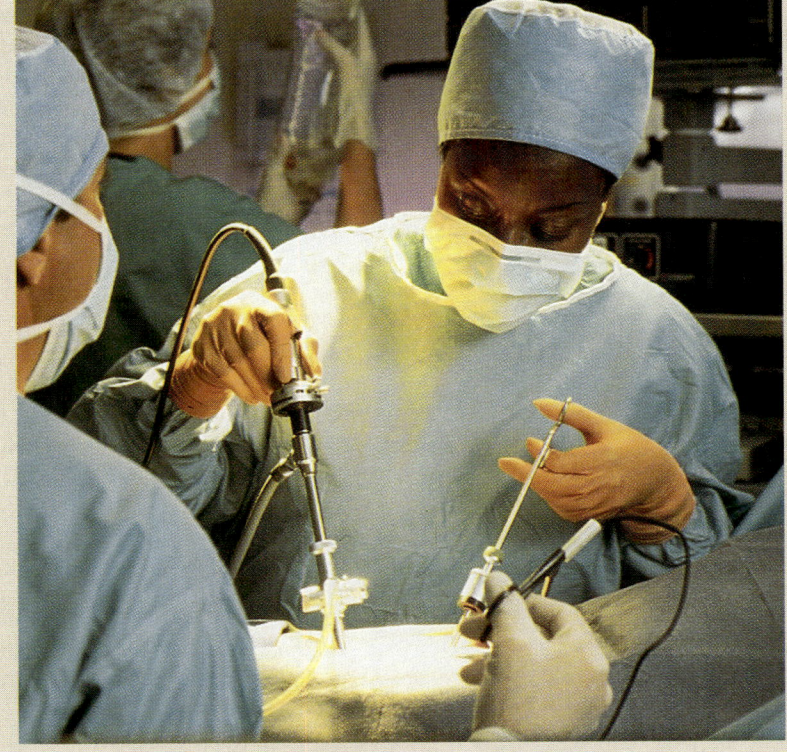

Medical costs in many areas, including surgical operations, are under increasing scrutiny. Hospital administrators are now examining how different cost allocation methods affect the resources used at the hospital as well as the amount of any cost reimbursement paid by insurers and the government.

Cost allocation is an inescapable problem in nearly every organization and in nearly every facet of accounting. How should the costs of shared services be allocated among departments? How should university costs be allocated among undergraduate programs, graduate programs, and research? How should the costs of expensive medical equipment, facilities, and staff be allocated in a hospital? How should manufacturing overhead be allocated to individual products in a multiproduct company?

Finding answers to cost-allocation questions is difficult. The answers are seldom clearly right or clearly wrong. Nevertheless, we will try to obtain some insight into cost allocation and to understand the dimensions of the questions, even if the answers seem elusive. You will undoubtedly be faced with many cost allocation questions in your career.

THE TERMINOLOGY OF COST ALLOCATION

Objective 1
Describe how a costing system can have multiple cost objects

The terms related to cost allocation vary throughout the literature and among organizations. Be sure that you and the people with whom you are dealing agree on the meaning of key terms when you confront allocation problems in practice. Important terms used in this chapter (and these are familiar to you from earlier chapters) include the following:

- *Cost object:* anything for which a separate measurement of costs is desired.
- *Direct costs of a cost object:* costs that are related to the cost object and can be traced to it in an economically feasible way. The term *cost tracing* describes assigning direct costs to the chosen cost object.
- *Indirect costs of a cost object:* costs that are related to the cost object but cannot be traced to it in an economically feasible way. The term *cost allocation* describes assigning indirect costs to the chosen cost object.

Examples of direct costs and indirect costs for a product and for an activity area follow:

Cost Object	Example of a Direct Cost	Example of an Indirect Cost
Product: Microwave oven manufactured by a home-	Materials assembled in making the microwave	Rent for manufacturing plant, which manufactures 200 different products.
Activity area: Document-photocopying area of a law firm	Paper and liquids used in photocopying machine	Electricity used to run machine. Electricity metered to firm but not to individual machines.

Organizations differ in how they classify costs. A direct cost item in one organization, such as assembly labor or energy, can be an indirect cost item in another organization. Moreover, changes in classification can occur over time. For example, advances in information-gathering technology, such as bar coding and optical scanning, now enable organizations to change the classification of some items from indirect costs to direct costs. For example, bar codes on individual cutting tools enable some machining shops to trace the costs of these cutting tools directly to the products that used them rather than having to allocate their total costs to products on the basis of, say, machine-hours.

PURPOSES OF COST ALLOCATION

Objective 2
Outline four purposes for allocating costs to cost objects

Indirect costs are often a sizable percentage of the costs assigned to cost objects such as products, distribution channels, and customers. Exhibit 14-1 outlines and illustrates four purposes for allocating costs to such cost objects:

EXHIBIT 14-1

Purposes of Cost Allocation

Purpose	Illustrations
1. To provide information for economic decisions	◆ To decide whether to add a new airline flight ◆ To decide whether to make a component part of a television set or to purchase it from an outside vendor ◆ To decide what selling price to charge for a customized product or service
2. To motivate managers and employees	◆ To encourage the design of products that are simpler to manufacture or less costly to service ◆ To encourage sales representatives to push high-margin products or services
3. To justify costs or compute reimbursement	◆ To cost products at a "fair" price, often done with defense contracting ◆ To compute reimbursement for a consulting firm that is paid a percentage of the cost savings resulting from its recommendations
4. To measure income and assets for reporting to external parties	◆ To cost inventories for financial reporting to stockholders, bondholders, and so on. (Under generally accepted accounting principles, inventoriable costs include manufacturing costs but exclude research and development, marketing, distribution, and customer-service costs.) ◆ To cost inventories for reporting to tax authorities

1. to provide information for economic decisions
2. to motivate managers and employees
3. to justify costs or compute reimbursement
4. to measure income and assets for reporting to external parties

The allocation of one particular cost need not satisfy all purposes simultaneously. Consider the salary of an aerospace scientist in a central research department of Boeing or Airbus. This salary cost may be allocated as part of central research costs to satisfy purpose 1 (economic decisions); it may or may not be allocated to satisfy purpose 2 (motivation); it may or may not be allocated to a government contract to justify a cost to be reimbursed to satisfy purpose 3 (cost reimbursement); and, it must not be allocated (under generally accepted accounting principles) to inventory to satisfy purpose 4 (income and asset measurement).

Different costs for different purposes is a major theme of this book. Exhibit 14-1 emphasizes this theme. Consider product costs of the following business functions in the value chain:

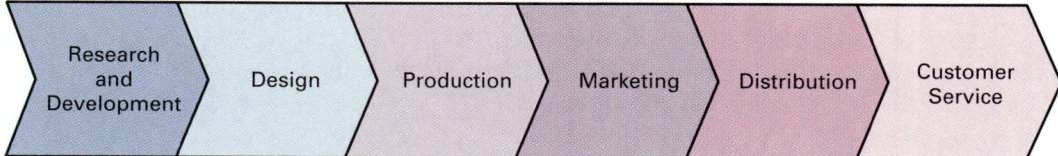

For the economic-decision purpose (for example, product pricing), the costs in all six functions may be included. For the motivation purpose, costs from more than one function are often included to emphasize to managers how costs in different functions are related to each other. For example, some Japanese companies require product designers to incorporate downstream costs such as distribution and customer service, as well as manufacturing, into their product cost estimates. The aim is to focus attention on how different product design options affect the total costs of the organization. For the cost-reimbursement purpose, the particular contract often will stipulate whether all six of the business function areas or only a subset are to be reimbursed. For instance, some U.S. government contracts explicitly exclude marketing costs. For the purpose of income and asset measurement for reporting to external parties, inventoriable costs under generally accepted accounting principles include only manufacturing costs (and product design costs in some cases). Research and development costs are expensed to the period in which they are incurred, as are marketing, distribution, and customer-service costs.

CRITERIA TO GUIDE COST-ALLOCATION DECISIONS

Role of Dominant Criteria

Objective 3
Describe alternative criteria used to guide decisions related to cost allocations

Exhibit 14-2 outlines and illustrates four criteria used to guide decisions related to cost allocations. Managers must first choose the primary purpose that a particular cost allocation is to fulfill and then select the appropriate criterion in implementing the allocation. This text emphasizes the superiority of the cause-and-effect and the benefits-received criteria, especially when the purpose for cost allocation is related to the economic decision or motivation purposes.

The benefits-received criterion and fairness-or-equity criterion are sometimes cited in regulations governing U.S. federal government procurement. The Federal Acquisition Regulation (FAR) includes the following definition of "allocability" (in FAR 31.201-4):

A cost [is] allocable if it is assignable or chargeable to one or more cost objectives in accordance with the relative benefits received or other equitable relationship. Subject to the foregoing, a cost is allocable to a government contract if it:

EXHIBIT 14-2

Criteria to Guide Cost-Allocation Decisions

1. CAUSE AND EFFECT. Using this criterion, managers identify the variable or variables that cause resources to be used. For example, managers may use hours of testing as the causal variable when allocating the costs of a quality-testing area to products. Cost allocations based on the cause-and-effect criterion are likely to be the most credible to operating personnel.

2. BENEFITS RECEIVED. Using this criterion, managers identify the beneficiaries of the outputs of the cost object. The costs of the cost object are allocated among the beneficiaries in proportion to the benefits each receives. For example, consider a corporatewide advertising program that promotes the general image of the corporation rather than any individual product. The costs of this program may be allocated on the basis of division sales; the higher the sales, the higher the division's share of the advertising costs. The rationale behind this allocation is the belief that divisions with higher sales levels apparently benefited from the advertising more than did divisions with lower sales levels and therefore ought to be allocated more of the advertising costs.

3. FAIRNESS OR EQUITY. This criterion is often cited in government contracting when cost allocations are the means for establishing a price satisfactory to the government and its suppliers. The allocation here is viewed as a "reasonable" or "fair" means of establishing a selling price in the minds of the contracting parties. For most allocation decisions, fairness is a lofty objective rather than an operational criterion.

4. ABILITY TO BEAR. This criterion advocates allocating costs in proportion to the cost object's ability to bear them. An example is the allocation of corporate executive salaries on the basis of divisional operating income; the presumption is that the more profitable divisions have a greater ability to absorb corporate headquarters' costs.

♦ Is incurred specifically for the contract;
♦ Benefits both the contract and other work, . . . and can be distributed to them in reasonable proportion to the benefits received; or
♦ Is necessary to the overall operation of the business, although a direct relationship to any particular cost objective cannot be shown.[1]

Further discussion of the contract reimbursement purpose of cost allocation is presented in Chapter 15.

The Cost-Benefit Consideration

Many companies place great importance on cost-benefit considerations when designing their cost-allocation systems. Companies incur costs not only in gathering data but also in taking the time necessary to educate management about the chosen system. The more sophisticated the system, in general, the higher these education costs.

The costs of designing and implementing sophisticated cost-allocation systems are highly visible, and most companies work to reduce them. In contrast, the benefits from using a well-designed cost-allocation system—being able to make more-informed make-or-buy decisions, pricing decisions, cost-control decisions, and so on—are difficult to measure and are frequently less visible. Still, designers of cost allocation systems should consider these benefits as well as costs.

---

[1]F. Alston, M. Worthington, and L. Goldsman, *Contracting with the Federal Government*, 3d ed. (New York: Wiley, 1993, p. 136). This book contains extensive discussion of the use of cost data in government contracting.

Spurred by rapid reductions in the costs of collecting and processing information, organizations today are moving toward more-detailed cost-allocation systems. Several companies have developed manufacturing or distribution overhead costing systems that use more than 10 different cost-allocation bases. Also, some businesses have state-of-the-art information technology already in place for operating their plants or distribution networks. Applying this existing technology to the development and operation of a cost allocation system is less expensive—and thus more inviting—than starting up such a system from scratch.

COST TRACING AND COST ALLOCATION

Exhibit 14-3 presents an overview of the costing system at the St. Louis assembly plant of the MicroComputer Division of Computer Horizons. This plant assembles two product lines: (1) the Plum line, used mostly by students, in homes, and in small offices, and (2) the Plum Laptop line, the portable version of the Plum. The area within the box in Exhibit 14-3 shows a costing overview for the Model A version of the Plum. This costing overview is similar to that presented in earlier chapters. See, for example, Exhibit 4-3 (p. 102), Exhibit 5-1 (p. 141), and Exhibit 8-1 (p. 269).

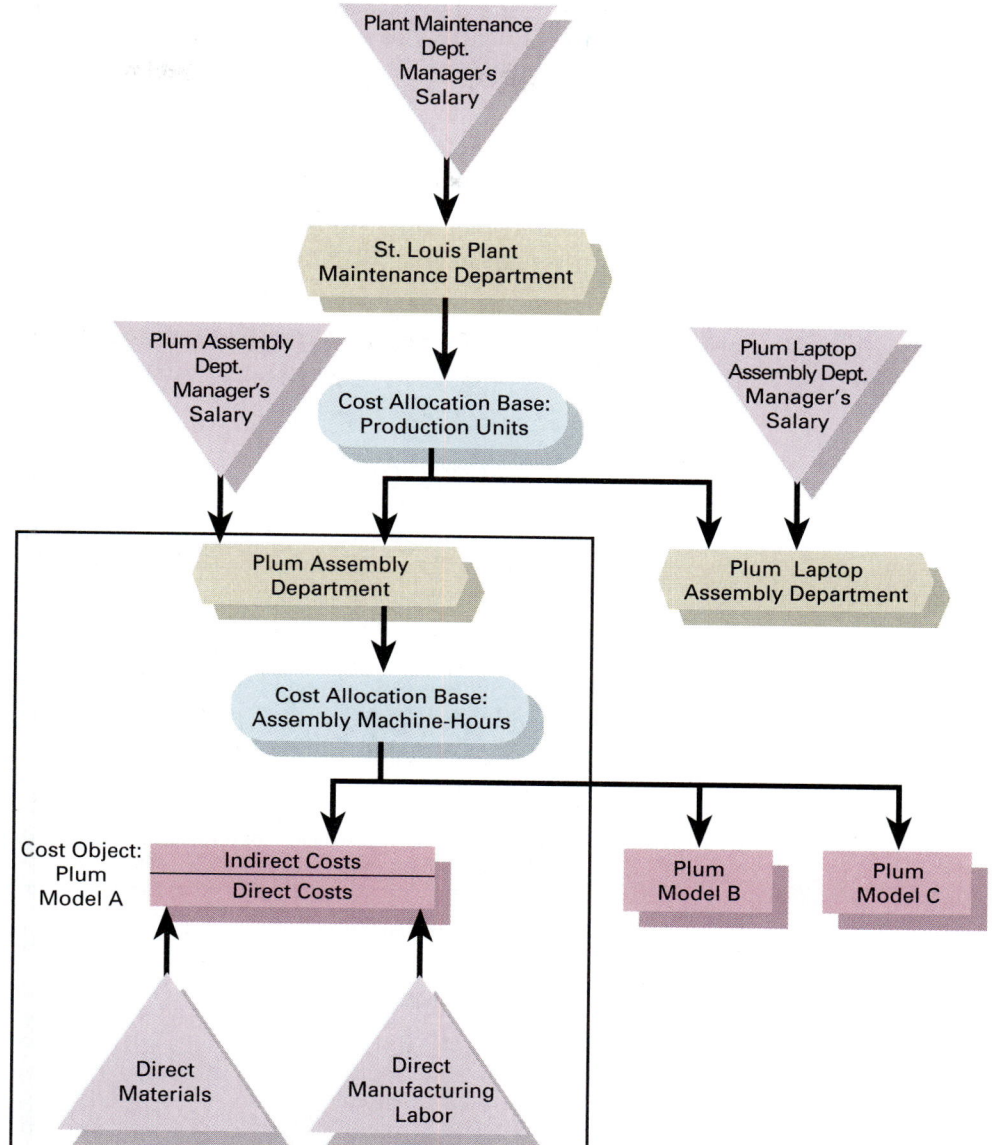

The product costing overviews presented in earlier chapters are typically only parts of larger costing systems. This larger costing system can be for a plant, a division, or even a whole company with multiple plants and divisions in many countries. Computer Horizons has eight manufacturing plants, located in the United States, Canada, Singapore, and the United Kingdom. It has marketing operations in more than 20 countries. Every month it consolidates accounting information from each of its operations to use in its planning and control decisions.

The costing system for the St. Louis plant portrayed in Exhibit 14-3 highlights two important points:

◆ There are multiple cost objects in most costing systems. Examples in Exhibit 14-3 in the St. Louis plant include the Plant Maintenance Department, the Plum Assembly Department, the Plum Laptop Assembly Department, and the separate products in the Plum Assembly Department—for example, Plum Model A, Plum Model B, and Plum Model

C. Note, however, that Exhibit 14-3 presents only a small subset of the separate cost objects at the St. Louis plant. Other examples include the procurement department, the energy department, and the various Plum Laptop products.

◆ An individual cost item can be simultaneously a direct cost of one cost object and an indirect cost of another cost object. Consider the salary of the Plant Maintenance Department manager. This is a direct cost traced to the Plant Maintenance Department. Computer Horizons then allocates the costs of this department to the two assembly departments at the St. Louis plant using units produced as the allocation base. In turn, the costs of the two assembly departments are allocated to individual products, such as the Plum Model A, using assembly machine-hours as the allocation base. Thus, the salary of the Plant Maintenance Department manager is both an indirect cost of each computer assembled at the plant and a direct cost of the Plant Maintenance Department.

COST POOLS AND COST ALLOCATION

Objective 4
Discuss key decisions faced when collecting costs in indirect cost pools

A *cost pool* is a grouping of individual cost items. This section discusses issues related to indirect cost pools when designing the cost-allocation component of an accounting system.

Examples of Cost Pools

Let's return to Computer Horizons. It has two manufacturing divisions. The Microcomputer Division manufactures the Plum and the Plum Laptop line of products. The Peripheral Equipment Division manufactures printers and other items that can be used with the Plum and Plum Laptop products. Exhibit 14-4 illustrates costs typically incurred at the corporate-headquarters level and at the manufacturing-plant level. Exhibit 14-4 focuses on indirect costs with respect to the products of the manufacturing divisions. Direct product costs, such as direct materials and direct manufacturing labor, are not included in Exhibit 14-4.

EXHIBIT 14-4
Indirect Cost Pools (with Respect to Individual Products) of Computer Horizons

CORPORATE HEADQUARTERS	
1. Corporate Executive Salaries	7. Payroll Department
2. Central Computer Facility Department	8. Personnel Department
3. Controller's Department	9. Research and Development
4. Treasury Department	10. Government Relations
5. Legal Department	11. Investor Relations
6. Marketing Department	12. Public Relations

INDIRECT COSTS OF PRODUCTS ASSEMBLED IN MICROCOMPUTER MANUFACTURING DIVISION	INDIRECT COSTS OF PRODUCTS ASSEMBLED IN PERIPHERAL EQUIPMENT MANUFACTURING DIVISION
1. Indirect Labor	1. Indirect Labor
2. Indirect Materials	2. Indirect Materials
3. Machinery and Equipment-Related Costs	3. Machinery and Equipment-Related Costs
4. Building Costs	4. Building Costs
5. Energy Costs	5. Energy Costs
6. Product Development	6. Product Development
7. Process Development	7. Process Development

Questions in Collecting Costs

Assume that the cost object is the Microcomputer Manufacturing Division of Computer Horizons. Direct costs would be those costs that are related to the Microcomputer Division and can be traced to it in an economically feasible way. Examples include plant leasing costs of the Microcomputer Division and direct materials costs of products assembled in that division. Indirect costs of this division include costs incurred at Corporate Headquarters and possibly costs incurred in the Peripheral Equipment Division and other divisions. For example, if the Microcomputer Division uses the services of product designers in the Peripheral Equipment Division, some of the costs of these designers would be an indirect cost of the Microcomputer Division. Decisions required when accumulating and subsequently allocating the indirect costs of the Microcomputer Division include:

◆ Which cost categories from Corporate Headquarters and the other divisions should be included in the indirect costs of the Microcomputer Division? Should all 12 corporate cost pools in Exhibit 14-4 be allocated, or should only a subset be allocated? For example, some companies exclude corporate public relations from any corporate cost allocations to the divisions; division managers have little say in corporate public relations decisions and would object to allocations as "taxation without representation."

◆ How many cost pools should be used when allocating corporate costs to the Microcomputer Division? One extreme is to aggregate all corporate costs into a single cost pool. The other extreme is to have numerous individual corporate cost pools. The concept of homogeneity (described below) is important in making this decision.

◆ Which allocation base should be used for each of the corporate cost pools when allocating corporate costs to the Microcomputer Division? Examples follow:

Cost Pool	Possible Allocation Bases
Corporate executive salaries	Sales; assets employed; operating income
Treasury department	Sales; assets employed; estimated time or usage
Legal department	Estimated time or usage; sales; assets employed
Marketing department	Sales; number of sales personnel
Payroll department	Number of employees; payroll dollars
Personnel department	Number of employees; payroll dollars; number of new hires

These allocation bases are illustrative only. The base an organization prefers depends on the purpose served by the cost allocation (see Exhibit 14-1), the criteria used to guide the cost allocation (see Exhibit 14-2), and the costs of implementing the different allocation bases.

Homogeneity of Cost Pools

A cost pool is *homogeneous* if all the activities whose costs are included in the pool have the same or a similar cause-and-effect relationship or benefits-received relationship between the cost driver and the costs of the activity. A consequence of using a **homogeneous cost pool** is that the cost allocations using that pool will be the same as would be made if costs of each individual activity in the pool were allocated separately. The greater the degree of homogeneity, the fewer cost pools required to reflect accurately the differences in how products use resources of the organization.

Assume that Computer Horizons wishes to use the cause-and-effect criterion to guide cost-allocation decisions. The company should aggregate only those cost pools that have the same cause-and-effect relationship to the cost object.[2] For example, if the

---

[2]By the *same* cause-and-effect relationship to the cost object we mean that there are both the same cost driver *and* the same rate per unit of the cost driver.

number of employees in a division is the cause for incurring both corporate payroll department costs and corporate personnel department costs, the payroll cost pool and the personnel cost pool could be aggregated before determining the combined payroll and personnel cost rate per unit of the allocation base.

Allowability of Costs in Cost Pools

A given cost item or amount may be included or excluded from a cost pool depending on the purpose at hand. For example, a consulting firm, when pricing work done for a commercial client, may include the cost of meals and drinks of its partners if they are conducting business over lunch or dinner. In contrast, when billing the U.S. government under a contract, no cost amount for any alcoholic beverage is permitted to enter the cost pools from which costs are allocated to the government. The U.S. government requires all alcohol costs be excluded from the cost pools before any allocation is made.

ALLOCATING COSTS FROM ONE DEPARTMENT TO ANOTHER

Objective 5
Describe how the single-rate cost-allocation method differs from the dual-rate method

We now move from a general discussion of cost-allocation issues to a discussion of allocating costs from one department to another. Decisions related to three issues will be discussed:

Concepts in Action

HOW A QUART OF DISTILLED WATER COST CINDY CHASE $17[a]

Cost allocation is an inescapable aspect of life. Cindy Chase discovered this after breaking a leg while skiing in the Rocky Mountains. A four-day stay in Denver University Hospital cost her over $10,000. Cindy was covered by her health insurance plan, but she was still puzzled by how the bill for only four days could possibly reach to $10,000. The answer was cost allocation.

One item on the bill that caught Chase's eye was a $17 charge for a quart of distilled water. She discovered that the direct cost of the quart of distilled water was only $3.40. The overhead amount on this direct cost was $13.60 (or 400% of direct cost). A Denver University Hospital official provided the following breakdown of the $13.60 overhead charge:

$ 4.25	Salaries and equipment of people at hospital handling the distilled water
3.40	Malpractice insurance, teaching, and administrative costs
5.10	Cost of treating uninsured patients
0.85	Profit component
$13.60	

The $5.10 overhead charge for Denver University Hospital treatment of uninsured patients means that Chase is subsidizing those patients who have no insurance. A Denver University hospital administrator admitted that the cost-allocation rate involved cross-subsidization. He noted: "We are moving some of the costs of caring for uninsured patients onto people who do have insurance or who pay out of their own pocket. Nobody likes that. It is not rational. But it is just the way the system has evolved over the years."

Ability-to-bear is the cost allocation criterion that best explains Denver University Hospital's inclusion of the $5.10 amount in the $13.40 overhead allocated to Chase's $3.40 bottle of distilled water.

[a] Adapted from "ABC World News" (April 14, 1993).

- Whether to use a single-rate method or a dual-rate method. Under the **single-rate method**, all costs are grouped in one cost pool and allocated to cost objects using the same rate per unit of the single allocation base; there is no distinction between fixed costs and variable costs. Under a **dual-rate method,** costs are grouped in two separate cost pools, each of which has a different allocation rate and may have a different allocation base. Typically, the two cost pools are fixed costs and variable costs.
- Whether to use budgeted costs or actual costs when allocating the costs of the department.
- Whether to use budgeted quantities or actual quantities when allocating the costs of the department.

Single-Rate and Dual-Rate Methods

Consider the Central Computer Facility Department of Computer Horizons (see Exhibit 14-4). For simplicity, assume that the only users of this facility are the Microcomputer Division and the Peripheral Equipment Division. The following data apply to the coming budget year:

1. Fixed costs of operating the facility	$300,000
2. Total capacity available	1,500 hours
3. Budgeted long-term usage (quantity) in hours:	
Microcomputer Division	800
Peripheral Equipment Division	400
Total	1,200
4. Budgeted variable costs per hour in the 1,000-hour to 1,500-hour relevant range	$200 per hour

Under the single-rate method, the costs of the Central Computer Facility would be allocated as follows (assuming budgeted usage is the allocation base):

Total cost pool: $300,000 + (1,200 budgeted hours × $200)	$540,000
Budgeted usage	1,200 hours
Cost per hour: $540,000 ÷ 1,200	$450
Microcomputer Division	$450 per hour used
Peripheral Equipment Division	$450 per hour used

The $450 per hour rate differs sizably from the $200 budgeted variable cost per hour. This $450 rate includes an allocated $250 per hour amount ($300,000 ÷ 1,200 hours) for the fixed cost of operating the facility. These fixed costs will be incurred whether the machine runs its 1,500-hour capacity, its 1,200-hour budgeted use, or even, say, only 600 hours use.

Use of the $450 per hour single-rate method (combined with the budgeted usage allocation base) transforms what is a fixed cost to the Central Computer Facility (and to Computer Horizons) into a variable cost to users of that facility. This approach could lead internal users to purchase computer time outside the company. Consider an external vendor that charges less than $450 per hour but more than $200 per hour. A division that uses this vendor rather than the Central Computer Facility may decrease its own division costs, but the overall costs to Computer Horizons are increased. For example, suppose the Microcomputer Division uses an external vendor that charges $360 per hour when the Central Computer Facility has excess capacity. In the short-run, Computer Horizons incurs an extra $160 per hour because this external vendor is used ($360 external purchase price per hour minus the $200 internal variable cost per hour) and not its own Central Computer Facility.

When the dual-rate method is used, allocation bases for each different cost function must be chosen. Assume that the allocation bases are budgeted usage for fixed costs and actual usage for variable costs. The costs allocated to the Microcomputer Division would be:

Fixed-cost function: (800 ÷ 1,200) × $300,000 $200,000 per year
Variable-cost function $200 per hour used

The costs allocated to the Peripheral Equipment Division would be:

Fixed-cost function: (400 ÷ 1,200) × $300,000 $100,000 per year
Variable-cost function $200 per hour used

Assume now that during the coming year the Microcomputer Division actually uses 900 hours and the Peripheral Equipment Division actually uses 300 hours. The costs allocated to these two divisions would be computed as follows:
Under the single-rate method:

Microcomputer Division: 900 × $450 = $405,000
Peripheral Equipment Division: 300 × $450 = $135,000

Under the dual-rate method:

Microcomputer Division: $200,000 + (900 × $200) = $380,000
Peripheral Equipment Division: $100,000 + (300 × $200) = $160,000

The one obvious benefit of using a single-cost function is its low cost of implementation. It avoids the often expensive analysis necessary to classify the individual cost items of a department into fixed and variable categories. However, a single-cost function may lead divisions to take actions that appear to be in their own (division) best interest but are not in the best interest of the total organization.

An important benefit of the dual-rate method is that it signals to division managers how variable costs and fixed costs behave differently. This important information could steer division managers into making decisions that benefit the corporation as well as each division.

Budgeted versus Actual Costs

Objective 6
Explain how the choice of budgeted versus actual allocation rates changes the risks managers face

The decision whether to allocate budgeted versus actual costs affects the level of uncertainty user departments face. Budgeted costs let the user departments know the costs in advance. Users are then better equipped to determine the amount of the service to request and—if the option exists—whether to use the internal department source or an external vendor.

Budgeted costs also help motivate the manager of the support department to improve efficiency. During the budget period, the support department, not the user departments, bears the risk of any unfavorable cost variances. Why? Because the user department does not pay for any costs that exceed the budgeted costs.

Some organizations recognize that it may not always be best to impose all the risks of variances from budgeted amounts completely on the support department (as when costs are allocated using budgeted amounts) or completely on the user department (as when costs are allocated using actual amounts). For example, the two departments may agree to share the risk (through an explicit formula) of a large uncontrollable increase in the price of materials used by the support department.

Budgeted versus Actual Usage Allocation Bases

The choice between actual usage and budgeted usage for allocating departmental fixed costs also can affect a manager's behavior. Consider the budget of $300,000 fixed costs at the Central Computer Facility Department of Computer Horizons. Assume that actual and budgeted fixed costs are equal. Assume also that the actual usage by the Microcomputer Division is always equal to the budgeted usage. We now look at three different cases that might arise in allocating the $300,000 in total fixed costs.

- ◆ Case 1: Actual usage by the Peripheral Equipment Division *equals* budgeted usage.
- ◆ Case 2: Actual usage by the Peripheral Equipment Division *exceeds* budgeted usage.
- ◆ Case 3: Actual usage by the Peripheral Equipment Division *is less than* budgeted usage.

Recall that the budgeted usage is 800 hours for the Microcomputer Division and 400 hours for the Peripheral Equipment Division. Exhibit 14-5 presents the allocation of total fixed costs of $300,000 to each division for these three variations in actual usage in the Peripheral Equipment Division.

We will first consider how fixed-cost allocations can differ for the Microcomputer Division when actual usage is the allocation base. In case 1, the fixed-cost allocation equals the expected amount. In case 2, the fixed-cost allocation is $40,000 less than expected ($160,000 vs. $200,000). Case 3 fixed-cost allocation is $40,000 more than expected ($240,000 vs. $200,000). Let's consider now case 3. Why is there an increase of $40,000 even though the Microcomputer Division's actual usage is exactly what was budgeted? Because the fixed costs are spread over fewer output units. We see that variations in usage in another division affect the fixed costs allocated to the Microcomputer Division when fixed costs are allocated on the basis of actual usage. When actual usage is the allocation base, user divisions will not know what their allocated costs are until the end of the budget period.

When budgeted usage is the allocation base, user divisions will know their allocated costs in advance. This information helps the user divisions with both short-run and long-run planning. The main justification given for the use of budgeted usage to allocate fixed costs is related to long-run planning. Firms commit to infrastructure costs (such as the fixed costs of a support department) on the basis of a long-run planning horizon; the use of budgeted usage to allocate these fixed costs is consistent with this long-run horizon.

If fixed costs are allocated on the basis of long-run commitments or plans, some managers will be tempted to underestimate their planned usage. In this way, they will bear a lower fraction of the total costs. Top management can counteract this ploy by systematically comparing actual usage with planned usage. Some organizations offer rewards in the form of salary increases and promotions for managers who make accurate forecasts. Also, some cost allocation methods include penalties for underpredictions of budgeted usage. For instance, a higher variable rate may be charged after a division exceeds its budgeted usage.

EXHIBIT 14-5

Effect of Variations in Actual Quantity
of Allocation Base on Cost Allocations

	Actual Quantity		Budgeted Quantity as Allocation Base		Actual Quantity as Allocation Base	
Case	Micro. Div.	Periph. Div.	Micro. Div.	Periph. Div.	Micro. Div.	Periph. Div.
1	800 hours	400 hours	$200,000*	$100,000†	$200,000*	$100,000†
2	800 hours	700 hours	$200,000*	$100,000†	$160,000‡	$140,000‖
3	800 hours	200 hours	$200,000*	$100,000†	$240,000§	$ 60,000#

$$^*\frac{800}{(800 + 400)} \times \$300,000. \qquad ^{\ddagger}\frac{800}{(800 + 700)} \times \$300,000. \qquad ^{\|}\frac{700}{(800 + 700)} \times \$300,000.$$

$$^{\dagger}\frac{400}{(800 + 400)} \times \$300,000. \qquad ^{\S}\frac{800}{(800 + 200)} \times \$300,000. \qquad ^{\#}\frac{200}{(800 + 200)} \times \$300,000.$$

Allocations to Influence Behavior

Top managers sometimes resort to cost allocations that appear arbitrary. Why? In some cases, allocation is a means of getting subordinates to behave as top managers desire. Assume that Computer Horizons allocates to its operating divisions the costs incurred at the Research and Development Department at corporate headquarters. Cause-and-effect justifications may be weak, but top managers may believe that this allocation motivates division managers to take an interest in central research activities. In this way, the formal accounting system provides an important way of influencing behavior through an "arbitrary" cost-allocation scheme. This practice may be appropriate if the desired objective is reached without causing confusion in cost analysis and resentment about the methods of cost allocation.

Allocating costs is only one of several ways that top executives can influence managers' behavior. Other approaches include involving all affected managers in the budgeting process or simply issuing edicts. Organizations rarely ask which single approach for motivating managers is best. Rather, they view the challenge as designing a collective set of motivation mechanisms that reinforce each other.

ALLOCATING COSTS OF SUPPORT DEPARTMENTS

Operating Departments and Support Departments

Many organizations distinguish between operating departments and support departments. An **operating department** (also called a **production department** in a manufacturing organization) adds to a product or service value that is observable by a customer. A **support department** (also called a **service department**) provides the services that maintain other internal departments (operating departments and other support departments) in the organization. Support departments at Computer Horizons include the legal department and the personnel department.

Support departments create special accounting problems when they provide reciprocal services to each other as well as services to operating departments. An example of reciprocal services at Computer Horizons would be the legal department providing services to the personnel department (such as advice on compliance with labor laws) and the personnel department providing services to the legal department (such as advice about the hiring of attorneys and secretaries). A subsequent section of this chapter further discusses support departments.

Be cautious here. First, organizations differ in the departments located at the corporate and division levels. Some departments located at corporate headquarters of Computer Horizons (for example, Research and Development) are located at the division level in other organizations. Second, organizations differ in the definitions of *operating department* and *support department*. Always try to ascertain the precise meaning of these terms when analyzing data that include allocations of "operating department" costs and "support department" costs.

Support Department Cost-Allocation Methods

Objective 7
Distinguish among direct allocation, step-down, and reciprocal methods of allocating support department costs

We now examine three methods of allocating the costs of support departments: *direct*, *step-down*, and *reciprocal*. To focus on concepts, we use the single-rate method to allocate the costs of each support department. The Problem for Self-Study at the end of this chapter illustrates the use of the dual-rate method for allocating support department costs.

Consider Castleford Engineering, which manufactures engines used in electric power generating plants. Castleford has two support departments and two operating departments in its manufacturing facility:

	Support Departments	**Operating Departments**
	◆ Plant maintenance ◆ Information systems	◆ Machining ◆ Assembly

Costs are accumulated in each department for planning and control purposes. For inventory costing, however, the support department costs of Castleford must be allocated to the operating departments. The data for our example are in Exhibit 14-6.

Direct Allocation Method

The **direct allocation method** (often called the **direct method**) is the most widely used method of allocating support department costs. This method ignores any service rendered by one support department to another; it allocates each support department's total costs directly to the operating departments. Exhibit 14-7 illustrates this method using the data in Exhibit 14-6. Note how this method ignores the service rendered by the Plant Maintenance Department to the Information Systems Department and also ignores the service rendered by Information Systems to Plant Maintenance. For example, the base used to allocate Plant Maintenance is the budgeted total maintenance labor-hours worked in the operating departments: 2,400 + 4,000 = 6,400 hours. This amount excludes the 1,600 hours of service provided by Plant Maintenance to Information Systems. Similarly, the base used for allocation of Information Systems costs is 1,600 + 200 = 1,800 hours of computer time, which excludes the 200 hours of service provided by Information Systems to Plant Maintenance.

The benefit of the direct method is its simplicity. There is no need to predict the usage of support department resources by other support departments.

Step-Down Allocation Method

Some organizations use the **step-down allocation method** (sometimes called the **step,** or **sequential, allocation method**), which allows for *partial* recognition of services rendered by support departments to other support departments. This method is more complex than the direct method because a sequence of allocations must be

EXHIBIT 14-6

Data for Support Department Cost Allocation
at Castleford Engineering

	Support Departments		Operating Departments		
	Plant Maintenance	**Information Systems**	**Machining**	**Assembly**	**Total**
Budgeted manufacturing overhead costs before any interdepartmental cost allocations	$600,000	$116,000	$400,000	$200,000	$1,316,000
Service furnished:					
By Plant Maintenance					
Budgeted labor-hours	—	1,600	2,400	4,000	8,000
Proportion	—	0.2	0.3	0.5	1.0
By Information Systems					
Budgeted computer time	200	—	1,600	200	2,000
Proportion	0.1	—	0.8	0.1	1.0

EXHIBIT 14-7

Direct Method of Allocating
Support Department Costs at Castleford Engineering

	Support Departments		Operating Departments		
	Plant Maintenance	Information Systems	Machining	Assembly	Total
Budgeted manufacturing overhead costs before any interdepartmental cost allocations	$600,000	$116,000	$400,000	$200,000	$1,316,000
Allocation of Plant Maintenance: (3/8, 5/8)*	(600,000) $ 0		225,000	375,000	
Allocation of Information Systems: (8/9, 1/9)†		(116,000) $ 0	103,111	12,889	
Total budgeted manufacturing overhead of operating departments			$728,111	$587,889	$1,316,000

*Base is (2,400 + 4,000), or 6,400 hours; 2,400 ÷ 6,400 = 3/8; 4,000 ÷ 6,400 = 5/8.
†Base is (1,600 + 200), or 1,800 hours; 1,600 ÷ 1,800 = 8/9; 200 ÷ 1,800 = 1/9.

chosen. Many companies adopt the sequence that begins with the department that renders the highest percentage of its total services to other support departments. The sequence continues with the department that gives the next-highest percentage of its total services to other support departments, and so on, ending with the support department that renders the lowest percentage of its total services to other support departments.[3]

Let us continue with our Castleford Engineering example, in which we have only two service departments. In Exhibit 14-6, we see that the Plant Maintenance department supplies 2/10 of its total service to the other support department, Information Systems. Information Systems supplies 1/10 of its total service to Plant Maintenance. We begin, then, with the Plant Maintenance Department.

Exhibit 14-8 shows an allocation of $120,000 (2/10 of $600,000) of Plant Maintenance total service costs to Information Systems. The new total of Information Systems, $236,000, is then allocated to the two operating departments. The allocation base is the total service-hours provided to the departments to which the cost is to be allocated. *Once a support department's costs have been allocated, no subsequent support department costs are allocated or circulated back to it.*

Reciprocal Allocation Method

The **reciprocal allocation method** allocates costs by explicitly including the mutual services rendered among all support departments. Theoretically, the direct method and the step-down method are not accurate when support departments render ser-

---

[3]An alternative approach to selecting the sequence of allocations is to begin with the department that renders the highest dollar amount of services to other support departments. The sequence ends with the allocation of the costs of the department that renders the lowest dollar amount of services to other support departments.

EXHIBIT 14-8
Step-Down Method of Allocating
Support Department Costs at Castleford Engineering

	Support Departments		Operating Departments		
	Plant Maintenance	Information Systems	Machining	Assembly	Total
Budgeted manufacturing overhead costs before any interdepartmental cost allocations	$600,000	$116,000	$400,000	$200,000	$1,316,000
Allocation of Plant Maintenance: (2/10, 3/10, 5/10)*	(600,000) $\quad$ 0	120,000	180,000	300,000	
Allocation of Information Systems: (8/9, 1/9)†		(236,000) $\quad$ 0	209,778	26,222	
Total budgeted manufacturing overhead of operating departments			$789,778	$526,222	$1,316,000

*Base is (1,600 + 2,400 + 4,000), or 8,000 hours; 1,600 ÷ 8,000 = 2/10; 2,400 ÷ 8,000 = 3/10; 4,000 ÷ 8,000 = 5/10.
†Base is (1,600 + 200), or 1,800 hours; 1,600 ÷ 1,800 = 8/9; 200 ÷ 1,800 = 1/9.

vices to one another reciprocally. For example, the Plant Maintenance Department maintains all the computer equipment in the Information Systems Department. Similarly, the Information Systems provides data-base support for Plant Maintenance. The reciprocal allocation method enables us to incorporate interdepartmental relationships fully into the support department cost allocations. Implementing the reciprocal allocation method requires three steps.

<mark>STEP 1: *EXPRESS SUPPORT-DEPARTMENT COSTS AND SUPPORT-DEPARTMENT RECIPROCAL RELATIONSHIPS IN LINEAR EQUATION FORM.*</mark> Let *PM* be the complete reciprocated costs of Plant Maintenance and *IS* be the complete reciprocated costs of Information Systems. We then express the data in Exhibit 14-6 as follows:

$$(1)\quad PM = \$600,000 + 0.1IS$$
$$(2)\quad IS = \$116,000 + 0.2PM$$

The 0.1*IS* term in equation (1) is the proportion of the Information Systems work used by Plant Maintenance. The 0.2*PM* term in equation (2) is the proportion of the Plant Maintenance work used by Information Systems.

By **complete reciprocated cost** in equations (1) and (2), we mean the actual incurred cost of the support department plus a part of the costs of the other support departments that provide services to it. This complete reciprocated cost figure is sometimes called the **artificial cost** of the support department; it is always larger than the actual cost.

<mark>STEP 2: *SOLVE THE SYSTEM OF SIMULTANEOUS EQUATIONS TO OBTAIN THE COMPLETE RECIPROCATED COST OF EACH SUPPORT DEPARTMENT.*</mark>
Where there are two support departments, the following substitution approach can be used.

Substituting equation (2) into equation (1):

$$
\begin{aligned}
PM &= \$600,000 + [0.1(\$116,000 + 0.2PM)] \\
PM &= \$600,000 + \$11,600 + 0.02PM \\
0.98PM &= \$611,600 \\
PM &= \$624,082
\end{aligned}
$$

Substituting into equation (2):

$$
IS = \$116,000 + 0.2(\$624,082) = \$240,816
$$

Where there are more than two support departments with reciprocal relationships, computer programs can be used to calculate the complete reciprocated cost of each support department.

STEP 3: *ALLOCATE THE COMPLETE RECIPROCATED COST OF EACH SUP-PORT DEPARTMENT TO ALL OTHER DEPARTMENTS (BOTH SUPPORT DE-PARTMENTS AND OPERATING DEPARTMENTS) ON THE BASIS OF THE US-AGE PROPORTIONS (BASED ON TOTAL UNITS OF SERVICE PROVIDED TO ALL DEPARTMENTS).* Consider the Information Systems Department, which has a complete reciprocated cost of $240,816. This complete reciprocated cost would be allocated as follows:

To Plant Maintenance (1/10) × $240,816 = $ 24,082
To Machining (8/10) × $240,816   =   192,652
To Assembly (1/10) × $240,816   =   24,082
Total $240,816

Exhibit 14-9 presents summary data pertaining to the reciprocal method.

One source of confusion to some managers using the reciprocal cost allocation method is why the complete reciprocated costs of the support departments ($624,082 and $240,816 in Exhibit 14-9) exceed their budgeted amount of $716,000 ($600,000 and

EXHIBIT 14-9

Reciprocal Method of Allocating
Support Department Costs at Castleford Engineering

	Support Departments		Operating Departments		
	Plant Maintenance	Information Systems	Machining	Assembly	Total
Budgeted manufacturing overhead costs before any interdepartmental cost allocations	$600,000	$116,000	$400,000	$200,000	$1,316,000
Allocation of Plant Maintenance: (2/10, 3/10, 5/10)*	(624,082)	124,816	187,225	312,041	
Allocation of Information Systems: (1/10, 8/10, 1/10)†	24,082	(240,816)	192,652	24,082	
	$ 0	$ 0			
Total budgeted manufacturing overhead of operating departments			$779,877	$536,123	$1,316,000

*Base is (1,600 + 2,400 + 4,000), or 8,000 hours; 1,600 ÷ 8,000 = 2/10; 2,400 ÷ 8,000 = 3/10; 4,000 ÷ 8,000 = 5/10.
†Base is (200 + 1,600 + 200), or 2,000 hours; 200 ÷ 2,000 = 1/10; 1,600 ÷ 2,000 = 8/10; 200 ÷ 2,000 = 1/10.

$116,000). This excess of $148,898 ($24,082 for Plant Maintenance and $124,816 for Information Systems) is the total cost that is allocated from one support department to the other support department. The total costs allocated to the operating departments under the reciprocal cost allocation are still only $716,000.

Overview of Methods

Assume that the total budgeted overhead of each operating department in the example in Exhibits 14-7 to 14-9 is allocated to individual products on the basis of budgeted machine-hours for the machining department (4,000 hours) and budgeted direct labor-hours for the assembly department (3,000 hours). The budgeted overhead allocation rates associated with each support department allocation method are (rounded to the nearest dollar):

Support Department Cost-Allocation Method	Total Budgeted Overhead Costs after Allocation of All Support Department Costs		Budgeted Overhead Rate Per Hour for Product-Costing Purposes	
	Machining	Assembly	Machining (4,000 machine-hours)	Assembly (3,000 labor-hours)
Direct	$728,111	$587,889	$182	$196
Step-down	789,778	526,222	197	175
Reciprocal	779,877	536,123	195	179

These differences in budgeted overhead rates with alternative support department cost-allocation methods can be important to managers. For example, consider a cost-reimbursement contract. The cost-allocation method chosen may make a considerable difference in the amount of the reimbursement.

The reciprocal method, while theoretically the most defensible, is not widely used. The advantage of the direct and step-down methods is that they are relatively simple to compute and understand. However, with the ready availability of computer software to solve systems of simultaneous equations, the extra costs of using the reciprocal method will, in most cases, be minimal. The more likely roadblocks to the reciprocal method being widely adopted are (1) many managers find it difficult to understand, and (2) the numbers obtained by using the reciprocal method differ little, in some cases, from those obtained by using the direct or step-down method.

ALLOCATION OF COMMON COSTS

A **common cost** is a cost of operating a facility, operation, activity area, or like cost object that is shared by two or more users. For example, consider the *Financial Accounting Standards Board* (FASB) and the *Governmental Accounting Standards Board* (GASB). The FASB was set up in 1972 to issue standards governing financial reporting by profit-seeking corporations. The GASB was set up in 1984 to issue standards governing financial reporting by public-sector organizations. The FASB and GASB occupy the same building, and a single mailroom facility serves both. Variable costs of mailing are readily identifiable and kept in separate cost pools that are charged to the user. The fixed costs of the mailroom, however, cannot be identified with each individual user on the basis of a cause-and-effect relationship. How should these fixed costs be allocated between the two users?

Suppose the fixed costs of the mailroom facility for the next year are budgeted at $500,000. If the GASB did not use the mailroom, the fixed operating costs would be $480,000. An outside vendor offers to provide mailroom services to the FASB for a fixed fee of $600,000 plus variable costs of mailing. The same vendor offers to provide mailroom services to the GASB for a fixed fee of $200,000 plus variable costs of mailing. Two methods have been proposed for allocating common costs to individual users: the incremental method and the stand-alone method.

Incremental Common Cost Allocation Method

Objective 8
Distinguish between the incremental and stand-alone common cost allocation methods

The **incremental common cost allocation method** requires that one user be viewed as the primary party and the second user be viewed as the incremental party. The incremental party is allocated the extra common cost that arises from there being two users instead of only the primary user. For the mailroom example, the FASB would be allocated $480,000 of the fixed costs and the GASB $20,000. In the extreme case where the fixed costs remain unchanged when the incremental party uses the common facility, the incremental party would receive zero allocation of the common cost. Where there are more than two users of the common facility, the method requires the users to be ranked and the common costs allocated to those users in the ranked sequence.

Under the incremental method, the primary party typically receives the highest allocation of the common costs. Not surprisingly, most users in common cost allocation disputes propose themselves as the incremental party. In some cases, the incremental parties are newly formed organizations or new subparts of a corporation, such as a new product line or a new sales territory. Chances for their short-term survival may be enhanced if they bear a relatively low allocation of common costs.

Stand-Alone Common Cost Allocation Method

The **stand-alone common cost allocation method** allocates the common cost on the basis of each user's percentage of the total of the individual stand-alone costs. In the mailroom example, the total of the individual stand-alone costs is $800,000 ($600,000 for FASB and $200,000 for GASB as determined by the outside vendor). The common cost of the internally run mailroom facility is allocated as follows:

$$\text{FASB: } \frac{\$600,000}{\$800,000} \times \$500,000 = \$375,000$$

$$\text{GASB: } \frac{\$200,000}{\$800,000} \times \$500,000 = \$125,000$$

Advocates of the stand-alone method often emphasize an equity or fairness rationale; they argue that fairness requires that each user pay a percentage of actual common costs based on its share of the total stand-alone costs.

Several of the allocation bases described previously in this chapter have also been used in allocating common costs. For instance, relative revenues are sometimes adopted as an allocation base justified by the ability-to-bear criterion. Also, common fixed costs can be allocated on the basis of the user's estimates of long-run service requirements.

PROBLEM FOR SELF-STUDY

PROBLEM

This problem illustrates how support department cost allocation methods can be used in a setting different from the manufacturing example examined in the chapter (Exhibits 14-6 to 14-9). In this problem, the costs of central corporate support departments are allocated to operating divisions. The corporate departments provide services to each other as well as to the operating divisions. Also, this problem illustrates the use of the dual-rate method of allocating support department costs. (The dual-rate method can also be used in manufacturing support department cost allocations.)

Computer Horizons budgets the following amounts for its two central corporate support departments (legal and personnel) in servicing each other and the two manufacturing divisions—the Microcomputer Division (MCD) and the Peripheral Equipment Division (PED):

To Be Supplied By		Budgeted Capacity			
	Legal	Personnel	MCD	PED	Total
Legal (hours)	—	250	1,500	750	2,500
Legal (proportions)	—	0.10	0.60	0.30	1.00
Personnel (hours)	2,000	—	22,500	25,000	50,000
Personnel (proportions)	0.05	—	0.45	0.50	1.00

Details on actual usage:

To Be Supplied by		Actual Usage by:			
	Legal	Personnel	MCD	PED	Total
Legal (hours)	—	400	400	1,200	2,000
Legal (proportions)	—	0.20	0.20	0.60	1.00
Personnel (hours)	2,000	—	26,600	11,400	40,000
Personnel (proportions)	0.05	—	0.665	0.285	1.00

The actual costs were:

	Fixed	Variable
Legal	$360,000	$200,000
Personnel	$475,000	$600,000

Fixed costs are allocated on the basis of budgeted capacity. Variable costs are allocated on the basis of actual usage.

Required
What support department costs for legal and personnel will be allocated to MCD and PED using (a) the direct method, (b) the step-down method (allocating the legal department costs first), and (c) the reciprocal method?

SOLUTION

Exhibit 14-10 presents the computations for allocating the fixed and variable support department costs. A summary of these costs follows:

	Microcomputer Division (MCD)	Peripheral Equipment Division (PED)
A. Direct Method		
Fixed costs	$465,000	$370,000
Variable costs	470,000	330,000
	$935,000	$700,000
B. Step-Down Method		
Fixed costs	$458,053	$376,947
Variable costs	488,000	312,000
	$946,053	$688,947
C. Reciprocal Method		
Fixed costs	$462,513	$372,487
Variable costs	476,364	323,636
	$938,877	$696,123

EXHIBIT 14-10

Alternative Methods of Allocating Corporate Support Department Costs to Operating Divisions of Computer Horizons: Dual-Rate Method Illustrated

Allocation Method	Corporate Support Departments		Manufacturing Divisions	
	Legal	Personnel	MCD	PED
A. Direct Method				
Fixed costs	$360,000	$475,000		
Legal (6/9, 3/9)	(360,000)		$240,000	$120,000
Personnel (225/475, 250/475)	$ 0	(475,000)	225,000	250,000
		$ 0	$465,000	$370,000
Variable costs	$200,000	$600,000		
Legal (0.25, 0.75)	(200,000)		$ 50,000	$150,000
Personnel (0.7, 0.3)	$ 0	(600,000)	420,000	180,000
		$ 0	$470,000	$330,000
B. Step-Down Method				
(Legal Department first)				
Fixed costs	$360,000	$475,000		
Legal (0.10, 0.60, 0.30)	(360,000)	36,000	$216,000	$108,000
Personnel (225/475, 250/475)	$ 0	(511,000)	242,053	268,947
		$ 0	$458,053	$376,947
Variable costs	$200,000	$600,000		
Legal (0.20, 0.20, 0.60)	(200,000)	40,000	$ 40,000	$120,000
Personnel (0.7, 0.3)	$ 0	(640,000)	448,000	192,000
		$ 0	$488,000	$312,000
C. Reciprocal Method				
Fixed costs	$360,000	$475,000		
Legal (0.10, 0.60, 0.30)	(385,678)	38,568	$231,407	$115,703
Personnel (0.05, 0.45, 0.50)	25,678	(513,568)	231,106	256,784
	$ 0	$ 0	$462,513	$372,487
Variable costs	$200,000	$600,000		
Legal (0.20, 0.20, 0.60)	(232,323)	46,465	$ 46,465	$139,393
Personnel (0.05, 0.665, 0.285)	$32,323	(646,465)	429,899	184,243
	$ 0	$ 0	$476,364	$323,636

The simultaneous equations for the reciprocal method are:

Fixed Costs

$$L = \$360,000 + 0.05P$$
$$P = \$475,000 + 0.10L$$
$$L = \$360,000 + 0.05(\$475,000 + 0.10L) = \$385,678$$
$$P = \$475,000 + 0.10(\$385,678) = \$513,568$$

Variable Costs

$$L = \$200,000 + 0.05P$$
$$P = \$600,000 + 0.20L$$
$$L = \$200,000 + 0.05(\$600,000 + 0.20L) = \$232,323$$
$$P = \$600,000 + 0.20(\$232,323) = \$646,465$$

SUMMARY

The following points are linked to the chapter's learning objectives.

1. A *cost object* is anything for which a separate measurement of costs is desired. Costing systems in organizations have multiple cost objects, including departments, products, brands, and customers. These multiple cost objects mean that many individual costs are allocated and reallocated several times before becoming an indirect cost of a specific cost object.

2. The four purposes of cost allocation are to provide information for economic decisions, to motivate managers and employees, to justify costs or compute reimbursement, and to measure income and assets for reporting to external parties. Different cost allocations may be appropriate depending on the specific purpose.

3. The cause-and-effect and the benefits-received criteria guide most decisions related to cost allocations. Other criteria found in practice include fairness or equity and ability to bear.

4. A cost pool is a grouping of individual cost items. Two key decisions related to indirect cost pools are the number of indirect cost pools and the allowability of individual cost items to be included in those cost pools.

5. A single-rate cost allocation method pools all costs in one cost pool and allocates them to cost objects using the same rate per unit of the single allocation base. In the dual-rate method, costs are grouped in two separate cost pools, each of which has a different allocation rate and may have a different allocation base.

6. When cost allocations are made using budgeted rates, managers of divisions to which costs are allocated face no uncertainty about the rates to be used in that period. In contrast, when actual rates are used for cost allocation, managers do not know the rates to be used until the end of the accounting period.

7. The three main methods of allocating support department costs to operating departments are the direct, step-down, and reciprocal. The last is the most defensible, but the direct and step-down methods are more widely used.

8. Common costs are the costs of operating a facility, operation, or activity area that are shared by two or more users. Diverse allocation methods are used in practice to allocate common costs to individual users. These include the incremental cost and the stand-alone cost methods.

TERMS TO LEARN

This chapter and the Glossary at the end of the book contain definitions of the following important terms:

artificial cost *(p. 513)*　　common cost *(515)*　　complete reciprocated cost *(513)*
direct allocation method *(511)*　　direct method *(511)*　　dual-rate method *(507)*
homogeneous cost pool *(505)*　　incremental common cost allocation method *(516)*
operating department *(510)*　　production department *(510)*
reciprocal allocation method *(512)*　　service department *(510)*
single-rate method *(507)*　　sequential allocation method *(511)*
stand-alone common cost allocation method *(516)*　　step allocation method *(511)*
step-down allocation method *(511)*　　support department *(510)*

ASSIGNMENT MATERIAL

QUESTIONS

14-1 Why might the classification of a cost as a direct cost or an indirect cost of a cost object change over time?

14-2 How can an individual cost item, such as the salary of a plant security guard, be both a direct cost and an indirect cost?

14-3 "A given cost may be allocated for one or more purposes." List four purposes.

14-4 What criteria might be used to guide cost allocation decisions? Which criterion is the dominant one?

14-5 What is a cost pool? Give an example.

14-6 Name three decisions managers face when designing the cost allocation component of an accounting system.

14-7 What is a support department? What is another term for a support department?

14-8 Give examples of bases used to allocate corporate cost pools to the operating divisions of an organization.

14-9 What role does the homogeneity of cost pools play in decisions on the number of cost pools to use?

14-10 Why use a dual-rate cost-allocation method when it is less costly to use a single-rate method?

14-11 "Actual repair costs rather than budgeted repair costs should be used to allocate the motor vehicle repair and maintenance department costs to the sales division." Comment on this statement made by the manager of the repair maintenance department.

14-12 "It is essential to fully allocate all corporate costs to our divisions. This gives division managers the incentive to monitor any growth in corporate costs. Without this monitoring, corporate costs would mushroom." Do you agree?

14-13 Distinguish among the three methods of allocating the costs of service departments to production departments.

14-14 What is the theoretically most defensible method for allocating service department costs?

14-15 Distinguish between two methods of allocating common costs.

EXERCISES AND PROBLEMS

14-16 Cost allocation in hospitals, alternative allocation criteria. Dave Meltzer went to Lake Tahoe for his annual winter vacation. Unfortunately, he suffered a severe break in his ankle while skiing and had to spend two days at the Sierra University Hospital. Meltzer's insurance company received a $4,800 bill for his two-day stay. One item that caught Meltzer's eye was an $11.52 charge for a roll of cotton. Meltzer was a salesman for Johnson & Johnson and knew that the cost to the hospital of the roll of cotton would be in the $2.20 to $3.00 range. He asked for a breakdown of how the $11.52 charge was derived. The accounting office of the hospital sent him the following information:

1. Invoiced cost of cotton roll	$ 2.40
2. Processing of paperwork for purchase	0.60
3. Supplies room management fee	0.70
4. Operating-room and patient room handling charge	1.60
5. Administrative hospital costs	1.10
6. University teaching-related recoupment	.60
7. Malpractice insurance costs	1.20
8. Cost of treating uninsured patients	2.72
9. Profit component	0.60
Total	$11.52

Meltzer believes the overhead charge is obscene. He comments, "There was nothing I could do about it. When they come in and dab your stitches, it's not as if you can say, 'Keep your cotton roll. I brought my own.' "

Required

1. Compute the overhead rate Sierra University Hospital charged on the cotton roll.

2. What criteria might Sierra use to justify allocation of each of the overhead items numbered 2 to 9 in the list? Examine each item separately, and use the allocation criteria listed in Exhibit 14-2 (p. 501) in your answer.

3. What should Meltzer do about the $11.52 charge for the cotton roll?

14-17 Single-rate versus dual-rate cost allocation methods (W. Crum, adapted) Carolina Company has a power plant designed and built to serve its three factories. Data for 19_4 follow:

Factory	Usage in Kilowatt-Hours	
	Budget	Actual
Durham	100,000	80,000
Charlotte	60,000	120,000
Raleigh	40,000	40,000

Actual fixed costs of the power plant were $1 million in 19_4; actual variable costs, $2 million.

Required

1. Compute the amount of power costs that would be allocated to Charlotte using a single-rate method.

2. Compute the amount of power costs that would be allocated to Charlotte using a dual-rate method.

14-18 Single-rate versus dual-rate allocation methods, support department. The Chicago power plant that services all manufacturing departments of MidWest Engineering has a budget for the coming year. This budget has been expressed in the following terms on a monthly basis:

Manufacturing Departments	Needed at Practical Capacity Production Level* (Kilowatt-Hours)	Average Expected Monthly Usage (Kilowatt-Hours)
Rockford	10,000	8,000
Peoria	20,000	9,000
Hammond	12,000	7,000
Kankakee	8,000	6,000
Totals	50,000	30,000

*This factor was the most influential in planning the size of the power plant.

The expected monthly costs for operating the department during the budget year are $15,000: $6,000 variable and $9,000 fixed.

Required
1. Assume that a single-cost pool is used for the power plant costs. What dollar amounts will be allocated to each manufacturing department? Use (a) practical capacity and (b) average expected monthly usage as the allocation bases.
2. Assume a dual-rate method; separate cost pools for the variable and fixed costs are used. Variable costs are allocated on the basis of expected monthly usage. Fixed costs are allocated on the basis of practical capacity. What dollar amounts will be allocated to each manufacturing department? Why might you prefer the dual-rate method?

14-19 Allocation of travel costs. Joan Ernst, a graduating senior at a university near San Francisco, received an invitation to visit a prospective employer in New York. A few days later she received an invitation from a prospective employer in Chicago. She decided to combine her visits, traveling from San Francisco to New York, New York to Chicago, and Chicago to San Francisco.

Ernst received job offers from both companies. Upon her return, she decided to accept the offer in Chicago. She was puzzled about how to allocate her travel costs between the two employers. She gathered the following data:

Regular roundtrip fares with no stopovers:

San Francisco to New York	$1,400
San Francisco to Chicago	$1,100

Ernst paid $1,800 for her three-leg flight (San Francisco to New York, New York to Chicago, Chicago to San Francisco). In addition, she paid $30 for a limousine from her home to San Francisco Airport and another $30 for a limousine from San Francisco Airport to her home when she returned.

Required
1. How should Ernst allocate the $1,800 airfare between the employers in New York and Chicago? Show the actual amounts you would allocate, and give reasons for your allocations.
2. Repeat requirement 1 for the $60 limousine charges at the San Francisco end of her travels.

14-20 Allocation of common costs, splitting of shared hotel bill. Linda and Mark McGraw are planning a one-week vacation to celebrate their tenth wedding anniversary. For their last two vacations, Linda's sister, Rebecca Miller, has accompanied them. Two months ago, Rebecca graduated from a prestigious MBA program. She recently accepted a job on Wall Street as an investment banker. Linda was very proud of her sister, especially because Rebecca had borrowed money on her own account to put herself through the best (and most expensive) MBA program in the country.

Last year the three of them stayed at the Kokomo Vacation Resort in the Florida Keys for one week. They rented one room with two double beds. The room rate was $120 a night for a single, $140 for a double, and $150 for three people in a room. Linda put the $150-a-night bill on her credit card and allocated $140 a night to Mark and herself and $10 to Rebecca. Rebecca was happy with this arrangement and promptly paid her $70 share for the room for the week's vacation.

Linda has just heard that Rebecca would again like to join them on their vacation, under the same arrangement as last year. Rebecca was thrilled when Linda told her that Mark had made reservations at the Kokomo Resort and that the room rates were still $120 a night for a single, $140 for a double, and $150 for three people in a room. Mark was less than enthusiastic when Linda told him of the repeat of last year's "cast of characters." Among other complaints, he thought Rebecca could well afford a suite of her own, given her Wall Street salary.

Required
1. What justification is there for the current allocation scheme of $140 to Linda and Mark and $10 to Rebecca?
2. Propose an alternative allocation of the $150 a night rate. Defend your proposal.

14-21 Allocation of support department costs. Atherton Machining Company has one support department (electric power) and two operating departments in its plant. The flexible-budget formula for the support department costs for the next fiscal year is $6,000 monthly plus $0.40 per machine-hour in the operating departments. Fixed costs are allocated on a lump-sum capacity-available basis, 60% to Operating Department 1 and 40% to Operating Department 2. Variable costs are allocated to the operating departments at the budgeted unit rate of $0.40 per machine-hour.

Required

1. Assume that the actual costs coincided exactly with the flexible-budget amount. Departments 1 and 2 each worked at 4,000-hour levels. Tabulate the allocations of all costs.

2. Assume the same facts as in requirement 1 except that the fixed costs were allocated on the basis of actual hours rather than capacity available. Tabulate the allocations of all costs. As the manager of Department 2, would you prefer the method in requirement 1 or in requirement 2? Why?

3. Suppose the support department had inefficiencies at an 8,000-hour level, incurring variable costs that exceeded the flexible budget by $800. How would this change your answers in requirements 1 and 2? Be specific.

14-22 Departmental cost allocation, university computer-service center. A computer-service center of National University serves two major users, the College of Engineering and the College of Humanities and Sciences (H&S).

Required

1. When the computer equipment was initially installed, the procedure for cost allocation was straightforward. The actual monthly costs were compiled and were divided between the two colleges on the basis of the computer time used by each. In October, the costs were $100,000. H&S used 100 hours and Engineering used 100 hours. How much cost would be allocated to each college? Suppose costs were $110,000 because of various inefficiencies in the operation of the computer center. How much cost would be allocated? Does such an allocation seem justified? If not, what improvement would you suggest?

2. Use the same approach as in requirement 1. The actual cost-behavior pattern of the computer center was $80,000 fixed cost per month plus $100 variable cost per hour used. In November, H&S used 50 hours and Engineering used 100 hours. How much cost would be allocated to each college? Use a single-rate method.

3. As the computer-service center developed, a committee was formed that included representatives of H&S and Engineering. This committee determined the size and composition of the center's equipment. The committee based its planning on the long-run average utilization of 180 monthly hours for H&S and 120 monthly hours for Engineering. Suppose the $80,000 fixed costs are allocated through a budgeted monthly lump sum based on long-run average utilization. Variable costs are allocated through a budgeted unit rate of $100 per hour. How much cost would be allocated to each college? What are the advantages of this dual-rate allocation method over other methods?

4. What are the likely behavioral effects of lump-sum allocations of fixed costs? For example, if you were the representative of H&S on the facility planning committee, what would your biases be in predicting long-run usage? How would top management counteract the bias?

14-23 Allocation of administrative and marketing costs, revenue as an allocation basis. The Mideastern Transportation Company has had a long-standing policy of fully allocating all costs to its various divisions. Among the costs allocated are general and administrative costs in central headquarters, consisting of office salaries, executive salaries, travel expenses, accounting costs, office supplies, donations, rents, depreciation, postage, and similar items.

All these costs are difficult to trace directly to the individual divisions benefited, so they are allocated on the basis of the actual total revenue of each of the divisions. The same basis is used for allocating marketing costs. For example, in 19_3 the following allocations were made (in millions):

	Division			
	A	**B**	**C**	**Total**
Revenue	$50.0	$40.0	$10.0	$100.0
Costs allocated on the basis of revenue	6.0	4.8	1.2	12.0

Division A's revenue was expected to rise in 19_4, but the division encountered severe competition. Revenue remained at $50 million. In contrast, Division C enjoyed explosive growth in traffic because of the completion of several huge factories in its area. Its revenue rose to $30 million. Division B's revenue remained unchanged. Careful supervision kept the actual total costs allocated on the basis of revenue at $12 million.

Required
1. What costs will be allocated to each division for 19_4?
2. Using the results in requirement 1, comment on the limitations of using revenue as a basis for cost allocation.

14-24 Allocation of support department costs. For simplicity, suppose Advanced Technologies has one support department (maintenance) plus two operating departments (M and N). The flexible budget of the support department is $24,000 fixed costs monthly plus $1 per machine-hour. The expected "long-run" usage of the support department is 75% by Department M and 25% by Department N. Fixed costs are allocated using "long-run" usage.

Required
1. Indicate the dual rates for allocating the support department's costs to each operating department.
2. Assume that actual costs coincided exactly with the flexible-budget amount. Departments M and N each worked at 8,000-hour levels. Tabulate the allocation of all costs.
3. Assume the same facts as in requirement 2 except that the fixed costs were allocated on the basis of actual hours worked by each department rather than on the basis of long-run usage. Tabulate the allocation of all costs. As the manager of Department N, would you prefer the method in requirement 2 or in requirement 3? Why?
4. Suppose the support department had inefficiencies at a 16,000-hour level, incurring costs that exceeded the flexible budget by $6,000. How would this change your answers in requirements 2 and 3? Be specific.

14-25 Allocating costs of a central telephone reservation system. Helena Park, the chief operating officer of Happy Inns, faces a difficult problem. The manager of the most prestigious motel in the five-motel chain is upset about her proposed cost allocation method for the central telephone reservation system. Until one year ago, each motel handled its own reservations. Then Happy Inns leased a central telephone reservation system for $84,000 per year. Happy Inns pays the variable costs of operating the system. During the first year of operations, Happy Inns allocated none of the $84,000 fixed cost and none of the $56,000 variable costs of operating the system. Summary data for the first year of operating the central reservation system follow:

Motel	Total Number of Reservations Made via Central System	Total Number of Room Nights Available for Rental for Year	Average Room Nights Actually Rented during Year	Rate per Room Night	Total Room Revenues for Year
Vancouver	5,000 (12.5%)	28,000 (11.7%)	20,000 (10.8%)	$ 60	$ 1,200,000 (8.6%)
Halifax	4,000 (10.0)	31,000 (12.9)	25,000 (13.5)	76	1,900,000 (13.6)
Toronto	8,000 (20.0)	37,000 (15.4)	30,000 (16.2)	50	1,500,000 (10.7)
Jasper	3,000 (7.5)	32,000 (13.3)	30,000 (16.2)	100	3,000,000 (21.4)
Quebec City	20,000 (50.0)	112,000 (46.7)	80,000 (43.3)	80	6,400,000 (45.7)
	40,000	240,000	185,000		$14,000,000

The total number of rooms actually rented includes those rooms rented through the central reservation system, those rented through direct dial to the individual motel, and those rented by walk-in clientele. During the first year of operations, approximately 100,000 phone calls were made to the central reservation service.

Park decides to allocate the central telephone costs to each motel. She proposes that the fixed and variable costs be combined in a single pool and be allocated on the basis of actual total room revenues for the year.

Each Happy Inn manager receives a fixed salary plus a percentage of the operating income of the individual motel.

Required

1. What costs would have been allocated to each motel under Park's new proposal (the relative percentage of actual total room revenues) in the first year of operation of the central reservation system? Why might the manager of the Jasper Happy Inn oppose Park's proposal?

2. What limitations are associated with the proposed cost allocation method of Park in requirement 1?

3. Park considers alternative cost-allocation systems. She decides to retain a single-cost pool, but to consider alternative allocation bases. What costs would have been allocated to each motel under each of the following bases: (a) reservations made through the central system, (b) total number of room nights available for rental, and (c) total number of room nights actually rented during the year?

14-26 Allocation of corporate legal costs to product lines. (CMA, adapted) Spiral Laboratories Inc. manufactures and distributes generic pharmaceuticals for sale to customers through retail outlets. Spiral is organized along product lines, with profit-center responsibilities assigned to the unit manager of each product line. The unit managers receive incentive compensation based on product-line performance.

Currently, Spiral is reviewing its policies regarding the use of legal services. Historically, the company has used various outside legal firms, as needed by the unit managers and members of the corporate staff. The cost of legal services to the corporation has been increasing ("spiraling" is the term the president used). Recently, the decision was made to hire an in-house attorney to coordinate and assist in legal matters. The goal is to lower overall legal costs by reducing the use of outside services. However, the unit managers still have the option of using outside legal help. Legal services are anticipated to be an on-going need at the corporate level and a periodic need at the unit level.

The corporate controller has identified the following four alternatives for allocating the cost of corporate legal services provided by the in-house attorney.

A. An annual charge to each product-line unit based on the budgeted sales level of the unit. Corporate headquarters would also receive a fixed annual charge; this fixed charge would be based on the budgeted use of legal time by corporate headquarters relative to budgeted total use of legal time by all areas of Spiral Laboratories.

B. No charge to the product-line units, with corporate headquarters absorbing all costs for in-house legal services.

C. A fixed, hourly charge for the actual services required by each product-line unit determined on the basis of comparative costs for services obtained from external sources.

D. A fixed, hourly charge for actual services required by each product-line unit that is based on full recovery of the cost of the corporate legal services. Under- or over-allocation of in-house legal costs would be charged to the product-line units at year-end. ("Full recovery" means that the actual costs of the legal department are equal to the costs allocated to divisions and to corporate headquarters for legal services.)

Required

1. Describe the probable behavioral effects on the Spiral Laboratories product-line unit managers brought about by each of the four cost-allocation methods.

2. How might each of the four proposals affect the quality and availability of legal services?

3. How would the Spiral Laboratories' corporate attorney be likely to react to each of the four proposals?

14-27 Allocation of central corporate costs to divisions. Dusty Rhodes, the corporate controller of Richfield Oil Company, is about to make a presentation to the senior corporate executives and the top managers of its four divisions. These divisions are:

a. Oil & Gas Upstream—the exploration, production, and transportation of oil and gas

b. Oil & Gas Downstream—the refining and marketing of oil and gas

c. Chemical Products

d. Copper Mining

Under the existing internal accounting system, costs incurred at central corporate headquarters are collected in a single pool and allocated to each division on the basis of the actual revenues of each division. The central corporate costs for the most recent year are (in millions):

Interest on debt	$2,000
Corporate salaries	100
Accounting and control	100
General marketing	100
Legal	100
Research and development	200
Public affairs	208
Personnel and payroll	192
	$3,000

Public affairs includes the public relations staff, the lobbyists, and the sizable donations Richfield makes to numerous charities and nonprofit institutions.

Summary data related to the four divisions for the most recent year are (in millions):

	Oil & Gas Upstream	Oil & Gas Downstream	Chemical Products	Copper Mining	Total
Revenue	$ 7,000	$16,000	$4,000	$3,000	$30,000
Operating costs	$ 3,000	$15,000	$3,800	$3,200	$25,000
Operating income	$ 4,000	$ 1,000	$ 200	$ (200)	$ 5,000
Identifiable assets	$14,000	$ 6,000	$3,000	$2,000	$25,000
Number of employees	9,000	12,000	6,000	3,000	30,000

The top managers of each division share in a divisional income bonus pool. *Divisional income* is defined as operating income less allocated central corporate costs.

Rhodes is about to propose a change in the method used to allocate central corporate costs. He favors collecting these costs in four separate pools:

◆ *Cost Pool 1:* Allocated using identifiable assets of division
  Cost Item: Interest on debt
◆ *Cost Pool 2:* Allocated using revenue of division
  Cost Items: Corporate salaries, accounting and control, general marketing, legal, research and development
◆ *Cost Pool 3:* Allocated using operating income (if positive) of division, with only divisions with positive operating income included in the allocation base
  Cost Item: Public affairs
◆ *Cost Pool 4:* Allocated using number of employees in division
  Cost Item: Personnel and payroll

Required
1. What purposes might be served by the allocation of central corporate costs to each division at Richfield Oil?
2. Compute the divisional income of each of the four divisions when central corporate costs are allocated using revenue of each division.
3. Compute the divisional income of each of the four divisions when central corporate costs are allocated through the four cost pools.
4. What are the strengths and weaknesses of Rhodes's proposal relative to the existing single-cost pool method?

14-28 Division managers' reactions to the allocation of central corporate costs to divisions. (Continuation of 14-27) Dusty Rhodes presents his proposal for the use of four separate cost pools to allocate central corporate costs to the divisions. The comments of the top managers of each of the four divisions include the following:

a. By the top manager of Oil & Gas Upstream Division: "The multiple-pool method of Rhodes is absurd. We are the only division generating a substantial positive cash flow, and this is ignored in the proposed (and indeed the existing) system. We could pay off any debt very quickly if we were not a cash cow for the rest of the dog divisions in Richfield Oil."

b. By the top manager of Oil & Gas Downstream Division: "Rhodes's proposal is the first sign that the money we spend in the accounting and control function at corporate headquarters is justified. The proposal is fair and equitable."

c. By the top manager of Chemical Products Division: "I oppose any cost allocation method. Last year I was the only major player in the chemical industry to show a positive operating income. We are operating at the bare-bones level. Last year I saved $300,000 by making everyone travel economy class. This policy created a lot of dissatisfaction, but we finally managed to get it accepted. Then at the end of the year we get a charge of $400 million for corporate central costs. What's the point of our division economy drives when they get swamped by allocations of corporate fat?"

d. By the top manager of Copper Mining Division: "I should probably get concerned, but frankly I view it all as bookkeeping entries. If we were in the black, certain aspects would really infuriate me. For instance, why should corporate research and development costs be allocated to the Copper Division? The only research corporate does for us is how to best prepare our division for divestiture."

Required

How should Rhodes respond to these comments?

14-29 Cost allocation at a hospital, use of actual costs and quantities. Golden Life (GL) is a health-maintenance organization that operates six hospitals. In addition, it has an agreement with Houston University Hospital (HUH) to use up to 30% of its capacity for patient-days (a patient-day is defined as an overnight stay) for each of the next seven years. The contract between GL and HUH provides for GL to pay HUH each year the greater of:

a. $20 million per year, or

b. the sum of its share of (i) actual general-care variable costs, (ii) actual operating-room variable costs, and (iii) actual fixed hospital costs.

The contract provides the following details on computing these three costs:

◆ General-care variable costs: based on actual general-care variable costs per patient-day times the actual number of Golden Life patient-days used in that year

◆ Operating-room variable costs: based on actual operating-room variable costs per operating-hour times the actual number of operating-hours Golden Life patients use

◆ Fixed hospital costs: based on the ratio of Golden Year actual patient-days used to actual total patient-days times the actual fixed hospital costs

HUH is in charge of all major operating decisions at the hospital. GL's discretion over the costs HUH incurs is limited to decisions related to GL patients (such as number of patient-days). The following information describes costs for 19_5 (the third year of the contract):

Budgeted fixed hospital costs	$60,000,000
Budgeted general-care variable costs	$100 per patient-day
Budgeted operating-room variable costs	$500 per hour
Actual fixed hospital costs	$56,000,000
Actual general-care variable costs	$15,300,000
Actual operating-room variable costs	$39,600,000

Summary operating information for 19_5 is:

Total patient-day capacity	180,000 patient-days (500 beds × 360 days)
Total operating-room capacity	100,000 hours
Budgeted patient-days	162,000
Budgeted operating-room hours	88,000

Actual patient-days

HUH patients	90,000
GL patients	60,000
	150,000 patient-days

Actual operating-room hours

HUH patients	46,080
GL patients	25,920
	72,000 hours

Required

1. Compute the amount Golden Life will pay Houston University Hospital for 19_5.

2. As Controller of GL, evaluate three areas where you would be concerned about the level of costs at HUH that GL is required to pay.

3. What cost accounting arguments could the HUH controller use to justify the existing contract provisions?

14-30 Cost allocation at a hospital, use of budgeted costs and quantities. (Continuation of 14-29) The controller at GL proposes an alternative cost-sharing contract. The proposal is to change the way the three cost items are computed as follows:

◆ General-care variable costs: based on budgeted general-care variable costs per patient-day times the actual number of Golden Life patient-days used in that year

◆ Operating-room variable costs: based on budgeted operating-room variable costs per operating-hour times the actual number of operating-hours used by Golden Life patients

◆ Fixed hospital costs: 30% of budgeted fixed costs

Required

1. Compute the amount Golden Life will pay Houston University Hospital for 19_5 under the new proposal.

2. What cost accounting arguments could the GL controller use to justify the new proposal?

14-31 Cost allocation, choice of a denominator level. (Continuation of 14-29 and 14-30) The vice-president of HUH is less than impressed by the GL controller's proposal in Problem 14-30. One concern is related to fixed-cost allocation. He observes that the budgeted fixed cost of $60,000,000 for 19_5 was based on a budgeted 108,000 patient-days for the hospital and a budgeted 54,000 patient-days amount provided by GL. His concern is how to ensure that GL will give an unbiased estimate of its budgeted usage of HUH at the start of each year.

Required

1. Why might the vice-president of HUH be concerned about fixed-cost allocation using the GL controller's proposal?

2. Describe a fixed-cost allocation method that the vice-president might propose to do away with his concerns. What are its advantages and its disadvantages?

14-32 Allocating costs of support departments; step-down and direct methods. The Central Valley Company has prepared departmental overhead budgets for normal-volume levels before allocations, as follows:

Support departments:		
Building and grounds	$ 10,000	
Personnel	1,000	
General factory administration	26,090	
Cafeteria: operating loss	1,640	
Storeroom	2,670	
Total		$ 41,400
Operating departments:		
Machining	$ 34,700	
Assembly	48,900	
Total		83,600
Total for both departments		$125,000

Management has decided that the most sensible inventory costs are achieved by using individual departmental overhead rates. These rates are developed after appropriate support department costs are allocated to operating departments.

Bases for allocation are to be selected from the following:

Department	Direct Manufacturing Labor-Hours	Number of Employees	Square Feet of Floor Space Occupied	Total Manufacturing Labor-Hours	Number of Requisitions
Building and grounds	0	0	0	0	0
Personnel*	0	0	2,000	0	0
General plant administration	0	35	7,000	0	0
Cafeteria: operating loss	0	10	4,000	1,000	0
Storeroom	0	5	7,000	1,000	0
Machining	5,000	50	30,000	8,000	2,000
Assembly	15,000	100	50,000	17,000	1,000
Total	20,000	200	100,000	27,000	3,000

*Basis used is number of employees.

Required

1. Using a worksheet, allocate support department costs by the step-down method. Develop overhead rates per direct manufacturing labor-hour for machining and assembly. Allocate the support departments in the order given in this problem. Use the allocation base for each support department you think is most appropriate.

2. Using the direct method, rework requirement 1.

3. Based on the following information about two jobs, determine the total overhead costs for each job by using rates developed in requirements 1 and 2.

	Direct Manufacturing Labor-Hours	
	Machining	Assembly
Job 88	18	2
Job 89	3	17

14-33 Support department cost allocations; single-department cost pools; direct, step-down, and reciprocal methods. The Manes Company has two products. Product 1 is manufactured entirely in Department X. Product 2 is manufactured entirely in Department Y. To produce these two products, the Manes Company has two support departments: A (a materials-handling department) and B (a power-generating department).

An analysis of the work done by Departments A and B in a typical period follows:

	Used By			
Supplied By	A	B	X	Y
A	—	100	250	150
B	500	—	100	400

The work done in Department A is measured by the direct labor-hours of materials-handling time. The work done in Department B is measured by the kilowatt-hours of power.

The budgeted costs of the support departments for the coming year are:

	Department A	Department B
Variable indirect labor and indirect materials costs	$ 70,000	$10,000
Supervision	10,000	10,000
Depreciation	20,000	20,000
	$100,000	$40,000
	+Power costs	+Materials-handling costs

The budgeted costs of the operating departments for the coming year are $1,500,000 for Department X and $800,000 for Department Y.

Supervisory costs are salary costs. Depreciation in B is the straight-line depreciation of power-generation equipment in its nineteenth year of an estimated 25-year useful life; it is old but well-maintained equipment.

Required
1. What are the allocations of costs of support Departments A and B to operating Departments X and Y using the direct method, two different sequences of the step-down method, and the reciprocal method of reallocation?
2. The power company has offered to supply all the power needed by the Manes Company and to provide all the services of the present power department. The cost of this service will be $40 per kilowatt-hour of power. Should Manes accept? Explain.

 14-34 Allocating costs of support departments; dual rates; cost justification. Lindsay Transport Enterprises (LTE) operates an integrated transportation network including both rail operations and road operations. LTE has two support departments and two transportation departments:

Support Departments	Transportation Departments
Equipment and maintenance (EM)	Rail (train) operations
Information systems (IS)	Road (truck) operations

The *budgeted* level of service relationships at the start of the year was:

Supplied By	Used By			
	EM	IS	Rail	Road
EM	—	0.10	0.30	0.60
IS	0.20	—	0.50	0.30

The *actual* level of service relationships for the year was:

Supplied By	Used By			
	EM	IS	Rail	Road
EM	—	0.20	0.40	0.40
IS	0.25	—	0.55	0.20

LTE collects fixed costs and variable costs of each service department in separate cost pools. The actual costs in each pool for the year were (in thousands):

	Fixed-Cost Pool	Variable-Cost Pool
EM	$300	$540
IS	80	75

Fixed costs are allocated on the basis of the *budgeted* level of service. Variable costs are allocated on the basis of the *actual* level of service.

LTE monitors the cost per track-mile for the rail department and the cost per road-mile for the road department. These cost figures include costs allocated from the support departments to the transportation departments. During the year, the actual transportation miles were:

Rail operations	15,000,000 miles
Road operations	12,000,000 miles

Required
1. Allocate the support department costs to the two transportation departments using the following three methods:
   a. Direct method

b. Step-down method (allocate EM first)

c. Reciprocal method

Show full details of calculations. Present results in a format similar to Exhibit 14-10. Allocate separately the variable and fixed service department costs.

2. Compare the service department total costs per transportation mile for the rail and road operations under each of the three methods in requirement 1. (Round to four decimal places.)

3. The prices LTE charges for rail are regulated by a government agency and set on a full-cost basis. Full costs includes allocations of service department costs. The road rates LTE sets are unregulated, and competition among road transportation operators is intense. What advice would you give the government regulatory agency about how to minimize the ability of rail transportation operators to overstate the service department costs included in their submissions to the agency about the full costs of their rail operations? Be specific.

14-35 Allocating costs of support departments; dual rates; direct, step-down, and reciprocal methods. Magnum T.A. Inc. specializes in the assembly and installation of high-quality security systems for the home and business segments of the market. The four departments at its highly automated state-of-the-art assembly plant are:

Service Departments	Assembly Departments
Engineering Support	Home Security Systems
Information Systems Support	Business Security Systems

The *budgeted* level of service relationships at the start of the year was:

	Used By			
Supplied By	**Engineering Support**	**Information Systems Support**	**Home Security Systems**	**Business Security Systems**
Engineering Support	—	0.10	0.40	0.50
Information Systems Support	0.20	—	0.30	0.50

The *actual* level of service relationships for the year was:

	Used By			
Supplied By	**Engineering Support**	**Information Systems Support**	**Home Security Systems**	**Business Security Systems**
Engineering Support	—	0.15	0.30	0.55
Information Systems Support	0.25	—	0.15	0.60

Magnum collects fixed costs and variable costs of each department in separate cost pools. The actual costs in each pool for the year were (in thousands):

	Fixed-Cost Pool	Variable-Cost Pool
Engineering Support	$2,700	$8,500
Information Systems Support	8,000	3,750

Fixed costs are allocated on the basis of the budgeted level of service. Variable costs are allocated on the basis of the actual level of service.

The support department costs allocated to each assembly department are allocated to products on the basis of units assembled. The units assembled in each department during the year were:

Home Security Systems	7,950 units
Business Security Systems	3,750 units

Required

1. Allocate the support department costs to the assembly departments using a dual-rate system and (a) the direct method, (b) the step-down method (allocate Information Systems Support first), (c) the step-down method (allocate Engineering Support first), and (d) the reciprocal method. Present results in a format similar to Exhibit 14-10.

2. Compare the support department costs allocated to each Home Security Systems unit assembled and each Business Security Systems unit assembled under a, b, c, and d in requirement 1.

3. What factors might explain the very limited adoption of the reciprocal method by many organizations?

14-36 Common cost allocation, theater facility with multiple users.   The Downtown Theater is owned by the city of Los Angeles. It has seating capacity of 2,500. Two companies use the theater.

◆ *Civic Light Opera Company:* Agreement with the city enables it to use the facilities on Friday, Saturday, and Sunday nights 50 weeks a year. It has used the Downtown Theater each year for the last 10 years.

◆ *Experimental Drama Company:* Agreement with the city enables it to use the facilities on Tuesday, Wednesday, and Thursday nights 50 weeks a year. This company was organized last year and has used the Downtown Theater for one year.

Data for the most recent year are:

	Civic Light Opera	Experimental Drama
Nights theater available	150	150
Nights theater used	120	80
Average attendance per night	2,000	500
Average price per ticket	$15	$10
Revenues	$3,600,000	$400,000
Cost identifiable with each company	$2,200,000	$250,000

The common costs of the Downtown Theater are the $200,000 annual fixed rent payment to the city and the annual fixed operating costs of $800,000.

If the Civic Light Opera Company were the only user of the theater, the fixed costs of the Downtown Theater would be $875,000 ($200,000 fixed rent payment + $675,000 fixed operating costs). Both companies can use the facilities of other theaters, providing they sign a one-year lease contract. The Civic Light Opera Company could use the Santa Monica Theater at an annual rent payment of $900,000. The Experimental Drama Company could use the University of Southern California Theater facility for an annual rent payment of $300,000.

Required

1. How will the common cost of $1,000,000 for the Downtown Theater be allocated between the two companies using the following allocation bases: (a) relative capacity (nights theater available), (b) relative use (nights theater used), (c) relative revenues, and (d) relative identifiable costs? What criterion might be invoked to justify methods a, b, c, and d individually?

2. How will the common cost of $1,000,000 be allocated using (e) the incremental common cost allocation method (assuming Civic Light Opera is the primary user) and (f) the stand-alone common cost allocation method? What criterion might be invoked to justify methods e and f individually?

3. What is the operating income of each of the two companies under methods a, b, c, d, e, and f? *Operating income* is defined as revenues minus the sum of identifiable costs and common cost allocations.

4. The mayor of Los Angeles proposes to use the incremental method as calculated in requirement 2. The manager of the Civic Light Opera Company comments that if the city wants to promote experimental drama, it should provide a direct subsidy to the Experimental Drama Company rather than use a cost allocation method that is unfair to the Civic Light Opera Company. How should the mayor respond to the manager of the Civic Light Opera Company?

14-37 Cost allocation for all cost categories in the value chain, different costs for different purposes.   Laser Technologies develops, assembles, and sells two product lines:

◆ Product Line A: laser scanning systems
◆ Product Line B: laser cutting tools

Product Line A is sold exclusively to the Department of Defense under a cost-plus reimbursement contract. Product Line B is sold to commercial organizations.

Laser Technologies classifies costs in each of its six value-chain business functions into two cost pools; direct product-line costs (separately traced to Product Line A or B) and indirect product-line costs. The indirect product-line costs are grouped into a single cost pool for each of the six functions of the value-chain cost structure:

Value-Chain Indirect Product-Line Cost Function	Base for Allocating Indirect Costs to Each Product Line
1. Research and development	Hours of R&D time identifiable with each product line
2. Design	Number of new products
3. Production	Hours of machine assembly time
4. Marketing	Number of salespeople
5. Distribution	Number of shipments
6. Customer service	Number of customer visits

Summary data in 19_4 are:

	Product Line A: Direct Costs (millions)	Product Line B: Direct Costs (millions)	Total Indirect Costs (millions)	Allocation Base for Indirect Costs	Product Line A Units of Allocation Base	Product Line B Units of Allocation Base
Research and development	$10.0	$5.0	$20.0	Hours of R&D time	6,000 hours	2,000 hours
Design	2.0	3.0	6.0	Number of new products	8 new products	4 new products
Manufacturing/ Production	15.0	13.0	24.0	Hours of machine assembly time	70,000 hours	50,000 hours
Marketing	6.0	5.0	7.0	Number of salespeople	25 people	45 people
Distribution	2.0	3.0	2.0	Number of shipments	600 shipments	1,400 shipments
Customer service	5.0	3.0	1.0	Number of customer visits	1,000 visits	4,000 visits

Required

1. For product pricing on its Product Line B, Laser Technologies sets a preliminary selling price of 140% of full cost (made up of both direct costs and the allocated indirect costs for all six of the value-chain cost categories). What is the average full cost per unit of the 2,000 units of Product Line B produced in 19_4?

2. For motivating managers, Laser Technologies separately classifies costs into three groups:

◆ Upstream (R&D and product design)
◆ Manufacturing
◆ Downstream (marketing, distribution, and customer service)

Calculate the costs (direct and indirect) in each of these three groups for Product Lines A and B.

3. For the purpose of income and asset measurement for reporting to external parties, inventoriable costs under generally accepted accounting principles for Laser Technologies include manufacturing costs and product design costs (both direct and indirect costs of

each category). At the end of 19_4, what is the average inventoriable cost for the 300 units of Product Line B on hand? (Assume zero beginning inventories.)

4. The Department of Defense purchases all Product Line A units assembled by Laser Technologies. Laser is reimbursed 120% of allowable costs. *Allowable cost* is defined to include all direct and indirect costs in the research and development, product design, manufacturing, distribution, and customer-service functions. Laser Technologies employs a marketing staff that makes many visits to government officials, but the Department of Defense will not reimburse Laser for any marketing costs. What is the 19_4 allowable cost for Product Line A?

5. "Differences in the costs appropriate for different decisions, such as pricing and cost reimbursement, are so great that firms should have multiple accounting systems rather than a single accounting system." Do you agree?

14-38   Division cost allocation, R&D, ethics.   World Semiconductor (WS) has eight divisions. It has a central research and development group in San Diego that conducts contract research for each of these eight divisions. At the start of each year, each division estimates the hours of research-scientist time at the San Diego group it will use in the coming year. These estimates are added up for all divisions. Each division is charged for budgeted overhead costs incurred at the San Diego facility on the basis of its relative budgeted percentage use of research-scientist time in the coming year. Central R&D bears the risk of any overruns on overhead costs during the year. Each division also pays (in 19_4) the San Diego facility $100 per hour of research-scientist time and the actual costs of any materials used on the project.

Toni Goodwin is the controller of the Applied Semiconductor Division (ASD), which is based in Tuscon, Arizona. She notes that in the first nine months of 19_4, ASD was charged $12.597 million for contract research at the San Diego facility.

Research-scientist time	$ 2,564,000
Materials and other direct charges	2,883,000
Overhead cost charge	
(22% of $32,500,000)	7,150,000
	$12,597,000

It is now time to prepare the 19_5 budget. Goodwin estimates that ASD will have a 19_5 budget of 30,000 hours of research-scientist time at the San Diego facility. This estimate is based on detailed interviews she has had with operating managers at ASD and on a recent ASD retreat, at which the strategy and operations for 19_5 were finalized. Roy Masters, the new president of ASD, is less than pleased with the 30,000 budget number. Goodwin and Masters have the following conversation.

GOODWIN: But Roy, you were at the retreat where we all signed off on the 30,000 number.

MASTERS: I was there, but I think "signed off" is too strong a phrase. By all means use the 30,000 number in our internal planning and budgeting at ASD. However, I want you to tell San Diego that we are budgeting for only 25,000 hours in 19_5.

GOODWIN: But . . .

MASTERS: But nothing, Toni. Everyone plays games in this company. This is the fourth division of World Semiconductor I have worked in. I know for a fact that in all my three prior divisions, we deliberately understated budgeted usage of research scientists to the San Diego people at the start of each year. Anyway, San Diego always artificially inflates its estimate of overhead costs for the coming year. They do it every year. Anyone who thinks this is a level playing field is more naive than my dog.

GOODWIN: Roy, I have to think about this.

MASTERS: Don't think too long Toni. I want the senior managers on my team to be team players. The issue you face, Toni, is whether you want to remain on the team.

Required

1. Why might Masters want Goodwin to report 25,000 budgeted hours rather than 30,000 budgeted hours to San Diego?

2. What steps might San Diego take to reduce WS divisions' understating their budgeted usage of San Diego research-scientist time?

3. What should Goodwin do? In your answer, refer to "Standards of Ethical Conduct for Management Accountants" (p. 17).

15

Cost Allocation: II

- Cost tracing and cost allocation
- Choosing indirect cost pools and cost rates
- Evolving trends in cost assignment
- Used and unused capacity distinctions
- Cost justification and reimbursement
- Cost assignment and cost hierarchies

Companies are increasingly using cost drivers as their chosen allocation bases. The Saturn Corporation of General Motors uses six cost drivers when allocating manufacturing overhead costs to individual cars. These cost drivers include planned assembly time, load time, and total investment value of tooling.

Chapter 14 introduced four purposes of cost allocation and discussed the allocation of costs to departments. Chapter 15 continues our exploration of cost allocation. We now examine the allocation of costs to individual products, services, or jobs (such as a government contract). These costs can include the already allocated costs of corporate and other support departments. The final sections of Chapter 15 discuss the cost justification or reimbursement purpose of cost allocation and the role of cost allocations in cost hierarchies.

COST TRACING AND COST ALLOCATION

We will use the costing of a deluxe refrigerator model built by Consumer Appliances Inc. (CAI) to continue our illustration of cost-allocation issues. CAI assembles this refrigerator, along with eight other products, at its Windsor, Ontario, plant. It uses its own sales force to sell refrigerators to retail department stores. CAI employs the five-step approach to costing outlined in Chapter 4 (pp. 102–103) and Chapter 5 (pp. 141–142).

STEP 1: *IDENTIFY THE PRODUCT THAT IS THE CHOSEN COST OBJECT.* The cost object is a deluxe refrigerator model.

STEP 2: *IDENTIFY THE DIRECT COST CATEGORIES FOR THE PRODUCT.* CAI identifies three categories of direct costs. One direct cost is for customer warranty. CAI pays a third-party electrical goods repair company $75 per refrigerator to handle customer requests for service during the 24-month warranty period. At the Windsor manufacturing plant, there are two direct costs—direct materials and direct manufacturing labor. Amounts traced to each deluxe refrigerator model are:

Customer warranty	$ 75
Direct materials	140
Direct manufacturing labor	35
Total direct costs	$250

STEP 3: *IDENTIFY THE INDIRECT COST POOLS ASSOCIATED WITH THE PRODUCT.* CAI identifies six indirect cost pools associated with the manufacturing and sale of the deluxe refrigerator. These six pools are listed below in step 5.

STEP 4: *SELECT THE COST ALLOCATION BASE TO USE IN ASSIGNING EACH INDIRECT COST POOL TO THE PRODUCT.* The chosen cost-allocation bases are also listed below in step 5.

STEP 5: *DEVELOP THE RATE PER UNIT OF THE COST ALLOCATION BASE USED TO ALLOCATE INDIRECT COSTS TO THE PRODUCT.* The allocation base and rate for each indirect cost pool in the January to June 19_5 period is:

Indirect Cost Pool	Allocation Base	Allocation Rate
1. Procurement	Number of parts	$0.50 per part
2. Production: labor-paced assembly	Direct manufacturing labor-hours	$20 per hour
3. Production: machine-paced assembly	Machine-hours	$16 per hour
4. Production: quality testing	Testing-hours	$30 per hour
5. Distribution	Cubic feet	$2 per cubic foot
6. Marketing	Units sold	$70 per unit

These rates are revised every six months.

The allocation rate for each indirect cost pool is calculated as:

$$\frac{\text{Budgeted indirect}}{\text{cost rate}} = \frac{\text{Budgeted total costs in indirect cost pool}}{\text{Budgeted total quantity of cost-allocation base}}$$

For example, the procurement allocation rate of $0.50 per part is computed as follows:

$$\frac{\$2,000,000}{4,000,000 \text{ parts}} = \$0.50 \text{ per part}$$

NUMERATOR. The budgeted total costs for the procurement cost pool in the January to June 19_5 period are $2,000,000. This amount includes costs for labor in the procurement department, for their equipment (for example, computers), and for the handling and inspection of incoming materials.

DENOMINATOR. The budgeted total quantity of the allocation base is 4,000,000 parts. This figure is the budgeted number of parts for all products assembled at the Windsor plant in the January to June 19_5 period. It includes a budget of 252,000 parts for the deluxe refrigerator model (84 parts per refrigerator times the 3,000 budgeted production units of refrigerators). The remaining 3,748,000 parts included in the denominator are for other products.

An overview of the product-costing system for CAI appears in Exhibit 15-1. The allocation rate for each indirect cost pool is computed in a way similar to that outlined for the procurement indirect cost pool. Exhibit 15-2 presents the product-cost buildup for the deluxe refrigerator model. The full product cost is $530, consisting of $250 direct costs and $280 indirect costs. The $280 amount includes $150 indirect cost for distribution and marketing costs. In computations of the inventoriable product cost (for financial reporting to external parties) only manufacturing costs are included, so the $75 direct cost for customer warranty and the $150 indirect cost for distribution and marketing are excluded. Thus, the inventoriable product cost is $305 per deluxe refrigerator model ($530 − $75 − $150). Exhibit 15-2 reinforces the "different costs for different purposes" notion. The $530 figure captures the full set of cost categories CAI must cover in its pricing if it is to remain a profitable organization (see Chapter 12). For financial reporting, however, generally accepted accounting principles prohibit the inclusion of costs "downstream" to manufacturing (distribution, marketing, and customer service) in the inventoriable product cost figure. Note that this exclusion of downstream costs pertains to both direct costs and indirect costs.

EXHIBIT 15-1

Overview of Product Costing at Consumer Appliances, Inc.

INDIRECT COST POOLS	Procurement	Production: Labor-Paced Assembly	Production: Machine-Paced Assembly	Production: Quality Testing	Distribution	Marketing

COST ALLOCATION BASES	Number of Parts	Direct Manufacturing Labor-Hours	Machine-Hours	Testing-Hours	Cubic Feet	Units Sold

COST OBJECT: DELUXE REFRIGERATOR

INDIRECT PRODUCT COSTS

DIRECT PRODUCT COSTS

DIRECT COSTS

Direct Materials | Direct Manufacturing Labor | Customer Warranty

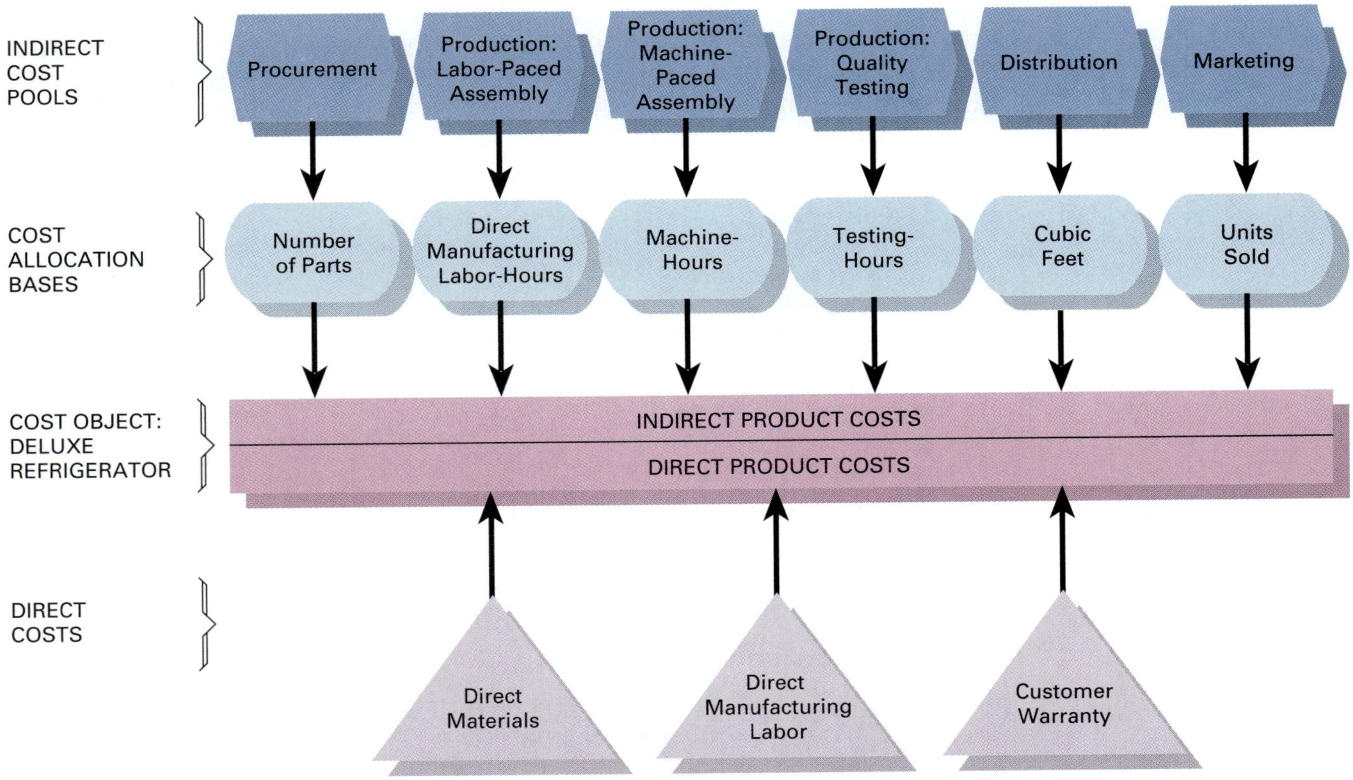

EXHIBIT 15-2

Costing of Deluxe Refrigerator Model of Consumer Appliances, Inc.

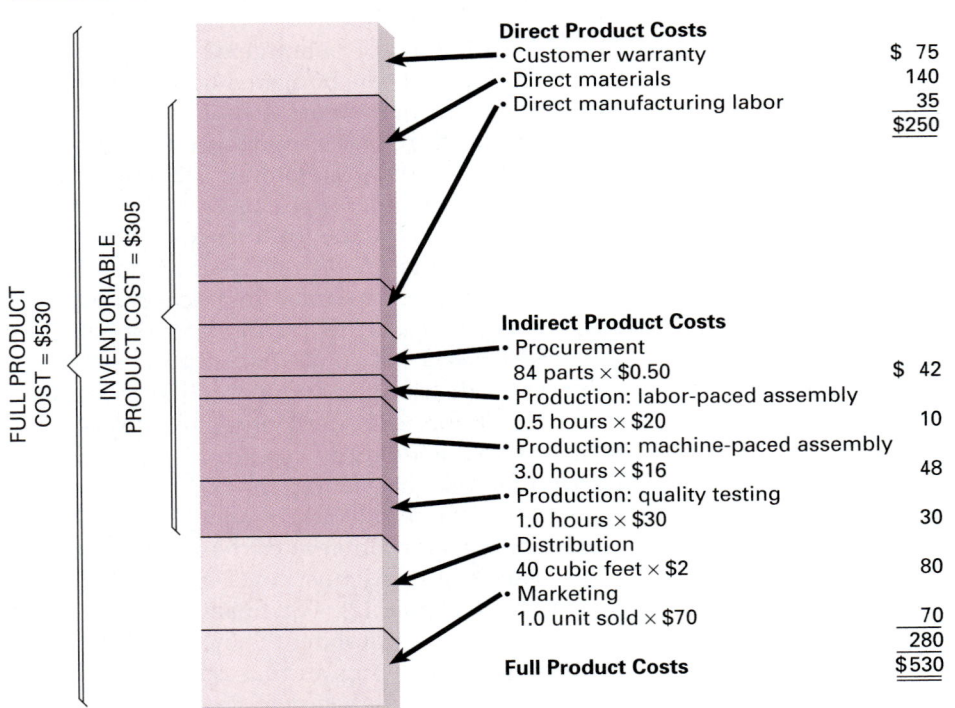

Direct Product Costs
- Customer warranty $ 75
- Direct materials 140
- Direct manufacturing labor 35
$250

Indirect Product Costs
- Procurement
  84 parts × $0.50 $ 42
- Production: labor-paced assembly
  0.5 hours × $20 10
- Production: machine-paced assembly
  3.0 hours × $16 48
- Production: quality testing
  1.0 hours × $30 30
- Distribution
  40 cubic feet × $2 80
- Marketing
  1.0 unit sold × $70 70
  280
Full Product Costs $530

FULL PRODUCT COST = $530

INVENTORIABLE PRODUCT COST = $305

CHOOSING INDIRECT COST POOLS AND COST RATES

Indirect costs allocated to individual products are a function of:

◆ the aggregate costs in each indirect cost pool,

◆ the allocation cost rate used for each indirect cost pool, and

◆ the allocation cost base used for each indirect cost pool.

This section discusses issues related to the choice of cost pools and cost rates. Subsequent sections discuss the choice of cost allocation bases.

The concept of *cost pool homogeneity*, discussed in Chapter 14, is central to the issues raised in this section. A cost pool is homogeneous if all activities whose costs are included in it have the same or a similar cause-and-effect or benefits-received relationship between the cost driver and the costs of the activity. Consider the cost of health benefits paid to employees in our Consumer Appliances example. Health-benefits costs are collected in a single cost pool and then allocated to each of the six indirect cost pools in Exhibit 15-1. The health-benefits cost pool includes health benefits paid to workers in the materials procurement, production, distribution, and marketing areas. Why? Because the same health-benefits rate is paid to a health-insurance group for every employee, the cost driver.

Plantwide Rates versus Department Rates

If a company produces many products, should it use a single plantwide manufacturing overhead rate or individual department manufacturing overhead rates? Two important questions arise:

1. Do individual departments *differ* in the cause-and-effect relationship of the department's manufacturing overhead costs and the driver of these costs?
2. Do individual products *differ* in the way they use individual departments in the plant?

If these differences are sizable, department overhead rates will provide more accurate product-cost figures than plantwide overhead rates. When these differences are minimal, department overhead rates will result in product-cost numbers similar to those computed using plantwide overhead rates.

Consider product costing at CAI's Windsor production plant. This production plant has three individual departments:

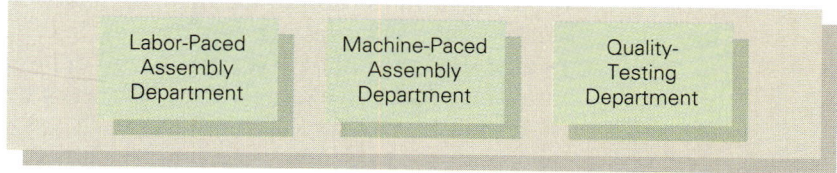

For many years, Consumer Appliances used a single overhead rate plantwide, based on direct manufacturing labor-hours in the labor-paced assembly area. When the plant was first built, labor-paced assembly was the largest area of the plant. In recent years, the size of the labor-paced assembly area has decreased as machine-paced assembly lines have grown, and CAI has responded by changing its costing system. The current product-costing approach has three department overhead rates for the production plant:

Production Cost Pool	Cost Allocation Rate
Labor-paced assembly	$20 per direct manufacturing labor-hour
Machine-paced assembly	$16 per machine-hour
Quality testing	$30 per testing-hour

The plantwide overhead rate for the current period would have been $100 per direct manufacturing labor-hour in the previous system.

The effect of using three department cost pools, compared with a single plantwide cost pool, can be illustrated with the example of two products—a deluxe refrigerator and a clothes dryer. These two products use the following resources from the three departments:

	Deluxe Refrigerator	Clothes Dryer
Labor-paced assembly department: direct manufacturing labor-hours	0.6 hours	0.8 hours
Machine-paced assembly department: machine-hours	4.0 hours	1.5 hours
Quality-testing department: testing-hours	3.0 hours	0.4 hours

The manufacturing overhead costs allocated to these two products using the plantwide rate and the department rates follow:

	Deluxe Refrigerator	Clothes Dryer
Plantwide manufacturing overhead rate		
0.6 direct manufacturing labor-hours × $100	$ 60	
0.8 direct manufacturing labor-hours × $100		$80
Department manufacturing overhead rates		
Labor-paced assembly department		
0.6 direct manufacturing labor-hours × $20	$ 12	
0.8 direct manufacturing labor-hours × $20		$16
Machine-paced assembly department		
4.0 machine assembly hours × $16	64	
1.5 machine assembly hours × $16		24
Quality-testing department		
3.0 testing-hours × $30	90	
0.4 testing-hours × $30		12
	$166	$52

The manufacturing overhead allocated to each deluxe refrigerator is $166 with department rates and $60 with a plantwide rate. Why the large difference? Because the department rates capture the deluxe's relatively high use of the machine-paced assembly and quality-testing departments. In contrast, the manufacturing overhead allocated to each clothes dryer is $52 with department rates and $80 with a plantwide rate. This product makes relatively low use of the machine-paced assembly and quality-testing areas. Note, however, that each clothes dryer makes relatively high use of the labor-paced assembly area, and recall that labor-hours was the allocation base for the plantwide rate. Indeed, one of the reasons CAI recently adopted department overhead rates was a complaint from the clothes-dryer products manager. She argued that the plantwide rate penalized her product line, making it appear that CAI was losing money on a product line she believed was a "winner." Top management accepted her argument and directed operating personnel to analyze the cause-and-effect cost relationships. The three different allocation bases in each department—manufacturing labor-hours, machine assembly hours, and testing-hours—resulted from that effort.

When based on the cause-and-effect criterion, department rates are preferable to a plantwide rate for allocating CAI's manufacturing overhead to products. The two products differ in their use of each of the three departments; use of a single cost-allocation base does not capture these differences. Many companies have observed similar differences among their manufacturing departments and their products. These differences are motivating the increased use of multiple rates rather than a single plantwide rate.

This discussion of plantwide versus department cost pools focused on manufacturing. However, it applies equally to other parts of the value chain, such as distribution, marketing, and customer service. The use of multiple indirect cost rates rather than a single cost rate will result in more accurate product costs when individual departments or activity areas represented by the multiple cost rates differ in cause-and-effect relationships and individual products differ in the way they use the individual departments or activity areas.

EVOLVING TRENDS IN COST ASSIGNMENT

Objective 4
Understand different trends occurring in cost assignment

When changes occur in information-gathering technology or in the underlying operations (such as an increase in automation or a change in the products manufactured), changes in the management accounting system should be considered. We now look at several trends in this area.

Interest in tracing direct costs is increasing. The use of more advanced information-gathering technologies is enabling companies to trace some cost items directly to products rather than allocating them using a cost-allocation base.

The number of separate indirect (overhead) cost pools is increasing. For example, some companies are now using a separate machine overhead cost pool when allocating machine costs to products. The CAI example discussed in this chapter illustrates a company adopting machine assembly hours as an allocation base for a department that previously used manufacturing labor-hours.

Machining costs include energy, lubricants, maintenance, repair, and depreciation. Depreciation is one of the largest cost items in a machining cost pool. Businesses routinely compute depreciation for income tax purposes. Tax-based depreciation estimates, however, may not reflect the economic decline in the service potential of machines. To refine estimates of a machine's useful life may require the expertise of industrial engineers, whose time may be a significant cost to the company.

One company making airplane blades and other metal parts uses multiple machining cost pools. The various machines used for cutting metal differ considerably in their capital costs, operating costs, and speed; the more expensive machines cut metal at a faster speed. Using three separate machine cost pools instead of a single machine cost pool, the company now allocates machining costs on the basis of a rate per machine-hour that better measures the cost and speed of the machine being used on a product. Different product costs are now reported for products that require the same total number of machine-hours but differ in their relative use of machines of different speeds.

A survey of U.K. companies reported that only 26% of companies used a single plantwide overhead rate; 31% used separate overhead rates for each department, and 38% used separate overhead rates for work centers within each department (5% used none of the above).[1] A work center is to a department as a department is to a plant. Different machine rates are an example of work-center overhead rates.

Companies in labor-paced work environments differ in their choice of allocation bases from companies in machine-paced work environments.

◆ **Labor-Paced Work Environment.** Worker dexterity and productivity determine the speed of production. Machines function as tools that aid production workers. In labor-paced environments, direct manufacturing labor costs or direct manufacturing labor-hours may still capture cause-and-effect relationships, even if operations are highly automated.

◆ **Machine-Paced Work Environment.** Machines conduct most (or all) phases of production, such as movement of materials to the production line, assembly and other activities on the production line, and shipment of finished goods to the delivery dock areas. Machine workers in such environments may simultaneously operate more than one machine. Workers focus their efforts on supervision of the production line and general trouble shooting rather than on

[1]C. Drury, S. Braund, P. Osborne, and M. Tayles, *A Survey of Management Accounting Practices in U.K. Manufacturing Companies* (London: Chartered Association of Certified Accountants, 1993).

the actual operation of the machines. Computer specialists and industrial engineers are the real "controllers" of the speed of production. In machine-paced environments, machine-hours likely will better capture cause-and-effect relationships than direct labor-hours.

A questionnaire on the use of manufacturing overhead allocation bases was sent to U.S. and Japanese companies. The results follow:[2]

Allocation Base	Labor-Paced Environments		Machine-Paced Environments	
	U.S.	Japan	U.S.	Japan
Standard labor-hours	32%	53%	10%	10%
Actual labor-hours	25	53	6	18
Standard labor costs	23	20	5	0
Actual labor costs	23	18	6	8
Standard machine-hours	7	0	24	38
Actual machine-hours	7	0	15	30
Time in work center	5	3	8	13

These percentages exceed 100% because many companies use multiple overhead allocation bases.

Increasingly, companies are turning to nonfinancial allocation bases. For example, several companies are now experimenting with manufacturing lead time as an allocation base. **Manufacturing lead time** is the time from when an order is ready to start on the production line to when it becomes a finished good. Any time delays, either at the start of production or during production, are included in the manufacturing lead time measure. The rationale for using this allocation base is that steps that lengthen manufacturing lead time frequently add to the indirect costs at the plant. For example, moving partly assembled products into and out of a work-in-process inventory area increases lead time and increases materials-handling costs. By using this allocation base, management signals to operating personnel that reported product costs can be reduced by shortening the lead time of products being assembled. The Problem for Self-Study at the end of this chapter illustrates use of the lead time allocation base for a medical instruments company.

Companies in the retail sector are also expanding their use of nonfinancial cost-allocation bases. For example, the cubic volume of a product is a driver of distribution and handling costs in the retail sector. Some retailers are now using cubic volume to allocate distribution and handling costs to individual products. Previously, product costs included only the purchase cost of goods sold. The inclusion of distribution and handling costs in product costs enables managers to gain more insight into the relative contribution to operating income of products with the same gross margin but different cubic volumes.[3]

Consequences of an Inappropriate Allocation Base

Objective 5

Outline the consequences of the inappropriate use of an allocation base

Cost figures play a key role in many important decisions. If these figures result from allocation bases that fail to capture cause-and-effect relationships, managers may make decisions that conflict with maximizing long-run company net income. Consider the use of direct manufacturing labor costs as an allocation base in machine-paced manufacturing settings. In this environment, indirect cost rates of 500% of direct manufacturing labor costs (or more) may be encountered. Thus, every $1 of indirect manufacturing labor costs has a $6 impact ($1 in direct costs + 500% per $1 in indirect costs). Possible negative consequences include the following points.

[2]National Association of Accountants Tokyo Affiliate, "Management Accounting in the Advanced Manufacturing Environment: Comparative Study on Survey in Japan and U.S.A.," (1988).

[3]See N. Coulthurst, "Management Accounting in Retail Organizations," in C. Drury (ed.), *Management Accounting Handbook* (London: Chartered Institute of Management Accountants, 1992).

1. Product managers may make excessive use of external vendors for parts with a high direct manufacturing labor content.

2. Manufacturing managers may pay excessive attention to controlling direct manufacturing labor-hours relative to the attention paid to controlling the more costly categories of materials and machining. By eliminating $1 of direct manufacturing labor costs, $6 of reported product cost can be eliminated. Managers can control much of the accounting amounts allocated by controlling direct labor use. However, this action does not control the actual incurrence of the larger materials and machining costs.

3. Managers may attempt to classify shop-floor personnel as indirect labor rather than as direct labor. As a result, part of these labor costs will be allocated (inappropriately) to other products.

4. Products may be undercosted or overcosted. The danger then arises that a company will push to gain market share on products that it believes are profitable when in fact the company loses money on them. Similarly, the company may neglect products that are profitable because it believes it is losing money on them.

Surveys of Company Practice

IS THE PRODUCT-COSTING SYSTEM BROKEN?

A viewer knows when a television set no longer works. A driver knows when a motor vehicle no longer starts. The breakdown of many products is easy to detect. The breakdown of a product-costing system is not. Nonetheless, guidelines for assessing whether a product-costing system is broken do exist. Robin Cooper of the Claremont Graduate School offers such a set of guidelines.[a] Although no one individual guideline is conclusive, collectively they can flag the need for a detailed review of an existing product-costing system. The following four questions focus on these guidelines. The responses are from a sample of Dutch companies:[b]

Guideline	Yes	No
1. Can managers easily explain changes in profit margins from one period to the next? (If they cannot, one explanation is that the existing system is broken.)	70%	30%
2. Can managers easily explain why their bids for business are successful or unsuccessful? (If they cannot, one explanation is that the costing system is broken.)	64	36
3. Does the costing system have a small number of cost pools, and are the items in each cost pool heterogeneous? (Reducing heterogeneity will require an increase in the number of cost pools.)	56	44
4. Are competitors pricing their high-volume products comparable to ours at prices substantially lower than our cost figure? (One explanation is that we are overcosting these products.)	54	46

The responses indicate a sizable percentage of companies saying no to questions 1 and 2 and yes to questions 3 and 4. These are the "red-flag" responses to the overall question of whether a product-costing system is broken. An individual company that gave a red-flag response to all four of these questions should quickly examine whether its existing costing system should be significantly changed.

[a]Cooper, "Does Your Company." [b]Boons and Roozen, "Symptoms." Full citations are in Appendix A.

Cost Drivers and Allocation Bases

When a cause-and-effect criterion is used, the chosen allocation bases are likely to be cost drivers. Because a change in a cost driver causes a change in the total cost of a related cost object, the use of cost drivers as allocation bases increases the accuracy of reported product costs. However, not all cost-allocation bases that organizations select are cost drivers. Consider the following reasons for using bases that are not cost drivers.

1. Improving the accuracy of individual product costs may be less important to an organization than other goals. Think about the goal of restraining the growth in headcount (the number of employees on a company's payroll). Several Japanese companies use direct manufacturing labor-hours as the cost-allocation base. These companies acknowledge that direct manufacturing labor is not the most important driver of their manufacturing overhead costs, but they want to send a clear signal to all managers that reduction in headcount is a key goal.[4]

Managers may prefer direct manufacturing labor-hours as an allocation base even when they are aware that direct manufacturing labor-hours do not drive short-run indirect costs. Why? To promote increased levels of automation. Using this allocation base, managers motivate product designers to decrease the direct manufacturing labor content of the products they design. Management may view increased automation as a strategic necessity to remain competitive in the long run.

2. Information about cost-driver variables may not be reliably measured on an ongoing basis. For example, managers often view the number of machine setups as a driver of indirect manufacturing costs, but some firms do not systematically record this information.

3. Accounting systems with many indirect cost pools and allocation bases are more expensive to use than systems with few cost pools and allocation bases. The investment required to develop and implement a system with many indirect cost pools—and to educate users about it—can be sizable. Unfortunately, some firms place a low priority on investments in their internal accounting systems, given that the benefits from such investments are frequently difficult to quantify.

USED AND UNUSED CAPACITY DISTINCTIONS

Budgeted indirect cost rates were computed in Chapters 4, 5, and 10 using the following formula:

$$\text{Budgeted indirect cost rate} = \frac{\text{Budgeted total costs in indirect cost pool}}{\text{Budgeted total quantity of cost-allocation base}}$$

Managers choose from among several alternative ways of computing the denominator in this formula. The following alternatives tie back into our discussion of two denominator-level concepts in Chapter 9 (pp. 322–24):

◆ Practical capacity: theoretical capacity adjusted for unavoidable operating interruptions, such as scheduled maintenance time, shutdowns for holidays, and so on.

◆ Master-budget utilization: the anticipated level of capacity utilization for the coming budget period.

In many cases, master-budget utilization is below the practical capacity. This section discusses how resource allocation in an organization can be affected by cost-allocation decisions related to capacity issues.

---

[4]See T. Hiromoto, "Another Hidden Edge—Japanese Management Accounting," *Harvard Business Review* (July–August 1988), pp. 22–26.

Capacity's Effect on Indirect Cost Rates

Rightway Foodmarkets is a national chain of supermarkets. For many years it has used Barton Transport, an independent company, to distribute refrigerated produce from its warehouses to each of its supermarkets. In 19_4 Barton Transport charged Rightway (and other comparable supermarket chains) $4 per ton-mile distributed. (A "ton-mile" is a unit measuring one ton moved one mile.) In July 19_4, Barton drivers went on strike for 3 weeks. One result was that Rightway had smaller amounts of produce delivered to each of its supermarkets.

Rightway decided to develop its own distribution network. It had two options:

1. develop its own distribution function (drivers, trucks, and so on) from the "ground up" or
2. acquire an existing trucking company.

Rightway opted for (2). In November 19_4 it acquired National Roadfreight, a superbly run trucking company.

Rightway's budgeted distribution requirements for 19_5 are 5,000,000 ton-miles. Rightway's National Roadfreight subsidiary has the practical capacity to transport 8,000,000 ton-miles in 19_5. Rightway expects to grow and so use the full 8,000,000 ton-miles capacity by 19_7. Budgeted 19_5 distribution costs for Rightway's National Roadfreight subsidiary are:

- ◆ Variable operating costs = $2.10 per ton-mile
- ◆ Fixed costs = $12,000,000

Each supermarket in the Rightway chain is charged for the cost of produce delivered to it. How should Rightway compute this cost? The main debate at Rightway is over the choice of the denominator to use for the fixed costs. The budgeted fixed costs are $2.40 per ton-mile if budgeted usage is used ($12,000,000 ÷ 5,000,000 ton-miles) and $1.50 per ton-mile if budgeted practical capacity is used as the denominator ($12,000,000 ÷ 8,000,000 ton-miles).

	Approach 1: Budgeted 19_5 Usage as Denominators	Approach 2: Budgeted 19_5 Practical Capacity as Denominator
Variable costs	$2.10	$2.10
Fixed costs	2.40	1.50
Total	$4.50	$3.60

The difference between approach 1 and approach 2 centers on the 3,000,000 ton-miles of budgeted unused capacity in 19_5.

In approach 1, this 3,000,000 ton-miles of unused capacity is excluded from the fixed costs per ton-mile computation. Hence, the fixed costs of $12,000,000 are fully charged out to the supermarkets in 19_5 as a cost of distribution. The manager of a supermarket would see the distribution cost increase from $4 per ton-mile in 19_4 (using Barton Transport) to $4.50 in 19_5 (using Rightway's transport company).

In approach 2, the 3,000,000 ton-mile unused capacity is included in the fixed costs per ton-mile computation. This approach results in $7,500,000 of 19_5 fixed costs—(5,000,000 ton-miles ÷ 8,000,000 capacity) × $12,000,000—being allocated to existing supermarkets as a distribution cost in 19_5. Because the remaining $4.5 million of fixed costs is excluded, the fixed costs that the Rightway supermarket manager faces drops to $3.60. Where does the remaining $4,500,000 of 19_5 fixed costs appear in the accounting system? One answer is to have a line item in the income statement for the fixed costs allocated to unused capacity:

Fixed costs:	
Used capacity	$ 7,500,000
Unused capacity	4,500,000
	$12,000,000

Senior managers of Rightway made the strategic decision to acquire a trucking company with capacity well above its 19_5 internal requirements. Reporting the $4,500,000 amount separately highlights that Rightway has this sizable excess capacity in its trucking subsidiary.

Many organizations have excess capacity in their manufacturing plants, their distribution networks, their sales forces, and elsewhere. Separate tracking of the costs of used capacity and unused capacity leads to a better understanding of cost-behavior patterns. Moreover, this separate tracking assists in evaluating managers at different levels who have different areas of responsibility. The decision to acquire Rightway was a strategic one, made with the full knowledge that 3,000,000 ton-miles of unused capacity would exist in 19_5. The separate tracking of this $4,500,000 unused-capacity cost might prompt senior managers to seek ways for using this capacity.

Unused Capacity Costs and the Downward Demand Spiral

Objective 6

Explain how attempts to recover fixed costs may lead to a downward demand spiral

Companies that use a cost-based approach to pricing face difficult issues when capacity costs are high and sizable excess capacity exists. Consider again our Rightway example. Assume now that each supermarket manager has the option to use the Rightway subsidiary (National Roadfreight) or to use Barton Transport. In 19_5 the Barton Transport rate is $4.10 per ton-mile. If Rightway computes its internal cost with budgeted usage as the denominator for fixed costs, the quoted price will be $4.50 per ton-mile. Suppose now that some supermarket managers decide to use Barton Transport. As a result, the budgeted denominator drops from 5,000,000 ton-miles to 4,000,000 ton-miles. The total cost per ton-mile increases from $4.50 to $5.10, as shown in the following table. This increase in total cost may spur more Rightway managers to shift to Barton, which has a $4.10 per ton-mile rate. Again the total cost per ton-mile increases as the denominator base (budgeted usage) is reduced, as shown by the following numbers.

Budgeted Denominator (Ton-Miles) (1)	Variable Costs per Ton-Mile (2)	Fixed Costs per Ton-Mile [$12,000,000 ÷ (1)] (3)	Total Costs per Ton-Mile (4)
8,000,000	$2.10	$1.50	$3.60
5,000,000	2.10	2.40	4.50
4,000,000	2.10	3.00	5.10
3,000,000	2.10	4.00	6.10
2,000,000	2.10	6.00	8.10

As Rightway revises its total costs per ton-mile to allocate the fixed costs over a smaller and smaller number of ton-miles, a higher revised costs per ton-mile occurs. More and more supermarkets likely would turn to Barton Transport rather than use the Rightway subsidiary.

The **downward demand spiral** (also called the **black hole demand spiral**) refers to the continuing reduction in demand that occurs when prices are raised and then raised again in an attempt to recover fixed costs from an ever-decreasing customer base (in our example, the Rightway managers of individual supermarkets).

COST JUSTIFICATION AND REIMBURSEMENT

Cost allocation issues arise in many contracting situations. Examples include:

1. A contract between the Department of Defense and a company designing and assembling a new fighter plane; the price paid is based on the contractor's cost plus a preset fixed fee.

2. A research contract between a university and a government agency; the university is reimbursed its direct costs plus an overhead rate that is a percentage of direct costs.

3. A contract between two oil companies in a joint venture; the costs of operating a shared oil-refining facility are allocated between the companies on the basis of expected usage of the refinery.

4. A contract between an energy-consulting firm and a hospital; the consulting firm receives a fixed fee plus a share of the energy-cost savings arising from the consulting firm's recommendations.

The areas of dispute between parties to such contracts can be reduced by making the "rules of the game" explicit (and preferably written) and well understood at the time the contract is signed. Such rules of the game include the definition of cost items allowed, the cost pools, and the permissible allocation bases.

Contracting with the U.S. Government

The U.S. government reimburses most contractors in one of two ways:[5]

1. *Contractor paid a preset price without analysis of actual contract cost data.* This approach is used, for example, where there is competitive bidding, where there is adequate price competition, or where there is an established catalog with prices quoted for items sold in substantial quantities to the general public.

2. *Contractor paid after analysis of actual contract cost data.* In some cases, the contract will explicitly state that reimbursement is based on actual allowable costs plus a fixed fee. This arrangement is a **cost-plus contract.** In other cases, the contractor is paid a preset fixed price, provided that a government contracting officer views this price as reasonable (that is, close to actual costs).

All contracts with any U.S. government agency must comply with cost accounting standards issued by the **Cost Accounting Standards Board (CASB).** The CASB was established in 1970 against a backdrop of systematic cost overruns by contractors on cost-reimbursement contracts and allegations that contractors were overstating their costs. In 1980, the CASB ceased operations because Congress refused to appropriate money for its annual budget. In 1988, the CASB was recreated as an independent board within the Office of Federal Procurement Policy. It has the exclusive authority to "make, promulgate, amend, and rescind cost accounting standards and interpretations thereof designed to achieve *uniformity* and *consistency* in the cost accounting standards governing measurement, assignment, and allocation of costs to contracts within the United States."

CASB Standards and Interpretations

Exhibit 15-3 lists the titles of selected accounting standards issued by the CASB that are incorporated into **Federal Acquisition Regulations (FARs).** FARs are procurement regulations with which companies contracting with the U.S. government must comply. Several CASB standards (such as 402) cover general issues related to the definition of cost items, consistency, and the prohibition of double-counting. **Double-counting** occurs when a cost item is included both as a direct cost item and as part of an indirect cost pool allocated to the contract using a budgeted rate. Other standards cover the allocation of indirect costs to government contracts.

Standard 403, Allocation of Home Office Expenses to Segments, illustrates the CASB approach to determining cost pools and the allocation bases for these pools. The thrust of the standard requires home-office expenses to be allocated on the basis of the beneficial or causal relationship between supporting and receiving activities. The standard sets forth the following hierarchy of allocation techniques for centralized services:

---

[5]For a detailed discussion of the issues in this section see F. Alston, M. Worthington, and L. Goldsman, *Contracting with the Federal Government*, 3d ed. (New York: Wiley, 1993).

EXHIBIT 15-3

Cost Accounting Standards of CASB Incorporated into Federal Acquisition Regulations

CAS 402	Consistency in Allocating Costs Incurred for the Same Purpose
CAS 403	Allocation of Home Office Expenses to Segments
CAS 404	Capitalization of Tangible Assets
CAS 405	Accounting for Unallowable Costs
CAS 406	Cost Accounting Period
CAS 409	Depreciation of Tangible Capital Assets
CAS 410	Allocation of Business Unit General and Administrative Expense to Final Cost Objectives
CAS 412	Composition and Measurement of Pension Costs
CAS 413	Adjustment and Allocation of Pension Cost
CAS 414	Cost of Money as an Element of the Cost of Facilities Capital
CAS 415	Accounting for the Cost of Deferred Compensation
CAS 416	Accounting for Insurance Costs
CAS 417	Cost of Money as an Element of the Cost of Capital Assets under Construction
CAS 418	Allocation of Direct and Indirect Costs
CAS 420	Accounting for Independent Research and Development Costs and Bid and Proposal Costs

Source: F. Alston, M. Worthington, and L. Goldsman, *Contracting with the Federal Government*, 3d ed. (New York: Wiley, 1993).

◆ Preferred: A measure of the activity of the organization performing the function. Supporting functions are usually labor-oriented, machine-oriented, or space-oriented.

◆ First Alternative: A measure of the output of the supporting function.

Exhibit 15-4 gives examples of the allocation bases provided in Standard 403 for individual home-office expense cost pools.

Fairness of Pricing

The entire field of negotiated government contracts is marked by the attempt to use *cost assignment*—which includes both the *cost-tracing* and the *cost-allocation* stages—as a means for establishing a mutually satisfactory *price*. A cost allocation may be difficult to defend on the basis of any cause-and-effect reasoning, but it may be a "reason-

EXHIBIT 15-4

Illustrative Allocation Bases Suggested by CASB for Centralized Home-Office Service Functions

Service Rendered	Cost-Allocation Bases
1. Personnel administration	1. Number of personnel, labor-hours, payroll, number of hires
2. Data-processing services	2. Machine time, number of reports
3. Centralized purchasing and subcontracting	3. Number of purchase orders, value of purchases, number of items
4. Centralized warehousing	4. Square footage, value of material, volume
5. Company aircraft service	5. Actual or standard rate per hour, mile, passenger mile, or similar unit
6. Central telephone service	6. Usage costs, number of telephones

able" or "fair" means to help establish a contract price in the minds of the appropriate parties. Various costs become "allowable," but others are "unallowable." An **allowable cost** is a cost that the parties to a contract agree to include in the costs to be reimbursed. Some contracts identify cost categories that are nonallowable. For example, the costs of lobbying activities and the costs of alcoholic beverages are not allowable costs on U.S. government contracts. Other contracts specify how allowable costs are to be determined. For example, only economy-class airfares are allowable for many contracts. Making the cost-assignment rules as explicit as possible (and in writing) reduces argument and litigation when costs are to be used for establishing a contract price.

COST ASSIGNMENT AND COST HIERARCHIES

Objective 9
Explain why managers may find cost hierarchy-based reports useful in their decisions

One extreme approach to cost assignment is to fully assign every cost to each individual unit of a product or service. There is growing interest in cost hierarchy systems that stop short of this full assignment of costs. A **cost hierarchy** is a categorization of costs into different cost pools on the basis of different classes of cost drivers, or different degrees of difficulty in determining cause-and-effect relationships. Not all costs in a cost hierarchy are driven by unit-level product or unit-level service-related variables. Chapter 10 (p. 355) introduced the cost hierarchy notion where the focal cost object was a product. The four hierarchial levels of costs discussed in Chapter 10 are:

- Output unit-level costs: cost driver is at the individual unit of product or service level
- Batch-level costs: cost driver is at the batch level, where a batch is a grouping of two or more units of a product or service
- Product-sustaining costs: cost driver is at the individual product or service level (irrespective of how units are produced)
- Facility-sustaining costs: costs do not fall into any of the above categories

The Concepts in Action box (p. 550) illustrates this four-tier cost hierarchy for a cattle ranch. Only output unit-level costs are allocated to individual products using this hierarchy.

This cost hierarchy is but one of several cost hierarchies that companies are considering when making changes to their management accounting system. We now discuss three additional cost hierarchies and how they result in different types of cost allocations: (1) an organization-structure cost hierarchy, (2) a customer cost hierarchy, and (3) a brand cost hierarchy.[6]

The examples presented in this section illustrate the diverse settings in which cost hierarchies can be used. Individual companies often tailor their own cost hierarchy to obtain the best information for decision making at all levels of their organization.

Organization-Structure Cost Hierarchy

A frequently encountered cost hierarchy classifies costs along the organization's structural lines (plants, divisions, corporate, and so on). We illustrate this approach with divisions and corporate. Within each division, costs are classified as variable (with respect to output units) or fixed. Finally, fixed costs are classified into those controllable within the division and those controllable outside the division. *Controllability* is the degree of influence that a specific manager has over the costs (revenues, or other items in question). A **segment** is an identifiable part of an organization.

---

[6]Many illustrations of product-based cost hierarchies are in R. Cooper, R. Kaplan, L. Maisel, E. Morrissey, and R. Oehm, *Implementing Activity-Based Cost Management* (Montvale, NJ: Institute of Management Accountants, 1992).

Concepts in Action

COST HIERARCHIES DOWN ON THE FARM

The Canadian Valley Cattle Ranch (CVCR) is a family-owned business in Seminole, Oklahoma. It owns 1,000 acres of land and over 600 head of registered Limousin cattle (so named because they were originally from the Limoges region of France). In its operations, CVCR makes decisions about when to sell individual cattle, about where to feed herds of cattle, about its Limousin cattle line, and about corporate planning. What would a product-based cost hierarchy model look like? The CVCR cost hierarchy follows, along with examples of the cost items at each level and their cost drivers.

Cost Hierarchy Level	Cost Items Included	Cost Drivers
Output unit-level costs	Semen costs	Number of artificial inseminations
	Feeding costs	Age and sex of animal
Batch-level costs	Seeding, spraying of pastures	Acreage of pastures
	Weaning costs	Number of labor-hours
Product-sustaining costs	Consultation on new Limousin strains	Number of visits by consultant
	Promotion at agriculture fairs	Number of shows attended
Facility sustaining costs	Property taxes	

The different groupings of costs in this cost hierarchy help CVCR managers make more informed decisions at all levels.

Source: C. Fulkerson, A. Lau, and H. Pourjalali, "Applying Activity-Based Costing to the Canadian Valley Cattle Ranch: A Case Study," *Advanced Management Accounting*, Vol. 3 (Greenwich, CT: JAI Press, 1994).

EXHIBIT 15-5

Monthly Income Statements of Newcastle Machining: Organization Division/Controllability-Based Cost Hierarchy

(all figures in thousands)		Total	Division A	Division B
Net revenues		$1,500	$500	$1,000
Variable manufacturing costs	$780		$250	$530
Variable marketing costs	220		100	120
Total variable costs		1,000	350	650
1. Contribution margin		500	150	350
Fixed division costs controllable by division managers*		190	140	50
2. Contribution controllable by division		310	10	300
Fixed division costs controllable outside division†		70	20	50
3. Contribution by divisions		240	$ (10)	$ 250
Corporate costs unallocated		135		
4. Operating income		$ 105		

*Examples include division-related advertising, sales promotion, salespersons' salaries, and engineering research.
†Examples include property taxes and the division manager's salary.

Exhibit 15-5 presents an income statement for Newcastle Machining, which has a corporate headquarters and two divisions (A and B). This income statement stresses cost-behavior patterns in relation to output levels at each of its two divisions. The assigning of revenues and variable costs to segments is usually the most straightforward approach. Line item 2 in Exhibit 15-5 is a measure of the managers' controllable contribution, and this item is often used as a measure of the managers' performance. Line item 3 describes the performance of the segment as an economic investment. For example, the Division A manager's objective for the forthcoming month may be to increase the division's contribution in line item 2 from $10,000 to $30,000. If this objective is attained, the manager may subsequently be judged successful by Newcastle Machining corporate management. However, using line item 3, top management may continue to regard the division as a miserable investment. Distinguishing between line items 2 and 3 can be difficult. There is often a gray area between cost items that segment managers control and those they do not control.

The income statement in Exhibit 15-5 shows that $135,000 is not allocated to divisions. Examples of unallocated costs in many organizations are corporate income taxes and interest on company debt.

Customer Cost Hierarchy

Chapter 4 introduced the important topic of customer costing. Given the importance of customer satisfaction in the new management approach, companies are very interested in refining customer-costing systems.

Exhibit 15-6 illustrates a customer cost hierarchy. Office Equipment groups its customers into two categories:

◆ Direct-sale customers: Typically, these are large companies that order in high volume and may request customized products for resale to other companies.

◆ Retail-outlet customers: These are retail stores that purchase the standard line of Office Equipment products for resale to other companies.

EXHIBIT 15-6
Customer Contribution Reports for Office Equipment (in thousands)

	Company as a Whole	Company Breakdown into Customer Lines		Breakdown of Direct Sales Line by Individual Customers		
		Direct Sale	Retail Outlets	Customer A	Customer B	...
Net revenues	$680	$240	$440	$27.2	$24.6	
Customer-specific costs*	392	150	242	17.8	15.1	
Customer-specific contribution	288	90	198	$ 9.4	$ 9.5	
Customer-line costs†	154	66	88			
Customer-line contribution	134	$ 24	$110			
Company-costs unallocated	66					
Operating income	$ 68					

*Includes customer-line and company costs allocated to individual customers.
†Includes company costs assigned to customer lines but not customer-line costs allocated to individual customers.

Its customer-based cost hierarchy has three levels:

1. Customer-specific costs are computed as the direct and indirect manufacturing costs of each product sold to the customer plus the direct and indirect customer-related costs assigned to each customer. An example is the cost of an express courier service for a customer who requests overnight delivery.

2. Customer-line categories are broken into the direct-sale and retail-outlet customer lines. An example of a direct-sale customer line item is the cost of mail catalogs sent to individual companies. Office Equipment does not allocate these costs to individual customers; management believes there is no cause-and-effect relationship between mailing catalogs and subsequent sales at the individual customer level, and with good reason. Less than 3% of potential customers who are sent catalogs make purchases in the following 12 months.

3. Company costs are not allocated to either customer lines or to individual customers. An example is the cost of advertisements to promote the brand names that Office Equipment sells to both its direct-sale and retail-outlet customers.

These three customer cost categories, and the related contribution line items, help Office Equipment managers make decisions that affect different levels of operations. For example, the company uses its individual customer contribution information to help guide it in selecting those customers to emphasize (those customers with a high contribution) and those customers to downplay (those with a minimal contribution). Information about the current profitability of its direct-sale and retail-outlet customer lines is a key input into forecasting future profitability, which in turn affects Office Equipment's decisions about the allocation of its marketing budget. (Many of the company's marketing decisions hinge on whether to promote direct-sale or retail-outlet customer lines rather than focusing on individual customers.)

Brand Cost Hierarchy

Brand names such as Coca-Cola, Nestlé, and Sony are valuable assets to the companies that own them. Managers in these companies frequently make decisions that focus on brand categories. Cost hierarchies have been developed to support these decisions. Consider the Swiss-based Nestlé Company. A brand cost hierarchy for its Nestlé brand would include:

1. Individual product-level costs. For Nestlé, these costs pertain to each unit of a single product, such as a one-pound block of milk chocolate sold under the Nestlé brand name.

2. Related-product-line costs. For Nestlé, these costs support the manufacture and marketing of a general product group (such as the Nestlé chocolate product line). The cost of a TV advertisement for a Nestlé chocolate product would fall into this category.

3. Brand-level costs. For Nestlé, these costs pertain to the general support of all products using the Nestlé brand. These costs are not assigned to individual product lines, individual products, or units of a product. The cost of a Nestlé-sponsored hot-air balloon that carries only the Nestlé name would fall into this category.

By making cost assignments at these different levels, a brand-level cost hierarchy can assist managers faced with challenging decisions at each level.

PROBLEM FOR SELF-STUDY

PROBLEM

Medical Instruments in 19_4 has changed the costing system at its manufacturing plant. The prior system had two direct product-cost categories (direct materials and direct manufacturing labor) and one indirect cost category (manufacturing overhead). Indirect costs were allocated to products on the basis of direct labor manufacturing costs.

The new costing system retains the same two direct product-cost categories. Now, however, indirect manufacturing costs are collected into two cost pools:

1. Materials-handling overhead—allocated on the basis of the budgeted number of parts in a product. (When the individual parts in a product are all different, the number of parts and the number of individual parts in the product will be equal. When the same part number is used multiple times in a product, the number of parts will exceed the number of separate parts in that product.)

2. Production overhead—allocated on the basis of the budgeted manufacturing lead time for each product. Manufacturing lead time is the time from when a product is ready to start on the production line to when it becomes a finished good.

Management made the following assumptions in developing the 19_4 budgeted indirect cost allocation rates:

Materials-Handling Overhead

Budgeted total materials-handling overhead costs	$8,000,000
Budgeted number of separate part numbers	5,000
Budgeted average usage per separate part number	800
Budgeted total number of parts (5,000 × 800)	4,000,000

$$\text{Budgeted materials-handling overhead cost allocation rate} = \frac{\$8,000,000}{4,000,000 \text{ parts}}$$

$$= \$2 \text{ per part}$$

Production Overhead

Budgeted total production overhead costs	$12,000,000
Budgeted number of individual products	400
Budgeted average production output per product	100 units
Budgeted average manufacturing lead time per product	6 hours
Budgeted total manufacturing lead time (400 × 100 × 6 hours)	240,000 hours

$$\text{Budgeted production overhead cost allocation rate} = \frac{\$12,000,000}{240,000 \text{ hours}}$$

$$= \$50 \text{ per hour}$$

Curt Henning is examining how the new costing system affects the reported costs of three products. Details of these products in 19_4 follow:

	Product A	Product B	Product C
Direct materials costs	$1,680	$1,250	$2,070
Direct manufacturing labor-hours	7.2	4.3	6.1
Number of parts	128	86	260
Manufacturing lead time in hours	4.8	3.9	18.5

The direct manufacturing labor rate in 19_4 is $30 per hour. Under the prior product-costing system (with one indirect cost category), an indirect cost allocation rate of 300% of direct manufacturing labor costs would have been used in 19_4.

Required

1. What characteristics of a product will lead to its having a much higher cost under the 19_4 costing system than it would have had under the prior costing system?
2. Compute the manufacturing costs of products A, B, and C using (a) the prior product-costing system and (b) the product-costing system introduced in 19_4.
3. Why might there be a cause-and-effect relationship between actual manufacturing lead time and production overhead costs?

SOLUTION

1. The characteristics of a product that will lead to its having a much higher cost under the new costing system are (a) low direct manufacturing labor cost content, (b) high number of parts, and (c) long manufacturing lead time.

2. a.

	Product A	Product B	Product C
Direct manufacturing unit cost			
Direct materials	$1,680	$1,250	$2,070
Direct manufacturing labor			
(7.2; 4.3; 6.1 × $30)	216	129	183
	$1,896	$1,379	$2,253
Indirect manufacturing unit costs			
($216; $129; $183 × 300%)	648	387	549
Total manufacturing unit costs	$2,544	$1,766	$2,802

b.

	Product A	Product B	Product C
Direct manufacturing unit cost			
Direct materials	$1,680	$1,250	$2,070
Direct manufacturing labor			
(7.2; 4.3; 6.1 × $30)	216	129	183
	$1,896	$1,379	$2,253
Indirect manufacturing unit cost			
Materials handling			
(128; 86; 260 × $2)	$ 256	$ 172	$ 520
Production			
(4.8; 3.9; 18.5 × $50)	240	195	925
	$ 496	$ 367	$1,445
Total manufacturing unit cost	$2,392	$1,746	$3,698

3. The actions required to reduce manufacturing lead time will probably reduce the activities that drive production overhead costs. For example, many firms achieving dramatic reductions in manufacturing lead times also achieve

◆ lower inventory levels (meaning lower materials-handling costs),
◆ higher quality levels (meaning reduced rework activities), and
◆ reduced complexity in scheduling (meaning lower manufacturing administrative costs).

SUMMARY

The following points are linked to the chapter's learning objectives.

1. The purpose for computing costs guides the costs to be assigned to a chosen cost object. For example, many marketing costs cannot be allocated to U.S. government contracts. In contrast, the costing of products sold to other companies should include any marketing costs associated with doing business with those companies.

2. The indirect costs allocated to individual products are a function of (a) the aggregate costs in each indirect cost pool, (b) the allocation base used for each indirect cost pool, and (c) the allocation rate used for each indirect cost pool.

3. Department overhead rates will provide more accurate product cost figures than plantwide overhead rates when the individual departments differ in their cost drivers and when the individual products differ in the way they use individual departments in the plant.

4. Companies are changing how they allocate costs in diverse ways. One change is to reduce the cost items that are allocated by increasing the cost items that are directly traced. Other changes are to increase the number of indirect cost pools and to increase the number of nonfinancial allocation bases.

5. The use of an inappropriate cost-allocation base can cause products to be manufactured less efficiently, management to be misfocused, and products to be mispriced in the marketplace.

6. Companies with high fixed costs and unused capacity may encounter ongoing and increasingly greater reductions in demand if they continue to raise selling prices to "fully recover" fixed costs from a declining sales base.

7. Contract disputes over amounts to be paid often can be reduced by making the cost assignment rules as explicit as possible (and in writing). These rules should include details such as the allowable cost items, the acceptable cost-allocation bases, and how differences between budgeted and actual costs are to be handled.

8. The U.S. government pays contractors in some cases with and in other cases without any analysis of their cost data. Even if the contract has a set price, frequently there is a provision that the set price must bear a reasonable relationship to the actual costs the contractor incurs.

9. There is growing interest in cost hierarchies, which are categorizations of costs into different cost pools based on different classes of cost drivers or different degrees of difficulty in determining cause-and-effect relationships. This chapter illustrated product, organization, customer, and brand-level cost hierarchies.

TERMS TO LEARN

This chapter and the Glossary at the end of the book contain definitions of the following important terms:

allowable cost (*p. 549*)   black-hole demand spiral (*546*)
Cost Accounting Standards Board (CASB) (*547*)   cost hierarchy (*549*)
cost-plus contract (*547*)   double-counting (*547*)   downward demand spiral (*546*)
Federal Acquisition Regulations (FARs) (*547*)   labor-paced work environment (*541*)
machine-paced work environment (*541*)   manufacturing lead time (*542*)
segment (*549*)

ASSIGNMENT MATERIAL

QUESTIONS

15-1 "Different costs for different purposes means that a cost allocated for one purpose is not allocated for another purpose." Do you agree?

15-2 Name three factors that affect the indirect costs allocated to individual products.

15-3 When are department overhead rates generally preferable to plantwide overhead rates?

15-4 "To obtain higher homogeneity, have more rather than fewer cost pools." Do you agree? Why?

15-5 "The traditional three cost categories of direct materials, direct manufacturing labor, and manufacturing overhead are outdated. It's time all firms recognized that the machining cost component of overhead should be reported separately as a fourth cost category." Do you agree? Why?

15-6 What is the most frequently used base for allocating manufacturing overhead costs to products in (a) labor-paced environments and (b) machine-paced environments in the United States and Japan?

15-7 "Manufacturing firms that make extensive use of machines should abandon the use of direct manufacturing labor cost as an allocation base." Do you agree? Why?

15-8 Describe two consequences of using direct manufacturing labor-hours as an allocation base in a machine-paced work environment.

15-9 Why might a firm not use a cost driver as the cost-allocation base?

15-10 Why might a company allocate fixed costs using budgeted total capacity available (theoretical or practical) rather than budgeted utilization in the coming period?

15-11 Explain why and how the downward demand spiral can occur when a company has a high level of fixed costs.

15-12 Name two ways that firms are reimbursed under government contracts.

15-13 What role can cost data play in the U.S. government's reimbursement of contractors?

15-14 Define double-counting. Give an example.

15-15 What is a cost hierarchy? Give two examples. Why might managers find the cost-hierarchy notion useful in analyzing costs?

EXERCISES AND PROBLEMS

15-16 Alternative allocation bases for a professional services firm. The Wolfson Group (WG) provides tax advice to multinational firms. WG charges clients for (a) direct professional time (at an hourly rate) and (b) support services (at 30% of the direct professional costs billed). The three professionals in WG and their rates per professional hour are:

Professional	Billing Rate per Hour
Myron Wolfson	$500 per hour
Ann Brown	120 per hour
John Anderson	80 per hour

WG has just prepared the May 19_7 bills for two clients. The hours of professional time spent on each client follow:

	Client	
Professional	Seattle Dominion	Tokyo Enterprises
Wolfson	15	2
Brown	3	8
Anderson	22	30
Total	40	40

Required

1. What amounts did WG bill to Seattle Dominion and Tokyo Enterprises for May 19_7?

2. Suppose support services were billed at $50 per professional labor-hour (instead of 30% of professional-labor costs). How would this change affect the amounts WG billed to the two clients for May 19_7? Comment on the differences between the amounts billed in requirements 1 and 2.

3. How would you determine whether professional labor costs or professional labor-hours is a more appropriate allocation base for WG's support services?

15-17 Cost allocation, use of a separate machining cost pool category. Mahitsu Motors is a manufacturer of motorcycles. Production and cost data for 19_4 follow:

	500 CC Brand	1,000 CC Brand
Units produced	10,000	20,000
Direct manufacturing labor-hours per unit	2	4
Machine-hours per unit	8	8

A single cost pool is used for manufacturing overhead. For 19_4, manufacturing overhead was $6,400,000. Mahitsu allocates manufacturing overhead costs to products on the basis of direct manufacturing labor-hours per unit.

Mahitsu's accountant now proposes that two separate pools be used for manufacturing overhead costs:

◆ Machining cost pool ($3,600,000 in 19_4)

◆ General plant overhead cost pool ($2,800,000 in 19_4)

Machining costs are to be allocated using machine-hours per unit. General plant overhead costs are to be allocated using direct manufacturing labor-hours per unit.

Required

1. Compute the overhead costs allocated per unit to each brand of motorcycle in 19_4 using the current single-cost-pool approach of Mahitsu.

2. Compute the machining costs and general plant overhead costs allocated per unit to each brand of motorcycle assuming that the accountant's proposal for two separate cost pools is used in 19_4.

3. What benefits might arise from the accountant's proposal for separate pools for machining costs and general plant costs?

15-18 Cost allocation with a nonfinancial variable, retailing. Best for Less is a retail chain of supermarkets. For many years, it has used gross margin (selling price minus cost of goods sold) to guide it in deciding on which products to emphasize or deemphasize. For many years it has not allocated any costs to products. Recently, it has changed its internal reporting system. Goods-handling costs are now allocated to individual products on the basis of cubic volume. (Most products are delivered to the shelves in cartons. A detailed study showed that cubic volume was the major driver of Best for Less's goods-handling costs. These costs make up over 30% of non-cost of goods sold costs of Best for Less.) The following data focus on four products in April 19_4:

Product	Revenue per Carton	Cost of Goods Purchased per Carton	Volume (cubic feet)
Breakfast cereal	$82	$56	24
Cheese product	64	52	12
Paper towels	36	26	24
Toothpaste	100	74	12

Each supermarket has a weekly report on "product contribution":

Revenue	$ R
Cost of goods sold	C
Gross margin (GM)	R – C
Goods-handling costs	D
Product contribution (PC)	$GM – D

The April 19_4 goods-handling cost allocation rate is $0.50 per cubic foot.

Required
1. Compute the gross margin of each of the four products. Rank these four products using their gross margin percentage.
2. Compute the product contribution of each of the four products. Rank these four products using the product contribution to revenue percentage.
3. Compare your ranking of products in requirement 2 with that in requirement 1. How is the requirement 2 analysis useful to Best for Less management?
4. The Best for Less controller wants to further increase the costs allocated to individual products. What other cost drivers might be considered for each supermarket?

15-19 Manufacturing cost allocation, use of a conversion cost pool category, automation. Medical Technology Products manufactures a wide range of medical instruments. Two testing instruments (101 and 201) are produced at its highly automated Quebec City plant. Data for December 19_4 follow:

	Instrument 101	Instrument 201
Direct materials	$100,000	$300,000
Direct manufacturing labor	$ 20,000	$ 10,000
Units produced	5,000	20,000
Actual direct labor-hours	1,000	500

Manufacturing overhead is allocated to each instrument product on the basis of actual direct manufacturing labor-hours per unit for that month. Manufacturing overhead cost for December 19_4 is $270,000. The production line at the Quebec City plant is a machine-paced one. Direct manufacturing labor is made up of costs paid to workers minimizing machine problems rather than actually operating the machines. The machines in this plant are operated by computer specialists and industrial engineers.

Required
1. Compute the cost per unit in December 19_4 for instrument 101 and instrument 201 under the existing cost accounting system.
2. The accountant at Medical Technology proposes combining direct manufacturing labor costs and manufacturing overhead costs into a single conversion costs pool. These conversion costs would be allocated to each unit of product on the basis of direct materials costs. Compute the cost per unit in December 19_4 for instrument 101 and instrument 201 under the accountant's proposal.

3. What are the benefits of combining direct manufacturing labor costs and manufacturing overhead costs into a single conversion costs pool?

15-20 Choice of cost pools and cost allocation bases. (CPA, adapted) You have been engaged to install a cost system for the Martin Company. Your investigation of the manufacturing operations of the business discloses these facts:

a. The company makes a line of lighting fixtures and lamps. The direct materials costs of any particular item ranges from 15% to 60% of total manufacturing costs, depending on the kind of metal and fabric used in making it.

b. The business is subject to wide cyclical fluctuations because the level of units sold follows new-housing construction.

c. About 60% of the manufacturing is normally done in the first quarter of the year.

d. For the whole plant, the wage rates range from $12.75 to $25.85 an hour. Within each of the eight individual departments, however, the spread between the high and low wage rates is less than 5%.

e. Each product uses all eight of the manufacturing departments but not proportionately.

f. Within the individual manufacturing departments, manufacturing overhead ranges from 30% to 80% of conversion costs.

Required

On the basis of the information above, write a letter to the president of the company explaining why in its costing system Martin Company should use each of the approaches below. Include the reasons supporting each of your three recommendations.

1. A denominator-level concept tied to normal utilization or master-budget utilization (see Chapter 9, pp. 322–24).

2. A plantwide overhead rate or departmental overhead rates.

3. A cost allocation base using direct manufacturing labor-hours, direct manufacturing labor costs, or prime costs (all direct manufacturing costs).

 15-21 Plantwide versus department overhead cost rates. (CGA, adapted) The Sayther Company manufactures and sells two products, A and B. Manufacturing overhead costs at its Portland plant are allocated to each product using a plantwide rate of $17 per direct manufacturing labor-hour. This rate is based on budgeted manufacturing overhead of $340,000 and 20,000 budgeted direct labor-hours:

Manufacturing Department	Budgeted Manufacturing Overhead	Budgeted Direct Manufacturing Labor-Hours
1	$240,000	10,000
2	100,000	10,000
Total	$340,000	20,000

The number of direct manufacturing labor-hours required to manufacture each product is:

Manufacturing Department	Product A	Product B
1	4	1
2	1	4
Total	5	5

Per-unit costs for the two categories of direct manufacturing costs are:

Direct Manufacturing Costs	Product A	Product B
Direct materials costs	$120	$150
Direct manufacturing labor costs	$ 80	$ 80

At the end of the year, there was no work in process. There were 200 finished units of product A and 600 finished units of product B on hand. Assume that the budgeted production level of the Portland plant was exactly attained.

Sayther sets the listed selling price of each product by adding 120% to its unit manufacturing costs—that is, if the unit manufacturing costs are $100, the listed selling price is $220 ($100 + $120). This 120% markup is designed to cover costs upstream to manufacturing (such as product design) and costs downstream from manufacturing (such as marketing and customer service) as well as to provide an operating income.

Required

1. What is the effect on the inventoriable costs for products A and B of using a plantwide overhead rate instead of department overhead rates?
2. What difference in the per-unit selling prices of product A and product B would result from the use of a plantwide overhead rate instead of department overhead rates?
3. Should Sayther Company prefer plantwide or department manufacturing overhead rates?

15-22 Plantwide versus department overhead cost rates. (CMA) MumsDay Corporation manufactures a complete line of fiberglass attaché cases and suitcases. MumsDay has three manufacturing departments—molding, component, and assembly—and two support departments—maintenance and power.

The sides of the cases are manufactured in the molding department. The frames, hinges, locks, and so on are manufactured in the component department. The cases are completed in the assembly department. Varying amounts of materials, time, and effort are required for each of the various cases. The maintenance department and power department provide services to the three manufacturing departments.

MumsDay has always used a plantwide overhead rate. Direct manufacturing labor-hours are used to allocate the overhead to each product. The budgeted rate is calculated by dividing the company's total budgeted overhead cost by the total budgeted direct labor-hours to be worked in the three manufacturing departments.

Whit Portlock, manager of Cost Accounting, has recommended that MumsDay use department overhead rates. Portlock has projected operating costs and production levels for the coming year. They are presented by department in the accompanying tables (thousands):

	Manufacturing Department		
	Molding	**Component**	**Assembly**
Department operating data			
Direct manufacturing labor-hours	500	2,000	1,500
Machine-hours	875	125	—
Department costs			
Direct manufacturing materials	$12,400	$30,000	$ 1,250
Direct manufacturing labor	3,500	20,000	12,000
Variable manufacturing overhead	3,500	10,000	16,500
Fixed manufacturing overhead	17,500	6,200	6,100
Total departmental costs	$36,900	$66,200	$35,850
Use of support departments			
Maintenance			
Estimated usage in labor-hours for coming year	90	25	10
Power (in kilowatt-hours)			
Estimated usage for coming year	360	320	120
Maximum allotted capacity	500	350	150

	Support Department	
	Maintenance	**Power**
Department operating data		
Maximum capacity	Adjustable	1,000 kW-h
Budgeted usage in coming year	125 hours	800 kW-h
Department costs		
Materials and supplies	$1,500	$ 5,000
Variable labor	2,250	1,400
Fixed overhead	250	12,000
Total support department costs	$4,000	$18,400

Required

1. Calculate the plantwide overhead rate for MumsDay Corporation for the coming year using the same method as used in the past.

2. Whit Portlock has been asked to develop department overhead rates for comparison with the plantwide rate. Follow these steps in developing the department rates:
   a. Allocate the maintenance department costs to the three manufacturing departments using a single rate.
   b. Allocate the power department costs to the three manufacturing departments using a dual-rate method; allocate fixed costs according to maximum capacity and variable costs according to budgeted usage in the coming year.
   c. Calculate department overhead rates for the three manufacturing departments using a machine-hour allocation base for the molding department and a direct manufacturing labor-hour allocation base for the component and assembly departments.

3. Should MumsDay Corporation use a plantwide rate or department rates to allocate overhead to its products? Explain your answer.

15-23 Plantwide versus department overhead cost rates, automation. (CMA) Rose Bach has recently been hired as controller of Empco Inc., a sheet-metal manufacturer. Empco has been in the sheet-metal business for many years and is currently investigating ways to modernize its manufacturing process. At the first staff meeting Bach attended, Bob Kelley, chief engineer, presented a proposal for automating the Drilling Department, Kelley recommended that Empco purchase two robots that would have the capability of replacing the eight direct manufacturing labor workers in the department. The cost savings in Kelley's proposal included elimination of direct manufacturing labor costs in the Drilling Department. Also, Empco charges manufacturing overhead on the basis of direct manufacturing labor costs using a plantwide rate, so manufacturing overhead costs in the department would be reduced to zero.

The president of Empco was puzzled by Kelley's explanation of cost savings, believing it made no sense. Bach agreed, explaining that as firms become more automated, they should rethink their manufacturing overhead systems. The president then asked Bach to look into the matter and prepare a report for the next staff meeting.

To refresh her knowledge, Bach reviewed articles on manufacturing overhead allocation for an automated plant and discussed the matter with some of her peers. Bach also gathered historical data on the manufacturing overhead rates Empco had used over the years. Bach also wanted to have some department data to present at the meeting and, using Empco's accounting records, was able to estimate averages for each manufacturing department in the 1990s.

Date	Annual Average Direct Manufacturing Labor Costs	Average Annual Manufacturing Overhead Costs	Average Manufacturing Overhead Allocation Rate
1950s	$1,000,000	$ 1,000,000	100%
1960s	1,200,000	3,000,000	250
1970s	2,000,000	7,000,000	350
1980s	3,000,000	12,000,000	400
1990s	4,000,000	20,000,000	500

	Averages for the Early 1990s		
	Cutting Department	Grinding Department	Drilling Department
Direct manufacturing labor	$ 2,000,000	$1,750,000	$ 250,000
Manufacturing overhead	11,000,000	7,000,000	2,000,000

Required

1. Disregarding the proposed use of robots in the Drilling Department, describe the shortcomings of the system for allocating manufacturing overhead that is currently used by Empco Inc.

2. Explain the misconceptions underlying Bob Kelley's statement that the manufacturing overhead costs in the Drilling Department would be reduced to zero if the automation proposal was implemented.

3. Recommend ways to improve Empco Inc.'s method for allocating manufacturing overhead by describing how it should revise its costing system
   a. in the Cutting and Grinding Departments and
   b. to accommodate the automation of the Drilling Department.

15-24 Allocation of robotic costs to products; alternative depreciation measures for a robotic cost pool. Consumer Electrics assembles three models of refrigerators (standard, deluxe, and supreme) at its Nashville plant. The plant has 20 robots on its assembly line. There are four plant cost pools: (1) direct materials, (2) direct manufacturing labor, (3) robotics-related costs, and (4) general plant overhead.

The robotics cost pool for July 19_5 totaled $900,000:

Depreciation	$380,000
Operating and maintenance	520,000
Total robotic costs	$900,000

Robotic costs are allocated to each refrigerator unit using the average actual robot-operating hours per unit for that model in the month times the average actual robotic costs per robot-operating hour in the month. Operating data for July 19_5 follow:

Model	Total Number of Robot Operating Hours Attributable to Refrigerator Model	Total Units of Model Produced
Standard	10,000	5,000
Deluxe	12,000	4,000
Supreme	8,000	2,000

The 20 robots were purchased 18 months ago for a total of $12 million. Under an income tax law designed to encourage investment, Consumer Electrics can claim 25% of this cost as depreciation in year 1, 38% as depreciation in year 2, and 37% as depreciation in year 3. These tax-based depreciation rates are also used for internal reporting purposes and product-cost accumulation at Consumer Electrics.

Required
1. Compute the total robotic cost allocated in July 19_5 to each unit of (a) standard, (b) deluxe, and (c) supreme refrigerator.
2. An industrial engineer at the plant observes that the robots are superbly maintained. Consequently, the three-year useful life underestimates their economic life to Consumer Electrics. She suggests that "true" depreciation is better measured using the straight-line method, a 5-year estimate of useful life (from the initial purchase date). Net disposal proceeds at the end of 5 years will be $3 million for the 20 robots. Compute how this measure of depreciation would affect the robotic costs allocated in July to each unit of (a) standard, (b) deluxe, and (c) supreme refrigerator. (Assume that the straight-line method had been used from the time the asset was first purchased.)
3. What are the strengths and weaknesses of relying on tax rules to measure the robotic depreciation cost that is allocated to an individual refrigerator unit?

15-25 Assignment of standard costs to products, automation. Photocopy Plus is a manufacturer and distributor of quality photocopying units. Consider its product line at the high-price range. It features color options and reduction/enlargement options. There are three brands in this line of products. The wholesale price of each brand and the units produced in the most recent month follow:

Brand Name	Wholesale Price	Units Produced (August)
Regal	$ 3,000	2,000
Royal	8,000	3,000
Monarch	30,000	5,000

These products are assembled at a highly automated plant in Biloxi, Mississippi.

For internal accounting purposes, Photocopy Plus uses a standard-costing system. Costs at the Biloxi plant are collected into four pools:

A. *Direct materials costs.* Standard costs are based on the standard materials allowed in assembling each photocopying unit times the standard cost for each part.

B. *Direct manufacturing labor costs.* Standard costs are based on standard direct manufacturing labor-hours allowed for each unit (from an engineering study) times the standard rate of $20 per manufacturing labor-hour.

C. *Machining costs.* Standard costs are based on standard direct machine-hours allowed for each unit (from an engineering study) times the standard rate of $50 per machine-hour.

D. *General plant overhead costs.* Standard rate is 80% of the total of the direct materials costs, direct manufacturing labor costs, and machining costs.

Standard unit production costs are calculated as the sum of A, B, C, and D.

The following information underlies the standard unit production costs for each photocopying unit:

	Regal	Royal	Monarch
Direct materials	$800	$2,000	$4,000
Direct manufacturing labor	5 hours	10 hours	30 hours
Machining	6 hours	20 hours	70 hours

Required
1. Compute the standard unit production costs for the Regal, Royal, and Monarch photocopying units.
2. What percentage is each cost pool of total standard production costs for August?
3. Wendy Reichter, the controller at the Biloxi plant, attends a conference on "Cost Accounting and the Automated Plant." A speaker at the conference proposes eliminating direct manufacturing labor costs as a separate cost pool and including all labor costs in general plant overhead costs. What factors should Reichter consider in evaluating this proposal for the Biloxi plant?

15-26 Effect of overhead cost allocation rate changes and product-design changes on reported product costs. (Extension of Problem for Self-Study, pp. 553–54) In 19_4, Medical Instruments adopted an accounting system with two direct product-cost categories (direct materials costs and direct manufacturing labor costs) and two indirect manufacturing product-cost categories:

a. Materials-handling overhead—allocated on the basis of budgeted number of parts in a product

b. Production overhead—allocated on the basis of budgeted average manufacturing lead time for each product

It is now 19_5. Curt Henning makes the following assumptions when developing the 19_5 cost-allocation rates for materials-handling overhead and production overhead:

Budgeted total materials-handling overhead costs	$7,695,000
Budgeted number of separate part numbers	4,500
Budgeted average usage per separate part number	900
Budgeted total production overhead costs	$12,240,000
Budgeted number of individual products	425
Budgeted average production output per product	120
Budgeted average manufacturing lead time per product	5 hours

Henning is now examining the reported costs of products A and C in 19_5. (Product B has been discontinued.) Product designers at Medical Instruments made several changes in these two products at the end of 19_4 that reduced both the direct manufacturing labor-hours content and the number of parts in each product. The 19_5 direct manufacturing labor rate is $32 per hour. Manufacturing has made substantial progress in reducing the manufacturing lead time for each product. Details of these two products in 19_4 and 19_5 follow:

	Product A		Product C	
	19_4	19_5	19_4	19_5
Direct materials dollars	$1,680	$1,618	$2,070	$2,027
Direct manufacturing labor-hours	7.2	6.9	6.1	5.2
Number of parts	128	116	260	224
Manufacturing lead time in hours	4.8	4.2	18.5	14.8

Required

1. Compute the 19_5 budgeted materials-handling overhead cost-allocation rate per part and the 19_5 budgeted production overhead cost-allocation rate per hour of manufacturing lead time.

2. Compute the 19_5 manufacturing costs of products A and C using the 19_5 indirect cost-allocation rates computed in requirement 1.

3. Compare the manufacturing cost figures for products A and C in 19_4 (see the Problem for Self-Study) and 19_5. Explain any differences between 19_4 and 19_5 costs for each product.

4. Assume that Medical Instruments uses actual rather than budgeted manufacturing lead time (at the budgeted rate per manufacturing lead time hour) when allocating production overhead costs to products. The plant is operating at full capacity. A special customer purchases 20 units of product C on the condition that they be "rushed" (say, at 10.8 hours per unit) through the production line. The customer will pay the listed price in the catalog. How should Medical Instruments compute the gross margin (selling price minus manufacturing costs) on this sale of 20 units of product C? Is the reported product cost of product C accurate for products sold to this customer?

15-27 Single versus multiple indirect cost pools, behavior change or accuracy in product costing. (Continuation of 15-26) Medical Instruments uses one indirect cost pool for production overhead. Several companies with production facilities similar to those at Medical Instruments use more than 10 separate production overhead cost pools, each with a different allocation base or rate. A manufacturing manager at Medical Instruments made the following observation:

> Our objective in using a single production overhead cost pool based on manufacturing lead time is to signal to a broad set of people (in product design, process engineering, and manufacturing) the strategic importance of Medical Instruments' reducing manufacturing lead time. The system is designed to cause behavioral change at Medical Instruments. A single production overhead rate based on manufacturing lead time sends a clear and unambiguous signal to reduce manufacturing lead time. Personally, I do not see the need to adopt a system using six or eight production overhead cost pools, each with their own allocation base. That would be overly complex and complete overkill.

Required

Comment on the manufacturing manager's rationale for allocating production overhead costs using a single indirect cost rate.

 15-28 Downward demand spiral, pricing, cost hierarchy. Francoise Le May is the new president (as of August 1, 19_4) of Sky Shuttle, a commuter airline that flies between San Francisco and Los Angeles. Sky Shuttle is a division of Global Shuttle, which operates more than 20 subsidiaries around the globe. Le May previously was president of Alliance Shuttle, which operates a commuter airline between Paris and Marseilles. Alliance Shuttle has the dominant market share on this air corridor because of its ownership of extensive gate slots at each airport.

The financial results of Sky Shuttle for July 19_4 were:

Revenues
  150 round-trip flights
  100 passengers per round-trip (average)
  $150 per round-trip passenger (average)
Cost data
  $20 variable costs per round-trip passenger
  $6,000 variable costs per round-trip flight
  $1,200,000 monthly fixed costs

Each flight has a maximum capacity of 160 passengers.

Le May is depressed by these July 19_4 results. She expresses concerns about the $150 round-trip price. She tells the director of marketing and the controller that a Paris-to-Marseilles round-trip ticket on Alliance Shuttle averaged $360. "How can you possibly recover all those fixed costs when you price at $150 per round-trip?" The controller agrees with her. Unfortunately, the marketing director is in a quandary. He was about to ask Le May to approve a special $120 per round-trip price in September to match a price promotion by Pacific West Airlines on its San Francisco-to-Los Angeles route.

Required

1. Compute the July 19_4 operating income of Sky Shuttle using a cost hierarchy that is based on cost variability at different levels of drivers. How might Le May find this cost hierarchy useful in her decision making?
2. Compute the operating income to Sky Shuttle if in August 19_4:
   a. Le May increases the round-trip price to $180 and the 150 scheduled flights average 90 passengers, and
   b. Le May increases the round-trip price to $200 and the 150 scheduled flights average 75 passengers.
   Does the (a) or (b) pricing strategy enable Le May to recover Sky Shuttle's fixed costs? Explain.
3. Should Le May approve the special $120 per round-trip price in September? Explain.

15-29 Cost allocation, downward demand spiral. Western Health Maintenance (WHM) operates a chain of 10 hospitals in the Los Angeles area. For many years it has operated a central food-catering facility in Santa Monica, which delivers meals to the 10 hospitals. The Santa Monica facility has the capacity to serve 3,650,000 meals a year (10,000 meals a day). It is currently operating at 2,920,000 meals a year (8,000 meals a day). The budgeted variable costs per meal in 19_4 are $3.80, which includes delivery to the hospital. Budgeted fixed costs for 19_4 are $4,380,000.

In July 19_4, the new WHM president announces that each hospital is to be a profit center. In addition, the head of each hospital can purchase services from outside WHM providing those services meet the WHM quality requirements. The president gives catering as an example. Roy Jenkins, the head of the Santa Monica catering facility, is less than pleased. This facility also will become a profit center (it has been a cost center for many years) under the reorganization.

Jenkins charged each hospital $5.30 per meal in 19_4—$3.80 variable cost + $1.50 allocation of fixed costs. Several hospitals complained about the $5.30 cost as well as about the quality of the food. (Jenkins sarcastically labels the quality complaints as "recycled mystery meat stories.") Indeed, the cost rose from $4.90 in 19_3 to $5.30 in 19_4. Jenkins defended the increase, claiming he needed to spread the same fixed costs over a smaller number of patient-days in 19_4. WHM experienced negative press on a local TV station in 19_3 and early 19_4, and local doctors are referring fewer patients to the WHM hospitals.

In October 19_4 Jenkins started to prepare the 19_5 budget, including the new cost to be charged per meal. He estimated that the total annual demand for meals at all 10 WHM hospitals will be 2,550,000. Then he learned that 3 of the 10 hospitals will use an outside canteen service, which reduces the 19_5 budgeted demand at the Santa Monica facility to 2,000,000 meals. No change in total fixed costs or variable costs per meal are expected in 19_5.

Required

1. How did Jenkins compute the budgeted fixed costs per meal in 19_4?
2. What alternative cost-per-meal figures might Jenkins compute for meals delivered to WHM hospitals in 19_5? Which cost figure should Jenkins use?
3. What factors should Jenkins consider in pricing meals the Santa Monica facility prepares for the WHM hospitals?

15-30 Product-line and territorial income statements. The Delvin Company shows the following results for the year 19_4:

Sales	$1,000,000	100.0%
Cost of goods sold	$ 675,000	67.5%
Marketing*	220,000	22.0%
Administrative (all fixed)	35,000	3.5%
Total costs	$ 930,000	93.0%
Operating income	$ 70,000	7.0%

*All fixed except for $40,000 freight-out cost.

The sales manager has asked you to prepare statements that will help him assess the company efforts by product line and by territories. You have gathered the following information:

	Product			Territory		
	A	B	C	North	Central	Eastern
Sales*	25%	40%	35%			
Product A				50%	20%	30%
Product B				15%	70%	15%
Product C				14/35	8/35	13/35
Variable manufacturing and packaging costs†	68%	55%	60%			
Fixed separable costs:						
Manufacturing	$15,000	$14,000	$21,000	(not allocated)		
Marketing	40,000	18,000	42,000	$48,000	$32,000	$40,000
Freight-out		(not allocated)		13,000	9,000	18,000

Note: All items not directly allocated were considered common costs.
*Percent of company sales.
†Percent of product sales.

Required
1. Prepare a product-line income statement, showing the results for the company as a whole in the first column, costs not allocated in the second column, and the results for the three products in adjoining columns. Show a contribution margin and a product margin, as well as operating income.
2. Repeat requirement 1 on a territorial basis. Show a contribution margin and a territory margin.
3. Should salespeople's commissions be based on contribution margins, product margins, territorial margins, operating income, or dollar sales? Explain.

15-31 Overhead disputes. (Suggested by Howard Wright) The Azure Ship Company works on U.S. Navy vessels and commercial vessels. General yard overhead (for example, the cost of the purchasing department) is allocated to the jobs on the basis of direct labor costs.

In 19_3, Azure's total $150 million of direct labor-cost consisted of $50 million navy and $100 million commercial. The general yard overhead was $30 million.

Navy auditors periodically examine the records of defense contractors. The auditors investigated a nuclear submarine contract, which was based on cost-plus-fixed-fee pricing. The auditors claimed that the Navy was entitled to a refund because of double-counting of overhead in 19_3.

The government contract included the following provision:

Par. 15-202. Direct Costs.

(a) A direct cost is any cost which can be identified specifically with a particular cost object. Direct costs are not limited to items which are incorporated in the end product as material or labor. Costs identified specifically with the contract are direct costs of the contract and are to be charged directly thereto. Costs identified specifically with other work of the contractor are direct costs of that work and are not to be charged to the contract directly or indirectly. When items ordinarily chargeable as indirect costs are charged to the contract as direct costs, the cost of like items applicable to other work must be eliminated from indirect costs allocated to the contract.

Azure had formed a special expediting purchasing group, the SE group, to join with the central purchasing group to obtain materials for the nuclear submarine only. Their direct costs, $5 million, had been included as direct labor of the nuclear work. Accordingly, overhead was allocated to the contracts in the usual manner. The SE costs of $5 million were not included in the general yard overhead. The auditors claimed that no overhead should have been allocated to these SE costs.

Required

1. Compute the amount of the refund that the Navy would claim.

2. Suppose the Navy also discovered that $4 million of general yard overhead was devoted exclusively to commercial engine-room purchasing activities. Compute the additional refund that the Navy would probably claim. (*Note*: This $4 million was never classified as "direct labor." Furthermore, the Navy would claim that it should be reclassified as a "direct cost" but not as "direct labor.")

15-32   Cost allocation, brand cost hierarchies.   The Heinz U.S.A. division of H.J. Heinz is currently re-examining its management accounting system. Donna Fargo, the controller, recently heard a presentation on product-cost hierarchies. The presenter spent a great deal of time outlining output unit-level, batch-level, product-sustaining-level, and facility-sustaining costs. The Heinz controller asked the presenter, "How can this cost hierarchy be applied to companies where the focus is on brands rather than individual products?" After much silence, the presenter said, "That's a good question. Let's talk about it later."

Fargo was intrigued by the cost hierarchy approach. The manager of the Heinz brand of products, Bob Rau, had long argued against fully allocating all the Heinz costs to every individual item of product sold. Currently all costs are allocated to individual items of product using revenues as the allocation base. Rau recently commented, "The only reason I can see to fully allocate all costs to individual products is to demonstrate that our people can divide and then add up."

Fargo decided to pursue the brand cost hierarchy notion herself. She developed the following hierarchy for the Heinz brand family of products:

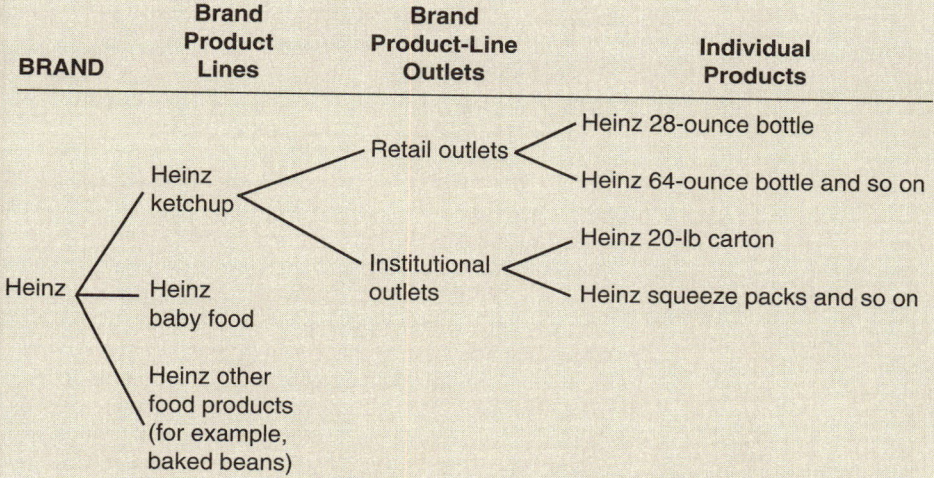

Required

1. Why might Rau be against fully allocating all the Heinz costs to every individual item of product sold?

2. Describe two specific cost items that would be included in the following categories:
    a. Individual product-level costs for the Heinz ketchup 28-ounce plastic bottle
    b. Related-product-line costs for the Heinz ketchup product line
    c. Brand-level costs for the Heinz brand name

3. How might Rau find a cost hierarchy based on (a), (b), and (c) in requirement 2 useful for decision making?

15-33   Fixed cost allocations, used and unused capacity.   The Integrated Testing (IT) Division of International Electronics (IE) provides product-testing services to its many operating divisions. A task force led by Nick Gagner, the president of the IT Division, recently had to decide which one of two new machines to purchase. It needs a new machine to test the engineer workstations assembled by IE's divisions.

1. The "Frontier" testing machine has the capacity to provide operating divisions with 8,000 annual hours of testing time. The yearly fixed costs are $2 million, with variable costs of $100 per hour.
2. The "Contemporary" testing machine has the capacity to provide operating divisions with 5,000 annual hours of testing time. The yearly fixed costs are $1.5 million, with variable costs of $120 per hour.

Both machines have expected useful lives of 5 years. The task force—two people from corporate and three from the operating divisions—recommends the Frontier.

International Electronics purchases the Frontier machine, even though the total budgeted demand for testing hours in 19_5 is only 4,000 hours. Top management at IE expects demand for testing time by its operating divisions will be 4,500 hours in 19_6, 5,250 hours in 19_7, 6,000 hours in 19_8, and 7,000 hours in 19_9.

It is October 15, 19_4. On December 1, 19_4, the IT Division will publish a schedule of the costs it will charge operating divisions for testing time in 19_5. Gagner's main concern is how to allocate the $2 million fixed costs of the Frontier machine when estimating the 19_5 per hour rate. The choice of the denominator level for 19_5 is his most difficult issue.

Current policy at IE is to prohibit outside sourcing of testing unless the IT Division is operating at full capacity. At the last company retreat, three operating division managers proposed that this policy be abandoned and that they be given complete discretion over choice of testing vendors. One manager commented, "We are told to behave in a decentralized way. However, we can't walk the talk. The IT Division charges excessive rates. The 19_4 rate of $540 per hour is well above the market rate. I have the greatest respect for their quality. However, I don't believe I should be forced to purchase testing time internally when they are not cost competitive." The IT Division is currently run as a cost center.

Required
1. What alternative denominator levels might Gagner use to compute a 19_5 fixed cost rate per testing hour? Give one pro and one con for each alternative.
2. Assume Gagner uses 8,000 hours of capacity as the denominator. How should the fixed costs of the Frontier machine that are not charged out to operating divisions be recognized in the accounting system?
3. Describe the likely behavioral consequences within IE of your recommendation in requirement 2.

15-34 Overhead charges, ethics. (Continuation of 15-33)   The IT Division charges out testing time to divisions at a per-hour rate plus an administrative fee. This fee is calculated each year on the basis of the budgeted overhead costs at the division (which include equipment maintenance, electricity, and personnel) divided by the budgeted equipment costs.

Rita Robinson, the controller of IT, is preparing the 19_5 overhead budget. The second biggest item in last year's overhead budget was $86,473 for equipment maintenance. The largest item was personnel costs. Robinson notes that the manufacturer of the Frontier equipment, to ensure the sale, agreed at the last moment to include free maintenance for 19_5 as part of the contract. This maintenance was budgeted to be $92,000 in 19_5. She distinctly remembers Gagner commenting that his negotiating skills yet again had helped reduce costs at the IT Division. She excludes the $92,000 amount from the budget when computing the 19_5 administrative fee to be charged to the division.

Gagner receives the 19_5 administrative markup proposal and is very angry with Robinson. He demands she put the $92,000 amount back in the budget. "We need that slack in our budget. I'm uncertain about several other cost items in the budget, and I want to have a safety net. Anyway, it was my negotiating skill that resulted in the IT Division's saving the $92,000. Why should the other divisions cream off all the gains via a reduced administrative charge? This is a one-year-only gain to IT, so it's not something you have to hide next year."

Required
1. What options does Robinson have in relation to Gagner's demand that the $92,000 be included in the 19_5 budget?
2. Which option would you recommend Robinson take? Refer to the "Standards of Ethical Conduct for Management Accountants" (p. 17) in your answer.

16

Cost Allocation: Joint Products and Byproducts

Meaning of terms

Why allocate joint costs?

Methods of allocating joint costs

Irrelevance of joint costs for decision making

Accounting for byproducts

*P*oultry farms seek revenue from all parts of their chicken and turkey products. Prestage Farms, at its Sampson, North Carolina, facility, obtains breasts, thighs, drumsticks, digest, feathermeal, and poultrymeal from each turkey.

Learning Objectives

When you have finished studying this chapter, you should be able to

1. Identify the splitoff point(s) in a joint-cost situation
2. Distinguish between joint products and byproducts
3. Provide several reasons for allocating joint costs to individual products
4. Distinguish alternative methods of allocating joint costs
5. Describe why the sales value at splitoff method is widely used
6. Describe the irrelevance of joint costs in deciding to sell or process further
7. Distinguish alternative methods of accounting for byproducts

J oint costs plague the accountant's work. This chapter examines methods for allocating joint costs to products and services. Many of the topics discussed in this chapter are related to issues already covered in Chapters 14 and 15. Before reading on, be sure you are comfortable with pages 498–502 of Chapter 14.

MEANING OF TERMS

Objective 1

Identify the splitoff points in a joint-cost situation

Consider a single process that yields two or more products (or services) simultaneously. The distillation of coal, for example, gives us coke, gas, and other products. A **joint cost** is the cost of a single process—distillation, for example—that yields multiple products simultaneously.

The juncture in the process when the products become separately identifiable—when the coal becomes coke, gas, and so on—is called the **splitoff point. Separable**

EXHIBIT 16-1

Examples of Joint-Cost Situations

Industry	Separable Products at the Splitoff Point

Agriculture
  Lamb
  Raw milk
  Turkey farm
	Lamb cuts, tripe, hides, bones, fat
	Cream, liquid skim
	Breast, wings, thighs, drumsticks, digest, feathermeal, and poultrymeal

Extractive industries
  Coal
  Copper ore
  Petroleum
  Salt
	Coke, gas, benzole, tar, ammonia
	Copper, silver, lead, zinc
	Crude oil, gas, raw LPG
	Hydrogen, chlorine, caustic soda

Chemical industries Raw LPG (liquefied petroleum gas)	Butane, ethane, propane
Semiconductor industry Fabrication of silicon-wafer chips	Memory chips of different quality (as to capacity), speed, life expectancy, and temperature tolerance

EXHIBIT 16-2

Joint Products, Main Product, Byproduct, and Scrap

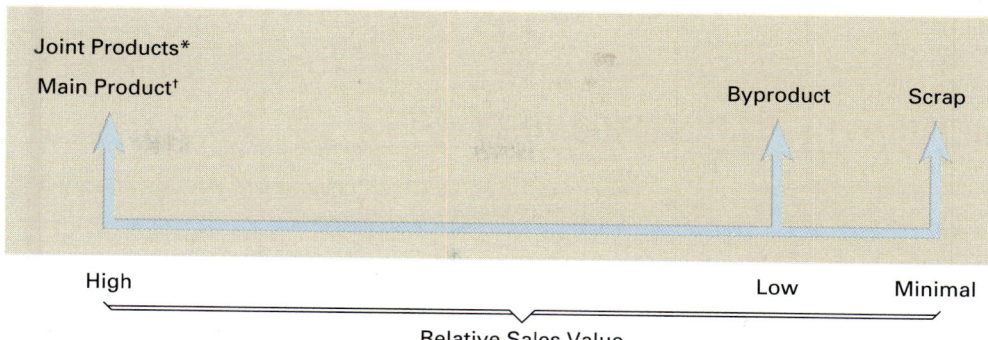

*If multiple products have relatively high sales values.
†If only one product has a relatively high sales value.

costs are costs incurred beyond the splitoff point that are assignable to individual products.

Industries in which single processes simultaneously yield two or more products abound. Exhibit 16-1 presents examples of joint-cost situations in diverse industries. In each example in Exhibit 16-1, no individual product can be produced without the accompanying products' appearing, although sometimes the proportions can be varied. A poultry farm cannot kill a turkey wing; it has to kill a whole turkey, which yields breasts, thighs, drumsticks, digest, feathermeal, and poultrymeal in addition to wings.

Generally accepted accounting principles for recording the costs and revenues from multiple products are often affected by their classification as joint or main products, byproducts, or scrap. The distinctions underlying these classifications are based on their relative sales values.

Joint products have relatively high sales value and are not separately identifiable as individual products until the splitoff point. When a single process yielding two or more products yields only one product with a relatively high sales value, that product is termed a **main product**. A **byproduct** has a low sales value compared with the sales value of the main or joint product(s). **Scrap** has a minimal (frequently zero) sales value. The classification of products as main, joint, byproduct, or scrap can change over time, especially for products (such as tin) whose market price can increase or decrease by, say, 30% or more in any one year.

Exhibit 16-2 is an overview of these definitions. Be careful. These distinctions are not firm in practice. The variety of terminology and accounting practice is bewildering. Always gain an understanding of the terms as used by the particular organization with which you are dealing.

Objective 2

Distinguish between joint products and byproducts

WHY ALLOCATE JOINT COSTS?

Objective 3

Provide several reasons for allocating joint costs to individual products

The purposes for allocating joint costs to individual products or services are similar to the purposes for cost allocation in general (see Exhibit 14-1, p. 499). They include:

1. Inventory cost and cost-of-goods-sold computations for external financial statements and reports for income tax authorities.
2. Inventory cost and cost-of-goods-sold computations for internal financial reporting. Such reports are used in division profitability analysis when determining compensation for division managers.
3. Cost reimbursement under contracts when only a portion of a business's products or services is sold or delivered to a single customer (such as a government agency).

4. Insurance settlement computations when damage claims made by businesses with joint products, main products, or byproducts are based on cost information.

5. Rate regulation when one or more of the jointly produced products or services are subject to price regulation.[1]

METHODS OF ALLOCATING JOINT COSTS

Objective 4
Distinguish alternative methods of allocating joint costs

There are three basic approaches to costing inventory (and computing cost of goods sold) in joint-cost situations:

◆ *Approach 1.* Allocate costs using market selling-price data. Three common methods are used in applying this approach:

The sales value at splitoff method
The estimated net realizable value (NRV) method
The constant gross-margin percentage NRV method

◆ *Approach 2.* Allocate costs using a physical measure.
◆ *Approach 3.* Do not allocate costs; use market selling-price data to guide inventory costing.

In the simplest situation, the joint products are sold at the splitoff point without further processing. We consider this case first. Then we consider situations involving processing beyond the splitoff point.

To highlight each joint-cost example, we make extensive use of exhibits in this chapter. We use the following notation:

Joint Product or Main Product   Byproduct or Scrap

EXAMPLE 1: Farmers' Dairy purchases raw milk from individual farms and processes it up to the splitoff point, where two products (cream and liquid skim) are obtained. These two products are sold to an independent company, which markets and distributes them to supermarkets and other retail outlets.

Exhibit 16-3 presents an overview of the basic relationships. Summary simplified data for the most recent month follow:

◆ Raw milk processed: 110 gallons (110 gallons of raw milk yield 100 gallons of good product with 10 gallons shrinkage)

◆ Production: cream   25 gallons
             liquid skim   75 gallons

◆ Sales: cream   20 gallons at $8 per gallon
             liquid skim   30 gallons at $4 per gallon

◆ | Product | Beginning Inventory | Ending Inventory |
  | --- | --- | --- |
  | Raw milk | 0 gallons | 0 gallons |
  | Cream | 0 gallons | 5 gallons |
  | Liquid skim | 0 gallons | 45 gallons |

◆ Cost of purchasing 110 gallons of raw milk and processing it up to the splitoff point to yield 25 gallons of cream and 75 gallons of liquid skim: $400

[1]See J. Crespi and J. Harris, "Joint Cost Allocation under the Natural Gas Act: An Historical Review," *Journal of Extractive Industries Accounting* 2, no. 2, pp. 133–142.

EXHIBIT 16-3

Farmers' Dairy: Example 1 Overview

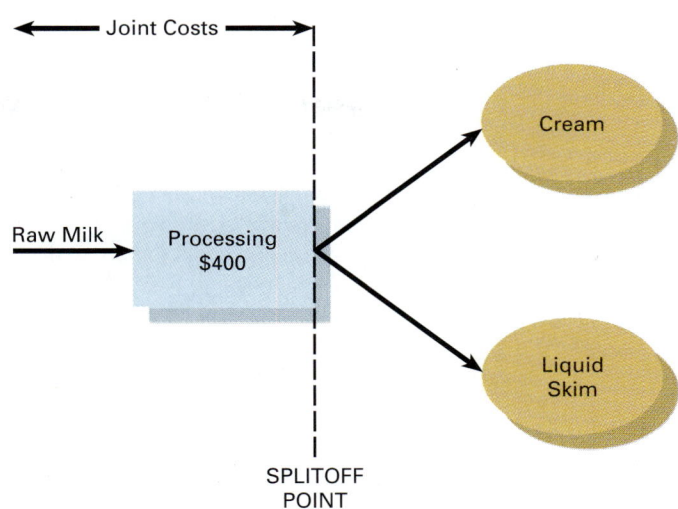

How should the accountant cost the ending inventory of 5 gallons of cream and 45 gallons of liquid skim? The $400 production costs cannot be uniquely identified with or traced to either product. Why? Because the products themselves were not separated before the splitoff point.

The example 1 data will be used to illustrate two of the joint-cost allocation methods: sales value at splitoff point and physical measure.

Sales Value at Splitoff Method

The **sales value at splitoff method** allocates joint costs on the basis of each product's relative sales value at the splitoff point. In example 1, the sales value at splitoff is $200 for cream and $300 for liquid skim. We then assign a weighted value to each product as a percentage of total sales value. Using this weighting, we allocate the joint costs to the individual products:

	Cream	Liquid Skim	Total
1. Sales value at splitoff point (cream, 25 gallons × $8; liquid skim, 75 gallons × $4)	$200	$300	$500
2. Weighting ($200 ÷ $500; $300 ÷ $500)	0.40	0.60	
3. Joint costs allocated (cream, 0.40 × $400; liquid skim, 0.60 × $400)	$160	$240	$400
4. Production costs per unit (cream, $160 ÷ 25 gallons; liquid skim, $240 ÷ 75 gallons)	$6.40	$3.20	

Objective 5

Describe why the sales value at splitoff method is widely used

Note that this method uses the sales value of the entire production of the period including the unsold portion, not just the actual sales of the period. Exhibit 16-4 presents the product-line income statement, using the sales value at splitoff method of joint-cost allocation. Both cream and liquid skim have gross-margin percentages of 20%.[2]

An advantage of the sales value at splitoff point method is its simplicity. The cost allocation base (sales value) is expressed in terms of a common denominator (dollars) that is systematically recorded in the accounting system. Many managers cite a second advantage: The costs are allocated in proportion to a measure of the relative revenue-generating power identifiable with the individual products.

---

[2]The equality of the gross-margin percentages for the two products is a mechanical result reached with the sales value at splitoff method when there are no beginning inventories and all products are sold at the splitoff point.

EXHIBIT 16-4

Product-Line Income Statement:
Joint Costs Allocated Using Sales Value at Splitoff Method

	Cream	Liquid Skim	Total
Sales (cream, 20 gallons × $8; liquid skim, 30 gallons × $4)	$160	$120	$280
Joint costs			
Production costs (cream, 0.4 × $400; liquid skim, 0.6 × $400)	160	240	400
Deduct ending inventory (cream, 5 gallons × $6.40; liquid skim, 45 gallons × $3.20)	32	144	176
Cost of goods sold	128	96	224
Gross margin	$ 32	$ 24	$ 56
Gross-margin percentage	20%	20%	20%

Physical Measure Method

The **physical measure method** allocates joint costs on the basis of their relative proportions at the splitoff point, using a common physical measure such as weight or volume. In example 1, the $400 joint costs produced 25 gallons of cream and 75 gallons of liquid skim. Joint costs are allocated as follows using these quantities:

	Cream	Liquid Skim	Total
1. Physical measure of production (gallons)	25	75	100
2. Weighting (25 gallons ÷ 100 gallons; 75 gallons ÷ 100 gallons)	0.25	0.75	
3. Joint costs allocated (cream, 0.25 × $400; liquid skim, 0.75 × $400)	$100	$300	$400
4. Production costs per unit (cream, $100 ÷ 25 gallons; liquid skim, $300 ÷ 75 gallons)	$4	$4	

Exhibit 16-5 presents the product-line income statement using this method of joint-cost allocation. The gross-margin percentages are 50% for cream and 0% for liquid skim.

The physical weights used for allocating joint costs may have no relationship to the revenue-producing power of the individual products. Consider a mine that extracts ore containing gold, silver, and lead. Use of a common physical measure (tons) would result in almost all the costs being allocated to the product that weighs the most—that product is lead, which has the lowest revenue-producing power. As a second example, if the joint cost of a hog were assigned to its various products on the basis of weight, center-cut pork chops would have the same cost per pound as pigs' feet, lard, bacon, ham, bones, and so forth. In a product-line income statement, products that have a high sales value per pound (for example, center-cut pork chops) would show a fabulous "profit," and products that have a low sales value per pound (for example, bones) would show consistent losses.

Physical measures are sometimes preferred to sales value methods in rate-regulation settings when the objective is to set a fair selling price. Why? Because it is circular reasoning to use selling prices as a basis for setting a selling price, as would be the case under the sales value at splitoff method.

EXAMPLE 2: Assume the same situation as in example 1 except that both cream and liquid skim can be further processed:

EXHIBIT 16-5

Product-Line Income Statement:
Joint Costs Allocated Using Physical Measure Method

	Cream	Liquid Skim	Total
Sales (cream, 20 gallons × $8; liquid skim, 30 gallons × $4)	$160	$120	$280
Joint costs			
Production costs (cream, 0.25 × $400; liquid skim, 0.75 × $400)	100	300	400
Deduct ending inventory (cream, 5 gallons × $4; liquid skim, 45 gallons × $4)	20	180	200
Cost of goods sold	80	120	200
Gross margin	$ 80	$ 0	$ 80
Gross-margin percentage	50%	0%	28.6%

◆ Cream → Butter cream: 25 gallons of cream are further processed to yield 20 gallons of butter cream at an additional processing (separable) cost of $280. Butter cream is sold for $25 per gallon.

◆ Liquid skim → Condensed milk: 75 gallons of liquid skim are further processed to yield 50 gallons of condensed milk at an additional processing cost of $520. Condensed milk is sold for $22 per gallon.

Exhibit 16-6 presents an overview of the basic relationships. Inventory information follows:

	Beginning Inventory	Ending Inventory
Raw milk	0 gallons	0 gallons
Cream	0 gallons	0 gallons
Liquid skim	0 gallons	0 gallons
Butter cream	0 gallons	8 gallons
Condensed milk	0 gallons	5 gallons

Sales during the period were 12 gallons of butter cream and 45 gallons of condensed milk.

EXHIBIT 16-6

Farmers' Dairy: Example 2 Overview

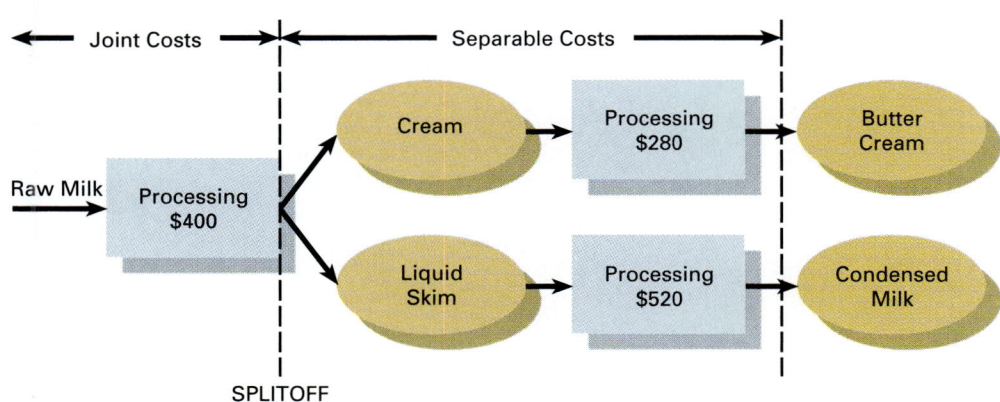

Example 2 will be used to illustrate the estimated net realizable value (NRV) method, the constant gross-margin percentage NRV method, and the no-allocation method.

Estimated Net Realizable Value (NRV) Method

The **estimated net realizable value (NRV) method** allocates joint costs on the basis of the *relative estimated net realizable value* (expected final sales value in the ordinary course of business minus the expected separable costs of production and marketing). Joint costs would be allocated as follows:

	Butter Cream	Condensed Milk	Total
1. Expected final sales value of production (butter cream, 20 gallons × $25; condensed milk, 50 gallons × $22)	$500	$1,100	$1,600
2. Deduct expected separable costs to complete and sell	280	520	800
3. Estimated net realizable value at splitoff point	$220	$ 580	$ 800
4. Weighting ($220 ÷ $800; $580 ÷ $800)	0.275	0.725	
5. Joint costs allocated (butter cream, 0.275 × $400; condensed milk, 0.725 × $400)	$110	$290	$400
6. Production costs per unit (butter cream, [$110 + $280] ÷ 20 gallons; condensed milk, [$290 + $520] ÷ 50 gallons)	$19.50	$16.20	

Note that *expected* final sales value of the *total production* of the period is used and not the *actual* final sales of the period. Exhibit 16-7 presents the product-line income statement using the estimated net realizable value method. The gross-margin percentages are 22.0% for butter cream and 26.4% for condensed milk.

Estimating the net realizable value of each product at the splitoff point requires information about the subsequent processing steps to be taken (and their expected

EXHIBIT 16-7

Product-Line Income Statement:
Joint Costs Allocated Using Estimated Net Realizable Value Method

	Butter Cream	Condensed Milk	Total
Sales (butter cream, 12 gallons × $25; condensed milk, 45 gallons × $22)	$300	$990	$1,290
Cost of goods sold			
Joint costs (butter cream, 0.275 × $400; condensed milk, 0.725 × $400)	110	290	400
Separable costs to complete	280	520	800
Cost of goods available for sale	390	810	1,200
Deduct ending inventory (butter cream, 8 gallons × $19.50; condensed milk, 5 gallons × $16.20)	156	81	237
Cost of goods sold	234	729	963
Gross margin	$ 66	$261	$ 327
Gross-margin percentage	22.0%	26.4%	25.3%

separable costs).[3] In some plants, there may be many possible subsequent steps. Firms may frequently change further processing to exploit fluctuations in the separable costs of each processing stage or in the selling prices of individual products. Under the estimated net realizable value method, each such change would affect the joint-cost allocation percentages. (In practice, a set of standard subsequent steps is assumed at the start of the accounting period when using this method.)

Constant Gross–Margin Percentage NRV Method

The **constant gross-margin percentage NRV method** allocates joint costs in such a way that the overall gross-margin percentage is identical for all the individual products. This method entails three steps:

◆ *Step 1*: Compute the overall gross-margin percentage.
◆ *Step 2*: Use the overall gross-margin percentage and deduct the gross margin from the final sales values to obtain the total costs that each product should bear.
◆ *Step 3*: Deduct the expected separable costs from the total costs to obtain the joint-cost allocation.

Exhibit 16-8 presents these three steps for allocating the $400 joint costs between butter cream and condensed milk. To determine the joint-cost allocation, Exhibit 16-8 uses the expected final sales value of the total production of the period ($1,600) and *not* the actual sales of the period.[4] The overall gross-margin percentage is 25%. A product-line income statement for the constant gross-margin percentage NRV method is presented in Exhibit 16-9.

---

[3]The estimated net realizable value method is clear-cut when there is only one splitoff point. When there are multiple splitoff points, however, additional allocations may be required if processes subsequent to the initial splitoff point remerge with each other to create a second joint-cost situation.

[4]The joint costs allocated to each product need not always be positive under the constant gross-margin percentage NRV method. Some products may receive negative allocations of joint costs to bring their gross-margin percentages up to the overall company average.

EXHIBIT 16-8

Joint Costs Allocated Using Constant Gross-Margin Percentage NRV Method

	Butter Cream	Condensed Milk	Total
Step 1			
Expected final sales value of production:			
(20 gallons × $25) + (50 gallons × $22)			$1,600
Deduct joint and separable costs ($400 + $280 +$520)			1,200
Gross margin			$ 400
Gross-margin percentage ($400 ÷ $1,600)			25%
Step 2			
Expected final sales value of production (butter cream, 20 gallons × $25; condensed milk, 50 gallons × $22)	$500	$1,100	$1,600
Deduct gross margin, using overall gross-margin percentage (25%)	125	275	400
Cost of goods sold	375	825	1,200
Step 3			
Deduct separable costs to complete and sell	280	520	800
Joint costs allocated	$ 95	$ 305	$ 400

EXHIBIT 16-9

Product-Line Income Statement:
Joint Costs Allocated Using Constant Gross-Margin Percentage NRV Method

	Butter Cream	Condensed Milk	Total
Sales (butter cream, 12 gallons × $25; condensed milk, 45 gallons × $22)	$300.0	$990.0	$1,290.0
Cost of goods sold			
Joint costs (see Exhibit 16-8)	95.0	305.0	400.0
Separable costs to complete and sell	280.0	520.0	800.0
Cost of goods available for sale	375.0	825.0	1,200.0
Deduct ending inventory (butter cream, 8 × $18.75*; condensed milk, 5 × $16.50†)	150.0	82.5	232.5
Cost of goods sold	225.0	742.5	967.5
Gross margin	$ 75.0	$247.5	$ 322.5
Gross-margin percentage	25%	25%	25%

*$375 ÷ 20 gallons = $18.75.
†$825 ÷ 50 gallons = $16.50.

The tenuous assumption underlying the constant gross-margin percentage NRV method is that all the products have the same ratio of cost to sales value. A constant ratio of cost to sales value across products is rarely seen in companies that produce multiple products but have no joint-cost situations.[5]

No Allocation of Joint Costs

All of these methods of allocating joint costs to individual products are subject to criticism. Some companies refrain from joint-cost allocation entirely. Instead, they carry all inventories at estimated net realizable value. Income on each product is recognized when production is completed. Industries that use variations of this approach include meatpacking, canning, and mining.

Accountants ordinarily criticize carrying inventories at estimated net realizable values. Why? Because income is recognized before sales are made. Partly in response to this criticism, some companies using this no-allocation approach carry their inventories at estimated net realizable values minus a normal profit margin.

Exhibit 16-10 presents the product-line income statement with no allocation of joint costs for example 2. The separable costs are assigned first, which highlights for managers the cause-and-effect relationship between individual products and the costs incurred on them. The joint costs are not allocated to butter cream and condensed milk as individual products.

Comparison of Methods

Which method of allocating joint costs should be chosen? Each one has weaknesses. Because the costs are joint in nature, managers cannot use the cause-and-effect criterion in making this choice. Managers cannot be sure what causes what cost when examining joint costs. The sales value at splitoff method is widely used when selling-price data are available (even if further processing is done). Reasons for this practice include the following:

---

[5]Another limitation of the constant gross-margin percentage NRV method is that it uses all separable costs (not just separable manufacturing costs) to compute the constant gross margin. The result is that inventory cost figures may implicitly include separable marketing and other "downstream" costs (contrary to what is permissible under generally accepted accounting principles).

EXHIBIT 16-10
Product-Line Income Statement:
No Allocation of Joint Costs

	Butter Cream	Condensed Milk	Total
Produced and sold (butter cream, 12 gallons × $25; condensed milk, 45 gallons × $22)	$300	$ 990	$1,290
Produced but not sold (butter cream, 8 gallons × $25; condensed milk, 5 gallons × $22)	200	110	310
Total sales value of production	500	1,100	1,600
Separable costs to complete	280	520	800
Contribution to joint costs and operating income	$220	$ 580	800
Joint costs			400
Gross margin			$ 400
Gross-margin percentage			25%

1. *No anticipation of subsequent management decisions.* The sales value at splitoff method does not presuppose an exact number of subsequent steps undertaken for further processing.

2. *Availability of a meaningful common denominator to compute the weighting factors.* The denominator of the sales value at splitoff method (dollars) is a meaningful one. In contrast, the physical measure method may lack a meaningful common denominator for all the separable products (for example, when some products are liquids and other products are solids).

3. *Simplicity.* The sales value at splitoff method is simple. In contrast, the estimated net realizable value method can be very complex in operations with multiple products and multiple splitoff points. The total sales value at splitoff is unaffected by any change in the production process after the splitoff point.

When sales values at the splitoff point are not available, other methods must be selected. Individual industries differ in terms of the complexity involved in applying the estimated net realizable value method and in the availability of an informative common physical measure. The same joint-cost allocation method is unlikely to pass the cost-benefit test in all industries.

IRRELEVANCE OF JOINT COSTS FOR DECISION MAKING

Objective 6
Describe the irrelevance of joint costs in deciding to sell or process further

The relevant-cost concepts introduced in Chapter 11 should guide management decisions regarding whether a product should be sold at the splitoff point or processed beyond splitoff. No technique for allocating joint-product costs should guide such management decisions. When a product is an inevitable result of a joint process, the decision to process further should not be influenced either by the size of the total joint costs or by the portion of the joint costs allocated to particular products.

Sell or Process Further

The decision to incur additional costs beyond splitoff should be based on the incremental operating income attainable beyond the splitoff point. Example 2 assumed that it was profitable for both cream and liquid skim to be separately processed into butter cream and condensed milk, respectively. The incremental analysis for these decisions to further process follows:

Further Processing Cream into Butter Cream

Incremental revenue ($500 – $200)	$300
Incremental costs	280
Incremental operating income	$ 20

Further Processing Liquid Skim into Condensed Milk

Incremental revenue ($1,100 – $300)	$800
Incremental costs	520
Incremental operating income	$280

The amount of joint costs incurred up to splitoff ($400)—and how they are allocated—are irrelevant in deciding whether to further process cream or liquid skim. Why? Because, the joint costs of $400 are the same whether or not further processing is done.

Many manufacturing companies constantly face the decision of whether to process a joint product further. Meat products may be sold as cut or may be smoked, cured, frozen, canned, and so forth. Petroleum refiners are perpetually trying to adjust to the most profitable product mix. The refining process necessitates separating all products from crude oil, even though only two or three may have high revenue potential. The refiner must decide what combination of processes to use to get the most profitable quantities of gasoline, lubricants, kerosene, naphtha, fuel oil, and the like.

In designing reports for executive decisions, the accountant must concentrate on incremental costs rather than on how historical joint costs are to be split among various products. The only relevant items are incremental revenue and incremental costs. This next example illustrates the importance of the incremental cost viewpoint.

EXAMPLE 3: Fragrance Inc. jointly processes a specialty chemical that yields two perfumes: 50 ounces of Mystique and 150 ounces of Passion. The sales values per ounce at splitoff are $6 for Mystique and $4 for Passion. The joint costs incurred before the splitoff point are $880. The manager has the option of further processing 150 ounces of Passion to yield 100 ounces of Romance. The total additional costs of converting Passion into Romance would be $160, and the selling price per ounce of Romance would be $8. Exhibit 16-11 summarizes the relationships in this example.

EXHIBIT 16-11
Fragrance Inc.: Example 3 Overview

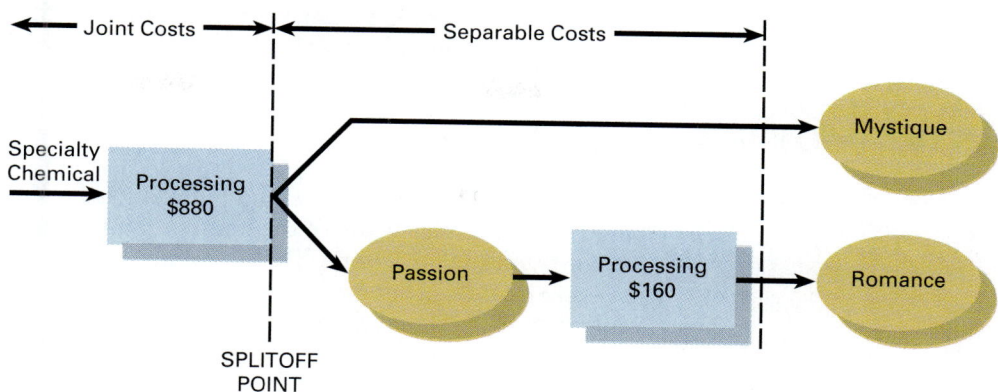

The correct approach in deciding whether to further process Passion into Romance is to compare the incremental revenue with the incremental costs:

Incremental revenue of Romance (100 × $8) − (150 × $4)	$200
Incremental costs of Romance, further processing	160
Incremental operating income from converting Passion into Romance	$ 40

A total income computation of each alternative follows:[6]

	Alternative 1: Sell Mystique and Passion		Alternative 2: Sell Mystique and Romance		Difference
Total revenues	($300 + $600)	$900	($300 + $800)	$1,100	$200
Total costs		880	($880 + $160)	1,040	160
Operating income		$ 20		$ 60	$ 40

In summary, it is profitable to extend processing and to incur additional costs on a joint product as long as the incremental revenue exceeds incremental costs.

Conventional methods of joint-cost allocation may mislead managers who rely on unit-cost data to guide their sell-or-process-further decisions. For example, allocating costs using the physical measure method (ounces in our example) would split the $880 joint costs as follows:

Product	Ounces Produced	Weighting	Allocation of Joint Costs
Mystique	50	50 ÷ 200 = 0.25	0.25 × $880 = $220
Passion	150	150 ÷ 200 = 0.75	0.75 × $880 = 660
	200		$880

The resulting product-line income statement for the alternative of selling Mystique and Romance would *erroneously* imply that the business would suffer a loss by selling Romance:

	Mystique	Romance
Revenues	$300	$800
Costs		
Joint costs allocated	220	660
Separable costs	—	160
Total cost of goods sold	220	820
Gross margin	$ 80	$(20)

---

[6]The revenues reported for each product are Mystique (50 ounces at $6 per ounce= $300), Passion (150 ounces at $4 per ounce = $600), and Romance (100 ounces at $8 per ounce = $800).

Incremental costs are those costs that differ between the alternatives considered (such as sell or process further). Do not assume that all separable costs in our joint-cost allocations for product-costing purposes are always incremental costs. For example, some separable costs may be allocated costs that do not differ between the specific alternatives being considered.

ACCOUNTING FOR BYPRODUCTS

Objective 7
Distinguish alternative methods of accounting for byproducts

Exhibit 16-2 (p. 571) illustrates that byproducts have relatively low sales value compared with the sales value of the main or joint product(s). We now discuss accounting for byproducts. To simplify the discussion, we deal with a two-product example—a main product and a byproduct. In more complex cases, there could be several joint products and several byproducts as well as scrap. (The accounting alternatives for scrap are discussed in Chapter 18.)[7]

EXAMPLE 4: The Meatworks Group processes meat from slaughterhouses. One of its departments cuts lamb shoulders and generates two products:

◆ Shoulder meat (the main product)—sold for $60 per pack
◆ Hock meat (the byproduct)—sold for $4 per pack

Both products are sold at the splitoff point without further processing, as Exhibit 16-12 shows. Data (number of packs) for this department's most recent accounting period follow:

	Production	Sales	Beginning Inventory	Ending Inventory
Shoulder meat	500	400	0	100
Hock meat	100	30	0	70

Total manufacturing costs of these products were $25,000.

Accounting methods for byproducts address two major questions:

◆ *When are byproducts first recognized in the general ledger?* The two main choices are (1) at the time of production and (2) at the time of sale.

---

[7]Further discussion on byproduct accounting methods is in C. Cheatham and M. Green, "Teaching Accounting for Byproducts," *Management Accounting News & Views* (Spring 1988), pp. 14–15; and D. Stout and D. Wygal, "Making By-Products a Main Product of Discussion: A Challenge to Accounting Educators," *Journal of Accounting Education* (1989), pp. 219–233.

EXHIBIT 16-12
Meatworks Group: Example 4 Overview

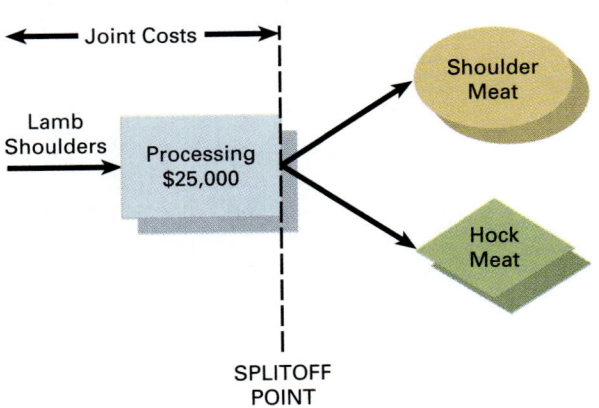

◆ *Where do byproduct revenues appear in the income statement?* The two main choices are (1) as a cost reduction of the main or joint product(s) and (2) as a separate item of revenue or other income.

Combining these two questions and choices gives four possible ways of accounting for byproducts:

Byproduct Accounting Method	When Byproducts Are Recognized in General Ledger	Where Byproduct Revenues Appear in Income Statement	Where Byproduct Inventories Appear on Balance Sheet
A	Production	Reduction of cost	Byproduct inventory reported at (unrealized) selling prices
B	Production	Revenue or other income item	
C	Sale	Reduction of cost	
D	Sale	Revenue or other income item	Byproduct inventory not recognized

Exhibit 16-13 presents the income statement figures and inventory figures that the Meatworks Group would report under each method. Methods A and B recognize the byproduct inventory at the point of production. Note, however, that byproduct inventories are reported on the balance sheet at selling prices rather than at a cost amount. (One variant of methods A and B is to report byproduct inventories at selling price minus a "normal profit margin." This variant avoids including unrealized gains as an offset to cost of goods sold in the period of production.)[8]

---

[8] One version of method A deducts the estimated net realizable value of the byproduct(s) from the joint costs before they are allocated to individual joint products. Another version of method A deducts the estimated net realizable value of the byproduct(s) from the total production costs (joint costs plus separable costs).

EXHIBIT 16-13
Meatworks Group: Income Statement

	Byproduct Accounting Method			
	A	**B**	**C**	**D**
When byproducts recognized in general ledger	At Production	At Production	At Sale	At Sale
Where byproduct revenues appear in income statement	Reduction of cost	Revenue item	Reduction of cost	Revenue item
Revenues				
Main product: shoulder meat (400 × $60)	$24,000	$24,000	$24,000	$24,000
Byproduct: hock meat (30 × $4)	—	120	—	120
Total revenue	$24,000	$24,120	$24,000	$24,120
Cost of goods sold				
Total manufacturing costs	$25,000	$25,000	$25,000	$25,000
Deduct byproduct net revenue (30 × $4)	120	—	120	—
Net manufacturing costs	$24,880	$25,000	$24,880	$25,000
Deduct main product inventory*	4,976	5,000	4,976	5,000
Deduct byproduct inventory (70 × $4)	280	280	—	—
Cost of goods sold	$19,624	$19,720	$19,904	$20,000
Gross margin	$4,376	$4,400	$4,096	$4,120
Gross-margin percentage	18.23%	18.24%	17.07%	17.08%
Inventory "cost" amount (end of period):				
Main product: Shoulder meat	$4,976	$5,000	$4,976	$5,000
Byproduct: Hock meat[†]	$ 280	$ 280	$ 0	$ 0

*(100 ÷ 500) × net manufacturing costs.
[†]Shown at selling prices.

Methods C and D are justified primarily on grounds of immateriality or on a cost-benefit criterion. Byproducts are not recognized in the general ledger until a sale occurs. Byproducts are viewed as incidental and therefore does not warrant costly accounting procedures (although some procedures may be adopted for control of physical quantities).

Separable Costs of Byproducts

The byproduct in example 4 requires no separable costs for processing, marketing, or disposal after the splitoff point. When separable costs are incurred, byproduct accounting becomes more complex. Conceptually, these separable costs should be assigned to each byproduct.

Byproduct accounting is an area where cost-benefit considerations often lead to choosing the expedient alternative. When firms recognize byproducts in their general ledger at the time of sale (methods C and D), the separable byproduct costs typically are expensed to the period in which they are incurred, a procedure justified by the product being immaterial. Although costs of main products are assigned to the production or marketing activities associated with those products, some firms find it too expensive to assign costs of byproducts in the same way.

PROBLEM FOR SELF-STUDY

PROBLEM

Inorganic Chemicals purchases salt and processes it into more-refined products such as caustic soda, chlorine, and PVC (polyvinyl chloride). In the most recent month (July), Inorganic Chemicals purchased salt for $40,000. Conversion costs of $60,000 were incurred up to the splitoff point, at which time two saleable products were produced: caustic soda and chlorine. Chlorine can be further processed into PVC.

The July production and sales information follows:

	Production	Sales	Sales Price per Ton
Caustic soda	1,200 tons	1,200 tons	$ 50
Chlorine	800 tons	—	—
PVC	500 tons	500 tons	$200

All 800 tons of chlorine were further processed, at an incremental cost of $20,000, to yield 500 tons of PVC. There were no byproducts or scrap from this further processing of chlorine. There were no beginning or ending inventories of caustic soda, chlorine, or PVC in July.

There is an active market for chlorine. Inorganic Chemicals could have sold all its July production of chlorine at $75 a ton.

Required

1. Calculate how the joint costs of $100,000 would be allocated between caustic soda and chlorine under each of the following methods: (a) sales value at splitoff, (b) physical measure (tons), and (c) estimated net realizable value.
2. What is the gross-margin percentage of (a) caustic soda and (b) PVC under the three methods cited in requirement 1?
3. Lifetime Swimming Pool Products offers to purchase 800 tons of chlorine in August at $75 a ton. This sale of chlorine would mean that no PVC would be produced in August. How would accepting this offer affect August operating income?

SOLUTION

1. a. **Sales value at splitoff method**

	Caustic Soda	Chlorine	Total
1. Sales value at splitoff (caustic, 1,200 × $50; chlorine, 800 × $75)	$60,000	$60,000	$120,000
2. Weighting ($60,000 ÷ $120,000; $60,000 ÷ $120,000)	0.5	0.5	
3. Joint costs allocated (caustic, 0.5 × $100,000; chlorine, 0.5 × $100,000)	$50,000	$50,000	$100,000

b. **Physical measure method**

	Caustic Soda	Chlorine	Total
1. Physical measure (tons)	1,200	800	2,000
2. Weighting (1,200 ÷ 2,000; 800 ÷ 2,000)	0.6	0.4	
3. Joint costs allocated (caustic, 0.6 × $100,000; chlorine, 0.4 × $100,000)	$60,000	$40,000	$100,000

c. **Estimated net realizable value method**

	Caustic Soda	Chlorine	Total
1. Expected final sales value of production (caustic, 1,200 × $50; PVC from chlorine, 500 × $200)	$60,000	$100,000	$160,000
2. Expected separable costs	—	20,000	20,000
3. Estimated net realizable value at splitoff point	$60,000	$ 80,000	$140,000
4. Weighting ($60,000 ÷ $140,000; $80,000 ÷ $140,000)	$\frac{3}{7}$	$\frac{4}{7}$	
5. Joint costs allocated (caustic, 3/7 × $100,000; chlorine, 4/7 × $100,000)	$42,857	$ 57,143	$100,000

2. a. **Caustic soda**

	Sales Value at Splitoff Point	Physical Measure	Estimated Net Realizable Value
Sales	$60,000	$60,000	$60,000
Joint costs	50,000	60,000	42,857
Gross margin	$10,000	$ 0	$17,143
Gross-margin percentage	16.67%	0%	28.57%

b. **PVC**

	Sales Value at Splitoff Point	Physical Measure	Estimated Net Realizable Value
Sales	$100,000	$100,000	$100,000
Joint costs	50,000	40,000	57,143
Separable costs	20,000	20,000	20,000
Gross margin	$ 30,000	$ 40,000	$ 22,857
Gross-margin percentage	30.00%	40.00%	22.86%

3. Incremental revenue from further processing of chlorine into PVC:

$(500 \times \$200) - (800 \times \$75)$	$\$40,000$
Incremental costs of further processing chlorine into PVC	20,000
Incremental operating income from further processing	$\$20,000$

The operating income of Inorganic Chemicals would be reduced by $20,000 if it sold 800 tons of chlorine to Lifetime Swimming Pool Products instead of further processing the chlorine into PVC for sale.

SUMMARY

The following points are linked to the chapter's learning objectives.

1. A joint cost is the cost of a single process that yields multiple products. The *splitoff point* is the juncture in the process when the products become separately identifiable.

2. Joint products have relatively high sales value and are not separately identifiable as individual products until the splitoff point. A byproduct has a low sales value compared with the sales value of a joint product. Individual products can change from being a byproduct or a joint product when their market prices move sizably in one direction.

3. The purposes for allocating joint costs to products include inventory costing for external financial reporting, internal financial reporting, cost reimbursement under contracts, insurance settlements, and rate regulation.

4. The accounting methods available for allocating joint costs include using market selling price (either sales value at splitoff or estimated net realizable value) or using a physical measure. Choosing not to allocate is also an option.

5. The sales value at splitoff method is widely used where selling price data are available because it does not anticipate subsequent management decisions on further processing, it uses a meaningful common denominator, and it is simple.

6. The relevant-cost analysis emphasized elsewhere in this book applies equally to joint-cost situations. No techniques for allocating joint-product costs should guide decisions about whether a product should be sold at the splitoff point or processed beyond splitoff because joint costs are irrelevant.

7. Byproduct accounting is an area where there is much inconsistency in practice and where some methods used are justified on the basis of expediency rather than theoretical soundness.

TERMS TO LEARN

This chapter and the Glossary at the end of the book contain definitions of the following important terms:

byproduct *(p. 571)*   constant gross-margin percentage NRV method *(577)*
estimated net realizable value (NRV) method *(576)*   joint cost *(570)*
joint products *(571)*   main product *(571)*   physical measure method *(574)*
sales value at splitoff method *(573)*   scrap *(571)*   separable costs *(570)*
splitoff point *(570)*

ASSIGNMENT MATERIAL

QUESTIONS

16-1   What is a *joint cost?*

16-2   Define *separable costs.*

16-3   Give two examples of industries in which joint costs are found. For each example, what are the individual products at the splitoff point?

16-4   Distinguish between a *joint product* and a *byproduct.*

16-5   Provide three reasons for allocating joint costs to individual products or services.

16-6   Name four methods of allocating joint costs to joint products.

16-7   Distinguish between the *sales value at splitoff method* and the *estimated net realizable value method.*

16-8   Many oil refineries do not allocate joint costs among their products. Explain why.

16-9   Give two limitations of the physical measure method of joint-cost allocation.

16-10   Give three reasons why firms like to use the *sales value at splitoff method.*

16-11   "Managers must decide whether a product should be sold at splitoff or processed further. The sales value at splitoff method of joint-cost allocation is the best method for generating the information managers need." Do you agree? Why?

16-12   "Managers should consider only additional revenues and separable costs when making decisions about selling now or processing further." Do you agree? Why?

16-13   Describe two major questions addressed by methods to account for byproducts.

16-14   Accounting methods used in practice for byproducts may be inconsistent with accounting methods for main or joint products. Give an example.

16-15   The marketing manager of a product available for sale at the first splitoff point commented: "This estimated net realizable value method is absurd. Every time they change the amount and mix of further processing at the processing plant, I see my costs change." Explain this statement. Do you agree with the marketing manager?

EXERCISES AND PROBLEMS

16-16   Estimated net realizable value method.   Illawara Inc. produces two joint products, cooking oil and soap oil, from a single vegetable oil refining process. In July 19_4, the joint costs of this process were $24,000,000. Separable processing costs beyond the splitoff point were: cooking oil, $30,000,000; soap oil, $7,500,000. Cooking oil sells for $50 per drum. Soap oil sells for $25 per drum. Illawara produced and sold 1,000,000 drums of cooking oil and 500,000 drums of soap oil. There are no beginning or ending inventories of cooking oil or soap oil.

Required
Allocate the $24,000,000 joint costs using the estimated net realizable value method.

16-17 Alternative joint-cost allocation methods, further process decision. The Wood Spirits Company produces two products, turpentine and methanol (wood alcohol), by a joint process. Joint costs amount to $120,000 per batch of output. Each batch totals 10,000 gallons: 25% methanol and 75% turpentine. Both products are processed further without gain or loss in volume. Separable processing costs: methanol, $3 per gallon; turpentine, $2 per gallon. Methanol sells for $21 per gallon; turpentine sells for $14 per gallon.

Required
1. What joint costs per batch should be allocated to the turpentine and methanol, assuming that joint costs are allocated on a physical measure (number of gallons at splitoff point) basis?
2. If joint costs are to be assigned on an estimated net realizable value basis, what amounts of joint cost should be assigned to the turpentine and to the methanol?
3. Prepare product-line income statements per batch for requirements 1 and 2. Assume no beginning or ending inventories.
4. The company has discovered an additional process by which the methanol (wood alcohol) can be made into a pleasant-tasting alcoholic beverage. The new selling price would be $60 a gallon. Additional processing would increase separable costs $9 (in addition to the $3 separable cost required to yield methanol). The company would have to pay excise taxes of 20% on the new selling price. Assuming no other changes in cost, what is the joint cost applicable to the wood alcohol (using the estimated net realizable value method)? Should the company use the new process?

16-18 Alternative joint-cost allocation methods, shrinkage during processing. Long Beach Refining operates a crude-oil refining plant. It purchases barrels of heavy crude oil at $20 per barrel. The refining process yields three saleable products—60% gasoline, 25% distillates, and 15% residual fuel—at a single splitoff point. Assume also that no reduction in liquid mass occurs at the refinery. Assume also that each product requires no additional processing beyond the splitoff point. Each barrel of heavy crude oil contains 42 gallons. Selling prices of the three saleable commodities are:

Gasoline	$0.80 per gallon
Distillates	$0.70 per gallon
Residual fuel	$0.60 per gallon

During April 19_5, Long Beach purchased and processed 100,000 barrels of heavy crude. There was no beginning or ending inventory. Operating costs of the refinery in April 19_5 were $600,000. Long Beach assigns the purchase cost of the crude and the operating costs to a single joint cost pool that is allocated to each product.

Required
1. What is the joint cost allocated to each product in April 19_5 using (a) the physical measure method and (b) the sales value at splitoff method?
2. Compute the gross-margin percentage of each product under the methods in (a) and (b) in requirement 1. Comment on the results.
3. Redo your answer to requirements 1 and 2 assuming that Long Beach Refining has a 10% reduction in liquid mass during refining—that is, 100 barrels of heavy crude yield the equivalent of 90 barrels of good output. Compare these results with the answers you computed assuming no shrinkage.

16-19 Alternative joint-cost allocation methods, effect of changing inputs on joint-cost allocations. (Continuation of 16-18). In May 19_5 Long Beach Refining purchased 100,000 barrels of light crude at $23 per barrel. The refining process yields three saleable products—70% gasoline, 21% distillates, and 9% residual fuel. The operating costs of the refinery in May 19_5 were $500,000. With light crude there is a 5% reduction in liquid mass during refining.

Required
1. Compute the gross-margin percentage of each product using (a) the physical measure method and (b) the sales value at splitoff method of joint-cost allocation.
2. Compare your answer with your response to requirement 3 in Problem 16-18. Comment on the results.

16-20 Alternative methods of joint-cost allocation, ending inventories. The Darl Company operates a simple chemical process to reduce a single material into three separate items, here referred to as X, Y, and Z. All three end products are separated simultaneously at a single splitoff point.

Products X and Y are ready for sale immediately upon splitoff without further processing or any other additional costs. Product Z, however, is processed further before being sold. There is no available market price for Z at the splitoff point.

The selling prices quoted below have not changed for three years, and no changes are foreseen for the coming year. During 19_5, the selling prices of the items and the total amounts sold were as follows:

- ◆ X—120 tons sold for $1,500 per ton
- ◆ Y—340 tons sold for $1,000 per ton
- ◆ Z—475 tons sold for $700 per ton

The total joint manufacturing costs for the year were $400,000. An additional $200,000 was spent in order to finish product Z.

There were no beginning inventories of X, Y, or Z. At the end of the year, the following inventories of completed units were on hand: X, 180 tons; Y, 60 tons; Z, 25 tons. There was no beginning or ending work in process.

Required
1. What will be the cost of inventories of X, Y, and Z for balance sheet purposes and what will be the cost of goods sold for income statement purposes as of December 31, 19_5, using (a) the estimated net realizable value method of joint-cost allocation and (b) the constant gross-margin percentage NRV method of joint-cost allocation?
2. Compare the gross-margin percentages for X, Y, and Z using the two methods given in requirement 1.

16-21 Net realizable value cost-allocation method, further process decision. (W. Crum) Tuscania Company crushes and refines mineral ore into three products in a joint-cost operation. Costs and production for 19_4 were as follows:

- ◆ *Department 1*: At initial joint costs of $420,000, produces 20,000 pounds of Alco, 60,000 pounds of Devo, 100,000 pounds of Holo
- ◆ *Department 2*: Processes Alco further at a cost of $100,000
- ◆ *Department 3*: Processes Devo further at a cost of $200,000

Results for 19_4:

- ◆ *Alco*: 20,000 pounds completed; 19,000 pounds sold for $20 per pound; ending inventory, 1,000 pounds
- ◆ *Devo*: 60,000 pounds completed; 59,000 pounds sold for $6 per pound; ending inventory, 1,000 pounds
- ◆ *Holo*: 100,000 pounds completed; 99,000 pounds sold for $1 per pound; ending inventory, 1,000 pounds; Holo required no further processing

Required
1. Use the estimated net realizable value method to allocate the joint costs of the three products.
2. Compute the total costs and unit costs of ending inventories.
3. Compute the individual gross-margin percentages of the three products.
4. Suppose Tuscania receives an offer to sell all of its Devo product for a price of $2 per pound at the splitoff point before going through Department 3, just as it comes off the production line in Department 1. Using last year's figures, would Tuscania be better off by selling Devo that way or processing it through Department 3 and selling it? Show computations to support your answer. Disregard all other factors not mentioned in the problem.

16-22 Estimated net realizable value cost allocation method, further process decision. (CPA) The Mikumi Manufacturing Company produces three products by a joint production process. Direct materials are put into production in Department A, and at the end of processing in this department three products appear. Product X is sold at the splitoff point, with no further processing. Products Y and Z require further processing before they are sold. Product Y is processed in Department B, and Product Z is processed in Department C. The company uses the estimated net realizable value method of allocating joint production costs. A summary of costs and other data for the year ended September 30, 19_4, follows.

	Product		
	X	Y	Z
Pounds sold	10,000	30,000	40,000
Pounds on hand at September 30, 19_4	20,000	0	20,000
Sales	$15,000	$81,000	$141,750

	Department		
	A	B	C
Direct materials costs	$56,000	$ 0	$ 0
Direct manufacturing labor costs	24,000	40,450	101,000
Manufacturing overhead	10,000	10,550	36,625

There were no inventories on hand at September 30, 19_3, and there were no direct materials on hand at September 30, 19_4. All the units of product on hand at September 30, 19_4, were fully complete as to processing.

Required
1. Determine the following amounts for each product: (a) estimated net realizable value as used for allocating joint costs, (b) joint costs allocated, (c) cost of goods sold, and (d) cost of finished goods inventory, September 30, 19_4.

2. Assume that the entire output of product X could be processed further at an additional cost of $2 per pound and then sold at a price of $4.30 per pound. What would be the effect on operating income if all the product X output for the year ended September 30, 19_4, had been processed further and sold rather than all sold at the splitoff point?

16-23 Alternative methods of joint-cost allocation, product-mix decisions. The Sunshine Oil Company buys crude vegetable oil. Refining this oil results in four products at the splitoff point: A, B, C, and D. Product C is fully processed at the splitoff point. Products A, B, and D can individually be further refined into Super A, Super B, and Super D. In the most recent month (December), the output at the splitoff point was:

Product A	300,000 gallons
Product B	100,000 gallons
Product C	50,000 gallons
Product D	50,000 gallons

The joint costs of purchasing the crude vegetable oil and processing it were $100,000.

Sunshine had no beginning or ending inventories. Sales of product C in December were $50,000. Total output of products A, B, and D was further refined and then sold. Data related to December are:

	Separable Processing Costs to Make Super Products	Sales
Super A	$200,000	$300,000
Super B	80,000	100,000
Super D	90,000	120,000

Sunshine had the option of selling products A, B, and D at the splitoff point. This alternative would have yielded the following sales for the December production:

Product A	$50,000
Product B	30,000
Product D	70,000

Required

1. What is the gross-margin percentage for each product sold in December, using the following methods for allocating the $100,000 joint costs: (a) sales value at splitoff, (b) physical measure, and (c) estimated net realizable value?

2. Could Sunshine have increased its December operating income by making different decisions about the further refining of products A, B, or D? Show the effect on operating income of any changes you recommend.

16-24 Comparison of alternative joint-cost allocation methods, further process decision, chocolate products. Roundtree Chocolates manufactures and distributes chocolate products. It purchases cocoa beans and processes them into two intermediate products:

♦ Chocolate-powder liquor base
♦ Milk-chocolate liquor base

These two intermediary products become separately identifiable at a single splitoff point. Every 500 pounds of cocoa beans yields 20 gallons of chocolate-powder liquor base and 30 gallons of milk-chocolate liquor base.

The chocolate-powder liquor base is further processed into chocolate powder. Every 20 gallons of chocolate-powder liquor base yields 200 pounds of chocolate powder. The milk-chocolate liquor base is further processed into milk chocolate. Every 30 gallons of milk-chocolate liquor base yields 340 pounds of milk chocolate.

An overview of the manufacturing operations at Roundtree Chocolates follows:

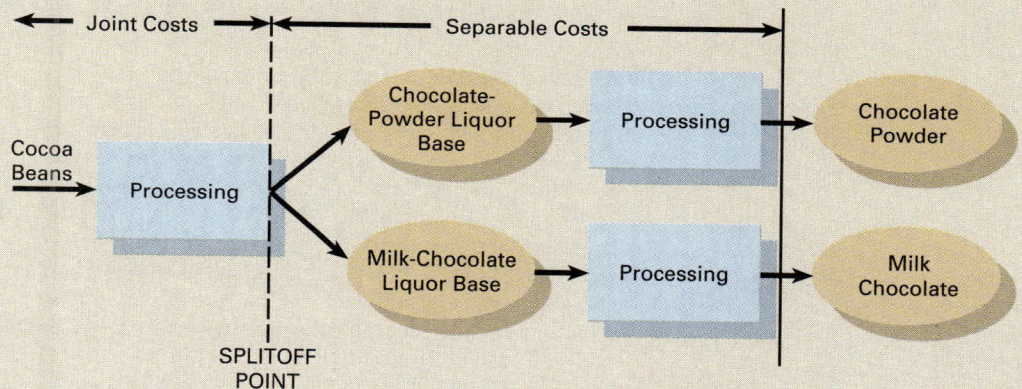

Production and sales data for August 19_4 are:

Cocoa beans processed, 5,000 pounds
Costs of processing cocoa beans to splitoff point (including purchase of beans) = $10,000

	Production	Sales	Unit Selling Price
Chocolate powder	2,000 pounds	2,000 pounds	$4 per pound
Milk chocolate	3,400 pounds	3,400 pounds	$5 per pound

The August 19_4 separable costs of processing chocolate-powder liquor base into chocolate powder are $4,250. The August 19_4 separable costs of processing milk-chocolate liquor base into milk chocolate are $8,750.

Roundtree fully processes both of its intermediate products into chocolate powder or milk chocolate. There is an active market for these intermediate products. In August 19_4,

Roundtree could have sold chocolate-powder liquor base for $21 a gallon and milk-chocolate liquor base for $26 a gallon.

Required

1. Calculate how the joint costs of $10,000 would be allocated between chocolate-powder liquor base and milk-chocolate liquor base under each of the following methods: (a) sales value at splitoff, (b) physical measure (gallons), (c) estimated net realizable value, and (d) constant gross-margin percentage NRV method.
2. What is the gross-margin percentage of chocolate-powder liquor base and milk-chocolate liquor base under methods (a), (b), (c), and (d) in requirement 1?
3. Could Roundtree Chocolates have increased its operating income by a change in its decision to fully process both of its intermediate products?

16-25 Alternative methods of joint-cost allocation, further process decision, memory chips. AMC is a semiconductor firm that specializes in memory chips. In the first stage of the manufacturing operation, raw silicon wafers are photolithographed and then baked at high temperatures. This process yields three individual products at a common splitoff point. For each batch of 1,600 raw silicon wafers, these products are:

1. 300 high-density (HD) memory chips
2. 900 low-density (LD) memory chips
3. 400 defective memory chips

The density of a memory chip is based on the number of good memory bits on each chip, with HD chips having more memory bits per chip than LD chips. The 400 defective memory chips from each batch have a zero disposal price. The joint costs of purchasing and processing the 1,600 raw silicon wafers up to the splitoff point are $5,000.

AMC has two options for each grade of good memory chip at the splitoff point:

1. Sell immediately. The selling price for each HD chip is $10. The selling price for each LD chip is $5.
2. Process further into extended-life memory chips. This processing step further exposes the chips to extreme temperatures, and the chips that survive are sold as extended-life memory chips. Data pertaining to this further processing stage follow.
   a. Extended-life high-density (EL-HD) chips: From a batch of 300 HD chips, the yield is 200 EL-HD chips. The 100 defective chips from this further processing step have a zero disposal price. The separable costs to further process the 300 HD chips are $1,000. The selling price for each EL-HD chip is $30.
   b. Extended-life low-density (EL-LD) chips: From a batch of 900 LD chips, the yield is 500 EL-LD chips. The 400 defective chips from this further processing step have a zero disposal price. The separable cost to further process the 900 LD chips is $3,000. The selling price for each EL-LD chip is $18.

AMC has consistently followed the policy of further processing the entire output of both the HD and LD chips into their EL-HD and EL-LD forms. Separable costs are equal to incremental costs.

Required

1. Compute how the joint costs of $5,000 would be allocated between HD and LD chips under each of the following methods: (a) sales value at splitoff, (b) physical measure (number of good chips at splitoff point), (c) estimated net realizable value, and (d) constant gross-margin percentage NRV. Assume that AMC has no beginning or ending inventories.
2. What is the gross-margin percentage of EL-HD and EL-LD chips under methods (a), (b), (c), and (d) in requirement 1?
3. Peach Computer Systems offers to buy 900 LD memory chips from AMC at $5 a chip. What would be the effect on operating income of accepting this offer rather than pursuing the current policy of further processing the LD chips into EL-LD form?

16-26 Estimated net realizable value method, byproducts. (CMA, adapted) Princess Corporation grows, processes, packages, and sells three joint apple products: (a) sliced apples that are used in frozen pies, (b) applesauce, and (c) apple juice. The outside skin of the apple, processed as animal feed, is treated as a byproduct. Princess uses the estimated net realizable value method to allocate costs of the joint process to its joint products. The byproduct is inventoried at its selling price when produced; the net realizable value of the byproduct is used to reduce the joint production costs before the splitoff point. Details of Princess's production process are presented below.

◆ The apples are washed and the outside skin is removed in the Cutting Department. The apples are then cored and trimmed for slicing. The three joint products and the byproduct are recognizable after processing in the Cutting Department. Each product is then transferred to a separate department for final processing.

◆ The trimmed apples are forwarded to the Slicing Department, where they are sliced and frozen. Any juice generated during the slicing operation is frozen with the slices.

◆ The pieces of apple trimmed from the fruit are processed into applesauce in the Crushing Department. The juice generated during this operation is used in the applesauce.

◆ The core and any surplus apple pieces generated from the Cutting Department are pulverized into a liquid in the Juicing Department. There is a loss equal to 8% of the weight of the *good* output produced in this department.

◆ The outside skin is chopped into animal feed and packaged in the Feed Department.

A total of 270,000 pounds of apples were entered into the Cutting Department during November. The schedule presented below shows the costs incurred in each department, the proportion by weight transferred to the four final processing departments, and the selling price of each end product.

Processing Data and Costs
November 19_4

Department	Costs Incurred	Proportion of Product by Weight Transferred to Departments	Selling Price per Pound of Final Product
Cutting	$60,000	—	—
Slicing	11,280	33%	$0.80
Crushing	8,550	30	0.55
Juicing	3,000	27	0.40
Feed	700	10	0.10
Total	$83,530	100%	

Required
1. Princess Corporation uses the estimated net realizable value method to determine inventory cost of its joint products; byproducts are reported on the balance sheet at their selling price when produced. For the month of November 19_4, calculate:
   a. the output for apple slices, applesauce, apple juice, and animal feed, in pounds.
   b. the estimated net realizable value at the splitoff point for each of the three joint products.
   c. the amount of the cost of the Cutting Department assigned to each of the three joint products and the amount assigned to the byproduct in accordance with corporate policy.
   d. the gross margins in dollars for each of the three joint products.
2. Comment on the significance to management of the gross-margin dollar information by joint product for planning and control purposes, as opposed to inventory costing purposes.

16-27   Accounting for a main product and a byproduct. (Cheatham and Green, adapted) Bill Dundee is the owner and operator of Louisiana Bottling, a bulk soft-drink producer. A single production process yields two bulk soft drinks—Rainbow Dew (the main product) and Resi-Dew (the byproduct). Both products are fully processed at the splitoff point, and there are no separable costs.

Summary data for September 19_5 are:

Cost of soft-drink operations = $120,000

Production and sales data:

	Production (in gallons)	Sales (in gallons)	Selling Price per Gallon
Main product: Rainbow Dew	10,000	8,000	$20.00
Byproduct: Resi-Dew	2,000	1,400	2.00

There were no beginning inventories on September 1, 19_5. An overview of operations follows.

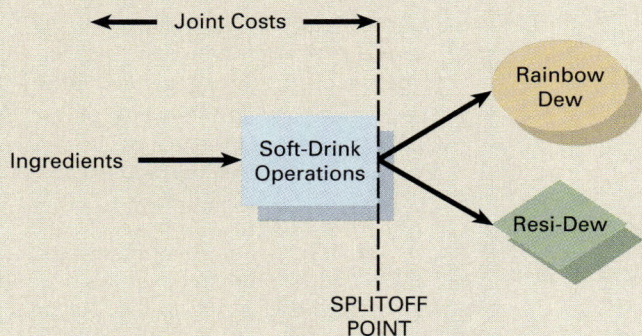

Required

1. What is the gross margin for Louisiana Bottling under methods A, B, C, and D of byproduct accounting described on p. 583 of this chapter?
2. What are the inventory amounts reported in the balance sheet on September 30, 19_5, for Rainbow Dew and Resi-Dew under each of the four methods of byproduct accounting cited in requirement 1?

16-28 Joint costs and byproducts. (W. Crum) Caldwell Company processes an ore in Department 1, out of which come three products, L, W, and X. Product L is processed further through Department 2. Product W is sold without further processing. Product X is considered a byproduct and is processed further through Department 3. Costs in Department 1 are $800,000 in total; Department 2 costs are $100,000; and Department 3 costs are $50,000. Processing 600,000 pounds in Department 1 results in 50,000 pounds of product L, 300,000 pounds of product W, and 100,000 pounds of product X.

Product L sells for $10 per pound. Product W sells for $2 per pound. Product X sells for $3 per pound. The company wants to make a gross margin of 10% of sales on product X and also allow 25% for marketing costs on product X.

Required

1. Compute unit costs per pound for products L, W, and X, treating X as a byproduct. Use the estimated net realizable value method for allocating joint costs. Deduct the estimated net realizable value of the byproduct produced from the joint cost of Products L and W.
2. Compute unit costs per pound for products L, W, and X, treating all three as joint products and allocating costs by the estimated net realizable value method.

16-29 Joint and byproducts, estimated net realizable value method. (CPA) The Harrison Corporation produces three products—Alpha, Beta, and Gamma. Alpha and Gamma are joint products, and Beta is a byproduct of Alpha. No joint costs are to be allocated to the byproduct. The production processes for a given year follow:

A. In Department 1, 110,000 pounds of direct material, Rho, are processed at a total cost of $120,000. After processing in Department 1, 60% of the units are transferred to Department 2, and 40% of the units (now Gamma) are transferred to Department 3.

B. In Department 2, the material is further processed at a total additional cost of $38,000. Seventy percent of the units (now Alpha) are transferred to Department 4 and 30% emerge as Beta, the byproduct, to be sold at $1.20 per pound. Separable marketing costs for Beta are $8,100.

C. In Department 4, Alpha is processed at a total additional cost of $23,660. After this processing, Alpha is ready for sale at $5 per pound.

D. In Department 3, Gamma is processed at a total additional cost of $165,000. In this department, a normal loss of units of Gamma occurs, which equals 10% of the good output of Gamma. The remaining good output of Gamma is then sold for $12 per pound.

Required

1. Prepare a schedule showing the allocation of the $120,000 joint costs between Alpha and Gamma using the estimated net realizable value method. The estimated net realizable value of Beta should be treated as an addition to the sales value of Alpha.

2. Independent of your answer to requirement 1, assume that $102,000 of total joint costs were appropriately allocated to Alpha. Assume also that there were 48,000 pounds of Alpha and 20,000 pounds of Beta available to sell. Prepare an income statement through gross margin for Alpha using the following facts:

   a. During the year, sales of Alpha were 80% of the pounds available for sale. There was no beginning inventory.
   b. The estimated net realizable value of Beta available for sale is to be deducted from the cost of producing Alpha. The ending inventory of Alpha is to be based on the net costs of production.
   c. All other cost and selling price data are listed in A through D above.

16-30 Joint products and byproducts. The Bulli Coal Company purchases coal at $30 a ton. Its coking plant produces three products—open-fire coke, closed-appliance coke, and tar. An overview of the operations in July 19_4 follows:

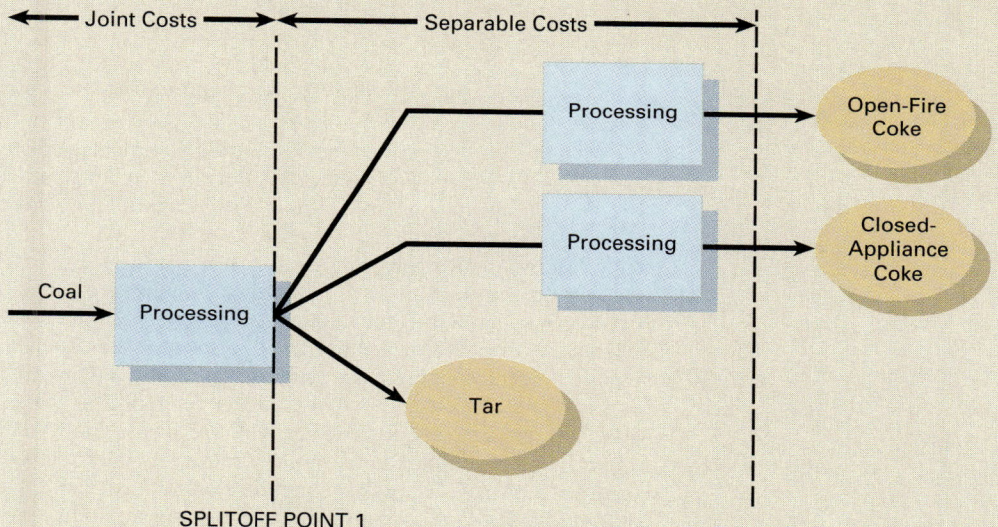

SPLITOFF POINT 1

The only saleable product at splitoff point 1 is tar. After further separable processing, open-fire coke and closed-appliance coke, the two additional products appear.

Each 100 tons of coal costs $2,000 to purchase and $1,000 to process up to splitoff point 1. The separable costs of further processing the 100 tons of coal beyond splitoff point 1 are $800 for open-fire coke and $600 for closed-appliance coke. The outputs from 100 tons of coal are:

Output	Tons	Selling Price per Ton
Open-fire coke	30	$100
Closed-appliance coke	40	80
Tar	10	20

In July 19_4, Bulli processes 10,000 tons of coal. It has no beginning inventory of coal or of finished product. On July 30, it has no ending inventory of coal, no work in process, and 300 tons of open-fire coke, 200 tons of closed-appliance coke, and 20 tons of tar in ending inventory.

Required

1. Bulli treats all three products as joint products. Calculate how the joint costs incurred up to splitoff point 1 would be allocated to open-fire coke, closed-appliance coke, and tar using: (a) the physical measure method and (b) the estimated net realizable value method.

2. What is the gross-margin percentage of each product using the two joint-cost allocation methods in requirement 1?

3. Should tar be viewed as a joint product or a byproduct? Explain.

4. Revisit requirement 1. What would be the implication of classifying tar as a byproduct (no extra calculations required)?

16-31 Joint products and byproducts, process-further decision, ethics. (Continuation of 16-30).

Bulli Coal Company is a wholly owned subsidiary of Pacific Resources. Ralph Adams is the manager of Bulli Coal Company. In his early career days, he was an accountant. Recently, Adams made the decision to further process all tar to road tar at the Bulli plant. Each 10 tons of tar yields 8 tons of road tar. The separable costs of further processing 10 tons of tar to 8 tons of road tar are $240. Bulli sells the road tar to Kembla Construction for $50 a ton.

Jim Murray is the management accountant at the Bulli Coal Company. He receives a request from Adams to change the accounting classification of tar from being a joint product to a byproduct. Adams noted that this change would help support his justification proposal to Pacific Resources for further processing the tar to road tar. Every other coal company in the region further processed tar to road tar, so Murray did not find Adams's further processing decision surprising. Adams's proposal includes the following gross margin percentage for each ton of road tar sold to Kembla Construction:

Revenues	$50
Processing costs	30
Gross margin	$20
Gross-margin percentage	40%

Adams noted that when tar is treated as a joint product, the costing system (which uses the physical measure for joint cost allocation) would show Bulli losing money on tar. He glowed with pride when he said that road tar would now be the product with the highest gross margin in the company.

Adams and Murray ran into each other at the coffee machine. After asking if Murray had received the proposal, Adams mentioned how he recently had become a board member of the Illawara Steelers, which for the last 2 years has been the nation's champion. Adams gave free tickets to Murray, his wife, and their two sons for the Illawara Steelers' football game. He even arranged for the captain of the Steelers (a national celebrity) to have his photo taken with Murray's two sons. This day, however, turned very sour. At the game Murray overheard a conversation about Kembla Construction. He discovered that Adams's father-in-law owns Kembla Construction and that Adams's wife is the chief operating officer. Murray has a sickening feeling in his stomach. He makes some discreet inquiries and learns that other road construction companies are paying $65 to $75 a ton for road tar comparable to what Bulli Coal is to sell to Kembla Construction.

Murray goes to work on Monday. On his desk he finds an 8″ by 10″ autographed glossy photo of the Steelers' captain and his two smiling sons. There is also a note from Adams saying how he hoped Murray and his wife enjoyed the day.

Required

1. Comment on Adams's gross-margin percentage computation for road tar. What assumptions underlie this computation?

2. Evaluate the decision to further process tar into road tar. (Assume separable processing costs are the same as incremental processing costs.)

3. What ethical and other issues does Murray face? Refer to the "Standards of Ethical Conduct for Management Accountants" (p. 17) in your answer.

4. What action should Murray take in relation to the issues you raise in requirement 3?

17

Process Costing Systems

- Process costing and equivalent unit computations
- Process costing with equivalent units: five key steps
- Weighted-average method
- First-in, first-out method
- Comparison of FIFO and weighted-average methods
- Standard costs and process costs
- Transferred-in costs in process costing
- Additional aspects of process costing

*R*aytheon's contract with the U.S. Navy for 529 Standard Missile-2s requires that all missiles be similar. Because all missiles pass through a uniform set of production stages, Raytheon uses process costing at its Bristol, Tennessee, plant when computing the unit cost of each missile assembled.

Process-costing systems are used for costing like or similar units of products, which are often mass produced. These units differ from the custom-made or unique products costed under job-costing systems. This chapter covers product costing using process-costing systems. It will be concerned only incidentally with *planning* and *control*, which are discussed in other chapters and are applicable to *all* product-costing systems regardless of whether process costing, job costing, or some hybrid system is used.

PROCESS COSTING AND EQUIVALENT UNIT COMPUTATIONS

An introduction to process costing appears in Chapter 5 (pp. 155–59). *Please review that material before proceeding.* Two cases are discussed in Chapter 5:

◆ Case 1: Process costing with all units fully completed at the end of the reporting period.
◆ Case 2: Process costing with some units incomplete at the end of the reporting period.

Objective 1
Explain when computing equivalent units makes a material difference

Both cases assumed no beginning inventory of work in process. Case 1 highlights the key feature of process costing, which is the averaging of costs over masses of like or similar units. Case 2 introduces the equivalent unit concept. This concept complicates process-costing computations but does not change its underlying averaging approach.

Case 1 is used in service industries in which a like or similar service (such as a bank processing deposits) is provided to many customers. By their very nature, service companies do not have inventory. Case 2 is used in manufacturing industries in which the following two conditions apply:

1. A significant percentage of total production during an accounting period is in work-in-process inventory. This condition applies, for example, to plants assembling helicopters that have long manufacturing lead times (lead time is the time from when a product is ready to start on the production line to when it becomes a finished good).

2. A material amount of unit cost can be assigned to a product before its final production point. This condition applies, for example, to the manufacture of some computer products (such as mainframes). In this case the final production process consists of testing fully assembled products over extended time periods. At this stage in the manufacture of these products all direct materials have been incorporated into the product. (Direct materials costs make up over 50% of the total manufacturing costs of many computer products.)

Either condition 1 or 2—and some companies experience them both—can result in a situation in which work-in-process inventory cost is a material percentage of total manufacturing costs.

Where work-in-process inventory cost is not a material percentage of total manufacturing cost (that is, neither condition 1 nor 2 holds), a manufacturer may still use a process-costing system, but it may not bother with the extra detail of computing equivalent units. As always, managers should consider the benefits and costs of making refinements when they choose specific costing methods.

Changes in operating conditions can affect those benefits and costs. For example, firms adopting just-in-time manufacturing practices (see Chapter 24) often markedly reduce their work-in-process inventories. These reductions can reduce the benefits gained by detailed tracking of equivalent units at each step in the production process.

We now discuss process costing when equivalent unit calculations are required.

PROCESS COSTING WITH EQUIVALENT UNITS: FIVE KEY STEPS

Objective 2
Describe five key steps of process costing using equivalent units

Our Chapter 5 introduction to process costing featured five key steps:

- Step 1: Summarize the flow of physical units of a product or output.
- Step 2: Compute output in terms of equivalent units.
- Step 3: Summarize the total costs to account for, which are the total of the costs charged to Work in Process (that is, the total debits to Work in Process).
- Step 4: Compute equivalent unit costs.
- Step 5: Assign total costs to units completed and to units in ending work in process.

Proceeding through these steps methodically helps minimize errors. The first two steps focus on what is occurring in physical or engineering terms. The dollar impact of the production process is measured in the final three steps. These steps are especially useful when there are beginning inventories.

The main objective of the five-step procedure is to compute the dollar amount of the credit to work in process and the corresponding debit to finished goods (or to work in process in the subsequent processing department) for work completed during the current period. We focus mainly on two alternative methods of process costing in this chapter—the weighted-average method and the first-in, first-out method. The difference between these two methods centers on how work done in the preceding period is taken into account in computing unit product costs. A third method of process costing—using standard costs—is also discussed in this chapter.

Data for Illustration

The easiest way to learn process costing is by example. Let us consider the following scenario.

EXAMPLE 1: Global Defense Inc. manufactures military equipment. It is best known for its Liberty Missile. Its product-costing system has a single direct cost category (direct materials) and a single indirect cost category (conversion costs). Each Liberty Missile passes through two departments—the Assembly Department and the Testing Department. Every effort is made to ensure that all Liberty Missiles are identical and meet many demanding performance specifications. Direct materials are added at the beginning of the process in Assembly. Additional direct materials are added at the end of processing in the Testing Department, and final assembly of each missile occurs there. Conversion costs are allocated evenly throughout both processes. When the Assembly Department finishes work on each missile, it is immediately transferred to Testing. When the Testing Department finishes work on each missile, it is immediately transferred to Finished Goods.

The following sketch summarizes these facts:
Data for the Assembly Department for March 19_4 are:

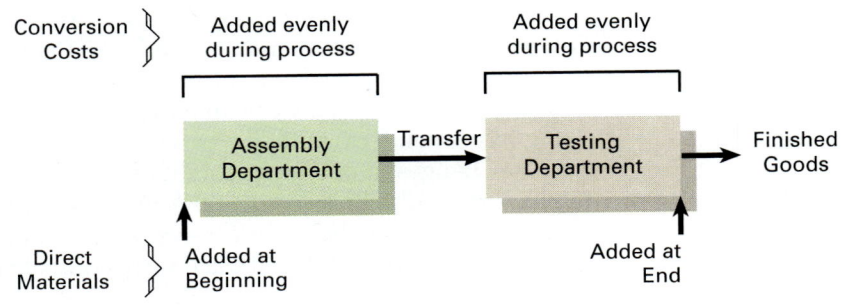

Physical Units for March 19_4

Work in process, beginning inventory (March 1)	100 units
Direct materials (100% complete)	
Conversion costs (40% complete)	
Started during March	400 units
Completed and transferred out during March	480 units
Work in process, ending inventory (March 31)	20 units
Direct materials (100% complete)	
Conversion costs (50% complete)	

Costs for March 19_4

Work in process, beginning inventory		
Direct materials	$4,000,000	
Conversion costs	1,110,000	$ 5,110,000
Direct materials costs added during March		22,000,000
Conversion costs added during March		18,000,000

EXHIBIT 17-1

Overview of Product Costing at Global Defense Inc.

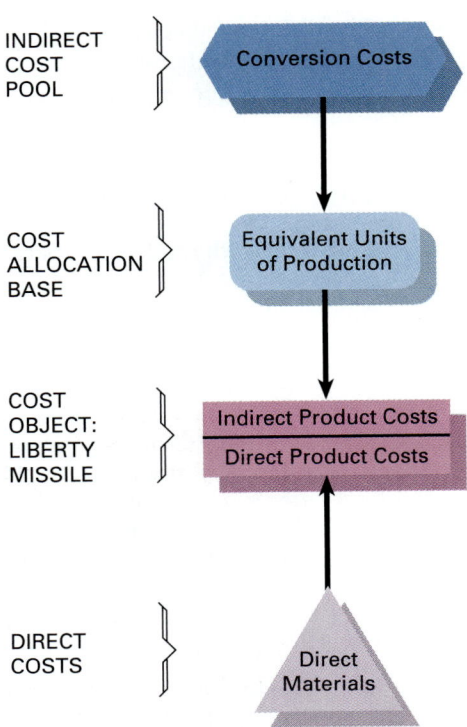

The per-unit cost amounts underlying example 1 are:

	February	March
Direct materials per unit	$40,000	$55,000
Conversion costs per unit	27,500	40,000
Total costs per unit	$67,500	$95,000

These per-unit increases from February to March, although larger than normal, are used to highlight differences among the process-costing methods we discuss.

For inventory-costing purposes, the accountant must

◆ Compute the cost of goods transferred out of a process or a department (in our example, Assembly).
◆ Compute the ending inventory of work in process.
◆ Prepare journal entries summarizing the transactions.

Exhibit 17-1 provides an overview of product costing at Global Defense for the Assembly Department. Note that conversion costs are usually accounted for as indirect costs and that the indirect cost-allocation base is usually expressed as equivalent units of production.

Step 1: Summarize the Flow of Physical Units

Step 1 traces the physical units of production. Where did the units come from? Where did they go? How many units are there to account for? How are they accounted for? It is usually helpful to draw flow charts as a preliminary step. For example:

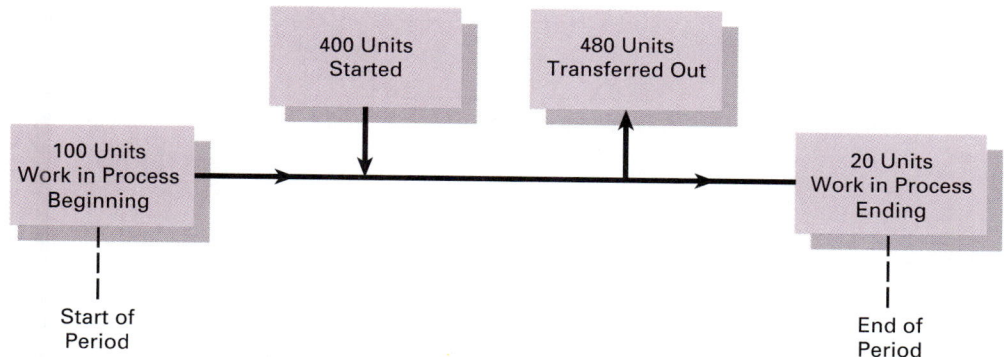

Exhibit 17-2 shows these relationships, which may also be expressed as an equation:

$$\left(\begin{array}{c} \text{Beginning} \\ \text{inventory} \end{array} \right) + \left(\begin{array}{c} \text{Units} \\ \text{started} \end{array} \right) = \left(\begin{array}{c} \text{Units} \\ \text{transferred out} \end{array} \right) + \left(\begin{array}{c} \text{Ending} \\ \text{inventory} \end{array} \right)$$

The total of the left side of the equation is shown in Exhibit 17-2 as the units to account for: 100 + 400 = 500. The total of the right side is shown in Exhibit 17-2 as the units accounted for: 100 + 380 + 20 = 500.

There are 100 physical units (missiles) in process at the start of the period. In addition, 400 units were begun during the current period. Of the total of 100 + 400 = 500 units to account for, 20 units remained in process at the end of the period. Therefore, 480 units (500 – 20) were completed and transferred out during the period, consisting of the 100 units from the beginning work in process plus 380 units from the 400 units started.

All costs incurred are debited to a department account that simultaneously serves as an inventory account, Work in Process—Assembly. As goods are completed

EXHIBIT 17-2

Step 1: Summarize the Flow of Physical Units:
Assembly Department of Global Defense Inc. for March 19_4

Flow of Production	Physical Units
Work in process, beginning*	100
Started during current period†	400
To account for	500
Completed and transferred out during current period:	
From beginning inventory	100
Started and completed	380
Work in process, ending‡	20
Accounted for	500

*Degree of completion: direct materials of this department, 100%; conversion costs of this department, 40%.

†"Current period" is used as a general term. In this example, the current period is one month, March.

‡Degree of completion: direct materials of this department, 100%; conversion costs of this department, 50%.

and transferred out, costs are shifted from Work in Process—Assembly to Work in Process—Testing , usually by monthly entries. For purposes of our example, these accounts will show physical units as well as dollars, which we will enter as we progress. Normally, only dollars are shown in these accounts. (The question marks in the following account are replaced by dollar amounts later in this chapter.)

Work in Process—Assembly

	Physical Units	Dollars		Physical Units	Dollars
Beginning inventory	100	?	Transferred out	480	?
Started	400	?	Ending inventory	20	?
To account for	500	?	Accounted for	500	?

WEIGHTED-AVERAGE METHOD

Objective 3
Demonstrate the weighted-average method of process costing

Step 1 is the same for both process-costing methods (weighted-average and first-in, first-out), but steps 2 through 5 are affected by the choice of method. This section describes the weighted-average method. A subsequent section discusses the first-in, first-out method.

The **weighted-average process-costing method** focuses on the total costs and total equivalent units completed to date; no distinction is made between work completed during the preceding period and work completed during the current period. Consequently, equivalent units include the work completed *before* the current period as well as the work done *during* the current period. Thus the stage of completion of the current period beginning work in process is *not* used in this computation.

Recall that *equivalent units* measure the output in terms of the quantities of each of the factors of production that have been used by the units. That is, an equivalent unit is a collection of production factors necessary to produce one complete physical unit of output.

Recall also that in process costing, costs are often divided into only two main classifications: direct materials and conversion costs. If manufacturing labor is a significant part of total manufacturing costs, an additional cost category (either direct manufacturing labor costs or conversion labor costs) may be used to separately assign these costs to products.

Step 2: Compute Output in Terms of Equivalent Units

Express the physical units in terms of equivalent units of work done. Because direct materials and conversion costs are usually assigned to production differently, the equivalent output is calculated separately for direct materials and conversion costs. Instead of thinking of output in terms of physical units, think of output in terms of direct materials doses of work and conversion-cost doses of work. *Disregard dollar amounts until equivalent units are computed.*

In the Assembly Department, direct materials are introduced at the beginning of the process. Therefore, both the physical units completed and the physical units in the ending work in process are "fully completed" in terms of equivalent units of work done regarding direct materials. Note especially that the direct materials component of work in process is fully completed as soon as work is started. Why? Because in our Liberty Missile example all direct materials added in the Assembly Department are kitted (packaged and assigned to a specific missile unit) at the initial stage of the process. Specific computations follow:

	Direct Materials
Completed and transferred out: 480 × 100%	480
Work in process, ending: 20 × 100%	20
Total equivalent units of work done	500

For conversion costs, the physical units completed and transferred out are fully completed. The physical units in ending work in process are 50% completed (on average, assuming continuous production). They will be weighted accordingly:

	Conversion Costs
Completed and transferred out: 480 × 100%	480
Work in process, ending: 20 × 50%	10
Total equivalent units of work done	490

Exhibit 17-3 combines steps 1 and 2. It summarizes the computation of output in terms of equivalent units.

EXHIBIT 17-3

Step 2: Compute Output in Terms of Equivalent Units: Weighted-Average Method of Process Costing, Assembly Department of Global Defense Inc. for March 19_4

	(Step 1)	(Step 2) Equivalent Units	
Flow of Production	**Physical Units**	**Direct Materials**	**Conversion Costs**
Work in process, beginning*	100	(work done before current period)	
Started during current period	400		
To account for	500		
Completed and transferred out during current period	480	480	480
Work in process, ending†	20	20‡	10§
Accounted for	500		
Work done to date		500	490

*Degree of completion: direct materials of this department, 100%; conversion costs of this department, 40%.
†Degree of completion: direct materials of this department, 100%; conversion costs of this department, 50%.
‡20 units × 100% complete = 20 equivalent units.
§20 units × 50% complete = 10 equivalent units.

Step 3: Summarize Total Costs to Account For

Exhibit 17-4 summarizes the total costs to account for—that is, the total debits in Work in Process. As given in our example data, the debits consist of the beginning balance, $5,110,000,[1] plus the costs added during March, $22,000,000 + $18,000,000 = $40,000,000.

Step 4: Compute Equivalent Unit Costs

Exhibit 17-5 shows the computation of equivalent unit costs. The weighted-average method has been called a "roll-back" method because the averaging of costs includes the work done in the preceding period(s) on the current period's beginning inventory of work in process. Thus, the total costs and the equivalent units combine some of the work begun in February with the work done during March. The equivalent units include all work done to date, including the work done on beginning work in process before the current period. The division of total costs by equivalent units yields equivalent unit costs.

---

[1]The beginning work-in-process cost of $5,110,000 comprises:

Direct materials:		
100 equivalent units × $40,000 per unit February costs		$4,000,000
Conversion costs:		
40 equivalent units × $27,500 per unit February costs		1,110,000
		$5,110,000

EXHIBIT 17-4
Step 3: Summarize Total Costs to Account for:
Weighted-Average Method of Process Costing,
Assembly Department of Global Defense Inc. for March 19_4

Work in Process—Assembly Department					
	Physical Units	**Dollars**		**Physical Units**	**Dollars**
Beginning inventory	100	$ 5,110,000	Transferred out	480	?
Started:	400		Ending inventory	20	?
Direct materials		22,000,000	Accounted for	500	?
Conversion costs		18,000,000			
To account for	500	$45,110,000			

EXHIBIT 17-5
Step 4: Compute Equivalent Unit Costs:
Weighted-Average Method of Process Costing,
Assembly Department of Global Defense Inc. for March 19_4

			Details	
	Totals	**Direct Materials**	**Conversion Costs**	**Equivalent Whole Unit**
Work in process, beginning	$ 5,110,000	$ 4,000,000	$ 1,110,000	
Costs added during current period	40,000,000	22,000,000	18,000,000	
Total costs to account for (step 3)	$45,110,000	$26,000,000	$19,110,000	
Divide by equivalent units (from Exhibit 17-3)		÷ 500	÷ 490	
Equivalent unit costs		$ 52,000	$ 39,000	$91,000

Step 5: Assign Total Costs to Units Completed and to Units in Ending Work in Process

Exhibit 17-6 shows how the equivalent unit costs computed in step 4 are the basis for assigning total costs to units completed and units in ending work in process. The 480 Liberty Missile units completed and transferred out of the Assembly Department are carried at a unit cost of $52,000 + $39,000 = $91,000. The 20 units in ending work in process consist of 20 equivalent units of direct materials at $52,000 and 10 equivalent units of conversion costs at $39,000. Note how the total costs accounted for can be checked against one another in steps 3 and 5. The $45,110,000 in Exhibit 17-4 agrees with the $45,110,000 in Exhibit 17-6.

A production-cost worksheet is a report of the units manufactured during a specified period and their costs (steps 3, 4, and 5). Such a worksheet may be highly summarized or extensively detailed. Supporting schedules are often provided. Exhibit 17-7 is a sample of a production-cost worksheet for the weighted-average method.

EXHIBIT 17-6

Step 5: Assign Total Costs to Units Completed and to Units in Ending Work in Process, Weighted-Average Method of Process Costing, Assembly Department of Global Defense Inc. for March 19_4

		Details	
	Totals	Direct Materials	Conversion Costs
Completed and transferred out (480 units)	$43,680,000	480 ($91,000)	
Work in process, ending (20 units)			
Direct materials	1,040,000	20 ($52,000)	
Conversion costs	390,000		10 ($39,000)
Total work in process	1,430,000		
Total costs accounted for	$45,110,000		

EXHIBIT 17-7

Production-Cost Worksheet:
Weighted-Average Method of Process Costing,
Assembly Department of Global Defense Inc. for March 19_4

			Details	
		Totals	Direct Materials	Conversion Costs
	Work in process, beginning	$ 5,110,000	$ 4,000,000	$ 1,110,000
	Costs added during current period	40,000,000	22,000,000	18,000,000
(Step 3)	Total costs to account for	$45,110,000	$26,000,000	$19,110,000
	Divide by equivalent units*		÷ 500	÷ 490
(Step 4)	Equivalent unit costs		$ 52,000	$ 39,000
(Step 5)	Assignment of costs:			
	Completed and transferred out (480 units)	$43,680,000	480 ($91,000†)	
	Work in process, ending (20 units):			
	Direct materials	1,040,000	20 ($52,000)	
	Conversion costs	390,000		10 ($39,000)
	Total work in process	1,430,000		
	Total costs accounted for	$45,110,000		

*For work done to date. For more details, see Exhibit 17-3.
†Cost per equivalent whole unit = $52,000 + $39,000 = $91,000.

EXHIBIT 17-8

Flow of Costs in a Process-Costing System: Weighted-Average Method of Process Costing, Assembly Department of Global Defense Inc. for March 19_4

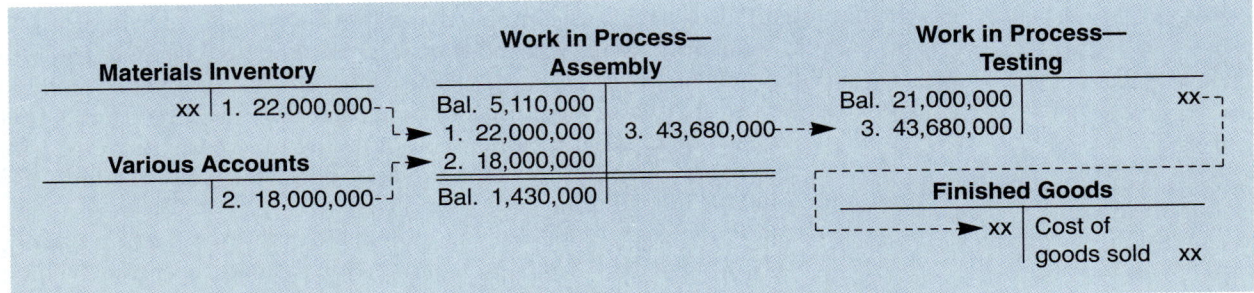

Journal Entries

Objective 4
Prepare journal entries for process-costing systems

Process-costing journal entries are basically like those made in the job-costing system. That is, direct materials and conversion costs are accounted for as in job systems. The main difference is that, in process costing, there is more than one Work-in-Process account.

The data in our weighted-average illustration would be journalized as follows:

1. Work in Process—Assembly $22,000,000
       Materials Inventory $22,000,000
   To record requisitions of direct materials for March.

2. Work in Process—Assembly $18,000,000
       Various accounts $18,000,000
   To record conversion costs for March. Examples of various accounts include manufacturing supplies, all manufacturing labor, and pertinent depreciation on plant and equipment.

3. Work in Process—Testing $43,680,000
       Work in Process—Assembly $43,680,000
   To record cost of goods completed and transferred from Assembly to Testing.

Exhibit 17-8 shows a general sketch of the flow of costs through the T-accounts. The key T-account, Work in Process—Assembly, shows an ending balance of $1,430,000:

Work in Process—Assembly

	Physical Units	Dollars		Physical Units	Dollars
Beginning inventory	100	$ 5,100,000	Transferred out	480	$43,680,000
Started:	400		Ending inventory	20	1,430,000
Direct materials		22,000,000	Accounted for	500	$45,110,000
Conversion costs		18,000,000			
To account for	500	$45,110,000			
Ending inventory	20	$ 1,430,000			

FIRST-IN, FIRST-OUT METHOD

Objective 5
Demonstrate the first-in, first-out (FIFO) method of process costing

The **first-in, first-out (FIFO) process-costing method** computes unit costs by confining equivalent units to work done during the current period only; costs of the current period are separately identified so that the unit costs are related only to the current period's work.

Steps 1 and 2: Physical Units and Equivalent Units

The first-in, first-out (FIFO) method regards the beginning inventory as if it were a batch of goods separate and distinct from the goods started *and* completed in a process during the current period. Step 1, the analysis of physical units, is the same under the two methods. However, subsequent steps taken under the FIFO method do differ from the steps taken under the weighted-average method.

Step 2, the computation of the output in terms of equivalent units, distinguishes between (1) the units carried over in beginning work in process and (2) the units started and completed in the current period. The beginning inventory was finished in March. All direct materials and 40% of conversion costs had been done in February. Hence, none of the direct materials and the remaining 60% of the conversion costs were done in March. Exhibit 17-9 presents the computations for step 2.

Steps 3, 4, and 5: Production-Cost Worksheet

Exhibit 17-10 is the production-cost worksheet for the FIFO method. It presents steps 3, 4, and 5. Concentrate for now on steps 3 and 4. The divisor for computing equivalent unit costs is confined to the work done during the current period. Therefore, the costs added during the current period (only) are divided by the equivalent units for work done during the current period (only). Thus, the $5,110,000 beginning inventory costs and the equivalent units of work done on the beginning inventory in the preceding period are *excluded* from the computation of unit costs.

The step 5 section of Exhibit 17-10 shows how the equivalent unit costs computed in step 4 are the basis for assigning costs to units completed and in ending work in process. First, compute the costs of the 480 units of goods completed and transferred out:

EXHIBIT 17-9

Step 2: Compute Output in Terms of Equivalent Units:
FIFO Method of Process Costing,
Assembly Department of Global Defense Inc. for March 19_4

Flow of Production	(Step 1) Physical Units	(Step 2) Equivalent Units	
		Direct Materials	Conversion Costs
Work in process, beginning*	100	(work done before current period)	
Started during current period	400		
To account for	500		
Completed and transferred out during current period:			
From beginning work in process	100	0	60†
Started and completed	380	380	380
Work in process, ending‡	20	20§	10‖
Accounted for	500		
Work done in current period only		400	450

*Degree of completion: direct costs of this department, 100%; conversion costs of this department, 40%.
†100 units × (100% − 40%) = 100 × 60% = 60 equivalent units.
‡Degree of completion: direct materials of this department, 100%; conversion costs of this department, 50%.
§20 units × 100% complete = 20 equivalent units.
‖20 units × 50% complete = 10 equivalent units.

Work in process, beginning, 100 units		$ 5,110,000
Costs added during the current period to complete the beginning inventory,		
60 equivalent units of conversion costs × $40,000		2,400,000
Total from beginning inventory		7,510,000
Started and completed during the current period:		
380 equivalent whole units × $95,000		36,100,000
Total costs transferred out		$43,610,000

Second, compute the cost of the 20 units in ending work in process:

Direct materials: 20 equivalent units × $55,000	$1,100,000
Conversion costs: 10 equivalent units × $40,000	400,000
Total costs of work in process, ending inventory	$1,500,000

The grand total of costs accounted for is:

Total costs transferred out	$43,610,000
Total costs of work in process, ending inventory	1,500,000
Total costs accounted for	$45,110,000

The average cost of units transferred out is $43,610,000 ÷ 480 units = $90,854 per Liberty Missile. Why does the $90,854 average cost of units transferred out differ from the $95,000 unit cost of units started and completed during March? The Assembly De-

EXHIBIT 17-10

Production-Cost Worksheet:
FIFO Method of Process Costing,
Assembly Department of Global Defense Inc. for March 19_4

		Totals	Details Direct Materials	Details Conversion Costs
	Work in process, beginning	$5,110,000	(costs of work done before current period)	
	Costs added during current period	40,000,000	$22,000,000	$18,000,000
(Step 3)	Total costs to account for	$45,110,000		
	Divide by equivalent units*		÷ 400	÷ 450
(Step 4)	Equivalent unit costs		$55,000	$40,000
(Step 5)	Assignment of costs:			
	Completed and transferred out (480 units):			
	Work in process, beginning (100 units):	$ 5,110,000		
	Costs added during current period:			
	Direct materials		—	
	Conversion costs	2,400,000		60* ($40,000)
	Total from beginning inventory	7,510,000		
	Started and completed (380 units)	36,100,000		380 ($95,000†)
	Total costs transferred out	43,610,000		
	Work in process, ending (20 units):			
	Direct materials	1,100,000	20* ($55,000)	
	Conversion costs	400,000		10* ($40,000)
	Total work in process	1,500,000		
	Total costs accounted for	$45,110,000		

*Equivalent units of work done. (See Exhibit 17-9.)
†Cost per equivalent whole unit = $55,000 + $40,000 = $95,000.

partment uses FIFO to distinguish between monthly batches of production. Succeeding departments, however, such as Testing and Final Assembly, "cost in" these units at the *one* average unit cost ($90,854 in this illustration). If the latter averaging were not done, the attempt to track costs on a pure FIFO basis throughout a series of processes would be cumbersome.

Only rarely is an application of pure FIFO ever encountered in process costing. It should really be called a *modified* or *departmental* FIFO method. Why? Because FIFO is applied within a department to compile the cost of units transferred *out*, but the units transferred *in* during a given period usually are carried at a single average unit cost as a matter of convenience.

COMPARISON OF FIFO AND WEIGHTED-AVERAGE METHODS

The key difference between the FIFO and weighted-average methods is how equivalent units are computed. Ponder the difference by comparing Exhibit 17-3 (p. 603) with Exhibit 17-9 (p. 607):

- ◆ Weighted-average—total work done to date (full weight given to all units completed and transferred out during the current period *plus* partial weight for work done on ending work in process).
- ◆ FIFO—work done during current period only (full weight given to all units started and completed currently *plus* partial weight for work done on beginning and ending work in process).

In turn, differences in equivalent units when coupled with differences in period-to-period unit costs generate differences in equivalent unit costs. Accordingly, there are differences in costs assigned to units completed and those still in process. In our example:

	Weighted Average[*]	FIFO[†]
Cost of units transferred out	$43,680,000	$43,610,000
Ending work in process	1,430,000	1,500,000
Total costs accounted for	$45,110,000	$45,110,000

[*]From Exhibit 17-7, p. 605.
[†]From Exhibit 17-10, p. 608.

In this example, the FIFO ending inventory is higher than the weighted-average ending inventory by $70,000, or 4.9% ($70,000 ÷ $1,430,000). The difference is attributable to variations in equivalent unit costs of direct materials and conversion costs in different months. The unit cost of the work done in March only was $55,000 + $40,000 = $95,000, as shown in Exhibit 17-10. In contrast, Exhibit 17-7 shows the weighted-average unit cost of $52,000 + $39,000 = $91,000. Therefore, the FIFO method results in a larger cost of work in process inventory, March 31, and a smaller March cost of units transferred out.

Unit costs can differ materially between the weighted-average and FIFO methods when (1) the direct materials or conversion costs per unit vary greatly from period to period and (2) the physical inventory levels of work in process are large in relation to the total number of units transferred out (or these work-in-process levels change greatly from period to period).[2]

---

[2] Recall that the per-unit cost changes from February 19_4 to March 19_4 were as follows:

	February	March
Direct materials	$40,000	$55,000
Conversion costs	$27,500	$40,000

During March, Liberty Missile transferred out 500 units. The beginning inventory (March 1) was 100 units, while the ending inventory (March 31) was 20 units.

The FIFO measurements of work done during the current period only are essential for judging current-period performance—March in this illustration. A major advantage of FIFO is that managers can judge the performance in the current period independently from the performance in the preceding period. In brief, the "work done during the current period" is vital information for planning and control purposes as well as for FIFO inventory costing purposes. Standard costing, which is described in the next section, uses "work done during the current period" as a basis for comparing actual costs of the current period with standard costs of the current period.

In practice, standard costing is especially popular among companies that do not have frequent changes in their basic products. For example, manufacturers of coffee, soap, candy bars, cereals, steel, and glass use standard costing for month-to-month planning, control, and inventory costing. However, FIFO and weighted-average process costing are also used, particularly in industries where short product life cycles reduce the attractions of installing a standard-cost system. Examples of such industries include computers and telecommunications.

STANDARD COSTS AND PROCESS COSTS

This section assumes that you have already studied Chapters 7 and 8. If you have not studied those chapters, proceed to the next major section, "Transferred-in Costs in Process Costing," page 614.

As we have mentioned, many companies that use process-costing systems produce numerous similar units of output. Setting standard quantities for inputs is often relatively straightforward in such companies. Standard costs per input unit may then be assigned to the physical standards to develop standard costs.

This section concentrates on the use of standard costs for inventory costing when a process-costing system is used. (Chapters 7 and 8 discussed how standard costs aid planning and control.) The intricacies and conflicts between weighted-average and FIFO historical costing methods are eliminated by using standard costs. Further, weighted-average and FIFO methods become very complicated when used in industries that produce a variety of products. Many observers have stressed that standard costing is especially useful where there are various combinations of materials, operations, and product sizes. For example, a steel-rolling mill uses various steel alloys and produces sheets of various sizes and of various finishes. The items of direct materials are not numerous; neither are the operations performed. But used in various combinations, they yield too great a variety of products to justify the broad averaging procedure of historical process costing. Similarly complex conditions are frequently found, for example, in plants that manufacture rubber products, textiles, ceramics, paints, and packaged food products.

Computations under Standard Costing

EXAMPLE 2: The facts here in example 2 about the Liberty Missile are basically the same as those we used for the Assembly Department in example 1 except that the following standard costs have been developed for the process. The same standard costs apply in February and March of 19_4.

Standard Costs for Assembly Department for March 19_4

Direct materials	$53,000 per unit
Conversion costs	37,000 per unit
Standard cost per unit	$90,000 per unit

The Work in Process, beginning inventory at standard costs is:

Direct materials	$5,300,000
Conversion costs	1,480,000
	$6,780,000

Data for the Assembly Department are:

Physical Units for March 19_4

Work in process beginning inventory (March 1)	100 units
Direct materials (100% complete)	
Conversion costs (40% complete)	
Started during March	400 units
Completed and transferred out during March	480 units
Work in process, ending inventory (March 31)	20 units
Direct materials (100% complete)	
Conversion costs (50% complete)	

For inventory-costing purposes, the accountant must

◆ Compute the standard cost of units completed and transferred out of a process or a department (in our example, Assembly).

Concepts in Action

PROCESS COSTING IN THE CERAMICS INDUSTRY

Ceramics Inc. produces ceramic products (such as multilayer packages for integrated circuits) in a batch flow manufacturing process. Forming and finishing are the two major production stages.

◆ Forming. Ceramic material is mixed, forced through an extruder, and sent to a dryer.

◆ Finishing. The products are fired in a kiln, cut, ground, and packaged.

For many years Ceramics Inc. has manufactured like or similar products in large production runs for industrial customers (termed "original equipment manufacturers," or OEMs) such as computer companies and defense companies.

Ceramic Inc. costs individual products using standard costs in a process-costing system. Cost data are accumulated and tracked in four departments: cutting, extruding, firing, finishing. Conversion costs are allocated using standard (scheduled) hours of production time in each department. Depreciation on plant and equipment is included in this conversion cost. The controller at Ceramics believes that this system "accurately measures the cost of manufacturing OEM products." These products are manufactured in large batches in a highly standardized way.

Ceramics Inc. recently added a "custom production line" at its plant. This line manufactures ceramic products that vary greatly in production volume and frequently are tailored to each individual customer's needs. For example, custom-designed nozzles used for pollution control are being manufactured for one customer who needs to rid its flue gas of sulfur. The controller is skeptical about the accuracy of product costs for these custom products based on the existing process-costing system. She believes that the costs of these products are driven by more variables than standard hours in production at each department.

For example, many custom jobs require specialized finishing steps that are undertaken in a job shop adjoining the main production area. Currently she is keeping a separate, largely manual job-costing system that uses some data from the main system and some separately maintained cost data.

The controller is now exploring ways to adapt the formal process-costing system to incorporate some elements of a job-costing system. Her point is that custom jobs put different demands on the resources of Ceramics Inc. than do the run-of-the-mill large production run jobs. For these custom jobs, a hybrid costing system with elements of both process costing and job costing may be appropriate.

Source: Adapted from U. Karmarkar, P. Lederer, and J. Zimmerman, "Choosing Manufacturing Production Control and Cost Accounting Systems," in R. Kaplan, *Measures For Manufacturing Excellence* (Boston, MA: Harvard Business School Press, 1990). Ceramics Inc. is a fictitious name for the actual company.

- ◆ Compute ending work in process.
- ◆ Compute total direct-materials variance and total conversion-cost variance. Actual direct materials costs were $22,000,000. Actual conversion costs were $18,000,000.
- ◆ Prepare journal entries summarizing the transactions.

The use of standard costs greatly simplifies process-costing computations. Like the FIFO method, standard costs use the "work done during current period only" — as shown in Exhibit 17-9, page 607, and repeated in Exhibit 17-11 below—as the basis for computing equivalent units.

EXHIBIT 17-11

Production-Cost Worksheet:
Use of Standard Costs in a Process-Costing System,
Assembly Department of Global Defense Inc. for March 19_4

	Equivalent Units		
Flow of Production	Physical Units	Direct Materials	Conversion Costs
Work in process, beginning*	100		
Started during current period	400		
To account for	500		
Units completed and transferred out during current period:			
From beginning inventory	100	0	60†
Started and completed	380	380	380
Work in process, ending‡	20	20§	10‖
Accounted for	500		
Work done in current period only		400	450

	Totals	Direct Materials	Conversion Costs
Standard cost per equivalent unit (given)		$ 53,000	$ 37,000
Multiply by work done in current period only		× 400	× 450
Costs assigned at standard prices during current period	$37,850,000	21,200,000	16,650,000
Work in process, beginning	6,780,000	5,300,000	1,480,000
Costs to account for	$44,630,000	$26,500,000	$18,130,000
Assignment of costs:			
Completed and transferred out (480 units)	$43,200,000	480 ($90,000#)	
Work in process, ending (20 units):			
Direct materials	1,060,000	20 ($53,000)	
Conversion costs	370,000		10 ($37,000)
Total work in process	1,430,000		
Total costs accounted for	$44,630,000		
Summary of variances for current performance:			
Current output in equivalent units		400	450
Current output at standard costs assigned		$21,200,000	$16,650,000
Actual costs incurred		22,000,000	18,000,000
Total unfavorable variance**		$ 800,000 U	$ 1,350,000 U

*Degree of completion: direct materials of this department, 100%; conversion costs of this department, 40%.
†100 units × (100% − 40%) = 100 × 60% = 60 equivalent units.
‡Degree of completion: direct materials of this department, 100%; conversion costs of this department, 50%.
§20 units × 100% complete = 20 equivalent units.
‖20 units × 50% complete = 10 equivalent units.
#Cost per equivalent whole unit = $53,000 + $37,000 = $90,000.
**These variances could be broken down further, depending upon available details.

EXHIBIT 17-12

Work-in-Process Account:
Use of Standard Costs in a Process-Costing System,
Assembly Department of Global Defense Inc. for March 19_4

Work in Process—Assembly (at standard costs)					
	Physical Units	Dollars		Physical Units	Dollars
Beginning inventory	100	$ 6,780,000*	Transferred out	480	$43,200,000§
Started:	400		Ending inventory	20	1,430,000‖
Direct materials		21,200,000†	Accounted for	500	$44,630,000
Conversion costs		16,650,000‡			
To account for	500	$44,630,000			
Ending inventory	20	$ 1,430,000			

*100 units × $53,000 $5,300,000
  100 units × 40% × $37,000   1,480,000
                       $6,780,000

†400 units × $53,000 = $21,200,000
‡450 units × $37,000 = $16,650,000
§480 units × ($53,000 + $37,000) = $43,200,000
‖20 units × $53,000 $1,060,000
  20 units × 50% × $37,000   370,000
                       $1,430,000

Steps 1 and 2 are the same for FIFO and standard costing. Exhibit 17-11 shows work done during the current period: direct materials, 400 equivalent units, and conversion costs, 450 equivalent units.

Steps 3, 4, and 5 are easier under standard costing than under the FIFO method. Why? Because the cost per equivalent unit does not have to be computed, as was done for the FIFO method. Instead, the cost per equivalent unit *is* the standard cost. The latter is the key to computing total costs to account for, costs completed and transferred out, and ending work-in-process inventory. Exhibit 17-11 illustrates a production-cost worksheet. Exhibit 17-12 summarizes all these calculations.

Accounting for Variances

Process-costing systems using standard costs usually accumulate actual costs separately from the inventory accounts. The following is an example. The actual data are recorded in the first two entries, and the variances are recorded in the next two entries. The final entry transfers the completed goods out at standard costs. Exhibit 17-13 shows how the costs flow through the accounts.

1. Assembly Department Cost Control (at actual) $22,000,000
        Materials Inventory $22,000,000
     To record requisitions of direct materials for March.
     This "cost control" account is debited with actual costs and credited later with standard costs assigned to the units worked on.

2. Assembly Department Cost Control (at actual) $18,000,000
        Various accounts $18,000,000
     To record conversion costs for March.

3. Work in Process—Assembly (at standard costs) $21,200,000
   Direct Materials Variances   800,000
        Assembly Department Cost Control $22,000,000
     To record inventory costs and direct materials variances.

EXHIBIT 17-13

Flow of Costs:
Use of Standard Costs in a Process-Costing System,
Assembly Department of Global Defense Inc. for March 19_4

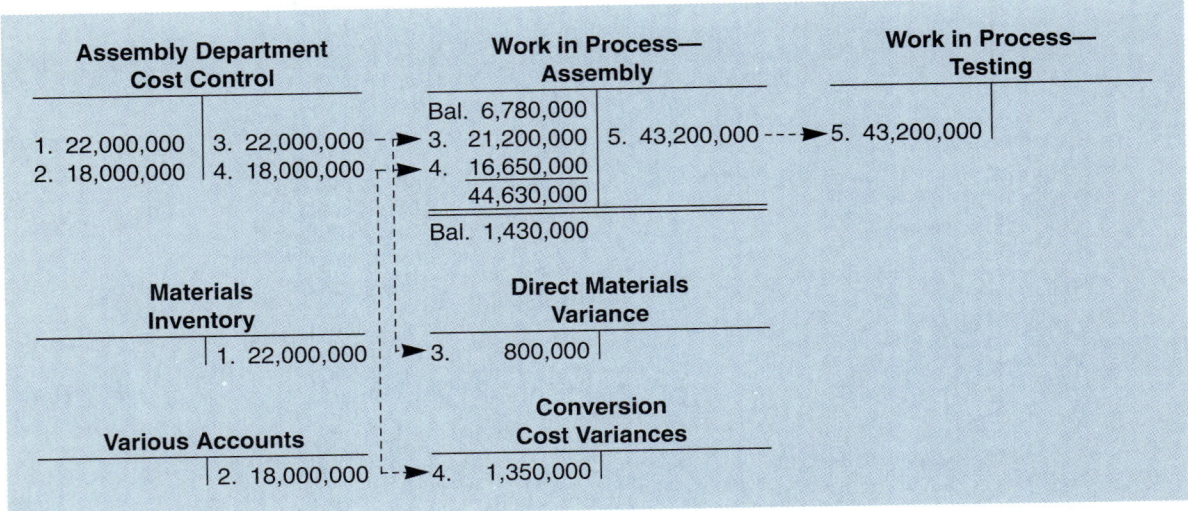

4. Work in Process—Assembly (at standard costs)	$16,650,000	
Conversion-Cost Variances	1,350,000	
Assembly Department Cost Control		$18,000,000
To record inventory costs and conversion-cost variances.		
5. Work in Process—Testing	$43,200,000	
Work in Process—Assembly		$43,200,000
To record cost of units completed and transferred at standard cost from Assembly to Testing.		

Variances can be measured and analyzed in little or great detail in the same manner as described in Chapters 7 and 8.

Standard costs not only eliminate the intricacies of weighted-average and FIFO inventory methods but also erase the need for burdensome computations of costs per equivalent unit. Standard costs *are* the costs per equivalent unit. In addition, a standard-cost approach helps control operations, as explained in Chapters 7 and 8.

Note how the equivalent units for the current period provide the key measures of the work accomplished during March. The analysis of the physical units and equivalent units in Exhibit 17-11 is vital for evaluating current-period performance. The equivalent units are used to cost inventories and to measure variances.

TRANSFERRED-IN COSTS IN PROCESS COSTING

Many process-costing systems have two or more departments or processes in the production cycle. Ordinarily, as units move from department to department, related costs are also transferred by monthly journal entries. If standard costs are used, the accounting for such transfers is relatively simple. However, if FIFO or weighted-average is used, the accounting can become more complex.

To make the ideas concrete, we now extend example 1 to encompass the Testing Department.

EXAMPLE 3: Recall that the Assembly Department of Global Defense transfers units (Liberty Missiles) to its Testing Department. Here the units receive additional direct materials at the *end* of the process. Conversion costs are allocated evenly throughout the Testing Department's process. As the process in Assembly is completed, units are immediately transferred to Testing; as units are completed in Testing, they are immediately transferred to Finished Goods.

Data for the Testing Department for the month of March 19_4 are:

Physical Units for March 19_4

Work in process, beginning inventory (March 1)	120 units
Transferred-in costs (100% complete)	
Direct materials (0% complete)	
Conversion costs (66 2/3% complete)	
Transferred-in during March	480 units
Completed during March	440 units
Work in process, ending inventory (March 31)	160 units
Transferred-in costs (100% complete)	
Direct materials (0% complete)	
Conversion costs (37.5% complete)	

Cost data assuming Global Defense uses the weighted-average method of process costing are:

Costs of Testing Department for March 19_4

Work in process, beginning inventory		
Transferred-in costs	$10,920,000	
Direct materials	0	
Conversion costs	10,080,000	$21,000,000
Transferred-in costs during March		43,680,000
Direct materials costs added during March		13,200,000
Conversion costs added during March		63,000,000

Data for the Testing Department for the month of March 19_4 using the FIFO method of process costing differs in only the amount of transferred-in costs:

Work in process, beginning inventory:	
Transferred-in costs	$10,905,000
Direct materials	0
Conversion costs	10,080,000
	$20,985,000
Transferred-in costs during March	$43,610,000

For inventory-costing purposes, the accountant must

◆ Compute cost of units completed and transferred out of a process or department (in our example, Testing).

◆ Compute ending inventory cost for the units remaining in the department, using the weighted-average method or first-in, first-out (FIFO) method.

◆ Prepare journal entries for transfers to the next processing department or to finished goods.

Transferred-in Costs and the Weighted-Average Method

Objective 7

Demonstrate how transferred-in costs affect weighted-average process costing

The five-step procedure described earlier (p. 599) also pertains when accounting for the costs of a subsequent department that has transferred-in costs. Exhibit 17-14 shows steps 1 and 2 for Testing, which compute physical units and compute equivalent units. Note that direct materials costs have no degree of completion regarding the ending work in process. Why? Because in Testing, direct materials are introduced at the *end* of the process. However, the equivalent units for transferred-in costs are, of course, fully completed because they are always introduced at the beginning of the process.

Treatment of transferred-in costs may become complex. **Transferred-in costs (or previous department costs)** are costs incurred in a previous department that have been charged to a subsequent department. As the units move from the first department to the second department, their costs move with them. Thus, Testing computations must include transferred-in costs, as well as any additional direct materials costs and conversion costs added in Testing.

EXHIBIT 17-14

Steps 1 and 2: Compute Output in Terms of Physical Units and Equivalent Units:
Weighted-Average Method of Process Costing,
Testing Department of Global Defense Inc. for March 19_4

	(Step 1) Physical Units	(Step 2) Equivalent Units		
Flow of Production		Transferred-in Costs	Direct Materials	Conversion Costs
Work in process, beginning*	120	(work done before current period)		
Transferred in during current period	480			
To account for	600			
Completed and transferred out during current period	440	440	440	440
Work in process, ending†	160	160	0‡	60§
Accounted for	600			
Work done to date		600	440	500

*Degree of completion: transferred-in costs, 100%; direct materials of this department, 0%; conversion costs of this department, 66⅔%
†Degree of completion: transferred-in costs, 100%; direct materials of this department, 0%; conversion costs of this department, 37.5%
‡160 units × 0% complete = 0 equivalent units
§160 units × 37.5% complete = 60 equivalent units

EXHIBIT 17-15

Production-Cost Worksheet:
Weighted-Average Method of Process Costing,
Testing Department of Global Defense Inc. for March 19_4

		Totals	Transferred-in Costs	Direct Materials	Conversion Costs
	Work in process, beginning	$ 21,000,000	$10,920,000	$ 0	$10,080,000
	Costs added during current period	119,880,000	43,680,000	13,200,000	63,000,000
(Step 3)	Total costs to account for	$140,880,000	$54,600,000	$13,200,000	$73,080,000
	Divide by equivalent units (from Step 2, Exhibit 17-14)		÷ 600	÷ 440	÷ 500
(Step 4)	Equivalent unit costs		$ 91,000	$ 30,000	$ 146,160
(Step 5)	Assignment of costs:				
	Completed and transferred out (440 units)	$117,550,400		440 ($267,160*)	
	Work in process, ending (160 units):				
	Transferred-in costs	14,560,000	160† ($91,000)		
	Direct materials	0		0	
	Conversion costs	8,769,600			60† ($146,160)
	Total work in process	23,329,600			
	Total cost accounted for	$140,880,000			

*Cost per equivalent whole unit = $91,000 + $30,000 + $146,160 = $267,160.
†Equivalent units of work done. See Exhibit 17-14 for details.

The essential difference between using the weighted-average method in the Assembly Department (Exhibit 17-3, p. 603) and using it in the Testing Department is the accounting for transferred-in costs in the Testing Department. In our example, steps 1 and 2 are unchanged, except for the addition of transferred-in units. Exhibit 17-14 displays the computations of output in equivalent units.

Exhibit 17-15 is a production-cost worksheet. It shows steps 3, 4, and 5. It presents the total costs to account for. Equivalent unit costs are computed and then assigned to the units completed and in ending work in process.

Examine Exhibit 17-15 closely, and note the following points: (1) Under the weighted-average method, unlike under the FIFO method, the costs of beginning work in process must be subdivided into their components. (2) These components are combined with the costs added in the current period to obtain a total amount incurred to date for each cost element: transferred-in costs, direct materials, and conversion costs. (3) Thus, the unit costs of each cost element will be weighted averages.

Note again that the total costs to account for in each of the three categories are divided by the equivalent units for the "work done to date." For instance, the divisor for conversion costs is 500 units. Exhibit 17-16 is a T-account portrayal of the total costs to account for (the total debits in Work in Process).

Production-cost worksheets may be presented in a briefer form than that shown in Exhibit 17-15. For example, the data in Exhibit 17-15 could be the supporting computations for a Production-Cost Report (see Exhibit 17-17).

EXHIBIT 17-16

Summary of Costs Accounted for:
Weighted-Average Method of Process Costing,
Testing Department of Global Defense Inc. for March 19_4

	Physical Units	**Dollars**		**Physical Units**	**Dollars**
		Work in Process—Testing			
Beginning inventory	210	$ 21,000,000	Transferred out	440	$117,550,400
Transferred in	480		Ending inventory	160	23,329,600
Transferred-in costs		43,680,000	Accounted for	600	$140,880,000
Direct materials		13,200,000			
Conversion costs		63,000,000			
To account for	600	$140,880,000			
Ending inventory	160	$ 23,329,600			

EXHIBIT 17-17

Production-Cost Report:
Weighted-Average Method of Process Costing,
Testing Department of Global Defense Inc. for March 19_4

Flow of Production	Physical Units	Total Costs
Work in process, beginning	120	$ 21,000,000
Add during current period	480	119,880,000
To account for	600	$140,880,000
Completed and transferred out	440	$117,550,400
Work in process, ending	160	23,329,600
Accounted for	600	$140,880,000

The journal entry for the transfer out to finished goods inventory would be:

Finished Goods Inventory	$117,550,400	
Work in Process—Testing		$117,550,400
To transfer units to finished goods.		

Sometimes a company may split the Work in Process account into Work in Process—Direct Materials and Work in Process—Conversion Costs. In these cases, the journal entries would contain this greater detail, even though the underlying reasoning and techniques would be unaffected.

Transferred-in Costs and the FIFO Method

Exhibit 17-18 shows the initial two steps for the FIFO method. Computation of equivalent units is basically the same as for FIFO in Assembly. However, transferred-in costs must now be considered.

Exhibit 17-19, the production-cost worksheet, shows steps 3, 4, and 5. Note again how the divisor equivalent units for FIFO differ from the divisor equivalent units for the weighted-average method. FIFO uses equivalent units for work done in the current period only.

Exhibit 17-20 summarizes the entries to Work in Process—Testing. As shown in Exhibit 17-19, the cost of units completed and transferred out would be $117,258,333 and the ending inventory would be $23,536,667.

Remember that in a series of interdepartmental transfers, each department is regarded as a distinct accounting entity. All costs transferred in during a given period are carried at one unit-cost figure, regardless of whether previous departments used the weighted-average or the FIFO method.

EXHIBIT 17-18
Steps 1 and 2: Compute Output in Terms of Physical Units and Equivalent Units:
FIFO Method of Process Costing,
Testing Department of Global Defense Inc. for March 19_4

	(Step 1)	(Step 2) Equivalent Units		
Flow of Production	**Physical Units**	**Transferred-in Costs**	**Direct Materials**	**Conversion Costs**
Work in process, beginning*	120	(work done before current period)		
Transferred in during current period	480			
To account for	600			
Completed and transferred out during current period:				
From beginning work in process	120	0	120	40†
Started and completed	320	320	320	320
Work in process, ending‡	160	160	0§	60‖
Accounted for	600			
Work done in current period only		480	440	420

*Degree of completion: transferred-in costs, 100%; direct materials of this department, 0%; conversion costs of this department, 66 ⅔%.
†120 units × (100% − 66⅔%) = 120 × 33⅓% = 40 equivalent units.
‡Degree of completion: transferred-in costs, 100%; direct materials of this department, 0%; conversion costs of this department, 37.5%.
§160 units × 0% complete = 0 equivalent units.
‖160 units × 37.5% complete = 60 equivalent units.

EXHIBIT 17-19

Production-Cost Worksheet: Physical Units and Equivalent Units,
FIFO Method of Processing Costing,
Testing Department of Global Defense Inc. for March 19_4

	Totals	Transferred-in Costs	Direct Materials	Conversion Costs
		Details		
Work in process, beginning	$ 20,985,000	(costs of work done before current period)		
Costs added during current period	119,810,000	$43,610,000	$13,200,000	$63,000,000
(Step 3) Total costs to account for	$140,795,000			
Divide by equivalent units		÷ 480	÷ 440	÷ 420
(Step 4) Equivalent unit costs		$90,854.167*	$ 30,000	$ 150,000
(Step 5) Assignment of costs:				
Completed and transferred out (440 units)				
Work in process, beginning (120 units)	$ 20,985,000			
Costs added during current period:				
Direct materials	3,600,000		120[†] ($30,000)	
Conversion costs	6,000,000			40 ($150,000[†])
Total from beginning inventory	$ 30,585,000			
Started and completed (320 units)				
	86,673,333	(320 × $270,854.167[‡])		
Total costs transferred out	$117,258,333			
Work in process, ending (160 units):				
Transferred-in costs	$ 14,536,667	160[†] ($90,854.167)		
Direct materials	0		—	
Conversion costs	9,000,000			60[†] ($150,000)
Total work in process	$ 23,536,667			
Total costs accounted for	$140,795,000			

*The unit costs are carried to several decimal places in these exhibits. Of course, they could be rounded with no harm done. However, small discrepancies in totals caused by rounding will occur.
[†]Equivalent units of work done. See Exhibit 17-18 for details.
[‡]Cost per equivalent whole unit = $90,854.167 + $30,000 + $150,000 = $270,854.167.

EXHIBIT 17-20

Summary of Total Costs to Account for (the Total Entries to Work in Process):
FIFO Method of Process Costing,
Testing Department of Global Defense Inc. for March 19_4

Work in Process—Testing					
	Physical Units	Dollars		Physical Units	Dollars
Beginning inventory	120	$ 20,985,000	Transferred out	440	$117,258,333
Transferred in:	480		Ending inventory	160	23,536,667
Transferred-in costs		43,610,000	Accounted for	600	$140,795,000
Direct materials		13,200,000			
Conversion costs		63,000,000			
To account for	600	$ 140,795,000			
Ending inventory	160	$ 23,536,667			

Common Mistakes

Avoid some common pitfalls when accounting for transferred-in costs:

1. Remember to include transferred-in costs from previous departments in your calculations. Such costs should be treated as if they were another kind of direct materials cost added at the beginning of the process. In other words, when successive departments are involved, transferred units from one department become all or a part of the direct materials of the next department, although they are called *transferred-in costs*, not direct materials costs.

2. In calculating costs to be transferred on a first-in, first-out basis, do not overlook the costs assigned at the beginning of the period to units that were in process but are now included in the units transferred. For example, do not overlook the $20,985,000 in Exhibit 17-19.

3. Unit costs may fluctuate between periods. Therefore, transferred goods may contain batches accumulated at different unit costs (see point 2). These goods, when transferred to the next department, are typically costed in the next department at *one* average unit cost.

4. Units may be expressed in terms of different units of measure in different departments. Consider each department separately. Unit costs could be based on kilograms in the first department and liters in the second. As units are received by the second department, they must be converted to the liter unit of measure.

ADDITIONAL ASPECTS OF PROCESS COSTING

This chapter's illustrations and almost all process-cost problems blithely mention various degrees of completion for inventories in process. The accuracy of these estimates depends on the care and skill of the estimator and the nature of the process. Estimating the degree of completion is usually easier for direct materials than for conversion costs. The conversion sequence usually consists of a number of basic operations or a specified number of hours, days, weeks, or months for mixing, heating, cooling, aging, curing, and so forth. Thus, the degree of completion for conversion costs depends on what proportion of the total effort needed to complete one unit or one batch has been devoted to units still in process. In industries in which no exact estimate is possible or, as in the textile industry, in which vast quantities in process prohibit costly physical estimates, all work in process in every department is assumed to be complete to some reasonable degree (for example, $\frac{1}{3}$, $\frac{1}{2}$, or $\frac{2}{3}$ complete).

Businesses use standard costing with process costing far more than they use actual costing (weighted-average or FIFO). The FIFO method is used the least.

PROBLEM FOR SELF-STUDY

PROBLEM

Allied Chemicals operates a thermo-assembly process at its plastics plant. Direct materials in thermo-assembly are added at the end of the process. The following data pertain to the Thermo-Assembly Department for June 19_4:

Work in process, beginning inventory	50,000 units
Transferred-in costs (100% complete)	
Direct materials (0% complete)	
Conversion costs (80% complete)	
Transferred in during current period	200,000 units
Completed and transferred out during current period	210,000 units
Work in process, ending inventory	? units
Transferred-in costs (100% complete)	
Direct materials (0% complete)	
Conversion costs (40% complete)	

Required

Compute the equivalent units for the current period using (1) the weighted-average method and (2) the FIFO method (or the standard-costing method, which would be the same as FIFO).

SOLUTION

1. Weighted-average

	(Step 1)	(Step 2) Equivalent Units		
Flow of Production	**Physical Units**	**Transferred-in Costs**	**Direct Materials**	**Conversion Costs**
Work in process, beginning*	50,000	(Work done before current period)		
Transferred in during current period	200,000			
To account for	250,000			
Completed and transferred out during current period	210,000	210,000	210,000	210,000
Work in process, ending†	40,000	40,000	0‡	16,000§
Accounted for	250,000			
Work done to date		250,000	210,000	226,000

*Degree of completion: transferred-in costs, 100%; direct materials of this department, 0%; conversion costs of this department, 80%.

†Degree of completion: transferred-in costs, 100%; direct materials of this department, 0%; conversion costs of this department, 40%.

‡40,000 units × 0% complete = 0 equivalent units.

§40,000 units × 40% complete = 16,000 equivalent units.

2. FIFO (or standard costing)

	(Step 1)	(Step 2) Equivalent Units		
Flow of Production	**Physical Units**	**Transferred-in Costs**	**Direct Materials**	**Conversion Costs**
Work in process, beginning*	50,000	(Work done before current period)		
Transferred in during current period	200,000			
To account for	250,000			
Completed and transferred out during current period				
From beginning work in process	50,000	0	50,000	10,000†
Started and completed during current period	160,000	160,000	160,000	160,000
Work in process, ending‡	40,000	40,000	0§	16,000‖
Accounted for	250,000			
Work done to date		200,000	210,000	186,000

*Degree of completion: transferred-in costs, 100%; direct materials of this department 0%; conversion costs of this department, 80%.

†50,000 units × (100% − 80%) = 50,000 × 20% = 10,000 equivalent units.

‡Degree of completion: transferred-in costs, 100%; direct materials of this department 0%; conversion costs of this department, 40%.

§40,000 units × 0% = 0 equivalent units.

‖40,000 units × 40% = 16,000 equivalent units.

SUMMARY

The following points are linked to the chapter's learning objectives.

1. Process-costing systems compute the unit costs of like or similar products or services. The key feature of process costing is the averaging of costs over a quantity (often large) of these like or similar units. Computing equivalent units in a process-costing system can make a material difference to reported inventory and income statement amounts when a significant percentage of total production in a period is in work-in-process inventory and a significant percentage of unit cost is assigned before the final production point of a product.

2. The five key steps in a process-costing system using equivalent units are: (a) summarize the physical flow, (b) compute output in equivalent units, (c) summarize total costs to account for, (d) compute equivalent unit costs, and (e) assign total costs to units completed and to units in ending work in process.

3. The weighted-average method of process costing computes unit costs by focusing on the total costs and the total equivalent units completed to date in a department; no distinction is made between work completed during the preceding period and work completed during the current period.

4. Journal entries in a process-costing system are similar to entries in a job-costing system. The main difference is that in a process-costing system, there is a separate work-in-process account for each department rather than for each product.

5. The first-in, first-out (FIFO) method of process costing computes unit costs by confining equivalent units to work done during the current period.

6. The use of standard costs in a process-costing system reduces the amount of detail found in weighted-average or FIFO process-costing systems.

7. Transferred-in costs in a weighted-average process-costing system must be subdivided into their components—transferred-in costs, direct materials costs, and conversion costs.

8. Transferred-in costs in a FIFO process-costing system can be retained as a single amount. FIFO uses only the costs of a period and equivalent units for work done in that period to determine the cost of each equivalent unit produced in that period.

TERMS TO LEARN

This chapter and the Glossary at the end of the book contain definitions of the following important terms:

first-in, first out (FIFO) process-costing method *(p. 606)*   previous department costs *(616)*
transferred-in costs *(616)*   weighted-average process-costing method *(602)*

ASSIGNMENT MATERIAL

QUESTIONS

17-1 State two conditions under which computing equivalent units will make a material difference to reported inventory amounts.

17-2 Name the five key steps in process costing when equivalent units are computed.

17-3 Name the three inventory methods commonly associated with process costing.

17-4 Name three tasks the accountant must perform for each department regarding inventory costing in process costing.

17-5 "Step 1 of the five key steps is the same for all inventory methods, but steps 2 through 5 are affected by the choice of method." Do you agree? Explain.

17-6 In process costing, into what two main classifications are costs often divided?

17-7 Explain why the stage of completion of the beginning inventory is not used when computing equivalent units under the weighted-average method.

17-8 Describe the distinctive characteristic of FIFO computations of equivalent units.

17-9 Why should the FIFO method be called a *modified* or *departmental* FIFO method?

17-10 Identify a major advantage of the FIFO method for purposes of planning and control.

17-11 "Standard-cost procedures are particularly applicable to process-costing situations." Do you agree? Why?

17-12 Why should the accountant distinguish between *transferred-in costs* and *additional direct materials costs* for a particular department?

17-13 "Previous department costs are those incurred in the preceding accounting period." Do you agree? Explain.

17-14 What problems might arise in estimating the degree of completion of an aircraft blade in a machining shop?

17-15 "There's no reason for me to get excited about the choice between the weighted-average and FIFO methods in my process-costing system. I have long-term contracts with my materials suppliers at fixed prices." State the conditions under which you would (a) agree and (b) disagree with this statement, made by a plant controller. Explain.

EXERCISES AND PROBLEMS

17-16 Weighted-average equivalent units. Consider the following data for West Virginia Steel:

Flow of Production	Physical Units (Tons)
Work in process, beginning*	20,000
Started in current period	70,000
To account for	90,000
Completed and transferred out during current period	?
Work in process, ending†	5,000
Accounted for	90,000

*Degree of completion: direct materials, 60%; conversion costs, 30%.
†Degree of completion: direct materials, 80%; conversion costs, 60%.

Required

Prepare a schedule of equivalent units for direct materials and conversion costs under the weighted-average method. Show physical units in the first column.

17-17 Weighted-average and FIFO equivalent units. Consider the following data for the satellite assembly division of Aerospatiale:

	Physical Units (Satellites)
Started in May 19_5	50
Completed in May 19_5	46
Ending work in process (May 31)	12
Beginning work in process (May 1)	8

The beginning work in process was 90% complete regarding direct materials and 40% complete regarding conversion costs. The ending work in process was 60% complete regarding direct materials and 30% complete regarding conversion costs.

Required

Prepare schedules of equivalent units for (1) the work done to date (for the weighted-average method) and (2) the work done during May 19_5 only (for the FIFO method).

17-18 Weighted-average and FIFO equivalent units. Consider these September data for physical units in the ThermoAssembly process of Bangkok Plastics: beginning work in process, 15,000; transferred-in from the Extruding Department during September, 9,000; ending work in process, 5,000. Direct materials are added when the process in the ThermoAssembly Department is 80% complete. Conversion costs are allocated evenly throughout the process. The beginning inventory was 60% completed as to conversion costs; the ending inventory was 20% completed as to conversion costs.

Required

Prepare schedules of equivalent units for (1) the work done to date (for the weighted-average method) and (2) the work done during September only (for the FIFO method). Include computations of equivalent units for transferred-in costs as well as for conversion costs and direct materials.

17-19 Weighted-average, FIFO. Kristina Company, which manufactures quality paint sold at premium prices, uses a single production department. Production begins with the blending of various chemicals, which are added at the beginning of the process, and ends with the canning of the paint. Canning occurs when the mixture reaches the 90% stage of completion. The gallon cans are then transferred to the Shipping Department for crating and shipment. Direct manufacturing labor and manufacturing overhead are added continuously throughout the process. Manufacturing overhead is allocated on the basis of direct manufacturing labor hours at the rate of $3 per hour.

Before May, work-in-process inventories were insignificant, but a change in the process has been implemented. This change enables greater production but results in material amounts of work in process for the first time. The company has always used the weighted-average method to determine equivalent production and unit costs. Now, production management is considering changing from the weighted-average method to the first-in, first-out method.

The following data are related to actual production during the month of May, 19_4:

Physical Units for May 19_4	
Work in process, beginning inventory	4,000 gallons
Direct materials: chemicals (?% complete)	
Direct materials: cans (0% complete)	
Direct manufacturing labor (25% complete)	
Manufacturing overhead (25% complete)	
Started in May	21,000 gallons
Sent to Shipping Department in May	20,000 gallons
Work in process, ending inventory	5,000 gallons
Direct materials: chemicals (100% complete)	
Direct materials: cans (?% complete)	
Direct manufacturing labor (80% complete)	
Manufacturing overhead (80% complete)	

Costs for May 19_4

Work in process, beginning inventory	
Direct materials: chemicals	$ 45,600
Direct manufacturing labor ($10 per hour)	6,250
Manufacturing overhead	1,875
May costs added	
Direct materials: chemicals	$228,400
Direct materials: cans	7,000
Direct manufacturing labor ($10 per hour)	35,000
Manufacturing overhead	10,500

Required

1. What is the percentage of completion for (a) direct materials: chemicals in work in process (beginning inventory) and (b) direct materials: cans in work in process (ending inventory)?

2. Prepare a schedule of equivalent units for each cost element for the month of May using (a) the weighted-average method and (b) the first-in, first-out method.

3. Calculate the cost (to the nearest cent) per equivalent unit for each cost element for the month of May using (a) the weighted-average method and (b) the first-in, first-out method.

4. Discuss the advantages and disadvantages of using the weighted-average method versus the first-in, first-out method.

17-20 Weighted-average method. Global Defense Inc. is a manufacturer of military equipment. Its Santa Fe plant manufactures the Interceptor Missile under contract to the U.S. government and "friendly" countries. All Interceptors go through an identical manufacturing process. Every effort is made to ensure that all Interceptors are identical and meet many demanding performance specifications. The product-costing system at the Santa Fe plant has a single direct cost category (direct materials) and a single indirect cost category (conversion costs). Each Interceptor passes through two departments—the Assembly Department and the Testing Department. Direct materials are added at the beginning of the process in Assembly. Conversion costs are allocated evenly throughout the two departments. When the Assembly Department finishes work on each Interceptor, it is immediately transferred to Testing.

Data for the Assembly Department for October 19_4 are:

Physical Units for October 19_4

Work in process, beginning inventory (October 1)	20 units
Direct materials (?% complete)	
Conversion costs (60% complete)	
Started during October	80 units
Completed and transferred out during October	90 units
Work in process, ending inventory (October 31)	10 units
Direct materials (?% complete)	
Conversion costs (70% complete)	

Costs for October 19_4

Work in process, beginning inventory		
Direct materials	$460,000	
Conversion costs	120,000	$ 580,000
Direct materials added during October		2,000,000
Conversion costs added during October		935,000

Global uses the weighted-average method of process costing.

Required

1. Prepare a schedule for October 19_4 of output in equivalent units for the Assembly Department.

2. Prepare a production-cost worksheet for the Assembly Department for October 19_4.

17-21 Journal entries. (Continuation of 17-20) Prepare a set of summarized journal entries for all October 19_4 transactions affecting Work in Process—Assembly Department. Set up a T-account for Work in Process—Assembly Department, and post the entries to it.

17-22 FIFO method. (Continuation of 17-20 and 17-21) Repeat Problem 17-20 using the FIFO method of process costing. Explain any difference between the cost per equivalent whole unit in the Assembly Department under the weighted-average method and the FIFO method.

17-23 Transferred-in costs, weighted average. (Related to 17-20 to 17-22)   Global Defense Inc., as you know, manufactures the Interceptor Missile at its Santa Fe plant. It has two departments—Assembly Department and Testing Department. This problem focuses on the Testing Department. (Problems 17-20 to 17-22 focused on the Assembly Department.) Direct materials are added at the end of the Testing Department. Conversion costs are allocated evenly throughout the Testing Department's process. As work in Assembly is completed, each unit is immediately transferred to Testing. As each unit is completed in Testing, it is immediately transferred to Finished Goods.

Data for the Testing Department for October 19_4 are:

Physical Units for October 19_4

Work in process, beginning inventory (October 1)	30 units
Transferred-in costs (?% complete)	
Direct materials (?% complete)	
Conversion costs (70% complete)	
Transferred-in during October	? units
Completed during October	105 units
Work in process, ending inventory (October 31)	15 units
Transferred-in costs (?% complete)	
Direct materials (?% complete)	
Conversion costs (60% complete)	

Cost of Testing Department for October 19_4

Work in process, beginning inventory		
Transferred-in costs	$985,800	
Direct materials	0	
Conversion costs	331,800	$1,317,600
Transferred-in costs during October		3,192,866
Direct materials costs added during October		3,885,000
Conversion costs added during October		1,581,000

Global Defense uses the weighted-average process costing method.

Required
1. What is the percentage of completion for (a) transferred-in costs and direct materials in work in process (beginning inventory) and (b) transferred-in costs and direct materials in work in process (ending inventory)?
2. Prepare a schedule of output in equivalent units. Prepare a production-cost worksheet for the Testing Department for October.
3. Prepare journal entries for October transfers from the Assembly Department to the Testing Department and from the Testing Department to Finished Goods.

17-24 Transferred-in costs, FIFO costing. (Continuation of 17-23). Using the FIFO process-costing method, repeat the requirements of Problem 17-23. The transferred-in costs from the Assembly Department for the beginning work in process on October 1 were $980,060. During October costs transferred in to the Testing Department were $3,188,000.

17-25 Weighted-average method. Star Toys manufactures wooden toy figures. It buys wood as its direct material for the Forming Department of its Madison plant. This plant processes one type of toy. The toys are transferred to the Finishing Department, where they are hand-shaped and metal is added to them.

Consider the following data for the Forming Department in April 19_5:

Physical Units for April 19_5

Work in process, beginning inventory (April 1)	300 units
Direct materials (100% complete)	
Conversion costs (40% complete)	
Started in April	2,200 units
Completed during April	2,000 units
Work in process, ending inventory (April 30)	500 units
Direct materials (100% complete)	
Conversion costs (25% complete)	

Costs for April 19_5

Work in process, beginning inventory		
Direct materials	$7,500	
Conversion costs	2,125	$ 9,625
Direct materials added during April		70,000
Conversion costs added during April		42,500
Total costs to account for		$122,125

Star Toys uses the weighted-average method of process costing.

Required

1. Prepare a schedule of output in equivalent units.
2. Prepare a production-cost worksheet for the Assembly Department for April. (For journal entries, see the next problem.)

17-26   Journal entries. (Continuation of 17-25)   Prepare a set of summarized journal entries for all April transactions affecting Work in Process—Forming Department. Set up a T-account for Work in Process—Forming Department, and post the entries to it.

17-27   FIFO computations. (Continuation of 17-25 and 17-26)   Repeat Problem 17-25, using FIFO and four decimal places for unit costs.

17-28   Transferred-in costs, weighted-average. (Related to 17-25 to 17-27)   Star Toys manufactures wooden toy figures at its Madison plant. It has two departments—a Forming Department and a Finishing Department. (Problems 17-25 to 17-27 focused on the Forming Department.) Consider now the Finishing Department, which processes the formed toys through hand shaping and the addition of metal. For simplicity here, suppose all additional direct materials are added at the end of the process.

The following is a summary of the April operations in the Finishing Department:

Physical Units for April 19_5	
Work in process, beginning inventory (April 1)	500 units
Transferred-in costs (100% complete)	
Direct materials (0% complete)	
Conversion costs (60% complete)	
Transferred-in during April	2,000 units
Completed during April	2,100 units
Work in process, ending inventory (April 30)	400 units
Transferred-in costs (100% complete)	
Direct materials (0% complete)	
Conversion costs (30% complete)	

Costs of Finishing Department for April 19_5		
Work in process, beginning inventory		
Transferred-in costs	$17,750	
Direct materials	0	
Conversion costs	7,250	$ 25,000
Transferred-in costs from Forming		
Department during April		104,000
Direct materials added during April		23,100
Conversion costs added during April		38,400
Total costs to account for		$190,500

Star Toys uses the weighted-average method of process costing.

Required

1. Prepare a schedule of output in equivalent units. Prepare a production-cost worksheet for the Finishing Department for April.
2. Prepare journal entries for April transfers from the Forming Department to the Finishing Department and from the Finishing Department to Finished Goods.

17-29   Transferred-in costs, FIFO costing. (Continuation of 17-28)   Using the FIFO process-costing method, repeat the requirements of Problem 17-28. The transferred-in costs from the Forming Department for the beginning work in process, April, were $17,520. During April the costs transferred in were $103,566.

17-30 Transferred-in costs, weighted-average and FIFO. Frito-Lay, Inc., manufactures convenience foods, including potato chips and corn chips. Production of corn chips occurs in four departments: cleaning, mixing, cooking, and drying and packaging. Consider the Drying and Packaging Department, where direct materials (packaging) is added at the end of the process. Conversion costs are allocated continuously during the period. Suppose the accounting records of a Frito-Lay plant provided the following information for corn chips in its Drying and Packaging Department during a weekly period (week 37):

Physical Units	
Work in process, beginning inventory	1,250 cases
Transferred in costs (100% complete)	
Direct materials (?% complete)	
Conversion costs (80% complete)	
Transferred in from the Cooking Department	
during week 37	5,000 cases
Completed during week 37	5,250 cases
Work in process, ending inventory	1,000 cases
Transferred-in costs (?% complete)	
Direct materials (?% complete)	
Conversion costs (40% complete)	

Costs (computed using weighted-average method)		
Work in process, beginning inventory		
Transferred-in costs	$29,000	
Direct materials	0	
Conversion costs	9,060	$ 38,060
Transferred-in costs from the Cooking Department		96,000
Direct materials added in week 37		25,200
Conversion costs added in week 37		38,400
Total costs to account for		$197,660

Required

1. Compute the equivalent units for transferred-in costs, direct materials, and conversion costs. Use (a) the weighted-average method and (b) the FIFO method.
2. Prepare a production-cost worksheet for the Drying and Packaging Department using its weighted-average method data.
3. Assume that the FIFO method is used for the Drying and Packaging Department. The transferred-in costs for work in process beginning inventory are $28,920. The transferred-in costs during the week from the Cooking Department were $94,000. Compute the unit costs for assigning the week's costs to products.

 17-31 FIFO process costing, transferred-in costs. Pepperell Mills, Inc., manufactures broadloom carpet in seven processes: spinning, dyeing, plying, spooling, tufting, latexing, and shearing.

First, fluff nylon purchased from a company such as DuPont or Monsanto is spun into yarn that is dyed the desired color. Then two or more threads of the yarn are joined together, or plied, for added strength. The plied yarn is spooled for use in the actual carpet making. Tufting is the process by which yarn is added to burlap backing. After the backing is latexed to hold it together and make it skid-resistant, the carpet is sheared to give it an even appearance and feel.

At March 31, before recording the transfer of costs from the Tufting Department to the Latexing Department, the Pepperell Mills general ledger included the following account for one of its lines of carpet:

Work in Process—Tufting Department

February 28 balance	33,900
Transferred in from Spooling	
Department	224,000
Direct materials	24,200
Conversion costs	28,150

Work in process inventory of the Tufting Department on February 28 consisted of 75 rolls that were 40% complete as to direct materials and 60% complete as to conversion costs. During March, 560 rolls were transferred in from the Spooling Department. The Tufting Department completed 500 rolls of the carpet in March, and 135 rolls were still in process on March 31. This ending inventory was 100% complete as to direct materials and 80% complete as to conversion costs.

Required

1. Using the FIFO method, compute the equivalent units of production for the Tufting Department for March.

2. Prepare a production-cost worksheet for March.

17-32 Journal entries. (Continuation of 17-31). Journalize all transactions affecting the Tufting Department during March, including those entries that have already been posted.

17-33 Standard costs. (Continuation of 17-31 and 17-32) Assume that standard costs per unit for the Tufting Department are $38 for direct materials and $47 for conversion costs. Compute the total March variances for direct materials and conversion costs.

17-34 Standard costing with beginning and ending work in process. The Victoria Corporation uses a standard-costing system for its manufacturing operations. Standard costs for the cooking process are $6 per unit for direct materials and $3 per unit for conversion costs. All direct materials are introduced at the beginning of the process, but conversion costs are allocated uniformly throughout the process. The operating summary for May 19_4 included the following data for the cooking process:

> Work-in-process inventories:
> May 1: 3,000 units*
> (direct materials $18,000; conversion costs $5,400)
> May 31: 5,000 units†
> Units started in May: 20,000
> Units completed and transferred out of cooking in May: 18,000
> Additional actual costs incurred for cooking during May:
> Direct materials: $125,000
> Conversion cost: $57,000

*Degree of completion: direct materials, 100%; conversion costs, 60%.
†Degree of completion: direct materials, 100%; conversion costs, 50%.

Required

1. Compute the total standard costs of units transferred out in May and the total standard costs of the May 31 inventory of work in process.

2. Compute the total May variances for direct materials and conversion costs.

17-35 Equivalent unit computations, benchmarking, ethics. Margaret Major is the Corporate Controller of Leisure Suits. Leisure Suits has 20 plants worldwide that manufacture basic suits for retail stores. Each plant uses a process-costing system. At the end of each month, each plant manager submits a production report and a production-cost report. The production report includes the plant manager's estimate of the percentage of completion of the ending work in process as to direct materials and conversion costs. Major uses these estimates to compute the equivalent work units done in each plant and the cost per equivalent work unit done for both direct materials and conversion costs. Plants are ranked from 1 to 20 in terms of (a) cost per equivalent unit of direct materials and (b) cost per equivalent unit of conversion costs. Each month Major publishes a report that she calls "Benchmarking for Efficiency Gains at Leisure Suits." The top three ranked plants on each category receive a bonus and are written up as the "best in their class" in the company newsletter.

Major has been pleased with the success of her benchmarking program. However, she has heard some disturbing news. She has received some unsigned letters stating that two plant managers have been manipulating their monthly estimates of percentage of completion in an attempt to obtain "best in class" status.

Required

1. How and why might plant managers "manipulate" their monthly estimates of percentage of completion?

2. Major's first reaction is to contact each plant controller and discuss the problem raised by the unsigned letters. Is that a good idea?

3. Assume that the plant controller's primary reporting responsibility is to the plant manager and that each plant controller receives the phone call from Major mentioned in requirement 2. What is the ethical responsibility of each plant controller (a) to Margaret Major and (b) to Leisure Suits in relation to the equivalent-unit information each plant provides for the "Benchmarking for Efficiency" report? Refer to the "Standards of Ethical Conduct for Management Accountants" (p. 17) in your answer.

4. How might Major gain some insight into whether the equivalent unit figures provided by particular plants are being manipulated?

18

Spoilage, Reworked Units, and Scrap

Management effort and control

Terminology

Spoilage in general

Process costing and spoilage

Job costing and spoilage

Reworked units

Accounting for scrap

Comparison of accounting for spoilage, rework, and scrap

Appendix: Inspection and spoilage at intermediate stages of completion in process costing

*C*ost management at carpentry plants requires efficient usage of timber materials. The De Vries window plant of The Rugby Group in Gorredijk, The Netherlands, monitors both financial and nonfinancial variables in its efforts to minimize spoilage, reworked units, and scrap.

In recent years, managers have paid more attention to the costs of spoilage, reworked units, and scrap. Accounting systems that record these costs in a timely and detailed way help managers make more-informed decisions, especially concerning production systems. For example, consider investments in cutting-edge production systems such as just-in-time (JIT) and computer-integrated manufacturing (CIM). Managers often cite reductions in costs of spoilage, reworked units, and scrap as a major justification for these investments.

This chapter concentrates on how spoilage, rework, and scrap are recorded in management accounting systems. Further discussion of these and other items appears in Chapter 23. Chapter 23 highlights how the costs discussed in this chapter are an important but far-from-complete estimate of the total costs to an organization of spoilage, rework, and scrap.

MANAGEMENT EFFORT AND CONTROL

Some amount of spoilage, rework, or scrap appears to be an inherent part of many production processes. One example is semiconductor manufacturing, where there is statistically significant variability over time in the number of good wafers produced. Another example is a mining company that processes batches of ore that contain a varying mix of valuable metals as well as scrap. The volume of this scrap cannot be controlled. This mining company, however, may employ production workers who pay inadequate attention to detail, and machines may be poorly maintained. Managers could take steps to improve those conditions and hence could better control quality.

Intensified competition in an expanding, more global marketplace has increasingly focused on improving quality. Executives have learned that a rate of defects regarded as normal in the past is no longer tolerable. Consider these words from a speech by the chief executive officer of Motorola, an electronics manufacturer:

> We want to improve our quality in everything we do by ten times in two years, by a hundred times in four years, and in six years . . . three and a half defects for every million operations, whether typing, manufacturing, or serving a customer.

TERMINOLOGY

The definitions and accounting used for spoilage, reworked units, and scrap vary considerably from organization to organization. This chapter uses the following definitions:

◆ **Spoilage.** Unacceptable units of production that are discarded or sold for net disposal proceeds. Partially completed or fully completed units may be spoiled. Net spoilage costs are the total of the costs assigned to the product up to the point of its rejection (plus its disposal costs or minus net disposal proceeds).

◆ **Reworked units.** Unacceptable units of production that are subsequently reworked and sold as acceptable finished goods. Such units may be sold through regular marketing channels or alternative channels, depending on the characteristics of the product and on the available alternatives.

◆ **Scrap.** Defined in Chapter 16 as a product that has minimal (frequently zero) sales value compared with the sales value of the main or joint product(s). Scrap may be either sold, disposed of, or reused. Examples are shavings and short lengths from woodworking operations and frayed and torn cloth and end cuts from suit-making operations.

SPOILAGE IN GENERAL

Two key objectives when accounting for spoilage, reworked units, and scrap are:

1. To spotlight for managers the magnitude of the costs of spoilage (the control purpose)
2. To identify the nature of the spoilage and distinguish between costs of normal spoilage and costs of abnormal spoilage when computing product costs (the product-costing purpose)

There is an unmistakable trend in manufacturing to increase quality. Why? Because managers have found that improved quality and intolerance for high spoilage have lowered overall costs and increased sales. The procedures described in this chapter help identify and make visible the costs of spoilage as special management problems. That is, spoilage costs are not ignored or buried as an unidentified part of the costs of good units produced.[1]

Normal Spoilage

Working within the selected set of production conditions, management must establish the rate of spoilage that is to be regarded as normal. **Normal spoilage** is spoilage that arises under efficient operating conditions; it is an inherent result of the particular process. Costs of normal spoilage are typically viewed as a part of the costs of *good* units produced, when attaining good units means the simultaneous appearance of spoiled units. In other words, normal spoilage is planned spoilage, in the sense that the choice of a given combination of factors of production entails a spoilage rate that management is willing to accept.

Normal spoilage rates should be computed using the total *good* outputs as the base, not the total *actual* units. Why? Because total actual inputs include the abnormal as well as the normal spoilage.

Abnormal Spoilage

Abnormal spoilage is spoilage that is not expected to arise under efficient operating conditions; it is not an inherent part of the selected production process. Most abnormal spoilage is usually regarded as controllable. Line operators and other plant personnel can generally increase efficiencies by minimizing machine breakdowns, accidents, and inferior materials. Costs of abnormal spoilage are the costs of inferior products that should be written off as losses of the accounting period in which detection occurs. For the most informative feedback, the Loss from Abnormal Spoilage account should appear in a detailed income statement as a separate line item and not be buried as an indistinguishable part of the cost of goods manufactured.

---

[1]The helpful suggestions of Samuel Laimon, University of Saskatchewan, are gratefully acknowledged.

Some companies adhere to a perfection standard as a part of their emphasis on total quality control. Their ideal goal is zero defects. Hence, all spoilage, rework, and even scrap would be treated as abnormal.

PROCESS COSTING AND SPOILAGE

Objective 2

Describe the general accounting procedures for normal and abnormal spoilage

Although this discussion of process costing emphasizes accounting for spoilage, the ideas here are equally applicable to waste, shrinkage, evaporation, or lost units. Again we must distinguish between cost control and product costing. For control, most companies use some version of budgeted or standard costs that incorporates an allowance for normal spoilage or shrinkage into the budgeted or standard amount. This section emphasizes product costing in process-costing systems.

Count All Spoilage

The two approaches to computing output units (or equivalent output units) in a process-costing system when spoilage (or shrinkage) occurs are to recognize spoiled units when computing output units (approach A) or to ignore spoiled units when computing output units (approach B). Approach A makes visible the costs associated with spoilage. Approach B spreads the spoilage costs over good units, potentially resulting in less accurate product costs.

Spoilage is typically assumed to occur at the stage of completion where inspection takes place. Why? Because spoilage is not detected until that point.

> EXAMPLE 1: Chipmakers Inc. manufactures computer chips for televisions. Consider the costing of direct materials in its process-costing system. All direct materials are added at the beginning of the chipmaking process. To highlight issues that arise with spoilage, we assume no beginning inventory. In May 19_4, $180,000 in direct materials are introduced. Production data for May indicate that 10,000 units were started, 5,000 good units were completed, 1,000 units were spoiled (all normal spoilage). Ending work in process has 4,000 units (each 100% complete as to direct materials costs). Spoilage is detected upon completion of the process.

The direct materials unit costs are computed and assigned using approach A and approach B as shown in Exhibit 18-1. Ignoring the equivalent units for spoilage decreases equivalent units; hence a higher unit cost of good units results. A $20 equivalent unit cost (instead of an $18 equivalent unit cost) is assigned to work in process that has not reached the inspection point. Simultaneously, the direct materials costs assigned to good units completed are too low ($100,000 instead of $108,000). Consequently, the 4,000 units in ending work in process contain costs of spoilage of $8,000 ($80,000 – $72,000) that do not pertain to those units and that, in fact, belong with the good units completed. The 4,000 units in ending work in process undoubtedly include some units that will be detected as spoiled in a subsequent accounting period. In effect, these units will bear two charges for spoilage. The ending work in process is being charged for spoilage in the current period, and it will be charged again when inspection occurs as the units are completed. Such cost distortions do not occur when spoiled units are recognized in the computation of equivalent units.

> EXAMPLE 2: Anzio Company manufactures a wooden recycling container in its Processing Department. Direct materials for this product are introduced at the beginning of the production cycle. At the start of production, all direct materials required to make one output unit are bundled together in a single kit. Conversion costs are incurred evenly throughout the cycle. Some units of this product are spoiled as a result of defects not detectable before inspection of finished units. Normally the spoiled units are 10% of the good output.

EXHIBIT 18-1

Effect of Recognizing Equivalent Units in Spoilage for Direct Materials Costs:
Chipmakers Inc. for May 19_4

	Approach A: Recognizing Spoiled Units When Computing Output in Equivalent Units	Approach B: Ignoring Spoiled Units When Computing Output in Equivalent Units
Costs to account for	$180,000	$180,000
Divide by equivalent units	÷10,000	÷9,000
Costs per equivalent unit	$ 18	$ 20
Assigned to:		
Good units completed: 5,000 × $18; 5,000 × $20	$ 90,000	$100,000*
Add normal spoilage: 1,000 × $18	18,000	0
Costs transferred out	$108,000	$100,000
Work in process, ending, 4,000 × $18; 4,000 × $20	72,000	80,000†
Costs accounted for	$180,000	$180,000

*5,000 × $20 = $100,000.
†4,000 × $20 = $ 80,000.

Summary data for January 19_4 are:

Physical Units for January

Work in process, beginning inventory (January 1)	2,000 units
Direct materials (100% complete)	
Conversion costs (80% complete)	
Started during January	8,000 units
Completed and transferred out in January	7,200 good units
Work in process, ending inventory (January 31)	1,500 units
Direct materials (100% complete)	
Conversion costs (66 2/3% complete)	

Costs for January

Work in process, beginning inventory		
Direct materials	$15,000	
Conversion costs	14,600	$29,600
Direct materials costs added during January		61,000
Conversion costs added during January		80,400

Computing Spoiled Units

The number of spoiled units is computed as follows:

$$\text{Spoiled units} = \left(\begin{array}{c}\text{Beginning units}\end{array} + \begin{array}{c}\text{Units started}\end{array}\right) - \left(\begin{array}{c}\text{Good units transferred out}\end{array} + \begin{array}{c}\text{Ending units}\end{array}\right)$$

$$= (2,000 + 8,000) - (7,200 + 1,500)$$
$$= 10,000 - 8,700$$
$$= 1,300 \text{ units}$$

Normal spoilage at Anzio is 10% of the 7,200 good output (the good units transferred out), or 720 units. Thus:

$$\text{Abnormal spoilage} = \text{Actual spoilage} - \text{Normal spoilage}$$
$$= 1{,}300 - 720$$
$$= 580 \text{ units}$$

We now illustrate how the weighted-average and FIFO methods of process costing discussed in Chapter 17 can incorporate both normal and abnormal spoilage in their computations. We will show calculations of:

◆ Equivalent units of production for January
◆ The dollar and unit amount of the abnormal spoilage during January
◆ Total inventoriable costs transferred to finished goods inventory
◆ The cost of work-in-process inventory at January 31
◆ Journal entries for units completed and transferred out of the department and for abnormal spoilage

The basic five-step approach used in Chapter 17 needs only slight modification to accommodate spoilage. The following observations pertain to both the weighted-average and FIFO methods:

◆ **Step 1:** *Summarize the flow of physical units of a product or output.* Identify both normal and abnormal spoilage.
◆ **Step 2:** *Compute output in terms of equivalent units.* Compute equivalent units for spoilage in the same way as for good units. Because Anzio inspects at the completion point, the same amount of work was done on each spoiled unit and each completed good unit.
◆ **Step 3:** *Summarize total costs to account for, which are all the costs charged to Work in Process (that is, the total debits to Work in Process).* The details of this step do not differ from those in Chapter 17.
◆ **Step 4:** *Compute equivalent unit costs.* Again, the details of this step do not differ from those in Chapter 17. We assume that spoiled units are included in the computation of output units.
◆ **Step 5:** *Assign costs to units completed and to units in ending work in process.* This step now includes computation of the cost of spoiled units and the cost of good units.

Exhibit 18-2 is an overview of product costing for process costing with spoilage at Anzio Company.

Weighted-Average Method and Spoilage

Objective 3
Account for spoilage in process costing using the weighted-average method

Exhibit 18-3 (p. 638) is a production-cost worksheet based on the weighted-average method. The costs of abnormal spoilage are assigned to the Loss from Abnormal Spoilage account, 580 units × $17.60 = $10,208. The costs of normal spoilage, $12,672, are added to the costs of their related good units. This illustration assumes inspection upon completion. In contrast, inspection may take place at some other stage—say, at the halfway point in the production cycle. In such a case, normal spoilage costs would be added to completed goods and to the units in process that are more than half completed.

Managers should try to have inspection done as early in the production process as technically possible. This timing will reduce the direct materials and conversion costs assigned to spoiled units. Thus, in the Exhibit 18-3 example, if inspection can occur when units are 80% complete as to conversion costs and 100% complete as to direct materials, the company can avoid incurring the final 20% of conversion costs on the spoiled units.

Objective 4
Account for spoilage in process costing using the first-in, first-out method

FIFO Method and Spoilage

Exhibit 18-4 presents the results of all five steps using the FIFO method. All spoilage costs are assumed to be related to units completed during this period, using the unit

EXHIBIT 18-2
Overview of Product Costing at Anzio Company

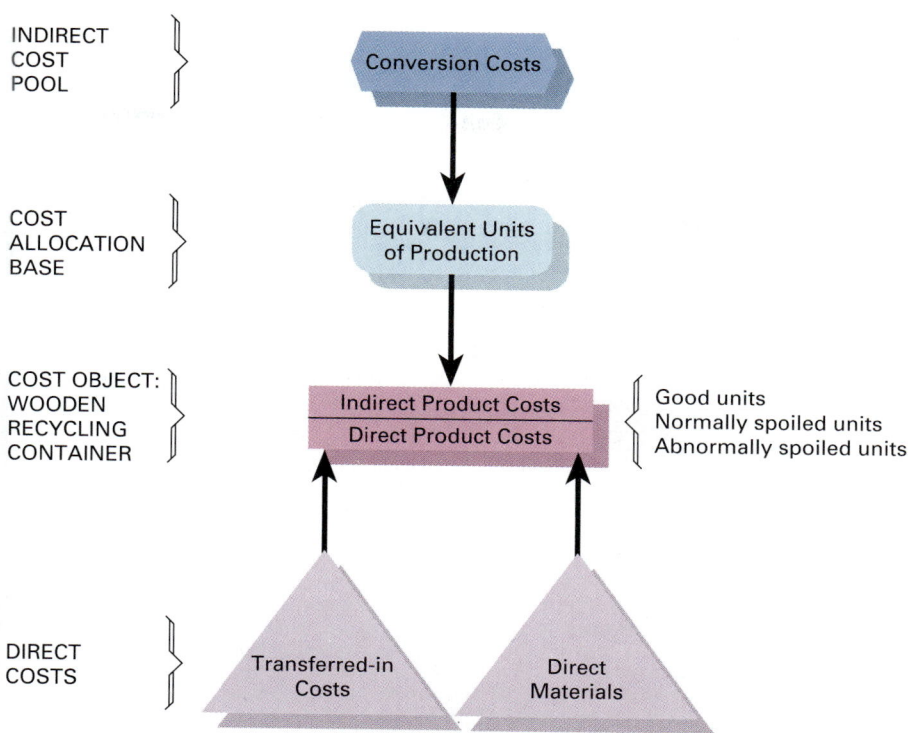

costs of the current period.[2] With the exception of accounting for spoilage, the FIFO method is the same as presented in Chapter 17.

Journal Entries

A production-cost report is shown in Exhibit 18-5 (p. 640). It is confined to physical units and the total costs that flow through the Work in Process account. If managers prefer, the supporting computations could be embodied in a more elaborate production-cost report like the production-cost worksheets in Exhibits 18-3 and 18-4.

The information in Exhibit 18-5 supports the following journal entries:

	Weighted Average		FIFO	
	Debit	**Credit**	**Debit**	**Credit**
Finished Goods	$139,392		$139,060	
Work in Process—Processing Department		$139,392		$139,060
To transfer good units completed in January.				
Loss from Abnormal Spoilage	$10,208		$10,325	
Work in Process—Processing Department		$10,208		$10,325
To recognize abnormal spoilage in January.				

---

[2]If the FIFO method were used in its purest form, normal spoilage costs would be split between the goods started and completed during the current period and those completed from beginning work in process—using the appropriate unit costs of the period in which the units were worked on. The simpler, modified FIFO method, as illustrated in Exhibit 18-4, in effect uses the unit costs of the current period for assigning normal spoilage costs to the goods completed from beginning work in process. This modified method assumes that all normal spoilage traceable to the beginning work in process was begun and completed during the current period, an obvious contradiction to a pure FIFO method.

EXHIBIT 18-3
Production-Cost Worksheet:
Weighted-Average Method of Process Costing,
Processing Department of Anzio Company for January 19_4

		(Step 1) Physical Units	(Step 2) Equivalent Units	
			Direct Materials	Conversion Costs
	Flow of Production			
	Work in process, beginning*	2,000		
	Started during current period	8,000		
	To account for	10,000		
	Good units completed and transferred out during current period	7,200	7,200	7,200
	Normal spoilage†	720	720	720
	Abnormal spoilage‡	580	580	580
	Work in process, ending§	1,500	1,500	1,000
	Accounted for	10,000		
	Total work done to date		10,000	9,500

		Totals	Details Direct Materials	Details Conversion Costs
	Work in process, beginning	$ 29,600	$15,000	$14,600
	Costs added during current period	141,400	61,000	80,400
(Step 3)	Total costs to account for	$171,000	$76,000	$95,000
	Divide by equivalent units		÷10,000	÷9,500
(Step 4)	Equivalent unit costs		$ 7.60	$ 10.00
(Step 5)	Assignment of costs:			
	(A) Abnormal spoilage (580 units)‡	$ 10,208	580 ($17.60)‖	
	Completed and transferred out (7,200 units):			
	Costs before adding normal spoilage	126,720	7,200 ($17.60)‖	
	Normal spoilage (720 units)†	12,672	720 ($17.60)‖	
	(B) Total costs transferred out	139,392		
	Work in process, ending (1,500 units):			
	Direct materials#	11,400	1,500 ($7.60)	
	Conversion costs**	10,000		1,000 ($10.00)
	(C) Total work in process	21,400		
(A) + (B) + (C)	Total costs accounted for	$171,000		

*Degree of completion: direct materials, 100%; conversion costs, 80%.
†Normal spoilage is 10% of good units transferred out: 10% × 7,200 = 720 units.
‡Abnormal spoilage = actual spoilage − normal spoilage = 1,300 − 720 = 580 units.
§Degree of completion: direct materials, 100%; conversion costs, 66 2/3%.
‖Cost per equivalent whole unit = $7.60 + $10.00 = $17.60.
#1,500 units × 100% complete = 1,500 equivalent units.
**1,500 units × 66 2/3% complete = 1,000 equivalent units.

EXHIBIT 18-4

Production-Cost Worksheet:
FIFO Method of Process Costing,
Processing Department of Anzio Company for January 19_4

		(Step 1) Physical Units	(Step 2) Equivalent Units	
Flow of Production			Direct Materials	Conversion Costs
Work in process, beginning*		2,000	(work done before current period)	
Started during current period		8,000		
To account for		10,000		
Good units completed and transferred out during current period:				
From beginning work in process		2,000	—	400
Started and completed		5,200	5,200	5,200
Normal spoilage†		720	720	720
Abnormal spoilage‡		580	580	580
Work in process, ending§		1,500	1,500	1,000
Accounted for		10,000		
Work done in current period only			8,000	7,900

		Totals	Details	
			Direct Materials	Conversion Costs
Work in process, beginning		$ 29,600	(costs of work done before current period)	
	Costs added during current period	141,400	$61,000	$ 80,400
(Step 3)	Total costs to account for	$171,000		
	Divide by equivalent units		÷8,000	÷7,900
(Step 4)	Equivalent unit costs		$7.6250	$10.1772
(Step 5)	Assignment of costs:			
(A)	Abnormal spoilage (580 units)‡	$ 10,325	580 ($17.8022)‖	
	Units completed (7,200 units)			
	Work in process, beginning (2,000 units)	29,600		
	Costs added during current period	4,071		400 ($10.1772)
	Total from beginning inventory before normal spoilage	33,671		
	Started and completed before normal spoilage (5,200 units)	92,571		5,200 ($17.8022)‖
	Normal spoilage (720 units)†	12,818		720 ($17.8022)‖
(B)	Total costs transferred out	139,060		
	Work in process, ending (1,500 units):			
	Direct materials#	11,438	1,500 ($7.625)	
	Conversion costs**	10,177		1,000 ($10.1772)
(C)	Total work in process	21,615		
(A) + (B) + (C)	Total costs accounted for	$171,000		

*Degree of completion: direct materials, 100%; conversion costs, 80%.
†Normal spoilage is 10% of good units transferred out: 10% × 7,200 = 720 units.
‡Abnormal spoilage = actual spoilage − normal spoilage = 1,300 − 720 = 580 units.
§Degree of completion: direct materials, 100%; conversion costs, 66 2/3%.
‖Cost per equivalent whole unit = $7.6250 + $10.1772 = $17.8022.
#1,500 units × 100% complete = 1,500 equivalent units.
**1,500 units × 66 2/3% complete = 1,000 equivalent units.

EXHIBIT 18-5

Production-Cost Report:
Weighted-Average and FIFO Methods of Process Costing,
Process Department of Anzio Company for January 19_4

| | | Total Costs | |
Flow of Production	Physical Units	Weighted-Average Method	FIFO
Work in process, beginning	2,000	$ 29,600	$ 29,600
Started during current period	8,000	141,400	141,400
To account for	10,000	$171,000	$171,000
Good units completed and transferred out	7,200 }	$139,392	$139,060
Normal spoilage	720 }		
Abnormal spoilage	580	10,208	10,325
Work in process, ending	1,500	21,400	21,615
Accounted for	10,000	$171,000	$171,000

Assumptions for Allocating Normal Spoilage

Spoilage might actually *occur* at various points or stages of the production cycle but spoilage is typically not *detected* until one or more specific points of inspection. The cost of spoiled units is assumed to be all costs incurred prior to inspection. The unit costs of abnormal and normal spoilage are the same when the two are detected simultaneously. However, situations might arise when abnormal spoilage is detected at a different point than normal spoilage. In such cases, the unit cost of abnormal spoilage would differ from the unit cost of normal spoilage.

Various assumptions prevail regarding how to allocate the cost of normal spoilage between completed units and the ending inventory of work in process. A popular approach (applicable to both the weighted-average and FIFO methods) follows.

When normal spoilage is presumed to occur at a specific point in the production cycle, allocate its cost (a unit cost of $17.60 in Exhibit 18-3) over all units that have passed that point. In our example, spoilage is assumed to occur upon completion, so no cost of normal spoilage is allocated to the ending work in process.

Whether the cost of normal spoilage is allocated to the units in ending work in process inventory depends strictly on whether they have passed the point of inspection. For example, if the inspection point is presumed to be the halfway stage of the production cycle, work in process that is 70% completed would be allocated a full measure of normal spoilage cost. But work in process that is 40% completed would not be allocated any normal spoilage cost.

So the point of inspection is the key to the application of spoilage costs. Avoid the idea that normal spoilage costs are assigned solely to units completed and transferred out. Why? Because if units in ending work in process have passed inspection, they should have normal spoilage costs added to them. The Appendix to this chapter contains additional discussion concerning various assumptions about spoilage.

JOB COSTING AND SPOILAGE

Objective 5
Account for spoilage in job costing

The concepts of normal spoilage and abnormal spoilage are equally applicable to both job-costing systems and process-costing systems. When spoiled goods have a disposal value, the net cost of spoilage is computed by deducting disposal value from the costs of the spoiled goods accumulated to the point of inspection.

In some cases, spoilage may be considered a normal characteristic of a given production cycle. For example, machines might malfunction at random. The spoilage, then, is not attributable to any specific job being worked on. Hence, it is inappropriate to charge the job that happens to be under way at the time the spoilage occurs with the cost of the spoilage. In this context, the cost of spoilage is allocated to work done on all jobs. The budgeted manufacturing overhead rate includes a provision for normal spoilage cost. Therefore, normal spoilage cost is spread, through overhead allocation, over all jobs rather than loaded on particular jobs only.

> EXAMPLE 3: In the Hull Machine Shop, 5 pieces out of a job lot of 50 aircraft parts are normally spoiled. Costs assigned to the point of inspection are $100 per unit. The current disposal price of the spoiled pieces is estimated to be $30 per unit. As the normal spoilage is detected, the spoiled goods are inventoried ($30 per unit), the Manufacturing Department Overhead Control account is debited for the net cost of spoilage ($70 per unit), and Work in Process is credited for the costs assigned to the goods up to the point of inspection ($100 per unit).[3]

Materials Control (spoiled goods at current disposal price): 5 × $30	$150	
Manufacturing Department Overhead Control (normal spoilage): 5 × $70	350	
Work in Process Control (particular job): 5 × $100		$500

Items in parentheses indicate subsidiary postings.

When the spoilage occurs because of the exacting specifications of a particular job, that job may be credited with the costs of that spoilage. The journal entry, for the same data just used, follows:

Materials Control (spoiled goods at current disposal price)	$150	
Work in Process Control (particular job)		$150

REWORKED UNITS

Objective 6
Account for reworked units

Reworked units are unacceptable units of production that are subsequently reworked and sold. Reworked units are (or should be) indistinguishable from non-reworked good units when completed. As a result, most companies assign equivalent costs to all good output units (whether they have been reworked or not).

Two approaches used to account for the costs of reworked units are:

1. To charge the costs of rework to the current period as a separate expense item. This approach highlights to managers how a reduction in rework costs can directly benefit reported operating income.

2. To charge the costs of rework to manufacturing overhead. We now illustrate this approach.

Consider the Hull Machine Shop data (example 3 above). Assume that the five spoiled pieces used in our Hull Machine Shop illustration are reworked at a total cost of $190 and sold through regular marketing channels. Journal entries follow:

Original cost assignment:	Work in Process Control	$500	
	Materials Control		$200
	Wages Payable Control		200
	Manufacturing Overhead Allocated		100
Rework (figures assumed):	Manufacturing Department Overhead Control (rework)	$190	
	Materials Control		$ 40
	Wages Payable Control		100
	Manufacturing Overhead Allocated		50
Transfer to finished goods:	Finished Goods Control	$500	
	Work in Process Control		$500

---

[3]Conceptually, the prevailing treatment just described can be criticized, primarily because *costs already assigned to products* are being charged back to Manufacturing Department Overhead Control, which logically should accumulate only *costs incurred*, not both costs incurred and costs assigned.

ACCOUNTING FOR SCRAP

Objective 7
Account for scrap

Scrap was defined in Chapter 16 in conjunction with the discussion of byproducts. Scrap is a product that has minimal (frequently zero) sales value compared with the sales value of the main or joint product(s).

There are two major aspects of accounting for scrap:

1. Planning and control, including physical tracking
2. Inventory costing, including when and how to affect operating income

Detailed records of physical quantities of scrap are often kept at all stages of production. For example, a survey of manufacturers shows that 71% of the responding companies tracked scrap by specific materials.[4] Scrap records not only help measure efficiency but also often focus on a tempting source for theft.

[4]Price Waterhouse, *Survey of the Cost Management Practices of Selected Midwest Manufacturers* (Cleveland: Price Waterhouse, 1989), p. 21.

Surveys of Company Practice

REJECTION IN THE ELECTRONICS INDUSTRY

From country to country and from industry to industry, the rates of rejected and reworked units vary tremendously. The data in the following table focus on different segments of the U.S. electronics industry. The data are drawn from companies that are members of the American Electronics Association.[a] The reject rate is the number of rejects per 1,000 items checked by quality control. The rework rate is the percentage of items returned for rework. Also reported is the median operating income to net sales figure for each segment of the electronics industry.

Segment of Electronics Industry	Reject Rate (no. per 1,000)	Rework Rate (% rework)	Oper. Inc. Net sales
1. Semiconductors	12.5	2.0%	4.2%
2. Components—includes capacitors, amplifiers, oscillators, and wire and cable	100.0	5.0	4.7
3. Computers—includes mainframes, minicomputers, and microcomputers	45.0	6.0	4.8
4. Peripherals—includes disc drives, printers, and keyboards	30.0	4.0	1.1
5. Office and consumer products—includes word processors and electronic games	67.0	7.0	6.7
6. Instruments—includes testing and analytical equipment for medicine	76.0	5.0	5.0
7. Telecommunications—includes telephone, radio, and TV apparatus	60.0	4.0	3.6
8. Software—includes prepackaged and custom software	10.0	N/A	2.2

The reject rate for components (100 ÷ 1,000 = 10%) is eight times as great as that for semiconductors (12.5 ÷ 1,000 = 1.25%). Semiconductors show the lowest percentage of rework (in part because rework is not always possible when defects arise). The operating income to net sales ratio ranges from 1.1% for peripherals to 6.7% for office and consumer products. Given these profitability percentages, substantial reductions in reject rates can markedly increase the profitability of many companies in the electronics industry.

[a]Adapted from American Electronics Association, *Operating Ratios Survey.* Full citation is in Appendix A.

Initial entries to scrap records are most often in nonfinancial terms such as in pounds or units. In various industries, items such as metal chips, filings, and wood shavings are quantified by weighing, counting, or some other expedient means. Excessive scrap indicates inefficiency. Costs are determined later after shop-floor production reports have been compiled. Scrap reports are prepared as source documents for periodic summaries of the amount of actual scrap compared with budgeted norms or standards. Scrap is either sold quickly or disposed of or stored in some routine way for later sale, disposal, or for reuse.

The tracking of scrap often extends into the financial records. For example, in one survey, 60% of the firms maintained a distinct cost for scrap somewhere in their cost accounting system.[5] The issues here are similar to those discussed in Chapter 16 regarding the accounting for byproducts:

1. When should any value of scrap be recognized in the accounting records: at the time of production or at the time of sale?
2. How should revenue from scrap be accounted for?

In our Hull example, assume that scrap related to the job lot in question has a total sales value of $45. Many companies will track only the physical amount of scrap and indicate its presence in storage awaiting sale.

When scrap is sold, the simplest accounting is to regard scrap sales as a separate line item of other revenues:

| Sale of scrap: | Cash or Accounts Receivable | $45 | |
| | Sales of Scrap | | $45 |

However, many companies account for the sales as offsets against manufacturing overhead:

Sale of scrap:	Cash or Accounts Receivable	$45	
	Manufacturing Department Overhead Control		$45
	Posting made to subsidiary record—"Sales of Scrap" column on department cost record.		

This method is both simple and accurate enough in theory to justify its wide use. A normal amount of scrap is often an inevitable result of production operations. Basically, this method does not link scrap with any particular physical product; instead, all products bear regular production costs without any credit for scrap sales except in an indirect manner. What really happens in such situations is that sales of scrap are considered when budgeted manufacturing overhead rates are being set. Thus, the budgeted overhead rate is lower than it would be if no credit for scrap sales were allowed in the overhead budget. This accounting for scrap is used in both process-costing and job-costing systems.

In job-costing systems, sometimes sales of scrap are traced to the jobs that yielded the scrap. This method is used only when feasible and economically desirable. For example, the Hull Machine Shop and particular customers, such as the U.S. Department of Defense, may reach an agreement that provides for charging specific, difficult jobs with all rework or spoilage costs and crediting such jobs with all scrap sales that arise from that job. Journal entries follow:

Scrap returned to storeroom:	No journal entry.		
	[Memo of quantity received and related job is entered in the perpetual record.]		
Sale of scrap:	Cash or Accounts Receivable	$45	
	Work in Process Control		$45
	Posting made to specific job record.		

In these illustrations, we assume that scrap returned to the storeroom is not assigned an inventory cost figure. Scrap, however, sometimes has a significant value,

---

[5]Ibid., p. 10.

and the time between storing it and selling or reusing it may be quite long. Under these conditions, the company is justified in inventorying scrap at some conservative estimate of net realizable value so that production costs and related scrap recovery may be recognized in the same accounting period. Some companies tend to delay sales of scrap until the price is most attractive. Volatile price fluctuations are typical for scrap metal. If scrap inventory becomes significant, it should be inventoried at some "reasonable" value — a difficult task in the face of volatile market prices.

Scrap is sometimes reused as direct materials rather than sold as scrap. Then it should be debited to Materials Control as a class of direct materials and carried at its estimated net realizable value.

Scrap returned to storeroom:	Materials Control	$45	
	Manufacturing Department Overhead Control		$45
Reuse of scrap:	Work in Process Control	$45	
	Materials Control		$45

COMPARISON OF ACCOUNTING FOR SPOILAGE, REWORK, AND SCRAP

The basic approach to accounting for spoilage, rework, and scrap distinguishes:

A. The normal amount common to all jobs
B. The normal amount attributable to specific jobs
C. Any abnormal amount

The following entries recapitulate the preceding examples.

Spoilage Cost (net $350 from p. 641)

A. Normal (common to all jobs):	Materials Control	$150	
	Manufacturing Department Overhead Control	350	
	Work in Process Control		$500
B. Normal (peculiar to specific jobs):	Materials Control	$150	
	Work in Process Control		$150
C. Abnormal:	Materials Control	$150	
	Loss from Abnormal Spoilage	350	
	Work in Process Control		$500

Rework Cost ($190 from p. 641)

A. Normal (common to all jobs):	Manufacturing Department Overhead Control	$190	
	Materials Control		$ 40
	Wages Payable Control		100
	Manufacturing Overhead Allocated		50
B. Normal (peculiar to specific jobs):	Same as preceding entry except that the debit of $190 would be to Work in Process Control		
C. Abnormal:	Same as preceding entry except that the debit of $190 would be to Loss from Abnormal Spoilage		

Scrap Value Recovered ($45 from p. 643)

A. Normal (common to all jobs):	Materials Control or Cash or Accounts Receivable	$ 45	
	Manufacturing Department Overhead Control		$ 45
B. and C. Normal/abnormal (peculiar to specific jobs):	Same as preceding entry except that the credit would be to Work in Process Control		

Although these journal entries are based on a job-order costing system, similar entries are made in a process-costing system. Of course, process costing, by definition, has no specific jobs for the identification of costs. Practices vary considerably from company to company.

PROBLEM FOR SELF-STUDY

PROBLEM

Burlington Textiles has some spoiled goods that had an assigned cost of $4,000.

Required

Prepare a journal entry for each of the following conditions under both (a) process costing (Department A) and (b) job costing:

1. Abnormal spoilage of $4,000
2. Normal spoilage of $4,000 related to general plant operations
3. Normal spoilage of $4,000 related to specifications of a particular job

SOLUTION

(a) Process Costing			(b) Job Costing		
1. Loss from Abnormal			Loss from Abnormal		
Spoilage	$4,000		Spoilage	$4,000	
Work in Process—			Work in Process		
Department A		$4,000	Control (job)		$4,000
2. No entry until units are transferred. Then the normal spoilage costs are transferred along with the other costs:			Manufacturing Department		
			Overhead Control	$4,000	
			Work in Process		
			Control (job)		$4,000
Work in Process—					
Department B	$4,000				
Work in Process—					
Department A		$4,000			
3. Not applicable			No entry. Spoilage cost remains in Work in Process Control (job).		

SUMMARY

The following points are linked to the chapter's learning objectives.

1. Spoilage is unacceptable units of production that are discarded or sold for net disposal proceeds. Reworked units are unacceptable units that are subsequently reworked and sold as acceptable finished goods. Scrap is product that has minimal sales value compared with the sales value of the main or joint product(s).

2. Normal spoilage is spoilage that arises under efficient operating conditions. Abnormal spoilage is spoilage that is not expected to arise under efficient operating conditions. Many accounting systems explicitly recognize both forms of spoilage when computing output units. Normal spoilage typically is included in the cost of good output units, while abnormal spoilage is recorded as a period cost.

3. Under the weighted-average method of process costing, costs in beginning inventory are pooled with costs in the current period when determining the costs of good units (which includes a normal spoilage amount) and the costs of abnormal spoilage.

4. Under the FIFO method of process costing, costs in beginning inventory are kept separate from the costs in the current period when determining the cost of good units (which includes a normal spoilage amount). The cost of abnormal spoilage typically is kept separate from the cost of good units.

5. With a job-costing system, companies can decide to assign spoilage to specific jobs. Alternatively, they can allocate spoilage to all jobs as part of manufacturing overhead. Different accounting treatments may be made for normal and abnormal spoilage.

6. Reworked units should be indistinguishable from nonreworked good units when completed and hence the two are assigned the same costs. The costs of extra materials (labor and so on) associated with reworked units are generally included as part of manufacturing overhead.

7. Accounting for scrap is similar to the accounting for byproducts discussed in Chapter 16. Companies differ as to both when and how scrap is recognized in the accounting records.

APPENDIX: INSPECTION AND SPOILAGE AT INTERMEDIATE STAGES OF COMPLETION IN PROCESS COSTING

Consider how the timing of an inspection affects the amount of normal and abnormal spoilage. What happens when the inspection occurs when production has reached various stages of completion for conversion costs? Assume that normal spoilage is 10% of the good units passing inspection in the Forging Department of Dana Corporation, a manufacturer of automobile parts. Direct materials are added at the start of production in the Forging Department. Conversion costs are allocated evenly throughout the process.

Suppose inspection had occurred at the 20%, the 50%, or the 100% conversion stage. A total of 8,000 units are spoiled in all cases. Note how the number of units of normal spoilage and abnormal spoilage change. Data are for January.

	Physical Units: Inspection at Stage of Completion		
Flow of Production	**at 20%**	**at 50%**	**at 100%**
Work in process, beginning (25%)*	11,000	11,000	11,000
Started during January	74,000	74,000	74,000
To account for	85,000	85,000	85,000
Good units completed and transferred out			
(85,000 − 8,000 spoiled − 16,000 ending)	61,000	61,000	61,000
Normal spoilage	6,600†	7,700‡	6,100§
Abnormal spoilage (8,000 − normal spoilage)	1,400	300	1,900
Work in process, ending (75%)*	16,000	16,000	16,000
Accounted for	85,000	85,000	85,000

*Degree of completion for conversion costs of this department at the dates of the work in process inventories.
†10% × (50,000 + 16,000).
‡10% × (11,000 beginning + 50,000 started and completed during current period + 16,000 ending).
§10% × (11,000 + 50,000).

Exhibit 18-6 shows the computation of equivalent units under the weighted-average method and assuming inspection at the 50% conversion stage. The calculations depend on how much direct materials and conversion costs were incurred to get the units to the point of inspection. In Exhibit 18-6, the spoiled units have a full measure of direct materials and a 50% measure of conversion costs.

The key computations are those of equivalent units. The computations of equivalent unit costs and the assignments of total costs to units completed and in ending work in process would be similar to those in previous illustrations. In this example, however, note that ending work in process has passed the inspection point. Therefore, these units would bear a normal spoilage cost, just like the units that have been completed and transferred out.

EXHIBIT 18-6

Steps 1 and 2: Computing Equivalent Units with Spoilage:
Weighted-Average Method of Process Costing,
Forging Department of Dana Corporation for January 19_4

Flow of Production	(Step 1) Physical Units	(Step 2) Equivalent Units Direct Materials	Conversion Costs
Work in process, beginning*	11,000		
Started during current period	74,000		
To account for	85,000		
Good units completed and transferred out	61,000	61,000	61,000
Normal spoilage	7,700	7,700	3,850
Abnormal spoilage	300	300	150
Work in process, ending†	16,000	16,000	12,000
Accounted for	85,000		
Total work done to date		85,000	77,000

*Degree of completion: direct materials, 100%; conversion costs, 25%.
†Degree of completion: direct materials, 100%; conversion costs, 75%.

TERMS TO LEARN

This chapter and the Glossary at the end of the book contain definitions of the following important terms:

abnormal spoilage *(p. 633)*   normal spoilage *(633)*   reworked units *(633)*
spoilage *(633)*

ASSIGNMENT MATERIAL

QUESTIONS

18-1   Why is there an unmistakable trend in manufacturing to improve quality?

18-2   Distinguish among spoilage, reworked units, and scrap.

18-3   "Normal spoilage is planned spoilage." Discuss.

18-4   "Costs of abnormal spoilage are lost costs." Explain.

18-5   "What has been regarded as normal spoilage in the past is not necessarily acceptable as normal in the present or future." Explain.

18-6   "Abnormal units are inferred rather than identified." Explain.

18-7   "In accounting for spoiled goods, we are dealing with cost assignment rather than cost incurrence." Explain.

18-8   "Total input includes abnormal as well as normal spoilage and is therefore irrational as a basis for computing normal spoilage." Do you agree? Why?

18-9   "The point of inspection is the key to the allocation of spoilage costs." Do you agree? Explain.

18-10   "The unit cost of normal spoilage is the same as the unit cost of abnormal spoilage." Do you agree? Explain.

18-11   "In job-order costing, the costs of specific normal spoilage are charged to specific jobs." Do you agree? Explain.

18-12   "The costs of reworking defective units are always charged to the specific jobs where the defects were originally discovered." Do you agree? Explain.

18-13   "Abnormal rework costs should be charged to a loss account, not to manufacturing overhead." Do you agree? Explain.

18-14   When is a company justified in inventorying scrap?

18-15   Prepare a journal entry appropriate for the placing of scrap in a storeroom for future use as direct materials. Assume that its estimated net realizable value is $2,000.

EXERCISES AND PROBLEMS

18-16   Normal and abnormal spoilage in units.   The following data, in physical units, describe a grinding process for January:

Work in process, beginning	19,000
Started during current period	150,000
To account for	169,000
Spoiled units	12,000
Good units completed and transferred out	132,000
Work in process, ending	25,000
Accounted for	169,000

Inspection occurs at the 100% conversion stage. Normal spoilage is 5% of the good units passing inspection.

Required

1. Compute the normal and abnormal spoilage in units.

2. Assume that the equivalent-unit cost of a spoiled unit is $10. Compute the amount of potential savings if all spoilage were eliminated, assuming that all other costs would be unaffected. Comment on your answer.

18-17   Normal and abnormal spoilage in units.   Data for a milling process for November follow: work in process, beginning, 20,000; good units completed and transferred out during current period, 90,000; work in process, ending, 17,000. Inspection is at the 100% stage of completion regarding conversion costs, which are incurred evenly throughout the process. Total spoilage is 7,000 units. Normal spoilage is 4% of the good units passing inspection.

Required

1. Compute the normal and abnormal spoilage in units.

2. Assume that the equivalent unit cost of a spoiled unit is $1,000. Compute the amount of potential savings if all spoilage were eliminated, assuming that all other costs would be unaffected. Comment on your answer.

18-18   Weighted-average method, spoilage.   Superchip specializes in the manufacture of microchips for aircraft. Direct materials are added at the start of the production process. Conversion costs are allocated evenly throughout the process. Some units of this product are spoiled as a result of defects not detectable before inspection of finished goods. Normally the spoiled units are 15% of the good units transferred out. Spoiled units are disposed of at zero net disposal price.

Summary data for September 19_4 are:

Physical Units for September 19_4	
Work in process, beginning inventory	400 units
Direct materials (100% complete)	
Conversion costs (30% complete)	
Started during September	1,700 units
Completed and transferred out during September	1,400 good units
Work in process, ending inventory	300 units
Direct materials (100% complete)	
Conversion costs (40% complete)	

Costs for September 19_4		
Work in process, beginning inventory		
Direct materials	$64,000	
Conversion costs	10,200	$ 74,200
Direct materials costs added during September		378,000
Conversion costs added during September		153,600

Required

Using the weighted-average method, prepare a production-cost worksheet for September. Distinguish between normal and abnormal spoilage.

18-19   FIFO method, spoilage. (Continuation of 18-18)   Using the FIFO method, prepare a production-cost worksheet for September. Distinguish between normal and abnormal spoilage.

18-20   Normal and abnormal spoilage.   The Van Brocklin Company manufactures one style of long, tapered wax candle, which is used on festive occasions. Each candle requires a two-foot-long wick and one pound of a specially prepared wax. The wick and melted wax are placed in molds and allowed to harden for 24 hours. Upon removal from the molds, the candles are immediately dipped in a special coloring mixture that gives them a glossy lacquer finish. Dried candles are inspected, and all defective ones are pulled out. Because the coloring mixture penetrates into the wax itself, the defective candles cannot be salvaged for reuse. They are destroyed in an incinerator. Normal spoilage is regarded as 3% of the number of candles that pass inspection.

Cost and production statistics for a particular week follow:

Direct materials requisitioned (including wicks and wax)	$3,340.00
Conversion costs (allocated at a uniform rate during the hardening process)	1,219.50
Total costs	$4,559.50

During the week, 7,800 candles were completed; 7,500 passed inspection, and the remainder were defective. At the end of the week, 550 candles were still in the molds; they were considered 60% complete. There was no beginning inventory. Van Brocklin uses the weighted-average method of process costing.

Required

1. The cost of the 7,500 candles that passed inspection was (choose one and show computations): (a) $4,333.73, (b) $4,125.00, (c) $4,217.85, (d) $4,290.00, (e) $4,248.75.

2. The cost of the candles still in the molds at the end of the week was (choose one and show computations): (a) $393.25, (b) $269.50, (c) $185.13, (d) $274.72, (e) $300.30.

18-21 Weighted-average method, spoilage. Consider the following data for a cooking department for the month of January:

	Physical Units
Work in process, beginning inventory*	11,000
Started during current period	74,000
To account for	85,000
Good units completed and transferred out during current period:	
From beginning work in process	11,000
Started and completed	50,000
Good units completed	61,000
Spoiled units	8,000
Work in process, ending inventory†	16,000
Accounted for	85,000

* Direct materials, 100% complete; conversion costs, 25% complete.
† Direct materials, 100% complete; conversion costs, 75% complete.

Inspection occurs when production is 100% completed. Normal spoilage is 11% of good units completed and transferred out during the current period.
The following cost data are available:

Work in process, beginning inventory:		
Direct materials	$220,000	
Conversion costs	30,000	$ 250,000
Costs added during current period:		
Direct materials		1,480,000
Conversion costs		942,000
Costs to account for		$2,672,000

Required
Prepare a detailed production-cost worksheet. Use the weighted-average method. Distinguish between normal and abnormal spoilage.

18-22 FIFO method, spoilage. Consider the data in Problem 18-21. Using the FIFO method, prepare a detailed production-cost worksheet. Distinguish between normal and abnormal spoilage.

18-23 Weighted-average method, spoilage. The Alston Company operates under a process-costing system. It has two departments, Cleaning and Milling. For both departments, conversion costs are allocated in proportion to the stage of completion. But direct materials are added at the *beginning* of the process in the Cleaning Department, and additional direct materials are added at the *end* of the milling process. The costs and unit production statistics for May follow. All unfinished work at the *end* of May is 25% completed as to conversion costs. The beginning inventory (May 1) was 80% completed as to conversion costs as of May 1. All completed work is transferred to the next department.

	Cleaning	Milling
Beginning Inventories		
Cleaning: $1,000 direct materials, $800 conversion costs	$1,800	
Milling: $6,450 previous department cost		
(transferred-in cost) and $2,450 conversion costs		$8,900
Costs Added during Current Period		
Direct materials	$9,000	$ 640
Conversion costs	$8,000	$4,950
Physical Units		
Units in beginning inventory	1,000	3,000
Units started this month	9,000	7,400
Good units completed and transferred out	7,400	6,000
Normal spoilage	740[*]	300[†]
Abnormal spoilage	260	100

[*]Normal spoilage in Cleaning Department is 10% of good units completed and transferred out.
[†]Normal spoilage in Milling Department is 5% of good units completed and transferred out.

Additional Information

1. Spoilage is assumed to occur at the end of each of the two processes when the units are inspected.
2. Assume that there is no shrinkage, evaporation, or abnormal spoilage other than that indicated in the information given above.
3. Carry unit-cost calculations to three decimal places where necessary. Calculate final totals to the nearest dollar.

Required
Using the weighted-average method, prepare a production-cost worksheet for the Cleaning Department. (Problem 18-25 explores additional facets of this problem.)

18-24 FIFO. Redo Problem 18-23, using FIFO. (Problem 18-26 explores additional facets of this problem.)

18-25 Continuation of 18-23, Second Department. Using the facts in Problem 18-23 and the weighted-average method, prepare a production-cost worksheet for the Milling Department.

18-26 Continuation of 18-24, Second Department. Using the facts in Problem 18-23 and the FIFO method, prepare a production-cost worksheet for the Milling Department.

18-27 Process costs, weighted-average, two departments, spoilage. (SMA) The Quebec Manufacturing Company produces a single product. There are two producing departments, 1 and 2.

There were no work-in-process inventories at the beginning of the year. In January, direct materials for 1,000 units were issued to production in Department 1 at a cost of $5,000. Conversion costs for the month were $2,700. During the month, 800 good units were completed and transferred to Department 2.

The work-in-process inventory at the end of the month contained 200 units, complete as to direct materials and one-half complete as to conversion costs. Normally there is no spoilage in Department 1.

Conversion costs in Department 2 amounted to $6,250 in January. During the month, 500 good units were completed and transferred to finished goods. At the end of the month, 200 units remained in process, one-quarter complete. Normally, spoilage in Department 2 is 8% of the good units completed and transferred out. Spoilage in Department 2 is recognized upon inspection at the end of the process. However, in January there was an abnormal loss (due to an industrial accident) of 50 units when they were one-half complete. The effect of this abnormal loss is not to be included in ending inventory. No materials are added in Department 2.

Required

Prepare a detailed production-cost worksheet for each department for the month of January. Use the weighted-average method.

18-28 Reworked units, costs of rework. White Goods assembles washing machines at its Auburn plant. In February 19_5 there was a major problem with 60 tumbler units that cost $44 each from a new supplier. All 60 units were defective and had to be disposed of at zero disposal price. White Goods was able to rework all 60 washing machines by substituting new tumbler units. Unfortunately, the new supplier had already gone bankrupt. White Goods purchased the replacement tumblers from one of its existing suppliers. Each replacement tumbler cost $50.

Required

1. What alternative approaches are there to account for the materials costs of reworked units?
2. Should White Goods use the $44 or $50 amount as the costs of materials reworked? Explain.
3. What other costs might White Goods include in its analysis of the total costs of rework due to the tumbler units purchased from the (now) bankrupt supplier?

18-29 Job-cost spoilage and scrap. (F. Mayne) Santa Cruz Metal Fabricators, Inc., has a large job, no. 2734, that calls for producing various ore bins, chutes, and metal boxes for enlarging a copper concentrator. The following charges were made to the job in November 19_4:

Direct materials	$26,951
Direct manufacturing labor	15,076
Manufacturing overhead	7,538

The contract with the customer called for the total price to be based on a cost-plus approach. The contract defined *cost* to include direct materials, direct manufacturing labor costs, and manufacturing overhead to be allocated at 50% of direct manufacturing labor costs. The contract also provided that the total costs of all work spoiled were to be removed from the billable cost of the job and that the benefits from scrap sales were to reduce the billable cost of the job.

Required

1. In accordance with the stated terms of the contract, prepare journal entries for the following two items:
   a. A cutting error was made in production. The up-to-date job-cost record for the batch of work involved showed materials of $650, direct manufacturing labor of $500, and allocated overhead of $250. Because fairly large pieces of metal were recoverable, the company believed that the scrap value was $600 and that the materials recovered could be used on other jobs. The spoiled work was sent to the warehouse.
   b. Small pieces of metal cuttings and scrap in November 19_4 amounted to $1,250, which was the price quoted by a scrap dealer. No journal entries have been made with regard to the scrap until the price was quoted by the scrap dealer. The scrap dealer's offer was immediately accepted.
2. Consider normal and abnormal spoilage. Suppose the contract described above had contained the clause "a normal spoilage allowance of 1% of the job costs will be included in the billable costs of the job."
   a. Is this clause specific enough to define exactly how much spoilage is normal and how much is abnormal? Explain.
   b. Repeat requirement 1(a) with this "normal spoilage of 1%" clause in mind. You should be able to provide two slightly different journal entries.

18-30 Spoilage and job costing. (L. Bamber) Bamber Kitchens produces a variety of items in accordance with special job orders of hospitals, plant cafeterias, and university dormitories. An order for 2,500 cases of mixed vegetables costs $6 per case: direct materials, $3; direct manufacturing labor, $2; and manufacturing overhead allocated, $1. The manufacturing-overhead rate includes a provision for normal spoilage.

Required

1. Assume that a laborer dropped and totally destroyed 200 cases. Prepare a journal entry to record this event. What is the unit cost of the remaining 2,300 cases?
2. Reconsider requirement 1. Suppose part of the 200 cases could be sold to a nearby prison for $200 cash. How would your answer to requirement 1 change?

3. Refer to the original data. (The facts given in requirement 1 are not pertinent to requirement 3.) Tasters at the company reject 200 of the 2,500 cases, and the 200 rejected cases are destroyed. Assume that this rejection rate is considered normal. Prepare a journal entry to record this event, and calculate the unit cost if

   a. The rejection is attributable to exacting specifications of this particular job.

   b. The rejection is characteristic of the production process and is not attributable to this specific job.

4. Reconsider requirement 3. Suppose part of the 200 cases could be sold to a nearby welfare agency for $400. How would your answers to requirement 3 change?

5. Refer to the original data. Tasters rejected 200 cases that had insufficient salt. The product can be placed in a vat, salt added, and reprocessed into jars. This operation will cost $200; this step is taken regularly because of the difficulty in seasoning. Prepare a journal entry to record this event. What is the average unit cost of all the cases?

6. How would your answer to requirement 5 change if the rejection occurred because of the exacting specifications of this particular job?

7. How would your answer to requirement 5 change if the rejection occurred because a worker simply forgot to add the salt?

18-31 Process versus job costing, spoilage, ethics.   Chip Contractors assembles high-performance microchips for two customers—Prestige Jets and the U.S. Air Force. Chip views the U.S. Air Force as a customer that Chip might use to impress the market. Indeed, several potential customers of Chip Contractors are especially interested in how the Air Force views the quality of its chips.

    Kelly Gillis, the president of Chip Contractors, decides to use a double-testing procedure for chips sold to the U.S. Air Force. Tom Sanders, the controller of Chip Contractors, asks Gillis how she wants the cost of this extra testing shown in the accounting records. Each month Sanders has to compile a production-cost report that he sends to Prestige Jets. Prestige has a contract under which it pays Chip the actual production cost per chip purchased plus a 20% profit margin. Sanders tells Gillis that last month (April 19_4) the spoilage rate was 9% of good output (7½% normal and 1½% abnormal). He was concerned about the impact of extra testing on the spoilage rate and hence on the reported actual production costs. The U.S. Air Force contract is for a fixed price amount per good chip. The Air Force is unlikely to be concerned if Chip Contractors has a higher cost per good chip for chips the Air Force buys than for chips sold to other customers, especially if the chips have passed a more demanding set of quality inspections.

    Gillis tells Sanders to follow the normal practice of computing the cost of good chips at the total plant level rather than for each customer. Sanders strongly objects to this approach. He believes it will result in Chip Contractors overcharging Prestige Jets. Gillis believes that they should wait until Prestige Jets complains before they make any adjustments. She is adamant that no internal spoilage data should be shown to Prestige Jets.

    Sanders decides to privately track cost and spoilage data in May 19_4 for Chip's two product lines. This tracking is relatively easy because a separate production line is used for each customer:

	Prestige Jets	U.S. Air Force	Total Plant
Physical Units for May			
Work in process, beginning inventory	0	0	0
Started during May	6,000	4,000	10,000
Good units completed and transferred out during May	4,000	2,000	6,000
Work in process, ending inventory	1,500	1,100	2,600
Direct materials (both 100% complete)			
Conversion costs (both 70% complete)			
Normal spoilage (% of good units transferred out)	7½%	25%	?
Costs for May			
Work in process, beginning inventory	$ 0	$ 0	$ 0
Direct materials added during May	12,750,000	8,500,000	21,250,000
Conversion costs added during May	7,834,000	6,327,080	14,161,080

All direct materials are introduced at the beginning of the production process. Conversion costs are allocated evenly throughout the process. Inspection occurs at the end of the production process. Spoiled chips are disposed of at zero disposal price. Chip Contractors uses the weighted-average method of process costing.

Required

1. Compute the actual cost per good chip transferred out in May for (a) Prestige Jets, (b) the U.S. Air Force, and (c) the total plant. Chip Contractors includes both normal and abnormal spoilage costs when computing actual costs for Prestige Jets.
2. Explain any differences between (a) and (b) in requirement 1.
3. What cost number per good chip sold in May 19_4 should Sanders report to Prestige Jets? Explain any ethical dilemmas that he might face.
4. Suppose the purchasing officer of Prestige Jets receives the May 19_4 bill based on the plant unit cost per good output unit. He immediately requests that Sanders provide full details on how the May unit-cost amount was determined. What should Sanders do?

Coverage of Chapter Appendix

18-32 Physical units, inspection at various stages of completion. (Study the chapter Appendix.) Normal spoilage is 6% of the good units passing inspection in a forging process. In March, a total of 10,000 units were spoiled. Other data follow: units started during March, 120,000; work in process, beginning, 14,000 units (20% completed for conversion costs); work in process, ending, 11,000 units (70% completed for conversion costs).

Required

In columnar form, compute the normal and abnormal spoilage in units, assuming inspection at 15%, 40%, and 100% stages of completion.

18-33 Weighted-average, inspection at 80% completion. (A. Atkinson) (Study the chapter Appendix.) Ottawa Manufacturing produces a plastic toy in a two-stage manufacturing operation. The company uses a weighted-average process-costing system. During the month of June, the following data were recorded for the Finishing Department:

Units of beginning inventory	10,000
Completion of beginning units (%)	25
Cost of direct materials in beginning work in process	$ 0
Units started	70,000
Units completed	50,000
Units in ending inventory	20,000
Completion of ending units (%)	95
Spoiled units	10,000
Costs added during current period:	
Direct materials	$655,200
Direct manufacturing labor	$635,600
Manufacturing overhead	$616,000
Work in process, beginning:	
Conversion costs	$ 42,000
Transferred-in costs	$ 82,900
Cost of units transferred in during current period	$647,500

Conversion costs are incurred evenly throughout the process. Direct materials costs are incurred when production is 90% complete. Inspection occurs when production is 80% complete. Normal spoilage is 10% of all good units that pass inspection.

Required

Prepare a production-cost worksheet for the month of June. Show supporting computations.

Operation Costing, Backflush Costing, and Project Control

PART ONE: OPERATION COSTING AND BACKFLUSH COSTING SYSTEMS

Hybrid costing systems

Operation costing

Just-in-time (JIT) production systems

Backflush costing

PART TWO: CONTROL OF PROJECTS

Project features

Similarity of control of jobs and projects

*C*hanges by companies in their production systems are leading to changes in their management control systems. Sikorsky has shifted to a just-in-time production system when assembling helicopters. Sikorsky's reduced inventory levels and shorter manufacturing lead times have enabled it to reduce the complexity of its cost accounting system.

This chapter has two major parts. Part One focuses on operation costing and backflush costing systems. We illustrate how underlying production systems affect the design of product-costing systems. Backflush costing is often found when a just-in-time (JIT) production system is used. Before discussing backflush costing, we present a brief overview of JIT. (Further discussion of JIT appears in Chapter 24.) Part Two, which can be studied separately, examines the control of projects.

◆ PART ONE:
OPERATION COSTING AND BACKFLUSH COSTING SYSTEMS

HYBRID COSTING SYSTEMS

Objective 1
Explain how hybrid costing systems develop in relation to production systems

Hybrid costing systems are blends of characteristics from both job-costing systems and process-costing systems. Recall that job-costing and process-costing systems are best viewed as ends of a continuum:

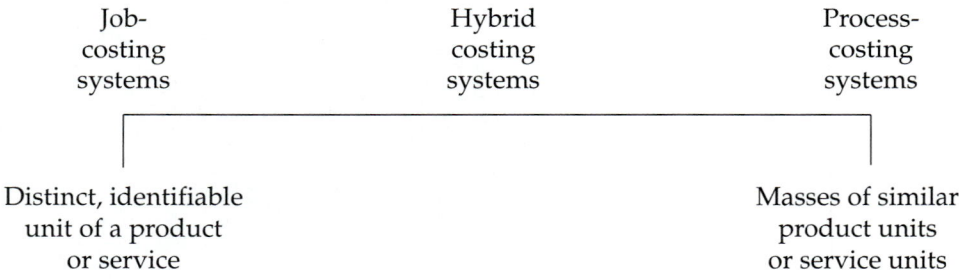

Job-costing systems	Hybrid costing systems	Process-costing systems

Distinct, identifiable unit of a product or service

Masses of similar product units or service units

As we have seen, job costing usually accompanies the custom-order manufacturing of relatively heterogeneous (different) products or services (for example, printing posters, hand-tailoring a suit, constructing a homeowner's patio). At its extreme, custom-order manufacturing entails issuing a specific job order that entails direct materials and a differing set of production steps.

In contrast, process costing usually accompanies the mass production and continuous-flow manufacturing of homogeneous (uniform) products (for example, printing rolls of wallpaper, stitching shirts, and bagging cement). Continuous-flow manufacturing usually entails mass production of standardized products.

Obviously, a product-costing system should be tailored to the underlying production system. Hybrid costing systems develop because there are hybrid production systems, which are blends of custom-order manufacturing and mass-production manufacturing. Manufacturers of a relatively wide variety of closely related standardized products tend to use a hybrid system. Consider Ford automobiles. They may be manufactured in a continuous flow, but each automobile may be customized with a special combination of motor, transmission, radio, and so on. Many manufacturing companies that produce generic products sold by other companies under a "private label" use an operation costing system. Examples of such products include canned soups, vegetables, and soft drinks. Each company develops its own hybrid costing system to meet its individual needs.

We will now explore two costing systems in some detail. The first, operation costing, is a hybrid that has existed for many years. The second, backflush costing, is a simplified budgeted or standard costing system that began to receive attention in the 1980s.

OPERATION COSTING

Overview of Operation Costing

Operation costing is a hybrid costing system applied to batches of similar products. Each batch of products uses the same resources to the same extent as all other batches. That is, a single batch of products proceeds through a sequence of selected activities or operations. Within each operation, all product units are treated exactly alike, using identical amounts of the operation's resources. Batches are also termed *production runs*.

Each batch is often a variation of a single design and requires a particular sequence of standardized operations. Consider a business that makes suits. Management may select a single basic design for every suit that the company manufactures. Depending on specifications, batches of suits vary from each other. One batch may use wool; another batch, cotton. One batch may require special hand stitching; another batch, machine stitching. Other products that are likewise often manufactured in batches are semiconductors, textiles, and shoes.

An **operation** is defined as a standardized method or technique that is performed repetitively regardless of the distinguishing features of the finished good. Operations are usually conducted within departments. For instance, a suit maker may have a cutting operation and a hemming operation within a single department. The term *operation*, however, is often used loosely. It may be a synonym for a *department* or *process*. For example, some companies may call their finishing department a finishing process or a finishing operation.

In operation costing, work orders initiate production. Product costs are compiled for each work order, which, in an operation costing system, will be made up of two or more units of a product. (Recall that in a job-costing system, each individual unit of a product has its own work order.) Direct materials that are different in work orders are specifically identified with the appropriate order. Conversion costs are compiled for each operation and then allocated to all units passing through the operation. A single average unit conversion cost is used in each operation. Typical allocation bases are the number of units worked on and the minutes used to complete the individual operation. As we will illustrate, the method for allocating conversion costs to products is similar to allocating manufacturing overhead. (Our examples assume only two cost categories, direct materials and conversion costs. Of course, operation costing can have more than two cost categories.)

As always, cost management requires attention to physical processes. For example, in the manufacturing of clothing, managers are concerned with fabric waste, the number of fabric layers that can be cut at one time, and so on. The cost-accounting system captures the financial impact of the control of physical processes. Feedback provided by the cost-accounting system helps managers improve the physical processes.

Illustration of Operation Costing

An operation costing system uses work orders that specify the needed direct materials and step-by-step operations. Work orders for each batch differ as to the combinations of direct materials and operations to be undergone.

Consider Baltimore Company, a clothing manufacturer, which produces two lines of blazers for department stores. Wool blazers use better-quality materials and undergo more operations than do polyester blazers. Let's look at the following (condensed) operations in 19_5:

	Work Order 423	Work Order 424
Direct materials	Wool	Polyester
	Full satin lining	Partial rayon lining
	Bone buttons	Plastic buttons
Operations		
1. Cutting cloth	Use	Use
2. Checking edges	Use	Do not use
3. Sewing body	Use	Use
4. Checking seams	Use	Do not use
5. Sewing collars and lapels by machine	Do not use	Use
6. Sewing collars and lapels by hand	Use	Do not use

Suppose work order 423 is for 50 wool blazers and work order 424 is for 100 polyester blazers. The following costs are assumed for these two work orders, which were started and completed in March 19_5:

	Work Order 423	Work Order 424
Number of blazers	50	100
Direct materials costs	$ 6,000	$3,000
Conversion costs allocated:		
Operation 1	580	1,400
Operation 2	400	—
Operation 3	1,900	3,560
Operation 4	500	—
Operation 5	—	875
Operation 6	700	—
Total manufacturing costs	$10,080	$8,835

This operation costing system uses a budgeted rate to allocate manufacturing overhead to each operation. For example, the costs of operation 5 might be budgeted as follows (amounts assumed):

$$\begin{aligned} \text{Operation 5 budgeted} \\ \text{conversion cost-} \\ \text{allocation rate in 19\_5} \end{aligned} = \frac{\begin{aligned}\text{Operation 5 budgeted} \\ \text{conversion costs in 19\_5}\end{aligned}}{\begin{aligned}\text{Operation 5 budgeted} \\ \text{machine-hours in 19\_5}\end{aligned}}$$

$$= \frac{\$1,225,000}{35,000 \text{ machine-hours}}$$

$$= \$35 \text{ per machine-hour}$$

The budgeted conversion costs of operation 5 include labor, power, repairs, supplies, depreciation, and other overhead of this operation.

As goods are manufactured, conversion costs are allocated to the work orders by multiplying the $35 hourly allocation rate times the number of machine-hours used in operation 5. Suppose the 100 polyester blazers require 25 machine-hours in operation 5. The conversion costs of operation 5 for 100 polyester blazers are 25 × $35 = $875.

All product units in a work order are assumed to consume identical amounts of resources of a particular operation. Consider operation 3. Work order 424, containing 100 units, has twice the total costs of the 50 units in work order 423 ($3,800 compared with $1,900). If work order 424 contained 75 units, its total costs in operation 3 would be 150% rather than 200% of the cost of work order 423.

Journal Entries

Objective 3
Prepare journal entries for an operation costing system

Summary journal entries for assigning costs to the polyester blazers follow. Entries for the wool blazers would be similar.

The journal entry for the requisition of direct materials, which are traced directly to particular batches, for the 100 polyester blazers is:

1. Work in Process, Operation 1 $3,000
        Materials Inventory Control $3,000

Actual conversion costs (assumed in March 19_5 to be $24,400 for operation 1) are entered into a Conversion Costs account:

2. Conversion Costs $24,400
        Various accounts (such as Wages Payable Control and
        Accumulated Depreciation) $24,400

The allocation of conversion costs to products in operation costing uses the budgeted allocation rate. Assume that the amount allocated to work order 424 for operation 1 is the $1,400 (40 machine-hours × $35) in the tabulation on p. 658:

3. Work in Process, Operation 1 $1,400
        Conversion Costs Allocated $1,400

The transfer of the polyester blazers from operation 1 to operation 3 (recall that the polyester blazers do not go through operation 2) would be journalized as follows:

4. Work in Process, Operation 3 $4,400
        Work in Process, Operation 1 $4,400

After posting, Work in Process, Operation 1, appears as follows:

Work in Process, Operation 1			
1. Direct materials	3,000	4. Transferred to Operation 3	4,400
3. Conversion costs allocated	1,400		

The costs of the blazers are transferred through the pertinent operations and then to finished goods in the usual manner. Costs are added throughout the year in the accounts Conversion Costs and Conversion Costs Allocated. Any overallocation or underallocation of conversion costs is disposed of in the same way as overallocated or underallocated manufacturing overhead in a job-costing system. (See pp. 152–55 for discussion.)

In sum, operation costing is a hybrid of job costing and process costing:

◆ Its job-costing feature is the specific assignment of direct materials costs to individual work orders.

◆ Its process-costing features are that conversion costs are tracked to each operation and that within each operation all product units are treated exactly alike, using identical amounts of the operation's resources.

EXHIBIT 19-1

Variety of Cost-Accounting Systems

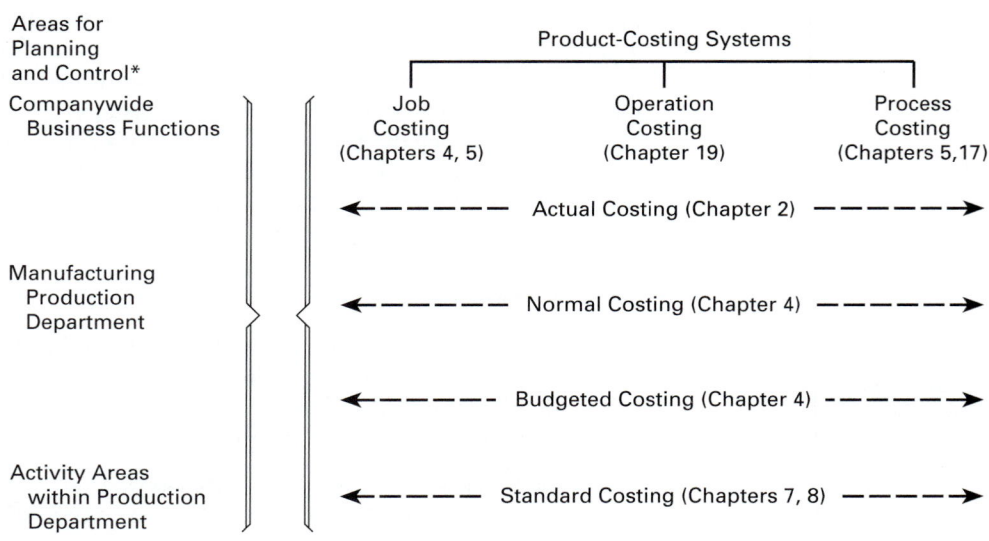

*See Exhibit 5-9 (p. 166) for elaboration of areas for planning and control.

Operation Costing and Activity-Based Accounting Systems

Exhibit 19-1 presents an overview of the variety of costing systems described in this book. While operation costing is placed on the continuum of product costing between job costing and process costing, it does not necessarily have to be exactly in the middle. Below the continuum, we see how these three product-costing systems can be combined with actual costing, normal costing, budgeted costing, and standard costing (which were compared in Exhibit 8-6, p. 286). Our operation-costing example above used normal costing, but actual costing, budgeted costing, or standard costing could also have been used.

The left-hand vertical classification, which is a condensed version of a similar presentation in Exhibit 5-9, page 166, displays how an accounting system can accumulate costs for planning and control. Costs can be identified with business functions alone or in more detailed ways with departments or activity areas. Costing at each activity area provides the most detailed or finely granulated accounting information. As Chapters 4 and 5 explain, activity-based costing is regarded as the strongest link connecting accounting and intelligent cost management. Using several different cost drivers leads to the most accurate information for evaluating the performance of activity areas and for building up product costs. Activity-based costing may be combined with any form or combination of product-costing systems.

Exhibit 19-1 emphasizes the variety of costing-system choices available to management. The choice of a costing system depends on the underlying production system and managers' desires regarding cost management and accuracy in product costing.

JUST-IN-TIME (JIT) PRODUCTION SYSTEMS

Objective 4

Describe the key features of just-in-time production systems

Companies adopting just-in-time are making major changes to their production areas. We now briefly note these changes as a lead-in to our discussion of backflush costing.

Just-in-time (JIT) production is a system in which each component on a production line is produced immediately as needed by the next step in the production line. In a JIT production line, manufacturing activity at any particular workstation is prompted by the use of that station's output at downstream stations. Ideally, the sale of a unit of finished goods triggers the completion of a unit in final assembly, and so forth, backward in the sequence of manufacturing steps all the way to the delivery of materials. This characteristic is often called the "demand-pull" feature of a JIT production line. At Hewlett-Packard (HP), workers and managers frequently quote the following demand-pull slogan: "Never build nothing, nowhere, for nobody, unless they *ask* you for it."

Major Features of JIT

There are four key features in a JIT production system:

◆ Inventory is regarded as an evil because of its "nonvalue-added" nature. A nonvalue-added activity is one that a customer does not perceive as increasing the utility (usefulness) of the product.

◆ Production activities are simplified. Nonvalue-added activities are spotlighted and then reduced or eliminated.

◆ Emphasis is placed on reducing the manufacturing lead time, which is the time from when an order is ready to start on the production line to when it becomes a finished good. Reduced lead time enables a company to respond better to changes in customer demand. It also means there is less work in process at any one point in time.

◆ Production stops if parts are absent or defects are discovered. Emphasis is placed on eliminating the root causes of defects as quickly as possible. In contrast, in many traditional production systems extra parts and subassemblies are held at workstations in anticipation of shortages or production breakdowns. Typically, there is less rework or spoilage in JIT systems.

Financial Benefits of JIT

JIT should lead to many financial benefits, including

1. Lower investment in inventories
2. Reductions in carrying and handling costs of inventories
3. Reductions in risk of obsolescence of inventories
4. Lower investment in plant space for inventories and production
5. Reductions in total manufacturing costs
6. Reductions in paperwork

Reductions in paperwork can be dramatic. For example, a Taiwanese manufacturing plant was processing 10,000 purchase orders, receiving reports, and materials requisitions per week. After switching to JIT, a simpler accounting system emerged. In some JIT plants, there are no individual purchase orders, no receiving reports, no materials requisitions, no work orders, and no tracking of manufacturing labor as a separate direct cost category.

JIT tends to focus broadly on the control of *total manufacturing costs* instead of individual costs such as direct manufacturing labor. For example, idle time may rise because production lines are starved for materials more frequently than before. Nevertheless, many manufacturing overhead costs will decline, such as the costs of materials handling and the costs of special quality control inspectors.

BACKFLUSH COSTING

Objective 5
Prepare journal entries for three typical versions of backflush costing systems

Simplified Budgeted or Standard Costing

The traditional budgeted and standard costing systems (discussed in Chapters 4, 7 and 8) use **sequential tracking** (also called **synchronous tracking**), which is any product-costing method in which the timing of entries recorded in the accounting system is synchronized with the physical sequences of purchases and production. The traditional budgeted or standard costing system tracks costs sequentially with the passage of the products from direct materials, through work in process, to finished goods. Sequential tracking is often expensive, especially if attempts are made to track direct materials requisitions and time tickets to individual operations and products.

An alternative to the sequential tracking approach in many costing systems is to delay the recording of journal entries until after the physical sequences have occurred. An extreme form of delay is to wait until sale of a finished good has occurred. The term **backflush costing** (also called **delayed costing, endpoint costing,** or **post-deduct costing**) describes a costing system that delays recording changes in a product being produced until good finished units appear; it then uses budgeted or standard costs to work backward to assign manufacturing costs to units produced. Typically no record of work in process appears in the accounting system. Backflush costing often accompanies JIT production systems, although backflush can be coupled with any production system. The term backflush probably arose because the trigger point for inventory costing entries can be delayed until as late as the time of sale. Costs are then finally flushed back through the accounting system.

Companies that adopt backflush costing often meet three conditions:

1. Management wants a simple accounting system. No detailed tracking of actual amounts of direct materials costs or direct manufacturing labor costs through a series of operations, step by step to the point of completion is thought to be necessary.
2. Each product has a set of budgeted or standard costs.
3. Backflush costing yields approximately the same financial results as sequential tracking would generate.

If inventories are low, the vast bulk of manufacturing costs will flow directly into Cost of Goods Sold and not be delayed in inventory. Hence, managers may not believe it worthwhile to spend resources tracking costs through Work in Process, Finished Goods, and Cost of Goods Sold. Backflush costing, therefore, is especially attractive in companies that have low inventories resulting from JIT. Even when inventory levels are high, if they are relatively stable, sequential tracking and backflush costing should produce approximately the same results. Constant amounts of costs will be deferred in inventory each period.

The following three examples demonstrate backflush costing. To underscore basic concepts, we assume no direct materials variances in any of the examples. (We do, however, discuss variances in a separate section following example 1.) The three examples differ in the number and placement of trigger points for making journal entries in the accounting system:

	Number of Journal Entry Trigger Points	Location of Journal Entry Trigger Points
Example 1	2	1. Purchase of raw materials (direct materials) and components 2. Completion of good finished units of product
Example 2	2	1. Purchase of raw materials and components 2. Sale of good finished units of product
Example 3	1	1. Completion of good finished units of product

Example 1: Trigger Points Are Materials Purchase and Finished Goods

This example illustrates how backflushing can eliminate the need for a separate Work in Process account. A hypothetical company, Silicon Valley Computer (SVC), has two trigger points, which are places in their operations at which journal entries are made in the accounting system. SVC produces keyboards for personal computers. For April, there were no beginning inventories of raw materials. Moreover, there is zero beginning and ending work in process.

SVC has only one direct manufacturing cost category (direct materials, which are simply called "raw") and one indirect manufacturing cost category (conversion costs). All labor costs at the manufacturing facility are included in conversion costs. The April standard direct materials costs per keyboard unit are $19; the standard conversion costs are $12.

SVC has two inventory accounts:

Type	Account Title
Combined materials and work in process	Inventory: Raw and In-Process Control
Finished goods	Finished Goods Control

Trigger point 1 occurs when materials are purchased. These costs are charged to Inventory: Raw and In-Process Control.

Actual conversion costs are recorded as incurred under backflush costing, just as in other costing systems. Conversion costs are allocated to products at trigger point 2—the transfer of units to Finished Goods. This example assumes that underallocated conversion costs are carried forward and not disposed of until year-end.

SVC takes the following steps when assigning costs to units sold and to inventories.

STEP 1: *RECORD THE DIRECT MATERIALS PURCHASED IN THE REPORTING PERIOD.* Assume April purchases of $1,950,000:

Entry (a) Inventory: Raw and In-Process Control	$1,950,000	
Accounts Payable Control		$1,950,000

STEP 2: *RECORD THE INCURRENCE OF CONVERSION COSTS DURING THE REPORTING PERIOD.* Assume conversion costs were $1,260,000:

Entry (b) Conversion Costs Control	$1,260,000	
Various accounts (such as Accounts Payable Control and Wages Payable Control)		$1,260,000

STEP 3: *DETERMINE THE NUMBER OF FINISHED UNITS MANUFACTURED DURING THE REPORTING PERIOD.* Assume that 100,000 keyboard units were manufactured in April.

STEP 4: *COMPUTE THE BUDGETED OR STANDARD COST OF EACH FINISHED UNIT.* This step normally requires a bill of materials (description of the types and quantities of materials) and an operations list (description of operations to be undergone) or similar records. For SVC, the standard cost is $31 ($19 direct materials + $12 conversion costs) per unit.

STEP 5: *RECORD THE COST OF FINISHED GOODS COMPLETED IN THE REPORTING PERIOD:* (100,000 units × $31 = $3,100,000).

This step gives backflushing its name. Up to this point in the operations, the costs have not been recorded synchronously with the flow of product along its production route. Instead, the output trigger reaches back and pulls costs from Inventory: Raw and In-process.

Entry (c) Finished Goods Control	$3,100,000	
Inventory: Raw and In-Process Control		$1,900,000
Conversion Costs Allocated		1,200,000

STEP 6: *RECORD THE COST OF GOODS SOLD IN THE REPORTING PERIOD.*
Assume that 99,000 units were sold in April (99,000 units × $31 = $3,069,000).

| Entry (d) Cost of Goods Sold | $3,069,000 | |
| Finished Goods Control | | $3,069,000 |

The April ending inventory balances are:

Inventory: Raw and In-Process	$50,000
Finished Goods, 1,000 units × $31	31,000
Total inventories	$81,000

Exhibit 19-2 provides an overview of this version of backflush costing. The elimination of the typical Work in Process account reduces the amount of detail in the accounting system. Units on the production line may still be tracked in physical terms, but there is no "costs-attach" tracking to specific work orders. In fact, there are no work orders or time tickets in the accounting system. By "costs-attach" we mean costs that adhere like barnacles to the physical products as they flow along the production cycle.

Accounting for Variances

The accounting for variances between actual costs incurred and budgeted or standard costs allowed and the disposition of variances is basically the same under all standard costing systems. The procedures are described in Chapters 7 and 8. In our example 1, suppose the direct materials purchased had an unfavorable price variance of $42,000. Entry (a) would then be:

Inventory: Raw and In-Process Control	$1,950,000	
Raw Materials Price Variance	42,000	
Accounts Payable Control		$1,992,000

Direct materials are often a large proportion of total manufacturing costs, sometimes over 80%. Consequently, many companies will at least measure direct materi-

EXHIBIT 19-2
Overview of Backflush Costing, Example 1

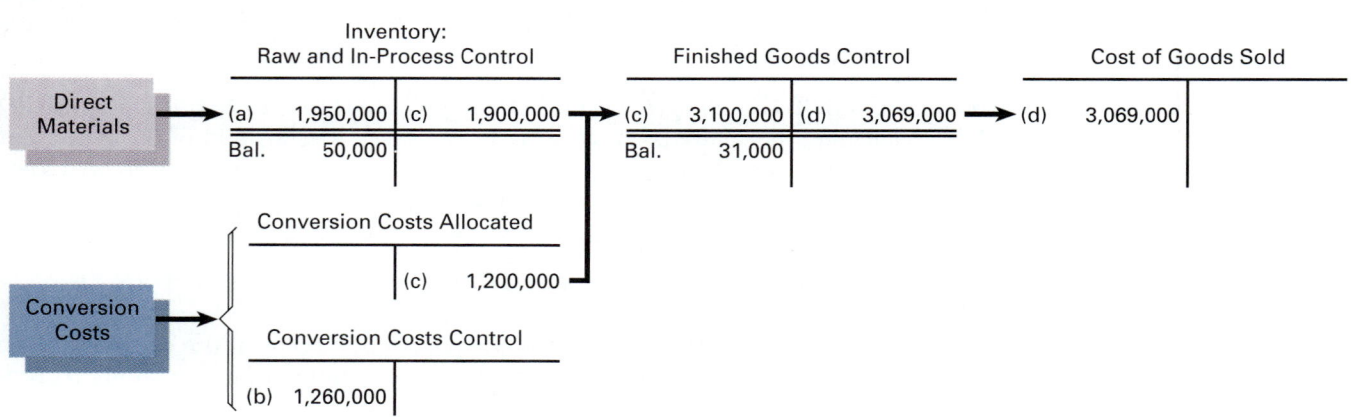

als efficiency variances in total by physically comparing what remains in direct materials inventory against what should be remaining, given the output of finished goods for the accounting period. In our example, suppose such a comparison showed an unfavorable efficiency variance of $90,000. The journal entry would be:

Raw Materials Efficiency Variance	$90,000	
Inventory: Raw and In-Process Control		$90,000

Actual conversion costs may be underallocated or overallocated in any given accounting period. Chapter 5 (pp. 152–55) discussed various ways to account for underallocated or overallocated manufacturing overhead costs. Many companies write off underallocations or overallocations only at year-end; other companies do so monthly. Suppose SVC's policy is to write off underallocated or overallocated conversion costs to cost of goods sold monthly. The journal entry for this difference between actual conversion costs incurred and standard conversion costs allocated (assumed to be $60,000) would be:

Conversion Costs Allocated	$1,200,000	
Cost of Goods Sold	60,000	
Conversion Costs Control		$1,260,000

Alternatively, if SVC's policy were to dispose of underallocated or overallocated conversion costs at year-end, the balances would accumulate throughout the year in Conversion Costs and Conversion Costs Allocated. At year-end, an entry like the immediately preceding one would be necessary.

Any direct materials variances and underallocated or overallocated conversion costs are usually written off to Cost of Goods Sold instead of being prorated. Companies that use backflush costing typically have low inventories, so proration is less often necessary.

Example 2: Trigger Points Are Materials Purchase and Sale

Example 2, also based on SVC and using the same data, presents a backflush costing system that is a more dramatic departure from a sequential tracking inventory costing system than example 1. The first trigger point in example 2 is the same as the first trigger point in example 1, but the second trigger point is the sale—not the completed manufacture—of finished units. Toyota's cost accounting at its Kentucky plant is similar to example 2. There are two justifications for this accounting system:

◆ To remove the incentive for managers to produce for inventory. Under the typical costs-attach tracking assumption, managers can bolster operating income by producing units not sold. Having trigger point 2 as the sale instead of the completion of production, however, reduces the attractiveness of this action by recording conversion costs as period costs instead of capitalizing them as inventoriable costs.

◆ To increase the focus of managers on a plantwide goal (selling units) rather than on an individual subunit goal (for example, increasing machine utilization at an individual work center).

This variation of backflush costing is sometimes called "super-variable costing" or "throughput costing." Why? Because it has the same effect on operating income as the immediate expensing of conversion costs. Allied Signal Limited, Skelmersdale, uses this approach (see Concepts in Action, p. 818).

The inventory account in example 2 is confined solely to direct materials (whether they are in storerooms, in process, or in finished goods). There is only one inventory account:

Type	Account Title
Combined direct materials inventory and any direct materials in work in process and finished goods	Inventory Control

Entry (a) is prompted by the same trigger point 1 as in example 1, the purchase of materials and components:

Entry (a) Inventory Control $1,950,000
  Accounts Payable Control $1,950,000

The incurrence of conversion costs is recorded in an identical manner as in example 1:

Entry (b) Conversion Costs Control $1,260,000
  Various accounts $1,260,000

Trigger point 2 is the *sale* of good finished units (not their *production*, as in example 1): 99,000 units sold × $31 = $3,069,000, consisting of direct materials (99,000 × $19 = $1,881,000) and conversion costs allocated (99,000 × $12 = $1,188,000).

Entry (c) Cost of Goods Sold
  (99,000 units × $31) $3,069,000
  Inventory Control
   (99,000 × $19) $1,881,000
  Conversion Costs Allocated
   (99,000 × $12)   1,188,000

No conversion costs are inventoried. Either the $1,260,000 in Conversion Costs is allocated to the units sold at standard ($1,188,000), or Conversion Costs is underallocated by $72,000 ($1,260,000 − $1,188,000). Suppose SVC, like many companies, writes these underallocated costs off monthly as additions to cost of goods sold:

Entry (d) Conversion Costs Allocated $1,188,000
  Cost of Goods Sold   72,000
   Conversion Costs Control $1,260,000

The April ending balance of Inventory is $69,000 ($50,000 direct materials still on hand + $19,000 direct materials embodied in the 1,000 units manufactured but not sold during the period). Exhibit 19-3 provides an overview of this version of backflush costing.

Example 3: Trigger Point Is Finished Goods

This example presents the most extreme and the simplest version of backflush costing. It has only one trigger point for making journal entries to inventory. Conversion Costs would be debited as actual costs are incurred. The trigger point is SVC's production of finished units. Using the same data as in examples 1 and 2, the summary entry is:

EXHIBIT 19-3
Overview of Backflush Costing, Example 2

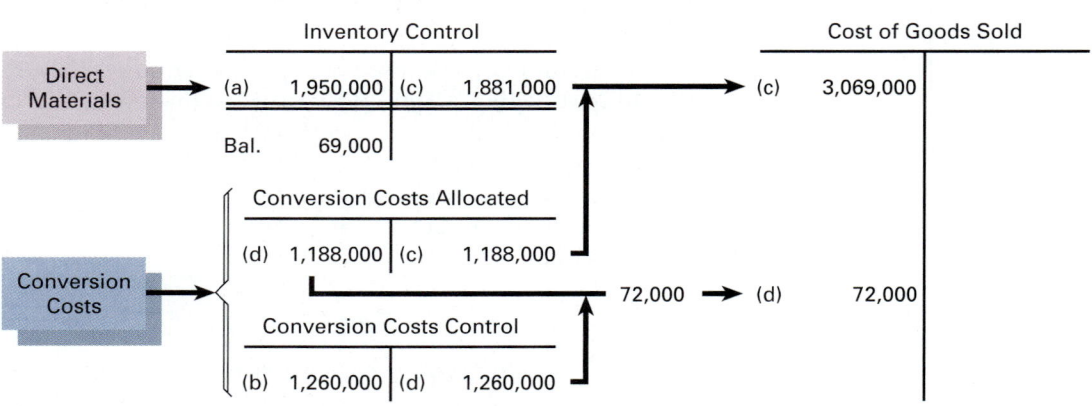

Finished Goods Control	$3,100,000	
Accounts Payable Control (for materials)		$1,900,000
Conversion Costs Allocated		1,200,000

Compare this entry with entries (a) and (c) in example 1 on pp. 663–64. The simpler version in example 3 ignores the $1,950,000 purchases of direct materials (entry [a] of example 1, p. 663). At the end of April, the $50,000 of direct materials purchased has not yet been placed into production ($1,950,000 − $1,900,000 = $50,000), nor has it been entered into the inventory costing system. The example 3 variation of backflush costing is suitable for a JIT production system with virtually no direct materials inventory and minimal work-in-process inventories. It is less feasible otherwise.

Difficulties of Backflush Costing

The accounting illustrated in examples 2 and 3 does not strictly adhere to generally accepted accounting principles of external reporting. Work in process (an asset) exists but is not recognized in the accounting system. Advocates of backflushing, however,

Surveys of Company Practice

ADOPT JIT AND SIMPLIFY YOUR LIFE AS A MANAGEMENT ACCOUNTANT

Many companies adopting JIT production systems are reporting reductions in the complexity and detail of their management accounting systems. The results of a survey of 22 U.S. manufacturing companies implementing JIT support these claims. These 22 manufacturers were made up of 11 machinery, 7 transportation, 2 computer, and 2 consumer products companies. On average, the companies had begun converting to JIT four years prior to the survey. At the time of the survey, they had converted an average 63% of their plant operations and inventory systems to JIT.

Key results from the survey follow[a]:

A. Mean reductions following JIT in four key areas:
- ◆ Number of vendors reduced by 67%
- ◆ Quantity of rework and scrap reduced by 44%
- ◆ Setup time for product changes on machines reduced by 47%
- ◆ Quantity in total inventory reduced by 46%

B. Types of cost accounting system in use:

	Before JIT	After JIT
Job costing	70%	30%
Process costing	20	60
Hybrid costing	10	10

C. Complexity of cost accounting systems following implementation of JIT:
- ◆ Less complex 72.7%
- ◆ More complex 27.3

D. Eight of the 22 companies adopted backflush costing after implementing JIT, eliminating accounting transactions for the movement of materials to work in process.

E. Complexity of performance measurement system following implementation of JIT:
- ◆ Less complex 77.3%
- ◆ More complex 22.7

[a]Adapted from Swenson and Cassidy, "The Effect of JIT." Full citation is in Appendix A.

cite the materiality concept in support of these versions of backflush accounting. They claim that if inventories are low or not subject to significant change from one accounting period to the next, operating income and inventory costs developed in a backflush costing system will not differ materially from the results generated by a system that does adhere to generally accepted accounting principles.

Suppose material differences in operating income and inventories do exist between the results of a backflush costing system and of a conventional standard costing system. An adjustment can be recorded to make the backflush numbers satisfy external reporting requirements. For example, the backflush entries in example 2 would result in expensing all conversion costs as a part of Cost of Goods Sold ($1,188,000 at standard + $72,000 write-off of underallocated conversion costs = $1,260,000). But suppose conversion costs were regarded as sufficiently material in amount to be included in Inventory. Then entry (d), closing the Conversion Costs accounts, would change as follows:

Original entry (d) Conversion Costs Allocated	$1,188,000	
Cost of Goods Sold	72,000	
Conversion Costs Control		$1,260,000
Revised entry (d) Conversion Costs Allocated	$1,188,000	
Inventory (1,000 units × $12)	12,000	
Cost of Goods Sold	60,000	
Conversion Costs Control		$1,260,000

A chief attraction of backflush costing is its simplicity. Simple systems, however, generally do not yield as much information as do more complex systems. Criticisms of backflush costing focus mainly on the absence of audit trails—the ability of the accounting system to pinpoint the uses of resources at each step of the production process. Managers, however, keep track of operations by personal observations, computer monitoring, and nonfinancial measures. In addition, actual materials quantities, conversion costs, and scrap can be identified with individual departments and activity areas. The good units are produced and recorded. Their budgeted or standard costs allowed are also measured. Variances are computed at the department or activity-area level at least monthly and sometimes daily.

◆ PART TWO:
CONTROL OF PROJECTS

Part One covered product costing for products that are ordinarily manufactured within hours or days. Part Two provides an overview of the control of entire projects, which would generally last much longer, sometimes even years.

PROJECT FEATURES

Objective 6
Identify the four critical success factors of project costing

What is a job? What is a project? Distinctions are fuzzy. An accounting firm may use "engagements" to describe its audits for various clients. The audit staff may regard the audit of a local country club as a "job" but the audit of a multinational company such as Exxon as a "project." In general, a **project** is a complex job that often takes months or years to complete and requires the work of many different departments, divisions, or subcontractors. Examples of projects arise in such fields as construction (for example, bridges, shopping centers, power plants); developing and introducing a new product model (for example, automobiles, computers, space vehicles); and conducting complex lawsuits (for example, asbestos litigation).

The planning and controlling of jobs and projects have common characteristics. Projects, however, are more challenging than jobs. Projects are unique and nonrepetitive, have more uncertainties, involve more skills and specialties, and require more coordination over a longer run. Managers use special control techniques for long-run projects. The projects are often subdivided into a series of work packages, each having its individual time schedules.

Managers' control of projects generally focuses on four critical success factors, the major keys to both profitability and customer satisfaction: (a) scope, (b) quality, (c) time schedule, and (d) costs. *Scope* is the technical description of the final product. Many projects are subjected to engineering change orders as the work proceeds, whereby the final product has features different from those originally planned. Obviously, changes in scope usually also affect quality, time schedule, and costs.

In the 1990s, companies are becoming increasingly sensitive to time-based competition (see Chapter 23). For example, automobile companies now set faster schedules for developing new products. Electronics companies aim at continually reducing schedules for research and development, design, production, marketing, and distribution of a new product. Time is a key in project control.

Chapter 12 described life-cycle product budgeting and costing. The focus was on the costs attributable to each product from its initial research and development to its final customer servicing and support in the marketplace. Life-cycle reports for each product require tracking costs and revenues on a product-by-product basis over several calendar periods. Reports for project costing likewise require tracking over long periods of time.

Project Variances

Objective 7
Compute project variances

The U.S. Department of Defense (DOD) spends billions of dollars each year on projects having varying schedules. The DOD requires that these expenditures be monitored by cost performance reporting (CPR). Exhibit 19-4 shows a CPR for an engineering project named Delphi, undertaken by Structural Design Associates. This project has a time horizon of 10 months (July 19_4 to April 19_5). Three variables are shown on Exhibit 19-4:

EXHIBIT 19-4
Cost and Schedule Performance Report

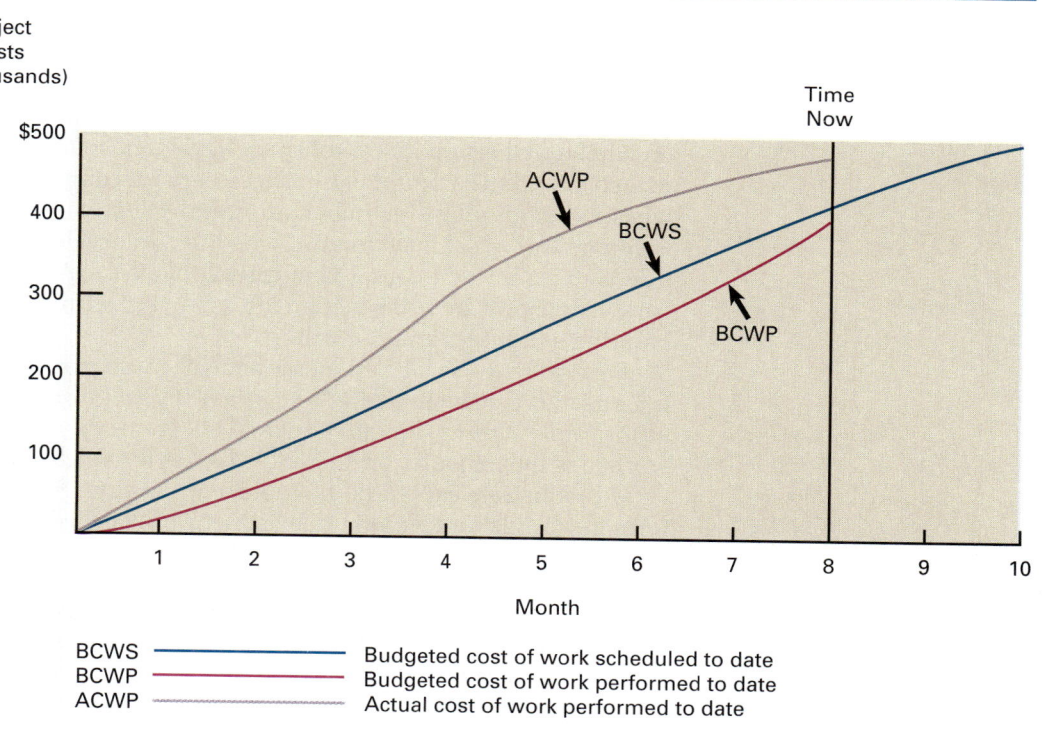

BCWS	Budgeted cost of work scheduled to date
BCWP	Budgeted cost of work performed to date
ACWP	Actual cost of work performed to date

- ◆ BCWS—Budgeted cost of work *scheduled* to date. The Delphi project should be 85% complete by February 28, 19_5 (month 8 in project time), with a budgeted cost of $425,000.
- ◆ BCWP—Budgeted cost of work *performed* to date. By February 28, 19_5, the Delphi project is only 80% complete. The budgeted cost of the 80% completed project is $400,000.
- ◆ ACWP—Actual cost of work performed to date, regardless of any budgets or schedules. By February 28, 19_5, costs assigned to the Delphi project total $480,000.

These three variables can be analyzed as follows:

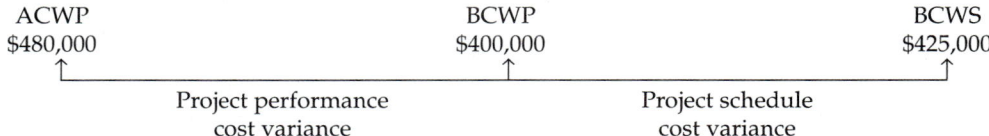

The **project performance cost variance** is the difference between the actual cost of work performed to date and the budgeted costs of the work performed to date. It is a measure of the project cost overrun (or underrun), controlling for the actual degree of completion to date on the project:

$$\text{ACWP} - \text{BCWP} = \$480,000 - \$400,000 = \$80,000 \text{ U}$$

The Delphi project has a cost overrun of $80,000 (20% above the $400,000 budget). This is a serious (unfavorable) concern regarding cost management on the project.

The **project schedule cost variance** is the difference between the budgeted cost of work performed to date and the budgeted cost of work scheduled to date:

$$\text{BCWP} - \text{BCWS} = \$400,000 - \$425,000 = \$25,000 \text{ U}$$

The Delphi project is behind schedule (viewed as unfavorable). It is at only the 80% completion stage at February 28, 19_5, instead of the budgeted 85% stage. The project schedule cost variance of $25,000 captures the budgeted cost difference between the 80% and 85% project completion stages.

These two major variances underscore that the cost performance report system would be more accurately labeled as a *cost and schedule performance report system*. The manager can use the unfolding information to predict final costs and completion dates, as we now discuss.

Focus on the Future

The graph in Exhibit 19-4 also illustrates another key aspect of project control: a focus on what remains to be accomplished. Managers do not want surprises concerning future costs and the amount of time necessary for completing a project. The "time now" label on the graph is the time of appraisal: Where has the project been, and where it is going? Appraisal may occur at scheduled intervals, such as weeks or months, or at designated stages of completion called *milestones*. An example of a milestone is the completion of a building's foundation; a second milestone is the completion of a building's framing, and so forth.

Exhibit 19-4 is labeled a cost and schedule performance report, but many contractors and consultants refer to it as a *performance report*, an *earned value report*, or an *earned hours report*. The latter two labels are often used when control of the hours worked is the major determinant of success (or at least of project billing).

To illustrate earned hours reporting, consider Structural Design Associates. Information on labor costs for an existing project called Gamma, follows:

- ◆ Original time budget to complete, 1,500 hours
- ◆ BCWS—Budgeted cost (hours) of work scheduled to date, 1,000 hours.
- ◆ BCWP—Budgeted cost (hours) of work performed to date, 700 hours. This amount is sometimes called earned value or earned hours.
- ◆ ACWP—Actual cost (hours) of work performed to date, 600 hours at $80 per hour

The performance report could be expressed in hours, in dollars, or in both terms. Suppose the actual labor cost is also $80 per hour. The variances in dollars would be:

◆ Project performance cost variance:

$$ACWP - BCWP = (600 \times \$80) - (700 \times \$80)$$
$$= 100 \times \$80 = \$8,000 \text{ F}$$

The Gamma project is under budget for labor costs, given the current stage of completion.

◆ Project schedule cost variance:

$$BCWP - BCWS = (700 \times \$80) - (1,000 \times \$80)$$
$$= -300 \times \$80 = -\$24,000 \text{ U}$$

The Gamma project is sizably behind schedule in terms of labor-hours worked. The project schedule cost variance shows a budgeted cost difference equal to 300 (24,000 ÷ $80 per hour) hours behind schedule.

We now relax the assumption that the actual and budgeted labor rates ($80 per hour) are the same. Suppose the actual rate was $85 per hour. The project performance cost variance would then have both an efficiency component and a price component (as explained on p. 238 of Chapter 7):

Efficiency variance, as calculated above	$8,000 F
Price variance, actual hours x difference in price per hour, or 600 hours x $5	3,000 U
Project performance cost variance	$5,000 F*

*Or ($85 × 600) − ($80 × 700) = $51,000 − $56,000 = $5,000 F.

In view of these variances, the project manager might predict a late completion date and offer revised budgeted total final costs of $115,000 (or less):

Original budgeted costs, 1,500 hours × $80	$120,000
Deduct: Favorable project cost variance to date	5,000
Subtotal	115,000
Deduct: Additional favorable project cost variance (prediction is needed)	?
Revised budget, final costs	$?

SIMILARITY OF CONTROL OF JOBS AND PROJECTS

The control of jobs is like the control of projects, although on a smaller, simpler scale. Moreover, jobs are usually more repetitive. Consider the partner in charge of the audit engagement of a local country club. She must monitor progress and adjust her predictions of actual hours compared with budgeted hours. Perhaps little can be done to alter performance or profitability of the current job. The information gathered on this year's job, however, may be vital to the budgeting and negotiating of fees on next year's job.

Projects such as the development of weapons systems and finding cures for diseases are often undertaken despite high uncertainties and likely changes as work progresses. Should interim performance reports compare progress against the original budget or against a revised budget (much like a flexible budget)? Ideally, management should be provided with both comparisons. In this way, the performance of managers as planners can be assessed by comparing the original budget against the revised budget. Similarly, the performance of managers regarding control of operations can be assessed by comparing the actual results against the revised budget.

PROBLEM FOR SELF-STUDY

PROBLEM

A Dallas manufacturing company uses standard costs and an operation costing system. It has a storeroom and several buffer stocks of parts and in-process inventories at various work centers along the production lines. No separate major cost category for direct manufacturing labor exists; all manufacturing labor is a part of conversion costs. For simplicity, assume that there are no beginning inventories and no standard cost variances of materials.

Required

1. Prepare summary journal entries (without disposing of underallocated or overallocated conversion costs) based on the following data for a given month (in thousands):

Raw materials purchased	$35,000
Raw materials used	30,000
Conversion costs incurred	22,000
Conversion costs allocated	20,000
Costs transferred to finished goods	47,500
Cost of goods sold	40,000

   For simplicity, you are not given the data to prepare journal entries for each underlying transfer from, say, Work in Process, Operation 1 to Work in Process, Operation 2 to Work in Process, Operation 3. Instead, assume that there is only a single account, Work in Process Control, that is supported by subsidiary Work in Process accounts for each operation.

2. Post the entries in requirement 1 to T-accounts for Inventories (Materials, Work in Process, and Finished Goods), Conversion Costs Control, Conversion Costs Allocated, and Cost of Goods Sold.

3. The plant adopts a JIT production system and a backflush costing system with two trigger points for journal entries: the purchase of materials and the completion of good finished units. Prepare summary journal entries based on the same data as in requirement 1. Note, however, that the raw materials used and the conversion costs assigned would be affected by the goods completed, not the work in process. Assume that 95% of the work placed in process is completed.

4. Post the entries in requirement 3 to T-accounts for Inventories (Raw and In-Process, and Finished Goods), Conversion Costs Control, Conversion Costs Allocated, and Cost of Goods Sold.

5. Compare the applicable inventory balances in requirements 2 and 4. Explain any differences.

SOLUTION

1. a. Materials Inventory Control	$35,000		
Accounts Payable Control		$35,000	
b. Conversion Costs Control	$22,000		
Various Accounts		$22,000	
c. Work in Process Control	$30,000		
Materials Inventory Control		$30,000	
d. Work in Process Control	$20,000		
Conversion Costs Allocated		$20,000	
e. Finished Goods Control	$47,500		
Work in Process Control		$47,500	
f. Cost of Goods Sold	$40,000		
Finished Goods Control		$40,000	

2.

Materials Inventory Control		Work in Process Control		Finished Goods Control		Cost of Goods Sold	
(a) 35,000	(c) 30,000	(c) 30,000	(e) 47,500	(e) 47,500	(f) 40,000	(f) 40,000	
Bal. 5,000		(d) 20,000		Bal. 7,500			
		Bal. 2,500					

Conversion Costs Control	Conversion Costs Allocated	
(b) 22,000		(d) 20,000

3. a. Inventory: Raw and In-Process Control $35,000
     Accounts Payable Control $35,000
b. Conversion Costs Control $22,000
     Various Accounts $22,000
c. Finished Goods Control $47,500
     Inventory: Raw and In-Process Control
        (0.95 × $30,000) $28,500
     Conversion Costs Allocated
        (0.95 × $20,000) $19,000
d. Cost of Goods Sold $40,000
     Finished Goods Control $40,000

4.

Inventory: Raw and In-Process Control		Finished Goods Control		Cost of Goods Sold	
(a) 35,000	(c) 28,500	(c) 47,500	(d) 40,000	(d) 40,000	
Bal. 6,500		Bal. 7,500			

Conversion Costs Control	Conversion Costs Allocated	
(b) 22,000		(c) 19,000

5.

	Sequential (Req. 2)	Backflush (Req. 4)
Materials inventory	$ 5,000	$ 0
Inventory: raw and in-process	0	6,500
Work in process	2,500	0
Subtotal	7,500	6,500
Finished goods	7,500	7,500
Total inventories	$15,000	$14,000
Underallocated conversion costs	$ 2,000	$ 3,000
Cost of goods sold (at standard)	$40,000	$40,000

The $1,000 difference in inventories is explained by the accounting for conversion costs. Conversion costs assigned in operation costing were $1,000 greater than conversion costs allocated under backflush costing because the trigger point for assignment is work in process, not finished goods.

To ease comparisons between the journal entries for operation costing and backflush costing, the numbers used for the various inventory balances are identical except for the $1,000 difference just explained. Advocates of JIT using backflush costing, however, maintain that inventories would decline substantially under JIT, particularly inventories of raw materials and work in process.

SUMMARY

The following points are linked to the chapter's learning objectives:

1. Each output unit can, in some extreme cases, be classified either (a) as a distinct, identifiable unit of a product or service or (b) as identical to numerous other units from that production system. A job-costing system is well suited to a production system yielding the first type of products, while a process-costing system is well suited to the second type of products. Most production systems have elements of both (a) and (b), the mix of which changes over time. Hybrid costing systems, such as operation costing, have developed so that the costing system captures the key features of the particular production system whose products are being costed.

2. Operation costing is a hybrid costing system applied to batches of homogeneous products. Production costs are compiled for each work order that will be made up of two or more similar (in many cases, homogeneous) units of a product.

3. Journal entries in an operation costing system focus on a work-in-process account for a batch of similar products. As the batch proceeds through various operations, direct costs (typically direct materials) and conversion costs are assigned to the work in process, which is similar to how costs are assigned in a job-costing system. Within each operation, all product units are treated exactly alike, which is similar to a process-costing system.

4. The key features of a just-in-time production system are: (a) inventory is regarded as an evil because of its nonvalue-added nature, (b) focus is centered on simplifying all production activities, (c) emphasis is placed on reducing manufacturing lead time, and (d) production is stopped if parts are absent or defects are discovered.

5. Journal entries in a backflush costing system are not made sequentially to match the flow of a product in a plant. Journal entries for work in process are not made. The accounting system does not record details of the changes in a product being produced until, at the soonest, finished units appear. Budgeted or standard costs are used to work backward to assign manufacturing costs to products.

6. A project is a complex job that often takes months or years to complete and requires the work of many different departments, divisions, or subcontractors. The four key success factors of a project are scope, quality, time schedule, and costs.

7. Two variances of specific interest in project costing are the project performance cost variance (a comparison of actual costs of work performed to date with the budgeted costs of work performed to date) and project schedule cost variance (a comparison of budgeted costs of work scheduled to date with the budgeted costs of work performed to date).

TERMS TO LEARN

This chapter and the Glossary at the end of this book contain definitions of the following important terms:

backflush costing *(p. 662)*   delayed costed *(662)*   endpoint costing *(662)*
hybrid costing system *(656)*   just-in-time (JIT) production *(660)*   operation *(657)*
operation costing *(657)*   post-deduct costing *(662)*   project *(668)*
project performance cost variance *(670)*   project schedule cost variance *(670)*
sequential tracking *(661)*   synchronous tracking *(661)*

ASSIGNMENT MATERIAL

QUESTIONS

19-1 Hybrid production systems develop because there are hybrid costing systems. Do you agree? Explain.

19-2 "Operation costing means that *conversion costs* and *manufacturing overhead* are synonyms." Do you agree? Explain.

19-3 Give three examples of industries that are likely to use operation costing.

19-4 Identify (a) the major job-costing feature of operation costing and (b) the major process-costing feature of operation costing.

19-5 Identify two allocation bases that are commonly used to assign conversion costs to products in operation costing systems.

19-6 List four major features of JIT production systems.

19-7 Give two examples of nonvalue-added activities that are lessened in JIT production systems.

19-8 Distinguish between the sequential tracking approach used in traditional budgeted or standard costing systems and the approach used in backflush costing.

19-9 Describe the essence of backflush costing.

19-10 "Companies adopting backflush costing often meet three conditions." Describe these three conditions.

19-11 Outline how three different versions of backflush costing can differ.

19-12 "Backflush accounting should be prohibited. It is not in conformity with GAAP." Do you agree? Explain.

19-13 "Projects are more challenging than jobs." Explain.

19-14 "Managers' control of projects generally focuses on four critical success factors." Identify those factors.

19-15 What is the relationship between cost and schedule performance reporting and earned hours reporting?

EXERCISES AND PROBLEMS

Coverage of Part One of the Chapter

19-16 **Operation costing journal entries.** The Omaha Desk Company specializes in making desks of varying shapes, materials, and sizes. It has a cutting operation, an assembly operation, and a staining operation. Some goods are sold unstained, so they do not go through the staining operation. Consider the following data for November.

Direct materials requisitioned by cutting	$200,000
Direct materials requisitioned by staining	20,000
Conversion costs of all operations (actual)	150,000
Conversion costs allocated:	
Cutting	50,000
Assembly	110,000
Staining	30,000

Required
Prepare journal entries. Assume that there is no beginning or ending work in process and that all goods are transferred to Finished Goods. The total manufacturing costs of the products not undergoing staining was $60,000.

19-17 Operation costing. Penske Company manufactures a variety of plastic products. The company has an extrusion operation and subsequent operations to form, trim, and finish parts such as buckets, covers, and automotive interior components. Plastic sheets are produced by the extrusion operation. Many of these sheets are sold as finished goods directly to other manufacturers. Additional direct materials (chemicals and coloring) are added in the finishing operation.

The company's manufacturing costs assigned to products for October were:

	Extrude	Form	Trim	Finish	Totals
Direct materials	$650,000	$ —	$ —	$ 80,000	$ 730,000
Direct manufacturing labor	55,000	30,000	20,000	40,000	145,000
Manufacturing overhead	270,000	90,000	40,000	60,000	460,000
	$975,000	$120,000	$60,000	$180,000	$1,335,000

In addition to plastic sheets, two types of automotive products (firewalls and dashboards) were produced:

	Units	Plastic Sheet Direct Materials	Additional Direct Materials
Plastic sheets, sold after extrusion	10,000	$500,000	$ —
Firewalls, sold after trimming	1,000	50,000	—
Dashboards, sold after finishing	2,000	100,000	80,000
	13,000	$650,000	$80,000

For simplicity, assume that all of the items and units produced received the same steps within each operation.

Required
1. Tabulate the conversion costs of each operation, the total units produced, and the conversion costs per unit.
2. Tabulate the total costs, the units produced, and the costs per unit. Be sure to account for all the total costs.

19-18 Operation costing, equivalent units. (CMA, adapted) Gregg Industries manufactures a variety of plastic products, including a series of molded chairs. The three models of molded chairs, which are all variations of the same design, are Standard (can be stacked), Deluxe (with arms), and Executive (with arms and padding). The company uses batch manufacturing and has an operation costing system.

Gregg has an extrusion operation and subsequent operations to form, trim, and finish the chairs. Plastic sheets are produced by the extrusion operation, some of which are sold directly to other manufacturers. During the forming operation, the remaining plastic sheets are molded into chair seats and the legs are added. The Standard model is sold after this operation. During the trim operation, the arms are added to the Deluxe and Executive models and the chair edges are smoothed. Only the Executive model enters the finish operation, where the padding is added. All of the units produced receive the same steps within each operation.

The May production run had a total manufacturing cost of $898,000. The units of production and direct materials costs incurred follow:

	Units Produced	Extrusion Materials	Form Materials	Trim Materials	Finish Materials
Plastic sheets	5,000	$ 60,000	$ 0	$ 0	$ 0
Standard model	6,000	72,000	24,000	0	0
Deluxe model	3,000	36,000	12,000	9,000	0
Executive model	2,000	24,000	8,000	6,000	12,000
	16,000	$192,000	$44,000	$15,000	$12,000

Manufacturing costs assigned during the month of May were:

	Extrusion Operation	Form Operation	Trim Operation	Finish Operation
Direct manufacturing labor	$152,000	$60,000	$ 30,000	$ 18,000
Manufacturing overhead	240,000	72,000	39,000	24,000

Required
1. For each product produced by Gregg Industries during the month of May, determine (a) the unit cost and (b) the total cost. Be sure to account for all costs incurred during the month, and support your answer with appropriate calculations.
2. Without considering your answer in requirement 1, assume that 1,000 units of the Deluxe model remained in work-in-process at the end of the month. These units were 100% complete as to materials costs and 60% complete in the trim operation. Determine the cost of the 1,000 units of the Deluxe model in the work-in-process inventory at the end of May.

19-19 Operation costing, ending inventory, equivalent units. Gilhooley Products uses three operations in sequence to manufacture an assortment of picnic baskets. In each operation, the same procedures, time, and costs are used to perform that operation for a given quantity of baskets, regardless of the basket style being produced.

During April, a batch of materials for 1,100 baskets of style X was put through the first operation. This batch was followed in turn by separate batches of materials for 400 baskets of style Y and 1,300 of style Z. All the materials for a batch are introduced at the beginning of the operation for that batch. The costs as shown below were incurred in April for the first operation:

Direct manufacturing labor	$29,600
Manufacturing overhead	13,135
Direct materials:	
Style X	18,700
Style Y	8,000
Style Z	13,000

All the units started in April were completed during the month and transferred out to the next operation except 350 units of style Z, which were only partially completed at April 30. These were 40% completed as to conversion costs and 100% completed as to direct materials costs. There were no work-in-process inventories at the beginning of the month.

Required
1. For each basket style, compute the total cost of work completed and transferred out to the second operation.
2. Compute the total cost of the ending inventory in process.

19-20 Operation costing with ending work in process and journal entries. Galvez Co. produces two models of video recorders. The deluxe units undergo two operations. The super-deluxe units undergo three operations. Consider the following:

	Production Orders	
	For 1,000 Deluxe Units	**For 500 Super-Deluxe Units**
Direct materials (actual costs)	$50,000	$54,000
Conversion costs (allocated on the basis of machine-hours used):		
Operation 1	20,000	10,000
Operation 2	?	?
Operation 3	—	5,000
Total manufacturing costs assigned	?	?

Required

1. Operation 2 is highly automated. Product manufacturing costs depend on a budgeted allocation rate for conversion costs based on budgeted machine-hours per unit. The budgeted costs for 19_2 were $100,000 direct manufacturing labor and $440,000 manufacturing overhead. Budgeted machine-hours were 18,000. Each product unit had a budgeted 6 minutes of time in operation 2. Compute the total costs of processing the deluxe products and super-deluxe products in operation 2.

2. Compute the total manufacturing costs and the unit costs of the deluxe and super-deluxe products.

3. Suppose that at the end of the year, 500 deluxe units were in process through operation 1 only and 300 super-deluxe units were in process through operation 2 only. Compute the cost of the ending work-in-process inventory. Assume that no direct materials are charged in operation 2 but that $4,000 of additional direct materials are to be charged to the 500 units processed in operation 3. (Ignore end-of-year variances.)

4. Prepare journal entries that track the costs of all 1,000 deluxe units through operations to Finished Goods.

 19-21 Operation costing, journal entries. Pafko Company, a small manufacturer, makes a variety of tool boxes. The company's manufacturing operations and their costs for November were:

	Cutting	Assembly	Finishing	Totals
Direct manufacturing labor	$2,600	$16,500	$ 4,800	$23,900
Manufacturing overhead	3,000	22,900	3,300	29,200
	$5,600	$39,400	$ 8,100	$53,100

Three styles of boxes were produced in November. The quantities and direct materials costs were:

Style	Quantity	Direct Materials
Standard	1,200	$18,000
Home	600	6,660
Industrial	200	5,400
		$30,060

The company uses actual costing. It tracks direct materials to each style of box. It combines direct manufacturing labor and manufacturing overhead and allocates the conversion costs on the basis of all product units passing through an operation. All product units are assumed to receive an identical amount of time and effort in each operation. The Industrial style, however, does not go through the Finishing operation.

Required

1. Compute the total cost and unit cost of each style produced.

2. Prepare summary journal entries for each operation. For simplicity, assume that all direct materials are introduced at the beginning of the cutting operation. Also, assume that all

units were transferred to Finished Goods when completed and that there was no beginning or ending work in process. Prepare one summary entry for all conversion costs incurred, but prepare a separate entry for allocating conversion costs in each operation.

19-22 Backflush journal entries and JIT production. The Lee Company has a plant that manufactures transistor radios. The production time is only a few minutes per unit. The company uses a just-in-time production system and a backflush costing system with two trigger points for journal entries:

◆ Purchase of raw materials and components
◆ Completion of good finished units of product

There are no beginning inventories. The following data pertain to April manufacturing (in thousands):

Raw materials purchased	$ 8,800
Raw materials used	8,500
Conversion costs incurred	4,220
Allocation of conversion costs	4,000
Costs transferred to finished goods	12,500
Cost of goods sold	11,900

Required

1. Prepare summary journal entries for April (without disposing of underallocated or overallocated conversion costs).
2. Post the entries in requirement 1 to T-accounts for Inventories, Conversion Costs Control, Conversion Costs Allocated, and Cost of Goods Sold.
3. Under an ideal JIT production system, how would the amounts in your journal entries differ from those in requirement 1?

19-23 Backflush costing and JIT production. Ronowski Company produces telephones. For June, there were no beginning inventories of raw materials and no beginning and ending work in process. Ronowski uses a JIT production system and backflush costing with two trigger points for making entries in its accounting system:

◆ Purchase of raw materials and components
◆ Completion of good finished units of product

Ronowski's June standard cost per unit of telephone product is: direct materials, $26; conversion costs, $15. There are two inventory accounts:

◆ Inventory: Raw and In-process Control
◆ Finished Goods Control

The following data apply to June manufacturing:

Raw materials and components purchased	$5,300,000
Conversion costs incurred	$3,080,000
Number of finished units manufactured	200,000
Number of finished units sold	192,000

Required

1. Prepare summary journal entries for June (without disposing of underallocated or overallocated conversion costs). Assume no direct materials variances.
2. Post the entries in requirement 1 to T-accounts for applicable Inventories Control, Conversion Costs Control, Conversion Costs Allocated, and Cost of Goods Sold.

19-24 Backflush, second trigger is sale. (Continuation of 19-23) Assume that the second trigger point for Ronowski Company is the sale—rather than the production—of finished units. Also, the inventory account is confined solely to direct materials, whether they would be in a storeroom, in work in process, or in finished goods.

No conversion costs are inventoried. They are allocated to the units sold at standard. Any underallocated or overallocated conversion costs are written off monthly to Cost of Goods Sold.

Required

1. Prepare summary journal entries for June, including the disposition of underallocated or overallocated conversion costs. Assume no direct materials variances.

2. Post the entries in requirement 1 to T-accounts for applicable Inventory Control, Conversion Costs Control, Conversion Costs Allocated, and Cost of Goods Sold. Explain the composition of the ending balance of Inventory.

3. Suppose conversion costs were sufficiently material in amount to be included in Inventory. Using a backflush system, show how your journal entries would be changed in requirement 1. Explain.

19-25 Backflush, one trigger point. Assume the same facts as in Problem 19-23. Now assume, however, that there is only one trigger point, the completion of good finished units.

Required

1. Prepare the journal entry at the trigger point.

2. Compare this entry with your entries (a) and (c) in example 1, pp. 663–64. Explain any differences in the results.

19-26 Accounting for variances. Assume the same facts as in Problem 19-23. Suppose the same quantity of raw materials had additional costs as follows:

Unfavorable price variance	$30,000
Unfavorable efficiency variance	70,000

Required

1. Prepare summary journal entries (without explanations) to record the raw-material variances.

2. Assume that underallocated or overallocated conversion costs are written off monthly to Cost of Goods Sold. Prepare the pertinent summary journal entry.

Coverage of Part Two of the Chapter

19-27 Basic project cost control. The marketing, manufacturing, and research and development (R&D) departments of the Perry Company have worked together in deciding which new products and components to develop. The R&D department has launched a project to develop a new compressor for the Perry line of Frosty Refrigerators.

Consider the following data:

◆ Original budget to complete	30,000 hours
◆ Budgeted average cost per hour	$ 120
◆ Price variance to date	$ 0
◆ Budgeted hours for work scheduled to date	22,100 hours
◆ Budgeted hours for work performed to date (earned hours)	20,000 hours
◆ Actual hours of work performed to date	21,200 hours
◆ Price variance expected	$39,000 unfavorable

Required

1. Compute the project performance cost variance and the project schedule cost variance.

2. Prepare a revised budget that is designed to predict actual final costs. Assume that the project performance cost variance will continue as unfavorable at the same rate as shown to date.

19-28 Basic project cost control. Phoenix Company, a defense subcontractor, has a project to design and produce a subsystem for guided missiles. The original budget to complete is 20,000 hours. The budgeted hours for the work scheduled to date are 12,000 hours. The budgeted hours for the work performed to date (earned hours) were 11,000. The actual hours of work performed to date were 11,440. The budgeted cost is $70 per hour. There was no price variance.

Required

1. Compute the project performance cost variance and the project schedule cost variance.
2. Prepare a revised budget that is designed to predict actual final costs. Assume that the project performance cost variance will continue to be unfavorable at the same rate as shown to date.

19-29 Project cost control. Consider a Grumman research project on a new wing design for a fighter aircraft. The original budget to complete was 30,000 hours at an average cost of $120 per hour. The budgeted hours for the work scheduled to date are 21,000 hours. The budgeted hours for the work performed to date (earned hours) were 22,000 hours. The actual hours of work performed to date were 21,500. Actual costs were $2,795,000.

Required

1. Compute in dollars: the project performance cost variance, efficiency variance, price variance, and project schedule cost variance.
2. As the manager, how would you interpret these variances? If the project performance cost variance persists at the same rate as shown to date, what are the expected actual final costs?

19-30 Project cost control. Wechtel Company, an engineering consulting firm, has a large project: the design and testing of a production control system for an airline's maintenance base. The following data pertain to the project:

Original budgeted costs		$600,000
Add (deduct) cost variances to date:		
Unfavorable price variances	$42,000	
Favorable efficiency variances	(60,000)	(18,000)
Subtotal		582,000
Add (deduct) additional expected variances		?
Revised budget, final costs		$?

The budgeted cost per hour is $100. The budgeted hours for work performed to date were 3,000. The budgeted hours for work scheduled to date were 3,500.

Required

1. Compute the project performance cost variance and the project schedule cost variance for the work done to date.
2. Assume that there will be no further price variances. Assume also that the same relative efficiency will persist throughout the remainder of the project. Prepare an estimate of the revised budget of final costs. Will the manager of the consulting firm be pleased? Explain.

19-31 Project cost analysis, project cancellation. Anthony Oxford is the president of Northern Power and Light (NPL). NPL owns and operates three coal-fueled power plants in the north of England. Four years ago, NPL signed a contract with Industrial Construction (IC) for the assembly of a nuclear power plant at Oldham. Work started on July 19_1. The contract included a clause that stated: "In the event of cancellation by NPL, NPL will pay IC for all costs attributable to the work done to date plus an additional 10% of these costs."

In September 19_5, NPL decided to cancel completion of the Oldham nuclear power plant. Ambiguity existed about how to determine "all costs attributable to the work done to date." IC argued that the work was 100% complete as to planning and engineering costs, 60% complete as to materials costs, and 40% complete as to on-site conversion costs. Oxford requested information about the:

♦ BCWS (budgeted cost of work scheduled to date)
♦ BCWP (budgeted cost of work performed to date)
♦ ACWP (actual cost of work performed to date)

Richard Ayers, the head of the Nuclear Construction Division of IC, provided NPL with the following numbers (in millions):

Project Area	BCWS	BCWP	ACWP
Planning and engineering costs	£104.7	£104.7	£141.7
Materials costs	161.1	164.6	159.8
On-site conversion costs	68.3	73.1	69.3
	£334.1	£342.4	£370.8

Ayers noted that IC had been able to accomplish the work related to materials costs and on-site conversion costs under budget because of some major changes in its planning and engineering division. He expressed great disappointment at NPL canceling the contract, noting that he was confident of completing the total project 5% under budget and on time.

Oxford approached Naveed Hasan, the controller of NPL, to seek her advice. His first comment was one of disbelief. "They say they have spent £370.800 million to build that partially finished eyesore. The total claim is for £407.880 million" (£370.800 + 10% of £370.800). Oxford told Hasan to assume that "the claim is a complete fraud" and "to leave no stone unturned in reducing it." He believed that £150 million was the maximum due. Even that amount would nearly bankrupt NPL!

Required

1. Using the numbers provided by IC, evaluate the project cost management skills of IC regarding construction of the Oldham nuclear plant.

2. What issues should Hasan examine when analyzing the validity of the IC claim for £407.880 million?

19-32 Project cost analysis, ethics. (Continuation of 19-31) Naveed Hasan, the controller of Northern Power and Light (NPL), made a visit to the Oldham plant. She also met financial and technical personnel in the Nuclear Plant Construction Division of Industrial Construction (IC). Hasan found the morale of IC's personnel to be very low, with many people worried about keeping their jobs because there were no new construction contracts.

Two weeks later, Hasan received an anonymous letter from a person apparently just laid off by IC in its most recent reduction of staff in the Nuclear Plant Construction Division. The letter was very detailed. It alleged that IC was engaging in multiple efforts to boost the cancellation cost claim. These efforts included double counting of cost items (including them in both direct and indirect cost pools) and overstatement of engineer time records. The letter also noted that IC, experiencing declining demand for nuclear plant construction, was now attempting to recover from NPL some planning and engineering costs that were initially budgeted to be spent on other contract bids. These bids were unsuccessful or were canceled at a very early stage.

Hasan is sure she knows who wrote the letter. She can match the handwriting with that on several documents IC previously sent to NPL. She shows the letter to Oxford. He suggests Hasan "have dinner with the disgruntled ex-employee. If necessary, hold out the prospect of obtaining a good job with NPL if he can provide more details about the £407.880 million claim. We are talking about economic survival of our company. That claim will bankrupt us."

Required

1. Discuss whether Hasan should have shown the letter to Oxford.

2. How should Hasan handle Oxford's heavy pressure to meet with the disgruntled ex-IC employee and solicit further information?

3. How should NPL use the information provided by the disgruntled employee in subsquent negotiations with IC?

20

Capital Budgeting
and Cost Analysis

- Contrast in purposes of cost analysis
- Definition and stages of capital budgeting
- The discounted cash-flow methods
- Sensitivity analysis
- Analysis of selected items using discounted cash flow
- The payback method
- The accrual accounting rate-of-return method
- Complexities in capital-budgeting applications

*C*apital budgeting decisions for new plant construction require forecasts of revenues, costs, and investment outlays over many future periods. Asea Brown Boveri used discounted cash-flow methods when making investment decisions for a new polyethylene facility in Venezuela.

Should we invest in a new research and development project, a computer-aided design system for new product development, or a different distribution network for our products? Should a city invest in a large multipurpose hospital or a network of smaller special-purpose hospitals? Such decisions frequently involve large dollar amounts and have long-lasting effects and uncertain actual outcomes. Once a project is implemented, project costs cannot be changed for long periods of time. Careful long-range planning, rather than day-to-day monitoring, is the key to managing long-run costs.

Capital budgeting is the principal accounting technique for analyzing and controlling project costs. *Capital budgeting* helps managers make long-range decisions. This chapter examines the role of accounting data in the popular decision models for capital budgeting. We also study the relationship of the decision models to the performance-evaluation models that managers use to help judge the results of capital-budgeting decisions.

At this stage, we again focus on purpose. Both income determination and the planning and controlling of routine operations focus primarily on the *current time period*. Special decisions and long-range planning focus primarily on the *project* or *program* with a much longer time span.

CONTRAST IN PURPOSES OF COST ANALYSIS

Objective 1

Differentiate between project-by-project orientation of capital budgeting and period-by-period orientation of accrual accounting

Exhibit 20-1 illustrates two different dimensions of cost analysis: (1) a project dimension, represented by the vertical axis, and (2) a time dimension, represented by the horizontal axis.

Each project is represented in Exhibit 20-1 as a distinct horizontal rectangle. The life of each project is longer than one year. We focus in this chapter on an individual project throughout its life, recognizing the time value of money. The time value of money takes into account that a dollar (or any other monetary unit) received today is worth more than a dollar received tomorrow. The reason is that a dollar received today can be invested to start earning interest, and so it is worth more than a dollar re-

EXHIBIT 20-1
The Project and Time Dimensions of Capital Budgeting

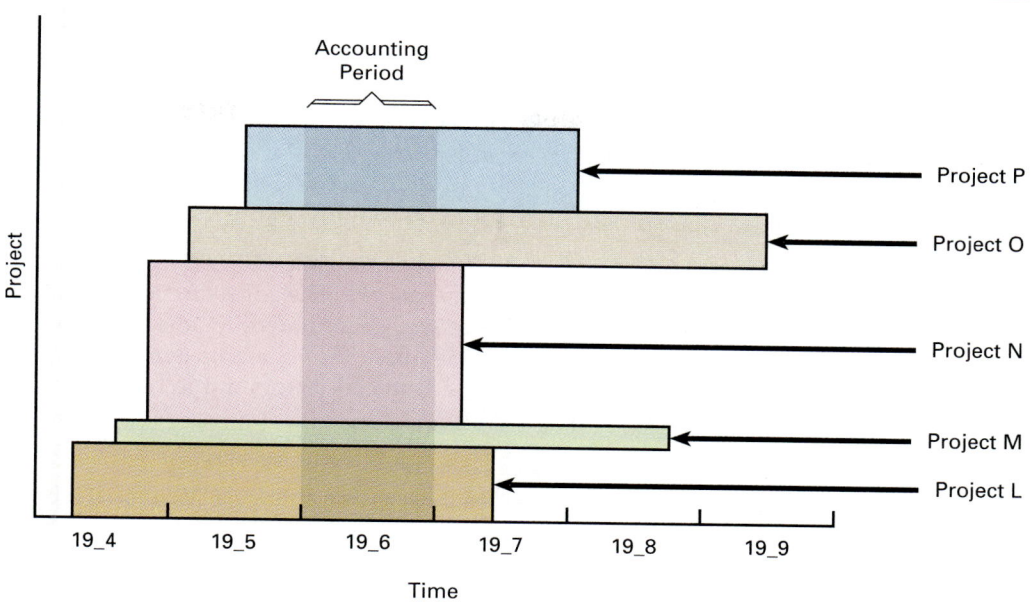

Objective 2
Explain the time value of money and opportunity costs

ceived tomorrow. The time value of money is the opportunity cost (interest forgone) from not having the money today.

The shaded vertical rectangle in Exhibit 20-1 illustrates the accounting period focus on income determination and routine planning and control. This cross section emphasizes the company's performance for a year (in our example, 19_6). Because the accounting period is a year or less, the time value of money usually is not a primary concern during that relatively short time span.

Recall a central theme of this book: different costs for different purposes. Capital-budgeting decisions focus on the project. The revenues and costs specifically associated with the project should be measured over the life of the project. There is a great danger in basing capital-budgeting decisions on an indiscriminate use of the data routinely reported in the current accounting period's income statements. We illustrate this point later in the chapter through several examples.

The accounting system that corresponds to the project dimension in Exhibit 20-1 is termed life-cycle costing. This system, described in Chapter 12, accumulates revenues and costs on a project-by-project basis. For example, a life-cycle costing statement for a new-car project at Ford Motor Company could encompass a 10-year period and would accumulate costs along the length of the value chain, in the research and development, design, production, marketing, distribution, and customer-service categories. This accumulation, covering many years, differs from the accrual accounting system that measures income on a period-by-period basis.

DEFINITION AND STAGES OF CAPITAL BUDGETING

Objective 3
Identify the six stages of capital budgeting for a project and its predicted outcomes

Capital budgeting is the making of long-term planning decisions for investments. It is a six-stage process. We illustrate these stages using a decision facing the manager of Lifetime Care Hospital (a nonprofit hospital) on whether to buy a new state-of-the-art X-ray machine. The new machine would reduce labor costs of operating the X-ray equipment. Also, its improved technology would mean that patients would need fewer X-rays.

The six stages in capital budgeting are:

STAGE 1: *IDENTIFICATION STAGE.* *To distinguish which capital expenditure projects will accomplish organizational objectives.* In the Lifetime Care example, the goal is to improve the productivity of the X-ray department, and the capital expenditure project identified is to purchase a new X-ray machine.

STAGE 2: *SEARCH STAGE.* *To explore several potential capital expenditure investments that will achieve organizational goals.* Lifetime Care can consider several alternative X-ray machines to improve productivity. Some alternatives are discarded early. Others are evaluated, more thoroughly, in the information-acquisition stage.

STAGE 3: *INFORMATION-ACQUISITION STAGE.* *To consider the consequences of alternative capital investments.* The consequences of capital expenditures can be quantitative and qualitative. Quantitative factors are outcomes that are measured in numerical terms. Some quantitative factors can be expressed in financial terms. Examples of quantitative financial factors in the Lifetime Care capital budgeting decision are:

   a. The cash paid to buy the X-ray machine
   b. The lower labor costs of operating the X-ray machine
   c. The lower costs resulting from having to take fewer X-rays per patient.

Other quantitative factors are nonfinancial—that is, they can be measured numerically but they are difficult to express in financial terms. In the Lifetime Care example, the amount of time a patient is exposed to radiation is an example of a quantitative, nonfinancial factor.

Qualitative factors are outcomes that cannot be measured in numerical terms and consequently are also difficult to express in financial terms. Examples of qualitative factors in the Lifetime Care capital investment decision are:

   a. The quality of X-rays. Higher-quality X-rays may lead to improved diagnoses and better patient treatment.
   b. The safety of employees and patients. Because fewer X-rays would have to be taken with the new machine, employees and patients would undergo less exposure to the possibly harmful effects of X-rays.
   c. Lifetime's reputation as a first-rate research hospital. Lifetime's ability to attract better physicians and radiologists may improve if Lifetime can provide its doctors with the latest equipment.

Capital budgeting emphasizes quantitative, financial factors. But nonfinancial quantitative factors and qualitative factors are also important.[1] Organizations are increasingly attempting to cast nonfinancial and qualitative factors in a financial light. Lifetime might search its records for cases where employee injury resulted from overexposure to X-rays. Employee safety is generally considered a qualitative factor, but the benefits of fewer employee injuries can potentially be quantified by measuring medical costs of treating the worker, the costs of work stoppages, and the costs of hiring replacement workers. In general, though, expressing nonfinancial quantitative factors and qualitative factors in financial terms is a difficult task.

STAGE 4: *SELECTION STAGE.* *To choose projects for implementation.* This chapter discusses the following methods for analyzing predicted outcomes of capital budgeting decisions that can be quantified in financial terms:

---

[1]Surveys indicate that U.S. and Japanese managers regard both financial and nonfinancial factors as important in capital investment decisions. The five most important factors cited by U.S. managers were sales-quantity forecasts, unit variable manufacturing costs, and contribution margin (all financial factors) and improved quality and improved delivery time (both nonfinancial factors). See A. C. Sullivan and K. Smith, "Capital Investment Justification for U.S. Factory Automation Projects," *Journal of the Midwest Finance Association* (1994) and P. Scarbrough, A. Nanni, and M. Sakurai, "Japanese Management Accounting Practices and the Effects of Assembly and Process Automation," *Management Accounting Research* 2 (1991).

♦ Discounted cash flow (both net present value and internal rate of return)
♦ Payback (both traditional and bailout)
♦ Accrual accounting rate of return

Chapter 23 discusses the breakeven time method, which focuses on the importance of time-to-market in the development of new products.

Our discussion emphasizes discounted cash-flow methods. Both the net present value and internal rate-of-return methods explicitly recognize the time value of money as a critical factor in decisions affecting long time spans.

Only predicted outcomes quantified in financial terms are formally included in the analysis. Conclusions reached on the basis of quantitative analyses are then reevaluated, taking into account qualitative considerations. Because qualitative factors are not part of the formal analysis, however, managers must exercise their own judgment about how much weight they will place on qualitative factors in making capital-budgeting decisions.

STAGE 5: *FINANCING STAGE.* *To obtain project funding.* Sources of financing include internally (within the organization) generated cash and the capital market (equity and debt securities). Financing is often the responsibility of the treasury function of an organization.

STAGE 6: *IMPLEMENTATION AND CONTROL STAGE.* *To put the project in motion and monitor performance.* In some cases, monitoring may include a postdecision audit, in which the predictions made at the time a project was selected are compared with the actual results. To provide effective feedback, a cost system must track project revenues and costs across periods. Costing systems that have an exclusive period-by-period focus (see Exhibit 20-1) are often unable to identify project costs over multiple periods.

This chapter focuses on the financial investment-decision aspects of capital budgeting. Beyond the numbers, however, the ability of individual managers to "sell" their own projects to senior management is often pivotal in the acceptance or rejection of projects.

The relevant cost and revenue framework outlined in Chapter 11 should guide decisions about what items to include in capital-budgeting analysis. Consider only future costs and revenues that differ among the alternatives under investigation. Do not assume that only those costs classified as variable in the short run are relevant. Most capital-budgeting projects have a multiyear time horizon. Costs that are fixed for short time periods can usually change over longer period horizons.

THE DISCOUNTED CASH-FLOW METHODS

Objective 4

Describe the two main discounted cash-flow (DCF) methods, the net present value (NPV) method, and the internal rate-of-return (IRR) method.

We now consider the first of three methods for analyzing the financial aspects of projects. **Discounted cash flow (DCF)** measures the cash inflows and outflows of a project as if they occurred at a single point in time so that they can be compared in an appropriate way. The *discounted cash-flow* method recognizes that the use of money has an opportunity cost—interest forgone. Because the discounted cash flow method explicitly and routinely weighs the time value of money, it is usually the best (most comprehensive) method to use for long-run decisions.

DCF focuses on *cash* inflows and outflows rather than on *operating income* as used in conventional accrual accounting. Cash is invested now with the expectation of receiving a greater amount of cash in the future. Try to avoid injecting accrual concepts of accounting into DCF analysis.

The compound interest tables and formulas used in DCF analysis are included in Appendix C at the back of the text. **APPENDIX C WILL BE USED FREQUENTLY IN CHAPTERS 20 AND 21.**

There are two main DCF methods:

1. Net present value (NPV)
2. Internal rate of return (IRR)

The **required rate of return (RRR),** which is the minimum acceptable rate of return on an investment, is used to calculate NPV and is used as a point of comparison in working with IRR. The RRR is the return that the organization could expect to receive elsewhere for an investment of comparable risk. This rate is also called the **discount rate, hurdle rate,** or **(opportunity) cost of capital.** Chapter 21 discusses issues encountered in estimating this rate.

For simplicity, this chapter (and Chapter 21) assumes that the cash outflows and cash inflows occur at the beginning or end of each period. We use the following information from Lifetime Care to illustrate various capital-budgeting decision methods.

EXAMPLE: As you recall, the manager of Lifetime Care Hospital is considering the purchase of a new state-of-the-art X-ray machine to replace an existing X-ray machine. The existing X-ray machine can operate for another five years and will have a zero terminal disposal price at the end of five years. The required net initial investment for the new machine is $379,100. The initial investment consists of the cost of the new machine—$372,890—plus an additional cash investment in working capital (supplies and spare parts for the new machine) of $10,000 minus cash of $3,790 obtained from the disposal of the existing machine ($372,890 + $10,000 − $3,790 = $379,100).

The manager expects the new machine to have a five-year useful life and a zero terminal disposal price at the end of five years. The new machine will reduce the average number of X-rays taken per patient and will decrease labor costs. The manager expects the investment to result in annual cash inflows of $100,000. These cash flows will generally occur throughout the year. For simplicity, we assume that the cash flows occur at the end of each year. (The cash inflows will come from cash savings in operating costs of $100,000 for each of the first four years and $90,000 in year 5. The manager also expects a cash inflow of $10,000 when the working capital investment is recovered [disposed of] at the end of year 5.) The hospital is not subject to income taxes. There are two alternatives:

1. Continue operation of the X-ray department without change, or
2. Buy the new X-ray machine and replace the existing X-ray machine.

Regardless of the decision, revenue will not change. Lifetime Care charges a fixed rate for a particular diagnosis, regardless of the number of X-rays taken. Government reimbursements for Medicare/Medicaid patients depend on the specific diseases treated and not on the number or type of X-rays taken. The only relevant financial benefit in evaluating Lifetime's decision to purchase the X-ray machines is the cash savings in operating costs.

Compute the net present value of the project. Assume that the required rate of return, or discount rate, for the project is 8%. (This relatively low discount rate is not unusual for nonprofit institutions, which can borrow funds at low rates because lenders pay no income taxes on interest received from nonprofit institutions.) Also, compute the internal rate of return on the project.

Net Present Value Method

Net present value (NPV) is a DCF method of calculating the expected net monetary gain or loss from a project by discounting all expected future cash inflows and outflows to the present point in time, using the required rate of return. Only projects with a positive net present value are acceptable. Why? Because the return from these projects exceeds the cost of capital (the return available by investing the capital elsewhere). Projects with higher NPVs are preferred to projects with lower NPVs, if all other things are equal. Using the NPV method entails the following steps:

STEP 1: *DRAW A SKETCH OF RELEVANT CASH INFLOWS AND OUTFLOWS.* The right side of Exhibit 20-2 shows how these cash flows are portrayed. Outflows ap-

EXHIBIT 20-2

Net Present Value Method for Lifetime Care Hospital

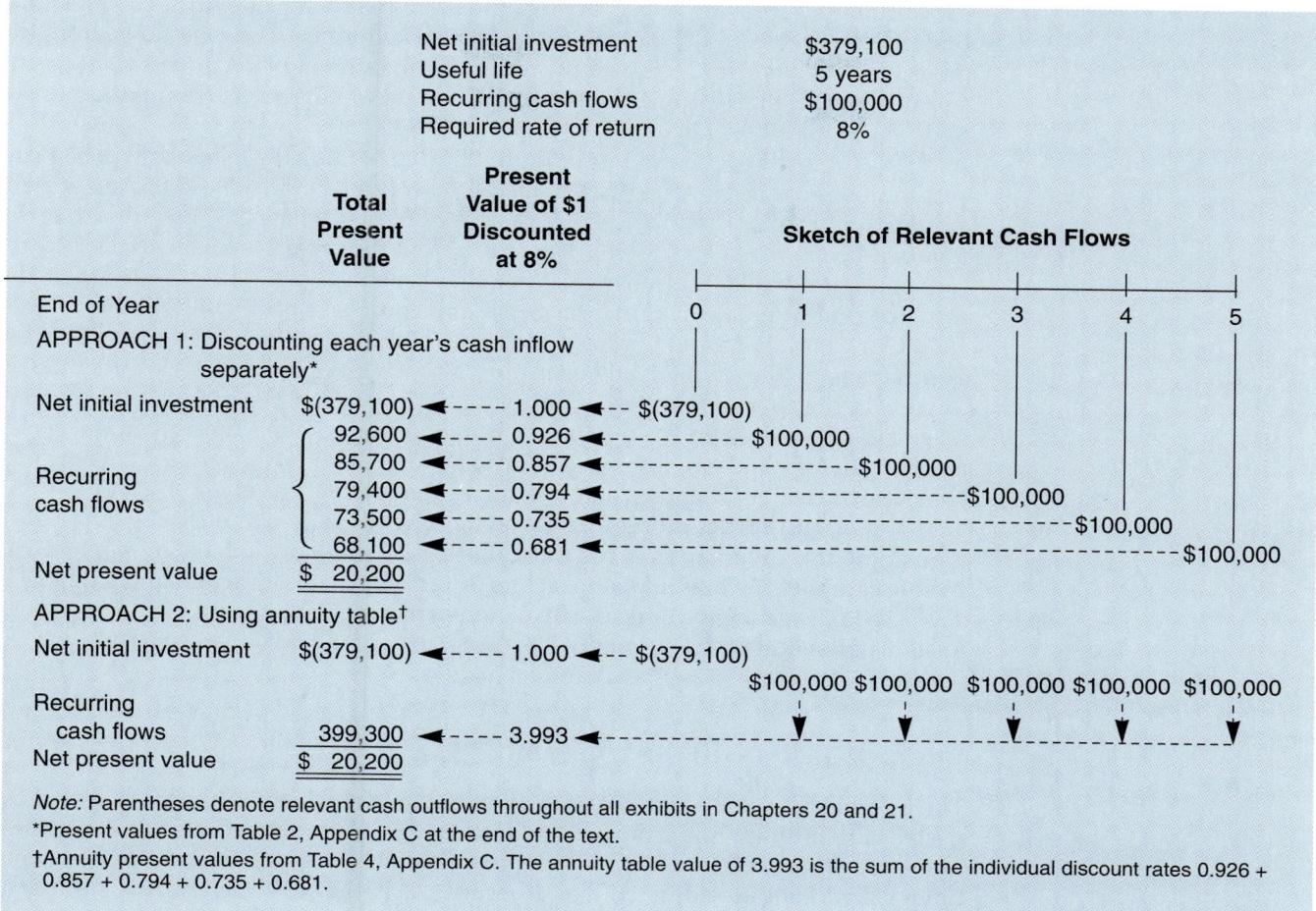

	Net initial investment	$379,100
	Useful life	5 years
	Recurring cash flows	$100,000
	Required rate of return	8%

Note: Parentheses denote relevant cash outflows throughout all exhibits in Chapters 20 and 21.

*Present values from Table 2, Appendix C at the end of the text.

†Annuity present values from Table 4, Appendix C. The annuity table value of 3.993 is the sum of the individual discount rates 0.926 + 0.857 + 0.794 + 0.735 + 0.681.

pear in parentheses. Although such a sketch is not absolutely necessary, it clarifies relationships. It helps the decision maker organize the data in a systematic way. Note that Exhibit 20-2 includes the outflow for the new machine at year 0, the time of the acquisition. The NPV method focuses only on cash flows. NPV analysis is indifferent to where the cash flows come from (operations, purchase or sale of equipment, or investment or recovery of working capital) and to the accounting treatments of individual cash flow items (for example, depreciation costs on equipment purchases).

STEP 2: *CHOOSE THE CORRECT COMPOUND INTEREST TABLE FROM APPENDIX C.* In our example, we can discount each year's cash flow separately using Table 2, or we can compute the present value of an annuity using Table 4. If we use Table 2 (App. C), we find the discount factors for periods 1 through 5 under the 8% column. Approach 1 in Exhibit 20-2 presents the five discount factors. If we use Table 4 (App. C), we find the discount factor for 5 periods under the 8% column. Approach 2 in Exhibit 20-2 shows that this discount factor is 3.993 (3.993 is the sum of the five discount factors used in Approach 1). To obtain the present values, multiply the discount factors by the appropriate cash amounts in the sketch in Exhibit 20-2.

STEP 3: *SUM THE PRESENT VALUES TO DETERMINE THE NET PRESENT VALUE.* If the sum is zero or positive, the financial aspect of the project indicates that it should be accepted. Its expected rate of return equals or exceeds the required rate of return. If the total is negative, the financial aspect indicates that the project is undesirable. Its expected rate of return is below the required rate of return.

Exhibit 20-2 indicates a NPV of $20,200 at the required rate of return of 8%; the expected return from the project exceeds the 8% required rate of return. Therefore, the project is desirable on the basis of financial considerations. The cash flows from the project are adequate to (1) recover the net initial investment in the project and (2) earn a return greater than 8% on the investment tied up in the project from period to period. The manager will likely buy the machine.

The higher the required rate of return, the less willing the manager will be to invest in this project. At a rate of 12%, the NPV would be –$18,600 (3.605, the present value annuity factor from Table 4, × $100,000 = $360,500, which is $18,600 less than the required investment of $379,100). The return from the project is less than the 12% return on capital available elsewhere. When the required rate of return is 12%, then, the machine is undesirable in a financial sense at its purchase price of $379,100.

Our example has emphasized the financial aspect of Lifetime Care's decision. The manager of the hospital, however, must also weigh other factors. Consider the reduction in the average number of individual X-rays taken per patient with the new machine. This reduction is a qualitative benefit of the new machine given the health risks to patients (and employees) of X-rays. Other qualitative benefits of the new machine are the better diagnoses and treatments that patients receive.

Important: Do not proceed until you thoroughly understand Exhibit 20-2. Compare approach 1 with approach 2 in Exhibit 20-2 to see how Table 4 in Appendix C merely compiles the present value factors of Table 2. That is, the fundamental table is Table 2; Table 4 merely reduces calculations when there is an annuity—a series of *equal* cash flows at equal intervals.

Internal Rate-of-Return Method

The **internal rate of return (IRR)** is the discount rate at which the present value of expected cash inflows from a project equals the present value of expected cash outflows of the project. That is, the IRR is the discount rate that makes NPV = $0. IRR is sometimes called the **time-adjusted rate of return.** As in the NPV method, the sources of cash flows and the accrual accounting treatment of individual cash flows are irrelevant to the IRR calculations. We illustrate the computation of the IRR using the X-ray machine project of Lifetime Care. Exhibit 20-3 presents the cash flows and shows the calculation of the NPV using a 10% discount rate. At a 10% discount rate the NPV of the project is zero. Therefore, the IRR for the project is 10%.

How do we determine the 10% discount rate that yields NPV = $0? In most cases, people solving capital-budgeting problems have a calculator or computer programmed to provide the internal rate of return. Without a calculator or computer program, a trial-and-error approach can yield the answer.

◆ *Step 1:* Try a discount rate and calculate the NPV of the project using that discount rate.
◆ *Step 2:* If the NPV is less than zero, try a lower discount rate. (A lower discount rate will increase the NPV; remember that we are trying to find a discount rate for which NPV = $0. If the NPV is greater than zero, try a higher discount rate to lower the NPV. Keep adjusting the discount rate until NPV = $0. In the Lifetime Care example, a discount rate of 8% yields NPV of + $20,200 (see Exhibit 20-2). A discount rate of 12% yields NPV of – $18,600 (see above). Therefore, the discount rate that makes NPV = $0 must lie between 8% and 12%. We try 10% and get NPV = $0. Hence, the IRR is 10%.

The step-by-step computations of an internal rate of return are easier when the cash flows are equal, as is the case in our example. Exhibit 20–3 uses the following equation:

$379,100 = Present value of annuity of $100,000 at X% for 5 years

Or, what factor F in Table 4 (App. C) will satisfy the following equation:

EXHIBIT 20-3
Internal Rate of Return Method for Lifetime Care Hospital

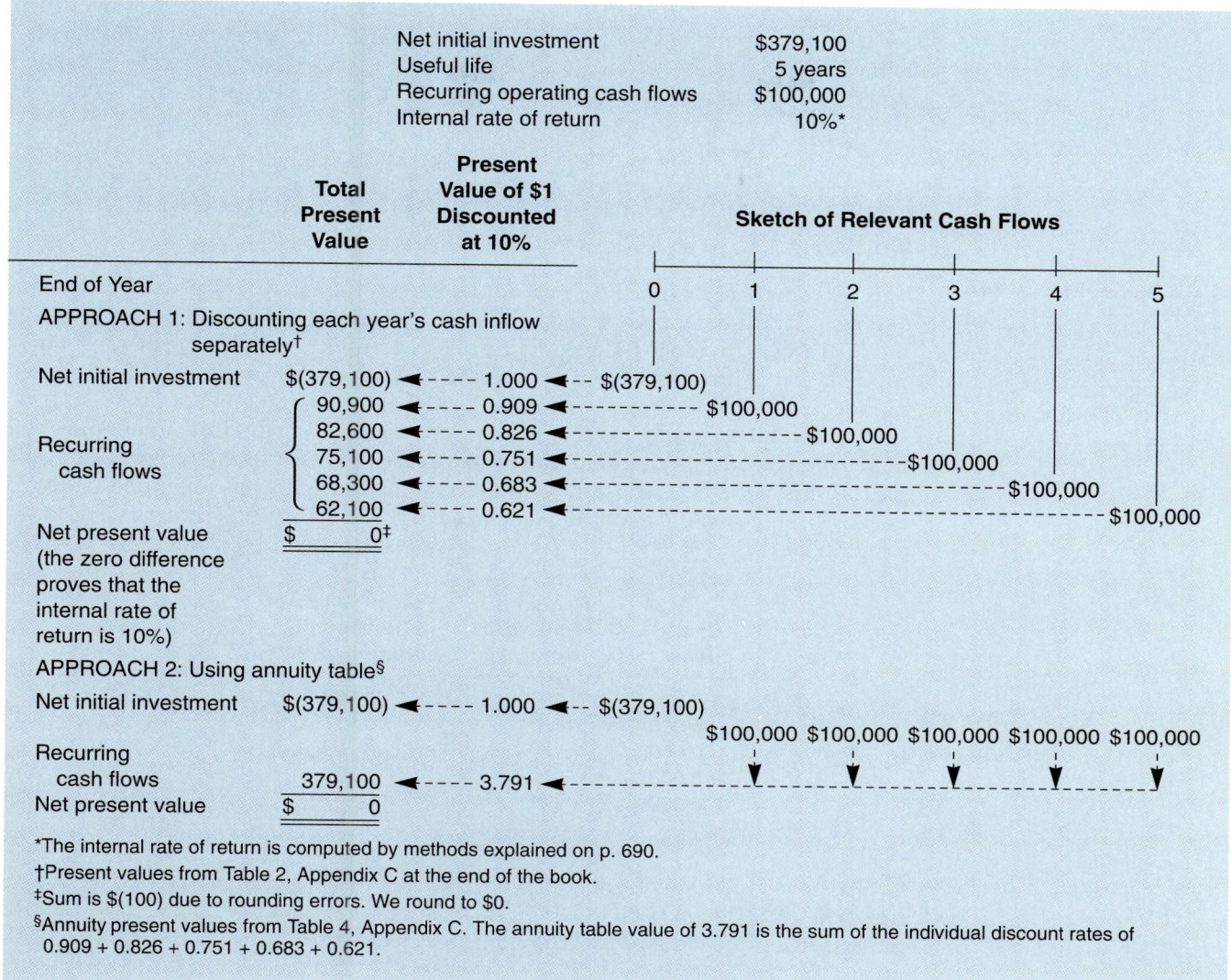

	Net initial investment	$379,100
	Useful life	5 years
	Recurring operating cash flows	$100,000
	Internal rate of return	10%*

*The internal rate of return is computed by methods explained on p. 690.
†Present values from Table 2, Appendix C at the end of the book.
‡Sum is $(100) due to rounding errors. We round to $0.
§Annuity present values from Table 4, Appendix C. The annuity table value of 3.791 is the sum of the individual discount rates of 0.909 + 0.826 + 0.751 + 0.683 + 0.621.

$$\$379{,}100 = \$100{,}000F$$
$$F = 3.791$$

On the 5-period line of Table 4, find the percentage column that is closest to 3.791. It is exactly 10%. If the factor (F) falls between the factors in two columns, straight-line interpolation is used to approximate the IRR. (For an illustration of interpolation, see requirement 1 of the Problem for Self-Study, pp. 705–6.)

A project is accepted if the internal rate of return exceeds the required rate of return (the opportunity cost of capital). In the Lifetime Care example, the X-ray machine has an IRR of 10%, which is greater than the required rate of return of 8%. Lifetime Care should invest in the new machine. If the IRR exceeds the opportunity cost of capital, then the project has a positive NPV when project cash flows are discounted at the opportunity cost of capital. If the IRR equals opportunity cost of capital, NPV = $0. If the IRR is less than the opportunity cost of capital, NPV is negative. Projects with higher IRRs are preferred to projects with lower IRRs, if all other things are equal.

The IRR of 10% means that the cash inflows from the project are adequate to (1) recover the net initial investment in the project and (2) earn a return of exactly 10% on investment tied up in the project from year to year. Note that the amount invested in the new X-ray machine changes from year to year. Consider, for example, the first year. At the beginning of the year, Lifetime Care's investment in the new machine is $379,100. A 10% return on the investment is equal to $37,910. The operating cash flow (cash savings) from the machine in the first year is $100,000 (see p. 688). The cash inflow in excess of $37,910 ($100,000 − $37,910 = $62,090) is the amount of investment recovered at the end of the first year. Lifetime's remaining investment in the machine at the beginning of the second year is $379,100 − $62,090 = $317,010. At the end of five years, the cash inflow exactly recovers the initial investment and earns a return of 10% on the unrecovered investment.

Comparison of Net Present Value and Internal Rate-of-Return Methods

Objective 5

Explain how the two main discounted cash-flow methods (NPV and IRR) differ

This text emphasizes the NPV method, which has the important advantage that the end result of the computations is dollars, not a percentage. Therefore, we can add up the NPVs of individual independent projects to estimate the effect of accepting a combination of projects. In contrast, the internal rates of return of individual projects cannot be added or averaged to derive the IRR of the combination of projects.

A second advantage is that we can use the NPV method in situations where there is not a constant required rate of return for each year of the project. For example, assume in the X-ray machine example that Lifetime Care has a required rate of return of 8% in years 1, 2, and 3 and 12% in years 4 and 5. The total present value of the cash inflows is:

Year	Cash Inflow	Required Rate of Return	Present Value of $1 Discounted at Required Rate	Total Present Value of Cash Inflows
1	$100,000	8%	0.926	$ 92,600
2	100,000	8	0.857	85,700
3	100,000	8	0.794	79,400
4	100,000	12	0.636	63,600
5	100,000	12	0.567	56,700
Total				$378,000

Given the net initial investment of $379,100, the project is unattractive: It has a negative NPV of $1,100 ($379,100 − $378,000). However, it is not possible to use the IRR method to infer that the project should be rejected. The existence of different required rates of return in different years (8% for years 1, 2, and 3 versus 12% for years 4 and 5) means there is not a single hurdle rate that the IRR (a single figure) must exceed for the project to be acceptable.

A third advantage of the NPV method over the IRR method is that the NPV method always yields a unique answer. Certain patterns of investments and cash flows have multiple IRRs.[2]

SENSITIVITY ANALYSIS

To highlight the basic differences among various decision methods, we have assumed that the expected values of cash flows will occur for certain. Obviously, managers know that predictions are imperfect. The Appendix to Chapter 3 explains how to cope with uncertainty in decision making. For now, we concentrate on sensitivity analysis, which was also introduced in Chapter 3. *Sensitivity analysis* is a "what-if" technique

---

[2]See, for example, R. Brealey and S. Myers, *Principles of Corporate Finance* (New York: McGraw-Hill, 1991), Chapter 5, p. 83.

that examines how a result will change if the original predicted data are not achieved or if an underlying assumption changes.

Sensitivity analysis can take various forms. For instance, management may want to know how far annual cash inflows must fall to reach the point of indifference (when the NPV equals zero). For the data in Exhibit 20-2, let ACI = annual cash inflows and let NPV = $0. The net initial investment is $379,100, and the present value factor at the 8% required rate for a five-year annuity of $1 is 3.993. Then:

$$\begin{aligned} \text{NPV} &= \$0 \\ 3.993 \times \text{ACI} - \$379,100 &= \$0 \\ 3.993 \times \text{ACI} &= \$379,100 \\ \text{ACI} &= \$94,941 \end{aligned}$$

Thus, at the discount rate of 8%, annual cash inflows can decrease by $100,000 − $94,941 = $5,059 before reaching the point of indifference regarding the investment (that is, before NPV falls to zero).

Computer spreadsheets enable managers to conduct systematic, efficient sensitivity analysis. Exhibit 20-4 shows how the net present value of the X-ray machine project is affected by variations in (1) the annual cash inflows and (2) the required rate of return. Sensitivity analysis provides an immediate measure of the financial effect of differences between forecasts and actual outcomes. It helps managers focus on those decisions that may be very sensitive indeed, and it eases the manager's mind about those decisions that are not so sensitive. For the X-ray machine project, Exhibit 20-4 shows that variations in either the annual cash inflows or the required rate of return have sizable effects on NPV.

All the NPV calculations in Exhibit 20-4 assume a five-year useful life. NPVs also vary with the useful life of a project. Suppose annual cash inflows from investing in the new machine were $100,000 for four years rather than five years. At a discount rate of 8%, the NPV would be −$47,900 (3.312, present value of the annuity fac-

EXHIBIT 20-4

Spreadsheet Analysis of Sensitivity of Net Present Value for Lifetime Care Hospital

Useful Life (in years)	Annual Cash Inflows	Required Rate of Return	Net Present Value
5	$ 80,000	6%	$(42,140)
5	80,000	8	(59,660)
5	80,000	10	(75,820)
5	90,000	6	(20)
5	90,000	8	(19,730)
5	90,000	10	(37,910)
5	100,000	6	42,100
5	100,000	8	20,200
5	100,000	10	0
5	110,000	6	84,220
5	110,000	8	60,130
5	110,000	10	37,910
5	120,000	6	126,340
5	120,000	8	100,060
5	120,000	10	75,820

tor for 4 years at 8% from Table 4 × $100,000 = $331,200 minus net initial investment of $379,100). The negative NPV would indicate that the machine would be an unwise investment.

ANALYSIS OF SELECTED ITEMS USING DISCOUNTED CASH FLOW

The key point in capital budgeting is to recognize that discounted cash-flow methods focus exclusively on *incremental* future cash flows. DCF techniques ignore all accrual accounting concepts such as accrued revenues and accrued costs. Only cash inflows and cash outflows matter. For example, pension costs incurred but not paid are not considered in discounted cash flow analysis. Only actual cash payments of pensions are. Why? Because it is only when cash is paid that the firm sacrifices alternative uses of cash and hence needs to recognize the time value of money. All cash flows are treated the same, whether they arise from operations, purchase or sale of equipment, or investment in or recovery of working capital. The opportunity cost and the time value of money are tied to the cash flowing in or out of the company, not to the source of the cash.

We discuss below several issues and problems that arise when predicting relevant cash inflows and cash outflows in capital-budgeting decisions. Relevant cash inflows and cash outflows are (1) expected future cash flows that (2) differ among the alternatives. In the Lifetime Care example, the alternatives are either to continue to use the old X-ray machine or to replace it with the new machine. The relevant cash inflows and cash outflows are the *differences* in cash flows between continuing to use the old machine and purchasing the new one. *When reading this section, focus on identifying incremental future cash flows and differences in cash flows.*

Capital investment projects typically have five major categories of cash flows: (1) initial investment in machines and working capital, (2) cash flow from current disposal of the old machine, (3) recurring operating cash flows, (4) cash flow from terminal disposal of machines and recovery of working capital, and (5) income tax cash savings due to depreciation deductions. We discuss the first four categories below, using Lifetime Care's purchase decision of the X-ray machine as an illustration. Income tax benefits are described in Chapter 21.

1. INITIAL INVESTMENT. Two components of investment cash flows are (a) the cash outflow to purchase the machine and (b) the working-capital cash outflows.

  a. **Initial machine investment.** These outflows, made for purchasing the plant and equipment, occur in the early periods of the project's life and include cash outflows for transporting and installing the item. In the Lifetime Care example, the $372,890 cost of the X-ray equipment is an outflow in year 0.

  b. **Initial working capital investment.** Investments in plant, equipment, and machines and in the sales promotions for product lines are invariably accompanied by incremental investments of working capital. These take the form of current assets, such as receivables and inventories (supplies and spare parts in the Lifetime Care example) minus current liabilities, such as accounts payable. Working capital investments are similar to machine investments. In each case, available cash is tied up in noncash assets.

The Lifetime Care example assumes an incremental $10,000 investment in working capital (supplies and spare parts inventory) if the new machine is acquired. The incremental working capital investment is the difference between the working capital required to operate the new machine (say, $15,000) and the working capital investment required to operate the old machine (say, $5,000). The $10,000 additional investment in working capital is a cash outflow in year 0.

Discounted cash flow methods record all investments (in equipment and working capital) made in year 0 as cash outflows in year 0, regardless of how they may be accounted for (capitalized or expensed) under accrual accounting. For example, the investment in the new machine will appear as a cash outflow in year 0 irrespective of how the machine is depreciated over its five-year life.

Note that both equipment and working capital investment cash flows are relevant for the capital-budgeting decision to acquire the new X-ray machine. Why? Because they are expected future cash flows that will differ between the alternatives of investing and not investing in the new machine. These cash flows are incurred only if Lifetime decides to purchase the new machine.

2. CURRENT DISPOSAL PRICE OF OLD MACHINE. In the decision of whether to keep or to replace equipment, how should the current disposal price of the existing machine affect the NPV computations? Any cash received from disposal of the old machine is a relevant cash inflow (in year 0) because it is an expected future cash flow that differs between the alternatives of investing and not investing in the new project. If Lifetime Care invests in the new X-ray machine, it will be able to dispose of its old machine for $3,790. These proceeds are included as cash inflow in year 0.

Recall from Chapter 11 that the book value (original cost minus accumulated depreciation) of the old equipment is irrelevant. It is a past cost. Nothing can change what has already been spent or what has already happened. Lifetime Care, as a nonprofit organization, is exempt from income taxes, and, consequently, disposal has no tax implication. Firms subject to tax must account for cash flows due to tax payments or tax savings on the disposal. Chapter 21 explains the effect of income taxes.

As we have seen, the net initial investment for the new X-ray machine, $379,100, is the initial machine investment plus the initial working capital investment minus current disposal price of the old machine: $372,890 + $10,000 − $3,790 = $379,100.

3. RECURRING OPERATING CASH FLOWS. This category includes all recurring operating cash flows that differ among the alternatives. Firms make capital investments to generate cash inflows in the future. These inflows may result from producing and selling additional goods or services, or, as in the Lifetime Care example, from savings in operating cash costs. Focus on operating *cash* flows, not on revenues and costs.

To underscore this point, consider the following additional facts about the Lifetime Care X-ray machine example:

1. Acquiring the new machine will save cash outflows of $10,000 in materials each year. It will also save cash outflows for operating and maintenance labor of $90,000 for each of the first four years and $80,000 for the fifth year.
2. Total X-ray department overhead costs will not change whether the new machine is purchased or the old machine is kept. The X-ray department overhead costs are allocated to individual X-ray machines—Lifetime has several—on the basis of the labor costs for operating each machine. Because the new X-ray machine will have lower labor costs, overhead allocated to it will be $30,000 less than the amount allocated to the machine it is replacing.
3. Depreciation on the new X-ray machine using the straight-line method is $74,578 ([original cost, $372,890 − expected terminal disposal price, $0] ÷ useful life, 5 years).

The materials costs and labor cost savings are clearly relevant operating cash flows because they are expected future cash flows that will differ between the alternatives of investing and not investing in the new machine. What about the decrease in allocated overhead costs of $30,000? What about depreciation of $74,578?

a. **Overhead costs.** The key question is, Do total overhead cash flows decrease? In our example, they do not. Total X-ray department overhead costs remain the same whether or not the new machine is acquired. Only the overhead allocated to individual machines changes. The overhead costs allocated to the new machine are $30,000 less. These $30,000 will be allocated to *other* machines. No cash flow savings in total overhead result. Therefore the $30,000 should not be included as part of recurring operating cash inflows.

b. **Book value and depreciation.** In using DCF methods when there are no income tax considerations, book value and depreciation are irrelevant. Why? Because DCF is based on inflows and outflows of *cash*, not on the *accrual* accounting concepts of revenues and costs. No adjustments should be made to cash flows for depreciation, because depreciation is not a cash flow. In DCF methods, the initial cost of an asset is usually regarded as a *lump-sum* outflow of cash at year 0. For example, the IRR of 10% shown in Exhibit 20-3 represents a return of 10% on investment after recovery of the net initial investment of $379,100. Deducting depreciation from operating cash inflows would be counting the lump-sum amount twice. (Chapter 21 explains how book value and depreciation affect *after-tax* cash flows from operations, as distinguished from *before-tax* cash flows from operations.)

4. TERMINAL DISPOSAL PRICE OF INVESTMENT. *The disposal of the investment at the date of termination of a project generally increases cash inflow in the year of disposal.* Errors in forecasting the terminal disposal price are seldom crucial because the present value for amounts to be received in the distant future is usually small. Two components of the terminal disposal price of investment are (a) the terminal disposal price of the machine and (b) recovery of working capital.

    a. Terminal disposal price of machine. At the end of the useful life of the project, the initial machine investment may not be recovered at all, or it may be only partially recovered in the amount of the terminal disposal price.

        The relevant cash inflow is the difference in expected terminal disposal prices under the two alternatives—the terminal disposal price of the new machine (zero in the case of Lifetime Care) minus the terminal disposal price of the old machine (also zero in the Lifetime Care example).

    b. Recovery of working capital. The initial investment in working capital is usually fully recouped when the project is terminated. At that time, inventories and receivables needed to support the project are no longer needed. The relevant cash inflow is the difference in the expected working capital recovered under the two alternatives. If the new X-ray machine is purchased, Lifetime Care will recover $15,000 of working capital in year 5. If the new machine is not acquired, Lifetime will recover $5,000 of working capital in year 5, at the end of the useful life of the old machine. The relevant cash inflow if Lifetime invests in the new machine is $10,000 ($15,000 – $5,000).

The difference between the initial working capital investment and the present value of its recovery is the present value of the cost of using working capital in the project. Assume a required rate of return of 8%. The present value of $10,000 at the end of five years is $6,810: 0.681, the present value factor at 8% for 5 years from Table 2, × $10,000. Thus, the present value of the cost of using the $10,000 working capital for the life of the project is $3,190 ($10,000 – $6,810).

Some capital investments reduce working capital. For example, plant automation projects frequently reduce work in process inventories. In this case, the timing of the cash inflows and outflows is reversed. Assume that a computer-integrated manufacturing project with a seven-year life will reduce working capital by $20 million from, say, $50 million to $30 million. This reduction will be represented as a $20 million cash *inflow* for the project at year 0 and a $20 million cash *outflow* for the project when it ends in seven years.

5. INCOME TAXES. In practice, when an organization is subject to taxes, comparison between investment alternatives is best made after considering income-tax effects because the tax impact may alter the picture. (Chapter 21 explains the effects of income taxes.)

Lifetime Care's analysis assumed that the new machine has a useful life of five years and that the existing machine has a remaining useful life of five years. Suppose instead the existing machine has a remaining useful life of only two years. Management could choose to evaluate the replacement decision over the useful life of either the longer-lived or the shorter-lived machine. Suppose management chooses a five-year analysis horizon, equal to the life of the longer-lived machine. The decision, then, is whether to invest in the new machine now or to invest in a new machine two years later. To evaluate the alternatives of buying the new machine in two years, Lifetime's management would have to predict the terminal disposal price of the new machine when it has been used for three years.

Exhibit 20-5 presents the relevant cash inflows and outflows for Lifetime Care's decision to purchase the new machine as described in items 1 through 4 above. The

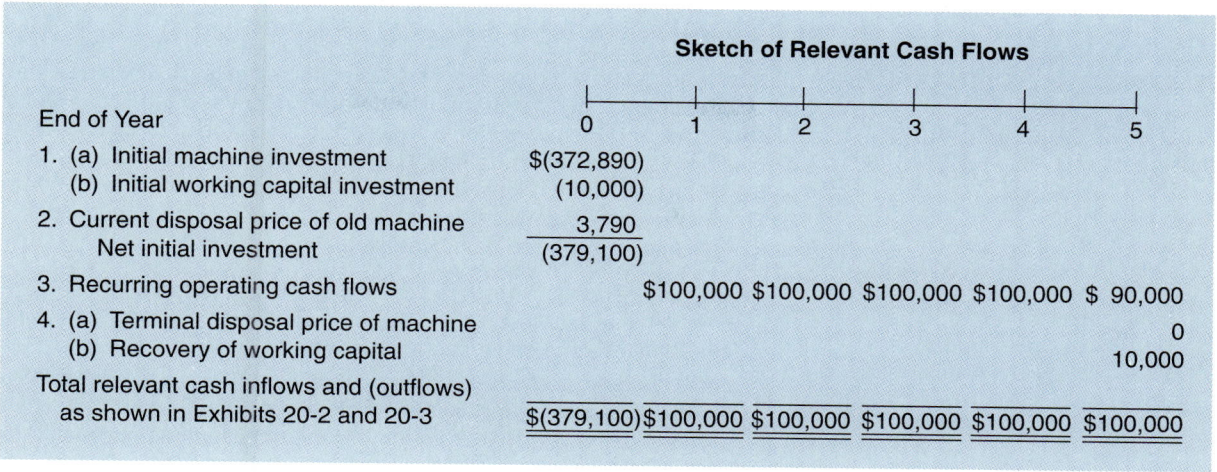

total relevant cash flows for each year are precisely the same as the relevant cash flows used in Exhibits 20-2 and 20-3 to illustrate the NPV and IRR methods.

THE PAYBACK METHOD

Objective 7
Describe the payback method

We now consider the second of three leading methods for analyzing the financial aspects of projects. The **payback** method measures the time it will take to recoup, in the form of net cash inflows, the net dollars invested in a project. Like NPV and IRR, the payback method does not distinguish the sources of cash inflows (operations, disposal of equipment, or recovery of working capital). In the Lifetime Care example, the X-ray machine costs $379,100, has a five-year expected useful life, and generates a $100,000 uniform cash inflow each year. The payback calculations follow:

$$\text{Payback} = \frac{\text{Net initial investment}}{\text{Uniform increase in annual cash flow}}$$

$$\text{Payback} = \frac{\$379,100}{\$100,000} = 3.791 \text{ years}$$

Under the payback method, firms often choose a cutoff period. Projects with a payback period less than the cutoff period are acceptable. Those with a payback period beyond the cutoff period are rejected. If Lifetime's cutoff period under the payback rule is three years, Lifetime will reject the new machine. If Lifetime uses a cutoff period of four years, Lifetime will consider the new machine to be acceptable.

The payback method highlights liquidity, which is often an important factor in business decisions. Managers prefer projects with shorter paybacks (more liquid) to projects with longer paybacks, if all other things are equal. Projects with shorter payback periods give the firm more flexibility because funds for other projects become available sooner. Also, managers are less confident about cash flow predictions that stretch far into the future. The faster the payback, the more certain managers can feel that their forecasts are on target.

The payback method is easy to understand. Like the DCF methods described previously, the payback method is not affected by accrual accounting conventions such as depreciation. Advocates of payback argue that it is a handy measure (1) when

Concepts in Action Box

ENVIRONMENTAL COSTS AND CAPITAL BUDGETING[a]

How relevant are environmental costs to capital budgeting? Environmental costs are becoming increasingly important. Recent amendments to the Clean Water Act, the Clean Air Act, the Resource Conservation Recovery Act, and the Superfund Amendments and Reauthorization Act have two targets: to reduce the quantity and toxicity of pollution, and to increase penalties and fines for violating environmental standards. Environment-related costs—in terms of their relevance to capital-budgeting decisions—fall into one of three broad categories.

1. *Regulatory costs,* such as the costs of training appropriate personnel, the costs of monitoring and testing for pollution, and the costs of notifying and reporting emission levels to government agencies.

2. *Liability costs,* such as penalties and fines for failing to comply with environmental laws and legal claims, awards, and settlements for personal injury and property damage arising from the release of hazardous material. For example, United Technologies was fined $3 million for dumping industrial solvents and polluted material on its property. Union Carbide Corporation paid several million dollars in settling claims arising from a chemical leak in Bhopal, India. Estimating potential liability costs requires an understanding of the detailed workings of many different pieces of pollution-control equipment.

3. *Less tangible costs,* such as the costs (or, in some cases, the decreased revenues) resulting from customer backlash against a company's environmental policies. These costs, though significant, may be difficult to express in financial terms. And how about the environment of the workplace itself? The company certainly incurs costs from the injuries that arise in an unsafe workplace. Moreover, if workers perceive their jobs to be physically risky, employee-management relations may deteriorate, and a drop-off in worker performance may follow. Quantifying this cost is extremely hard to do.

What makes all these environmental costs relevant? Consider choosing a new piece of equipment from among several alternatives. Each piece of equipment carries a different—and therefore relevant—set of "environmental costs." A company that keeps environmental costs in mind may choose one piece of equipment. A company that ignores these costs may choose a different piece of equipment. Also, proper recognition of environmental costs and accounting for them are particularly important when investing in actual pollution-control equipment.

[a]Adapted from P. Bailey, "Full Cost Accounting for Life Cycle Costs," *Environmental Finance* (Spring 1991); and L. Spencer, "Designated Inmates," *Forbes* (October 26, 1992).

precision in estimates of profitability is not crucial and preliminary screening of many proposals is necessary and (2) when the predicted cash flows in later years are highly uncertain.

Two major weaknesses of the payback method are (1) it neglects the time value of money and (2) it neglects project profitability. Consider an alternative to the $379,100 X-ray machine mentioned earlier. Assume that another X-ray machine, with a three-year useful life, requires only a $300,000 net initial investment and will also result in cash inflows of $100,000 per year ($100,000 in operating cash inflows in the first two years, and $90,000 in operating cash inflows and $10,000 in working capital recovery in the third year). Lifetime Care's required rate of return on investments is 8%. First, compare the two payback periods:

$$\text{Machine 1: } \frac{\$379,100}{\$100,000} = 3.791 \text{ years}$$

$$\text{Machine 2:} \frac{\$300,000}{\$100,000} = 3.000 \text{ years}$$

The payback criterion would favor buying the $300,000 machine, because it has a shorter payback. In fact, if the cutoff period is three years, then Lifetime Care would not acquire machine 1, which would not meet the payback criterion. Consider next the NPV of the two investment options. The second machine has a negative NPV. At a discount rate of 8%, the NPV of machine 2 is –$42,300 (2.577, the present value annuity factor for three years at 8% from Table 4 × $100,000 = $257,700 minus the net initial investment of $300,000). Machine 1, as we know, has a positive net present value of $20,200 (from Exhibit 20-2). The NPV criterion suggests that Lifetime Care should acquire machine 1. Machine 2 would fail the NPV criterion.

Why do the payback method and the NPV method yield different answers? For two reasons: (1) The payback method does not consider cash flows after the payback date, and (2) the payback method considers but does not discount cash flows prior to the payback date.

An added problem with the payback method is that choosing too short a cutoff date for project acceptance may promote the selection of only short-lived projects. The firm will tend to reject long-term, positive NPV projects.

Nonuniform Cash Flows

The payback formula presented on page 697 is designed for *uniform* annual cash inflows. When annual cash inflows are *not uniform*, the payback computation takes a cumulative form. The years' net cash inflows are accumulated until the amount of the net initial investment has been recovered. Assume that a law firm is considering purchase of a $1,500 fax machine for electronically transmitting documents to its clients. This machine is expected to produce a total cash savings of $3,200 over the next five years (primarily due to a reduction in the use of courier services). The cash savings occur evenly within each year but at a nonuniform rate across years. Payback occurs between the second and third years:

Year	Cash Savings	Cumulative Cash Savings	Net Initial Investment Yet to Be Recovered at End of Year
0	—	—	$1,500
1	$500	$ 500	1,000
2	600	1,100	400
3	800	1,900	—
4	700	2,600	—
5	600	3,200	—

Straight-line interpolation within the third year, which has cash savings of $800, reveals that the final $400 needed to recover the $1,500 investment (that is, $1,500 – $1,100 recovered by the end of year 2) will be achieved halfway through year 3:

$$\text{Payback} = 2 \text{ years} + \left(\frac{400}{800} \times 1 \text{ year} \right) = 2.5 \text{ years}$$

The X-ray machine example (p. 688) has a net initial investment of $379,100 at year 0. Where a project has multiple cash outflows occurring at different points in time, these outflows are added to derive a total cash outflow figure for the project. No adjustment is made for the time value of money when adding these cash outflows in computing the payback period. (The breakeven time method, discussed in Chapter 23, has several similarities to the payback method but does consider the time value of money.)

The Bailout Factor: An Extension of Payback

The payback computation discussed above tries to answer the question, How soon can the net initial investment for a project be recouped *if operations proceed as planned?* An additional question to ask is, Which of the competing projects has the best protection if things go wrong? The **bailout payback** method measures the time it will take for the cumulative cash flows from operations plus the disposal price of equipment plus the recovery of working capital at the end of a particular year to equal the net initial investment. The bailout payback examines the downside protection if the project is disbanded. Projects with shorter bailout payback periods are preferable to those with longer bailout payback periods, if all other things are equal.

Disposal prices of general-purpose equipment frequently far exceed those of special-purpose equipment because general-purpose equipment has a wider-ranging market. Consider again Lifetime Care's choice between the two X-ray machines. Machine 1, which costs $379,100, is a multipurpose machine. Machine 2, which costs $300,000, is a specialized machine that takes only certain types of X-rays.

	Multipurpose X-Ray Machine	Specialized X-Ray Machine
Net initial investment	$379,100	$300,000
Uniform annual cash inflows	$100,000	$100,000
Useful life of machine	5 years	3 years

Management expects the disposal price of the multipurpose X-ray machine to be $225,100 at the end of year 1 and to decline by $56,000 annually thereafter. The specialized X-ray machine has a disposal price of $150,000 at the end of year 1 and declines by $75,000 annually thereafter. Whichever machine is acquired, working capital can be recovered, at any time, for $10,000.

Exhibit 20-6 compares the two X-ray machines using the traditional payback method and the bailout payback method. With the traditional payback method, the specialized machine has a shorter payback period (3 years) than the multipurpose machine (3.791 years). In contrast, the multipurpose machine has a shorter bailout

EXHIBIT 20-6
Comparison of Traditional Payback and
Bailout Payback Methods for Lifetime Care Hospital

Machine	Traditional Payback			Bailout Payback			
	Net Initial Investment	Uniform Annual Cash Inflows	Payback	At End of	Cumulative Cash Inflows	Disposal Price of Equipment Plus Recovery of Working Captial	Cumulative Total
Specialized X-ray machine	$300,000	$100,000	3 years	Year 1	$100,000	+ $160,000 =	$260,000
				Year 2	200,000	+ 85,000* =	285,000
				Year 3	290,000	+ 10,000 =	300,000
				Bailout payback is 3 years.			
Multi-purpose X-ray machine	$379,100	$100,000	3.791 years	Year 1	$100,000	+ $235,100 =	$335,100
				Year 2	200,000	+ 179,100 =	379,100
				Bailout payback is 2 years.			

*$85,000 = disposal price of machine, $75,000 + recovery of working capital, $10,000.

payback period (2 years) than does the specialized machine (3 years). The specialized machine suffers substantial declines in its disposal price each year. Its bailout payback period is the same as the traditional payback period. The bailout payback period for the multipurpose machine is only two years because the disposal price of the machine declines only gradually each year. Like the traditional payback method, the bailout payback method neglects the time value of money and the profitability of the project.

THE ACCRUAL ACCOUNTING RATE-OF-RETURN METHOD

Objective 8

Explain the accrual accounting rate-of-return (AARR) method.

We now consider the last of the three methods for analyzing the financial aspects of capital budgeting projects. The **accrual accounting rate of return (AARR)** is an accounting measure of income divided by an accounting measure of investment. It is also called **accounting rate of return.** Advocates of this method prefer projects with higher, rather than lower, accrual accounting rates of return, if all other things are equal.

The denominator is often the project's net initial investment:

$$\text{Accrual acounting rate of return} = \frac{\text{Increase in expected average annual operating income}}{\text{Net initial investment}}$$

If Lifetime Care purchases the new X-ray machine, the increase in expected average annual operating savings will be $98,000: This amount is the total operating savings of $490,000 ($100,000 for four years and $90,000 in year 5) ÷ 5. The new machine has a zero terminal disposal price. Straight-line depreciation on the new machine is $372,890 ÷ 5 = $74,578. The net initial investment is $379,100 (initial machine investment, $372,890, minus current disposal price of old machine, $3,790, plus initial working capital investment, $10,000). The accrual accounting rate of return is equal to:

$$\text{AARR} = \frac{\$98,000 - \$74,578}{\$379,100}$$

$$= \frac{\$23,422}{\$379,100}$$

$$= 6.18\%$$

Sometimes the denominator is expressed as the average increase in investment, rather than the net initial increase. In the Lifetime Care example, the investment remaining at the end of the project is $10,000 (expected terminal disposal price of the new machine, $0, plus recovery of working capital investment, $10,000). The following table summarizes the net initial investment and the investment remaining at the end of the project

	Net Initial Investment	Investment Remaining at the End of the Project
Initial machine investment minus current disposal price of old machine	$369,100	$ 0
Initial working capital investment	10,000	10,000
Total	$379,100	$10,000

Under the average investment method, the investment base for the AARR will be $194,550 [(net initial investment of $379,100 + ending investment of $10,000) ÷ 2]. The numerator is the same as before, $23,422. The AARR will increase to $23,422 ÷ $194,550 = 12.04%.[3]

The label *accrual accounting rate of return method* is not universal. Synonyms are *book-value method, rate-of-return-on-assets method,* and *accounting return-on-investment method.* The AARR computations dovetail most closely with how operating income and required investment are calculated under accrual accounting.

The AARR is simple and easy to understand. Unlike the payback method, the AARR method considers profitability. Unlike the NPV method, however, the AARR method does not discount future dollars at a higher rate than present dollars. The AARR ignores the time value of money. It treats future dollars as equal to present dollars.

Computing the Investment Base

Practice is not uniform. Should the net initial investment or the average increase in investment be used in the denominator of the AARR? Neither measure is technically superior to the other. Some companies prefer the use of the net initial investment because it does not require estimates of the investment remaining at the end of the project. The key is to apply the chosen method consistently to facilitate comparisons of accrual accounting rates of return on project-to-project, plant-to-plant, division-to-division, and year-to-year bases.

In most cases, the rankings of competing projects based on accrual accounting rate of return will not differ whether the initial or average investment base is the denominator. Using the average base will show substantially higher rates of return; therefore, the desirable rate of return used as a basis for accepting projects should be proportionately higher.

Evaluation of Performance

The use of the accrual accounting rate of return for evaluating performance is often a stumbling block to the implementation of the DCF method for capital-budgeting decisions. Consider Mary Conley, the manager of a flour-milling division in a multi-division company. Mary has just completed courses in management accounting and finance in an executive education program. She became convinced that the NPV method for capital budgeting would lead to decisions that would better achieve the long-range operating income goals of the company.

Upon returning to her company, Mary was more frustrated than ever. Top management used the AARR for judging her division's performance. Each year division operating income was divided by average division assets to obtain the rate of return on investment (ROI).[4] Such a measure usually inhibits some investments in plant and equipment that would be attractive using the DCF methods. Why? Because a huge investment often boosts depreciation inordinately in the early years under accelerated depreciation methods, thus reducing the numerator (accounting income) in the ROI computation. Also, the denominator (net initial investment) is increased substantially

---

[3]The average cash inflow in the example is $98,000 a year. The amount of average annual depreciation is $372,890 ÷ 5 = $74,578. The increase in expected average annual operating income is $98,000 − $74,578 = $23,422.

However, if the project has a nonzero terminal disposal price and straight-line depreciation is used, the *average* investment is computed by *adding* the sum of the terminal disposal price and the recovery of working capital to the net initial investment and dividing by 2. For example, assume that the X-ray machine in Exhibits 20-2 and 20-3 has a terminal disposal price of $60,000 instead of zero. The annual depreciation would be ($372,890 − $60,000) ÷ 5 = $62,578. The average investment is ($379,100 + $70,000) ÷ 2 = $224,550. The accrual accounting rate of return with the denominator calculated as the average investment is

$$\text{AARR} = (\$98,000 - \$62,578) \div \$224,550$$
$$\text{AARR} = \$35,422 \div \$224,550 = 15.77\%$$

[4]Chapter 26 discusses ROI calculations and limitations in more detail.

INTERNATIONAL COMPARISON OF CAPITAL BUDGETING METHODS

What methods do companies around the world use for analyzing capital investment decisions? The percentages in the table below indicate how frequently particular capital-budgeting methods are used in eight countries. The reported percentages exceed 100% because many companies surveyed use more than one capital-budgeting method.

	United States[a]	Australia[b]	Canada[c]	Ireland[d]	Japan[b]	Scotland[e]	South Korea[f]	United Kingdom[b]
Payback	59%	61%	50%	84%	52%	78%	75%	76%
Internal rate of return (IRR)	52%	37%	62%	84%	4%	58%	75%	39%
Net present value (NPV)	28%	45%	41%		6%	48%	60%	38%
Accrual actg. rate of return (AARR)	13%	24%	17%	24%	36%	31%	68%	28%
Other	44%	7%	8%	—	5%	—	—	7%

We make several observations:

1. Companies in the United States, Australia, Canada, Ireland, Scotland, South Korea, and the United Kingdom tend to use two methods, on average, to evaluate capital investments. (The sum of the capital-budgeting percentages in the columns for each of these countries is approximately 200%.)

2. Japanese companies tend to use only one method. (The sum of the capital-budgeting percentages for Japan is approximately 100%.)

3. The payback method is a very popular method among companies in all countries. Japanese companies use the payback method as the primary method analysis in their capital-budgeting decisions. Companies in the United States, Australia, Canada, Ireland, Scotland, South Korea, and the United Kingdom use discounted cash flow (DCF) methods (IRR and NPV) extensively.

4. The accrual accounting rate-of-return (AARR) method lags behind DCF methods in the United States, Australia, Canada, Ireland, Scotland, and the United Kingdom. It is on par with DCF methods in South Korea, and it is very much preferred to DCF methods in Japan.

---

[a]Adapted from Smith and Sullivan, "Survey of Cost." [b]Blayney and Yokoyama, "Comparative Analysis." [c]Jog and Srivastava, "Capital Formation." [d]Clarke, "Management Accounting." [e]Sangster, "Capital Investment." [f]Kim and Song, "Accounting Practices." Full citations are in Appendix A.

by the initial cost of the new assets. As Conley said, "Top management is always giving me hell about my new flour mill, even though I know it is the most efficient we've got regardless of what the figures say."

Obviously, there is an inconsistency between citing DCF methods as being best for capital-budgeting decisions and then using different concepts to evaluate subsequent performance. As long as such practice continues, managers will be tempted to make decisions on the basis of accrual accounting rates of return, even though such decisions are not in the best interests of the firm. Such temptations become more pronounced if managers are frequently transferred (or promoted), or if annual accounting income is important in their evaluations and their compensation plans. Why? Because the manager's performance is being evaluated over short time horizons. The manager has no motivation to do a DCF analysis to take into account cash flows that will occur in the distant future. Those cash flows will not influence the manager's performance evaluation.

COMPLEXITIES IN CAPITAL-BUDGETING APPLICATIONS

The most challenging aspects of capital budgeting are identifying the project and predicting their outcomes. Textbook examples typically ignore much of the complexity associated with the identification and information-acquisition stages in capital budgeting. This section illustrates some of the complexity often found in practice.

Consider a firm deciding whether to invest in computer-integrated manufacturing (CIM) technology. In CIM plants, computers give instructions that automatically set up and run equipment. Computers monitor the product and directly control the process to ensure defect-free, high-quality output. Applying CIM to its full extent can result in a highly automated plant, where the role of manufacturing labor is largely restricted to computer programming, engineering support, and maintenance of the robotic machinery. The amounts at stake in CIM decisions can be huge—in the billions of dollars for such companies as General Motors and Toyota. Two important factors when evaluating CIM investments are (1) recognizing the full set of benefits and costs and (2) recognizing the full time horizon of the project.

Recognizing the Full Set of Benefits and Costs

Objective 10

Recognize the impact of nonfinancial and qualitative factors in capital-budgeting decisions

The factors that firms consider in making CIM decisions are far broader than costs alone. The reasons for introducing CIM technology—faster response time, higher product quality, and greater flexibility in meeting changes in customer preferences—are often to increase revenues and contribution margins. Ignoring revenue effects underestimates the financial benefits of CIM investments. As we describe below, however, revenue benefits of technology investments are often difficult to quantify in financial terms. Nevertheless, competitive and revenue advantages are important managerial considerations.

Exhibit 20-7 presents examples of the broader set of factors that U.S., Australian, Japanese, and U.K. firms weigh in evaluating CIM technology. The benefits include the following points, which are often difficult to quantify in financial terms.[5]

1. Faster response to market changes. An automated plant can, for example, make major design modifications (such as switching from a two-door to a four-door car) relatively quickly. To quantify this benefit requires some notion of consumer demand changes that may occur many years in the future and the manufacturing technology choices made by competitors.

   For example, a firm may view itself as gaining significant competitive advantage over a rival firm by being the first to invest in CIM technology. This conclusion is based on the presumption that the rival firm will continue to use conventional technology. In fact, the rival firm may respond by itself investing in CIM technology. As a result, the benefits to the first firm from investing in CIM technology may be significantly smaller than predicted. The key point: In determining outcomes of an investment decision, a firm must consider how a rival firm will respond to the firm's decision.

2. Increased learning by workers about automation. If workers have a positive experience with CIM, the company can implement other automation projects more quickly and more successfully. Quantifying this benefit requires a prediction of the company's subsequent automation plans. Survey evidence emphasizes the importance of linking CIM decisions to the overall competitive strategies of a firm.

---

[5]C. Sullivan and K. Smith, "Capital Investment Justification for U.S. Factory Automation Projects." M. Freeman and G. Hobbes, "Capital Budgeting: Theory versus Practice" *Australian Accountant* (September 1991); C. Drury, S. Braund, P. Osborne, and M. Tayles, *A Survey of Management Accounting Practices in U.K. Manufacturing Companies* (London: Certified Accountants Educational Trust, 1993); M. Sakurai, "The Change in Cost Management Systems in the Age of CIM," Working paper, Senshu University, 1992.

Category	Examples
Financial outcomes	Lower direct manufacturing labor costs
	Lower hourly support labor costs
	Less scrap and rework
	Increase in software costs
	Retraining of personnel costs
Nonfinancial and qualitative outcomes	Reduction in manufacturing lead time
	Increase in manufacturing flexibility
	Lower product defect rate
	Increase in business risk due to higher fixed-cost structure
	Improved product delivery and service
	Improved competitive position
	Reduction in new product development time
	Faster response to market changes
	Increased learning by workers about automation

Recognizing the full set of costs also presents problems. Three classes of costs are difficult to measure and often are underestimated:

1. Costs associated with a reduced competitive position in the industry. If other firms in the industry are investing in CIM, a firm not investing in CIM may likely suffer a decline in market share. Several firms in the machine-tool industry that continued to use a conventional manufacturing approach experienced rapid drops in market share after their competitors introduced CIM.

2. Costs of retraining the operating and maintenance personnel to handle the automated facilities.

3. Costs of developing and maintaining the software and maintenance programs to operate the automated manufacturing activities.

Recognizing the Full Time Horizon of the Project

The time horizon of CIM projects can stretch well beyond 10 years. Many of the costs are incurred and are highly visible in the early years of adopting CIM. In contrast, important benefits may not be realized until many years after the adoption of CIM. A long time horizon should be considered when evaluating CIM investments.

PROBLEM FOR SELF-STUDY

PROBLEM

Revisit the Lifetime Care X-ray machine project. Assume that the expected annual cash inflow is $130,000 instead of $100,000. All other facts are unchanged: a $379,100 net initial investment, a five-year useful life, a zero terminal disposal price, and an 8% required rate of return. Year 5 cash inflow includes $10,000 recovery of working capital. Compute the following:

1. Discounted cash flow:
   a. Net present value
   b. Internal rate of return
2. Payback period
3. Accrual accounting rate of return on net initial investment

Assume (for calculation purposes) that cash outflows and cash inflows occur at the beginning or end of each period (as specified in the problem).

SOLUTION

1. a. NPV = ($130,000 × 3.993) − $379,100
      = $519,090 − $379,100 = $139,990

   b. There are several approaches to computing the IRR. One is to use a calculator with an IRR function; this gives a 21.18% IRR. An alternative approach is to use the tables in Appendix C at the end of the text:

$$\$379,100 = \$130,000F$$

$$F = \frac{\$379,100}{\$130,000} = 2.916$$

On the 5-period line of Table 4 (Appendix C), the column closest to 2.916 is 22%. A more accurate number can be obtained by using interpolation:

	Present Value	**Factors**
20%	2.991	2.991
IRR	—	2.916
22%	2.864	—
Difference	0.127	0.075

$$\text{Internal rate of return} = 20\% + \frac{0.075}{0.127}(2\%) = 21.18\%$$

2. $$\text{Payback} = \frac{\text{Net initial investment}}{\text{Uniform increase in annual cash inflow}}$$
    $$= \frac{\$379,100}{\$130,000} = 2.92 \text{ years}$$

3. $$\text{Accrual accounting ROR} = \frac{\text{Increase in expected average annual operating income}}{\text{Net initial investment}}$$

Increase in expected average annual operating savings = [($130,000 × 4) + $120,000] ÷ 5

= $128,000

Average annual depreciation = $372,890 ÷ 5 = $74,578

Increase in expected average annual operating income = $128,000 − $74,578 = $53,422

$$\text{Accrual accounting ROR} = \frac{\$53,422}{\$379,100} = 14.09\%$$

SUMMARY

The following points are linked to the chapter's learning objectives.

1. Capital budgeting is long-term planning for proposed capital projects. The life of a project is usually longer than one year, so capital budgeting decisions consider revenues and costs over relatively long periods. In contrast, accrual accounting measures income on a year-by-year basis.

2. The time value of money takes into account this fact: A dollar received today can be invested to start earning interest, so it is worth more than a dollar received tomorrow. The time value of money is the opportunity cost (interest forgone) from not having the money today.

3. Capital budgeting is a six-stage process: (a) the identification stage, (b) the search stage, (c) the information-acquisition stage, (d) the selection stage, (e) the financing stage, and (f) the implementation and control stage.

4. Discounted cash-flow (DCF) methods explicitly include all project cash flows and the time value of money in capital-budgeting decisions. Two DCF methods are the net present value (NPV) method and the internal rate of return (IRR) method. The NPV method calculates the expected net monetary gain or loss from a project by discounting all expected future cash inflows and outflows to the present point in time, using the required rate of return. A project is acceptable if it has a positive NPV. The IRR method computes the rate of interest (discount rate) at which the present value of expected cash inflows from a project equals the present value of expected cash outflows from a project. A project is acceptable if its IRR exceeds the required rate of return.

5. The NPV method has two advantages over the IRR method: (a) NPVs of individual projects can be added together to obtain a valid estimate of accepting a combination of projects, and (b) the NPV method accommodates different required rates of return across different years of the project.

6. Relevant cash inflows and outflows are (a) expected future cash flows that (b) differ among the alternatives. Only cash inflows and outflows matter. Accrual accounting concepts such as accrued revenues and accrued expenses are irrelevant for the discounted cash-flow methods.

7. The payback method measures the time it will take to recoup, in the form of cash inflows, the total amount invested in a project. The payback method neglects profitability and the time value of money.

8. The accrual accounting rate of return (AARR) is an accounting measure of operating income divided by an accounting measure of investment. Different ways of calculating AARR appear in practice. In most cases, the AARR considers profitability but ignores the time value of money.

9. The widespread use of accrual accounting for evaluating the performance of a manager or division impedes the adoption of DCF methods in capital budgeting. Frequently, the optimal decision made using a DCF method will not report good "operating income" results in the project's early years on the basis of accrual accounting methods, so managers are tempted to ignore DCF methods even though the decisions that stem from them would be optimal for the company over the long run.

10. Nonfinancial and qualitative factors such as the effects of investment decisions on employee learning and on the ability to respond faster to market changes are often not explicitly considered in capital-budgeting decisions. But nonfinancial and qualitative factors can be extremely important. In reaching conclusions, managers must at times give more weight to nonfinancial and qualitative factors than to financial factors.

TERMS TO LEARN

This chapter and the Glossary at the end of the book contain definitions of the following important terms:

accounting rate of return *(p. 701)*   accrual accounting rate of return (AARR) *(701)*
bailout payback *(700)*   capital budgeting *(685)*   discounted cash flow (DCF) *(687)*
discount rate *(688)*   hurdle rate *(688)*   internal rate of return (IRR) *(690)*
net present value (NPV) *(688)*   (opportunity) cost of capital *(688)*   payback *(697)*
required rate of return (RRR) *(688)*   time-adjusted rate of return *(690)*

ASSIGNMENT MATERIAL

QUESTIONS

20-1 Capital budgeting has the same focus as accrual accounting. Do you agree? Explain.

20-2 List and briefly describe each of the six stages in capital budgeting.

20-3 What is the essence of the discounted cash-flow method?

20-4 "Only quantitative outcomes are relevant in capital-budgeting analyses." Do you agree? Explain.

20-5 List three methods of analyzing the quantitative and financial aspects of a capital-budgeting project.

20-6 What are the two main discounted cash-flow methods? How do they differ?

20-7 What is the payback method? What are its main strengths and weaknesses?

20-8 How is the bailout payback method different from the payback method?

20-9 Describe the accrual accounting rate of return method. What are its main strengths and weaknesses?

20-10 "The trouble with discounted cash-flow techniques is that they ignore depreciation costs." Do you agree? Explain.

20-11 "Let's be more practical. DCF is not the gospel. Managers should not become so enchanted with DCF that strategic considerations are overlooked." Do you agree? Explain.

20-12 "The net present value method is the preferred method for capital-budgeting decisions. Therefore managers will always use it." Do you agree? Explain.

20-13 "All overhead costs are relevant in NPV analysis." Do you agree? Explain.

20-14 What conflicts can arise in using DCF methods for capital-budgeting decisions and accrual accounting for performance evaluation?

20-15 Bill Watts, president of Western Publications, accepts a capital-budgeting project advocated by Division X. This is the division in which the president spent his first 10 years with the company. On the same day, the president rejects a capital-budgeting project proposal from Division Y. The manager of Division Y is incensed. She believes that the Division Y project has an internal rate of return at least 10 percentage points above the Division X project. She comments, "What is the point of all our detailed DCF analysis? If Watts is panting over a project, he can arrange to have the proponents of that project "massage" the numbers so that it looks like a winner." What advice would you give the manager of Division Y?

EXERCISES AND PROBLEMS

Throughout the assignment material, ignore the effects of income taxes.

20-16 Exercises in compound interest. To be sure that you understand how to use the tables in Appendix C at the end of this book, solve the following exercises. Ignore income-tax considerations. The correct answers, rounded to the nearest dollar, appear on pages 717–18.

Required

1. You have just won $5,000. How much money will you have at the end of 10 years if you invest it at 6% compounded annually? At 14%?

2. Ten years from now, the unpaid principal of the mortgage on your house will be $89,550. How much do you have to invest *today* at 6% interest compounded annually to accumulate the $89,550 in 10 years?

3. If the unpaid mortgage on your house in 10 years will be $89,550, how much money do you have to invest *annually* at 6% to have exactly this amount on hand at the end of the tenth year?

4. You plan to save $5,000 of your earnings at the end of each year for the next 10 years. How much money will you have at the end of the tenth year if you invest your savings compounded at 12% per year?

5. You hold an endowment insurance policy that will pay you a lump sum of $200,000 at age 65. If you invest the sum at 6%, how much money can you withdraw from your account in equal amounts each year so that at the end of 10 years (age 75) there will be nothing left?

6. You have estimated that for the first 10 years after you retire you will need an annual cash inflow of $50,000. How much money must you invest at 6% at your retirement age to obtain this annual cash inflow? At 20%?

7. The following table shows two schedules of prospective operating cash inflows, each of which requires the same net initial investment of $10,000 now:

	Annual Cash Inflows	
Year	Plan A	Plan B
1	$ 1,000	$ 5,000
2	2,000	4,000
3	3,000	3,000
4	4,000	2,000
5	5,000	1,000
Total	$15,000	$15,000

The required rate of return is 6% compounded annually. All cash inflows occur at the end of each year. In terms of net present value, which plan is more desirable? Show computations.

20-17 Basic nature of present value. Santa Ynez Products is considering investing in a project with a two-year life and a zero terminal disposal price. Operating cash inflows will be equal payments of $4,000 at the end of each of the two years. How much would the company be willing to invest to earn an internal rate of return of 8%? Use Table 2, Appendix C, to find your answer. Prepare a tabular analysis of each payment as follows:

Year (1)	Investment at Beginning of Year (2)	Operating Cash Inflows (3)	Return at 8% per Year on Investment at Beginning of Year (4) = 8% × (2)	Amount of Investment Recovered at End of Year (5) = (3) − (4)	Unrecovered Investment at End of Year (6) = (2) − (5)
1	?	$4,000	?	?	?
2	?	$4,000	?	?	?

Does Santa Ynez Products recover all of its net initial investment by the end of year 2?

20-18 Comparison of approaches to capital budgeting. The Building Distributors Group is thinking of buying, at a cost of $220,000, some new packaging equipment that is expected to save $50,000 in cash-operating costs per year. Its estimated useful life is 10 years, and it will have zero terminal disposal price. The required rate of return is 16%.

Required

Compute:

1. Net present value
2. Payback period

3. Internal rate of return

4. Accrual accounting rate of return based on (a) net initial investment and (b) average investment. Assume straight-line depreciation.

20-19 Comparison of approaches to capital budgeting. City Hospital, a nontaxable institution, estimates that it can save $28,000 a year in cash-operating costs for the next 10 years if it buys a special-purpose machine at a cost of $110,000. A zero terminal disposal price is expected. City Hospital's required rate of return is 14%.

Required
Compute:
1. Net present value
2. Payback period
3. Internal rate of return
4. Accrual accounting rate of return based on (a) net initial investment and (b) average investment. Assume straight-line depreciation.

20-20 Payback, bailout payback, sensitivity analysis. University Food Service is choosing between two machines for producing doughnuts. The food service now buys doughnuts from an outside supplier at $0.25 per doughnut. The food service estimates sales of 50,000 doughnuts each year for the next five years.

	Semiautomatic Doughnut Machine	Automatic Doughnut Machine
Initial machine investment	$5,000	$7,500
Useful life of machine	5 years	5 years
Variable costs (all cash) per doughnut	$ 0.15	$ 0.12
Annual fixed cash costs to run machine	$3,750	$4,500

The semiautomatic machine is the more popular model. Management expects its disposal price to be $3,150 at the end of year 1 and to decline $650 annually thereafter. Management expects the disposal price of the automatic machine to be $3,000 at end of year 1 and to decline $750 annually thereafter.

Required
1. Compute the payback period for each project. If payback were the sole criterion for the decision, which project would you choose?
2. Compute the bailout payback period for each project. On the basis of the bailout payback criterion, which project would you choose?
3. Compare your answers in requirements 1 and 2. Are they the same? Explain.
4. Compute the payback period for each project if University Food Service can sell (a) 40,000 doughnuts a year (b) 60,000 doughnuts a year.

20-21 Capital budgeting with uneven cash flows. Southern Cola is considering the purchase of a special-purpose bottling machine for $28,000. It is expected to have a useful life of seven years with a zero terminal disposal price. The plant manager estimates the following savings in cash-operating costs:

Year	Amount
1	$10,000
2	8,000
3	6,000
4	5,000
5	4,000
6	3,000
7	3,000
Total	$39,000

Southern Cola uses a required rate of return of 16% in its capital-budgeting decisions.

Required

Compute:

1. Net present value
2. Payback period
3. Internal rate of return
4. Accrual accounting rate of return based on (a) net initial investment and (b) average investment. Assume straight-line depreciation. Use the average annual savings in cash-operating costs when computing the numerator of the accrual accounting rate of return.

20-22 Equipment purchase for customer-service unit, net present value and payback. Solar Energy Inc. manufactures and markets solar panels to heat water for swimming pools. It currently owns a service van used by its customer-service representatives when making repair and service visits. The van was purchased three years ago for $56,000. The van has a current book value of $35,000 and a remaining useful life of five years but will require a $10,000 overhaul two years from now. Its current disposal price is $20,000; in five years its terminal disposal price is expected to be $8,000, assuming that the $10,000 overhaul is done on schedule. The cash-operating costs of the van are expected to be $40,000 annually.

A salesperson has offered a new van for $51,000, or for $31,000 plus the old van as a trade-in. The new van would reduce annual cash-operating costs by $10,000, would not require any overhauls, would have a useful life of five years, and would have a terminal disposal price of $3,000.

Solar Energy has a required rate of return of 14% in its capital-budgeting decisions.

Required

1. Using a net present value criterion, should Solar Energy Inc. purchase the new van?
2. Compute the payback period for Solar Energy Inc. if it purchases the new van.

20-23 Sporting contract, net present value. Milano Capri is an Italian soccer team with a long tradition of winning. However, the last three years have been traumatic. The team has not won a major championship, and attendance at games has dropped considerably. Bennetelo Company is the major corporate sponsor of Milano Capri. Rocky Balboa, the president of Bennetelo, is also the president of Milano Capri. Balboa proposes that the team purchase the services of Andreas Brehme. Brehme would create great excitement for Milano's fans and sponsors. Brehme's agent notifies Balboa that terms for the superstar's signing with Milano Capri are a bonus of $3 million payable now (start of 19_5) plus the following four-year contract (assume all amounts are in millions and are paid at the end of each year):

	19_5	19_6	19_7	19_8
Salary	$4.500	$5.000	$6.000	$6.500
Living and other costs	1.000	1.200	1.300	1.400

Balboa's initial reaction is horror. As president of Bennetelo, he has never earned more than $800,000 a year. However, he swallows his pride and decides to examine the expected additions to Milano Capri's cash inflows if Brehme is signed for the four-year contract (assume all cash inflows are in millions and are received at the end of each year):

	19_5	19_6	19_7	19_8
Net gate receipts	$2.000	$3.000	$3.000	$3.000
Corporate sponsorship	3.000	3.500	4.000	4.000
Television royalties	0.000	1.200	1.400	2.000
Merchandise income (net of costs)	0.600	0.600	0.700	0.700

Balboa believes that a 12% required rate of return is appropriate for investments by Milano Capri.

Required

1. For Brehme's proposed four-year contract, compute (a) the net present value and (b) the payback period.
2. What other factors should Balboa consider when deciding whether to sign Brehme to the four-year contract?

20-24 DCF, accrual accounting rate of return, working capital, evaluation of performance. Hammerlink Company has been offered a special-purpose metal-cutting machine for $110,000. The machine is expected to have a useful life of eight years with a terminal disposal price of $30,000. Savings in cash-operating costs are expected to be $25,000 per year. However, additional working capital is needed to keep the machine running efficiently and without stoppages. Working capital includes such items as filters, lubricants, bearings, abrasives, flexible exhaust pipes, and belts. These items must continually be replaced, so that an investment of $8,000 must be maintained in them at all times, but this investment is fully recoverable (will be "cashed in") at the end of the useful life. Hammerlink's required rate of return is 14%.

Required

1. Compute the net present value.
2. Compute the internal rate of return.
3. Compute the accrual accounting rate of return based on (a) the net initial investment and (b) the average investment. Assume straight-line depreciation.
4. You have the authority to make the purchase decision. Why might you be reluctant to base your decision on the DCF model?

20-25 DCF, accrual accounting rate of return, working capital, evaluation of performance. Jana Wendt is the manager of the local Country West Department Store. She is considering whether to renovate some space in the expectation of increasing sales quantities. She could just continue to use the current display fixtures and equipment. Existing fixtures and equipment are expected to last for six years and will have a terminal disposal price of zero at the end of six years.

If she does renovate the space, new display fixtures and equipment will cost $73,000, with an expected useful life of six years and a terminal disposal price of $4,000. Existing fixtures and equipment that have a current book value of $15,000 can be sold immediately for $3,000. Additional cash-operating inflows are expected to be $25,000 per year.

However, experience has shown that in order to sustain the higher sales volume, similar renovations have required additional investments in current assets, such as merchandise inventories and accounts receivable. An initial investment of $40,000 is needed to finance or "carry" this working capital, and this level must be maintained continuously. If and when she decides to terminate this plan or to use the store space for other purposes, the inventories and receivables can soon be recovered or "cashed in."

Required

1. Compute (a) net present value, using a required rate of return of 12%; (b) internal rate of return; (c) accrual accounting rate of return based on the net initial investment; and (d) accrual accounting rate of return on the average investment. Assume straight-line depreciation.
2. As the store manager, which capital-budgeting method would you prefer for the purposes of making this decision and for evaluating subsequent performance? Give reasons, and compare the methods.

20-26 Relevant costs, replacement decisions. George Handley, general manager of the Coronado Company, is contemplating replacing the existing assembly-line equipment in the Assembly Department with automated assembly equipment. Production output and revenues will be unaffected by the replacement decision. Transactions related to the capital investment are cash transactions that would occur today.

	Existing Assembly Equipment	New Automated Assembly Equipment
Initial equipment investment	$1,100,000	$1,200,000
Useful life in years	11	5
Current age in years	6	0
Useful life remaining in years	5	5
Accumulated depreciation	$ 600,000	$ 0
Book value	$ 500,000	Not acquired yet
Current disposal price (in cash)	$ 200,000	Not acquired yet
Terminal disposal price (in cash, in 5 years)	$ 0	$ 0
Average working capital needed	$ 120,000	$ 70,000

Current annual Assembly Department costs follow:

Direct materials	$ 600,000
Direct manuf. labor	400,000
Depreciation	100,000
Maintenance and repairs	150,000
Other operating costs	50,000
Supervision (allocated as 10% of direct manuf. labor costs)	40,000
Allocated rent (based on space used)	40,000
Allocated corporate overhead (based on direct manuf. labor costs)	120,000
Total	$1,500,000

Additional information:

a. Coronado uses straight-line depreciation calculated on the difference between the initial equipment investment and the terminal disposal price of the equipment.

b. The new equipment will produce output more swiftly. Therefore, the average working capital investment, if the new equipment is purchased, will decrease.

c. Of the total direct materials costs, $120,000 is waste and scrap. The new equipment is expected to reduce scrap costs to $20,000.

d. The new equipment is expected to reduce direct manufacturing labor costs by $150,000 each year.

e. Maintenance and repairs on the old equipment have been excessive. If the new equipment is acquired, maintenance and repair costs are expected to decrease to $100,000.

f. Coronado collects all supervision costs for all manufacturing departments in the plant into one cost pool. These costs are then allocated to departments on the basis of direct manufacturing labor costs. The assembly department has only one supervisor currently. The supervisor will continue in her current position if the new equipment is purchased.

g. The new equipment will reduce the space required for assembly operations by 20%, reducing allocated rent by $8,000. Coronado Company has no alternative uses for this extra space.

h. Corporate overhead costs are allocated to each department at 30% of direct manufacturing labor costs of each department.

Handley estimates a required rate of return of 12% for this project.

Required

1. On the basis of the net present value method, should Handley replace the existing assembly equipment?

2. Suppose that next year is the last year Coronado will offer the attractive bonus plan currently in place. Handley's bonus hinges on short-run accrual accounting income for that year. Will Handley be inclined to replace the assembly department equipment? Provide quantitative support for your answer.

3. What nonfinancial and qualitative factors should Handley consider in coming to a decision?

 20-27 Equipment replacement, relevant costs, sensitivity analysis. A toy manufacturer that specializes in making fad items has just developed a $50,000 molding machine for automatically producing a special toy. The machine has been used to produce only one unit so far. The company will depreciate the $50,000 initial machine investment evenly over four years, after which time production of the toy will be stopped. The company's expected annual costs will be: direct materials, $10,000; direct manufacturing labor, $20,000; and variable manufacturing overhead, $15,000. Variable manufacturing overhead varies with direct manufacturing labor costs. Fixed manufacturing overhead, exclusive of depreciation, is $7,500 annually, and fixed marketing and administrative costs are $12,000 annually.

Suddenly a machine salesperson appears. He has a new machine that is ideally suited for producing this toy. His automatic machine is distinctly superior. It reduces the cost of direct materials by 10% and produces twice as many units per hour. It will cost $44,000 and will have a zero terminal disposal price at the end of four years.

Production and sales of 25,000 units per year (sales of $100,000) will be the same whether the company uses the molding machine or the automatic machine. The current disposal price of the toy company's molding machine is $5,000. Its terminal disposal price in four years will be $2,600.

Required

1. Assume that the required rate of return is 18%. Using the net present value method, show whether the new machine should be purchased. What is the role of the book value of the old machine in the analysis?

2. What is the payback period for the new machine?

3. As the manager who developed the $50,000 molding machine, you are trying to justify not buying the new $44,000 machine. You question the accuracy of the expected cash operating savings. By how much must these cash savings fall before the point of indifference—the point where the net present value of the project is zero—is reached?

20-28   Relevant costs, outsourcing, capital budgeting.   The Strubel Company currently makes as many units of Part No. 789 as it needs. David Lin, general manager of the Strubel Company, has received a bid from another company for making Part No. 789. Gabriella will supply 1,000 units of Part No. 789 per year at $50 a unit. Gabriella can begin supplying on January 1, 19_5, and continue for five years, after which time Strubel will not need the part. Gabriella can accommodate any change in Strubel's demand for the part and will supply it for $50 regardless of quantity.

Jack Tyson, the controller of the Strubel Company, reports the following costs for manufacturing 1,000 units of Part No. 789.

Direct materials	$22,000
Direct manufacturing labor	11,000
Variable manufacturing overhead	7,000
Depreciation on machine	10,000
Product and process engineering	4,000
Rent	2,000
Allocation of general plant overhead costs	5,000
Total costs	$61,000

The following additional information is available.

a. Part No. 789 is made on a machine used exclusively for the manufacture of Part No. 789. The machine was acquired on January 1, 19_4, at a cost of $60,000. The machine has a useful life of six years and zero terminal disposal price. Depreciation is calculated on the straight-line method.

b. The machine could be sold today for $15,000.

c. Product and process engineering costs are incurred to ensure that the manufacturing process for Part No. 789 works smoothly. Although these costs are fixed in the short run, with respect to units of part No. 789 produced, they can be saved in the long run if Part No. 789 is no longer produced. If Part No. 789 is outsourced, product and process engineering costs of $4,000 will be incurred for 19_5 but not thereafter.

d. Rent costs of $2,000 are allocated to products on the basis of the floor space used for manufacturing the product. If Part No. 789 is discontinued, the space currently used to manufacture it would become available. The company could then use the space for storage purposes and save $1,000 currently paid for outside storage.

e. General plant overhead costs are allocated to each department on the basis of direct manufacturing labor dollars. These costs will not change in total. But no general plant overhead will be allocated to Part No. 789 if the part is outsourced.

Assume that Strubel requires a 12% rate of return for this project.

Required

1. Should David Lin outsource Part No. 789? Prepare a quantitative analysis.

2. Describe any sensitivity analysis that seems advisable, but you need not perform any sensitivity calculations.

3. What other factors should Lin consider in making a decision.

4. Lin is particularly concerned about his bonus for 19_5. The bonus is based on accounting income of Strubel. What decision will Lin make if he wants to maximize his bonus in 19_5?

20-29   Capital budgeting and relevant costs, sensitivity analysis.   The city of Los Angeles has been operating a cafeteria for its employees, but it is considering a conversion from this form of food service to a completely automated set of vending machines. If the change is made, the old equipment would be sold now for whatever cash it might bring.

The vending machines would be purchased immediately for cash. A catering firm would take complete responsibility for servicing and replenishing the vending machines and would pay the city a predetermined percentage of the gross vending receipts.

The present cafeteria equipment has 10 years of remaining useful life. The new vending machines have a 10-year useful life. The following data are available (in thousands):

Cafeteria cash revenues per year	$ 120
Cafeteria cash costs per year	$ 124
Present cafeteria equipment	
Net book value	$ 84
Annual depreciation cost	$ 6
Current disposal price	$ 4
Terminal disposal price (10 years from now)	$ 0
New vending machines	
Initial machine investment	$ 64
Terminal disposal price	$ 5
Expected annual gross receipts	$ 80
City's percentage share of receipts	10%
Expected annual cash costs (negligible)	
Present values at 14%	
$1 due in 10 years	$0.27
Annuity of $1 a year for 10 years	$5.20

The city of Los Angeles has a 14% required rate of return.

Required

For the two alternatives, compute the following:

1. Expected increase in net annual operating cash inflows
2. Payback period
3. Net present value
4. Point of indifference (zero NPV) in terms of annual gross vending machine receipts

20-30 Relevant costs, capital budgeting. (D. Solomons, adapted) The Rainier Company sells a range of high-grade chemical products that must be carefully packaged. The company's containers have a special patented lining made from a material known as GHJ, and the firm operates a special department to maintain its containers in good condition and to make new ones to replace those beyond repair.

Rona Wood, the general manager, has received a two-part quotation from Closure Inc. (a) Closure is prepared to supply all 15,000 new containers Rainier requires each year for $300,000 per year. The contract would run for five years. If the number of containers required decreased, Rainier would still pay $300,000 per year. If the number of containers required increased, the contract price would be raised proportionately. (b) Whether or not the above contract is agreed to, Closure has offered to carry out just the maintenance work on containers for a sum of $70,000 per year for five years.

Brian Wrend, the controller, provides the following up-to-date statement of the annual costs of operating the container department.

Direct materials		$100,000
Direct manufacturing labor		175,000
Departmental overhead		
Manager's salary	$50,000	
Rent	6,000	
Depreciation on machinery	30,000	
Maintenance of machinery	7,000	
Other costs	31,000	124,000
		399,000
Allocation of general administrative overhead		33,000
Total costs		$432,000

Wood asks the manager of the department, Skip Spencer, for his comments. She mentions that even if his department were closed, Spencer would be appointed to another management position soon becoming vacant, without loss of pay or prospects. If Spencer were not available, Wood would have to hire an outsider at a cost of $52,000 per year.

Spencer responds "Let me raise two issues. First, the machinery cost us $240,000 four years ago but could be sold for only $40,000 now, even though it is good for at least another five years. Second, the inventory of GHJ we bought a year ago cost $120,000—800 tons at $150 per ton—and will last another three years. We used up a quarter of it last year. Wrend's figure of $100,000 for materials includes $30,000 for GHJ. Today you'd have to pay $180 per ton, but if we sold it we would get only $120 per ton net of all handling costs."

Wrend adds, "We are paying $10,000 a year in rent for warehouse space that other departments of the company use. If we closed Spencer's department, we'd have all the warehouse space we need without renting. Also, we should not forget that if the department is closed, termination costs for the oldest employees released will amount to $14,000 annually for five years, but general administrative overhead costs will be unaffected. We don't expect to fire anyone in the general office if Spencer's department is closed."

"Well," says Wood, "what are your thoughts on doing the maintenance work ourselves?"

"Let's see," says Spencer. "We wouldn't need any machinery. The supervision could be handled by a new supervisor instead of the manager, which would save $30,000 a year. And we would need one-fifth of the workers. If we retained the oldest ones, we would not have to pay any termination costs. We would not save any space, so I suppose the rent would be the same. Other costs would be about $13,000 a year. We use $10,000 per year of materials on maintenance, and GHJ is not used for maintenance."

Required

1. Assume that the required rate of return is 10%. What action should Wood take? Prepare a financial analysis.

2. Describe any sensitivity analysis that seems particularly advisable. Without doing calculations, predict the effect of the sensitivity analysis on your final computations. For example, what would happen to your two best alternatives if the required rate of return were 20% instead of 10 percent?

3. What nonfinancial or qualitative factors should Wood consider in coming to a decision?

20-31 Capital-budgeting approaches, computer-integrated manufacturing.

Craig Young, the production manager of Brittania Tools, is concerned about Brittania's ability to maintain its competitive position in the industrial machine-tool market. A recent surge of imports is priced 30% below Brittania's full product cost. A major domestic competitor recently switched to a computer-integrated manufacturing (CIM) operation for its machine-tool plant.

Young attends a trade exhibition titled "Automate, Emigrate, or Evaporate" and starts negotiations with a vendor of CIM equipment. The vendor will provide the necessary machines and associated equipment for a cost of $80 million. Young estimates the following annual cash-flow effects from implementing CIM:

a. Reduction in rental payments due to reduced floor space requirements: $ 4.0 million
b. Lower number of product defects and reduced reworking of products: $14.0 million
c. Additional revenues minus cash operating costs from CIM investment: $8.0 million.

Another benefit of CIM is reduced levels of working capital. The average combined level of inventories and accounts receivable (working capital) for Brittania Tools at present is $14 million. Young estimates that after implementation of CIM, the average combined level of working capital would be $4 million. (For simplicity, assume that this reduction in working capital occurs instantaneously at the time the investment in CIM is made.)

The one-time internal costs of implementing CIM are estimated to be $40 million. These costs include the retraining of operating and maintenance personnel plus any lost production during the changeover. For internal reporting purposes, the $40 million internal costs are capitalized along with the $80 million purchase price when determining the initial equipment investment required for the CIM proposal. (For simplicity, assume that these $40 million implementation costs are incurred at the same time the $80 million equipment purchase is made.) The annual cash costs of maintaining the software programs and of the CIM hardware equipment and machinery are estimated to be $8 million. The vendor of the CIM equipment indicates that, if the equipment is properly maintained, a 20-year useful life may be expected.

The estimated disposal price of the CIM equipment is $30 million at the end of 10 years and $10 million at the end of 20 years. Brittania uses a required rate of return of 14%. The maximum time period Brittania currently considers for any investment proposal is 10 years.

Required
1. Compute the payback period for the CIM proposal.
2. Compute the net present value of the CIM proposal. Should Brittania adopt CIM, given its existing investment criteria?
3. Compute the accrual accounting rate of return based on (a) net initial investment and (b) average investment. Assume straight-line depreciation and a 10-year useful life for the investment.
4. Young reads an article arguing that many companies are rejecting CIM proposals because either (a) the discount rate used in DCF analysis is too high or (b) the time period over which the benefits are considered is too short. He believes that Brittania should use an 8% discount rate and consider benefits for a 20-year period in its CIM analysis. Prepare a report for Young on the effects of making these changes in the DCF calculations.
5. What nonfinancial or qualitative factors would you recommend that Brittania Tools consider in deciding whether to adopt the CIM proposal?

20-32 Ethics, capital budgeting. (CMA, adapted) Evans Company must expand its manufacturing capabilities to meet the growing demand for its products. The first alternative is to expand its current manufacturing facility, which is located next to a vacant lot in the heart of the city. The second alternative is to convert a warehouse, already owned by Evans, located 20 miles outside the city. Evan's controller, George Watson, assigns Helen Dodge, assistant controller, to use net present value computations to evaluate both proposals. On completing her analysis, Dodge reports to Watson that the proposal to expand the current manufacturing facility has a slightly positive net present value. The proposal to convert the warehouse has a large negative net present value.

Watson is upset at Dodge's conclusions. He returns the proposal to her with the comment, "You must have made an error. The warehouse proposal should look better and have a positive net present value. Work on the projections and estimates."

Dodge suspects that Watson is anxious to have the warehouse proposal selected because the choice of this location would eliminate his long commute into the city. Feeling some pressure, she checks her calculations but finds no errors. Dodge reviews her projections and estimates. These too are quite reasonable. Even so, she replaces some of her original estimates with new estimates that are more favorable to the warehouse proposal, although these new estimates are less likely to occur. The revised proposal still has a negative net present value. Dodge is confused about what she should do.

Required
1. Referring to the "Standards of Ethical Conduct for Management Accountants" described in Chapter 1 (p. 17), explain:
   a. Whether George Watson's conduct was unethical when he gave Helen Dodge specific instructions on revising the proposal.
   b. Whether Helen Dodge's revised proposal for the warehouse conversion was unethical.
2. Identify the steps that Helen Dodge should follow in attempting to resolve this situation.

Answers to Exercises in Compound Interest (Problem 20-16)

The general approach to these exercises centers on a key question: Which of the four basic tables in Appendix C should be used? No computations should be made until after this basic question has been answered with confidence.

1. From Table 1. The $5,000 is a present value. The value 10 years hence is an amount of future worth.

$$S = P(1 + r)^n$$

The conversion factor, $(1 + r)^n$, is on line 10 of Table 1.

$$\text{Substituting at 6\%: } S = 5{,}000(1.791) = \$8{,}955$$

$$\text{Substituting at 14\%: } S = 5{,}000(3.707) = \$18{,}535$$

2. From Table 2. The $89,550 is an *amount of future worth*. You want the present value of that amount. $P = S \div (1 + r)^n$; the conversion factor, $1 \div (1 + r)^n$, is on line 10 of Table 2. Substituting:

$$P = \$89{,}550(0.558) = \$49{,}969$$

3. From Table 3. The $89,550 is *future worth*. You are seeking the uniform amount (annuity) to set aside annually. Note that $1 invested each year for 10 years at 6% has a future worth of $13.181 after 10 years, from line 10 of Table 3.

$$S_n = \text{Annual deposit}(F)$$
$$\$89{,}550 = \text{Annual deposit}(13.181)$$
$$\text{Annual deposit} = \frac{\$89{,}550}{13.181} = \$6{,}794$$

4. From Table 3. You are seeking the *amount of future worth* of an annuity of $5,000 per year. Note that $1 invested each year for 10 years at 12% has a future worth of $17.549 after 10 years.

$$S_n = \$5{,}000F, \text{ where } F \text{ is the conversion factor}$$
$$S_n = \$5{,}000(17.549) = \$87{,}745$$

5. From Table 4. When you reach age 65, you will get $200,000, a present value at that time. You must find the annuity that will exactly exhaust the invested principal in 10 years. To pay yourself $1 each year for 10 years when the interest rate is 6% requires you to have $7.360 today, from line 10 of Table 4.

$$P_n = \text{Annual withdrawal}(F)$$
$$\$200{,}000 = \text{Annual withdrawal}(7.360)$$
$$\text{Annual withdrawal} = \frac{\$200{,}000}{7.360} = \$27{,}174$$

6. From Table 4. You need to find the present value of an annuity for 10 years.

At 6%:
$$\begin{cases} P_n = \text{Annual withdrawal}(F) \\ P_n = \$50{,}000(7.360) \\ P_n = \$368{,}000 \end{cases}$$

At 20%:
$$\begin{cases} P_n = \$50{,}000(4.192) \\ P_n = \$209{,}600, \text{ a much lower figure} \end{cases}$$

7. Plan B is preferable. The net present value of Plan B exceeds that of Plan A by $980 ($3,126 − $2,146):

		Plan A		Plan B	
Year	PV Factor at 6%	Cash Inflows	PV of Cash Inflows	Cash Inflows	PV of Cash Inflows
0	1.000	$(10,000)	$(10,000)	$(10,000)	$(10,000)
1	0.943	1,000	943	5,000	4,715
2	0.890	2,000	1,780	4,000	3,560
3	0.840	3,000	2,520	3,000	2,520
4	0.792	4,000	3,168	2,000	1,584
5	0.747	5,000	3,735	1,000	747
			$2,146		$3,126

21

Capital Budgeting:

A Closer Look

Income tax factors

Effects of income taxes on cash flow

Capital budgeting and inflation

Required rate of return

Applicability to nonprofit organizations

Implementing the net present value decision rule

Implementing the internal rate-of-return decision rule

Appendix: Modified Accelerated Cost Recovery System

*T*axes and inflation are key aspects of many infrastructure investments. Harvest States considered both taxes and inflation when making a discounted cash-flow analysis for new rail car facilities at its Winona, Minnesota, grain-handling terminal.

Benjamin Franklin said that two things in life are certain: death and taxes. We might add a third: changing prices. This chapter examines how managers analyze income taxes and changing prices in capital budgeting. (We also recognize death in this chapter, although only of projects, not of the individuals who select them.) This chapter also covers estimation of the required rate of return, capital budgeting in nonprofit organizations, and some issues in using the net present value decision rule and the internal rate-of-return decision rule.

No matter where a company does business, income taxes have a major impact on two things:

◆ the amounts of cash inflows and outflows and
◆ the timing of those cash flows

Cash flows generally occur throughout the year. For simplicity, in this chapter, as in Chapter 20, we assume that all cash outflows and cash inflows occur at the beginning or end of each year (as specified).

Objective 1
Identify two major ways that income taxes affect business decisions

INCOME TAX FACTORS

The Importance of Income Taxes

Income taxes often have a tremendous influence on decisions. For example, income taxes can sizably reduce the net cash inflows from individual projects and so change their relative desirability.

We concentrate on a general approach to understanding income taxes that applies globally. In taking this approach, we focus on income tax provisions affecting depreciation and confine our discussion to corporations (excluding partnerships and individuals).[1]

Many tax rules regarding income measurement are the same as the financial rules regarding reporting to stockholders. Other rules differ. For example, income tax

---

[1]A general framework for examining income tax factors in business decisions is presented in M. Scholes and M. Wolfson, *Taxes and Corporate Financial Strategy: A Global Planning Approach* (Englewood Cliffs, NJ: Prentice Hall, 1991).

rules frequently allow taxpayers to use shorter useful lives for depreciation than generally accepted accounting principles permit.

Treatment of Depreciation for Tax Purposes

Objective 2

Identify three factors that influence the amount of depreciation claimed as a tax deduction

Tax rules for depreciation vary considerably among countries. Even within a single country, marked changes can occur over short periods of time. However, tax rules typically cover three factors that influence depreciation deductions: the amount allowable for depreciation, the time period over which the asset is to be depreciated, and the pattern of allowable depreciation.

Amount Allowable for Depreciation.   In most cases, the amount allowable for depreciation is the original cost of (initial investment in) the asset. However, sometimes the amount allowable for depreciation exceeds or is less than the original cost. In countries where corporations have the option of claiming an investment tax credit,[2] the amount allowable for depreciation may be reduced below the original cost of the asset acquired. Some countries permit corporations to write off more than the original cost (as measured by nominal monetary units) for depreciation purposes. For example, when determining the amount allowable for periodic depreciation, Brazilian corporations may use inflation indexes to write up the cost of assets. Allowance for inflation is important in a country such as Brazil, which has seen average annual inflation rates of more than 100% in some recent years. As we shall see, without indexing, the dollar value of the depreciation deductions would be insignificant compared with the dollar value to acquire the asset.

Time Period over Which the Asset Is to Be Depreciated.   Throughout the years and in various countries, tax authorities have permitted three main methods of determining the depreciation time period:

1. The taxpayer estimates the useful life.
2. The tax authority estimates the useful life.
3. Tax legislation specifies a table of allowable lives. An example is the property-class life categories (recovery periods) used in the Modified Accelerated Cost Recovery System that is applicable in the United States as of this writing.

Other things being equal, the shorter the allowable (depreciable) life, the higher is the project's net present value. A short allowable life means that depreciation deductions can be claimed early in the project's life. For profitable companies (which is the case we assume throughout this chapter), this means cash savings from tax deductions occur in the early periods of the project, when cash savings have higher present value.

Pattern of Allowable Depreciation (for a Given Time Period). Tax authorities allow three main depreciation methods:

1. **Straight-line depreciation (SL),** in which an equal amount of depreciation is taken each year.
2. Accelerated depreciation procedures, such as the double-declining balance (DDB) method.[3] **Accelerated depreciation** is any pattern of depreciation that writes off depreciable assets more quickly than does straight-line depreciation.
3. Depreciation using a table of allowable percentage write-offs as specified by tax law.

---

[2]An **investment tax credit (ITC)** is a direct reduction of income taxes payable arising from the acquisition of depreciable assets. Governments use the ITC to stimulate investments in specific assets or in specific industries. To illustrate: If a firm purchases an asset costing $100,000 and there is a 4% ITC, the firm obtains an immediate tax credit of $4,000; this credit increases the net present value of an asset purchase by $4,000. The depreciable amount of the asset would be $96,000 (the net cost, $100,000 − $4,000) or $100,000, depending on the specific tax law. In the United States, the ITC option has been made available (and then subsequently withdrawn) several times since 1962.

[3]**Double-declining balance (DDB) depreciation** is a form of accelerated depreciation in which first-year depreciation is twice the amount of straight-line depreciation when a zero terminal disposal price is assumed. The DDB method is illustrated on p. 728.

All other things being equal, the more accelerated the pattern of depreciation, the higher the project's net present value. Accelerated depreciation means that more depreciation occurs in the earlier years of a project, when the tax savings from depreciation tax deductions are in dollars with a relatively high present value.

EXAMPLE: Martina Enterprises, a newly formed corporation, is considering purchasing its first machine. The following tax laws apply:

- *Amount allowable for depreciation*—original cost of the machine minus any predicted terminal disposal price.
- *Time period over which the asset is to be depreciated*—based on an estimate of useful life made by the taxpayer (Martina).
- *Pattern of allowable depreciation*—only the straight-line depreciation method is permitted.

The original cost of the machine is $90,000 payable in cash immediately. Martina predicts that the machine will have an expected useful life of five years and a terminal disposal price of $0. No change in working capital is required if the machine is purchased. The corporate income tax rate of 40% will apply each year. The company uses a required rate of return of 12% to discount after-tax cash flows. Martina Enterprises estimates sales of $100,000 and operating costs other than depreciation of $62,000 in its first year of operation. For simplicity, assume that all sales are cash sales and that all operating costs other than depreciation are paid in cash.

The pertinent data follow:

- *Amount allowable for depreciation*—$90,000 (the original cost of $90,000 minus the terminal disposal price of $0).
- *Time period over which the asset is to be depreciated*—5 years.
- *Pattern of allowable depreciation*—straight-line method, $18,000 [($90,000 − 0) ÷ 5 years].

Panel A of Exhibit 21-1 shows the calculation of cash flow from operations, net of income taxes, using two different methods.

1. $S - E - T = \$100,000 - \$62,000 - \$8,000 = \$30,000$
2. $NI + D = \$12,000 + \$18,000 = \$30,000$

where S = sales, $100,000 per year
E = costs excluding depreciation (assumed all paid in cash), $62,000 per year
D = depreciation, $18,000 per year
T = income taxes, $8,000 per year
NI = net income, $12,000 per year
t = income tax rate, 40% per year

The first method calculates cash flow from operations, net of income taxes, by subtracting *cash* operating costs and taxes paid from cash sales. The second method starts with net income and adds back depreciation because depreciation is an operating cost that reduces net income but does not reduce cash outflow.

Panel B of Exhibit 21-1 describes a third method, frequently used to compute cash flow from operations, net of income taxes in capital-budgeting situations. Recall from Panel A of Exhibit 21-1 that

$$T = t \times OI$$
$$\text{where } OI = S - E - D$$

Therefore,

$$T = t(S - E - D)$$

The third method follows from the first.

$$
\begin{aligned}
\text{Cash flow from operations, net of income taxes} &= S - E - T \\
&= S - E - [t(S - E - D)] \\
&= S - E - [t(S - E)] + tD \\
&= [(S - E)(1 - t)] + tD \\
&= S - tS - E + tE + tD \qquad \text{Equation (1)}
\end{aligned}
$$

PANEL A: METHODS BASED ON THE INCOME STATEMENT

(S)	Sales		$100,000
	Costs		
(E)	Costs, excluding depreciation*	$62,000	
(D)	Depreciation (straight-line of $90,000 ÷ 5 years)	18,000	
	Total costs		80,000
(OI)	Operating income		20,000
(T)	Income taxes (40% × OI) = 40% × $20,000		8,000
(NI)	Net income		$ 12,000

Total cash flow from operations, net of income taxes, is:

1. $S - E - T = \$100,000 - \$62,000 - \$8,000 = \$30,000$

or

2. $NI + D = \$12,000 + \$18,000 = \$30,000$

PANEL B: ITEM-BY-ITEM METHOD

	Effect of cash-operating flows:	
(S − E)	Cash inflow from operations: $100,000 − $62,000	$38,000
	Income tax cash outflow at 40%	15,200
	After-tax cash-operating flows	22,800
	Effect of depreciation	
(D)	Straight-line depreciation: $90,000 ÷ 5 = $18,000	
	Income tax savings at 40% × $18,000	7,200
	Total cash flow from operations, net of income taxes	$30,000

Total cash flow from operations, net of income taxes,
can be computed as (letting t be the income tax rate):

$$[(1 - t) \times (S - E)] + (t \times D) = [(1 - 0.4) \times (\$100,000 - \$62,000)] + (0.4 \times \$18,000)$$
$$= \$22,800 + \$7,200 = \$30,000$$

or $S - (t \times S) - E + (t \times E) + (t \times D) = \$100,000 - (0.4 \times \$100,000) - \$62,000$
$$+ (0.4 \times \$62,000) + (0.4 \times \$18,000)$$
$$= \$100,000 - \$40,000 - \$62,000 + \$24,800 + \$7,200$$
$$= \$30,000$$

*All costs, other than depreciation, are assumed to be paid in cash (that is, depreciation is the only accrual cost item).

That is,

Cash flow from operations, net of income taxes

$$= \$100,000 - (0.40 \times \$100,000) - \$62,000 + (0.40 \times \$62,000) + (0.40 \times \$18,000)$$
$$= \$100,000 - \$40,000 - \$62,000 + \$24,800 + \$7,200 = \$30,000$$

The easiest way to interpret the last equation is to think of the government as a 40% (equal to the tax rate) partner in Martina Enterprises. Every time Martina earns $1, the government claims $0.40 by way of income taxes, leaving $0.60 ($1 − $0.40) as Martina's share. In other words, every time Martina earns revenue of $S, it pays 40% of that revenue (0.40S) in taxes (see the first two terms in equation (1), p. 722). Similarly, every time Martina incurs a cost (E or D), it saves 40% of that cost (0.40E or 0.40D) in taxes. The tax savings are cash savings and appear as cash inflows (see the last two terms in equation (1), p. 722). The examples in this chapter combine the cash-operating flows (those operating items that actually entail an inflow or outflow of cash) and the tax effects of those flows into a single term, *after-tax cash-operating flows*, equal to $(1 − t)(S − E)$. Depreciation is considered separately. Depreciation cost itself does not affect cash flow from operations and is not part of equation (1) because depreciation is a noncash cost. But depreciation reduces tax payments by (tD) and in this way it increases the firm's cash flow.

Objective 3
Explain why depreciation deductions are an important source of tax savings but are not themselves inputs into DCF computations

The present value of the tax savings from depreciation for a company such as Martina Enterprises, which is expected to have operating income, is calculated as follows:

Year	Income Tax Deduction for Depreciation	Income Tax Savings at 40%	12% Discount Factor	Present Value at 12%
1	$18,000	$ 7,200	0.893	$ 6,430
2	18,000	7,200	0.797	5,738
3	18,000	7,200	0.712	5,126
4	18,000	7,200	0.636	4,579
5	18,000	7,200	0.567	4,082
Total	$90,000	$36,000		$25,955

The tax benefit of depreciation deserves further clarification. In our example, purchasing and using the machine result in a net cash outflow of $90,000 over five years (original cost of $90,000 minus terminal disposal price of $0). The total depreciation cost also equals $90,000 ($18,000 per year over five years). Tax savings from depreciation deductions are 40% of $90,000 spread over five years. The tax savings occur as depreciation is recognized, not when the cash of $90,000 is paid to purchase the equipment. The present value of the tax savings of $25,955 (shown in preceding table) offsets the cost of purchasing and using the machine. The after-tax cost of using the machine is $90,000 – $25,955 = $64,045. The present value of the tax savings depends on the depreciation method used, the applicable tax rate, and the interest rate used for discounting future cash flows.

Suppose Martina Enterprises had operating losses before depreciation in year 1. Martina would have no tax liability, and so depreciation deductions would yield no tax savings in year 1. Tax rules allow depreciation deductions that cannot be taken advantage of in a particular year to be carried forward and used to reduce tax payments in subsequent years. If Martina's operating income in year 2, before depreciation, is $36,000 or larger, Martina could reduce its tax payment in year 2 using $36,000 of depreciation deductions ($18,000 from year 2 and $18,000 from year 1 that had not been utilized in year 1). In this case, of course, the present value of the year 1 depreciation deduction of $18,000 would be lower, because the tax saving from the depreciation deduction would save cash in year 2, not in year 1.

EFFECTS OF INCOME TAXES ON CASH FLOW

We turn now to a fuller discussion of how income tax can affect cash inflows and outflows and also how it influences managers' decisions. Our detailed example highlights the effect of the tax deductibility of depreciation on the net present value of a project.

EXAMPLE: Potato Supreme processes potato products for sale to supermarkets and other retail outlets. It is considering replacing an old packaging machine (purchased six years ago) with a new, more efficient packaging machine. The new machine is less labor-intensive and can process a higher volume of potato-chip packs per hour than can the old machine. For simplicity, we assume that:

a. All cash outflows or inflows occur at the start or end of the year (although cash-operating costs generally occur throughout the year)

b. The tax effects of cash inflows and outflows occur at the same time that the inflows and outflows occur

c. The income tax rate is 30% each year

d. Gains or losses on the sale of depreciable assets are taxed at the same rate as ordinary income[4]

e. Both the old and the new machine have the same working capital requirements

f. Potato Supreme is a profitable company. Tax savings from depreciation deductions occur in the year in which depreciation becomes available

The following income tax rules apply to Potato Supreme for the old machine and the new machine:

◆ *Amount allowable for depreciation.* Original cost is the basis for depreciation computations. No account is taken of terminal disposal price when computing depreciation for either the old or the new machine. When an existing asset is sold, any difference between the disposal price and the book value (original cost minus accumulated depreciation at the time of the sale) is treated as ordinary income (or loss) for tax purposes.

◆ *Time period over which the asset is to be depreciated.* A table of allowable lives specified in tax legislation determines the period for depreciation. The old machine is depreciated over 10 years. It is six years old, so its remaining useful life is four years. The new machine is depreciated over four years.

◆ *Pattern of allowable depreciation.* The old machine is depreciated using straight-line depreciation. The new machine, however, would qualify for a special tax provision permitting use of the double-declining balance (DDB) depreciation method.

Summary data for the two machines follow:

	Old Machine	New Machine
Original cost	$125,000	$200,000
Accumulated depreciation	$ 75,000	—
Current book value	$ 50,000	—
Current disposal price	$ 26,000	—
Terminal disposal price, 4 years from now	$ 6,000	$ 20,000
Annual cash-operating costs in potato-chip packing area	$250,000	$150,000
Remaining useful life	4 years	4 years
After-tax required rate of return	10%	10%

Objective 4

Distinguish between the total project approach and the incremental approach in capital-budgeting decisions

Potato Supreme wants to evaluate whether it should replace the old packaging machine with the new packaging machine. Consider two approaches:

◆ **Total project approach** calculates the present value of *all* cash inflows and outflows under each alternative. This is a two-step approach. (1) Calculate the present value of all cash inflows and outflows if Potato Supreme continues to use the old packaging machine for the next four years (the remaining useful life of the old machine). (2) Separately calculate the present value of all cash outflows and inflows if Potato Supreme replaces the old machine with the new machine for the next four years (the useful life of the new machine).

◆ **Incremental approach** analyzes only those cash outflows and inflows that differ between the alternatives of using the old machine and replacing the old machine.

Five categories of cash flows are considered in both approaches:

1. Initial machine investment

2. After-tax cash flow from current disposal of old machine

3. Recurring after-tax cash-operating flows (exclusive of depreciation effects)

Objective 5

Give examples of five categories of cash flows considered in capital-budgeting analyses

---

[4]The example assumes that if the new machine is purchased, the old machine is sold outright for cash. When the old machine is traded in for a new machine of like kind, no gain or loss is recognized for tax purposes in the year of the transaction. Rather, the new machine is capitalized at the book value of the old machine plus the cash payment. As a result, any gain or loss is spread over the life of the new machine through the new depreciation charges.

When the old machine is sold outright for cash, under U.S. tax rules, part of the gain may be taxed at ordinary income tax rates and part at capital-gains rates. The chapter Appendix describes the manner in which gains and losses on sale of assets are taxed under U.S. tax rules. For simplicity, this chapter assumes that gains on disposal are taxed at ordinary rates.

4. Income tax cash savings from depreciation deductions
5. After-tax cash-flow from terminal (at the end of the project) disposal of machine

Total Project Approach

STEP 1. CALCULATE THE PRESENT VALUE OF TOTAL CASH FLOWS OF KEEPING THE OLD PACKAGING MACHINE. Under this alternative, cash flow categories that specifically pertain to the new machine are not relevant.

1. *Initial machine investment.* This amount is zero because we are examining the case when Potato Supreme keeps the old packaging machine and makes no new investment. Exhibit 21-2, item 1, shows an initial machine investment of $0 in year 0.

2. *After-tax cash flow from current disposal of old machine.* This amount is also zero because we are considering the case when Potato Supreme keeps rather than disposes of the old packaging equipment. Exhibit 21-2, item 2, shows after-tax cash flow from current disposal of old machine of $0 in year 0.

3. *Recurring after-tax cash-operating flows (exclusive of depreciation effects).*

Recurring cash-operating flows (costs) for the old machine	$(250,000)
Deduct income tax savings at 30% of $250,000	75,000
Recurring after-tax cash-operating flows	$(175,000)

After-tax cash-operating flows of $(175,000) each year for four years appear as relevant cash outflows in Exhibit 21-2, item 3.

4. *Income tax cash savings from depreciation deductions.* The old machine is depreciated on a straight-line basis. The original cost of the machine is $125,000 and the allowable life is 10 years. Depreciation cost is $125,000 ÷ 10 = $12,500 per year. The accumulated depreciation to date is $12,500 × 6 years = $75,000; book value of the old machine is $50,000 ($125,000 − $75,000). The following table illustrates the savings in taxes resulting from depreciation deductions in future years.

Year (1)	Book Value at Start of Year (2)	Income Tax Deduction for Depreciation (3)	Income Tax Rate (4)	Income Tax Cash Savings (5) = (3) × (4)	Book Value at End of Year (6) = (2) − (3)
1	$50,000	$12,500	30%	$3,750	$37,500
2	37,500	12,500	30%	3,750	25,000
3	25,000	12,500	30%	3,750	12,500
4	12,500	12,500	30%	3,750	0

Note that depreciation on the old machine is itself irrelevant because it is a non-cash cost. Depreciation deductions, however, decrease taxable income. In turn, income tax payments decrease and so Potato Supreme's overall cash flow increases. Exhibit 21-2, item 4, shows the income-tax cash savings from depreciation deductions.

Our example assumes that Potato Supreme's income tax rate is 30% each year. When this assumption is not appropriate, analysts must predict the tax rate applicable for each year of a project.

5. *After-tax cash flow from terminal disposal of old machine.*

Terminal disposal price of old machine at end of year 4 (given, see p. 725)	$ 6,000
Deduct book value of old machine at end of year 4	0
Gain on disposal of old machine	$ 6,000
Terminal disposal price of old machine at end of year 4	$ 6,000
Deduct taxes on gain (30% of $6,000)	(1,800)
After-tax cash flow from terminal disposal of old machine	$ 4,200

EXHIBIT 21-2
Total Project Approach for Potato Supreme: After-Tax Analysis of Keeping Old Machine

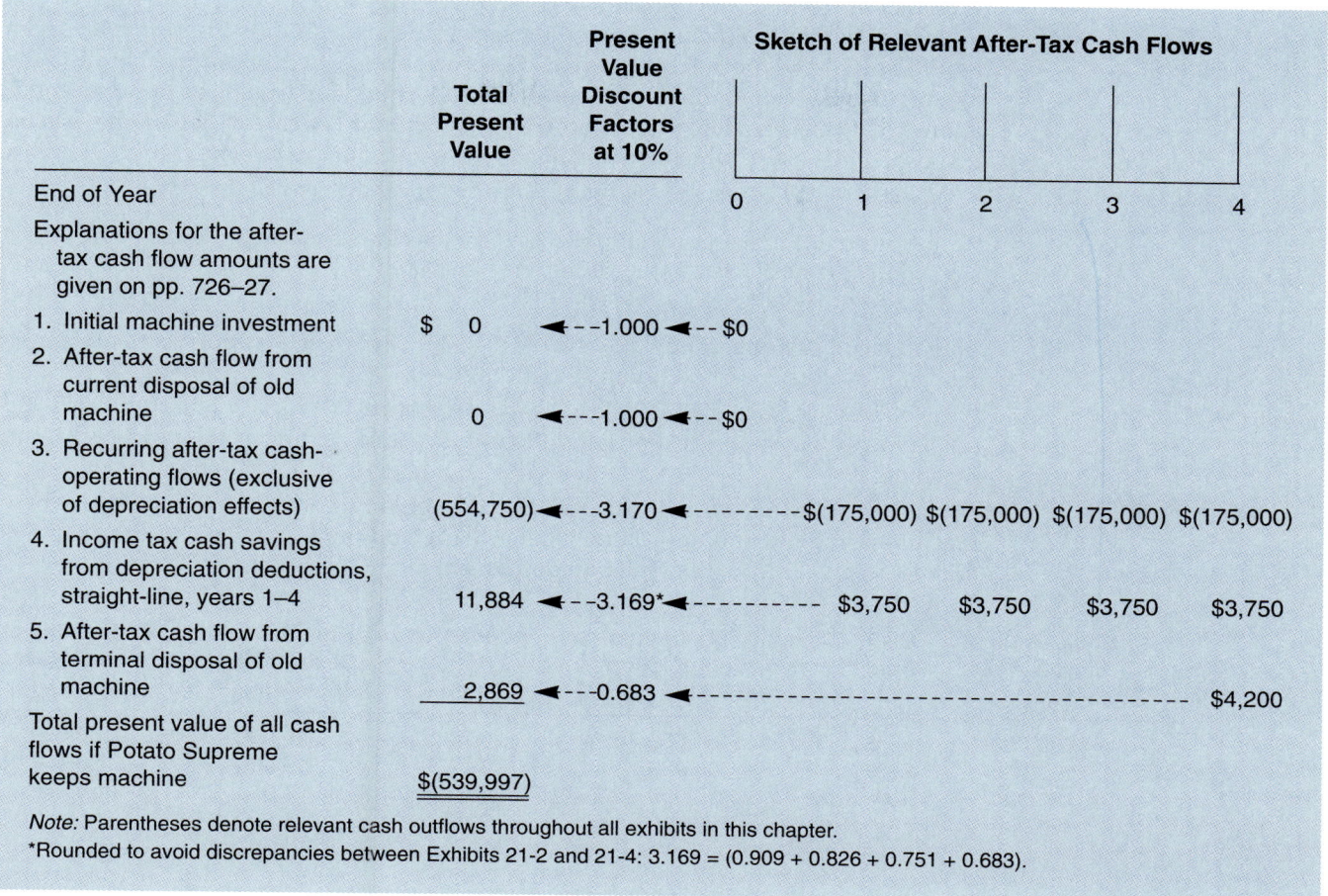

| | Total Present Value | Present Value Discount Factors at 10% | Sketch of Relevant After-Tax Cash Flows |

End of Year — 0 1 2 3 4

Explanations for the after-tax cash flow amounts are given on pp. 726–27.

1. Initial machine investment — $ 0 ◄- -1.000 ◄-- $0

2. After-tax cash flow from current disposal of old machine — 0 ◄---1.000 ◄-- $0

3. Recurring after-tax cash-operating flows (exclusive of depreciation effects) — (554,750) ◄---3.170 ◄--------$(175,000) $(175,000) $(175,000) $(175,000)

4. Income tax cash savings from depreciation deductions, straight-line, years 1–4 — 11,884 ◄--3.169* ◄---------- $3,750 $3,750 $3,750 $3,750

5. After-tax cash flow from terminal disposal of old machine — 2,869 ◄---0.683 ◄-- $4,200

Total present value of all cash flows if Potato Supreme keeps machine — $(539,997)

Note: Parentheses denote relevant cash outflows throughout all exhibits in this chapter.
*Rounded to avoid discrepancies between Exhibits 21-2 and 21-4: 3.169 = (0.909 + 0.826 + 0.751 + 0.683).

The after-tax cash flow from terminal disposal of the old machine appears as a cash inflow in year 4 of Exhibit 21-2, item 5.

Exhibit 21-2 presents all after-tax cash flows that would arise if Potato Supreme continued to use the old packaging machine. Each cash flow is multiplied by its corresponding present value discount factor to give the present value of each cash flow. The total present value is $(539,997).

STEP 2. *CALCULATE THE PRESENT VALUE OF TOTAL CASH FLOWS OF RE-PLACING THE OLD PACKAGING MACHINE.*

1. *Initial machine investment.* The original cost of the new packaging machine is $200,000. This amount appears as a cash outflow in year 0 in Exhibit 21-3, item 1.

2. *After-tax cash flow from current disposal of old machine.*

Current disposal price of old machine (given, see p. 725)	$ 26,000
Deduct current book value of old machine (given, see p. 725)	50,000
Loss on disposal of machine	$(24,000)
Current disposal price of old machine	$ 26,000
Add savings in taxes (30% of loss of $24,000)	7,200
After-tax cash flow from current disposal of old machine	$ 33,200

The after-tax cash flow of $33,200 from the disposal of the old machine appears as a cash inflow in year 0 of Exhibit 21-3, item 2. The initial machine investment, $200,000, minus the after-tax cash inflow from disposal of the old machine, $33,200, is the net initial investment of $166,800, shown as a cash outflow in Exhibit 21-3.

Review what is included in the present value analysis. It is the *cash inflow* from asset disposal and the *cash savings* in taxes. The book value of the old machine and the loss on disposal do not themselves affect cash flow and are irrelevant in capital-budgeting analysis. The book value, however, enters into the calculation of the loss on disposal of the asset, which in turn affects the income tax payments.

3. *Recurring after-tax cash-operating flows (exclusive of depreciation effects).*

Recurring cash-operating flows (costs) for the new machine	$(150,000)
Deduct income tax savings at 30% of $150,000	45,000
Recurring after-tax cash-operating flows	$(105,000)

The after-tax cash-operating flows of $(105,000) each year for four years appear as relevant cash outflows in Exhibit 21-3, item 3.

4. *Income tax cash flows from depreciation deductions.* Depreciation deductions yield tax savings that, in effect, partially offset the cost of acquiring the new packaging machine.

The following table illustrates (1) the calculation of depreciation on the new machine using the double-declining balance method[5] and (2) the income tax savings arising each year from depreciation deductions.

Year (1)	Book Value at Start of Year (2)	DDB Rate (3)	Income Tax Deduction for Depreciation (4) = (2) × (3)	Income Tax Rate (5)	Income Tax Cash Savings (6) = (4) × (5)	Book Value at End of Year (7) = (2) − (4)
1	$200,000	50%	$100,000	30%	$30,000	$100,000
2	100,000	50%	50,000	30%	15,000	50,000
3	50,000	50%	25,000	30%	7,500	25,000
4	25,000	—	25,000	30%	7,500	0

Exhibit 21-3, item 4, shows the income tax cash savings resulting from depreciation deductions.

Note that the initial investment in the asset of $200,000 is included in the capital-budgeting analysis as a *lump-sum* cash outflow in year 0 (see Exhibit 21-3, item 1). Depreciation on the new machine, as it was on the old machine, is in and of itself excluded in the capital-budgeting analysis because it is a noncash cost. Depreciation deductions, however, decrease income tax payments and so increase Potato Supreme's overall cash flow.

---

[5] The DDB depreciation pattern is calculated as follows:

a. Compute the rate by dividing 100% by the years of useful life. Then double the rate. In the Potato Supreme example, 100% ÷ 4 years = 25%. The DDB rate would be 2 × 25%, or 50%.

b. To compute the depreciation for any year, multiply the beginning book value at the start of the year (original cost minus any accumulated depreciaton) by the DDB rate (ignoring the terminal disposal price). In the Potato Supreme example, depreciation for the second year is 50% of $100,000 (book value of the packaging machine at the beginning of the second year) = $50,000. Unmodified, this method would never fully depreciate the existing book value. In the Potato Supreme example, for simplicity, we assume that the depreciation in the fourth (last) year is the book value at the start of that year.

An alternative method of calculating DDB depreciation follows: (a) Each year compare depreciation charged using DDB method with straight-line depreciation applied to the undepreciated value of the asset over the remaining useful life of the asset. (b) When straight-line depreciaton first exceeds the amount in the DDB schedule, switch to straight-line depreciation for the remaining useful life of the asset. In the Potato Supreme example, the DDB depreciation at the start of year 4 is equal to 50% of $25,000, or $12,500. A switch to straight-line would yield $25,000 (undepreciated value of asset of $25,000 divided by remaining useful life of 1 year). Therefore, depreciation in year 4 equals $25,000 since straight-line depreciation yields a higher depreciation charge.

EXHIBIT 21-3

Total Project Approach for Potato Supreme: After-Tax Analysis of Purchasing New Machine

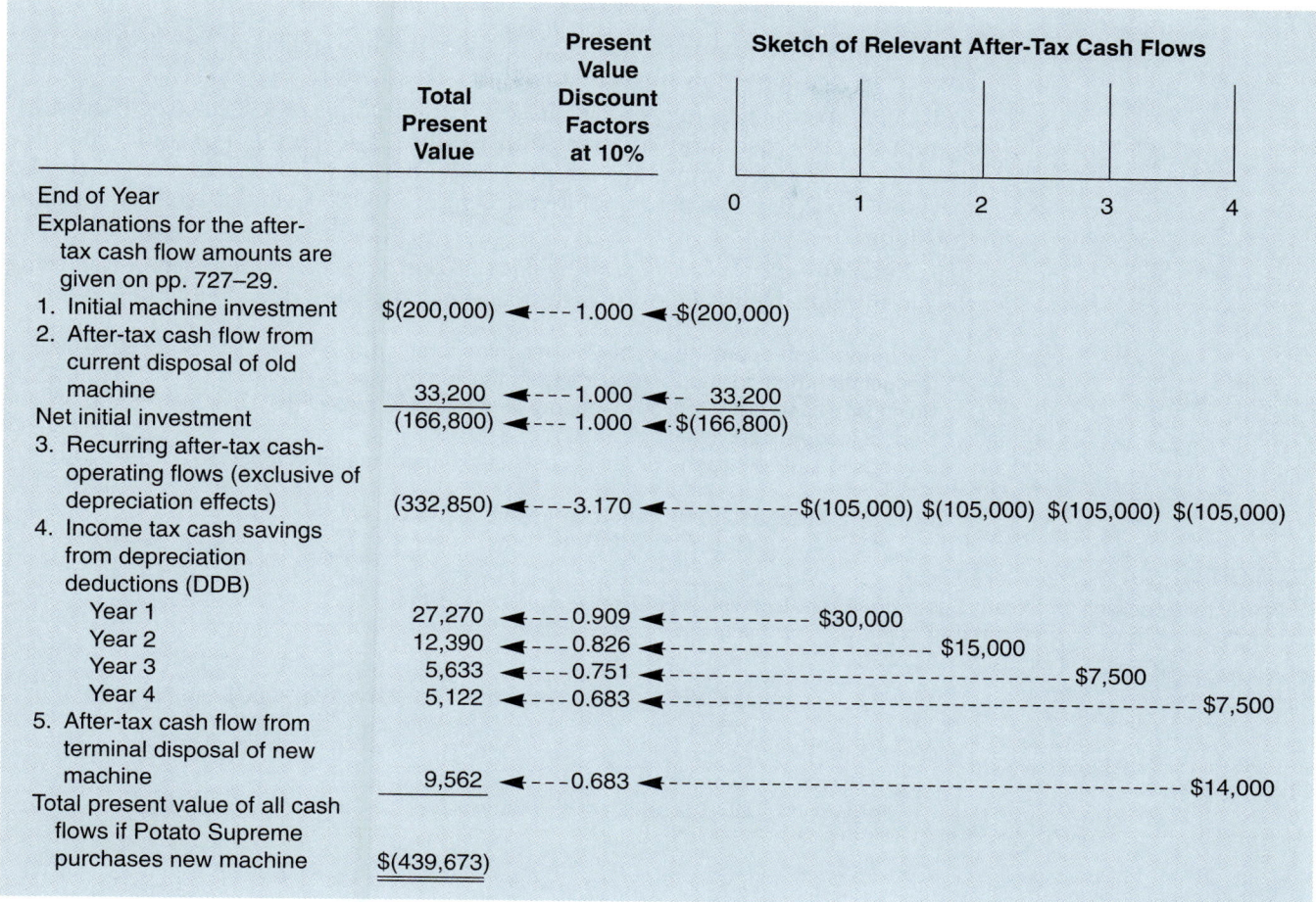

5. *After-tax cash flow from terminal disposal of new machine.*

Terminal disposal price of new machine at end of year 4 (given, see p. 725)	$20,000
Deduct book value of new machine at end of year 4	0
Gain on disposal of new machine	$20,000
Terminal disposal price of new machine at end of year 4	$20,000
Deduct taxes on gain (30% of $20,000)	6,000
After-tax cash flow from terminal disposal of new machine	$14,000

The after-tax cash flow of $14,000 from terminal disposal of the new machine appears as a cash inflow in year 4 of Exhibit 21-3, item 5.

Exhibit 21-3 summarizes the relevant after-tax cash flows that would occur if Potato Supreme replaced its old packaging machine. Present values are derived by multiplying cash flows by the corresponding present value discount factors. The total present value of cash flows equals $(439,673). Recall from Exhibit 21-2 that the present value of after-tax cash flows of continuing to use the old packaging machine is $(539,997). The alternative of using the new packaging machine has a lower present value of cash outflows and is therefore preferred. The decision to replace the old machine with the new machine has a net present value of $100,324 ($539,997 – $439,673).

Incremental Approach

Unlike the two-step total project approach, the incremental approach is a one-step method that includes only those cash inflows and outflows that *differ* between the two alternatives. The incremental approach compares (1) the cash outflows arising from replacing the old machine with (2) the *savings* in future cash outflows resulting from using the new machine rather than the old machine. We used the incremental approach when presenting the Lifetime Care example in Chapter 20.

1. *Initial machine investment* of $200,000 for the new machine (see Exhibit 21-3) appears as a cash outflow in year 0 in Exhibit 21-4, item 1.

2. *After-tax cash flow from current disposal of old machine* of $33,200 (see Exhibit 21-3) appears as a cash inflow in year 0 in Exhibit 21-4, item 2. The initial machine investment, $200,000, minus the after-tax cash flow from current disposal of the old machine, $33,200, is the net initial investment of $166,800, shown as a cash outflow in Exhibit 21-4.

3. *Recurring after-tax cash-operating flows (exclusive of depreciation effects)*. Replacing the old machine results in lower cash-operating costs, described below:

Recurring cash-operating costs if old machine retained	$250,000
Deduct recurring cash-operating costs if machine replaced	150,000
Savings in cash-operating costs each year if machine replaced	$100,000
Deduct income tax on savings at 30% of $100,000	(30,000)
Savings in after-tax cash-operating costs each year if machine replaced	$ 70,000

Exhibit 21-4, item 3, shows this increase in cash flows.

EXHIBIT 21-4
Incremental Approach for Potato Supreme: After-Tax Analysis of Replacing Old Machine

	Total Present Value	Present Value Discount Factors at 10%	Sketch of Relevant After-Tax Cash Flows				
			0	1	2	3	4
End of Year							
Explanations for the after-tax cash inflow amounts are given on pp. 730–31.							
1. Initial machine investment	$(200,000) ◀---1.000 ◀-$(200,000)						
2. After-tax cash flow from current disposal of old machine	33,200 ◀---1.000 ◀- 33,200						
Net initial investment	(166,800) ◀---1.000 ◀-$(166,800)						
3. Recurring after-tax cash-operating flows (exclusive of depreciation effects)	221,900 ◀---3.170 ◀----------			$70,000	$70,000	$70,000	$70,000
4. Income tax cash savings from depreciation deductions							
Year 1	23,861 ◀---0.909 ◀----------			$26,250			
Year 2	9,293 ◀---0.826 ◀-------------				$11,250		
Year 3	2,816 ◀---0.751 ◀-------------					$3,750	
Year 4	2,561 ◀--- 0.683 ◀-------------						$3,750
5. After-tax cash flow from terminal disposal of machines	6,693 ◀---0.683 ◀-------------						$9.800
Net present value if new machine is purchased	$100,324						

4. *Income tax savings from depreciation deductions.* Larger depreciation deductions are available on the new machine than on the old machine. The following table describes the additional tax savings resulting from the higher depreciation deductions allowed on the new machine.

Year (1)	Income Tax Deduction for Depreciation when Machine Replaced (see table, p. 728) (2)	Income Tax Deduction for Depreciation If Old Machine Kept (see table, p. 726) (3)	Increase in Income Tax Deduction If Machine Replaced (4) = (2) − (3)	Tax Rate (5)	Income Tax Savings (6) = (4) × (5)
1	$100,000	$12,500	$87,500	30%	$26,250
2	50,000	12,500	37,500	30%	11,250
3	25,000	12,500	12,500	30%	3,750
4	25,000	12,500	12,500	30%	3,750

Exhibit 21-4, item 4, shows the increase in cash flow resulting from depreciation-related incremental income tax savings.

5. *After-tax cash inflow from terminal disposal of machines.*

After-tax cash inflow from terminal (at end of year 4) disposal of new machine (see Exhibit 21-3)	$14,000
Deduct after-tax cash inflow from terminal (at end of year 4) disposal of old machine (see Exhibit 21-2)	4,200
Increase in after-tax cash inflow from terminal disposal of machine if machine is replaced	$ 9,800

Exhibit 21-4, item 5, shows this relevant, incremental after-tax cash inflow from the disposal of the machines at the end of year 4.

Both the total project approach (Exhibits 21-2 and 21-3) and the incremental approach (Exhibit 21-4) result in a net present value of $100,324 in favor of replacing the old packaging machine with the new one. Where there are only two alternatives, the incremental approach is faster. However, it rapidly becomes unwieldy when there are more than two alternatives.

U.S. Taxation Rules

We have concentrated on a general approach to analyzing income tax effects in capital budgeting, an approach that is applicable around the globe. The tax rules in the United States change almost every year. The rules in effect at the time of this writing are called the **Modified Accelerated Cost Recovery System (MACRS).** MACRS is a federal income tax regulation that classifies every depreciable asset into one of several categories. MACRS specifies (1) the number of years over which assets in each category can be depreciated and (2) the pattern of allowable depreciation (double-declining balance, 150% declining balance, or straight-line) for each category. MACRS is a modification of tax laws first introduced in 1981. Its specific provisions have been revised several times since then. The two depreciation methods illustrated in the Potato Supreme example—straight-line and double-declining balance—and in the exercises and problems for this chapter are the main alternatives available in the version of MACRS applicable at the time this text was written.[6] The Appendix to this chapter summarizes some key provisions of U.S. tax rules for depreciable assets. The Appendix can be read now without any interruption in the flow of this chapter.

---

[6]The exact application of MACRS is slightly different because of the half-year convention.

CAPITAL BUDGETING AND INFLATION

Inflation can be defined as the decline in the general purchasing power of the monetary unit. Some countries—for example, Brazil, Israel, Mexico, and Russia—have experienced annual inflation rates of 15% or more. Even an annual inflation rate of 5% over, say, a five-year period can result in sizable losses in purchasing power over that time. We now examine how inflation can be explicitly recognized in capital-budgeting analysis.

Real and Nominal Rates of Return

Objective 6

Distinguish between the nominal rate of return and the real rate of return

When analyzing inflation, distinguish between the real rate of return and the nominal rate of return:

1. **Real rate of return** is the rate of return demanded to cover investment risk (with no inflation). This rate is made up of two elements: (a) a risk-free element, the "pure" rate of return on risk-free long-term government bonds when there is no expected inflation, and (b) a business-risk element, the "risk" premium above the pure rate that is demanded for bearing risks.

2. **Nominal rate of return** is also made up of two elements: (a) the real rate of return and (b) an inflation element—the premium above the real rate that is demanded for the anticipated decline in the general purchasing power of the monetary unit. It is the rate of return demanded to cover both investment risk and any anticipated decline in general purchasing power.

Assume that the real rate of return for investments in high-risk cellular data-transmission equipment at Network Communications is 20% and that the expected inflation rate is 10%. The nominal rate of return is:

$$\text{Nominal rate} = [(1 + \text{Real rate})(1 + \text{Inflation rate})] - 1$$
$$= [(1 + 0.20)(1 + 0.10)] - 1$$
$$= [(1.20)(1.10)] - 1 = 1.32 - 1 = 0.32$$

An alternative approach to deriving the nominal rate of return from the real rate of return and the inflation rate is:

Real rate of return	0.20
Inflation rate	0.10
Combination (0.20 × 0.10)	0.02
Nominal rate of return	0.32

Note especially that the formula does not allow a mere addition of the real rate of return (0.20) to the inflation rate (0.10) to determine the nominal rate of return. For example, the nominal rate is *not* 0.20 + 0.10 = 0.30. The combination term recognizes that inflation also decreases the purchasing power of the real rate of return earned during the year. As the formula indicates, the relationship between real rates, inflation rates, and nominal rates is multiplicative, not additive.[7] Let us now look at applying the nominal rate and the real rate to the net present value method.

Net Present Value Method and Inflation

The watchwords when incorporating inflation into net present value (NPV) analysis are *internal consistency*. There are two internally consistent approaches:

---

[7] The real rate of return can be derived from the nominal rate of return as follows:

$$\text{Real rate} = \frac{(1 + \text{Nominal rate})}{(1 + \text{Inflation rate})} - 1$$
$$= \frac{(1 + 0.32)}{(1 + 0.10)} - 1 = 0.20$$

	Nominal Approach:	Predict cash inflows and outflows in nominal monetary units *and* use a nominal discount rate.
	Real Approach:	Predict cash inflows and outflows in real monetary units *and* use a real discount rate.

Many managers find the nominal approach easier to understand. This approach uses the same type of dollar figures that will be recorded in the accounting system—dollars that include the impact of inflation.

Whether a manager uses the nominal rate or the real rate, his or her choice must be applied consistently. Let's revisit Network Communications, which is deciding whether to invest in equipment to make a cellular data-transmission product the company would sell. The equipment would cost $750,000 immediately. It is expected to have a four-year useful life with a zero terminal disposal price. An annual inflation rate of 10% is expected over this four-year period.

The predicted net cash inflows from the equipment over the next four years (excluding the $750,000 investment in the equipment and before any income tax payments) are:

Year (1)	Real Dollars (2)	Cumulative Inflation Rate Factor* (3)	Nominal Dollars (4) = (2) × (3)
1	$500,000	$(1.10)^1 = 1.1000$	$550,000
2	600,000	$(1.10)^2 = 1.2100$	726,000
3	600,000	$(1.10)^3 = 1.3310$	798,600
4	300,000	$(1.10)^4 = 1.4641$	439,230

*1.10 = 1.00 + 0.10 inflation rate.

Network Communications requires an after-tax real rate of return of 20% from this project. The after-tax nominal rate of return is:

$$\text{Nominal rate} = [(1 + \text{Real rate})(1 + \text{Inflation rate})] - 1$$
$$= [(1 + 0.20)(1 + 0.10)] - 1 = 0.32$$

Objective 7

Demonstrate the equivalence of the nominal approach and the real approach to incorporating inflation into capital budgeting

The corporate income tax rate is 40%. For tax purposes, the equipment will be depreciated using the double-declining balance (DDB) method.[8]

Exhibit 21-5 presents the capital-budgeting approach for predicting cash flows in nominal dollars and using a nominal discount rate.[9] Exhibit 21-6 presents the approach of predicting cash flows in real terms and using a real discount rate.[10] (The

[8]Given a four-year useful life, the DDB factor is 0.5 (2 × 0.25). Depreciation each year of the equipment will be:

Year	Beginning Book Value	DDB Factor	Annual Depreciation
1	$750,000	0.5	$375,000
2	375,000	0.5	187,500
3	187,500	0.5	93,750
4	93,750	—	93,750

The income-tax deduction for depreciation in year 4 is the book value at the start of that year.

[9]The present value discount factors in the example are calculated using six-decimal digits to eliminate doubt about the equivalence of the two approaches. In practice, the present value discount factors (to three-decimal digits) can be obtained using Table 2 (Present value of $1) of Appendix C at the end of the text. The Problem for Self-Study at the end of this chapter uses Table 2.

[10]The inflation factors used in Exhibit 21-6 to compute the tax savings in real dollars, given an inflation rate of 10%, are:

Year	Inflation Factor Formula	Inflation Factor
1	$1 \div (1.10)^1$	0.909091
2	$1 \div (1.10)^2$	0.826446
3	$1 \div (1.10)^3$	0.751315
4	$1 \div (1.10)^4$	0.683013

EXHIBIT 21-5

Nominal Approach to Inflation for Network Communications: Predict Cash Inflows and Outflows in Nominal Dollars and Use a Nominal Discount Rate*

Sketch of Relevant After-Tax Cash Flows

	Total Present Value	Present Value Discount Factors at 32%†	0	1	2	3	4

1. Initial equipment investment

Year	Investment Outflows							
0	$(750,000)							

$(750,000) ◄--- 1.000000 ◄--- $(750,000)

2. Recurring after-tax cash-operating flows

Year	Cash-Operating Inflows	Income Tax Outflows (40%)	After-Tax Cash-Operating Inflows
1	$550,000	$220,000	$330,000
2	726,000	290,400	435,600
3	798,600	319,440	479,160
4	439,230	175,692	263,538

250,000	0.757576		$330,000
250,000	0.573921		$435,600
208,333	0.434789		$479,160
86,805	0.329385		$263,538
795,138			

3. Income tax cash savings from depreciation deductions

Year	Depreciation	Tax Savings (40%)
1	$375,000	$150,000
2	187,500	75,000
3	93,750	37,500
4	93,750	37,500

113,636	0.757576	$150,000
43,044	0.573921	$75,000
16,305	0.434789	$37,500
12,352	0.329385	$37,500
185,337		

Net present value $230,475

*The nominal discount rate of 32% is made up of the real rate of interest of 20% and the inflation rate of 10%: $[(1 + 0.20)(1 + 0.10)] - 1 = 0.32$.
†Present value factors shown to six decimal digits to emphasize that the Exhibit 21-5 and Exhibit 21-6 approaches to inflation are equivalent.

1. Initial equipment investment

Year	Investment Outflows	Total Present Value	Present Value Discount Factors at 20%*
0	$(750,000)	$(750,000) ◄─ 1.000000 ◄─ $(750,000)	

2. Recurring after-tax cash-operating flows

Year	Cash-Operating Inflows	Income Tax Outflows (40%)	After-Tax Cash-Operating Inflows	Total Present Value	Present Value Discount Factors at 20%*
1	$500,000	$200,000	$300,000	250,000 ◄ 0.833333	
2	600,000	240,000	360,000	250,000 ◄ 0.694444	
3	600,000	240,000	360,000	208,333 ◄ 0.578704	
4	300,000	120,000	180,000	86,805 ◄ 0.482253	
				795,138	

3. Income tax cash savings from depreciation deductions

Year	Depreciation	Tax Savings (40%) in Nominal Dollars	Inflation Factor 10%†	Tax Savings in Real Dollars	Total Present Value	Present Value Discount Factors at 20%*
1	$375,000	$150,000	0.909091	$136,364	113,636 ◄ 0.833333	
2	187,500	75,000	0.826446	61,983	43,044 ◄ 0.694444	
3	93,750	37,500	0.751315	28,174	16,305 ◄ 0.578704	
4	93,750	37,500	0.683013	25,613	12,352 ◄ 0.482253	
					185,337	

Net present value $230,475

Sketch of Relevant After-Tax Cash Flows

	0	1	2	3	4
	$(750,000)				
		$300,000			
			$360,000		
				$360,000	
					$180,000
		$136,364			
			$61,983		
				$28,174	
					$25,613

*Present value factors shown to six decimal digits to emphasize that the Exhibit 21-5 and Exhibit 21-6 approaches to inflation are equivalent.
†The formula on Table 2 of Appendix C is used to compute the inflation factor.

present value factors in Exhibits 21-5 and 21-6 have six-decimal digits to show the equivalence of the two approaches.) *Both approaches indicate that the project has a net present value of $230,475.* Using a net present value criterion, Network Communications should develop the product.

An often overlooked adjustment to the real approach is necessary in countries (such as the United States) where tax rules restrict the amount allowed for depreciation to the asset's original cost *in nominal dollars.* In these cases, the tax savings each year will be in nominal dollars and will have to be adjusted for inflation before they are discounted in real terms. (See Exhibit 21-6, item 3, which shows the conversion of tax savings in nominal dollars into tax savings in real dollars.)

The most frequently encountered error when accounting for inflation in capital budgeting is keeping cash inflows and outflows in real terms and using a nominal discount rate. This approach is internally inconsistent. The cash flows are adjusted for inflation by considering real rather than nominal flows. The discount rate is again adjusted for inflation by using nominal rather than real rates. Inflation has been adjusted for twice! This error understates the discounted present value of cash flows that occur in the future. The error would create a bias against the adoption of many worthwhile capital investment projects.

REQUIRED RATE OF RETURN

The *required rate of return (RRR)* is a critical variable in discounted cash-flow analysis. It is the rate of return that the firm forgoes by investing in a particular project rather than investing in an alternative of comparable risk. *Risk* here refers to the business risk of the project, *independent* of the specific manner in which the project is financed—whether with debt or with equity. Here is a safe generalization: The higher the risk, the higher the required rate of return. Alternative names for RRR include the *discount rate,* the *hurdle rate,* and the *(opportunity) cost of capital.*

The RRR used in DCF analysis should be internally consistent with the approach applied to predict cash inflows and outflows. The options include various combinations of (1) the real rate and the nominal rate and (2) the pretax rate and the after-tax rate. The numerical magnitude of differences among these rates can be sizable, given estimates of inflation that may exceed 10% and corporate tax rates of 30% or more.

Organizations typically use at least one of the following approaches in dealing with the risk factor of projects:

Objective 8
Describe alternative approaches used to recognize the degree of risk in capital-budgeting projects

1. *Varying the Payback Time.* Companies that use payback as a project-selection criterion vary the required payback to reflect differences in project risk. The higher the risk, the shorter the required payback time. Alternatively, they compute the bailout payback period to evaluate their downside protection if the project is disbanded.[11]

2. *Adjusting the Required Rate of Return.* The higher the risk, the higher the required rate of return. Estimating a precise risk factor for each project is difficult. Some organizations simplify the task by having three or four general-risk categories (for example, very high, high, average, and low). Each potential project is assigned to a specific category. Management uses a predetermined discount rate, assigned to each category, as the required rate of return for projects in that category.

3. *Adjusting the Estimated Future Cash Inflows.* This approach systematically reduces the estimated future cash inflows of riskier projects. One company has a policy of systematically reducing the predicted cash inflows of very-high-risk projects by 30%, high-risk projects by 20%, and average-risk projects by 10%. It makes no change to the projected cash inflows of low-risk projects. This approach is called

[11]See J. Grinyer and N. Daing, "The Use of Abandonment Values in Capital Budgeting—A Research Note," *Management Accounting Research,* 4 (1993).

the *certainty equivalent approach.* It discounts these risk-adjusted cash flows by the RRR for low-risk projects. Using risk-adjusted RRRs for discounting, as in approach 2 above, would double-count risk.

4. *Sensitivity ("What-If?") Analysis.* This approach involves examining the consequences of changing key assumptions underlying a capital-budgeting project. For example, a copper-mining company might examine what changes would occur if the world price of copper were to increase (decrease) by 10%, 20%, and 30%. How would these changes affect the economic attractiveness of an investment in a new copper mine?

5. *Estimating the Probability Distribution of Future Cash Inflows and Outflows for Each Project.* The Appendix to Chapter 3 discussed this approach to uncertainty. This approach gives due weight to all possible outcomes, favorable and unfavorable, to arrive at an *expected* cash flow, and then discounts the expected cash flow at the risk-adjusted required rate of return for the investment. Estimating these distributions is difficult. But a practical guideline is not to include so comprehensive a set of outcomes that it makes the analysis overwhelming. The manager should always focus on the most important issues. There are other benefits of estimating the proba-

Surveys of Company Practice

RISK ADJUSTMENT METHODS IN CAPITAL BUDGETING

How do companies around the globe adjust for risk when evaluating capital investments? The percentages in the table below indicate how frequently particular risk-adjustment methods are used in capital budgeting in four countries. The reported percentages exceed 100% because some companies use more than one risk-adjustment method.

	United States[a]	Australia[b]	Canada[c]	United Kingdom[d]
Sensitivity analysis	29%	57%	59%	63%
Increase the required rate of return	18%	—	31%	42%
Shorten payback period	17%	—	24%	34%
Estimate probability distribution of future cash flows	12%	11%	18%	15%
Compare optimistic and pessimistic forecasts	—	63%	—	—
Make subjective, nonquantitative assessment	54%	37%	29%	22%
Make no adjustments	37%	—	—	10%

Dashes indicate information was not disclosed in survey.

The surveys indicate that the specific methods managers use vary among countries. A common feature, however, is that managers appear to favor simpler methods (such as sensitivity analysis, analysis of optimistic and pessimistic forecasts, and subjective, nonquantitative assessments) rather than more sophisticated techniques (such as estimating the probability distribution of future cash flows).

———
[a]Adapted from Sullivan and Smith, "Capital Investment Justification." [b]Freeman and Hobbes, "Capital Budgeting." [c]Jog and Srivastava, "Capital Formation." [d]Ho and Pike, "Risk Analysis." Full citations are in Appendix A.

bility distribution of future cash inflows and outflows. For example, suppose a project has a 60% likelihood of very high cash inflows in its early years and a 40% likelihood of minimal cash inflows in its early years. This 40% probability may prompt managers to establish lines of credit with a bank. If the low outcome occurs, these lines of credit would enable the company to avoid a short-run cash-flow crisis.

APPLICABILITY TO NONPROFIT ORGANIZATIONS

Discounted cash-flow analysis applies to both profit-seeking and nonprofit organizations. Almost all organizations must decide which fixed assets will accomplish various tasks at the least cost. U.S. federal agencies use a 7% required rate of return in capital budgeting for water projects (dams, irrigation, and so on) and 10% for all other projects. These lower rates exist, in part, because nonprofit organizations such as governments, universities, and certain hospitals are not subject to income taxes.

Studies of the capital-budgeting practices of government agencies at various levels (federal, state, and local) and in several countries report these findings:

1. Urgency is an important factor when allocating funds. For example, capital budgeting for roads often starts with a list of the physical deficiencies in an existing highway rather than with a systematic analysis of whether it would be preferable to build an alternative highway.
2. Project estimates are often systematically biased. For example, studies of irrigation projects by the U.S. Bureau of Reclamation report overestimates of the benefits, underestimates of the costs, and underestimates of the time it takes to construct dams and other irrigation infrastructure.
3. A tendency to cut capital budget projects first when there is a strong push to balance a budget or reduce a deficit.

The national attention on containing health-care costs in the United States has forced hospitals to evaluate more carefully their investments in new equipment. Consider, for example, the effect of changes in Medicare, a government-run program that finances hospital care for the elderly. Medicare had reimbursed hospitals that treated patients covered under the program for 100% of certain capital equipment expenditures. Recent policy changes, however, have limited reimbursement to 85%.[12] As a result of these changes and the increased emphasis on controlling hospital prices through competition and regulation, hospitals are increasingly using analytical capital-budgeting methods (such as discounted cash-flow methods) and more carefully auditing the effect of capital investments.

IMPLEMENTING THE NET PRESENT VALUE DECISION RULE

This section discusses problems in using the net present value method when there is a restriction on the total funds available for capital spending. Executives must frequently work within an overall capital-budgeting limit. In nonprofit enterprises, restrictions on the total funds available for capital spending are the norm. For example, an annual government budget typically will provide an upper limit on the funds available to each of the individual government departments.

The **excess present value index** (sometimes called the *profitability index*) is the total present value of future net cash inflows of a project divided by the total present value of the net initial investment. Consider this index for two software graphics packages—Superdraw and Masterdraw—that Business Systems is evaluating. The writers of each package require that Business Systems market only one software graphics package, so accepting one software package automatically means rejecting the other—that is, the packages are mutually exclusive. Summary financial data from the capital budget proposal of each package are:

---

[12]S. Finkler, "Analytical Capital Budgeting," *Hospital Cost Management and Accounting* (March 1992).

Project (1)	Present Value at 10% RRR (2)	Net Initial Investment (3)	Excess Present Value Index (4) = (2) ÷ (3)	Net Present Value (5) = (2) − (3)
Superdraw	$1,400,000	$1,000,000	140%	$400,000
Masterdraw	3,900,000	3,000,000	130%	900,000

Superdraw has an excess present value index of 140% compared with 130% for Masterdraw. (Projects with excess present value indexes of less than 100% have negative net present values.) If all other things, such as risk and alternative use of funds, are equal, Superdraw is the preferred project. But "all other things" are rarely equal.

Assume that Business Systems has a total capital budget limit of $5,000,000 for the coming year. It is considering investing in Superdraw or Masterdraw and in any one or more of eight other projects (coded B, C, . . . , H, I). Exhibit 21-7 presents two alternative combinations of these projects. Note that the project portfolio in alternative 2 is superior to alternative 1, despite the greater profitability per dollar obtained by investing in Superdraw compared with Masterdraw. Why? Because the $2,000,000 incremental investment in Masterdraw has an incremental net present value (NPV) of $500,000. The $2,000,000 would otherwise be invested in projects E and B, which have a lower combined incremental NPV of $256,000:

	Present Value	Net Initial Investment	Increase In Net Present Value
Masterdraw	$3,900,000	$3,000,000	
Superdraw	1,400,000	1,000,000	
Increment	$2,500,000	$2,000,000	$500,000
Project E	$ 912,000	$ 800,000	
Project B	1,344,000	1,200,000	
Total	$2,256,000	$2,000,000	$256,000

EXHIBIT 21-7

Allocation of $5,000,000 Capital Budget: Comparison of Two Alternatives for Business Systems

	Alternative 1				Alternative 2			
Project	Net Initial Investment	Excess Present Value Index	Total Present Value at 10%	Project	Net Initial Investment	Excess Present Value Index	Total Present Value at 10%
C	$ 600,000	167%	$1,002,000	C	$ 600,000	167%	$1,002,000
Superdraw	1,000,000	140%	1,400,000				
D	400,000	132%	528,000	D	400,000	132%	528,000
				Masterdraw	3,000,000	130%	3,900,000
F	1,000,000	115%	1,150,000	F	1,000,000	115%	1,150,000
					$5,000,000*		$6,580,000‡
E	800,000	114%	912,000	E	$ 800,000	114%	Reject
B	1,200,000	112%	1,344,000	B	1,200,000	112%	Reject
	$5,000,000*		$6,336,000†				
H	$ 550,000	105%	Reject	H	550,000	105%	Reject
G	450,000	101%	Reject	G	450,000	101%	Reject
I	1,000,000	90%	Reject	I	1,000,000	90%	Reject

*Total budget constraint.
†Net present value, $1,336,000.
‡Net present value, $1,580,000.

This example illustrates that managers cannot base decisions involving mutually exclusive investments of different sizes on the excess present value index. The net present value method is the best general guide.

IMPLEMENTING THE INTERNAL RATE-OF-RETURN DECISION RULE

Objective 9

Explain why the internal rate-of-return and the net present value decision rules may rank projects differently

The NPV method always indicates the project (or set of projects) that maximizes the NPV of future cash flows. However, surveys of practice report widespread use of the internal rate-of-return (IRR) method. Why? Probably because managers find that method easier to understand and because, in most instances, their decisions would be unaffected by using one method or the other. In some cases, however, the two methods will not indicate the same decision.

Where mutually exclusive projects have unequal lives or unequal investments, the IRR method can rank projects differently from the NPV method. Consider Exhibit 21-8.[13] The ranking by the IRR method favors project X, while the NPV method favors project Z. The projects ranked in Exhibit 21-8 differ in both life (5, 10, and 15 years) and net initial investment ($286,400, $419,200, and $509,200).

Managers using the IRR method implicitly assume that the reinvestment rate is equal to the indicated rate of return for the shortest-lived project. Managers using the NPV method implicitly assume that the funds obtainable from competing projects can be reinvested only at the company's required rate of return.

Corporate finance texts cover, in great detail, the problems of ranking projects with unequal lives or unequal investments. Ideally, there should be a common terminal date for all projects with explicit assumptions as to the appropriate reinvestment rates of funds. The practical difficulties of predicting future profitability on reinvestment are greater than those of predicting profitability of immediate projects. But reinvestment opportunities should be considered when they can be foreseen and measured.

---

[13]Exhibit 21-8 concentrates on differences in project lives. Similar conflicting results can occur when the terminal dates are the same but the sizes of the net initial investments differ.

EXHIBIT 21-8

Ranking of Projects Using Internal Rate of Return and Net Present Value

| Project | Life | Net Initial Investment | Annual Net After-Tax Cash Flows | Ranking by Internal Rate of Return | | Ranking by Net Present Value | | |
				IRR	Ranking	PV of Annual Net After-Tax Cash Flows at 10% RRR	NPV	Ranking
X	5	$286,400	$100,000	22%	1	$379,100	$ 92,700	3
Y	10	419,200	100,000	20	2	614,500	195,300	2
Z	15	509,200	100,000	18	3	760,600	251,400	1

PROBLEM FOR SELF-STUDY

This is a comprehensive review problem. It illustrates both income tax factors and capital budgeting with inflation.

PROBLEM

Stone Aggregates operates 92 plants across the country. Each plant produces crushed stone used in many construction projects. Transportation is a major cost item. A scale clerk prepares a delivery ticket using a customer master file. The clerk weighs the products and records details of the product shipped: its weight, its freight charges, and whether or not it is taxed.

Stone Aggregates is considering a proposal to use a computerized ticket-writing system for all of its 92 plants. One plant has been a pilot site for the past 12 months, generating cash-operating cost savings (before taxes) of $300,000. These savings arose mainly from a reduction in operating costs at the plant and from a reduction in overshipments (amounts shipped in excess of amounts ordered) to customers. The cost analyst estimates that if the computerized ticket system had been operating at all of the company's plants for the past year, cost savings would have been $25 million (expressed in today's dollars). This cost-savings estimate takes into account the estimated $5 million cash-operating costs that would have been incurred in operating the computerized ticket-writing system at all of the company's plants in the past year.

The cost of the equipment for all 92 plants is $45 million, payable immediately. This equipment has an expected useful life of four years and a terminal disposal price of $10 million (expressed in today's dollars). Income tax rules applying to Stone Aggregates are:

- *Amount allowable for depreciation.* Original cost of the equipment is the basis for depreciation computations. No account is taken of predicted terminal disposal price when computing depreciation. The gain on disposal (in nominal dollars) will be taxed at the ordinary income tax rate in the year the disposal is made.
- *Time period over which the asset is to be depreciated.* Under a tax law designed to encourage investment, Stone Aggregates can use a three-year write-off period for the asset.
- *Pattern of allowable depreciation.* Straight-line depreciation is required. Given an original cost of $45 million and a three-year write-off period, annual depreciation is $15 million.

Stone Aggregates expects a 30% income tax rate in each of the next four years.

Required

1. Does the automated delivery ticket-writing proposal meet Stone Aggregates' 16% after-tax required rate-of-return criterion? This 16% required rate of return includes an 8% inflation component. (The real rate of interest is 7.4%; recall that nominal rate of interest = $[(1 + 0.074)(1 + 0.08)] - 1 = 0.16$.) This 8% inflation prediction applies to both the cost savings and the terminal disposal price of the equipment. Prepare a net present value analysis using nominal dollars and a nominal required rate of return.

2. What other factors would you recommend that Stone Aggregates consider in more detail when evaluating the automated delivery ticket-writing proposal?

SOLUTION

1. Exhibit 21-9 shows the net present value analysis. To illustrate an alternative presentation found in practice, the format of Exhibit 21-9 differs from that of Exhibits 21-2, 21-3, and 21-4. The proposal for an automated delivery ticket-writing system has a net present value of $29.086 million, indicating that—on the basis of financial factors—it is an attractive investment. Note especially how the tax law enables Stone Aggregates to fully depreciate the equipment by the end of the third year. No depreciation occurs in year 4.

2. The analysis in Exhibit 21-9 assumes that the cost of operating the system is $5 million each year. However, costs in the year of changeover to the computerized system would probably be much higher. Many companies find that actual implementation and operat-

EXHIBIT 21-9

Net Present Value Analysis of Computerized Ticket-Writing System for Stone Aggregates
(in millions of dollars; n.d. = nominal dollars)

	Total Present Value	End of Year 1	End of Year 2	End of Year 3	End of Year 4
Operating Savings					
1. Cash-operating savings (real dollars)	—	$25.000	$25.000	$25.000	$25.000
2. Cumulative inflation factor (from Table 1, App. C for 8%)	—	1.080	1.166	1.260	1.360
3. Cash-operating savings (n.d.): 1 × 2	—	$27.000	$29.150	$31.500	$34.000
4. Tax Payment: 30% × 3	—	$ 8.100	$ 8.745	$ 9.450	$10.200
5. After-tax cash-operating savings (n.d.): 3 − 4	—	$18.900	$20.405	$22.050	$23.800
6. Present value factor (16%)	—	0.862	0.743	0.641	0.552
7. PV of after-tax cash-operating savings (n.d.): 5 × 6	$58.725	$16.292	$15.161	$14.134	$13.138
Depreciation and Disposal Proceeds					
8. Depreciation deductions	—	$15.000	$15.000	$15.000	—
9. Income tax cash savings (30%)	—	$ 4.500	$ 4.500	$ 4.500	—
10. After-tax cash flow from terminal disposal of equipment*	—	—	—	—	9.520
11. Total income tax cash savings and disposal proceeds: 9 + 10	—	$ 4.500	$ 4.500	$ 4.500	$ 9.520
12. Present value factor (16%)	—	0.862	0.743	0.641	0.552
13. PV of income tax savings + disposal proceeds: 11 × 12	15.361	$ 3.879	$ 3.343	$ 2.884	$ 5.255
PV of total cash inflows: 7 + 13	74.086	$20.171	$18.504	$17.018	$18.393
PV of initial equipment investment	(45.000)				
Net Present Value	$29.086				

*Terminal disposal price of $10 million in year 0 dollars will be $13.600 million ($10 million × 1.360) in year 4 nominal dollars. At a 30% tax rate, the after-tax cash flows from terminal disposal of equipment in nominal dollars are $9.520 million ($13.600 million × [1 − 30%]).

Sketch of Relevant After-Tax Cash Flows

	0	1	2	3	4
		$18.900	$20.405	$22.050	$23.800
		$ 4.500	$ 4.500	$ 4.500	
					$9.520
	$(45.000)				

ing costs in the changeover year exceed the operating costs in subsequent years by a factor of 200% or more.

We wrote this problem after reading an article describing Vulcan Materials' adoption of a computerized ticket-writing system.[14] The benefits Vulcan reported included the following:

◆ "Hauler productivity increased because the new system improved the movement of product across the scales."

◆ "Scale clerks benefited through job enrichment and a much more predictable work load."

◆ "Communication between the scale-house microcomputers and the division office computers has afforded Vulcan several benefits. The first was the opportunity to eliminate costs associated with transporting the punch cards required under the old system. The second was the elimination of transferring transaction data from punch cards to computer tape. Vulcan also has benefited markedly from the reduction of errors contained in the transaction data and reduced costs associated with making corrections. In addition to improving accuracy, the system has accelerated the issuance of Vulcan's invoices."

◆ "Overshipments are eliminated because the up-to-date status of a customer's sales order is available at the plant."

◆ "Cash on delivery customers are identified by the system to prevent unintentional credit sales."

◆ "Faster flow of data has narrowed the time lag in detecting and correcting problems."

Stone Aggregates should learn from the experience of Vulcan Materials and include analysis of the preceding benefits that will possibly occur with the purchase of a computerized ticket-writing system.

SUMMARY

The following points are linked to the chapter's learning objectives.

1. Income tax factors almost always play an important role in decision making. Income taxes have a major impact on the amounts of cash inflow and outflow and on their timing.

2. Three factors influence the amount of depreciation claimed as a tax deduction: (a) the amount allowable for depreciation, (b) the time period over which the asset is to be depreciated, and (c) the pattern of allowable depreciation.

3. Depreciation is a noncash cost and therefore does not itself affect operating cash flows. But depreciation is a deductible cost for tax purposes. The taxes saved as a result of depreciation deductions increase cash flows in discounted cash-flow (DCF) computations.

4. The total project approach calculates the present value of all cash inflows and outflows under each alternative. The incremental approach includes only those cash inflows and outflows that differ among the alternatives.

5. Five categories of cash flows considered in capital-budgeting analysis involving a machine are (a) initial machine investment, (b) after-tax cash flow from current disposal of old machine, (c) recurring after-tax cash-operating flows, (d) income tax cash savings from depreciation deductions, and (e) after-tax cash flow from terminal disposal of machine.

6. The real rate of return is the rate of return demanded to cover investment risk absent inflation. The nominal rate of return is the rate of return demanded to cover both investment risk and the anticipated decline in the general purchasing power of money due to inflation.

[14]J. Bush and R. Stewart, "Vulcan Materials Automates Delivery Ticket Writing," *Management Accounting* (August 1985).

7. Two internally consistent ways to account for inflation in capital budgeting are (a) to predict cash inflows and outflows in nominal terms and to use a nominal discount rate and (b) to predict cash inflows and outflows in real terms and to use a real discount rate. The nominal and real approaches are equivalent: Both yield the same net present value.

8. The higher the risk, the higher the required rate of return on an investment. Alternative approaches to recognizing project risk in capital-budgeting decisions are (a) reducing the required payback time, (b) increasing the required rate of return, (c) reducing estimated future cash inflows, (d) performing sensitivity analysis, and (e) estimating the probability distribution of future cash inflows and outflows.

9. The internal rate of return and net present value methods make different assumptions about the rate at which project cash inflows are reinvested. Consequently, the two methods may rank projects differently.

APPENDIX: MODIFIED ACCELERATED COST RECOVERY SYSTEM

The tax rules governing depreciation in the United States are collectively called the Modified Accelerated Cost Recovery System (MACRS). Changes in this set of rules are made periodically, as they are in most tax regulations. MACRS is a modification of the tax laws first introduced in 1981, termed the Accelerated Cost Recovery System (ACRS). For most depreciable assets placed in service in the 1981–1986 period, the ACRS system applies. Assets acquired since 1987 are subject to MACRS. Both ACRS and MACRS have more accelerated depreciation schedules than existed with the prior tax rules. Some highlights of the current version of MACRS follow.

AMOUNT ALLOWABLE FOR DEPRECIATION. In general, the amount allowable for depreciation is the original cost of the asset. MACRS uses the phrase "cost recovery" to describe the amount allowable each year as a "depreciation" deduction. Estimates of future disposal prices are ignored under MACRS.

For proceeds on disposal of an asset up to the original cost of the asset, the difference between the proceeds and the asset's tax book value is taxed at the same rate as ordinary income or loss (35% at the time of this writing). Proceeds greater than the original asset cost are taxed at special capital gains tax rates (28% at the time of this writing).

TIME PERIOD OVER WHICH THE ASSET IS TO BE DEPRECIATED. The time period is specified in a "table of allowable lives" (termed *recovery periods*). Eight different recovery periods are possible: 3, 5, 7, 10, 15, 20, 27.5, and 31.5 years. These recovery periods do not necessarily reflect the estimated useful life of the assets included in each category.

PATTERN OF ALLOWABLE DEPRECIATION (FOR A GIVEN TIME PERIOD). The depreciation method specified is a function of the recovery period. The eight different recovery periods and their depreciation methods are:

Recovery Period	Depreciation Method
3, 5, 7, 10 years	Double-declining balance (also called 200% declining balance)
15, 20 years	150% declining balance
27.5, 31.5 years	Straight-line

Many other tax considerations are not discussed in this chapter. Some examples are investment tax credits, loss carrybacks and carryforwards, state income taxes, short-term and long-term capital gains, and distinctions between how capital assets and other assets affect income taxes.

TERMS TO LEARN

This chapter and the Glossary at the end of the book contain definitions of the following important terms:

accelerated depreciation *(p. 721)*   double-declining balance (DDB) depreciation *(721)*
excess present value index *(738)*   incremental approach *(725)*   inflation *(732)*
investment tax credit (ITC) *(721)*
Modified Accelerated Cost Recovery System (MACRS) *(731)*
nominal rate of return *(732)*   real rate of return *(732)*
straight-line depreciation *(721)*   total project approach *(725)*

ASSIGNMENT MATERIAL

QUESTIONS

21-1 What are the two major impacts that income taxes have on capital-budgeting decisions?

21-2 Describe three factors that influence the amount claimed as a depreciation deduction for tax purposes.

21-3 "It doesn't matter what depreciation method is used. The total dollar tax bills are the same." Do you agree? Explain.

21-4 "Accelerated depreciation provides higher cash flows in the early years than does straight-line depreciation." Do you agree? Explain.

21-5 Distinguish between the *total* project approach and the *incremental* approach to choosing between two capital-budgeting projects.

21-6 "Depreciation is an irrelevant factor in deciding whether to replace an existing delivery vehicle with a more energy-efficient vehicle." Do you agree? Explain.

21-7 Give examples of three categories of cash flows considered in capital-budgeting analyses.

21-8 What are the main depreciation methods permitted under the Modified Accelerated Cost Recovery System (MACRS)?

21-9 Distinguish between the *nominal* rate of return and the *real* rate of return.

21-10 What are the two internally consistent approaches to incorporating inflation into DCF analysis?

21-11 What approaches might be used to recognize risk in capital budgeting?

21-12 "Discounted cash flow techniques are relevant only to profit-seeking organizations." Do you agree? Explain.

21-13 "In practice there is no single rate that a given company can use as a guide for sifting among all projects." Do you agree? Explain.

21-14 How do the reinvestment assumptions implicit in the use of the internal rate of return method and the net present value method differ when comparing projects with lives of different lengths?

21-15 "The net present value method and the internal rate of return method always rank different projects identically." Do you agree? Explain.

EXERCISES AND PROBLEMS

21-16 Recapitulation of role of depreciation in Chapters 11, 20, and 21. Antonio Inoki, president of Yokohoma Steel, remarked, "I've read three chapters that have included discussions of depreciation in relation to decisions regarding the replacement of equipment. I'm confused. Chapter 11 said that depreciation on old equipment is irrelevant but that depreciation on new equipment is relevant. Chapter 20 said that depreciation was irrelevant in relation to discounted cash-flow models, but Chapter 21 indicated that depreciation was indeed relevant."

Required
Prepare a clear explanation for the president that would minimize his confusion.

21-17 New equipment purchase, taxation, straight-line and DDB depreciation. Presentation Graphics prepares slides and other aids for individuals making group presentations. It estimates it can save $35,000 a year in cash-operating costs for the next five years if it buys a special-purpose color-slide workstation at a cost of $75,000. The workstation will have a zero terminal disposal price at the end of year 5. Presentation Graphics has a 12% after-tax required rate of return. Its income tax rate is 40% each year for the next five years.

Required
1. Assume that Presentation Graphics uses straight-line depreciation on its tax return. Compute (a) net present value, (b) payback period, and (c) internal rate of return.
2. Assume that Presentation Graphics uses the double-declining balance method on its tax return with depreciation for the fifth year being the book value at the start of the fifth year. Compute (a) net present value, (b) payback period, and (c) internal rate of return.

21-18 Multiple choice, including straight-line depreciation. (CPA, adapted) The Apex Company is evaluating a capital-budgeting proposal for the current year. The relevant data follow:

Year	Present Value of an Annuity of $1 in Arrears at 15%
1	$0.870
2	1.626
3	2.284
4	2.856
5	3.353
6	3.785

The initial equipment investment would be $30,000. Apex would depreciate the equipment for tax purposes on a straight-line basis over six years with a zero terminal disposal price. The before-tax annual cash inflow arising from this investment is $10,000. The income tax rate is 40%, and income tax is paid the same year as incurred. The after-tax required rate of return is 15%. Choose the best answer for each question and show computations.

1. What is the after-tax accrual accounting rate of return on Apex's initial equipment investment?
   (a) 10%, (b) 16⅔%, (c) 26⅔%, (d) 33⅓%.
2. What is the after-tax payback period (in years) for Apex's capital-budgeting proposal?
   (a) 5, (b) 3. 75, (c) 3, (d) 2.
3. What is the net present value of Apex's capital-budgeting proposal?
   (a) $(7,290), (b) $280, (c) $7,850, (d) $11,760.
4. How much would Apex have had to invest five years ago at 15% compounded annually to have $30,000 now?
   (a) $12,960, (b) $14,910, (c) $17,160, (d) cannot be determined from the information given.

 21-19 Capital budgeting, income taxes, DDB depreciation, ATM network. Rural Bank is considering whether to invest in an automatic teller machine (ATM) network. Its main competitor (Security Bank) has started installing ATMs at several of its branches. Rick Sterling, a retail banking financial analyst at Rural Bank, estimates the following:

◆ Initial equipment and software investment = $30 million. The total $30 million amount would be invested immediately and would be depreciated over a five-year period using the double-declining balance (DDB) method, with depreciation for the fifth year being the book value of the equipment at the start of the fifth year.

◆ Annual cash costs of operating the ATMs (for example, equipment and software maintenance) = $3 million.

◆ Annual reduction in labor and related cash costs from using the ATMs as compared with the current labor-intensive teller system = $10 million.

Sterling estimates that the useful life of the ATMs is six years. The terminal disposal price of the ATMs at the end of six years is $3 million. For tax purposes, the terminal disposal price is included as an income item in the year of disposal.

Rural Bank has a 14% after-tax required rate of return for retail-banking investments. Rural Bank has a 30% income tax rate, which is expected to remain constant over the foreseeable investment horizon.

Required
1. Sketch the after-tax cash inflows and outflows from Rural Bank's investing in the ATMs. Show all your calculations.
2. Compute the payback period and the net present value of the ATM investment.
3. What questions should Rural Bank consider in deciding whether to invest in an ATM network?

21-20 Comprehensive equipment-replacement decision, income taxes, straight-line depreciation. A manufacturer of automobile parts acquired a special-purpose shaping machine for automatically producing a particular part. The machine has been used for one year. It will have no useful economic life after three more years. The machine is being depreciated on a straight-line basis for income tax purposes. It cost $88,000, has a current disposal price of $29,000, and has a terminal disposal price of $6,000. However, a terminal disposal price of zero was assumed in computing straight-line depreciation for tax purposes.

A new machine has become available and is far more efficient than the present machine. It would cost $63,000, would cut annual cash-operating costs from $60,000 to $40,000, and would have zero terminal disposal price at the end of its useful life of three years. Straight-line depreciation would be used for tax purposes. The applicable income tax is 30%. The after-tax required rate of return is 14%.

Required
Using the net present value method, show whether the new machine should be purchased (a) under a total-project approach and (b) under an incremental approach.

21-21 Automated materials-handling capital project, income taxes, double-declining balance depreciation. Ontime Distributors operates a large distribution network for health-related products. It is considering an automated materials-handling (AMH) proposal for its major warehouse. The before-tax cash-operating savings from the automation are estimated to be $2.5 million a year. These savings arise from reduced storage space requirements, increased labor productivity, less product damage, and higher inventory record-keeping accuracy. This $2.5 million annual savings is calculated as $3.5 million gross cash-operating savings minus the $1.0 million cash costs of operating and maintaining the AMH equipment. The AMH equipment will cost $6 million, payable immediately. The equipment has a useful life of four years and zero terminal disposal price. The lease on the warehouse expires in four years and is not expected to be renewed. The company has an income tax rate of 40% and after-tax required rate of return of 12%. Under existing tax laws, the $6 million equipment cost qualifies for use of the double-declining balance depreciation method with a four-year useful life, with depreciation for the fourth year being the book value of the equipment at the start of the fourth year. The terminal disposal price of the equipment is included as a taxable income item in the year of its disposal.

Required
1. Compute (a) the net present value and (b) the payback period on the automated materials-handling project.
2. Assume that the AMH equipment has a $1 million terminal disposal price at the end of the fourth year instead of a zero terminal disposal price. How will this $1 million terminal disposal price affect the net present value?
3. What other factors should Ontime Distributors consider in its decision?

21-22 Equipment replacement, income taxes. (CMA, adapted) VacuTech manufactures testing instruments for microcircuits. These instruments sell for $3,500 each. VacuTech incurs cash-operating costs of $2,450 to manufacture these instruments. On January 1, 19_1, VacuTech bought a vacuum pump for $400,000. VacuTech is considering the purchase of a new, more efficient pump on January 1, 19_5 (four years later). The new pump costs $620,000. Under the income tax code, the original cost of the pump would be depreciated as follows: 19_5, 33%; 19_6, 45%; 19_7, 15%; 19_8, 7%. The new pump is expected to have a terminal disposal price of $80,000 at the end of four years. At current rates of production, the new pump's greater efficiency will result in annual cash savings of $125,000.

The old pump will be fully depreciated by December 31, 19_4, but it can still be used for another four years. It has a current disposal price of $50,000. If it is used for another four years, the pump's terminal disposal price will be zero.

VacuTech is able to sell all the testing instruments it produces. Because of the increased speed of the new pump, output is expected to increase by 30 units in 19_5, 50 units in 19_6 and 19_7, and 70 units in 19_8. Over and above the annual cash savings at current production levels, VacuTech's cash manufacturing costs will decrease by $150 per unit on all *additional* units produced.

VacuTech is subject to a 40% tax rate. VacuTech's after-tax required rate of return is 16%.

Required
1. Determine whether VacuTech should purchase the new pump by calculating the net present value at January 1, 19_5, of the estimated after-tax cash flows that would result from the acquisition.
2. Describe nonfinancial and qualitative factors that VacuTech should consider before making the pump replacement decision.

21-23 Replacement of a machine, income taxes, straight-line depreciation. (CMA, adapted) WRL Company operates a snack-food center at the Hartsfield Airport. On January 2, 19_3, WRL purchased a special cookie-cutting machine, which has been used for three years. WRL is considering purchasing a newer, more efficient machine. If purchased, the new machine would be acquired on January 2, 19_6. WRL expects to sell 300,000 cookies in each of the next four years. The selling price of each cookie is expected to average $0.50.

WRL has two options: (1) continue to operate the old machine or (2) sell the old machine and purchase the new machine. The seller of the new machine offered no trade-in. The following information has been assembled to help management decide which option is more desirable.

	Old Machine	New Machine
Initial machine investment	$80,000	$120,000
Terminal disposal price at the end of useful life assumed for depreciation purposes	$10,000	$20,000
Useful life from date of acquisition	7 years	4 years
Expected annual cash-operating costs		
Variable costs per cookie	$0.20	$0.14
Total fixed costs	$15,000	$14,000
Depreciation method used for tax purposes	Straight-line	Straight-line
Estimated disposal prices of machines:		
January 2, 19_6	$40,000	$120,000
December 31, 19_9	$7,000	$20,000

WRL has a 40% income tax rate. Assume that any gain or loss on the sale of machinery is treated as an ordinary tax item and will affect the taxes paid by WRL in the year in which it occurs. WRL has an after-tax required rate of return of 16%.

Required
1. Use the net present value method to determine whether WRL should retain the old machine or acquire the new machine.
2. Assume that the financial differences between the net present values of the two options are so slight that WRL is indifferent between the two proposals. Identify and discuss the nonfinancial and qualitative factors that WRL should consider.

21-24 Capital budgeting, make versus buy, income taxes, double-declining balance depreciation, relevant costs. (CMA, adapted)   Jonfran Company manufactures three different models of paper shredders. Each has a waste container. Jonfran estimates the following number of waste containers needed over the next five years: 19_4, 50,000; 19_5, 50,000; 19_6, 52,000; 19_7, 55,000; 19_8, 55,000.

The equipment used to manufacture waste containers must be replaced because it has broken. The old equipment is fully depreciated and has a current disposal price of $1,500. The new equipment would cost $960,000. The equipment would go into service on January 1, 19_4, and would have a five-year useful life. Under the prevailing tax laws, depreciation is calculated on the double-declining balance method over the five years, with depreciation in the fifth year being the book value of the equipment at the start of that year. The double-declining balance method assumes zero terminal disposal price at the end of five years, but actual disposal price would be $12,000.

Jonfran's current manufacturing costs for waste containers follows.

Direct materials		$10.00
Direct manufacturing labor		8.00
Variable manufacturing overhead		4.00
Fixed overhead		
Supervision	$2.00	
Depreciation on old equipment	3.00	
General administrative overhead	6.00	11.00
Total manufacturing cost per unit		$33.00

An outside supplier has offered to supply all the containers that Jonfran needs over the next five years at a fixed price of $29 per container. If the supplier's offer is accepted, Jonfran would not need to replace the equipment.

If the waste containers are purchased outside, the salary and benefits of one supervisor, included in the fixed overhead at $45,000, would be eliminated. There would, however, be no change in general administrative overhead. Jonfran has no alternative use for the extra space that would become available if the containers were purchased from outside. Working capital requirements are approximately the same whether the containers are made or purchased.

Jonfran has a 40% income tax rate. Its after-tax required rate of return on new equipment is 12%.

Required

1. Use a net present value analysis to determine whether Jonfran should purchase the waste containers from the outside supplier or purchase the new equipment.

2. What nonfinancial and qualitative factors should Jonfran consider before coming to a decision?

21-25 Capital budgeting, inflation, income taxes, book depreciation different from tax depreciation, lease versus buy. (S. Huddart, adapted)   Home Electronics manufactures electrical controls for residences. It has recently acquired rights to produce and sell a newly developed thermostat. The following information is available.

a. Net initial investment

Plant and equipment	$500,000
Working capital	300,000
	$800,000

b. The plant and equipment have an estimated useful life of four years and zero terminal disposal price. Home Electronics recognizes depreciation on a straight-line basis on its own books. Depreciation is claimed for tax purposes as follows: year 1, 33%; year 2, 45%; year 3, 15%; year 4, 7%.

c. Home Electronics expects to sell 100,000 units each year for the next four years.

d. Revenue and cost information for the new thermostat follows:

Selling price	$ 15.00
Variable costs per unit	$ 10.00
Additional fixed costs	$175,000

Additional fixed costs include straight-line depreciation on new plant and equipment. Assume that all of Home Electronics' transactions are cash transactions that, except for the initial investment, occur at the end of each year.

e. Home Electronics is subject to a 40% tax rate. Taxes are paid for taxable income in year 1 at the end of year 1, taxes for year 2 at the end of year 2, and so on.

f. Management expects inflation to be 10% per year, affecting both revenues and costs. A 10% increase in working capital is needed at the end of each year.

g. Home Electronics' after-tax nominal required rate of return is 14%.

Required

1. Determine whether Home Electronics should undertake the project by calculating its net present value. Conduct your analysis in nominal terms.

2. Suppose Home Electronics could lease the new plant and equipment for four years for a fixed annual payment of $180,000. Is the lease option better or worse for Home Electronics than purchasing the plant and equipment?

21-26 Decision to go to college, net present value, income taxes, inflation. Angela Avila is in her final year of high school. Her best friend, Mary Smith, has decided not to go to college. She has accepted a $16,000-a-year job with a local bank. Mary argues that she will be able to save a fortune by the time Angela has recouped her college tuition fees. Angela, who has never previously considered taking a job straight out of high school, seeks your advice on the financial benefits of going to college. She views herself as similar to Mary in ability, family support, and intelligence. A time period of 20 years from when Angela and Mary leave high school is chosen for analysis. You decide to examine the following two scenarios:

a. Angela too joins the bank at $16,000 per year. Her salary increases in real terms at 3.7736% per year.

b. Angela goes to college for four years. The college will cost $12,000 a year for fees, books, and other items (assume paid at the *end* of each year). After college, Angela joins an accounting firm at $30,000 per year. Her salary increases in real terms at 5.7692% per year.

The required real rate of return is 10% per year. (Use 10% in your answers to all four requirements.) Hint: Note that $1.037736 \div 1.10 = 1 \div 1.06$; $1.057692 \div 1.10 = 1 \div 1.04$.

Required

1. Assume zero inflation and zero income taxes. Compare the net present value for the 20-year period of (a) Angela joining the bank and (b) Angela going to college and then joining an accounting firm. Comment on your results.

2. Assume zero inflation. Repeat requirement 1 assuming a 28% income tax rate on all income. There are no tax deductions allowed for education costs. Comment on your results.

3. How might a 12% inflation rate affect your answers to requirements 1 and 2? (No computations are required.)

4. What other factors would you recommend that Angela consider in her decision on whether to go to college?

21-27 Capital budgeting for product design, inflation, taxation, straight-line depreciation. Cila Black, president of the Liverpool Product Design Group, is considering an investment in computer-aided design equipment. The equipment will cost £110,000 and will have a five-year useful economic life. It has a zero terminal disposal price and will generate annual cash-operating savings of £36,000 (before income taxes), using 19_4 prices. It is December 31, 19_4.
    The after-tax required rate of return is 18% per year.

Required

1. Compute the net present value of the project. Assume a 40% tax rate and straight-line depreciation.

2. Black is wondering if the method in requirement 1 provides a correct analysis of the effects of inflation. The 18% required rate of return incorporates an element attributable to anticipated inflation. For purposes of her analysis, she assumes that the existing rate of inflation, 10% annually, will persist over the next five years. Repeat requirement 1, adjusting the cash-operating savings upward in accordance with the 10% inflation rate.

3. What generalizations about the effects of inflation on capital-budgeting methods and decisions can you develop?

 21-28 Capital budgeting, inflation, taxation, straight-line depreciation. (J. Fellingham, adapted) Abbie Young is manager of the customer-service division of an electrical appliance store. Abbie is considering buying a repairing machine that costs $10,000. The machine will last five years. Abbie estimates that the incremental pretax cash savings from using the machine will be $3,000 annually. The $3,000 is measured at current prices and will be received at the end of each year. For tax purposes, she will depreciate the machine straight-line, assuming zero terminal disposal price. Abbie requires a 10% real rate of return (that is, the rate of return is 10% when all cash flows are denominated in 19_4 dollars). Use the 10% rate in your answers to all four requirements.

Required
Treat each of the following cases independently.

1. Abbie lives in a world without income taxes and without inflation. What is the net present value of the machine in this world?
2. Abbie lives in a world without inflation, but there is an income tax rate of 40%. What is the net present value of the machine in this world?
3. There are no income taxes, but the annual inflation rate is 20%. What is the net present value of the machine? The cash savings each year will be increased to reflect the inflation rate.
4. The annual inflation rate is 20%, and the income tax rate is 40%. What is the net present value of the machine?

21-29 Robotics capital project, inflation, income taxes, double-declining balance depreciation. Rustbelt America Inc. purchases secondhand pipeline equipment and "rehabilitates" it for resale. A major problem in its plant is the spot-welding activity. There have been many industrial accidents involving workers at this activity. Rustbelt looks into the possibility of investing in robots. The investment in robots will cost $10 million payable immediately and will reduce labor costs, worker insurance costs, and materials usage costs by a total of $7 million (in 1/1/19_4 dollars) a year. The robots require an addition to annual cash-operating costs of $3 million (in 1/1/19_4 dollars) a year. Hence the net cash-operating savings from using the robots will be $4 million annually (in 1/1/19_4 dollars). Rustbelt believes that using the robots will eliminate industrial accidents involving workers at the spot-welding activity.

The robots have a four-year useful life with a terminal disposal price of $1 million (in 1/1/19_4 dollars). The robots qualify for a four-year recovery period using the double-declining balance depreciation method. Any terminal disposal price of the robots is treated as taxable income in the year of the disposal. Rustbelt anticipates inflation in its operating costs and in the terminal disposal price of the robots of 20% per year. It uses a 10% after-tax required rate of return for investments expressed in real dollars. Rustbelt's income tax rate is 40%.

Required
1. What is the nominal after-tax required rate of return of Rustbelt America for investments expressed in nominal dollars?
2. What is the net present value of the $10 million investment in robots? Use the approach of predicting cash inflows and outflows in nominal dollars *and* using a nominal discount rate.
3. What are the advantages of the approach to capital budgeting for inflation in requirement 2 relative to the approach of predicting real cash inflows and outflows *and* using a real discount rate?
4. What factors other than the net present value figure in requirement 2 should Rustbelt America consider in deciding whether to invest in robots?

21-30 Comparison of projects with unequal lives. The manager of the Robin Hood Company is considering two investment projects that are mutually exclusive.

The after-tax required rate of return of this company is 10%, and the anticipated cash flows are as follows:

Project No.	Investment Required Now	Cash Inflows			
		Year 1	Year 2	Year 3	Year 4
1	$10,000	$12,000	$0	$0	$ 0
2	10,000	0	0	0	17,500

Required

1. Compute the internal rate of return of both projects. Which project is preferable?
2. Compute the net present value of both projects. Which project is preferable?
3. Comment briefly on the results in requirements 1 and 2. Be specific in your comparisons.

21-31   Ranking projects. (Adapted from *NAA Research Report No. 35*, pp. 83–85)   Assume that six projects, A through F in the table that follows, have been submitted for inclusion in the coming year's budget for capital expenditures:

		Project					
	Year	A	B	C	D	E	F
Investment	0	$(100,000)	$(100,000)	$(200,000)	$(200,000)	$(200,000)	$(50,000)
	1	0	20,000	70,000	0	5,000	23,000
	2	10,000	20,000	70,000	0	15,000	20,000
	3	20,000	20,000	70,000	0	30,000	10,000
	4	20,000	20,000	70,000	0	50,000	10,000
	5	20,000	20,000	70,000	0	50,000	
Per year	6–9	20,000	20,000		200,000	50,000	
	10	20,000	20,000			50,000	
Per year	11–15	20,000					
Internal rate of return		14%	?	?	?	12.6%	12.0%

Required

1. Compute internal rates of return (to the nearest half percent) for projects B, C, and D. Rank all projects in descending order in terms of internal rate of return. Show computations.
2. Based on your answer in requirement 1, state which projects you would select, assuming a 10% required rate of return (a) if $500,000 is the limit to be spent, (b) if $550,000 is the limit, and (c) if $650,000 is the limit.
3. Assuming a 16% required rate of return and using the net present value method, compute the net present values and rank all the projects. Which project is more desirable, C or D? Compare your answer with your ranking in requirement 2.
4. What factors other than those considered in requirements 1 through 3 would influence your project rankings? Be specific.

21-32   Ethics, capital budgeting. (CMA, adapted)   Instant Dinners Inc. (IDI) makes microwaveable frozen foods. The company is considering purchasing an automated materials-movement system (AMMS) for its Western Plant. Bill Rolland, IDI's chief financial officer, has asked Lealand Forrest, assistant controller, to prepare a net present value analysis for the proposal.

Rolland was instrumental in convincing the Board of Directors to open the Western Plant. Now, unless significant improvements in cost control and production efficiency are achieved, the Western Plant may be sold. Rolland is anxious to have the Western Plant continue to operate to maintain his credibility with the Board and also to help Western's production manager, an old friend of Rolland.

Rolland has given Forrest the information for preparing the net present value analysis. When Forrest completed his initial analysis, the proposed project had a positive net present value. After further investigation, Forrest discovered that the estimated terminal disposal price of the AMMS should be $100,000, not $850,000, and that the useful life of the system was expected to be 8 years, not 10 years. Forrest prepared a revised, second analysis based on this new information. On seeing the second analysis, Rolland told Forrest to discard the revised analysis and not to discuss it with anyone at IDI or with the Board of Directors.

Required

Referring to the "Standards of Ethical Conduct for Management Accountants" described in Chapter 1, explain how Lealand Forrest, a management accountant, should evaluate Bill Rolland's directives to conceal the revised analysis. Identify the specific steps Lealand Forrest should take to resolve this situation.

22

Measuring Mix, Yield, and Productivity

PART ONE: SALES VARIANCES

Sales-volume variances

Sales-quantity and sales-mix variances

Market-size and market-share variances

PART TWO: INPUT VARIANCES

Direct materials yield and direct materials mix variances

Direct labor yield and direct labor mix variances

PART THREE: PRODUCTIVITY MEASUREMENT

What is productivity?

Partial productivity measures

Total factor productivity

Analysis of annual cost changes

Service-sector productivity

*S*upermarkets around the globe, such as in Japan, have as goals both increasing total sales and increasing the mix of higher-margin items sold. Sales-quantity and sales-mix variances help managers monitor their progress in achieving these two goals.

Comparing actual results with budgets can help managers evaluate operations. Managers can see the impact of key variables on the actual results and focus on areas that deserve more attention. Chapters 7 and 8 illustrated various uses of variance information relating to direct materials, direct manufacturing labor, direct marketing labor, manufacturing overhead, and marketing overhead.

This chapter extends our look at some major areas in the analysis of accounting variances that have been covered only briefly in earlier chapters and develops a new topic—productivity measurement. Three subjects are discussed. Because each subject can be studied independently, the chapter is divided into three major parts.

1. Part One: *Sales variances.* We examine five variances: sales volume, sales quantity, sales mix, market size, and market share.
2. Part Two: *Input variances.* For illustrative purposes, we present yield and mix variances for each of two inputs—direct materials (to illustrate manufacturing input variances) and direct service labor (to illustrate input variances in the service sector). The variances presented can readily be adapted to other inputs, such as energy.
3. Part Three: *Productivity measurement.* We illustrate partial productivity and total factor productivity measures using two inputs—direct materials and direct manufacturing labor.

The examples presented in this chapter involve no more than two products, two direct materials inputs, or two direct service labor inputs. When there are many products sold or many direct materials or direct labor inputs, computer software packages are essential in variance analysis.

◆ PART ONE: SALES VARIANCES

Chapter 7 showed how price, efficiency, and sales-volume variances assist managers in single-product situations. Most organizations, though, produce more than one product or service. Budgets for total sales usually specify the quantity of each product to be sold. *Sales mix* is the relative combination of quantities of products that constitute total sales. We now explore complications that arise in multiple-product situations when the sales quantity or the sales mix or both differ from the budget.

The literature on variance analysis in multiple-product situations is sizable and bewildering. Be on guard whenever you see the terms *quantity variance, volume variance, activity variance, mix variance,* or *yield variance.* Organizations use these terms differently. In any discussion, be sure that all parties agree on definitions.

SALES-VOLUME VARIANCES

Objective 1

Describe the insight gained from dividing the sales-volume variance into the sales-quantity variance and the sales-mix variance

Chapter 7 introduced flexible budgets and standards. Exhibit 7-9 (p. 244) provided an overview of variances. Please pause and review that exhibit before reading further. Chapter 7 focused on *price variances* and *efficiency variances.* We now examine *sales-volume* variances.

The *sales-volume variance* is the difference between the flexible-budget amount and the static-budget amount. Budgeted unit selling prices, budgeted unit variable costs, and budgeted fixed costs are held constant. We measure this variance in terms of contribution margin because budgeted fixed costs are the same in a flexible budget and a static budget.

EXAMPLE: Party Wholesaler has an exclusive contract with Bordeaux Winery to import into the United States and sell two brands of wine—regular wine and premium wine. Party Wholesaler sells to liquor stores and other retail outlets. The monthly budget for the Bordeaux Winery contract is based on a combination of last year's performance, a forecast of general industry sales, and the company's expected share of the U.S. market for imported wine.

Assume the following budgeted and actual data for October 19_5 in our example:

	Budgeted		Actual	
	Regular Wine	Premium Wine	Regular Wine	Premium Wine
Selling price per bottle	$ 5	$16	$5.50	$15.50
Variable costs per bottle	4	9	4.30	9.50
Contribution margin per bottle	$ 1	$ 7	$1.20	$ 6.00
Sales (in bottles)	1,200	400	1,100	500

Budgeted fixed costs for October 19_5 are $3,000. Actual fixed costs for October 19_5 are $3,050.

Party Wholesaler prepared the following static budget for October 19_5 for its contract with Bordeaux Winery:

Static Budget

Revenues	
(Regular, 1,200 × $5 + Premium, 400 × $16)	$12,400
Variable costs	
(Regular, 1,200 × $4 + Premium, 400 × $9)	8,400
Contribution margin	$ 4,000
Fixed costs	3,000
Operating income	$ 1,000

The budgeted average contribution margin per bottle is $2.50 ($4,000 ÷ 1,600 total number of bottles budgeted to be sold).

Exhibit 22-1 compares the actual results for October 19_5 with the budgeted results. (Exhibit 22-1 is a Level 2 analysis in the terminology used in Chapter 7.) The favorable static-budget variance of $270 is made up of an unfavorable flexible-budget variance of $330 and a favorable sales-volume variance of $600.

EXHIBIT 22-1

Flexible Budget–Based Variance Analysis for Party Wholesaler, October 19_5

	Actual Results (1)	Flexible-Budget Variances (2) = (1) − (3)	Flexible Budget* (3)	Sales-Volume Variances (4) = (3) − (5)	Static Budget† (5)
Units sold	1,600	—	1,600	—	1,600
Revenues (sales)	$13,800	$300 F	$13,500	$1,100 F	$12,400
Variable costs	9,480	580 U	8,900	500 U	8,400
Contribution margin	$ 4,320	$280 U	$ 4,600	$ 600 F	$ 4,000
Fixed costs	3,050	50 U	3,000	—	3,000
Operating income	$ 1,270	$330 U	$ 1,600	$ 600 F	$ 1,000

$330 U
Total flexible-budget variance

$600 F
Total sales-volume variance

$270 F
Total static-budget variance

F = favorable effect on operating income; U = unfavorable effect on operating income.
*Flexible budget based on actual units sold and budgeted selling price, variable costs per output unit, and fixed costs.
†Static budget based on budgeted units sold and budgeted selling price, variable costs per output unit, and fixed costs.

The Effects of More Than One Product

Exhibit 22-1 provides a useful overview of the variances, but most managers want more detail. Consider a puzzle regarding the final three columns. How can the flexible-budget amounts differ from the static-budget amounts if the number of bottles actually sold equals the 1,600 bottles originally budgeted? The reason, as explained in this section, is that this is a multiple-product company. In a single-product company, this puzzle would not arise.

In a multiple-product company, each product has its own flexible budget. If the *mix* of products sold changes, the company's overall flexible budget is affected because it merely sums up the flexible budgets prepared for the individual products sold.

Exhibit 22-2 shows that the original budgeted contribution margin of $4,000 is the sum of the budgeted contribution margins for 1,200 bottles of regular wine and 400 bottles of premium wine. But 1,100 bottles of regular wine and 500 bottles of premium wine were actually sold. The flexible budget based on actual sales of 1,600 bottles differs from the static budget because the sales mix used in the static budget differs from the sales mix that actually occurred. It is only by coincidence that the number of bottles budgeted—1,600—was also the number of bottles actually sold.

Exhibit 22-2 provides details of the sales-volume variance. It shows how the flexible budget based on actual units sold is the sum of individual flexible budgets for regular wine and premium wine. The total sales-volume variance is $600 favorable:

$$\text{Sales-volume variance} = \left(\begin{array}{c} \text{Actual sales} \\ \text{quantity in units} \end{array} - \begin{array}{c} \text{Budgeted sales} \\ \text{quantity in units} \end{array} \right) \times \begin{array}{c} \text{Budgeted individual product} \\ \text{contribution margin per unit} \end{array}$$

	Regular Wine			Premium Wine			Total		
	Flexible Budget* (1)	Sales-Volume Variances (2)	Static Budget† (3)	Flexible Budget* (4)	Sales-Volume Variances (5)	Static Budget† (6)	Flexible Budget* (7) = (1) + (4)	Sales-Volume Variances (8) = (2) + (5)	Static Budget† (9) = (3) + (6)
Units sold	1,100	100 U	1,200	500	100 F	400	1,600	—	1,600
Revenues (sales)‡	$5,500	$500 U	$6,000	$8,000	$1,600 F	$6,400	$13,500	$1,100 F	$12,400
Variable costs§	4,400	400 F	4,800	4,500	900 U	3,600	8,900	500 U	8,400
Contribution margin‖	$1,100	$100 U	$1,200	$3,500	$ 700 F	$2,800	4,600	600 F	4,000
Fixed costs							3,000	—	3,000
Operating income							$ 1,600	$ 600 F	$ 1,000

F = favorable effect on operating income; U = unfavorable effect on operating income.
*Flexible budget based on actual units sold and budgeted selling prices, variable costs per output unit, and fixed costs.
†Static budget based on budgeted units sold and budgeted selling prices, variable costs per output unit, and fixed costs.
‡Regular wine $5 per unit, premium wine $16 per unit.
§Regular wine $4 per unit, premium wine $9 per unit.
‖Regular wine $1 per unit, premium wine $7 per unit.

$$\begin{aligned}\text{Regular wine} &= (1{,}100 - 1{,}200) \times \$1 = \$100 \text{ U} \\ \text{Premium wine} &= (500 - 400) \times \$7 \quad = \underline{\$700} \text{ F} \\ \text{Total} && \underline{\$600} \text{ F}\end{aligned}$$

Other things being equal, the failure to sell the originally expected 1,200 bottles of regular wine would reduce operating income by 100 units × $1 = $100. Selling 100 more bottles of premium wine than originally expected, however, would increase operating income by 100 bottles × $7 = $700. Because the sales-volume variance depends on how the market reacts to a product, some people prefer to call the sales-volume variance the marketing variance.

How Managers Use Sales-Variance Information

The manager in charge of the Bordeaux Winery contract at Party Wholesaler may initially concentrate on the Total columns in Exhibit 22-2 (columns 7, 8, and 9). However, these columns provide only a beginning. An inquiring manager would dig more deeply. The manager should keep the following three points in mind when analyzing the information underlying Exhibit 22-2.

First, the budgeted and actual *aggregated* units are equal. But is this information helpful? Adding units of different products together may yield a meaningless amount. (This is often called the "apples and oranges problem.") As we move beyond a single-product analysis, aggregated units fail to provide a common denominator for measuring overall sales quantities. To be useful to a manager, units are converted into monetary equivalents such as sales dollars or contribution margins. But we should not forget the fundamental inputs and outputs of organizations—their products or services.

Second, the aggregated dollars give an overall picture of the product line, but managers must often focus on individual products. For example, a manager must decide how best to allocate the advertising budget. Should Party Wholesaler push the premium wine or the regular wine? Without information on the contribution margins of the individual products that make up the product line, a manager cannot make an informed decision.

Third, there can be confusion regarding the sales-volume variances of $1,100 F, $500 U, and $600 F in column 8 of Exhibit 22-2. Taken together, *the total sales volume in units* was unchanged, but the original proportions—the sales mix—changed. The actual quantities of *individual* products deviated from what was expected. If the original budget in total units is attained and a larger proportion of products with a higher budgeted contribution margin are sold than was specified in the original mix, higher operating income will result.

The general manager of Party Wholesaler usually receives summary monetary figures rather than a flood of detail. For example, she reviews only the dollar amounts for each contract with a winery or other supplier. Party Wholesaler simply deals with too many outside businesses for the general manager to review every detail of each contract. In our example, the report she receives would present only the information in the Total columns of Exhibit 22-2.

Rapid advances in information technology are making more information more readily available to more people. Marketing managers at Pepsi-Cola, for example, can link into a computer network to seek detailed explanations for individual amounts in their summary sales-volume-variance reports. Indeed, they may be able to access on line the same detailed product-by-product variance information typically examined only by individual product managers.

SALES-QUANTITY AND SALES-MIX VARIANCES

Many managers favor probing the sales-volume variances further. The analysis in Exhibit 22-2 reveals how two major factors affect total sales-volume variances: (1) the actual quantity of units sold and (2) the relative proportions of products with different contribution margins. The sales-quantity variance captures factor 1, and the sales-mix variance captures factor 2. To simplify the computations, we focus solely on contribution margins, even though these variances could be computed separately for revenues and variable costs.

Given that the budgeted sales mix is unchanged, the **total sales-quantity variance** is the difference between two amounts: (1) the budgeted contribution margin based on actual quantities sold of all products and (2) the contribution margin in the static budget based on the budgeted quantities to be sold of all products. Budgeted selling prices and budgeted unit variable costs, and hence budgeted unit contribution margins for each product are held constant. This variance is measured in terms of contribution margin because total fixed costs are the same for different total quantities sold of all products.

Given the total actual quantities sold of all products, the **total sales-mix variance** is the difference between two amounts: (1) the budgeted contribution margin for the actual sales mix (flexible-budget contribution margin) and (2) the budgeted contribution margin if the budgeted sales mix had been unchanged. Budgeted selling prices for all products, budgeted unit variable costs, and hence budgeted unit contribution margins are held constant. This variance is measured in terms of contribution margin because total fixed costs are unchanged for different mixes of products sold.

Exhibit 22-3 presents the sales-quantity variances and the sales-mix variances in columnar form.[1]

- ◆ Ponder the column titles in Exhibit 22-3.
- ◆ Observe that the difference between the flexible budget (column 1) and the static budget (column 3) is the *sales-volume variance* for regular wine (Panel A), for premium wine (Panel B), and for all wines (Panel C).

---

[1]The presentations in the remainder of Part One draw (in part) from teaching notes prepared by John Harris.

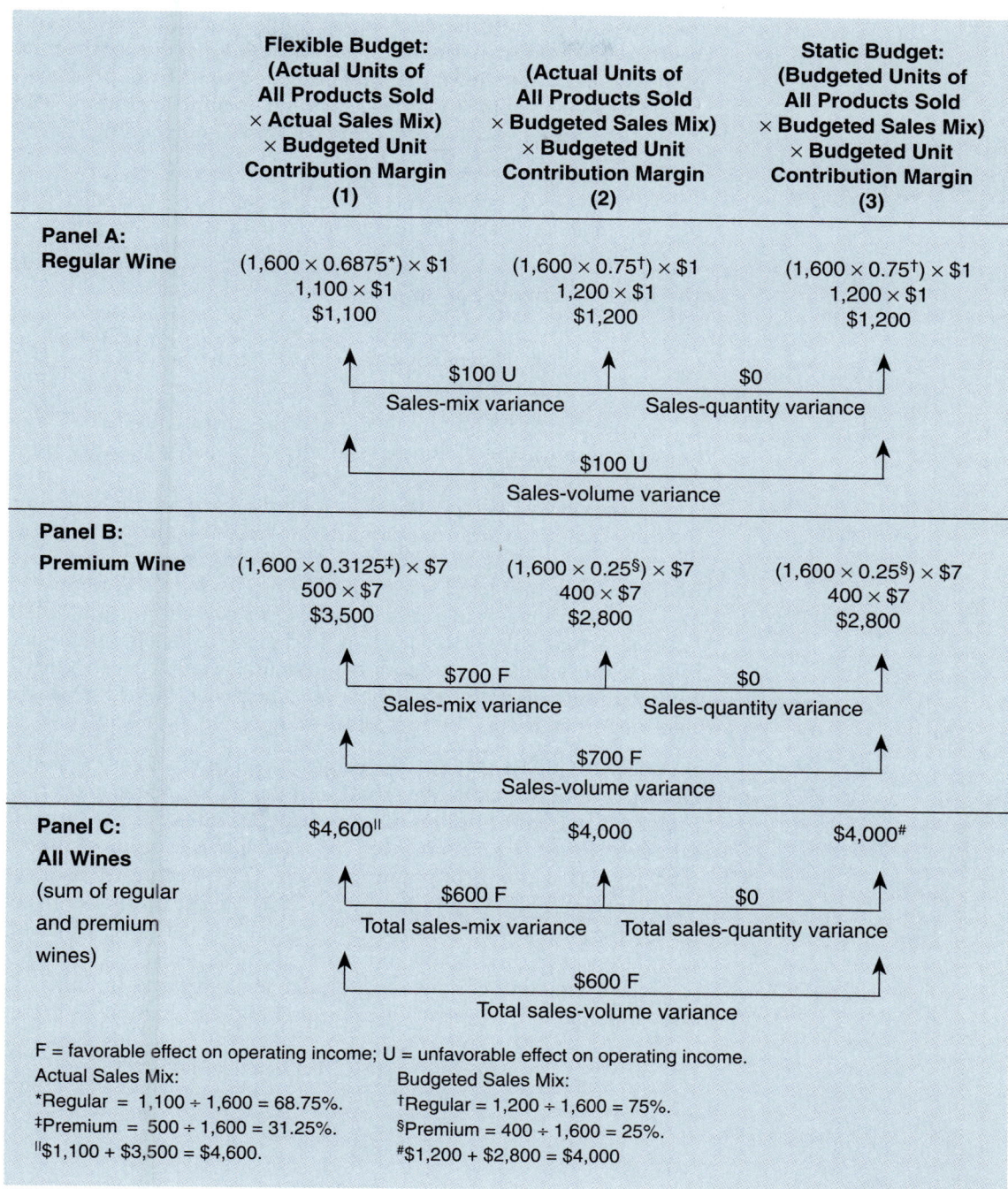

	Flexible Budget: (Actual Units of All Products Sold × Actual Sales Mix) × Budgeted Unit Contribution Margin (1)	(Actual Units of All Products Sold × Budgeted Sales Mix) × Budgeted Unit Contribution Margin (2)	Static Budget: (Budgeted Units of All Products Sold × Budgeted Sales Mix) × Budgeted Unit Contribution Margin (3)
Panel A: **Regular Wine**	$(1,600 \times 0.6875^*) \times \$1$ $1,100 \times \$1$ $\$1,100$	$(1,600 \times 0.75^\dagger) \times \$1$ $1,200 \times \$1$ $\$1,200$	$(1,600 \times 0.75^\dagger) \times \$1$ $1,200 \times \$1$ $\$1,200$

$100 U Sales-mix variance $0 Sales-quantity variance

$100 U Sales-volume variance

Panel B: **Premium Wine**	$(1,600 \times 0.3125^\ddagger) \times \$7$ $500 \times \$7$ $\$3,500$	$(1,600 \times 0.25^\S) \times \$7$ $400 \times \$7$ $\$2,800$	$(1,600 \times 0.25^\S) \times \$7$ $400 \times \$7$ $\$2,800$

$700 F Sales-mix variance $0 Sales-quantity variance

$700 F Sales-volume variance

Panel C: **All Wines** (sum of regular and premium wines)	$\$4,600^{\parallel}$	$\$4,000$	$\$4,000^{\#}$

$600 F Total sales-mix variance $0 Total sales-quantity variance

$600 F Total sales-volume variance

F = favorable effect on operating income; U = unfavorable effect on operating income.

Actual Sales Mix: Budgeted Sales Mix:
*Regular = 1,100 ÷ 1,600 = 68.75%.   †Regular = 1,200 ÷ 1,600 = 75%.
‡Premium = 500 ÷ 1,600 = 31.25%.   §Premium = 400 ÷ 1,600 = 25%.
‖$1,100 + $3,500 = $4,600.   #$1,200 + $2,800 = $4,000

◆ In each panel, the *sales-volume variance* is divided into two variances, the *sales-mix variance* and the *sales-quantity variance* calculated as the difference between two adjacent columns. The difference between column 1 and column 2 is the sales-mix variance; the difference between column 2 and column 3, the sales-quantity variance.

It is easiest to start with the static budget and work our way to the flexible budget, so we start with column 3, describe the sales-quantity variance, and then the sales-mix variance.

SALES-QUANTITY VARIANCE. Focus on column 3 and column 2. Compare the descriptions of the two columns. *The only difference* between the two columns is that column 3 *uses budgeted units of all products sold* while column 2 uses the *actual units of all products sold*. Both columns use the budgeted sales mix (regular wine, 75%; premium wine, 25%) and the budgeted unit contribution margins (regular wine, $1; premium wine, $7). The difference in contribution margin between column 2 and column 3 is the sales-quantity variance, which arises *solely* because the total actual units of all products sold differs from the total budgeted units of all products sold.

The sales-quantity variance for regular wine appears in Panel A and for premium wine in Panel B. In comparing column 2 and column 3 of Exhibit 22-3, for each type of wine

$$
\begin{pmatrix} \text{Sales-} \\ \text{quantity} \\ \text{variance} \end{pmatrix} = \begin{pmatrix} \text{Actual units} & \text{Budgeted units} \\ \text{of all products} - \text{of all products} \\ \text{sold} & \text{sold} \end{pmatrix} \times \begin{array}{c} \text{Budgeted} \\ \text{sales-mix} \\ \text{percentage} \end{array} \times \begin{array}{c} \text{Budgeted unit} \\ \text{contribution} \\ \text{margin} \end{array}
$$

The sales-quantity variances are:

Regular wine   = (1,600 – 1,600) × 0.75 × $1 = $0
Premium wine = (1,600 – 1,600) × 0.25 × $7 =   0
All wines $0

Note that in both Exhibit 22-3 and the formula described above, the total sales-quantity variance for all wines is the sum of the sales-quantity variances for each type of wine.

How do we interpret the sales-quantity variance for each product? We calculate the sales-quantity variance assuming the sales-mix proportion does not change from the budget and using the budgeted contribution margin per unit. An increase of actual sales over budgeted sales for all wines leads to a favorable sales-quantity variance for each wine (remember that the sales-mix proportion does not change). Similarly, if actual sales fall short of budgeted sales for all wines, an unfavorable sales-quantity variance for each wine occurs. Managers can study how the budgeted contribution margin for each type of wine and in total will fluctuate if total actual sales depart from total budgeted sales.

SALES-MIX VARIANCE. Focus next on columns 2 and 1. Both columns compute contribution margins using actual units of all products sold (1,600 bottles) and the budgeted unit contribution margin (regular wine, $1; premium wine, $7). *Only the sales-mix percentages differ between the two columns.* Column 2 uses the budgeted sales mix (regular wine, 75%; premium wine, 25%). Column 1 uses the actual sales mix (regular wine, 68.75%; premium wine 31.25%). The difference in contribution margin between column 1 and column 2 is the sales-mix variance, which arises *solely* because of differences in the actual and budgeted sales mixes.

The sales-mix variance for regular wine appears in Panel A and for premium wine in Panel B. In comparing columns 1 and 2 of Exhibit 22-3, for each type of wine

$$
\begin{pmatrix} \text{Sales-} \\ \text{mix} \\ \text{variance} \end{pmatrix} = \begin{pmatrix} \text{Actual sales-} & \text{Budgeted sales-} \\ \text{mix percentage} - \text{mix percentage} \end{pmatrix} \times \begin{array}{c} \text{Actual units} \\ \text{of all products} \\ \text{sold} \end{array} \times \begin{array}{c} \text{Budgeted unit} \\ \text{contribution} \\ \text{margin} \end{array}
$$

The sales-mix variances are:

Regular wine   = (0.6875 – 0.75) × 1,600 × $1 = $100 U
Premium wine = (0.3125 – 0.25) × 1,600 × $7 =   700 F
All wines $600 F

Note that in both Exhibit 22-3 and the formula described above, the sales-mix variance for all wines is the sum of the sales-mix variances for each type of wine.

A favorable sales-mix variance arises in the case of premium wine because the actual sales-mix percentage of premium wine (31.25%) is greater than the budgeted sales-mix percentage (25%). An unfavorable sales-mix variance occurs in the case of regular wine because the actual sales-mix percentage of regular wine (68.75%) is less than the budgeted sales-mix percentage (75%).

The total sales-mix variance is favorable ($600 F), indicating the overall effect of the differences between actual and budgeted sales mixes. It focuses on the total, not on individual products. In the case of Party Wholesaler, the total sales-mix variance is favorable because the actual mix of wine bottles sold has shifted in favor of the high-contribution-margin premium wine. The total sales-mix variance helps managers understand how, for a given total actual sales of all products, the budgeted contribution margin fluctuates as the actual product mix fluctuates from the budgeted product mix.

The following table summarizes the computations for regular wine and premium wine described in Exhibit 22-3.

	Sales-Quantity Variance	Sales-Mix Variance	Sales-Volume Variance
Regular wine	$0	$100 U	$100 U
Premium wine	0	700 F	700 F
Total	$0	$600 F	$600 F

The table indicates that the sales-volume variance of $600 F was due to a shift in the mix of products produced in favor of the product with the higher contribution margin rather than an increase in the total quantity of products sold.

Suppose the original sales mix of 75% of regular wine and 25% of premium wine and the original budgeted quantity of all wines sold (1,600 bottles) was maintained but total actual quantity of wine sold increased by 5%. We have assumed two changes from the previous example: (1) the total actual quantity of all wines sold has increased and (2) the actual sales mix is as budgeted. As we might expect, if we repeat our earlier analysis, there will be no sales-mix variance (no change in sales mix relative to the budgeted mix has occurred), only a sales-quantity variance (actual quantity of all wines sold exceeds the budgeted quantity). The actual sales of regular wine would be 1,260 ($1,600 \times 1.05 \times 0.75$) bottles, and actual sales of premium wine would be 420 ($1,600 \times 1.05 \times 0.25$) bottles. The sales-quantity variance would be:

$$
\begin{aligned}
\text{Regular wine} &= (1{,}260 - 1{,}200) \times \$1 &=& \ \$ \ 60 \ \text{F} \\
\text{Premium wine} &= (420 - 400) \times \$7 &=& \ \underline{\ 140 \ \text{F}} \\
\text{Total} & & & \ \underline{\$200 \ \text{F}}
\end{aligned}
$$

The result is consistent with the idea that, given an unchanging sales mix, a 5% increase in actual unit volume should produce a 5% increase in budgeted sales dollars and a 5% increase in budgeted contribution margin. The budgeted contribution margin would increase from $4,000 to $4,200, which is the $200 increase calculated above.

MARKET-SIZE AND MARKET-SHARE VARIANCES

Objective 2

Explain how market-size and market-share variances provide different explanations for a sales-quantity variance

Sales depend on overall demand for the industry's products and the company's ability to maintain its share of the market. Statistics are readily available for some industries—for example, automobiles, television sets, and soft drinks—so those companies can easily monitor their market shares. In other industries—for example, management consulting and restaurants—these statistics are difficult to obtain in a timely fashion. Sometimes, defining the market is difficult. For example, in the case of television sets, should the market be defined in terms of all television sets or television sets of specific styles and sizes (19-inch television sets, 27-inch television sets, or flat-screen television sets)?

Assume that the following information applies to our Party Wholesaler illustration:

	Actual	Budget
Total U.S. market for all imported premium and regular wine (units)	25,000	20,000
Bordeaux Winery's share of total U.S. market for imported premium and regular wine	6.4%	8%

Party Wholesaler makes all sales of Bordeaux Winery's products in the United States. Panel C of Exhibit 22-3 shows that the total sales-quantity variance equals $0. The *total sales-quantity variance* can be divided into the *market-size variance* and the *market-share variance*.[2] Observe that we compute the market-size and market-share variances for all products together and not for each product individually.

Given that the budgeted mix of products is unchanged, the **market-size variance** is the difference between two amounts: (1) the budgeted contribution margin based on the *actual market size in units* times the budgeted market share and (2) the contribution margin in the static budget based on the *budgeted market size in units* times the budgeted market share. Given that the budgeted mix of products is unchanged, the **market-share variance** is the difference between two amounts: (1) the budgeted contribution margin based on the actual market size in units times the *actual market share* and (2) the budgeted contribution margin based on actual market size in units times the *budgeted market share*.

[2]See J. Shank and V. Govindarajan, *Strategic Cost Management: Three Key Themes for Managing Costs Effectively"* The Free Press, (New York: 1993), Chapter 7.

EXHIBIT 22-4
Columnar Presentation of Market-Size and Market-Share Variance Analysis for Party Wholesaler, October 19_5

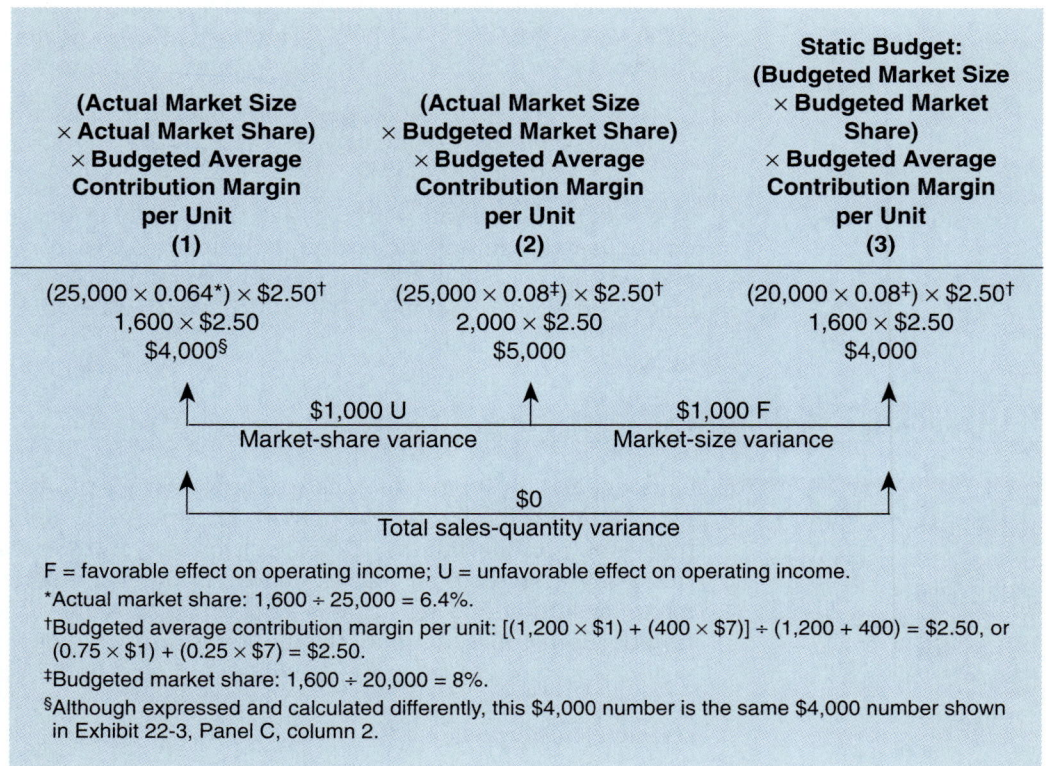

(Actual Market Size × Actual Market Share) × Budgeted Average Contribution Margin per Unit (1)	(Actual Market Size × Budgeted Market Share) × Budgeted Average Contribution Margin per Unit (2)	Static Budget: (Budgeted Market Size × Budgeted Market Share) × Budgeted Average Contribution Margin per Unit (3)
(25,000 × 0.064*) × $2.50† 1,600 × $2.50 $4,000§	(25,000 × 0.08‡) × $2.50† 2,000 × $2.50 $5,000	(20,000 × 0.08‡) × $2.50† 1,600 × $2.50 $4,000

$1,000 U
Market-share variance
$1,000 F
Market-size variance

$0
Total sales-quantity variance

F = favorable effect on operating income; U = unfavorable effect on operating income.
*Actual market share: 1,600 ÷ 25,000 = 6.4%.
†Budgeted average contribution margin per unit: [(1,200 × $1) + (400 × $7)] ÷ (1,200 + 400) = $2.50, or (0.75 × $1) + (0.25 × $7) = $2.50.
‡Budgeted market share: 1,600 ÷ 20,000 = 8%.
§Although expressed and calculated differently, this $4,000 number is the same $4,000 number shown in Exhibit 22-3, Panel C, column 2.

Exhibit 22-4 presents the market-size and market-share variances in columnar form. Each column uses the budgeted average contribution margin per unit, Party Wholesaler's total budgeted contribution divided by the budgeted units of all wines sold: ($[1,200 \times \$1] + [400 \times \$7]) \div 1,600 = \$2.50$. Budgeted average contribution margin per unit indicates the rate at which the total budgeted contribution changes as the *total* actual quantity of wine sold changes relative to the budget. We start with column 3 of Exhibit 22-4 and work our way to column 1.

MARKET-SIZE VARIANCE. Focus on columns 3 and 2. Both columns use the budgeted market shares (8%) and the budgeted average contribution margin per unit ($2.50). *Only the market size differs between the two columns.* Column 3 uses budgeted market size (20,000 units), and column 2 uses actual market size (25,000 units). Hence the difference in contribution margins between the two columns is the market-size variance.

$$\begin{matrix} \text{Market-} \\ \text{size} \\ \text{variance} \end{matrix} = \left(\begin{matrix} \text{Actual} \\ \text{market size} \\ \text{in units} \end{matrix} - \begin{matrix} \text{Budgeted} \\ \text{market size} \\ \text{in units} \end{matrix} \right) \times \begin{matrix} \text{Budgeted} \\ \text{market} \\ \text{share} \end{matrix} \times \begin{matrix} \text{Budgeted average} \\ \text{contribution margin} \\ \text{per unit} \end{matrix}$$

$$= (25,000 - 20,000) \times 0.08 \times \$2.50 = \$1,000 \text{ F}$$

MARKET-SHARE VARIANCE. Compare columns 2 and 1. Both columns use the actual market size and the budgeted average contribution margin per unit. *The only difference: column 2 uses budgeted market share (8%) while column 1 uses the actual market share (6.4%).* Hence the difference in contribution margins between the two columns is the market-share variance.

$$\begin{matrix} \text{Market-} \\ \text{share} \\ \text{variance} \end{matrix} = \begin{matrix} \text{Actual} \\ \text{market size} \\ \text{in units} \end{matrix} \times \left(\begin{matrix} \text{Actual} \\ \text{market} \\ \text{share} \end{matrix} - \begin{matrix} \text{Budgeted} \\ \text{market} \\ \text{share} \end{matrix} \right) \times \begin{matrix} \text{Budgeted average} \\ \text{contribution margin} \\ \text{per unit} \end{matrix}$$

$$= 25,000 \times (0.064 - 0.08) \times \$2.50 = \$1,000 \text{ U}$$

To summarize,

Market-size variance	$1,000 F
Market-share variance	1,000 U
Total sales quantity variance	$ 0

Managers are likely to gain some insights from these variances. The market-size variance measures the additional contribution margin, $1,000, that would be expected because the market size expanded. Unfortunately, the company attained only 6.4% of the industry market. This failure to maintain budgeted market share—the drop from 8% to 6.4%—created an equal, offsetting market-share variance.

Is dividing the sales-quantity variance into market-size and market-share variances useful for evaluating the marketing manager's performance? Suppose market size and the demand for an industry's products are largely influenced by factors such as growth and interest rates in the economy. Then, the market-size variance does not tell us much about the marketing manager's performance because it is largely determined by factors outside the manager's control. Top management may, then, put greater weight on the market-share variance when evaluating the marketing manager. In the Party Wholesaler's example, at first glance a zero sales-quantity variance would indicate an acceptable level of performance. But the unfavorable market-share variance may suggest a somewhat inferior performance and serve as a reason for further investigation.

Exhibit 22-5 presents a helpful overview of the variances we have computed in this section. Always consider possible interdependencies among these individual variances.

The Problem for Self-Study at the end of the chapter covers sales-volume, sales-quantity, and sales-mix variances.

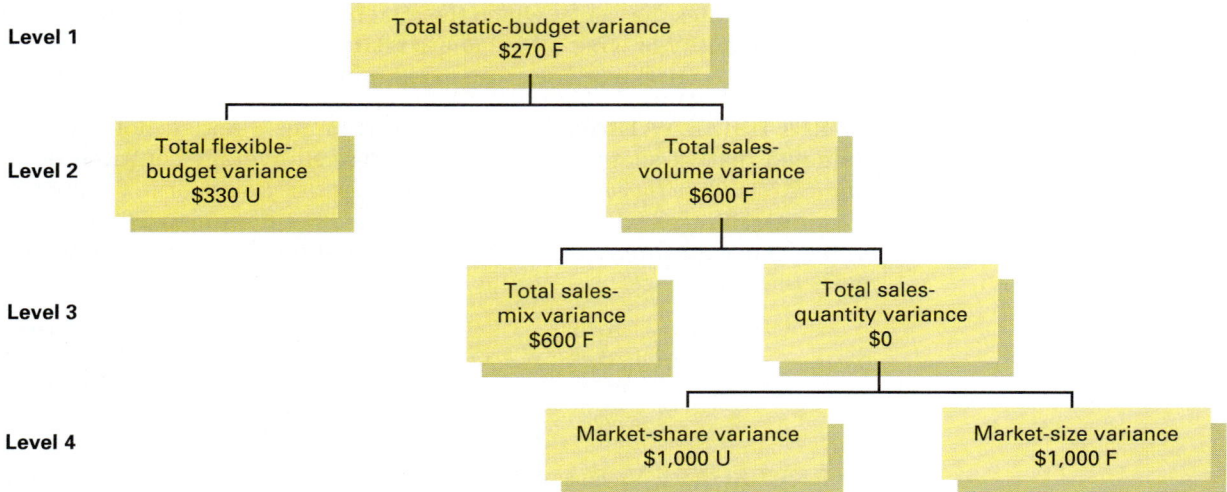

◆ PART TWO: INPUT VARIANCES

This part of the chapter focuses on variance analysis for inputs in manufacturing, merchandising, and service organizations. In the case of manufacturing processes, input variances are also called production variances. Manufacturing processes often require that a number of different direct materials and a number of different direct manufacturing labor skills be combined to obtain a unit of finished product. For some products, this combination must be exact. The manager of a plant that assembles portable computers has a prespecified list of parts to include in each computer. It is not possible to substitute one part or component for another without altering the final product. We refer to these materials as *nonsubstitutable* materials. For other products, a manufacturer has some leeway in combining the materials. The manager of an agricultural fertilizer plant can combine materials (such as elemental phosphorus and acids) in varying proportions. Elemental phosphorus and acids are *substitutable* materials.

What do the terms *yield* and *mix* mean in a production setting? *Yield* refers to the quantity of finished output produced from a budgeted or standard mix of inputs. *Mix* refers to the relative proportion or combination of the various inputs required to produce the quantity of finished output. Yield and mix variances are useful when examining direct manufacturing labor inputs or direct service labor inputs. Recall from Chapter 7 that a *variance* is the difference between an actual result and a budgeted amount, when that budgeted amount is a financial variable reported by the accounting system. The sources of the budgeted figures discussed in this chapter include:

- ◆ Internally generated actual costs from the most recent reporting period
- ◆ Internally generated actual costs of the most recent period adjusted for expected improvement
- ◆ Internally generated *standard costs* based on perfection standards
- ◆ Internally generated *standard costs* based on currently attainable standards
- ◆ Externally generated *target cost* numbers based on an analysis of the cost structure of the leading competitor in an industry

The interpretation of a variance will depend on the source of the budgeted figure(s) used in computing the variance.

DIRECT MATERIALS YIELD AND DIRECT MATERIALS MIX VARIANCES

Objective 3
Distinguish between variance analysis procedures when inputs cannot be substituted for one another and variance analysis procedures when inputs can be so substituted

When we initially examined material and labor variances in Chapter 7, we saw that managers sometimes make trade-offs between price and efficiency variances. For example, an orange-juice bottler may use oranges whose juice content is lower than budgeted if their price is significantly lower than the price of oranges with the budgeted juice content. The yield and mix variances computed in this section provide insight into the effect that such decisions have on operating income.

The Familiar Analysis: Direct Materials Efficiency and Direct Materials Price Variances

Consider a specific example of multiple direct materials inputs and a single product output. Delpino Corporation makes tomato ketchup. To get the correct ketchup color and taste, Delpino mixes two types of tomatoes grown in two different regions—California tomatoes (called Caltoms) and Florida tomatoes (called Flotoms). Delpino has the following direct materials input standards to produce 1 ton of ketchup.

1.2 tons of Caltoms at $80 per ton	$ 96
0.4 tons of Flotoms at $90 per ton	36
Total standard cost of 1.6 tons of tomatoes	$132

Delpino's production standards specify 1.6 tons of tomatoes to produce 1 ton of ketchup. Therefore, the standard cost of tomatoes required to make 1 ton of ketchup is $132.

Because Delpino uses fresh tomatoes to make ketchup, no inventories of direct materials are kept. Purchases are made as needed, so all price variances are related to direct materials used. Actual results show that a total of 6,500 tons of tomatoes were used in a recent period to produce 4,000 tons of ketchup:

5,200 tons of Caltoms at actual cost of $82 per ton	$426,400
1,300 tons of Flotoms at actual cost of $97 per ton	126,100
6,500 tons of tomatoes	552,500
4,000 tons of ketchup were produced at a standard cost of $132 per ton	528,000
Total variance to be explained	$ 24,500 U

Given the standard that 1.2 tons of Caltoms and 0.4 tons of Flotoms are required to produce 1 ton of ketchup, the budgeted input of tomatoes required to produce 4,000 tons of ketchup are:

Caltoms: $1.20 \times 4,000 = 4,800$ tons
Flotoms: $0.40 \times 4,000 = 1,600$ tons

Exhibit 22-6 presents the familiar approach to analyzing the flexible budget direct materials variance discussed in Chapter 7. The direct materials efficiency variance and the direct materials price variance are calculated separately for each input material and then added together.

The analysis in Exhibit 22-6 may suffice when the two direct materials used are not substitutes. Managers control each individual input, and no discretion is permitted regarding the substitution of materials inputs. For example, there is often a specified mix of parts needed for the assembly of motor vehicles, radios, and washing machines. In these cases, all deviations from the input-output relationships are due to efficient or inefficient usage of individual direct materials; the price and efficiency variances individually computed for each material typically provide the information necessary for decisions.

EXHIBIT 22-6
Columnar Presentation of Direct Materials Price and Efficiency Variances for Delpino Corporation

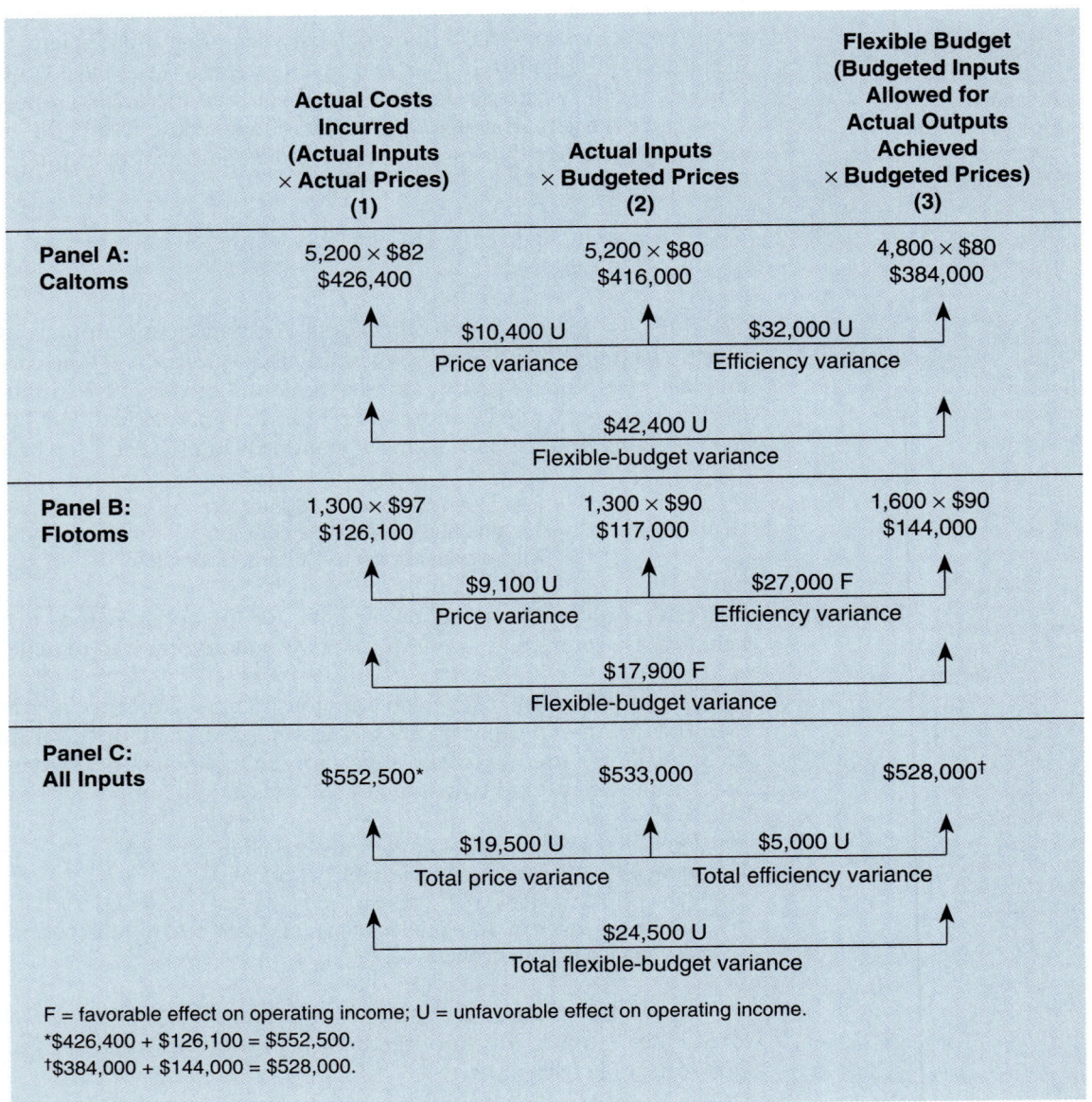

	Actual Costs Incurred (Actual Inputs × Actual Prices) (1)	Actual Inputs × Budgeted Prices (2)	Flexible Budget (Budgeted Inputs Allowed for Actual Outputs Achieved × Budgeted Prices) (3)
Panel A: **Caltoms**	5,200 × $82 $426,400	5,200 × $80 $416,000	4,800 × $80 $384,000
	↑ $10,400 U Price variance ↑	$32,000 U Efficiency variance ↑	
	↑	$42,400 U Flexible-budget variance	↑
Panel B: **Flotoms**	1,300 × $97 $126,100	1,300 × $90 $117,000	1,600 × $90 $144,000
	↑ $9,100 U Price variance ↑	$27,000 F Efficiency variance ↑	
	↑	$17,900 F Flexible-budget variance	↑
Panel C: **All Inputs**	$552,500*	$533,000	$528,000†
	↑ $19,500 U Total price variance ↑	$5,000 U Total efficiency variance ↑	
	↑	$24,500 U Total flexible-budget variance	↑

F = favorable effect on operating income; U = unfavorable effect on operating income.
*$426,400 + $126,100 = $552,500.
†$384,000 + $144,000 = $528,000.

The Role of Direct Materials Yield and Direct Materials Mix Variances

Managers sometimes do have discretion to substitute one material for another. For example, the manager of Delpino's ketchup plant has some leeway in combining Caltoms and Flotoms to produce the ketchup without affecting quality. We assume that to maintain quality at least 75% and no more than 80% of the tomatoes used must be Caltoms. That is, of the 1.6 tons of tomatoes required to produce 1 ton of ketchup, at least 1.2 (1.6 × 75%) tons but no more than 1.28 (1.6 × 80%) tons must be Caltoms. When inputs are substitutable, direct materials efficiency improvement can come from two sources: (1) using less input to achieve a given output and (2) using a cheaper mix to produce a given output. The direct materials yield variance and the direct materials mix variance divide the efficiency variance into two variances: the yield variance focusing on total inputs used and the mix variance on how the inputs are combined.

Using the framework in Chapter 7, the direct materials yield variance and the direct materials mix variance are Level 4 variances. They provide insight into the direct materials efficiency variance, a Level 3 variance as diagrammed below:

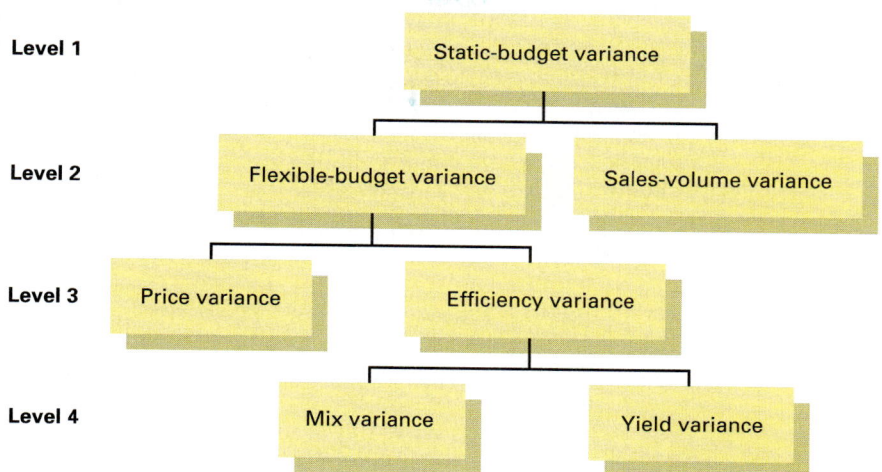

<table>
<tr><td>Level 1</td><td colspan="2" align="center">Static-budget variance</td></tr>
<tr><td>Level 2</td><td align="center">Flexible-budget variance</td><td align="center">Sales-volume variance</td></tr>
<tr><td>Level 3</td><td align="center">Price variance</td><td align="center">Efficiency variance</td></tr>
<tr><td>Level 4</td><td align="center">Mix variance</td><td align="center">Yield variance</td></tr>
</table>

Objective 4

Understand how materials yield and materials mix variances highlight trade-offs among material inputs

Given that the budgeted input mix is unchanged, the **total direct materials yield variance** is the difference between two amounts: (1) the budgeted cost of direct materials based on actual quantities of all direct materials inputs used and (2) the flexible-budget cost of direct materials based on the budgeted quantities of direct materials inputs for the actual output achieved. Given that the actual quantities of all direct materials inputs used are unchanged, the **total direct materials mix variance** is the difference between two amounts: (1) the budgeted cost for the actual direct materials input mix and (2) the budgeted cost if the budgeted direct materials input mix had been unchanged. The analysis of the direct materials yield variance and the direct materials mix variance is conceptually very similar to the analysis of the sales-quantity variance and the sales-mix variance described in Part One of this chapter.

Exhibit 22-7 presents the *direct materials yield variance* and *direct materials mix variance* for Delpino Corporation. We start with column 3 and work our way to column 1.

DIRECT MATERIALS YIELD VARIANCE. Compare columns 3 and 2. Column 3 calculates the flexible-budget cost based on the quantities of Caltoms (4,800 tons) and Flotoms (1,600 tons) required to produce the actual output (4,000 tons of ketchup). Column 3 can be restated as the budgeted cost of the budgeted quantities of all inputs used (6,400 tons) times the budgeted input mix (Caltoms, 75%; Flotoms, 25%). The *only difference in the two columns is that column 3 uses budgeted quantities of all inputs used, and column 2 uses actual quantities of all inputs used (6,500 tons).* Hence the difference in cost between the two columns is the direct materials yield variance, due solely to differences in input yields.

$$
\begin{array}{ccc}
\begin{array}{c}\text{Direct}\\\text{materials}\\\text{yield variance}\end{array} =
&
\left(
\begin{array}{cc}
\begin{array}{c}\text{Actual quantity}\\\text{of all direct materials}\\\text{inputs used}\\\times \text{budgeted input mix}\end{array} -
&
\begin{array}{c}\text{Budgeted quantity}\\\text{of direct materials}\\\text{inputs allowed for}\\\text{actual output units achieved}\end{array}
\end{array}
\right)
&
\times
\begin{array}{c}\text{Budgeted}\\\text{price of}\\\text{direct materials}\\\text{inputs}\end{array}
\end{array}
$$

The direct materials yield variances are:

Caltoms: $[(6,500 \times 0.75) - (4,000 \times 1.20)] \times \$80 = (4,875 - 4,800) \times \$80 = \$6,000$ U
Flotoms: $[(6,500 \times 0.25) - (4,000 \times 0.40)] \times \$90 = (1,625 - 1,600) \times \$90 = \underline{2,250}$ U
All inputs: $\underline{\underline{\$8,250}}$ U

EXHIBIT 22-7
Columnar Presentation of Direct Materials Yield and Mix Variances for Delpino Corporation

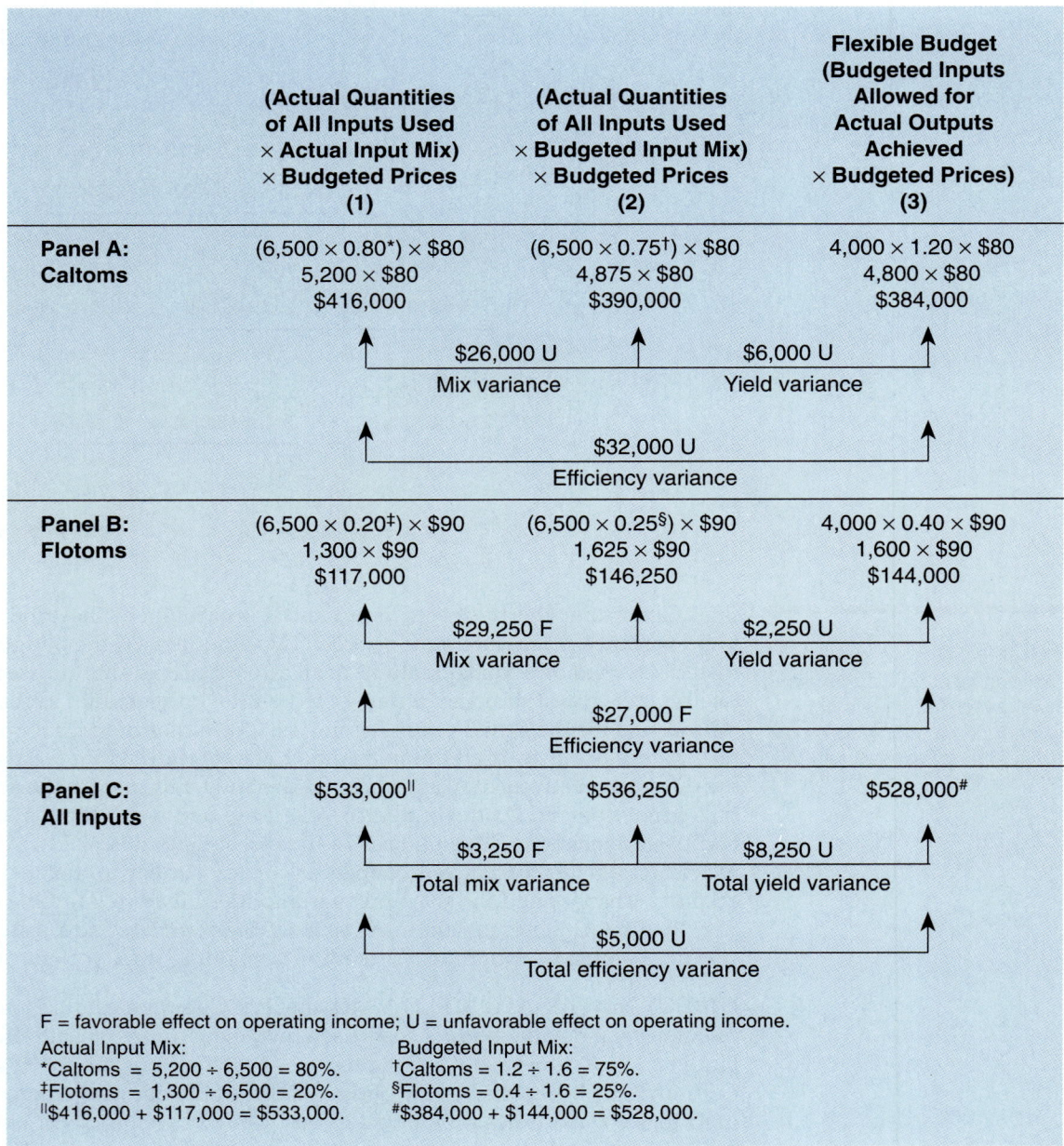

	(Actual Quantities of All Inputs Used × Actual Input Mix) × Budgeted Prices (1)	(Actual Quantities of All Inputs Used × Budgeted Input Mix) × Budgeted Prices (2)	Flexible Budget (Budgeted Inputs Allowed for Actual Outputs Achieved × Budgeted Prices) (3)
Panel A: Caltoms	$(6,500 \times 0.80^*) \times \$80$ $5,200 \times \$80$ $\$416,000$	$(6,500 \times 0.75^\dagger) \times \$80$ $4,875 \times \$80$ $\$390,000$	$4,000 \times 1.20 \times \$80$ $4,800 \times \$80$ $\$384,000$
	⬆ $26,000 U Mix variance ⬆	$6,000 U Yield variance ⬆	
	⬆ $32,000 U Efficiency variance ⬆		
Panel B: Flotoms	$(6,500 \times 0.20^\ddagger) \times \$90$ $1,300 \times \$90$ $\$117,000$	$(6,500 \times 0.25^\S) \times \$90$ $1,625 \times \$90$ $\$146,250$	$4,000 \times 0.40 \times \$90$ $1,600 \times \$90$ $\$144,000$
	⬆ $29,250 F Mix variance ⬆	$2,250 U Yield variance ⬆	
	⬆ $27,000 F Efficiency variance ⬆		
Panel C: All Inputs	$533,000‖	$536,250	$528,000#
	⬆ $3,250 F Total mix variance ⬆	$8,250 U Total yield variance ⬆	
	⬆ $5,000 U Total efficiency variance ⬆		

F = favorable effect on operating income; U = unfavorable effect on operating income.

Actual Input Mix:	Budgeted Input Mix:
*Caltoms = 5,200 ÷ 6,500 = 80%.	†Caltoms = 1.2 ÷ 1.6 = 75%.
‡Flotoms = 1,300 ÷ 6,500 = 20%.	§Flotoms = 0.4 ÷ 1.6 = 25%.
‖$416,000 + $117,000 = $533,000.	#$384,000 + $144,000 = $528,000.

Note from both Exhibit 22-7 and the formula that the total direct materials yield variance is the sum of the direct materials yield variances for each input.

To interpret the direct materials yield variance, recall that we calculated the direct materials yield variance assuming that the direct materials mix does not change from the budget and using budgeted prices per ton. An increase in the actual quantities over the budgeted quantities of all direct materials inputs leads to an unfavorable direct materials yield variance for each direct materials input—Caltoms and Flotoms in this case. (Remember that the direct materials mix proportion does not change.) Similarly, if actual quantities fall short of budgeted quantities of all direct materials inputs, a favorable direct materials yield variance for each input occurs. Managers can study how the budgeted quantities and costs for each direct materials input and in total will fluctuate if total actual inputs depart from total budgeted inputs.

DIRECT MATERIALS MIX VARIANCE. Compare columns 2 and 1 in Exhibit 22-7. Both columns compute costs using actual quantities of all inputs used (6,500 tons) and budgeted input prices (Caltoms, $80; Flotoms, $90). *The only difference is that column 2 uses budgeted input mix (Caltoms, 75%; Flotoms, 25%), and column 1 uses actual input mix (Caltoms, 80%; Flotoms, 20%).* The difference in costs between the two columns is the direct materials mix variance, which arises solely due to differences in the mix of inputs used.

As the word *mix* implies, the direct materials mix variance in Exhibit 22-7 concentrates on the changes in the percentages (proportions) of the individual inputs used. Comparing columns 1 and 2 of Exhibit 22-7, note that:

$$
\begin{array}{l}
\text{Direct} \\
\text{materials} \\
\text{mix} \\
\text{variance}
\end{array}
=
\left(
\begin{array}{l}
\text{Actual} \\
\text{direct materials} \\
\text{input mix} \\
\text{percentage}
\end{array}
-
\begin{array}{l}
\text{Budgeted} \\
\text{direct materials} \\
\text{input mix} \\
\text{percentage}
\end{array}
\right)
\times
\begin{array}{l}
\text{Actual quantities} \\
\text{of all} \\
\text{direct materials} \\
\text{inputs used}
\end{array}
\times
\begin{array}{l}
\text{Budgeted} \\
\text{price} \\
\text{of direct materials} \\
\text{inputs}
\end{array}
$$

The direct materials mix variances are:

$$
\begin{array}{llll}
\text{Caltoms} & = (0.80 - 0.75) \times 6,500 \times \$80 & = 0.05 \times 6,500 \times \$80 & = \$26,000 \text{ U} \\
\text{Flotoms} & = (0.20 - 0.25) \times 6,500 \times \$90 & = (-0.05) \times 6,500 \times \$90 & = \underline{\ 29,250 \text{ F}} \\
\text{All inputs} & & & \underline{\$\ \ 3,250 \text{ F}}
\end{array}
$$

An unfavorable direct materials mix variance occurs with Caltoms because Delpino used a greater percentage of Caltoms in its direct materials mix (80%) than budgeted (75%). Flotoms shows a favorable direct materials mix variance because the actual direct materials mix percentage of Flotoms (20%) is less than the budgeted direct materials mix percentage (25%).

The total direct materials mix variance indicates that the overall effect of the change in actual direct materials mix from the budgeted mix is favorable ($3,250 F). The actual mix of direct materials inputs had a greater proportion of the cheaper input (Caltoms) than the budgeted mix had. The total direct materials mix variance helps managers understand how total budgeted costs vary as the actual direct materials mix varies from the budgeted mix.

How should we interpret the analysis in Exhibit 22-7? The total direct materials yield variance in Exhibit 22-7 is $8,250 U, and the total direct materials mix variance is $3,250 F. There was a trade-off among ingredients that reduced the cost of the mix of inputs used but hurt yield. The net effect is that the budgeted average unit cost of the overall direct materials inputs is higher than expected, and a net unfavorable variance of $5,000 occurs. The direct materials yield and direct materials mix variances provide additional insights for managers comparing actual results and budgeted amounts.

A summary of the direct materials variances computed in this section follows:

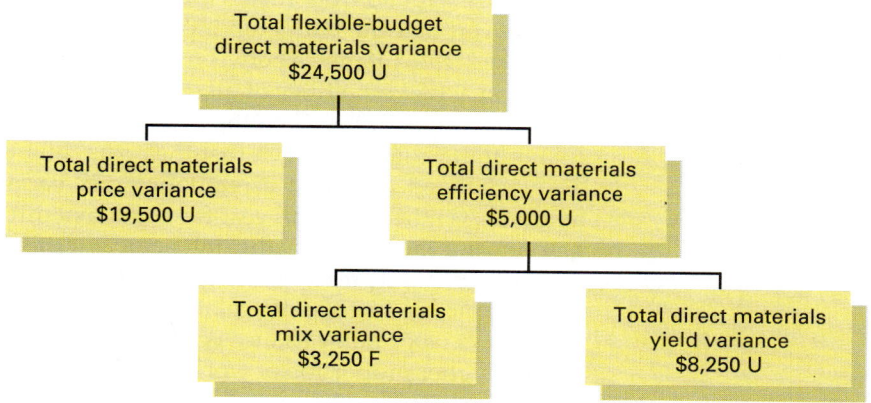

DIRECT LABOR YIELD AND DIRECT LABOR MIX VARIANCES

Objective 5

Explain the direct labor yield and direct labor mix variances

As we have just seen, managers frequently develop the standard costs of products by using specified combinations of direct materials with different individual prices. Managers often use the same approach in developing direct manufacturing labor or direct service labor standards. We illustrate direct labor yield and mix variances for a service company, Lee and Associates, a firm of architects. These variances closely parallel the direct materials yield and mix variances described in the preceding section. Lee and Associates has two levels of professional staff: senior architects, who are responsible for the main designs, and junior architects, who provide technical support for the senior architects. Budgeted costs for five architectural jobs done over a particular period follow:

2,400 hours of senior architects' time at $75 per hour	$180,000
3,600 hours of junior architects' time at $25 per hour	90,000
6,000 total hours	$270,000

The budgeted average cost per hour is $45 ($270,000 ÷ 6,000 total hours). Actual results show that Lee and Associates completed the five jobs in 5,800 hours:

2,610 hours of senior architects' time at $80 per hour	$208,800
3,190 hours of junior architects' time at $20 per hour	63,800
5,800 total hours	272,600
Budgeted costs of the jobs completed: 6,000 hours at $45 per hour	270,000
Total variance to be explained	$ 2,600U

The budgeted mix and actual mix of hours of senior architects and junior architects are:

	Budgeted Mix	**Actual Mix**
Senior architect labor	0.40 (2,400/6,000)	0.45 (2,610/5,800)
Junior architect labor	0.60 (3,600/6,000)	0.55 (3,190/5,800)
	1.00	1.00

Exhibit 22-8 presents the direct labor price and the direct labor efficiency variances for each employee category and for the two combined. The total price variance is favorable ($2,900 F) because of the lower wage rates paid to junior architects, but the total efficiency variance is unfavorable ($5,500 U). The unfavorable total direct labor efficiency variance of $5,500 may be divided into yield and mix variances in the same way that we divided the direct materials efficiency variance in the preceding section. Because we are focusing on the direct labor efficiency variance, all price variances are excluded from these calculations.

Given that the budgeted input mix is unchanged, the **total direct labor yield variance** is the difference between two amounts: (1) the budgeted cost of direct labor based on actual quantities of all direct labor inputs used and (2) the flexible budget cost of direct labor based on the budgeted quantities of direct labor inputs for the actual output achieved. Given that the actual quantities of all direct labor inputs used are unchanged, the **total direct labor mix variance** is the difference between two amounts: (1) the budgeted cost for the actual direct labor input mix and (2) the budgeted cost if the budgeted direct labor input mix had been unchanged.

Exhibit 22-9 presents the computations for the direct labor yield and direct labor mix variances. The total direct labor yield variance in Exhibit 22-9 is $9,000 F, and the total direct labor mix variance is $14,500 U. The unfavorable mix variance occurs because a higher proportion of work was done by senior architects. (Senior architects accounted for 45% of the total actual direct labor-hours spent on the various jobs but had been budgeted to handle only 40%). The favorable yield variance indicates that

EXHIBIT 22-8

EXHIBIT 22-8
Columnar Presentation of Direct Labor Price and Efficiency Variances for Lee and Associates

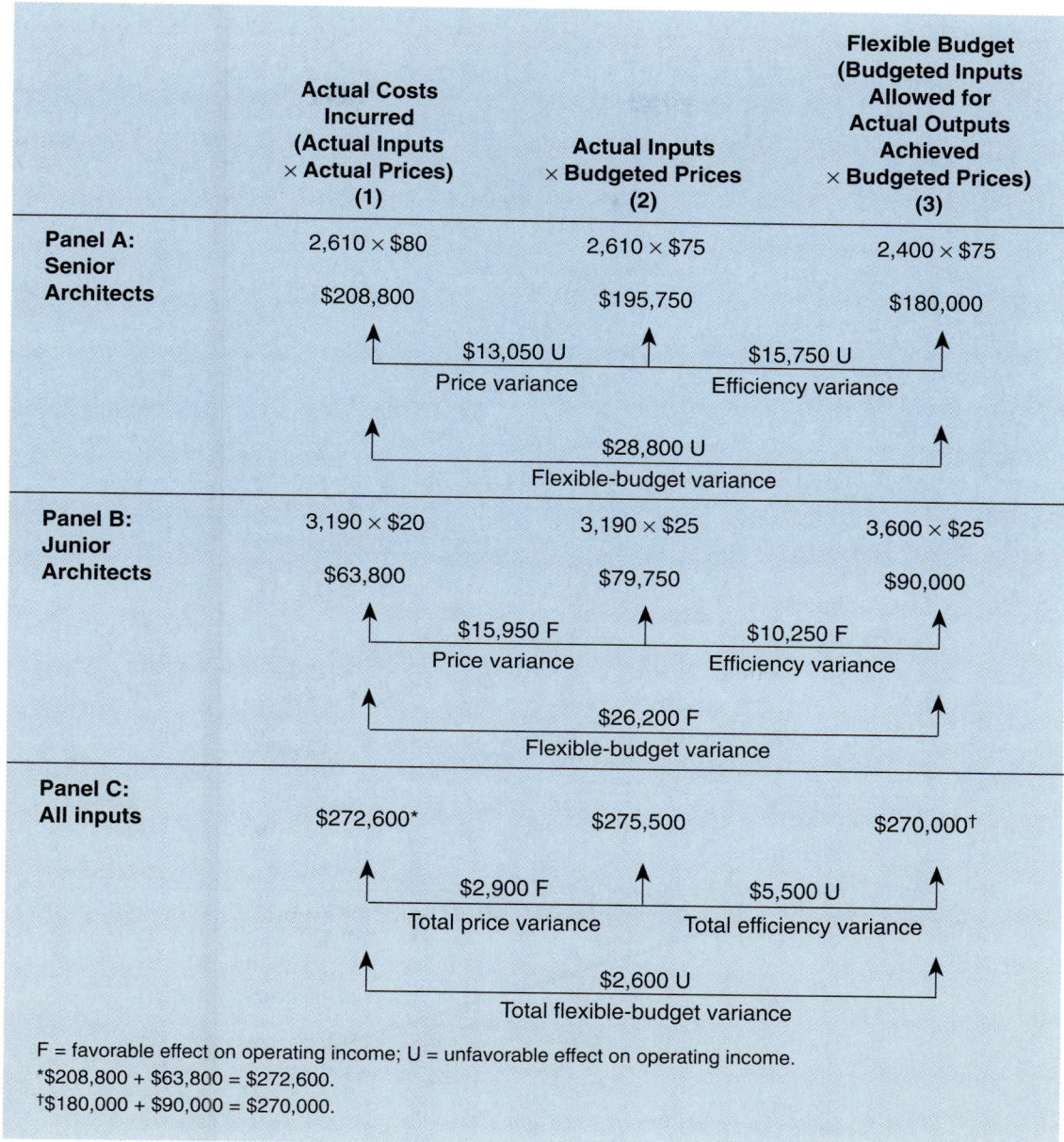

F = favorable effect on operating income; U = unfavorable effect on operating income.
*$208,800 + $63,800 = $272,600.
†$180,000 + $90,000 = $270,000.

the work was completed faster: in 5,800 total actual hours compared with 6,000 total budgeted hours. Perhaps this result was due to the extra time spent by the senior architects on the job. But was the mix-versus-yield trade-off worth it? No, because the overall direct labor efficiency decreased.

◆ PART THREE: PRODUCTIVITY MEASUREMENT

"Most people don't know a lot about productivity—exactly what it is; what you do to raise it—but they do know it's a good thing. It's the economist's answer to motherhood. Everyone's for it; no-one's against it.

EXHIBIT 22-9

Columnar Presentation of Direct Labor
Yield and Mix Variances for Lee and Associates

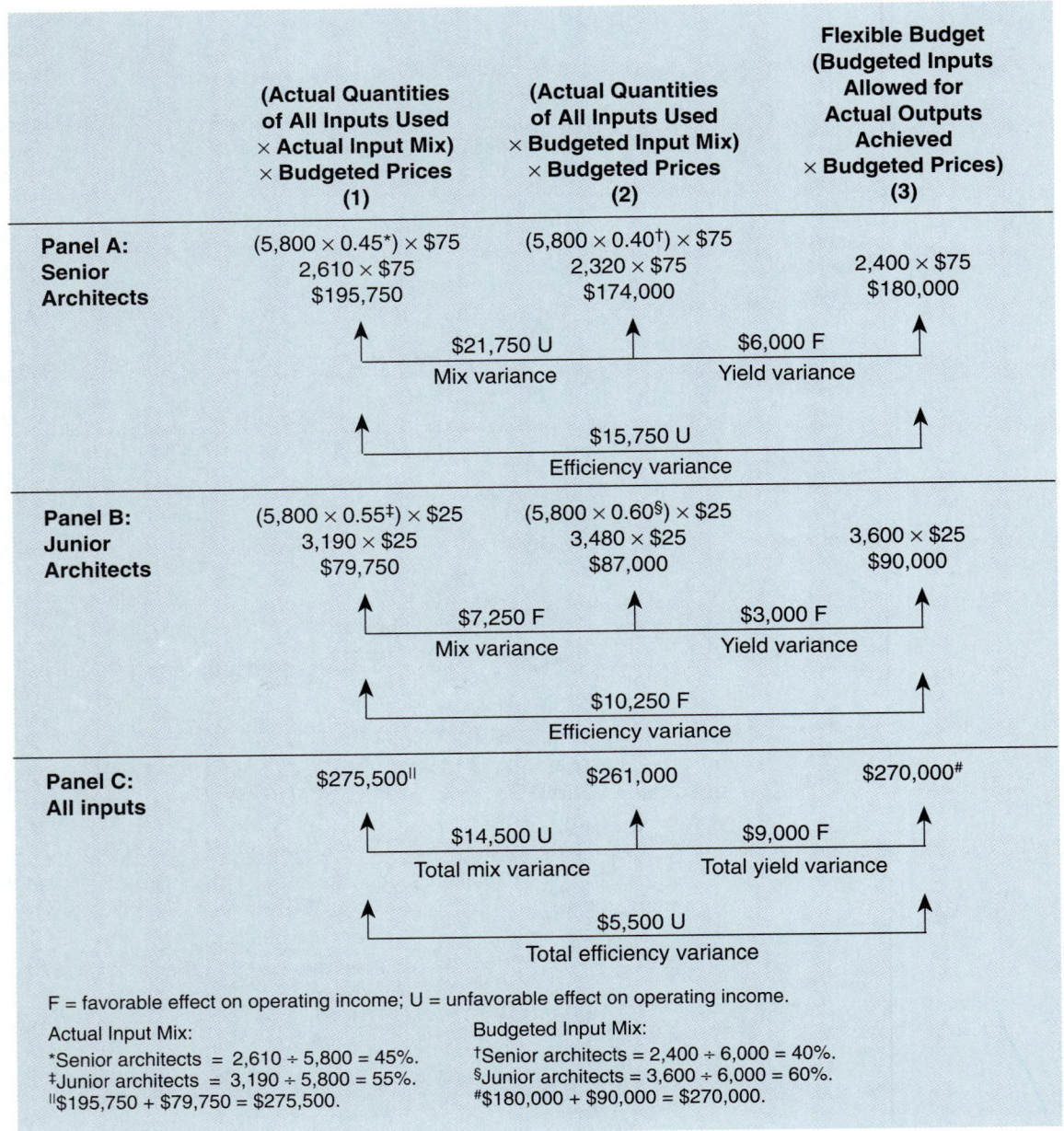

	(Actual Quantities of All Inputs Used × Actual Input Mix) × Budgeted Prices (1)	(Actual Quantities of All Inputs Used × Budgeted Input Mix) × Budgeted Prices (2)	Flexible Budget (Budgeted Inputs Allowed for Actual Outputs Achieved × Budgeted Prices) (3)
Panel A: Senior Architects	(5,800 × 0.45*) × $75 2,610 × $75 $195,750	(5,800 × 0.40†) × $75 2,320 × $75 $174,000	2,400 × $75 $180,000

$21,750 U → Mix variance $6,000 F → Yield variance

$15,750 U → Efficiency variance

Panel B: Junior Architects	(5,800 × 0.55‡) × $25 3,190 × $25 $79,750	(5,800 × 0.60§) × $25 3,480 × $25 $87,000	3,600 × $25 $90,000

$7,250 F → Mix variance $3,000 F → Yield variance

$10,250 F → Efficiency variance

Panel C: All inputs	$275,500‖	$261,000	$270,000#

$14,500 U → Total mix variance $9,000 F → Total yield variance

$5,500 U → Total efficiency variance

F = favorable effect on operating income; U = unfavorable effect on operating income.

Actual Input Mix:
*Senior architects = 2,610 ÷ 5,800 = 45%.
‡Junior architects = 3,190 ÷ 5,800 = 55%.
‖$195,750 + $79,750 = $275,500.

Budgeted Input Mix:
†Senior architects = 2,400 ÷ 6,000 = 40%.
§Junior architects = 3,600 ÷ 6,000 = 60%.
#$180,000 + $90,000 = $270,000.

And everyone's right. Increased productivity is the key to increased material prosperity. 'Doing more with less'—the simplest view of what it means makes us wealthier."[3]

Countries and companies pay great attention to productivity. Economists reason that productivity gains are the engine for improvements in standards of living. Of the major industrialized nations, Japan has experienced the highest *growth* in labor productivity over the last 30 years, while U.S. labor productivity has grown at the

[3]R. Gittins, "Enterprise Bargaining, the Snake Oil for the National Malaise," *The Sydney Morning Herald*, April 27, 1991.

slowest rate. U.S. workers, however, still lead the world in labor productivity, although other nations have, over the years, narrowed the lead. Chief executive officers (CEOs) consistently cite productivity improvement and cost containment as among the most important issues requiring their attention.

WHAT IS PRODUCTIVITY?

Objective 6
Describe productivity and productivity measures

Productivity measures the relationship between actual inputs used and actual outputs achieved; the lower the inputs for a given set of outputs or the higher the outputs for a given set of inputs, the higher the level of productivity. Productivity measurement focuses on two aspects of the relationship between inputs and outputs. It evaluates (1) whether more inputs than necessary have been used to produce output and (2) whether the best mix of inputs has been used to produce output.

We have described techniques to evaluate both these issues. Efficiency variances discussed in Chapters 7 and 8 evaluate points 1 and 2 together. As described in Part Two of this chapter, yield variances address point 1 and mix variances examine point 2. Recall, however, that the yield and mix variance calculations require that substitutions can be made within a given category of inputs: direct materials inputs (Caltoms and Flotoms) in the Delpino Corporation example, and direct labor inputs (senior and junior architects) in the Lee and Associates example.

In many contexts, substitutions can occur between direct materials and direct manufacturing labor. Garment manufacturers, for example, can substitute fasteners for buttons and so save the labor costs they would incur in making buttonholes. Substitutions may also occur between labor and capital. Manufacturing companies, for example, must decide on the extent of capital investment in automation that in turn reduces labor costs. Productivity measures must confront these more general substitutions among different types of inputs.

Chapter 7 illustrated variance analysis and the computations of efficiency and price variances using information from a standard costing system. A productivity measure compares the relationship between actual inputs and actual outputs over time. It indicates, for example, if an organization is using fewer and fewer inputs to produce the same quantity of output. This chapter describes how the Chapter 7 framework can be used to analyze productivity changes. Productivity comparisons provide an effective summary of an organization's effort at continuous improvement. An organization could also compare its productivity to the productivity of another similar organization. For the most part, however, managers use productivity to describe the actual performance of an organization over time, which is the approach we take in this chapter.

We illustrate productivity measures using data from Ramona Inc. Ramona makes wooden door handles. For simplicity, we focus on only two inputs, direct materials and direct manufacturing labor, which are partial substitutes. Carving handles out of wooden boards creates trims and edges (the portions of the board that remain after the handles have been carved out). These trims and edges could be reused, but they would require more care and attention and consequently more direct manufacturing labor time. Alternatively, trims and edges can be thrown away, saving direct manufacturing labor time but increasing direct materials usage.

Ramona provides the following information for the years 19_4 and 19_5.

	19_4	19_5
Number of wooden handles produced and sold	425,000	510,000
Direct manufacturing labor-hours used	34,000	37,400
Wages per hour	$14	$15
Direct materials (in square feet) used	170,000	219,500
Direct materials costs per square foot	$2.05	$2.00
Direct manufacturing labor costs: (direct manufacturing labor hours × wages per hour)	$476,000	$561,000

| | Direct materials costs: (direct materials used × direct materials costs per square foot) | $348,500 | $439,000 |
| | Total input costs: (direct manufacturing labor costs + direct materials costs) | $824,500 | $1,000,000 |

Ramona's total costs for direct materials and direct manufacturing labor increased from $824,500 in 19_4 to $1,000,000 in 19_5. Many reasons can account for the change in costs—a change in the quantity of outputs produced, a change in the prices of inputs used, or a change in productivity (resulting from a change in the way inputs are converted into output). The goal is to isolate the effect of any productivity change on costs. Management can use this information to evaluate actions aimed at improving productivity and reducing costs. These actions may include training workers and building better supplier relationships, among many others. To understand productivity changes, we consider both partial productivity measures and total factor productivity.

PARTIAL PRODUCTIVITY MEASURES

Objective 7

Discuss the benefits and drawbacks of partial productivity measures

Partial productivity compares the quantity of output produced with the quantity of a *single* input used. In its most common form, partial productivity is expressed as a ratio:

$$\text{Partial productivity} = \frac{\text{Quantity of output produced}}{\text{Quantity of input used}}$$

The higher the ratio, the greater the productivity. Partial productivity measures ignore the price of inputs.

Consider direct manufacturing labor productivity at Ramona in 19_4:

$$\frac{\text{Direct manufacturing}}{\text{labor partial productivity}} = \frac{\text{Quantity of wooden handles produced during 19\_4}}{\text{Direct manufacturing labor-hours used to produce handles in 19\_4}}$$

$$= \frac{425,000 \text{ units}}{34,000 \text{ labor hours}}$$

$$= 12.5 \text{ handles per direct manufacturing labor-hour}$$

Observe that the direct manufacturing labor partial productivity measure is, indeed, a *partial* productivity measure. Ramona uses two inputs, direct manufacturing labor-hours and direct materials to make its wooden handles. The direct manufacturing labor partial productivity measure considers only one of those inputs.

We can similarly define the direct materials partial productivity at Ramona for 19_4 as:

$$\frac{\text{Direct materials}}{\text{partial productivity}} = \frac{\text{Quantity of wooden handles produced during 19\_4}}{\text{Direct materials used to produce handles in 19\_4}}$$

$$= \frac{425,000 \text{ units}}{170,000 \text{ square feet}}$$

$$= 2.5 \text{ wooden handles per square foot of direct materials}$$

By itself, a partial productivity measure has little meaning. It gains meaning only when comparisons are made that examine productivity changes:

◆ Over time
◆ Among several facilities
◆ Relative to a benchmark

The following table presents Ramona's direct manufacturing labor partial productivity and direct materials partial productivity for the years 19_4 and 19_5.

Partial Productivities

	19_4	19_5
Direct manufacturing labor partial productivity	$\dfrac{425{,}000}{34{,}000} = 12.50$	$\dfrac{510{,}000}{37{,}400} = 13.64$
Direct materials partial productivity	$\dfrac{425{,}000}{170{,}000} = 2.50$	$\dfrac{510{,}000}{219{,}500} = 2.32$

Comparing Changes in Partial Productivities and Efficiency Variances

How do we compare partial productivities across periods? By using information about the production function or production technology that describes how inputs get converted into outputs. A *production technology* or *production function* describes the relationship between different quantities of inputs used and the quantities of output produced. The production technology indicates the effect of changes in input quantities on the output produced. A particular form of the production technology is the *constant returns to scale technology,* which describes the following relationship between inputs and output: increasing each input by a particular percentage increases output by the same percentage. We assume that Ramona's production technology is a constant returns to scale technology. That is, if Ramona increases the quantities of direct manufacturing labor and direct materials by 10% each in 19_4, the output quantity will also increase by 10%. If Ramona increases the quantity of each input by 20% in 19_4, output quantity will increase by 20%, and so on. This assumption is similar to the one made in earlier chapters of the book that output increases linearly with direct manufacturing labor and direct materials (in the relevant range).

Ramona produced 510,000 handles in 19_5, which is 20% more than the output produced in 19_4 (510,000 = 425,000 × 1.20). If Ramona had produced 510,000 handles in 19_4, how many direct manufacturing labor-hours and how much direct materials would Ramona be expected to use? Assuming constant returns to scale, Ramona would require 40,800 (34,000 × 1.20) direct manufacturing labor-hours and 204,000 (170,000 × 1.20) square feet of direct materials.

How would producing more output have affected Ramona's partial productivity measures in 19_4?

$$\text{Direct manufacturing labor partial productivity} = \frac{510{,}000}{40{,}800} = \text{12.5 handles per direct manufacturing labor-hour}$$

$$\text{Direct materials partial productivity} = \frac{510{,}000}{204{,}000} = \text{2.5 handles per square foot of direct materials}$$

Partial productivity measures would be unaffected if Ramona had produced 510,000 handles. Exhibit 22-10 indicates why this is the case. The numerator and denominator in the partial productivity measures are both being multiplied by the same number (1.20). More generally stated, under a constant returns to scale technology, partial productivity measures for 19_4 would be unchanged whatever the quantity of output produced in 19_4. This means that it is valid to directly compare the partial productivities in the two years, even though the output quantity has increased in 19_5 relative to 19_4. The table of partial productivities above indicates that

- ◆ Direct manufacturing labor productivity increased by 9.12% [(13.64 − 12.5) ÷ 12.5]
- ◆ Direct materials productivity decreases by 7.20% [(2.32 − 2.5) ÷ 2.5]

To see why, recall that to produce 510,000 handles Ramona used 37,400 direct manufacturing labor-hours in 19_5. It would have used 40,800 direct manufacturing labor-hours in 19_4 (a decrease of 3,400 direct manufacturing labor-hours in 19_5). Conversely, Ramona used 219,500 square feet of wood in 19_5. It would have used 204,000 square feet in 19_4 (an increase of 15,500 square feet of wood in 19_5).

EXHIBIT 22-10

Partial Productivities for Ramona in 19_4 under Constant Returns to Scale If Inputs
Increased by 20%

	19_4 Actuals		Output and Inputs If Each Input Increased by 20%
Number of wooden handles produced	425,000	× 1.20 →	510,000
Direct manufacturing labor-hours used	34,000	× 1.20 →	40,800
Direct materials used (square feet)	170,000	× 1.20 →	204,000
Direct manufacturing labor partial productivity	$\frac{425,000}{34,000} = 12.5$		$\frac{510,000}{40,800} = 12.5$
Direct materials partial productivity	$\frac{425,000}{170,000} = 2.5$		$\frac{510,000}{204,000} = 2.5$

Chapter 7 stressed price and efficiency variances. Exhibit 22-11 illustrates that even though direct manufacturing labor partial productivity increased from 19_4 to 19_5, the direct manufacturing labor efficiency variance (calculated on the basis of a standard costing system) was favorable in 19_4 and unfavorable in 19_5. Ramona's workers have not performed up to the standards set for 19_5 although they have outperformed the standards for 19_4 ($0.084 \times 510,000 = 42,840$ hours). The reason, of course, is that the standards for 19_5 were tighter than the standards for 19_4; perhaps improvements were expected to occur. Ramona's workers did not make the targeted improvements, although the higher productivity indicates that their 19_5 performance did improve over their 19_4 actual performance.

Contrast the efficiency variances of Chapter 7 with the productivity measures here:

EXHIBIT 22-11

Comparison of Direct Manufacturing Labor Productivity and
Direct Manufacturing Labor Efficiencies Based on a Standard Costing System

Year (1)	Standard Direct Manufacturing Labor-Hours per Output Unit (numbers assumed) (2)	Output Produced (3)	Standard Direct Manufacturing Labor-Hours for Actual Output (4)	Actual Direct Manufacturing Labor Hours (5)	Standard Price Assumed to Equal Actual Price (6)	Direct Manufacturing Labor Efficiency Variance (7) = [(5) − (4)] × (6)	Direct Manufacturing Labor Partial Productivity (8) = (3) ÷ (5)
19_4	0.084	425,000	35,700	34,000	$14	(34,000 − 35,700) × $14 = $23,800 F	12.5
19_5	0.072	510,000	36,720	37,400	$15	(37,400 − 36,720) × $15 = $10,200 U	13.64

- Efficiency variances generally focus on whether *currently specified standard* relationships between inputs and outputs have been achieved.
- Annual productivity measures focus on whether the *actual* relationship between inputs and outputs has changed from one year to the next.

Thus, a company may have an unfavorable efficiency variance in a particular year when partial productivity has improved. The differences derive from the source of the benchmark used—an efficiency standard in one case and last year's actual performance in the other.

Usefulness of the Partial Productivity Measures

Partial productivity measures are important tools:

1. Partial productivity measures inform managers whether the productivity of individual inputs is increasing or decreasing. For example, Ford Motor Company's assembly productivity was 4.71 workers per vehicle in 1979 and 3.01 workers per vehicle in 1992. The corresponding numbers for General Motors are 5.12 in 1979 and 4.55 in 1992.[4]

2. Partial productivity measures focus on a single input and are simple to calculate and easily understood at the operations level. Operations personnel can more easily understand the physical number of wooden products made per direct manufacturing labor-hour than they can understand financial numbers.

Drawbacks of Partial Productivity Measures

For all their advantages, partial productivity measures also have some serious drawbacks. Partial productivity does not focus on the productivity of all inputs simultaneously. It fails to consider the effect of substituting one input for another. Direct manufacturing labor productivity improved in 19_5 relative to 19_4 but direct materials productivity declined. Is declining direct materials productivity a concern? Not necessarily. Rigidly interpreting partial productivity measures can be misleading. The key is to recognize that direct manufacturing labor and direct materials are substitutes (for example, throwing away edges and trims increases direct materials consumption but decreases direct manufacturing labor-hours). Ramona can produce any given level of output (510,000 units in 19_5, for example) using various combinations of direct manufacturing labor-hours and direct materials. Each possible combination uses a different mix of direct manufacturing labor-hours and direct materials. The specific substitutions that will lead to the lowest costs depend on the price of each input resource. For example, in 19_5, Ramona may have used more direct materials and fewer direct manufacturing labor-hours to take advantage of the lower price of direct materials. But partial productivity measures ignore the possibilities and advantages of input substitution.

This discussion leads to a conclusion that may contradict commonly held views about productivity. We often think of productivity as a physical measure, how many units of output produced per unit of input, lacking financial content. This is not the case. Productivity measurement is intricately tied to minimizing total cost—a financial objective.

TOTAL FACTOR PRODUCTIVITY

Objective 8
Describe total factor productivity and its advantages and disadvantages

Total factor productivity (TFP), or total productivity, measures the combined productivity of all inputs used to produce output by taking into account the relative prices of inputs. **Total factor productivity** is the ratio of the quantity of output produced to the

---

[4]*Manufacturing Engineering* (April 1993), p. 16.

quantity of *all* inputs used, where the inputs are combined on the basis of current period prices.

$$\text{Total factor productivity} = \frac{\text{Quantity of output produced}}{\text{Costs of all inputs used}}$$

TFP considers all inputs simultaneously and also considers the trade-offs across inputs based on input prices. Our goal is to measure changes in total factor productivity (TFP) from one period to the next.

Calculating Total Factor Productivity and Change in Total Factor Productivity

We first calculate Ramona's total factor productivity (TFP) in 19_5, using 19_5 prices and quantity of output produced = 510,000 units.

$$
\begin{pmatrix} \text{Costs of} \\ \text{inputs used} \\ \text{in 19\_5} \\ \text{based on} \\ \text{19\_5 prices} \end{pmatrix} = \begin{pmatrix} \text{Direct} & & \text{Direct} \\ \text{manufacturing} & \times & \text{manufacturing} \\ \text{labor-hours used} & & \text{labor rate} \\ \text{in 19\_5} & & \text{in 19\_5} \end{pmatrix} + \begin{pmatrix} \text{Direct} & & \text{Direct} \\ \text{materials} & \times & \text{materials} \\ \text{used in} & & \text{prices} \\ \text{19\_5} & & \text{in 19\_5} \end{pmatrix}
$$

= (37,400 × $15) + (219,500 × $2)

= $561,000 + $439,000 = $1,000,000

$$
\begin{aligned}
\text{Total factor} \atop \text{productivity} &= \frac{\text{Quantity of output produced}}{\text{Costs of inputs used in 19\_5 based on 19\_5 prices}} \\
&= \frac{510,000}{\$1,000,000}
\end{aligned}
$$

= 0.51 units of output per dollar of inputs

By itself, the 19_5 TFP of 0.51 handles per dollar of input is not particularly helpful. We cannot tell if the manager used fewer physical inputs to produce output, or if the manager wisely substituted one input for another to take advantage of changes in input prices in 19_5 relative to 19_4. We need something to compare the 19_5 TFP against. We use, as a comparison, the units of output per dollar of inputs (TFP) that the manager would have achieved in 19_4 on the basis of the inputs used in 19_4 to produce 425,000 units of output, calculated using 19_5 prices.

$$
\begin{pmatrix} \text{Costs of} \\ \text{inputs used} \\ \text{in 19\_4} \\ \text{based on} \\ \text{19\_5 prices} \end{pmatrix} = \begin{pmatrix} \text{Direct} & & \text{Direct} \\ \text{manufacturing} & \times & \text{manufacturing} \\ \text{labor-hours used} & & \text{labor rate} \\ \text{in 19\_4} & & \text{in 19\_5} \end{pmatrix} + \begin{pmatrix} \text{Direct} & & \text{Direct} \\ \text{materials} & \times & \text{materials} \\ \text{used in} & & \text{prices} \\ \text{19\_4} & & \text{in 19\_5} \end{pmatrix}
$$

= (34,000 × $15) + (170,000 × $2)

= $510,000 + $340,000

= $850,000

$$
\begin{aligned}
\text{Total factor} \atop \text{productivity} &= \frac{\text{Quantity of output produced}}{\text{Costs of inputs used in 19\_4 based on 19\_5 prices}} \\
&= \frac{425,000}{\$850,000}
\end{aligned}
$$

= 0.50 units of output per dollar of inputs

Using 19_5 prices, total factor productivity increases from 0.50 units of output per dollar of inputs in 19_4 to 0.51 units of output per dollar of inputs in 19_5. The increase in total factor productivity from 19_4 to 19_5 is 2% ([0.51 − 0.50] ÷ 0.50). The decrease in partial productivity of direct materials was more than offset by productivity gains in direct manufacturing labor.

Why did TFP increase by 2% from 19_4 to 19_5? To understand the reasons for the change in TFP, we follow the same basic structure that we used to understand the reasons for partial productivity changes. Using 19_5 prices, we compare the input costs that Ramona incurred in 19_5 to manufacture 510,000 handles with the input costs that Ramona would have incurred in 19_4 to produce the same output. The steps in the analysis follow:

STEP 1. Determine the amount of direct manufacturing labor-hours and direct materials that Ramona would have used if it had produced 510,000 units of output in 19_4. From our previous computations (see Exhibit 22-10), we know that Ramona would have used 40,800 direct manufacturing labor-hours and 204,000 square feet of direct materials.

STEP 2. Cost all the inputs computed in step 1 using 19_5 prices. Why? Because by using 19_5 prices, we are able to compute what it would have cost Ramona in 19_5 to produce 510,000 handles on the basis of the mix and quantities of inputs it used in 19_4. This cost equals $1,020,000 [(40,800 × $15) + (204,000 × $2)].

STEP 3. Calculate the actual costs incurred by Ramona in 19_5 to produce the 510,000 handles. This cost equals $1,000,000 [(37,400 × $15) + (219,500 × $2)], $20,000 less than what it would have cost on the basis of the quantities and mix of inputs used in 19_4.

Ramona shows a gain in total factor productivity because, on the basis of 19_5 prices, it uses a mix of inputs that is less costly than the mix of inputs it would have used in 19_4 to produce 510,000 handles. That is, the actual output per dollar of inputs increases in 19_5. Note that the $20,000 difference in costs between step 2 and step 3 is not due to input price differences; we used 19_5 prices to evaluate the input mix in both 19_4 and 19_5.

Two sources account for gains in TFP from one period to the next. First, TFP increases if a company uses fewer physical quantities of all inputs in the second period to produce the same quantity of output that it produced in the first period. Second, TFP increases if a company uses a mix of inputs to produce output in the second period that is cheaper than the mix of inputs it used in the first period.

TFP measures the net effect of the two factors—the physical inputs used and the mix of inputs used. For example, the 37,400 hours of direct manufacturing labor-hours in 19_5 is the net effect of taking less (or more) time to do the same tasks and eliminating some labor tasks altogether at the cost of increasing direct materials usage.

Advantages and Disadvantages of Total Factor Productivity

A major *advantage* of total factor productivity is that it measures the combined productivity of all inputs to produce output and therefore explicitly evaluates substitution among inputs. TFP recognizes that productivity gains can come from using fewer inputs to produce a given level of output or by altering the mix of inputs used to produce that output. However, TFP has two main *disadvantages*.

1. Operations personnel find physical rather than financial numbers more useful in performing their tasks. Physical measures provide direct feedback. Productivity-based bonuses, therefore, are often tied to partial labor productivity. But this situation creates incentives for workers to substitute materials (and capital) for labor, which improves their own productivity measure though possibly decreasing overall productivity of the company as measured by TFP. To overcome the incentive problems that partial productivity measures might create, some companies—for example, TRW, Eaton, and Whirlpool—explicitly adjust bonuses based on partial labor productivity for the effects of other factors such as investments in new equipment and higher levels of scrap.

2. TFP measures are difficult to link across multiple periods. Consider the Ramona example. Our analysis used 19_5 prices to evaluate changes in total factor productivity between 19_4 and 19_5. To study the change in TFP between 19_5 and 19_6, we would use 19_6 prices. Because TFP calculations in each year are calculated using the prices in that year, we would compute the change in TFP from 19_4 to 19_6 using 19_6 prices only. Now the change in TFP between 19_4 and 19_6 will not be the same as the product of (a) the change in TFP from 19_4 to 19_5 and (b) the change in TFP from 19_5 to 19_6. Why? Because the change in TFP for 19_5 is evaluated using 19_5 prices, while the change in TFP for 19_6 is evaluated using 19_6 prices.

ANALYSIS OF ANNUAL COST CHANGES

> **Objective 9**
> Explain how productivity changes can help explain cost changes from one period to the next

We next explore productivity's role in explaining the change in costs from 19_4 to 19_5. Exhibit 22-12 describes three components—the output adjustment component, the input price change component, and the productivity change component—that account for the cost changes. Each component is calculated as the difference between two adjacent columns. It is easiest to start with 19_4 actual costs and work our way toward 19_5 actual costs, so we start at the right side of the exhibit with column 4. The description of each component follows.

1. OUTPUT ADJUSTMENT COMPONENT. Focus on columns 4 and 3. Compare the descriptions of the two columns. Only the quantities of *output produced in 19_4 (425,000 handles) and in 19_5 (510,000 handles) differ between the two columns,* and hence the corresponding amounts of inputs that would have been used to produce the 19_5 output quantity in 19_4. Both columns use 19_4 prices. We label the cost difference between the two columns as an output adjustment. Why? Because the cost increase of $164,900 (for all inputs) arises solely on account of differences in the output produced in 19_4 and the output produced in 19_5.

2. INPUT PRICE CHANGE COMPONENT. Next focus on columns 3 and 2. Each column computes costs using the inputs that would have been used in 19_4 to produce the 510,000 units of output that were produced in 19_5—40,800 direct manufacturing labor-hours and 204,000 square feet of direct materials. *The only difference between the columns is the use of 19_4 actual prices in column 3 and 19_5 actual prices in column 2.* The increase in costs between columns 3 and 2 of $30,600 (for all inputs) is due solely to net increases in input prices in 19_5 over 19_4.

3. PRODUCTIVITY CHANGE COMPONENT. Finally focus on columns 2 and 1. Both columns use actual 19_5 input prices. Both columns calculate input resources used to produce the actual quantity of 510,000 handles in 19_5. The difference between the two columns arises solely because of differences in the quantities and mix of resources used in 19_4 (40,800 direct manufacturing labor-hours and 204,000 square feet of direct materials) and the quantities and mix of input resources used in 19_5 (37,400 direct manufacturing labor-hours and 219,500 square feet of direct materials). Ramona shows cost savings of $20,000 (for all inputs) as a result of productivity gains.

The difference between Ramona's actual costs in 19_4 and Ramona's actual costs in 19_5 can be explained as:

Total change in costs	=	Change in costs due to output adjustment	+	Change in costs due to input price changes	+	Change in costs due to productivity change
$175,500 U	=	$164,900 U	+	$30,600 U	+	$20,000 F

Note the following:

1. In calculating the cost difference due to output adjustment, we hold input prices and the quantity and mix of inputs constant.

EXHIBIT 22-12
Analysis of Change in Actual Costs from 19_4 to 19_5

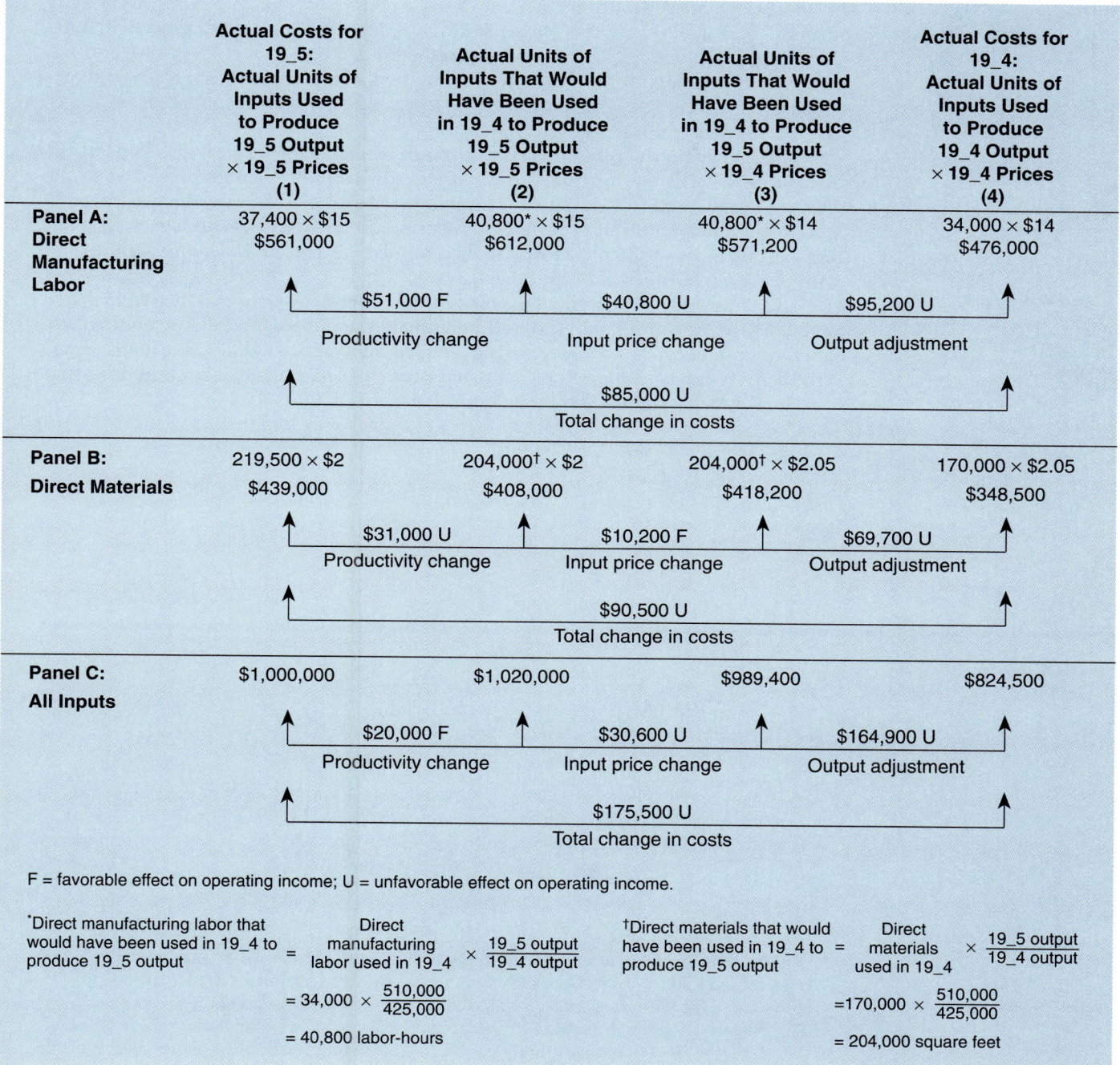

	Actual Costs for 19_5: Actual Units of Inputs Used to Produce 19_5 Output × 19_5 Prices (1)	Actual Units of Inputs That Would Have Been Used in 19_4 to Produce 19_5 Output × 19_5 Prices (2)	Actual Units of Inputs That Would Have Been Used in 19_4 to Produce 19_5 Output × 19_4 Prices (3)	Actual Costs for 19_4: Actual Units of Inputs Used to Produce 19_4 Output × 19_4 Prices (4)
Panel A: Direct Manufacturing Labor	37,400 × $15 $561,000	40,800* × $15 $612,000	40,800* × $14 $571,200	34,000 × $14 $476,000
	↑ $51,000 F Productivity change	↑ $40,800 U Input price change	↑ $95,200 U Output adjustment	↑
	↑	$85,000 U Total change in costs		↑
Panel B: Direct Materials	219,500 × $2 $439,000	204,000† × $2 $408,000	204,000† × $2.05 $418,200	170,000 × $2.05 $348,500
	↑ $31,000 U Productivity change	↑ $10,200 F Input price change	↑ $69,700 U Output adjustment	↑
	↑	$90,500 U Total change in costs		↑
Panel C: All Inputs	$1,000,000	$1,020,000	$989,400	$824,500
	↑ $20,000 F Productivity change	↑ $30,600 U Input price change	↑ $164,900 U Output adjustment	↑
	↑	$175,500 U Total change in costs		↑

F = favorable effect on operating income; U = unfavorable effect on operating income.

*Direct manufacturing labor that would have been used in 19_4 to produce 19_5 output $=$ Direct manufacturing labor used in 19_4 $\times \dfrac{\text{19\_5 output}}{\text{19\_4 output}}$

$= 34,000 \times \dfrac{510,000}{425,000}$

$= 40,800$ labor-hours

†Direct materials that would have been used in 19_4 to produce 19_5 output $=$ Direct materials used in 19_4 $\times \dfrac{\text{19\_5 output}}{\text{19\_4 output}}$

$=170,000 \times \dfrac{510,000}{425,000}$

$= 204,000$ square feet

2. In calculating the cost difference due to input price changes, we hold output quantity and the quantities and mix of inputs constant.

3. In computing the cost difference resulting from productivity changes, we hold quantity of output and the input prices constant.

Exhibit 22-12 also presents how much of the cost change in each component is due to direct manufacturing labor and to direct materials. This additional information provides managers with more details about the underlying sources of the cost changes.

SERVICE-SECTOR PRODUCTIVITY

The service sector employs well over 60% of the workforce in the United States and in most developed countries. In most of these countries, service-sector productivity growth (0.2% per year in the United States over 25 years) has lagged growth in manufacturing productivity (2.6% per year in the United States over 25 years), and has been a major reason for the slow rates of growth in overall productivity.[5] Overall productivity growth rates will continue to be low unless improvements occur in service or "white collar" productivity.

The basic productivity measures used in the service sector are the same as the measures used in the manufacturing sector—the ratio of outputs produced to the costs of inputs used to produce the output. In some service activities such as research and development, output is difficult to measure. In other service industries, measuring output is less difficult. Hospitals use number of patient-days and airlines use number of flight-miles as output measures.

There are several potential ways to improve service-sector productivity.[6] A key step is to define carefully all tasks and eliminate the unnecessary. Bank lending productivity can be enhanced by reducing detailed analysis for small-value loans. Retail productivity can be enhanced by using bar codes to electronically scan purchases and doing away with manually totaling customers' charges.

---

[5]R. Schmidt, "Services: A Future of Low Productivity Growth?" *Federal Reserve Bank of San Francisco Weekly Letter*, Febuary 14, 1992.

[6]See P. Drucker, "The New Productivity Challenge," *Harvard Business Review* (November–December 1991), pp. 69–79; and H. D. Sherman, *Service Organization Productivity Management* (Hamilton, Ontario: The Society of Management Accountants of Canada, November 1988).

PROBLEM FOR SELF-STUDY

PROBLEM

Assume the same budget for Party Wholesaler as in Exhibit 22-2 (p. 757). Suppose, however, that actual sales were 1,440 units of regular wine and 360 units of premium wine. Compute the sales-volume variances for each product and in total. Subdivide the total sales-volume variance into sales-quantity and sales-mix variances. Using your numerical results, comment on the meaning of each variance.

SOLUTION

Exhibit 22-13 presents the variances for sales volume, sales quantity, and sales mix for the revised Party Wholesaler problem. A favorable sales-quantity variance of $500 represents the increase in contribution that results, given an unchanging mix, from exceeding the budgeted unit quantity of 1,600 units by 200 units. The actual sales mix is 80% of regular wine (75% in the static budget) and 20% of premium wine (25% in the static budget). The unfavorable sales-mix variance occurs because the actual mix of wine sold had a greater proportion of the regular wine, with its lower budgeted contribution margin, than the budgeted mix had.

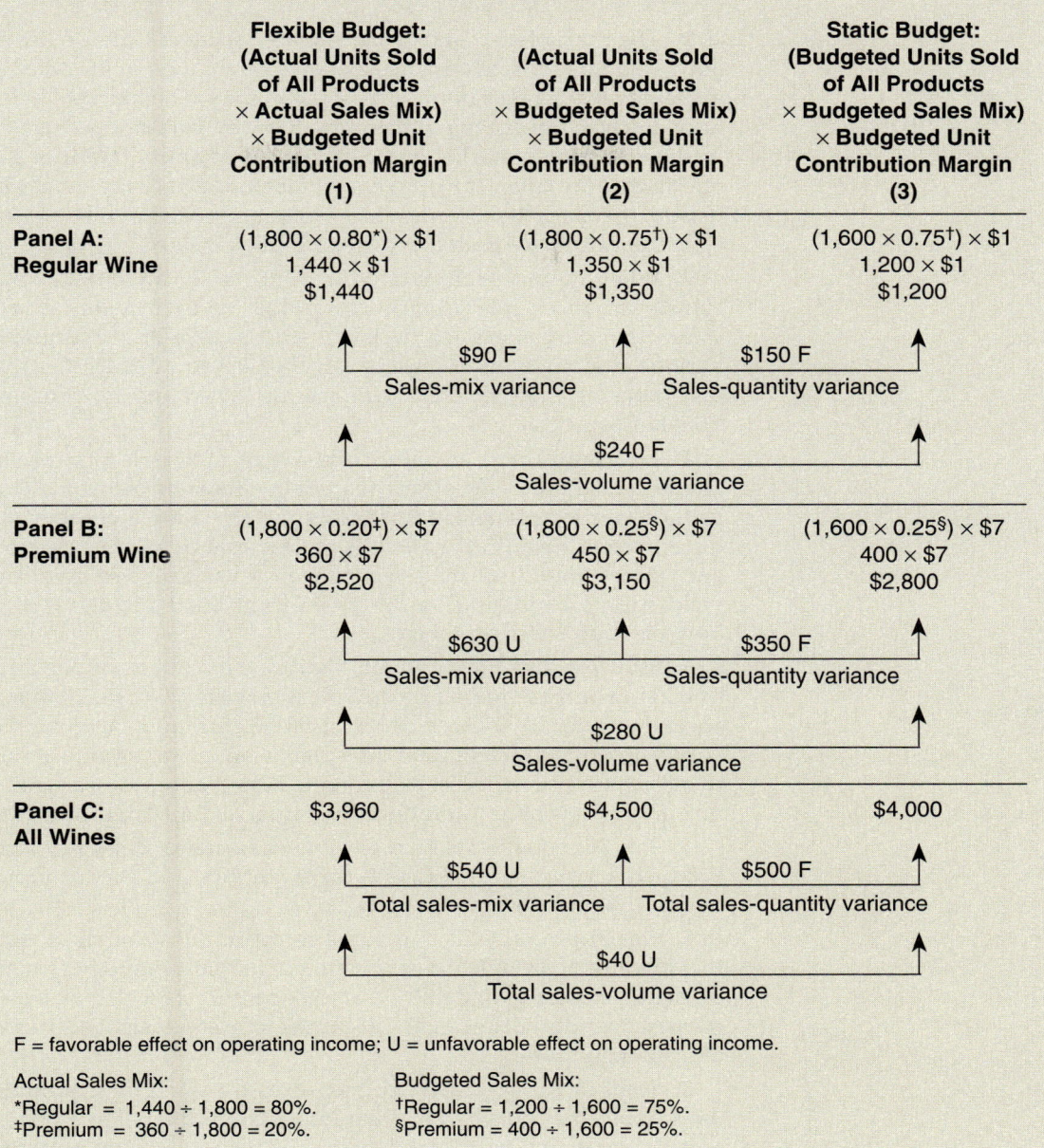

EXHIBIT 22-13
Columnar Presentation of Sales-Quantity and Sales-Mix Variances
for Party Wholesaler, October 19_5

	Flexible Budget: (Actual Units Sold of All Products × Actual Sales Mix) × Budgeted Unit Contribution Margin (1)	(Actual Units Sold of All Products × Budgeted Sales Mix) × Budgeted Unit Contribution Margin (2)	Static Budget: (Budgeted Units Sold of All Products × Budgeted Sales Mix) × Budgeted Unit Contribution Margin (3)
Panel A: **Regular Wine**	$(1,800 \times 0.80^*) \times \$1$ $1,440 \times \$1$ $\$1,440$	$(1,800 \times 0.75^\dagger) \times \$1$ $1,350 \times \$1$ $\$1,350$	$(1,600 \times 0.75^\dagger) \times \$1$ $1,200 \times \$1$ $\$1,200$

$90 F ← Sales-mix variance $150 F ← Sales-quantity variance

$240 F ← Sales-volume variance

Panel B: **Premium Wine**	$(1,800 \times 0.20^\ddagger) \times \$7$ $360 \times \$7$ $\$2,520$	$(1,800 \times 0.25^\S) \times \$7$ $450 \times \$7$ $\$3,150$	$(1,600 \times 0.25^\S) \times \$7$ $400 \times \$7$ $\$2,800$

$630 U ← Sales-mix variance $350 F ← Sales-quantity variance

$280 U ← Sales-volume variance

Panel C: **All Wines**	$3,960	$4,500	$4,000

$540 U ← Total sales-mix variance $500 F ← Total sales-quantity variance

$40 U ← Total sales-volume variance

F = favorable effect on operating income; U = unfavorable effect on operating income.

Actual Sales Mix:
*Regular = 1,440 ÷ 1,800 = 80%.
‡Premium = 360 ÷ 1,800 = 20%.

Budgeted Sales Mix:
†Regular = 1,200 ÷ 1,600 = 75%.
§Premium = 400 ÷ 1,600 = 25%.

SUMMARY

The following points are linked to the chapter's learning objectives.

1. The sales-volume variance is the difference between the flexible-budget amount and the static-budget amount (budgeted unit selling prices, budgeted unit variable costs, and budgeted fixed costs are held constant). When a company sells multiple products, two major factors affect the sales-volume variance: (a) the total actual quantity of units sold and (b) the relative proportions of products bearing different contribution margins. The sales-quantity variance captures point 1. The sales mix variance captures point 2. Separating the sales-volume variance into its quantity and mix components is especially useful when managers have discretion over the mix of products to sell.

2. The market-size and market-share variances divide the sales-quantity variance. The market-size variance explains how much of the sales-quantity variance is explained by changes in the size of the market. The market-share variance measures how much of the change in the sales-quantity variance occurred because of changes in the company's market share. The sales-quantity variance includes market-size effects that are often the outcome of factors outside the manager's control. For this reason, the market-share variance is often regarded as a better indicator of the manager's performance than the sales-quantity variance.

3. When inputs, such as two direct materials, are not substitutes, price and efficiency variances individually computed for each material typically provide the information necessary for decisions. In the case of substitutable inputs, however, various combinations of inputs can be used to produce the same output. Further decomposing the efficiency variance into yield and mix variances provides additional information.

4. Many products use multiple direct materials that are substitutes for one another. In these cases, direct materials efficiency can come from two sources: (a) using fewer inputs of both materials and (b) using a cheaper mix of inputs to produce output. The direct materials yield variance and the direct materials mix variance divide the direct materials efficiency variance into two components, with the yield variance focusing on the total inputs used and the mix variance evaluating how the inputs are combined.

5. Multiple direct labor inputs that are substitutes for one another are often used to manufacture a product or provide a service. The direct labor yield and mix variances indicate the sources of direct labor efficiency. A favorable direct labor yield variance results if fewer total direct labor-hours are used to produce a given quantity of products or services. A favorable direct labor mix variance results if a cheaper mix of inputs is used to produce the actual quantity of products or services.

6. Productivity measures the relationship between inputs and outputs. Productivity measures compare the quantities and mix of inputs used to produce output over two or more periods.

7. Partial productivity measures compare the quantity of output produced with the quantity of a single input consumed. Partial productivity measures are simple to calculate and easily understood at the operations level. However, partial productivity measures do not focus on all inputs simultaneously and cannot evaluate trade-offs among inputs.

8. Total factor productivity measures the combined productivity of all inputs used to produce output by taking into account the relative prices of inputs. A major advantage of total factor productivity is that it explicitly evaluates substitution possibilities among inputs. Two disadvantages are that total factor productivity measures are more difficult to understand at the operations level and that they are more difficult to link across multiple periods than are partial productivity measures.

9. Cost changes from one period to the next can be broken down into changes due to output adjustment, changes due to input prices, and changes due to productivity. The productivity change compares the cost to produce output in a particular period with what it would have cost to produce the same level of output in a previous period at current period input prices.

TERMS TO LEARN

This chapter and the Glossary at the end of the book contain definitions of the following important terms:

market-share variance (p. 762)   market-size variance (762)   partial productivity (774)
productivity (773)   total direct labor mix variance (770)
total direct labor yield variance (770)   total direct materials mix variance (767)
total direct materials yield variance (767)   total factor productivity (TFP) (777)
total sales-mix variance (758)   total sales-quantity variance (758)

ASSIGNMENT MATERIAL

QUESTIONS

22-1   "The sales-volume variance might be better labeled as a marketing variance." Explain why such a comment might be made.

22-2   Distinguish between a *sales-quantity variance* and a *sales-mix variance*.

22-3   Describe how a favorable sales-mix variance can arise in a company selling two products.

22-4   Why might a manager not compute market-size variances and market-share variances?

22-5   "A company can sell exactly the same number of units as specified in the static budget and still have a flexible budget with numbers that differ from those in the static budget." Do you agree? Explain.

22-6   Name three sources of the standards used in the direct materials yield variance and the direct materials mix variance.

22-7   Distinguish between a *direct materials yield variance* and a *direct materials mix variance*.

22-8   The manager of a highly automated plant that assembles desktop computers commented, "Yield and mix variance information is irrelevant to my cost management decisions." Give two possible reasons for the manager's making this statement.

22-9   Describe how an unfavorable direct manufacturing labor mix variance can arise in a manufacturing plant using two categories of direct manufacturing labor.

22-10   Define productivity.

22-11   What is a partial productivity measure?

22-12   What is total factor productivity?

22-13   What information does total factor productivity provide that partial productivities do not?

22-14   Give one advantage and one limitation of total factor productivity.

22-15   "We are already measuring total factor productivity. Measuring partial productivities would be of no value." Do you agree? Comment briefly.

EXERCISES AND PROBLEMS

Coverage of Part One of the Chapter

22-16 Sales-volume, sales-quantity, and sales-mix variances.   The Detroit Penguins play in the American Ice Hockey League. The Penguins play in the Downtown Arena, which has a capacity of 30,000 seats (10,000 lower-tier seats and 20,000 upper-tier seats). All tickets are sold by the Reservation Network. The Penguins' budgeted contribution margin for each type of ticket in 19_6 is computed as follows:

	Lower-Tier Tickets	Upper-Tier Tickets
Selling price	$35	$14
Downtown Arena fee	10	6
Reservation Network fee	5	3
Contribution margin	$20	$ 5

The budgeted and actual average attendance figures per game in the 19_6 season are:

	Budgeted Seats Sold	Actual Seats Sold
Lower-tier	8,000	6,600
Upper-tier	12,000	15,400
Total	20,000	22,000

There was no difference between the budgeted and actual contribution margins for lower-tier or upper-tier seats.

The manager of the Penguins was delighted that actual attendance was 10% above budgeted attendance per game, especially given the depressed state of the local economy in the past six months.

Required
1. Compute the total sales-volume variance for the Detroit Penguins in 19_6.
2. Compute the budgeted average contribution margin per unit (seat sold) in 19_6.
3. Compute the budgeted and actual sales-mix percentages for the lower-tier and the upper-tier seats.
4. Compute the total sales-quantity variance and the total sales-mix variance in 19_6.
5. Present a summary of the variances in requirements 1 and 4. Comment on the results.

22-17 Sales-volume, sales-quantity, and sales-mix variances. Debbie's Delight Inc. operates a chain of cookie stores. Budgeted and actual operating data of the downtown store for August 19_5 follow:

Budget for August

	Selling Price per Pound	Variable Costs per Pound	Contribution Margin per Pound	Sales Volume in Pounds
Chocolate chip	$4.50	$2.50	$2.00	45,000
Oatmeal raisin	5.00	2.70	2.30	25,000
Coconut	5.50	2.90	2.60	10,000
White chocolate	6.00	3.00	3.00	5,000
Macadamia nut	6.50	3.40	3.10	15,000
				100,000

Actual for August

	Selling Price per Pound	Variable Costs per Pound	Contribution Margin per Pound	Sales Volume in Pounds
Chocolate chip	$4.50	$2.60	$1.90	57,600
Oatmeal raisin	5.20	2.90	2.30	18,000
Coconut	5.50	2.80	2.70	9,600
White chocolate	6.00	3.40	2.60	13,200
Macadamia nut	7.00	4.00	3.00	21,600
				120,000

Required
1. Compute the individual product and total sales-volume variances for August 19_5.
2. Compute the individual product and total sales-quantity variances for August 19_5.

3. Compute the individual product and total sales-mix variances for August 19_5.
4. Comment on your results in requirements 1, 2, and 3.

22-18 Sales-volume, sales-quantity, and sales-mix variances. Computer Horizons manufactures and sells three related microcomputer products:

1. Plum—sold mostly to college students
2. Portable Plum—smaller version of the Plum that can be carried in a briefcase
3. Super Plum—has a larger memory and more capabilities than the Plum and is targeted at the business market

Budgeted and actual operating data for 19_6 follow:

Budget for 19_6

	Selling Price per Unit	Variable Costs per Unit	Contribution Margin per Unit	Sales Volume in Units
Plum	$1,200	$ 700	$ 500	700,000
Portable Plum	800	500	300	100,000
Super Plum	5,000	3,000	2,000	200,000
				1,000,000

Actual for 19_6

	Selling Price per Unit	Variable Costs per Unit	Contribution Margin per Unit	Sales Volume in Units
Plum	$1,100	$ 500	$ 600	825,000
Portable Plum	650	400	250	165,000
Super Plum	3,500	2,500	1,000	110,000
				1,100,000

During 19_6, competition sparked by overseas suppliers drove the cost of computer chips down, allowing Computer Horizons to buy key components at bargain prices. Computer Horizons had budgeted for a major expansion into the lucrative microcomputer business market in 19_6. Unfortunately, it underestimated the marketing power of its rival, the Big Blue Company.

Required
1. Compute the individual product and total sales-volume variances for Computer Horizons in 19_6.
2. Compute the individual product and total sales-quantity variances for 19_6.
3. Compute the individual product and total sales-mix variances for 19_6.
4. Comment on your results in requirements 1, 2, and 3.

22-19 Market-size and market-share variances. (Extension of 22-18) Computer Horizons derived its total unit sales budget for 19_6 from an internal management estimate of a 20% market share and an industry sales forecast by Micro-Information Services of 5,000,000 units. At the end of 19_6, Micro-Information reported actual industry sales of 6,875,000 units.

Required
Compute the market-size and market-share variances for Computer Horizons.

22-20 Sales-volume, sales-quantity, and sales-mix variances; working backward. Jinwa Corporation sells two brands of wine glasses—Plain and Chic. Jinwa provides the following information for sales in the month of June 19_6:

Static budget total contribution margin	$5,600
Budgeted units to be sold of all glasses in June 19_6	2,000 units
Budgeted contribution margin per unit of Plain	$2 per unit
Budgeted contribution margin per unit of Chic	$6 per unit
Total sales-quantity variance	$1,400 U
Actual sales-mix percentage of Plain	60%

Required

1. Calculate the sales-quantity variances for each product for June 19_6.
2. Calculate the individual product and total sales-mix variances for June 19_6.
3. Calculate the individual product and total sales-volume variances for June 19_6.
4. Briefly describe the conclusions you would draw from the variances.

Coverage of Part Two of the Chapter

 22-21 Direct materials price and efficiency variances, direct materials yield and mix variances, food processing. Tropical Fruits Inc. processes tropical fruit into fruit salad mix, which it sells to a food service company. Tropical Fruits has in its budget the following standards for the direct materials inputs to produce 80 pounds of tropical fruit salad:

50 pounds of pineapple at $1.00 per pound	$50
30 pounds of watermelon at $0.50 per pound	15
20 pounds of strawberries at $0.75 per pound	15
100	$80

Note that 100 pounds of input quantities are required to produce 80 pounds of fruit salad. No inventories of direct materials are kept. Purchases are made as needed, so all price variances are related to direct materials used. The actual direct materials inputs used to produce 54,000 pounds of tropical fruit salad for the month of October were:

36,400 pounds of pineapple at $0.90 per pound	$32,760
18,200 pounds of watermelon at $0.60 per pound	10,920
15,400 pounds of strawberries at $0.70 per pound	10,780
70,000	$54,460

Required

1. Compute the direct materials price and direct materials efficiency variances for each product and for the total output of tropical fruit salad in October.
2. Compute the individual product and total direct materials yield variances for October.
3. Compute the individual product and total direct materials mix variances for October.
4. Comment on your results in requirements 1, 2, and 3.
5. Why might direct materials yield and direct materials mix variances be especially informative to the management of Tropical Fruits Inc.?

22-22 Direct materials efficiency variance, mix and yield variances; working backward. Agrichem Enterprises manufactures and sells fertilizers. Agrichem uses the following standard direct materials costs to produce 1 ton of fertilizer:

75% of the input materials is Base at $400 per ton	$360
25% of the input materials is Grade at $200 per ton	60
Total standard cost of 1.2 tons of inputs	$420

Note that 1.2 tons of input quantities are required to produce 1 ton of fertilizer. No inventories of direct materials are kept. Purchases are made as needed, so all price variances are related to direct materials used. Agrichem produced 2,000 tons of fertilizer in a particular period. The total direct materials yield variance for the period was $35,000 unfavorable. The actual input mix for the period was 50% of Base and 50% of Grade.

Required

1. Calculate the individual direct materials yield variances for the period.
2. Calculate the individual and total direct materials mix variances for the period.
3. Calculate the individual and total direct materials efficiency variances for the period.
4. Briefly describe the conclusions you would draw from the total variances.

22-23 Direct manufacturing labor price and efficiency variances, direct manufacturing labor yield and mix variances. A supervisor in a sheet metal operation of Midwest Industries has the following direct manufacturing labor standard:

(a) Direct manufacturing labor price per hour:

2 artisans × $22	$44
3 helpers × $12	36
Total cost of standard combination of direct manufacturing labor	$80
Average cost per direct manufacturing labor hour: $80 ÷ 5 =	$16

(b) Standard direct manufacturing labor cost per unit of output
    at 8 units per hour: $16 ÷ 8 = $ 2

(c) Standard direct manufacturing labor cost of 20,000 units
    of output, 20,000 × $2, or 2,500 × $16 = $40,000

(d Actual inputs, 2,900 hours consisting of:

900 hours of artisans × $23	$20,700
2,000 hours of helpers × $11	22,000
	$42,700

The supervisor had to pay a higher average wage rate to the artisans as the result of a bargained agreement. He tried to save some costs by using more helpers per artisan than usual.

Required

1. Compute the direct manufacturing labor price and efficiency variances.
2. Divide the direct manufacturing labor efficiency variance into yield and mix components.
3. What would the actual costs probably have been if the standard direct manufacturing labor mix had been held constant even though the actual direct manufacturing labor prices were incurred?

22-24 Direct distribution labor price, efficiency variances, direct distribution labor yield and mix variances. (CMA, adapted)   The Memphis distribution center receives and distributes products for the Landeau Manufacturing Company. An analysis comparing the actual results with a flexible budget is prepared monthly.

The standard direct distribution labor rates in effect for the fiscal year ending June 30, 19_5, and the standard hours allowed for the April output follow. The labor classes reflect different skill levels and different jobs: forklift operator, manual handler, and helper.

	Standard Direct Distribution Labor Rate per Hour	Standard Direct Distribution Labor-Hours Allowed for Actual Products Distributed
Distribution labor class III	$8.00	500
Distribution labor class II	$7.00	500
Distribution labor class I	$5.00	500

The actual direct distribution labor-hours worked and the actual direct distribution labor rates per hour for April follow:

	Actual Direct Distribution Labor Rate per Hour	Actual Direct Distribution Labor-Hours
Distribution labor class III	$8.50	550
Distribution labor class II	$7.50	650
Distribution labor class 1	$5.40	375
		1,575

Required

1. Calculate the price and efficiency variances of each class of distribution labor and in total for April.
2. Calculate the mix and yield variances of each class of distribution labor and in total for April.

Coverage of Part Three of the Chapter

22-25 Partial productivity measurement. Warren Company produces 20,000 units of printed circuit boards on a contract basis using labor and machine time as inputs. The cost per direct manufacturing labor-hour is $15 per unit. The cost per machine-hour is $20 per unit. Warren's engineers believe that the output can be produced using either of the following two combinations of inputs:

	Approach 1	Approach 2
Direct manufacturing labor-hours	10,000	15,000
Machine-hours	20,000	16,000

Required
1. Compute the partial productivity ratios for each input under approach 1.
2. Compute the partial productivity ratios for each input under approach 2.
3. On the basis of your answers to requirements 1 and 2, should Warren use approach 1 or approach 2 to manufacture the product? Use other information given in the problem to advise Warren as to which approach is preferable.

22-26 Partial productivity measurement. Hanover Corporation makes small parts from steel alloy sheets. Hanover's management has some ability to substitute direct materials for direct manufacturing labor. If workers cut the steel carefully, Hanover can manufacture more parts out of a metal sheet, but this will require more direct manufacturing labor-hours. Alternatively, Hanover can use fewer direct manufacturing labor-hours if it is willing to tolerate more waste of direct materials. Hanover provides the following information for the years 19_5 and 19_6.

	19_5	19_6
Output units	400,000	520,000
Direct manufacturing labor-hours used	10,000	13,875
Wages per hour	$26	$25
Direct materials used	160 tons	190 tons
Direct materials cost per ton	$3,187.50	$3,437.50

Required
1. Compute the partial productivity ratios for 19_5 and 19_6.
2. On the basis of the partial productivity ratios alone, can you conclude whether productivity improved overall in 19_6 relative to 19_5? Explain.

22-27 Total factor productivity and its comparison between two time periods. Use the data given for Hanover Corporation in Problem 22-26 for solving this problem.

Required
1. Calculate Hanover Corporation's total factor productivity in 19_6.
2. Compare Hanover Corporation's total factor productivity performance in 19_6 relative to 19_5.
3. What does total factor productivity tell you that partial productivity measures do not?

22-28 Analysis of cost changes. Use the data given for Hanover Corporation in Problem 22-26 for solving this problem.

Required
1. Calculate the actual costs incurred by Hanover Corporation in 19_5 and 19_6.
2. Calculate how much of the change in costs is due to the output adjustment component, the input price change component, and the productivity change component.
3. Interpret your answers in requirement 2 to explain why Hanover Corporation's costs changed from 19_5 to 19_6.

22-29 Partial productivity measurement. Pittsburgh Industries makes chemical products using direct materials and direct manufacturing labor as substitutable inputs. It reports the following data for the last two years of operations:

	19_4	19_5
Output units	375,000	525,000
Direct manufacturing labor-hours used	7,500	9,500
Wages per hour	$20	$25
Direct materials used, in kilograms	450,000	610,000
Direct materials cost per kilogram	$1.20	$1.25

Required

1. Compute the partial productivity ratios for 19_4 and 19_5.
2. On the basis of the partial productivity ratios alone, can you conclude whether productivity improved overall in 19_5 relative to 19_4? Explain.

22-30 Total factor productivity and its comparison between two time periods. Use the data given for Pittsburgh Industries in Problem 22-29 for solving this problem.

Required

1. Compute Pittsburgh Industries' total factor productivity in 19_5.
2. Compare Pittsburgh Industries' total factor productivity performance in 19_5 relative to 19_4.
3. Why did Pittsburgh Industries' total factor productivity change from 19_4 to 19_5?

22-31 Analysis of cost changes. Use the data given for Pittsburgh Industries in Problem 22-29 for solving this problem.

Required

1. Calculate the actual costs incurred by Pittsburgh Industries in 19_4 and 19_5.
2. Calculate how much of the change in cost is due to the output adjustment component, the input price change component, and the productivity change component.
3. Interpret your answers in requirement 2 to explain why Pittsburgh Industries' costs changed from 19_4 to 19_5.

22-32 Partial productivity, estimation of inputs. Eva Lyon, the office manager at Rapid Copiers, was trying to forecast the number of units of toner-refill cartridges and hours of maintenance labor needed for 19_5. In 19_4, machines were used to make 1,800,000 copies. Eva predicts there will be a 10% increase in the number of copies made in 19_5. In 19_4, a toner-refill cartridge was required after every 60,000 copies, and one hour of maintenance time was needed after every 6,000 copies. In 19_5, Eva expects to use new robust machines; one hour of maintenance will be required only every 12,000 copies. New toner-refill cartridges will last for 99,000 copies before they need to be replaced.

Required

1. What is the partial productivity of toner and of maintenance labor in 19_4 and 19_5?
2. Estimate the number of toner-refill cartridges and maintenance labor-hours required in 19_5.

22-33 Comparing efficiency and partial productivity measures. Vander Investments invests in stocks on behalf of its clients. Many of the transactions are done over the telephone. In 19_5, Vander expects its investment representatives to handle 10 calls per hour. In December 19_5, Vander buys a new client-information system that it expects will enable the representatives to handle 12 calls per hour. The standard and actual wages paid to investment representatives is $15 per hour. Vander obtains the following information about the performance of its representatives in June 19_5 and June 19_6.

◆ Representatives took 16,600 calls in 1,600 hours in June 19_5.
◆ Representatives took 18,360 calls in 1,600 hours in June 19_6.

Required

1. Calculate the direct labor partial productivity in June 19_5 and June 19_6.
2. Calculate direct labor efficiency variances for June 19_5 and June 19_6.
3. Have both direct labor partial productivity and direct labor efficiency measures improved in June 19_6 compared with June 19_5? Compare and comment on your answers to requirements 1 and 2.
4. What do you think of Vander's plan to set specific targets for its investment representatives to achieve?

22-34 Productivity and ethics. Donovan Inc. manufactures valves for automobiles. In the latest round of negotiations, the union has reluctantly agreed to give up regular pay increases and accept instead a productivity-based gain-sharing program. Under the gain-sharing program bonuses paid to workers are tied to improvements in labor productivity. Still, the labor union leaders continue to be skeptical about the gain-sharing program. They fear that management will try to avoid making bonus payments by never disclosing the correct numbers. Jerry Mason is the cost analyst in charge of maintaining productivity records and the productivity data that determine the productivity-based bonus.

Donovan has just finished its first year under the gain-sharing program. The results are not good. Ken Hunter, the controller at Donovan, approaches Jerry Mason with the following request, "I think we better show something positive in the productivity numbers. The union already mistrusts our motives. If we don't pay at least some bonus, we are never going to get this productivity program off the ground. One approach is to make a smaller adjustment for all the new capital that we put in and increase the amount of productivity gain associated with the worker's efforts." Mason is confused about what he should do.

Required
Refer to the "Standards of Ethical Conduct for Management Accountants" described in Chapter 1 (p. 17). Evaluate whether Ken Hunter's suggestion to Mason to modify the productivity numbers is unethical. Would it be unethical for Mason to modify his analysis to enable a small bonus payment to be made to the workers? What steps should Mason take to resolve this situation?

23

Cost Management:

Quality and Time

Quality as a competitive weapon

Costs of quality

Methods used to identify quality problems

Relevant costs and benefits of quality improvement

Quality and customer-satisfaction measures

Quality and internal performance measures

Evaluating quality performance

Time as a competitive weapon

New product development time

Breakeven time for new products

Operational measures of time

Costs of time

Theory of constraints and throughput contribution analysis

Customers are demanding ever-increasing levels of quality at competitive prices in the products and services they purchase. Solectron monitors a combination of financial and nonfinancial variables to ensure printed circuit boards assembled at its plants in France, Malaysia, and the United States meet demanding quality levels at competitive prices.

Learning Objectives

When you have finished studying this chapter, you should be able to

1. Explain four cost categories in a costs of quality program
2. Describe three methods that companies use to identify quality problems
3. Identify the relevant costs and benefits of quality improvements
4. Provide examples of nonfinancial quality measures of customer satisfaction and internal performance
5. Understand why companies use both financial and nonfinancial measures of quality
6. Explain breakeven time
7. Describe customer-response time, and explain the reasons for and the costs of lines and delays
8. Define the three main measurements in the theory of constraints
9. Describe four steps in managing constraints

Quality and time are two important areas in which companies compete in the marketplace. This chapter first examines how management accounting can assist managers in taking initiatives in the quality area. It then describes management accounting issues that are related more specifically to time-based competition.

QUALITY AS A COMPETITIVE WEAPON

Many executives regard total quality management (TQM) as among the most important issues of the 1990s. Several prestigious, high-profile awards, for example, the Malcolm Baldrige Quality Award in the United States and the Deming Prize in Japan, are given to companies for quality. International quality standards have emerged. For example, ISO 9000, developed by the International Organization for Standardization, is a set of five international standards for quality management adopted by more than 50 countries. ISO 9000 was created to help companies effectively document their quality system elements.

Why this growing emphasis on quality? Because quality costs are significant. Quality costs range from 15% to 20% of sales revenue for many organizations. Quality improvement programs can result in substantial savings and higher revenues. Motorola, the telecommunications and electronics manufacturer, estimates that it saves $2.2 billion annually from quality programs. This amounts to savings of 16.5% on annual revenues of $13.3 billion in 1992. Motorola's 1992 operating income was $576 million. Without the savings from its quality programs, Motorola claims its losses would be over $1.5 billion.[1]

Many definitions of quality have been proposed over the years. Each definition focuses on a different aspect of quality—fitness for use, the degree to which a product satisfies the wants of a customer, and the degree to which a product conforms to design specification and engineering requirements.[2] We discuss two basic aspects of quality—*quality of design* and *conformance quality*.[3]

---

[1]*New York Times,* January 21, 1993.

[2]The American Society for Quality Control defines *quality* as the totality of features and characteristics of a product made or a service performed according to specifications, to satisfy customers at the time of purchase and during use. ANSI/ASQC A3-1978, *Quality Systems Terminology* (Milwaukee, WI: American Society for Quality Control, 1978).

[3]See R. DeVor, T. Chang, and J. Sutherland, *Statistical Quality Design and Control* (New York : Macmillan Publishing Company, 1992); J. Evans and W. Lindsay, *The Management and Control of Quality* (St. Paul: West Publishing Company, 1993).

Quality of design measures how closely the characteristics of products or services match the needs and wants of customers. Suppose customers of photocopying machines want copiers that combine copying, faxing, scanning, and electronic printing. Photocopying machines that fail to meet these customer needs fail in the quality of their design.

Conformance quality is making the product according to design, engineering, and manufacturing specifications. For example, if a photocopying machine mishandles paper or breaks down, it will have failed to satisfy conformance quality. Products not conforming to specifications need to be repaired, reworked, or scrapped at an additional cost to the organization. If nonconformance errors are not corrected within the plant and the product breaks down at the customer's site, even greater repair costs as well as loss of customer goodwill may result.

The following diagram illustrates our framework. Conformance quality failure and quality of design failure point to the shortfall between actual performance and customer satisfaction that needs to be bridged through continuous improvement.

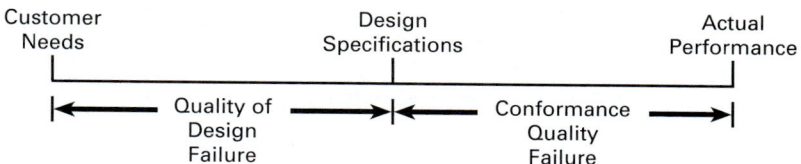

Any failure a company encounters as it travels along this route from customer needs to actual performance exacts a cost, both financially and nonfinancially. We first examine financial measures of quality, in particular costs of quality. We next describe nonfinancial measures of customer satisfaction. Finally, we consider nonfinancial measures to evaluate quality improvement activities within the company.

COSTS OF QUALITY

The **costs of quality** are those costs incurred to prevent poor quality from occurring or those costs incurred because poor quality has occurred. Costs of quality (COQ) encompass costs incurred throughout the firm.

Four categories of costs are often distinguished:

1. **Prevention costs:** costs incurred in preventing the production of products that do not conform to specifications

2. **Appraisal costs:** costs incurred in detecting which of the individual units of products do not conform to specifications

3. **Internal failure costs:** costs incurred when a nonconforming product is detected *before* it is shipped to customers

4. **External failure costs:** costs incurred when a nonconforming product is detected *after* it is shipped to customers

Exhibit 23-1 presents examples of individual line items in each of the prevention, appraisal, internal failure, and external failure categories of quality costs. Category A lists those costs of quality items generally reported on COQ reports. Category B presents costs of poor quality items often not reported on COQ reports, such as forgone contribution margin from lost sales and from lower prices of products sold. Category B items are opportunity costs, which are difficult to estimate and generally not recorded in accounting systems. Nevertheless, the forgone contribution margin from poor quality can be substantial, and Category B items can be important driving forces in quality improvement programs.

Note that the items included in Category A come from all value-chain business functions. Most existing accounting systems satisfy multiple objectives and are not

EXHIBIT 23-1
Items Pertaining to Costs of Quality Reports

Prevention Costs	Appraisal Costs	Internal Failure Costs	External Failure Costs
Category A Items Quality engineering Supplier evaluations Equipment maintenance Manufacturing process engineering Design engineering Quality training New materials	**Category A Items** Inspection of materials received Technical services laboratory Product testing Product acceptance (finished goods inspection) Manufacturing product inspection (inspecting products on line to evaluate if the manufacturing process is operating correctly)	**Category A Items** Scrap Rework Rescheduling, retesting, and reinspection Manufacturing and process engineering on internal failure **Category B Items** Lost contribution margin from poor-quality production	**Category A Items** Distribution costs of returned products Marketing costs on external failure Manufacturing and process engineering on external failure Repair Travel costs related to quality problems Warranty claims Liability claims **Category B Items** Lost contribution margin from lower sales, declining market share, and lower price realizations

Note: Category A items: costs of quality items generally included in costs of quality reports.
Category B items: costs of quality items generally not included in costs of quality reports (mainly lost contribution margins resulting from poor quality).

designed to focus comprehensively on costs of quality for the organization as a whole. They report only a subset of the cost items in Category A of Exhibit 23-1 (such as the costs of rework; see Chapter 18) and underestimate the benefits of quality improvements.

We illustrate the computation and use of costs of quality reports for Photon Corporation. Photon Corporation makes multiple products. Our presentation focuses on Photon's photocopying machines, which earned operating income of $24 million on sales of $300 million (20,000 copiers) in 19_6. Photon determines Category A costs of quality using an activity-based approach with five steps.

STEP 1: *IDENTIFY ALL QUALITY-RELATED ACTIVITIES AND ACTIVITY COST POOLS.* Inspecting (including testing) the photocopying machines is an example.

STEP 2: *DETERMINE THE QUANTITY OF THE COST ALLOCATION BASE (OR COST DRIVER) FOR EACH QUALITY-RELATED ACTIVITY.* Photon identifies inspection hours as the cost-allocation base (cost driver) of the inspection activity. Assume that photocopying machines use 240,000 hours (12 hours per copier × 20,000 copiers) of the cost-allocation base.

STEP 3: *DETERMINE THE RATE PER UNIT OF EACH COST-ALLOCATION BASE.* In the Photon example, the total (fixed and variable) costs of inspection are $40 per hour.

STEP 4: *COMPUTE THE COSTS OF EACH QUALITY-RELATED ACTIVITY FOR PHOTOCOPYING MACHINES BY MULTIPLYING THE QUANTITY OF THE COST-ALLOCATION BASE COMPUTED IN STEP 2 BY THE RATE PER UNIT OF THE COST-ALLOCATION BASE COMPUTED IN STEP 3.* Quality-related inspection costs are $9,600,000 (240,000 hours × $40 per hour).

STEP 5: *OBTAIN THE TOTAL COSTS OF QUALITY BY ADDING THE COSTS OF ALL QUALITY-RELATED ACTIVITIES FOR PHOTOCOPYING MACHINES IN ALL VALUE-CHAIN BUSINESS FUNCTIONS.* Exhibit 23-2, Panel A, presents Photon Corporation's analysis of all Category A costs of quality items. The exhibit classifies costs into prevention, appraisal, internal failure, and external failure categories. Column 2 indicates that quality costs occur in all value-chain functions. Column 3 describes the allocation bases and cost drivers. If you are familiar with the hierarchy of costs described in Chapter 10 (pp. 355–57), you will recognize design engineering and process engineering costs as product-sustaining costs. All other costs are output unit-

EXHIBIT 23-2
Activity-Based Costs of Quality Analysis for Photon Corporation

PANEL A: CATEGORY A COSTS OF QUALITY ITEMS GENERALLY RECORDED IN COQ REPORTS

Costs of Quality Category (1)	Value-Chain Cost Category Affected (2)	Quantity of Allocation Base or Cost Driver (3)	Allocation or Cost Driver Rate (Number Assumed) (4)	Total Costs (5) = (3) × (4)	Percentage of Sales (6) = (5) ÷ $300,000,000
Prevention costs					
Design engineering	R&D/Design	40,000* hours	$80 per hour	$ 3,200,000	1.07%
Process engineering	R&D/Design	45,000* hours	$60 per hour	2,700,000	0.90
Total prevention costs				5,900,000	1.97
Appraisal costs					
Inspection	Manufacturing	240,000† hours	$40 per hour	9,600,000	3.20
Total appraisal costs				9,600,000	3.20
Internal failure costs					
Rework	Manufacturing	2,500‡ copiers reworked	$4,000 per copier reworked	10,000,000	3.33
Total internal failure costs				10,000,000	3.33
External failure costs					
Customer support	Marketing	3,000§ copiers repaired	$200 per copier repaired	600,000	0.20
Returning and replacing parts	Distribution	3,000 copiers repaired	$240 per copier repaired	720,000	0.24
Warranty repair	Customer service	3,000 copiers repaired	$4,400 per copier repaired	13,200,000	4.40
Total external failure costs				14,520,000	4.84
Total costs of quality				$40,020,000	13.34%

PANEL B: CATEGORY B COSTS OF QUALITY ITEMS GENERALLY NOT RECORDED IN COQ REPORTS

Costs of Quality Category (1)	Revenue and Cost Items Affected (2)	Quantity of Lost Sales (3)	Contribution per Copier (Number Assumed) (4)	Total Lost Contribution Margin (5) = (3) × (4)	Percentage of Sales (6) = (5) ÷ $300,000,000
External failure costs					
Estimated forgone contribution margin on lost sales	Revenue and all cost categories	2,000‖	$6,000	$12,000,000	4.00%
Total costs of quality				$12,000,000	4.00%

*Based on special studies.
†12 hours per copier × 20,000 copiers.
‡12.5% of 20,000 copiers manufactured require rework.
§15% of 20,000 copiers manufactured require warranty repair service.
‖Estimated by Photon.

level costs. For example, rework costs are expected to vary with each additional copier produced.

To avoid clutter, we do not provide details of the calculations.[4] The allocation rates include both fixed and variable cost components. Exhibit 23-2, Panel A, shows Photon's costs of quality for photocopying machines at $40.02 million, of which the largest categories are $14.52 million in total external failure costs and $10 million in total internal failure costs—a sum of $24.52 million. Total Category A costs of quality are 13.34% of current sales.

Exhibit 23-2, Panel B, presents the analysis of the opportunity costs of poor quality—the Category B items generally not reported on COQ reports. Photon Corporation estimates lost sales of 2,000 photocopying machines because of external failures. The forgone contribution of $12 million measures the financial costs from dissatisfied customers who have returned machines to Photon and from sales lost because of Photon's bad reputation for quality. Total costs of quality (including lost contribution margin) equal $52.02 million (Panel A, $40.02 million + Panel B, $12 million), which equals 17.34% of current sales. Failure-related costs and failure-related lost contribution margin are $36.52 million (Panel A, $24.52 million + Panel B, $12 million).

The COQ report highlights Photon's high internal and external failure costs. To reduce costs of quality, Photon must reduce failures. How? The following section briefly describes some of the methods that organizations use to identify quality problems—control charts, Pareto diagrams, and cause-and-effect diagrams.

METHODS USED TO IDENTIFY QUALITY PROBLEMS

Control Charts

Objective 2
Describe three methods that companies use to identify quality problems

Statistical quality control (SQC) or statistical process control (SPC) is a formal means of distinguishing between random variation and nonrandom variation in an operating process. A key tool in SQC is a control chart. A **control chart** is a graph of a series of successive observations of a particular step, procedure, or operation taken at regular intervals of time. Each observation is plotted relative to specified ranges that represent the expected distribution. Only those observations outside the specified limits are ordinarily regarded as nonrandom and worth investigating.

Exhibit 23-3 presents control charts for the defect rates observed at Photon's three production lines. Defect rates in the prior 60 days for each plant were assumed to provide a good basis from which to calculate the distribution of daily defect rates. The arithmetic mean (μ, read "mu") and standard deviation (σ, read "sigma") are the two parameters of the distribution that are used in the control charts in Exhibit 23-3.

---

[4]Allocation rates are computed using methods described in Chapter 4 (pp. 107–9), Chapter 5 (pp. 144–45), and Chapter 15 (pp. 536–38).

EXHIBIT 23-3

Statistical Quality Control Charts: Daily Defect Rate at Three Production Lines of Photon Corporation

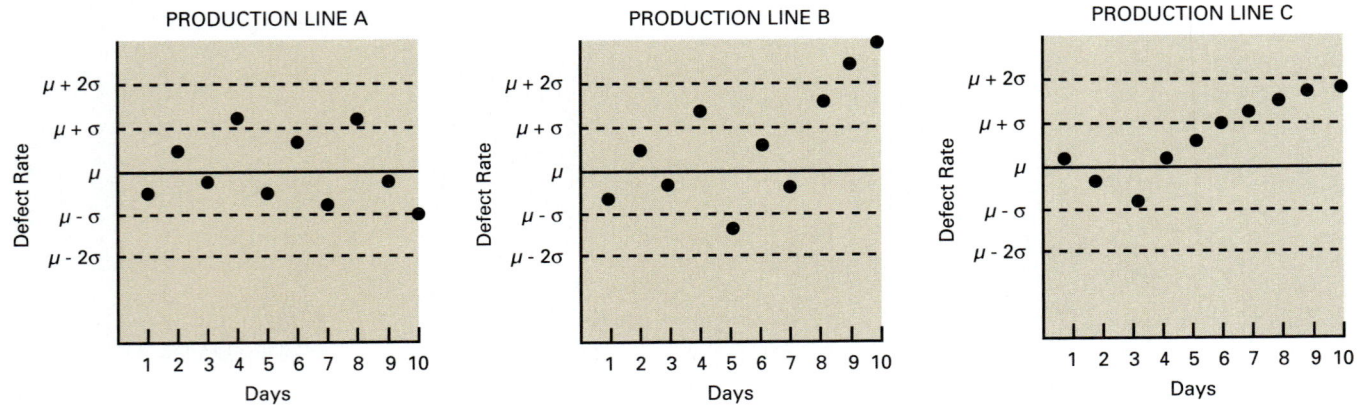

On the basis of experience, the firm decides that any observation outside the $\pm 2\sigma$ range should be investigated.

For production line A in Exhibit 23-3, all observations are within the range of $\pm 2\sigma$ from the mean. Management, then, believes no investigation is necessary. For production line B, the last two observations signal a possible out-of-control occurrence. Given the $\pm 2\sigma$ rule, both observations would lead to an investigation. Production line C illustrates a process that would not prompt an investigation under the $\pm 2\sigma$ rule but may well be out of control. Note that the last eight observations show a clear direction and that the direction by day 5 (the third point in the last eight) is away from the mean. Statistical procedures have been developed that consider the trend as well as the level of the variable in question.

Pareto Diagrams

A **Pareto diagram** indicates how frequently each type of failure (defect) occurs. Exhibit 23-4 presents a Pareto diagram for Photon's failures. Fuzzy and unclear copies are the most frequently occurring problem.

The fuzzy copy problem results in high rework costs because, in many cases, Photon discovers the fuzzy image problem only after the copier has been built. Sometimes fuzzy images occur at customer sites, resulting in high warranty and repair costs.

Cause-and-Effect Diagrams

The **cause-and-effect diagram** helps to identify potential causes of failure. As a first step, Photon analyzes the causes of the most frequently occurring failure, "fuzzy and unclear copies." Exhibit 23-5 presents the cause-and-effect diagram for the problem of "fuzzy and unclear copies." The exhibit identifies four major categories of potential causes of failure—human factors, methods and design factors, machine-related factors, and materials and components factors. As additional arrows are added for each cause, the general appearance of the diagram begins to resemble a fishbone (hence, cause-and-effect diagrams are also called *fishbone diagrams*).[5]

---

[5]Managers in U.S. electronics companies consider the following factors as contributing to improvements in quality (ranked in order of importance with 1 = most important):

1. Better product design
2. Improved process design
3. Improved training of operators
4. Improved products from suppliers
5. Investments in technology and equipment

See G. Foster and L. Sjoblom, "Survey of Quality Practices in the U.S. Electronics Industry," Working Paper, Stanford University, 1993.

EXHIBIT 23-4

Pareto Diagram for Photon Corporation

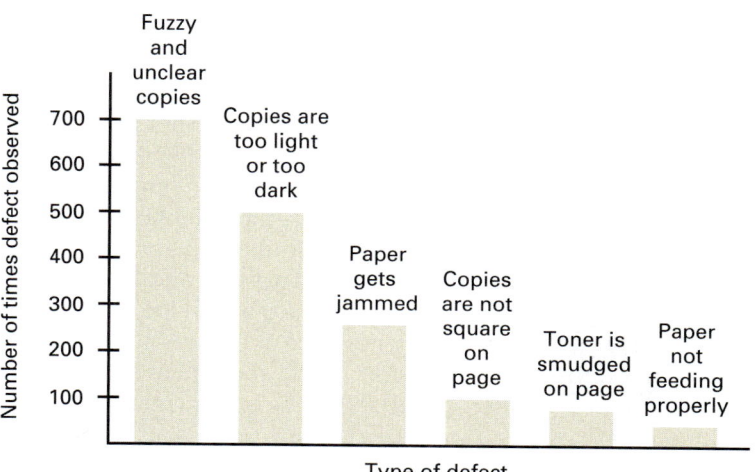

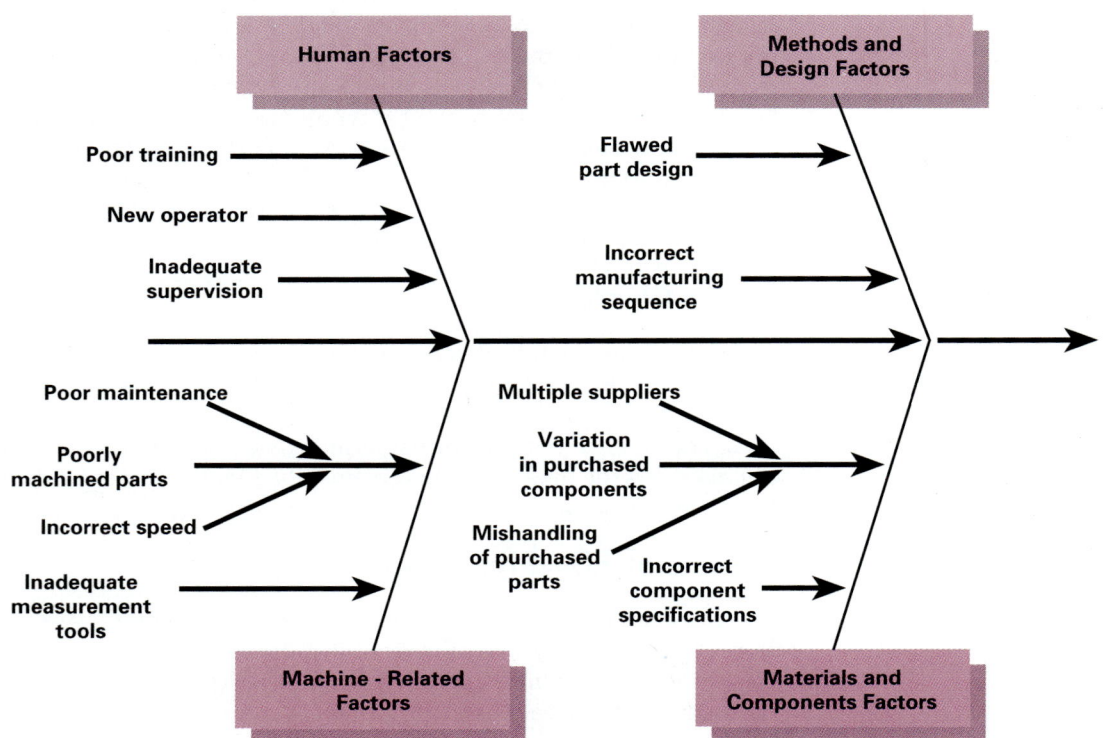

RELEVANT COSTS AND BENEFITS OF QUALITY IMPROVEMENT

Careful study and analysis of the cause-and-effect diagram reveals that the steel frame (or chassis) of the copier is often mishandled as it travels from the supplier's warehouse to Photon's plant. The frame must satisfy very precise specifications and tolerances; otherwise, various copier components (such as drums, mirrors, and lenses) attached to the frame will be improperly aligned. Mishandling causes the dimensions of the frame to vary from specifications, resulting in fuzzy images.

The team of engineers working toward solving the fuzzy image problem offers two alternative solutions. The first alternative is to improve the inspection of the frame immediately upon delivery. Photon could use various gauges to ensure that the frame satisfies specifications before assembling components on the frame. Some companies use sophisticated cameras to measure and check different dimensions of incoming materials and components. A second alternative is to redesign and strengthen the frame and the containers used to transport them to better withstand mishandling during transportation.

Objective 3

Identify the relevant costs and benefits of quality improvements

What choice should Photon make: Photon can either inspect incoming frames more carefully or redesign them and their containers. Management estimates that *additional inspection* will cost $400,000 ($40 per hour × 10,000 hours). Redesign will cost an *additional* $460,000 (design engineering: $80 per hour × 2,000 hours; process engineering: $60 per hour × 5,000 hours). The potential benefits of incurring these costs are lower internal and external failure costs. The key question at this point is, What are the relevant cost savings and other relevant benefits? Photon considers only a one-year time horizon for analyzing this decision because Photon plans to introduce a completely new line of copiers at the end of the year.

Consider first, the internal failure costs of rework. Photon reworks 2,500 (12.5% × 20,000) of the copiers manufactured. Analysis of rework costs indicates that of the total costs ($4,000) incurred to rework a copier, variable costs (direct materials, direct rework labor, and supplies) equal $1,600, and allocated fixed costs (equipment, space, and allocated overhead) average $2,400. If Photon chooses to inspect the frame, it expects to eliminate rework on 600 copiers and to save the variable costs of $960,000 ($1,600 × 600) in rework. See Exhibit 23-6, column 2. Photon believes that fixed rework costs will be unaffected. If Photon chooses the redesign alternative, it expects to eliminate rework on 800 copiers, saving $1,280,000 ($1,600 × 800). See Exhibit 23-6, column 3.

Next consider external failure costs. What effect will more closely inspecting incoming frames or redesigning the frames have on the number of copiers that will require repair in the field? As fewer copiers fail at customer sites and Photon's reputation for quality improves, how many additional copiers can Photon expect to sell?

Photon currently repairs 3,000 copiers at customer sites. If incoming frames are inspected, Photon estimates that 500 fewer copiers will require warranty repair and that it will be able to sell 250 additional copiers. If the frame is redesigned, Photon estimates that 700 fewer copiers will require warranty repair and that it will be able to sell 300 additional copiers.

Variable and fixed cost components of individual external failure costs of quality items described in Exhibit 23-2 column 4 follow:

EXHIBIT 23-6

Estimated Effect of Quality Improvement Actions on Costs of Quality for Photon Corporation

Description (1)	Incremental Costs and Benefits of Inspecting Incoming Frame (2)	Incremental Costs and Benefits of Redesigning Frame (3)
Category A Costs of Quality Items		
Additional design engineering costs		
$80 × 2,000 hours		$ 160,000
Additional process engineering costs		
$60 × 5,000 hours	—	300,000
Additional inspection and testing costs		
$40 × 10,000 hours	$ 400,000	—
Savings in rework costs		
$1,600 × 600 fewer copiers reworked	(960,000)	
$1,600 × 800 fewer copiers reworked		(1,280,000)
Savings in customer-support costs		
$80 × 500 fewer copiers repaired	(40,000)	
$80 × 700 fewer copiers repaired		(56,000)
Savings in costs to return and replace parts		
$180 × 500 fewer copiers repaired	(90,000)	
$180 × 700 fewer copiers repaired		(126,000)
Savings in warranty repair costs		
$1,800 × 500 fewer copiers repaired	(900,000)	
$1,800 × 700 fewer copiers repaired		(1,260,000)
Category B Costs of Quality Items		
Contribution margin from increased sales		
$6,000 × 250 additional copiers sold	(1,500,000)	
$6,000 × 300 additional copiers sold		(1,800,000)
Net cost savings and additional contribution margin	$(3,090,000)	$(4,062,000)
Cost savings in favor of redesigning frame	$972,000	

	Variable Costs per Copier Repaired	Fixed Costs per Copier Repaired	Total Costs per Copier Repaired
Customer-support costs	$ 80	$ 120	$ 200
Costs of returning and replacing parts	180	60	240
Warranty repair costs	1,800	2,600	4,400

As Photon eliminates repair work on copiers, it expects to save only the variable costs of warranty repair, customer support, and returning and replacing parts. Fixed costs will be unaffected by fewer repairs.

Exhibit 23-6 presents Photon's analysis of the estimated incremental effect of either (1) inspecting or (2) redesigning the frame on all cost of quality items described in Exhibit 23-2. Note that the savings per copier in rework costs, customer-support costs, costs to return and replace parts, and warranty repair costs differ from the costs per copier for each of these items in Exhibit 23-2. Why? Because Exhibit 23-6 shows only the variable costs that Photon expects to save. Exhibit 23-2 shows the *total* (fixed and variable) costs of each of these items. Also note that Exhibit 23-6 includes the incremental contribution margin from the estimated increases in sales due to the improved quality and performance of Photon's copiers.

Exhibit 23-6 indicates that the net estimated cost savings are $972,000 greater under the "redesign the frame" alternative. Photon estimates that redesigning for quality will reduce internal and external failure costs more than would inspecting for quality. Photon incurs the costs of a poorly designed frame in the form of higher manufacturing, marketing, distribution, and customer-service costs, as internal and external failures begin to mount. But these costs are locked in when the frame is designed. Thus, it is not surprising that redesign will yield significant savings.

In the Photon example, lost contribution margin occurs because Photon's repeated external failures damage its reputation for quality, resulting in lost sales. Lost contribution margin can also occur as a result of internal failures. Suppose Photon's manufacturing capacity is fully used. In this case, rework uses up valuable manufacturing capacity and causes the company to forgo contribution margin from producing and selling additional copiers. Suppose Photon could produce (and subsequently) sell an additional 600 copiers by improving quality and reducing rework. The costs of internal failure would then include lost contribution margin of $3,600,000 ($6,000 contribution margin per copier × 600 copiers). This $3,600,000 is the opportunity cost of poor quality.

Consider a general effect that quality has on revenues and hence on contribution margin. If competitors are improving quality, then a company that does not invest in quality improvement will likely suffer a decline in its market share and revenues. In this case, the benefit of better quality is in preventing lower revenues, not in generating higher revenues.

Quality improvement also has nonfinancial and qualitative effects. For example, managers and workers focusing on quality gain expertise about product and process. This knowledge may lead to lower costs in the future. Also, manufacturing a product of high quality enhances a company's reputation and increases customer goodwill, which may lead to higher future revenues.

Photon can use its costs of quality report to examine interdependencies across the four categories of quality-related costs. In our example, redesigning the frame increases costs of prevention activities (design and process engineering), decreases costs of internal failure (rework), and decreases costs of external failure (warranty repairs). Costs of quality yield more insight when managers compare trends over time. In successful quality programs, the costs of quality as a percentage of sales and the costs of internal and external failure as a percentage of total costs of quality should decrease over time.

The Photon Corporation example shows large initial net gains from reducing defects. The key question is, Will Photon continue to realize these net gains until it

Concepts in Action

CRYSEL WINS PREMIO NACIONAL DE CALIDAD—MEXICO'S PREMIER QUALITY AWARD[a]

Crysel, a member of the Mexican industrial group CYDSA (*Celulosa y Derivados Society Anonimos*), is the largest producer of acrylic fiber in Latin America and among the top-10 producers of acrylic fiber in the world. In 1991, Crysel was awarded the Premio Nacional de Calidad, Mexico's equivalent of the Malcolm Baldrige quality award.

One element of the Premio Nacional de Calidad's evaluation criteria is a company's costs of quality reporting. Companies vary as to which items they include in their COQ reports, choosing those cost categories that management feels warrant the greatest emphasis. Crysel classifies its costs of failure into six main classes:

1. *Consumption factors*: Excess or wasted direct materials, steam, or energy
2. *Maintenance:* Costs of repairing machines that break down
3. *Human resources:* Costs of extra workers and staff, such as rework crew employed to correct quality problems
4. *Accounts receivable:* Finance costs of not receiving money from customers on time
5. *Substandard quality:* Contribution margin lost from selling inferior-grade rather than top-grade fiber
6. *Sales volume:* Contribution margin lost from selling less than the available plant capacity because of quality problems

The first three classes appear in the costs of quality reports that most companies prepare. Crysel's innovation in quality reporting is to include the last three classes: accounts receivable, substandard quality, and sales volume. Each of these three classes measures an opportunity cost of poor quality, a cost not generally found in costs of quality reports. The following table indicates that these opportunity costs are a significant percentage of Crysel's total costs of failure. Including them in the cost of quality reports signals to all employees that top management believes these classes deserve close attention.

Costs of Quality as a Percentage of Sales

	1985	1989	1992
Items Generally Recorded in COQ Reports			
Consumption factors	3.8%	4.1%	2.6%
Maintenance	0.8	0.9	0.8
Human resources	0.6	0.5	0.4
Total	5.2%	5.5%	3.8%
Items Generally Not Recorded in COQ Reports			
Accounts receivable	3.7	0.9	0.5
Substandard quality	1.0	0.4	0.8
Sales volume	8.5	2.4	1.6
Total	13.2%	3.7%	2.9%
Total costs of quality as a percentage of sales	18.4%	9.2%	6.7%

An important component of the Premio Nacional de Calidad is a company's safety record. Crysel's safety index, measured by the number of accidents per million labor-hours, declined from 6.3 in 1986 to 3.0 in 1992. Crysel eliminated accidents by redesigning machines, training operators in safety practices, and implementing safe operating procedures.

---

[a] Based on a presentation by Raul Gil Dufoo, Director General of Crysel, and discussions with company management.

achieves zero defects, or will there come a point when prevention and appraisal costs increase faster than the decrease in failure costs, so that some level of defects is optimal? A precise answer to whether a zero-defect policy is optimal depends on correctly measuring the costs of failure on the one hand, and the costs of prevention and appraisal on the other hand. Many managers believe the costs of failure are sufficiently high and the costs of effective and efficient prevention and appraisal are sufficiently low that a zero-defect policy could be optimal.

QUALITY AND CUSTOMER-SATISFACTION MEASURES

Customer satisfaction is an important element of quality programs. Producing a defect-free, high-quality product is profitable only if it also satisfies customers. Companies that emphasize quality, for example, Mitsubishi and Toyota in Japan and Ford, Xerox, Motorola, and Federal Express in the United States give special attention to customer satisfaction when choosing the design and features of their products.

Motorola describes its program of total customer satisfaction as:

◆ Giving the customer product-performance features that are perceived by the customer as providing fair value
◆ Delivering the product when promised
◆ Delivering the product with no defects
◆ Ensuring that the product will not experience early failure
◆ Ensuring that the product will not fail excessively in service

To evaluate and track how well it is doing, Motorola and other companies track customer-satisfaction trends over time. Customer satisfaction is difficult to measure precisely, but companies can choose from among many indicators in their search for answers.

Financial Measures of Customer Satisfaction

Costs of external failures—such as warranty repair costs, liability claims, forgone contribution margin on lost sales, and lower prices for products sold—are all financial indicators of poor customer satisfaction. But financial measures do not indicate the specific areas that need improvement, nor do they reveal the future needs and preferences of customers. For these reasons, most companies also use nonfinancial measures.

Nonfinancial Measures of Customer Satisfaction

Nonfinancial measures of customer satisfaction include:

◆ The number of defective units shipped to customers as a percentage of total units of products shipped
◆ The number of customer complaints. Companies estimate that for every customer who actually complains, there are 10 to 20 others who have had bad experiences with the product but have not complained.
◆ Excess customer-response time (the difference between scheduled delivery date and date requested by customer)
◆ On-time delivery (percentage of shipments made on or before the scheduled delivery date)

Management will step in and investigate if these numbers deteriorate over time.

In addition to these routine nonfinancial measures, many companies conduct surveys to measure customer satisfaction. Surveys serve two objectives. First, they provide a deeper perspective into customer experiences and preferences. Second, they provide a glimpse into features that customers would like future products to have.

Xerox Corporation conducts several surveys each year to measure customer satisfaction with its products.[6] In its periodic surveys, which Xerox mails to 40,000 randomly selected customers each month, the company measures satisfaction in several dimensions. Are customers pleased with Xerox overall? Have the products proven to be reliable? If a problem arose, how quickly did Xerox technical support arrive to solve it?

Soon after Xerox installs a machine, it sends out post-installation surveys. It questions customers about product performance (copy quality, ease of use, jamming of paper), responsiveness of the sales representative, order process (whether the product was what the customer wanted), delivery process (on-time), installation process, and support activities (user training, documentation, responsiveness to questions).

Surveys are not the only source of customer feedback. Consumer products firms, such as Procter & Gamble, obtain information on customer satisfaction through in-depth personal interviews and meetings with consumer focus groups. Customer feedback alerts design and operations staff to problems that company teams then seek to correct.

QUALITY AND INTERNAL PERFORMANCE MEASURES

Prevention costs, appraisal costs, and internal failure costs are examples of financial measures of quality performance inside the company. Most companies monitor both financial and nonfinancial measures of internal quality.

What nonfinancial measures might a business use? Analog Devices, a semiconductor manufacturer, follows trends in these gauges of quality:

◆ The number of defects for each product line
◆ Process yield (ratio of good output to total output)
◆ Manufacturing lead time (the time taken to convert direct materials into finished output)
◆ Employee turnover (ratio of number of employees who left the company to the total number of employees)

By themselves, nonfinancial measures of quality have limited meaning. They are more informative when management examines trends over time. Analog Devices uses an innovative method called the half-life method to set targets for rates of improvement. The *half-life method* is the amount of time it takes for a defect rate to be cut in half. Suppose Analog Devices sets target defect levels per million units produced as follows:

Date (1)	Target Defects per Million Units Produced (2)	Percentage Defects (3) = (2) ÷ 1,000,000 units
January 1, 1995	80,000 units	8%
July 1, 1995	40,000 units	4%
January 1, 1996	20,000 units	2%
July 1, 1996	10,000 units	1%

The half-life in this example is six months. Given an initial defect rate of 8%, the company expects to take six months to cut the defect rate in half, to 4%, and another six months to cut the defect rate in half again, to 2% (half of 4%), and so on.

Half-lives measure how quickly an organization perceives, analyzes, and solves problems. Half-lives are estimated for each type of defect identified—for example process losses, lead time, and defective products. Half-lives for different types of defects vary from less than a month to several years, depending on the complexity of the

[6]Xerox Corporation: The Customer Satisfaction Program, Case 9-591-055 (Boston, MA.: Harvard Business School).

problem and the number of different organizational units or departments that need to work together to solve it. The half-life concept can also be applied to nonfinancial measures of customer satisfaction, such as on-time deliveries.

EVALUATING QUALITY PERFORMANCE

Objective 5
Understand why companies use both financial and nonfinancial measures of quality

Exhibit 23-7 summarizes our description of financial and nonfinancial measures of customer satisfaction and internal performance. Measuring the financial costs of quality and the nonfinancial aspects of quality have distinctly different advantages.

Advantages of the Costs of Quality (COQ) Measure

1. COQ focuses attention on how costly poor quality can be, although COQ measures sometimes exclude important but difficult-to-measure costs such as the impact of poor quality on customer goodwill.
2. Financial COQ measures are a useful way of comparing different quality-improvement projects and setting priorities for maximum cost reduction.
3. Financial COQ measures serve as a common denominator for evaluating trade-offs among prevention and failure costs. COQ provides a single, summary measure of quality performance.

Advantages of Nonfinancial Measures of Quality

1. Nonfinancial measures of quality are easy to quantify and easy to understand.
2. Nonfinancial measures direct attention to physical processes and hence focus attention on the precise problem areas that need improvement.
3. Nonfinancial measures provide immediate short-run feedback on whether quality improvement efforts have in fact succeeded in improving quality.

Many of the advantages cited for COQ are disadvantages of nonfinancial measures and vice versa. Most organizations use both financial and nonfinancial quality measures to measure quality performance.

Some companies (for example, Analog Devices, Milliken, and Octel) present both financial and nonfinancial measures of quality performance in a single report, sometimes called a *balanced scorecard*. The balanced scorecard helps top management evaluate whether lower-level managers have improved one area at the expense of others. For example, a manager at risk of not meeting operating income goals may start to ship high-margin products and delay deliveries of low-margin products. The balanced scorecard will recognize the improvement in financial performance but will also reveal that operating income targets were achieved by sacrificing on-time per-

EXHIBIT 23-7
Financial and Nonfinancial Measures of Quality

	Measures of Quality	
	Financial Measures	**Nonfinancial Measures**
Quality Measures For Customer Satisfaction	Costs of external failures	Physical measures of customer satisfaction (for example, customer response times, on-time performance)
Quality Measures For Internal Performance	Costs of prevention, appraisal, and internal failures	Physical measures of internal performance (for example, defect rates, manufacturing lead times)

formance. Moreover, corporations often use multiple measures of quality in managers' bonus plans. For example, Motorola explicitly ties employee bonuses to improvements in customer satisfaction, yield, and cycle time.

As corporations' responsibilities toward the environment grow, many companies are directing managers to apply total quality management principles to problems of air pollution, waste water, oil and chemical spills, hazardous waste, and waste management. The costs of environmental damage (failure costs) can be extremely high to corporations under the 1990 amendments to the Clean Air Act. Companies can be charged multimillion dollar fines. For example, Exxon paid $125 million in fines and restitution on top of $1 billion in civil payments for the *Exxon Valdez* oil spill, which harmed the Alaskan coast.

TIME AS A COMPETITIVE WEAPON

Companies increasingly view time as a key variable in competition.[7] We consider time from two different perspectives:

1. *New product development times*, which measure how long it takes companies to develop new products and bring them to market
2. *Operational measures of time*, which reveal how quickly companies respond to customers' demands for their existing products

NEW PRODUCT DEVELOPMENT TIME

Successful new product ventures are essential to most companies. IBM introduces a new computer, AT&T introduces a new cordless telephone, and Honda introduces a new car. Bringing new products to market faster than competitors enables a company to gain market share. The first company to come out with a new product preempts the competition, outracing its rivals to the customers. Perhaps customers are anxious to try a product featuring a promised new technology, and the quicker a company gets to market, the more likely it is to gain those customers. Whatever the market's desire or need, a company that responds quickly stands a better chance of succeeding than does a company that comes out with the product too late. Also, the company that brings new products to market faster generally incurs lower development costs. **New product development time** is the amount of time from when the initial concept for a new product is approved by management to its market introduction.

The increased emphasis on reducing new product development time has come about because of shorter product life cycles. The *product life cycle* is the time from initial research and development to the time at which support to customers is withdrawn as the product dies. Because of shrinking product life cycles, products become obsolete more quickly. Successful companies have responded by introducing new products more quickly.

Many companies have reported spectacular reductions in new product development times. Deere and Co. reduced the time needed to develop new construction equipment from seven years to four years. Honeywell developed its latest thermostat in less than a year, whereas previously its new products had taken four years to evolve. NCR reduced development time from four years to two years on its checkout terminals. Toyota and Honda indicate that they now take three years (or less) to develop a new car. General Motors, Ford, and Chrysler report new product development times of closer to four years, down from the previous five years.

The management accountant's task is to develop consistent measurements for new product development time. Several decisions need to be made. For example,

---

[7] See G. Stalk and T. Hout, *Competing Against Time* (New York: Free Press, 1990).

what date should be considered the start date for measuring new product development time? Is it when the need for a new product is first identified and approved by management or when the product development team is first put together? What is the end date? Is it when the first unit of the product is produced or when the first unit is sold? The choices made should be consistently applied and used to monitor trends in new product development time. Analyzing trends helps managers understand the factors that cause new product development time to increase or decrease.

BREAKEVEN TIME FOR NEW PRODUCTS

Objective 6
Explain breakeven time

Missing from our discussion of new product development time thus far is a consideration of financial success after the product is introduced. For example, new product development time does not address how quickly a company recovers its cash investment in the new product. Decreasing new product development time is not an end in itself. The goal of introducing products to market faster is to increase the chances of success. But reduced new product development time does not guarantee faster and higher cash flows. Hence the need to measure new product development in financial terms. This section introduces one such measure—*breakeven time*.

Breakeven time (BET) is the amount of time from when the initial concept for a new product is approved by management until the time when the cumulative present value of net cash inflows from the project equals the cumulative present value of net investment outflows. BET is the length of time it takes to recover the investment made in new product development. New-product proposals with shorter BETs are preferred to new-product proposals with longer BETs, if all other things are equal.

Hewlett-Packard (HP) is an enthusiastic advocate of BET:

> One of HP's major objectives is to remain a leader in product development. . . . Some traditional finance methods, e.g., net present value and internal rate of return, were considered for use as HP's primary product-development-process metric. However, within the context of HP's high-technology business, time-to-market is considered to be of utmost importance, and so BET, with its emphasis on how fast these technologies and products are brought to market, has been chosen to be HP's focal product-development metric.[8]

Example of BET Computations

Consider a new medical instrument (SP-108) developed by the Stanford Park division of Soltron. Soltron puts together the project investigation team on December 31, 19_4. Because of other urgent commitments, the project is put on hold until 19_5. In December 19_5, Soltron starts to work intensively on the new instrument. Information on the project follows.

1. The initial investment for the project consists of:
   a. Research and development costs for developing the instrument
   b. Costs of designing the product and process, including the costs of designing, building, and testing product prototypes and acquiring new tools
   c. Manufacturing investments in special machines and additional working capital
   d. Marketing costs for market research and advertising

   Soltron's initial investment is $12 million. Assume for simplicity that the investment occurs on December 31, 19_5. The new instrument has a product life cycle of four years.

2. Soltron's annual cash inflows from sales of SP-108 occur throughout the year, but to ease computations we assume they occur at the end of each year: 19_6, $18 million; 19_7, $33 million; 19_8, $40 million; 19_9, $14 million.

3. To generate sales, Soltron incurs incremental annual cash outflows of direct materials, direct manufacturing labor, manufacturing overhead, marketing, distribution and cus-

[8]Corporate Engineering, Hewlett-Packard, *Breakeven Time at Hewlett-Packard*, 1990, p. 2.

tomer-service costs. Cash outflows occur throughout the year, but to ease computations, we again assume they take place at the end of each year: 19_6, $13 million; 19_7, $24 million; 19_8, $30 million; 19_9, $11 million.

4. Cash inflow from recovery of initial investment at the end of 19_9 is $2 million.

Assume a 14% required rate of return for discounting cash flows on a before-tax basis. Ignore income taxes and inflation effects.

Exhibit 23-8, Panel A, presents the computations necessary to calculate BET for SP-108. BET calculations cover the period of time that begins with the concept approval by management on December 31, 19_4, not the period of time that begins when investment cash outflows start. Why? Because the goal of BET is to evaluate how quickly new ideas are converted into profitable products.

How long did it take Soltron to recover the $10.524 million cumulative present value of investment cash outflows? In present value terms, Soltron recovers $9.920 million of its investment through product cash inflows by December 31, 19_7, and $15.840 million by December 31, 19_8. This means that Soltron recovers the present value of its initial investment cash outflows of $10.524 million sometime in 19_8. Soltron's present value of product cash inflows during 19_8 is $5.920 million, and it needs $0.604 million

EXHIBIT 23-8
Breakeven Time Computations for SP-108 at Soltron (in millions)

PANEL A: INVESTMENT AND PRODUCT CASH FLOWS

Year (1)	PV Discount Factor at 14% (2)	Investment Cash Outflows (3)	PV of Investment Cash Outflows* (4) = (2) × (3)	Cumulative PV of Investment Cash Outflows* (5)	Product Cash Inflows† (6)	PV of Product Cash Inflows* (7) = (2) × (6)	Cumulative PV of Product Cash Inflows* (8)
19_4	1.000	—	—	—	—	—	—
19_5	0.877	$(12.000)	$(10.524)	$(10.524)			
19_6	0.769				$ 5.000	$3.845	$ 3.845
19_7	0.675				9.000	6.075	9.920
19_8	0.592				10.000	5.920	15.840
19_9	0.519				3.000	1.557	17.397
	0.519				2.000‡	1.038	18.435

PANEL B: CASH OUTFLOWS AND INFLOWS

Year (1)	PV Discount Factor at 14% (2)	Cash Inflows (3)	Cash Outflows (4)	Net Cash Flows (5) = (3) − (4)	PV of Net Cash Flows* (6) = (2) × (5)	Cumulative PV of Net Cash Flows* (7)
19_4	1.000	—	—	—	—	—
19_5	0.877	—	$12.000	$(12.000)	$(10.524)	$(10.524)
19_6	0.769	$18.000	13.000	5.000	3.845	(6.679)
19_7	0.675	33.000	24.000	9.000	6.075	(0.604)
19_8	0.592	40.000	30.000	10.000	5.920	5.316
19_9	0.519	16.000	11.000	5.000	2.595	7.911

*At December 31, 19_4.

†Product cash inflows = sales cash inflows − cash outflows incurred to generate sales. Cash inflows from disposal of investment on December 31, 19_9 are also included under this column.

‡Cash inflow from disposal of investment.

($10.524 million – $9.920 million) to recover its investment. Soltron takes 3.10 years to recover the cumulative present value of investment cash outflows.

$$3 \text{ years (up to December 31, 19\_7)} + \frac{\$0.604 \text{ million}}{\$5.920 \text{ million}} = 3.10 \text{ years}$$

After 3.10 years, the cumulative present value of *all* project cash outflows equals the cumulative present value of *all* project cash inflows.[9]

An alternative approach to calculate BET is to *ignore* all distinctions between investment and product cash flows and to compute the time it takes for the present value of net cash flows to equal zero. Exhibit 23-8, Panel B, presents the computations to calculate BET for SP-108 using this approach. Observe that the present value of cash flows becomes positive between December 31, 19_7 and December 31, 19_8. Present value of product cash inflows during 19_8 is $5.920 million, and Soltron needs to recover $0.604 million. Therefore as in Exhibit 23-8, Panel A,

$$\text{BET} = 3 \text{ years (up to December 31, 19\_7)} + \frac{\$0.604 \text{ million}}{\$5.920 \text{ million}} = 3.10 \text{ years}$$

What are the relevant cash flows in BET analysis? The incremental future cash inflows and future cash outflows that will change as a result of introducing the new product. The focus is on cash flows, not on accrued and deferred items, such as depreciation. Overhead costs are irrelevant if adding a new product changes only the amount of overhead costs allocated to each product without changing the total cash outflow for overhead costs. BET analysis should also include both positive and negative cash-flow effects that the new product may have on sales of existing products. For example, introducing a new razor may increase overall sales of blades but decrease sales of the old-model razors.

Our example illustrates the calculation of BET based on Soltron's actual experience with SP-108. This is the performance evaluation role of BET—management compares actual BET with the estimates presented when the project was approved. Calculating actual BETs for different projects also allows management to examine how BET has changed over time. Take considerable care, however, when interpreting differences in BETs across time periods and across projects. Shorter BETs could result from choosing simpler projects rather than from improving product development.

BET can also be used as a decision tool for choosing among design options and for choosing among new development projects. For example, management may decide to consider only projects with a projected BET of, say, less than four years. In this case, managers must forecast investment and product cash flows to calculate BET.

The BET computations illustrated in this section consider only the present value of net cash flows until the cumulative present value of all project cash outflows equals the cumulative present value of all project cash inflows. As computed here, BET measures how fast a new product recovers the investment made in its development. An alternative approach is to consider how long it takes to recover the present value of the total cash outflows for the product over its life cycle. Because this approach includes more cash outflows (all cash outflows over the project's life) it will result in longer BETs for new products than the approach presented in this section.

BET versus the Payback Method

BET differs from the payback method (see Chapter 20, p. 697) in two ways. First, BET, unlike payback, starts counting time when management approves a project, rather than from the time that cash outflows occur. Second, BET considers the time value of

---

[9]A literal interpretation of the assumption that cash flows only occur at the end of each year would imply a BET of four years because Soltron will recover the present value of its investment only when cash inflows occur at the end of the fourth year. The calculations shown in the chapter, however, better approximate Soltron's actual BET computed on the basis of uniform cash flows and continuous compounding.

money by discounting cash inflows and outflows. The payback method ignores the time value of money.

Limitations of BET

As useful as BET is, the approach does have limitations:

1. The BET method neglects project profitability because it ignores all project cash flows after the breakeven time. A project with high projected cash inflows in later years may be rejected in favor of a less profitable project with moderate cash flows in early years if the latter project has a shorter BET.
2. Critics of BET say that it deemphasizes long-range thinking. An emphasis on BET for performance evaluation may encourage managers to undertake short-run incremental projects rather than truly innovative, long-run projects based on new technology that generally have longer BETs.
3. The BET measure has a strictly financial focus and does not consider strategic and nonfinancial reasons for product development.
4. BET is not comparable among businesses. BET can vary greatly from business to business because of differences in product life cycles and investment needs.

OPERATIONAL MEASURES OF TIME

Time has many components. In addition to developing and bringing new products to market quickly, organizations compete on the time it takes to respond to customer requests and the reliability with which scheduled delivery dates are met. Operational measures of time indicate the speed and reliability with which organizations supply products and services to customers. Two common operational measures of time are customer-response time and on-time performance.

Customer-Response Time

Objective 7
Describe customer-response time, and explain the reasons for and the costs of lines and delays

Customer-response time is the amount of time from when a customer places an order for a product or requests service to when the product or service is delivered to the customer. The time to respond to customer requests is a key competitive factor in many industries. For example, a retail store able to deliver a washing machine of a given make and color in two days will probably attract more business than a store with a two-week delivery time. Similarly, a grocery store that averages 5 minutes per customer in a check-out line will probably attract more business than a store that averages 20 minutes per customer. The time a customer spends in line is critical in many other industries, such as banking, car-rental, and fast-food.

Why do lines—often called queues—form and delays occur? Consider a fast-food restaurant. Customers who arrive at the counter when all waiters are busy serving other customers must wait. Likewise, in manufacturing, delays can occur when products that require work at a particular machine arrive at a time when that machine is processing products that arrived earlier.

A **time driver** is any factor that causes a change in the speed with which an activity is undertaken when the factor itself changes. What are the drivers of time? We consider two of the most important. (1) Uncertainty about when customers will order products or services. For example, the more randomly customers arrive at a fast-food restaurant, the more likely that lines will form and delays will occur. (2) Limited capacity and bottlenecks. A **bottleneck** is an operation where the work required to be performed approaches or exceeds the available capacity. For example, a bottleneck is created when products that need to be processed at a machine arrive while the machine is busy processing other products. The demand for time on the machine exceeds available capacity, and delays begin to occur.

Effects of Uncertainty and Bottlenecks on Delays

Consider the following numerical example. Falcon Works (FW) uses one turning machine to convert steel bars into component parts. FW makes one specialty component, A22. FW makes this component only after FW's customers order the component.

FW does not know exactly how many orders of A22 it will receive during the year. FW expects it will receive 30 orders, but it could actually receive 10 orders, 20 orders, or 50 orders. Each order is for 1,000 units of A22. Each order takes 100 hours of manufacturing time (8 hours of setup time to clean and prepare the machine, and 92 hours of processing time). The annual capacity of the machine is 4,000 hours. If FW receives the number of orders it expects, the total amount of manufacturing time required on the machine will be 3,000 (100 × 30) hours, which is within the available machine capacity of 4,000 hours. Even though capacity limits are not strained, queues and delays will still occur. Why? Because of uncertainty about the actual dates when FW's customers will place the orders. FW may receive an order from one or more customers while the machine is processing another customer's order. The second order must wait in line for the machine to be free.

Under certain assumptions about the pattern of customer orders and how orders will be processed,[10] we can calculate the *average waiting time*. The **average waiting time** is the average amount of time that an order will wait in line before it is set up and processed. In our example, the average waiting time (in hours) before manufacturing an order commences is calculated as follows:[11]

$$
= \frac{\left(\begin{array}{c}\text{Average number} \\ \text{of orders of A22}\end{array}\right) \times \left(\begin{array}{c}\text{Manufacturing} \\ \text{time for A22}\end{array}\right)^2}{2 \times \left\{\begin{array}{c}\text{Annual machine} \\ \text{capacity}\end{array} - \left[\left(\begin{array}{c}\text{Average number} \\ \text{of orders of A22}\end{array}\right) \times \left(\begin{array}{c}\text{Manufacturing} \\ \text{time for A22}\end{array}\right)\right]\right\}}
$$

$$
= \frac{30 \times (100)^2}{2 \times [4,000 - (30 \times 100)]} = \frac{30 \times 10,000}{2 \times (4,000 - 3,000)} = \frac{300,000}{2 \times 1,000} = \frac{300,000}{2,000} = 150 \text{ hours}
$$

On average, any order received will have to wait for 150 hours before it is set up and processed on the machine. Note that the formula describes only the *average* waiting time. A particular order may happen to arrive when the machine is free, in which case manufacturing will start immediately. In other situations, FW may receive an order while two other orders are waiting to be processed. In this case, the delay will be longer than 150 hours.

Manufacturing lead (or cycle) time for an order is any waiting time plus manufacturing time for the order. It is the time from when the order is ready to start on the production line (ready to be set up) to when it becomes a finished good. In our example, average manufacturing lead time for an order of A22 is 250 hours (150 hours of average waiting time + 100 hours of manufacturing time). Throughout this section we use manufacturing lead time to refer to manufacturing lead time for an order.

FW is considering whether to introduce a new product, C33. FW expects to receive 10 orders of C33 (each order is for 800 units) in the coming year. Each order takes 50 hours of manufacturing time (4 hours of setup time and 46 hours of processing time). The expected demand for A22 will be unaffected whether or not FW introduces C33.

If FW introduces C33, and receives the number of orders it expects to receive,

---

[10]The precise technical assumptions are that customer orders for the product follow a Poisson distribution with a mean equal to the expected number of orders (30 in the case of A22) and that orders are processed on a first-in first-out (FIFO) basis. The Poisson arrival pattern for customer orders has been found to be reasonable in many real-world settings. The FIFO assumption can be modified. The basic queuing and delay effects will still occur, but the precise formulas will be different.

[11]$\left(\begin{array}{c}\text{Manufacturing time} \\ \text{required for A22}\end{array}\right)^2 = \left(\begin{array}{c}\text{Manufacturing time} \\ \text{required for A22}\end{array}\right) \times \left(\begin{array}{c}\text{Manufacturing time} \\ \text{required for A22}\end{array}\right)$

the total amount of manufacturing time required on the machine will be 3,500 hours [(100 × 30) for A22 + (50 × 10) for C33], which is still less than the available machine capacity of 4,000 hours. What is the effect of introducing C33 on waiting times and manufacturing lead times? The average waiting time before an order is set up and processed is given by the following formula, which is an extension of the formula described earlier for the single-product case.

$$\frac{\left[\left(\begin{matrix}\text{Average number}\\\text{of orders of A22}\end{matrix}\right)\times\left(\begin{matrix}\text{Manufacturing}\\\text{time for A22}\end{matrix}\right)^2\right]+\left[\left(\begin{matrix}\text{Average number}\\\text{of orders of C33}\end{matrix}\right)\times\left(\begin{matrix}\text{Manufacturing}\\\text{time for C33}\end{matrix}\right)^2\right]}{2\times\left\{\begin{matrix}\text{Annual machine}\\\text{capacity}\end{matrix}-\left[\left(\begin{matrix}\text{Average number}\\\text{of orders of A22}\end{matrix}\right)\times\left(\begin{matrix}\text{Manufacturing}\\\text{time for A22}\end{matrix}\right)\right]-\left[\left(\begin{matrix}\text{Average number}\\\text{of orders of C33}\end{matrix}\right)\times\left(\begin{matrix}\text{Manufacturing}\\\text{time for C33}\end{matrix}\right)\right]\right\}}$$

$$=\frac{[30\times(100)^2]+[10\times(50)^2]}{2\times[4,000-(30\times100)-(10\times50)]}=\frac{(30\times10,000)+(10\times2,500)}{2\times(4,000-3,000-500)}$$

$$=\frac{(300,000+25,000)}{2\times500}=\frac{325,000}{1,000}=325\text{ hours}$$

Average manufacturing lead time for A22 is 425 hours (325 hours of average waiting time + 100 hours of manufacturing time), and for C33 it is 375 hours (325 hours of average waiting time + 50 hours of manufacturing time).

Introducing C33 causes a dramatic increase in average waiting time, more than doubling the 150 hours for A22 alone to 325 hours for both A22 and C33. Why so large an increase in average waiting time? Because choosing to manufacture C33 increases the average capacity utilization of the machine from 75% (3,000 ÷ 4,000) to 87.5% (3,500 ÷ 4,000). Think of excess capacity as a cushion for absorbing the shocks of variability and uncertainty in the arrival of customer orders. As excess capacity shrinks, the chances increase that, at any instant in time, new orders will arrive while existing orders are being manufactured. At the time the orders come in, there are more orders to be worked on at the machine than the capacity available. The resulting bottleneck creates lines and delays.

Also observe the ratio of manufacturing time (the time to set up and process an order) to average manufacturing lead time as average capacity utilization increases. For C33 the ratio is 13.33% (50 ÷ 375). On average, C33 spends 86.67% (100% − 13.33%) of its manufacturing lead time (the time it spends in FW's plant) just waiting for manufacturing to start!

Manufacturing lead time is just one component of customer response time, as the following figure describes.

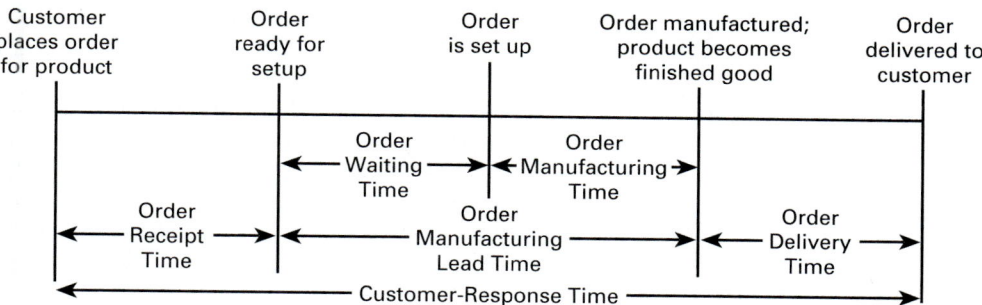

In addition to manufacturing lead time, customer-response time includes the time it takes marketing to place an order with manufacturing (order receipt time) and the time it takes distribution to deliver the product to the customer (order delivery time). In the FW example, we assume that order receipt time and order delivery time are minimal.

On-Time Performance

On-time performance refers to situations in which the product or service is actually delivered at the time it is scheduled to be delivered. Customer orders for products or services generally specify quantity, cost, and delivery time. For example, Federal Express specifies a price per package and a next-day delivery time for its overnight courier service. Federal Express's on-time performance measures evaluate how often the company meets its stated delivery times. On-time performance in the airline industry is becoming increasingly prominent. For example, Southwest Airlines reported that it was rated number one for on-time performance by the Department of Transportation in October 1992 (93.4% of its daily flights arrived within 15 minutes of the scheduled times).

Although beyond the scope of this book, bottlenecks and uncertainty about when customers will order products or services also affect on-time performance. There is, however, one additional trade-off between customer-response time and on-time performance. Scheduling longer customer-response times improves on-time performance. For example, a supermarket announcing that it will serve its customers in five minutes is more likely to meet its claim for on-time performance than is a supermarket promising that it will serve its customers in two minutes.

COSTS OF TIME

Relevant Revenues and Relevant Costs

Should FW introduce product C33? Consider the following information:

Product	Average Number of Orders	Average Selling Price per Order If Average Manufacturing Lead Time Less Than 300 Hours	Average Selling Price per Order If Average Manufacturing Lead Time More Than 300 Hours	Variable Costs per Order	Inventory Carrying Costs per Order per Hour
A22	30	$22,000	$21,500	$16,000	$1.00
C33	10	10,000	9,600	8,000	0.50

Note that manufacturing lead times affect both revenues and costs in our example. Revenues are affected because customers are willing to pay a small premium for faster delivery. Costs are affected through inventory carrying costs. *Inventory carrying costs* usually consist of the opportunity costs of investment tied up in inventory (see Chapter 11, p. 399) and the relevant costs of storage, such as space rental, spoilage, deterioration, and materials handling. (Chapter 24 describes the computation of relevant inventory carrying costs in more detail.) Companies usually calculate inventory carrying costs on a per order per year basis. To simplify computations, we express inventory carrying costs on a per order per hour basis. FW incurs inventory carrying costs for the duration of the wait time and manufacturing time.

Exhibit 23-9 presents relevant revenues and relevant costs for this decision. The preferred alternative is not to introduce C33. Note that C33 is rejected despite having a positive contribution margin of at least $1,600 ($9,600 – $8,000) per order. Recall, too, that FW's machine has the capacity to process C33 because the machine will, on average, use only 3,500 of the available 4,000 hours. Why is C33 rejected? *The key is to recognize the negative effects of C33 on the existing product A22.* The following table presents the opportunity costs of using up extra capacity on the machine to manufacture C33.

**Expected Loss in Revenues and Expected Increase in Costs
from Introducing C33**

Product (1)	Expected Loss in Revenues for A22 from Increasing Average Manufacturing Lead Times (2)	Expected Increase in Carrying Costs for All Products from Increasing Average Manufacturing Lead Times (3)	Expected Loss in Revenues Plus Expected Increase in Costs of Introducing C33 (4) = (2) + (3)
A22	$15,000*	$5,250†	$20,250
C33	—	1,875‡	1,875
Total	$15,000	$7,125	$22,125

*($22,000 − $21,500) × 30 orders.
†(425 hours − 250 hours) × $1.00 × 30 orders.
‡(375 hours − 0) × $0.50 × 10 orders.

Introducing C33 causes the average manufacturing lead time of A22 to increase from 250 hours to 425 hours. This increases inventory carrying costs. It also causes A22's revenues to decrease because it would, on average, take more than 300 hours to manufacture A22. The expected costs of introducing C33 equal $22,125, which exceeds C33's expected contribution margin of $16,000 ($1,600 per order × 10 orders). FW chooses not to produce C33.

Assume that FW's customers want further reductions in the current average manufacturing lead time of 250 hours for A22. In response, FW is considering cutting its order size in half, from 1,000 units an order to 500 units an order. FW expects the number of units sold to remain the same, so cutting order size in half would double the number of orders that FW expects to receive, from 30 to 60. How long will it take to manufacture each order of 500 units? Each order will still require 8 hours of setup but only 46 (92 ÷ 2) hours of processing. Total manufacturing time per order will be 54 hours.

What is the effect on average waiting time? Using our formula, we get an average waiting time (rounded to the nearest hour) of

EXHIBIT 23-9

Determining Expected Relevant Revenues and Expected
Relevant Costs for Falcon Works' Decision to Introduce C33

	Alternative 1: Introduce C33 (1)	Alternative 2: Do Not Introduce C33 (2)	Relevant Revenues and Relevant Costs (3) = (1) − (2)
Expected revenues	$741,000*	$660,000†	$81,000
Expected variable costs	560,000‡	480,000§	80,000
Expected carrying costs	14,625‖	7,500#	7,125
Expected variable and carrying costs	574,625	487,500	87,125
Expected revenues minus expected costs	$166,375	$172,500	$ (6,125)

*($21,500 × 30) + ($9,600 × 10); average manufacturing lead times will be more than 300 hours.
†$22,000 × 30; average manufacturing lead times will be less than 300 hours.
‡($16,000 × 30) + ($8,000 × 10).
§$16,000 × 30.
‖(A22's average manufacturing lead time, 425 hours x A22's unit carrying costs, $1.00 x 30 expected orders) + (C33's average manufacturing lead time, 375 hours x C33's unit carrying costs, $0.50 x 10 expected orders).
#A22's average manufacturing lead time, 250 hours x A22's unit carrying costs, $1.00 x 30 expected orders.

$$\frac{60 \times (54)^2}{2 \times [4{,}000 - (60 \times 54)]} = \frac{60 \times 2{,}916}{2 \times (4{,}000 - 3{,}240)} = \frac{174{,}960}{2 \times 760} = \frac{174{,}960}{1{,}520} = 115 \text{ hours}$$

A22's average manufacturing lead time will decrease from 250 hours to 169 hours (115 + 54), because smaller order sizes require less time on the machine, which reduces waiting time. Waiting time is also reduced because there is less chance that the machine will be busy when an order is received. Will average manufacturing lead times decrease further if we cut order size in half again? Not in our example. Why? Because too much time would then be wasted on setups. To further reduce manufacturing lead time, FW must reduce setup time.

We have described a simple setting to explain the effects of uncertainty and capacity constraints and the relevant revenues and relevant costs of time.[12] How can delays be reduced? Increasing the capacity of the bottleneck resource can reduce lines, delays, and inventories. When demand uncertainty is high, *some* excess capacity is desirable. Increasing bottleneck capacity does not necessarily require capital investment. Available physical capacity increases as the time it takes to set up and process existing products decreases. Delays can also be reduced through careful scheduling of orders on machines—for example, by batching similar jobs together for processing.

Choice of Cost Allocation Bases for Product Costing and Performance Evaluation

Several firms have adopted manufacturing lead time as the base for allocating indirect manufacturing costs to products. The rationale for doing so is best expressed by the controller at Zytec Corporation, a manufacturer of computer equipment:

> We wanted to pick [a cost allocation base] that was meaningful to the people on the floor. We wanted the [chosen allocation base] to capture the essence of our drive for continuous improvement. In particular, we were convinced that the cost system could become a potent tool for behavior modification.[13]

Zytec believes that using manufacturing lead time motivates managers to reduce the time taken to manufacture products. In turn, total overhead costs decrease and operating income rises.

Companies that emphasize customer-response time and on-time performance use these measures to evaluate managers. Bonuses paid to regional managers of several overnight courier services include a component based on on-time delivery performance. Drivers for courier companies are given very clear signals that on-time delivery is their top priority each day.

THEORY OF CONSTRAINTS AND THROUGHPUT CONTRIBUTION ANALYSIS

Objective 8
Define the three main measurements in the theory of constraints

We now expand the discussion of the previous section by considering products that are made from multiple parts and processed on different machines. With multiple parts in a product, dependencies arise among operations; some operations cannot be started until parts from a previous operation are available. Waiting times now appear for two reasons: (1) parts that require processing at a bottleneck machine must wait in line until the bottleneck machine is free, and (2) parts made on nonbottleneck machines must wait until parts coming off the bottleneck machine arrive.

---

[12]Other complexities such as analyzing a network of machines, priority scheduling, and allowing for uncertainty in processing times are beyond the scope of this book. In these cases, the basic queuing and delay effects persist, but the precise formulas are different.

[13]R. Cooper and R. P. Turney, "Internally Focused Activity-Based Cost Systems," in *Performance Excellence in Manufacturing and Service Organizations,* ed. P. Turney (Sarasota, FL: American Accounting Association, 1990), p. 95.

The theory of constraints (TOC) focuses on revenue and cost management when faced with bottlenecks.[14] It defines three measurements:

1. **Throughput contribution,** equal to sales dollars minus direct materials costs.
2. *Investments (inventory),* equal to the sum of materials costs of direct materials inventory, work-in-process inventory, and finished goods inventory; research and development costs; and costs of equipment and buildings.
3. *Other operating costs,* equal to all operating costs (other than direct materials) incurred to earn throughput contribution. Other operating costs include salaries and wages, rent, utilities, and depreciation.

Objective 9

Describe four steps in managing constraints

The objective of TOC is to increase throughput contribution while decreasing investments and operating costs. *The theory of constraints considers short-run time horizons and assumes other current operating costs to be fixed costs.* The key steps in managing bottleneck resources follow:

STEP 1. Recognize that the bottleneck resource determines throughput contribution of the plant as a whole.

STEP 2. Search and find the bottleneck resource by identifying resources with large quantities of inventory waiting to be worked on.

STEP 3. Subordinate all nonbottleneck resources to the bottleneck resource. That is, the needs of the bottleneck resource determine the production schedule of nonbottleneck resources.

Step 3 represents a key notion described in Chapter 11: To maximize overall contribution margin, the plant must maximize contribution margin (in this case throughput contribution) of the constrained resource (see p. 394). Step 3 implies that workers at nonbottleneck machines should not be motivated to improve their productivity if the additional output cannot be processed by the bottleneck machine. Producing more nonbottleneck output only creates more inventory; it does not increase throughput contribution. The preferred course of action is for the bottleneck machine to set the pace for the nonbottleneck machine.

STEP 4. Take actions to increase bottleneck efficiency and capacity—the objective is to increase throughput contribution minus the incremental costs of taking such actions.

Cardinal Industries (CI) manufactures car doors in two operations—stamping and pressing. Additional information follows:

	Stamping	Pressing
Capacity per hour	20 units	15 units
Annual capacity (6,000 hours of capacity available in each of stamping and pressing)	120,000 units	90,000 units
Annual production	90,000 units	90,000 units
Operating costs other than direct materials (fixed)	$720,000	$1,080,000
Other fixed operating costs per unit produced ($720,000 ÷ 90,000; $1,080,000 ÷ 90,000)	$8 per unit	$12 per unit

Each door sells for $100 and has direct materials costs of $40. Variable costs in other functions of the value chain—research and development, design of products and processes, marketing, distribution, and customer service—are negligible. CI's output is constrained by the capacity of 90,000 units at the pressing operation. What can CI do to relieve the bottleneck constraint at the pressing operation?

1. *Eliminate idle time (time when the machine is neither being set up to process a part nor actually processing a part) on the bottleneck operation.* CI is considering permanent-

---

[14]See E. Goldratt and J. Cox, *The Goal* (New York: North River Press, 1986); and E. Goldratt, *The Theory of Constraints* (New York : North River Press, 1990).

Concepts in Action

THROUGHPUT ACCOUNTING AT ALLIED-SIGNAL SKELMERSDALE, U.K.[a]

Allied-Signal Skelmersdale, U.K., manufactures turbochargers for the automotive industry. In the late 1980s and early 1990s, the Skelmersdale plant was forced to change from producing few products in large quantities to producing many products in small quantities in a very competitive market. The plant also had to cope with frequent changes in its sales mix. The plant often missed delivery dates and incurred high transportation costs to ship via air those parts urgently needed by its automotive customers. John Darlington, the controller of the Skelmersdale plant, recognized the important role finance and accounting could play in this environment, but "we were just not supporting, communicating with, and complementing shop-floor management—not until we began emphasizing throughput contributions."

The format designed by the Allied-Signal accountants for the throughput contribution–based operating income statement follows:

Throughput Operating Income Statement for 19_4
(in thousands)

Sales revenues		£50,000
Direct materials costs		28,500
Throughput contribution		21,500
Other operating costs		
Direct manufacturing labor	£ 4,275	
Engineering costs	1,767	
Other manufacturing costs	11,585	
Marketing costs	1,873	
Total other operating costs		19,500
Operating income		£ 2,000

The Skelmersdale management viewed operating costs, other than direct materials costs, as fixed in the short run. The key to improving profitability was maximizing throughput contribution by identifying and optimizing the use of bottleneck resources. Management reduced the load on the bottleneck machines by shifting operations performed there onto other machines. New investments to improve efficiency at nonbottleneck machines were turned down because greater efficiency at nonbottleneck machines did nothing to improve throughput contribution. Instead Allied-Signal made additional investments to increase bottleneck capacity.

To motivate workers to improve throughput, Allied-Signal managers designed new performance measures. Instead of measuring localized efficiency such as direct labor efficiency at various operations, management introduced "adherence to schedule" as the key performance measure. Workers at nonbottleneck operations were asked not to produce more than what was required according to the bottleneck schedule. In the surplus time available to these workers, they received training in total quality management practices and in improving operator skills. The Skelmersdale plant also introduced four other performance measures—costs of quality, customer due-date delivery, days inventory on hand, and manufacturing lead time—all with the objective of satisfying customers and maximizing throughput contribution. Over a four-year period, the Skelmersdale plant showed dramatic increases in each of these measures and in profitability, cash flow, and return on investment.

[a] Adapted from J. Darlington, J. Innes, F. Mitchell, and J. Woodward, "Throughput Accounting: The Garrett Automotive Experience," *Management Accounting* (April 1992); P. Coughlan and J. Darlington, "As Fast as the Slowest Operation: The Theory of Constraints," *Management Accounting* (June 1993); and discussions with Allied-Signal, Skelmersdale management.

ly positioning two workers at the pressing operation. Their sole responsibility would be to unload finished parts as soon as one batch of parts is processed and to set up the machine to process the next batch. Suppose the annual cost of this action is $48,000 and the effect of this action is to increase bottleneck output by 1,000 units per year. Should CI incur the additional costs? Yes, because CI's relevant throughput contribution increases by $60,000 (1,000 units × [selling price, $100 – direct materials costs, $40]), which exceeds the additional cost of $48,000. All other costs are irrelevant.

2. *Process only those parts or products that increase sales and throughput contribution, not parts or products that remain in finished goods or spare parts inventory.* Building products that sit in inventory does not increase throughput contribution.

3. *Shift parts that do not have to be made on the bottleneck machine to nonbottleneck machines or to outside facilities.* Suppose Spartan Corporation, an outside contractor, offers to press 1,500 doors from direct materials that CI supplies at $15 per door. Spartan's quoted price is greater than CI's own operating cost of $12 per door. Should CI accept the offer? Yes, because pressing is a bottleneck operation. Getting additional doors pressed from outside increases throughput contribution by $90,000 ([$100 – $40] × 1,500 doors), while relevant costs increase by $22,500 ($15 × 1,500). The fact that CI's unit cost is less than Spartan's quoted price is irrelevant.

Suppose Gemini Industries, another outside contractor, offers to stamp 2,000 doors from direct materials that CI supplies at $6 per door. Gemini's price is lower than CI's cost of $8 per door. Should CI accept the offer? Other operating costs are fixed costs, so CI will not save any costs by subcontracting the stamping operations. Total costs will be greater by $12,000 ($6 × 2,000) under the subcontracting alternative. Stamping more doors will not increase throughput contribution, which is constrained by pressing capacity. CI should not accept Gemini's offer.

4. *Reduce setup time and processing time at bottleneck operations (for example, by simplifying the design or reducing the number of parts in the product).* Suppose CI can reduce setup time at the pressing operation by incurring additional costs of $55,000 a year. Suppose further that reducing setup time enables CI to press 2,500 more doors a year. Should CI incur the costs to reduce setup time? Yes, because throughput contribution increases by $150,000 ([$100 – $40] × 2,500), which exceeds the additional costs incurred of $55,000. Will CI find it worthwhile to incur costs to reduce machining time at the stamping operation? No. Other operating costs will increase, but throughput contribution will remain unaffected. Throughput contribution increases only by increasing bottleneck output; increasing nonbottleneck output has no effect.

5. *Improve the quality of parts manufactured at the bottleneck operation.* Poor quality is often more costly at a bottleneck operation than it is at a nonbottleneck operation. The cost of poor quality at a nonbottleneck operation is the cost of materials wasted. If CI produces 1,000 defective doors at the stamping operation, the cost of poor quality is $40,000 (direct materials cost per unit, $40 × 1,000 doors). No throughput contribution is forgone because stamping has surplus capacity. Despite the defective production, stamping can produce and transfer 90,000 doors to the pressing operation. At a bottleneck operation, the cost of poor quality is the cost of materials wasted *plus* the opportunity cost of lost throughput contribution. Bottleneck capacity not wasted in producing defective parts could be used to generate additional sales and throughput contribution. If CI produces 1,000 defective units at the pressing operation, the cost of poor quality is $100,000: direct materials cost of $40,000 (direct materials cost per unit, $40 × 1,000 units) plus forgone throughput contribution of $60,000 ([$100 – $40] × 1,000 doors).

The high costs of poor quality at the bottleneck operation means that bottleneck time should not be wasted processing parts that are defective. Also, quality improvement programs should focus on ensuring that bottlenecks produce minimal defects.

PROBLEM FOR SELF STUDY

PROBLEM

Revisit the Falcon Works (FW) example. FW has convinced all its customers to place orders for A22 in order sizes of 500 units. FW expects to receive and manufacture 60 orders of A22. Each order will take 54 hours of manufacturing time (8 hours of setup time plus 46 hours of processing time). Assume the following with respect to A22: Average selling price per order, if average manufacturing lead time is less than 300 hours, is $11,000; average selling price per order, if average manufacturing lead time is greater than 300 hours, is $10,750; variable costs per order are $8,000; and inventory carrying costs per order are $0.50 per order per hour. Assume the same data for C33 as in the chapter illustration. Given this new information, FW is reconsidering whether it should introduce C33. What should FW do?

SOLUTION

Average waiting time for A22, if C33 is not introduced = 115 hours (see p. 816).
Average waiting time for A22 if C33 is introduced using the formula from p. 813 is

$$\frac{[60 \times (54)^2] + [10 \times (50)^2]}{2 \times [4,000 - (60 \times 54) - (10 \times 50)]} = \frac{(60 \times 2,916) + (10 \times 2,500)}{2 \times (4,000 - 3,240 - 500)}$$

$$= \frac{174,960 + 25,000}{2 \times 260} = \frac{199,960}{520} = 385 \text{ hours}$$

Average manufacturing lead time for A22 = 385 + 54 = 439 hours
Average manufacturing lead time for C33 = 385 + 50 = 435 hours

The following table describes the expected loss in revenues and expected increase in costs from introducing C33.

Expected Loss in Revenues and Expected Increase in Costs from Introducing C33

Product	Expected Loss in Revenues for A22 from Increasing Average Manufacturing Lead Times	Expected Increase in Carrying Costs for All Products from Increasing Average Manufacturing Lead Times	Expected Loss in Revenues Plus Expected Increase in Costs of Introducing C33
A22	$15,000*	$ 9,720†	$24,720
C33	—	2,175‡	2,175
Total	$15,000	$11,895	$26,895

* ($11,000 − $10,750) × 60 orders.
† (439 hours − 115 hours) × $0.50 × 60 orders.
‡ (435 hours − 0) × $0.50 × 10 orders.

FW is better off not introducing C33. The additional costs of $26,895 exceed C33's expected contribution of $16,000 ($1,600 per order × 10 orders).

SUMMARY

The following points are linked to the chapter's learning objectives.

1. Four cost categories in a costs of quality program are *prevention costs* (costs incurred in preventing the manufacture of products that do not conform to specifications), *appraisal costs* (costs incurred in detecting which of individual products produced do not conform to specifications), *internal failure costs* (costs incurred when a nonconforming product is detected before its shipment to customers), and *external failure costs* (costs incurred when a nonconforming product is detected after its shipment to customers).

2. Three methods that companies use to improve quality are *control charts*, to distinguish random variations from other sources of variation in an operating process, *Pareto diagrams*, which indicate how frequently each type of failure occurs, and *cause-and-effect diagrams*, which identify potential factors or causes of failure.

3. The relevant costs of quality improvement are costs incurred to implement the quality program. The relevant benefits are the savings in total costs and the estimated increase in contribution margin from the higher sales that result from the quality improvements.

4. Nonfinancial measures of customer satisfaction include the number of customer complaints, the on-time delivery rate, and the customer-response time. Nonfinancial measures of internal performance include product defect levels, process yields, and manufacturing lead times.

5. Financial measures are helpful to evaluate trade-offs among prevention and failure costs. They focus attention on how costly poor quality can be. Nonfinancial measures help focus attention on the precise problem areas that need attention.

6. *Breakeven time (BET)* is the time from when the initial concept for a new product is approved by management until the time when the cumulative present value of net cash inflows from the project equals the cumulative present value of net investment outflows. BET measures the amount of time it takes to recover the investment in the new product.

7. *Customer-response time* is the amount of time from when a customer places an order for a product or requests service to when the product or service is delivered to the customer. Lines and delays occur because of (a) uncertainty about when customers will order products or services and (b) limited capacity and bottlenecks. Bottlenecks are operations at which the work to be performed approaches or exceeds the available capacity. The costs of lines and delays include lower revenues and increased inventory carrying costs.

8. The three main measurements in the theory of constraints are throughput contribution (equal to sales dollars minus direct materials costs), investments or inventory (equal to sum of materials costs of direct materials inventory, work-in-process inventory and finished goods inventory; research and development costs; and costs of equipment and buildings), and other operating costs (equal to all operating costs other than direct materials costs incurred to earn throughput contribution).

9. The four steps in managing bottlenecks are (a) recognize that the bottleneck operation determines throughput, (b) search for and find the bottleneck, (c) subordinate all nonbottleneck operations to the bottleneck operation, and (d) increase bottleneck efficiency and capacity.

TERMS TO LEARN

This chapter and the Glossary at the end of the book contain definitions of the following important terms:

appraisal costs *(p. 795)*   average waiting time *(812)*   bottleneck *(811)*
breakeven time (BET) *(808)*   cause-and-effect diagram *(799)*   control chart *(798)*
costs of quality *(795)*   customer-response time *(811)*   external failure costs *(795)*
internal failure costs *(795)*   new product development time *(807)*
on-time performance *(814)*   Pareto diagram *(799)*   prevention costs *(795)*
throughput contribution *(817)*   time driver *(811)*

ASSIGNMENT MATERIAL

QUESTIONS

23-1   How does conformance quality differ from quality of design? Explain.

23-2   Name two items classified as prevention costs.

23-3   Distinguish between internal failure costs and external failure costs.

23-4   Describe three methods that companies use to identify quality problems.

23-5   "Companies should focus on financial measures of quality because these are the only measures of quality that can be linked to bottom-line performance." Do you agree? Explain.

23-6   Give two examples of nonfinancial measures of customer satisfaction.

23-7   Give two examples of nonfinancial measures of internal performance.

23-8   Why is there increased interest in reducing new product development time?

23-9   What is the advantage of breakeven time over new product development time?

23-10   "When selecting new product development projects, companies should always choose the projects that have the shortest breakeven time." Do you agree? Explain.

23-11   Distinguish between customer-response time and manufacturing lead time.

23-12   Give two reasons why waiting lines and delays occur.

23-13   "Companies should always make and sell all products whose selling prices exceed variable costs." Do you agree? Explain.

23-14   Describe the three main measures used in the theory of constraints.

23-15   Describe three ways to improve the performance of a bottleneck operation.

PROBLEMS AND EXERCISES

23-16   Costs of quality analysis, nonfinancial quality measures.   Hartono Corporation manufactures and sells industrial grinders. The following table presents financial information pertaining to quality in 19_6 and 19_7.

	19_7 (in thousands)	19_6 (in thousands)
Sales	$12,500	$10,000
Costs		
Line inspection	85	110
Scrap	175	250
Design engineering	240	100
Returned goods	45	60
Product testing equipment	50	50
Customer problems and complaints	30	40
Rework	135	160
Preventive maintenance	90	35
Product liability claims	125	200
Incoming materials inspection	40	20
Breakdown maintenance	40	90
Product testing labor	75	220
Training	120	45
Warranty repair	200	300
Supplier evaluations	50	20

Required

1. Classify the cost items in the table into prevention, appraisal, internal failure, or external failure categories.

2. Calculate the ratio of each costs of quality category to sales in 19_6 and 19_7.

3. Comment on the trends in costs of quality between 19_6 and 19_7.

4. Give two examples of nonfinancial quality measures that Hartono Corporation could monitor as part of a total quality control effort.

23-17 Costs of quality analysis, nonfinancial quality measures. Ontario Industries manufactures two types of refrigerators, Olivia and Solta. Information on each refrigerator follows:

	Olivia	Solta
Units manufactured and sold	10,000 units	5,000 units
Selling price	$2,000	$1,500
Variable costs per unit	$1,200	$800
Hours spent on design engineering	6,000	1,000
Testing and inspection hours per unit	1	0.5
Percentage of units reworked in plant	5%	10%
Rework costs per refrigerator	$500	$400
Percentage of units repaired at customer site	4%	8%
Repair costs per refrigerator	$600	$450
Estimated lost sales from poor quality	—	300 units

The labor rates per hour for various activities follow:

Design	$75 per hour
Testing and inspection	$40 per hour

Required

1. Calculate the costs of quality for Olivia and Solta classified into prevention, appraisal, internal failure, and external failure categories.

2. For each type of refrigerator, calculate the ratio of each cost of quality item as a percentage of sales.

3. Compare and comment on the costs of quality for Olivia and Solta.

4. Give two examples of nonfinancial quality measures that Ontario Industries could monitor as part of a total quality control effort.

23-18 Nonfinancial quality measures, half-life method. Transco Retailers operates a general merchandise retail store. The average wait for a customer at the check-out counter is 20 minutes. Transco wants to reduce the waiting time to 2½ minutes. It estimates the half-life of reducing the waiting time at nine months.

Required
Estimate how long it will take Transco to reduce time in the line to 2½ minutes.

23-19 Quality improvement, relevant costs, and relevant revenues. Thomas Corporation sells 300,000 V262 valves to the automobile and truck industry. Thomas has a capacity of 110,000 machine-hours and can produce 3 valves per hour. V262's contribution margin per unit is $8. Thomas sells only 300,000 valves because 30,000 valves (10% of the good valves) need to be reworked. It takes 1 hour to rework 3 valves so that 10,000 hours of capacity are lost in the rework process. Thomas's rework costs are $210,000.
　　　Rework costs consist of:

Direct materials and direct rework labor (variable costs)	$3 per unit
Fixed costs of equipment, rent, and overhead allocation	$4 per unit

　　　Thomas's process designers have come up with a modification that would maintain the speed of the process and would ensure 100% quality and no rework. The new process would cost $315,000 per year. The following additional information is available.

◆ The demand for Thomas's V262 valves is 370,000 per year.
◆ Jackson Corporation has asked Thomas to supply 22,000 T971 valves if Thomas implements the new design. The contribution margin per T971 valve is $10. Thomas can make two T971 valves per hour on the existing machine with 100% quality and no rework.

Required
1. Suppose Thomas's designers implemented the new design. Should Thomas accept Jackson's order for 22,000 T971 valves? Explain.
2. Should Thomas Corporation implement the new design?
3. What nonfinancial and qualitative factors should Thomas Corporation consider in deciding whether to implement the new design?

23-20 Quality improvement, relevant costs, and relevant revenues. Tan Corporation makes multicolor plastic lamps in two operations, molding and welding. The molding operation has a capacity of 200,000 units per year; welding has a capacity of 300,000 units per year. Costs of quality information recorded by Tan follows:

Design of product and process costs	$240,000
Inspection and testing costs	170,000
Scrap costs (all in the molding department)	750,000

The demand for lamps is very strong. Tan will be able to sell whatever output quantities it can produce at $40 per lamp.
　　　Tan can start only 200,000 units into production in the molding department because of capacity constraints on the molding machines. If a defective unit is produced at the molding operation, it must be scrapped, and the scrap yields no revenue. Of the 200,000 units started at the molding operation, 30,000 units (15%) are scrapped. Scrap costs, based on total (fixed and variable) manufacturing costs incurred up to the molding operation, equal $25 per unit as follows:

Direct materials (variable)	$16 per unit
Direct manufacturing labor, setup labor, and materials-handling labor (variable)	3 per unit
Equipment, rent, and other allocated overhead including inspection and testing costs on scrapped parts (fixed)	6 per unit
	$25 per unit

The good units from the molding department are sent to the welding department. Variable manufacturing costs at the welding department are $2.50 per unit. There is no scrap in the welding department. Therefore, Tan's total sales quantity equals the molding department's output. Tan incurs no other variable costs.

Tan's designers have determined that adding a different type of material to the existing direct materials would reduce scrap to zero, but it would increase the variable costs per unit in the molding department by $3. Recall that only 200,000 units can be started each year.

Required

1. What is the additional direct materials cost of implementing the new method?
2. What is the additional benefit to Tan from using the new material and improving quality?
3. Should Tan use the new material?
4. What other nonfinancial and qualitative factors should Tan consider in making a decision?

23-21 Statistical quality control, airline operations. People's Skyway operates daily round-trip flights on the London–New York route using a fleet of three 747s, the *Spirit of Birmingham*, the *Spirit of Glasgow*, and the *Spirit of Manchester*. The budgeted quantity of fuel for each round-trip flight is the mean (average) fuel usage. Over the last 12 months, the average fuel usage per round-trip is 100 gallon-units with a standard deviation of 10 gallon-units. A gallon-unit is 1,000 gallons.

Cilla Black, the operations manager of People's Skyway, uses a statistical quality control (SQC) approach in deciding whether to investigate fuel usage per round-trip flight. She investigates those flights with fuel usage greater than two standard deviations from the mean.

Black receives the following report for round-trip fuel usage in October by the three planes operating on the London–New York route:

Flight	Spirit of Birmingham (gallon-units)	Spirit of Glasgow (gallon-units)	Spirit of Manchester (gallon-units)
1	104	103	97
2	94	94	104
3	97	96	111
4	101	107	104
5	105	92	122
6	107	113	118
7	111	99	126
8	112	106	114
9	115	101	117
10	119	93	123

Required

1. Using the ±2σ rule, what variance investigation decisions would be made?
2. Present SQC charts for round-trip fuel usage for each of the three 747s in October. What inferences can you draw from them?
3. Some managers propose that People's Skyway present its SQC charts in monetary terms rather than in physical quantity terms (gallon-units). What are the advantages and disadvantages of using monetary fuel costs rather than gallon-units in the SQC charts?

23-22 Compensation linked with profitability, on-time delivery, and external quality performance measures; balanced scorecard. Pacific-Dunlop supplies tires to major automotive companies. It has two tire plants in North America, in Detroit and Los Angeles. The quarterly bonus plan for each plant manager has three components:

1. Profitability performance: Add 2% of operating income.
2. On-time delivery performance: Add $10,000 if on-time delivery performance to the 10 most important customers is 98% or better. If on-time performance is below 98%, add nothing.
3. Product quality performance: Deduct 50% of cost of sales returns from the 10 most important customers.

Quarterly data for 19_7 on the Detroit and Los Angeles plants follow:

	Jan.–Mar.	Apr.–June	July–Sept.	Oct.–Dec.
DETROIT				
Operating income	$ 800,000	$ 850,000	$ 700,000	$ 900,000
On-time delivery	98.4%	98.6%	97.1%	97.9%
Cost of sales returns	$ 18,000	$ 26,000	$ 10,000	$ 25,000
LOS ANGELES				
Operating income	$1,600,000	$1,500,000	$1,800,000	$1,900,000
On-time delivery	95.6%	97.1%	97.9%	98.4%
Cost of sales returns	$ 35,000	$ 34,000	$ 28,000	$ 22,000

Required
1. Compute the bonus paid each quarter of 19_7 to the plant manager of the Detroit plant and to the plant manager of the Los Angeles plant.
2. Discuss the three components of the bonus plan as measures of profitability, on-time delivery, and product quality.
3. Why would you want to evaluate plant managers on the basis of both operating income and on-time delivery?
4. Give one example of what might happen if on-time delivery were dropped as a performance evaluation measure.

23-23 Quality improvement, Pareto charts, fishbone diagrams. Murray Corporation manufactures, sells, and installs photocopying machines. Murray has placed heavy emphasis on reducing defects and failures in its production operations. Murray wants to apply the same total quality management (TQM) principles to managing its accounts receivables.

Required
1. On the basis of your knowledge and experience, what would you classify as failures in accounts receivables?
2. Give examples of prevention activities that could reduce failures in accounts receivables.
3. Draw a Pareto diagram of the types of failures in accounts receivables and a fishbone diagram of possible causes of one type of failure in accounts receivables.

23-24 New product proposal, breakeven time. Brooks Corporation is considering developing a new scientific instrument, MX-505. Management is expected to approve the project and form the project development team on December 31, 19_3, but work on the new product is not expected to start until very late in 19_4. The following cash flows are projected (in millions).

Year	Investment Cash Outflows	Product Cash Inflows[*]
19_4	$165	
19_5		$30
19_6		66
19_7		73
19_8		80
19_9		88

[*] Product cash inflow includes cash inflows of $20 million from disposal of investment on December 31, 19_9.

Brooks uses a 10% required rate of return for this investment. Ignore income taxes.

Required
1. Calculate the breakeven time for MX-505.
2. How could Brooks reduce BET?
3. Calculate the payback period for MX-505.

23-25 New product proposal, breakeven time, relevant costs. Harbor Marine is going to develop a new navigational device, N701. Management is expected to approve the project and form the project development team on December 31, 19_3. Owing to lack of funding, the project team is not expected to start working on the project until the end of 19_4. The following cash flows are projected (in millions).

Year Ended	Initial Investment Outflow	Cash Revenues	Cash Outflows for Manufacturing, Marketing Distribution and Customer Service
19_4	$2.50		
19_5		$1.50	$1.00
19_6		2.00	1.20
19_7		4.00	2.20
19_8		3.50	2.00
19_9		3.00	1.80

Additional Information

a. All of the initial investment will be capitalized and depreciated on a straight-line basis over five years at the rate of $0.5 million a year, assuming a zero terminal disposal price for the investment.

b. Each year the chief executive officer's salary is allocated to products at the rate of 1% of product revenues. For example, in 19_5, 1% of $1.50 million ($0.015 million) is expected to be allocated to N701.

c. As a result of introducing the new navigational device, Harbor Marine estimates it will lose some sales of its existing navigational device, R265, as follows: 19_5, $0.40 million; 19_6, $0.60 million; 19_7, $0.70 million; 19_8, $0.70 million; 19_9, $0.70 million.

d. The cash contribution margin percentage on sales of R265 is 30%.

e. Harbor Marine uses a 12% required rate of return on this project. Ignore income taxes.

Required
1. Calculate the breakeven time for N701.
2. Why might Harbor Marine be interested in reducing the breakeven time on N701? What steps can Harbor Marine take to reduce breakeven time on the N701 project?

23-26 New product proposal, breakeven time, working backward. Dallas Instruments is considering manufacturing and selling a new memory chip, the P-Chip. The new product development committee will not fund a new product proposal if it has a breakeven time of greater than four years. The cash investments to make the P-Chip will be made on January 1, 19_5 soon after the committee approves the project. The projected cash sales of the P-Chip are expected to be $10 million each year in 19_5, 19_6, 19_7, 19_8, and 19_9. The cash costs of manufacturing, marketing, distribution, and customer service to support P-Chip sales are expected to be $6 million each year. The required rate of return on the investment is 14%.

Required
1. What is the maximum cash investment that the new product development committee will agree to fund for the P-Chip project?
2. Why might Dallas Instruments specify a policy that it will not fund new product proposals with an estimated BET greater than four years?

23-27 New-product proposal, breakeven time. The product development committee of Detroit Motors is examining a capital-budgeting proposal from its new product development group. The new car, code named Project Nirvana, requires research and development and product design investments in the years 1994, 1995, and 1996. Management approval of the project is expected at the end of 1993. Sales revenues will begin in 1997. The following cash outflows and cash inflows (in millions) are projected. Ignore income taxes.

	1994	1995	1996	1997	1998	1999	2000
Cash Outflows							
Research and development	$20	$40	$ 5	$ 0	$ 0	$ 0	$ 0
Product design	10	25	50	0	0	0	0
Manufacturing	0	0	0	25	300	200	45
Marketing	0	0	0	60	140	80	30
Distribution	0	0	0	2	60	40	10
Customer service	0	0	0	0	15	60	40
Cash Inflows							
Revenues	0	0	0	80	760	540	160

Detroit Motors uses a 12% required rate of return.

Required
1. Compute the breakeven time of Project Nirvana.
2. What steps can Detroit Motors take to reduce breakeven time on Project Nirvana?

23-28   Waiting time, banks.   Regal Bank has a small branch in Orillia, Canada. The counter is staffed by one teller. The counter is open for 5 hours (300 minutes) each day. It takes 5 minutes to serve a customer. The Orillia branch expects to receive 40 customers each day.

Required
1. How long, on average, will a customer wait in line before being served?
2. How long, on average, will a customer wait in line if the branch expects 50 customers each day?
3. The bank is considering ways to reduce waiting time. How long will customers have to wait on average, if the time to serve a customer is reduced to 4 minutes and the bank expects to serve 50 customers each day?

23-29   Waiting time, relevant costs, and relevant revenues; working backward. (Continuation of 23-28)   The Orillia branch is thinking of offering additional services to its customers. With new services, the bank expects to serve an average of 60 customers each day instead of the 40 customers it currently averages. It will take four minutes to serve each customer regardless of whether or not the new services are offered.

Required
1. How long, on average, will a customer wait in line before being served?
2. Regal Bank's policy is that the average waiting time in the line should not exceed five minutes. The bank cannot reduce the time to serve a customer below four minutes without significantly affecting quality. To reduce waiting time, the bank would have to keep the counter open for more hours each day. How many minutes must the counter be kept open so that the average waiting time that a customer spends in line is five minutes?
3. The bank expects to generate, on average, $30 in additional operating income each day as a result of offering the new services. The teller is paid $10 per hour and is employed in increments of an hour (that is, the teller can be employed for 5 hours, 6 hours, 7 hours, and so on, but not for a fraction of an hour). If the bank wants average waiting time to be no more than five minutes, should the bank offer the new services?

23-30   Waiting times, manufacturing lead times.   SRG Corporation uses an injection molding machine to make a plastic product, Z39. SRG makes products only after receiving firm orders from its customers. SRG estimates that it will receive 50 orders for Z39 (each order is for 1,000 units) during the coming year. It takes 80 hours to manufacture each order of Z39 (4 hours to clean and prepare the machine, called setup, and 76 hours to process the order). The annual capacity of the machine is 5,000 hours.

Required
1. What percentage of the total available machine capacity does SRG expect to use during the coming year?
2. Calculate the average amount of time that an order for Z39 will wait in line before it is processed.
3. Calculate the average manufacturing lead time for Z39.
4. SRG is considering introducing a new product, Y28. SRG estimates that, on average, it will receive 25 orders of Y28 (each order for 200 units) in the coming year. It takes 20 hours to manufacture each order of Y28 (2 hours to clean and prepare the machine, and 18 hours to process the order). The average demand for Z39 will be unaffected by the introduction of Y28. Calculate the average waiting time for an order received and the average manufacturing lead time for each product, if SRG introduces Y28.
5. If SRG introduces Y28, on average what fraction of the total manufacturing lead time will Y28 spend just waiting to be processed?
6. Briefly describe why delays occur in the processing of Z39 and Y28.

23-31 Waiting times, relevant revenues and relevant costs. (Continuation of 23-30) SRG is still deciding whether or not it should introduce and sell Y28. The table below provides information on selling prices, variable costs, and inventory carrying costs for Z39 and Y28. SRG will incur additional variable costs and inventory carrying costs for Y28 only if it introduces Y28. Fixed costs equal to 40% of variable costs are allocated to all products produced and sold during the year.

Product	Average Number of Orders	Average Selling Price per Order If Average Manufacturing Lead Time Less Than 320 Hours	Average Selling Price per Order If Average Manufacturing Lead Time More Than 320 Hours	Variable Costs per Order	Inventory Carrying Costs per Order per Hour
Z39	50	$27,000	$26,500	$15,000	$0.75
Y28	25	8,400	8,000	5,000	0.25

Required

1. Should SRG manufacture and sell Y28? Show all your computations.
2. What is the cutoff price above which SRG should manufacture and sell Y28 and below which SRG should choose not to manufacture and sell Y28?

23-32 Theory of constraints, throughput contribution, relevant costs. Colorado Industries manufactures electronic testing equipment. Colorado also installs the equipment at the customer's site and ensures that it functions smoothly. Additional information on the manufacturing and installation departments follow. Capacities are expressed in terms of the number of units of equipment.

	Equipment Manufactured	Equipment Installed
Annual capacity	400 units per year	300 units per year
Equipment manufactured and installed	300 units per year	300 units per year

Colorado manufactures only 300 units per year because the installation department has only enough capacity to install 300 units. The equipment sells for $40,000 per unit and has direct materials costs of $15,000. All costs other than direct materials costs are fixed. The following requirements refer only to the data given above; there is *no connection* between the situations.

Required

1. Colorado's engineers have found a way to reduce equipment manufacturing time. The new method would cost an additional $50 per unit and allow Colorado to manufacture 20 additional units a year. Should Colorado implement the new method?
2. Colorado's designers have proposed a change in the direct materials that would increase direct materials costs by $2,000 per unit. This change would enable Colorado to install 320 units of equipment each year. If Colorado makes the change, it will implement the new design on all equipment sold. Should Colorado use the new design?
3. A new installation technique has been developed that will enable Colorado's engineers to install 10 additional units of equipment a year. The new method will increase installation costs by $50,000 each year. Should Colorado implement the new technique?
4. Colorado is considering how to motivate workers to improve their productivity (output per hour). One proposal is to evaluate and compensate workers in manufacturing and installation departments on the basis of their productivities. Do you think the new proposal is a good idea? Explain briefly.

23-33 Theory of constraints, throughput contribution, quality, relevant costs. Aardee Industries manufactures pharmaceutical products in two departments—mixing and tablet making. Additional information on the two departments follows. Each tablet contains 0.5 grams of direct materials.

	Mixing	Tablet Making
Capacity per hour	150 grams	200 tablets
Monthly capacity (2,000 hours of capacity available in mixing and tablet making)	300,000 grams	400,000 tablets
Monthly production	200,000 grams	390,000 tablets
Other operating costs (fixed)	$16,000	$39,000
Other operating costs per unit ($16,000 ÷ 200,000; $39,000 ÷ 390,000)	$0.08 per gram	$0.10 per tablet

The mixing department makes 200,000 grams of direct materials mixture (enough to make 400,000 tablets) because the tablet-making department has only enough capacity to process 400,000 tablets. All direct materials costs are incurred in the mixing department. Aardee incurs $156,000 in direct materials costs. The tablet-making department manufactures only 390,000 tablets from the 200,000 grams of mixture processed; 2.5% of the direct materials mixture is lost in the tablet-making process. Each tablet sells for $1. All costs other than direct materials costs are fixed costs. The following requirements refer only to the data given above; there is *no connection* between the situations.

Required
1. An outside contractor makes the following offer: If Aardee will supply the contractor with 10,000 grams of mixture, the contractor will manufacture 19,500 tablets for Aardee (allowing for the normal 2.5% loss during the tablet-making process) at $0.12 per tablet. Should Aardee accept the contractor's offer?

2. Another firm offers to prepare 20,000 grams of mixture a month from direct materials Aardee supplies. The company will charge $0.07 per gram of mixture. Should Aardee accept the company's offer?

3. Aardee's engineers have devised a method that would improve quality in the tablet-making operation. They estimate that the 10,000 tablets currently being lost would be saved. The modification would cost $7,000 a month. Should Aardee implement the new method?

4. Suppose that Aardee also loses 10,000 grams of mixture in its mixing operation. These losses can be reduced to zero if the company is willing to spend $9,000 per month in quality improvement methods. Should Aardee adopt the quality improvement method?

5. What are the benefits of improving quality at the mixing operation compared with the benefits of improving quality at the tablet-making operation?

23-34 Ethics and quality. John Emerson, the controller at Grant Semiconductors, called the assistant controller, Mary Hughes, into his office. "Our plant manager, Harry Davis, is quite upset with the recent costs of quality and nonfinancial measures of quality reports that you prepared. He feels his workers have made progress in improving quality at the plant but that our reports are just not picking up this fact. He wants to apply for various quality awards that would bring a lot of prestige to Grant, but he obviously cannot do so on the basis of the numbers we are reporting. Can you look over these quality numbers, and see what you can do? I think Harry has a point. Nobody wants Grant to miss out on all the wonderful press we'd get if we won one of these quality awards." Hughes is quite certain that her numbers are correct. Yet she would very much like Grant to win these prestigious quality awards. She is confused about how to handle Emerson's request.

Required
Refer to the "Standards of Ethical Conduct for Management Accountants" described in Chapter 1 (p. 17). Is John Emerson's suggestion to Hughes to recalculate her quality numbers unethical? Would it be unethical for Hughes to modify her analysis? What steps should Hughes take to resolve this situation?

24

Inventory Management and Just-in-Time

- Managing goods for sale in retail organizations
- Difficulties with accounting data for managing goods for sale
- Just-in-time purchasing
- Managing inventories in manufacturing organizations

Merchandisers are demanding more-frequent deliveries with shorter purchase-order lead times from their suppliers. American Greetings is investing in new information systems and technology to provide better service to retailers who resell their greeting card products. These new approaches are helping both merchandisers and their suppliers to keep inventory levels low.

Objective 1

Explain why cost management of materials is pivotal in many organizations

The costs of materials account for more than 50% of total costs in manufacturing companies and over 70% of total costs in retail companies. Not surprisingly, managers are paying increasing attention to inventory management. The 1990s have seen General Motors (GM) demand double-digit price cuts from its suppliers. General Electric (GE) announced "Target 10," an initiative that aims to reduce suppliers' costs by 10%. To achieve these goals, GM and GE would assist suppliers in reducing costs, and those suppliers that did not cooperate would probably lose business.

Inventory management is the planning, organizing, and control activities focused on the flow of inventory into, through, and from the organization. Many decisions fall under the inventory management umbrella. When is the best time to purchase materials or merchandise? How should purchasing arrangements be structured? Which is the best way to handle materials or merchandise inventories, once they are received? These questions are among the many that inventory management seeks to answer.

Accounting information can play a key role in inventory management. This chapter illustrates the importance of accounting information in two areas:

1. The management of goods for sale in retail organizations
2. The management of materials, work in process, and finished goods in organizations with manufacturing operations

MANAGING GOODS FOR SALE IN RETAIL ORGANIZATIONS

Cost of goods sold constitutes the largest single cost item for most retailers. For example, a survey of retail food chains reports the following breakdown of operations.[1]

Sales		100.0%
Deduct costs:		
Cost of goods sold	75.4%	
Payroll (including fringe benefits)	12.6	
Other operating costs and taxes	11.0	
Total costs		99.0
Net income		1.0%

Net income for the surveyed firms averaged only 1% of sales, but cost of goods sold averaged 75.4%. This paper-thin net income percentage means that better decisions

---

[1]*Marketing Costs* (Washington, D.C.: Food Marketing Institute, 1992), p. 1.

regarding the purchasing and managing of goods for sale can cause dramatic percentage increases in net income.

Two decisions are central to managing goods for sale in a retail organization: How much to order (the economic-order-quantity, EOQ, decision), and when to order (the reorder decision). This chapter discusses EOQ models that can help managers make these decisions.

Costs Associated with Goods for Sale

Objective 2

Identify five categories of costs associated with goods for sale

In managing goods for sale in a retail organization, the following cost categories are important.

1. *Purchase costs*: **Purchase costs** consist of the costs of goods acquired from suppliers including freight and transportation costs. These costs usually make up the largest single cost category. Discounts for different purchase-order sizes and supplier credit terms affect purchasing costs.

2. *Ordering costs*: **Ordering costs** consist of the costs of preparing and issuing a purchase order. Related to the number of orders processed are special processing, receiving, and inspection costs.

3. *Carrying costs*: **Carrying costs** arise when a business holds inventories of goods for sale. These costs include the opportunity cost of the investment tied up in inventory (see Chapter 11, pp. 399–400) and the costs associated with storage, such as space rental and insurance.

4. *Stockout costs*: A **stockout** arises when a customer demands a unit of product and that unit is not readily available. A company may respond to the shortfall by expediting an order from a supplier. Expediting costs include the additional ordering costs plus any associated transportation costs. Alternatively, the company may lose a sale due to the stockout. In this case, stockout costs include the lost contribution margin on the sale plus any customer ill-will generated by the stockout that hurts future sales.

5. *Quality costs*: The *quality* of a product or service is its conformance with a preannounced or prespecified standard. As described in Chapter 23, four categories of costs of quality are often distinguished. (a) *Prevention costs* are costs incurred to avoid the sale of products that do not conform to specification—for example, the costs of using high-quality vendors to deliver reliable component parts. (b) *Appraisal costs* are costs incurred to detect which individual product units do not conform to specification—for example, inspection costs. (c) *Internal failure costs* are costs that arise when nonconforming products are detected *before* they are shipped to customers—for example, costs of defective parts. (d) *External failure costs* are costs that occur when nonconforming products are detected *after* they have been shipped to customers—for example, costs of replacing products.

Advances in information-gathering technology are increasing the reliability and timeliness of data in these five cost categories. Do not assume, however, that all the relevant data are available in existing accounting systems. Opportunity costs, which are not typically recorded in accounting systems, are an important component in several of these cost categories.

Economic-Order-Quantity Decision Model

Objective 3

Explain the economic-order-quantity (EOQ) decision model and how it balances ordering costs and carrying costs

The **economic-order-quantity (EOQ)** decision model calculates the optimal quantity of inventory to order. The simplest version of this model incorporates only ordering costs and carrying costs into the calculation. It assumes:

1. The same fixed quantity is ordered at each reorder point.

2. Demand, ordering costs, and carrying costs are certain. The **purchase-order lead time**—the time between the placement of an order and its delivery—is also certain.

3. Purchasing costs per unit are unaffected by the quantity ordered. This assumption makes purchasing costs irrelevant to determining the optimal EOQ size.

4. No stockouts occur. One justification for this assumption is that the costs of a stockout are prohibitively high. We assume that to avoid these potential costs, management always maintains adequate inventory so that no stockout can occur.

5. In deciding the size of the purchase order, management considers the costs of quality only to the extent that these costs affect ordering costs or carrying costs.

Given these assumptions, the purchase costs, stockout costs, and quality costs can be safely ignored. The total relevant costs for determining the optimal EOQ are the sum of the ordering costs and the carrying costs:

Total relevant costs = Total relevant ordering costs + Total relevant carrying costs

EXAMPLE: Video Galore sells packages of blank VCR tapes to its customers; it also rents out tapes of movies and sporting events. It purchases packages of VCR tapes from Sontek, a distributor, at $20 a package. Sontek pays all incoming freight. No incoming inspection is necessary, as Sontek has a superb reputation for delivering quality merchandise. Annual demand is 10,400 packages, at a rate of 200 packages per week. Video Galore earns 12% on its cash investments. The purchase-order lead time is two weeks. The following cost data are available:

Relevant ordering costs per purchase order		$62.50
Relevant carrying costs per package per year:		
Required annual return on investment, 12% × $20	$2.40	
Relevant insurance, materials handling, breakage, and so on, per year	2.80	5.20

What is the economic order quantity of packages of VCR tapes?

Study Exhibit 24-1 carefully. It tabulates the total annual relevant costs of ordering and carrying inventory under various order sizes, and illustrates the trade-off between the two types of costs. The larger the order quantity, the higher the annual carrying costs, but the lower the annual ordering costs. The smaller the order quantity, the lower the annual carrying costs, but the higher the annual ordering costs. Exhibit 24-2 analyzes the behavior of these two cost functions graphically. The total annual relevant costs will be at a minimum where total ordering costs and total carrying costs

EXHIBIT 24-1
Annualized Ordering Costs and Carrying Costs at Various Order Quantities for Video Galore

D: Demand in units: (10,400 units per year)	10,400	10,400	10,400	10,400	10,400	10,400	10,400
Q: Order quantity in units	50	100	400	500	600	1,000	10,400
$Q/2$: Average inventory in units	25	50	200	250	300	500	5,200
D/Q: Number of purchase orders	208	104	26	20.80	17.33	10.40	1
$(D/Q) \times P^*$: Annual ordering costs for purchase orders	$13,000	$6,500	$1,625	$1,300	$1,083	$ 650	$ 62
$(Q/2) \times C^\dagger$: Annual carrying costs	130	260	1,040	1,300	1,560	2,600	27,040
Total relevant costs of ordering and carrying inventory	$13,130	$6,760	$2,665	$2,600	$2,643	$3,250	$27,102

Minimum cost

*P = ordering cost per purchase order = $62.50.
†C = carrying cost of one unit in stock = $5.20 per year.

EXHIBIT 24-2

Graphic Analysis of Ordering Costs and Carrying Costs for Video Galore

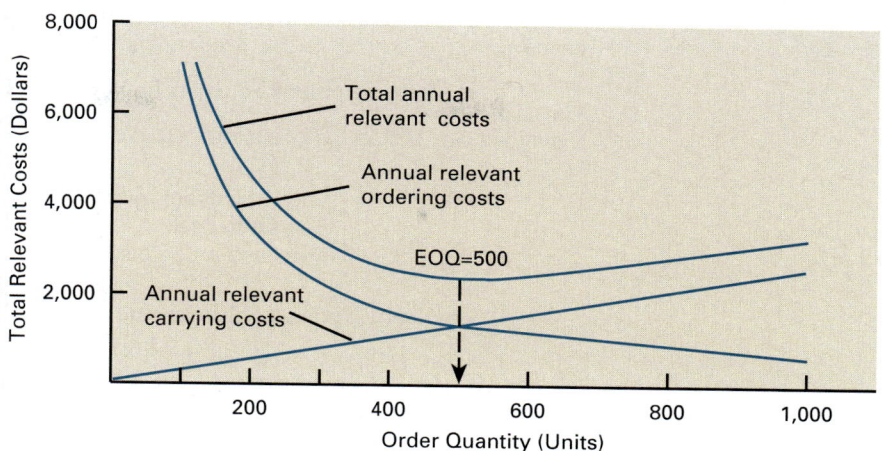

are equal. The EOQ that minimizes the total annual relevant costs in Exhibits 24-1 and 24-2 is 500 packages. The EOQ is most easily found by using the formula given in the next section.

EOQ Formula

The formula underlying the EOQ model is

$$EOQ = \sqrt{\frac{2DP}{C}}$$

where

EOQ = economic order quantity
D = demand in units for a specified time period
P = ordering costs per purchase order
C = costs of carrying one unit in stock for the time period used for D (one year in this example)

We illustrate this formula using the data for Video Galore:

$$EOQ = \sqrt{\frac{2 \times 10{,}400 \times \$62.50}{\$5.20}} = \sqrt{250{,}000} = 500 \text{ packages}$$

The total relevant costs (TRC) can be calculated using the following formula where Q = the order quantity:

$$TRC = \frac{DP}{Q} + \frac{QC}{2}$$

(In this formula, Q can be any order quantity, not just the EOQ.)
When $Q = 500$ units:

$$TRC = \frac{10{,}400 \times \$62.50}{500} + \frac{500 \times \$5.20}{2}$$

$$= \$1{,}300 + \$1{,}300 = \$2{,}600$$

The number of deliveries each time period (in our example, one year) is

$$\frac{D}{EOQ} = \frac{10{,}400}{500} = 20.8 \text{ deliveries}$$

When to Order, Assuming Certainty

The **reorder point** is the quantity level of the inventory on hand that triggers a new order. The reorder point is simplest to compute when both demand and lead time are certain:

$$\text{Reorder point} = \text{Sales per unit of time} \times \text{Purchase-order lead time}$$

Consider our Video Galore example. We choose a week as the unit of time:

Economic order quantity	500 packages
Sales per week	200 packages
Purchase-order lead time	2 weeks

$$\text{Reorder point} = \text{Sales per unit of time} \times \text{Purchase-order lead time}$$
$$= 200 \times 2 = 400 \text{ packages}$$

The graph in Exhibit 24-3 presents the behavior of the inventory level of VCR packages, assuming demand occurs uniformly throughout each week.[2] If the purchase-order lead time is two weeks, a new order will be placed when the inventory level reaches 400 VCR packages.

Safety Stock

Exhibit 24-3 assumes that demand and purchase-order lead time are certain. When retailers are uncertain about the demand, the lead time, or the amount that suppliers can provide, they often hold safety stock. **Safety stock** is the buffer inventory held as a cushion against unexpected increases in demand or lead time and unexpected unavailability of stock from suppliers. Video Galore's expected demand is 200 packages per week, but its managers feel that a maximum demand of 320 packages per week may occur. If Video Galore's managers decide that the costs of stockout are prohibitive, they may decide to hold safety stock of 240 packages. This amount is the maximum excess demand of 120 per week for the two weeks of purchase-order lead time.

The computation of safety stock hinges on demand forecasts. In our Video Galore example with uncertain demand, we used 240 VCR packages as a safety stock. The 240 amount is based on projected demand. Managers will have some notion—usually based on experience—of the range of weekly demand.

The major relevant costs in maintaining safety stock are the carrying costs and the stockout costs. Suppose a customer calls Video Galore to buy five packages of VCR tapes. The store has none in stock at the moment, but it can supply them within 24 hours at an extra payment to the supplier of $4 per package. The customer is willing to wait the 24 hours. The stockout costs in this case are a minimum of $4 per package. The optimal safety stock level is the quantity of safety stock where the costs of carrying an extra unit are exactly counterbalanced by the expected stockout costs. This would be the level that minimizes the total annual stockout costs and annual carrying costs.

A frequency distribution based on prior daily or weekly levels of demand provides data for computing the associated costs of maintaining safety stock. Assume that one of seven different levels of demand for two weeks will occur at Video Galore.

---

[2]This handy but special formula does not apply when the receipt of the order fails to increase inventory to the reorder-point quantity (for example, when the lead time is three weeks and the order is a one-week supply). In these cases, orders will overlap. The reorder point will be average usage during lead time plus safety stock minus orders placed but not yet received. *Safety stock* is the buffer inventory held as a cushion against unexpected increases in demand or lead time and unexpected unavailability of stock from suppliers.

EXHIBIT 24-3

Inventory Level of VCR Packages for Video Galore*

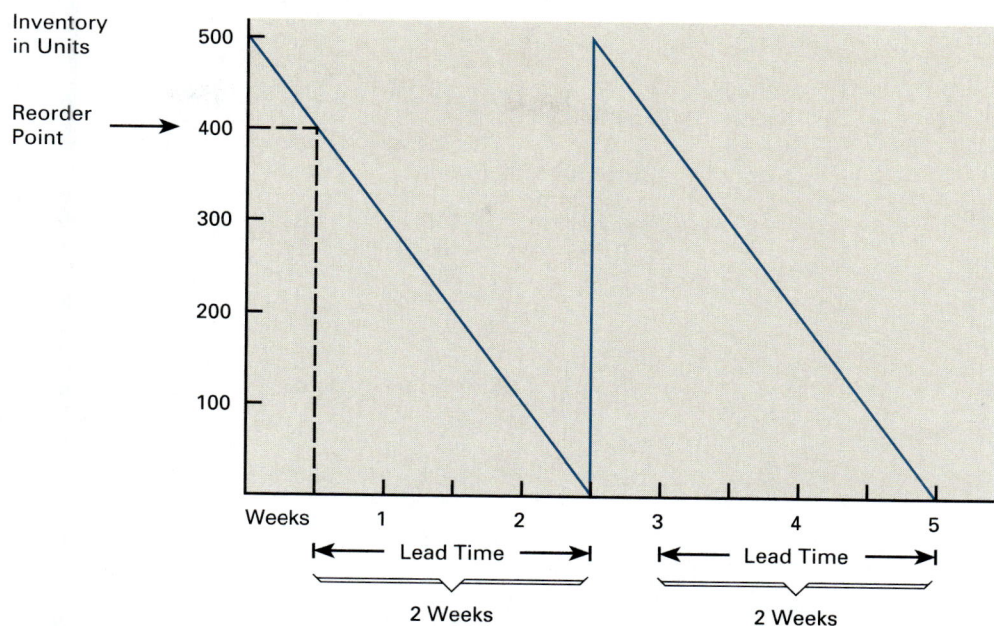

*Exhibit assumes that demand and purchase-order lead time are certain.

Total demand for 2 weeks	160 Units	240 Units	320 Units	400 Units	480 Units	560 Units	640 Units
Probability (sums to 1.00)	0.04	0.06	0.20	0.40	0.20	0.06	0.04

We see that 400 is the most likely level of demand for two weeks because it is assigned the highest of the probabilities in the chart. We see also that there is a 0.30 probability that demand will be between 480 and 640 packages (0.20 + 0.06 + 0.04 = 0.30).

If we decided to use a reorder point equal to 400 packages, stockouts would occur only if demand were for 480, 560, or 640 units. The following table shows the probability of a stockout with a reorder point of 400 units and safety stocks of 0, 80, 160, and 240 units. For example, suppose actual demand is 640. Subtracting the demand provided for (400 units) gives us a stockout of 240 units if the business decides not to keep any safety stock, a stockout of 160 units if 80 units of safety stock are held, and so on.

Probability of Stockout	0.40	0.20	0.06	0.04
Total actual demand during lead time	400	480	560	640
Deduct demand provided for during lead time	400	400	400	400
Stockout if provision is also made for safety stock of:				
0 units	0	80	160	240
80 units	0	0	80	160
160 units	0	0	0	80
240 units	0	0	0	0

Exhibit 24-4 presents the total annual stockout costs and carrying costs assuming the stockout costs are $4 per package and the carrying costs are $5.20 per unit per year. Of the safety stock levels presented in Exhibit 24-4, the total annual stockout costs and annual carrying costs are minimized at $1,098, when a safety stock of 160 packages is maintained.

EXHIBIT 24-4
Computation of Safety Stock for Video Galore

Safety Stock Level in Units	Probability of Stockout	Stockout Costs				Carrying Costs§	Total Costs
		Stockout in Units	Stockout Costs*	Orders per Year†	Expected Stockout Costs‡		
0	0.20	80	$320	20.8	$1,331		
	0.06	160	640	20.8	799		
	0.04	240	960	20.8	799		
					$2,929	$ 0	$2,929
80	0.06	80	$320	20.8	$ 399		
	0.04	160	640	20.8	532		
					$ 931	$ 416	$1,347
160	0.04	80	$320	20.8	$ 266	$ 832	$1,098
240	0.00	0	$ 0	20.8	$ 0	$1,248	$1,248

*Stockout units × stockout costs of $4 per unit.
†Annual demand 10,400 ÷ 500 EOQ = 20.8 orders per year.
‡Stockout costs × probability × number of orders per year.
§Safety stock × annual carrying costs of $5.20 per unit (assumes that safety stock is on hand at all times and that there is no overstocking caused by decreases in expected usage).

DIFFICULTIES WITH ACCOUNTING DATA FOR MANAGING GOODS FOR SALE

Estimating Cost Parameters

Obtaining accurate estimates of the cost parameters used in the EOQ decision model is a challenging task. For example, the relevant annual carrying costs of inventory consist of *outlay costs* plus the *opportunity cost* of capital.

What are the relevant outlay costs of carrying inventory? Only those costs that vary with the quantity of inventory held—for example, insurance, property taxes, and costs of breakage. Consider for example, the salaries paid to clerks, storekeepers, and materials handlers. These costs are irrelevant if they are unaffected by changes in inventory levels. Suppose, however, that as inventories decrease, these salary costs also decrease as the clerks, storekeepers, and materials handlers are transferred to other activities. In this case, the salaries paid to these people are relevant outlay costs of carrying inventory. Similarly, the costs of storage space owned that cannot be used for other profitable purposes as inventories decrease are irrelevant. But if the space has other profitable uses, or if there is a rental cost geared to the amount of space occupied, storage costs are relevant outlay costs of carrying inventory.

What is the relevant opportunity cost of capital? It is the interest forgone by investing capital in inventory. It is calculated as the required rate of return multiplied by those costs per unit that vary with the number of units purchased and that are incurred at the time the units are received. (Examples of these costs per unit are purchase price, freight-in, and incoming inspection.) An opportunity cost must be charged because interest income is forgone when money is tied up in inventory rather than being invested elsewhere (say, in a bank). Interest would not be forgone if the units could be acquired at the instant of their use or sale. Most internal reporting systems do not formally record opportunity costs. Users of EOQ models cannot, then, rely exclusively on the accounting system for all the components included in the costs of carrying inventory.

Opportunity costs are also relevant to estimating stockout costs. When a sale is forgone because an item is out of stock, the opportunity cost includes the lost contribution margin on that sale as well as the lost contribution margin on potential future sales that will not be made to the disgruntled customer. Indeed, the opportunity costs may be even higher if the customer informs other customers and potential customers of his or her experience.

Cost of a Prediction Error

Objective 5

Discuss why EOQ models are rarely sensitive to minor variations in cost predictions

Predicting costs is difficult. Intelligent managers understand that their projections will seldom be perfectly on target. What is the cost of an incorrect prediction when actual costs are different from predicted costs?

Suppose Video Galore's ordering costs per purchase order are $90 instead of the $62.50 prediction used in Exhibits 24-1 and 24-2. We can calculate the cost of this prediction error with a three-step approach.

STEP 1: Compute the monetary outcome from the best action that could have been taken, given the actual amount of the cost input. The appropriate inputs are $D = 10,400$ units, $P = \$90$, and $C = \$5.20$. The economic order quantity size is

$$EOQ = \sqrt{\frac{2DP}{C}}$$

$$= \sqrt{\frac{2 \times 10,400 \times \$90}{\$5.20}} = \sqrt{\frac{\$1,872,000}{\$5.20}} = \sqrt{360,000}$$

$$= 600 \text{ packages}$$

The total annual relevant costs when EOQ = 600 is

$$TRC = \frac{DP}{Q} + \frac{QC}{2}$$

$$= \frac{10,400 \times \$90}{600} + \frac{600 \times \$5.20}{2}$$

$$= \$1,560 + \$1,560$$

$$= \$3,120$$

STEP 2: Compute the monetary outcome from the best action based on the incorrect amount of the predicted cost input. The planned action when the ordering costs per purchase order are predicted to be $62.50 is to purchase quantities of 500 packages. The total annual relevant costs using this order quantity when $D = 10,400$ units, $P = \$90$, and $C = \$5.20$ are

$$TRC = \frac{10,400 \times \$90}{500} + \frac{500 \times \$5.20}{2}$$

$$= \$1,872 + \$1,300$$

$$= \$3,172$$

STEP 3: Compute the difference between the monetary outcomes from step 1 and step 2.

	Monetary Outcome
Step 1	$3,120
Step 2	3,172
Difference	$ (52)

The cost of the prediction error is only $52. The total annual relevant costs curve in Exhibit 24-2 is relatively flat over the range of order quantities from 400 to 600. *An important feature of the EOQ model is that the total relevant costs are rarely sensitive to minor variations in cost predictions. The square root in the EOQ model reduces the sensitivity of the decision to errors in predicting its inputs.*

Goal-Congruence Issues

Objective 6

Describe the potential conflicts that can arise between EOQ decision models and models used for performance evaluation

Goal-congruence issues can arise when there is an inconsistency between the decision model and the model used to evaluate the performance of the person implementing the model. Consider opportunity costs. Opportunity costs are an important component of the total relevant costs when making decisions with an EOQ model. The absence of recorded opportunity costs in conventional accounting systems, however, raises the possibility of a conflict between the order quantity that the EOQ model indicates as optimal and the order quantity that the purchasing manager regards as optimal.

If annual carrying costs are not included in the measures used to evaluate the performance of store managers, the managers may favor purchasing a larger order quantity than the EOQ decision model implies is optimal. One way to resolve this conflict is to design the performance evaluation system so that the carrying costs are charged to the appropriate manager. For example, an "imputed interest" charge can be levied against managers for the inventories under their responsibility. This practice inhibits managers from overbuying inventories, a common temptation. The chairman of the Coca-Cola Company summed up Coca-Cola's practice of charging imputed interest as follows: "I learned that when you start charging people for their capital, all sorts of things happen. All of a sudden inventories get under control. You don't have three months' [inventory] sitting around for an emergency."

The inventory area provides another illustration of why accountants and managers must be alert to the motivational implications of failing to dovetail a performance evaluation model with the decision model that top management favors.

JUST-IN-TIME PURCHASING

Objective 7

Describe how just-in-time (JIT) purchasing reduces costs

Organizations are giving increased attention to the potential gains from (1) making smaller and more frequent purchase orders and (2) restructuring their relationship with suppliers. Both items 1 and 2 are related to the heightened interest in just-in-time purchasing systems. **Just-in-time (JIT) purchasing** is the purchase of goods or materials such that delivery immediately precedes demand or use. In the extreme, no inventories (goods for sale of a retailer; materials, work in process, or finished goods for a manufacturer) would be held.

Organizations using JIT purchasing often stress the hidden costs associated with holding high inventory levels. These hidden costs include larger amounts of inventory storage space and sizable amounts of spoilage. Retailers have long recognized these costs in relation to perishable goods. For example, bread and milk have been delivered daily to supermarkets for many years. Retailers using JIT purchasing are now attempting to extend daily deliveries to as many areas as possible so that the goods spend less time in warehouses or on store shelves before they are sold to customers.

Some vendors are highly cooperative in a business's attempts to adopt JIT purchasing. For example, consider the Frito-Lay company, which has a large market share in potato chips and other snack foods. Frito-Lay makes more frequent deliveries than many of its competitors. Its corporate strategy emphasizes service to retailers and consistency of product quality delivered. Companies moving toward JIT purchasing often argue that the "full cost" of carrying inventories (including the opportunity and other costs not recorded in the accounting system) has been dramatically underestimated in the past.

EOQ Implications of Just-in-Time Purchasing

Exhibit 24-5 uses sensitivity analysis of the Video Galore example in Exhibit 24-1 to illustrate the economics of smaller and more frequent purchase orders. Reductions in the ordering costs of each purchase order and increases in the carrying costs of inventories reduce the economic order quantity. The rows of Exhibit 24-5 report how reductions in the ordering costs of Video Galore's placing a purchase order (P) reduce EOQ. Consider the case with $D = 10,400$ and $C = \$5.20$; the EOQ is 500 packages when $P = \$62.50$, 447 packages when $P = \$50$, and 346 when $P = \$30$. The columns of Exhibit 24-5 report how increases in the costs of Video Galore's carrying a unit of inventory are associated with reductions in the EOQ. Consider the case with $D = 10,400$ and $P = \$62.50$; the EOQ is 500 when $C = \$5.20$, 431 when $C = \$7$, and 294 when $C = \$15$. The analysis presented in Exhibit 24-5 supports JIT purchasing—that is, having a smaller economic order quantity—as carrying costs increase and ordering costs per purchase order decrease.

Do not assume that a JIT purchasing policy will always be guided by an EOQ decision model. At a minimum, the EOQ model assumes a constant order quantity. If demand fluctuates, a JIT purchasing policy may well require a different quantity for each order. Further, the EOQ model considers only carrying and ordering costs. As described earlier in this chapter, inventory management extends beyond ordering and carrying costs to include purchase costs, stockout costs, and quality costs (p. 833). The quality of materials and goods and timely deliveries are important motivations for using JIT purchasing, and stockout costs are an important concern. To understand the full costs and benefits of JIT purchasing, we move outside the confines of the EOQ model.

The next section considers relevant purchase costs and stockout costs in JIT purchasing decisions. We then discuss relevant costs of quality and timeliness of deliveries.

Relevant Benefits and Relevant Costs of JIT Purchasing

Revisit the Video Galore example, and consider the following information. Video Galore has recently established an electronic hookup to Sontek. A purchase order is now triggered by a single computer entry that makes ordering costs negligible. Video Galore is negotiating with Sontek to have that company deliver 100 packages of VCR tapes twice each week instead of delivering 500 packages every 2½ weeks (52 weeks ÷ 20.8 deliveries each year) as calculated in Exhibit 24-1. Sontek is willing to make these frequent deliveries but it would tack on a small additional price of $0.02 per VCR package. Video Galore's required return on investment remains 12%. Assume relevant outlay carrying costs of insurance, materials handling, breakage, and so on per package per year also remain at $2.80 per year.

EXHIBIT 24-5
Sensitivity of EOQ to Variations in Ordering Costs and Carrying Costs*

Carrying Costs per Package per Year (C)	Ordering Costs per Purchase Order (P)			
	$62.50	$50.00	$40.00	$30.00
$ 5.20	EOQ = 500	EOQ = 447	EOQ = 400	EOQ = 346
$ 7.00	431	385	345	299
$10.00	361	322	288	250
$15.00	294	263	236	204

*All cells in the matrix assume that annual demand (D) is 10,400 units.

Video Galore's one concern is that lower inventory levels from implementing JIT purchasing will lead to more stockouts. Assume that Video Galore incurs no stockout costs under its current purchasing policy. By shifting to a JIT purchasing policy, Video Galore expects to incur stockout costs on five VCR packages each month. In the event of a stockout, Video Galore will have to rush-order VCR packages at a cost of $4 per package. Should Video Galore implement JIT purchasing?

Exhibit 24-6 compares (1) the incremental costs Video Galore incurs when it purchases VCR tapes from Sontek under its current purchasing policy with (2) the incremental costs Video Galore would incur if Sontek supplied VCR tapes under a JIT policy. The difference in the two incremental costs is the net relevant savings of JIT purchasing. An alternative approach would compare the two alternatives directly and include only those costs that differ between the two alternatives. Exhibit 24-6 shows net cost savings of $591.88 per year by shifting to a JIT purchasing policy. Note that ordering costs are irrelevant and hence excluded from our analysis. Why? Because Video Galore will implement the new electronic technology to reduce its ordering costs whether or not it shifts to JIT purchasing. Exhibit 24-6 focuses only on the incremental benefits and costs of lower inventories.

Relevant Costs of Quality and Timely Deliveries

To emphasize quality and delivery, retailing and manufacturing companies calculate the total costs of materials. Total costs include quality costs of materials (for example, incoming materials inspection costs, returns, scrap costs, and rework costs), costs of late deliveries (for example, expediting costs, idle time on machines, and forgone contribution margin on lost sales), and costs of early deliveries (carrying costs). Texas Instruments (TI) reports purchase costs of $0.55 per unit for its electrical connectors and ordering and carrying costs of $0.45 per unit. If a connector is found to be defective,

EXHIBIT 24-6
Annual Relevant Costs of Current Purchasing Policy and JIT Purchasing Policy for Video Galore

	Incremental Costs under Current Purchasing Policy	Incremental Costs under JIT Purchasing Policy
Purchasing costs		
$20.00 per unit x 10,400 units per year	$208,000.00	
$20.02 per unit x 10,400 units per year		$208,208.00
Required return on investment		
12% per year x $20.00 cost per unit x 250* units of average inventory per year	600.00	
12% per year x $20.02 cost per unit x 50† units of average inventory per year		120.12
Outlay carrying costs (insurance, materials handling, breakage, and so on)		
$2.80 per unit per year x 250 units of average inventory per year	700.00	
$2.80 per unit per year x 50 units of average inventory per year		140.00
Stockout costs		
No stockouts	0	
$4 per unit x 5 units per month x 12 months		240.00
Total annual incremental costs	$209,300.00	$208,708.12
Difference in favor of JIT purchasing (annually)	↑ $591.88 ↑	

*Order quantity ÷ 2 = 500 ÷ 2.
†Order quantity ÷ 2 = 100 ÷ 2.

however, it costs TI $15 to replace during manufacturing, $57 to replace during final testing, and $97 to replace at the customer's site.[3]

Costs of quality and timely deliveries are particularly crucial in JIT purchasing environments. Defective materials and late deliveries often bring the whole plant to a halt, resulting in lost contribution. Companies that implement JIT purchasing choose their suppliers carefully and pay special attention to developing long-run supplier partnerships (see Chapter 7, p. 243). Price is only one component in evaluating suppliers.

What are the relevant costs when choosing suppliers? Consider again our Video Galore example. Denton and Company also distributes VCR tapes. It offers to supply all of Video Galore's VCR tape needs at a price of $19.50 per package (less than Sontek's price of $20.02) under the same JIT delivery terms Sontek offers. Video Galore's relevant outlay carrying costs of insurance, materials handling, breakage, and so on per package per year is $2.80 whether it purchases VCR tapes from Sontek or from Denton. Should Video Galore accept Denton's offer? Not before considering the relevant costs of quality and also the costs that a failure in delivery would create.

Video Galore has used Sontek in the past and knows that Sontek fully deserves its reputation for delivering quality merchandise on time. In fact, Video Galore does not find it necessary to inspect the VCR packages that Sontek supplies. Denton, however, does not enjoy so sterling a reputation for quality. Video Galore anticipates the following negative aspects of using Denton:

- Video Galore would incur inspection costs of $0.05 per package.
- Average stockouts of 25 VCR packages each month, largely resulting from late deliveries. Video Galore anticipates lost contribution margin per unit of $10 from stockouts. Denton cannot rush-order VCR packages to Video Galore at short notice.
- Customers would likely return 2% of all packages sold owing to poor quality of the tapes. Video Galore estimates it would cost $40 to handle each returned package.

Exhibit 24-7 presents the incremental costs of purchasing (1) from Sontek and (2) from Denton. Even though Denton is offering a lower price per package, the *total* costs of purchasing goods from Sontek are lower by $6,048.88. Exhibit 24-7 includes incremental carrying costs of insurance, materials handling, breakage, and so on under each alternative. These costs could be ignored in the relevant cost analysis since these costs are the same whether Video Galore purchases from Sontek or from Denton. Selling high-quality merchandise also has nonfinancial and qualitative benefits. For example, offering Sontek's high-quality tapes enhances Video Galore's reputation and increases customer goodwill, which may lead to higher future revenues.

Cost Management

Companies are now paying more attention to better management of the costs associated with inventories. Organizations that have moved toward JIT purchasing have made substantial changes in their purchasing practices:

- A reduction in the number of suppliers for each item, with an associated reduction in negotiation time. For example, a division of Xerox reduced its number of suppliers from 5,000 to 300.
- The use of long-run contracts with suppliers, with minimal paperwork involved in each individual transaction. A purchase transaction may involve only a single telephone call or a single computer entry.
- Minimal checking by purchasers of the quantity and quality of goods received. In the initial negotiations, suppliers are made aware of the premium placed on the on-time delivery of high-quality goods in the exact quantity ordered.
- Payment to suppliers made for batches of deliveries rather than for each individual delivery. For example, the materials-receiving group at one Hewlett-Packard plant receives documents on a daily basis but sends them to Accounts Payable only on a weekly basis.

---

[3]L. Carr and C. Ittner, "Measuring the Cost of Ownership," *Journal of Cost Management* (Fall 1992).

EXHIBIT 24-7

Annual Relevant Costs of JIT Purchasing from Sontek and Denton

	Incremental Costs of Purchasing from Sontek	Incremental Costs of Purchasing from Denton
Purchasing costs		
$20.02 per unit × 10,400 units per year	$208,208.00	
$19.50 per unit × 10,400 units per year		$202,800.00
Inspection costs		
No inspection necessary	0	
$0.05 per unit × 10,400 units		520.00
Required return on investment		
12% per year × $20.02 × 50* units of average inventory per year	120.12	
12% per year × $19.50 × 50* units of average inventory per year		117.00
Outlay carrying costs (insurance, materials handling, breakage, and so on)		
$2.80 per unit per year × 50 units of average inventory per year	140.00	
$2.80 per unit per year × 50 units of average inventory per year		140.00
Stockout costs		
$4 per unit × 5 units per month × 12 months	240.00	
$10 per unit × 25 units per month × 12 months		3,000.00
Customer returns costs		
No customer returns	0	
$40 per unit returned × 2% × 10,400 units returned		8,320.00
Total annual incremental costs	$208,708.12	$214,897.00
Difference in favor of Sontek (annually)	↑ $6,188.88 ↑	

*Order quantity ÷ 2 = 100 ÷ 2.

Computer software programs "match" each receiving document with the purchase-order number. A computer program then sums all amounts due each supplier, and a single check is written to each supplier.

These changes in purchasing practices can substantially reduce the cost of placing each individual purchase order.

Retailers are often highly aggressive in their cost-management activities. For example, Kaufmann's, a department store, requires suppliers to use their own salespeople to provide in-store services such as setting up promotion displays and picking up damaged merchandise in some of their departments. Wal-Mart "fines" suppliers for late deliveries and incomplete orders.

MANAGING INVENTORIES IN MANUFACTURING ORGANIZATIONS

Objective 8

Explain the roles an accountant can play in a materials requirements planning (MRP) system

Managers in companies with manufacturing facilities face the challenging task of producing high-quality products at competitive cost levels. Numerous systems have been developed to help managers plan and implement activities.

We distinguish between two basic types of systems, a centralized "push-through" system (which emphasizes demand forecasts) and a decentralized "demand-pull" system (which responds to actual customer demand). A push-through system, often described as a *materials requirement planning (MRP)* system focuses first on the forecasted amount and timing of finished goods demand and then determines the demand for materials, components, and subassemblies at each of the prior stages of production.[4] Once scheduled production starts, the output of each department is pushed through the system to the next department for processing or into inventory to be retrieved later.

A demand-pull system, often described as a just-in-time (JIT) production system, emphasizes simplicity and close coordination among work centers. The company adapts final goods schedules as demands for products change. In response to customer demand, final assembly places production orders for the needed subassemblies and components. Only those subassemblies and components required by final assembly are pulled through the system.

The following section outlines the key features of MRP and JIT production and the role of accounting information in each of these manufacturing systems. Our descriptions of MRP and JIT are necessarily brief and simple. Be careful when examining the manufacturing systems of individual firms. Businesses vary in the labels they use to describe their systems.

Materials Requirements Planning

Exhibit 24-8 is a production structure diagram of an MRP system that offers an overview of the production process and its materials requirements for three end products (FG 1, FG 2, and FG 3). Working backward, each end product is sequentially exploded (that is, separated) into its necessary components and materials. Four direct materials (DM 1, DM 2, DM 3, and DM 4) are purchased for finished goods. For both FG 1 and FG 3, the materials are used to manufacture the components that are assembled into the end product. For FG 2, no intermediary components are produced.

Key elements of an MRP system include the following:

1. Master production schedule, which specifies both the quantity and the timing of each item to be produced.
2. Bill of materials file, which outlines the materials, components, and subassemblies for each end product. MRP distinguishes outside purchases from components derived from prior steps in the production process.

---

[4]MRP is sometimes used to describe *manufacturing resources planning*, which is a planning system that considers such items as capacity planning and labor scheduling as well as materials planning. Common terminology is MRP I for materials requirements planning and MRP II for manufacturing resource planning.

EXHIBIT 24-8
Production Structure Diagram Underlying an MRP System

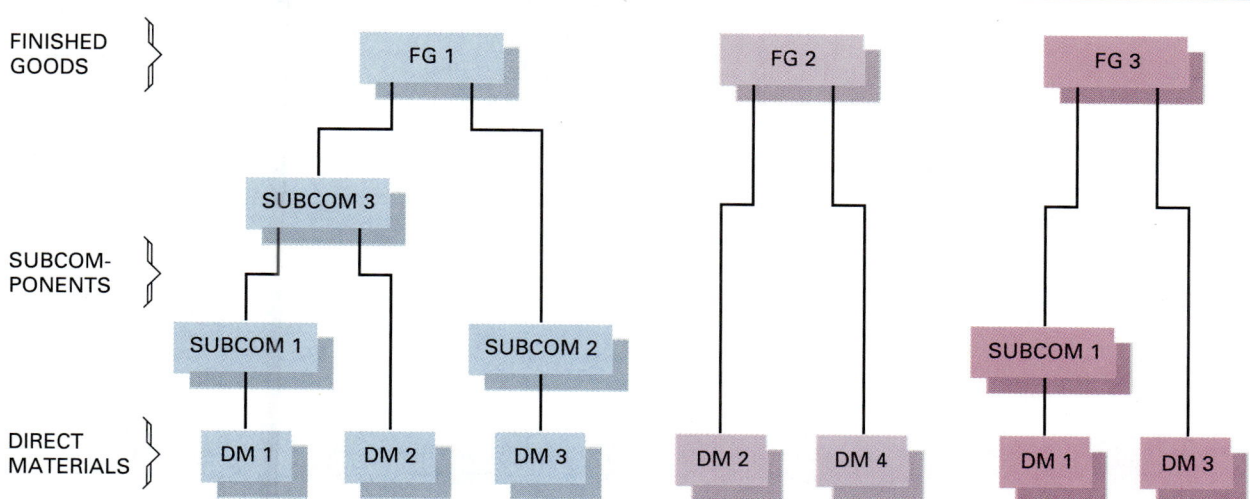

3. Inventory report of each part, component, or subassembly, in which each such item is carried in a separate computer file with details on the number of items on hand and the arrival times and quantities of items scheduled to be received.

4. Lead times of all items to be purchased and standard construction times for all components and subassemblies produced internally.

The management accountant can play several important roles in the operation of an MRP system. First, MRP requires accurate and timely information pertaining to materials, work in process, and finished goods. A major cause of unsuccessful attempts to implement MRP has been the problem of collecting and updating inventory records.

A second role of the management accountant consists of providing estimates of the costs of setting up each production run at a plant, the costs of downtime, and the costs of holding inventory. Costs of setting up the machine are analogous to ordering costs in the EOQ model. When the costs of setting up machines or sections of the production line are high (for example, as with a blast furnace in an integrated steel mill), processing larger batches of materials and incurring larger inventory carrying costs is the optimal approach, because it reduces the number of times the machine must be set up. When setup costs are small, processing smaller batches is optimal because it reduces carrying costs. Similarly, when the costs of downtime are high, there can be sizable benefits from maintaining continuous production.

Just-in-Time Production

Just-in-time (JIT) production is a system in which each component on a production line is produced immediately as needed by the next step in the production line.[5] JIT production includes three key features:

1. The production line is run on a demand-pull basis, so that activity at each workstation is authorized by the demand of downstream workstations. There are many ways to implement this demand-pull feature, but perhaps the most common is a *Kanban* system. *Kanban* is the Japanese term for a visual record or card.

In the simplest *Kanban* system, one operation uses a *Kanban* card to authorize another operation to produce a specified quantity of a particular part. For example, suppose the assembly department of a muffler manufacturer receives an order for 10 mufflers. The assembly department triggers production of the 10 metal pipes it needs to make the 10 mufflers by sending a *Kanban* card to the machining department. On receiving the *Kanban* card, the machining department begins production of the pipes. When production is complete, the machining department attaches the *Kanban* card to the box containing the metal pipes and ships the package downstream to the assembly department. The assembly department starts the cycle over again when it receives the next customer order.

2. Emphasis is placed on minimizing the setup time and the manufacturing lead time for each unit. *Manufacturing lead time* is the time from when a product is ready to start on the production line to when it becomes a finished good. Producing to demand often means manufacturing small quantities of the product. Producing small batches is economical only if setup times are small.

3. The production line is stopped if parts are absent or defective work is discovered. Stoppage creates an urgency about correcting problems that cause defective units. Each employee puts a premium on minimizing the potential sources of stoppages (such as defective material parts). In contrast, under traditional methods the

---

[5]Detailed discussion of JIT production management is in M. Schniederjans, *Topics in Just-in-Time Management* (Needham Heights, MA: Allyn and Bacon, 1992). For case studies, see R. M. Lindsay and S. Kalagnanam, *The Adoption of Just-in-Time Production Systems in Canada and Their Association with Management Control Practices* (Hamilton, Canada: Society of Management Accountants, 1993).

inventory of parts and work in process is large enough to enable workers to set aside defective parts and continue their normal operations.

Exhibit 24-9 summarizes the effects Hewlett-Packard reported from adopting JIT at several of its production plants. Early advocates of JIT emphasized the benefits associated with lower inventories. *Most firms adopting JIT report even greater benefits from the heightened emphasis on eliminating the root causes of rework and waste and on reducing the manufacturing lead time of their products.*

In computing the relevant benefits and relevant costs of implementing JIT systems, the cost analyst must consider all benefits. Consider Hudson Corporation, a manufacturer of brass taps. Hudson is considering implementing a JIT production system. Suppose that to implement JIT production, Hudson must incur $100,000 in annual tooling costs to reduce setup times. Suppose further that JIT will reduce average inventory by $500,000. Also, relevant costs of insurance, space, materials handling, and setup will decline by $30,000 per year. The company's required rate of return on inventory investments is 10% per year. Should Hudson implement JIT? On the basis of the numbers provided, we would be tempted to say no. Cost savings amount to $80,000 ([10% of $500,000] + $30,000), which is less than the incremental costs of $100,000.

Our analysis, however, has not considered other benefits of JIT. For example, Hudson estimates that implementation of JIT will reduce rework on 500 units, a savings of $50 a unit. Also better quality and faster delivery will allow Hudson to charge $2 more per unit on each of the 20,000 units that Hudson sells. The quality and deliv-

EXHIBIT 24-9
The Effects of JIT Production

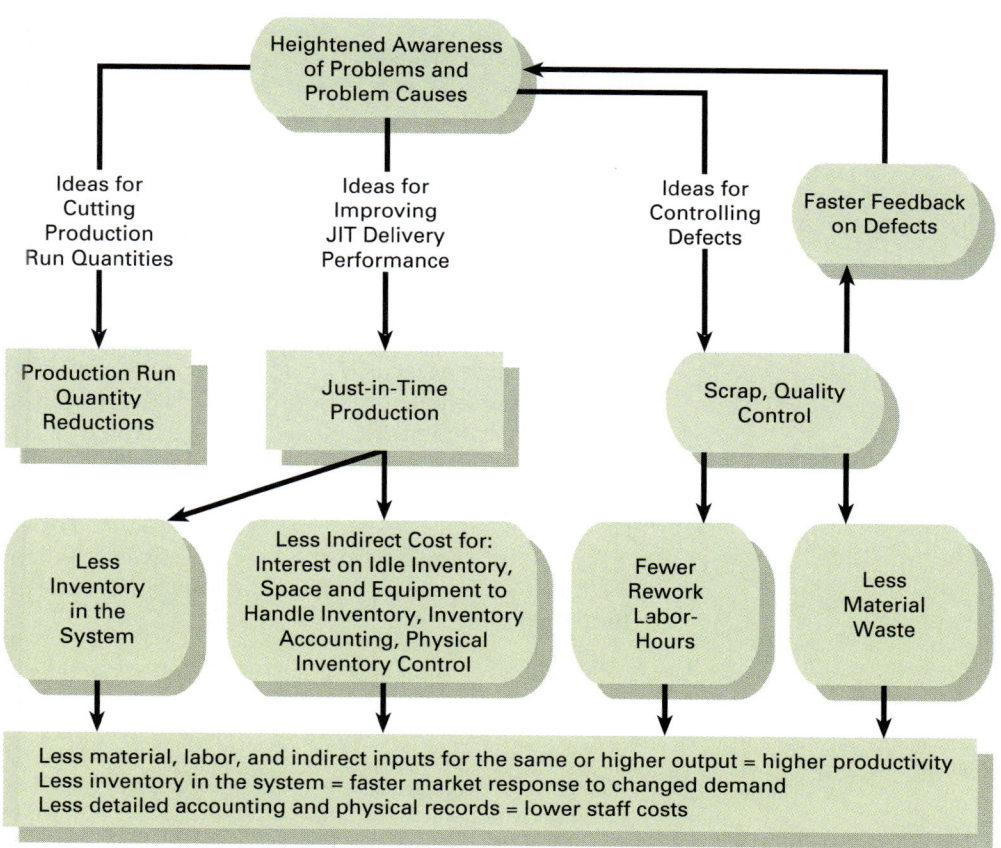

Source: Hewlett-Packard, adapted.

JIT PERFORMANCE MEASURES AROUND THE GLOBE

What performance measures do companies around the globe use to evaluate their JIT systems? The table below ranks in order of importance (1 = most important) the performance measures companies in four countries apply. The rankings also indicate the relative importance of the different reasons that motivated the companies to implement JIT in the first place.

	United States[a]	Canada[b]	Ireland[c]	United Kingdom[a]
Inventory reduction	1	1	3	1
On-time delivery performance	2	4	1	2
Improved quality	3	2	4	3
Decreased manufacturing lead time	4	3	2	5
Increased labor productivity	5	—	—	4
Reduced space utilization	6	5	—	6

A pattern emerges. The most important reasons for JIT implementation are reducing inventory cost, getting deliveries on time, and improving quality. To a lesser extent, companies also view reducing manufacturing lead time as important.

A survey of Italian companies[d] also reveals extensive use of these performance measures. The survey, however, provides no information on the relative importance of individual measures.

One survey[c] also found distinct differences between management control systems in JIT and non-JIT companies. JIT companies are characterized by greater decentralization, more frequent and timely reporting, and increased worker responsibility and autonomy for starting and stopping production to ensure quality.

Adapted from [a]Billesbach, Harrison, and Croom-Morgan, "Just-in-Time." [b]Lindsay and Kalagnanam, "The Adoption." [c]Clarke and O'Dea, "Management Accounting." [d]Bartezzaghi, Turco, and Spina, "The Impact." Full citations are in Appendix A.

ery benefits from JIT equal $65,000 (rework savings, $50 × 500 + additional contribution margin, $2 × 20,000). *Total* benefits and cost savings equal $145,000 ($80,000 + $65,000), which exceeds JIT implementation costs of $100,000.

The underlying philosophy of JIT is the simplification of the production process so that only essential activities that add value to the product are conducted. Several JIT adopters have extended this simplification to their internal accounting system. For example, the product-costing records in several JIT plants have only two entries (based on standard costs). The first is made when direct materials are used at the start of the production line, and the second is made when finished goods leave the production line. The reasons for using simplified product-costing records with JIT include the following:

1. Materials control in JIT plants can best be accomplished by a manager's personal observations. The absence of large amounts of materials and work-in-process inventory means that managers can spend more time observing and monitoring the existing materials and work in process.

2. Work in process constitutes a lower percentage of the total costs of production because of the reductions in lead time and the demand-pull feature of production.

3. There is a reduction in detailed record keeping associated with rework. Rework can occur at many different stages in a production process. Detailed records are necessary if the costs of rework are to be traced to individual jobs. Reducing the percentage of jobs with rework reduces the need for keeping detailed records.

BOEING GOES "JUST IN TIME" JUST IN TIME[a]

Facing deferral of orders from some of its customers, Boeing has begun implementing changes to make manufacturing and assembly of its airplanes cheaper and more efficient. The key components of Boeing's strategy: just-in-time production and just-in-time purchasing.

Boeing's ambitious performance targets are to cut manufacturing lead time on all its models (for example, Boeing aims to reduce lead time from 13 months to 6 months on its 737 model), to pare the time necessary to build completely new planes from 38 months to 28 months, and to lower costs by 25% to 50%, all by the year 2000. Boeing has already made impressive strides toward these goals. Assembly times (a component of manufacturing lead times) of 45–60 days have been sliced by 5 days for 737 and 747 aircraft and by 9 days on the 757 aircraft. In the plant where Boeing builds its wing structures, reductions in setup time have enabled Boeing to decrease average batch sizes from 16 to 2, thereby cutting manufacturing lead times.

To get its efforts at JIT purchasing airborne, Boeing is working closely with its largest suppliers—such as the three engine manufacturers General Electric, Pratt & Whitney, and Rolls Royce—to have parts delivered just before they are needed. At some of Boeing's plants, direct materials inventories have decreased from six months' to six weeks' worth of parts, resulting in substantial savings in inventory costs.

[a]Adapted from D. Yang and A. Rothman, "Reinventing Boeing, Radical Changes Amid Crisis," *Business Week*, March 1, 1993, and presentations by Boeing Company executives.

Additional examples of simplifications in accounting systems are described in Chapter 19.

Companies adopting JIT production make heavy use of nonfinancial performance measures of manufacturing lead time, inventory, setup time, and quality in the day-to-day control of operations at the plant level.[6] Examples of such measures are:

- Manufacturing lead time
- Units produced per hour
- Days inventory on hand
- $\dfrac{\text{Total setup time for machines}}{\text{Total production time}}$
- $\dfrac{\text{Number of units requiring rework or scrap}}{\text{Total number of units started and completed}}$

[6]See M. DeLuzio, "Management Accounting in a Just-in-Time Environment," *Journal of Cost Management* (Winter 1993).

PROBLEMS FOR SELF-STUDY

PROBLEM 1

The Complete Gardener (CG) is deciding on the economic order quantity for two brands of lawn fertilizer: Super Grow and Nature's Own. The following information is collected:

	Super Grow	Nature's Own
Annual demand	2,000 bags	1,280 bags
Ordering costs per purchase order	$30	$35
Annual carrying costs per bag	$12	$14

Required

1. Compute the EOQ for Super Grow and Nature's Own.
2. What is the sum of the total annual ordering costs and total annual carrying costs for Super Grow and Nature's Own?
3. Compute the number of deliveries per year for Super Grow and Nature's Own.

SOLUTION 1

1.

Super Grow	**Nature's Own**
$EOQ = \sqrt{\dfrac{2(2,000)(\$30)}{\$12}}$	$EOQ = \sqrt{\dfrac{2(1,280)(\$35)}{\$14}}$
$= 100$ bags	$= 80$ bags

2.

Super Grow	**Nature's Own**
$TRC = \dfrac{2,000(\$30)}{100} + \dfrac{100(\$12)}{2}$	$TRC = \dfrac{1,280(\$35)}{80} + \dfrac{80(\$14)}{2}$
$= \$1,200$	$= \$1,120$

3.

Super Grow	**Nature's Own**
$\dfrac{2,000}{100} = 20$ deliveries	$\dfrac{1,280}{80} = 16$ deliveries

PROBLEM 2

The Complete Gardener (CG) signs a long-term contract with the Super Grow distributor, and they set up a new procedure for placing purchase orders. A single entry is made into a computer network operated by the Super Grow distributor. CG will make no incoming inspection of bags; the distributor has "guaranteed" to maintain a 100% product quality level in return for CG's signing the long-term contract. CG's new ordering costs per purchase order will be $4.50. CG reexamined its materials-handling costs and revised upward its annual carrying cost per bag to $28.80.

Required

1. Consider the new ordering costs per purchase order and the new carrying costs per bag. Compute CG's economic order quantity and the number of deliveries per year for Super Grow.
2. How might your answers to requirement 1 provide insight into a just-in-time purchasing policy?

SOLUTION 2

1. For Super Grow, $D = 2,000$, $P = \$4.50$, and $C = \$28.80$:

$$EOQ = \sqrt{\frac{2(2,000)(\$4.50)}{\$28.80}}$$

$$= 25 \text{ bags}$$

$$\frac{D}{EOQ} = \frac{2,000}{25} = 80 \text{ deliveries}$$

2. A just-in-time purchasing policy involves the purchase of goods or materials such that delivery immediately precedes demand. The decrease in the EOQ for Super Grow from 100 bags to 25 bags increases the number of deliveries from 20 to 80. By restructuring relationships with its supplier, CG has dramatically reduced its ordering costs. This pattern is a familiar one for companies adopting JIT purchasing.

SUMMARY

The following points are linked to the chapter's learning objectives.

1. Managing costs of materials and other inventories is pivotal in many organizations because cost of goods sold constitutes the largest single cost item for most retail and manufacturing companies.

2. Five categories of costs associated with goods for sale are purchase costs, ordering costs (costs of preparing a purchase order and receiving goods), carrying costs (costs of holding inventory of goods for sale), stockout costs (costs arising when a customer demands a unit of product and that unit is not readily available), and quality costs (prevention costs, appraisal costs, internal failure costs, and external failure costs).

3. The economic-order-quantity (EOQ) decision model calculates the optimal quantity of inventory to order. The larger the order quantity, the higher the annual carrying costs and the lower the annual ordering costs. The smaller the order quantity, the lower the annual carrying costs and the higher the annual ordering costs. The EOQ model includes those transactions routinely recorded in the accounting system and opportunity costs not routinely recorded.

4. The reorder point is the quantity level of inventory that triggers a new order. It equals the sales per unit of time multiplied by the purchase-order lead time. Safety stock is the buffer inventory held as a cushion against unexpected unavailability of stock from suppliers.

5. The EOQ model is not sensitive to minor variations in cost predictions. The square root in the EOQ model reduces the sensitivity of the decision to errors in predicting its inputs.

6. The EOQ model exemplifies the potential conflict between decision models and performance evaluation models. The opportunity cost of the forgone interest on the investment in inventory plays a key role in the EOQ decision model. Still, many organizations measure management performance on the basis of reported costs or total reported operating income, without providing for any opportunity cost of forgone interest.

7. Just-in-time (JIT) purchasing is the purchase of goods or materials such that delivery immediately precedes demand or use. Cost savings arise in a JIT purchasing system because, for example, the costs of inventory carrying, materials handling, and breakage are reduced. Savings also occur because quality is higher. A potential cost of JIT purchasing is stockout costs.

8. A push-through system, often described as a *materials requirement planning (MRP)* system, focuses first on the forecasted amount and timing of finished goods and then determines the demand for materials, components, and subassemblies at each of the prior stages of production. The accountant's role in an MRP system is to provide information on quantities of inventories and costs of setups and downtime.

9. Just-in-time production is a system in which each component on a production line is produced immediately as needed by the next step in the production line. Measurements in JIT production systems emphasize manufacturing lead time, inventory levels, setup time, and quality.

TERMS TO LEARN

This chapter and the Glossary at the end of the book contain definitions of the following important terms:

carrying costs *(p. 833)*   economic order quantity (EOQ) *(833)*
just-in-time (JIT) purchasing *(840)*   inventory management *(832)*   ordering costs *(833)*
purchase costs *(833)*   purchase-order lead time *(833)*   reorder point *(836)*
safety stock *(836)*   stockout *(833)*

ASSIGNMENT MATERIAL

QUESTIONS

24-1 Give two examples of decisions that fall under the inventory management umbrella.

24-2 Why do better decisions regarding the purchasing and managing of goods for sale frequently cause dramatic percentage increases in net income?

24-3 Name two decisions that are central to the management of goods for sale in a retail organization.

24-4 Name five cost categories that are important in managing goods for sale in a retail organization.

24-5 What assumptions are made when using the simplest version of the economic-order-quantity (EOQ) decision model?

24-6 Give examples of costs included in annual carrying costs of inventory when using the EOQ decision model.

24-7 Give three examples of opportunity costs that typically are not recorded in accounting systems, although they are relevant to the EOQ model.

24-8 What are the steps in computing the cost of a prediction error when using the EOQ decision model?

24-9 "The practical approach to determining economic order quantity is concerned with locating a minimum cost range rather than a minimum cost point." Explain.

24-10 Why might goal-congruence issues arise when an EOQ model is used to guide decisions on how much to order?

24-11 Name two cost factors that can explain why an organization finds it cost-effective to make smaller and more frequent purchase orders.

24-12 "Accountants have placed inventories on the wrong side of the balance sheet. They are a liability, not an asset." Comment on this statement by a plant manager.

24-13 Give two examples of changes that organizations that moved toward just-in-time purchasing have made in their purchasing practices.

24-14 What roles can the accountant play in the operation of a materials requirements planning system?

24-15 Describe key features of just-in-time (JIT) production.

EXERCISES AND PROBLEMS

24-16 Economic order quantity for retailer. The Cloth Center sells fabrics to a wide range of industrial and consumer users. One of its products is denim cloth, used in the manufacture of jeans and carrying bags. The supplier for the denim cloth pays all incoming freight. No incoming inspection of the denim is necessary because the supplier has a track record of delivering high-quality merchandise. The purchasing officer of the Cloth Center has collected the following information:

Annual demand for denim cloth	20,000 yards
Ordering costs per purchase order	$160
Carrying costs per year	20% of purchase cost
Safety stock requirements	None
Cost of denim cloth	$8 per yard

The purchasing lead time is two weeks. The Cloth Center is open 250 days a year (50 weeks for 5 days a week).

Required

1. Calculate the economic order quantity for denim cloth.
2. Calculate the number of orders that will be placed each year.
3. Calculate the reorder point for denim cloth.

24-17 Economic order quantity for retailer. Tru-Value Hardware (TVH) is deciding the purchase order quantity for its standard line of neon light fittings. Annual demand is 18,000 units. Ordering costs per purchase order are $16. Carrying costs per light-fitting unit are $1.50 per year. TVH uses an economic-order-quantity model in its purchasing decisions. The store is open 360 days a year.

Required

1. Calculate TVH's economic order quantity for neon light fittings.
2. Assume that demand is known with certainty and the purchasing lead time is five days. Calculate TVH's reorder point for neon light fittings.

24-18 Effect of different order quantities on ordering costs and carrying costs, EOQ. Koala Blue retails a broad line of Australian merchandise at its Santa Monica store. It sells 26,000 Ken Done linen bedroom packages (two sheets and two pillow cases) each year. Koala Blue pays Ken Done Merchandise Inc. $104 per package. Its ordering costs per purchase order are $72. The carrying costs per package are $10.40 per year.

Liv Carrol, manager of the Santa Monica store, seeks your advice on how ordering costs and carrying costs vary with different order quantities. Ken Done Merchandise Inc. guarantees the $104 purchase cost per package for the 26,000 units budgeted to be purchased in the coming year.

Required

1. For purchase order quantities of 300, 500, 600, 700, and 900 compute the annual ordering costs, the annual carrying costs, and their sum. Use the format of Exhibit 24-1 (p. 834) to present your results. What is the economic order quantity? Comment on your results.
2. Assume that Ken Done Merchandise Inc. introduces a computerized ordering network for its customers. Liv Carrol estimates that Koala Blue's ordering costs will be reduced to $40 per purchase order. How will this reduction in ordering costs affect Koala Blue's economic order quantity for linen bedroom packages?

24-19 Effect of different order quantities on ordering costs and carrying costs, EOQ. Ritchway Supplies retails office equipment. It purchases 10,000 units a year of the Ultimate Executive Desk from Lumber Products at $1,000 per desk. Ritchway's ordering costs per purchase order to Lumber Products are $960. The carrying costs per Ultimate Executive Desk are $120 per year.

David Watson, president of Ritchway Supplies, seeks your advice on the economics of purchasing in different order quantities. Lumber Products has guaranteed that the purchase price for the 10,000 units will be $1,000 per desk.

Required

1. For purchase order quantities of 100, 300, 400, 500, and 700 compute the annual ordering costs, the annual carrying costs, and their sum. Use the format of Exhibit 24-1 (p. 834) to present your results. What is the economic order quantity? Comment on your results.
2. Assume that Ritchway Supplies is able to streamline its relationships with Lumber Products. The ordering costs per purchase order drop to $240. How will this reduction in ordering costs affect Ritchway Supplies' economic order quantity for Ultimate Executive Desks?

24-20 Purchase order size for retailer, EOQ, just-in-time purchasing. The 24 Hour Mart operates a chain of supermarkets. Its best-selling soft drink is Fruitslice. Demand in April 19_5 for Fruitslice at its Memphis supermarket is estimated to be 6,000 cases (24 cans in each case). In March 19_5, the Memphis supermarket estimated the ordering costs per purchase order (P) for Fruitslice to be $30. The carrying costs (C) of each case of Fruitslice in inventory for a month were estimated to be $1. At the end of March 19_5, the Memphis 24-Hour Mart reestimated its carrying costs to be $1.50 per case per month to take into account an increase in warehouse-related costs.

During March 19_5, 24-Hour Mart restructured its relationship with suppliers. It reduced the number of suppliers from 600 to 180. Long-term contracts were signed only with those suppliers that agreed to make product-quality checks before shipping. Each purchase order would be made by linking into the supplier's computer network. The Memphis 24-Hour Mart estimated that these changes would reduce the ordering costs per purchase order to $5. The 24-Hour Mart is open 30 days in April 19_5.

Required
1. Calculate the economic order quantity in April 19_5 for Fruitslice. Use the EOQ model, and assume in turn:
   a. $D = 6,000; P = $30; C = 1
   b. $D = 6,000; P = $30; C = 1.50
   c. $D = 6,000; P = $5; C = 1.50
2. How does your answer to requirement 1 give insight into the retailer's movement toward just-in-time purchasing policies?

24-21 Purchase order size for retailer, EOQ, just-in-time purchasing. The Family Discount Store (FDS) operates a chain of retail discount stores. Its best-selling brand of baby diapers is Baby Care. Demand in October 19_4 for Baby Care at its Quebec City store is 800 cases (100 one-pound packages of Baby Care are in each case). In October 19_4, the Quebec City store estimated the ordering costs per purchase order (P) for Baby Care to be $40. The carrying costs (C) of each case of Baby Care diapers in inventory for a month were estimated to be $6. At the end of September 19_4, the Quebec City store reestimated its carrying costs to be $10 per case to take into account an increase in warehouse-related costs.

During September, FDS restructured its relationship with suppliers. It reduced the number of suppliers from 950 to 370. Only those suppliers that agreed to ship in the exact quantities ordered, and with quality-control checks made before shipment, were given long-term supply contracts. Each individual purchase order involved minimal paperwork, and FDS was to make no quality checks of deliveries. The Quebec City store estimated its ordering costs per purchase order to be $4.20 after these changes were made. FDS is open 31 days in October 19_4.

Required
1. Calculate the economic order quantity in October 19_4 for Baby Care. Use the EOQ model and assume in turn:
   a. $D = 800; P = $40; C = 6
   b. $D = 800; P = $40; C = 10
   c. $D = 800; P = $4.20; C = 10
2. How does your answer to requirement 1 give insight into the retailer's movement toward just-in-time purchasing policies?

24-22 EOQ, cost of prediction error. Ralph Menard is the owner of a truck repair shop. He uses an economic-order-quantity model for each of his truck parts. He initially predicts the annual demand for heavy-duty tires to be 2,000. Each tire has a purchase price of $50. The incremental ordering costs per purchase order are $40. The incremental carrying costs per year are $4 per unit plus 10% of the supplier's purchase price.

Required
1. Calculate the EOQ for heavy-duty tires, along with the total sum of ordering costs and carrying costs.
2. Suppose Menard is precisely correct in all his predictions except the purchase price. He ignored a new law that abolished tariff duties on imported heavy-duty tires, which led to lower prices from foreign competitors. If he had been a faultless predictor, he would have foreseen that the purchase price would drop to $30 at the beginning of the year and would be unchanged throughout the year. What is the cost of the prediction error?

24-23 EOQ, uncertainty, safety stock, reorder point. (CMA, adapted) The Starr Company distributes a wide range of electrical products. One of its best-selling items is a standard electric motor. The management of Starr Company uses the economic-order-quantity (EOQ) decision model to determine the optimum number of motors to order. Management now wants to determine how much safety stock to hold.

Starr Company estimates annual demand (300 working days) to be 30,000 electric motors. Using the EOQ decision model, the company orders 3,000 motors at a time. The lead time for an order is five days. The annual carrying costs of one motor in safety stock are $10. Management has also estimated that the stockout costs are $20 for each motor they are short.

Starr Company has analyzed the demand during 200 past reorder periods. The records indicate the following patterns:

Demand during Lead Time	Number of Times Quantity Was Demanded
440	6
460	12
480	16
500	130
520	20
540	10
560	6
	200

Required
1. Determine the level of safety stock for electric motors that Starr Company should maintain in order to minimize expected stockout costs and carrying costs. When computing carrying costs, assume that the safety stock is on hand at all times and that there is no overstocking caused by decreases in expected demand. (Consider safety stock levels of 0, 20, 40, and 60 units.)
2. What would be Starr Company's new reorder point?
3. What factors should Starr Company have considered in estimating the stockout costs?

24-24 Just-in-time purchasing, relevant benefits, relevant costs. (CMA, adapted) AgriCorp sells farm equipment. AgriCorp's Service Division provides spare parts to various repair centers that repair AgriCorp's equipment. In an effort to reduce inventory costs, the Service Division implemented a just-in-time inventory program on January 1, 19_4. On January 1, 19_5, Janice Grady, the Service Division controller, decides to evaluate the effect the program has had on the Service Division's financial performance. Grady documents the following results.

◆ The Service Division's average inventory declined from $550,000 to $150,000.

◆ Projected annual insurance costs of $80,000 declined 60% owing to the lower average inventory.

◆ A leased 8,000-square-foot warehouse, previously used for materials storage, was not used at all during the year. The division paid $11,200 annual rent for the warehouse and was able to sublet 75% of the building to several tenants at $2.50 per square foot. The balance of the space remained idle.

◆ Two warehouse employees whose services were no longer needed were transferred on January 1, 19_4, to the purchasing department to assist in the coordination of the just-in-time program. The annual salary costs for these two employees totaled $38,000 and continued to be charged to the indirect manufacturing labor portion of fixed manufacturing costs.

◆ Despite the use of overtime to manufacture 7,500 spare parts, lost sales due to stockouts totaled 3,800 spare parts. The overtime premium incurred amounted to $5.60 per part manufactured. The use of overtime to fill spare parts orders was immaterial prior to January 1, 19_4.

Before the decision to implement the just-in-time inventory program, AgriCorp's Service Division had completed its 19_4 budget. The division's budgeted income statement, without any adjustments for just-in-time inventory, is presented below. AgriCorp's required rate of return for investment in inventory is 15% per year.

AgriCorp Service Division Budgeted Income Statement
for the Year Ended December 31, 19_4
(in thousands)

Revenues (280,000 spare parts)		$6,160
Cost of goods sold		
Variable manufacturing costs	$2,660	
Fixed manufacturing costs	1,120	3,780
Gross margin		2,380
Marketing and distribution costs		
Variable marketing and distribution costs	700	
Fixed marketing and distribution costs	555	1,255
Operating income		$1,125

Required

1. Calculate the cash savings (loss) of AgriCorp's Service Division for 19_4 that resulted from the adoption of the just-in-time inventory program.

2. Identify and explain the factors, other than financial, that should be considered before a company implements a just-in-time program.

24-25 Choosing suppliers for just-in-time purchasing. Jeffrey Chang runs a print shop. Chang requires 100,000 boxes of printing paper each year. He wants his suppliers to deliver the boxes on a JIT basis in order quantities of 400 boxes. Savoy Corporation currently supplies the paper to Chang. Savoy charges $100 per box and has a superb reputation for quality and timely delivery. Chang reports the following revenue and cost information for a typical print job:

Revenues	$100,000
Costs of printing paper ($100 per box × 400 boxes)	40,000
Other direct materials (ink and so on)	2,000
Variable printing overhead	3,000
Fixed printing overhead	25,000
Variable marketing and distribution overhead	1,000
Fixed marketing and distribution overhead	12,000

Bond Corporation has approached Chang with a proposal to supply all 100,000 boxes to Chang at a price of $95 a box. The savings in purchase costs are substantial, and Chang is tempted to accept Bond's offer, but before doing so Chang decides to check on Bond's reputation for quality and timely delivery. The information Chang gathers is not all positive. Chang estimates that late deliveries from Bond would lead to his incurring overtime and subcontracting costs of $30,000 per job on 10 jobs during the coming year. Chang also recognizes that Bond's paper quality would not be uniformly high, and the ink would sometimes smudge after printing. Chang expects that smudging would occur on five jobs during the year. Chang would then have to buy paper in the open market at $110 per box and rerun the job. Chang does not expect both delivery problems and quality problems to occur on the same jobs. Chang requires a rate of return of 15% per year on investments in inventory.

Required

1. Calculate Chang's costs if he purchases paper from Savoy and if he purchases paper from Bond. Which supplier should Chang choose on the basis of only the financial numbers given in the problem?

2. What other factors should Chang consider before choosing a supplier?

24-26 Production batch size, EOQ. (CMA, adapted) Clyde Peterson, general manager for Adam Furniture Company, is upset because the company exhausted its finished goods inventory of Style 103—Modern Desk twice during the previous month. These stockouts led to customer complaints and disrupted the normal flow of operations.

"We should plan better," declared Peterson. "Our annual sales demand is 18,000 units for this model or an average of 75 desks per day based upon our 240-day work year. Unfortunately, the sales pattern is not uniform. Our daily demand on that model varies considerably. When we run out of units, we cannot convert immediately because we would disrupt the production of our other products and cause cost increases. The setup process for this model costs $600. Once we get the line up, we can produce 200 units per day. I would prefer to have several planned runs of a uniform quantity rather than short unplanned runs often required to meet unfilled customer orders."

The manager of the Cost Accounting Department has suggested that an EOQ model be adopted to determine optimal production runs and then a safety stock established to guard against stockouts. The cost data for the Modern Desk, which sells for $110.00, is readily available from the accounting records. The manufacturing costs follow.

Direct materials	$30.00
Direct manufacturing labor (1 direct manufacturing labor hour—DMLH × $14.00)	14.00
Variable manufacturing overhead (1 DMLH × $6.00)	6.00
Fixed manufacturing overhead (1 DMLH × $10.00)	10.00
Total manufacturing costs	$60.00

The Cost Accounting Department estimates that the company's carrying costs are 10.8% per year of the incremental cash manufacturing costs.

Required
1. Explain which costs the company would be attempting to balance if it adopted the EOQ model for its production runs.
2. Calculate Adam's optimal quantity for each production run of Style 103—Modern Desk.
3. Calculate the number of production runs of Modern Desks that Adam Furniture Company would schedule during the year on the basis of the optimal quantity calculated in requirement 2.

24-27 Just-in-time production, relevant benefits, relevant costs. Evans Corporation manufactures cordless telephones. Evans is planning to implement a just-in-time production system, which requires annual tooling costs of $150,000. Evans estimates that the following annual benefits would arise from JIT production.

a. Average inventory will decline by $700,000, from $900,000 to $200,000.
b. Insurance, space, materials handling, and setup costs, which currently total $200,000, would decline by 30%.
c. The emphasis on quality inherent in JIT systems would reduce rework costs by 20%. Evans currently incurs $350,000 on rework.
d. Better quality would enable Evans to raise the prices of its products by $3 per unit. Evans sells 30,000 units each year. Evans's required rate of return on inventory investment is 12% per year.

Required
1. Calculate the net benefit or cost to Evans Corporation from implementing a JIT production system.
2. What other nonfinancial and qualitative factors should Evans consider before deciding on whether it should implement a JIT system?

24-28 Just-in-time production, operating efficiency. The Mannheim Group is a major manufacturer of metal-cutting machines. It has plants in Frankfurt and Stuttgart. The managers of these two plants have different manufacturing philosophies.

Richard Stehle, the recently appointed manager of the Frankfurt plant, is a convert to just-in-time production. In September 19_4, he commenced a four-month phase-in period for JIT. By January 19_5, JIT had been fully implemented.

Frank Kohl, manager of the Stuttgart plant, has adopted a wait-and-see approach to JIT. He commented to Stehle: "In my time, I have forgotten more manufacturing acronyms than you have read about in your five-year career. In two years' time, JIT will join the manufacturing buzzword scrapheap." Kohl continues with his "well-honed" traditional approach to manufacturing at the Stuttgart plant.

Summary operating data for the two plants in 19_5 follow:

	Jan.–Mar.	Apr.–June	July–Sept.	Oct.–Dec.
Manufacturing lead time (days)				
Frankfurt	9.2	8.7	7.4	6.2
Stuttgart	8.3	8.2	8.4	8.1
Total setup time for machines				
Total production time				
Frankfurt	52.1%	49.6%	43.8%	39.2%
Stuttgart	47.6	48.1	46.7	47.5
Number of units requiring rework				
Total number of units started and completed				
Frankfurt	64.7%	59.6%	52.1%	35.6%
Stuttgart	53.8	56.2	51.6	52.7

Required

1. What are the key features of JIT production?
2. Compare the operating performance of the Frankfurt and Stuttgart plants in 19_5. Comment on any differences you observe.
3. Stehle is concerned about the level of detail on the job-cost records for the cutting machines manufactured at the Frankfurt plant during 19_6. What reasons might lead Stehle to simplify the job-cost records?

24-29 Inventory management, ethics. (CMA, adapted) Belco Manufacturing builds and distributes industrial storage racks. Belco's earnings increased sharply in 19_5, and earnings-based bonuses were paid to the management staff for the first time in several years.

Jim Kern, vice-president of finance, was pleased with Belco's 19_5 earnings and thought that the pressure from top management to show improved financial results would ease. However, Ellen North, Belco's president, told Kern that she saw every reason for earnings to continue growing, even to the point that the 19_6 bonuses would be double those of 19_5. As a result, Kern felt pressure to increase reported earnings in order to ensure increased bonuses.

Kern met with Bill Keller of Pristeel Inc., a primary vendor of Belco's manufacturing supplies and equipment. Kern and Keller have been close business contacts for many years. Kern asked Keller to invoice *all* of Belco's purchases (equipment and supplies) as equipment. The reason Kern gave for his request was that Belco's president had imposed stringent budget constraints on operating costs but not on capital expenditures. Kern planned to capitalize the purchase of supplies and defer the expense recognition for these items. This procedure would increase reported earnings, leading to increased bonuses. Keller agreed to do as Kern asked.

While analyzing the second-quarter financial statements, Gary Wood, Belco's controller, noticed a large decrease in supplies expense from one year ago. Wood reviewed the Supplies Expense account and noticed that only equipment and no supplies had been purchased from Pristeel, a major source for supplies. Wood, who reports to Kern, immediately brought this matter to Kern's attention.

Kern told Wood of President North's high expectations and of the arrangement made with Bill Keller of Pristeel. Wood told Kern that his action was an improper accounting treatment for the supplies purchased from Pristeel. Wood requested that he be allowed to correct the accounts and urged that the arrangement with Pristeel be discontinued. Kern refused the request and told Wood not to become involved in the Pristeel arrangement.

After thinking about the matter for a while, Wood arranged to meet with North, and he disclosed the arrangement Kern had made with Pristeel.

Required

1. Is Gary Wood, Belco's controller, correct in saying that the supplies purchased from Pristeel Inc. were accounted for improperly? Explain your answer.
2. Refer to the "Standards of Ethical Conduct for Management Accountants" described in Chapter 1 (p. 17). Explain why the use of the alternative accounting method to manipulate reported earnings is unethical.
3. Without prejudice to your answers to requirements 1 and 2, assume that Jim Kern's arrangement with Pristeel Inc. was in violation of the Standards of Ethical Conduct for Management Accountants. Discuss whether Wood's actions were appropriate or inappropriate.

25

Systems Choice:

Decentralization

and Transfer Pricing

- Organizational structure and decentralization
- Choices about responsibility centers
- Transfer pricing
- Illustration of transfer pricing
- Tax considerations
- Multinational transfer pricing
- Market-based transfer prices
- Cost-based transfer prices
- A general guideline for transfer-pricing situations

Transfer-pricing policy is of much interest to tax officials in different countries and to division managers of multinational companies. Westvaco considers tax as well as other factors—such as goal congruence, incentives, and autonomy—when determining its worldwide policy on how wood and paper-related products are priced for transactions between internal divisions based in different countries.

Which company has the better management control system: Ford Motor Company or Toyota Motor Company? Michelin or Pirelli? What role can accounting information play in management control systems? Should products exchanged between individual profit centers or investment centers of an organization be transferred at market price, full cost, or variable cost? Senior managers often ask such questions. This chapter and Chapter 26 cover important topics in this area of systems choice. This chapter discusses the benefits and costs of centralized and decentralized organizations and looks at the pricing of products or services transferred between subunits of the same organization. Chapter 26 examines performance evaluation and compensation.

Chapters 25 and 26 contain a blend of cost accounting (narrowly conceived), strategic management, economics, and organizational behavior. This material is "softer" than that in many other chapters—that is, it presents relatively few numbers. Nevertheless, the concepts in these chapters are important.

ORGANIZATIONAL STRUCTURE AND DECENTRALIZATION

As we discuss the issues of decentralization, we use the term subunit to refer to any part of an organization. In practice, a subunit may be a large division (for example, the Chevrolet Division of General Motors) or a small group (for example, the two-person advertising department of a local clothing boutique).

Top management makes decisions about decentralization that affect day-to-day operations at all levels of the organization. The essence of **decentralization** is the freedom for managers at lower levels (subunits) of the organization to make decisions. But why is this freedom given? Among the most important reasons is *information asymmetry*. **Information asymmetry** exists when one particular manager (for example, a subunit manager) has more and better information about factors that affect the performance of his or her job than do other managers (for example, top management or managers of other subunits). A division manager, for example, has more and better information about her customers, her competitors, and ways to decrease cost and improve quality of products produced in her division than does top management.

Decentralization is a matter of degree. Total decentralization means minimum constraints and maximum freedom for managers to make decisions at the lowest levels of an or-

ganization. Total centralization means maximum constraints and minimum freedom for managers at the lowest levels. In most organizations, we find a degree of centralization and a degree of decentralization.

Benefits of Decentralization

Objective 1

Describe the benefits and costs of decentralization

How should top managers decide how much decentralization is optimal? Conceptually, they try to choose the degree of decentralization that maximizes the excess of benefits over costs. From a practical standpoint, top managers can seldom quantify either the benefits or the costs. Still, the cost-benefit approach helps them focus on the central issues.

Advocates of decentralizing decision making and granting responsibilities to managers of subunits claim the following benefits:

1. *Creates greater responsiveness to local needs.* Compared with top managers, subunit managers are better informed about their customers, competitors, suppliers, and employees. Subunit managers who have decision-making authority can use their information to be responsive to the immediate demands of their customers, suppliers, and employees. Eastman Kodak reports that one advantage of decentralization is an "increase in the company's knowledge of the marketplace and improved service to customers." Interlake, a manufacturer of materials-handling equipment, notes this important benefit of increased decentralization: "We have distributed decision-making powers more broadly to the cutting edge of product and market opportunity."

2. *Leads to quicker decision making.* An organization that gives lower-level managers the responsibility for making decisions can make decisions quickly, creating a competitive advantage over organizations that drag their heels by sending the decision-making responsibility upward through layer after management layer. Lower-level managers have the flexibility to act quickly and decisively, a benefit we call *flexibility gain.* Managers at Minnesota Manufacturing and Mining Co. (3M) believe that quicker decision making is paramount in today's competitive environments.

3. *Increases motivation.* Subunit managers are usually more highly motivated when they can exercise greater individual initiative. Johnson & Johnson, a highly decentralized company, argues that "Decentralization = Creativity = Productivity."

4. *Aids management development and learning.* Giving managers more responsibility promotes the development of an experienced pool of management talent—a pool that the company can draw from to fill higher-level management positions. The company also learns which people are not management material. Tektronix, an instruments company, expressed this benefit as follows: "Decentralized units provide a training ground for general managers, and a visible field of combat where product champions may fight for their ideas."

5. *Sharpens the focus of managers.* In a decentralized setting, the manager of a small subunit has a concentrated focus. A small subunit is more flexible and nimble than a larger subunit and better able to adapt itself quickly to a fast-opening market opportunity. Also, top management, relieved of the burden of day-to-day operating decisions, can focus more time and energy on strategic planning for the entire organization.

Costs of Decentralization

Advocates of more centralized decision making point out the following costs of decentralizing decision making:

1. *Leads to **suboptimal** (also called **dysfunctional) decision making,** which arises when a decision's benefit to one subunit is more than offset by the costs or loss of benefits to the organization as a whole.* This cost arises because top management has given up some control over decision making, a cost we call *control loss.*

   Suboptimal decision making may occur (a) when there is lack of harmony or congruence among the overall organization goals, the subunit goals, and the individual goals of decision makers or (b) when no guidance is given to subunit managers concerning the effects of their decisions on other parts of the organization.

Suboptimal decision making is most likely to occur when the subunits in the organization are highly interdependent—that is, when decisions affecting one subunit influence the decisions and performance of another subunit. Examples of interdependencies are:

◆ Subunits competing with each other for the same input factors (such as direct materials) or for the same customers
◆ Subunits that are vertically integrated so that the end product of one subunit is the direct material of another subunit

2. *Allows duplication of activities.* Several individual subunits of the organization may undertake the same activity separately. For example, there may be a duplication of staff functions (such as accounting, employee relations, and legal) if an organization is highly decentralized.

3. *Decreases loyalty toward the organization as a whole.* Individual subunit managers may regard the managers of other subunits in the same organization as external parties. Consequently, managers may be unwilling to share significant information or to assist when another subunit faces an emergency situation.

4. *Increases costs of gathering information.* Managers may spend too much time negotiating the prices for internal products or services transferred among subunits.

Sourcing Restrictions

Some organizations restrict what products or services their subunits can purchase from external suppliers when internal subunits can provide the needed products or services. For example, John Deere requires that when excess capacity is available at one particular subunit, other subunits should give it business, provided that internal incremental costs are less than the outside bids. Likewise, some organizations restrict the products or services their subunits sell to external customers when internal subunits demand those particular products or services. Such restrictions mean that some degree of centralization exists in the organization. Various rationales are given for restricting sources:

1. The attempt to formulate long-run production plans; subunits can make capital-budgeting and operating decisions with greater predictability.
2. The attempt to improve long-run supply reliability.
3. The protection of an "infant" subunit. A subunit may be in its infancy, and the parent wants to protect it from "market forces" in its developmental stage by providing a guaranteed market for some of its outputs.
4. The belief that products or services produced internally are of higher quality and will be delivered quickly and on time.

Comparison of Benefits and Costs

Top managers must compare the benefits and costs of decentralization, often on a function-by-function basis. For example, the controller's function may be highly decentralized for many attention-directing and problem-solving purposes (such as preparing operating budgets and performance reports) but highly centralized for other purposes (such as processing accounts receivables and developing income tax strategies). Organizations are rarely totally centralized or totally decentralized.

Surveys of U.S. and European firms report that the decisions made most frequently at the decentralized level and least frequently at the corporate level are related to sources of supply, products to manufacture, and product advertising. Decisions related to the type and source of long-term financing are made least frequently at the decentralized level and most frequently at the corporate level.[1] Decentralized

[1] *Evaluating the Performance of International Operations* (New York: Business International, 1989), p. 4; and *Managing the Global Finance Function* (London: Business International, 1992), p. 31.

companies are generally large and unregulated, face great uncertainties in their environments, require detailed local knowledge for performing various jobs, and have few interdependencies among divisions.[2]

CHOICES ABOUT RESPONSIBILITY CENTERS

Objective 2
Recognize that the same type of responsibility center can be found in both centralized and decentralized organizations

An organization, whether centralized or decentralized, can measure the performance of its subunits by using one of the four types of responsibility centers presented in Chapter 6:

◆ Cost center—manager accountable for costs only
◆ Revenue center—manager accountable for revenues only
◆ Profit center—manager accountable for revenues and costs
◆ Investment center—manager accountable for investments, revenues, and costs

Centralization or decentralization is not mentioned in these descriptions. Why? Because each of these responsibility units can be found in the extremes of centralized and decentralized organizations. It is not the case that certain types of responsibility centers are found only in centralized organizations and that other types of responsibility centers are found only in decentralized organizations.

A common misconception is that the term *profit center* (and in some cases *investment center*) is a synonym for a decentralized subunit and that *cost center* is a synonym for a centralized subunit. *Profit centers can be coupled with a highly centralized organization, and cost centers can be coupled with a highly decentralized organization.* For example, managers in a division organized as a profit center may have little leeway in making decisions. They may need to obtain approval from corporate headquarters for every expenditure over, say, $10,000 and may be forced to accept central-staff "advice." In another company, divisions may be organized as cost centers, but their managers may have great latitude on capital expenditures and on where to purchase materials and services. *In short, the labels "profit center" and "cost center" can be independent of the degree of decentralization in an organization.*

Whatever the type of responsibility center it uses, a company may divide itself into reporting units—that is, units that report back to headquarters—in several possible ways. Four frequently encountered alternatives are geographic, organizational or functional, product-line, and customer.

A geographic reporting unit is a country, a state, a province, a city, or the like. An organizational reporting unit is an R&D division, a manufacturing division, a marketing department, or the like. A product-line reporting unit is the breakfast cereals product line, the candy product line, or the like. A customer reporting unit is the direct sales customer group, the retail outlet customer group, or the like. Given four types of responsibility centers and four types of reporting units for each type of responsibility center, 16 possible reporting combinations exist.

TRANSFER PRICING

Consider News Ltd., which has a Paper Division and a Printing Division. Each division is a profit center. The Paper Division produces rolls of paper that are either transferred to the Printing Division of News Ltd. or sold to other companies that print books, magazines, and newspapers. The Printing Division of News Ltd. prints magazines that are sold in supermarkets and newsstands and through subscriptions. The paper it uses comes from the Paper Division or from an outside source.

---

[2] See A. Christie, M. Joye, and R. Watts, "Decentralization of the Firm: Theory and Evidence." Working Paper, University of Rochester, April 1991.

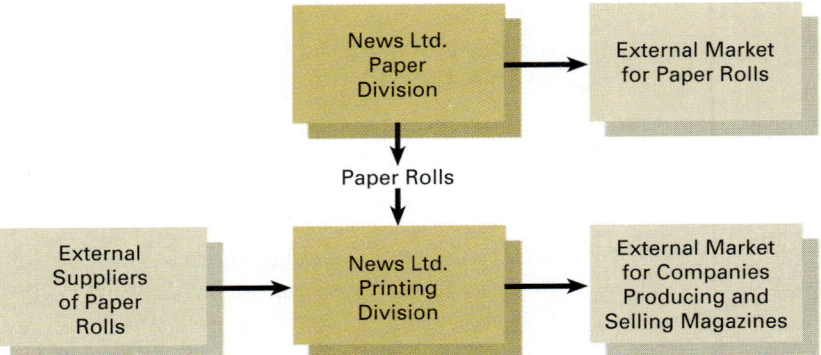

An **intermediate product** is a product transferred from one subunit to another subunit of the same organization. This product is further processed and sold to an external customer. A **transfer price** is the price one subunit (segment, department, division, and so on) of an organization charges for a product or service supplied to another subunit of the same organization. Paper rolls are the intermediate product in our News Ltd. example. The price at which paper rolls are transferred from the Paper Division to the Printing Division is termed the *transfer price.*

The transfer price creates revenue for the Paper Division and a cost for the Printing Division, yielding operating income numbers for both divisions. The operating incomes can be used to evaluate the performance of each division and to motivate managers. Division operating incomes also have tax consequences and, in the case of multinational corporations, restrict the amount of dividends a foreign division or subsidiary can transfer to its parent corporation.

News Ltd. must make decisions related to both the sourcing and the pricing of paper rolls. Sourcing decisions raise issues such as, Should the Paper Division be able to sell paper rolls to external customers when the paper requirements of the Printing Division have not been met? Should the Printing Division be able to buy paper rolls from external suppliers when the output of the Paper Division is not yet fully sold? Transfer-pricing decisions focus on questions such as, What price(s) will be set for internal transfers of paper rolls from the Paper Division to the Printing Division? Subsequent sections of this chapter discuss issues that News Ltd. and other companies should consider in their sourcing and pricing decisions.

Ideally, the chosen transfer-pricing method should lead each subunit manager to make optimal decisions for the organization as a whole. Two specific criteria can help in choosing a transfer-pricing method:

Objective 3

Specify three criteria that can help in choosing a transfer-pricing method

1. *Promotion of goal congruence.* Goal congruence exists when each subunit manager acting in his or her own best interest takes actions that automatically result in achieving the organization goals established by top management.

2. *Promotion of a sustained high level of management effort.* Effort is defined as exertion toward a goal—for example, sellers are motivated to hold down costs of supplying a product or service, and buyers are motivated to acquire and use inputs efficiently.

If top management favors a high degree of decentralization, a third criterion is also appropriate:

3. *Promotion of a high level of subunit autonomy* in decision making. **Autonomy** is the degree of freedom to make decisions.

These criteria are also explored in Chapter 13 (pp. 468–70).

Alternative Transfer-Pricing Methods

Objective 4

Identify three general methods for determining transfer prices

There are three general methods for determining transfer prices:

1. *Market-based transfer prices.* A company may choose to use the price of a similar product or service publicly listed in, say, a trade journal. Also, a company may select for its internal price the external price that a subunit charges to outside customers.

2. *Cost-based transfer prices.* Examples include variable manufacturing costs, full manufacturing (absorption) costs, and full product costs. "Full product costs" include all production costs as well as costs from other business functions (R&D, design, marketing, distribution, and customer service). The costs can be actual costs or budgeted costs.

3. *Negotiated transfer prices.* In some cases, the subunits of a company are free to negotiate the transfer price between themselves. Subunits may use information about costs and market prices in these negotiations, but there is no requirement that the chosen transfer price have any specific relationship to either cost or market-price data. Negotiated transfer prices are often employed when market prices are volatile and change constantly. The negotiated transfer price is the outcome of a bargaining process between the "selling" and the "buying" divisions.

The next section illustrates the use of market-based and cost-based transfer-pricing methods in a petroleum company that has production, transportation, and refining divisions. This example shows how the choice of a transfer-pricing method can dramatically affect the operating income of individual divisions. A description of tax implications follows. Subsequent sections of the chapter then discuss how the criteria of goal congruence, management effort, and subunit autonomy affect the choice of transfer-pricing methods within organizations.

ILLUSTRATION OF TRANSFER PRICING

Objective 5
Understand how a transfer-pricing method can affect the operating income of individual subunits

Horizon Petroleum has three divisions. Each operates as a profit center. The Production Division manages the production of crude oil from a petroleum field near Tulsa, Oklahoma. The Transportation Division manages the operation of a pipeline that transports crude oil from the Tulsa area to Houston, Texas. The Refining Division manages a refinery at Houston that processes crude oil into gasoline. (For simplicity, assume that gasoline is the only saleable product the refinery makes and that it takes two barrels of crude oil to yield one barrel of gasoline.)

The data in Exhibit 25-1 summarize the variable and fixed costs. Variable costs in each division are assumed to be variable with respect to a single cost driver: barrels of crude oil produced by the Production Division, barrels of crude oil transported by the Transportation Division, and barrels of gasoline produced by the Refining Di-

EXHIBIT 25-1
Operating Data for Horizon Petroleum

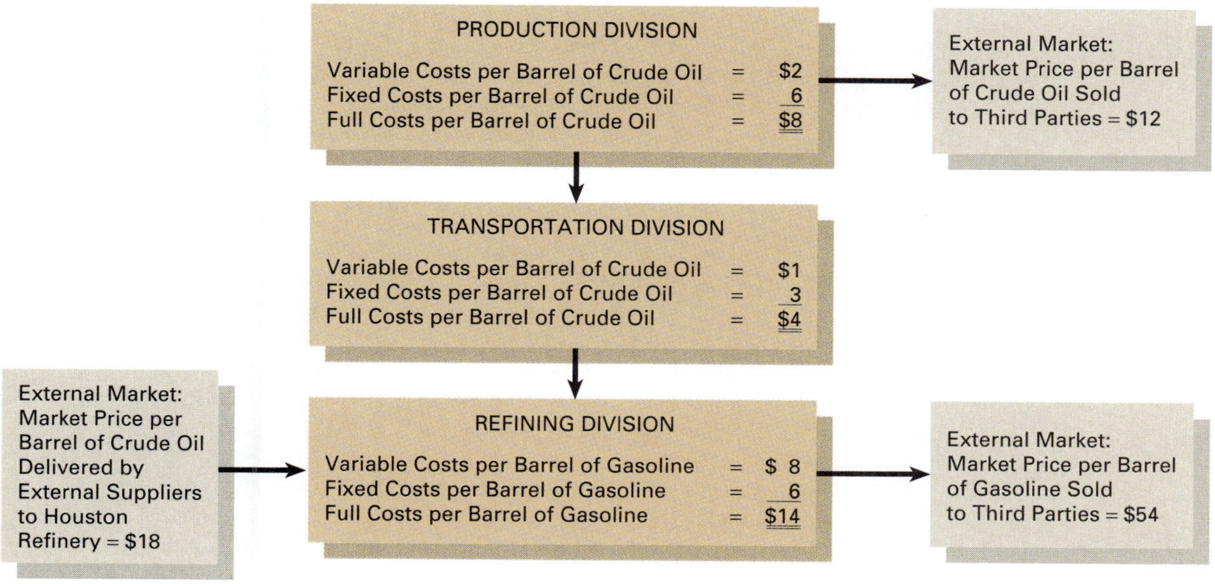

vision. The fixed costs per unit are based on the budgeted annual output of crude oil to be produced and transported and the amount of gasoline to be produced.

- The Production Division can sell crude oil to other parties in the Tulsa area at $12 per barrel.
- The Transportation Division "buys" crude oil from the Production Division, transports it to Houston, and then "sells" it to the Refining Division. The pipeline from Tulsa to Houston can carry 40,000 barrels of crude oil per day.
- The Refining Division has been operating *at capacity*, 30,000 barrels of crude oil a day, using oil from Horizon's Production Division (an average of 10,000 barrels per day) and oil bought from other producers and delivered to the Houston Refinery (an average of 20,000 barrels per day, at $18 per barrel).

Exhibit 25-2 presents division operating income resulting from three transfer-pricing methods applied to a series of transactions involving 100 barrels of crude oil produced by Horizon's Production Division.

EXHIBIT 25-2

Division Operating Income of Horizon Petroleum for 100 Barrels of Crude Oil under Alternative Transfer-Pricing Methods

	Method A Internal Transfers at 150% of Variable Costs	Method B Internal Transfers at 125% of Full Costs	Method C Internal Transfers at Market Price
1. PRODUCTION DIVISION			
Revenues:			
$3, $10, $12, × 100 barrels crude oil	$ 300	$1,000	$1,200
Deduct:			
Division variable costs:			
$2 × 100 barrels crude oil	200	200	200
Division fixed costs:			
$6 × 100 barrels crude oil	600	600	600
Division operating income	$ (500)	$ 200	$ 400
2. TRANSPORTATION DIVISION			
Revenues:			
$6, $17.50, $18, × 100 barrels crude oil	$ 600	$1,750	$1,800
Deduct:			
Transferred-in costs:			
$3, $10, $12, × 100 barrels crude oil	300	1,000	1,200
Division variable costs:			
$1 × 100 barrels crude oil	100	100	100
Division fixed costs:			
$3 × 100 barrels crude oil	300	300	300
Division operating income	$ (100)	$ 350	$ 200
3. REFINING DIVISION			
Revenues:			
$54 × 50 barrels gasoline	$2,700	$2,700	$2,700
Deduct:			
Transferred-in costs:			
$6, $17.50, $18, × 100 barrels crude oil	600	1,750	1,800
Division variable costs:			
$8 × 50 barrels gasoline	400	400	400
Division fixed costs:			
$6 × 50 barrels gasoline	300	300	300
Division operating income	$1,400	$ 250	$ 200

♦ Method A: 150% of variable costs, where variable costs are the cost of the transferred-in product plus the division's own variable costs

♦ Method B: 125% of full costs, where full costs are the cost of the transferred-in product plus the division's own variable and fixed costs

♦ Method C: Market price

The transfer prices per barrel of crude oil under each method follow. The transferred-in cost component in methods A and B is denoted by the symbol*.

♦ **Method A: 150% of Variable Costs**
Production Division to Transportation Division = 1.5($2) = $3
Transportation Division to Refining Division = 1.5($3* + $1) = $6

♦ **Method B: 125% of Full Costs**
Production Division to Transportation Division = 1.25($2 + $6) = $10
Transportation Division to Refining Division = 1.25($10* + $1 + $3) = $17.50

♦ **Method C: Market Price**
Production Division to Transportation Division = $12
Transportation Division to Refining Division = $18

The $12 and $18 amounts in method C are market prices per third-party transactions reported in industry newsletters.

The division operating income per 100 barrels of crude oil reported under each transfer-pricing method (see Exhibit 25-2) is:

Division	150% of Variable Costs	125% of Full Costs	Market Price
Production	$(500)	$200	$400
Transportation	(100)	350	200
Refining	1,400	250	200
Horizon Petroleum	$ 800	$800	$800

The total operating income to Horizon Petroleum from producing, transporting, and refining the 100 barrels of crude oil is $800, regardless of the set of internal transfer prices used. The division operating incomes, however, differ dramatically under the three methods, as Exhibit 25-2 shows. Note that the operating income amounts span a $900 range in the Production Division: $(500) to $400; a $450 range in the Transportation Division: $(100) to $350; and a $1,200 range in the Refining Division: $200 to $1,400. Note also that each division would choose a different method if its sole criterion were to maximize its own division operating income. Little wonder that division managers take considerable interest in the setting of transfer prices, especially those managers whose compensation or promotion directly depends on division operating income.

Exhibit 25-2 illustrates that the choice of a transfer-pricing method can affect how the companywide operating income pie is divided among individual divisions. Subsequent sections of this chapter illustrate that the choice of a transfer-pricing method also affects the size of the operating income pie itself.

TAX CONSIDERATIONS

Managers should always consider the tax implications of alternative transfer-pricing methods. Tax factors include not only income taxes but also payroll taxes, tariffs, sales taxes, value-added taxes, environment-related taxes, and other government levies on organizations. Full consideration of tax aspects of transfer-pricing decisions is beyond the scope of this book. Our aim here is to highlight tax factors as an important consideration in transfer-pricing decisions.

Consider the Horizon Petroleum data in Exhibit 25-2. Assume that the Production Division has to pay the state of Oklahoma income tax (at 5% of operating income) and that both the Transportation and Refining Divisions operate in Texas and do not pay state income tax. Horizon Petroleum would minimize its state income tax payments with the 150% of variable cost transfer-pricing method:

Transfer-Pricing Method (1)	Operating Income of Production Division for 100 Barrels of Crude Oil (2)	State Income Tax on 100 Barrels of Crude Oil (3) = 0.05 × (2)
A. 150% of variable costs	$(500)	$ 0
B. 125% of full costs	200	10
C. Market price	400	20

The state of Oklahoma is fully aware of Horizon Petroleum's incentives to minimize income tax. Typically, detailed tax rules constrain the options available to companies when selecting a transfer-pricing method (and other accounting methods). Some companies keep one set of accounting records for tax reporting and a second set for internal management reporting; many companies do not. Using the same transfer-pricing method for both tax reporting and management reporting can help support a company's claim that it is not "manipulating" its reported taxable income to avoid tax payments. International tax issues arise when a multinational corporation transfers goods between divisions located in different countries.

MULTINATIONAL TRANSFER PRICING

Many organizations have divisions located in different countries. Determining a set of transfer prices for exchanges between these divisions requires consideration of additional factors.

1. *International taxation.* Tax considerations, similar to those described in the previous section, apply in the case of international transfers. Countries have different tax rates, and they enforce their tax codes to different degrees. Also, different countries allow different deductions. Consider a transfer of motor vehicle parts by the Irish division of World Motors in Ireland to the Sentel division of World Motors in the United States. Taking advantage of tax and other incentives offered by the Republic of Ireland, the Irish division pays no tax on its income in Ireland, but Sentel pays state and federal income taxes at a 40% rate in the United States. World Motors has an incentive to use a high transfer price for transfers from Ireland to the United States to maximize the income reported in Ireland, which has no income tax. The U.S. tax authorities are fully aware of this incentive and will make efforts to maximize the U.S. taxes paid by the Sentel Division of World Motors.

Transfer of intangible property—such as patents, licenses, technology, and brand names—poses special problems. A pharmaceutical company, for example, could license its subsidiary in Puerto Rico (which is not subject to tax) to manufacture and sell drugs developed in the United States in exchange for a small royalty on sales. With this arrangement, most of the income would be earned in Puerto Rico and escape tax. U.S. tax authorities would contend that the transfer price for royalties set by the U.S. corporation, in exchange for the license, was not properly negotiated and would attempt to levy higher taxes on the U.S. company.

Section 482 of the Internal Revenue Code governs taxation of multinational transfer pricing. Section 482 requires that transfer prices for both tangible and intangible property between a company and its foreign division or subsidiary be set to equal the price that would be charged by an unrelated third party in an arm's-length transaction.

In January 1993, the IRS developed important new regulations, too detailed for our consideration, describing allowable methods for determining arm's-length prices. The regulations present methods for handling both tangible and intangible property. The key point: Transfer prices charged should be in line with prices charged, costs incurred, or taxable income earned by comparable companies. The IRS has applied similar criteria to determine arm's-length transfer prices in cases involving Eli Lilly and Co. (technology transfer to Puerto Rico) and Bausch and Lomb (manufacture and sale of contact lenses from Ireland).[3]

Implementing section 482 entails the difficult tasks of identifying comparable companies, obtaining relevant data, and adjusting for differences in accounting procedures and estimates. Disputes in applying these methods are expected to arise between taxpayers and the IRS because of difficulties in determining and agreeing on costs, comparable prices, and margins.

2. *Income or dividend payment restrictions.* Some countries restrict the payment of income or dividends to parties outside their national borders. By increasing the prices of goods or services transferred into divisions in these countries, companies can increase the funds paid out of these countries without "appearing to violate" income or dividend restrictions.

Additional factors that arise in multinational transfer pricing include tariffs, customs duties, and risks associated with movements in foreign-currency exchange rates.

MARKET-BASED TRANSFER PRICES

Perfectly Competitive Market Case

Objective 6

Illustrate how market-based transfer prices generally promote goal congruence in perfectly competitive markets

When the intermediate market is perfectly competitive,[4] interdependencies of subunits are minimal, and there are no additional costs or benefits to the corporation as a whole in using the market instead of transacting internally, transferring products or services at market prices generally leads to optimal decisions. In using market prices with perfectly competitive markets, a company can meet the criteria of goal congruence, management effort, and subunit autonomy (if desired).

Reconsider the Horizon Petroleum example in Exhibits 25-1 and 25-2. Assume that there is a perfectly competitive market for crude oil in the Tulsa area. The Production Division can sell and the Transportation Division can buy as much crude oil as each wants at $12 per barrel. Horizon would like its managers to buy or sell crude oil internally. Consider the decisions that Horizon's division managers would make if each had the option to sell or buy crude oil externally. If the transfer price between Horizon's Production Division and Transportation Division is set below $12, the manager of the Production Division will be motivated to sell all production to outside buyers at $12 per barrel. If the transfer price is set above $12, the manager of the Transportation Division will be motivated to purchase all its crude oil requirements from outside suppliers. A transfer price of $12 will motivate the Production Division and the Transportation Division to buy and sell internally.

Each division manager will exert effort to maximize his or her own division operating income. The Production Division will sell (either internally or externally) as much crude oil as it can profitably sell, and the Transportation Division will buy (either internally or externally) as much crude oil as it can profitably transport. The actions that maximize division operating income are also the actions that maximize Horizon Petroleum's corporate operating income.

---

[3]Business International Corporation, *International Transfer Pricing* (New York, 1991); A. King, "The IRS's New Neutron Bomb," *Management Accounting* (December 1992); and Coopers and Lybrand, *Tax Topics Advisory* (January 21, 1993).

[4]A **perfectly competitive market** *exists when there is a homogeneous product with equivalent buying and selling prices and no individual buyers or sellers can affect those prices by their own actions.*

Distress Prices

When an industry has excess capacity, market prices may drop sizably below their historical average. If the drop in price is expected to be temporary, these low market prices are sometimes called "distress prices." Deciding whether a current market price is a distress price is often difficult. The market prices of several agricultural commodities and minerals have stayed for many years at what observers initially believed were temporary distress levels.

What transfer-pricing method should be used for judging performance if distress prices prevail? Some companies use the distress prices, but others use long-run average prices, or "normal" market prices. In the short run, the manager of the supplier division should meet the distress price as long as it exceeds the incremental costs of supplying the product or service. In the long run, if the manager forecasts that prices will remain low, the manager must decide whether to dispose of some manufacturing facilities. If long-run average market price is used, forcing managers to buy internally at prices above current market prices will hurt the short-run performance of the buying division.

COST-BASED TRANSFER PRICES

Sometimes market prices are unavailable, inappropriate, or too costly to obtain for transfer pricing. The product may be specialized or unique, price lists may not be widely available, or the internal product may be different from products available externally in terms of quality and service. In these cases, many organizations use cost-based measures as transfer prices. This section explores examples of the uses and limitations of these measures.

Full-Cost Bases

In practice, many companies use transfer prices based on full costs. These prices, however, can lead to suboptimal decisions. Assume that Horizon Petroleum (see Exhibits 25-1 and 25-2) makes internal transfers at 125% of full cost. The Houston Refining Division purchases, on average, 20,000 barrels of crude oil per day from a local Houston supplier, who delivers the crude oil to the Refinery. Purchase and delivery cost $18 per barrel. To reduce crude oil costs, the Refining Division has located an independent producer in Tulsa who is willing to sell 20,000 barrels of crude oil per day at $13 per barrel, delivered to Horizon's pipeline in Tulsa. Given Horizon's organizational structure, the Transportation Division would purchase the 20,000 barrels of crude oil in Tulsa, ship it to Houston, and then sell it to Horizon's Refining Division. The pipeline has excess capacity and can ship the 20,000 barrels at its variable costs of $1 per barrel without affecting the shipment of crude oil from Horizon's own Production Division. Will Horizon Petroleum incur lower costs by purchasing crude oil from the independent producer in Tulsa or by purchasing crude oil from the Houston supplier? Will the Refining Division show lower crude oil purchasing costs by using oil from the Tulsa producer or by using its current Houston supplier?

The following analysis shows that Horizon Petroleum would prefer to purchase oil from the independent Tulsa producer.

◆ *Alternative 1:* Buy 20,000 barrels from Houston supplier at $18 per barrel.
  Total cost to Horizon Petroleum = 20,000 × $18 = $360,000

◆ *Alternative 2:* Buy 20,000 barrels in Tulsa at $13 per barrel and transport it to Houston at $1 per barrel variable cost.
  Total cost to Horizon Petroleum = 20,000 × ($13 + $1) = $280,000

There is a reduction in total costs to Horizon Petroleum of $80,000 by using the independent producer in Tulsa.

In turn, suppose the Transportation Division's transfer price to the Refining Division is 125% of full cost. The Refining Division will see its reported division costs increase if the crude oil is purchased from the independent producer in Tulsa:

$$\text{Transfer price} = 1.25 \times \left(\begin{array}{ccc} \text{Purchase price} & \text{Unit variable costs} & \text{Unit fixed costs} \\ \text{from Tulsa} & + \text{ of Transportation} & + \text{ of Transportation} \\ \text{producer} & \text{Division} & \text{Division} \end{array} \right)$$

$$= 1.25 \times (\$13 + \$1 + \$3) = 1.25 \times \$17 = \$21.25$$

- ◆ *Alternative 1*: Buy 20,000 barrels from Houston supplier at $18 per barrel.
  Total cost to Refining Division = 20,000 × $18 = $360,000
- ◆ *Alternative 2*: Buy 20,000 barrels from the Transportation Division of Horizon Petroleum that are purchased from the independent producer in Tulsa.
  Total cost to Refining Division = 20,000 × $21.25 = $425,000

As a profit center, the Refining Division can maximize its short-run division operating income by purchasing from the Houston supplier. If instead it chooses the alternative that maximizes the operating income of the company as a whole, the Refining Division will increase its own division operating costs by $425,000 − $360,000 = $65,000.

This situation illustrates the goal incongruence that is induced by a transfer price based on full cost. *The transfer-pricing method has led the Refining Division to regard the fixed cost (and the 25% markup) of the Transportation Division as a variable cost.* From the viewpoint of the company as a whole, the full-cost–based transfer price can lead to suboptimal decisions in the short run.

Evidence from Company Practice

Objective 7
Explain why many companies use full-cost–based transfer prices although they can lead to goal-congruence problems

Objective 8
Describe factors executives consider important when determining transfer prices

Survey evidence (see Surveys of Company Practice, p. 872) indicates that among firms that use cost-based transfer prices, full-cost transfer prices are by far the most common. But full-cost transfer prices are clearly inadequate and lead to goal incongruence for decisions that require knowledge of short-run variable costs. Why, then, do so many companies use full-cost transfer pricing? Companies often prefer to use one transfer price to serve several purposes simultaneously. One explanation for companies' using full-cost transfer pricing is that it yields relevant costs for long-run decisions even though full-cost allocations may lead to poor short-run decisions.

The surveys indicate that management uses transfer prices for performance evaluation, for pricing decisions, to take advantage of differences in tax rates and customs duties, and to transfer income and dividends from foreign countries. A full-cost transfer price, for example, will be generally preferred when transferring products from a division operating in a low-tax area to a division operating in a high-tax area. In other cases, tax authorities may insist that products be transferred at full cost.

Chapter 12 indicated that managers prefer to use full-product costs as their cost base for pricing decisions. (See Surveys of Company Practice, p. 447). The advantages managers cite include fixed-cost recovery and price stability (there is a reduced temptation to engage in long-run price cutting). To facilitate full-product cost pricing, managers transfer products from one division to another using full-cost transfer prices.

Using full-cost transfer prices that include an allocation of fixed overhead costs raises other issues. How are indirect costs of manufacturing and marketing allocated to products? Have the correct activities and cost drivers been identified? Full-cost–based transfer prices calculated using activity-based cost drivers can provide more refined allocation bases for identifying costs with products.

Surveys of Company Practice

DOMESTIC AND MULTINATIONAL TRANSFER-PRICING PRACTICES

What transfer-pricing practices do companies around the world use? The percentages in the table below indicate how frequently particular transfer-pricing methods are used in different countries.

A. Domestic Transfer-Pricing Methods

Method	United States[a]	Australia[b]	Canada[c]	Japan[d]	India[e]	United Kingdom[f]
1. Market-price-based	37%	13%	34%	34%	47%	26%
2. Cost-based:						
Variable costs	4	—	6	2	6	10
Full costs (absorption costs)	41	—	37	44	47	38
Other	1	—	3	0	0	1
Total	46%	65%	46%	46%	53%	49%
3. Negotiated	16%	11%	18%	19%	0%	24%
4. Other	1%	11%	2%	1%	0%	1%
	100%	100%	100%	100%	100%	100%

B. Multinational Transfer-Pricing Methods

Method	United States[a]	Australia[b]	Canada[c]	Japan[d]	India[e]	United Kingdom[g]
1. Market-price-based	46%	—	37%	37%	—	31%
2. Cost-based:						
Variable costs	3	—	5	3	—	5
Full costs (absorption costs)	37	—	26	38	—	28
Other	1	—	2	0	—	5
Total	41%	—	33%	41%	—	38%
3. Negotiated	13%	—	26%	22%	—	20%
4. Other	0%	—	4%	0%	—	11%
	100%	—	100%	100%	—	100%

Note: Dashes indicate information was not disclosed in survey.

The surveys indicate that managers in all countries use cost-based transfer prices more frequently than market-price-based transfer prices for domestic transfer pricing. For multinational transfer pricing, managers use market-price-based and cost-based methods equally frequently.

What factors do executives consider important in decisions on domestic transfer pricing? Survey evidence indicates the following (in order of importance): (1) performance evaluation, (2) management motivation, (3) pricing and product emphasis, and (4) external market recognition.[h]

Factors cited as important in decisions on multinational transfer-pricing policy are (in order of importance): (1) overall income of the company, (2) income tax rate and other tax differences among countries, (3) income or dividend repatriation restrictions, and (4) competitive position of subsidiaries in their respective markets.[a]

[a]Adapted from Tang, "Transfer Pricing." [b]Joye and Blayney, "Cost and Management Accounting." [c]Tang, "Canadian Transfer." [d]Tang, Walter, and Raymond, "Transfer Pricing." [e]Govindarajan and Ramamurthy, "Transfer Pricing." [f]Drury, Braund, Osborne, and Tayles, "A Survey of Management Accounting." [g]Mostafa, Sharp, and Howard, "Transfer Pricing." [h]Price Waterhouse, *Transfer Pricing Practices.* Full citations are in Appendix A.

The following situation at Bellcore,[5] the research arm of AT&T, demonstates the negative effects of incorrect allocations in determining transfer prices: In the 1980s, Bellcore discovered that its researchers and engineers were spending large amounts of time typing documents and making overhead slides, or alternatively subcontracting this work to outside vendors rather than giving the work to Bellcore's own in-house staff. The reason: Prices charged by AT&T's in-house word-processing, graphics, technical publications, and secretarial services departments were way too high. As the internal demand for these services reduced, the costs of these services increased further, as the fixed costs of operating these support departments now had to be spread over fewer units of work. At one point, it cost $50 to type a single page!

The cause of the problem was that a disproportionate share of fixed company overhead was being allocated to these support departments. For example, rent, taxes, utilities, and general administration costs were allocated to departments based on the square footage that each department used. The different costs of the different *types* of spaces being used were not recognized. Typists, for example, use open space. The cost per square foot of open space is considerably less than the costs per square foot of laboratories and computer room space, which require more security, greater climate control, and multiple underground cables. Square footage is not an appropriate cost driver. It systematically overcosted typing department costs and undercosted laboratory and computer room costs.

Bellcore implemented a "fairer" way to allocate fixed costs. The changes reduced overhead costs allocated to word processing, graphics, technical publications, and secretarial services departments. As costs decreased and efficiency improved, the demand for these departments' services steadily increased.

Prorating the Overall Contribution

Consider again the example of Horizon Petroleum purchasing crude oil from the independent producer in Tulsa at $13 per barrel. The variable transportation costs from Tulsa to Houston are $1 per barrel. Each division acts autonomously. What transfer price will promote goal congruence and maximum management effort for both the Transportation Division and the Refining Division? Recall that the company as a whole benefits from purchasing the crude oil from the independent Tulsa producer. The minimum transfer price is $14 per barrel; a transfer price below $14 does not provide the Transportation Division with an incentive to purchase crude oil from the independent producer in Tulsa. The maximum transfer price is $18 per barrel; a transfer price above $18 will not provide an incentive for the Refining Division to purchase crude oil from the Transportation Division.

Companies sometimes impose a variable-cost transfer price and credit each division for a prorated share of the contribution to companywide operating income. In the case of Horizon Petroleum, the contribution to companywide operating income is $4, the difference between the variable-cost transfer price of $14 per barrel and the variable costs of $18 per barrel that Horizon Petroleum would have incurred if the Refining Division had purchased crude oil from the outside supplier. The proration of this contribution can be negotiated in many ways. Suppose the proration is made on the basis of budgeted variable costs incurred by the respective divisions to transport and refine a given quantity of crude oil. In our example, Horizon Petroleum would prorate the $4 difference ($18 − $14) between the Transportation Division and the Refining Division on the basis of data drawn from Exhibit 25-2 (p. 866):

---

[5] See E. Kovac and H. Troy, "Getting Transfer Prices Right: What Bellcore Did," *Harvard Business Review* (September–October 1989).

Variable costs of Transportation Division to transport 100 barrels of crude oil	$100
Variable costs of Refining Division to refine 100 barrels of crude oil	400
	$500

The $4-per-barrel overall contribution would be allocated as follows:

To Transportation Division $\frac{\$100}{\$500} \times \$4.00 = \$0.80$

To Refining Division $\frac{\$400}{\$500} \times \$4.00 = \$3.20$

The transfer price between the Transportation Division and the Refining Division would be $14.80 per barrel of crude oil ($13 purchase cost + $1 variable costs + $0.80 contribution allocated to Transportation Division). With this transfer price, both divisions will increase their own reported division operating income by purchasing crude oil from the independent producer in Tulsa. Essentially, this approach is a budgeted-variable-cost plus transfer-pricing system; the "plus" is a portion of the overall contribution to corporate operating income.

To decide on the $0.80 and $3.20 allocation of the $4.00 contribution to total corporate operating income per barrel, the divisions must share information about their variable costs. In effect, each division does not operate (at least for this transaction) in a totally decentralized manner. Because most organizations are hybrids of centralization and decentralization anyway, this approach deserves serious consideration when transfers are significant. Note, however, that each division has incentives to overstate its variable costs in order to capture a larger share of the total contribution.

Dual Pricing

There is seldom a *single* transfer price that simultaneously meets the criteria of goal congruence, management effort, and subunit autonomy. Some companies turn to **dual pricing**, using two separate transfer-pricing methods to price each interdivision transaction. An example of dual pricing arises when the selling division receives a full-cost plus markup-based price and the buying division pays the market price for the internally transferred products. Assume that Horizon Petroleum purchases crude oil from the independent producer in Tulsa at $13 per barrel. One way of recording the journal entry for the transfer between the Transportation Division and the Refining Division is:

1. Credit the Transportation Division (the selling division) with the 125%-of-full-cost transfer price of $21.25 per barrel of crude oil.
2. Debit the Refining Division (the buying division) with the market-based transfer price of $18 per barrel of crude oil.
3. Debit a corporate account for the $3.25 ($21.25 – $18.00) difference between the two transfer prices.

The dual pricing method promotes goal congruence because it makes the Refining Division no worse off if it purchases the crude oil from the Transportation Division rather than from the outside supplier. In either case, the Refining Division's cost is $18 per barrel of crude oil. This dual-pricing system essentially gives the Transportation Division a corporate subsidy. The results of dual pricing? The operating income for Horizon Petroleum as a whole is less than the sum of the operating incomes of the divisions.

Dual pricing is not widely used in practice even though it reduces the goal-congruence problems associated with a pure cost-plus–based transfer-pricing method. One concern of top management is that the manager of the supplying division does not have sufficient incentives to control costs with a dual-price system. A second concern is that the dual-price system does not provide clear signals to division managers about the level of decentralization top management seeks. Above all, dual-pricing tends to insulate managers from the frictions of the marketplace. Managers should know as much as possible about their subunits' buying and selling markets, and dual pricing reduces the incentives to gain this knowledge.

A GENERAL GUIDELINE FOR TRANSFER-PRICING SITUATIONS

Objective 9
Present a general guideline for determining a minimum transfer price in transfer-pricing situations

Is there an all-pervasive rule for transfer pricing that leads toward optimal decisions for the organization as a whole? No. Why? Because the three criteria of goal congruence, management effort, and subunit autonomy must all be considered simultaneously. The following general guideline, however, has proven to be a helpful *first step* in setting a minimum transfer price in many specific situations:

$$\begin{matrix} \text{Minimum} \\ \text{transfer price} \end{matrix} = \begin{pmatrix} \text{Additional } outlay \text{ costs} \\ \text{per unit incurred} \\ \text{to the point of transfer} \end{pmatrix} + \begin{pmatrix} Opportunity \text{ costs} \\ \text{per unit to the} \\ \text{supplying division} \end{pmatrix}$$

The term *outlay costs* in this context represents the cash outflows that are directly associated with the production and transfer of the products or services. *Opportunity costs* are defined here as the maximum contribution forgone by the supplying division if the products or services are transferred internally. We distinguish outlay costs from opportunity costs because the accounting system typically records outlay costs but not opportunity costs. Opportunity costs may be forgone contribution margins in some instances or even forgone net proceeds from the sale of facilities in other instances. Consider the application of the general guideline to some specific situations.

1. ***A competitive intermediate market exists, and the supplying division has no idle capacity.*** Consider the Production Division and the Transportation Division of Horizon Petroleum. If there is no idle capacity and if the market for crude oil is perfectly competitive, the Production Division can sell all the crude oil it produces to the external market at $12 per barrel. The Production Division's outlay costs (see Exhibit 25-1, p. 865) are $2 per barrel of crude oil. The Production Division's opportunity cost of transferring the oil internally is the contribution margin per barrel of $10 (market price, $12 – variable cost, $2) forgone by not selling the crude oil in the external market. In this case,

Minimum transfer price per barrel = Outlay costs per barrel + Opportunity costs per barrel
= $2 + $10 = $12 = Market price per barrel

2. ***A competitive intermediate market exists, and the supplying division has idle capacity.*** In this case, the Production Division's opportunity cost of transferring the oil internally is zero because the Division does not forgo any external sales and hence does not forgo any contribution margin from internal transfers.

Minimum transfer price per barrel = Outlay costs per barrel = $2 per barrel

Note that there is a range of transfer prices that can achieve the objective of having the Production Division sell its crude oil to the Tranportation Division. To calculate the range, first observe that the Production Division will not accept a transfer price of less than $2 per barrel (it will choose not to produce crude oil if the transfer price is below its own outlay costs). The Transportation Division will not be willing to pay more than $12 per barrel, the price at which it can buy crude oil in the external market. But any transfer price between $2 and $12 motivates the Production Division to produce and sell crude oil to the Transportation Division and the Transportation Division to buy crude oil from the Production Division. The exact price between $2 and $12 will depend on the bargaining strengths of the two divisions.

3. No market exists for the intermediate product. Here, the opportunity cost of supplying crude oil internally is zero because no contribution margin is forgone. If no market exists, there is no opportunity for the supplying division to sell outside. At the Production Division of Horizon Petroleum, the minimum transfer price under the general guideline would be the outlay costs per barrel ($2). At this transfer price, of course, the Production Division would never record positive operating income. One approach to overcoming this problem is to have the Transportation Division make a lump-sum payment to cover fixed costs and generate some operating income for the Production Division while the Production Division continues to make transfers at outlay costs.

4. No competitive intermediate market exists, and the supplying division has idle capacity. The opportunity cost of each potential internal transfer of products or services is difficult to measure using our general guideline. Is the opportunity cost zero? Probably not. This division might reduce the price to increase demand, hoping to increase overall operating income. But measuring the effect of the price cut is difficult to do, so measuring the opportunity costs is also difficult. Clearly, the transfer price depends on constantly changing levels of supply and demand. There is not just one transfer price; rather, a transfer-price function yields the transfer price for various quantities supplied and demanded, depending on the outlay costs and opportunity costs of the units transferred.

PROBLEM FOR SELF-STUDY

PROBLEM

Pillercat Corporation is a highly decentralized company. Each division head has full authority for sourcing decisions and sales decisions. The Tractor Division can purchase a key component—the crankshaft—from the Machining Division of Pillercat or from external suppliers:

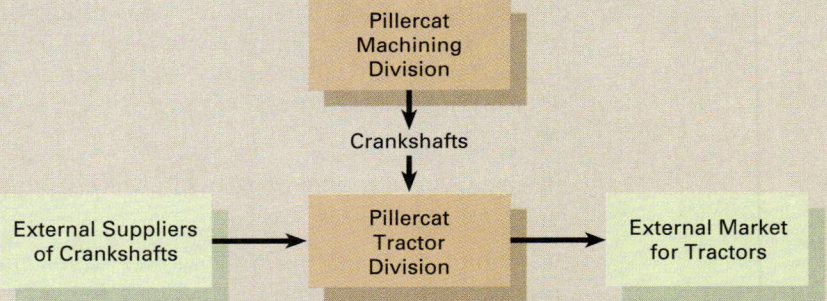

The Machining Division of Pillercat has been the major supplier of crankshafts to the Tractor Division in recent years. The Tractor Division, however, has just announced that it will purchase all its crankshafts in the forthcoming year from two external suppliers at $200 per crankshaft. The Machining Division of Pillercat recently increased its unit price for the forthcoming year to $220 (from $200 in the current year).

Juan Gomez, manager of the Machining Division, felt that the price increase was fully justified. The Machining Division recently purchased some specialized equipment to manufacture crankshafts. The resulting higher depreciation charge as well as an increase in labor costs led to the 10% price increase. Gomez met with the president of Pillercat Corporation and requested that the Tractor Division be directed to buy all its crankshafts from the Machining Division at the $220 price. Gomez supplied the following cost information for the Machining Division: variable costs per crankshaft, $190; fixed costs per crankshaft, $20.

The additional outlay costs per unit that Pillercat incurs to produce each crankshaft is the Machining Division's variable cost of $190. The Tractor Division purchases 2,000 crankshafts per month.

Required

1. Compute the advantage or disadvantage (in terms of monthly operating income) to Pillercat as a whole if the Tractor Division buys crankshafts internally from the Machining Division under each of the following three cases.

    a. Machining Division has no alternative use for the facilities used to manufacture crankshafts.

    b. Machining Division can use the facilities for other production operations, which will result in monthly cash-operating savings of $29,000.

    c. Machining Division has no alternative use for the facilities, and the external supplier drops its price to $185 per crankshaft.

2. As the president of Pillercat, how would you respond to Juan Gomez's request to order the Tractor Division to purchase all of its crankshafts from the Machining Division? Would your response differ according to the scenarios described in (a), (b), and (c) of requirement 1? Why?

SOLUTION

1.

	(a)	(b)	(c)
Total purchase costs if buying from an external supplier:			
2,000 x $200, $200, $185	$400,000	$400,000	$370,000
Total outlay costs if buying from the Machining Division: 2,000 × $190	380,000	380,000	380,000
Total opportunity costs of the Machining Division	—	29,000	—
Total relevant costs	380,000	409,000	380,000
Monthly operating income advantage (disadvantage) to Pillercat if buying from the Machining Division	$ 20,000	$ (9,000)	$(10,000)

2. Pillercat Corporation is a highly decentralized company. If no forced transfer were made, the Tractor Division would use an external supplier, resulting in an optimal decision for the company as a whole in parts (b) and (c) of requirement 1 but not in part (a).

    Suppose that in requirement 1(a), the Machining Division refuses to meet the price of $200. This decision means that the company will be $20,000 worse off in the short-run. Should top management interfere and force a transfer at $200? This interference would undercut the philosophy of decentralization. Many top managements would not interfere because they would view the $20,000 as the cost of potential suboptimal decisions that occasionally occur under decentralization. But how high must this cost go before the temptation to interfere would be irresistible? $30,000? $40,000?

    Any superstructure that interferes with lower-level decision making weakens decentralization. Of course, such interference may occasionally be necessary to prevent costly blunders. But recurring interference and constraints simply transform a decentralized organization into a centralized organization.

    The "general guideline" that was introduced in the chapter as a first step in setting a transfer price can be used to highlight the alternatives in requirement 1:

Case	Additional Outlay Costs per Unit Incurred to Point of Transfer	+	Opportunity Costs per Unit to the Supplying Division	=	Transfer Price	External Market Price
(a)	$190	+	$0	=	$190	$200
(b)	$190	+	$14.50 ($29,000 ÷ 2,000)	=	$204.50	$200
(c)	$190	+	$0	=	$190	$185

    The Tractor Division will maximize monthly operating income of Pillercat as a whole by purchasing from the Machining Division in case (a) and by purchasing from the external supplier in cases (b) and (c).

SUMMARY

The following points are linked to the chapter's learning objectives.

1. Decentralization's benefits include: (a) greater responsiveness to local needs, (b) flexibility gain from quicker decision making, (c) increased motivation, (d) greater management development and learning, and (e) sharper management focus. Decentralization's costs include (a) dysfunctional decision making (control loss), (b) duplication of activities, (c) decreased loyalty toward the organization, and (d) increased costs of information gathering.

2. Responsibility units such as profit centers can exist with or without a high degree of decentralization.

3. A transfer-pricing system should be judged in relation to its impact on (a) goal congruence, (b) management effort, and (c) subunit autonomy.

4. Transfer prices can be (a) market-based, (b) cost-based, or (c) negotiated.

5. Different transfer-pricing methods produce different revenues and costs for individual subunits, and hence different operating incomes.

6. In perfectly competitive markets with no idle capacity, division managers can buy and sell as much as they want to at the market price. Setting the transfer price at the market price motivates division managers to transact internally and to take exactly the same actions as they would if they were transacting in the external market.

7. Companies find full-cost transfer pricing useful for pricing decisions, to take advantage of differences in tax rates and customs duties, and to transfer income and dividends from foreign countries. Full-cost transfer prices are clearly inadequate and lead to goal incongruence for decisions that require knowledge of total variable costs incurred.

8. Executives regard performance evaluation, taxes, restrictions on transfer of dividends and income from foreign countries, and product pricing as among the most important considerations when determining transfer prices.

9. The general guideline for transfer pricing states that the minimum transfer price equals the outlay costs incurred to the point of transfer *plus* the opportunity costs per unit to the supplying division resulting from transferring products or services internally.

TERMS TO LEARN

This chapter and the Glossary at the end of the book contain definitions of the following important terms:

autonomy *(p. 864)*   decentralization *(860)*   dual pricing *(874)*
dysfunctional decision making *(861)*   information asymmetry *(860)*
intermediate product *(864)*   perfectly competitive market *(869)*
suboptimal decision making *(861)*   transfer price *(864)*

ASSIGNMENT MATERIAL

QUESTIONS

25-1   Name three benefits of decentralization.

25-2   Name two costs of decentralization.

25-3   "Organizations typically adopt a consistent decentralization or centralization philosophy across all their business functions." Do you agree? Explain.

25-4   Give an example of each of the following four reporting units: geographic, organizational or functional, product-line, and customer.

25-5   Distinguish between *sourcing* restrictions and *transfer-pricing* restrictions.

25-6   "Transfer pricing is confined to profit centers." Do you agree? Why?

25-7   Describe three criteria useful in choosing a transfer-pricing method.

25-8   What are the three general methods for determining transfer prices?

25-9   "All transfer pricing methods give the same division operating income." Do you agree? Explain.

25-10   Why should managers consider income tax issues when choosing a transfer-pricing method?

25-11   Under what conditions is a market-based transfer price optimal?

25-12   What is one major limitation of full-cost–based transfer prices?

25-13   In transfer pricing, what is a common conflict between a division and the company as a whole?

25-14   Give two reasons why a dual-pricing approach to transfer pricing is not widely used.

25-15   "Under the general transfer-pricing guideline, the minimum transfer price will vary depending on whether the supplying division has idle capacity or not." Do you agree? Explain.

EXERCISES AND PROBLEMS

25-16   Transfer-pricing methods, goal congruence.   British Columbia Lumber has a Raw Lumber Division and Finished Lumber Division. The variable costs are:

◆ Raw Lumber Division: $100 per 100 board-feet of raw lumber
◆ Finished Lumber Division: $125 per 100 board-feet of finished lumber

Assume that there is no board-feet loss in processing raw lumber into finished lumber. Raw lumber can be sold at $200 per 100 board-feet. Finished lumber can be sold at $275 per 100 board-feet.

Required
1. Should British Columbia Lumber process raw lumber into its finished form?
2. Assume that internal transfers are made at 110% of variable costs. Will each division maximize its division operating income contribution by adopting the action that is in the best interests of British Columbia Lumber?

3. Assume that internal transfers are made at market prices. Will each division maximize its division operating income contribution by adopting the action that is in the best interests of British Columbia Lumber?

25-17 Responding to a transfer-pricing question. The Plastics Company has a separate division that produces a special molding powder (MP). For the past three years, about two-thirds of the output (measured in pounds) has been sold to another division within the company. The remainder has been sold to outsiders. Last year's operating data follow:

	To Other Division		To Outsiders	
Sales	10,000 MP × $70* =	$700,000	5,000 MP × $100 =	$500,000
Variable costs: $50 per pound		500,000		250,000
Fixed costs		150,000		75,000
Total costs		650,000		325,000
Operating income		$ 50,000		$175,000

*The $70 price is ordinarily determined by the outside selling price minus marketing and administrative costs wholly applicable to outside business.

The manager of the buying division has a chance to get a firm contract with an outside supplier at $65 per pound for the coming year.

Required
Assume that the manager of the molding powder division asserts that no operating income can be earned if sales are made at $65 per pound. As the manager of the buying division, write a short reply. Assume that the molding powder division cannot sell the 10,000 pounds to other customers.

25-18 Transfer-pricing dispute. Allison-Chambers Corp., manufacturer of tractors and other heavy farm equipment, is organized along decentralized lines, with each manufacturing division operating as a separate profit center. Each division manager has been delegated full authority on all decisions involving the sale of that division's output both to outsiders and to other divisions of Allison-Chambers. Division C has in the past always purchased its requirement of a particular tractor-engine component from Division A. However, when informed that Division A was increasing its selling price to $150, Division C's manager decided to purchase the engine component from outside suppliers.

Division C can purchase the component for $135 on the open market. Division A insisted that, because of the recent installation of some highly specialized equipment and the resulting high depreciation charges, A would not be able to make an adequate return on its investment unless it raised its price. A's manager appealed to top management of Allison-Chambers for support in the dispute with C and supplied the following operating data:

C's annual purchases of tractor-engine component	1,000 units
A's variable costs per unit of tractor-engine component	$120
A's fixed costs per unit of tractor-engine component	$ 20

Required
1. Assume that there are no alternative uses for internal facilities. Determine whether the company as a whole will benefit if Division C purchases the component from outside suppliers for $135 per unit.
2. Assume that internal facilities of A would not otherwise be idle. By not producing the 1,000 units for C, A's equipment and other facilities would be used for other production operations that would result in annual cash-operating savings of $18,000. Should C purchase from outside suppliers?
3. Assume that there are no alternative uses for A's internal facilities and that the price from outsiders drops $20. Should C purchase from outside suppliers?

25-19 Transfer-pricing problem. Refer to Problem 25-18. Assume that Division A could sell the 1,000 units to other customers at $155 per unit with variable marketing costs of $5 per unit. If this were the case, determine whether Allison-Chambers would benefit if C purchased the 1,000 components from outside suppliers at $135 per unit.

25-20 Transfer pricing, goal congruence. (I.M.A., adapted) Nogo Motors, Inc., has several regional divisions that often purchase component parts from each other. The company is fully decentralized, each division buying and selling to other divisions or in outside markets. Each division makes its decision on where to buy and sell in conformity with divisional goals. Igo Division purchases most of its airbags from Letgo Division. The managers of these two divisions are currently negotiating a transfer price for the airbags for next year, when the airbag will be standard equipment on all Igo vehicles. Letgo Division prepared the following financial information for negotiating purposes:

Costs of airbag as manufactured by Letgo:	
Direct materials costs	$ 40
Direct manufacturing labor costs	55
Variable manufacturing overhead costs	10
Fixed manufacturing overhead costs	25
Variable marketing costs	5
Fixed marketing costs	15
Fixed administrative costs	10
Total costs	$160

Letgo Division is currently working at 80% of its capacity. Letgo's policy is to achieve an operating income of 20% of sales.

There has been a drop in price for airbags. The current market price is $130 per unit.

Required
Consider each of the requirements independently.

1. If Letgo Division desires to achieve its operating income goal of 20% of sales, what should be the transfer price?
2. Assume that Letgo Division wants to maximize its operating income. What transfer price would you recommend that the Letgo Division negotiate?
3. What is the transfer price that you believe Letgo Division should charge if overall company operating income is to be maximized?

25-21 Pertinent transfer price. Europa Inc., has two divisions, A and B, which manufacture expensive bicycles. Division A produces the bicycle frame, and Division B assembles the rest of the bicycle onto the frame. There is a market for both the subassembly and the final product. Each division has been designated as a profit center. The transfer price for the subassembly has been set at the long-run average market price.

The following data are available to each division:

Estimated selling price for final product	$300
Long-run average selling price for intermediate product	200
Outlay costs for completion in Division B	150
Outlay costs in Division A	120

The manager of Division B has made the following calculation:

Selling price—final product		$300
Transferred-in costs (market)	$200	
Outlay costs for completion	150	350
Contribution (loss) on product		$ (50)

Required
1. Should transfers be made to Division B if there is no excess capacity in Division A? Is the market price the correct transfer price?
2. Assume that Division A's maximum capacity for this product is 1,000 units per month and sales to the intermediate market are now 800 units. Should 200 units be transferred to Division B? At what transfer price? Assume that for a variety of reasons A will maintain the $200 selling-price indefinitely; that is, A is not considering cutting the price to outsiders even if idle capacity exists.

3. Suppose A quoted a transfer price of $150 for up to 200 units. What would be the contribution to the company as a whole if the transfer were made? As manager of B, would you be inclined to buy at $150?

25-22 Pricing in imperfect markets. Refer to Problem 25-21.

Required

1. Suppose the manager of Division A has the option of (a) cutting the external price to $195 with the certainty that sales will rise to 1,000 units or (b) maintaining the outside price of $200 for the 800 units and transferring the 200 units to Division B at some price that would produce the same operating income for Division A. What transfer price would produce the same operating income for Division A? Does that price coincide with that produced by the "general guideline" in the chapter, so that the desirable decision for the company as a whole would result?

2. Suppose that if the selling price for the intermediate product is dropped to $195, outside sales can be increased to 900 units. Division B wants to acquire as many as 200 units if the transfer price is acceptable. For simplicity assume that there is no outside market for the final 100 units of Division A capacity.

   a. Using the "general guideline," what is (are) the transfer price(s) that should lead to the correct economic decision? Ignore performance evaluation considerations.

   b. Compare the total contributions under the alternatives to show why the transfer price(s) recommended lead(s) to the optimal economic decision.

25-23 Effect of alternative transfer-pricing methods on division operating income. Oceanic Products is a tuna-fishing company based in San Diego. It has three divisions:

1. Tuna Harvesting: operates a fleet of 20 trawling vessels
2. Tuna Processing: processes the raw tuna into tuna fillets
3. Tuna Marketing: packages tuna fillets in two-pound packets that are sold to wholesale distributors at $12 each

The Tuna Processing Division has a yield of 500 pounds of processed tuna fillets from 1,000 pounds of raw tuna provided by the Tuna Harvesting Division. The Tuna Marketing Division has a yield of 300 two-pound packets from every 500 pounds of processed tuna fillets provided by the Tuna Processing Division. (The weight of the packaging material is included in the two-pound weight.) Cost data for each division follow:

Tuna Harvesting Division	
Variable costs per pound of raw tuna	$0.20
Fixed costs per pound of raw tuna	$0.40
Tuna Processing Division	
Variable costs per pound of processed tuna	$0.80
Fixed costs per pound of processed tuna	$0.60
Tuna Marketing Division	
Variable costs per 2-pound packet	$0.30
Fixed costs per 2-pound packet	$0.70

Fixed costs per unit are based on the estimated quantity of raw tuna, processed tuna, and two-pound packets to be produced during the current fishing season.

Oceanic Products has chosen to process internally all raw tuna brought in by the Tuna Harvesting Division. Other tuna processors in San Diego purchase raw tuna from boat operators at $1 per pound. Oceanic Products has also chosen to process internally all tuna fillets into the two-pound packets sold by the Tuna Marketing Division. Several fish-marketing companies in San Diego purchase tuna fillets at $5 per pound.

Required

1. Compute the overall operating income to Oceanic Products of harvesting 1,000 pounds of raw tuna, processing it into tuna fillets, and then selling it in two-pound packets.

2. Compute the transfer prices that will be used for internal transfers (i) from the Tuna Harvesting Division to the Tuna Processing Division and (ii) from the Tuna Processing Division to the Tuna Marketing Division under each of the following transfer-pricing methods:

   a. 200% of variable costs: Variable costs are the costs of the transferred-in product (if any) plus the division's own variable costs.

b. 150% of full costs: Full costs are the costs of the transferred-in product (if any) plus the division's own variable and fixed costs.

c. Market price.

3. Oceanic rewards each division manager with a bonus, calculated as 10% of division operating income (if positive). What is the amount of the bonus that will be paid to each division manager under each of the three transfer-pricing methods in requirement 2? Which transfer-pricing method will each division manager prefer to use?

25-24   Goal congruence problems with cost-plus transfer-pricing methods, dual-pricing methods. (Extension of Problem 25-23)   Assume that Oceanic Products uses a transfer price of 150% of full cost. Pat Forgione, the company president, attends a seminar on the virtues of decentralization. Forgione decides to implement decentralization at Oceanic Products. A memorandum is sent to all division managers: "Starting immediately, each division of Oceanic Products is free to make its own decisions regarding the purchase of its direct materials and the sale of its finished product."

Required

1. Give two examples of goal congruence problems that may arise if Oceanic continues to use the 150% of full costs transfer-pricing method and a policy of decentralization is adopted.

2. Forgione is investigating whether a dual transfer pricing policy will reduce goal congruence problems at Oceanic Products. Transfers out of each selling division will be made at 150% of full cost; transfers into each buying division will be made at market price. Using this dual transfer pricing policy, compute the operating income of each division for a harvest of 1,000 pounds of raw tuna that is further processed and marketed by Oceanic Products.

3. Compute the sum of the division operating incomes in requirement 2. Why might this sum not equal the overall corporate operating income from the harvesting of 1,000 pounds of raw tuna and its further processing and marketing?

4. What problems may arise if Oceanic Products uses the dual transfer pricing system described in requirement 2?

25-25   Multinational transfer pricing, global tax minimization.   Industrial Diamonds Inc., based in Los Angeles, has two divisions:

1. Philippine Mining Division—operates a mine containing a rich body of raw diamonds

2. U.S. Processing Division—processes the raw diamonds into polished diamonds used in industrial applications

The costs of the Philippine Mining Division are:

◆ Variable costs, 2,000 pesos per pound of raw industrial diamonds

◆ Fixed costs, 4,000 pesos per pound of raw industrial diamonds

Industrial Diamonds Inc. has a corporate policy of further processing in Los Angeles all raw diamonds mined in the Philippines. Several diamond-polishing companies in the Philippines buy raw diamonds from other local mining companies at 8,000 pesos per pound. Assume that the current foreign exchange rate is 20 pesos = $1 U.S.

The costs of the U.S. Processing Division are:

◆ Variable costs, $200 per pound of polished industrial diamonds

◆ Fixed costs, $600 per pound of polished industrial diamonds

Assume that it takes two pounds of raw industrial diamonds to yield one pound of polished industrial diamonds. Polished diamonds sell for $4,000 per pound.

Required

1. Compute the transfer price (in $U.S.) for one pound of raw industrial diamonds transferred from the Philippine Mining Division to the U.S. Processing Division under two methods: (a) 300% of full costs, (b) market price.

2. Assume a world of no income taxes. One thousand pounds of raw industrial diamonds are mined by the Philippine Division and then processed and sold by the U.S. Processing Division. Compute the operating income (in $U.S.) for each division of Industrial Diamonds Inc. under each transfer-pricing method in requirement 1.

3. Assume that the corporate income tax rate is 20% in the Philippines and 35% in the United States. Compute the after-tax operating income (in $U.S.) for each division under each transfer-pricing method in requirement 1. (Income taxes are not included in the computation of the cost-based transfer price. Industrial Diamonds does not pay U.S. taxes on income already taxed in the Philippines.)

4. Which transfer-pricing method in requirement 1 will maximize the total after-tax operating income of Industrial Diamonds?

5. What factors, in addition to global tax minimization, might Industrial Diamonds consider in choosing a transfer-pricing method for transfers between its two divisions?

25-26 Multinational transfer pricing and taxation. (Richard Lambert, adapted)   Anita Corporation, headquartered in the U.S., manufactures state-of-the-art milling machines. It has two marketing subsidiaries, one in Brazil and one in Switzerland, that sell its products. Anita is considering building one new machine, at a cost of $500,000. There is no market for the equipment in the United States. The equipment can be sold in Brazil for $1,000,000, but the Brazilian subsidiary would incur transportation and modification costs of $200,000. Alternatively, the equipment can be sold in Switzerland for $950,000, but the Swiss subsidiary would incur transportation and modification costs of $250,000. The U.S. company can sell the equipment either to its Brazilian subsidiary or to its Swiss subsidiary but not to both. Anita Corporation and its subsidiaries operate in a very decentralized manner. Managers in each company have considerable autonomy, with each division manager interested in maximizing his or her own division's income.

Required
1. From the viewpoint of Anita and its subsidiaries taken together, should Anita Corporation manufacture the equipment? If it does, where should it sell the equipment to maximize corporate operating income? What would the operating income for Anita and its subsidiaries be from the sale? Ignore any income tax effects.

2. What range of transfer prices will result in achieving the actions determined to be optimal in requirement 1? Explain your answer.

3. The effective income tax rates for this transaction follow: 40% in the United States, 60% in Brazil, and 15% in Switzerland. The tax authorities in the three countries are uncertain about the cost of the intermediate product and will allow any transfer price between $500,000 and $700,000. If Anita and its subsidiaries want to maximize after-tax operating income, (a) should the equipment be manufactured and (b) where and at what price should it be transferred?

4. Now suppose each manager acts autonomously to maximize his or her own subsidiary's after-tax operating income. The tax authorities will allow transfer prices only between $500,000 and $700,000. Which subsidiary will get the product and at what price? Is your answer the same as your answer in requirement 3? Explain why or why not.

25-27 Transfer price, goal congruence. (Nahum Melumad, adapted)   The Cheap Shot Company has three divisions (A, B, and C), organized as decentralized profit centers. Division A produces the basic chemical Aldon (in multiples of 1,000 pounds) and transfers it to Divisions B and C. Division B processes Aldon into the final chemical product Baxon, and Division C processes Aldon into the final chemical product Calmite. No material is lost during processing.
Division A's costs follow:

Fixed costs per pound	$ 0
Variable costs per pound of Aldon	$0.18

Division A has a capacity limit of 10,000 pounds; Divisions B and C have capacity limits of 4,000 and 6,000 pounds, respectively. Given the high cost of storing Aldon, Baxon, and Calmite, Cheap Shot's divisions produce no more than the quantities they plan to sell. Divisions B and C sell their final product in separate markets.

The total revenues minus processing costs (net revenues) for each division are summarized in the following tables. Observe that the net revenues change for each incremental 1,000 pounds of Aldon converted to Baxon or to Calmite.

DIVISION B

Pounds of Aldon Processed in B	Revenues – Processing Costs from Selling Baxon
1,000	$ 500
2,000	850
3,000	1,100
4,000	1,200

DIVISION C

Pounds of Aldon Processed in C	Revenues – Processing Costs from Selling Calmite
1,000	$ 600
2,000	1,200
3,000	1,800
4,000	2,100
5,000	2,250
6,000	2,350

Required

1. Suppose there is no external market for Aldon. What quantity of Aldon should Cheap Shot produce to maximize Cheap Shot's operating income? How should this quantity be allocated between the two processing divisions?

2. What range of transfer prices will motivate Divisions B and C to demand the quantities that maximize Cheap Shot's operating income as determined in requirement 1, as well as motivate Division A to produce the sum of those quantities?

25-28 Transfer prices, goal congruence, external markets, capacity constraints.   Refer to the information in Problem 25-27.

Required

1. Suppose that Division A has the option of selling as much Aldon as it wants in a perfectly competitive market for $0.33 per pound. To maximize Cheap Shot's operating income, how many pounds of Aldon should Division A transfer to Division B and to Division C, and how many pounds should it sell in the external market?

2. What range of transfer prices will result in Divisions A, B, and C taking the actions determined as optimal in requirement 1? Explain your answer.

3. Assume the same scenario as in requirement 1, except that Division A's capacity constraint limits production to only 4,000 pounds of Aldon. To maximize Cheap Shot's operating income, how many pounds of Aldon should Division A transfer to Division B and to Division C, and how many pounds should it sell in the external market?

4. What range of transfer prices will result in Divisions A, B, and C taking the actions determined as optimal in requirement 3? Explain your answer.

25-29 Transfer prices, goal congruence.   Seville Corporation has several operating divisions. Each division manager is compensated on the basis of his or her own division's operating income.

The Television (TV) Division manufactures and sells television sets. The TV Division's budgeted income statement for 19_5 is on page 886. The division anticipates sales of 20,000 units.

The TV Division currently purchases all its TV screens from an outside supplier. The Screen Division of Seville itself manufactures and sells 35,000 TV screens to outside customers at a price of $120 per screen. The Screen Division has the capacity to make 50,000 screens. The Screen Division has variable manufacturing costs of $50 per screen and variable marketing costs of $4 per screen. Its fixed manufacturing costs are $500,000.

The Television Division manager approaches the Screen Division manager with a proposal to purchase screens from the Screen Division. The Screen Division would incur variable manufacturing costs of $45 per screen to make the screens the Television Division requires. Further, the Screen Division would need to incur no variable marketing costs if it supplied the screens internally to the Television Division.

TELEVISION DIVISION
Budgeted Income Statement for 19_5

	Per Unit	Total (in thousands)
Sales revenue	$300	$6,000
Manufacturing costs		
TV screen	80	1,600
Other variable costs	70	1,400
Fixed costs	40	800
Total manufacturing costs	190	3,800
Gross margin	110	2,200
Marketing costs:		
Variable	15	300
Fixed marketing	20	400
Total marketing costs	35	700
Operating income before taxes	$ 75	$1,500

Required

1. The Television Division manager proposes to buy exactly 20,000 screens from the Screen Division at a total price of $1,200,000 ($60 × 20,000). If the Screen Division manager has the discretion to accept or reject this proposal, should he or she accept or reject it? (Recall how division managers are compensated.)

2. Suppose that the Television Division is willing to consider sourcing some of its screens from an outside supplier and some from the Screen Division. From the viewpoint of Seville Corporation as a whole, how many screens should be sold (1) to outsiders and (2) to TV Division?

3. For what range of transfer prices will the manager of the Screen Division agree that the screen manufacturing and selling plan identified in your answer to requirement 2 is best for his or her division?

4. For what range of transfer prices will the manager of the Television Division agree that buying screens from the Screen Division is better than buying from an outsider?

25-30 Transfer prices, goal congruence, taxes. (Continuation of 25-29) Refer to the information in Problem 25-29. Suppose the Screen Division is located in a state that imposes a 10% tax on income earned within its boundaries, while the Television Division is located in a state that imposes no tax on income earned within its boundaries.

Required

1. What transfer price would be chosen by Seville Corporation to minimize tax payments for the company as a whole? Assume that only transfer prices that are greater than or equal to unit manufacturing costs and less than or equal to the market price of "substantially similar" screens are acceptable to the taxing authorities.

2. Suppose Seville Corporation announces the transfer price computed in requirement 1 to price all transfers between the Screen and Television Divisions. Each division manager then acts autonomously to maximize his or her own division's operating income. Will division managers acting in a decentralized manner achieve the optimal production and sourcing plans identified in requirement 2 of Problem 25-29? Why or why not?

25-31 Paper company. (Copyright 1986 by the President and Fellows of Harvard College. Reproduced by permission.) "If I were to price these boxes any lower than $480 a thousand," said Mr. Brunner, manager of Birch Paper Company's Thompson Division, "I'd be countermanding my order of last month for our sales representatives to stop shaving their bids and to bid full-cost quotations. I've been trying for weeks to improve the quality of our business, and if I turn around now and accept this job at $430 or $450 or something less than $480 I'll be tearing down this program I've been working so hard to build up. The division can't very well show an operating income by putting in bids that don't even cover a fair share of overhead costs, let alone give us operating income."

Birch Paper Company was a medium-sized, partly integrated paper company, producing white and kraft papers and paperboard. A portion of its paperboard output was converted into corrugated boxes by the Thompson Division, which also printed and colored the outside surface of the boxes. Including Thompson, the company had four producing divisions and a timberland division, which supplied part of the company's pulp requirements.

For several years each division had been judged independently on the basis of its operating income and return on investment. Top management had been working to gain effective results from a policy of decentralizing responsibility and authority for all decisions except those relating to overall company policy. The company's top officials felt that in the past few years the concept of decentralization had been successfully applied and that the company's operating income and competitive position had definitely improved.

In early 19_7 the Northern Division designed a special display box for one of its papers in conjunction with the Thompson Division, which was equipped to make the box. Thompson's package design and development staff spent several months perfecting the design, production methods, and materials that were to be used. Because of the unusual color and shape these aspects were far from standard. According to an agreement between the two divisions, the Thompson Division was reimbursed by the Northern Division for the cost of its design and development work.

When the specifications were all prepared, the Northern Division asked for bids on the box from the Thompson Division and from two outside companies. Each division manager was normally free to buy from whichever supplier he wished, and, even on sales within the company, divisions were expected to meet the going market price if they wanted the business.

In early 19_7 the operating income margins of converters such as the Thompson Division were being squeezed. Thompson, as did many other similar converters, bought its board, liner, or paper, and its function was to print, cut, and shape it into boxes. Though it bought most of its materials from other Birch divisions, most of Thompson's sales were to outside customers. If Thompson got the business, it would probably buy the linerboard and corrugating medium from the Southern Division of Birch. The walls of a corrugated box consist of outside and inside sheets of linerboard sandwiching the fluted corrugating medium. About 70% of Thompson's out-of-pocket costs of $400 represented the cost of linerboard and corrugating medium. Though Southern had been running below capacity and had excess inventory, it quoted the market price, which had not noticeably weakened as a result of the oversupply. Its out-of-pocket costs on both liner and corrugating medium were about 60% of the selling price.

The Northern Division received bids on the boxes of $480 a thousand from the Thompson Division, $430 a thousand from West Paper Company, and $432 a thousand from Erie Papers, Ltd. Erie Papers offered to buy from Birch the outside linerboard with the special printing already on it but would supply its own inside liner and corrugating medium. The outside liner would be supplied by the Southern Division at a price equivalent of $90 per thousand boxes and would be printed for $30 a thousand by the Thompson Division. Of the $30, about $25 would be out-of-pocket costs.

Because this situation appeared a little unusual, William Kenton, manager of the Northern Division, discussed the wide discrepancy of bids with Birch's commercial vice-president. He told the vice-president: "We sell in a very competitive market, where higher costs cannot be passed on. How can we be expected to show a decent operating income and return on investment if we have to buy our supplies at more than 10% over the going market?"

Knowing that Mr. Brunner had on occasion in the past few months been unable to operate the Thompson Division at capacity, the vice-president found it odd that Mr. Brunner would add the full 20% overhead and operating income charge to his out-of-pocket costs. When asked about this, Mr. Brunner's answer was the statement that appears at the beginning of this problem. He went on to say that having done the development work on the box, and having received no operating income on that, he felt entitled to a good markup on the production of the box itself.

The vice-president explored further the cost structures of the various divisions. He remembered a comment that the controller had made at a meeting last week to the effect that costs that were variable for one division could be largely fixed for the company as a whole. He knew that in the absence of specific orders from top management, Mr. Kenton would accept the lowest bid, which was that of the West Paper Company for $430. However, it would be possible for top management to order the acceptance of another bid if the situation warranted such action. And though the quantity represented by the transactions in question was less than 5% of the quantity of any of the divisions involved, other transactions could conceivably raise similar problems later.

Required

1. In the controversy described, which alternative seems best for the company as a whole? Prepare an analysis of the cash flows under each alternative.

2. As the commercial vice-president, what action would you take? Explain.

25-32 Ethics, transfer pricing. The Winchester Division of Boston Industries makes two component parts, X23 and Y99. It supplies X23 to the Dearborn Division to be used in the manufacture of car engines and supplies Y99 to the Flint Division to be used in the manufacture of car transmissions. The Winchester Division is the only supplier of these specialized components. When transfers are made in-house, Boston Industries uses a full-cost transfer-pricing policy. X23 is transferred out at $40 per unit, and Y99 at $20 per unit. The Dearborn Division feels that the price for X23 is too high and has told Winchester that it is trying to locate an outside vendor to supply the part at a lower price.

Cliff Malone, Winchester's management accountant, calls Sam Fraser, his assistant, into his office. "We can't afford to lose the Dearborn Division business. Our fixed costs won't go away even if we stop supplying Dearborn, and this means that the costs of supplying Y99 to Flint will increase. Then they'll start wanting to buy from outside. We're seriously looking at possibly shutting down the entire division if we lose the Dearborn business. See if you can find a different method of allocating fixed costs that will decrease X23's transfer price by $5. I think Flint will be willing to pay a somewhat higher price for Y99."

Fraser is uncomfortable making any changes because he knows that any other allocation method would violate corporate guidelines on overhead cost allocation. Still, he strongly believes that changing the fixed cost allocations is in the best interests of Boston Industries. Fraser is confused about what he should do.

Required

Refer to the "Standards of Ethical Conduct for Management Accountants" described in Chapter 1 (p. 17). Evaluate whether Cliff Malone's suggestion to Fraser to change the fixed cost allocations is ethical. Would it be ethical for Fraser to revise the fixed cost allocations at his boss's urging? What steps should Fraser take to resolve this situation?

26

Systems Choice:

Performance Measurement

and Compensation

Financial and nonfinancial performance measures

Designing an accounting-based performance measure

Different performance measures

Alternative definitions of investment

Measurement alternatives for assets

Goal congruence and performance measures

Distinction between managers and organization units

Performance measures at the individual activity level

Performance measures at the total organization level

Environmental and ethical responsibilities

*D*epartment stores aim to maximize return on investment by increasing their income margin on each sale and by increasing their investment turnover. Hudson's Bay Company promotes merchandise from around the globe to increase return on investment at its flagship Toronto store.

Performance measures are a central component of a management control system. This chapter examines issues in designing performance measures for different levels of an organization and for managers at these different levels. We discuss both financial and nonfinancial performance measures.

Performance measurement of an organization's subunits should be a prerequisite for allocating resources within that organization. When a subunit undertakes new activities, it projects revenues, costs, and investments. Periodic comparisons of the actual revenues, costs, and investments with the budgeted amounts can help guide top management's decisions about future allocations.

Performance measurement of managers is used in decisions about their salaries, bonuses, future assignments, and career advancement. Moreover, the very act of measuring their performance can motivate managers to strive for the goals used in their evaluation.

FINANCIAL AND NONFINANCIAL PERFORMANCE MEASURES

Objective 1

Provide examples of financial and nonfinancial measures of performance

Chapter 13 noted how the information used in a management control system can be financial or nonfinancial and can be based on internal or external measurements. Many common performance measures such as return on investment (ROI) rely on internal financial and accounting information. Increasingly, companies are supplementing internal financial measures with measures based on external financial information (for example, stock prices), internal nonfinancial information (such as manufacturing lead time), and external nonfinancial information (such as customer satisfaction).

Some companies present financial and nonfinancial performance measures for various organization subunits in a single report called the *balanced scorecard* (see Chapter 23). Different companies stress various elements in the scorecard, but most scorecards include (1) profitability measures, (2) customer-satisfaction measures, (3) innovation measures, and (4) internal measures of efficiency, quality, and time.[1]

---

[1] See R. Kaplan and D. Norton, "The Balanced Scorecard—Measures That Drive Performance," *Harvard Business Review* (January–February 1992) and S. Hronec, *Vital Signs* (New York: American Management Association, 1993).

Conner Peripherals, a manufacturer of hard disk drives, identifies the following measures within each category:

◆ *Profitability measures:* Operating income and revenue growth.
◆ *Customer-satisfaction measures:* Market share, customer-response time, on-time performance, and product reliability.
◆ *Innovation measures:* Number of new patents, number of new product launches, and time-to-market.
◆ *Efficiency, quality, and time measures:* Direct materials efficiency variance, defects, yield, manufacturing lead time, head count, and inventory.

The balanced scorecard highlights trade-offs that the manager may have made. For example, it indicates if improvements in financial performance resulted from sacrificing investments in new products or on-time delivery. The specific nonfinancial measures chosen signal to employees the areas that top management views as critical to the company's success.

Some performance measures, such as the number of new patents developed, have a long-run time horizon. Other measures, such as direct materials efficiency variance and yield, have a short-run time horizon.

We now examine accounting-based measures used to evaluate performance over an intermediate to long-run time horizon. The discussion concentrates on performance measures at the total organization level, at the division level (which may have multiple individual facilities), and at the individual activity level. Chapters 7 and 8 discuss accounting performance measures with a short-run time horizon (such as variances in direct materials and direct labor).

DESIGNING AN ACCOUNTING-BASED PERFORMANCE MEASURE

Objective 2

Describe the steps in designing an accounting-based performance measure

Designing an accounting-based performance measure requires the following steps:

STEP 1: *CHOOSING A VARIABLE(S) THAT REPRESENTS TOP MANAGEMENT'S FINANCIAL GOAL(S).*   Does operating income, net income, return on investment, or revenues, for example, best measure a division's performance?

STEP 2: *CHOOSING DEFINITIONS OF THE ITEMS INCLUDED IN THE VARIABLE(S) IN STEP 1* (such as operating income, net income, and investment). For example, should investment be defined as total assets or total assets minus liabilities?

STEP 3: *CHOOSING MEASURES FOR THE ITEMS INCLUDED IN THE VARIABLE(S) IN STEP 1.*   For example, should assets be measured at historical cost, current cost, or present value?

STEP 4: *CHOOSING A TARGET AGAINST WHICH TO GAUGE PERFORMANCE.* For example, should all divisions have as a target the same required rate of return on investment?

STEP 5: *CHOOSING THE TIMING OF FEEDBACK.*   Should reports on the performance of a division be sent to corporate headquarters weekly, monthly, or quarterly?

We discuss the first three steps in subsequent sections of this chapter. The fourth step is discussed in Chapter 21, where we point out that different required rates of return should be used for different divisions. The fifth step, the timing of feedback, has been discussed in many places in this book. Timing depends largely on the specific level of management that is receiving the feedback and on the sophistication of the organization's information technology.

These five steps need not be taken sequentially. The issues considered in each step are interdependent, and a decision maker will often proceed through these steps several times before deciding on an accounting-based performance measure(s). The answers to the questions raised at each step depend on top management's beliefs

about how each alternative fulfills, in a cost-effective manner, the behavioral criteria of goal congruence, management effort, and subunit autonomy discussed in Chapter 25.

DIFFERENT PERFORMANCE MEASURES

Hospitality Inns owns and operates three motels, one each in San Francisco, Chicago, and New Orleans. Exhibit 26-1 summarizes data for each of the three motels and for corporate headquarters (located in San Francisco) for the most recent year (19_8). At present, Hospitality Inns does not allocate to the three separate motels the operating costs and assets of corporate headquarters or the total long-term debt of the company. Exhibit 26-1 indicates that the New Orleans motel generates the highest operating income, $480,000. The Chicago motel generates $300,000; the San Francisco motel, $240,000. But is this comparison appropriate? Is the New Orleans motel the most "successful"? Actually, the comparison of operating income ignores potential differences in the *size* of the investments in the different motels.

Two approaches incorporate the amount of investment into a performance measure: return on investment (ROI) and residual income.

Return on Investment

Objective 3
Understand the duPont method of profitability analysis

Return on investment (ROI) is income divided by investment.

$$\text{Return on investment} = \frac{\text{Income}}{\text{Investment}}$$

	San Francisco Motel (1)	Chicago Motel (2)	New Orleans Motel (3)	Corporate Headquarters (4)	Total (1) + (2) + (3) + (4)
Motel revenues (sales)	$1,200	$2,000	$3,000	—	$6,200
Motel variable costs	540	750	840	—	2,130
Motel fixed costs	420	950	1,680	—	3,050
Motel operating income	$ 240	$ 300	$ 480	—	1,020
Central corporate costs	—	—	—	$ 200	200
Interest on long-term debt	—	—	—	400	400
Income before income taxes	—	—	—	—	420
Income taxes	—	—	—	—	150
Net income	—	—	—	—	$ 270
Average Book Values for 19_8:					
Current assets	$ 400	$ 500	$ 600	$ 200	$1,700
Long-term assets	600	1,500	2,400	300	4,800
Total assets	$1,000	$2,000	$3,000	$ 500	$6,500
Current liabilities	$ 230	$ 320	$ 350	$ 100	$1,000
Long-term liabilities	—	—	—	4,000	4,000
Stockholders' equity	—	—	—	—	1,500
Total liabilities and stockholders' equity					$6,500

ROI is the most popular approach to incorporating the investment base into a performance measure. ROI appeals conceptually because it blends all the major ingredients of profitability (revenues, costs, and investment) into a single number. ROI can be compared with opportunities elsewhere, inside or outside the company. Like any single performance measure, however, ROI should be used cautiously and in conjunction with other performance measures.

ROI is often called the accounting rate of return or the accrual accounting rate of return. Companies vary in the way they define both the numerator and the denominator of the ROI. For example, some firms use operating income for the numerator. Other firms use net income.

ROI can often provide more insight into performance when it is divided into the following components:

$$\left(\begin{array}{c}\text{Investment}\\\text{turnover}\end{array}\right) \times \left(\begin{array}{c}\text{Income}\\\text{margin}\end{array}\right) = \text{Return on investment}$$

$$\frac{\text{Revenue}}{\text{Investment}} \times \frac{\text{Income}}{\text{Revenue}} = \frac{\text{Income}}{\text{Investment}}$$

This approach is widely known as the *duPont method of profitability analysis*. The components of the duPont method lead to the following generalization: ROI is increased by any action that:

1. Increases revenue
2. Decreases costs
3. Decreases investment

while holding the other two factors constant. Put another way, there are two basic ingredients in profit making: investment turnover and income margin. An improvement in either ingredient without changing the other increases return on investment.

Consider the ROI of each of the three Hospitality Inns motels in Exhibit 26-1. Our definitions here are operating income of each motel for the numerator and total assets of each motel for the denominator:

Motel	Operating Income	÷	Total Assets	=	ROI
San Francisco	$240	÷	$1,000	=	24%
Chicago	300	÷	2,000	=	15%
New Orleans	480	÷	3,000	=	16%

Using these ROI figures, the San Francisco motel appears to make the best use of its total assets.

Assume that the top management at Hospitality Inns adopts a 30% target ROI for the San Francisco motel. How can this return be attained? The duPont method illustrates the present situation and three alternatives (in thousands):

	$\dfrac{\text{Revenues}}{\text{Total Assets}}$	×	$\dfrac{\text{Operating Income}}{\text{Revenues}}$	=	$\dfrac{\text{Operating Income}}{\text{Total Assets}}$
Present Situation	$\dfrac{\$1,200}{\$1,000}$	×	$\dfrac{\$240}{\$1,200}$	=	$\dfrac{\$240}{\$1,000}$ or 24%
Alternatives					
A. Increase operating income by increasing revenues	$\dfrac{\$1,500}{\$1,000}$	×	$\dfrac{\$300}{\$1,500}$	=	$\dfrac{\$300}{\$1,000}$ or 30%
B. Increase operating income by decreasing costs	$\dfrac{\$1,200}{\$1,000}$	×	$\dfrac{\$300}{\$1,200}$	=	$\dfrac{\$300}{\$1,000}$ or 30%
C. Decrease total assets	$\dfrac{\$1,200}{\$800}$	×	$\dfrac{\$240}{\$1,200}$	=	$\dfrac{\$240}{\$800}$ or 30%

Alternative A can be achieved by increasing the selling price per unit (the room rate) or by increasing the number of units sold (rooms rented) or both. Alternative B demonstrates the cost-reduction approach by, say, reducing cleaning, maintenance, or front-office costs. Alternative C shows that reducing total assets (such as accounts receivable or supplies inventories) improves ROI.

ROI highlights the benefits that managers can obtain by reducing their investments in current or fixed assets. Some managers are conscious of the need to boost revenues or to control costs but give less attention to reducing their investment base. Investments in cash, accounts receivable, inventory, and fixed assets should be managed carefully at all levels of performance. This approach means investing idle cash, managing credit judiciously, determining proper inventory levels, and spending carefully on fixed assets.

Residual Income

Objective 4
Describe the motivation for using residual income measurement

Residual income is income minus an imputed interest charge for the investment base.

$$\text{Residual income} = \text{Income} - \text{Imputed interest charge for investment}$$

Imputed interest is not regularly recognized by usual accounting procedures. Imputed interest in the residual income calculation represents the interest forgone by Hospitality Inns as a result of tying up cash in various investments. Assume that Hospitality Inns defines residual income for each motel as motel operating income minus an imputed interest charge of 10% of the total assets of the motel:

Motel	Operating Income	−	Imputed Interest Charge	=	Residual Income
San Francisco	$240	−	$100 (10% of $1,000)	=	$140
Chicago	$300	−	$200 (10% of $2,000)	=	$100
New Orleans	$480	−	$300 (10% of $3,000)	=	$180

Given the 10% imputed interest charge, the New Orleans motel is performing best in terms of residual income.

The objective of maximizing residual income assumes that as long as the division earns a rate in excess of the imputed charge for investments, the division should expand. Some firms favor the residual-income approach because managers will concentrate on maximizing an absolute amount (dollars of residual income) rather than a percentage (return on investment).

The objective of maximizing ROI may induce managers of highly profitable divisions to reject projects that, from the viewpoint of the organization as a whole, should be accepted. Assume that Hospitality Inns' required rate of return on investment is 10%. Assume also that an expansion of the San Francisco motel will increase its operating income by $160,000 and increase its total assets by $800,000. The ROI for the expansion is 20% ($160,000 ÷ $800,000), which makes it attractive to Hospitality Inns. By making this expansion, however, the San Francisco manager will see the motel's ROI decrease :

$$\text{Preexpansion ROI} = \frac{\$240}{\$1,000} = 24\%$$

$$\text{Postexpansion ROI} = \frac{(\$240 + \$160)}{(\$1,000 + \$800)} = \frac{\$400}{\$1,800} = 22.2\%$$

The annual bonus paid to the San Francisco manager may decrease if ROI is a key factor in the bonus calculation and the expansion option is selected. In contrast, if the annual bonus is a function of residual income, the San Francisco manager will view the expansion favorably:

$$\text{Preexpansion residual income} = \$240 - (10\% \times \$1,000) = \$140$$
$$\text{Postexpansion residual income} = \$400 - (10\% \times \$1,800) = \$220$$

Goal congruence is more likely to be promoted by using residual income rather than ROI as a division manager's performance measure.

Both ROI and residual income represent the results for a single time period (such as a year). Managers can take actions that cause short-run increases in ROI or residual income but are in conflict with the long-run interests of the organization. For example, managers may curtail research and development and plant maintenance in the last three months of a fiscal year to achieve a target level of annual operating income.

Return on Sales

The income-to-revenue (sales) ratio—often called return on sales (ROS)—is a frequently used key performance measure. ROS is one component of ROI in the duPont method of profitability analysis. The ROS of each motel in the Hospitality Inns chain is:

Motel	Operating Income	÷	Revenue (Sales)	=	ROS
San Francisco	$240	÷	$1,200	=	20%
Chicago	$300	÷	$2,000	=	15%
New Orleans	$480	÷	$3,000	=	16%

The San Francisco motel has a higher ROS than either the Chicago or the New Orleans motel.

The following table summarizes the performance and ranking of each motel under each of the three performance measures.

Motel	ROI (Rank)	Residual Income (Rank)	ROS (Rank)
San Francisco	24% (1)	140 (2)	20% (1)
Chicago	15% (3)	100 (3)	15% (3)
New Orleans	16% (2)	180 (1)	16% (2)

The residual income rankings differ somewhat from the ROI and ROS rankings. Consider the ROI and residual income rankings for the San Francisco and New Orleans motels. The New Orleans motel has a smaller ROI. Although its income is twice that of the San Francisco motel ($480,000 versus $240,000), its total assets are three times as large ($3,000,000 versus $1,000,000). The return on assets invested in the New Orleans motel is not as high as the return on assets invested in the San Francisco motel. The New Orleans motel has a higher residual income because it earns a higher operating income after covering the 10% interest charge on investment. The motels have similar ROI and ROS rankings because the revenue-to-investment ratios are similar across motels. However, ROI and ROS do not always yield similar rankings. Is any one method superior to the others? No, each evaluates a slightly different aspect of performance.

Surveys of U.S. and Japanese companies indicate extensive use of net income as a performance measure. Next to net income measures, U.S. companies favor ROI over ROS, while Japanese companies use ROS more than ROI. These differences are also consistent with differences in pricing practices in the two countries. Japanese companies emphasize sales margins, while U.S. companies emphasize return on investment.[2] Some researchers speculate that Japanese managers favor ROS because it is easier to calculate and because achieving a sufficient sales margin will be likely to benefit ROI sooner or later. Deemphasizing ROI has other advantages. Managers are not induced to delay investment in equipment because of the negative effects it might have on ROI in the short-run.

Exhibit 26-2 presents the key financial performance measures used by nine companies. Note the diversity in the use of income measures, ROS, and ROI.

[2]See K. Smith and C. Sullivan, "Survey of Cost Management Systems in Manufacturing," Working paper, Purdue University, 1990; and P. Scarbrough, A. Nanni, and M. Sakurai, "Japanese Management Accounting Practices and the Effects of Assembly and Process Automation," *Management Accounting Research* 2 (1991).

EXHIBIT 26-2

Key Financial Performance Measures of Nine Companies

Company Name	Country Headquarters	Product	Key Financial Performance Measure(s)
Dow Chemical	U.S.	Chemicals	Income (profit)
Xerox	U.S.	Photocopiers	ROS and ROI
Ford Motor	U.S.	Automotive	ROS and ROI
Guinness	U.K.	Consumer products	Income and ROS
Krones	Germany	Machinery/equipment	Sales and income
Mayne Nickless	Australia	Security/transportation	ROI and ROS
Mitsui	Japan	Trading	Sales and income
Pirelli	Italy	Tires/manufacturing	Income and cash flow
Swedish Match	Sweden	Consumer products	ROI

Source: Business International Corporation, *Evaluating the Performance of International Operations* (New York, 1989); and Business International Corporation, *101 More Checklists for Global Financial Management* (New York, 1992).

ALTERNATIVE DEFINITIONS OF INVESTMENT

Different companies use different definitions of investment. The definitions include the following:

1. *Total assets available.* Includes all business assets, regardless of their particular purpose.
2. *Total assets employed.* Defined as total assets available minus idle assets and minus assets purchased for future expansion. For example, if the New Orleans motel in Exhibit 26-1 has unused land set aside for potential expansion, the total assets employed by the motel would exclude the cost of that land.
3. *Working capital (current assets minus current liabilities) plus other assets.* This definition excludes that portion of current assets financed by short-term creditors.
4. *Stockholders' equity.* Use of this definition for each individual motel in Exhibit 26-1 requires allocation of the long-term liabilities of Hospitality Inns to the three motels, which would then be deducted from the total assets of each motel.

Most firms that employ ROI for performance measurement use either total assets available or working capital plus other assets as the definition of the ROI denominator. However, when top management directs a division manager to carry extra assets, total assets employed can be more informative than total assets available. The most common rationale for using working capital plus other assets is that the division manager often influences decisions on the short-term debt the division utilizes.

MEASUREMENT ALTERNATIVES FOR ASSETS

Objective 5
Distinguish between present value, current cost, and historical cost asset-measurement methods

How should the assets included in the investment base be measured? At present value, current cost, current disposal price, or historical cost? Should gross book value or net book value be used for depreciable assets?

Present Value

Chapters 20 and 21 discuss the relevance of discounted cash-flow (DCF) analysis for both asset acquisition and disposal decisions. **Present value** is the asset measure based on DCF estimates. Consider an existing motel with an expected useful life of 10 years, expected net cash inflows of $1,200,000 each year, and an expected terminal disposal price of $2,000,000. The required rate of return is 12%. The present value of the motel would be $7.424 million (see Tables 2 and 4, Appendix C):

Present value of annuity of $1,200,000 for 10 years discounted at 12%:	
$1,200,000 x 5.650	$6,780,000
Present value of $2,000,000 10 years later discounted at 12%:	
$2,000,000 x 0.322	644,000
Present value of motel	$7,424,000

Few firms systematically incorporate total present value (and disposal price) information into their routine accounting reports. Nevertheless, managers make periodic attempts to approximate present values and current disposal prices in judging the desirability of investment in assets. Otherwise managers may overlook investment or disinvestment opportunities that will improve the value of the company. The decision rule for disinvestment compares the total present value with the current disposal price. If the current disposal price exceeds the present value, the asset(s) should be sold now.

Current Cost

Current cost is the cost today of purchasing an asset identical to the one currently held. It is the cost of purchasing the services provided by that asset if an identical asset cannot currently be purchased. Consider the following comment by American

Standard, a manufacturer of household fixtures and an enthusiastic advocate of current-cost measures when evaluating the performance of both its activities and its managers:

> [For some years] American Standard has been modifying its financial measurements to eliminate the distortions and inequities of historical cost accounting. A considerable amount of the difference in returns on book assets between units can just be a reflection of when a unit acquired its fixed assets. The unit that acquired its assets some time ago at deflated dollars in today's terms can look very good compared with a unit that acquired its fixed assets recently. It may be that the converse relation is actually true in terms of economic returns on new investments. Then too, the units with high returns may be inhibited from making necessary fresh investments because the investments will depress the historical return on book assets. The internal accounting system should have correct incentives for management to make economic capital investment decisions. By trying to maintain unrealistically high book returns, management can be inhibited from investing in the business. Book returns [under historical cost accounting] can lead management into a trap of "milking the business" and shying away from necessary replacements with new and better equipment and facilities.[3]

The American Standard argument can be illustrated using the Hospitality Inns example discussed earlier in this chapter (see Exhibit 26-1). Assume the following information about the fixed assets of each motel (dollar amounts are in thousands):

	San Francisco	Chicago	New Orleans
Age of facility (at end of 19_8)	8 years	4 years	2 years
Gross book value	$1,400	$2,100	$2,800
Accumulated depreciation	$ 800	$ 600	$ 400
Net book value (at end of 19_8)	$ 600	$1,500	$2,400
Depreciation for 19_8	$ 100	$ 150	$ 200

Hospitality Inns assumes a 14-year useful life, assumes no terminal disposal price for the physical facilities, and calculates depreciation on a straight-line basis.

An index of construction costs for the eight-year period that Hospitality Inns has been operating (19_0 year-end = 100) is as follows:

Year	19_1	19_2	19_3	19_4	19_5	19_6	19_7	19_8
Construction cost index	110	122	136	144	152	160	174	180

Earlier in this chapter we computed an ROI of 24% for San Francisco, 15% for Chicago, and 16% for New Orleans (see p. 894). One possible explanation of the high ROI for San Francisco is that this motel's fixed assets are expressed in terms of 19_0 construction price levels and that the fixed assets for the Chicago and New Orleans motels are expressed in terms of more recent construction price levels.

Exhibit 26-3 illustrates a step-by-step approach for incorporating current-cost estimates for fixed assets and depreciation into the ROI calculation. The current-cost adjustment dramatically reduces the ROI of the San Francisco motel.

	ROI: Historical Cost	ROI: Current Cost
San Francisco	24%	10.81%
Chicago	15%	11.05%
New Orleans	16%	13.79%

---

[3]From an internal document ("Management Uses of Inflation-Adjusted Accounting Data: The American Standard Approach") by K. Todd.

EXHIBIT 26-3

ROI for Hospitality Inns: Computed Using Current-Cost Estimates
as of the End of 19_8 for Depreciation and Fixed Assets

Step 1: Restate fixed assets from gross book value at historical cost to current cost as of the end of 19_8.

$$\text{Gross book value} \atop \text{at historical cost} \quad \times \quad \frac{\text{Construction cost index in 19\_8}}{\text{Construction cost index in year of construction}}$$

San Francisco $1,400 × (180 ÷ 100) = $2,520
Chicago $2,100 × (180 ÷ 144) = $2,625
New Orleans $2,800 × (180 ÷ 160) = $3,150

Step 2: Derive net book value of fixed assets at current cost as of the end of 19_8. (The estimated total useful life of each motel is 14 years.)

$$\text{Gross book value} \atop \text{at current cost in 19\_8} \quad \times \quad \frac{\text{Estimated useful life remaining}}{\text{Estimated total useful life}}$$

San Francisco $2,520 × (6 ÷ 14) = $1,080
Chicago $2,625 × (10 ÷ 14) = $1,875
New Orleans $3,150 × (12 ÷ 14) = $2,700

Step 3: Compute current cost of total assets in 19_8. (Assume current assets of each motel expressed in 19_8 dollars.)

$$\text{Current assets} \atop \text{(from Exhibit 26-1)} \quad + \quad \text{Fixed Assets} \atop \text{(from step 2 above)}$$

San Francisco $400 + $1,080 = $1,480
Chicago $500 + $1,875 = $2,375
New Orleans $600 + $2,700 = $3,300

Step 4: Compute current-cost depreciation expense in 19_8 dollars.

Gross book value at current cost in 19_8 × (1 ÷ 14)

San Francisco $2,520 × (1 ÷ 14) = $180
Chicago $2,625 × (1 ÷ 14) = $187.50
New Orleans $3,150 × (1 ÷ 14) = $225

Step 5: Compute 19_8 current-cost operating income using 19_8 current-cost depreciation.

$$\text{Historical-cost} \atop \text{operating income} \quad - \quad \left(\text{Current-cost} \atop \text{depreciation} \quad - \quad \text{Historical-cost} \atop \text{depreciation} \right)$$

San Francisco $240 − ($180 − $100)   = $160
Chicago $300 − ($187.50 − $150) = $262.50
New Orleans $480 − ($225 − $200)   = $455

Step 6: Compute ROI using current-cost estimates for fixed assets and depreciation.

$$\frac{\text{Current-cost operating income (from step 5)}}{\text{Current-cost of total assets (from step 3)}}$$

San Francisco $160 ÷ $1,480   = 10.81%
Chicago $262.50 ÷ $2,375 = 11.05%
New Orleans $455 ÷ $3,300   = 13.79%

Thus, the 24% ROI of the San Francisco motel can give a highly misleading picture of the returns that Hospitality Inns should expect from subsequent investments in motel facilities.

Obtaining current-cost estimates for some assets can be difficult.[4] The exact assets currently held by an organization may no longer be traded or manufactured. When making current-cost estimates, the company should focus on the expected cash flows from the assets held, not on their precise physical or technological features. The aim is to approximate how much it would cost to obtain similar assets that would produce the same expected operating cash inflows as do the assets currently held. Managers rarely replace assets with new assets having identical operating and economic characteristics.

Plant and Equipment: Gross or Net Book Value?

Because historical-cost investment measures are used often in practice, there has been much discussion about the relative merits of using gross book value (original cost) or net book value (original cost minus accumulated depreciation). Those who favor using gross book value claim that it helps comparisons among plants and divisions. If income decreases as a plant ages, the decline in earning power will be made evident. In contrast, if net book value is used, the constantly decreasing base can show a higher ROI in later years; this higher rate may mislead decision makers.

The proponents of using net book value as a base maintain that it is less confusing because (1) it is consistent with the total assets shown on the conventional balance sheet, and (2) it is consistent with net income computations that include deductions for depreciation. If net book value is used in a manner that is consistent with the planning model, it can be useful for auditing past decisions. Surveys of practice report net book value to be the dominant asset measure used by companies in their internal performance evaluations.

GOAL CONGRUENCE AND PERFORMANCE MEASURES

Objective 6

Explain goal-congruence problems arising from accrual accounting performance measures

Individual components of a management control system should be consistent and mutually reinforcing. Goal-congruence problems can arise when the measures used to evaluate a manager's performance conflict with the decision models advocated by top management.

Limitations of Accrual Accounting Performance Measures

As we have discussed, capital investment decisions in many companies are based on discounted cash-flow (DCF) models. These same companies, however, often use accrual accounting models for evaluating the performance of managers. Managers, therefore, may reject capital projects that are justified on a DCF basis because these projects produce dismally low accrual accounting ROI numbers in the first year or two after the initial investment.

Several different approaches promote consistency between the DCF model used to select capital-budgeting projects and the model used for performance evaluation. One approach is to compare directly the cash-flow predictions made in the DCF analysis with the actual cash flows that occur. A second approach is to predict the accrual accounting rate of return for each year of the project at the time the project is selected. The actual accrual accounting rate of return in each year could then be compared with the predicted rate of return.[5]

---

[4] When a specific cost index, such as the construction cost index, is not available, companies use a general index, such as the consumer price index, to approximate current costs.

[5] A third approach is to use the *compound-interest depreciation method*. This method calculates depreciation in such a manner that the year-by-year accounting rate of return on beginning-period investment is equal to the DCF internal rate of return on the investment. This depreciation method is rarely adopted in practice.

Comparability of Historical-Cost Measures

Despite the problems with historical cost-based accounting measures for evaluating economic returns on new investments and the disincentives historical cost-based accounting measures sometimes create for new expansion, historical-cost ROIs *can* be used for evaluating current performance. Consider our Hospitality Inns example. The key is to recognize that the motels were built at different times, which in turn means they were built at different costs. Top management could adjust the target ROIs accordingly, perhaps setting San Francisco's ROI at 26%, Chicago's at 18%, and New Orleans' at 16%.

This alternative is frequently overlooked in the literature. Critics of historical cost have indicated how high rates of return on old assets may erroneously induce a manager not to replace assets. Regardless, the manager's mandate is often "Go forth and attain the budgeted results." The budget, then, should be carefully negotiated with full knowledge of historical-cost accounting pitfalls. The desirability of tailoring a budget to a particular manager and a particular accounting system cannot be overemphasized. For example, many problems of asset valuation and income measurement (whether based on historical cost or current cost) can be satisfactorily solved if top management gets everybody to focus on what is attainable in the forthcoming budget period—regardless of whether the financial measures are based on historical costs or some other measure, such as replacement costs.

DISTINCTION BETWEEN MANAGERS AND ORGANIZATION UNITS[6]

As noted in several earlier chapters, performance evaluation of a manager should be distinguished from performance evaluation of an organization subunit, such as a division of a company. For example, historical cost-based ROIs can be used to evaluate a manager's performance even though historical cost ROIs may be unsatisfactory for evaluating economic returns earned by the organization subunit.

Consider another example. Companies often put the most skillful division manager in charge of the weakest division in an attempt to change its fortunes. Such an effort may take years, not months. Furthermore, the manager's efforts may result merely in bringing the division up to a minimum acceptable ROI. The division may continue to be a poor profit performer in comparison with other divisions. But it would be a mistake to conclude from the poor performance of the division that the manager is performing poorly.

This section focuses on developing basic principles for evaluating the performance of a division manager of an individual facility. The concepts we discuss apply, however, to all organization levels. Later sections consider specific examples at the individual activity level and the total organization level. For specificity, we use the ROI performance measure throughout.

The Basic Trade-off: Creating Incentives versus Imposing Risk

Objective 7

Recognize the trade-off between providing incentives and imposing risk

Compensation arrangements for managers and employees run the range from a flat salary with no direct performance-based bonus (as, for example, in the case of some government officials) to rewards based only on performance (as, for example, in the case of managers of real estate agencies). Most often, however, a manager's total compensation includes some combination of salary and a performance-based bonus. An important consideration in designing compensation arrangements is the trade-off between creating incentives and imposing risk. We illustrate this trade-off in the context of our Hospitality Inns example.

---

[6]The presentations here draw (in part) from teaching notes prepared by S. Huddart, N. Melumad and S. Reichelstein.

Sally Fonda owns the Hospitality Inns chain of motels. Roger Brett manages the Hospitality Inns, San Francisco (HISF). Assume that Fonda uses ROI to measure performance. To achieve good results as measured by ROI, Fonda would like Brett to keep the motel clean, reduce waste, control costs, provide prompt and courteous service, and reduce receivables and supplies inventories. But even if Brett did all those things, good results are by no means guaranteed. HISF's ROI is affected by many factors outside Fonda's and Brett's control, such as recession in the San Francisco economy, fewer travelers, violence in the city, or road and building construction near the motel that might negatively affect HISF. Alternatively, noncontrollable factors might have a positive influence on HISF's ROI. Noncontrollable factors make HISF's profitability uncertain and risky.

Fonda is an entrepreneur and does not mind bearing risk; she is risk neutral. But Brett has chosen to be a manager rather than an entrepreneur, in part because he does not like being subject to risk; he is risk averse. One way of insuring Brett against risk is to pay Brett a flat salary, regardless of the actual outcome. All the risk would then be borne by Fonda, who has a greater tolerance for it. Excessive reliance on salary, however, with no performance-based compensation will provide Brett with no incentives to work harder or undertake extra physical and mental effort. The issue here is not that Brett will put in no effort. To retain his job or to uphold his own personal values, Brett will put in some work. The question is whether Brett can be motivated to work even harder.

Moral hazard describes contexts in which an employee is tempted to put in less effort (or report distorted information) because the employee's interests differ from the owner's and because the employee's effort cannot be accurately monitored and enforced. In the employer-employee context, moral hazard[7] has three important aspects:

1. There are two parties and the actions of one party (Brett, in our example) affect the benefits received by the other (Fonda).
2. The two parties have conflicting interests. Fonda would like Brett to put in more effort and increase HISF's ROI. Brett prefers to get more money for less effort.
3. The affected party (Fonda) cannot monitor and enforce the other party's (Brett's) actions. In some repetitive jobs—for example, in electronic assembly—a supervisor can monitor the workers' actions, and the moral hazard problem does not arise. The manager's job is often to gather information and exercise judgment on the basis of the information obtained. Hence, monitoring a manager's effort is considerably more difficult.

Paying no salary and rewarding Brett *only* on the basis of some performance measure—ROI, in our example—raises different concerns. Brett would now be motivated to put in effort to increase ROI because his rewards would increase with increases in ROI. But compensating Brett on ROI also subjects Brett to risk. Why? Because HISF's ROI depends not only on Brett's effort but *also* on random factors over which Brett has no control.

To compensate Brett for taking on uncontrollable risk, Fonda must pay Brett some *extra* compensation within the structure of the ROI-based arrangement. Thus, using performance-based incentives will cost Fonda more money, *on average*, than paying Brett a flat salary. Why "on average"? Because Fonda's compensation payment to Brett will vary with ROI outcomes. When averaged over these outcomes, the ROI-based compensation will cost Fonda more than would paying Brett a flat salary. The motivation for having some salary and some performance-based bonus in compensation arrangements is to balance the benefits of incentives against the extra costs of imposing uncontrollable risk on the manager.

---

[7]The term *moral hazard* originated in insurance contracts to represent situations where insurance coverage caused insured parties to take less care of their properties than they might otherwise. One response to moral hazard in insurance contracts is the system of deductibles (that is, the insured pays for damages below a specified amount).

Intensity of Incentives and Accounting Measurements

What dictates the intensity of the incentives? That is, how large should the incentive component be relative to salary? A key question is, *How well does the performance measure capture the manager's ability to influence the desired results?*

Good performance measures change significantly with the manager's performance and not very much with changes in factors that are beyond the manager's control. As a result, good performance measures limit the manager's exposure to uncontrollable risk and hence reduce the cost of providing incentives to get the manager to accept the incentive program. Lower costs favor the use of performance-based incentives because the benefits from incentives will, more than likely, outweigh the costs.

The absence of good performance measures restricts the owner's ability to motivate a manager through performance-based incentives. From the owner's standpoint, compensating a manager on the basis of poor performance measures adds uncontrollable risk and is more costly to implement. From the manager's standpoint, poor performance measures do not capture performance and hence do not induce improvement.

Suppose Brett has no authority for determining investments. Further suppose revenue is determined largely by external factors such as the local economy. Brett's actions influence only costs. Using ROI as a performance measure in these circumstances puts Brett's bonus at risk because two components of the performance measure (investments and revenues) are unrelated to his actions. ROI fails to capture Brett's performance. Using ROI-based performance measures is costly and ineffective. The management accountant might suggest that Fonda consider using a different performance measure—perhaps HISF's costs—that more closely captures Brett's effort. Measuring costs alone (that is, defining HISF as a cost center) reduces risk in the performance measure. Fonda can then effectively motivate Brett using cost-based performance incentives. Note that in this case, ROI may be a perfectly good measure of the economic viability of the division but it is not a good measure of Brett's performance.

The benefits of tying performance measures more closely to a manager's efforts encourage the use of nonfinancial measures in management control systems. Consider two possible measures for evaluating the manager of the housekeeping department at Hospitality Inns —the costs of the housekeeping department and the average time taken by the housekeeping staff to clean a room. Suppose housekeeping costs are affected by factors such as wage rates, which the housekeeping manager does not determine. In this case, using housekeeping department costs adds uncontrollable risk to the manager's performance evaluation measure. The average time taken to clean a room may more sharply capture the manager's performance.

The salary component of compensation dominates when only weak measures of performance are available (as in the case of some corporate staff and government officials). This is not to say that incentives are completely absent; promotions and salary increases do depend on some overall measure of performance, but the incentives are less direct and much weaker. Employers give stronger incentives when good measures of performance are available and when monitoring the manager's effort is very difficult (as in the case of real estate agencies).

Benchmarks and Relative Performance Evaluation

Benchmarking is the continuous process of measuring products, services, and activities against the best levels of performance, which can be found either inside or outside the organization. Revisit our Hospitality Inns example, and suppose Brett has authority over revenues, costs, and investments. Fonda can benchmark Brett's performance by comparing HISF's ROI against other motels within the Hospitality Inns chain or against external competitors. In evaluating Brett's performance, the benchmark that is most helpful is the performance of other similar motels influenced by the same noncontrollable factors that affect HISF. In effect, this benchmarking, also called *relative performance evaluation,* "cancels" the common noncontrollable factors and provides better information about Brett's performance.

Can the performance of two managers running similar operations within a company be benchmarked against one another? Yes, but one problem is that the use of these benchmarks may reduce incentives for these managers to help one another. A manager's performance evaluation measure improves if he or she does a better job or if he or she makes other managers look bad.

PERFORMANCE MEASURES AT THE INDIVIDUAL ACTIVITY LEVEL

Objective 9

Describe the incentive problems that could arise when employees perform multiple tasks as part of their jobs

This section focuses on incentive issues that arise in the context of individual activities. The principles described here, however, can be applied at all levels of the organization.

Performing Multiple Tasks

Most employees perform more than one task as part of their jobs. Marketing representatives sell products, provide customer support, and gather market information. Manufacturing workers are responsible for both the quantity and quality of their products. Employers want employees to intelligently allocate time and effort among various tasks.

Consider, for example, mechanics at an auto repair shop. Their jobs have at least two distinct and important aspects. The first aspect is the repair work. Performing more repair work would generate more revenue for the shop. The second aspect is customer satisfaction. The higher the quality of the job, the more likely the customer will be pleased. If the employer wants an employee to focus on both these aspects, then the employer must measure and compensate performance on both.

Suppose the employer can easily measure the quantity of auto repairs but not their quality. If the employer rewards workers on a piece-rate system—which pays workers only on the basis of the number of repairs actually performed—mechanics will likely increase the number of repairs they make at the expense of quality. Management could consider taking the following steps to motivate workers to maintain a balance between quantity and quality: (1) Management could drop the piece-rate system and pay mechanics an hourly salary, a step that would deemphasize repair quantity. Mechanics' promotions and pay increases could then be determined on the basis of management's assessment of each mechanic's overall performance with respect to quantity and quality of repairs. (2) Management could develop explicit measures for quality and customer satisfaction such as the number of dissatisfied customers or the number of customer complaints. Management could go so far as to gather customer satisfaction data in surveys (Chapter 23). Management could use these data to create incentives for mechanics to maintain high-quality work. (3) Management could also employ independent staff to randomly monitor whether the repairs performed were of high quality.

Note that nonfinancial measures (such as customer-satisfaction measures) play a central role in motivating mechanics to emphasize both quantity and quality. The goal is to measure both aspects of the mechanics' jobs and to balance incentives so that one aspect is not overemphasized.

Team-Based Compensation Arrangements

Many manufacturing, marketing, and design problems require employees with multiple skills, experiences, and judgments to pool their talents. In these situations a team of employees achieves better results than individuals acting on their own.[8] Compa-

---

[8]J. Katzenbach and D. Smith, *The Wisdom of Teams* (Boston: The Harvard Business School Press, 1993).

nies give incentives and bonuses to individuals on the basis of team performance. Team incentives encourage cooperation, with individuals helping one another as they strive toward a common goal. The blend of knowledge and skills needed to change methods and improve efficiency puts a team in a better position than a lone individual to respond to the incentives management presents. Eaton, TRW, Whirlpool, Monsanto, Dana, and Analog Devices in the United States and Nissan and Nippon Steel in Japan are examples of companies that use some form of team-based incentives.

Whether team-based compensation is desirable depends, to a great extent, on the culture and management style of a particular organization. One criticism of teams, especially in the United States, is that individual incentives to excel are dampened, harming overall performance.

PERFORMANCE MEASURES AT THE TOTAL ORGANIZATION LEVEL

Objective 10
Describe proposed changes in the accounting for stock options and compensation disclosures required by the Securities and Exchange Commission.

The principles of performance evaluation described in the previous sections also apply to executive compensation plans at the total organization level. Executive compensation plans are based on both financial and nonfinancial information and consist

Concepts in Action

SEARS AUTO REPAIR SHOPS: WHAT PRICE QUALITY?[a]

In 1990, Sears, Roebuck and Co. introduced commissions and by-the-job rates for its mechanics. Mechanics who had been paid a wage of $15 per hour were put on base wages of $12 per hour plus piece-work incentives, which depended on the type of job done and the standard time required to do the job. For example, a front-end alignment estimated to take one hour had a piece rate of $3 per alignment.

What behavior did this incentive policy encourage? Because the quality of work was not carefully monitored, the piece-rate structure created incentives to do a greater quantity of repair with little attention paid to *quality*. A mechanic who could rush through two front-end alignment jobs in an hour would see his wages increase by $6 an hour.

Sears compounded the problems that its incentive structure created by establishing quotas for parts (shock absorbers and struts) and services (number of oil changes) for every eight-hour shift. If mechanics failed to meet quotas, their hours were reduced or they were transferred out.

As quality declined, consumer complaints increased, and the California Department of Consumer Affairs (CDCA) began an investigation of Sears's repair and billing practices. CDCA found that its agents were overcharged at Sears repair shops. Sears's mechanics were recommending repair jobs when no repairs were needed. CDCA agents were charged for front and rear springs and for new alternators and batteries when none were needed.

Sears responded by changing its incentive arrangements to promote customer satisfaction. In a full-page advertisement in *The Wall Street Journal* of June 25, 1992, Edward Brennan, Chairman and CEO of Sears, stated, "We have concluded that our incentive compensation and goal-setting program inadvertently created an environment in which mistakes have occurred. We are moving quickly and aggressively to eliminate that environment."

Sears cut all incentive compensation and goal-setting rewards for its automotive service staff that were linked to the number of jobs done or quantity of parts sold. Compensation would now be based on customer-satisfaction measures and quality assessments made by independent staff. Sears also settled the class-action suit filed against it in California at an estimated cost of $8 million.

[a]Adapted from, T. Yin, "Sears Is Accused of Billing Fraud at Auto Centers," *The Wall Street Journal*, June 12, 1992; K. Kelly and E. Schine, "How Did Sears Blow This Gasket?" *Business Week*, June 29, 1992, J. Quinn; "Repair Job," *Incentive* (October 1992).

of a mix of (1) base salary, (2) annual incentives (for example, cash bonus based on annual reported income), (3) long-term incentives (for example, stock options [described later in this section] based on achieving a set level of ROI by the end of a five-year period), and (4) fringe benefits (for example, life insurance, an office with a view, and a personal secretary). Three factors that designers of executive compensation plans emphasize are achievement of organization goals, administrative ease, and the likelihood that managers affected by the plan believe it is fair.

Owners and stockholders can steer executives to a shorter-run or a longer-run perspective by carefully choosing the time horizon over which to measure their performance. For example, evaluating performance on the basis of annual ROI would sharpen an executive's short-term focus. Using ROI over, say, five years would lengthen that executive's view. To achieve a perspective balanced over time, owners use both short-run and long-run measures to evaluate executive performance.

In the early 1990s some stockholders, academics, and regulators strongly criticized the executive compensation practices at some of the largest U.S. corporations.[9] At issue are the large pay packages given to some chief executive officers (CEOs). The average pay at the largest U.S. corporations in 1992 equaled $3.8 million, pay at times unrelated to company performance. The criticism, in large part, centers on the failure of corporate boards of directors to design compensation plans in a fair and unbiased manner. Two factors that critics believe have contributed to excessive compensation are (1) improper accounting for stock options granted to executives (stock options often constitute a significant percentage of executive pay) and (2) inadequate disclosure to stockholders of the details of executive compensation plans.

Stock options give executives the *right* to buy company stock at a specified price (called the exercise price) within a specified period. Suppose on July 1, 1993, Marriott Corporation gave its CEO the option to buy 200,000 shares of Marriott stock at any time before June 30, 1998, at the July 1, 1993, market price of $25 per share. If Marriott's stock price rises to, say, $40 per share on March 24, 1996, and the CEO chooses to exercise his option on all 200,000 shares, he will earn $3 million. (The CEO would exercise his right to buy Marriott stock from the company on March 24, 1996, for $25 per share and sell it in the market at $40 per share, earning $15 per share on 200,000 shares.) If Marriott's stock price stays below $25 the entire period, the CEO will simply forgo his right to buy the shares.

As we write this book, Accounting Principles Board (APB) Opinion 25 governs the accounting for options granted to executives. APB 25 does not require a company to record compensation cost in its income statement for any options granted if the exercise price of the option equals or exceeds the market price of the stock on the day the options are granted. The company recognizes no cost even though the company has sacrificed something of value. This value stems from the potentially large payments that the executive will receive if the stock price increases. In June 1993, the Financial Accounting Standards Board (FASB) agreed in principle to change the accounting for stock options granted to executives. It proposed that companies record as a cost in their income statements the fair market value of the options on the date the options are given.

In October 1992, the Securities and Exchange Commission (SEC) issued new rules requiring more detailed disclosures of the compensation arrangements of top-level executives. In complying with these rules in April 1993, Marriott Corporation disclosed a summary compensation table showing the salary, bonus, stock options, other stock awards, and other compensation earned by its top five officers during the 1990, 1991, and 1992 fiscal years. The SEC rules also require companies to disclose the principles underlying their executive compensation plans and the performance criteria—such as profitability, sales growth, and market share—used in determining compensation. In its annual report, Marriott Corporation described these principles as "building a strong correlation between stockholder return and executive compensation, offering incentives which encourage attainment of near and long-term business

---

[9]"Executive Pay: Reform Is Coming, But You Wouldn't Know It," *Business Week*, April 26, 1993.

goals, and providing a total level of pay which is commensurate with performance." Marriott uses cash flow, earnings per share, and guest satisfaction as performance criteria to determine annual cash incentives for its executives. The SEC rules further required Marriott to disclose how well its stock performed relative to the stocks of other motels and hotels over a five-year period.

ENVIRONMENTAL AND ETHICAL RESPONSIBILITIES

Managers in all organizations shoulder environmental and ethical responsibilities. Environmental violations (such as water and air pollution) and unethical practices (such as bribery and corruption) carry heavy fines and are jailable offenses under the laws of the United States and other countries. But environmental responsibilities and ethical conduct extend beyond legal requirements.

Socially responsible companies are increasingly setting targets to reduce pollution and measuring their environmental management performance against these targets. Some companies, such as Lockheed, make environmental performance a line item in every employee's salary appraisal sheet. Duke Power Company's employee performance appraisal includes an evaluation of the employee's performance in reducing solid waste, cutting emissions and discharges, and implementing the company's environmental plans. Has explicitly evaluating and rewarding employees on environmental performance had any effect? Apparently so. Duke Power has met every environmental goal management has set.

Many foreign countries also emphasize environmental responsibilities. Companies operating in Germany, Switzerland, the Netherlands, and the Scandinavian countries report on environmental performance as part of a larger set of social responsibility disclosures (which include employee welfare and community development information).

Chapter 1 concluded with a discussion of professional ethics for management accountants. The same guidelines apply to managers. In particular, the numbers that subunit managers report should not be tainted by "cooking the books." That is, managers are responsible for reporting reliable figures uncontaminated by, for example, padded assets, understated liabilities, fictitious sales, and understated costs.

In turn, senior management should unequivocally communicate a positive "tone from the top"—an absolute intolerance for manipulation of accounting transactions and reports. These communications should be both oral and written and repeated often. A culture of integrity can permeate an organization only if senior managers speak and act as good examples.

Codes of business conduct are circulated in some organizations to signal appropriate and inappropriate individual behavior. Caterpillar Tractor's "Code of Worldwide Business Conduct and Operating Principles" follows:

> The law is a floor. Ethical business conduct should normally exist at a level well above the minimum required by law. . . . Caterpillar employees shall not accept costly entertainment or gifts (excepting mementos and novelties of nominal value) from dealers, suppliers and others with whom we do business. And we won't tolerate circumstances that produce, or reasonably appear to produce, conflict between personal interests of an employee and interests of the company.

Division managers often cite enormous pressures "to make the budget" as excuses or rationalizations for not adhering to ethical accounting policies and procedures. The exertion of pressure from senior managers to meet budget targets may have positive motivational effects. A healthy amount of pressure is not bad by itself—as long as the tone from the top simultaneously communicates the absolute need for all managers to behave ethically at all times. Management should promptly and severely reprimand unethical conduct irrespective of the benefits that accrue to the company from such actions. Some companies such as Lockheed emphasize ethical behavior by routinely evaluating employees against a business code of ethics.

PROBLEM FOR SELF-STUDY

PROBLEM

Budgeted data of the baseball manufacturing division of Home Run Sports for February 19_4 follow:

Current assets	$ 400,000
Long-term assets	600,000
Total assets	$1,000,000

Production output: 200,000 baseballs per month
Target ROI (operating income ÷ total assets): 30%
Fixed costs: $400,000 per month
Variable costs: $4 per baseball

Required

1. Compute the selling price per unit necessary to achieve the 30% target ROI.
2. Using the selling price from requirement 1, compute profitability under the duPont method.
3. Pamela Stephenson, division manager, receives 15% of the monthly residual income of the baseball manufacturing division as a bonus. What is her bonus for February 19_4, using the selling price from requirement 1? Home Run Sports uses a 12% required rate of return on total division assets when computing division residual income.

SOLUTION

1. Target operating income = 30% of $1,000,000
   = $300,000

   Let P = Selling price
   Sales – Variable costs – Fixed costs = Operating income
   $200,000P - (200,000 \times \$4) - \$400,000 = \$300,000$
   $200,000P = \$1,500,000$
   $P = \$7.50$

Proof:		
Sales: 200,000 × $7.50		$1,500,000
Variable costs: 200,000 × $4		800,000
Contribution margin		700,000
Fixed costs		400,000
Operating income		$ 300,000

2. $$\frac{\text{Revenue}}{\text{Investment}} \times \frac{\text{Income}}{\text{Revenue}} = \frac{\text{Income}}{\text{Investment}}$$

   $$\frac{\$1,500,000}{\$1,000,000} \times \frac{\$300,000}{\$1,500,000} = \frac{\$300,000}{\$1,000,000}$$

   $$1.5 \times 0.2 = 0.30, \text{ or } 30\%$$

3. Residual income = Operating income – Imputed interest charge for investment
   = $300,000 – (0.12 × $1,000,000)
   = $300,000 – $120,000
   = $180,000

Stephenson's bonus is $27,000 (15% of $180,000).

The following points are linked to the chapter's learning objectives.

1. Financial measures such as ROI and residual income can capture important aspects of both manager performance and organization-subunit performance. In many cases, however, financial measures need to be supplemented with nonfinancial measures of performance, such as those relating to customer service, product quality, and productivity.

2. The steps in designing an accounting-based performance measure are (a) choosing variables to include in the performance measure, (b) defining the terms, (c) measuring the items included in the variables, (d) choosing a target for performance, and (e) choosing the timing of feedback.

3. The duPont method describes return on investment (ROI) as the product of two components: investment turnover (revenue ÷ investment) and income margin (income ÷ revenue). This method highlights three ways to increase ROI—increase revenue, decrease costs, and decrease investment.

4. Residual income is income minus an imputed interest charge for the investment base. Residual income was designed to overcome some of the limitations of ROI. For example, residual income is more likely than ROI to promote goal congruence—actions that are in the best interests of the organization maximize residual income. The objective of maximizing ROI, conversely, may induce managers of highly profitable divisions to reject projects that, from the viewpoint of the organization as a whole, should be accepted.

5. Present value of an asset is the present value of the discounted cash flows (DCF) from the asset. Current cost of an asset is the cost today of purchasing an identical asset. Historical cost asset measurement methods consider the cost at which the asset was originally acquired, adjusted for depreciation.

6. Goal-congruence problems arise when firms base investment decisions on DCF methods while using accrual accounting models for evaluating the performance of managers. Managers may reject capital projects that are justified on a DCF basis because these projects produce very low accrual accounting ROI numbers in the first year or two after the initial investment.

7. Organizations create incentives by rewarding managers on the basis of performance. But the manager may face risks because random factors beyond the manager's control may also affect performance. Owners choose compensation arrangements to trade off the incentive benefit against the cost of imposing risk.

8. Good measures of employee performance are critical for implementing strong incentives. Many management accounting practices, such as the design of responsibility centers and the establishment of financial and nonfinancial measures, have as their goal better performance evaluation.

9. Most employees perform multiple tasks as part of their jobs. In some situations, one aspect of a job is easily measured (for example, the quantity of work done), while another aspect is not (for example, the quality of work done). Creating incentives to promote that aspect of the job that is easily measured (quantity) may cause workers to ignore an aspect of their job that is more difficult to measure (quality).

10. Managers around the world are facing increasing regulatory demands for greater disclosure of compensation arrangements for top-level executives and of the performance criteria used in determining compensation.

TERMS TO LEARN

This chapter and the Glossary at the end of the book contain definitions of the following important terms:

current cost *(p. 897)*   moral hazard *(902)*   present value *(897)*
residual income *(894)*   return on investment (ROI) *(892)*

ASSIGNMENT MATERIAL

QUESTIONS

26-1   Give two examples of financial performance measures and two examples of nonfinancial performance measures.

26-2   What are the five steps in designing an accounting-based performance measure?

26-3   What factors affecting ROI does the duPont method highlight?

26-4   "Residual income is not identical to ROI although both measures incorporate income and investment into their computations." Do you agree? Explain.

26-5   Give three definitions of investment used in practice when computing ROI.

26-6   Distinguish between measuring assets based on present value, current-cost, and historical-cost.

26-7   What approaches can promote consistency between the DCF model used to select capital-budgeting projects and the model used for performance evaluation?

26-8   Why is it important to distinguish between the performance of a manager and the performance of the organization subunit for which the manager is responsible? Give examples.

26-9   Describe moral hazard.

26-10   "Managers should be rewarded only on the basis of their performance measures. They should be paid no salary." Do you agree? Explain.

26-11   Explain the management accountant's role in helping organizations design stronger incentive systems for their employees.

26-12   Explain the role of benchmarking in evaluating managers.

26-13   Explain the incentive problems that can arise when employees have to perform multiple tasks as part of their jobs.

26-14   Describe two disclosures required by the SEC with respect to executive compensation and the change in accounting for executive stock options proposed by the FASB.

26-15   Peta Milano made the following comment when her friend Jake Ali was made manager of the Nuclear Construction Division of General Projects: "It was like putting a new captain on the bridge of the *Titanic* after it had hit the iceberg. That division has no prospect of ever being profitable. It is a dinosaur that should have died a long time ago." How can Jake Ali be motivated to make a sustained effort to achieve General Projects' goal of maximizing company-wide ROI?

EXERCISES AND PROBLEMS

26-16   Return on investment; comparisons of three companies. (IMA, adapted)   Return on investment is often expressed as follows:

$$\frac{\text{Income}}{\text{Investment}} = \frac{\text{Revenue}}{\text{Investment}} \times \frac{\text{Income}}{\text{Revenue}}$$

Required

1. What advantages are there in the breakdown of the computation into two separate components?

2. Fill in the blanks:

	Companies in Same Industry		
	A	**B**	**C**
Revenue	$1,000,000	$500,000	—
Income	100,000	50,000	—
Investment	500,000	—	$5,000,000
Income as a percentage of revenue	—	—	0.5%
Investment turnover	—	—	2
Return on investment	—	1%	—

After filling in the blanks, comment on the relative performance of these companies as thoroughly as the data permit.

26-17 Analysis of return on invested assets, comparison of three divisions. Quality Products, Inc., is a soft-drink and food-products company. It has three divisions: soft drinks, snack foods, and family restaurants. Results for the past three years follow (in millions):

	Soft-Drink Division	Snack-Foods Division	Restaurant Division	Quality Products, Inc.
Operating Revenues				
19_4	$2,800	$2,000	$1,050	$5,850
19_5	3,000	2,400	1,250	6,650
19_6	3,600	2,600	1,530	7,730
Operating Income				
19_4	120	360	105	585
19_5	160	400	114	674
19_6	240	420	100	760
Total Assets				
19_4	1,200	1,240	800	3,240
19_5	1,250	1,400	1,000	3,650
19_6	1,400	1,430	1,300	4,130

Required
Use the duPont method to explain changes in the operating income to total assets ratio over the 19_4 to 19_6 period for each division. Comment on the results.

26-18 ROI and residual income. (D. Solomons, adapted)   Consider the following data for General Electric Company (in thousands):

	Division A	Division B
Total assets	$1,000	$5,000
Operating income	$ 200	$ 750
Return on investment	20%	15%

Required
1. Which is the more successful division? Why?
2. General Electric has used residual income as a measure of management success, the variable it wants a manager to maximize. Using this criterion, what is the residual income for each division if the imputed interest rate is (a) 12%, (b) 14%, (c) 18%? Which division is more successful under each of these imputed interest rates?

26-19 ROI and residual income. (D. Kleespie)   The Gaul Company produces and distributes a wide variety of recreational products. One of its divisions, the Goscinny Division, manufactures and sells "menhirs," which are very popular with cross-country skiers. The demand for these menhirs is relatively insensitive to price changes. The Goscinny Division is considered to be an investment center and in recent years has averaged a return on investment of 20%. The following data are available for the Goscinny Division and its product:

Total annual fixed costs	$1,000,000
Variable costs per menhir	$300
Average number of menhirs sold each year	10,000
Average operating assets invested in the division	$1,600,000

Required
1. What is the minimum selling price per unit that the Goscinny Division could charge in order for Mary Obelix, the division manager, to get a favorable performance rating? Management considers an ROI below 20% to be unfavorable.
2. Assume that the Gaul Company judges the performance of its investment center managers on the basis of residual income rather than ROI, as was assumed in requirement 1. The company's required rate of return is considered to be 15%. What is the minimum selling price per unit that the Goscinny Division should charge for Obelix to receive a favorable performance rating?

26-20 Pricing and return on investment. Hardy Inc. assembles motorcycles and uses long-run (defined as 3 to 5 years) average demand to set the budgeted production level and costs for pricing. Prices are then adjusted only for large changes in assembly wage rates or direct materials prices.

You are given the following data:

Direct materials, assembly wages, and other variable costs	$1,320 per unit
Fixed costs	$300,000,000 per year
Target return on investment	20%
Normal utilization of capacity (average output)	1,000,000 units
Investment (total assets)	$900,000,000

Required
1. What operating income percentage on revenues is needed to attain the target return on investment of 20%?
2. What rate of return on investment will be earned if Hardy assembles and sells 1,500,000 units? 500,000 units?
3. The company has a management bonus plan based on yearly division performance. Assume that Hardy assembled and sold 1,000,000, 1,500,000, and 500,000 units in three successive years. Each of three people served as division manager for one year before being killed in an automobile accident. As the principal heir of the third manager, comment on the bonus plan.

26-21 ROI, residual income, investment decisions. The Media Group has three major divisions:

a. Newspapers—owns leading newspapers on four continents
b. Television—owns major television networks on three continents
c. Film studios—owns one of the five largest film studios in the world

Summary financial data (in millions) for 19_6 and 19_7 follow:

	Operating Income		Revenues		Total Assets	
	19_6	19_7	19_6	19_7	19_6	19_7
Newspapers	$900	$1,100	$4,500	$4,600	$4,400	$4,900
Television	130	160	6,000	6,400	2,700	3,000
Film studios	220	200	1,600	1,650	2,500	2,600

The manager of each division has an annual bonus plan based on division return on investment (ROI). ROI is defined as operating income divided by total assets. Senior executives from divisions reporting increases in ROI from the prior year are automatically eligible for a bonus. Senior executives of divisions reporting a decline in the division ROI have to provide persuasive explanations for the decline to be eligible for any bonus, and they are limited to 50% of the bonus paid to the division managers reporting an increase in ROI.

Ken Kearney, manager of the Newspapers Division, is considering a proposal to invest $200 million in fast-speed printing presses with color-print options. The estimated increment to 19_8 operating income would be $30 million. The Media Group has a 12% required rate of return for investments in all three divisions.

Required
1. Use the duPont method to explain differences among the three divisions in their 19_7 division ROI. Use 19_7 total assets as the denominator.
2. Why might Kearney be less than enthusiastic about the fast-speed printing press investment proposal?
3. Rupert Prince, chairman of the Media Group, receives a proposal to base senior executive compensation at each division on division residual income. Compute the residual income of each division in 19_7.
4. Would adoption of a residual income measure reduce Kearney's reluctance to adopt the fast-speed printing press investment proposal?

26-22 Division manager's compensation. (Continuation of 26-21). Rupert Prince seeks your advice on revising the existing bonus plan for division managers of the Media Group. He is considering three ideas:

◆ Reduce the salary of each division manager and make most, if not all, of the division manager's compensation depend on division ROI.
◆ Reduce the salary of each division manager and make most, if not all, of the division manager's compensation depend on companywide (the Media Group) ROI.
◆ Use benchmarking, and evaluate each division manager on the basis of his or her own division's ROI minus the average ROI of the other two divisions.

Required
Evaluate each of the three ideas Prince has put forth using performance evaluation concepts described in the chapter. Indicate the positive and negative features of each proposal.

26-23 Various measures of profitability. When the Coronet Company formed three divisions a year ago, the president told the division managers that an annual bonus would be paid to the most profitable division. However, absolute division operating income as conventionally computed would not be used. Instead, the ranking would be affected by the relative investments in the three divisions. Options available include ROI and residual income. Investment can be measured using gross book value or net book value. Each manager has now written a memorandum claiming entitlement to the bonus. The following data are available:

Division	Gross Book Value of Division Assets	Division Operating Income
Mastex	$400,000	$47,500
Banjo	380,000	46,000
Randal	250,000	30,800

All the assets are fixed assets that were purchased 10 years ago and have 10 years of useful life remaining. A zero terminal disposal price is predicted. Coronet's required return on investment used for computing the imputed interest charge in residual income is 10% of investment.

Required
Which method for computing profitability did each manager choose? Make your description specific and brief. Show supporting computations. Where applicable, assume straight-line depreciation.

26-24 Evaluating managers, ROI, value-chain analysis of cost structure. User Friendly Computer is one of the largest personal computer companies in the world. The board of directors was recently (March 19_5) informed that User Friendly's president, Brian Clay, was resigning to "pursue other interests." An executive search firm recommends that the board consider appointing Peter Diamond (current president of Computer Power) or Norma Provan (current president of Peach Computer). You collect the following financial information on Computer Power and Peach Computer for 19_3 and 19_4 (in millions):

	Computer Power		Peach Computer	
	19_3	19_4	19_3	19_4
Total assets	$360.0	$340.0	$160.0	$240.0
Revenues	$400.0	$320.0	$200.0	$350.0
Costs				
Research and development	36.0	16.8	18.0	43.5
Design	15.0	8.4	3.6	11.6
Production	102.0	112.0	82.8	98.6
Marketing	75.0	92.4	36.0	66.7
Distribution	27.0	22.4	18.0	23.2
Customer service	45.0	28.0	21.6	46.4
Total costs	300.0	280.0	180.0	290.0
Operating income	$100.0	$ 40.0	$ 20.0	$ 60.0

In early 19_5, a computer magazine gave Peach Computer's main product five stars (its highest rating on a 5-point scale). Computer Power's main product was given three stars, down from five stars a year ago because of customer-service problems. The computer magazine also ran an article on new-product introductions in the personal computer industry. Peach Computer received high marks for new products in 19_4. Computer Power's performance was called "mediocre." One "unnamed insider" of Computer Power commented: "Our new-product cupboard is empty."

Required
1. Use the duPont method to analyze the ROI of Computer Power and Peach Computer in 19_3 and 19_4. Comment on the results.
2. Compute the percentage of costs in each of the six business-function cost categories for Computer Power and Peach Computer in 19_3 and 19_4. Comment on the results.
3. Rank Diamond and Provan as potential candidates for president of User Friendly Computer.

26-25 ROI, residual income, performance evaluation. (CMA, adapted)   Lawton Industries has manufactured prefabricated houses for over 20 years. Lawton expanded into the pre-cut housing market in 19_1 when it acquired Presser Company, one of its suppliers. In this market, various types of lumber are precut to the appropriate lengths, banded into packages, and shipped to customers' lots for assembly. Lawton decided to maintain Presser's separate identity and, thus, established the Presser Division as an investment center of Lawton.

Lawton uses return on investment (ROI) as a performance measure with investment defined as average total assets employed. Management bonuses are based in part on ROI. All investments in assets are expected to earn a minimum return of 15% before income taxes.

Presser's ROI has ranged from 19.3 to 22.1% since it was acquired in 19_1. Presser had an investment opportunity in 19_5 that had an estimated ROI of 18%. Presser's management decided against the investment because it believed the investment would decrease the division's overall ROI.

The 19_5 income statement for Presser Division is presented below. The division's total assets employed were $12,600,000 at the end of 19_5, a 5% increase over the 19_4 year-end balance.

Presser Division Operating Income Statement
for the Year Ended December 31, 19_5
(in thousands)

Revenue		$24,000
Cost of goods sold		15,800
Gross margin		8,200
Other operating costs		
Administrative	$2,140	
Marketing	3,600	5,740
Operating income		$ 2,460

Required

1. Calculate the return on investment in average total assets employed (ROI) for 19_5 for the Presser Division.

2. Compute Presser Division's residual income on the basis of average total assets employed.

3. Would the management of Presser Division have been more likely to accept the investment opportunity it had in 19_5 if residual income were used as a performance measure instead of ROI? Explain your answer.

4. The Presser Division is a separate investment center within Lawton Industries. Identify the items Presser must control if it is to be evaluated fairly by either the ROI or residual income performance measures.

26-26   ROI, residual income, management incentives. (CMA, adapted)   Jump-Start Co. (JSC), a subsidiary of Mason Industries, manufactures go-carts and other recreational vehicles. Family recreational centers, featuring go-cart tracks and miniature golf, batting cages, and arcade games have increased in popularity. As a result, JSC has been receiving some pressure from the Mason management to diversify into some of these other recreational areas. Recreational Leasing Inc. (RLI), one of the largest companies that leases arcade games to these family recreational centers, is looking for a friendly buyer. Mason's top management believes that RLI's assets could be acquired for an investment of $3 million and has strongly urged Bill Grieco, division manager of JSC, to consider acquiring RLI.

Grieco has reviewed RLI's financial statements with his controller, Marie Donnelly, and they believe that the acquisition may not be in JSC's best interest. "If we decide not to do this, the Mason people are not going to be happy," said Grieco. "If we could convince them to base our bonuses on something other than return on investment, maybe this acquisition would look more attractive. How would we do if the bonuses were based on residual income using the company's 15% required rate of return on investment?"

Mason has traditionally evaluated all of its divisions on the basis of return on investment, which is defined as the ratio of operating income to total assets; the desired rate of return for each division is 20%. The management team of any division reporting an annual increase in the return on investment is automatically eligible for a bonus. The management team of any division reporting a decline in the return on investment must provide convincing explanations for the decline to be eligible for a limited bonus.

Presented below are condensed financial statements for both JSC and RLI for the fiscal year ended May 31, 19_5.

	JSC	RLI
Revenue	$10,500,000	—
Leasing revenue	—	$2,800,000
Variable costs	7,000,000	1,000,000
Fixed costs	1,500,000	1,200,000
Operating income	$ 2,000,000	$ 600,000
Current assets	$ 2,300,000	$1,900,000
Long-term assets	5,700,000	1,100,000
Total assets	$ 8,000,000	$3,000,000
Current liabilities	$ 1,400,000	$ 850,000
Long-term liabilities	3,800,000	1,200,000
Stockholders' equity	2,800,000	950,000
Total liabilities and stockholders' equity	$ 8,000,000	$3,000,000

Required

1. If Mason Industries continues to use return on investment as the sole measure of division performance, explain why Jump-Start Co. (JSC) would be reluctant to acquire Recreational Leasing Inc. (RLI). Be sure to support your answer with appropriate calculations.

2. If Mason Industries could be persuaded to use residual income to measure the performance of JSC, explain why JSC would be more willing to acquire RLI. Be sure to support your answer with appropriate calculations.

3. Discuss how the behavior of division managers is likely to be affected by the use of
   a. return on investment as a performance measure
   b. residual income as a performance measure

26-27 Alternative measures for the investment base of gasoline stations. ARCO is having trouble in deciding whether to continue to use its old gasoline stations and in evaluating the performance of these stations and their managers in terms of return on investment. Top management has explored various ways of measuring investment for the stations:

a. *Historical cost:* original cost of land and buildings minus accumulated depreciation (sometimes called net book value)

b. *Current cost:* cost to currently replace the operating cash inflows provided by the existing gasoline station

c. *Current disposal price:* the net proceeds from selling the gasoline station to another company

Information on three gasoline stations was collected to help clarify the issues:

	Fresno Station	Las Vegas Station	Modesto Station
Operating income	$100,000	$ 120,000	$ 60,000
Historical cost of investment	$400,000	$ 200,000	$260,000
Current cost of investment	$640,000	$ 480,000	$290,000
Current disposal price of investment	$600,000	$2,500,000	$300,000
Age	6 years old	15 years old	2 years old

The Las Vegas station is located next to the largest casino on the Las Vegas Strip and was purchased before the current boom in casinos. The current-cost estimate of the Las Vegas station is for a site one mile away from the existing site. The new site would generate the same amount of operating income as the old site. The current-cost estimates for the Fresno and Modesto stations are for the same site as the existing station in each city.

Required

1. Which of the three measures of investment is relevant for deciding whether to dispose of any one (or more) of the gasoline stations? Why?

2. Compute the ratio of operating income to investment for the Fresno, Las Vegas, and Modesto stations under each of the three measures of investment.

3. Which of the three measures is applicable for judging the performance of a gasoline station as an investment activity?

4. Which of the three measures is applicable for judging the performance of the manager of a gasoline station? Is your answer the same as, or different from, your answers in requirements 1 and 3?

5. What measures of performance, in addition to ROI, might be used to evaluate the performance of a manager of a gasoline station?

26-28 ROI performance measures based on historical cost and current cost. Mineral Waters Ltd. operates three divisions that process and bottle sparkling mineral water. The historical-cost accounting system reports the following data for 1995 (in thousands):

	Calistoga Division	Alpine Springs Division	Rocky Mountains Division
Revenues	$500	$ 700	$1,100
Operating costs (excluding depreciation)	300	380	600
Depreciation	70	100	120
Operating income	$130	$ 220	$ 380
Current assets	$200	$ 250	$ 300
Fixed assets—plant	140	900	1,320
Total assets	$340	$1,150	$1,620

Mineral Waters estimates the useful life of each plant to be 12 years with zero terminal disposal price. The straight-line depreciation method is used. The respective age of each plant at the end of 1995 is Calistoga (10 years old), Alpine Springs (3 years old), and Rocky Mountains (1 year old).

An index of construction costs of plants for mineral water production for the 10-year period that Mineral Waters has been operating (1985 year-end = 100) is:

1985	1992	1994	1995
100	136	160	170

Given the high turnover of current assets, management believes that the historical-cost and current-cost measures of current assets are approximately the same.

Required
1. Compute the ROI (operating income to total assets) ratio of each division using historical-cost measures. Comment on the results.
2. Use the approach in Exhibit 26-3 (p. 899) to compute the ROI of each division, incorporating current-cost estimates as of 1995 for depreciation and fixed assets. Comment on the results.
3. What advantages might arise from using current-cost asset measures as compared with historical-cost measures for evaluating the performance of the managers of the three divisions?

26-29 Relevant costs, performance evaluation, goal congruence. Pike Enterprises has three operating divisions. The managers of these divisions are evaluated on their divisional operating income, a figure that includes an allocation of corporate overhead *proportional to the revenues of each division*. The operating income statement (in thousands) for the first quarter of 19_5 appears below:

	Andorian Division	Orion Division	Tribble Division	Total
Revenues	$2,000	$1,200	$1,600	$4,800
Cost of goods sold	1,050	540	640	2,230
Gross margin	950	660	960	2,570
Division overhead	250	125	160	535
Corporate overhead	400	240	320	960
Division operating income	$ 300	$ 295	$ 480	$1,075

The manager of the Andorian Division is unhappy that his profitability is about the same as the Orion Division's and is much less than the Tribble Division's, even though his revenues are much higher than either of these other two divisions. The manager knows that he is carrying one line of products with very low profitability. He was going to replace this line of business as soon as more profitable product opportunities became available, but he has kept it because the line is marginally profitable and uses facilities that would otherwise be idle. That manager now realizes, however, that the sales from this product line are attracting a fair amount of corporate overhead because of the allocation procedure, and maybe the line is already unprofitable for him. This low-margin line of products had the following characteristics for the most recent quarter (in thousands):

Revenues	$800
Cost of goods sold	600
Avoidable division overhead	100

Required
1. Prepare the operating income statement for Pike Enterprises for the second quarter of 19_5. Assume that revenues and operating results are identical to the first quarter except that the manager of the Andorian Division has dropped the low-margin product line entirely from his product group.
2. Is Pike Enterprises better off from this action?
3. Is the Andorian Division manager better off from this action?
4. Suggest changes for Pike's system of division reporting and evaluation that will motivate division managers to make decisions that are in the best interest of Pike Enterprises as a whole. Discuss any potential disadvantages of your proposal.

26-30 Risk sharing, incentives, multiple tasks. Gastek Incorporated makes rubber gaskets for the machine tool industry. Gastek has access to as much input as it needs and can currently sell all the gaskets it manufactures. Ralph Hogan is the manufacturing manager. He manufactures the gaskets and gives them to the owner, Sally Bates, who then sells them. Sally is also responsible for all purchases and for setting wages and salaries. Hogan does not like being exposed to risk. Bates is indifferent toward risk. The number of gaskets that Hogan produces depends partly on the care and effort that he takes and partly on random events beyond his control. Bates cannot observe or monitor Hogan's effort. Assume that Gastek's investment is fixed.

Required
1. How should Bates compensate Hogan:
   a. Pay Hogan a flat salary.
   b. Reward Hogan only on the basis of Gastek's revenues minus manufacturing costs before any bonus payments.
   c. Reward Hogan on the basis of a flat salary and a bonus based on Gastek's revenues minus manufacturing costs before any bonus payments.
   d. Reward Hogan only on the basis of the quantity of gaskets manufactured.
   e. Reward Hogan on the basis of a flat salary and a bonus based on the quantity of gaskets manufactured.

   Briefly explain why you chose the compensation measure that you did. Assume that the quality of gaskets is not a concern.
2. Suppose, for the purposes of this question, that Hogan's job has two aspects. He needs to produce as many gaskets as he can, but he also needs to pay attention to quality. Assume that Bates cannot determine the quality of the gaskets. Would your answer to requirement 1 change? Explain briefly.
3. What other steps would you recommend that Bates take if she wants to emphasize both the quality and quantity of gaskets produced?

26-31 Risk sharing, incentives, benchmarking, multiple tasks. The Dexter Division of AMCO sells car batteries. AMCO's corporate management gives Dexter management considerable operating and investment autonomy in running the division. AMCO is considering how it should compensate Jim Marks, the general manager of the Dexter Division. Proposal 1 calls for paying Marks a fixed salary. Proposal 2 calls for paying Marks no salary and compensating him only on the basis of the division's ROI (calculated based on operating income before any bonus payments). Proposal 3 calls for paying Marks some salary and some bonus based on ROI. Assume that Marks does not like bearing risk.

Required
1. Evaluate each of the three proposals, specifying the advantages and disadvantages of each.
2. Suppose that AMCO competes against Tiara Industries in the car battery business. Tiara is roughly the same size and operates in a business environment that is very similar to Dexter's. The senior management of AMCO is considering evaluating Marks on the basis of Dexter's ROI minus Tiara's ROI. Marks complains that this approach is unfair because the performance of another firm, over which he has no control, is included in his performance evaluation measure. Is Marks's complaint valid? Why or why not?
3. Suppose that Marks has no authority for making capital investment decisions. Corporate management makes these decisions. Under this condition, is return on investment a good performance evaluation measure for Marks? Explain.
4. Dexter's salespersons are responsible for selling and providing customer service and support. Sales are easy to measure. Although customer service is very important to Dexter in the long-run, it has not yet implemented customer-service measures. Marks wants to compensate his salesforce only on the basis of sales commissions paid for each unit of product sold. He cites two advantages to this plan: (a) It creates very strong incentives for the salesforce to work hard; and (b) the company pays salespersons only when the company itself is earning revenues and has cash. Do you like Marks's plan? Why or why not?

 26-32 Companywide versus division bonus plans, compensation plans. Andersen, Price and Young (APY) has three divisions: the Auditing Division, Consulting Division, and Tax Division. Gary Schofield, managing director of APY, has just heard that Jonathan Davies and five other senior partners in APY's Consulting Division are about to resign and set up the Davies Consulting Group. The possible resignation of Davies came as no surprise to Schofield, but the departure of the other five partners did. Schofield and Davies had openly clashed in APY partner meetings during the last six months over the current bonus plan. The current bonus per partner is based on 25% of total APY net income divided by the total number of APY partners. All partners receive the same bonus. The salary paid to each partner is based on guidelines set by an executive committee. Currently, the highest-paid partner (Schofield) can receive no more than 300% of the salary of the lowest-paid partner. The financial press recently ran a series of articles on internal personality problems at APY. Both Schofield and Davies were described as having "elephant-sized egos."

Exhibit 26-4 presents the revenues and net income of each division over the 19_1 to 19_5 period. In 19_4, APY had a $100 million out-of-court settlement cost resulting from a banking audit client (Western Bank) going bankrupt in 19_2. Stockholders of the Western Bank sued APY for $800 million. The suit alleged that APY had failed to use appropriate auditing procedures and had failed to detect gross irregularities in loan procedures at Western Bank. The 19_4 Auditing Division net income was based on a $49 million operating income and the $100 million settlement. (APY's insurance company also settled out of court for an additional $200 million beyond the $100 million paid by APY.)

Schofield is in his late fifties. He joined APY 30 years ago. He worked exclusively in the Auditing Division prior to his promotion to APY managing partner. Davies is in his late thirties and does not have an accounting certification (such as a CMA or CPA certificate). He is the author of two books as well as six articles in the *Harvard Business Review*. Davies has an MBA degree and is on a first-name basis with the senior executives of many corporations. A sizable part of APY's consulting business comes from corporations implementing the strategic frameworks that Davies developed.

The executive committee of APY makes the key decisions concerning compensation structure and resource allocation. There are 11 members of this committee—six from auditing, two from consulting, and three from tax. Davies is not a member of the executive committee because the APY articles of association require all executive committee members to have an accounting certification.

Required

1. Compute the bonus paid to each of the APY partners in each year of the 19_1 to 19_5 period.
2. Discuss two advantages and two disadvantages of including the $100 million out-of-court settlement as a 19_4 period cost when computing the total bonus pool for APY.
3. Assume that partner bonuses are calculated separately for each division. All partners in the same division would receive the same bonus. For example, the Consulting Division bonus would be based on 25% of the net income of the Consulting Division divided by the number of partners in the Consulting Division. Compute the bonus paid to each of the (a) auditing partners, (b) consulting partners, and (c) tax partners of APY in each year of the 19_1 to 19_5 period under this revised bonus plan. Comment on the differences between the bonuses paid to partners under this approach and the bonuses computed in requirement 1.

26-33 Ethics, cooking the books, division managers, internal control. PepsiCo's three main divisions are beverages (main product is the Pepsi soft drink), food products (main prod-

EXHIBIT 26-4

Andersen, Price and Young: Summary Data

	19_1	19_2	19_3	19_4	19_5
Revenues (in millions)					
Auditing Division	$601	$610	$ 625	$ 632	$ 648
Consulting Division	110	151	205	278	385
Taxation Division	193	212	248	265	294
Total	$904	$973	$1,078	$1,175	$1,327
Net Income (in millions)					
Auditing Division	$ 60	$ 55	$ 56	$ (51)*	$ 38
Consulting Division	36	45	71	89	121
Taxation Division	38	40	41	42	49
Total	$134	$140	$ 168	$ 80	$ 208
Number of Partners					
Auditing Division	624	614	636	648	628
Consulting Division	68	92	128	170	182
Taxation Division	176	180	188	202	208
Total	868	886	952	1,020	1,018

*Includes $100 million out-of-court settlement cost.

ucts are Frito-Lay chips and other snack foods), and food service (mainly Pizza Hut and Taco Bell).

Over the 19_1 to 19_5 period, the beverage division was organized into Pepsi-Cola Company (U.S. operations) and PepsiCo International. PepsiCo International bottled soft drinks in more than 600 foreign plants.

In November 19_5, PepsiCo issued the following in a press release:

> Internal auditors at PepsiCo, Inc. recently discovered significant accounting irregularities in certain company-owned foreign bottling operations of its International division. The foreign subsidiaries involved accounted for less than 5% of PepsiCo's operating income in 19_4.
>
> It appears that these irregularities involve the overstatement of assets and understatement of costs over several years, going back at least to 19_1.
>
> The company's investigation, being conducted by a task force that includes special legal counsel and public accountants retained for this purpose, has shown that accounts were falsified by managers of these foreign subsidiaries, principally in Mexico and the Philippines, to improve the apparent performance of their operations. Extensive collusion, creation of false documentation, and the evasion of company internal controls combined to make these misstatements possible. It does not presently appear that these misrepresentations were designed to divert company funds to personal, improper, or illegal use.
>
> PepsiCo is terminating and replacing the appropriate individuals, including the U.S.-based manager of the bottling unit of the International division.

In a December 19_5 press release, PepsiCo reported the following details on a restatement of earnings for the 19_1 to 19_5 period based on the investigation (in millions):

	19_1	19_2	19_3	19_4	19_5
Net income as reported	$225.8	$264.9	$291.8	$333.5	$273.0
Net reduction resulting from restatement	2.6	14.5	31.1	36.0	8.1
Net income as restated	$223.2	$250.4	$260.7	$297.5	$264.9

> A former SEC commissioner stated that "numerous techniques were used [by the foreign subsidiaries of PepsiCo] to report false operating income, including falsifying costs, failing to write off broken or unusable bottles and uncollectible accounts receivable, and writing up bottle inventories above cost. To further these schemes, certain individuals made false statements to PepsiCo and its independent auditors concerning the financial condition of certain subsidiaries and, at various times, participated in or were aware of the falsifications of the books and records of the foreign beverage operations."
>
> The operating income of PepsiCo's three main divisions over the 19_1 to 19_5 period, before any adjustments for the practices disclosed in the investigation, were (in millions):

Division	19_1	19_2	19_3	19_4	19_5
Beverage	$227.0	$254.0	$274.7	$281.9	$217.7
Food Products	158.2	195.4	245.8	298.5	326.4
Food Service	64.1	49.9	59.5	81.9	119.3

Required
1. For each year, compute the percentage that the "restatement of net income" is (a) of the net income PepsiCo originally reported and (b) of the operating income of the beverage division. Comment on the results.
2. What factors may have motivated the Mexico and Philippine senior managers to engage in the unethical practices?
3. One press commentator described the practices reported by the investigation as "business as usual in a company with a high pressure to perform. It is the proverbial storm in a teacup. I don't think the issue is a serious one for PepsiCo's top management." Do you agree? Explain your answer.

APPENDIX A

Surveys of Company Practice

This appendix provides the full citations to the individual publications cited in the many Surveys of Company Practice boxes included in the text.

AMERICAN ELECTRONICS ASSOCIATION, *Operating Ratios Survey 1991-92*, (Santa Clara, CA: American Electronics Association, 1991)—cited in Chapters 8 and 18.

ARMITAGE, H., AND R. NICHOLSON, "Activity-Based Costing: A Survey of Canadian Practice," forthcoming in Supplement to *CMA Magazine* (1993)—cited in Chapter 5.

ASADA, T., J. BAILES, AND M. AMANO, "An Empirical Study of Japanese and American Budget Planning and Control Systems," (Working Paper, Tsukuba University and Oregon State University, 1989)—cited in Chapter 6.

ASK, U., AND C. AX, "Trends in the Development of Product Costing Practices and Techniques—A Survey of the Swedish Manufacturing Industry," (Working Paper, Gothenburg School of Economics, Gothenburg, Sweden, 1992)—cited in Chapters 7 and 9.

ATKINSON, A., *Intrafirm Cost and Resource Allocations: Theory and Practice*, (Hamilton, Canada: Society of Management Accountants of Canada and Canadian Academic Accounting Association Research Monograph, 1987)—cited in Chapter 14.

BARTEZZAGHI, E., F. TURCO, AND G. SPINA, "The Impact of the Just-in-Time Approach on Production System Performance: A Survey of Italian Industry," *International Journal of Operations & Production Management* (Vol. 12, No. 1, 1992)—cited in Chapter 24.

BILLESBACH, T., A. HARRISON, AND S. CROOM-MORGAN, "Just-in-Time: A United States–United Kingdom Comparison," *International Journal of Operations & Production Management* (Vol. 11, No. 10, 1991)—cited in Chapter 24.

BLAYNEY, P., AND I. YOKOYAMA, "Comparative Analysis of Japanese and Australian Cost Accounting and Management Practices," (Working Paper, The University of Sydney, Sydney, Australia, 1991)—cited in Chapters 3, 5, 6, 9, 12, 14, 20, and 26.

BOONS, A., AND F. ROOZEN, "Symptoms of Dysfunctional Cost Information Systems: Some Preliminary Evidence from the Netherlands," (Working Paper, Erasmus Universiteit, Rotterdam, Netherlands, 1992)—cited in Chapter 15.

BRIGHT, J., R. DAVIES, C. DOWNES, AND R. SWEETING, "U.K. National" published as "The Deployment of Costing Techniques and Practices: A UK Study," *Management Accounting Research* (September 1992)—cited in Chapter 5.

BUSINESS INTERNATIONAL, *Managing the Global Finance Function*, (London, U.K.: Business International, 1992)—cited in Chapter 1.

CLARKE, P., "Management Accounting Practices and Techniques in Irish Manufacturing Firms: A Pilot Study," (Working Paper, Trinity College, Dublin, Ireland, 1992)—cited in Chapters 5, 7, 20, and 26.

CLARKE, P., AND T. O'DEA, "Management Accounting Systems: Some Field Evidence from Sixteen Multinational Companies in Ireland," (Working Paper, Trinity College, Dublin, Ireland, 1993)—cited in Chapter 24.

COHEN, J., AND L. PAQUETTE, "Management Accounting Practices: Perceptions of Controllers," *Journal of Cost Management* (Fall 1991)—cited in Chapter 5.

COOPER, R., "Does Your Company Need a New Cost System?" *Journal of Cost Management* (Spring 1987)—cited in Chapter 15.

CORNICK, M., W. COOPER, AND S. WILSON, "How Do Companies Analyze Overhead," *Management Accounting* (June 1988)—cited in Chapters 7 and 12.

DEAN, G., M. JOYE, AND P. BLAYNEY, *Strategic Management Accounting Survey*, (Sydney, Australia: The University of Sydney, 1991)—cited in Chapter 14.

DE WITH, E., AND E. IJSKES, "Current Budgeting Practices in Dutch Companies," (Working Paper, Vrije Universiteit, 1992, Amsterdam, Netherlands)—cited in Chapter 6.

DRURY, C., S. BRAUND, P. OSBORNE, AND M. TAYLES, *A Survey of Management Accounting Practices in UK Manufacturing Companies*, (London, U.K.: Chartered Association of Certified Accountants, 1993)—cited in Chapters 7, 12, and 25.

FREEMAN, M., AND G. HOBBES, "Capital Budgeting: Theory versus Practice," *Australian Accountant* (September 1991)—cited in Chapter 21.

FREMGEN, J., AND S. LIAO, *The Allocation of Corporate Indirect Costs* (New York: National Association of Accountants, 1981)—cited in Chapter 14.

GAUMNITZ, B., AND F. KOLLARITSCH, "Manufacturing Variances: Current Practice and Trends," *Journal of Cost Management* (Spring 1991)—cited in Chapter 7.

GOVINDARAJAN, V., AND B. RAMAMURTHY, "Transfer Pricing Policies in Indian Companies: A Survey," *Chartered Accountant* (November 1983)—cited in Chapter 25.

HO, S., AND R. PIKE, "Risk Analysis in Capital Budgeting Contexts: Simple or Sophisticated?" *Accounting and Business Research* (Vol. 21, No. 83, 1991)—cited in Chapter 21.

INOUE, S., "A Comparative Study of Recent Development of Cost Management Problems in U.S.A., U.K., Canada, and Japan," *Kagawa University Economic Review* (June 1988)—cited in Chapters 7 and 8.

JOG, V., AND A. SRIVASTAVA, "Capital Formation and Corporate Financial Decision Making in Canada," (Working Paper, Carleton University, Ottawa, Canada 1991)—cited in Chapters 20 and 21.

JOYE, M., AND P. BLAYNEY, "Cost and Management Accounting Practices in Australian Manufacturing Companies: Survey Results," (Accounting Research Centre, The University of Sydney, 1991)—cited in Chapters 10 and 25.

KIM, I., AND J. SONG, "U.S., Korea, and Japan: Accounting Practices in Three Countries," *Management Accounting* (August 1990)—cited in Chapters 20 and 21.

KPMG, *Building Global Profitability and Competitiveness: The New Role of Finance* (Montvale, NJ: KPMG Peat Marwick, 1989)—cited in Chapter 1.

LINDSAY, R., AND S. KALAGNANAM, *The Adoption of Just-in-Time Production Systems in Canada and Their Association with Management Control Practices*, (Hamilton, Canada:

Society of Management Accountants of Canada, 1993)—cited in Chapter 24.

MANAGEMENT ACCOUNTING RESEARCH GROUP, "Investigation into the Actual State of Target Costing, Corporate Accounting," (Working Paper, Kobe University, Japan, May 1992)—cited in Chaper 12.

MILLER, J., A. DE MEYER, AND J. NAKANE, *Benchmarking Global Manufacturing* (Homewood, IL: Business One Irwin, 1992)—cited in Chapter 2.

MILLS, R., AND C. SWEETING, "Pricing Decisions in Practice: How Are They Made in U.K. Manufacturing and Service Companies?" (London, U.K.: Chartered Institute of Management Accountants, Occasional Paper, 1988)—cited in Chapter 12.

MOSTAFA, A., J. SHARP, AND K. HOWARD, "Transfer Pricing—A Survey Using Discriminant Analysis," *Omega*, (Vol. 12, No. 5, 1984)—cited in Chapter 25.

MOWEN, M., *Accounting for Costs as Fixed and Variable* (National Association of Accountants: Montvale, NJ, 1986)—cited in Chapter 3.

NAA TOKYO AFFILIATE, "Management Accounting in the Advanced Manufacturing Surrounding: Comparative Study on Survey in Japan and U.S.A.," (Tokyo, Japan, 1988)—cited in Chapter 10.

PRICE WATERHOUSE, *Transfer Pricing Practices of American Industry* (New York: Price Waterhouse, 1984)—cited in Chapter 25.

RAMADAN, S., "The Rationale for Cost Allocation: A Study of UK Divisionalised Companies," *Accounting and Business Research* (Winter 1989)—cited in Chapter 14.

SALOWSKY, H., "Labor Costs in Twenty Industrialized Countries 1970-1991" (Working Paper, Institute of the German Economy in Cologne, Germany, 1992)—cited in Chapter 2.

SANGSTER, A., "Capital Investment Appraisal Techniques: A Survey of Current Usage," *Journal of Business Finance & Accounting* (April 1993)—cited in Chapter 20.

SCARBROUGH, P., A. NANNI, AND M. SAKURAI, "Japanese Management Accounting Practices and the Effects of Assembly and Process Automation," *Management Accounting Research* (March 1991)—cited in Chapter 7.

SCHIFF, J., "ABC on the Rise," *Cost Management Update Issue No. 24* (February 1993)—cited in Chapter 5.

SLATER, K., AND C. WOOTON, *A Study of Joint and By-Product Costing in the UK* (London, U.K.: Institute of Cost and Management Accountants, 1984)—cited in Chapter 16.

SMITH, K., AND C. SULLIVAN, "Survey of Cost Management Systems in Manufacturing," (Working Paper, Purdue University, West Lafayette, Indiana, 1990)—cited in Chapters 20 and 26.

SULLIVAN, C., AND K. SMITH, "Capital Investment Justification for U.S. Factory Automation Projects," *Journal of the Midwest Finance Association* (1994)—cited in Chapter 21.

SWENSON, D., AND J. CASSIDY, "The Effect of JIT on Management Accounting," *Journal of Cost Management* (Spring 1993)—cited in Chapter 19.

TANG, R., "Environmental Variables of Multinational Transfer Pricing: A U.K. Perspective," *Journal of Business Finance & Accounting* (Summer 1982)—cited in Chapter 25.

TANG, R., "Canadian Transfer Pricing in the 1990s." *Management Accounting* (February 1992)—cited in Chapter 25.

TANG, R., C. WALTER, AND R. RAYMOND, "Transfer Pricing–Japanese vs. American Style," *Management Accounting* (January 1979)—cited in Chapter 25.

THORNTON, GRANT, *Survey of American Manufacturers*, (New York: Grant Thornton, 1992)—cited in Chapter 12.

APPENDIX B

Recommended Readings

The literature on cost accounting and related areas is vast and varied. The following books illustrate recent publications that capture current developments:

BRIMSON, J., *Activity Accounting: An Activity-Based Costing Approach*. New York: Wiley, 1991.

COOPER, R., AND R. KAPLAN, *The Design of Cost Management Systems*. Englewood Cliffs, NJ: Prentice-Hall, 1991.

ERNST & YOUNG, *International Transfer Pricing*. London, U.K.: Business International, 1991.

HRONEC, S., *Vital Signs*. New York: American Management Association, 1993.

JOHNSON, T., *Relevance Regained*. New York: Free Press, 1992.

SCHWEITZER, M., E. TROSSMANN, AND G. LAWSON, *Break-even Analyses: Basic Model, Variants, Extensions*. Chichester, U. K.: Wiley, 1992.

SHANK, J., AND V. GOVINDARAJAN, *Strategic Management Accounting*. New York: The Free Press, 1993.

TURNEY, P., *Common Cents: The ABC Performance Breakthrough*. Hillsboro, OR: Cost Technology, 1991.

Books of readings related to cost or management accounting include:

ASHTON, D., T. HOPPER, AND R. SCAPENS, EDS., *Issues in Management Accounting*. Hemel Hempstead, U. K.: Prentice Hall International, 1991).

BRINKER, B., ED., *Emerging Practices in Cost Management*. Boston, MA: Warren, Gorham, and Lamont, 1992.

RATNATUNGA, J., J. MILLER, N. MUDALIGE, AND A. SOHALLED, EDS., *Issues in Strategic Management Accounting*. Sydney, Australia: Harcourt Brace Jovanovich, 1993.

The Harvard Business School series in accounting and control offers important contributions to the cost accounting literature, including:

ANTHONY, R., *The Management Control Function*. Boston. Harvard Business School Press, 1988.

BERLINER, C., AND J. BRIMSON, EDS., *Cost Management for Today's Advanced Manufacturing: The CAM-I Conceptual Design*. Boston: Harvard Business School Press, 1988.

BRUNS, W., ED., *Performance Measurement, Evaluation, and Incentives*. Boston: Harvard Business School Press, 1992.

BRUNS, W., AND R. KAPLAN, EDS., *Accounting and Management: Field Study Perspectives*. Boston: Harvard Business School Press, 1987.

JOHNSON, H., AND R. KAPLAN, *Relevance Lost: The Rise and Fall of Management Accounting*. Boston: Harvard Business School Press, 1987.

KAPLAN, R., ED., *Measures for Manufacturing Excellence*. Boston: Harvard Business School Press, 1990.

MERCHANT, K. A., *Rewarding Results: Motivating Profit Center Managers*. Boston: Harvard Business School Press, 1989.

Productivity Press publishes many books with a global focus on cost and management accounting, including:

KANUTSU, T., *TQC for Accounting: A New Role in Companywide Improvement*. Cambridge, MA: Productivity Press, 1990.

MONDEN, Y., *Cost Management in the New Manufacturing Age: Innovations in the Japanese Automotive Industry*. Cambridge, MA: Productivity Press, 1992.

The Institute of Management Accountants publishes monographs and books covering cost accounting topics, such as:

COOPER, R., R. KAPLAN, L. MAISEL, E. MORRISSEY, AND R. OEHM, *Implementing Activity-Based Cost Management: Moving from Analysis to Action*. Montvale, NJ: Institute of Management Accountants, 1992.

KLAMMER, T., *Managing Strategic and Capital Investment Decisions*. Burr Ridge, IL: Irwin & IMA, 1994.

The Financial Executives Research Foundation publishes monographs and books concerning topics of interest to financial executives, such as:

HOWELL, R., J. SHANK, S. SOUCY, AND J. FISHER, *Cost Management for Tomorrow: Seeking the Competitive Edge*. Morristown, NJ: Financial Executives Research Foundation, 1992.

KEATING, P., AND S. JABLONSKY, *Changing Roles of Financial Management*. Morristown, NJ: Financial Executives Research Foundation, 1990.

The Chartered Institute of Management Accountants publishes monographs and books, including:

DRURY, C., ED., *Management Accounting Handbook*. London, U.K.: Butterworth Heinemann and Chartered Institute of Management Accountants, 1992.

WARD, K., *Strategic Management Accounting*. Oxford, U.K.: Butterworth and Chartered Institute of Management Accountants, 1992.

The following are detailed annotated bibliographies of the cost and management accounting research literatures:

CLANCY, D., *Annotated Management Accounting Readings*. Management Accounting Section of the American Accounting Association, 1986.

DEAKIN, E., M. MAHER, AND J. CAPPEL, *Contemporary Literature in Cost Accounting*. Homewood, IL: Richard D. Irwin, 1988.

KLEMSTINE, C., AND M. MAHER, *Management Accounting Research: 1926–1983*. New York: Garland Publishing, 1984.

The *Journal of Cost Management for the Manufacturing Industry* contains numerous articles on modern management accounting. It is published by Warren, Gorham and Lamont, 210 South Street, Boston, MA 02111.

Two journals bearing on management accounting are published by sections of the American Accounting Association, 5717 Bessie Drive, Sarasota, FL 34233: *Journal of Management Accounting Research* and *Behavioral Research in Accounting*.

Professional associations that specialize in serving members with cost and management accounting interests include:

◆ *Institute of Management Accountants*, 10 Paragon Drive, P.O. Box 433, Montvale, NJ 07645. Publishes the *Management Accountant* journal.

◆ *Financial Executives Institute*, 10 Madison Avenue, P.O. Box 1938, Morristown, NJ 07960. Publishes *Financial Executive*.

◆ *Society of Cost Estimating and Analysis*, 101 South Whiting Street, Suite 313, Alexandria, VA 22304. Publishes the *Journal of Cost Analysis* and monographs related to cost estimation and price analysis in government and industry.

◆ *The Institute of Internal Auditors*, 249 Maitland Avenue, Altamonte Springs, FL 32701. Publishes *The Internal Auditor* journal. Also publishes monographs on topics related to internal control.

◆ *Society of Management Accountants of Canada*, 154 Main Street East, MPO Box 176, Hamilton, Ontario, L8N 3C3. Publishes the *CMA Magazine*.

◆ *The Chartered Institute of Management Accountants*, 63 Portland Place, London, WIN 4AB. Publishes the *Management Accounting* journal. Also publishes monographs covering cost and managerial accounting topics.

In many countries, individuals with cost and management accounting interests belong to professional bodies that serve members with financial reporting and taxation, as well as cost and management accounting, interests. An example is the Australian Society of Accountants.

APPENDIX C

Notes on Compound Interest and Interest Tables

Interest is the cost of using money. It is the rental charge for funds, just as renting a building and equipment entails a rental charge. When the funds are used for a period of time, it is necessary to recognize interest as a cost of using the borrowed ("rented") funds. This requirement applies even if the funds represent ownership capital and if interest does not entail an outlay of cash. Why must interest be considered? Because the selection of one alternative automatically commits a given amount of funds that could otherwise be invested in some other alternative.

Interest is generally important, even when short-term projects are under consideration. Interest looms correspondingly larger when long-run plans are studied. The rate of interest has significant enough impact to influence decisions regarding borrowing and investing funds. For example, $100,000 invested now and compounded annually for ten years at 8% will accumulate to $215,900; at 20%, the $100,000 will accumulate to $619,200.

INTEREST TABLES

Many computer programs and pocket calculators are available that handle computations involving the time value of money. You may also turn to the following four basic tables to compute interest.

Table 1—Future Amount of $1

Table 1 shows how much $1 invested now will accumulate in a given number of periods at a given compounded interest rate per period. Consider investing $1,000 now for three years at 8% compound interest. A tabular presentation of how this $1,000 would accumulate to $1,259.70 follows:

Year	Interest per Year	Cumulative Interest Called Compound Interest	Total at End of Year
0	$ —	$ —	$1,000.00
1	80.00	80.00	1,080.00
2	86.40	166.40	1,166.40
3	93.30	259.70	1,259.70

This tabular presentation is a series of computations that could appear as follows:

$$S_1 = \$1,000(1.08)^1$$
$$S_2 = \$1,000(1.08)^2$$
$$S_3 = \$1,000(1.08)^3$$

The formula for the "amount of 1," often called the "future value of $1" or "future amount of $1," can be written:

$$S = P(1 + r)^n$$
$$S = \$1,000(1 + .08)^3 = \$1,259.70$$

S is the future value amount; P is the present value, $1,000 in this case; r is the rate of interest; and n is the number of time periods.

Fortunately, tables make key computations readily available. A facility in selecting the *proper* table will minimize computations. Check the accuracy of the answer above using Table 1, page 931.

Table 2—Present Value of $1

In the previous example, if $1,000 compounded at 8% per year will accumulate to $1,259.70 in three years, then $1,000 must be the present value of $1,259.70 due at the end of three years. The formula for the present value can be derived by reversing the process of *accumulation* (finding the future amount) that we just finished.

$$S = P(1 + r)^n$$

If
$$P = \frac{S}{(1 + r)^n}$$

then
$$P = \frac{\$1,259.70}{(1.08)^3} = \$1,000$$

Use Table 2, page 932, to check this calculation.

When accumulating, we advance or roll forward in time. The difference between our original amount and our accumulated amount is called *compound interest*. When discounting, we retreat or roll back in time. The difference between the future amount and the present value is called *compound discount*. Note the following formulas (where $P = \$1,000$):

$$\text{Compound interest} = P\,[(1 + r)^n - 1] = \$259.70$$
$$\text{Compound discount} = S\left[1 - \frac{1}{(1 + r)^n}\right] = \$259.70$$

Table 3—Amount of Annuity of $1

An (ordinary) *annuity* is a series of equal payments (receipts) to be paid (or received) at the *end* of successive periods of equal length. Assume that $1,000 is invested at the end of each of three years at 8%:

End of Year	Amount
1st payment	$1,000.00 → $1,080.00 → $1,166.40, which is $1,000(1.08)^2
2nd payment	$1,000.00 → 1,080.00, which is $1,000(1.08)^1
3rd payment	1,000.00
Accumulation (future amount)	$3,246.40

The arithmetic shown above may be expressed algebraically as the amount of an ordinary annuity of $1,000 for three years $= \$1,000(1 + r)^2 + \$1,000(1 + r)^1 + \$1,000$.

We can develop the general formula for S_n, the amount of an ordinary annuity of $1, by using the example above as a basis:

1. $S_n = 1 + (1 + r)^1 + (1 + r)^2$

2. Substitute: $S_n = 1 + (1.08)^1 + (1.08)^2$

3. Multiply (2) by $(1 + r)$: $(1.08)S_n = (1.08)^1 + (1.08)^2 + (1.08)^3$

4. Subtract (2) from (3): $1.08S_n - S_n \, (1.08)^3 - 1$

   Note that all terms on the right-hand side are removed except $(1.08)^3$ in equation (3) and 1 in equation (2).

5. Factor (4): $S_n(1.08 - 1) = (1.08)^3 - 1$

6. Divide (5) by $(1.08 - 1)$: $S_n = \dfrac{(1.08)^3 - 1}{1.08 - 1} = \dfrac{(1.08)^3 - 1}{.08}$

7. The general formula for the amount of an ordinary annuity of $1

   becomes: $S_n = \dfrac{(1 + r)^n - 1}{r}$ or $\dfrac{\text{Compound interest}}{\text{Rate}}$

This formula is the basis for Table 3, page 933. Look at Table 3 or use the formula itself to check the calculations.

Table 4—Present Value of an Ordinary Annuity of $1

Using the same example as for Table 3, we can show how the formula of P_n, *the present value of an ordinary annuity*, is developed.

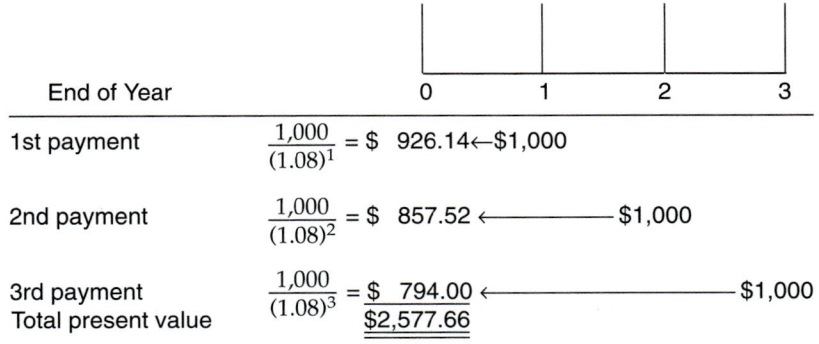

End of Year		
1st payment	$\dfrac{1,000}{(1.08)^1} = \$\ 926.14$ ← $1,000	
2nd payment	$\dfrac{1,000}{(1.08)^2} = \$\ 857.52$ ←	$1,000
3rd payment	$\dfrac{1,000}{(1.08)^3} = \$\ 794.00$ ←	$1,000
Total present value	$\$2,577.66$	

For the general case, the present value of an ordinary annuity of $1 may be expressed:

1.
$$P_n = \frac{1}{1+r} + \frac{1}{(1+r)^2} + \frac{1}{(1+r)^3}$$

2. Substitute
$$P_n = \frac{1}{1.08} + \frac{1}{(1.08)^2} + \frac{1}{(1.08)^3}$$

3. Multiply by $\frac{1}{1.08}$:
$$P_n \frac{1}{1.08} = \frac{1}{(1.08)^2} + \frac{1}{(1.08)^3} + \frac{1}{(1.08)^4}$$

4. Subtract (3) from (2):
$$P_n - P_n \frac{1}{1.08} = \frac{1}{1.08} - \frac{1}{(1.08)^4}$$

5. Factor:
$$P_n \left(1 - \frac{1}{(1.08)}\right) = \frac{1}{1.08}\left[1 - \frac{1}{(1.08)^3}\right]$$

6. or
$$P_n \left(\frac{.08}{1.08}\right) = \frac{1}{1.08}\left[1 - \frac{1}{(1.08)^3}\right]$$

7. Multiply by $\frac{1.08}{.08}$:
$$P_n = \frac{1}{.08}\left[1 - \frac{1}{(1.08)^3}\right]$$

The general formula for the present value of an annuity of $1.00 is:

$$P_n = \frac{1}{r}\left[1 - \frac{1}{(1+r)^n}\right] = \frac{\text{Compound discount}}{\text{Rate}}$$

Solving,

$$P_n = \frac{.2062}{.08} = 2.577$$

The formula is the basis for Table 4, page 934. Check the answer in the table. The present value tables, Tables 2 and 4, are used most frequently in capital budgeting.

The tables for annuities are not essential. With Tables 1 and 2, compound interest and compound discount can readily be computed. It is simply a matter of dividing either of these by the rate to get values equivalent to those shown in Tables 3 and 4.

TABLE 1

Compound Amount of $1.00 (The Future Value of $1.00)

$S = P(1 + r)^n$. In this table $P = \$1.00$.

PERIODS	2%	4%	6%	8%	10%	12%	14%	16%	18%	20%	22%	24%	26%	28%	30%	32%	40%	PERIODS
1	1.020	1.040	1.060	1.080	1.100	1.120	1.140	1.160	1.180	1.200	1.220	1.240	1.260	1.280	1.300	1.320	1.400	1
2	1.040	1.082	1.124	1.166	1.210	1.254	1.300	1.346	1.392	1.440	1.488	1.538	1.588	1.638	1.690	1.742	1.960	2
3	1.061	1.125	1.191	1.260	1.331	1.405	1.482	1.561	1.643	1.728	1.816	1.907	2.000	2.097	2.197	2.300	2.744	3
4	1.082	1.170	1.262	1.360	1.464	1.574	1.689	1.811	1.939	2.074	2.215	2.364	2.520	2.684	2.856	3.036	3.842	4
5	1.104	1.217	1.338	1.469	1.611	1.762	1.925	2.100	2.288	2.488	2.703	2.932	3.176	3.436	3.713	4.007	5.378	5
6	1.126	1.265	1.419	1.587	1.772	1.974	2.195	2.436	2.700	2.986	3.297	3.635	4.002	4.398	4.827	5.290	7.530	6
7	1.149	1.316	1.504	1.714	1.949	2.211	2.502	2.826	3.185	3.583	4.023	4.508	5.042	5.629	6.275	6.983	10.541	7
8	1.172	1.369	1.594	1.851	2.144	2.476	2.853	3.278	3.759	4.300	4.908	5.590	6.353	7.206	8.157	9.217	14.758	8
9	1.195	1.423	1.689	1.999	2.358	2.773	3.252	3.803	4.435	5.160	5.987	6.931	8.005	9.223	10.604	12.166	20.661	9
10	1.219	1.480	1.791	2.159	2.594	3.106	3.707	4.411	5.234	6.192	7.305	8.594	10.086	11.806	13.786	16.060	28.925	10
11	1.243	1.539	1.898	2.332	2.853	3.479	4.226	5.117	6.176	7.430	8.912	10.657	12.708	15.112	17.922	21.199	40.496	11
12	1.268	1.601	2.012	2.518	3.138	3.896	4.818	5.936	7.288	8.916	10.872	13.215	16.012	19.343	23.298	27.983	56.694	12
13	1.294	1.665	2.133	2.720	3.452	4.363	5.492	6.886	8.599	10.699	13.264	16.386	20.175	24.759	30.288	36.937	79.371	13
14	1.319	1.732	2.261	2.937	3.797	4.887	6.261	7.988	10.147	12.839	16.182	20.319	25.421	31.691	39.374	48.757	111.120	14
15	1.346	1.801	2.397	3.172	4.177	5.474	7.138	9.266	11.974	15.407	19.742	25.196	32.030	40.565	51.186	64.359	155.568	15
16	1.373	1.873	2.540	3.426	4.595	6.130	8.137	10.748	14.129	18.488	24.086	31.243	40.358	51.923	66.542	84.954	217.795	16
17	1.400	1.948	2.693	3.700	5.054	6.866	9.276	12.468	16.672	22.186	29.384	38.741	50.851	66.461	86.504	112.139	304.913	17
18	1.428	2.026	2.854	3.996	5.560	7.690	10.575	14.463	19.673	26.623	35.849	48.039	64.072	85.071	112.455	148.024	426.879	18
19	1.457	2.107	3.026	4.316	6.116	8.613	12.056	16.777	23.214	31.948	43.736	59.568	80.731	108.890	146.192	195.391	597.630	19
20	1.486	2.191	3.207	4.661	6.727	9.646	13.743	19.461	27.393	38.338	53.358	73.864	101.721	139.380	190.050	257.916	836.683	20
21	1.516	2.279	3.400	5.034	7.400	10.804	15.668	22.574	32.324	46.005	65.096	91.592	128.169	178.406	247.065	340.449	1171.356	21
22	1.546	2.370	3.604	5.437	8.140	12.100	17.861	26.186	38.142	55.206	79.418	113.574	161.492	228.360	321.184	449.393	1639.898	22
23	1.577	2.465	3.820	5.871	8.954	13.552	20.362	30.376	45.008	66.247	96.889	140.831	203.480	292.300	417.539	593.199	2295.857	23
24	1.608	2.563	4.049	6.341	9.850	15.179	23.212	35.236	53.109	79.497	118.205	174.631	256.385	374.144	542.801	783.023	3214.200	24
25	1.641	2.666	4.292	6.848	10.835	17.000	26.462	40.874	62.669	95.396	144.210	216.542	323.045	478.905	705.641	1033.590	4499.880	25
26	1.673	2.772	4.549	7.396	11.918	19.040	30.167	47.414	73.949	114.475	175.936	268.512	407.037	612.998	917.333	1364.339	6299.831	26
27	1.707	2.883	4.822	7.988	13.110	21.325	34.390	55.000	87.260	137.371	214.642	332.955	512.867	784.638	1192.533	1800.927	8819.764	27
28	1.741	2.999	5.112	8.627	14.421	23.884	39.204	63.800	102.967	164.845	261.864	412.864	646.212	1004.336	1550.293	2377.224	12347.670	28
29	1.776	3.119	5.418	9.317	15.863	26.750	44.693	74.009	121.501	197.814	319.474	511.952	814.228	1285.550	2015.381	3137.935	17286.737	29
30	1.811	3.243	5.743	10.063	17.449	29.960	50.950	85.850	143.371	237.376	389.758	634.820	1025.927	1645.505	2619.996	4142.075	24201.432	30
35	2.000	3.946	7.686	14.785	28.102	52.800	98.100	180.314	327.997	590.668	1053.402	1861.054	3258.135	5653.911	9727.860	16599.217	130161.112	35
40	2.208	4.801	10.286	21.725	45.259	93.051	188.884	378.721	750.378	1469.772	2847.038	5455.913	10347.175	19426.689	36118.865	66520.767	700037.697	40

TABLE 2 *(Place a clip on this page for easy reference.)*

Present Value of $1.00.

$$P = \frac{S}{(1+r)^n}. \text{ In this table } S = \$1.00.$$

PERIODS	2%	4%	6%	8%	10%	12%	14%	16%	18%	20%	22%	24%	26%	28%	30%	32%	40%	PERIODS
1	0.980	0.962	0.943	0.926	0.909	0.893	0.877	0.862	0.847	0.833	0.820	0.806	0.794	0.781	0.769	0.758	0.714	1
2	0.961	0.925	0.890	0.857	0.826	0.797	0.769	0.743	0.718	0.694	0.672	0.650	0.630	0.610	0.592	0.574	0.510	2
3	0.942	0.889	0.840	0.794	0.751	0.712	0.675	0.641	0.609	0.579	0.551	0.524	0.500	0.477	0.455	0.435	0.364	3
4	0.924	0.855	0.792	0.735	0.683	0.636	0.592	0.552	0.516	0.482	0.451	0.423	0.397	0.373	0.350	0.329	0.260	4
5	0.906	0.822	0.747	0.681	0.621	0.567	0.519	0.476	0.437	0.402	0.370	0.341	0.315	0.291	0.269	0.250	0.186	5
6	0.888	0.790	0.705	0.630	0.564	0.507	0.456	0.410	0.370	0.335	0.303	0.275	0.250	0.227	0.207	0.189	0.133	6
7	0.871	0.760	0.665	0.583	0.513	0.452	0.400	0.354	0.314	0.279	0.249	0.222	0.198	0.178	0.159	0.143	0.095	7
8	0.853	0.731	0.627	0.540	0.467	0.404	0.351	0.305	0.266	0.233	0.204	0.179	0.157	0.139	0.123	0.108	0.068	8
9	0.837	0.703	0.592	0.500	0.424	0.361	0.308	0.263	0.225	0.194	0.167	0.144	0.125	0.108	0.094	0.082	0.048	9
10	0.820	0.676	0.558	0.463	0.386	0.322	0.270	0.227	0.191	0.162	0.137	0.116	0.099	0.085	0.073	0.062	0.035	10
11	0.804	0.650	0.527	0.429	0.350	0.287	0.237	0.195	0.162	0.135	0.112	0.094	0.079	0.066	0.056	0.047	0.025	11
12	0.788	0.625	0.497	0.397	0.319	0.257	0.208	0.168	0.137	0.112	0.092	0.076	0.062	0.052	0.043	0.036	0.018	12
13	0.773	0.601	0.469	0.368	0.290	0.229	0.182	0.145	0.116	0.093	0.075	0.061	0.050	0.040	0.033	0.027	0.013	13
14	0.758	0.577	0.442	0.340	0.263	0.205	0.160	0.125	0.099	0.078	0.062	0.049	0.039	0.032	0.025	0.021	0.009	14
15	0.743	0.555	0.417	0.315	0.239	0.183	0.140	0.108	0.084	0.065	0.051	0.040	0.031	0.025	0.020	0.016	0.006	15
16	0.728	0.534	0.394	0.292	0.218	0.163	0.123	0.093	0.071	0.054	0.042	0.032	0.025	0.019	0.015	0.012	0.005	16
17	0.714	0.513	0.371	0.270	0.198	0.146	0.108	0.080	0.060	0.045	0.034	0.026	0.020	0.015	0.012	0.009	0.003	17
18	0.700	0.494	0.350	0.250	0.180	0.130	0.095	0.069	0.051	0.038	0.028	0.021	0.016	0.012	0.009	0.007	0.002	18
19	0.686	0.475	0.331	0.232	0.164	0.116	0.083	0.060	0.043	0.031	0.023	0.017	0.012	0.009	0.007	0.005	0.002	19
20	0.673	0.456	0.312	0.215	0.149	0.104	0.073	0.051	0.037	0.026	0.019	0.014	0.010	0.007	0.005	0.004	0.001	20
21	0.660	0.439	0.294	0.199	0.135	0.093	0.064	0.044	0.031	0.022	0.015	0.011	0.008	0.006	0.004	0.003	0.001	21
22	0.647	0.422	0.278	0.184	0.123	0.083	0.056	0.038	0.026	0.018	0.013	0.009	0.006	0.004	0.003	0.002	0.001	22
23	0.634	0.406	0.262	0.170	0.112	0.074	0.049	0.033	0.022	0.015	0.010	0.007	0.005	0.003	0.002	0.002	0.000	23
24	0.622	0.390	0.247	0.158	0.102	0.066	0.043	0.028	0.019	0.013	0.008	0.006	0.004	0.003	0.002	0.001	0.000	24
25	0.610	0.375	0.233	0.146	0.092	0.059	0.038	0.024	0.016	0.010	0.007	0.005	0.003	0.002	0.001	0.001	0.000	25
26	0.598	0.361	0.220	0.135	0.084	0.053	0.033	0.021	0.014	0.009	0.006	0.004	0.002	0.002	0.001	0.001	0.000	26
27	0.586	0.347	0.207	0.125	0.076	0.047	0.029	0.018	0.011	0.007	0.005	0.003	0.002	0.001	0.001	0.001	0.000	27
28	0.574	0.333	0.196	0.116	0.069	0.042	0.026	0.016	0.010	0.006	0.004	0.002	0.002	0.001	0.001	0.000	0.000	28
29	0.563	0.321	0.185	0.107	0.063	0.037	0.022	0.014	0.008	0.005	0.003	0.002	0.001	0.001	0.000	0.000	0.000	29
30	0.552	0.308	0.174	0.099	0.057	0.033	0.020	0.012	0.007	0.004	0.003	0.002	0.001	0.001	0.000	0.000	0.000	30
35	0.500	0.253	0.130	0.068	0.036	0.019	0.010	0.006	0.003	0.002	0.001	0.001	0.000	0.000	0.000	0.000	0.000	35
40	0.453	0.208	0.097	0.046	0.022	0.011	0.005	0.003	0.001	0.001	0.000	0.000	0.000	0.000	0.000	0.000	0.000	40

TABLE 3

Compound Amount of Annuity of $1.00 in Arrears* (Future Value of Annuity)

$$S_n = \frac{(1+r)^n - 1}{r}$$

PERIODS	2%	4%	6%	8%	10%	12%	14%	16%	18%	20%	22%	24%	26%	28%	30%	32%	40%	PERIODS
1	1.000	1.000	1.000	1.000	1.000	1.000	1.000	1.000	1.000	1.000	1.000	1.000	1.000	1.000	1.000	1.000	1.000	1
2	2.020	2.040	2.060	2.080	2.100	2.120	2.140	2.160	2.180	2.200	2.220	2.240	2.260	2.280	2.300	2.320	2.400	2
3	3.060	3.122	3.184	3.246	3.310	3.374	3.440	3.506	3.572	3.640	3.708	3.778	3.848	3.918	3.990	4.062	4.360	3
4	4.122	4.246	4.375	4.506	4.641	4.779	4.921	5.066	5.215	5.368	5.524	5.684	5.848	6.016	6.187	6.362	7.104	4
5	5.204	5.416	5.637	5.867	6.105	6.353	6.610	6.877	7.154	7.442	7.740	8.048	8.368	8.700	9.043	9.398	10.946	5
6	6.308	6.633	6.975	7.336	7.716	8.115	8.536	8.977	9.442	9.930	10.442	10.980	11.544	12.136	12.756	13.406	16.324	6
7	7.434	7.898	8.394	8.923	9.487	10.089	10.730	11.414	12.142	12.916	13.740	14.615	15.546	16.534	17.583	18.696	23.853	7
8	8.583	9.214	9.897	10.637	11.436	12.300	13.233	14.240	15.327	16.499	17.762	19.123	20.588	22.163	23.858	25.678	34.395	8
9	9.755	10.583	11.491	12.488	13.579	14.776	16.085	17.519	19.086	20.799	22.670	24.712	26.940	29.369	32.015	34.895	49.153	9
10	10.950	12.006	13.181	14.487	15.937	17.549	19.337	21.321	23.521	25.959	28.657	31.643	34.945	38.593	42.619	47.062	69.814	10
11	12.169	13.486	14.972	16.645	18.531	20.655	23.045	25.733	28.755	32.150	35.962	40.238	45.031	50.398	56.405	63.122	98.739	11
12	13.412	15.026	16.870	18.977	21.384	24.133	27.271	30.850	34.931	39.581	44.874	50.895	57.739	65.510	74.327	84.320	139.235	12
13	14.680	16.627	18.882	21.495	24.523	28.029	32.089	36.786	42.219	48.497	55.746	64.110	73.751	84.853	97.625	112.303	195.929	13
14	15.974	18.292	21.015	24.215	27.975	32.393	37.581	43.672	50.818	59.196	69.010	80.496	93.926	109.612	127.913	149.240	275.300	14
15	17.293	20.024	23.276	27.152	31.772	37.280	43.842	51.660	60.965	72.035	85.192	100.815	119.347	141.303	167.286	197.997	386.420	15
16	18.639	21.825	25.673	30.324	35.950	42.753	50.980	60.925	72.939	87.442	104.935	126.011	151.377	181.868	218.472	262.356	541.988	16
17	20.012	23.698	28.213	33.750	40.545	48.884	59.118	71.673	87.068	105.931	129.020	157.253	191.735	233.791	285.014	347.309	759.784	17
18	21.412	25.645	30.906	37.450	45.599	55.750	68.394	84.141	103.740	128.117	158.405	195.994	242.585	300.252	371.518	459.449	1064.697	18
19	22.841	27.671	33.760	41.446	51.159	63.440	78.969	98.603	123.414	154.740	194.254	244.033	306.658	385.323	483.973	607.472	1491.576	19
20	24.297	29.778	36.786	45.762	57.275	72.052	91.025	115.380	146.628	186.688	237.989	303.601	387.389	494.213	630.165	802.863	2089.206	20
21	25.783	31.969	39.993	50.423	64.002	81.699	104.768	134.841	174.021	225.026	291.347	377.465	489.110	633.593	820.215	1060.779	2925.889	21
22	27.299	34.248	43.392	55.457	71.403	92.503	120.436	157.415	206.345	271.031	356.443	469.056	617.278	811.999	1067.280	1401.229	4097.245	22
23	28.845	36.618	46.996	60.893	79.543	104.603	138.297	183.601	244.487	326.237	435.861	582.630	778.771	1040.358	1388.464	1850.622	5737.142	23
24	30.422	39.083	50.816	66.765	88.497	118.155	158.659	213.978	289.494	392.484	532.750	723.461	982.251	1332.659	1806.003	2443.821	8032.999	24
25	32.030	41.646	54.865	73.106	98.347	133.334	181.871	249.214	342.603	471.981	650.955	898.092	1238.636	1706.803	2348.803	3226.844	11247.199	25
26	33.671	44.312	59.156	79.954	109.182	150.334	208.333	290.088	405.272	567.377	795.165	1114.634	1561.682	2185.708	3054.444	4260.434	15747.079	26
27	35.344	47.084	63.706	87.351	121.100	169.374	238.499	337.502	479.221	681.853	971.102	1383.146	1968.719	2798.706	3971.778	5624.772	22046.910	27
28	37.051	49.968	68.528	95.339	134.210	190.699	272.889	392.503	566.481	819.223	1185.744	1716.101	2481.586	3583.344	5164.311	7425.699	30866.674	28
29	38.792	52.966	73.640	103.966	148.631	214.583	312.094	456.303	669.447	984.068	1447.608	2128.965	3127.798	4587.680	6714.604	9802.923	43214.343	29
30	40.568	56.085	79.058	113.283	164.494	241.333	356.787	530.312	790.948	1181.882	1767.081	2640.916	3942.026	5873.231	8729.985	12940.859	60501.081	30
35	49.994	73.652	111.435	172.317	271.024	431.663	693.573	1120.713	1816.652	2948.341	4783.645	7750.225	12527.442	20188.966	32422.868	51869.427	325400.279	35
40	60.402	95.026	154.762	259.057	442.593	767.091	1342.025	2360.757	4163.213	7343.858	12936.535	22728.803	39792.982	69377.460	120392.883	207874.272	1750091.741	40

*Payments (or receipts) at the end of each period.

TABLE 4 *(Place a clip on this page for easy reference.)*

Present Value of Annuity $1.00 in Arrears*.

$$P_n = \frac{1}{r}\left[1 - \frac{1}{(1+r)^n}\right]$$

PERIODS	2%	4%	6%	8%	10%	12%	14%	16%	18%	20%	22%	24%	26%	28%	30%	32%	40%	PERIODS
1	0.980	0.962	0.943	0.926	0.909	0.893	0.877	0.862	0.847	0.833	0.820	0.806	0.794	0.781	0.769	0.758	0.714	1
2	1.942	1.886	1.833	1.783	1.736	1.690	1.647	1.605	1.566	1.528	1.492	1.457	1.424	1.392	1.361	1.331	1.224	2
3	2.884	2.775	2.673	2.577	2.487	2.402	2.322	2.246	2.174	2.106	2.042	1.981	1.923	1.868	1.816	1.766	1.589	3
4	3.808	3.630	3.465	3.312	3.170	3.037	2.914	2.798	2.690	2.589	2.494	2.404	2.320	2.241	2.166	2.096	1.849	4
5	4.713	4.452	4.212	3.993	3.791	3.605	3.433	3.274	3.127	2.991	2.864	2.745	2.635	2.532	2.436	2.345	2.035	5
6	5.601	5.242	4.917	4.623	4.355	4.111	3.889	3.685	3.498	3.326	3.167	3.020	2.885	2.759	2.643	2.534	2.168	6
7	6.472	6.002	5.582	5.206	4.868	4.564	4.288	4.039	3.812	3.605	3.416	3.242	3.083	2.937	2.802	2.677	2.263	7
8	7.325	6.733	6.210	5.747	5.335	4.968	4.639	4.344	4.078	3.837	3.619	3.421	3.241	3.076	2.925	2.786	2.331	8
9	8.162	7.435	6.802	6.247	5.759	5.328	4.946	4.607	4.303	4.031	3.786	3.566	3.366	3.184	3.019	2.868	2.379	9
10	8.983	8.111	7.360	6.710	6.145	5.650	5.216	4.833	4.494	4.192	3.923	3.682	3.465	3.269	3.092	2.930	2.414	10
11	9.787	8.760	7.887	7.139	6.495	5.938	5.453	5.029	4.656	4.327	4.035	3.776	3.543	3.335	3.147	2.978	2.438	11
12	10.575	9.385	8.384	7.536	6.814	6.194	5.660	5.197	4.793	4.439	4.127	3.851	3.606	3.387	3.190	3.013	2.456	12
13	11.348	9.986	8.853	7.904	7.103	6.424	5.842	5.342	4.910	4.533	4.203	3.912	3.656	3.427	3.223	3.040	2.469	13
14	12.106	10.563	9.295	8.244	7.367	6.628	6.002	5.468	5.008	4.611	4.265	3.962	3.695	3.459	3.249	3.061	2.478	14
15	12.849	11.118	9.712	8.559	7.606	6.811	6.142	5.575	5.092	4.675	4.315	4.001	3.726	3.483	3.268	3.076	2.484	15
16	13.578	11.652	10.106	8.851	7.824	6.974	6.265	5.668	5.162	4.730	4.357	4.033	3.751	3.503	3.283	3.088	2.489	16
17	14.292	12.166	10.477	9.122	8.022	7.120	6.373	5.749	5.222	4.775	4.391	4.059	3.771	3.518	3.295	3.097	2.492	17
18	14.992	12.659	10.828	9.372	8.201	7.250	6.467	5.818	5.273	4.812	4.419	4.080	3.786	3.529	3.304	3.104	2.494	18
19	15.678	13.134	11.158	9.604	8.365	7.366	6.550	5.877	5.316	4.843	4.442	4.097	3.799	3.539	3.311	3.109	2.496	19
20	16.351	13.590	11.470	9.818	8.514	7.469	6.623	5.929	5.353	4.870	4.460	4.110	3.808	3.546	3.316	3.113	2.497	20
21	17.011	14.029	11.764	10.017	8.649	7.562	6.687	5.973	5.384	4.891	4.476	4.121	3.816	3.551	3.320	3.116	2.498	21
22	17.658	14.451	12.042	10.201	8.772	7.645	6.743	6.011	5.410	4.909	4.488	4.130	3.822	3.556	3.323	3.118	2.498	22
23	18.292	14.857	12.303	10.371	8.883	7.718	6.792	6.044	5.432	4.925	4.499	4.137	3.827	3.559	3.325	3.120	2.499	23
24	18.914	15.247	12.550	10.529	8.985	7.784	6.835	6.073	5.451	4.937	4.507	4.143	3.831	3.562	3.327	3.121	2.499	24
25	19.523	15.622	12.783	10.675	9.077	7.843	6.873	6.097	5.467	4.948	4.514	4.147	3.834	3.564	3.329	3.122	2.499	25
26	20.121	15.983	13.003	10.810	9.161	7.896	6.906	6.118	5.480	4.956	4.520	4.151	3.837	3.566	3.330	3.123	2.500	26
27	20.707	16.330	13.211	10.935	9.237	7.943	6.935	6.136	5.492	4.964	4.524	4.154	3.839	3.567	3.331	3.123	2.500	27
28	21.281	16.663	13.406	11.051	9.307	7.984	6.961	6.152	5.502	4.970	4.528	4.157	3.840	3.568	3.331	3.124	2.500	28
29	21.844	16.984	13.591	11.158	9.370	8.022	6.983	6.166	5.510	4.975	4.531	4.159	3.841	3.569	3.332	3.124	2.500	29
30	22.396	17.292	13.765	11.258	9.427	8.055	7.003	6.177	5.517	4.979	4.534	4.160	3.842	3.569	3.332	3.124	2.500	30
35	24.999	18.665	14.498	11.655	9.644	8.176	7.070	6.215	5.539	4.992	4.541	4.164	3.845	3.571	3.333	3.125	2.500	35
40	27.355	19.793	15.046	11.925	9.779	8.244	7.105	6.233	5.548	4.997	4.544	4.166	3.846	3.571	3.333	3.125	2.500	40

*Payments (or receipts) at the end of each period.

APPENDIX D

Cost Accounting in

Professional Examinations

This appendix describes the role of cost accounting in professional examinations. We use professional examinations in the United States, Canada, Australia, Japan, and the United Kingdom to illustrate this role[1]. A conscientious reader who has solved a representative sample of the problems at the end of the chapters will be well prepared for the professional examination questions dealing with cost accounting. This appendix aims to provide perspective, instill confidence, and encourage readers to take the examinations.

AMERICAN PROFESSIONAL EXAMINATIONS

CPA and CMA Designations

Many American readers may eventually take the Certified Public Accountant (CPA) examination or the Certified Management Accountant (CMA) examination. Certification is important to professional accountants for many reasons, such as:

1. Recognition of achievement and technical competence by fellow accountants and by users of accounting services
2. Increased self-confidence in one's professional abilities
3. Membership in professional organizations offering programs of career-long education
4. Enhancement of career opportunities
5. Personal satisfaction

The CPA certificate is issued by individual states; it is necessary for obtaining a state's license to practice as a Certified Public Accountant. A prominent feature of public accounting is the use of independent (external) auditors to give assurance about the re-

---

[1]We appreciate help from Hadassah Baum (U.S.A.), Lisa Anguish (Canada), Jan Chamberlain (Australia), Takeo Tanaka and Michi Sakurai (Japan) and Andrea Jeffries (United Kingdom) in preparing this appendix.

liability of the financial statements supplied by managers. These auditors are called Certified Public Accountants in the United States and Chartered Accountants in many other English-speaking nations. The major U.S. professional association in the private sector that regulates the quality of external auditing is the American Institute of Certified Public Accountants (AICPA).

The CMA designation is offered by the Institute of Management Accountants (IMA). The IMA is the largest association of management accountants in the world.[2] The major objective of the CMA certification is to enhance the development of the management accounting profession. In particular, focus is placed on the modern role of the management accountant as an active contributor to and a participant in management. The CMA designation is gaining increased stature in the business community as a credential parallel to the CPA designation.

The CMA examination consists of four parts taken during two days (16 hours):

- Part 1: Economics, finance, and management
- Part 2: Financial accounting and reporting
- Part 3: Management reporting, analysis, and behavioral issues
- Part 4: Decision analysis and information systems

Questions regarding ethical issues will appear on any part of the examination. A person who has successfully completed the U.S. CPA examination is exempt from Part 2.

Cost/management accounting questions are prominent in the CMA examination. The CPA examination also includes such questions, although they are less extensive than questions regarding financial accounting, auditing, and business law. On the average, cost/managerial accounting represents 35–40% of the CMA examination and 5% of the CPA examination. This book includes many questions and problems used in past CMA and CPA examinations. In addition, a supplement to this book, *Student Guide and Review Manual* (John K. Harris and Dudley W. Curry [Englewood Cliffs, NJ: Prentice Hall, 1994]), contains over one hundred CMA and CPA questions and explanatory answers. Careful study of appropriate topics in this book will give candidates sufficient background for succeeding in the cost accounting portions of the professional examinations.

The IMA publishes *Management Accounting* monthly. Each issue includes advertisements for courses that help students prepare for the CMA examination.[3]

CANADIAN PROFESSIONAL EXAMINATIONS

Three professional accounting designations are available in Canada:

Designation	Sponsoring Organization
Certified Management Accountant (CMA)	Society of Management Accountants (SMA)
Certified General Accountant (CGA)	Certified General Accountants' Association (CGA)
Chartered Accountant (CA)	Canadian Institute of Chartered Accountants

The SMA represents over 24,000 certified management accountants employed throughout Canadian business, industry, and government.

---

[2]The IMA has a wide range of activities driven by many committees. For example, the Management Accounting Practices Committee issues statements on both financial accounting and management accounting. The IMA also has an extensive continuing-education program.

[3]Other U.S. professional associations also require detailed knowledge of cost accounting. For example, the Certified Cost Estimator/Analyst (CCEA) program is administered by the Society of Cost Estimating and Analysis, 101 South Whiting Street, Suite 313, Alexandria, VA 22304. The society's primary purpose is to improve the effectiveness of cost estimation and price analysis. Special attention is given to contract cost estimation.

The CMA Entrance Examination is a two-day examination, divided into three broad categories:

1. Management accounting area   50-60%
2. Financial accounting area   20-30%
3. Management studies   15-25%

Objective questions comprise 50–60% and cases 50–40% of the exam. Topics covered on recent examinations in the management accounting area include relevant costing, transfer pricing, capital budgeting, performance measures, activity-based costing, cost allocation, and productivity.

The Society of Management Accountants publishes *CMA: The Management Accounting Magazine* monthly. This magazine includes details of courses that assist students in preparing for the CMA examination.

AUSTRALIAN PROFESSIONAL EXAMINATIONS

The Australian Society of Certified Practicing Accountants is the largest body representing accountants in Australia. Their professional designation is termed a CPA (Certified Practicing Accountant). The basic entry requirements for Associate membership of the Society are having an approved Bachelor's degree and passing the CPA program. There are two compulsory core segments in the exam. Core I covers the practical application of the more common accounting standards and ethics, while more technical standards (such as foreign currency translation) are covered in the Core II segment. Candidates are then required to take three segments from seven areas. These areas are: (1) external reporting, (2) insolvency and reconstruction, (3) management accounting, (4) management of information technology, (5) auditing, (6) treasury, and (7) taxation.

The management accounting segment topics include:

1. Defining the management accounting function
2. Cost accumulation and product costing (including activity-based costing)
3. Feedback control I: Long-term planning (including capital budgeting)
4. Feedback control II: Short-term planning (including CVP and cost estimation)
5. Feedback control (including responsibility accounting and performance measures)

The Australian Accountant, published each month by the society, includes advertisements for courses that help students prepare for the CPA examination.

JAPANESE PROFESSIONAL EXAMINATIONS

The Japan Industrial Management and Accounting Institute (JIMAI) promotes the knowledge of finance, accounting, and management for business enterprises and public utilities throughout Japan. JIMAI has more than 600 corporate members comprising leading industrial and business organizations of the country. It is the oldest, the largest, and the most authoritative accounting organization of its kind in Japan.

JIMAI directs a School of Cost Control and a School of Corporate Tax Accounting. There are two courses in the School of Cost Control—Preparatory Course and Cost Control Course. These courses are taught by university professors and executives from member corporations. There are more than 12,000 graduates of this school since it was organized in 1942.

JIMAI sponsors specialist committees to study and research problems including the Committee on Management Accounting and the Committee on Corporate Finance.

Financial and Cost Accounting is published monthly by JIMAI.

The Chartered Institute of Management Accountants (CIMA) is the recognized professional management accounting body in the United Kingdom. CIMA provides a wide range of services to members in commerce, education, government, and the accounting profession.

The syllabus for the CIMA examination consists of four stages:

1. Preparation for business and accounting (including "foundation costing")
2. The tools of management accounting (including "operational cost accounting")
3. The rules of a profession (including "management accounting applications")
4. The application of knowledge to business management and finance (including "strategic management accounting" and "management accounting control systems")

Management Accounting, published monthly by CIMA, includes details of courses assisting students in preparing for their examinations.

Glossary

Abnormal spoilage. Spoilage that is not expected to arise under efficient operating conditions; it is not an inherent part of the selected production process. (633)

Absorption costing. Method of inventory costing in which all direct manufacturing costs and all manufacturing overhead costs—both variable and fixed—are considered as inventoriable costs. (309)

Accelerated depreciation. Depreciation method in which the pattern of depreciation writes off depreciable assets more quickly than does straight-line depreciation. (721)

Account analysis method. Approach to cost estimation that classifies cost accounts in the ledger as variable, fixed, or mixed with respect to the cost driver. Typically, qualitative rather than quantitative analysis is used in making these classification decisions. (345)

Accounting rate of return. See *accrual accounting rate of return (AARR)*.

Accrual accounting rate of return (AARR). Accounting measure of income divided by an accounting measure of investment. Also called *accounting rate of return*. (701)

Activity-based budgeting. Approach to budgeting that focuses on the costs of activities necessary to produce and sell products and services. (199)

Activity-based costing (ABC). Approach to costing that focuses on activities as the fundamental cost objects. It uses the cost of these activities as the basis for assigning costs to other cost objects such as products, services, or customers. (115)

Actual costing. A costing method that traces direct costs to a cost object by using the actual direct cost rate(s) times the actual quantity and allocates indirect costs based on the actual indirect cost rate(s) times the actual quantity. (109)

Actual costs. Costs incurred (historical costs), as distinguished from predicted or forecasted costs. (27)

Allowable cost. Cost that the parties to a contract agree to include in the costs to be reimbursed. (549)

Appraisal costs. Costs incurred in detecting which of the individual units of products do not conform to specifications. (795)

Artificial cost. See *complete reciprocated cost*.

Autocorrelation. See *serial correlation*.

Autonomy. The degree of freedom to make decisions. (864)

Average cost. See *Unit cost*.

Average waiting time. The average amount of time that an order will wait in line before it is set up and processed. (812)

Backflush costing. Costing system that delays recording changes in a product being produced until good finished units appear; it then uses budgeted or standard costs to work backward to assign manufacturing costs to units produced. Also called *delayed costing, endpoint costing,* or *post-deduct costing*. (662)

Bailout payback. Capital budgeting method that measures the time it will take for the cumulative cash flows from operations plus the disposal price of equipment plus the recovery of working capital at the end of a particular year to equal the net initial investment. (700)

Batch-level costs. The costs of resources sacrificed on activities that are related to a group of units of products or services rather than to each individual unit of product or service. (356)

Benchmarking. The continuous process of measuring products, services, or activities against the best levels of performance that may be found either inside or outside the organization. (7)

Black hole demand spiral. See *downward demand spiral*.

Book value. The original cost minus accumulated depreciation of an asset. (403)

Bottleneck. An operation where the work required to be performed approaches or exceeds the available capacity. (811)

Breakeven point. Quantity of output where total revenues and total costs are equal; that is where the operating income is zero. (62)

Breakeven time (BET). The amount of time from when the initial concept for a new product is approved by management until the time when the cumulative present value of net cash inflows from the project equals the cumulative present value of net investment outflows. (808)

Budget. The quantitative expression of a plan of action and an aid to the coordination and implementation of the plan. (8)

Budgetary slack. See *padding.*

Budgeted costing. A costing method that traces direct costs to a cost object by using the budgeted direct cost rate(s) times the actual quantity and allocates indirect costs based on the budgeted indirect cost rate(s) times the actual quantity. (109)

Business function costs. The sum of all the costs in a particular business function. (393)

Byproduct. Product from a joint process that has a low sales value compared with the sales value of the main or joint product(s). (571)

Capital budgeting. The making of long-term planning decisions for investment. (685)

Capitalized costs. Costs that are first recorded as an asset and then subsequently become an expense. (37)

Carrying costs. Costs that arise when a business holds inventories of goods for sale. (833)

Cash budget. Schedule of expected cash receipts and disbursements. (210)

Cash cycle. See *self-liquidating cycle.*

Cause-and-effect diagram. Diagram that helps identify the potential causes of failure. Four major categories of potential causes of failure are identified—human factors, methods and design factors, machine-related factors, and materials and components factors. Also called a *fishbone diagram.* (799)

Certified Management Accountant (CMA). The professional designation for management accountants and financial executives in the United States. (16)

Chief financial officer (CFO). The senior officer empowered with oversight of the finance operations of an organization. Also called *finance director.* (12)

Choice criterion. Objective that can be quantified in a decision model. Also called *objective function.* (77)

Coefficient of determination (r^2). Measures the percentage of variation in a dependent variable explained by one or more independent variables. (352)

Collusive pricing. Companies in an industry conspiring in their pricing and output decisions to achieve a price above the competitive price. (452)

Common cost. The cost of operating a facility, operation, activity area, or like cost object that is shared by two or more users. (515)

Complete reciprocated cost. The actual cost incurred by the service department plus a part of the costs of the other service departments that provide services to it; it is always larger than the actual cost. Also called *artificial cost* of the service department. (513)

Conference method. Approach to cost estimation that develops cost estimates on the basis of analysis and opinions gathered from various departments of an organization (purchasing, process engineering, manufacturing, employee relations, and so on). (344)

Constant. The intercept term (g) in a cost estimation model—the component of total costs that, within the relevant range, does not vary with changes in the level of the cost driver. Also called *intercept.* (341)

Constant gross-margin percentage NRV method. Joint cost allocation method that allocates joint costs in such a way that the overall gross-margin percentage is identical for all the individual products. (577)

Constraint. A mathematical inequality or equality that must be satisfied by the variables in a mathematical model. (411)

Continuous improvement standard cost. Standard cost that is successively reduced over succeeding time periods. Also called *moving cost reduction standard cost.* (246)

Contribution income statement. Income statement that groups individual line items to highlight the contribution margin, which is the difference between revenues and all costs that vary with respect to an output-related driver. (63)

Contribution margin. Revenues minus all costs that vary with respect to an output-related cost driver. (62)

Contribution-margin percentage. Total contribution margin divided by revenues. (74)

Control. Covers both action that implements the planning decision and performance evaluation of its personnel and its operations. (8)

Control chart. Graph of a series of successive observations of a particular step, procedure, or operation taken at regular intervals of time. Each observation is plotted relative to specified ranges that represent the expected distribution. (798)

Control-factor unit. Measure of workload used in work measurement. (479)

Controllable cost. Any cost that is primarily subject to the influence of a given manager of a given responsibility center for a given time span. (206)

Controllability. The degree of influence that a specific manager has over costs, revenues, or other items in question. (206)

Controller. The financial executive primarily responsible for both management accounting and financial accounting. (14)

Conversion costs. All manufacturing costs other than direct materials costs. (43)

Correlation. The general relationship (co-movement) between two variables; in a cost estimation model, the general relationship between the dependent variable and the cost driver. (348)

Cost. Resource sacrificed or forgone to achieve a specific objective. (26)

Cost accounting. Management accounting plus a part of financial accounting—to the extent that cost accounting provides information that helps the requirements of external reporting. (4)

Cost Accounting Standards Board (CASB). Independent board within the Office of Federal Procurement Policy of the U.S. government. Has the exclusive authority to make, promulgate, amend, and rescind cost accounting standards used to guide contracts with U.S. government. (547)

Cost accounting system. Means by which cost accounting information is reported. Also called *costing system.* (4)

Cost accumulation. The collection of cost data in some organized way through an accounting system. (27)

Cost allocation. The assigning of indirect costs to the chosen cost object. (29)

Cost allocation base. A factor that is the common denominator for systematically linking an indirect cost or group of indirect costs to a cost object. (99)

Cost assignment. General term that encompasses both (1) tracing accumulated costs to a cost object and (2) allocating accumulated costs to a cost object. (27)

Cost-benefit approach. Primary criterion for choosing among alternative accounting systems, which is how each system achieves organization goals in relation to the cost of each system. (10)

Cost center. A responsibility center in which a manager is accountable for costs only. (203)

Cost driver. Any factor that affects costs. That is, a change in the cost driver will cause a change in the total cost of a related cost object. (29)

Cost estimation. The measurement of past cost relationships. (340)

Cost hierarchy. Categorization of costs into different cost pools on the basis of different classes of cost drivers, or different degrees of difficulty in determining cause-and-effect relationships. (549)

Cost incurrence. Occurs when resources are actually sacrificed. (439)

Cost management. The actions by managers to satisfy customers while continuously reducing and controlling costs. (5)

Cost object. Anything for which a separate measurement of costs is desired. (26)

Costs of quality. Costs incurred to prevent poor quality from occurring or those costs incurred because poor quality has occurred. (795)

Cost-plus contract. Contract in which reimbursement is based on actual allowable cost plus a fixed fee. (547)

Cost pool. A grouping of individual cost items. (99)

Cost predictions. Forecast of future costs. (340)

Cost tracing. The assigning of direct costs to the chosen cost object. (29)

Costing system. See *cost accounting system.*

Costing system refinement. Making changes to an existing costing system that result in a better measure of the way that jobs, products, customers, and so on, differentially use the resources of the organization. (110)

Cost-volume-profit (CVP). Analyzes the behavior of total costs, total revenues, and operating income as changes occur in the output level, selling price, variable costs, or fixed costs; a single revenue driver and a single cost driver are used in this analysis. (61)

Cumulative average-time learning model. Learning curve model in which the cumulative average time per unit declines by a constant percentage each time the cumulative quantity of units produced is doubled. (358)

Current cost. Asset measure based on the cost today of purchasing an asset identical to the one currently held. It is the cost of purchasing the services provided by that asset if an identical asset cannot currntly be purchased. (897)

Currently attainable standards. A good level of performance taking into account normal spoilage, waste, and the like. (234)

Customer costing system. Costing system that reports cost numbers that reflect the way that customers differentially use the resources of a company. (119)

Customer life-cycle costs. Focuses on the total costs to a customer of acquiring and using a product or service until it is replaced. (451)

Customer-response time. Amount of time from when a customer places an order for a product or requests service to when the product or service is delivered to the customer. (811)

Customer service. The support activities provided to customers. (3)

Customer-support costs. Costs incurred at the individual customer level, regardless of the quantity or mix of individual services or products that the customer buys. (122)

Decentralization. The freedom for managers at lower levels (subunits) of the organization to make decisions. (860)

Decision model. Formal model for making a choice under uncertainty, frequently involving quantitative analysis. (77)

Decision table. Summary of the contemplated actions, events, outcomes, and probabilities of events in a decision. Also called *payoff table* or *payoff matrix.* (80)

Delayed costing. See *backflush costing.*

Denominator level. Preselected level of the cost allocation base used to set a budgeted fixed overhead rate for allocating fixed overhead costs to a cost object. In manufacturing settings, often termed the *production denominator level* or *production denominator volume.* (276)

Dependent variable. The cost variable to be predicted in a cost estimation or prediction model. (347)

Design of products, services, or processes. The detailed planning and engineering of products, services, or processes. (3)

Designed-in costs. See *locked-in costs.*

Differential cost. See *incremental cost.*

Direct allocation method. Method of support cost allocation that ignores any service rendered by one support department to another; it allocates each support department's total costs directly to the operating departments. Also called *direct method.* (511)

Direct costing. See *variable costing.*

Direct costs of a cost object. Costs that are related to the cost object and can be traced to it in an economically feasible way. (27)

Direct manufacturing labor costs. Compensation of all manufacturing labor that is considered to be part of the cost object (say, units finished or in process) and that can be traced to the cost object in an economically feasible way. (41)

Direct materials costs. The acquisition costs of all materials that eventually become part of the cost object (say, units finished or in process) and that can be traced to that cost object in an economically feasible way. (41)

Direct materials inventory. Direct materials in stock and awaiting use in the manufacturing process. (35)

Direct method. See *direct allocation method.*

Discounted cash flow (DCF). Capital budgeting method that measures the cash inflows and outflows of a project as if they occurred at a single point in time so that they can be compared in an appropriate way. (687)

Discount rate. See *required rate of return.*

Discretionary costs. Arise from periodic (usually yearly) decisions regarding the maximum outlay to be incurred. They are not tied to a clear cause-and-effect relationship between inputs and outputs. (475)

Distribution. The mechanism by which products or services are delivered to the customer. (3)

Disturbance term. See *residual term.*

Double-counting. Occurs when a cost item is included in a contract reimbursement report both as a direct cost item and as part of an indirect cost pool allocated to the contract using a budgeted rate. (547)

Double-declining balance (DDB) depreciation. Accelerated depreciation method in which the first-year depreciation is twice the amount of straight-line depreciation when a zero terminal disposal price is assumed. (721)

Downstream costs. See *post manufacturing costs.*

Downward demand spiral. Continuing reduction in demand that occurs when prices are raised and then raised again in an attempt to recover fixed costs from an ever-decreasing customer base. Also called *black hole demand spiral.* (546)

Dual pricing. Approach to transfer pricing using two separate transfer-pricing methods to price each interdivision transaction. (874)

Dual-rate method. Cost allocation method in which costs are grouped in two separate cost pools, each of which has a different allocation rate and may have a different allocation base. (507)

Dumping. Under U.S. laws, occurs when a non-U.S.–based company sells a product in the U.S. at a price below the market value in the country of its creation, and this action materially injures or threatens to materially injure an industry in the U.S. (452)

Dysfunctional decision making. See *suboptimal decison making.*

Economic order quantity (EOQ). Decision model that calculates the optimal quantity of inventory to order. Simplest model incorporates only ordering costs and carrying costs. (833)

Effectiveness. The degree to which a predetermined objective or target is met. (237)

Efficiency. The amount of inputs used to achieve a given level of output. (237)

Efficiency variance. The difference between the actual quantity of input used (such as yards of materials) and the budgeted quantity of input that should have been used, multiplied by the budgeted price. Also called *quantity variance* or *usage variance.* (238)

Effort. Exertion toward a goal. (468)

Endpoint costing. See *backflush costing.*

Engineered costs. Costs that result specifically from a clear cause-and-effect relationship between inputs and outputs. (475)

Equivalent units. Measure of the output in terms of the quantities of each of the factors of production that have been consumed by the units. It is the collection of inputs necessary to produce one fully complete unit of product or output. (157)

Error term. See *residual term.*

Estimated net realizable value (NRV) method. Joint cost allocation method that allocates joint costs on the basis of the relative estimated net realizable value (expected final sales value in the ordinary course of business minus the expected separable costs of production and marketing). (576)

Events. A possible occurrence in a decision model. Also called *states* or *states of nature.* (77)

Excess present value index. Capital budgeting measure in which the total present value of future net cash inflows of a project is divided by the total present value of the net initial investment. (738)

Expected monetary value. See *expected value.*

Expected value. Weighted average of the outcomes of a decision with the probability of each outcome serving as the weight. Also called *expected monetary value.* (78)

Experience curve. Function that shows how full product costs per unit (including manufacturing, distribution, marketing, and so on) decline as units of output increase. (358)

Extended value chain. Focuses on all the business functions related to a product or service from its cradle to grave (womb to tomb), irrespective of whether those functions occur in the same organization or in a set of legally independent organizations. (7)

External failure costs. Costs incurred when a nonconforming product is detected *after* it is shipped to customers. (795)

Facility-sustaining costs. The costs of resources sacrificed on activities that cannot be traced to specific products or services but support the organization as a whole. (356)

Factory burden costs. See *manufacturing overhead costs.*

Factory overhead costs. See *manufacturing overhead costs.*

Favorable variance. Variance that increases operating income relative to the budgeted amount. Denoted F. (228)

Federal Acquisition Regulations (FARs). Procurement regulations with which companies contracting with the U.S. government must comply. (547)

Finance director. See *chief financial officer (CFO).*

Financial accounting. Focuses on external reporting through financial statements to investors, government authorities, and other outside parties. (4)

Financial budget. That part of the master budget that comprises the capital budget, cash budget, budgeted balance sheet, and budgeted statement of cash flows. (188)

Financial planning models. Mathematical statements of the relationships among all operating activities, financial activities, and the other major internal and external factors that may affect decisions. (200)

Finished goods inventory. Goods fully completed but not yet sold. (35)

First-in, first-out (FIFO) process costing method. Method of process costing that computes unit costs by confining equivalent units to work done during the current period only; costs of the current period are separately identified so that the unit costs are related only to the current period's work. (606)

Fishbone diagram. See *cause-and-effect diagram.*

Fixed cost. Cost that does not change *in total* despite changes of a cost driver. (30)

Flexible budget. Budget that is developed using budgeted revenue or cost amounts: it is adjusted (flexed) to the actual level of output achieved or expected to be achieved during the budget period. (228)

Flexible-budget variance. Difference between the actual result and the flexible budget amount for the actual output achieved. (230)

Full product costs. The sum of all the costs in all the business functions—R&D, design, production, marketing, distribution, and customer service. (393)

Goal congruence. Exists when individuals and groups work toward the organization goals that top management desires. (468)

Goods in process inventory. See *work in process inventory*.

Gross margin. Revenues minus cost of goods sold. (73)

High-low method. Method used to estimate a cost function that entails using only the highest and lowest observed values of the cost driver within the relevant range. (348)

Homogeneous cost pool. Cost pool in which all costs have a similar cause-and-effect or benefits-received relationship between the cost driver and the costs of the activity. (505)

Hurdle rate. See *required rate of return*.

Hybrid costing systems. Blends of characteristics from both job costing systems and process costing systems. (656)

Imputed costs. Costs recognized in particular situations that are not regularly recognized by accrual accounting procedures. (400)

Incremental approach. Approach to decision making that analyzes only those cash outflows and inflows that differ between the alternatives. (725)

Incremental common cost allocation method. Common cost allocation method requiring that one user be viewed as the primary party and the second user be viewed as the incremental party. (516)

Incremental cost. Difference in total cost between two alternatives. Also called *differential cost* and *net relevant cost*. (389)

Incremental unit-time learning model. Learning curve model in which the incremental unit time (the time needed to produce the last unit) declines by a constant percentage each time the cumulative quantity of units produced is doubled. (358)

Indirect costs of a cost object. Costs that are related to the cost object but cannot be traced to it in an economically feasible way. (28)

Indirect manufacturing costs. See *manufacturing overhead costs*.

Industrial-engineering method. Approach to cost estimation that first analyzes the relationship between inputs and outputs in physical terms. Also called *work measurement method*. (344)

Inflation. The decline in the general purchasing power of the monetary unit. (732)

Information asymmetry. Exists when one particular manager (for example, a subunit manager) has more and better information about factors that affect the performance of his or her job than do other managers (for example, top management or managers of other subunits). (860)

Infrastructure costs. Costs that arise from having property, plant, equipment, and a functioning organization; little can be done in the short run to change infrastructure costs. (476)

Insourcing. Process of producing goods or providing services within the firm rather than purchasing those same goods or services from outside vendors. (394)

Institute of Management Accountants (IMA). The largest association of internal accountants in the United States. Formerly called the National Association of Accountants. (16)

Intercept. See *constant*.

Intermediate product. Product transferred from one subunit to another subunit of the organization. This product is further processed and sold to an external customer. (864)

Internal auditing. The function responsible for reviewing and analyzing the financial and other records of an organization to attest to the integrity of its financial statements. (16)

Internal failure costs. Costs incurred when a nonconforming product is detected *before* it is shipped to customers. (795)

Internal rate of return (IRR). Discount rate at which the present value of expected cash inflows from a project equals the present value of expected cash outflows of the project. The IRR is the discount rate that makes NPV = $0. Also called the *time-adjusted rate of return*. (690)

Inventoriable costs. Specific type of capitalized cost. They are costs associated with the purchase of goods for resale (in the case of merchandise inventory) or costs associated with the acquisition and conversion of materials and all other manufacturing inputs into goods for sale (in the case of manufacturing inventories). (37)

Inventory management. The planning, organizing, and control activities focused on the flow of materials into, through, and from the organization. (832)

Investment center. A responsibility center in which a manager is accountable for investments, revenues, and costs. (203)

Investment tax credit (ITC). A direct reduction of income taxes payable arising from the acquisition of depreciable assets. (721)

Job cost record. Record used to accumulate the costs of a job. Also called *job order sheet* or *job cost sheet*. (142)

Job costing system. Costing system in which the cost of a product or service is obtained by assigning costs to a distinct, identifiable product or service. (99)

Job cost sheet. See *job cost record*.

Job order sheet. See *job cost record*.

Joint cost. Cost of a single process that yields multiple products simultaneously. (570)

Joint products. Products from a joint process that have relatively high sales value and are not separately identifiable as individual products until the splitoff point. (571)

Just-in-time (JIT) production. Production system in which each component on a production line is produced immediately as needed by the next step in the production line. (660)

Just-in-time (JIT) purchasing. The purchase of goods or

materials such that delivery immediately precedes demand or use. In the extreme, no inventories would be held. (840)

***Kaizen* budgeting.** Budgeting approach that projects costs on the basis of future improvements rather than current practices and methods. (200)

Key success factors. Factors that directly affect customer satisfaction, such as cost, quality, time, and innovative products and services. (6)

Labor-paced work environment. Worker dexterity and productivity determine the speed of production. (541)

Labor time record. Record used to charge departments and job cost records for labor time used. (143)

Learning curve. Function that shows how labor-hours per unit decline as units of output increase. (358)

Life-cycle budgeting. Budget that incorporates the revenues and costs attributable to each product from its initial research and development to its final customer servicing and support in the market place. (448)

Life-cycle costing. Budget that tracks and accumulates the actual costs attributable to each product from start to finish. (448)

Line management. Managers directly responsible for attaining the objectives of the organization. (12)

Linear cost function. Cost function in which a single constant and a single slope coefficient describe the behavior of total costs for all changes in the level of the cost driver within the relevant range. (341)

Locked-in costs. Costs that have not yet been incurred but that will be incurred in the future on the basis of decisions that have already been made. Also called *designed-in costs* (439)

Machine-paced work environment. Machines conduct most (or all) phases of production, such as movement of materials to the production line, assembly and other activities on the production line, and shipment of finished goods to the delivery dock areas. (541)

Main product. When a single process yielding two or more products yields only one product with a relatively high sales value, that product is termed a main product. (571)

Management accounting. Focuses on internal customers; it measures and reports financial and other information that assists managers in fulfilling goals of the organization. (4)

Management by exception. The practice of concentrating on areas that deserve attention and placing less attention on areas operating as expected. (8)

Management by objectives (MBO). Subordinate and his or her manager jointly formulate the subordinate's set of objectives and the plans for attaining those objectives for a subsequent period. (207)

Management control system. Means of gathering information to aid and coordinate the process of making planning and control decisions throughout the organization. (466)

Manufacturing lead time. Time from when an order is ready to start on the production line to when it becomes a finished good. (542)

Manufacturing overhead allocated. All manufacturing costs that are assigned to a product (or service) using a cost allocation base because they cannot be traced to a product (or service) in an economically feasible way. (151)

Manufacturing overhead costs. All manufacturing costs that are considered to be part of the cost object (say, units finished or in process) but that cannot be traced to that cost object in an economically feasible way. Also called *indirect manufacturing costs, factory overhead costs,* and *factory burden costs.* (42)

Manufacturing-sector companies. Provide to their customers tangible products that have been converted to a different basic form from the materials purchased from suppliers. (35)

Margin of safety. Excess of budgeted revenues over the breakeven revenues. (68)

Marketing. The process by which individuals or groups (a) learn about and value the attributes of products or services and (b) purchase those products or services. (3)

Market-share variance. The difference between the budgeted contribution margin based on the actual market size in units times the actual market share and the budgeted contribution margin based on actual market size in units times the budgeted market share. (762)

Market-size variance. The difference between the budgeted contribution margin based on the actual market size in units times the budgeted market share and the contribution margin in the static budget based on the budgeted market size in units times the budgeted market share. (762)

Master budget. Budget that summarizes the financial projections of all the organization's budgets and plans. It describes the financial plans for all value-chain functions. (183)

Master-budget utilization. The denominator-level concept based on the anticipated level of capacity utilization for the coming budget period. (323)

Materials requisition record. Record used to charge departments and job cost records for the cost of the materials used on a specific job. (143)

Merchandising-sector companies. Provide to their customers tangible products they have previously purchased in the same basic form from suppliers. (35)

Mixed cost. A cost that has both fixed and variable elements. Also called a *semivariable cost.* (342)

Modified Accelerated Cost Recovery System (MACRS). Federal income tax regulation that classifies every depreciable asset into one of several categories. MACRS specifies (1) the number of years over which assets in each category can be depreciated, and (2) the pattern of allowable depreciation for each category. (731)

Moral hazard. Describes contexts in which an employee is tempted to put in less effort (or report distorted information) because the employee's interests differ from the owner's and because the employee's effort cannot be accurately monitored and enforced. (902)

Motivation. The desire to attain a selected goal (the goal-congruence aspect) combined with the resulting drive or pursuit toward that goal (the effort aspect). (469)

Moving cost reduction standard cost. See *continuous improvement standard cost.*

Multicollinearity. Exists when two or more independent variables in a regression model are highly correlated with each other. (372)

Multiple regression. Regression model that uses more than one independent variable to estimate the dependent variable. (350)

Negotiated static budget. Budget in which a fixed amount of costs is established through negotiations before the start of the budget period. (478)

Net income. Operating income plus nonoperating revenues (such as interest revenues) minus nonoperating costs (such as interest costs) minus income taxes. (61)

Net present value method. Discounted cash-flow method of calculating the expected net monetary gain or loss from a project by discounting all expected future cash inflows and outflows to the present point in time, using the required rate of return. (688)

Net relevant cost. See *incremental cost.*

New product development time. Amount of time from when the initial concept for a new product is approved by management to its market introduction. (807)

Nominal rate of return. Return made up of two elements: (1) the real rate of return and (2) an inflation element. The rate of return demanded to cover both investment risk and any anticipated decline in general purchasing power. (732)

Nonlinear cost function. Cost function in which a single constant (*g*) and a single slope coefficient (*w*) do not adequately describe the behavior of costs for all changes in the level of the cost driver. (357)

Non-value-added cost. A cost that customers perceive as not adding value to a product or service. (270)

Normal costing. A costing method that traces direct costs to a cost object by using the actual direct cost rate(s) times the actual quantity and allocates indirect costs based on the budgeted indirect cost rate(s) times the actual quantity. (109)

Normal spoilage. Spoilage that arises under efficient operating conditions; it is an inherent result of the particular process. (633)

Normal utilization. The denominator-level concept based on the level of capacity utilization that satisfies average customer demand over a period (say, two or three years) that includes seasonal, cyclical, or other trend factors. (323)

Objective function. Expresses the objective to be maximized (for example, operating income) or minimized (for example, operating costs) in a decision model. Also called *choice criterion.* (77)

On-time performance. Situation in which the product or service is actually delivered at the time it is scheduled to be delivered. (814)

Operating budget. The budgeted income statement and its supporting schedules. (188)

Operating cycle. See *self-liquidating cycle.*

Operating department. A department that adds to a product or service value that is observable by a customer. Also called a *production department* in manufacturing organizations. (510)

Operating income. Operating revenues for the accounting period minus all operating costs, including cost of goods sold. (61)

Operation. A standardized method or technique that is performed repetitively regardless of the distinguishing features of the finished good. (657)

Operation costing. Hybrid costing system applied to batches of similar products. Each batch of products uses the same resources to the same extent as all other batches. (657)

Opportunity cost. The contribution to income that is forgone (rejected) by not using a limited resource in its best alternative use. (399)

Opportunity cost of capital. See *required rate of return.*

Ordering costs. Costs of preparing and issuing a purchase order. (833)

Ordinary incremental budget. Budget based on the budget of the previous period and actual results. (478)

Organizational structure. The arrangement of lines of responsibility within the entity. (202)

Outcomes. Predicted consequences of the various possible combinations of actions and events in a decision model. Also called *payoffs.* (77)

Outlay costs. Current or near-future cash disbursements made to meet costs incurred because of a specific decision. Also called *out-of-pocket costs.* (392)

Out-of-pocket costs. Current or near-future cash disbursements made to meet costs incurred because of a specific decision. Also called *outlay costs.* (392)

Output level overhead variance. Difference between budgeted fixed overhead and the fixed overhead allocated to actual output units achieved. Also called a *production volume variance* or *a production level variance* in a manufacturing setting. (278)

Output unit-level costs. The costs of resources sacrificed on activities performed on each unit of product or service. (355)

Outsourcing. Process of purchasing goods and services from outside vendors rather than producing the same goods or providing the same services within the firm. (394)

Overabsorbed indirect (overhead) costs. See *overallocated indirect (overhead) costs.*

Overallocated indirect (overhead) costs. Allocated amount of indirect costs in an accounting period is greater than the actual (incurred) amount in that period. Also called *overapplied indirect (overhead) costs* and *overabsorbed indirect (overhead) costs.* (152)

Overapplied indirect (overhead) costs. See *overallocated indirect (overhead) costs.*

Padding. The practice of underestimating budgeted revenues (or overestimating budgeted costs) in order to make budgeted targets more easily achievable. Also called *budgetary slack.* (199)

Parameter. A constant or a coefficient in a model equation. (341)

Pareto diagram. Diagram that indicates how frequently each type of failure (defect) occurs. (799)

Partial productivity. Measures the quantity of output produced with the quantity of a single input used. (774)

Payback method. Capital budgeting method that measures the time it will take to recoup, in the form of net cash inflows, the net dollars invested in a project. (697)

Payoffs. See *outcomes.*

Payoff matrix. See *decision table.*

Payoff table. See *decision table.*

Peak-load pricing. Practice of charging a higher price for the same product or service when demand approaches physical capacity limits. (446)

Peanut-butter costing. A costing approach that uniformly assigns ("spreads" or "smoothes out") the cost of resources to cost objects (such as products, services, or customers) although the individual products, services, or customers in fact use those resources in a nonuniform way. (114)

Perfection standards. The best level of performance under the best conceivable condition. No provision is made for spoilage, waste, and the like. (234)

Perfectly competitive market. Exists when there is a homogeneous product with equivalent buying and selling prices and no individual buyers or sellers can affect those prices by their own actions. (869)

Performance gap. Difference between an actual result and a benchmark amount. (227)

Period costs. Costs that are reported as expenses of the period in question. (37)

Periodic inventory method. Inventory method that does not maintain a continuous record of inventory changes. (41)

Perpetual inventory method. Inventory method that maintains a continuous ("real-time") record of additions to and reductions from inventory. (40)

Physical measure method. Joint cost allocation method that allocates joint costs on the basis of their relative proportions at the splitoff point, using a common physical measure such as weight or volume. (574)

Planning. Choosing goals, predicting results under various alternative ways of achieving those goals, and then deciding how to attain the desired goals. (7)

Post-deduct costing. See *backflush costing.*

Postmanufacturing costs. All costs incurred in the value-chain business functions of marketing, distribution, and customer service. Also called *downstream costs.* (429)

Practical capacity. The denominator-level concept that reduces theoretical capacity for unavoidable operating interruptions such as scheduled maintenance time, shutdowns for holidays and other days, and so on. (323)

Predatory pricing. Company deliberately prices below its costs in an effort to drive out competitors and restrict supply and then raises prices rather than enlarge demand or meet competition. (451)

Premanufacturing costs. All costs incurred in the value-chain business functions of research and development and the design of products or processes. Also called *upstream costs.* (429)

Present value. An asset measure based on DCF estimates. (897)

Prevention costs. Costs incurred in preventing the production of products that do not conform to specifications. (795)

Previous department costs. See *transferred-in costs.*

Price discrimination. Practice of charging some customers a higher price than is charged to other customers. (446)

Price variance. The difference between actual price and budgeted price multiplied by the actual quantity of input in question. Also called *rate variances,* especially when those variances are for direct labor categories. (238)

Prime costs. All direct manufacturing costs. (43)

Priority incremental budget. Similar to an ordinary incremental budget, with the inclusion of a description of what incremental activities or changes would occur (1) if the budget were increased by, say, 10% and (2) if the budget were decreased by the same percentage. (478)

Probability. Likelihood of occurrence of an event in a decision model. (77)

Probability distribution. Describes the likelihood (or probability) of each of the mutually exclusive and collectively exhaustive sets of events. (78)

Process costing system. Costing system in which the cost of a product or service is obtained by assigning costs to masses of similar units and then computing unit costs on an average basis. (99)

Product cost. Sum of the costs assigned to a product for a specific purpose. (45)

Product-cost cross-subsidization. Costing outcome where at least one miscosted product results in the miscosting of other products in the organization. (114)

Product life cycle. Spans the time from initial research and development to the time at which support to customers is withdrawn. (448)

Product line. A grouping of similar products. (116)

Product overcosting. A product consumes a relatively low level of resources but is reported to have a relatively high cost. (114)

Product-sustaining costs. The costs of resources sacrificed on activities undertaken to support specific products or services. (356)

Product undercosting. A product consumes a relatively high level of resources but is reported to have a relatively low cost. (114)

Production. The coordination and assembly of resources to produce a product or deliver a service. (3)

Production denominator level. See *denominator level.*

Production denominator volume. See *denominator level.*

Production department. See *operating department.*

Production level variance. See *output level overhead variance.*

Production volume variance. See *output level overhead variance.*

Productivity. Measures the relationship between actual inputs used and actual outputs achieved; the lower the inputs for a given set of outputs or the higher the outputs for a given set of inputs, the higher the level of productivity. (773)

Profit center. A responsibility center in which a manager is accountable for revenues and costs. (203)

Pro forma statements. Budgeted financial statements of an organization. (187)

Project. Complex job that often takes months or years to complete and requires the work of many different departments, divisions, or subcontractors. (668)

Project performance cost variance. Difference between

the actual cost of work performed to date and the budgeted cost of the work performed to date. (670)

Project schedule cost variance. Difference between the budgeted cost of work performed to date and the budgeted cost of work scheduled to date. (670)

Proration. The spreading of underallocated or overallocated overhead among ending work in process, finished goods, and cost of goods sold. (153)

Purchase costs. Cost of goods acquired from suppliers, including freight and transportation costs. (833)

Purchase-order lead time. Amount of time between the placement of an order and its delivery. (833)

PV chart. Shows the impact on operating income (profit) of changes in the output level (volume). (68)

Qualitative factors. Outcomes that cannot be measured in numerical terms. (389)

Quality. The conformance of a product or service with a preannounced or prespecified standard. (6)

Quantitative factors. Outcomes that are measured in numerical terms. (389)

Quantity variance. See *efficiency variance.*

Rate variance. See *price variance.*

Real rate of return. The rate of return demanded to cover investment risk (with no inflation). This rate is made up of two elements: (1) a risk-free element, and (2) a business-risk element. (732)

Reciprocal allocation method. Method of support-cost allocation that explicitly includes the mutual services rendered among all support departments. (512)

Regression analysis. Statistical model that measures the average amount of change in the dependent variable that is associated with a unit change in one or more independent variables. (350)

Relevant costs. Expected future costs that differ among alternative courses of action. (388)

Relevant range. Range of the cost driver in which a specific relationship between cost and the driver is valid. (31)

Relevant revenues. Expected future revenues that differ among alternative courses of action. (388)

Reorder point. The quantity level of the inventory on hand that triggers a new order. (836)

Required rate of return (RRR). The minimum acceptable rate of return on an investment; the return that the organization could expect to receive elsewhere for an investment of comparable risk. Also called *discount rate, hurdle rate,* and *opportunity cost of capital.* (688)

Research and development. The generation of, and experimentation with, ideas related to new products, services, or processes. (3)

Residual income. Income minus an imputed interest charge for the investment base. (894)

Residual term. The difference between the actual and predicted amount of a dependent variable (such as a cost) in a regression model. Also called the *disturbance term* or *error term.* (351)

Responsibility accounting. System that measures the plans (by budgets) and actions (by actual results) of each responsibility center. (202)

Responsibility center. A part, segment, or subunit of an organization whose manager is accountable for a specified set of activities. (p. 202)

Return on investment (ROI). Income divided by investment. (892)

Revenue center. A responsibility center in which a manager is accountable for revenues only. (203)

Revenue driver. Any factor, such as units sold, that affects revenues. (60)

Reworked units. Unacceptable units of production that are subsequently reworked and sold as acceptable finished goods. (633)

Rolling budget. Budget or plan that is always available for a specified future period by adding a month, quarter, or year in the future as the month, quarter, or year just ended is dropped. (186)

Safety stock. The buffer inventory held as a cushion against unexpected increases in demand or lead time and unexpected unavailability of stock from suppliers. (836)

Sales mix. Relative combination of quantities of products or services that constitutes total sales. (69)

Sales value at splitoff method. Joint cost allocation method that allocates joint costs on the basis of each product's relative sales value at the splitoff point. (573)

Sales-volume variance. Difference between the flexible-budget amount and the static-budget amount; unit selling prices, unit variable costs, and fixed costs are held constant. (230)

Scrap. Product that has a minimal (frequently zero) sales value. (571)

Segment. Identifiable part of an organization. (549)

Self-liquidating cycle. Movement from cash to inventories to receivables and back to cash. Also called *working-capital cycle, cash cycle,* or *operating cycle.* (211)

Semivariable costs. See *mixed cost.*

Sensitivity analysis. A "what-if" technique that examines how a result will change if the original predicted data are not achieved or if an underlying assumption changes. (67)

Separable costs. Costs incurred beyond the splitoff point that are assignable to individual products. (570)

Sequential allocation method. See *step-down allocation method.*

Sequential tracking. Product-costing method in which the timing of entries recorded in the accounting system is synchronized with the physical sequences of purchases and production. Also called *synchronous tracking.* (661)

Serial correlation. Systematic pattern in the sequence of residuals of a regression such that the residual in period t conveys information about the residuals in periods $t+1$, $t+2$, and so on. Also called *autocorrelation.* (368)

Service department. See *support department.*

Service-sector companies. Provide services or intangible products to their customers—for example, legal advice or an audit. (35)

Service sustaining costs. The costs of resources sacrificed on activities undertaken to support specific services. (356)

Shrinkage. The difference between goods purchased for sale and goods actually sold (after making inventory adjustments). This difference can arise from breakages before a sale, theft, perishability, and so on. (117)

Simple regression. Regression model that uses only one independent variable to estimate the dependent variable. (350)

Single-rate method. Cost allocation method in which all costs are grouped in one cost pool and allocated to cost objects using the same rate per unit of the single allocation base. (507)

Slope coefficient. Coefficient term in a cost estimation model—indicates how much total costs change for each unit change in the cost driver within the relevant range. (341)

Specification analysis. Testing of the assumptions of regression analysis. (353)

Splitoff point. Juncture in the process when the products become separately identifiable. (570)

Spoilage. Unacceptable units of production that are discarded or sold for net disposal proceeds. (633)

Staff management. Managers who provide advice and assistance to line management. (12)

Stand-alone common cost allocation method. Common cost allocation method that allocates the common cost on the basis of each user's percentage of the total of the individual stand-alone costs. (516)

Standard. Represents a good or best level of performance. Usually developed from a careful study of the specific operations and expressed on a per-unit basis. (226)

Standard cost. The per-unit cost of output for good or best level of performance. Most frequently found in the manufacturing and distribution parts of the value chain. (227)

Standard costing. Costing method that traces direct costs to a cost object by using the standard price(s) or rate(s) times the standard inputs allowed for actual outputs achieved and allocates indirect costs on the basis of the budgeted indirect rate(s) times the standard inputs allowed for the actual outputs achieved. (286)

Standard error of the estimated coefficient. Regression statistic that indicates how much the estimated value is likely to be affected by random factors. (366)

Standard input. Allowed quantity of inputs (such as hours of labor time or pounds of materials) for one unit of output, given a good level or the best level of performance. (226)

States. See *events*.

States of nature. See *events*.

Static budget. Budget that is based on one level of output and is not adjusted or altered after it is finalized. (228)

Step allocation method. See *step-down allocation method*.

Step-down allocation method. Method of support cost allocation that allows for partial recognition of services rendered by support departments to other support departments. Also called *step* or *sequential allocation method*. (511)

Stockout. A stockout arises when a customer demands a unit of product and that unit is not readily available. (833)

Straight-line depreciation (SL). Depreciation method in which an equal amount of depreciation is taken each year. (721)

Strategy. Selection of overall objectives of an organization. (184)

Strategy and administration. Business function that includes senior executives charged with the overall responsibility for the organization. (3)

Suboptimal decision making. Decisions in which the benefit to one subunit is more than offset by the costs or loss of benefits to the organization as a whole. Also called *dysfunctional decision making*. (861)

Sunk costs. Past costs that are unavoidable because they cannot be changed no matter what action is taken. (405)

Support department. A department that provides the services that maintain other internal departments (operating departments and other support departments) in the organization. Also called a *service department*. (510)

Synchronous tracking. See *sequential tracking*.

Tactics. General means used by an organization to attain its strategic goals. (184)

Target cost. Estimated long-run cost of a product (or service) that when sold enables the company to achieve the targeted income. Target cost is derived by subtracting the target profit margin from the target price. (438)

Target price. Estimated price for a product (or service) that potential customers will be willing to pay. (438)

Target return on investment. The target operating income that an organization expects to earn divided by invested capital. (443)

Theoretical capacity. The denominator-level concept that is based on the production of output at maximum efficiency for 100% of the time. (323)

Throughput contribution. Sales dollars minus direct materials costs. (817)

Time-adjusted rate of return. See *internal rate of return (IRR)*.

Time driver. Any factor that causes a change in the speed with which an activity is undertaken when the factor itself changes. (811)

Total direct labor mix variance. The difference between the budgeted cost for the actual direct labor input mix and the budgeted cost if the budgeted direct labor input mix had been unchanged. (770)

Total direct labor yield variance. The difference between the budgeted cost of direct labor based on actual quantities of all direct labor inputs used and the flexible budget cost of direct labor based on the budgeted quantities of direct labor inputs for the actual output achieved. (770)

Total direct materials mix variance. The difference between the budgeted cost for the actual direct materials input mix and the budgeted cost if the budgeted direct materials input mix had been unchanged. (767)

Total direct materials yield variance. The difference between the budgeted cost of direct materials based on actual quantities of all direct materials inputs used and the flexible-budget cost of direct materials based on the budgeted quantities of direct materials inputs for the actual output achieved. (767)

Total factor productivity. The ratio of the quantity of output produced to the quantity of all inputs used, where the inputs are combined on the basis of current period prices. (777)

Total project approach. Approach to decision making that incorporates all relevant revenues and relevant costs under each alternative. In capital budgeting decisions, incorporates the present value of all cash inflows and outflows of each alternative. (725)

Total sales-mix variance. The difference between the budgeted contribution margin for the actual sales mix (flexible-budget contribution margin) and the budgeted contribution margin if the budgeted sales mix had been unchanged. (758)

Total sales-quantity variance. The difference between the budgeted contribution margin based on actual quantities sold of all products and the contribution margin in the static budget based on the budgeted quantities to be sold of all products. (758)

Transfer price. Price one subunit (segment, department, division, and so on) of an organization charges for a product or service supplied to another subunit of the same organization. (864)

Transferred-in costs. Costs incurred in a previous department that have been charged to a subsequent department. Also called *previous department costs*. (616)

Treasurer. The financial executive primarily responsible for obtaining investment capital and managing cash. (14)

Uncertainty. The possibility that an actual amount will deviate from an expected amount. (67)

Underabsorbed indirect (overhead) costs. See *underallocated indirect (overhead) costs*.

Underallocated indirect (overhead) costs. Allocated amount of indirect costs in an accounting period is less than the actual (incurred) amount in that period. Also called *underapplied indirect (overhead) costs* or *underabsorbed indirect (overhead) costs*. (152)

Underapplied indirect (overhead) costs. See *underallocated indirect (overhead) costs*.

Unfavorable variance. Variance that decreases operating income relative to the budgeted amount. Denoted U. (228)

Unit cost. Computed by dividing some total cost (the numerator) by some number of units (the denominator). Also called *average cost*. (33)

Upstream costs. See *premanufacturing costs*.

Usage variance. See *efficiency variance*.

Value-added activities. Activities that customers perceive as adding utility (usefulness) to the products or services they purchase. (29)

Value-added cost. A cost that customers perceive as adding value to a produce or service. (270)

Value chain. The sequence of business functions in which utility (usefulness) is added to the products or services of an organization. (2)

Value engineering. Systematic evaluation of all aspects of research and development, design of products and processes, production, marketing, distribution, and customer service, with the objective of reducing costs while satisfying customer needs. (438)

Variable cost. Cost that changes *in total* in proportion to changes of a cost driver. (29)

Variable costing. Method of inventory costing in which all direct manufacturing costs and variable manufacturing overhead costs are included as inventoriable costs; fixed manufacturing overhead costs are excluded from the inventoriable costs and are costs of the period in which they are incurred. Also called *direct costing*. (309)

Variable cost percentage. Total variable costs (with respect to an output-related factor) divided by revenues. (74)

Variable overhead efficiency variance. The difference between the actual and budgeted quantity of the variable overhead cost allocation base and the budgeted variable overhead cost allocation rate. (274)

Variable overhead spending variance. The difference between the actual amount of variable overhead incurred and the budgeted amount allowed for actual output achieved. (274)

Variance. Difference between an actual result and a budgeted amount when that budgeted amount is a financial variable reported by the accounting system. (227)

Weighted-average process costing method. Method of process costing that focuses on the total costs and total equivalent units completed to date; no distinction is made between work completed during the preceding period and work completed during the current period. (602)

Work in process inventory. Goods partially worked on but not yet fully completed. Also called *work in progress inventory* or *goods in process inventory*. (35)

Work in progress inventory. See *work in process inventory*.

Working capital cycle. See *self-liquidating cycle*.

Work measurement. Careful analysis of a task, its size, the method used in its performance, and the efficiency with which it is performed. (479)

Work-measurement method. See *industrial-engineering method*.

Zero-base budgeting (ZBB). Budgeting from the ground up, as though the budget were being prepared for the first time. Every proposed expenditure comes under review. (478)

Photo Credits

Author Index

Alston, F., 501*n*, 547*n*, 548*n*
Amano, M., 187*n*
American Electronics Association, 642*n*
Anguish, Lisa, 935*n*
Areeda, P., 451, 451*n*
Armitage, H., 161*n*
Armstrong, R., 15*n*
Asada, T., 187*n*
Ask, U., 318*n*
Atkinson, A., 353*n*, 358*n*, 502*n*
Ax, C., 318*n*

Bailes, J., 187*n*
Bailey, P., 698*n*
Barkman, A., 451*n*
Bartezzaghi, E., 848*n*
Baum, Hadassah, 935*n*
Berndt, E., 352*n*
Billesbach, T., 848*n*
Blayney, P., 73*n*, 145*n*, 187*n*, 318*n*, 345*n*, 447*n*, 502*n*, 516*n*, 703*n*, 872*n*, 892*n*
Blevins, L., 479*n*
Braund, S., 232*n*, 447*n*, 541*n*, 704*n*, 872*n*
Brealey, R., 692*n*
Bright, J., 161*n*
Bush, J., 743*n*

Carr, L., 243*n*, 843*n*
Cassidy, J., 667*n*
Chamberlain, Jan, 935*n*
Chang, T., 794*n*
Cheatham, C., 582*n*
Chen, J., 358*n*
Christie, A., 863*n*
Clarke, P., 145*n*, 199*n*, 232*n*, 395*n*, 703*n*, 848*n*, 892*n*
Cohen, J., 145*n*
Cooper, W., 232*n*, 447*n*
Cooper, R., 355*n*, 543*n*, 549*n*, 816*n*
Coopers and Lybrand, 869*n*
Cornick, M., 232*n*, 447*n*

Coughlan, P., 818*n*
Coulthurst, N., 542*n*
Cox, J., 817*n*
Crespi, J., 572*n*
Croom-Morgan, S., 848*n*
Curry, D. W., 191*n*, 936

Daing, N., 736*n*
Darlington, J., 818*n*
Dean, G., 502*n*
DeLuzio, M., 849*n*
De Meyer, A., 46*n*
DeVor, R., 794*n*
de With, E., 199*n*
Drucker, P., 782*n*
Drury, C., 232*n*, 447*n*, 541*n*, 542*n*, 704*n*, 872*n*
Dufoo, R. G., 803*n*

Eppen, G., 410*n*
Evans, J., 794*n*

Failing, R., 479*n*
Finkler, S., 738*n*
Fitzgerald, C., 472*n*
Foster, G., 799*n*
Freeman, M., 704*n*, 737*n*
Fremgen, J., 502*n*
Fulkerson, C., 550*n*
Fuller, S., 123*n*

Gaumnitz, B., 255*n*
Gietzmann, M., 184*n*
Gittins, R., 772*n*
Goldratt, E., 817*n*
Goldsman, L., 501*n*, 547*n*, 548*n*
Gould, F., 410*n*
Govindarajan, V., 7*n*, 184*n*, 449*n*, 762*n*, 872*n*
Granelli, J., 343*n*
Green, M., 582*n*
Greer, D., 451*n*
Grinyer, J., 736*n*

Guest, N., 9*n*

Harrington, J., 451*n*
Harris, J. K., 191*n*, 572*n*, 758*n*, 936
Harrison, A., 848*n*
Hiromoto, T., 544*n*
Ho, S., 737*n*
Hobbes, G., 704*n*, 737*n*
Hout, T., 807*n*
Howard, K., 872*n*
Hronec, S., 466*n*, 890*n*
Huddart, S., 901*n*

Ijskes, E., 199*n*
Innes, J., 818*n*
Inoue, S., 184*n*, 318*n*
Itami, H., 477*n*
Ittner, C., 243*n*, 843*n*

Janzen, J., 479*n*
Jeffries, Andrea, 935*n*
Jog, V., 703*n*, 737*n*
Johnson, H. T., 473*n*
Johnston, J., 352*n*
Jolley, J., 451*n*
Joye, M., 345*n*, 502*n*, 863*n*, 872*n*

Kalagnanam, S., 846*n*, 848*n*
Kaplan, R., 355*n*, 473*n*, 549*n*, 611*n*, 890*n*
Karmarkar, U., 611*n*
Katzenbach, J., 904*n*
Kelly, K., 905*n*
Kim, I., 703*n*
King, A., 869*n*
Knight, C. F., 202*n*
Kollaritsch, F., 255*n*
Kovac, E., 873*n*

Laimon, S., 633*n*
Lau, A., 550*n*
Lawson, G., 70*n*
Leader, C., 611*n*

Liao, S., 502n
Lindsay, R. M., 846n, 848n
Lindsay, W., 794n

Maisel, L., 549n
Manes, R., 358n
Melumad, N., 901n
Merchant, K., 206n
Miller, J., 46n, 476n
Mills, R., 447n
Mitchell, F., 818n
Morrissey, E., 549n
Mostafa, A., 872n
Mowen, M. M., 345n
Mudalige, N., 476n
Myers, S., 692n

Nahmias, S., 410n
Nakane, J., 46n
Nanni, A., 232n, 686n, 896n
Nicholson, R., 161n
Norton, D., 890n

Ochsenchlager, T., 313n
O'Dea, T., 848n
Oehm, R., 549n
O'Halloran, D., 250n
Osborne, P., 232n, 447n, 541n, 704n,
    872n

Paquette, L., 145n
Pender, K., 430n
Petruno, T., 343n
Pike, R., 737n
Pourjalali, H., 550n
Price Waterhouse, 642n

Quinn, J., 905n

Ramamurthy, B., 872n
Ratnatunga, J., 476n

Raymond, R., 872n
Reichelstein, S., 199n, 901n
Rothman, A., 849n

Sakurai, M., 201n, 232n, 686n, 704n,
    896n, 935n
Salowsky, H., 44n
Sangster, A., 703n
Saparito, B., 202n
Scarbrough, P., 232n, 686n, 896n
Scherer, F., 451n
Schiff, J., 16n, 161n
Schine, E., 905n
Schmenner, R., 322n
Schmidt, C., 410n
Schmidt, R., 782n
Schniederjans, M., 846n
Scholes, M., 720n
Schweitzer, M., 70n
Shank, J., 7n, 184n, 449n, 762n
Sharp, J., 872n
Sherman, H. D., 782n
Sjoblom, L., 799n
Slater, K., 580n
Smith, D., 904n
Smith, K., 686n, 703n, 704n, 737n, 892n,
    896n
Sohal, A., 476n
Song, J., 703n
Spencer, L., 698n
Spina, G., 848n
Srivastava, A., 703n, 737n
Stalk, G., 807n
Stewart, R., 743n
Stout, D., 582n
Studness, C. M., 401n
Sullivan, C., 686n, 703n, 704n, 737n,
    892n, 896n
Sutherland, J., 794n
Sweeting, C., 447n, 935n
Swenson, D., 667n

Tanaka, T., 200n, 201n, 443n
Tang, R., 872n
Tayles, M., 232n, 447n, 541n, 704n, 872n
Templeman, J., 443n
Teresko, J., 397n
Todd, K., 898n
Trossman, E., 70n
Troy, H., 873n
Turco, F., 848n
Turk, W., 42n
Turner, D., 451n, 451n
Turney, R. P., 816n

Vernon, J., 451n
Villers, R., 203n
Viscusi, W., 451n

Walleck, A., 250n
Walter, C., 872n
Watts, R., 863n
Williamson, R., 79n
Wilson, S., 232n, 447n
Winchell, W., 345n
Wolfson, M., 720n
Woodward, J., 818n
Wooton, C., 580n
Worthington, M., 501n, 547n, 548n
Worthy, F., 430n
Wygal, D., 582n

Yang, D., 849n
Yin, T., 905n
Yokoyama, I., 73n, 145n, 187n, 318n,
    447n, 516n, 703n, 892n

Zimmerman, J., 611n

Note: For further listing of authors, see Appendix B.

Company Index

Adjustor's Replace-a-Car, 451–52
Agency Rent-a-Car, 451–52
Allied Signal Limited, 665, 818
Aluminum Company of America, 15
American Airlines, 452
American & Efird, 267
American Greetings, 831
American Standard, 897–98
Analog Devices, 805, 806, 905
Apple Computer, 184, 273, 438
Arizona Public Service Company, 186
Arthur Andersen & Company, 98
Asea Brown Boveri, 683
AT&T, 456, 873

Bausch and Lomb, 869
Beijing Engineering, 396
Bell Canada, 2
Ben & Jerry's, 472
Boeing, 849
British Petroleum, 187

Cadbury Schweppes, 2, 219
California, State of, 10
Canadian Valley Cattle Ranch, 550
Caterpillar Tractor, 907
Chrysler, 807
Clark-Hurth, 165
Clorox Company, 13, 15, 16
Coca-Cola Company, 5, 840
Compaq, 438
Conner Peripherals, 891
Continental Airlines, 452
Crysel, 803

Daihatsu, 438
Dana, 905
Dayton-Hudson, 479
Deere, John, and Co., 807, 862
Delta Airlines, 452
Denver University Hospital, 506
Digital Equipment Company, 123
Duke Power Company, 907

Eastman Kodak, 199, 397
Eaton, 779, 905
Emerson Electric, 202
Emery Air Freight, 66–67
Episcopal Diocese of Newark, 207
Equitable Resources, 1
Exxon, 807

Federal Express Corporation, 98, 226,
    804, 814
Ford Motor Company, 27, 28, 98, 427,
    469, 777, 804, 807, 860, 896
Frito-Lay, Inc., 628, 840, 920
Fujitsu, 6–7

General Electric, 10, 473–74, 832
General Motors Corporation, 11, 71,
    535, 704, 777, 807, 832, 860
Guinness, 896

Harley-Davidson, 10, 42
Harvest States, 719
Hewlett-Packard, 16, 187, 273, 390, 661,
    808, 843, 847
Hitachi, 273
Home Depot, 465–67, 469
Honda, 11, 807
Honeywell, 807
Hudson's Bay Company, 889

ICI, 468
Illinois, University of, 10
Intel, 430
Interlake, 861
International Business Machines
    (IBM), 199, 807
Isuzu Motors, 430

John Deere and Co. 807, 862
Johnson & Johnson, 9, 861

Kaufmann's, 844
Kodak Corporation, 199, 397
Krones, 896

Lever Brothers, 117, 475
Lilly, Eli, and Co., 869
Lindsay Brothers, 385
Lockheed, 10, 907

Macy's, 98
Market Insights, 482
Marriott Corporation, 430, 906–7
Matsushita, 438
Mayne Nickless, 896
MCI Communications, 446
Mercedes Benz, 438, 443, 451
Michelin, 860
Microsoft, 10, 119
Midway Airline, 452
Milliken, 806
Minnesota Manufacturing and Mining
    Co. (3M), 861
Mitsubishi, 804
Mitsui, 896
Monsanto, 905
Motorola, 632, 794, 804, 807

NCR, 807
NEC Corporation, 186
Neptune Plastics Manufacturing Com-
    pany, 353
Nestlé Company, 552
Nestlé-Rowntree, 225
Nike, 2, 478
Nippon Steel, 905
Nissan Motor Company, 2, 187, 438,
    451, 905
Northeast Utilities, 401
Northrop Aircraft Division, 243
Northwest Airlines, 452

Octel, 806
Owens-Corning, 187

Pacific Bell, 430
PanAm, 452
Panasonic, 438
PepsiCo, 2, 6, 27, 758
Philips, 273
Pirelli, 860
Pizza Hut, 181, 203–05, 207
Pratt & Whitney, 849
Prestage Farms, 569
Procter & Gamble, 117, 202, 805
Public Service Company of New
        Hampshire, 401

Raychem, 430
Raytheon, 339, 597
Reebok, 478
Robert Half International, 2n
Rolls Royce, 849
Rugby Group, 631

Sandoz, 472
Sara Lee, 313
Saturn Corporation, 535

Sears Corporation, 472, 905
Sharp, 438
Shell Canada, 202
Siemens, 273
Sikorsky, 655
Snapple Beverage Corp., 307
Solectron, 793
Southwest Airlines, 814
Sun Microsystems, 59
Swedish Match, 896
Swissair, 16

Taco Bell, 920
Takeda Chemicals, 139
Tandem Computers, 430
Target, 97
Tektronix, 861
Texas Instruments, 842
3M (Minnesota Manufacturing and
        Mining Co.), 861
Toshiba, 273, 438
Toyota Motor Company, 7, 52, 200, 201,
        438, 443, 451, 665, 704, 804, 807,
        860

Toys "Я" Us, 187
TRW, 25, 779, 905
TWA, 452

Union Carbide Corporation, 698
Union Pacific Railroad, 196
United Airlines, 227, 452
United Technologies, 698
U.S. Air, 452

Wal-Mart, 844
Westvaco, 859
Whirlpool, 779, 905

Xerox Corporation, 804, 805, 843, 896

Yamazaki Mazak, 249, 260
Yoplait Company, 15

Zytec Corporation, 816

Subject Index

Ability to bear
  as cost-allocation criterion, 501
  and hospital cross-subsidization, 506
Abnormal spoilage, 633–34, 644
Absorption costing, 194, 309, 314–19
  breakeven points in, 320–21
  denominator-level concepts in, 322–26
  income effects of, 320
  in inventory-costing system, 312
  and output level, 390–92
  and performance measures, 321–22
  and special order, 432–34
Accelerated Cost Recovery System (ACRS), 744
Accelerated depreciation, 721–22
  double-declining balance, 721n, 728, 728n
Account analysis method, for cost estimation, 345–46
Accounting
  importance of, 3
  and opportunity costs, 399, 833
  types of
    cost, 2–3, 4
    financial, 4–5
    management, 2, 3, 4, 9, 98
  See also specific types
Accounting-based performance measures, 891–92
Accounting Principles Board (APB) *Opinion 25*, 906
Accounting rate of return, 701–4
Accounting return-on-investment method, 702
Accounting systems
  change in, 43, 98
  choice of, 10–11
  simplicity desirable in, 110
Accounting terms, defining of, 43–45
Accrual accounting rate of return (AARR), 687, 701–4
  in international comparison, 703
Accumulation, cost, 27–28
Actions, vs. events, 77

Activity area, direct and indirect costs for, 499
Activity-based budgeting, 199–200
Activity-based costing (ABC), 115, 660
  and decision making, 397
  example of, 165
  growing interest in, 161
  in manufacturing sector, 159–66
  merchandising sector application of, 116–18
Activity variance, 755
Actual costs and costing, 27, 109, 286
  and interdepartmental cost allocation, 508
  in inventory-costing system, 312
  in manufacturing, 145–46
  vs. budgeted costs, 105–6, 108–9
Actual usage allocation bases, 508–9
Adjustments, end-of-period, 153, 249
Administration, cost drivers in, 30
Administration costs budget, 197
Advertising, money wasted on, 476
After-tax cash-operating flows, 723
Aggregated amounts, 757–58
Allocation bases. *See* Cost allocation bases
Allocation of resources, and performance measures, 890
Allowable costs, 549
  and cost pools, 506
American Institute of Certified Public Accountants (AICPA), 936
Annual cost changes, analysis of, 780–81
Annuity
  future value of, 928–29, 933
  present value of, 929–30, 934
Antitrust laws, 451–52
  and price justification, 445
Apples-and-oranges problem, 757
Appraisal costs, 795, 796, 833
Artificial cost, 513
Assets, measurement alternatives for, 897–900
Assignment, cost. *See* Cost assignment

Attention-directing function, 15–16
Auditing, internal, 16
Audit trails, and backflush costing, 668
Australian professional examinations, 937
Autocorrelation, 368
Automation, benefits and costs of, 704–5
Autonomy, and transfer pricing, 864
Average cost. *See* Unit cost
Average waiting time, 812

B

Backflush costing method, 35n, 656, 661–62
  difficulties of, 667–68
  examples of, 663–67
Bailout payback, 700–701
Balanced scorecard, 806, 890
Balance sheet, and byproducts, 583
Baldrige Quality Award, 794
Bar codes, 28, 782
Batch-level costs, 356
Benchmark(ing), 7, 227, 249, 903–4
  at Australian company, 250
  and cost management, 481–83
  financial and nonfinancial, 250
  and performance gaps, 285
  three types of, 227–28
Black boxes, 477
Black hole demand spiral, 546
"Blame game," 206, 247
Bonuses
  and quality, 807
  and total factor productivity, 779
Book value, 403
  gross vs. net, 900
Book-value method, 702
Bottleneck, 811–13
  and theory of constraints, 816–19
Brand cost hierarchy, 552
Breakeven analysis, 64. *See also* Cost-volume-profit analysis
Breakeven point, 62

Breakeven point, *(cont.)*
    methods for determining, 62–65
    in variable and absorption costing,
        320–21
Breakeven time, 687, 808–11
Budgetary slack, 199
Budget(s) and budgeting, 8, 182–84
    activity-based, 199–200
    advantages of, 184–86
    capital, 684, 685–87 (*see also* Capital
        budgeting)
    cash, 209–13
    classifications of, 187–88
    and control, 671
    for discretionary costs, 478–79
    at Emerson Electric, 202
    engineered- vs. discretionary-cost
        approaches to, 479–81
    and ethical accounting policies, 907
    financial, 188, 209
    flexible, 226–38
    historical-cost pitfalls in, 901
    human aspects of, 208
    international comparisons on, 187
    life-cycle, 448–49
    master, 183, 188–99, 323–24, 544
    paperless (Union Pacific), 196
    at Pizza Hut, 181
    and standard costs, 226–27
    time coverage of, 186–87
Budgeted balance sheet, 209
Budgeted costs or costing, 109, 286
    indirect allocation rates for, 107–9
    and interdepartmental cost alloca-
        tion, 508
    in inventory-costing system, 312
    in manufacturing, 145–46
    vs. actual costs, 105–6, 108–9
Budgeted income statement, 197–98
Budgeted statement of cash flows, 209
Budgeted usage allocation bases, 508–9
Budgeted variable overhead rates,
    270–71
Business function costs, 393
Byproducts, 571, 582–84
    from poultry farms, 569, 571

C

Canadian Institute of Chartered Ac-
    countants, 936
Canadian professional examinations,
    936–37
Capacity, theoretical and practical, 323
Capacity issues, 544–46
Capital, opportunity cost of, 688, 736,
    838
Capital budgeting, 684, 685–87
    accrual accounting rate-of-return
        method of, 687, 701–4
    complexities in application of, 704–5
    discounted cash flow method of,
        687–90 (*see also* Discounted cash
        flow method)
    and environmental costs, 698
    and income taxes, 719, 720–31
    and inflation, 732–36

international comparison of, 703
    and net present value decision rule,
        738–40
    for new plant construction, 683
    and nonprofit organizations, 738
    payback method of, 687, 697–701,
        703, 810–11
    risk adjustment methods in (survey),
        737
Capital investment, 694
Capitalized costs, 37–40
Carpentry plants, cost management at,
    631
Carrying costs, 833
    of inventory, 399–400
Cash budget, 209–13
Cash cycle, 211
Cash flows
    and income taxes, 724–31
    and risk, 736–37
    See also Discounted cash flow
        method
Cause and effect, 342–43
    as cost-allocation criterion, 501
Cause-and-effect diagram, 799–800
Ceramics industry, process costing in,
    611
Certainty equivalent approach, 737
Certified Cost Estimator/Analyst
    (CCEA) program, 936*n*
Certified General Accountant (CGA),
    936
Certified General Accountants' Associ-
    ation (CGA), 936
Certified Management Accountant
    (CMA), 16, 936
Certified Management Accountant
    (CMA) examination, 935–36
Certified Public Accountant (CPA) ex-
    amination, 935–36
Chartered Accountant (CA), 936
Cherry picking of production line, 322
Chief financial officer (CFO), 12–14
Choice criterion, 77
CIM. *See* Computer-integrated manu-
    facturing
Clayton Act, 451
*CMA: The Management Accounting
    Magazine*, 937
Coefficient of determination, 352
Collusive pricing, 452
Common costs, 515
    allocation of, 515–17
Communication, 185–86
Compensation
    executive compensation plans, 905–7
    and incentive, 901–3
    and multiple tasks, 904
    and risk, 206
    term-based, 904–5
    See also Incentives and incentive
        plans
Competence, as obligation, 17
Competitor analysis, 430
Competitors, and pricing, 428
Complete reciprocated cost, 513
Compound interest. *See* Tables, interest

Compound-interest depreciation
    method, 900*n*
Computer-based financial planning
    models, 200–202
Computer-integrated manufacturing
    (CIM)
    and capital budgeting, 704–5
    as investment, 632
Computers, and job cost information,
    104
Conference method, for cost estima-
    tion, 344–45
Confidentiality, as obligation, 17
Constant, 341
Constant gross-margin percentage
    NRV method of joint cost allo-
    cation, 577–79
Constant returns to scale technology,
    775
Constant variance of residuals, 367
Constraint, 411
Constraints, theory of, 817–19
Continuous improvement standard
    cost, 246
Contracting situations, cost allocation
    issues in, 45, 546–49
Contribution income statement, 63
Contribution margin, 62, 73–74
Contribution margin method
    for computing breakeven point,
        62–63
    and output level, 390–92
    and relevant-cost analysis, 396–97
Contribution-margin percentage, 74
Control, 8–10
    benchmarks in, 285
    and costing systems, 166
    and discretionary- vs. engineered-
        cost approach, 480–81
    and job cost records, 144
    of projects, 668–71
    of projects and jobs, 671
Control chart, 798–99
Control-factor unit, 479
Controllability, 206, 549
    and responsibility, 206–8
Controllable cost, 206
Controller, 14–15
Control systems, evolution of, 182
Conversion costs, 43, 156
    estimating degree of completion for,
        620
Cooking the books, 907
Coordination, 185–86
Correlation, 348
    vs. causation, 342–43
Cost(s), 26
    of capital, 688, 736
    classification of, 46–47
    environment-related, 698
    hierarchy of, 355–57
    interrelated, 467–68
    linear, 31
    and pricing, 429
    of quality, 794, 795–98, 806, 833,
        842–43
    relevant, 388–89, 841–43

as success factor, 6
of time, 814–16
transferred-in (process costing), 614–20
Cost(s), major types of
actual (historical), 27, 109, 286 (see also Actual costs and costing)
budgeted, 105–6, 107–9, 145–46, 286, 312, 508
capitalized, 37–40
carrying, 399–400, 833
conversion, 43, 156, 620
direct, 27–29, 32–33, 98, 104–6, 286, 441–42, 498–99
direct manufacturing labor, 41–42
direct materials, 41, 620
engineered, discretionary, and infra-structure, 475–76
fixed, 30–31, 32–33 (see also Fixed costs)
hidden ("buried"), 450
imputed, 400
indirect, 28–29, 32–33, 99 (see also Indirect costs)
inventoriable, 37–40, 194, 308, 309, 314
manufacturing, 41–43 (see also Manufacturing costs)
normal, 109, 145–46, 286, 312
opportunity, 397, 399, 833 (see also Opportunity cost)
ordering, 833
outlay, 392, 399, 875
overhead, 268 (see also Overhead costs)
past (sunk), 403–6
period, 37–40, 309, 314
prime, 43
purchase, 833
standard, 227, 286 (see also Standard cost)
unit vs. total, 33–34 (see also Unit cost)
value-added vs. non-value-added, 270
variable, 29–31, 32–33, 309, 314–19 (see also Variable cost)
See also Costing
Cost accounting, 4
as management accounting, 2
and value-chain, 2–3 (see also Value chain)
Cost Accounting Standards Board (CASB), 547–48
Cost accounting system, 4. See also Costing system
Cost accumulation, 27–28
Cost allocation, 27, 29, 99, 498, 502–4
in ABC system, 161, 164
of common costs, 515–17
in contract situations, 546–49
and cost pools, 504–6 (see also Cost pools)
criteria to guide, 500–502
and cross–subsidization, 115
in customer costing, 119–23

department-to-department, 506–10
fairness of, 548–49
in hospitals, 497, 506
in job costing
for manufacturing organizations, 141–42, 144–45
for service or merchandising organizations, 103, 107–9
of joint costs, 571–79 (see also Joint costs)
in product costing, 503, 536–38
purposes of, 499–500, 502, 510
and support departments, 510–15, 873
terminology of, 498–99
Cost allocation bases, 99
in budgeted variable overhead rate, 270
and cost drivers, 544
in cost system refinement, 110
inappropriate, 542–43
in job costing examples, 103, 141–42, 144, 161
and manufacturing lead time, 816
for manufacturing overhead, 145
nonfinancial, 542
in traditional vs. ABC approach, 164
U.S.-Japan comparison on, 542
Cost analysis, 684–85
Cost assignment, 27–29, 99, 548
and cost hierarchies, 549–52
evolving trends in, 541–44
See also Cost allocation; Costing systems; Cost tracing
Cost-based transfer prices, 865, 870–75
Cost-benefit approach, 10–11, 98
in cost allocation, 501–2
and indirect-cost-rate revision, 108
and inventory categories, 35n
for management control system, 468
and variance investigation, 255
Cost center, 203, 863
Cost classification
international comparison of, 345
for value chain, 476–78
Cost driver(s), 29, 60, 340
and allocation bases, 544
and cause-effect criterion, 342–43
examples of for business function areas, 30
and fixed vs. variable costs, 30, 31
hierarchy of, 355–57
identifying of, 348
nonoutput-related, 61
in regression model, 353
relevant ranges for, 31
at Saturn Corporation, 535
Cost-effectiveness, 28
Cost estimation, 340
approaches to
account analysis, 345–46
conference, 344–45
industrial-engineering, 344
quantitative analysis, 346–47
assumptions in, 340
high-low method of, 348–50

at Neptune Plastics Manufacturing, 353
regression analysis in, 350–52
steps in, 347–48
Cost function(s), 340–41
evaluating and choosing among, 352–53, 369–71
and learning curve, 358–61
linear, 340, 341–42
nonlinear, 357
Cost of goods manufactured, 38n
Cost of goods sold (COGS)
generally accepted accounting principles on, 295
and under/overallocated overhead, 153
Cost of goods sold budget, 195
Cost hierarchy, 355–57, 549–52
Cost incurrence, 439–40
Costing
absorption, 309, 314–19 (see also Absorption costing)
actual vs. normal vs. budgeted, 109, 145–46
actual vs. normal vs. budgeted vs. standard, 286
five-step approach to, 536–38
life-cycle, 448, 451, 685
for long run, 434–37
peanut-butter, 114–15
for short run, 431–34
super-variable (throughput), 313, 322, 665
target, 438–39, 440–42, 443, 446, 450–51
See also Cost(s)
Costing down
budgeting for, 183–84
and continuous improvement, 429
and cost drivers, 445
Costing system(s), 4, 98, 503, 656
and activity-based costing, 115–18, 159–66, 397, 660
backflush, 35n, 661–68
choice of, 660
customer costing, 118–23, 124
hybrid, 656–57
job, 99–100, 140, 656, 660 (see also Job costing system)
in manufacturing sector, 140
multipurpose nature of, 166
operation, 657–60
process, 99–100, 140, 598–99, 656, 660 (see also Process costing system)
varieties of, 660
Costing system refinement, 110–14
with activity-based costing, 160
Cost justification, 499, 546–49
Cost management, 5, 29
in automotive industry, 427
and benchmarking, 481–83
at carpentry plants, 631
international comparison of, 447
and just-in-time purchasing, 843–44
in pharmaceutical industry, 139
See also Spoilage

Cost object, 26–27, 98, 340, 498–99
  choice of, 343
  and classification of costs, 29
  direct costs of, 498–99 (see also Direct costs)
  indirect costs of, 498–99 (see also Indirect costs)
  and variable vs. fixed costs, 30
Cost planning, 65–67
Cost-plus contract, 547
Cost-plus pricing, 443–46
Cost-pool homogeneity, 505–6, 539
Cost pools, 99, 504–6
  allowability of costs in, 506
  in budgeted variable overhead rate, 270
  homogeneous, 110, 505–6
  indirect (overhead)
    choice of (plantwide vs. department), 539–41
    in example, 504
    increased number of as refinement, 111–14
    increasing number of, 541
    in job costing, 102–3, 141, 144, 160
    in product costing, 536–37
    in traditional vs. ABC approach, 164
  plantwide vs. department, 539–41
  and repair costs, 352
  questions on, 505
Cost predictions, 340
Cost rates, choice of, 539–41
Costs. See Cost(s)
Costs-attach tracking, 664, 665
Cost and schedule performance report, 669, 670
Costs of quality, 794, 795–98, 833
  relevant, 842–43
Costs of quality measure, 806
Cost tracing, 27, 29, 99, 498, 502–4, 548, 536
  increasing interest in, 541
  in service organization, 104–6
  See also Cost allocation
Cost variances, overhead, 280–82
  fixed, 277–78
  integrated analysis of, 280–81
  in nonmanufacturing settings, 281–82
  variable, 271–75
Cost-volume-profit (CVP) analysis, 60, 61
  assumptions in, 65
  and breakeven point, 62–65
  in cost planning, 65–67
  and income taxes, 70–72
  key terms in, 61–62
  and nonprofit institutions (revenue), 72
  and profit-volume chart, 68–69
  and sales mix, 69–70
  and target operating income, 64–65
  and uncertainty/sensitivity analysis, 67–68
  and variable-costing operating income, 318–19

vs. "breakeven analysis", 64
Cross-sectional data, 346–47
Cross-subsidization, 114–15
  at hospitals, 506
Cumulative average-time learning model, 358, 359
Current cost, 897–900
Currently attainable standards, 234
Customer cost hierarchy, 551–52
Customer costing system, 118–23
  at DEC-Europe, 123
  value chain in, 124
Customer life-cycle costs, 451
Customer profitability, and relevant costs, 400–403
Customer reporting unit, 863
Customer-response time, 811
Customer satisfaction, 3, 5, 6, 118
  and discretionary-cost centers, 486
  as Johnson & Johnson goal, 9
  measures of, 804–5
  and pricing, 428
  and quality, 793
  and value-adding, 29, 473
Customer-satisfaction measures, 891
Customer service, 3
  cost drivers in, 30
Customer-service costs budget, 197
Customer-support costs, 122
Custom products, and product costs (ceramics firm), 611
Cycle, self-liquidating, 211

D

Data
  plotting of, 348
  problems encountered with, 354–55
  time-series vs. cross-sectional, 346–47
Data base, 4
  ideal characteristics for, 354
Decentralization, 860–63
  and responsibility centers, 863
Decision process or model, 386
  and good decisions vs. good outcomes, 81
  ideal process of, 397
  illustrative examples of, 26
  information in, 386–88
  joint costs as irrelevant for, 579–82
  and learning curve, 361
  and performance-evaluation model, 406–7
  quantitative vs. qualitative factors in, 389–90
  relevant considerations in, 388 (see also at Relevant)
  suboptimal, 861–62
  total-alternatives approach to, 397–99
  and uncertainty, 77–81, 253
Decision table, 80
Decision tree, 80–81
Degrees of freedom, 366n
Delayed costing, 662. See also Backflush costing method

Demand cycle, and forecasting of normal utilization, 324
Demand-pull feature of JIT, 661, 845, 846
Demand spiral, downward, 546
Demand-supply relation, and prices, 429
Deming Prize, 794
Denominator level, 276
  in absorption costing, 322–26
  and capacity distinctions, 544
  at Snapple Beverage, 307
Denominator reason, 108, 152
Department costing, 142
Department of Defense (DOD), and project variances, 669
Department-to-department cost allocation, 506–10
Department rates
  and transfer prices, 873
  vs. plantwide rates, 539–41
Department stores, and return on investment, 889
Dependent variable, 347–48, 350
Depreciation
  in cash budget example, 210, 211, 212
  compound-interest, 900n
  as fixed cost, 344
  and income tax, 724, 725, 726, 728
  and inventoriable vs. period costs, 314
  in machining cost pool, 541
  with MACRS, 731, 744
  and taxes, 721–24, 731
Design, 3
  of accounting systems, 98
  cost drivers in, 30
  quality of, 795
Designed-in costs, 439
"Different costs for different purposes," 4, 45
Differential cost, 389
Direct allocation method (direct method), 511
Direct costing, 311. See also Variable cost(ing)
Direct costs, 27–29, 32–33, 98
  actual, normal, budgeted and standard, 286
  and cost object, 498–99
  manufacturing
    in inventory-costing system, 312
    labor, 41–42
  materials, 41, 620
  product, 429–30, 499
  in target costing, 441–42
  tracing of, 29, 111 (see also Cost tracing)
Direct labor variances
  investigating of (survey), 255
  mix, 770–71
  yield, 770–71
Direct manufacturing labor budget, 194
Direct manufacturing labor variances, proration of, 291–93

Direct materials inventory, 35
Direct materials purchases budget, 192–93
Direct materials usage budget, 192–93
Direct materials variances
    investigating of (survey), 255
    mix, 765–69
    proration of, 293–95
    yield, 765–69
Direct method, 511
Discounted cash flow method, 687–88
    and accrual accounting performance measures, 900
    at Harvest States, 719
    internal rate-of-return, 690–92
    in international comparison, 703
    issues and problems in, 694–97
    net present value, 688–90, 692, 722, 732–36
    and nonprofit organizations, 738
    required rate of return in, 688, 736–38
Discount rate, 688, 736
Discretionary-cost centers, 485–86
Discretionary costs, 475–76, 477
    as budgeting approach, 480–81
    and negotiated static budget, 478
Disposal price, and discounted cash-flow method, 695, 696
Distress prices, 870
Distribution, 3
    cost drivers in, 30
Distribution costs budget, 197
Disturbance term, 367
Documents, source, 104, 142–44
Double-counting, 547
Double-declining balance (DDB) depreciation, 721n, 728, 728n
Downstream costs, 429
Downward demand spiral, 546
Dual internal/external focus, 6, 7, 249
Dual pricing, 874–75
Dual-rate method of cost allocation, 507–8
Dumping, 452
DuPont method of profitability analysis, 893–94
Dysfunctional decision making, 861–62

E

Earned hours report, 670
Earned value report, 670
Economic feasibility, 28
Economic-order-quantity (EOQ) decision model, 833–35
    cost parameters used in, 838–40
    and goal-congruence issues, 840
    and just-in-time purchasing, 841
Economic plausibility, 343, 352, 370
Effectiveness, 237, 478
    in discretionary-cost centers, 485–86
Efficiency, 237, 478
    in discretionary-cost centers, 485–86
Efficiency, quality, and time measures, 891
Efficiency variances, 238, 240–43

variable overhead, 274–75
Effort, 468–69
Electronics industry, rejection rates in, 642
Ending inventory budget, 194–95
End-of-period account(s) approach, to under/overallocated overhead, 153–54, 155, 249
End-of-period adjustments, 153, 249
Endpoint costing, 662. See also Backflush costing method
Engineered costs, 475, 476, 477
    as budgeting approach, 480
Engineering change orders (ECO), 473, 473n
Environmental costs and responsibilities, 435, 907
    and capital budgeting, 698
    and integrated project planning, 468
    and management control systems, 472
    and total quality management, 807
EOQ. See Economic-order-quantity decision model
Equation method, for computing breakeven point, 62
Equity
    as cost-allocation criterion, 501
    and stand-alone method, 517
Equivalent units, 157
    in process costing, 599–609
Errors, in recording costs, 354
Error term, 367
Estimated net realizable value (NRV) method of joint cost allocation, 576–77
Estimation, cost. See Cost estimation
Ethics, professional, 16–18, 907
Evaluation. See Performance evaluation or measures
Events, 77
    vs. actions, 77
Examinations. See Professional examinations
Excess capacity, 813
Excess present value index, 738
Executive compensation plans, 905–7
Executive judgment, group, 199
Expectations operator, 341
Expected monetary value, 78–79
Expected value, 78–79
"Expenses," 42
Experience curve, 358
Extended value chain, 6–7
External failure costs, 795, 796, 833
External reporting, 4
    and cost allocation, 499
    responsibility for, 18
    variable costing for, 312
Exxon Valdez oil spill, 807

F

Facility-sustaining costs, 356–57
Factory burden costs, 42
Factory overhead costs, 42. See also Manufacturing overhead costs

Failure costs, 795, 796, 833
Fairness
    as cost-allocation criterion, 501
    of pricing, 548–49
    and stand-alone method, 517
Favorable variance, 228
Feasibility, economic, 28
Federal Acquisition Regulations (FARs), 500–501, 547
Federal Energy Regulatory Commission (FERC), 401
Federal Trade Commission Act, 451
Feedback, 10, 387–88
    and responsibility, 206
FIFO. See First-in, first-out process-costing method
Finance director, 12
Financial accounting, 4–5
Financial Accounting Standards Board (FASB), 515
    on executive stock options, 906
Financial benchmarks, 250
Financial budget, 188, 209
Financial planning models, computer-based, 200–202
Financial reporting
    external, 4, 18, 312, 499
    internal, 4, 5, 18, 312–13
    inventory costing for, 166
    life-cycle, 450–51
    and product cost, 45
    responsibility for, 18
Financial statements
    and denominator-level concept, 324–26
    and opportunity costs, 399
    and proration, 295
    for service-, merchandising-, and manufacturing-sector companies, 34–37, 38
Finished goods inventory, 35
First-in, first-out (FIFO) process-costing method, 606–9
    and spoilage, 636–37, 637n
    and transferred-in costs, 618–19
    use of, 620
    vs. weighted-average method, 609–10, 617
Fishbone diagrams, 799
Fixed costs, 30–31, 32–33
    and cost object, 343
    overhead, 270, 276
    problems encountered with, 354
    relevant range for, 31–32, 344
    and support departments, 873
    and time span, 343
Fixed manufacturing overhead costs, 276–77, 279, 282–83
    and denominator-level concept, 325
    and internal reporting (international comparison), 318
    in inventory-costing system, 312
    for planning/control vs. inventory costing, 268, 283
    and variable vs. absorption costing, 308–14 (see also Absorption costing; Variable costing)

Fixed overhead cost variances, 277–78
Fixed manufacturing overhead in ending inventory (FMOH:EI), 317
Fixed overhead rates, 276–77
Flexibility gain, 861
Flexible budgets, 226, 228
    and standard costs, 226–27, 233–38
    without standard costs, 228–33
    and variable overhead manufacturing costs, 271–73
Flexible-budget variances, 230–31, 237–38
Forecasting, sales, 198–99
4–, 3–, 2–, 1–variance analysis, 280–81
"Fraction of the action" programs, 486
Franklin, Benjamin, on death and taxes, 720
Fringe benefits, 44
Full-cost bases, for transfer prices, 870–71
Full-cost-plus basis, 282
Full product costs, 393, 436, 865
    and pricing decisions, 445
Future, and decision making, 388
Future value table, 927–28, 931
    for annuity, 928–29, 933

G

General ledger
    and byproducts, 582–84
    in manufacturing job costing system, 146–52, 153
    and standard costs, 248–49
Generally accepted accounting principles, 4
    and actual vs. standard costs, 295
    on inventoriable costs, 37, 281, 314
Geographic reporting unit, 863
Goal congruence, 468–69
    and EOQ model, 840
    and moral hazard, 902, 902n
    and performance measures, 900–901
    and residual income, 895
    and transfer pricing, 864
    See also Performance-evaluation model
Goodness of fit, 352, 365–66, 370
Goods in process, 35
Goods for sale, managing of, 832–40
Government
    antitrust laws, 445, 451–52
    contracting with, 45, 547
    environmental laws, 435
Governmental Accounting Standards Board (GASB), 515
Graphic solution approach, 413
Graph method, for computing breakeven point, 63–64
Gross book value, 900
Gross margin, 73–74
Group executive judgment, in sales forecasting, 199

H

Half-life method, 805

Heteroscedasticity, 367
Hidden (buried) cost categories, 450
Hierarchy of costs and cost drivers, 355–57, 549–52
High-low method, for cost estimation, 348–50
Historical costs, 27. *See also* Actual costs and costing
Historical data, and performance, 185
Homogeneous cost pool, 110, 505–6
Homoscedasticity, 367
Hospitals
    benchmark reports for, 482
    and capital budgeting, 738
    cost allocation in, 497, 506
Hurdle rate, 688, 736
Hybrid costing systems, 656–57

I

Idle facilities, 395–96
IMA. *See* Institute of Management Accountants
Improvement
    continuous, 7, 246, 429
    and *kaizen* budgeting, 200
    and perfection standards, 234
    and pricing, 429
Imputed costs, 400
Imputed interest charge, 840, 894
Incentives and incentive plans, 470–72
    and discretionary-cost centers, 486
    in executive compensation, 906–7
    intensity of, 903
    vs. risk, 901–2
    See also Compensation
Income statement
    budgeted, 197–98
    and byproducts, 583
    contribution, 63
    and denominator-level concepts, 326
    for manufacturers, 37, 38
    for merchandisers, 36–37
    for service sector, 36
    for special order, 432, 433–34
    under variable and absorption costing, 315–16
Income taxes, 720–21
    and cash flow, 724–31
    in CVP, 70–72
    and denominator-level concept, 325
    and depreciation, 721–24, 731
    and discounted cash-flow method, 696–97, 719
    relevant aspects of, 389
    See also Taxes
Incremental approach, 725, 729–31
Incremental common cost allocation method, 516
Incremental costs, 389, 582
    and decision making, 580–82
Incremental revenue, and decision making, 580–82
Incremental unit-time learning model, 358, 360
Independent variable, 350
    significance of, 352–53, 366

Indirect cost(s), 28–29, 32–33, 99
    actual, normal, budgeted and standard, 286
    allocation of, 29, 498 (*see also* Cost allocation)
    and cost object, 498–99
    factors in, 539
    manufacturing, 42
    product, 430, 499
    under/overallocation of, 152–55
Indirect cost pools
    choice of (plantwide vs. department), 539–41
    in example, 504
    increased number of as refinement, 111–14
    increasing number of, 541
    in job costing, 102–3, 141, 144, 160
    in product costing, 536–37
    in traditional vs. ABC approach, 164
Indirect cost rates, capacity as affecting, 545–46
Individual activities, performance measures for, 904–5
Industrial-engineering method, 344
Inflation, 732
    and capital budgeting, 732–36
    and cost estimation, 355
    and depreciation, 721
    and discounted cash-flow analysis, 719
Information
    and decision process, 386–88
    for management control system, 466–67 (*see also* Management control systems)
Information asymmetry, 860
Infrastructure costs, 476
Innovation, as success factor, 6
Innovation measures, 891
Input variances, 764
    direct labor yield and direct labor mix, 770–71
    direct materials yield and direct materials mix, 765–69
Insourcing, 394–95
In-sourcing initiative, at GM, 71
Inspection, and spoilage, 634, 640, 646–47
Institute of Certified Management Accountants (IMA), 936
Institute of Management Accountants (IMA), 16, 45n
    "Standards of Ethical Conduct for Management Accountants" of, 17
Intangible property, international transfer of, 868
Integrity, as obligation, 17
Intercept, 341
Interest, 927
    in cash budget example, 211, 212
    imputed, 840, 894
    and time value of money, 389, 684–85
Interest tables. *See* Tables, interest
Interim performance reports, 671

Intermediate product, 864
Internal auditing, 16
Internal failure costs, 795, 796, 833
Internal performance measures, and quality, 805–6
Internal rate of return (IRR), 690
Internal rate-of-return decision rule, 740
Internal rate-of-return method, 690–92
    vs. net present value, 692
Internal reporting, 4, 5
    and cost method, 312–13
    responsibility for, 18
Internal Revenue Service, 313, 314, 325, 869. *See also* Income taxes
International comparisons and trends
    on ABC approach, 161
    of budget practices, 187
    of capital budgeting methods, 703
    of cost allocation bases, 145
    on cost assignment, 541, 542
    of cost classification, 345
    on JIT performance measures, 848
    on labor fringe benefits, 44
    on nonfinancial measures of performance, 892
    of pricing and cost management, 447
    of risk adjustment methods, 737
    on standard costs, 232
    on support department cost allocation, 516
    of variable vs. absorption costing, 318
    on variable vs. fixed costs, 73
International taxation, 868
Inventoriable costs, 37–40, 308, 314
    and absorption costing, 194
    and overhead costs, 281
    and variable vs. absorption costing, 309
    see also Absorption costing; Variable costing
Inventory(ies)
    carrying costs of, 399–400
    manufacturing
        direct materials, 35
        finished goods, 35
        work in process, 35
    merchandise, 35
    in no-allocation approach to joint cost, 578
    and service companies, 35, 37
    undesirable buildup of, 321–22
    and variances, 247–48
Inventory carrying costs, 814
Inventory costing, 308
    and backflush costing, 668
    comparison of methods for, 312
    and costing systems, 166
    generally accepted accounting principles on, 295
    in operating budget, 194–95
    for process costing, 610–12, 614
Inventory management, 832
    at American Greetings, 831
    and just-in-time system, 661, 662, 840–44 (*see also at* Just-in-time)

in manufacturing organizations, 844–49
in retail organization, 832–40
Inventory methods
    periodic, 40, 41
    perpetual, 40–41
Inventory shrinkage, 117
Investment
    alternative definitions of, 897
    capital, 694
    and department stores, 889
    JIT or CIM as, 632
Investment base, computing of, 702
Investment center, 203, 863
Investment tax credit (ITC), 721, 721*n*

J

Japanese professional examinations, 937
Job(s)
    control of, 671
    vs. project, 668
Job costing system, 99–100, 140, 656, 660
    and activity-based costing, 115, 159, 160–62
    in manufacturing, 140–44
        and actual vs. normal vs. budgeted costing, 145–46
        and allocation of indirect costs, 144–45
        illustration of, 146–52
        and under/overallocation, 152–55
    in service organizations, 101–4
        and activity-based costing, 115–18
        and actual vs. normal vs. budgeted costing, 109
        and allocation of indirect costs to jobs, 107–9
        and costing system refinemement, 110–14
        and customer costing, 118–23
        and peanut-butter costing, 114–15
        and tracing of direct costs to jobs, 104–6
    and spoilage, 640–41, 645
    value chain in, 124
Job cost record (job order or job cost sheet), 142
Jobs bank, at GM, 71
Joint costs, 570
    allocation of, 571–79
    and byproducts, 582–84
    and decision making, 579–82
Joint products, 571
    from poultry farms, 569, 571
Journal entries
    for operation costing, 659
    for overhead costs and variances, 283–85
    and process costing, 159, 606
    and spoilage, 637–40
Justification of costs, 499, 546–49
Just-in time (JIT) production system, 660–61, 845, 846–48
    and backflush costing, 656, 662

as investment, 632
at Sikorsky, 655
survey results on, 667
Just-in-time (JIT) purchasing, 840–44

K

Kaizan budgeting or management, 7, 200
    at Toyota, 201
Kanban system, 846
Key financial performance measures, 896
Key success factors, 6

L

Labor costs
    direct manufacturing, 41–42
    fringe benefits in, 44
    in law firms, 104
    in service organizations, 104
Labor-paced work environment, 541
Labor time record, 143–44
Labor unions, and labor cost, 71
Law firms, labor compensation costs in, 104
Laws
    antitrust, 445, 451–52
    environmental, 435
    and ethical business conduct, 907
Leadership, and discretionary-cost centers, 486
Lead time, manufacturing, 542
Learning curve, 358
    and cost function, 358–61
    and decision making, 361
Ledgers. *See* General ledger; Subsidiary ledgers
Level of uncertainty, 477–78
Life-cycle, product, 448, 807
Life-cycle budgeting, 448–49
Life-cycle costing, 448, 685
    and customer life-cycle costs, 451
Life-cycle reports, 450–51
Linear cost function, 340, 341–42
Linear costs, 31
Linear programming (LP), 410–13
Line management (vs. staff), 12
Locked-in (designed-in) costs, 439
    and life-cycle costs, 450–51
Long run, costing and pricing for, 434–37

M

Machine-paced work environment, 541
Main product, 571
Make-or-buy decisions, 395
Malcolm Baldrige Quality Award, 794
Management accounting, 2, 3, 4
    choice of system for, 98
    customer-driven changes in, 9
Management Accounting, 936
Management Accounting (U.K.), 938
Management by exception, 8, 204

Management by objectives, 207
Management control systems, 466–67
  and benchmarking in cost management, 481–83
  and budgeting for discretionary costs, 478–79
  and cost categories in value chain, 476–78
  and discretionary-cost centers, 485–86
  and engineered- vs. discretionary-cost budgeting approaches, 479–81
  and engineered vs. discretionary vs. infrastructure costs, 475–76
  and environmental concerns, 472
  evaluating of, 468–70
  and feedback, 10
  formal and informal aspects of, 469
  at Home Depot, 465
  and interrelated cost, 467–68
  and motivation, 468–72
    and risk, 469–70
    and sales compensation plans, 470–72
  planning and control in, 7–10
  and value-added analysis, 473–75
Managerial behavior, and performance-evaluation model, 406–7
Managers, incentives for. See Incentives and incentive plans; Motivation
Manufacturing costs, 41–43
  direct, 41, 312 (see also at Direct)
  in example, 435–37
  overhead, 42, 273 (see also Manufacturing overhead costs; Overhead costs)
  and variable costing, 311
"Manufacturing expenses," 42
Manufacturing lead time, 542, 812–13
  Boeing's attempt to cut, 849
  as cost allocation base, 816
  and costs of time, 814–16
  and just-in-time production, 846
Manufacturing organizations, inventory management in, 844–49
Manufacturing overhead allocated, 151
Manufacturing overhead (MOH) costs, 42, 273
  cost allocation bases for, 145
  fixed, 276–77, 279, 282–83, 308–14 (see also Fixed manufacturing overhead costs)
  as inventoriable, 281
  and inventoriable vs. period costs, 314
  in target costing, 441–42
  for thread manufacturers, 267
  variable, 279, 282–83, 312
  variation in, 435
Manufacturing overhead costs budget, 194
Manufacturing rates, plantwide vs. department, 539–41
Manufacturing resources planning, 845n

Manufacturing-sector companies, 35
  activity-based costing in, 159–66
  capitalized, inventoriable, and period costs for, 37–40
  and contribution margin vs. gross margin, 74
  financial statements for, 34–37, 38
  inventoriable costs for, 314
  job costing systems in, 140–44
    and actual vs. normal vs. budgeted costing, 145–46
    and cost allocation, 141–42, 144–45
    illustration of, 146–52
    and under/overallocation, 152–55
  process costing systems in, 155–59
Manufacturing variances, proration of, 290–95. See also Variances
Margin of safety, 68
Market, defining of, 761
Market-based transfer prices, 869–70
Marketing, 3
  cost drivers in, 30
Marketing costs budget, 196
Market-share variance, 762–64
Market-size variance, 762–64
Markup, 445
Master budget, 183
  illustration of, 188–99
Master-budget utilization, 323–24, 544
Materiality concept, and backflush costing, 667–68
Materials, substitutable and nonsubstitutable, 764
Materials management, 832. See also Inventory management
Materials requirement planning (MRP), 844–46
Materials requisition record, 143
Maximum capacity, 323
Measures and measurement
  of assets, 897–900
  nonfinancial, 322, 804–5, 806, 890–91, 892
  of output, 62
  of performance, 890 (see also Performance evaluation or measures)
Medical costs, 497. See also Hospitals
Menu-based costing, 114
Merchandising-sector companies, 35
  activity-based costing in, 116–18
  capitalized, inventoriable, and period costs for, 37–40
  and contribution margin vs. gross margin, 74
  financial statements for, 34–37
Milestones, 670
Mix, 764
  product, 393–94
  sales, 69–70, 754
Mixed cost, 342
Mix variances, 753, 755, 758–61, 765–71
Modified Accelerated Cost Recovery System (MACRS), 731, 744
Moral hazard, 902, 902n
Motivation, 468–69
  through cost allocation, 499, 502, 510
  through decentralization, 861

and risk sharing, 469–70
  through sales compensation plans, 470–72
Moving cost reduction standard cost, 246
Mu (μ), 798
Multicollinearity, 372
Multinational transfer pricing, 859, 868–69, 872
Multiple jobs, in customer costing, 120–23
Multiple products, and variance analysis, 755, 756–57
Multiple regression, 350, 371–72
Multiple tasks, 904

N

National Association of Accountants, 16, 45n
Negotiated static budget, 478–79
Net book value, 900
Net income, 61
Net present value decision rule, 738–40
Net present value (NPV) method, 688–90
  and depreciation, 722
  and inflation, 732–36
  vs. internal rate-of-return, 692
Net relevant cost, 389
New product development time, 807–8
Nominal rate of return, 732
Non-cost considerations, in pricing, 446–48
Nonfinancial allocation bases, 542
Nonfinancial benchmarks, 250
Nonfinancial factors
  and capital budgeting, 686
  and CIM investment, 705
  environmental, 698
Nonfinancial measures
  of customer satisfaction, 804–5
  of internal quality, 805
  of performance, 322, 890–91, 892
  of quality, 806
Nonlinear cost function, 357
Nonmanufacturing settings, 281–82
Nonprofit institutions
  and cost-volume-revenue analysis, 72
  discounted cash-flow analysis for, 738
Nonsubstitutable materials, 764
Nonvalue-added activity, 661
Non-value-added cost, 270
Normal costing, 109, 286
  in inventory-costing system, 312
  in manufacturing, 145–46
Normal spoilage, 633, 640
  common vs. specific-job, 644
Normal utilization, 323
Numerator reason, 107–8, 152

O

Objective function, 77, 410
Objectivity, as obligation, 17

Office of Federal Procurement Policy, 547
One-time-only special orders, 390–92, 431–34
On-time performance, 814, 816
Operating budget, 188
Operating costs, of manufacturer, 38n
Operating cycle, 211
Operating department, 510
Operating income, 61
  under absorption and variable costing, 317–19
  and prorating of manufacturing variances, 294–95
Operation, 657
Operation costing, 657, 660
  illustration of, 658–59
  journal entries for, 659
Opportunity cost, 397, 399, 833
  of capital, 688, 736, 838
  in energy transmission, 401
  and managing goods for sale, 833
  and stockout costs, 839
  and transfer pricing, 875
Ordinary incremental budgets, 478
Organizational culture, and discretionary-cost centers, 486
Organizational reporting unit, 863
Organizational structure, 11–16, 202
  and decentralization, 860–63
  and responsibility accounting, 202–6
Organization-structure cost hierarchy, 549–51
Outcomes, 77
  vs. decisions, 81
Outlay (out-of-pocket), costs, 875, 392, 399
Output level overhead variance, 278–80
Output levels, choice of, 390–93
Output measures, 62
Output unit-level costs, 355
Outsourcing, 394–95
  at Eastman Kodak, 397
Overabsorbed indirect (overhead) costs, 152–55
Overallocated indirect (overhead) costs, 152–55
Overapplied indirect (overhead) costs, 152–55
Overcosting, product, 114
Overhead costs, 268
  fixed, 270, 276
  journal entries for, 283–85
  manufacturing, 42 (see also Manufacturing overhead costs)
  planning and control of, 269–70
  variable, 270, 282–83, 312
  See also Indirect costs
Overhead rates
  fixed, 276–77
  variable, 270–71
Overhead variances
  cost
    fixed, 277–78
    integrated analysis of, 280–81

  in nonmanufacturing settings, 281–82
  variable, 271–75
  efficiency variance (variable), 274–75
  journal entries for, 283–85
  output level (production volume), 278–80
  spending (variable), 274

P

Padding, 199
Paperless budget, at Union Pacific, 196
Parameter, 341
Pareto diagram, 799
Partial productivity, 774–77
Past, and decision making, 388
Past costs, irrelevance of, 403–6
Payback method, 687, 697–99
  bailout, 700–701
  in international comparison, 703
  vs. BET, 810–11
Payoff matrix, 80
Payoffs, 77
Payoff table, 80
Pay-for-performance plans, 470–71
Peak-load pricing, 446
Peanut-butter costing, 114–15
Perfection standards, 234
Perfectly competitive market, 869, 869n
Performance evaluation or measures, 890
  and absorption costing, 321–22
  accounting-based, 891–92
  and alternative definitions of investment, 897
  and asset measurement, 897–900
  and budgeting, 185
  and capital-budget decisions, 702–4
  care needed in, 285–86
  cost allocation bases for, 816
  and decision model, 78
  and goal congruence, 900–901
  at individual activity level, 904–5
  and managerial incentives, 901–4
  nonfinancial, 322, 890–91, 892
  proposals for revising, 322
  and quality, 805–7
  standards in, 245–47
  and throughput accounting (Allied-Signal), 818
  at total organizational level, 905–7
  types of
    residual income, 894–95
    return on investment, 892–94, 898–900
    return on sales, 895–96
  variance analysis in, 206
Performance-evaluation model, 406–7
Performance gap, 227, 285. See also Variance analysis
Performance report, 670
  interim, 671
Period costs, 37–40
  and reporting, 314
  and variable vs. absorption costing, 309

Periodic inventory method, 40, 41
Perpetual inventory method, 40–41
Pharmaceutical products, and cost management, 139
Physical measure method of joint cost allocation, 574–75
Planning, 7–8
  benchmarks in, 285
  and budgeting, 184–85
  and costing systems, 166
  of variable and fixed overhead costs, 269–70
Plant and equipment, gross or net book value for, 900
Plantwide rates, vs. department rates, 539–41
Poisson arrival pattern, for customer orders, 812n
Post-deduct costing, 662. See also Backflush costing method
Postmanufacturing (downstream) costs, 429
Poultry farms, joint products from, 569, 571
Practical capacity, 323, 544
Predatory pricing, 451–52
  and price justification, 445
Predictions
  cost, 340
  and models, 386–87
Premanufacturing (upstream) costs, 429
Premio Nacional de Calidad, 803
Present value, 897
Present value table, 928, 932
  for annuity, 929–30, 934
Prevention costs, 795, 796, 833
Previous department (transferred-in) costs, 616
  and process costing, 614–20
Price discrimination, 446
Price justification, 445
Price (rate) variance, 238–40
  direct materials, 293–94
  information required for, 242–43
  and performance measurement, 245
  presentation of, 241–42
Pricing decisions, 428, 429
  alternative approaches to, 437–38
  and antitrust laws, 451
  cost-plus, 443–46
  fairness of, 548–49
  international comparison of, 447
  and life-cycle budgeting, 448–49
  for long run, 434–37
  major influences on, 428–29
  non-cost considerations in, 446–48
  predatory, 445, 451–52
  and product cost, 45
  profitability included in, 428n
  for short run, 431–34
  and target cost, 438–39
  See also Transfer pricing
Prime costs, 43
Priority incremental budgets, 478
Probability(ies), 77
  and variance investigatioon, 254–55

Probability distribution, 78
Problem-solving function, 15–16
Process costing system, 99–100, 140,
    598–99, 660
    and activity-based costing, 115, 159
    in ceramics industry, 611
    with equivalent units, 599–602
        first-in, first-out (FIFO) method,
            606–10, 617, 618–19, 620, 636–37,
            637n
        weighted-average method, 602–6,
            609–10, 615–18, 636
    estimating degree of completion for,
        620
    in manufacturing, 155–59
    at Raytheon, 597
    and spoilage, 634–40, 645
        and inspection timing, 646–47
    and standard costs, 610–14, 620
    transferred-in costs in, 614–20
Product cost(s), 45
    calculating of, 157–59, 434–37
    categories of, 429–30
    direct vs. indirect, 429–30, 499
Product-cost cross-subsidization,
    114–15
Product costing, 660
    cost allocation bases for, 816
    five-step approach to, 536–38
    guidelines for judging, 543
    and larger systems, 503
Product emphasis, and product cost, 45
Production, 3
    cost drivers in, 30
Production budget, 192
Production-cost worksheet, 607, 608,
    638, 639
Production denominator level, 276
Production denominator volume, 276
Production department, 510
Production level variance, 278–80
Production technology (production)
    function, 775
Production volume variance, 278–80
Productivity, 773–74
    and analysis of annual cost changes,
        780–81
    improvement of, 771–73
    partial measures of, 774–77
    service-sector, 782
    total factor, 777–80
Product life cycle, 448, 807
Product-life-cycle reporting, 450
Product line, 116
Product-line reporting unit, 863
Product mix decisions, 393–94
    and joint products, 580
Product overcosting, 114
Product parts, multiple, 816
Product-sustaining (service-sustaining)
    costs, 356, 437
Product undercosting, 114
Professional ethics. *See* Ethics, profes-
    sional
Professional examinations
    American, 935–36
    Australian, 937

Canadian, 936–37
Japanese, 937
United Kingdom, 938
Profitability
    and nonprofit organizations, 738
    and pricing decision, 428n
Profitability index, 738
Profitability measures, 891
Profit center, 203, 863
    and cost allocation, 502
Profit plan, budgeting as, 187
Profit-volume (PV) chart, 68–69. *See
    also* Cost-volume-profit analysis
Pro forma statements, 187
Project, 668
    and capital-budgeting decisions, 685
        and accrual-accounting rate-of-re-
            turn method, 701–4
        and discounted cash-flow method,
            687–92, 694–97
        and payback method, 697–701
        planning and control of, 668–71
        risk factor of, 736–37
Project performance cost variance, 670
Project schedule cost variance, 670
Proration
    of contribution to companywide op-
        erating income, 873
    of manufacturing variances, 290–95
    of under/overallocated overhead,
        153, 154–55
Purchase costs, 833
Purchase-order lead time, 833
Purpose(s)
    of cost allocation, 499–500, 502, 510
    "different costs for different pur-
        poses," 4, 45
    multipurpose nature of costing sys-
        tems, 166
    of product costing, 45
PV (profit-volume) chart, 68–69. *See
    also* Cost-volume-profit analysis

Q

Qualitative factors, 389
    and capital budgeting, 686
    and CIM investment, 705
Quality, 6, 794, 794n
    conformance, 795
    costs of, 794, 795–98, 806, 833, 842–43
    costs and benefits of improvement
        in, 800–804
    customers' demand for, 793
    and defect rate, 632
    of design, 795
    evaluating performance of, 806–7
    measures of, 891
        customer satisfaction, 804–5
        internal, 805–6
    and problem identification, 798–800
    as success factor, 6
    total management of (TQM), 794,
        807
Quantitative analysis of cost relation-
    ships, 346–47
Quantitative factors, 389

Quantity variances, 238, 755
Queues, 811

R

Rate of return
    nominal, 732
    real, 732
Rate-of-return-on-assets method, 702
Rate variance, 238@I2:
Real rate of return, 732
Reciprocal allocation method, 512–15
Refinement of costing system, 110–14
    with activity-based costing, 160
Regression analysis, 350–52, 364–68
    multiple regression, 371–72
    in sales forecasting, 199
Rejection (defect) rate
    in electronics industry, 642
    and JIT, 661
    need to improve, 632
    See also Spoilage
Relative performance evaluation,
    903–4
Relevant benefits, of JIT purchasing,
    841–42
Relevant cost(s), 388–89
    of JIT purchasing, 841–42
    of quality and timely deliveries,
        842–43
Relevant-cost analysis
    and contribution-margin analysis,
        396–97
    and customer profitability, 400–403
    and income taxes, 389
    and output level, 390–93
    and past costs, 403–6
    pitfalls in, 392
    and product mix decisions, 393–94
    and qualitative vs. quantitative fac-
        tors, 389–90
    and time value of money, 389
Relevant range, 31–32, 344
    linearity within, 367
Relevant revenues, 388–89
Reorder point, 836
Reporting relationships, 16
Reporting units, 863
Required rate of return (RRR), 688,
    736–38
Research and development, 3
    cost drivers in, 30
Research and development costs bud-
    get, 196
Residual income, 894–95
Residuals (residual term), 351, 367
    constant variance of, 367
    independence of, 368
    normality of, 368
Resource allocation, and performance
    measures, 890
Responsibility
    and controllability, 206–8
    for financial reports, 18
    and job cost records, 144
    and variances, 206, 246–47
Responsibility accounting, 202–6

Responsibility centers, 202, 863
Restated allocation rate approach, to under/overallocated overhead, 153, 155, 249
Retail organizations, managing goods for sale in, 832–40
Retail sector, and nonfinancial cost-allocation bases, 542
Return on investment (ROI), 890, 892–94, 898–900
  and historical cost, 901
Return on sales, 895–96
Revenue budget, 191–92
Revenue center, 203, 863
Revenue driver, 60
Revenues, relevant, 388–89
Reverse engineering, 430
Reworked units, 633, 641, 644
  and JIT, 661
Risk
  approaches to, 736–37
  and compensation, 206, 901–2
  and management control system, 469–70
  and required rate of return, 736
Robinson-Patman Act, 451
Robotics, 25
Rolling budget, 186

S

Safety and environment (S&E) concerns, and management control systems, 472
Safety margin, 68
Safety stock, 836–38
Sales budget, 191–92
  vs. forecast, 198
Sales forecasting, 198–99
Sales mix, 69–70, 754
Sales-mix variances, 753, 758–61
Sales quantity variances, 753, 758–61
Sales value at splitoff method of joint cost allocation, 573–74, 578–79
Sales variances, 754–55
  market-size and market-share, 761–64
  quantity and mix, 758–61
  volume, 230, 232–33, 237, 755–58
Scope, 669
Scorecard, balanced, 806, 890
Scorekeeping function, 15
Scrap, 571, 633, 642–44
Securities and Exchange Commission (SEC), 906
Segment, 549
Self-liquidating cycle, 211
Semivariable cost, 342
Sensitivity ("what-if") analysis, 67–68, 200, 692–94, 737
Separable costs, 570–71
  of byproducts, 584
Sequential allocation method, 511
Sequential tracking, 661–62
Serial correlation, 368
Service department, 510
Service-sector companies, 35

capitalized, inventoriable, and period costs for, 37–40
  financial statements for, 34–37
  and inventory, 35, 37
  job costing in, 101–4
    and activity-based costing, 115
    and actual vs. normal vs. budgeted costing, 109
    and cost allocation, 103, 107–9
    and costing system refinement, 110–14
    and customer costing, 118–23
    and peanut-butter costing, 114–15
    and tracing of direct costs to jobs, 104–6
  productivity of, 782
Service-sustaining costs, 356
Setup costs, and short-run pricing decisions, 432–34
Sherman Act, 451
Short run, pricing decisions for, 431–34
Shrinkage, 117
Sigma (Å), 798
Simple regression, 350
Simplex method, 413n
Single-rate method of cost allocation, 507–8
Slack, budgetary, 199
Slope coefficient, 341
Society of Cost Estimating and Analysis, 936n
Society of Management Accountants (SMA), 936–37
Source documents, 104
  in job costing systems, 104, 142–44
Sourcing restrictions, 862
Special-order decisions, 390–92, 431–34
  by trucking firm, 385
Specification analysis, 353, 366–68, 370
Spending variance, variable overhead, 274
Splitoff point, 570
Spoilage, 633
  abnormal, 633–34, 644
  and JIT, 661
  and job costing, 640–41, 645
  normal, 633, 640, 644
  and process costing, 634–40, 645
    and inspection timing, 646–47
Spreadsheets, and sensitivity analysis, 67–68, 693
Staff management, 12
Stand-alone common cost allocation method, 516–17
Standard(s), 226
  perfection vs. currently-attainable, 234
  in performance measurement, 245–47
  as range of outcomes, 247
Standard cost(ing), 227, 286
  flexible budgeting with, 226–27, 233–38
  and general ledger, 248–49
  international use of, 232
  in inventory-costing system, 312
  and process costs, 610–14, 620
  and proration of manufacturing

variances, 290–95
Standard error of estimated coefficient, 366
Standard input, 226–27
Statements on Management Accounting, 45
States (of nature), 77
Static budget, 228
  and variable overhead manufacturing costs, 271–73
Static-budget variance, 228–29
Statistical approaches, in sales forecasting, 199
Statistical process control (SPC), 798–99
Statistical quality control (SQC), 798–99
Step-down allocation method (step or sequential allocation method), 511–12
Step-fixed cost function, 357
Step functions, 357
Stockholders' equity, 897
Stock options, for executives, 906
Stockout, 833
Straight-line (SL) depreciation, 721
Strategy, 184
  cost drivers in, 30
Strategy and administration function, 3
Structure, organizational. *See* Organizational structure
Suboptimal (dysfunctional) decision making, 861–62
Subsidiary ledgers, in manufacturing job costing system, 146–52, 153
Substitutable materials, 764
Success factors, 6
Sunk costs, 405
Super-absorption costing, 313
Supermarkets, 753
Super-variable costing, 313, 322, 665
Suppliers, evaluation of (Northrup), 243
Support department, 510
  and cost allocation, 510–15, 873
  and transfer prices, 873
Surveys of company practice
  on activity-based costing, 161
  on budget practices, 187
  on capital budgeting, 703
  on CFO responsibilities, 14
  on cost allocation, 145, 502, 516
    for manufacturing overhead, 273
  on cost classification, 345
  on decision to investigate variances, 255
  on distinction between fixed and variable costs, 73
  on JIT, 667, 848
  on joint-cost allocation (U.K.), 580
  on labor compensation costs (law firms), 104
  on labor fringe benefits, 44
  on manufacturing cost categories, 46
  on nonfinancial measures of performance, 892
  on pricing practices and cost man-

Surveys of company practice, (cont.)
    agement, 447
    on product-costing system, 543
    on rejection rates (electronics), 642
    on risk adjustment methods in capi-
        tal budgeting, 737
    on standard costs, 232
    on transfer prices (world-wide), 872
Synchronous tracking, 661–62
Systems design, 98

T

Tables, interest
    future value, 927–28, 931
        of annuity, 928–29, 933
    present value, 928, 932
        of annuity, 929–30, 934
Tactics, 184
Target cost(ing), 438–39, 440–42
    by Japanese firms, 446
    and life-cycle costs, 450–51
    at Mercedes Benz and Toyota, 443
Targeting, budgeting as, 187
Target operating income, 64–65
    and income taxes, 70–72
Target price(ing), 438–39, 443
Target return on investment, 443
Taxes
    and breakeven point, 72n
    and discounted cash-flow analysis,
        719
    and MACRS, 731, 744
    and transfer pricing, 867–68
    See also Income taxes
Term-based compensation arrange-
        ments, 904–5
Theoretical capacity, 323
Theory of constraints, 817–19
Throughput contribution, 313, 816–19
    at Allied-Signal, 818
Throughput (super-variable) costing,
        313, 322, 665
Time
    breakeven, 808–11
    costs of, 814–16
    operational measures of, 811–14
    as success factor, 6
Time-adjusted rate of return, 690
Time driver, 811
Time period (time horizon, time span)
    of budgets, 186–87
    and capital budgeting projects, 687
    and executive performance, 906
    and indirect cost rates, 107–8
    and pricing decision, 430
    problems encountered with, 354
    of project, 705
    and variable vs. fixed costs, 67, 343
    See also Long run; Short run
Time-series data, 346–47
Time value of money, 684–85
    and discounted cash-flow methods,
        687
    relevant aspects of, 389
Total-alternatives approach to decision
        making, 397–99

Total assets available, 897
Total assets employed, 897
Total costs, vs. unit costs, 33–34
Total direct labor mix variance, 770
Total direct labor yield variance, 770
Total direct materials mix variance, 767
Total direct materials yield variance, 767
Total factor productivity, 777–80
Total manufacturing costs, and JIT, 661
Total project approach, 725, 726–29
Total quality management (TQM), 794,
        807
Total sales-mix variance, 758
Total sales-quantity variance, 758
Total value-chain analysis, 6–7
Trade fairs, 59
Trails, audit, and backflush costing,
        668
Transfer price(ing), 863–65
    cost-based, 865, 870–75
    guideline for, 875–76
    illustration of, 865–67
    market-based, 869–70
    multinational, 859, 868–69, 872
    survey on, 872
    tax considerations in, 867–68
Transferred-in costs (previous depart-
        ment costs), 616
    and process costing, 614–20
Treasurer, 14
Trend analysis, in sales forecasting, 199
Trial-and-error solution approach,
        411–13

U

Uncertainty, 67, 77–79
    decision making model for, 79–81
    and delays, 812–13
    level of, 477–78
    and responsibility, 207
    and variance investigation decisions,
        253–55
Underabsorbed indirect (overhead)
        costs, 152–55
Underallocated indirect (overhead)
        costs, 152–55
Underapplied indirect (overhead)
        costs, 152–55
Undercosting, product, 114
Unfavorable variance, 228
Unit cost (average cost), 33–34
    assigned before final production
        point, 598
    as misleading, 392–93
    and output level, 390
    in pricing decisions, 444–45
United Auto Workers (UAW), 71
United Kingdom professional exami-
        nations, 938
Unit fixed manufacturing overhead
        costs, 276–77
Unit variable costs, and learning-curve
        effects, 339
Upstream costs, 429
Usage variances, 238
U.S. government

antitrust laws, 445, 451–52
contracting with, 45, 547
environmental laws, 435
Utilization, normal and master-budget,
        323–24

V

Value-added activities, 29, 473–75
Value-added cost, 270
Value chain, 2–3, 6–7, 124
    cost categories in, 476–78
    and life-cycle budgeting, 448–49
Value engineering, 438
    in example, 439
    and life-cycle costs, 450–51
Variable(s), dependent and indepen-
        dent, 347–48, 350, 352–53, 366
Variable cost(ing), 29–31, 32–33, 309,
        314–19
    breakeven points in, 320–21
    and cost object, 343
    and "direct costing," 311
    for external reporting, 312
    income effects of, 320
    for internal reporting, 312–13
    international usage of, 318
    in inventory-costing system, 312
    overhead, 270, 282–83, 312
    and relevant range, 31–32, 344
    and special order, 432–34
    and time period, 67, 343
Variable-cost percentage, 74
Variable manufacturing overhead
        costs, 279, 282–83
    in inventory-costing system, 312
    for planning/control and inventory
        costing, 283
Variable overhead rates, 270–71
Variable overhead (VOH) variances
    cost, 271–75
    efficiency, 274–75
    spending, 274
Variance analysis, 227, 245
    columnar presentation of, 279
    4–, 3–, 2–, 1–, 280–81
    guidelines for investigation, 247
    illustration of, 244, 245
    and process-costing systems,
        613–14
    and responsibility, 206, 246–47
    survey on, 255
    and uncertainty, 253–55
Variances, 8, 227, 285
    and backflush costing system,
        664–65, 668
    and effectiveness or efficiency,
        237–38
    efficiency, 238, 240–43, 274–75
    favorable, 228
    fixed overhead cost, 277–78
    flexible-budget, 230–31, 237–38
    and flexible vs. static budget, 228
    input, 764
        direct labor yield and direct labor
            mix, 770–71
        direct materials yield and direct

materials mix, 765–69
interdependencies among, 281
and inventories, 247–48
overhead, 271–75, 277–82, 283–85
price (rate), 238–40, 241–43, 245
  direct materials, 293–94
project, 669–70
proration of, 290–95
sales, 754–55
  market-size and market-share, 761–64
  sales-quantity and sales-mix, 758–61
  sales-volume, 230, 232–33, 237, 755–58
static-budget, 228–29
unfavorable, 228
volume, 755
yield, 755, 765–71

W

Waiting time, average, 812
Waste stream, cost of, 472
Weighted-average process-costing method, 602–6
  and spoilage, 636
  and transferred-in costs, 615–18
  vs. FIFO, 609–10, 617
"What-if" (sensitivity) analysis, 67–68, 200, 692–93, 737
White collar productivity, 782
Work environments, labor-paced and machine-paced, 541–42
Working capital, 897
Working-capital cycle, 211
Work measurement, 479, 481
Work-measurement method of cost estimation, 344

Work in process inventory (work in progress, goods in process), 35, 35n
and backflush costing, 662
and inventory-costing systems, 312
Work in progress, 35
Worksheet, production-cost, 607, 608, 638, 639
Write-offs, and denominator-level concept, 325

Y

Yield, 764
Yield variance, 755, 765–71

Z

Zero-based budgeting (ZBB), 478–79